The Biographical Encyclopedia of Astronomers

The Biographical Encyclopedia of Astronomers

Volume II
M–Z

Editor-in-Chief

Thomas Hockey

Senior Editors

Virginia Trimble
Thomas R. Williams

Editors

Katherine Bracher
Richard A. Jarrell
Jordan D. Marché, II
F. Jamil Ragep

Associate Editor

JoAnn Palmeri

Assistant Editor

Marvin Bolt

Dr. Thomas Hockey
Professor of Astronomy
University of Northern Iowa
Department of Earth Science
Office: Latham 112
Cedar Falls
IA 50614
USA

ISBN 13: 978-0-387-31022-0
The electronic version of the whole set will be available under ISBN-13: 978-0-387-30400-7.
The print and electronic bundle of the whole set will be available under ISBN-13: 978-0-387-33628-2.

springer.com

Printed on acid-free paper SPIN: 11494034 2109 — 5 4 3 2 1 0

To my teachers

Aldrich Syverson
Joseph Freimeyer
Connie Mitchell
John Miller
Paul Coke
Peggy Hudson
Irwin Shapiro
John Lewis
Reta Beebe
Herbert Beebe
William Eamon
Clyde Tombaugh

Preface

Like that of any human activity, the history of astronomy has been played out under the influence of myriad cultural, institutional, political, sociological, technological, and natural forces. Any history that focuses only on the greatest participants in a field likely misses a great deal of interest and historical value. Inasmuch as astronomy is undertaken by and for human beings, therefore, its history cannot be limited to the lives and achievements of a narrow group.

Here we analyze the lives of people who, in our view, produced some substantial contribution to the field of astronomy, were involved in some important astronomical event, or were in some other manner important to the discipline. In doing so we do not discount the work of countless other journeyman astronomers without whom the science would not have progressed as it has.

Scope

Biographical Encyclopedia of Astronomers [*BEA*] entries presented here do not pretend to illuminate all aspects of a given person's vita. Moreover, some figures included are better known for their enterprises outside of astronomy. In these situations, their astronomical contributions are emphasized.

For many of our entries, the length is limited to something substantially less than 1,000 words due to the lack of available information. There is, of course, an inclination to write a great deal more about persons for whom there is a significant literature already available, *e. g.*, Copernicus, Kepler, Newton, William Herschel, or Einstein. Many such individuals are covered in other standard resources, and we have not felt compelled to repeat all that is already published in those cases. In fact, we look at our entries as a guide to recent scholarship and a brief summary of the important facts about the lives involved. On the other hand, two-thirds of the entries in this encyclopedia are about individuals for whom there is no readily available standard source. In those cases, the length of the article may be longer than might be expected in comparison with those of better known astronomers, and reflects the fact that an entry offers the first (and perhaps only) easily available information about the astronomer involved: It is not difficult to find sources on "Greats" such as Galileo Galilei; however, it *is* hard to find information on Galilei's acolyte, Mario Guiducci.

Citations within the text have been avoided to enhance readability. Nearly all articles end with a list of selected references. The reader is thus presented with opportunities for further research; no article is intended to be a dead end. Toward that end, if we do not provide additional resources for an entry, the subject will be cross-referenced within other articles for which we do provide selected references.

In compiling the selected references, we have tried to include difficult-to-identify secondary sources. At the same time we have largely excluded standard reference works and include only some of the latest canonical works covering the best-known figures in astronomy.

The *BEA* documents individuals born from Antiquity to approximately mid-1918. Subjects may be living or dead. While some ancient figures have become legendary, we have tried to avoid clearly mythological ones. For example, while the royal Chinese astronomers Ho and Hsi (supposedly third millennium BCE) appear in nearly every history of eclipses, they warrant no entry here.

This terminal birth date assures that the subjects written about have completed most of their careers, and that sufficient time likely has elapsed since their featured accomplishments that a historical perspective on their work is possible. Note that almost all of our subjects began their careers before the watershed transformation of astronomy brought about by the events of World War II. It is also true that the number of astronomers significantly increased after this time. Our youngest subject is Gérard de Vaucouleurs; our oldest is Homer.

Inclusion Parameters

Our entry selection embraces a broad definition of the word "astronomer." In modern science, little differentiation is made between the words "astronomy" and "astrophysics"; we do not use such a distinction here. For example, our definition includes astrometrists, cosmologists, and planetologists. These three fields were considered separate and self-contained for most of human history. Cosmology, especially, requires the inclusion of many philosophers and theologians.

Early astronomers often also were astrologers. If they performed astronomical pursuits in addition to simple divination, we include them. Likewise, no distinction is made between the professional and the contributing amateur.

With the exception of a few important cases, instrument makers are included only if they pursued astronomical work with their instruments. Surveyors and cartographers are included if their study of the stars went beyond mere reference for terrestrial mapmaking. Lastly, a select group of authors, editors of astronomical journals, founders of astronomical societies, observatory builders and directors, astronomy historians, and patrons of astronomy are included.

A common pitfall in the history of science is to make the story of a discipline appear to be a single ladder ascending toward modern theory. Instead, it is a tree with many branches, only some of which have led to our current understanding of the Universe. Indeed, seemingly dead branches may become reanimated later in time. And branches may merge as ideas once considered unrelated are brought together. A better metaphor may be a vine, one with many grafts.

Scientists who contributed theories no longer held salient, or who made observations now considered suspect, nonetheless are included on our list if their effort was considered scientifically useful in its time, and the basis for further inquiry. At the same time, scientists whose ideas or techniques are now considered prescient, but who were unrecognized in their lifetimes, may appear as well.

The contributions of persons selected for entries in this work were weighed in the context of their times. Thus, while a contribution made by a medieval scholar might seem small by today's standards, it was significant for its era. We are especially proud of our inclusion of "non-western" figures who often have been given little treatment in histories of astronomy. Finally, we have included numerous entries of fewer than 100 words, some just a sentence or two, to introduce their names and place them in context within the broader vistas of astronomy.

Construction of the subject list was done by the editor-in-chief in consultation with the content editors. Well-known historian of astronomy Owen Gingerich generously volunteered his time to comment upon draft lists. Still, while an earnest attempt was made to make an objective selection of our more than 1,500 entries, responsibility for omissions must rest with the editor-in-chief. Most vulnerable to omission were those born in the last century.

Project Staffing

Author solicitation was done by the editor-in-chief. Many of the shortest entries were crafted by the editor-in-chief; some but not most of these short entries were paraphrased from an unpublished typescript draft titled *Biographical Dictionary of Astronomers*, originally prepared by the historian Hector C. Macpherson in 1940. The standardized format of the articles was arrived at by consensus among the editors. Senior editor Thomas R. Williams's *Author Guidelines* proved indispensable.

Editors were invited to join the project by the editor-in-chief. This editorial board includes, more-or-less equally, individuals who entered history-of-astronomy scholarship with a background either in history of science or in astronomy. (Some have both.) Unlike many encyclopedists, we did not use our editorial role to eradicate the individual writing styles of the authors.

Each content editor was assigned a thematic editorial responsibility, though all were called upon, at one time or another, to edit articles outside of this specialty. The assignments were as follows:

Classical and Medieval Astronomers—Katherine Bracher
Renaissance and Enlightenment Astronomers—Richard A. Jarrell
Nineteenth Century Astronomers—Marvin Bolt
Twentieth Century Astronomers/Astrophysicists—Virginia Trimble
Astronomers of the Islamic World—Jamil Ragep
Nonvocational Astronomers—Thomas R. Williams
Astronomy Popularizers—Jordan D. Marché, II

All content editors also contributed articles to the *BEA*. JoAnn Palmeri edited the vital references for all entries. Additionally she served as our illustrations editor.

For *errata* information, e-mail us at HOCKEY@UNI.EDU

Thomas Hockey
October 2005

Acknowledgments

The *Biographical Encyclopedia of Astronomers* [*BEA*] is above all the product of its authors. These 410 contributors hail from 40 different countries. Nearly every article is an original piece of scholarship. In some cases, scholars about whom entries were written were themselves gracious enough to write articles for us on other subjects.

At the heart of this 6-year project has been its board of editors. Contrary to what the narrow definition of this job title might imply, these people have been actively providing aid, comfort, and advice to the project, since its inception. As to their editorial contribution specifically, this was often far greater, and more time consuming, than is commonly assumed.

The *BEA* was the idea of Peter Binfield (then Business Development at Kluwer). Dr. Binfield's assistant, Ms. Livia Iebba, also provided support "above and beyond." Dr. Harry Blom, Springer's Senior Editor for Astronomy and Astrophysics, traveled many kilometers to meet with the *BEA* editorial board and lend support on the long road to publication.

Usually unsung in a project of this nature are those individuals who did not write for us, but instead recommended other willing and qualified authors. Brevity permits me only two examples: Eva Isaksson of the University of Helsinki and Kevin Krisciunas of the Cerro Tololo Interamerican Observatory.

Brenda G. Corbin at the United States Naval Observatory kindly provided us with a manuscript copy of Hector Copland MacPherson's *Biographical Dictionary of Astronomers* (1940), which was never published. We hope that its use in assembling the *BEA* is similar to what Dr. MacPherson had wished to achieve. Many, though not most, of the shortest entries in the *BEA* were paraphrased from MacPherson's work.

Certain scholars consulted with us on astronomers of specific nationalities. We appreciate the assistance of Alexander A. Gurshtein (astronomers of the former USSR), Suzanne Débarbat (Francophone astronomers), Helge Kragh (Scandinavian astronomers), Robert Van Gent (Dutch), A. Vagiswari (Indian astronomers), Kevin D. Pang (Chinese astronomers), Jochi Shigeru (East Asian astronomers), and Rudi Paul Lindner (Byzantine astronomers).

The bibliographies of recent works in the history of astronomy published by Ruth Freitag (Library of Congress) were enormously useful. So was the Finding List of Obituary Notes of Astronomers (1900–1997) prepared by Hilmar Dürbeck and Beatrix Ott, with contributions by Wolfgang Dick. The Astrophysics Data System of the National Aeronautics and Space Administration was frequently accessed.

The effort of Daniel W. E. Green, Harvard-Smithsonian Center for Astrophysics and International Astronomical Union Center for Astronomical Telegrams, assured that the proper use of new International Astronomical Union comet and minor-planet nomenclatures was maintained.

H. Miller's Thryomanes font facilitated communicating Arabic text between editors. Yuliana Ivakh helped the editor-in-chief with Cyrillic.

Kari Aunan handled thousands of letters during the author-solicitation process. Wesley Even created and maintained the spreadsheet, so necessary for keeping track of the data and long lists generated by the project. Rachel Wiekhorst operated the document scanner. Jeff Guntren prepared the Table of Contents. I am proud to say that all did so while being undergraduate students at the University of Northern Iowa.

Ruby Hockey undertook the cumbersome filing process.

"Thank you" to the members of the Department of Earth Science, University of Northern Iowa [UNI], especially Lois Jerke. I relied on their infrastructure and good humor greatly. Generous, too, was the support of Dean Kichoon Yang, UNI College of Natural Sciences. Linda Berneking of the UNI Donald O. Rod Library, Interlibrary Loan, also deserves special mention.

Editor Marvin Bolt would like to thank the Adler Planetarium and Astronomy Museum and the Program in the History and Philosophy of Science at the University of Notre Dame for research support.

Editor Katherine Bracher would like to acknowledge the advice and support of Cynthia W. Shelmerdine, Professor of Classics at the University of Texas at Austin.

Editor Jordan D. Marché, II thanks the Department of Astronomy at the University of Wisconsin-Madison for its strong support, and especially the Woodman Astronomical Library. Concurrently, he acknowledges the other libraries of the University of Wisconsin-Madison system and the Wisconsin State Historical Society Library.

Editor Jamil Ragep wishes to acknowledge Sally P. Ragep for editorial work behind the scenes and also Julio Samsó for help with Andalusian/North African astronomers.

Editor Virginia Trimble wishes to acknowledge the assistance of Leon Mestel, George Herbig, Meinhard Mayer, Harry Lustig, M. G. Rodriguez, Adriaan Blaauw, and Dimitri Klimushkin.

Editor Thomas R. Williams would like to acknowledge Peter Hingley, librarian of the Royal Astronomical Society, and Richard McKim, as well as the staff of Fondren Library at Rice University for their assistance.

The editorial board is grateful for the aid received from the many other scholars and librarians, too many to list here, who assisted with facts, citations, and general comments on individual entries. This public support is echoed by officers of the International Astronomical Union Commission 41 (History of Astronomy)/Inter-Union Commission for History of Astronomy, Ileana Chinnici and Wayne Orchiston, who, in the *ICHA Newsletter* #3 (2002), wrote regarding the *Biographical Encyclopedia of Astronomy*: "While the formation of the ICHA came too late for it to be an active participant in the planning phase, we are happy to report that the ICHA Organizing Committee has given the project its whole-hearted support…"

Acknowledgments

Foreword

In the past four decades, the history of astronomy and cosmology has grown into a professional research area, complete with a journal (*Journal for the History of Astronomy*), sessions devoted to the subject at annual meetings of professional societies, and regular meetings of its own, such as the biennial meetings at the University of Notre Dame. Indeed, the field contains subspecialties, such as archaeoastronomy, that hold regular meetings of their own and have journals.

Astronomy is unique in several respects. First, although the research front in all sciences moves ever faster, constantly increasing the distance between the practitioner and the subject's history, in astronomy the time dimension plays a crucial role in current research (as opposed to, for instance, chemistry), and this means that past data, *e. g.*, of eclipse or sunspot observations, continue to play a role in astronomical research. The historian of astronomy is often the intermediary between the astronomer and these data, especially for earlier periods. Second, among the exact sciences, astronomy is the only field in which amateurs continue to play an active, if supporting, role: In a number of cases professional astronomers rely on the services of the amateurs, and many of the services delivered by these amateurs are very professional indeed. But the lines demarking astronomers from historians and professionals from amateurs are not cut–and-dried. There are museum curators and planetarium educators who are amateurs astronomers or do highly professional research on historical periods, and there are professional astronomers who have an abiding interest in the history of their field for various reasons. And lest we forget, there are very large numbers of readers and television viewers with a passive interest in the history of astronomy for whom the human dimension of the quest to understand the heavens is crucial.

Many of the standard histories of astronomy date from the 1930s and 1950s. But these single-volume histories, which once served both as teaching tools and reference works, have become obsolete in the past few decades. More recent single-volume histories of astronomy can serve only as teaching tools and works of general interest. There has, thus, been a growing need for reference works that cover the results of research into the history of astronomy published in the past half century. Recently, two encyclopedias have been published, *History of Astronomy: an Encyclopedia*, edited by John Lankford, and *Encyclopedia of Cosmology*, edited by Norriss S. Hetherington. Concepts and issues are central in these works. The *Biographical Encyclopedia of Astronomers* is a reference work that focuses on individuals; it adds the human dimension without which no science, or its history, can come to life.

Albert van Helden
Utrecht, September 2005

Contributors

Victor K. Abalakin
Pulkovo Observatory

Mohammed Abattouy
Fez University

Leonard B. Abbey
Independent Scholar

Helmut A. Abt
Kitt Peak National Observatory

Narahari Achar
University of Memphis

Meltem Akbas
Istanbul University

Durruty Jesús de Alba Martinez
Universidad de Guadalajara

Roberto de Andrade Martins
Universidade da Campinas

S. M. Razaullah Ansari
Aligarh Muslim University

Adam Jared Apt
Independent Scholar

Stuart Atkinson
Independent Scholar

David Aubin
Université Pierre-et-Marie Curie

Salim Aydüz
Fatih University

Ennio Badolati
Università delgi Studi del Molise

Mohammad Bagheri
Encyclopaedia Islamica Foundation

Yuri V. Balashov
University of Georgia

Sallie Baliunas
Harvard-Smithsonian Center for Astrophysics

Alan Baragona
Virginia Military Institute

Edward Baron
University of Oklahoma

Raymonde Barthalot
Observatoire de la Cote d'Azur

Alan H. Batten
National Research Council (Canada)

Richard Baum
Independent Scholar

Anthony F. Beavers
University of Evansville

Herbert Beebe
New Mexico State University

Martin Beech
University of Regina

Ari Belenkiy
Hebrew University

Trudy E. Bell
Independent Scholar

Isaac Benguigui
Universitat Geneva

J. Len Berggren
Simon Fraser University

Giuseppe Bezza
Independent Scholar

Charlotte Bigg
Max-Planck-Institut für Wissenschafts Geschichte

Albert Bijaoui
Observatoire de Nice

Adriaan Blaauw
Rijksuniversiteit Groningen

Nicolaas Bloembergen
Harvard University

Thomas J. Bogdan
University of Colorado

Karl-Heinz Bohm
University of Washington

Marvin Bolt
Adler Planetarium

Patrick J. Boner
University of Florida

Fabrizio Bònoli
Berrera Osservatorio

Alan J. Bowden
Liverpool Museum

Alan C. Bowen
Princeton University

Katherine Bracher
Whitman College

Raffaello Braga
Independent Scholar

Ronald Brashear
Chemical Heritage Foundation

Sonja Brentjes
Aga Khan University

Peter Broughton
Independent Scholar

C. Brown-Syed
Wayne State University

Mary T. Brück
University of Edinburgh

Charles Burnett
Warburg Institute

Paul L. Butzer
Rheinisch-Westfalische Technische Hochschule

Chris K. Caldwell
University of Tennessee

Emilia Calvo
Universitad de Barcelona

Gary L. Cameron
Iowa State University

Nicholas Campion
Bath Spa University College

Juan Casanovas
Vatican Observatory

Josep Casulleras
Universitad de Barcelona

Patrick A. Catt
Independent Scholar

Roger Cayrel
Observatoire de Paris

Davide Cenadelli
Osservatorio di Brera

Michelle Chapront-Touzé
Observatoire de Paris

Paul Charbonneau
University of Colorado

François Charette
Ludwig-Maximilian University

Ileana Chinnici
Palermo Osservatorio

J. S. R. Chisholm
University of Kent

Grant Christie
Aukland Observatory

George W. Clark
Smithsonian Institution

Donald D. Clayton
Clemson University

Mercè Comes
Universitad de Barcelona

Glen M. Cooper
Brigham Young University

Brenda G. Corbin
United States Naval Observatory

Alan D. Corré
University of Wisconsin

Paul Couteau
Observatoire de Nice

George V. Coyne
Vatican Observatory

Mary Croarken
University of Warwick

Michael J. Crowe
University of Notre Dame

David Cunning
Northern Illinois University

Clifford J. Cunningham
Star Lab Press

Martijn P. Cuypers
Universiteit Leiden

Alex Dalgarno
Harvard University

Dennis Danielson
University of British Columbia

A. Clive Davenhall
University of Edinburgh

Suzanne Débarbat
Observatoire de Paris

Robert K. DeKosky
University of Kansas

Deng Kehui
Inner Mongolian Normal University

David DeVorkin
Smithsonian Institution

Jozef T. Devreese
Universiteit Antwerpen

David W. Dewhirst
Cambridge University

Gregg DeYoung
American University in Cairo

Alnoor Dhanani
Institute of Ismaili Studies

Dimitris Dialetis
University of Athens

Steven J. Dick
National Aeronautics and Space Administration (USA)

Richard R. Didick
Independent Scholar

Thomas A. Dobbins
Independent Scholar

John W. Docktor
Independent Scholar

Audouin Dollfus
Observatoire de Paris

Emmanuel Dormy
Institute de Physique du Globe de Paris

Matthew F. Dowd
University of Notre Dame

Ellen Tan Drake
Independent Scholar

Simone Dumont
Observatoire de Meudon

Wolcott B. Dunham, Jr.
Fund for Astrophysics Research, Incorporated

Storm Dunlop
Sussex University

Sven Dupré
Universiteit Ghent

Ian T. Durham
University of Saint Andrews

Suvendra Nath Dutta
Harvard University

James Dye
Northern Illinois University

Frank K. Edmondson
Indiana University

Philip Edwards
Institute of Space and Astronautical Science (UK)

Yuri N. Efremov
Moscow State University

Alv Egeland
Universitet Oslo

Arthur J. Ehlmann
Texas Christian University

Ian Elliott
Dunsink Observatory

David S. Evans
University of Texas

Glenn S. Everett
Stonehill College

Peter S. Excell
University of Bradford

Carl-Gunne Fälthammar
Alfvénlaboratoriet

İhsan Fazlıoğlu
Istanbul University

Fernando B. Figueiredo
Instituto Politécnico de Tomar

Maurice A. Finocchiaro
University of Nevada

Ronald Florence
Independent Scholar

Miquel Forcada
Universitad de Barcelona

Kenneth W. Ford
National Aeronautics and Space Administration (USA)

Malcolm R. Forster
University of Wisconsin

Michael Fosmire
Purdue University

Harmut Frommert
Independent Scholar

Michael Frost
Independent Scholar

Patrick Fuentes
Independent Scholar

George Gale
University of Misourri

Karl Galle
Universität Göttingen

Robert A. Garfinkle
Independent Scholar

Leonardo Gariboldi
Università delgi Studi di Milan

Roy H. Garstang
University of Colorado

Stephen Gaukroger
University of Sydney

Steven J. Gibson
Arecibo Observatory

Henry L. Giclas
Lowell Observatory

Adam Gilles
Observatoire de Lyon

Owen Gingerich
Harvard-Smithsonian Center for Astrophysics

M. Colleen Gino
Dudley Observatory

Ian S. Glass
South African Astronomical Observatory

André Goddu
Stonehill College

Gunther Görz
Universität Erlangen-Nürnberg

Daniel W. E. Green
Harvard-Smithsonian Center for Astrophysics

Solange Grillot
Observatoire de Paris

Monique Gros
Observatoire de Paris & Université Pierre-et-Marie Curie

Jiří Grygar
Akademie Ved, Ceská Republika

Françoise le Guet Tully
Observatoire de la Côte d'Azur

Alastair G. Gunn
University of Manchester

Guo Shirong
Inner Mongolian Normal University

Alexander A. Gurshtein
Russian Academy of Sciences

Fathi Habashi
Laval University

Peter Habison
Kuffner Sternwarte

Margherita Hack
Osservatorio Trieste

Petr Hadrava
Akademie Ved, Ceská Republika

Alena Hadravová
Akademie Ved, Ceská Republika

Graham Hall
University of Aberdeen

Fernand Hallyn
Universiteit Ghent

Jürgen Hamel
Museum für Astronomie und Technikgeschichte (Germany)

Truls Lynn Hansen
Universitet Tromsø

Katherine Haramundanis
Hewlett Packard Company

Behnaz Hashemipour
Isfahan University of Technology

Robert Alan Hatch
University of Florida

Christian E. Hauer, Jr.
Westminster College

John Hearnshaw
University of Canterbury

Klaus Hentschel
Universität Göttingen

Dieter B. Herrmann
Independent Scholar

Norriss S. Hetherington
Independent Scholar

Donald W. Hillger
Colorado State University

John Hilton
University of Natal

Alan W. Hirshfeld
University of Massachusetts

Thomas Hockey
University of Northern Iowa

Laurent Hodges
Iowa State University

Dorrit Hoffleit
Yale University

Julian Holland
University of Sydney

Gustav Holmberg
Lunds Universitet

Gerald Holton
Harvard University

Elliott Horch
Rensselaer Polytechnic Institute

Léo Houziaux
Académie Royale (Belgium)

Mark Hurn
Cambridge University

Robert J. Hurry
Calvert Marine Museum

Gary Huss
University of Hawaii

Roger D. Hutchins
Oxford University

Siek Hyung
Bohyunsan Optical Astronomy Observatory

Saori Ihara
Kochi University

Satoru Ikeuchi
Nagoya University

Setsuro Ikeyama
Independent Scholar

Balthasar Indermühle
Independent Scholar

Francine Jackson
University of Rhode Island

Richard A. Jarrell
York University

David Jefferies
University of Surrey

Derek Jensen
University of California at San Diego

Mihkel Joeveer
Tartu Astrophuusika Observatoorium

J. Bryn Jones
University of Nottingham

Mustafa Kaçar
Istanbul University

Horst Kant
Max-Planck-Institut für Wissenschafts Geschichte

Hannu Karttunen
Independent Scholar

Katalin Kèri
Janus Ponnonius University

Paul T. Keyser
Cornell University

Elaheh Kheirandish
Harvard University

Kevin J. Kilburn
University of Manchester

Stamatios Kimigis
Johns Hopkins University

David A. King
Johann Wolfgang Göthe Universität

Thomas D. Kinman
Kitt Peak National Observatory

Gyula Klima
Fordham University

Thomas Klöti
Universität Bern

Gillian Knapp
University of Washington

Oliver Knill
Harvard University

Wolfgang Kokott
Universität München

Daniel Kolak
William Paterson University

Nicholas Kollerstrom
University College of London

Anne J. Kox
Universiteit Amsterdam

Yoshihide Kozai
National Astronomical Obervatory of Japan

Helge Kragh
Universitet Aarhus

John Kraus
Ohio State University

Henk Kubbinga
Rijksuniversiteit Groningen

Suhasini Kumar
University of Toledo

Paul Kunitzsch
Ludwig-Maximilian Universität

Takanori Kusuba
Osaka University

Alistair Kwan
University of Melbourne

Claud H. Lacy
University of Arkansas

Keith R. Lafortune
University of Notre Dame

Edgar Laird
Southwest Texas State University

Cindy Lammens
Universiteit Ghent

Jérôme Lamy
Observatoire de Paris

Harry G. Lang
Rochester Institute of Technology

Y. Tzvi Langermann
Bar Ilan University

James M. Lattis
University of Wisconsin at Madison

Françoise Launay
Observatoire de Meudon

Raimo Lehti
Tekniska Högskolan

Jacques M. Lévy
Observatoire de Paris

Li Di
Inner Mongolian Normal University

Kurt Liffman
Commonwealth Science and Industrial Research Organization (Australia)

Rudi Paul Lindner
University of Michigan

Jean-Pierre Luminet
Observatoire de Paris

Gene M. Lutz
University of Northern Iowa

Kirsten Lutz
Independent Scholar

Brian Luzum
United States Naval Observatory

Joseph F. MacDonnell
Holycross University

H. Clark Maddux
Indiana University at Kokomo

Jordan D. Marché, II
University of Wisconsin at Madison

Theresa Marché
University of Kutztown Pennsylvania

Tapio Markkanen
Tekniska Högskolan

Brian G. Marsden
Harvard-Smithsonian Center for Astrophysics

M. J. Martres
Observatoire de Paris

Ursula B. Marvin
Harvard-Smithsonian Center for Astrophysics

Sergei Maslikov
Tomsk State University

Kenneth Mayers
Universitet Bergen

Dennis D. McCarthy
United States Naval Observatory

John McFarland
Armagh Observatory

Robert D. McGown
Independent Scholar

Donald J. McGraw
University of San Diego

John M. McMahon
Lemoyne College

Marjorie Steele Meinel
National Aeronautics and Space Administration (USA)

John Menzies
South African Astronomical Observatory

Michael Meo
Independent Scholar

Raymond Mercier
Independent Scholar

Mark D. Meyerson
United States Naval Academy

Michael E. Mickelson
Denison University

Jan Mietelski
Universitas Iagellonica Cracoviensis

Cirilo Flórez Miguel
Universitad de Salamanca

Eugene F. Milone
University of Calgary

Kristian Peder Moesgaard
Steno Museet

Patrick Moore
British Broadcasting Company

Nidia Irene Morrell
Universidad Nacional de La Plata

James Morrison
University of Waterloo

Robert Morrison
Whitman College

Adam Mosley
Cambridge University

George S. Mumford
Tufts University

Marco Murara
Independent Scholar

Paul Murdin
Cambridge Institute of Astronomy

Negar Naderi
Encyclopaedia Islamica Foundation

Victor Navarro-Brotóns
Universidad de Valencia

Davide Neri
Università di Bologna

Claudia Netz
Independent Scholar

Christian Nitschelm
Universiteit Antwerpen

Peter Nockolds
Independent Scholar

Marilyn Bailey Ogilvie
University of Oklahoma

Takeshi Oka
University of Chicago

Timothy O'Keefe
University of Minnesota

Ednilson Oliveira
Universidade de Sao Paolo

Wayne Orchiston
Anglo-Australian Observatory

JoAnn Palmeri
University of Oklahoma

Kevin D. Pang
California Institute of Technology

Jay M. Pasachoff
Williams College

Naomi Pasachoff
Independent Scholar

Stuart F. Pawsey
Independent Scholar

Mariafortuna Pietroluongo
Università di Molise

Luisa Pigatto
Osservatorio Padova

Christof A. Plicht
Independent Scholar

Kim Plofker
Universität Utrecht

Patrick Poitevin
Independent Scholar

Roser Puig
Universitad de Barcelona

F. Jamil Ragep
McGill University

Sally P. Ragep
University of Oklahoma

Steven L. Renshaw
Kochi University

Michael Rich
University of California at Los Angeles

Lutz Richter-Bernburg
Universität Tübingen

Michael S. Reidy
University of Wisconsin

Peter Riley
University of Texas

Mònica Rius
Universitad de Barcelona

Leif J. Robinson
Sky and Telescope

Nadia Robotti
Università degli Studi di Genova

John Rogers
Cambridge University

Stanislaw Rokita
Planetarium Wladyslawa Dziewulskiego

Philipp W. Roseman
University of Dallas

Eckehard Rothenberg
Archenhold-Sternwarte

Marc Rothenberg
Smithsonian Institution

Tamar M. Rudavsky
Ohio State University

M. Eugene Rudd
University of Nebraska

Steven Ruskin
University of Notre Dame

David M. Rust
Johns Hopkins University

John J. Saccoman
Seaton Hall University

K. Sakurai
Kanagawa University

Michael Saladyga
American Association of Variable Star Observers

Julio Samsó
Universitad de Barcelona

Voula Saridakis
Virginia Technological University

Hüseyin Sarıoğlu
Istanbul University

Ke Ve Sarma
SSES Research Centre (India)

Gilbert E. Satterthwaite
Imperial College (UK)

Peggy Huss Schaller
Collections Research for Museums

Petra G. Schmidl
Johann Wolfgang Göthe Universität

Anneliese Schnell
Universität Wiena

Paul A. Schons
University of Saint Thomas

Ronald A. Schorn
University of Texas

Douglas Scott
University of British Columbia

Mary Woods Scott
Ohio State University

R. W. Sharples
University College of London

Stephen Shectman
Carnegie Observatories

William Sheehan
Independent Scholar

Steven N. Shore
Università di Pisa

Edward Sion
Villanova University

Lucas Siorvanes
King's College of London

Lorenzo Smerillo
Biblioteca Nazionale Protocenobio Sublacense

Charles H. Smith
Western Kentucky University

Horace A. Smith
Michigan State University

Laura Ackerman Smoller
University of Arkansas

Keith Snedegar
Utah Valley State College

Stephen D. Snobelen
University of King's College

Martin Solc
Univerzita Karlova

Kerstin Springsfeld
Rheinisch-Westfalische Technische Hochschule

Frieda A. Stahl
California State University at Los Angeles

Matthew Stanley
Iowa State University

Donn R. Starkey
Independent Scholar

David Strauss
Kalamazoo College

David J. Sturdy
University of Ulster

Woodruff T. Sullivan, III
University of Washington

Raghini S. Suresh
Kent State University

Jeff Suzuki
Brooklyn College

László Szabados
Konkoly Obszervatórium

Richard J. Taibi
Independent Scholar

Hidemi Takahashi
Johann Wolfgang Göthe Universität

Scott W. Teare
Mount Wilson Observatory

Pekka Teerikorpi
Turku University

Antonio E. Ten
Universidad de Valencia

Joseph S. Tenn
Sonoma State University

Antonella Testa
Università di Milan

Christian Theis
Universität Kiel

William Tobin
University of Canterbury

Hüseyin Gazi Topdemir
Ankara University

Roberto Torretti
University of Puerto Rico

Tim Trachet
Zenit

Virginia Trimble
University of California at Irvine & Las Cumbres Observatory

Jean-Louis Trudel
Université du Quebec

Giancarlo Truffa
Independent Scholar

Milcho Tsvetkov
Bulgarian Academy of Sciences

Pasquale Tucci
Università di Milan

Steven Turner
Smithsonian Institution

Arthur Upgren
Wesleyan University

A. Vagiṣwari
Indian Institute of Astrophysics

Ezio Vailati
Southern Illinois University

David Valls-Gabard
Observatoire de Paris

Glen Van Brummelen
Bennington College

Benno van Dalen
Johann Wolfgang Göthe Universität

Guido Van den Berghe
Universiteit Ghent

Petra Van der Heijden
Universiteit Leiden

Frans van Lunteren
Universiteit Utrecht

Steven M. van Roode
Independent Scholar

Ilan Vardi
California Institute of Technology

Yatendra P. Varshni
University of Ottawa

Gerald P. Verbrugghe
Rutgers University

Andreas Verdun
Universität Bern

Graziella Vescovini
Università di Firenze

Živa Vesel
Centre National de la Recherche Scientifique (France)

Jan Vondrák
Observatória na Skalnatom Plese

Bert G. Wachsmuth
Seaton Hall University

Christoffel Waelkens
Universiteit Leuven

Craig B. Waff
Independent Scholar

Glenn A. Walsh
Independent Scholar

Alun Ward
Independent Scholar

Gary A. Wegner
Dartmouth College

Gerald White
Independent Scholar

Raymond E. White
University of Arizona

Patricia S. Whitesell
University of Michigan

Sven Widmalm
Uppsala Universitet

Roland Wielen
Astronomisches Rechen-Institut

Christian Wildberg
Princeton University

Richard P. Wilds
Independent Scholar

Thomas R. Williams
Rice University

Thomas Nelson Winter
University of Nebraska

Peter Wlasuk
Florida International University

Bernd Wöbke
Max-Planck-Institut für Aeronomie

Lodewijk Woltjer
Observatoire de Saint Michel

Shin Yabushita
Nara Sangyo University

Keiji Yamamoto
Kyoto Sangyo University

Michio Yano
Kyoto Sangyo University

Hamid-Reza Giahi Yazdi
Encyclopaedia Islamica Foundation

Donald K. Yeomans
National Aeronautics and Space Administration (USA)

Robinson M. Yost
Iowa State University

Miloslav Zejda
Práce Hvezdárny a Planetária Mikuláše Koperníka

Endre Zsoldos
Konkoly Obszervatórium

Table of Entries

Names preceded by an article or preposition are alphabetized by the next word in the name. There are two exceptions: One is the Dutch "Van," "Van de," "Van den," and "Van der." Another is "Warren De La Rue" (alphabetized under D). (Arabic names are alphabetized under the shortened version of the name.)

If a name within the text appears in **bold**, there exists an entry on that astronomer elsewhere in the encyclopedia.

Introduction

History is the essence of innumerable biographies.
Thomas Carlyle, *Essays*, "On History"

Astronomy has a long and rich tradition, and as the record shows, the history of that tradition is tied closely to collective biography.[1] The present volumes represent a modern attempt to provide a comprehensive biographical encyclopedia of astronomers. The purpose of these volumes is twofold. First, as ready reference, they are designed to provide easy access to biographical information in the history of astronomy. Cutting across space and time, biographical entries are international in scope and cover the period from classical Antiquity to the late 20th century. Second, drawing on a variety of specialized scholars, these volumes aim to serve as an "access point" for continuing research. While individual entries "stand alone" as ready reference, taken collectively, they offer a map of the complex communities that gave science shape.[2] The following introduction has two purposes: first, to sketch the origins of collective biography and its place in the history of astronomy; second, to illustrate the design and use of collective biographies as reference and research tools.

Biography And History

There is properly no history, only biography.
Ralph Waldo Emerson, *Essays*, "History"

History—here I mean historical writing—traces its origins to classical Antiquity, to the celebration of heroes and the lives of great men. Although *lives* were written before Plutarch's aptly titled classic, the modern sense of biography—a fair-minded history of a particular life—took mature form only in the 19th century.[3] The history of writing lives challenges the boundaries that currently separate history, biography, literature, rhetoric, and political commentary. While the roots of modern biography can be traced to the Renaissance (including early examples of science biography), sharp distinctions between "history and biography" are difficult to sustain, not only because the categories continue to overlap but because both share a common ancestor—what we now call collective biography.[4] As background to the present volumes, the following historiographic essay sketches these changing relations.[5]

The origins of *biography* (literally, *life writing*) are found in classical Antiquity as part of a long tradition dedicated to the celebration of heroes.[6] For two millennia, what we now know as *history* was often viewed as philosophy teaching by example. A brief glance at early writers suggests that biography and collective biography share a complex evolution. While Damascius (sixth century) was the first writer to use the Latin term *biographia*, John Dryden was the first to use *biography* in print (1683), this in reference to Plutarch's *Lives*. Words are important but much more was at work. Viewed over time, historical writing included what is now known as history, biography, and collective biography, as well as elements from other branches of the humanities and social sciences.

Biography has served many masters. Between Antiquity and the Renaissance, its main role was to tell the lives of statesmen, philosophers, and saints. As a display of literary and rhetorical skill, its principal aim was to instruct and inspire. Among ancient Greek and Latin authors, the biographical art is evident in the *Lives* of Critias, the *Memorabilia* of Xenophon, the *Lives of the Philosophers* by Diogenes

[1] I wish to thank the *BEA* Editorial Board for the invitation to write the Introduction. While I have contributed several articles in these volumes, I have had no role in designing or editing the present work.

[2] Collective biography invites the reader to explore the interplay of individuals, ideas, and groups. One scholar went further: "In group biography, one becomes defined by the many. The group biography in fact becomes a protest against the erosion of a viable communal life and marks the socialization of biography as it incorporates several lives, not a single life." Nadel, Ira Bruce (1984) *Biography: Fiction, Fact & Form,* New York, p. 192.

[3] See *Telling Lives: The Biographer's Art,* Marc Pachter, ed., Philadelphia, 1979; *Telling Lives in Science: Essays on Scientific Biography,* Eds. M. Shortland and M. Yeo, Cambridge, 1996; Edmund Gosse, "Biography," in *Encyclopaedia Britannica,* 11th Edition (New York, 1910) Vol. 3: 952–954; Virginia Woolf, "The Art of Biography," *The Atlantic Monthly* 163 (1939): 506–510; and Sidney Lee, "Principles of Biography." *Elizabethan and Other Essays.* Oxford, 1927: 31–57.

[4] Collective biography—short sketches of individual lives representing a group—is a recent term that might be applied to earlier traditions. Collective biography is sometimes associated with prosopography, a method used by social scientists and social historians based on data from collective biography. For an overview, see Helge Kragh, "Prosopography," *An Introduction to the Historiography of Science,* Cambridge, 1987, pp. 174–181. As an example of trends in a specific historical field, see *Fifty Years of Prosopography: The Later Roman Empire, Byzantium and Beyond,* Ed. Averil Cameron, Oxford, 2003.

[5] Historiography—the history of historical writing—suggests that history, biography, and collective biography share common roots. For background, see Herbert Butterfield, "Historiography," *Dictionary of the History of Ideas,* Vols. 2, (New York, 1973): 464–498; for history of science, see John R. R. Christie, "The Development of the Historiography of Science," *Companion to the History of Modern Science,* London and New York, 1990, pp. 5–22, and Helge Kragh, *An Introduction to the Historiography of Science,* Cambridge, 1987.

[6] Over time, biography seized on the individual character of virtue and vice; collective biography celebrated group achievement by virtue of vocation. A counter example is *Catalogus Hereticorum* (1522?) by Bernardus de Lutzenburg, which devotes two chapters to heretics and their errors.

Laertius, Plutarch's *Parallel Lives*, and Suetonius's *Lives of the Twelve Caesars*.[7] It should be noted that these authors are often not identified as historians, but as scholars, poets, or letter writers. When we consider the best-known early historians—from Herodotus (*circa* 480–*circa* 430 BCE) and Thucydides (*circa* 460–400 BCE) to noted writers such as Pliny (23–79), Livy (59 BCE-17), and Vespasiano (1421–1498)—short biography was an essential element in their annals and accounts.[8]

Origins of Modern Biography

The origins of modern biography—the first sustained attempts to write the life of a single individual—can be traced to the Renaissance. The earliest examples were literary. William Roper (1496–1578) wrote the life of Sir Thomas More, George Cavendish (1500–1561?), the life of Cardinal Wolseÿ later, Izaak Walton published a series of biographies, including the life of John Donne (1640).[9] Collective biography also found favor as poets, artists, and scholars joined ranks with statesmen, saints, and kings.[10] Thomas Fuller's *History of the Worthies of England* (1662) extended earlier traditions into more secular territory, while Aubrey's *Minutes of Lives* (its working title) is still widely read today. An early member of the Royal Society, John Aubrey (1626–1697) became interested in biography through his friend, Anthony à Wood (1632–1695), in researching the latter's *Athenae Oxonienses* (1691–1692), a "living and lasting history" of Oxford University based on group biography.[11] The more widely read work is now known as Aubrey's *Brief Lives*.[12] Although Wood judged him "credulous," Aubrey wrote vivid and often intimate biographical sketches, including a number of figures from the New Science—Robert Boyle, René Descartes, Edmond Halley, Thomas Hobbes, Robert Hooke, Nicolas Mercator, and Christopher Wren. Aubrey interviewed many of his subjects. In retrospect, a key problem was the scarcity of personal diaries and journals, as the publication of memoirs and letters was not yet fashionable.[13] Aubrey's contemporary, Thomas Sprat (1635–1713), wrote the *Life of Cowley* (1668) and his better-known *History of the Royal Society* (1667).[14] Drawing on institutional registers and journals, Sprat sprinkled his *History* with short biographies. His aim was to provide living proof of the "usefulness" of "true philosophy." Institutional histories have since used collective biography as a key component in their narratives.

Biography—indeed "science biography"—took recognizable form with the work of Pierre Gassendi (1592–1655). A noted philosopher and astronomer, Gassendi was among the first to write the lives of individual astronomers. An advocate of the New Science, Gassendi employed his knowledge of nature and the language skills of a classical scholar. According to his English translator, Gassendi was "comparable to any of the ancients."[15] His versatility served him well in telling the lives of Nicolaus Copernicus and Tycho Brahe, as well as Georg Peurbach and

[7] As one example of recent scholarly treatment of ancient biography, see Tomas Hägg and Philip Rousseau, Eds. *Greek Biography and Panegyric in Late Antiquity. The Transformation of the Classical Heritage*, 31. Berkeley, 2000. Examples from other periods include David J. Sturdy, *Science and Social Status: The Members of the Académie des sciences, 1666–1750*. Rochester, New York, 1995 and Frank A. Kafker, *The Encyclopedists as a Group: A Collective Biography of the Authors of the "Encyclopédie."* For an overview of key issues, see Clark A. Elliott, "Models of the American Scientist: A Look at Collective Biography." *Isis*, Vol. 73, No. 1 (March, 1982): 77–93.

[8] From preclassical times, the transition from oral traditions, epics, and story telling (understood as historical literature) was accompanied by the production of records. In addition to annals and chronologies, the earliest forms of government required dynastic lists, while legal considerations of inheritance (as one example of precedence) called for extended genealogies. Between Greek and Roman writers, early forms of historical writing would now be classified as political commentary, contemporary history, or history of the times. Cicero expresses the Roman ideal of the historian as a writer who seeks motives, portrays individual character, analyzes results, and who "supports the cause of virtue and moves the reader by literary artistry." (Herbert Butterfield, "Historiography." *Dictionary of the History of Ideas*, 5. Vols., New York, 1973, Vol. 2: 464–498, p. 470.) Butterfield summarizes the view of Tacitus: "the deeds of good men ought not to be forgotten and that evil men ought to be made to fear the judgment of posterity." "Historiography," p. 479.

[9] He also wrote biographies of Henry Wotton (1651), Richard Hooker (1665), George Herbert (1670), and Robert Saunderson (1678).

[10] A late 16th-century writer lamented: "For lives, I find it strange, when I think of it, that these our times have so little esteemed their own virtues, as that the commemoration and writings of the lives of those who have adorned our age should be no more frequent. For although there be but few sovereign kings or absolute commanders, and not many princes in free states (so many free states being now turned into monarchies), yet are there many worthy personages (even living under kings) that deserve better than dispersed report or dry and barren eulogy." Thomas Blundeville, *The True Order and Method of Writing and Reading Histories*, London, 1574 (no pagination), quoted in *Versions of History from Antiquity to the Enlightenment*, Ed. Donald R. Kelley, New Haven, 1991, 397–413, p. 407.

[11] Wood's *History*, prompted by his friend, Dr John Fell, dean of Christ Church, brought him much fame and notoriety. His grand project, the *Athenae Oxonienses*, was essentially a biographical dictionary mixing historical narrative, collective biography, and bio-bibliography. Assisted by Aubrey and Andrew Allam (neither adequately acknowledged), Wood drew on a variety of printed sources ranging from published works to institutional documents from libraries, archives, and governmental offices. John Fell, influential with the university press, assisted with publication. Wood was eventually sued for libel and removed from the university.

[12] Aubrey's *Lives*, written between 1669–1696, exists in four folio manuscript volumes. The public appearance of the *Lives* has a complicated publishing history. While early editions appeared in the late 18th century, an early standard edition appeared only in 1898. John Aubrey. *"Brief Lives," Chiefly Contemporaries, set down by John Aubrey, between the years 1669 & 1696*. Edited by Andrew Clark. 2 Vols. Oxford, 1898.

[13] Diaries and letters are critical resources for biographers and historians. The best known diaries of this period, published centuries later, include *The Diary of Robert Hooke* (Eds. H.W. Robinson and W. Adams, 1935); *The Diary of Samuel Pepys*, 11 Vols. (Eds. R. Latham and W. Matthews, 1970–1983); and *The Diary of John Evelyn*, 6 Vols. (Ed. E.S. de Beer, 1955–). Publication of personal and scholarly letters began in the 17th century. Early efforts include the letters of N-C Fabri de Peiresc, Galileo Galilei, Johannes Hevelius, and René Descartes, among others.

[14] Thomas Sprat. *The History of the Royal-Society of London, for the Improving of Natural Knowledge*. London, 1667. Sprat's polemic for the New Science is thematic, philosophical, and passionate. His use of biography is not central to his arguments but ever-present in illustrating his claims.

[15] Gassendi's *Vita*, discussed more fully below, was translated by William Rand and published as *The Mirrour of True Nobility & Gentility* (London, 1657).

Johannes Regiomontanus.[16] In retrospect, Gassendi's success was linked to an emerging biographical principle, to portray the "conjunction of life and mind."[17] Like other contemporaries, Gassendi used history to support his scientific claims while shedding light on the inner workings of science.[18] His most cited biography is a tribute to his friend and patron, Nicolas-Claude Fabri de Peiresc (1580–1637). A noted humanist scholar and amateur of science, Peiresc collaborated with Gassendi in astronomy and in conducting optical experiments. Gassendi's biography portrays Peiresc's motives for studying nature and the relation between his personality and worldview. One of the first biographies translated from Latin into English, Gassendi's *Mirrour of True Nobility* (W. Rand, trans., 1657; *Vita* 1641) has been favorably compared to a later classic biography, Boswell's *Life of Johnson* (1791). Gassendi met Boswell's strictest criteria: Boswell's masterpiece is an intimate and telling portrait; it clearly shows that the biographer and subject had "ate, drank, and communed."[19]

Boswell's *Life of Johnson* established biography as a legitimate form of historical writing. Importantly, Boswell's central interest in Johnson's life was to portray the "progress of his mind"—to tell his story accurately but not without passion. For Boswell, in "every picture there must be shade as well as light," and while not wishing "to cut his claws nor make a tiger a cat," his portrait of Johnson included all the "blotches and pimples."[20] Boswell transformed biography into a conventional and fashionable form of historical writing.

By the 19th century, biography gained maturity and great prestige. It was here, in the Century of Science, that a new genre appeared. It is now called "science biography." In the century that followed, particularly after World War II, numerous science biographies appeared. They celebrated traditional heroes as well as obscure figures. Classic studies of Isaac Newton, to take the oldest tradition, illustrate important shifts in the objectives of science biography. Since his death, Newton has been the subject of dozens of studies, from early hagiographic accounts to modern archive-based interpretations devoted to "Newton the Man."[21] Newton posed problems for biographers from the outset, particularly as unknown manuscripts came to light betraying his passion for alchemy, religion, and prophecy. Heralded as the "Splendid Ornament of Our Time" by Sir Edmond Halley, "High Priest of Science" by Sir David Brewster, and "Last of the Magicians" by Baron John Maynard Keynes, Newton's many faces continue to challenge traditional assumptions about the proper relation between science and biography. Despite differences and continuing debate, scholars agree that biography should leave readers less worshipful and more intrigued.[22]

The distinction between biography and history is a modern development. Although both share a common ancestor—and a strong family resemblance—each has a distinct physiognomy. To overstate a difference, biography stems from the belief that history is made by human beings, not by abstract ideas or impersonal forces. Equally overstated, history emphasizes the view that larger themes, trends, and movements account for change. In brief, if biography is a solo instrument, history is an orchestra. The limits of either perspective (assuming such distinctions can be sustained) are clear. In either case, authors assume a point of view. Biographers take the view that life is not encountered

[16] Latin versions appeared in several editions, the first in Paris (1654), the second in The Hague: Pierre Gassendi, *Tychonis Brahei, equitis Dani, astronomorum coryphaei, vita… Accessit Nicolai Copernici, Georgi Peurbachii, and Ioannis Regiomontani, astronomorum celebrium, vita*. Hagae Comitum (Vlacq) 1655.

[17] See Gassendi's introductory letter to Jean Chapelain in the Preface to Peurbach and Regiomontanus.

[18] Chronology was an important element in the New Science. Practitioners include not only Johannes Kepler and Issac Newton but an extraordinary group that mixed classical studies with advanced skills in astronomy, among them Joseph Scaliger, Wilhelm Schickard, Ismaël Boulliau, J-F Gronovius, John Greaves, Edward Bernard, Nicolas Heinsius, John Bainbridge, Sir Christopher Heydon, J-H Boecler, Henry Savile, James Ussher (archbishop of Armagh), Vincenzo Viviani, and Edmond Halley.

[19] Pierre Gassendi. *The Mirrour of True Nobility & Gentility, Being the Life of the Renowned Nicolaus Claudius Fabricius Lord of Peiresk, Senator of the Parliament at Aix*. Trans. W. Rand, London, 1657.

[20] The phrase "warts and all" biography (perhaps derived from Boswell's "blotches and pimples") resonates with Walt Whitman's charge to his biographer, "… do not prettify me: include all the hells and damns."

[21] The first full-scale biography of Isaac Newton was written by Sir David Brewster (1781–1868), the noted physicist and journalist. Brewster's first excursions in biography were popular. But as author of *The Life of Sir Isaac Newton* (1831) and *Martyrs of Science: Lives of Galileo, Tycho Brahe and Kepler* (1841), Brewster soon found himself defending his principal hero. In 1822, the French astronomer J-B Biot (1822) made claims that Isaac Newton was intellectually crippled by mental illness, and hinted at Newton's questionable moral behavior. A decade later, Francis Baily made much of Newton's unfairness in his *Account of the Revd John Flamsteed* (London, 1835). To defend Newton, Brewster gained access to little-known Newton manuscripts in the Portsmouth Collection (and Hurstbourne Collection). Much to his surprise, Brewster unearthed evidence that linked Newton to unorthodox religious and alchemical views. The result was Brewster's *Memoirs of the Life, Writings and Discoveries of Sir Isaac Newton* 2 Vols. (1855). On balance, Brewster did little to respond to the substance of the claims by Biot and Baily, essentially ignoring Newton's alchemy while denying Newton's illness of 1693. Some 80 years later, L.T. Trenchard More blasted Brewster's approach in his *Isaac Newton: A Biography* (1934). Charging him with playing the role of advocate to "The High Priest of Science," More claimed that Brewster made "almost no attempt to present Newton as a living man or to give a critical analysis of his character" (*Newton*, pp. vi–vii). Into this debate next came the noted economist, John Maynard Keynes (1883–1946). A wealthy collector of rare manuscripts, Keynes acquired hitherto unknown manuscripts of Isaac Newton on alchemy and religion. On the basis of these documents, Keynes famously proclaimed that "Newton was not the first of the age of reason. He was the last of the magicians" ("Newton the Man," 1947, *Newton Tercentenary Celebrations*, 1947, pp. 27–34). A generation later, the noted historian Frank Manuel published an important trilogy, *Isaac Newton, Historian* (1963), *The Religion of Isaac Newton* (1974), and *A Portrait of Isaac Newton* (1968)—a brilliant but controversial psycho-biographical study. Two decades later, a Newtonian synthesis of sorts appeared, *Never at Rest, A Biography of Isaac Newton* (Cambridge, 1980) by Richard S. Westfall. As Newton's biographer, Westfall aimed to "present his science, not as the finished product … but as the developing endeavor of a living man confronting it as problems still to be solved" (p. x). Westfall's credo captures the modern sense of science biography. Subsequent biographers have followed suit. In his *Isaac Newton, Adventurer in Thought* (London, 1992), A.R. Hall suggests the problem with earlier approaches was that the "mythical Newton, a new Adam born on Christmas Day and nourished by an apple from the tree of knowledge, came to obscure the real man who had worked in dynamics, astronomy, and optics" (p. xii). A number of important studies continue to appear. Although the biographical tradition surrounding Newton is longstanding, it shares important similarities with subsequent biographic traditions associated with Charles Sigmund Albert, Darwin, Freud, and Einstein.

[22] Thomas L. Hankins, "In Defence of Biography: The Use of Biography in the History of Science." *History of Science*, 17: 1–16. See also Helge Kragh, "The Biographical Approach," in H. Kragh, *An Introduction to the Historiography of Science*, Cambridge, 1987, 168–173.

as a category or theme. Although it focuses on an individual life, biography can be used as an historical lens to refract the full range of human experience—from individual aspirations to enduring achievements. Those who write "science biography" often aim to show how scientists go about their business, how ideas and theories emerge, and how life and work make a coherent whole. In the end, most readers recognize that biography can be honest without telling the whole truth.

Modern Collective Biography

A biography should either be as long as Boswell's or as short as Aubrey's.
Lytton Strachey

Collective biography—short sketches of individual lives representing a group—traces its roots to classical Antiquity, and since then it has been popularized, institutionalized, and widely embraced.[23] Collective biography has a long tradition of telling the story about science "in the making." Since the time of Aristotle, authors have taken pains to record the efforts of predecessors (if only to show how misguided their views) just as modern authors have summoned ancient authors to support new theories. Applied to astronomy, an important assumption of collective biography is that "astronomy" is not only a body of knowledge but a body of people. It addresses individual lives as well as forms of life. Taken collectively, most astronomers—observers, mathematicians, calculators, astrologers, speculative philosophers—were not heroic figures. While few historians doubt the significance of Newton, many are persuaded of the importance of minor figures.[24] Scholars continue to debate the appropriate balance between individuals and groups.

The history of astronomy—like other scholarly specialities—is inseparably linked to collective biography. Among the early pioneers in this genre, two deserve brief mention: Giovanni Battista Riccioli (1598–1671) and Edward Sherburne (1618–1702). Echoing tradition in his title, Riccioli's *Almagestum novum* (Bologna, 1651) was not the first work to use history as evidence for his cosmological views.[25] Engaged in the great debate over the Ptolemaic, Tychonic, and Copernican world systems, Riccioli used history to tip the scales in favor of an Earth-centered model. A Jesuit by training, Riccioli published his two-volume work in defense of charges leveled against Galileo Galilei (1616 and 1633). Riccioli heaped new observations on old theories to support the Tychonic model.[26] To counter Copernicus's claims, Riccioli marshaled an army of believers in the immobility of the Earth, and not surprisingly, the Copernicans were vastly outnumbered.[27] Working old arguments into a new narrative, Riccioli used history and biography in what amounted to a Copernican counter-reformation. Riccioli's collective biography contains some 400 astronomers from Antiquity to his own age. It fills 20 folio pages—in small type.[28]

Appearing several decades later, Edward Sherburne's *Sphere of Marcus Manilius* (1675) contains the first modern collective biography of astronomers.[29] Responding to wide-spread interest in the ancient astrologer Manilius (flourished 10), Edward Sherburne (1618–1702) presented the first English translation of Book One of the *Astronomicon*, and along with it, his remarkable "Catalogue of the Most Eminent Astronomers, Ancient & Modern." It was a model for future collective biographies. Following earlier traditions,[30] Sherburne's *Astronomical*

[23] As one recent scholar summarized, "Initially, the analytic life was a minority voice as large, multivolume biographies dominated Victorian lives. However, a tradition originating in short Latin lives, renewed by antiquaries of the 16th century, popularized by Aubrey's *Brief Lives* in the seventeenth, dignified by Johnson's *Lives of the Poets* in the eighteenth, and culminating in works like Strachey's *Portraits in Miniature* in the twentieth, reasserted the centrality of the brief life. In the 19th century, the form reached its apogee in collective lives, biographies in series and biographical dictionaries. Their extraordinary sales and continued influence is a measure of their importance." Ira Bruce Nadel, *Biography: Fiction, Fact & Form*, New York, 1984, p. 13.

[24] One reviewer of the *Dictionary of Scientific Biography* wrote, in some sense "obscure second-rate scientists are as important as, and probably even more significant than, scientific geniuses" given (in his view) that "the real subject matter of the history of science is not the individual scientist, but the scientific community as a whole." Jacques Roger, "The *DSB*: A Review Symposium," *Isis*, 71 (1980): 633–652, p. 650.

[25] Giovanni Battista Riccioli. *Almagestum novum, astronomiam veterem novamque complectens,* (2 Vols.) Bologna, 1651.

[26] The Tychonic model can be described as geocentric and geo-static, and more accurately as geo-heliocentric. A geo-heliocentric model has the planets to revolve around the Sun, but in turn, the Sun revolves annually around the central and stationary Earth. Geo-heliocentric models were in principle observationally equivalent to a heliocentric model. Viewed in context, they served as an intelligent alternative rather than as a "compromise" cosmology. See M.A. Hoskin and Christine Jones. "Problems in Late Renaissance Astronomy." *Le Soleil a la Renaissance*. Paris, 1965. Further details about the history and various mutations of the geo-heliocentric model can be found in Christine Schofield-Jones' doctoral dissertation.

[27] If theory selection is based on *Numerus, Mensura, Pondus*, historians have mused over the number, size, and weight of Riccioli's arguments. By one reckoning, J-B Delambre counted some 57 arguments against a moving Earth. For his part, Riccioli claims "40 new arguments in behalf of Copernicus and 77 against him." See J-B Delambre, *Histoire de l'Astronomie Moderne*, Vol. 1, Paris, 1821, pp. 672–681 and G-B-Riccioli, *Almagest novum*, 2 Vols., (Bologna, 1651). See Volume 2, Section 4, Ch. 1, pp. 290 *et seq.*, where Riccioli expands his list of Copernicans and non-Copernicans weighing arguments for and against a moving Earth; see also pp. 313–351. For Riccioli's reckoning of the number of arguments, see *Apologia pro Argumento Physicomathematico contra Systema Copernicanum adiecto contra illud Novo Argumento ex Reflexo motu Gravium Decidentium*. Venice, 1669; Dorothy Stimson, *The Gradual Acceptance of the Copernican Theory of the Universe*, New York, 1917, pp. 79–84, provides a general discussion.

[28] Riccioli. *Almagestum novum*, Pt I. Following a historical narrative, Riccioli offers a chronological outline of astronomy (xxvi–xxviii) followed by an alphabetical list of over 400 astronomers (xxviii–xlvii). Entry length varies from a few lines to nearly a full page in the case of Tycho Brahe. Though long and often laborious (over 1,500 pages), Riccioli's volumes provide one of the best introductions to the history of astronomy up to his time. Technically skilled and historically inclined, Riccioli provides useful perspectives on contemporary authors, including Copernicus, Brahe, Longomontanus, Kepler, Galilei, Boulliau, and others.

[29] Edward Sherburne, *The Sphere of Marcus Manilius made an English Poem with Annotations and an Astronomical Appendix* (London, 1675).

[30] The more noted early astronomer-historians include Schickard, Gassendi, Riccioli, Boulliau, Viviani, and eventually Halley.

Appendix (pp. 1–126) contains some 1,000 biographical entries, varying from several lines to several pages. Less polemical than Riccioli, Sherburne's purpose was no less passionate. He aimed to tell the story of the "origins and progress" of astronomy from the very beginning—literally, from Adam (5600 BCE). Sherburne's Catalogue contains detailed information about a large number of his friends and colleagues, and it remains useful for historians evaluating contemporary issues and reputations. Young Isaac Newton, as one example, receives a surprisingly short entry—easily dwarfed by those of Tycho and Hevelius.[31]

Collective biography came of age in the 17th century. Although writers continued to celebrate political and religious figures, a shift took place with the appearance of works on artists and scholars as well as advocates of the New Science. During the previous century, Konrad Gesner (1516–1565) published his pioneering *Bibliotheca Universalis* (Zürich, 1545–1549), Giorgio Vasari (1512–1574) his *Lives of the Artists*, and extending a long tradition, the *Acta Sanctorum* (1643 *et seq.*) swelled to 68 folio volumes. This monumental work gave new meaning to the word hagiography.[32] Toward the end of the century, men of learning again took center stage with the appearance of Charles Perrault's *Les hommes illustres*,[33] and soon thereafter, J-P Nicéron's *Mémoires pour servir à l'histoire des hommes dans la République des Lettres* (1729–1745, Paris). Both works included biographies of astronomers.[34]

The most comprehensive work of the century was published by Louis Moréri (1643–1680), *Le Grand Dictionnaire historique* (Lyon, 1671).[35] Unprecedented in scope and rigor, Moréri established new possibilities. For present purposes, while it contained biographies of all the major astronomers up to that day, Moréri's *Dictionnaire* represented unprecedented opportunities for combining history and biography.[36] First published in French, his *Dictionnarie* was soon translated into English, German, Italian, and Spanish, and within a century (1671–1759), some twenty editions appeared.[37] The success of Moréri's work was followed by an avalanche of encyclopedias and dictionaries that constituted an intellectual movement in itself. Less widely noted, the encyclopedia movement was paralleled by the publication of scholarly *Èloges*, most notably by Bernard de Fontenelle (1657–1757) and subsequent secretaries of the French Académie des sciences.[38] Certainly one of the most influential works of the century was the *Dictionnaire historique et critique* (4 Pts, 2 Vols., Rotterdam, 1697) of Pierre Bayle (1647–1706). Later called the "Arsenal of the Enlightenment," Bayle's *Dictionnaire* appeared in five editions over the next 50 years, not including an influential English translation (2nd Edition, 1734–1738).[39] Praised for its topical articles (particularly on reforming religion, philosophy, and politics), Bayle's *Dictionnaire* was less comprehensive than Moréri, and while prone to philosophical polemics, its influence was immense. Like Moréri, Bayle included important biographies on noted thinkers, many associated with the New Science, astronomy, and cosmology. By tradition, Bayle's *Dictionnaire* foreshadowed the *Encyclopédie*, an Enlightenment showcase designed by Denis Diderot (1713–1784), Jean D'Alembert (1717–1783), and other advocates of toleration and reform. The influence of the *Encyclopédie* in transforming political, social, and intellectual institutions would be difficult to overstate. Aided by dramatic increases in literacy, the explosive growth of the printing press, wider use of the vernacular, and the proliferation of learned journals, scholars joined the public sphere as never before, often pointing to Bacon, Galilei, and Descartes as models of free thinking and useful knowledge.[40] Historical evidence and philosophical principle soon became equal partners in political polemics. By the end of the century, collective works multiplied across national boundaries, among the most important, the *Encyclopaedia Britannica* (3 Vols., Edinburgh, 1771) and Chamber's *Cyclopaedia*

[31] Sherburne, *The Sphere*, Brahe, p. 63; Hevelius, pp, 110–111; Newton, p. 116.

[32] Hagiography can be described as a literary tradition devoted to telling the lives of ecclesiastical figures, notably martyrs and saints canonized by the Church of Rome. Hagiography has since gained a heroic connotation associated with "secular saints" such as Newton, Darwin, Freud, and Einstein.

[33] Charles Perrault. *Les hommes illustres qui ont paru en France pendant ce siècle avec leurs portraits au naturel*, 2 Volumes (1697 and 1700, Paris).

[34] Jean-Pierre Nicéron. *Mémoires pour servir à l'histoire des hommes dans la République des Lettres* (1729–1745, Paris).

[35] Louis Moréri. *Le Grand Dictionnaire historique, ou le mélange curieux de l'histoire sacrée et profane*, (Lyon, 1671 *et seq.*).

[36] The Moréri edition of 1759, for example, contains biographies of astronomers from Antiquity through the early 18th century, among them, Boulliau 2: 137; Copernicus 4: 105–106; Cunitz 4: 324; Descartes 4 (2): 115–119; Galilei 5 (2): 32–33; Kepler 6 (2): 17–18; Mersenne 7: 488; Brahe 10: 181–182; as well as Newton 8: 1001–1002 and other countrymen, Wallis 10: 756; and Ward 10: 764–765. Several articles are particularly noteworthy, for example, the early reception of Descartes's work in universities and subsequent controversies with church authorities is both thorough and unprecedented; the article on J-B Morin contains unique information and is nuanced in interpretation; and Newton is already showing signs of icon status, heralded as one of "the most learned men of our age." The Moréri edition is noteworthy for high standards; articles often quote from primary sources and occasionally from unpublished letters and manuscripts.

[37] Subsequent editions appeared under the editorship of C-P Goujet (1697–1767) and E-F Drouet (1715–1779).

[38] The impulse to publish these éloges (biographies of deceased men of learning) came from several directions. The *éloge* of the French Académie des sciences show similarities with earlier biographical traditions. As idealized portraits "extolling the moral virtues of the post-Renaissance sciences" (p. ix) they represent, as Charles B. Paul has argued, a classic form of collected scientific hagiography. Re-inventing an old tradition, Fontenelle (1657–1757) and his successors (Mairan, Fouchy, and Condorcet) published over 200 posthumous eulogies of Académie members during the 18th century. As commemorative pieces, they underscored societies' debt and popularized the belief that scientists were modest, dedicated, disinterested seekers after truth devoted to social improvement and human progress. See Charles B. Paul, *Science and Immortality: The* Èloges *of the Paris Academy of Sciences (1699–1791).* Berkeley, 1980.

[39] Pierre Bayle. *Dictionnaire historique et critique*, Rotterdam, 1697, fol. 2 Vols. Many editions followed: a second edition (3 Vols., Amsterdam, 1702); the fourth edition (4 Vols., Rotterdam, 1720), edited by Prosper Marchand; and a ninth edition in 10 Volumes appearing shortly thereafter. The second edition of the *Dictionnaire* was translated into English (4 Vols., London, 1709), and later the fifth edition (1730) was translated by Birch and Lockman (5 Vols., London, 1734–1740). Other editions with supplements and additional translations followed, among them a German translation (4 Vols., Leipzig, 1741–1744), with a preface by J.C. Gottsched. It is widely reported that Bayle undertook his *Dictionnaire* due to unacceptable errors and omissions found in Moréri. Later editions of Moréri show a remarkable level of scholarship.

[40] In his *Preliminary Discourse to the Encyclopedia of Diderot* (1751) d'Alembert rehearsed the "traditional litany" of heroes from the scientific revolution (traditionally Copernicus to Newton) explaining how "a few great men … prepared from afar the light which gradually, by imperceptible degrees, would illuminate the world" (Ed. R. Schwab, New York, 1963), p. 74. Voltaire echoed a similar view in his famous chapter on the "Academies" in his *Age of Louis XIV* (*Le Siècle de Louis XIV*, 1751).

(2 Vols., London, 1728).[41] By the end of the century, the publication of private letters of individuals—literary, political, philosophical—became fashionable as learned conversation and salon gossip found its way into print.

The 19th century saw an explosion of multivolume publications. Among them, a new tradition began to emerge with the publication of the complete works of individual scientists—*opera omnia*, collected papers, and published correspondence. Intellectuals increasingly entered the public sphere. One of the early landmarks reflecting the Republic of Letters was the *Biographie universelle ancienne et moderne* (52 Vols. Paris, 1810–1828), edited by J-F Michaud (1767–1839).[42] Spanning time and space, Michaud's *Biographie* remains one of the most enduring universal dictionaries of all time. Boasting high scholarly standards, it is composed of substantial articles signed by eminent authors. As one example, the article on Newton, written by the well-known physicist, Jean-Baptiste Biot (1774–1862), became a symbol of the international and increasingly controversial character of celebrity.[43] As local heroes gained international status, national reputations were hotly disputed. Astronomers were well represented.[44]

An extreme example—finally affecting reputations of both the living and the dead—involved the French mathematician, Michel Chasles (1793–1880), the noted Copley Medalist and Member of the Académie des sciences.[45] In 1867, Chasles claimed that his celebrated countryman, Blaise Pascal (1623–1662), had sent letters (hitherto unknown) to young Isaac Newton during the years 1654–1661. In effect, Chasles suggested that the French mathematician had handed over the secret of the Universe—the law of universal of gravitation—to an Englishman. The dispute that followed involved two years of public wrangling and scholarly exchanges between Newton and Galilei experts—finally followed by a trial and prison sentence. In the end, Chasles came to discover (along with an international audience) that his claims were based on false documents forged by one Vrain-Denis Lucas (1818- *circa* 1871).[46] Chasles eventually acknowledged that he had been duped, swindled, and humiliated.[47] The *Affaire Vrain Lucas* is an extreme example of historical celebrity and national pride gone awry, a dramatic reminder that biography, like other forms of historical writing, is always written from a perspective.

A watershed in collective biography came with specialized dictionaries devoted to individual countries.[48] These "national biographies" have since become showcases of scholarship and—increasingly—for international cooperation. Following a century of political conflict and upheaval, the great national biographies stemmed from a sense of pride and patriotism. First appearing in the early decades of the 19th century, major national biographies began to appear across Europe, from the great universal dictionary of Moréri in France (52 Vols., 1810–1828) to the national dictionaries of Sweden (23 Vols., 1835–1857); the Netherlands (24 Vols., 1852–1879); Austria, 35 Vols., (1856–1891); Belgium (35 Vols., 1866–); Germany (45 Vols., 1875–1900); Great Britain (63 Vols., 1882–1900); the United States (30 Vols., 1928–1936; 1994); France (19 Vols., 1933–); and Italy (59 Vols., 1960–).[49] Although defined geographically, national biographies can be an invaluable resource of information on astronomers, whether major or minor figures.

Among the national biographies that dominated 19th-century scholarly publication, the most eminent was the widely celebrated *Dictionary of National Biography* [*DNB*] (1882–1900). The *DNB* soon became a symbol of scholarly collaboration, not unlike the

[41] Ephraim Chambers, *Cyclopaedia; or an Universal Dictionary of Art and Sciences, containing an Explication of the Terms and an Account of the Things Signified thereby in the several Arts, Liberal and Mechanical, and the several Sciences, Human and Divine*, London, 1728, fol. 2 Vols. A noted example of publishing letters of the learned is Angelo Fabroni, *Lettre inedite di uomini illustri*, 2 Vols. Florence, 1773 and 1776.

[42] [Joseph-François] Michaud, *Biographie universelle ancienne et moderne*, 52 Vols., Paris, 1810–1828 (32 supplement Volumes); a good deal of the work was completed by his younger brother, Louis-Gabriel Michaud (1773–1858). A second revised edition appeared in 45 Volumes (Paris, 1843–1865).

[43] J-B Biot, "Isaac Newton," *Biographie Universelle*, Vol. 30: 366–404. As noted above, Biot raised important questions about Newton's mental illness—hinting at his beliefs in alchemy and religion—which later spurred a defense by Sir David Brewster as well as a growing tradition of scholarly debate.

[44] Michaud and subsequent editors enlisted the most noted scholars of the day as contributors. Several noted biographies of astronomers were written by J-B Delambre (Kepler; Boulliau; A-G Pingré) and by J-B Biot (Copernicus; Galilei; Newton).

[45] Articles by Chasles, and the many responses, are found in the *Comptes rendus des séances de l'Académie des sciences* beginning in July 1867 (Tome LXV). Consisting of hundreds of pages of text (involving extracts and complete transcriptions of "letters"), the appearance of these exchanges ran from roughly July 1867 to January 1868 (Tome LXVI). By this time, Sir David Brewster joined the fray, along with the English astronomer, Robert Grant. They were joined by scholars from Italy and France, Galileo scholars, among them Pietro Angelo Secchi and Paolo Volpicelli, and French specialists, among them the Pascal scholar, A-P Faugère. The *Affaire Vrain Lucas*, combined with the colossal theft of manuscripts by Guglielmo Libri (1802–1869), may have prompted European archivists to refine the inventories of their manuscript collections. This dramatic display of scholarly effort, fueled by scandal and the loss of national treasures, likely gave impetus to the publication of *Opera* and *Correspondence* of major figures. On the Libri Affair, see P.A. Maccioni Ruju and Marco Mostert, *The Life and Times of Guglielmo Libri (1802–1869), scientist, patriot, scholar, journalist and thief, A 19th century story*. Hilversum, 1995.

[46] On the Vrain-Lucas affair, see Henri Bordier and Émile Mabille, *Une fabrique de faux autographes, ou recit de l'Affaire Vrain Lucas*. Paris, 1870; *Le parfait secrétaire des grands hommes ou Les lettres de Sapho, Platon, Vercingétorix, Cléopâtre, Marie-Madeleine, Charlemagne, Jeanne d'Arc et autres personnages illustres*, Ed. Georges Girard, Paris, 2003; and Joseph Rosenblum, *Forging of False Autographs, Or, An Account Of The Affair Vrain Lucas*. New Castle, Delaware, 1998.

[47] Although Newton would have been 12 years old at the beginning of the exchange—and despite irregularities in other documents in his possession—Chasles persisted in publishing his views in the prestigious *Comptes rendus* of the Académie des sciences. Overall, Vrain Lucas forged some 27,000 documents, including letters purportedly written by Mary Magdalene, Aristotle, Alexander the Great, and Lazarus (both before and after his resurrection). Virtually all were written in French. Lucas was fond of the scientific revolution; among his favorite figures were Pascal, Galilei, Louis XIV, and Boulliau.

[48] Robert B. Slocum. *Biographical Dictionaries and Related Works; An International Bibliography of More than 16,000 Collected Biographies*, 2nd edition, 2 Vols., (Detroit, 1986) [First edition, 1967]. This volume lists major biographical dictionaries and encyclopedias according to standard categories, from national or area designations to vocation and related thematic distinctions.

[49] See Appendix for further bibliographic details.

Oxford English Dictionary and *Encyclopediae Britannica*.[50] Drawing on hundreds of contributors, the *DNB* contained some 30,000 entries, supplemented by 6,000 additions. The *DNB* was reprinted in 1908, and thereafter, future publication fell to Oxford University Press (1917). Significantly, the *DNB* was viewed not as a completed project but as an ongoing enterprise. That was a century ago. Jumping forward in time, plans were put in place in 1992 to publish the new *Oxford Dictionary of National Biography* [*ODNB*], which was completed in 2004.[51] This modern edition, the most comprehensive biographical dictionary of its kind, contains some 54,922 lives filling 60 volumes. Foreshadowing future efforts in collective biography, the *ODNB* has set new standards by providing electronic online access for subscribers, thus ensuring easy updates and unprecedented capacity for searching and comparing individuals across traditional categories.[52]

Since the Enlightenment

Since the Enlightenment, important developments have taken place in the theory and practice of historical writing. Like other specialized areas of research, the history of astronomy has benefited from increased access to manuscripts and primary sources, not to mention profound changes in educational institutions and dramatic increases in the availability of printed works. These ongoing and often parallel developments began to converge in the form of pioneering works in the history of science. Some of these early works are still available in print, several in the history of astronomy.

A classic example was published by the noted astronomer, J-B Delambre (1749–1822). His impressive multivolume study, *Histoire de l'Astronomie* (1817–1821; 1827) still shows exceptional talent as it moves across ancient, medieval, and modern astronomy.[53] Delambre's work combines the technical skills of an astronomer with the language skills of a classical scholar. Standing the test of time, his six-volume *Histoire* skillfully weaves technical analysis with biographical references—most memorable are entire pages filled with elegant equations. A work for specialists, Delambre's *Histoire* is based squarely on the analysis of published works. Today, his approach might be called "technical thick-description." Although his narrative sails boldly across difficult seas (observation, data reduction, mathematical procedures, and the calculation of tables), his travel-chart is organized around individuals, not concepts or historical periods.

But if Delambre's approach is not thematic, neither is it about *lives*.[54] While his chapter titles and subsections bear the names of individuals, Delambre tells the reader little about his subjects.[55] Instead of a biographical or historical narrative, he offers technical analysis of specific problems. For Delambre and his contemporaries, the use of a "thematic narrative" in the history of astronomy still lay in the future. For now, chronology, bibliography, and technical analysis ruled the day.[56] Delambre's mentor, Joseph-Jérôme de Lalande (1732–1807), echoes the point,[57] and a similar transitional approach is equally evident in the work of a learned contemporary, Alexandre-Guy Pingré

50 Known initially by the working title of *Biographia Britannica*, much of the early work was undertaken by the first editor, Sir Leslie Stephen (1824–1901); he was eventually replaced by Sir Sidney Lee (1859–1926). The first volume of the *DNB* appeared on 1 January 1885; the last, number 63, in 1900.

51 The *ODNB* has been widely reviewed by scholars, and was recently dubbed "the greatest reference work on earth" (*Daily Telegraph*). Stefan Collini, in "Our Island Story," *London Review of Books*, Vol. 27 (20 January, 2005) concludes his review suggesting that "In deeply unpropitious times, the *Oxford Dictionary of National Biography* has refreshed and fortified our sense of what can still be meant by the collective endeavour of 'scholarship.'"

52 Though widely discussed in recent decades, the advent of electronic texts and powerful search potential continue to change the scholarly landscape. After several minutes searching all the entries in the *ODNB*, I present the following purposely mixed findings: From 50,000 individuals, 3,267 are linked with *science*; within the entire *ODNB*, the word *revolutionary* appears 1,380 times; *child prodigy* 39 times; *intellectually brilliant* 7 times; *arrogant* 307 times; and *quite mad* 3 times. Overall, the *ODNB* contains biographies on 231 astronomers of whom six are women. Searching religious affiliation among the astronomers (selecting from 20 categories) yields two Lutherans (not further specified) and 33 Catholics (not refined here by seven subcategories). Electronic texts allow unprecedented capacities for linking words, concepts, and categories.

53 Jean-Baptiste Delambre, *Histoire de l'astronomie ancienne*. 2 Vols. (Paris, 1817); *Histoire de l'Astronomie du moyen age*. (Paris, 1819); *Histoire de l'astronomie moderne*. 2 Vols. (Paris, 1821); *Histoire de l'astronomie au XVIII siècle*. (Paris, 1827).

54 Delambre wrote a number of solid and lengthy biographical articles for the *Biographie universelle*, including articles on Hipparchus, Kepler, La Caille, Lalande, Ptolemy, and Picard. For an overview of Delambre's career, see the works of I. Bernard Cohen cited below.

55 Delambre's *Histoire de l'Astronomie Moderne*, which lacks a traditional table of contents, contains 16 books; each chapter title except the first (Réformation du Calendrier) is given a single individual name (Copernic, Tycho-Brahé, Képler, etc.) or the names of several individual astronomers ("Métius, Boulliaud, et Seth-Ward"). Minor figures, to Delambre's credit, receive substantial analysis.

56 A recent scholar suggested that Delambre's "six volume *Histoire* is the greatest full-scale technical history of any branch of science ever written by a single individual" further adding it "sets a standard very few historians of science may ever achieve." (I. Bernard Cohen, "Delambre," *Dictionary of Scientific Biography*. Vol. 4: 14–18, p. 17). Elsewhere Cohen explained that Delambre's approach was to go through "each chronological period by describing and analyzing first one treatise and then another [he] thereby avoids any attempt at a historical 'synthesis,' or generalization, largely confining himself to critical analyses and expositions of major and minor contributions within the rigid framework . . . ." "Introduction," J-B-J Delambre, *Historie de l'Astronomie Modern*, Reprint, New York, 1969, p. xvi.

57 Jérôme de Lalande (1732–1807) published a similarly impressive work—again, still useful today—that followed the tradition of linking units of information along a clean chronological line. It would now be known as annotated bibliography, *Bibliographie astronomique avec l'histoire de l'astronomie depuis 1781 jusqu'à 1802*. (Paris, 1803). Not a history but a reference tool, Lalande's *Bibliographie* lists every known astronomical work from *circa* 480 BCE to 1802. Containing some 660 pages, it was unrivaled as a chronological bibliography of the history of astronomy. By design, it also served as a chronological list of astronomers. At the end of his book, Lalande provided a concise "history of astronomy" (1781–1802), in effect, a calendar of astronomical events and activities similar to the annual publications of the *Académie des sciences*. A similar model was adopted by G. Bigourdan in publishing the work of A-G Pingré (see below).

(1711–1796).[58] But organizational approaches to historical writing were changing. At the close of the century, Adam Smith (1723–1790), the noted economist, developed a more thematic approach in his *Principles Which Lead and Direct Philosophical Enquiries; Illustrated by the History of Astronomy* (1795).[59] As the title suggests, Smith used history to explore the roots of human progress. As an ancient form of knowledge, astronomy provided Smith with an example that linked material and moral improvement.[60] Many of these early historical writings mixed technical analysis with bio-bibliography. In varying degrees, each shows a shift toward narrative, from chronicling events to evaluating themes. An important virtue of historical narrative is that it accommodates "time's arrow" along with traditional interests in analysis, biography, and bibliography.[61]

Since the Enlightenment, research and reference tools have appeared in growing numbers, and as philosophy and science have became more specialized, historical works have followed suit. In the history of science, the German physicist and bibliographer, Johann Christian Poggendorff (1796–1877) published a pioneering biographical handbook. Poggendorff's evolving multivolume *Biographisch-Literarisches Handwörterbuch der exakten Naturwissenschaften* (1863–1904, *et seq.*) initially contained some 8,400 biographical entries. It was the first comprehensive bio-bibliographical work of its kind. Although it emphasized the physical and exact sciences, it covered all countries and chronological periods.[62] Outside the physical sciences, William Munk (1816–1898) published his *Roll of the Royal College of Physicians* (3 Vols., 1878), one of many multivolume works showing increased specialization. An example: George Sarton (1884–1956), among the early founders of the discipline, provided a detailed roadmap to ancient science in his *Introduction to the History of Science* (1927–1948, Baltimore).[63] Continuing the journey (ancient to medieval) Pierre Duhem (1861–1916) published his monumental *Le système du monde*, 10 Vols. (1913–1959, Paris), providing a detailed study of the physical sciences, including the history of astronomy.[64] Similarly styled encyclopedic narratives appeared by Lynn Thorndike (1882–1965), *History of Magic and Experimental Science* (8 Vols., 1923–1958),[65] while R.T. Gunther's *Early Science in Oxford* (14 Vols. 1923–1945, Oxford) is more typical of institutional works. As pioneers, Sarton, Duhem, Thorndike, and Gunther represent a transitional encyclopedic tradition that joined bio-bibliography with a thin chronological narrative. Finally, a more recent trend in collective biography is evident in "Who's Who" publications. These works have helped fill biographical gaps left by other approaches, particularly in the professions. One of the most comprehensive works of collective science biography contains some 30,000 entries, *The World Who's Who in Science: A Biographical Dictionary of Notable Scientists, From Antiquity to the Present* (Chicago, 1968), edited by Alan Debus.[66]

[58] Pingré's *Annales céleste du dix-septième siècle* (1901), as the title suggests, is based on a year-by-year celestial calendar; it offers a treasure trove of detailed information about celestial events, observations, publications, and people. Like his predecessors, Pingré's skeletal structure was never fleshed out; there is no narrative theme and little life, although it sometimes offers exceptional biographical insight.

[59] Two early historians of astronomy, James Ferguson (1710–1776) and Robert Grant (1814–1892), followed similar strategies of mixing biography and historical narrative that echoed the interpretive themes of their day (Robert Grant, *History of Physical Astronomy, From the Earliest Ages to the Middle of the Nineteenth Century* (London, 1852)). Grant's title may be misleading. His 14-page introduction covers the period up to Newton; the following 13 chapters are devoted to the theory of gravitation, particularly the genesis and reception of the "immortal discoveries of Newton" (p. 20). Although occasional flourishes of whiggism may jar the modern reader, Grant's *History* remains impressive. On the solid basis of primary sources, it shows admirable technical mastery, historical rigor, and remarkable rectitude of judgment.

[60] Striking a more traditional note, Joseph Priestley (1733–1804), a Unitarian minister, echoed a similar theme. Priestly saw the natural philosopher as "something greater and better than another man" as his work involved the "contemplation of the works of God." Joseph Priestley, *The History and Present State of Electricity, with Original Experiments*. 2 Vols., 3rd ed. (London 1775): Vol. 1, p. xxiii.

[61] Earlier historians with interests in other areas had been emphasizing topical and thematic approaches since the beginning of the 17th century, notably John Selden (1584–1654) and the noted French historian, Jacques Auguste de Thou (1553–1617). In the nascent history of science, more thematic approaches are evident in William Whewell, *History of the Inductive Sciences* (1837). Voltaire, their contemporary, is widely noted for stretching historical narratives from political concerns to science, learning, and the arts. Although a trend toward historical narrative is evident in the history of science, two later classics, by Arthur Berry (1898) and J.L.E. Dreyer (1906), continued to entitle chapter headings (and many subsections) with the names of specific individuals. Biography remains an important organizational strategy in the history of astronomy.

[62] Johann Christian Poggendorff (1796–1877), Professor at the University of Berlin (1834), served as editor of *Annalen der Physik und Chemie* (1824–1877) and was a member of the Prussian Academy of Sciences (1839). Poggendorff's work first appeared in two volumes (1863) and gradually expanded into seven parts ("Band I" to "Band VII," 1863–1992; Part 8 was begun in 1999). Poggendorff is particularly strong for the physical sciences—astronomers, mathematicians, physicists, chemists, mineralogists, geologists, naturalists, and physicians. An electronic version of Poggendorff's work is now available in database format. It reportedly contains entries for some 29,000 scientists from ancient to modern times. The electronic edition (DVD) is under the auspices of Sächsische Akademie der Wissenschaften zu Leipzig. See Appendix for bibliographic details.

[63] George Sarton. *Introduction to the History of Science*. 3 Vols., Baltimore: Williams and Wilkins, 1927–1948.

[64] Pierre Duhem. *Le système du monde, Histoire des doctrines cosmologiques de Platon à Copernic*. The volumes include *I. La cosmologie hellénique; II. La cosmologie hellénique; III. L'astronomie latine au Môyen Age; IV. L'astronomie latine au Moyen Age; V. La crise de l'aristotélisme; VI. Le refus de l'aristotélisme; VII. La physique parisienne au XIVe siècle; VIII. La physique parisienne au XIVe siècle; IX. La physique parisienne au XIVe siècle; IX. La cosmologie de XVe siècle. Ecoles et universités.*

[65] Lynn Thorndike. *A History of Magic and Experimental Science* (8 Vols., New York, 1923–1958).

[66] Several thematic reference works have appeared in recent decades, notably the *Dictionary of the History of Ideas* (1974), now in a new edition; *Encyclopedia of Philosophy* (1967); *Companion to the History of Science* (1990); and particularly useful for identifying minor figures, the *Isis Cumulative Bibliography* (1971–).

An important scholarly tradition—which continues today—emerged in the 19th century with the publication of the complete works of noted scholars and scientists.[67] No discussion of science biography would be complete without mentioning the significance of these scholarly monuments. Among the oldest and most powerful research tools for historians of science, these works first appeared as *opera omnia, oeuvres complètes*, or as *Lettres* or *Complete Correspondence* of the traditional heroes of our discipline. Contemporary interest in heroic individuals reflects the philosophy of science at the time, not to mention nationalistic tendencies and expressions of local pride.[68] Challenging in scope and complexity, the extant body of letters and manuscripts of leading scientists required exceptional scholarship, collective effort, and substantial institutional support. Arguably, these requirements help define modern collective biography as well as the character of private, institutional, and national funding. Because these works have appeared over the course of several centuries, it is instructive to consider changing standards of scholarship.[69]

Heralded as "one of the most ambitious projects ever undertaken in studies of the history of science," the *Dictionary of Scientific Biography* (*DSB*) (1970–1980) occupies an important place at the end of this brief historical introduction.[70] The *DSB*, sponsored by the American Council of Learned Societies and supported by the National Science Foundation, has been identified as a collaborative work that at once asserted and affirmed the identity of a discipline.[71] Published with remarkable speed and regularity in the course of a decade (1970–1980), the original 16-volume set includes over 5,000 biographical entries in the history of science from Antiquity to the 20th century.[72] Overall, the scholarly response to the *DSB* was extremely positive. Some proclaimed it "magnificent" and "triumphantly executed," others offered detailed criticism and useful suggestions.[73] In the end, despite the unprecedented scope of a project this size, most reviewers returned to time-honored principles that define the design and use of collective biography—*inclusion criteria*, *entry length*, and issues of *coverage*. By tradition, key areas of concern turn on the relative importance of historical figures—their positive contributions, contemporary influence, subsequent significance, and their role in representing or typifying a group. Difficult decisions are involved. To suggest the size of the problem, what weight does a Leviathan like Isaac Newton have compared to a small fry like John Newton (a contemporary almanac writer)? Scholarly reviews of the *DSB* reconfirm a diversity of opinion—and sustained acceptance—of collective biography.[74] Classified by field, the *DSB* contains articles on some 750 astronomers, most from the modern period.[75]

[67] A selected list, considered chronologically, includes Pierre Gassendi, *Opera Omnia* (6 Vols., Lyon, 1658); Benedict de Spinoza, *Opera Posthuma* (Amsterdam 1677), Dutch edition, *Die nagelate Schriften van B. d. S.* (n.p., 1677); J. Bernoulli (1744); René Descartes (1824–1826 *et seq.*); Johannes Kepler (*Opera*, 1858–1871; *GW*, 1935–); A- L. Lavoisier (6 Vols., 1862–1893); C. F. Gauss (12 Vols., 1863–1933); J- L. Lagrange (14 Vols., 1867–1892); P-S Laplace (14 Vols., 1878–1912); A- L. Cauchy (26 Vols., 1882–1970); Christiaan Huygens (22 Vols., 1888–1950); René Descartes (12 Vols., 1897–1913); Galileo Galilei (20 Vols., 1890–1910); Blaise Pascal (14 Vols., 1904–1914; 1964–1992, *et seq.*); Leonard Euler (43; 72 Vols., 1909; 1911–1996); Tycho Brahe (15 Vols., 1913–1929); G-W Leibniz (1923–); Isaac Newton (7 Vols., 1959–1977); Nicolaus Copernicus (4 Vols., 1978–); Robert Boyle (1999–2000; 2001); and Albert Einstein (1987–). Similar volumes have recently appeared for Thomas Hobbes (1994), John Flamsteed (1995–2003), and John Wallis (2003 et seq.). Taken separately, less heroic figures have attracted scholarly interest, savants such as N-C Fabri de Peiresc (1888–1898; 1972), Marin Mersenne (1932–1986), and Henry Oldenburg (1965–1986). The *Discepoli di Galilei* (1975–1984) was designed to shed light not only on individuals but working groups. See Appendix for bibliographic details.

[68] On the title pages of one edition of Galilei's works, for example, one finds in over-sized colored type the name of Benito Mussolini. In France, Philippe Tamizey de Larroque, editor of the *Lettres* of N-C Fabri de Peiresc, was a enthusiastic but unrepentant promoter of his hero, the glory of Provence.

[69] As an example, Johannes Kepler has two major editions dedicated to his work. Christian Frisch edited the first major edition, *Joannis Kepleri opera omnia* 8 Vols. (Frankfort and Erlangen, 1858–1871); the more recent appeared as *Gesammelte Werke* (22 Vols., Munich, 1938–). The differences are notable. As an example, Frisch presents Kepler's letters unsystematically, sometimes appended to various parts of his relevant published works. The modern *Gesammelte Werke*, by contrast, supplies the complete text of all known correspondence organized and annotated in familiar modern format. A second example involves the *Lettres* of N-C Fabri de Peiresc. In more than one instance, the editor of Peiresc's letters, Tamizey de Larroque, combined various versions of letters (originals, drafts, copies) in a well-meaning effort to provide a more complete text—but alas, without alerting the reader. Larroque sometimes omitted portions of Peiresc's published letters (and on occasion entire letters) judging them "too scientific."

[70] Another reviewer proclaimed the *DSB* the "greatest contribution to scholarship in the history of science of the second half of the 20th century."

[71] The *DSB* was "designed to make available reliable information on the history of science through the medium of articles on the professional lives of scientists. All periods of science from classical Antiquity to modern times are represented, with the exception that there are no articles on the careers of living persons." (Preface). *DSB* entries are signed and usually include a bibliography; geographical coverage is international, although China, India, and the Far East are not treated as extensively as others.

[72] The *DSB* appeared in 16 Volumes during the years 1970–1980, followed by supplements. Entries provide the subject's birthplace and date, family information and background, education and intellectual development, treatment of growth and directions of the subject's scientific work and scientific personality in relation to predecessors, contemporaries, and successors. Inclusive across time and space, entry length was in three categories (300–700; 700–1300; and 1300–3600 words), reflecting the individual's contribution and influence.

[73] A brief survey suggests three principal concerns: thematic boundaries defining the group; inclusion criteria; and relative length of entries. As general principles, collective biography should be inclusive, symmetrical, authoritative, and where possible, based on primary sources. In practice, editors wisely supply contributors with an editorial "boiler plate" to ensure symmetry (date and place of birth and death; parents and siblings; birth order position; religion; education; publications; friends; students; appointments and honors; institutional affiliations; contemporary influence; personal finance; work habits; motives for pursuing science; etc.). One reviewer of the *DSB* suggested editors request "guideposts" to cue readers: "the subject's most significant work is X," or "a critical influence was Y." Editorial decisions are particularly acute when major collective biographies (such as the *DNB* and *DSB*) are reduced to a single comprehensive volume. The *Concise Dictionary of National Biography* (Pt. 1, Oxford, 1903; 2nd Ed. 1906) consists of entries one-fourteenth the number of words from the parent edition. Entries in the *Concise Dictionary of Scientific Biography* (New York, 1981) are 10 percent the length of those in parent volumes.

[74] The *DSB* is currently being revised and expanded to include individuals from the 20th century and those previously omitted. The new *DSB* will be in electronic format and fully searchable.

[75] The *Concise* DSB contains "Lists of Scientists By Field" (749–773) which facilitates this rough estimate; arguably, a more accurate reckoning would be 500 "astronomers."

Conclusion

Readers of the *BEA* will find a familiar format aimed at easy access. The only notable departure from tradition is that individual entry length shows less dramatic variation than in earlier works. With an eye toward supplying specialists and laymen with appropriate references, individual entries vary from 100 to 1500 words. Readers may note that entries for the likes of Newton and Einstein may be rivaled by less-known astronomers. The rationale is twofold: First, entry length helps rescue a number of astronomers from relative oblivion; second, it provides readers with scarce information not readily found in secondary works, sometimes not available in English or in modern languages. Major figures continue to receive substantial entries but with less lengthy largesse. This strategy also reflects the wider availability of source material for major figures.

As we look to the past, collective biography has not only proven adaptable to changes in historical writing, it has been central to the story from the start. Like other forms of scholarship, individual works of collective biography will continue to be judged by their rigor, utility, and scholarly merit. But while readers have come to expect increasingly higher levels of expertise, inclusion, and ease of access, most modern readers remain curiously consistent—even old fashioned—in their expectations about biography. As in the past, readers will continue to appreciate an appropriate anecdote, particularly if it puts a face on a thought or makes a life and career more coherent. In the end, the lives of scientists are human lives, and if *biography* is about an individual life, *collective biography* is about *forms of life*. Biography, like astronomy, has a long and rich tradition. It tells the story of forgotten constellations; it contemplates patterns of human acheivement and human aspiration. Those now distant worlds—puny and brief—seem no less majestic, no less alluring.

Robert Alan Hatch
University of Florida

Appendix

Reference and Research Resources

This list of biographical sources is suggestive, not exhaustive. It aims to provide selected sources that may be useful for identifying biographical sources in the history of astronomy and cosmology. Additional detailed research can be pursued by means of specialized scholarly studies found in the second section, which includes the complete works, correspondence, and cumulative bibliographies of noted figures. For further information on biographical reference sources, see Robert B. Slocum. *Biographical Dictionaries and Related Works: An International Bibliography of Approximately 16,000 Collective Biographies*, 2 Vols., 2nd ed., Detroit, 1986.

Selected Reference Sources

ADB (*Allgemeine Deutsche Biographie*). 56 Vols., Leipzig, 1875–1912; reprinted Berlin, 1967–1971.

ANB (*American National Biography*). 24 Vols., Oxford University Press, 1999.

AMWS (*American Men and Women of Science: A Biographical Directory*). New York, 1906–. (Prior to 12th edition (1971) entitled *American Men of Science*).

AO (*Athenae Oxonienses*), A New Edition. A facsimile of the London edition of 1813, Anthony Wood, 4 Vols., Reprint, New York and London, 1967.

B-DH (*Dictionnaire historique et critique*), Pierre Bayle, 4 Vols., Rotterdam, 1720.

BDAS (*Biographical Dictionary of American Science: The Seventeenth Through the Nineteenth Centuries.*), edited by Clark A. Elliott, Westport, 1979.

BDS (*Biographical Dictionary of Scientists*), 3rd ed., edited by Roy Porter and Marilyn Bailey Ogilvie, 2 Vols., New York, 2000.

BGA (*Bibliographie générale de l'astronomie*), edited by J.C. Houzeau de Lehaie and A.B.M. Lancaster, 3 Vols., Brussels, 1887–1889.

BK (*Bibliografia Kopernikowska 1509–1955*), edited by Henryk Baranowski, Reprint, New York, 1970.

BLH [P] (*Biographisch-literarisches Handworterbuch zur Geschichte der exakten Wissenschaften.*), edited by J. C. Poggendorff, Leipzig and Berlin, 1863–1926. Band VIIa -Supplement. Berlin, 1969.

BNB Académie Royale de Belgique. (*Biographie Nationale Belgique*), 20 Vols., Brussels, since 1866–.

BU (*Biographie Universelle, Ancienne et Moderne*) ou (*Histoire, par ordre alphabétique : de la vie publique et privée de tous les hommes qui se sont fait remarquer par leurs écrits, leurs actions, leurs talents, leurs vertus ou leurs crimes.*), J-F Michaud, 85 Vols., in 45 Vols. Paris: Michaud Frères, 1811–1862. *Second, revised edition.* (variants)

BWN (*Biographisch Woordenboek der Nederlanden*), 21 Vols., Haarlem,1852–1878.

CBD (*Chambers' General Biographical Dictionary*), 32 Vols., London, 1812–1817 (1984)

CA (*Alumni Cantabrigienses: A Biographical List of All Known Students, Graduates and Holders of Office at the University of Cambridge to 1900*), J. Venn, 10 Vols., Cambridge University Press, Cambridge, 1922–1954.

DAB (*Dictionary of American Biography*), 20 Vols., New York, 1928–1936; reprinted in 10 Vols. with supplements, New York.

DBF (*Dictionnaire de Biographie Française*), edited by J. Balteau *et al.*, with supplements, Paris, 1932–.

DBI (*Dizionario Biografico Degli Italiani*) (currently 59 Vols., Rome, 1960–).

DNB (*Dictionary of National Biography*), edited by Sir Leslie Stephen *et al.*, 72 Vols., 1885–1912 (1964); See **ODNB** below.

DSB (*Dictionary of Scientific Biography*). Charles Scribner's Sons, New York, edited by Charles Coulston Gillispie (Vols. I-XVI) and Frederic L. Holmes (Vols. 17–18). (Volumes I-XIV: 1970–1976; Volume XV: *Supplement I*, 1978; Volume 16: *Index*, 1980; Volumes 17–18: *Supplement II*, 1990.)

EC (*Encyclopedia of Cosmology*), edited by Norriss S. Hetherington, New York, 1993.

FS (*Les Femmes dans la Science*). Notes Recueillies by Alononse Rebiere, 2nd Edition, Paris, 1897.

G-HC (*A Historical Catalogue of Scientific Periodicals*) (*1665–1900*), New York, 1985.

HEA (*History of Astronomy: An Encyclopedia*), edited by John Lankford, New York, 1997.

ICB (*ISIS Cumulative Bibliography*). *A Bibliography of the History of Science formed from ISIS Critical Bibliographies 1–90, 1913–1965*, Vols., 1–2 (Personalities). London, 1971, *et seq.* (Critical Bibliographies 1–90 (1913–1965), 6 Vols.; 91–100 (1966–1975), 2 Vols.; 101–110 (1976–1985), 2 Vols.; (1986–1995), 4 Vols.

M (*Biographie universelle ancienne et moderne,* publiée par Michaud), Joseph-François Michaud, Paris, 1810–1828, 52 Vol. in-8, plus 32 Vols. supplément.

ML (Louis Moréri, *Le grand Dictionaire historique, ou le mélange curieux de l'histoire sacrée et profane*), Lyon, 1671 *et seq.*

N (Jean-Pierre Nicéron, *Mémoire pour servir a l'histoire des hommes illustres dans la République des Lettres, avec un catalogue raisonne de leurs ouvrages*), 43 Vols., Paris, 1727–1745.

NBG (*Nouvelle Biographie Générale, Depuis les temps les plus reculés jusqu'à nos jours*), 46 Vols. in 24, Paris: Firmin Didot, 1853–66, edited by F. Hoeffer, variants.

NBU (*Nouvelle Biographie Universelle*) (title variants) 46 Vols., Paris, 1852–1866; reprinted in 23 Vols., Copenhagen, 1963–1969.

NDB (*Neue Deutsche Biographie*), edited by Historischen Kommission of the Bayerischen Akademie der Wissenschaften, 7 Vols., *et seq.*, Berlin, 1953–.

ODNB (*Oxford Dictionary of National Biography*), 61 Vols., Oxford, 2004.

P-BLH (*Biographisch-literarisches Handworterbuch der exakten Naturwissenschaften*), Johann C. Poggendorff *et al.*, Leipzig: Barth, 1863–1904; Leipzig, 1925–1940; Berlin, 1955–. (Variant titles), Reprinted: Band 1–6, to 1931. Ann Arbor, 1945.

RS (Royal Society of London, *Catalogue of Scientific Papers, 1800–1900*). London, 1867–1902; Cambridge, 1914–1925, 19 Vols.

SBB (*Scientists since 1660: A Bibliography of Biographies*), edited by Leslie Howsam, Brookfield, Vermont, 1997.

SCB-I (*A Short-title Catalogue of Books printed in England . . . 1475–1640*), edited by A.W. Pollard and G.R. Redgrave, London, 1926.

SCB-2 (*Short-title Catalogue of Books printed in England . . . 1641–1700*), edited by D.G. Wing, 3 Vols., New York, 1945–1951.

W-BD (*The Biographical Dictionary of Women in Science*), edited by Marilyn Ogilvie and Joy Harvey, 2 Vols., New York and London, 2000.

WS (*Women in Science, Antiquity through the Nineteenth Century: A Biographical Dictionary with Annotated Bibliography*), edited by Marilyn Bailey Ogilvie. Boston, 1986.

WS-A (*American Women in Science: A Biographical Dictionary*), edited by Martha J. Bailey, Santa Barbara, 1994.

WSI (*Women Scientists From Antiquity to the Present: An Index*), edited by Caroline L. Herzenberg, West Cornwall, CT, 1986.

Selected Research Sources

AO (*Oeuvres complètes de d'Alembert*), Alembert, Jean Le Rond d', Paris, 1821–1822, Reprint 1967.

AOP (*Oeuvres philosophiques, historiques et littéraires de d'Alembert*), Alembert, Jean Le Rond d', 18 Vols., Paris, 1805.

BBO (*Jacobi Bernoulli, Basileensis, Opera*), Jacob Bernoulli, (1654–1705), 2 Vols., Geneva, 1744.

BF-W (*Works of Francis Bacon*), Francis Bacon, edited by J. Spedding, R.C. Ellis, and D.D. Heath, 14 Vols., London, 1857–1874.

BRC (*The Correspondence of Robert Boyle*), Robert Boyle, edited by Michael Hunter, Antonio Clericuzio, and Lawrence M. Principe, 6 Vols., London, 2001.

BRW (*The Works of Robert Boyle*), Robert Boyle, edited by Michael Hunter and Edward B. Davis, Pickering and Chatto Ltd, 14 Vols., London, 1999–2000.

BRW-B (*The Works of the Honourable Robert Boyle*), To which is prefixed The Life of the Author, Robert Boyle, edited by Thomas Birch, 5 Vols., in folio, London, 1744; "A New Edition," 6 Vols., London, 1772.

C (*Nicholas Copernicus' Complete Works*), Nicolas Copernicus, edited by Jerzy Dobrzycki, translation and commentary by Edward Rosen, 4 Vols., London and Basingstoke, 1978–.

CC (*Carteggio*), Bonaventura Cavalieri, edited by Giovanna Baroncelli, Florence, 1987.

COO (*Opera Omnia*), Girolamo Cardano, 10 Vols., Reprint, New York and London, 1967.

DC (*Correspondance*), René Descartes, edited by Charles Adam and Gaston Milhaud. 8 Vols., Paris, 1936–1963.

DGG (*Le Opere dei Discepoli di Galileo Galilei*), Carteggio, Edizione Nazionale, Vol. 1 (1642–1648), Vol. 2 (1649–1656), edited by Paolo Galluzzi and Maurizio Torrini, Florence, 1975, 1984.

DO (*Oeuvres de Descartes*), René Descartes, edited by Charles Adam and Paul T. Tannery, 13 Vols., 1897–1913.

DSP (*Scientific papers*), George Howard Darwin, Cambridge, 1907–1916.

EC (*Correspondance mathématique et physique de quelque célèbres géomètres du XVIIIeme siècle*), Leonard Euler, edited by P.H. Fuss, 2 Vols., St. Petersburg, 1843.

ECP (*The Collected Papers of Albert Einstein*), Princeton University Press, Princeton, 1987–.

EO (*Leonhardi Euleri Opera Omnia*), Leonard Euler, edited by Charles Blanc, Asot T. Grigorijan, Walter Habicht, Adolf P. Juskevic, Vladimir I. Smirnov, Ernst Trost, 3 Vols. Basil, 1975 (1911).

EO-2 (*Leonhardi Euleri Opera Omnia*), *Series prima* (*Opera mathematica*, 29 in 30 Vols.), *Series secunda* (*Opera mechanica et astronomica*, 31 in 32 Vols.), *Series tertia* (*Opera physica et Miscellanea*, 12 Vols.), *Series quarta A* (*Commercium epistolicum*, 9 Vols.), and *Series quarta B* (*Manuscripta*, approx. 7 Vols.), Basel, Birkhäuser, 1911–1996.

ESO (*Early Science in Oxford*), edited by R.T. Gunther, 14 Vols., Oxford, 1923–1945.

FGL (*The Gresham Lectures of John Flamsteed*), John Flamsteed, edited by Eric G. Forbes, London, 1975.

FO (*Oeuvres de Fermat*), Pierre Fermat, edited by Paul Tannery, Charles Henry, and Cornelis de Waard, 5 Vols., Paris, 1891–1922.

FOM (*Varia opera mathematica D. Petri de Fermat / accesserunt selectae quaedam ejusdem epistolae, vel ad ipsum a plerisque doctissimis viris Gallice, Latine, vel Italice, de rebus ad mathematicas disciplinas, aut physicam pertinentibus scriptae*), Pierre Fermat, Toulouse, 1679.

GAC (*Amici e corrispondenti di Galilei*), Galileo Galilei, edited by Antonio Favaro, with introductory notes by Paolo Galluzzi, 3 Vols., Florence (reprinted) 1983.

GGO (*Le Opere di Galileo Galilei*), Galileo Galilei, Edizione Nazionale, edited by Antonio Favaro, 20 Vols., Florence, 1890–1939.

GOO (*Petro Gassendi, Opera Omnia, hactenus edita auctor ante obit recensuit*), Pierre Gassendi, edited by H.L. Habert de Montmor and F. Henry, 6 Vols., Lyon, 1658–1675.

HC (*The Correspondence of Thomas Hobbes*), 2 Vols., Oxford, 1994.

HCP (*Correspondence and papers of Edmond Halley*), Edmond Halley, Oxford, 1932.

HD. (*The Diary of Robert Hooke MA., M.D., F.R.S. 1670–1680*), Robert Hooke, London, 1935.

HEW (*The English Works of Thomas Hobbes of Malmesbury*), Thomas Hobbes, edited by Sir William Molesworth, 11 Vols., London, 1839–1845.

HOC (*Oeuvres Complètes de Christiaan Huygens*), Christiaan Huygens, publiées par la Société Hollandaise des Sciences, 22 Vols., The Hague, 1888–1950.

HP (*The Hartlib Papers*), Samuel Hartlib, The Hartlib Project, directed by Michael Leslie, Mark Greengrass, Michael Hannon, Patrick Collinson, with assistance from Timothy Raylor, Judith Crawford and others, University of Sheffield. (CD-ROM edition)

IB (*Institut de France: index biographique des membres et correspondants de l'Académie des Sciences de 1666 a 1954*), Institute de France, Gauthier-Villars, Paris, 1954.

IBAC (*Académie des sciences. Index Biographique des Membres et Correspondants de l'Académie des Sciences*), Paris, 1968.

KA (*Joannis Kepleri astronomi opera omnia*), Johannes Kepler, edited by Christian Frisch, 8 Vols., Frankfurt, 1858–1871.

KGW (*Gesammelte Werke*), edited by Walther van Dyck, Max Caspar, and Franz Hammer. Munich, 1937–.

L (*The Correspondence of John Locke*), John Locke, edited by E.S. de Beer, 8 Vols., Oxford, 1976–1989.

L-CII (*Carteggio Linceo*), 3 parts, *Atti della Reale Accademia Nazionale dei Lincei, Memorie della Classe di Scienze Morali, Storiche e Filologiche* (*Part I anni* 1603–1609) pp 1–120, (*Part II, anni* 1610–1624, *Sezione I*, 1610–1615) Vol. 7, 1938 (XVI), pp 123–535; *Part II, Sezione II* (*anni* 1616–1624), pp 537–993; Part III (*anni* 1621–1630), pp 999–1446.

L-PG (*The Lives of the Professors of Gresham College*), John Ward, London, 1740; Reprint, New York and London, 1967.

LBO (*Bibliographie des Oeuvres de Leibniz*), edited by Emile Ravier, Hildesheim, 1966.

LCC (*Catalogue critique des manuscrits de Leibniz*), Gottfried Wilhelm Leibniz, edited by A. Rivaud, Poitiers, 1914–1924.

LMN (*Mathematischer Naturwissenschaftlicher und Technischer Briefwechsel*), Gottfried Wilhelm Leibniz, 2 Vols., (1663–1683) Berlin, 1976–1987.

LO. (*Oeuvres de Lagrange*), Joseph-Louis Lagrange, Paris, 1867–1892. Also, *Oeuvres*, Paris, 1973.

LOC (*Oeuvres complètes*), Pierre-Simon Laplace, 14 Vols., Paris, 1878–1912.

LR (*Register zu Gottfried Wilhelm Leibniz Mathematische Schriften und Der Briefwechsel mit Mathematikern*), Gottfried Wilhelm Leibniz, edited by Joseph Ehrenfried Hofman, Hildesheim and New York, 1977.

LSB (*Samtliche Schriften und Briefe*), Gottfried Wilhelm Leibniz, Damstadt, Leipsig, Berlin, 1923–.

LUI (*Lettre inedite di uomini illustri*), edited by Angelo Fabroni, 2 Vols., Florence, 1773 and 1776.

MAS (*Mémoires de l'Académie Royale des sciences depuis 1666 jusqu'à 1699*), 9 Vols., Paris, 1729–1732.

MC (*Correspondance du P. Marin Mersenne*), P. Marin Mersenne, edited by Paul Tannery, Cornelis de Waard, and Armand Beaulieu, 16 Vols., Paris, 1932–1986.

M-CL (*Collected letters of Colin MacLaurin*), Colin MacLaurin, Nantwich, Cheshire, England, 1982.

MCL (*Carteggio Magliabechi, Lettere di Borde, Arnaud e associati Lionesi ad Antonio Magliabechi (1661–1700)*), Antonio Magliabechi, edited by Salvatore Ussia, Florence.
MO (*Oeuvres de Malebranche*), Nicolas de Malebranche, Vols. 18–19, (*Correspondance actes et documents*), edited by André Robinet, Paris, 1978.
MP (*The Mathematical Practitioners of Tudor & Stuart England*), E.G.R. Taylor, Cambridge, 1954.
MP2 (*The Mathematical Practitioners of Hannoverian England*), E.G.R. Taylor, 1714–1840, Cambridge, 1966.
MPBS (*Manuscript Papers of British Scientists, 1600–1940*), London, 1982.
NC (*The Correspondence of Isaac Newton*), Isaac Newton, edited by H.W. Turnbull, J. F. Scott, and A. Rupert Hall, Cambridge, 7 Vols., 1959–1977.
NMP (*The Mathematical Papers of Isaac Newton*), Isaac Newton, edited by Derek T. Whiteside, 8 Vols., Cambridge, 1967–1981.
OC (*The Correspondence of Henry Oldenburg*), Henry Oldenburg, edited by. A. Rupert Hall and Marie Boas Hall, 9 Vols., Madison, 1965–1973; Vols., 10 and 11, Mansell, London, 1975–1977; Vols., 12–13, Taylor and Francis, 1986.
P-C. (*Les Correspondants de Peiresc, Lettres inédites*), Nicolas-Claude Fabri de Peiresc, 2 Vols., Reprint, Geneva, 1972.
P-L (*Lettres de Peiresc*), Nicolas-Claude Fabri de Peiresc, edited by Philippe Tamizey de Larroque, 7 Vols., Paris, 1888–1898.
PDC. (*Diary and Correspondence of Samuel Pepys, F.R.S.*), Samuel Pepys, edited by Richard Braybrooke, 4 Vols., London, 1848–1849.
PHI (*Les Hommes illustres qui ont paru en France pendant le XVIIe siècle*), Charles Perrault, 2 Vols., Paris, 1696–1700.
PO (*Oeuvres de Blaise Pascal*), Blaise Pascal, edited by Leon Brunschvicg, Pierre Boutroux, and Felix Gazier, 14 Vols., Paris, 1908–1914.
POC (*Oeuvres complètes*), Blaise Pascal, preface by Henri Gouhier, notes by Louis Lafuma, editions du Seuil, Paris, 1963.
PT (*Philosophical Transactions: giving some Accompt of the present Undertakings, Studies and Labours of the Ingenious in many considerable parts of the World*), edited by Henry Oldenburg, London and Oxford, 1665–1677.
S-C (*The Correspondence of Spinoza*), Benedict de Spinoza, edited and translated by Abraham Wolf, London, 1928.
S-OP (*Opera Posthuma*) Benedict de Spinoza, edited by J. Jellis, Amsterdam 1677; Dutch edition, *Die nagelate Schriften van B. d. S.* (n.p., 1677).
SS (*The Principal Works of Simon Stevin*), Simon Stevin, edited by E.J. Dijksterhuis, D. J. Struik, A. Pannekoek, Ernst Crone, and W.H. Schukking, 4 Vols., Amsterdam, 1955–1964.
TBO (*Tychonis Brahe Dani Opera Omnia*), Tycho Brahe, edited by J.L.E. Dreyer, 15 Vols., Copenhagen, 1913–1929.
TO (*Opere di Evangelista Torricelli*), Evangelista Torricelli, edited by Gino Loria and Giuseppe Vassura, 4 Vols., in 5 pts, Faenza, 1919–1944.

Geographical Place Names in Biography Headers

Birth and death places are given as [city], [country] when well known, *e. g.*, London, England and Rome, Italy. Lesser-known places are often accompanied by regional/provincial/county/state names, *e. g.*, Beverley, Humberside, England and Lusigny, Aube, France. States in the USA, Canadian provinces, and Australian states are included.

All place names are given as they are found on current maps. Where city names have changed historically, the modern version follows the original within parentheses, *e. g.*, Constantinople (Istanbul, Turkey) and Pitschen (Byczyna, Poland). In cases where cities have disappeared, the nearest modern place is given, *e. g.*, Colophon (near Selcuk, Turkey).

Regional/provincial/county/state names as well as country names are placed within parentheses if they did not exist at the time of the subject's birth or death. Place names are given in the original language except where common English versions exist, *e. g.*, Milan, Germany, Bavaria, Tuscany, Munich, *etc.*

Richard A. Jarrell

M

Maclaurin, Colin

Born **Kilmodan, Argyllshire, Scotland, February 1698**
Died **Edinburgh, Scotland, 14 June 1746**

Colin Maclaurin was, perhaps, the last of the great British mathematicians of the period following **Isaac Newton**. His geometrical methods influenced French work in celestial mechanics. An equilibrium shape of a rotating fluid body called a Maclaurin was thought for many years to be relevant to the formation of binary stars from single, rotating gas clouds.

Maclaurin's father, a parish Minister, died 6 weeks after Colin's birth, and his mother died when he was nine. His uncle, Daniel Maclaurin, Minister in Kilfinnan on Loch Fyne, took responsibility for him. In 1709, Maclaurin entered the University of Glasgow and, although a career in the church was originally intended, he was introduced to Euclid's *Elements* and turned his attention to mathematics and physics. After 4 years, he graduated MA with a thesis, "On the Power of Gravity."

In 1717, Maclaurin was appointed professor of mathematics at Marischal College, Aberdeen, aged only 19 (the youngest professor recorded at any university). Shortly afterwards, he became a good friend and disciple of Newton and, at this time, was elected to a fellowship of the Royal Society of London. Maclaurin travelled widely in Europe between 1722 and 1725. In 1724, he was awarded a prize from the Academy of Sciences in Paris for his work, "On the Percussion of Bodies."

Early in 1726, Maclaurin was appointed to the chair of mathematics at the University of Edinburgh, which he held for the remaining 20 years of his life. During this period, he was a major force in the world of mathematics. His contributions were not only to analysis and geometry but also to physics and astronomy.

During his time in Edinburgh, Maclaurin's efforts were important in the formation of scientific societies. A member of the Society for the Improvement of Medical Knowledge from 1731, he helped broaden its purview. The result was the new, more inclusive organization, the Philosophical Society of Edinburgh, founded in 1737. It was the latter society that, in 1783, was the catalyst for the foundation of the Royal Society of Edinburgh, which is still Scotland's premier learned society.

In 1733, Maclaurin married Anne Stewart, and this union produced 7 children.

An assessment of Maclaurin's place amongst the great mathematicians is obscured by, among other things, the difficulties in assigning priorities to certain mathematical discoveries. Nowadays, his name is known to all mathematicians because of Maclaurin's series. This result appeared in his book, *Treatise of Fluxions*, published in 1742, but the author acknowledged that it was a special case of an earlier result due to Brook Taylor in the latter's book, *Methodus Incrementorum* (1715). In any case, this result was, at least in some form, known to earlier workers, including **Jacob Bernoulli** and the Scottish mathematicians, **James Gregory** and James Stirling. On the other hand, Cramer's rule for systems of linear equations, Cauchy's integral test for convergence, and Bezout's theorem on the intersection of curves were developed by Maclaurin several years before those mathematicians whose names are now associated with them.

In the *Treatise of Fluxions*, Maclaurin gave a systematic account of Newton's theory of fluxions, mainly in response to an attack on these ideas by the Irish philosopher George Berkeley, in the latter's *Analyst* of 1734. Maclaurin's book also contained an account of his work on the gravitational stability of ellipsoids. This was accomplished by employing classical mechanics rather than fluxions. **Alexis Clairaut**, after reading Maclaurin's text, reverted to geometrical methods to attack the figure-of-the-earth problem.

Maclaurin also dealt with the tides (based on material that had previously been awarded a prize by the Paris Academy in 1740). In addition to the above, there were two other posthumous books, *Treatise of Algebra* and *An Account of Sir Isaac Newton's Philosophy.*

Maclaurin was also involved in some more practical branches of mathematics and in the organization of the defenses of Edinburgh against the Jacobite forces in 1745. When Edinburgh fell, Maclaurin fled to York. Although he soon returned, these exertions, together with a fall from a horse, seriously weakened his health, and he died shortly afterward.

Graham Hall

Alternate name

Cailean MacLabhruinn

Selected References

Sageng, Erik L. (1989). "Colin MacLaurin and the Foundations of the Method of Fluxions." Ph.D. diss., Princeton University.

Turnbull, H. W. (1951). *Bi-centenary of the Death of Colin Maclaurin (1698–1746), Mathematician and Philosopher, Professor of Mathematics in Marischal College, Aberdeen (1717–1725)*. Aberdeen: Aberdeen University Press.

Maclear, Thomas

Born **Newtownstewart, Co. Tyrone, (Northern Ireland), 17 March 1794**

Died **Mobray near Cape Town, (South Africa), 14 July 1879**

Thomas Maclear, the third director of the Royal Observatory at the Cape of Good Hope, established it as a leading astronomical observatory during his 36-year tenure. Thomas, eldest son of the Reverend James Thomas Maclear and Mary Magrath of Newtownstewart, was sent at age 15 to England to be cared for by his mother's brothers. Thomas was apprenticed to one of them, Dr T. Magrath, an eminent surgeon at Biggleswade, near Bedford, 50 miles north of London.

In 1814, Maclear went to London to study medicine at Guy's and Saint Thomas's Hospitals, and was admitted a member of the Royal College of Surgeons in 1815. He accepted the post of house surgeon at the Bedford Infirmary and, in 1823, joined his uncle's medical practice. In 1825, Maclear married Mary Pearse. He became acquainted with captain **William Smyth**, who established an observatory nearby. Maclear frequently visited Smyth's observatory, where he met Francis Beaufort and **John Herschel**, who became his lifelong friends.

In 1828, Maclear joined the Astronomical Society of London, setting up his own observatory at Biggleswade using a small transit instrument and a Dollond refractor on loan from the society. Maclear became expert in predicting lunar occultations, and calculated the occultations of Aldebaran (1829–1831) for ten European observatories. He corresponded about occultations with **Thomas Henderson**, then director of the Royal Observatory at the Cape of Good Hope. Maclear was elected a fellow of the Royal Society in December 1831.

After Henderson's resignation from the Royal Observatory at the Cape of Good Hope, the Admiralty offered the position to Maclear. His uncle disapproved; Maclear's annual salary as doctor brought in about £800, whereas the Cape post offered only £600, and Henderson had warned him of many difficulties. Nevertheless, Maclear decided to go, perhaps comforted by Herschel's plan to travel to the Cape to carry out a survey of the southern skies.

Maclear and his family, including his wife, five daughters, a governess, a nursemaid, and a manservant, arrived at the Cape on 7 January 1834 after a grueling 3-month voyage. His youngest daughter was suddenly taken ill and died within 2 days. The Herschels arrived at the Cape soon afterwards; Maclear and Herschel kept in close contact by letter.

When Maclear arrived at the Cape, he immediately inspected the Troughton mural circle, which had given trouble to his predecessors, **Fearon Fallows** and Henderson. Maclear confirmed their opinions that only by averaging the readings on all six microscopes could reliable results be achieved with the mural circle, a cumbersome process that delayed observational progress. His progress was further delayed when his one assistant, lt. William Meadows, proved unsatisfactory and left in December 1834. Meadows was eventually replaced by **Charles Piazzi Smyth**, the second son of captain Smyth. Maclear and Herschel observed the reappearance of Halley's comet (IP/Halley) in 1836, and made accurate observations of its path. After Herschel departed for England in March 1838, he reported favorably on Maclear's work, and recommended sending him more equipment and assistants.

One of Maclear's original instructions was to remeasure the meridian arc that **Nicolas de La Caille** had measured during

1751–1753. The arduous survey required most of the observatory's resources from 1837 to 1847. It showed that La Caille's measurements had been accurate but had not made allowance for the gravitational attractions of mountains on the plumb-bob.

The establishment of a magnetic observatory in 1841 was part of a global network of magnetic and meteorological observatories promoted by general **Edward Sabine** and supported by Herschel and others. Buildings and instruments were set up and run for 3 years by a detachment of the Royal Artillery, but when the magnetic observatory was transferred to the Admiralty in 1846, Maclear was instructed to add these observations to his duties. The magnetic work declined after 1857 and ceased on Maclear's retirement in 1869.

In 1833, Henderson had established a time service for vessels moored in Table Bay. Each night, a gun was fired to synchronize marine chronometers. Captain Robert Wauchope improved it with a time ball, which slid down a pole at an advertised time. His idea was adopted immediately by the Royal Observatory, Greenwich, in 1833 and by the Cape Observatory in 1836. A second time ball was erected at Simon's Town in 1857; they were connected to the observatory by electric telegraph in 1861. By August 1865, Maclear proudly declared that each day at one o'clock, the observatory ball, the Simon's Town ball, the Cape Town time gun, and the Port Elizabeth ball were discharged simultaneously.

In 1845, when Piazzi Smyth was appointed to succeed Henderson at Edinburgh, Maclear appointed William Mann as his first assistant. Mann helped erect and operate three new refractors: A 3-in. Dollond, a 3-in. Jones, and a 7-in. Mertz. A transit circle by Troughton and Simms was erected in the room formerly occupied by the "Mural Circle." Observations started in January 1855.

In 1859, Maclear spent a few months visiting England, Ireland, Paris, and Brussels. In June 1860, he was rewarded with a knighthood. The death of his wife in July 1861 caused him much grief, but his large family – six children were born in Cape Town – provided comfort to him. For his work on the extension of La caille's arc measurements, Maclear received the Lalande Prize of the French Academy of Sciences in 1867. In 1869, he was awarded the Gold Medal of the Royal Society. Other awards included membership of the Academy of Sciences of Palermo (1835), corresponding member of the Imperial Geological Institute and Geographical Society of Vienna (1858), and correspondent of the Institute of France (1863).

Maclear was greatly interested in exploration; with John Herschel he served on the Committee of the Association for Exploring Central Africa. In 1850, Maclear met David Livingstone and taught him how to use a sextant. They remained lifelong friends, and Livingstone sent his sextant and chronometer readings to Maclear for reduction. Maclear enthusiastically supported measures to improve the well-being of the Cape colony, taking particular interest in the provision of lighthouses, the standardization of weights and measures, and the improvement of hygiene.

After retiring from the observatory in 1870, Maclear went to live at Mobray, near Cape Town. He became blind in 1876. Half a dozen topological features in Cape Province carry his name; a 20-km-diameter lunar crater at 10°.5 N, 20°.1 E is named for him.

Ian Elliott

Selected References

Anon. (1880). "Sir Thomas Maclear." *Monthly, Notices of the Royal Astronomical, Society* 40: 200–204.

Evans, David S. (1967). "Historical Notes on Astronomy in South Africa." *Vistas in Astronomy* 9: 265–282.

Lastovica, Ethleen (1995) "Ardour in the Cause of Astronomy: Bibliography of the Publications of Sir Thomas Maclear with an Historical Introduction." *Transactions of the Royal Society of South Africa* 50: 65–77.

Warner, Brian (1979). "Thomas Maclear 1833–1870." In *Astronomers at the Cape Observatory, Cape of Good Hope*. Cape Town: A. A. Balkema.

——— (1995). "The Life and Astronomical Work of Sir Thomas Maclear." *Transactions of the Royal Society of South Africa* 50: 89–94.

Warner, Brian and N. Warner (1984). *Maclear and Herschel*. Cape Town: A.A. Balkema.

Macrobius, Ambrosius (Theodosius)

Flourished **5th century**

Neoplatonist Ambrosius Macrobius's diameter for the Sun (twice the diameter of the Earth) was cited frequently in the Middle Ages. The circuitous derivation of this figure can be read in his *Commentary on the Dream of Scipio*.

Selected Reference

Macrobius, Ambrosius Aurelius Theodosius (1990). *Commentarii in Somnium Scipionis* (Commentary on the dream of Scipio). Translated by William Harris Stahl. New York: Columbia University Press.

Mädler, Johann Heinrich von

Born **Berlin, (Germany), 29 May 1794**
Died **Hanover, Germany, 14 March 1874**

Johann von Mädler published a celebrated chart of the Moon and the first scientifically based selenography. He was also the author of the first map of Mars, made contributions to stellar astronomy, and compiled a useful history of astronomy.

Mädler was born into the family of a master tailor. Though frail as a child, at the age of 12 he was sent to the Friedrich-Werdersche Gymnasium in Berlin where he received a sound grounding in science and mathematics. His interest in astronomy was inspired by the Great Comet of 1811 (C/1811 F1). Although Mädler was an excellent scholar, he was unable to enter the university at the age of 19. An outbreak of typhus claimed both his parents, so instead of an academic career he considered it his duty to support four younger siblings. He enrolled in the tuition-free Kürstenschen seminary to prepare for a career as an elementary school teacher. At the same time, Mädler began giving lessons as a private teacher. In 1817, he got a job as a schoolmaster of calligraphy, and in 1819, Mädler founded a school for poor children. Meanwhile, he found time to attend lectures at the University of Berlin as an external student. Under the supervision of **Johann Bode**, **Johann Encke**, and Gustav Peter Lejeune Dirichlet (1805–1859), he studied astronomy and higher mathematics. With his enhanced education, Mädler was in a position to give private lessons at a higher level, a turning point in his life.

In 1824, **Alexander von Humboldt** introduced Mädler to the Berlin banker **Wilhelm Beer**, who applied to Mädler for lectures in higher mathematics and astronomy. Attracted by Mädler's lectures, Beer decided to set up his own observatory with Mädler as the main observer. A 97-mm refractor was installed in a small dome in the Tiergarten near Beer's home in 1828. There, Mädler and Beer began one of the more successful collaborations in the history of astronomy.

Mädler and Beer chose to map the surfaces of the Moon and Mars for their first projects. They observed Mars intently during that planet's perihelic opposition in September 1830, made drawings, and attempted to measure the coordinates of the most distinct spots. Their study left little doubt that the markings on Mars were permanent and disproved the previous belief that the spots on Mars were similar to the clouds of the Earth. In 1840 Mädler combined all the observations and drew the first map of Mars ever published. In the opinion of **Camille Flammarion,** Mädler and Beer deserve to be remembered as the true pioneers in this new conquest of Mars, a planet that had been the subject of intense study by "a phalanx of astronomers" for more than a century.

From 1830 to 1836, Mädler and Beer also observed the Moon. Mädler first measured the positions of a network of 106 reference points scattered across the lunar surface with a filar micrometer. Using these benchmarks, Mädler and Beer then measured the positions of 919 lunar formations, the heights of 1,095 mountains, and the diameters of 150 craters. On the basis of these measurements Mädler prepared the first scientifically designed lunar chart, *Mappa Selenographica*, which was published in four parts between 1834 and 1836. In 1837, a descriptive volume *Der Mond, nach zeinen kosmischen und individuellen Verhaltnissen oder allgemeine vergleichende selenography* (The Moon, concerning its cosmic and individual conditions or general comparative selenography) followed. In contrast to most of their predecessors, Mädler and Beer viewed the Moon as an airless, lifeless, and unchanging globe.

It is well known that both in the Mars project and in the lunar mapping and later in preparation of the *Selenograph* Mädler carried out most of the work, the observations, computation, map preparation, and writing. In the lunar-mapping project alone, Mädler spent 600 nights at the telescope. Although some observations were contributed by Beer, his role was primarily that of a patron who made the observatory available to Mädler.

In 1836, primarily because of the favorable reception of the lunar map, Encke employed Mädler as an observer at Berlin Observatory, a welcome relief from his previous occupation as a schoolteacher and part-time astronomer. Probably the best year in Mädler's life, however, was 1840 when he moved to Dorpat, Russia (now Tartu, Estonia) as the director of the observatory and professor of astronomy at the university, replacing **Friedrich Struve** when the latter left to found the Pulkovo Observatory. In the same year, Mädler married a poetess, Minna von Witte.

At Dorpat Observatory, Mädler used the 9-in. Fraunhofer refractor (the Great Dorpat refractor) for micrometric measurements of double stars from the catalog by Friedrich Struve. For 513 binaries he found the presence of orbital motions, for 15 binaries he calculated the orbit parameters. For 3,222 stars with positions observed by **James Bradley** from 1750 to 1762, Mädler found new positions on the basis of meridian observations at Dorpat Observatory and at other observatories, and calculated the proper motions. Subsequently these proper motions were used to study the motions in the stellar universe and to determine the solar motion parameters. Mädler correctly supposed that the motions of stars are governed by the collective gravitational field, but due to rather crude observational data of his time he was mistaken when he found that the center of our stellar system resides in the Pleiades cluster, not far from 180° from the true center of rotation in Sagittarius. In several papers and comments Mädler wrote about the sizes and periods of rotation of the planets.

In Dorpat, Mädler wrote popular books, read popular lectures, and actively contributed to local newspapers, besides doing the ordinary astronomer's work. The director's house near the observatory was a meeting place for literature for local friends. In 1865, Mädler retired from Dorpat University and went back to Germany to live in Wiesbaden, Bonn, and Hanover. In his retirement years, Mädler published an extensive and useful history of astronomy.

Mädler was a member of many scientific societies, the Madrid, Munich, and Wien academies and the Royal Astronomical Society among them. Nevertheless, he was not appointed to the Saint Petersburg academy because his relations with the influential academician Struve were not good. Struve also unsuccessfully opposed Mädler's appointment to the professorship at Dorpat University.

Mihkel Joeveer

Selected References

Both, Ernst E. (1961). *A History of Lunar Studies*. Buffalo, New York: Buffalo Museum of Science.

Eelsalu, Heino, Jürgen Hamel, and Dieter Hermann (1985). *Johann Heinrich Mädler, 1794–1874: Eine dokumentarische Biographie*. Berlin: Akademie-Verlag.

Mädler, Johann Heinrich. (Eight editions between 1841 and 1884). *Der Wunderbau des Weltalls, oder, Populäre Astronomie*.

——— (1846). *Astronomische Briefe*. Mitau.

——— (1846). *Die Centralsonne*. Dorpat.

——— (1847, 1848). *Untersuchungen über die Fixstern–Systeme*. Mitau.

——— (1856). *Die Eigenbewegungen der Fixsterne*. Dorpat.

——— (1862). *Astronomie zum Schulgebrauch*.

——— (1873). *Geschichte der Himmelskunde*. Braunschweig.

Sheehan, William P. and Thomas A. Dobbins (2001). *Epic Moon: A History of Lunar Exploration in the Age of the Telescope*. Richmond, Virginia: Willmann-Bell.

Whitaker, Ewen A. (1999). *Mapping and Naming the Moon: A History of Lunar Cartography and Nomenclature*. Cambridge: Cambridge University Press.

Magini, Giovanni Antonio

Born **Padua, (Italy), 13 June 1555**
Died **Bologna, (Italy), 11 February 1617**

Working at the cusp of Ptolemaic and Copernican astronomy, Giovanni Antonio Magini attempted to combine the best elements of both. Although Magini's theories have traditionally been identified as opposing **Galileo Galilei**'s, there is recent evidence that Magini

supported certain aspects of Galilei's work. Magini's contributions to mathematics and geography were also noteworthy.

Magini completed his early studies in Padua and then attended the University of Bologna, graduating in June 1579 with a degree in philosophy, although he had shown great interest in mathematics since childhood. After **Egnatio Danti** was transferred to Rome in 1587 the Bologna Senate called a competition for the chair of mathematics, which was assigned to Magini in August 1588. The Paduan scientist was chosen over the young Galilei, who had also applied for the chair, not only because of Magini's greater experience and notoriety at the time, but also because he had already published volumes of ephemerides and astronomical tables.

In addition to teaching Euclid, Magini also focused in particular on astrological and astronomical subjects, such as the theory of the planets, and on commentaries to **John of Holywood**'s *Sphaera Mundi* and **Ptolemy**'s astronomy. In 1597 he was given a lifetime teaching position and also received permission to go to Mantua. At the Gonzaga court, in 1599, Magini tutored the children of Duke Vincenzo, for whom he also wrote several astrological opinions. The sterling reputation he had earned among his peers was due also in part to his vast correspondence with all the illustrious scholars of his time, including Galilei, **Tycho Brahe**, and **Johannes Kepler**. Magini established an excellent rapport with Kepler, who in 1610 asked him to come to Prague to work with him on the new astronomical ephemerides. Magini did not accept Kepler's invitation, not only because he and the German astronomer had different viewpoints, but also because he did not want to leave the prestigious chair at Bologna.

According to 19th-century historian Antonio Favaro (editor of the complete edition of Galilei's works), Magini was one of Galilei's most dogged opponents. However, on the basis of recent studies by G. Betti, this view of Magini as Galilei's "enemy" seems exaggerated, particularly since Magini was probably the true author – or at least the direct inspiration – of the 1611 *Epistola Apologetica* against Martin Horky, written in Galilei's defense. Moreover, Magini's two best disciples, Cesare Marsili and Giovanni Antonio Roffeni, were Galilei's most ardent supporters in 17th-century Bologna.

Magini's attitude toward the Copernican system is intriguing. Although he was convinced that the Earth did not move, in some of his works he accepted **Nicolaus Copernicus**' theory as a working hypothesis. He justified this because it simplified calculations and yielded results that better matched observations, even though, in Magini's opinion, the theories were unlikely. Nonetheless, he never agreed with the concept of the Copernican system from a philosophical standpoint, replacing it with his own planetary model that combined the ideas of Copernicus and Ptolemy, and even added several new hypotheses. Magini claimed there was a need for a theory of planets that abandoned the model of the Alphonsine Tables in order to comply with recent observations, but rejected Copernicus' absurd hypotheses. Magini completely changed the Ptolemaic theories of the Sun and the Moon but adhered to the Ptolemaic system for the other five planets, albeit eliminating the equants. Furthermore, he accepted the idea that the stars and planets were pulled by their orbits or spheres and that they could not move independently. He also asserted that there had to be a ninth and tenth sphere between those of the fixed stars and the prime mover. For his theory of the Moon, Magini agreed with Copernicus in affirming that Ptolemaic theory did not comply with observation and experience. He later adopted the cosmological system of Brahe, with whom he established a rapport of both friendship and scientific collaboration. However, he modified the Tychonic system with elements from Kepler's astronomy. Magini defined his new theory in the *Tabulae novae iuxta Tychonis rationes elaborate,* but the work was unfinished when he died and was published posthumously in 1619. However, in 1623, 6 years after Magini's death, the Tribunal of the Holy Office ordered Magini's entire astrological library confiscated.

Magini was far more skilled at calculation than at theory, and the ephemerides he calculated for the years 1581 to 1630 are proof of this. He was a very talented instrument maker. He also wrote *Breve istruzione sopra le apparenze et mirabili effetti dello specchio concavo sferico* on concave mirrors. In 1592, Magini published *Tabula tetragonica sui quadrati dei numeri naturali*, which made it possible to determine the product of two factors, such as the difference between two squares. In 1609, he drew up accurate trigonometric tables in which he introduced new terms for the functions now known as cosine, cotangent, and cosecant. The nomenclature used by Magini attracted several followers and was adopted by **Bonaventura Cavalieri**. Magini also contributed to practical geometry with treatises on the sphere and on the application of trigonometry. He described the use of the quadrant and the astronomical square. Magini was the first person to suggest the use of the decimal point to separate the whole number from the decimals.

Magini was also active in medical astrology. He wrote a commentary on Galen's treatise, confirming that the stars govern the world of nature, and he recommended studying the annual recurrences of nativities and elections, essential for observing when the patient became ill, the critical days in the course of the illness, and the best times to administer medicine.

Magini's importance as a geographer and cartographer is undisputed. His edition of Ptolemy's *Guide to Geography,* which first appeared in Venice in 1596, is extremely important, not so much for Magini's careful descriptive comments but because he added 37 new maps to the 27 Ptolemaic maps, forming a true modern atlas.

However, the work to which Magini devoted most of the latter part of his life was an atlas of Italy, for which he prepared his own maps. Most of them were original and based on official surveys that various Italian governments had done. Because of this work, which Magini funded out of his own pocket, he was perennially in financial difficulty. The definitive compilation of the entire atlas, dedicated to Ferdinando Gonzaga, was published posthumously by Magini's son Fabio in 1620 with the title *Italia di Gio. Ant. Magini data in luce da Fabio suo figliolo*. The work, which had 61 tables and a brief commentary, enjoyed widespread and lasting fame.

Magini was buried in the Church of the Dominicans, with an epitaph dictated by his disciple Roffeni. His chair was offered to Kepler who, in a letter dated 15 May 1617 addressed to the rector of the University of Bologna, regretfully turned down the offer, fearful that as a Protestant he would feel ill at ease in a Catholic environment.

Fabrizio Bònoli

Selected References

Almagià, R. (1922). *L'Italia di Giovanni Antonio Magini e la cartografia dell'Italia nei secoli XVI e XVII*. Naples: F. Perrella.

Betti, G. (1997). "Il copernicanesimo nello Studio di Bologna." In *La diffusione del copernicanesimo in Italia (1543–1610)*. Florence: L. Olschki.

Bònoli, F. and E. Piliarvu (2001). *I lettori di astronomia presso lo Studio di Bologna dal XII al XX secolo*. Bologna: Clueb.

Favaro, A. (1886). *Carteggio inedito di Ticone Brahe, Giovanni Keplero e di altri celebri astronomi e matematici dei secoli XVI e XVII con Giovanni Antonio Magini*. Bologna: N. Zanichelli.

Mahendra Sūri

Flourished **(India), 14th century**

Mahendra Sūri's *Yantrarāja* (1370) helped to popularize Islamic astronomy in India.

Selected Reference

Plofker, Kim (2002). "Spherical Trigonometry and the Astronomy of the Medieval Kerala School." In *History of Oriental Astronomy*, edited by S. Razaullah Ansari, pp. 83–93. Dordrecht: Kluwer Academic Publishers.

Maimonides: Abū ʿImrān Mūsā [Moses] ibn ʿUbayd Allāh [Maymūn] al-Qurṭubī

Born **Cordova, (Spain), 1145 or 1148**
Died **Fusṭāṭ, (Egypt), 1204**

Maimonides, the renowned Jewish theologian and physician, also wrote on the relationship between Judaism and the sciences, astronomy in particular. He spent his formative years in Spain and North Africa. He eventually settled in Fusṭāṭ, near present-day Cairo where he achieved his great fame in learning and communal leadership. Nonetheless, Maimonides remained attached to the intellectual outlook of the western part of the Islamic world throughout his life, and this is especially true of his work in astronomy. In his youthful search for guidance, especially in matters of cosmography (which were later to be a major concern), he sought out the son of **Jābir ibn Aflaḥ** as well as some pupils of **Ibn Bājja**. Indeed, his career affords us one of the clearest examples of the distinctive features of the western Islamic astronomical tradition. Maimonides contributed to the Arabic astronomical literature by editing (*i. e.*, preparing corrected versions of texts that had become problematic) books written by two of his Andalusian predecessors, the above-mentioned Jābir and Ibn Hūd, ruler of Seville.

Astronomical issues are stressed at several places in Maimonides' great work of religious thought, the *Guide of the Perplexed*. The most detailed discussion is found in Part Two, Chapter 24, which is devoted entirely to a review of the state of what may be anachronistically called cosmology or celestial physics. Aristotelian physics had established by means of what were then taken to be irrefutable proofs that the motions of the heavenly bodies must be circular, with the Earth at the center. **Ptolemy**'s models clearly violate these principles. All of the solutions that had been offered to date were critically scrutinized and rejected; these included the proposals of **Thābit ibn Qurra** and Ibn Bājja, for which Maimonides remains our only source. Did Maimonides consider the problem insoluble, or to put it differently, did he think the "true configuration" to be beyond human ken? Opinions have differed sharply on this point. It is noteworthy, however, that Maimonides breaks away from some of the Andalusians in that he does not think the solution to lie in rediscovering **Aristotle**'s cosmology. Maimonides firmly believed that astronomy had advanced considerably since Aristotle's day. Although the Stagirite's proclamations in physics remain true, his teachings in astronomy can no longer be maintained. In this respect Maimonides' position is closer to that of the Egyptian **Ibn al-Haytham**.

Maimonides' sole contribution to mathematical astronomy is his procedure for determining the visibility of the lunar crescent, which takes up several chapters of his great law code, the *Mishneh Torah*. Before the calendar was fixed, Jewish law required that the beginning of each month be certified by the court at Jerusalem. No month can exceed 30 days. Hence, if the crescent is not seen on the eve of the 29th, the declaration of the new month is automatic. Maimonides' procedure is necessary only for those instances where witnesses do report a sighting on the eve of the 29th. Specifically, the members of the court need to know whether a sighting is possible, so that they may convene in the expectation of witnesses; and they need a few details about the appearance of the crescent for purposes of cross-examination. Conversely, the court needs to know when a sighting will be impossible, so as to be able to reject any purported sightings.

With these facts in mind, it will be readily understood why Maimonides presents his method in "cookbook" fashion. Solar and lunar parameters, listed by Maimonides, can be plugged in, and the computation is then carried out step-by-step. Eventually the result is a simple yes or no answer; if the answer is yes, some additional information about the appearance of the crescent can be obtained. Theoretical explanations or justifications are kept to a bare minimum. Certain parameters, for example the geographical latitude, are built in, since the computation is meant to be true only for Jerusalem and its environs. Maimonides states that he has allowed himself some approximations, but, he assures us, the round-off errors cancel each other out, so that there is no net effect on the computation.

Maimonides issued some critically important and repercussive statements on the relationship between Judaism and the sciences, astronomy in particular. He asserted that ancient Rabbinic views on the structure of the heavens have no privileged position. The tenets of astronomy can be proven or rejected by universal and invariant rules of logic; hence their source, or, as we might say, the cultural context out of which they emerge, is irrelevant. On the other hand, astronomy is by no means a "secular" science. Knowledge of God, the attainment of which is a primary religious obligation, can be approximated – Maimonides denies that it can be fully achieved – only by inference from creation. The stars are the most noble bodies in creation, and the study of their motions is one of the most religiously fulfilling activities at our disposal.

Y. Tzvi Langermann

Selected References

Freudenthal, Gad (1993). "Maimonides' Stance on Astrology in Context: Cosmology, Physics, Medicine, and Providence." In *Moses Maimonides: Physician, Scientist, and Philosopher*, edited by Fred Rosner and Samuel S. Kottek, pp. 77–90. Northvale, New Jersey: Jason Aronson. (Connects Maimonides's work in astronomy to his endeavors in other sciences and in philosophy.)

Langermann, Y. Tzvi (1991). "Maimonides' Repudiation of Astrology." *Maimonidean Studies* 2: 123–158. (Discusses the scientific, philosophical, and religious grounds of Maimonides' position.)

——— (1999). "Maimonides and Astronomy: Some Further Reflections." Essay IV in *The Jews and the Sciences in the Middle Ages*. Aldershot: Ashgate. (Study on Maimonides' views on cosmology, with references to earlier studies by the author and others.)

——— (2000). "Hebrew Astronomy: Deep Soundings from a Rich Tradition." In *Astronomy Across Cultures*, edited by Helaine Selin, pp. 555–584. Dordrecht: Kluwer Academic Publishers. (First section discusses Maimonides's privileging of scientific truth over traditional texts.)

Maimonides, Moses (1956). *Sanctification of the New Moon*, translated by Solomon Gandz. New Haven: Yale University Press. (English version of Maimonides's codification of the laws appertaining to the Jewish calendar, including his procedure for determining the visibility of the lunar crescent, which is clearly reformulated in modern language in an "Astronomical Appendix" by Otto Neugebauer.)

Mairan, Jean-Jacques

Born **Béziers, (Hérault), France, 26 November 1678**
Died **Paris, France, 20 February 1771**

French naturalist Jean-Jacques de Mairan shares credit (1721) with **Henry Cavendish** and John Dalton for accurately determining the height of the aurora. Mairan thought that the phenomenon was caused by zodiacal light particles falling into the Earth's atmosphere.

Alternate name

Dortous de Mairan, Jean-Jacques

Selected Reference

Akasofu, Syun-Ichi (2002). "Secrets of the Aurora Borealis." *Alaska Geographic* 29, No.1. Anchorage: Alaska Geographic Society.

Majrīṭī: Abū al-Qāsim Maslama ibn Aḥmad al-Ḥāsib al-Faraḍī al-Majrīṭī

Born **Madrid, (Spain), first half of the 10th century**
Died **Cordova, al-Andalus, (Spain), 1007**

Maslama al-Majrīṭī was considered by his Andalusian contemporaries as the foremost authority of his time in the field of astronomy. He traveled as a young man to Cordova, the capital of the Umayyad caliphate, where he studied and worked until his death. His achievements are mainly in the field of mathematical astronomy, although it is known that he wrote on commercial arithmetic (*muʿāmalāt*) and was also a renowned astrologer. Historians have at times misattributed to Majrīṭī works on magic and alchemy.

In addition to his own compositions, Majrīṭī's importance lies within the context of Andalusian science and his activity in scientific teaching. Majrīṭī was the founder of an original school of Andalusian astronomers in which the disciplines of arithmetic and geometry were also cultivated. Majrīṭī's disciples, who include outstanding figures like **Ibn al-Samḥ**, **Ibn al-Ṣaffār**, and Ibn Barghūth (died: 1052), spanned three generations and greatly influenced the development and expansion of the exact sciences throughout al-Andalus. Majrīṭī brought together for the first time in al-Andalus two distinct mathematical traditions, namely the tradition of *farāʾiḍ* (religiously based division of inheritances) and the tradition of mathematically based philosophical sciences, a category that included astronomy. Majrīṭī's combining of these two mathematical branches reflects the interests of his two known teachers: ʿAbd al-Ghāfir ibn Muḥammad al-Faraḍī, who wrote a treatise on *farāʾiḍ*, and ʿAlī ibn Muḥammad ibn Abī ʿĪsā al-Anṣārī, who is reported to have known astronomy.

In the field of astronomy, Majrīṭī was the first Andalusian to make his own astronomical observations. According to **Zarqālī**, he observed the star Regulus in the year 979 and found its ecliptical longitude to be 135° 40′. Starting from the determination of the longitude of this star, Majrīṭī was then able to determine the longitude for all fixed stars, thereby establishing a movement of precession of the equinoxes of 13° 10′ with respect to the epoch of compilation of the catalog of stars in **Ptolemy**'s *Almagest*.

The above value for the longitude of Regulus appears in the table of stars that accompanies Majrīṭī's commentary on Ptolemy's *Planisphaerium*, which is a treatise on the stereographic projection of the sphere (the basic technique for the construction of the standard astrolabe). Some historians mistakenly thought that Majrīṭī may have learned Greek and translated the *Planisphaerium* himself, but recent investigation has shown that he most likely revised an eastern Arabic translation of the work. Indeed, Majrīṭī's text contains several additions to the work of Ptolemy that considerably improved the procedures for tracing the fundamental lines of the astrolabe and for locating the fixed stars of its rete, or star map on the instrument, using several kinds of coordinates. In the second part of this work, Majrīṭī deals with a number of problems of spherical astronomy using the Theorem of Menelaus, which was the unique trigonometric tool employed in his time and upon which he had previously written several notes in another work.

Majrīṭī's major work in astronomy was the adaptation that he made, together with his disciple Ibn al-Ṣaffār, of **Khwārizmī**'s *Sindhind zīj*. This 9th century astronomical handbook with tables and explanatory text was based primarily on Indian methods, and thus differed from later Islamic astronomical material, which relied on planetary models laid out in the *Almagest*. Although Khwārizmī's original text appears to be lost, a Latin version by **Adelard of Bath** (12th century) of Majrīṭī's revision is extant. This text, which is referred to as the *zīj* of Khwārizmī-Maslama (Majrīṭī), contains tables derived from Khwārizmī's original *zīj* (which had material based upon Persian and Ptolemaic traditions in addition to Indian ones) as well as material and tables that were adaptations, additions, or replacements introduced by Majrīṭī and Ibn al-Ṣaffār. The aim of the Andalusian astronomers was to adapt the original tables to the time and place in which they were living. For example, the Persian solar calendar used in Khwārizmī's tables was replaced by the Muslim lunar calendar, and some tables that were observer-specific were adapted to the geographical coordinates of Cordova. Khwārizmī's mean motion tables were calculated for radix positions corresponding to the meridian of Arīn (the center of the world in the Indian systems). A significant outcome of using Cordova's longitude was that Majrīṭī provides the earliest evidence of an important correction to the size of the Mediterranean Sea to its actual size; this was preserved in most Andalusian geographical tables. On the whole, the transformations affected the tables for chronology, mean motions, mean conjunctions and oppositions, and visibility of the lunar crescent. They also involved the addition of new tables related to the astrological practices of equating the houses and projecting the rays. Moreover, the contents of the final version of the *zīj* suggest the redactors included some elements that, though not strictly necessary, were in use in contemporary Andalusia. This is the case of the two trigonometric tables that are extant in the Latin translation, one for the sine (based on a radius of 60 parts) and the other for the cotangent (shadow length), which presumably were not used in the original *Sindhind*. Other Andalusian contributions found in the *zīj* are the reference to the Hispanic era (38 BCE) in the chronological part, the use of the meridian and latitude of Cordova for certain tables, and improved calculation methods that were both accurate and easier to use.

As a professional astrologer, Majrīṭī was also interested in the conjunction of Saturn and Jupiter, which took place in 1006/1007; with it he foretold a change of dynasty, ruin, slaughter, and famine.

Josep Casulleras

Selected References

Balty-Guesdon, Marie Genevievè (1992). "Médecins et hommes de sciences en Espagne musulmane (IIe/VIIIe-Ve/XIe s.)." Ph.D. diss., Université de la Sorbonne Nouvelle – Paris III.

Comes, Mercè (1994). "The 'Meridian of Water' in the Tables of Geographical Coordinates of al-Andalus and North Africa." *Journal for the History of Arabic Science* 10: 41–51. (Reprinted in *The Formation of al-Andalus, Part 2: Language, Religion, Culture and the Sciences*, edited by Maribel Fierro and Julio Samsó, pp. 381–391. Aldershot: Ashgate, 1998.)

——— (2000). "Islamic Geographical, Coordinates: al-Andalus' Contribution to the Correct Measurement of the Size of the Mediterranean." In *Science in Islamic Civilisation: Proceedings of the International Symposium "Science Institutions in Islamic Civilisation" and "Science and Technology in the Turkish and Islamic World,"* edited by Ekmeleddin Ihsanoğlu and Feza Günergun, pp. 123–138. Istanbul: IRCICA.

Dalen, Benno van (1996). "al-Khwārizmī's Astronomical Tables Revisited: Analysis of the Equation of Time." In *From Baghdad to Barcelona: Essays on the History of the Islamic Exact Sciences in Honour of Prof. Juan Vernet*, edited by Josep Casulleras and Julio Samsó, Vol. 1, pp. 195–252. Barcelona: Instituto "Millás Vallicrosa" de Historia de la Ciencia árabe.

Fierro, Maribel (1986). "Bāṭinism in al-Andalus. Maslama b. Qāsim al-Qurṭubī (d. 353/964), Author of the *Rutbat al-Ḥakīm* and the *Ghāyat al-Ḥakīm (Picatrix)*." *Studia Islamica* 63: 87–112.

Hogendijk, Jan P. (1989). "The Mathematical Structure of Two Islamic Astrological Tables for 'Casting the Rays.'" *Centaurus* 32: 171–202.

Kunitzsch, Paul and Richard Lorch (1994). *Maslama's Notes on Ptolemy's* Planisphaerium *and Related Texts*. Munich: Bayerischen Akademie der Wissenschaften.

Samsó, Julio (1992). *Las ciencias de los antiguos en al-Andalus*. Madrid: Mapfre, pp. 80–110.

Vernet, Juan and María Asunción Catalá (1965). "Las obras matemáticas de Maslama de Madrid." *Al-Andalus* 30: 15–45. (English translation in *The Formation of al-Andalus, Part 2: Language, Religion, Culture and the Sciences*, edited by Maribel Fierro and Julio Samsó, pp. 359–379. Aldershot: Ashgate, 1998.)

Makaranda

Flourished **Kāśī (Vārānasī, Uttar Pradesh, India), 1478**

Makaranda, surnamed Ānandakanda, computed many tables of astronomical phenomena that he published in very useful forms. A reputed astronomer of Kāśī in North India, the hub of intellectual activity in India during medieval times, Makaranda was a follower of the Saura School, one of four principal schools of Hindu astronomy active during the classical period (late 5th to 12th centuries).

Makaranda's work, known simply as *Makaranda*, is an extensive treatise containing many astronomical tables that enable one to read the dates and times of different celestial phenomena. The tables span a large number of years after 1478, when they were commenced. The astronomical phenomena covered by Makaranda are *tithis* or lunar days, *nakṣatras*, or asterisms, *yogas* marking complementary positions of the Sun and Moon, *saṁkrāntis*, or the times of entry of the Sun into the zodiacal signs, the mean motions of the planets and their anomalies, the length of daylight on different days, weekdays, and times of eclipses. As a way of making access to the work easier, Makaranda provided, in certain cases, two sets of tables, one for single years and the other for groups of years.

The labor involved in the preparation of these tables must have been enormous and entailed much ingenuity. But this labor has benefited later generations of astronomers and astrologers by reducing their own time and effort. Makaranda's works are among the most popular in North India, especially at Bihar and Bengal. More than 20 commentaries were prepared on Makaranda's work by later astronomers that explained the principles of construction of the tables and their practical use, which attest to their popularity among the masses.

Ke Ve Sarma

Selected References

Pingree, David (1968). "Sanskrit Astronomical Tables in the United States." *Transactions of the American Philosophical Society* 58, no. 3: 1–77, see 39–46.

——— (1974). "Makaranda." In *Dictionary of Scientific Biography*, edited by Charles Coulston Gillispie. Vol. 9, p. 42. New York: Charles Scribner's Sons.

——— (1978). "History of Mathematical Astronomy in India." In *Dictionary of Scientific Biography*, edited by Charles Coulston Gillispie. Vol. 15 (Suppl. 1), pp. 533–633. New York: Charles Scribner's Sons.

Makemson, Maud Worcester

Born **Center Harbor, New Hampshire, USA, 16 September 1891**
Died **Weatherford, Texas, USA, 25 December 1977**

Maud Worcester Makemson, astronomy professor and director of the observatory at Vassar College, New York, worked in celestial mechanics and was a pioneering practitioner of archaeoastronomy (especially Polynesian and Mayan astronomy) and astrodynamics (the application of celestial mechanics to spacecraft).

Makemson took up astronomy late in life. Born Maud Lavon Worcester, daughter of Ira Eugene and Fannie Malvina (*née* Davisson) Worcester, she studied classics at the Boston Girls' Latin School (graduating in 1908) as well as at Radcliffe College; eventually she became conversant with Latin, Greek, French, German, Spanish, Italian, Japanese, and Chinese. After a year at Radcliffe, she began teaching in a rural one-room school in Sharon, Connecticut.

In 1911, Worcester moved with her family to a ranch in Pasadena, California. There she met farmer Thomas Emmet Makemson, whom she married the next year (1912), and moved with him to Arizona. By her fifth year of marriage, she had three children: a daughter Lavon (born: 1913) and two sons Donald (born: 1915) and Harris (born: 1917). She had also launched herself into a writing career, becoming a reporter for the *Arizona Gazette* in Phoenix in 1918, and eventually publishing two original plays. In 1919, she and Thomas divorced.

An astronomical spectacle in 1921, however, transformed the direction of Makemson's life. The night of a picnic in the desert north of Phoenix, she was dazzled by a remarkable aurora (widely witnessed throughout the United States on 14 and 15 May), its streamers so bright that they cast enough light to read. The next day, newspapers noted the display had coincided with the appearance of large sunspots. Her curiosity aroused, Makemson began devouring popular books on astronomy. That September, she resigned her newspaper job, moved her children to Palmdale, California (near her parents' home in Pasadena), and supported herself teaching grade school while taking correspondence courses in trigonometry and astronomy from the University of California. In August 1923, Makemson enrolled full-time at the University of California at Berkeley, receiving her BA in 1925 at the age of 34. Over the next 5 years, she earned her MA (1927) and Ph.D. (1930) in astronomy (for work in celestial mechanics under **Armin Leuschner**), receiving several fellowships and research assistantships.

After a year at the University of California as an instructor of astronomy (1930/1931) and another of teaching mathematics and astronomy at Rollins College in Winter Park, Florida (1931/1932), Makemson became an assistant professor of astronomy and navigation at Vassar College. For the next quarter century, Makemson remained at Vassar College , the first astronomy faculty member there not to have been a student of **Maria Mitchell**. By her retirement in 1957, she was director of the college's observatory (from 1936), chair of the astronomy department (from 1941), and a full professor (from 1944).

At Vassar College, Makemson did yeoman work in practical celestial mechanics, calculating the orbits of comets, asteroids, and double stars, as well as teaching astronomy, history of astronomy, and meteorology. She also introduced the heavens to thousands of school children, high-school students, and scout troops who viewed celestial wonders through the observatory's telescopes.

Moreover, Makemson began to spread her intellectual wings farther afield, making use of her knack for languages to pursue what would now be called archaeoastronomy (although her work is not widely cited by archaeoastronomers today). During the summer of 1935, she worked at the Bishop Museum in Honolulu, Hawaii, beginning research on Polynesian astronomy (gaining some contemporary notoriety by suggesting that the legendary star Kokoiki, which allegedly appeared just before the birth of Hawaii's first king, Kamehameha I, might have been Halley's comet (IP/Halley) in 1758). In 1941, Yale University Press published Makemson's book *The Morning Star Rises: An Account of Polynesian Astronomy*, based on the writings of missionaries, Polynesian historians, anthropologists, as well as her own astronomical research (which used her familiarity with Polynesian languages to verify translations of ancient chants).

In 1941 and 1942, Makemson held a John Simon Guggenheim Memorial Foundation Fellowship for the study of Mayan astronomy. Some of her findings were published in 1946 in *The Maya Correlation Problems* (on her attempt at a correlation between the ancient Mayan calendar and the Julian and Gregorian calendars) and in 1951 in *The Book of the Jaguar Priest* (her translation of the Mayan calendar from the original hieroglyphs). In 1953/1954 she was granted a Fulbright Fellowship to teach in Japan.

Upon her retirement from Vassar College in 1957, Makemson returned to California and entered yet her third career, this time in the booming new field of space technology. From 1959 to 1964, she was research astronomer and lecturer at the University of California at Los Angeles [UCLA] and consultant to Consolidated Lockheed-California (1961–1963).

Makemson generally taught positional (navigational) astronomy, and managed to make it seem perfectly natural that a 70-year-old woman should be doing this at a university with no other female faculty in astronomy, mathematics, or physics. Her touch was light, and a characteristic final-examination question said, "you are lost on a desert island with a sextant, chronometer, carrier pigeon and your copy of Smart's *Spherical Astronomy*. Explain how you will save yourself."

At UCLA, Makemson met Robert M. L. Baker Jr., who in 1958 had just received his Ph.D. in engineering, the first of its kind to be granted in the United States with the specialty in astronautics. Although a handful of schools then were teaching astrodynamics – a newly coined word referring to the application of celestial mechanics to spacecraft in orbit around the Earth or on trajectories to the Moon or planets – no textbook on the subject yet existed. Aware of Makemson's experience in celestial mechanics, astronomical history, and book publishing, Baker invited her to coauthor *An Introduction to Astrodynamics*. Over the next decade, their text stood as the only one in the field, going through two editions and multiple printings.

In 1965, in the heyday of the manned space program, Makemson moved to Texas and worked as a consultant to the National Aeronautics and Space Administration [NASA] through the Applied Research Laboratories of General Dynamics in Fort Worth. By 1971, she had devised a technique for the Apollo astronauts to determine their selenographic latitude and longitude by photographing the positions of stars through a zenith telescope, allowing them to navigate around the Moon's surface without radio or radar (although her method does not appear to have been used).

Intellectually active well into her 80s, Makemson's last project was translating *The Astronomical System of Philolaus*, originally published in Latin by **Ismaël Boulliau** in 1645. It was still incomplete at her death in a nursing home. She was survived by one son (who had legally changed his name to Donald Worcester), seven grandchildren, and eleven great-grandchildren.

In addition to her two books, articles by Makemson were published in an eclectic variety of scholarly journals and semipopular magazines, ranging from *American Anthropology* to the *Astronomical Journal* to *The Sky* to the *Bulletin of the Seismological Society of America*.

Trudy E. Bell

Selected References

Anon. "Mrs. Maud Worcester Makemson." Biographical File. Vassar College Archives, Poughkeepsie, New York.

Anon. (17 November 1935). "Vassar Professor May Upset Legend." *New York Times*.

Bailey, Martha J. (1994). "Makemson, Maud Worcester." In *American Women in Science: A Biographical Dictionary*, pp. 232–233. Denver: ABC-CLIO.

Baker, Jr., Robert M. L. and Maud W. Makemson (1960, 1967). *An Introduction to Astrodynamics*. New York: Academic.

Kiser, Helene Barker (1999). "Maud Worcester Makemson." In *Notable Women Scientists*, edited by Pamela Proffitt, pp. 350–351. Detroit: Gale Group.

Larsen, Kristine (2001). "Maud Worcester Makemson." In *Women in World History*, edited by Anne Commire, Vol. 10, pp. 119–120, Waterford, Connecticut: Yorkin Publishers.

Makemson, Maud W. (1936). "The Legend of Kokoiki and the Birthday of Kamehameha I." In *Forty-Fourth Annual Report of the Hawaiian Historical Society for the Year 1935*, pp. 44–50. Honolulu, Hawaii.

——— (1939). "South Sea Sailors Steer By The Stars." *Sky* 3: 3–5, 19–21.

Maksutov, Dmitry Dmitrievich

Born **Nikolayev (Mykolayiv, Ukraine), 11/23 April 1896**
Died **Leningrad (Saint Petersburg, Russia), 12 August 1964**

Dmitry Maksutov was a Soviet optician who is credited as a leading designer of astronomical optical instruments after World War II. A graduate (1913) of cadet school, he was enrolled in the Military Engineering College of Saint Petersburg, but his study was interrupted by World War I. Like other members of his family, he fought in that conflict, and after the 1917 Bolshevik Revolution, attempted to immigrate to the United States through China but failed. Maksutov continued his education at the Tomsk Technical Institute (Siberia) and worked in Saint Petersburg and Odessa, Ukraine, where in 1930 he was arrested but miraculously survived while every other randomly chosen "suspect" was shot. He was arrested for a second time during Stalin's Great Terror in 1937.

From 1930 to 1952, Maksutov worked in the State Optical Institute (Leningrad), where he founded and headed the laboratory for astronomical optics. After 1952, he worked at the Pulkovo Observatory. He was elected a corresponding member of the Soviet Academy of Sciences (1946). On two occasions (1941, 1946), Maksutov was awarded the highest scientific trophy of the USSR, the Stalin Prize, which was renamed the State Prize after Stalin's death. In spite of these honors and his international recognition, Soviet authorities never permitted him to travel abroad.

While not the first to consider melding the best attributes of a refractor and a parabolic reflector, in 1941 Maksutov proposed a meniscus optical system of exceptional performance (the Maksutov telescope). This compact type of a catadioptric telescope differs from the Schmidt design in that the correcting plate is a deeply curved, diverging meniscus lens. Since the primary mirror is also spherical, all three optical surfaces are simple to manufacture. The spherical aberration of the meniscus lens exactly balances that of the primary mirror, yielding a compact and well-corrected optical instrument, though stable and very precise alignment of the optical system is critical. Combining many advantages, Maksutov telescopes became very popular throughout the entire world, although they cannot have a large diameter. American optical engineer John Gregory popularized a variant of this system, foreseen by Maksutov, wherein an aluminized spot on the meniscus lens serves as a Cassegrain secondary. In the United States, a small coterie of amateur telescope makers founded the Maksutov Club, led by Allan Makintosh, and for many years published *Maksutov Club Circulars*.

During the era of slide rules and logarithm tables, Maksutov was actively involved in the design of many astronomical instruments, including the 6-m telescope (then the world's largest) called the Large Altazimuth Reflector [BTA] at the Special Astrophysical Observatory near Zelenchukskaya. He wrote several textbooks on astronomical optics. He is commemorated with a crater on the Moon's farside.

Alexander A. Gurshtein

Selected References

Kopal, Z. (ed.) (1956). *Proceedings of a Symposium on Astronomical Optics and Related Subjects Held in the University of Manchester April 19–22, 1955*. Amsterdam: North Holland.

Maksutov, D. D. (1954). *Technologie der astronomischen Optik*. Berlin: Verlag Technik.
——— (1984). *Manufacturing and Investigation of Astronomical Optics* (in Russian). 2nd ed. Moscow: Nauka.
Petrunin, Yuri and Eduard Trigubov (2001). "Dmitri Maksutov: The Man and His Telescopes." *Sky & Telescope* 102, no. 6: 52–58.

Malapert, Charles

Born **1581**
Died **1630**

French astronomer Charles Malapert published one of the earliest drawings of the Moon, as seen through a telescope (1619). A Jesuit, he defended geocentric cosmology against **Galileo Galilei** and Copernican heliocentricity. A lunar crater is named for him.

Selected Reference

Birkenmajer, Alexander (1967). "Alexius Sylvius Polonus (1593–ca. 1653), a Little-Known Maker of Astronomical Instruments." *Vistas in Astronomy* 9: 11–12.

Malebranche, Nicholas

Born **Paris, France, 6 August 1638**
Died **Paris, France, 13 October 1715**

Nicholas Malebranche, a prominent natural philosopher who wrote on the metaphysical nature of the Universe, developed a synthesis of Cartesian rationalism with accepted Christian dogma. His replacement of the Cartesian subtle matter with miniature elastic vortices received its most elaborate mathematical treatment in the work of **Johann** and **Daniel Bernoulli**, who attempted to make it consistent with the work of **Johannes Kepler**.

Malebranche was the youngest of 13 children born to the prosperous family of Nicholas Malebranche and Catherine de Lauzon. Because of a spinal condition, he was educated at home until the age of 16. He then moved to the Collège de la Marche, where he received the degree of *Maître ès Arts* in 1656. Malebranche studied theology at the Sorbonne University for the next 3 years and then in 1660 entered the oratory, a papally approved Augustinian order dedicated to reform the Catholic Church from within. He would remain in the order until his death. In 1664 there occurred two events of particular importance in Malebranche's life. He was ordained a priest, and had his first encounter with **René Descartes**' physics. His reading of Descartes' *Traité de l'homme* (*Treatise on man*) would contribute to his adoption of the view that all natural phenomena are to be explained in terms of matter and the laws that govern its motions. In 1669 Malebranche was elected to the Académie royale des sciences for his *Treatise on the Laws of the Communication of Movement*.

A popular move in 17th-century science was to attempt to subsume natural phenomena under general laws. Despite the great success of this in the work of figures like **Isaac Newton**, an issue that still bothered philosophers was the cause of the adherence of bodies to these laws. Some, like the Cambridge Platonists Henry More and Ralph Cudworth, argued that God created immaterial viceregents to keep bodies in line. Even Newton worried about this issue; he flirted with the latter view, also with the view that God Himself acts on bodies to make them adhere to laws. Famously, at the end of the day, Newton opted to "feign no hypotheses" on the issue. Where Newton did not like the metaphysical and theological consequences of the view that God does each and everything, Malebranche embraced them and so was a full-blown occasionalist.

Although he held that God does each and everything, Malebranche did not think that scientific explanations ought constantly to appeal to God's activity. Instead, he argued that they ought to be given in terms of the laws that God has instituted, the laws in accordance with which He constantly acts. Malebranche's view was that since the divine attributes include order and simplicity, God's constant activity is in accordance with general laws. He appears to have been committed to the view that even miracles are in accordance with God's general laws and that we call something a "miracle" when it is anomalous with respect to what we mistakenly take to be the laws in place. (An alternate interpretation has Malebranche committed to the view that God acts in accordance with general laws except when performing miracles.) Nonetheless, Malebranche thinks that in "explaining" a particular event, we should not say that God brings it about; we should instead appeal to the general laws under which it is subsumed. Science, for Malebranche, is the search to uncover these laws. He thus contributed to the tendency in 17th-century science to offer explanations in terms of matter and the laws that govern its motion.

An interesting wrinkle in Malebranche's view arises from a consideration of his deeper metaphysics. He adhered to a representational theory of perception, holding that in sensory perception we perceive objects indirectly *via* our mental representations of them. When we observe an object, we are really just having a mental perception that might or might not correspond to an actual object. Since a perception, like everything else, is caused by God, the question for Malebranche was not whether or not our perceptions correspond to their causes but whether or not there is a material reality that God has created to correspond to these perceptions. In fact, Malebranche held that we cannot know that there are any material objects, except by faith. Malebranche could even hold that the material Universe is perfectly harmonious and orderly. A common maneuver for people like **Johannes Kepler** was to insist on the mathematical order of the Universe even when the astronomical data suggested something less. Malebranche's system allowed him to not take empirical data so seriously. Our perceptions might sometimes be of anomalies and irregularities, but Malebranche could insist that these perceptions do not tell us all about the actual material reality that corresponds to them. Since God's Universe would be maximally perfect and harmonious, Malebranche could ignore unhappy sensory perceptions and hold that our best idealizations of the Universe describe it exactly.

David Cunning

Selected References

Malebranche, Nicholas (1992). *Treatise on Nature and Grace*, edited and translated by Patrick Riley. Oxford: Clarendon Press.

——— (1997). *Dialogues on Metaphysics and on Religion*, edited by Nicholas Jolley and translated by David Scott. Cambridge: Cambridge University Press.

McCracken, Charles (1998). "Knowledge of the Existence of Body." In *The Cambridge Companion to Seventeenth-Century Philosophy*, edited by Daniel Garber and Michael Ayers, Vol. 1, pp. 628–634. Cambridge: Cambridge University Press.

Nadler, Steven (ed.) (2000). *The Cambridge Companion to Malebranche.* Cambridge: Cambridge University Press.

Radner, Daisie (1993). "Occasionalism: Malebranche." In *The Renaissance and Seventeenth-century Rationalism*, edited by G. H. R. Parkinson, pp. 361–371. London: Routledge.

Mallius

➜ **Manilius [Manlius], Marcus**

Malmquist, Karl Gunnar

Born **Ystad, Sweden, 2 February 1893**
Died **Uppsala, Sweden, 27 June 1982**

Swedish astronomer (Karl) Gunnar Malmquist is eponymized in the Malmquist bias, the idea that a sample of distant objects will inevitably be dominated by the brightest ones, compared to a sample of nearby objects. He wrote down very useful equations for correcting this bias in the early 1920s, although the basic idea was already implicit in earlier work by **Jacobus Kapteyn**.

Malmquist was the son of Emil Vilhelm and Anne Alfrida (*née* Persson) Malmquist. By his first wife, Hanna Karola Gertrud Ingeborg (*née* Lundvall), he had two sons, Sten (a professor of statistics at Stockholm University) and Olle (a medical doctor). Hanna died in about 1951, and a second late marriage to Lisa Malmquist was childless.

Malmquist studied under **Carl Charlier** at the Lund Observatory where he developed methods of mathematical statistics for the analysis of astronomical data, receiving his Ph.D. in 1921. He moved to the Stockholm Observatory in 1931, participating in building the observatory at Saltsjöbaden. Malmquist was appointed professor at Uppsala University in 1939, where he continued his earlier theoretical work on observations of the Milky Way. In addition, he contributed to the founding of the Kvistaberg Observatory and the Uppsala Southern Station on Mount Stromlo in Australia. Malmquist was an active member of the Royal Academy of Sciences in Stockholm, the secretary of Royal Academy of Sciences in Uppsala from 1948 to 1963, and a member of numerous scientific societies.

Malmquist is best remembered for his description of the Malmquist bias, which plays a significant role in both stellar statistics and cosmology and about which an extensive literature exists. The classical Malmquist bias is that using a sample complete to some apparent brightness inevitably yields mean values for any measured quantity that are more and more dominated by the brightest objects at the largest distances. In general this is described by an integral equation incorporating several factors including the line-of-sight density and reddening distributions, the intrinsic properties of the observed sources, and the sensitivity of the detector. The equation has no simple solution. Malmquist showed, however, that under the simplifications of homogeneous space distribution, no absorption along the line-of-sight, and Gaussian distribution of absolute magnitudes with dispersion σ and intrinsic mean of M{o}, the mean value of M{m}, for any apparent magnitude m, is related to the intrinsic mean by his well known result:

$$M\{m\} = M\{o\} - 1.382\sigma^2.$$

Different forms of bias plague the measurement of many astronomical quantities, including number counts of sources, estimation of distances, and the motions of galaxies; properly correcting for biases is an important step in gaining knowledge about the Universe. Correct determination of the Hubble constant is particularly sensitive to Malmquist bias.

Other astronomical topics to which Malmquist made significant contributions include the large-scale inhomogeneities of the distribution of bright young stars in the galactic plane and the significance of interstellar absorption outside the plane. The asteroid (1527) Malmquista is named is his honor.

Gary A. Wegner

Selected References

Malmquist, K. G. (1920). *Meddelande från Lunds astronomiska observatorium*, ser. 2, no. 22: 1.

——— (1924). *Meddelanden från Lunds astronomiska observatorium*, ser. 2, 32: 64.

Teerikorpi, P. (1997). "Observational Selection Bias Affecting the Determination of the Extragalactic Distance Scale." *Annual Review of Astronomy and Astrophysics* 35: 101–136.
Westerlund, B. E. Private communication, 2002.

Ma'mūn: Abū al-ʿAbbās ʿAbdallāh ibn Hārūn al-Rashīd

Born **Baghdad, (Iraq), 14 September 786**
Died **near Tarsus, (Turkey), August 833**

Ma'mūn was the son of Caliph **Hārūn al-Rashīd**, a patron of the arts whose fame has come down to us in the tales of the *Thousand and One Nights*. Hārūn also supported a fine library in Baghdad, called "The Treasure House of Wisdom," as well as the translation of foreign works in various fields. So Ma'mūn, brought up in an educated environment, was not only learned in the traditional Muslim studies but also was aware of a wider world of foreign learning. When he came to the throne as the seventh caliph of the ʿAbbāsid Empire in 813, he was among the well-educated men of his time.

Ma'mūn spent his early years as caliph consolidating his reign and building internal unity in a diverse empire. It has been argued that part of that endeavor involved commissioning Arabic translations of important Persian documents, as part of a project of Arabicizing Persian learning. Since, in addition, many Persian intellectuals believed that Greek learning was in fact based upon older Persian learning, Ma'mūn commissioned translations of Greek material as well. Apart from these political considerations, however, there was undoubtedly a genuine interest on Ma'mūn's part in the learning of the Greeks. There is also a story of a dream in which Ma'mūn saw the Greek sage, **Aristotle**, reassuring him that religion and learning were not enemies and that Ma'mūn's support of foreign learning was not a threat to Islam.

Ma'mūn was zealous in his search for new material and sent the scholar Salm to Byzantine lands to buy manuscripts. (Salm also helped to improve an Arabic translation of **Ptolemy**'s astronomical classic, *The Almagest*.) According to some reports Ma'mūn founded, in the early 830s before his death, the Bayt al-Ḥikma, the House of Wisdom. However, some historians have argued that this was less a new foundation than an extension of the Treasury of Wisdom that was already in existence at the time of Hārūn. In any case we do know that Mā'mūn supported scholars of many nations and professing many faiths, who studied, translated, and disseminated wisdom and learning, particularly that of the Greeks.

In addition to his general interest in the learning of the ancients, part of Ma'mūn's support for astronomy was based on its utility for astrology, a subject with which it was to be closely associated for many centuries. Whatever the motives for his support, the result of these translation efforts was the translation into Arabic of a number of Greek astronomical works. These included the introductory treatises of **Theodosius,** Euclid, **Menelaus,** and **Aristarchus**, as well as all of Ptolemy's works.

In addition to supporting the intellectual climate in which this work could be done, Ma'mūn also sponsored two sets of observations. The first was done in Baghdad, in 828, in the Shammāsiyya area, by astronomers including **Yaḥya ibn Abī Manṣūr** and the noted mathematician **Khwārizmī**. (Two others were **Sanad ibn ʿAli** and **ʿAbbās al-Jawharī**.) The Shammāsiyya observations were conducted around the times of the solstices and equinoxes, and it appears that Ma'mūn took an active interest in them. **Bīrūnī** informs us in his *Taḥdīd* that Ma'mūn rejected the first set of observations of 828 because of the big difference between the values for the maximum and minimum altitudes of the Sun (at the summer and winter solstices, respectively) at those observations and at the latter ones.

Yaḥya died before Ma'mūn left on one of his campaigns against the Byzantines in the early 830s. After his death, Mā'mūn decided to do new observations at Dayr Murrān on a hill near Damascus. Accordingly, he charged **Khālid ibn ʿAbd al-Malik al-Marwarrūdhī** with the task of doing observations over the period of a year with a new set of instruments. The observations, done in two periods between 831 and 833, lasted more than a year. They pleased Ma'mūn sufficiently for him to order that astronomical tables be prepared on the basis of their results. Since the observations both in Damascus and Baghdad seem to focus entirely on the Sun and Moon, these tables must have reflected earlier material for planetary motions.

Quite apart from these undoubted contributions to astronomy, Ma'mūn furnished an example of the type of a ruler that found many echoes in medieval Islam. The result was the development of the observatory as a new scientific institution, a development directly inspired by Ma'mūn, and, more generally, a tradition of royal patronage of astronomy.

Len Berggren

Selected References

Ali, Jamil (trans.) (1967). *The Determination of the Coordinates of Cities: Al-Bīrūnī's* Taḥdīd al-Amākin. Beirut: American University of Beirut.
Gutas, Dimitri (1998). *Greek Thought, Arabic Culture: The Graeco-Arabic Translation Movement in Baghdad and Early ʿAbbāsid Society (2nd–4th/8th–10th centuries)*. London: Routledge.
Rekaya, M. (1991). "al-Ma'mūn, Abu 'l-ʿAbbās." In *Encyclopaedia of Islam*. 2nd edn. Vol. 6, pp. 331–339. Leiden: E. J. Brill.
Sayılı, Aydın (1960). *The Observatory in Islam*. Ankara: Turkish Historical Society.

Manfredi, Eustachio

Born **Bologna, (Italy), 20 September 1674**
Died **Bologna, (Italy), 15 February 1739**

Eustachio Manfredi, a skilled observer of the heavens, a geographer, and a geodesist, oversaw the restoration and continued development of astronomy in Italy following the departure of **Giovanni Cassini** to Paris.

The son of Alfonso Manfredi, a notary from Lugo di Romagna, and Anna Maria Fiorini, Eustachio was the eldest of a family of scholars devoted to science and mathematics. Manfredi completed his early studies at the Jesuit school in Bologna, focusing on philosophy. In 1692, he graduated with a degree in civil and canon law, but never practiced. At Bologna, Manfredi studied mathematics and hydraulics with Domenico Guglielmini, and together with his childhood friend Vittorio Francesco Stancari, he became interested in astronomy.

After Cassini left for Paris and **Geminiano Montanari** for Padua, Italian astronomy in universities faded. Lecturers focused mainly on hydraulics and the science of numbers, and few studied astronomy. Cassini's meridian line of San Petronio was no longer in use. Manfredi and Stancari—who were essentially self-taught—conducted observations with the meridian, and at Stancari's house they set up a small observatory with a sextant and several telescopes. From 1698 to 1702, they undertook systematic observations of the relative positions of stars; studied planetary movements; and observed lunar and solar eclipses, the eclipses of Jupiter's satellites, and lunar occultations—doing so to determine accurately Bologna's geographical position.

In 1690, Manfredi, Stancari, and others, including the famous physician Giovanni Battista Morgagni, founded the *Accademia degli Inquieti*, which became a driving force for Bolognese culture. The institution turned its attention to the physical sciences, studying new systems such as those of **René Descartes**, **Gottfried Leibniz**, and **Isaac Newton**, focusing in particular on experimental and observational reality. The Accademia contributed decisively to the establishment of the *Istituto delle Scienze*, founded by Count Luigi Ferdinando Marsili, a member of one of Bologna's most illustrious families. Marsili, a valiant general and a scientist, believed that scientific research was the cornerstone of technological progress. His military experiences gave him opportunities to collect scientific and documentary material in order to establish a center in Bologna. Astronomy played a predominant role in the Accademia, representing the basic element of reform for a scientific alternative to Aristotelian thought; as a result, in 1702 Marsili appointed Manfredi and Stancari to oversee the construction of an observatory in Bologna, in his own *palazzo*. Instruments were ordered from the Lusverg family in Rome: two movable quadrants, a mural semicircle (currently exhibited at the Astronomical Museum of the University of Bologna), and a 3-ft. telescope.

During this period, Manfredi also studied sunspots, noting that there were far fewer than those observed by earlier astronomers. This phenomenon is now referred to as the Maunder minimum. In 1703, Manfredi wrote a pamphlet entitled *Descrizione d'alcune macchie scoperte nel Sole*, publishing his observations, from which he also calculated the value of the inclination of the rotational axis of the Sun on the ecliptic. In 1699, he was appointed lecturer in mathematics, and in 1704 he was appointed rector of a Pontifical College and was named Superintendent of Waters, a position he held until his death.

In 1712, Marsili donated all his instruments and his collections to the Bologna Senate, and on 13 March 1714, with the financial support of Pope Clement XI, the Istituto delle Scienze was inaugurated in Palazzo Poggi (now the seat of the University of Bologna). The Istituto incorporated the Accademia, with the name Accademia delle Scienze, and the Accademia delle Belle Arti. Manfredi was one of the inspirers of the institute and, in drawing up its program, he looked to Cassini, with whom he corresponded. A new observatory alongside the Istituto delle Scienze was planned, and construction began in 1712 under Manfredi's supervision, but was not completed until 1725; the main instruments were installed in 1727.

In 1715 Manfredi compiled the *Bolognese ephemerides* for the years 1715–1725, based on Cassini's tables, completed in Paris and previously unpublished. They included tables of the planets' transit time across the meridian, the eclipses of Jupiter's satellites, and the lunar conjunctions, as well as maps of the regions on Earth where solar eclipses would take place. The *Ephemerides*, considered among the best in Europe for several decades, were accompanied by a valuable book of instructions, *Introductio in Ephemerides,* detailing their use. The ephemerides for the period of 1726–1750 were subsequently published in 1725.

At this time, with the help of several assistants, including his two sisters, Manfredi undertook systematic observations to verify if there were perceptible shifts in the positions of the stars. If observed, these displacements would allow him to measure stellar parallax and confirm the Earth's annual revolution around the Sun; the issue of heliocentric and geocentric systems was still being debated. The initial observations revealed small shifts in the positions of the stars, yet they were not attributable to the parallactic displacement. The results were published in 1729 in *De annuis inerrantium aberrationibus*. That year, **James Bradley** offered the correct explanation for this phenomenon, later called "annual aberration of starlight," after the title of Manfredi's publication. Two years later, in Tome I of the *Commentarii dell'Accademia delle Scienze*, Manfredi published a treatise entitled *De novissimis circa fixorum siderum errores observationibus*, adding other observations to those of Bradley. Manfredi was the first to confirm Bradley's hypotheses. He did not explicitly express an opinion that would link him too closely to Bradley's explanations, due to the local political and religious situation. (Bologna was part of the Papal States.) Nevertheless, it was the first evidence of the Earth's movement around the Sun.

In 1736 Manfredi published *De gnomone meridiano Bononiensi ad Divi Petronii*, in which he included the history and description of Cassini's meridian, as well as all observations made since the instrument was created in 1655. His analysis of nearly 80 years of observations revealed a progressive decrease of one second per year in the obliqueness of the ecliptic. Although the actual value is approximately half a second, this nevertheless revealed and measured – for the very first time – a process that if continues unchanged would abolish the seasons in less than 2,000 centuries.

The following year Manfredi oversaw the publication of **Francesco Bianchini**'s *Astronomicae ac geographicae observationes selectae*. He had also previously organized and completed Stancari's notes, published as *Schedae mathematicae et observationes astronomicae*. His university lectures were collected into a considerable work, *Instituzioni astronomiche*, published posthumously in 1749.

In 1738 Manfredi asked Jonathan Sisson to make a new set of instruments. However, the astronomer died 2 years before the instruments were delivered. Sisson's instruments, installed and used by Manfredi's successor **Eustachio Zanotti**, are also exhibited at the Astronomical Museum. Because of his scientific merit, Manfredi was honored as a member of the Paris Académie des sciences and London's Royal Society.

Manfredi's manuscripts and the astronomical logbooks are in the Historical Archive of the Department of Astronomy, University of Bologna.

Fabrizio Bònoli

Selected References

Baiada, E. and R. Merighi (1982). "The Revival of Solar Activity after Maunder Minimum in Reports and Observations of E. Manfredi." *Solar Physics* 77: 357–362.

Baiada, E., F. Bònoli and A. Braccesi (1985). "Astronomy in Bologna." *Museo della Specola – Catalog*. Bologna: Bologna University Press.
Bònoli, F. and E. Piliarvu (2001). *I lettori di astronomia presso lo Studio di Bologna dal XII al XX secolo*. Bologna: Clueb.

Manilius [Manlius], Marcus

Flourished **Rome, (Italy), 10**

Marcus Manilius, a citizen of Rome, authored *Astronomicon libri V*, the oldest and most widely cited work on ancient astrology. Nothing certain is known of his life, education, or related writings. Rediscovered in the Renaissance, the *Astronomica* soon developed a wide audience and an unparalleled tradition of scholarly editors, among them the foremost astronomers of their day. For the average reader, the *Astronomica* served as a literary introduction to the heavens and an advanced primer to astrology. Manilius' masterpiece, a Latin didactic poem in five books, unveils the cosmos in hexameter verse, explaining the celestial sphere and zodiac, "describing the stars, constellations, and planets," and above all, providing a Stoic vision of the celestial dance. It is not an introduction to astronomy – here even the basics are sometimes confused – and, indeed, astrological doctrine is often muddled. More significantly, the heavens for Manilius were a reminder to persist against fate, to trust divine reason. Influenced by **Ovid** and **Virgil**, Manilius opposed **Lucretius**' harsh view of civil society. For subsequent scholars, the *Astronomica* was an elegant and influential text, albeit beset with a bizarre style and complex history.

The traditional difficulty with the *Astronomica* stems from the absence of an archetype text and an abundance of corrupt copies. The longstanding debate concerning Manilius' manuscripts (and indeed the identity of the author) extends back to the 10th century, although it is now clear that Manilius lived during the reign of Augustus and Tiberius, and probably wrote the *Astronomica* between 14–27. The debate widened with the advent of the printing press (*circa* 1450), and since that time the *Astronomica* has had a remarkably rich publication history. The first notable printing (*circa* 1472) came from the celebrated Renaissance astronomer, **Johann Müller** (Regiomontanus), and by 1650 nearly a dozen printings had appeared. The first firm textual basis for the *Astronomica* was provided by the French scholar, J.-J. Scaliger (1540–1609), whose *Astronomica* (1579; Heidelberg 1590) was later reprinted by Johann Heinrich Boecler (1611–1672) in Strasbourg (1655) with commentary by the French astronomer **Ismaël Boulliau**.

Thereafter, interest in Manilius spread as the New Science took root in the Republic of Letters. Gerhard Vossius (1577–1649) and his son, Isaac Vossius (1618–1689), for example, corresponded with Boulliau (and others in his circle), each contributing to Manilius studies. This group included J.-F. Gronovius (1611–1671), Nicolaas Heinsius (1620–1681), Claude Saumaise (1588–1653), P.-D. Huet (1630–1721), and Edward Sherburne (1618–1702), professor of astronomy at Oxford. Sherburne's first English translation of Book I (*The Sphere*, London, 1675) was soon followed by the first complete English translation (Books I–V, London, 1697; 1700) published by Thomas Creech (1659–1700), also from Oxford. The Creech editions underwent several large printings and launched the modern popular tradition.

The modern scholarly tradition of the *Astronomica*, however, stems from Richard Bentley, the foremost classical scholar of his day, and here again classical studies and the New Science converged. Bentley's interest in Manilius began five decades before he published his *Astronomica* (London, 1739), and two years before he initiated his famous "Newton–Bentley Correspondence" (1692–1693). If that exchange epitomized Enlightenment it also echoed Antiquity. Ironically, Manilius' pagan concerns – Reason, Nature, Design – resonate throughout Bentley's "Confutation of Atheism" (1693), both texts claiming that the World was not a "fortuitous or causal concourse of atoms."

Two final editions of Manilius must be noted. As the Enlightenment drew to a close, the French astronomer **Alexandre Pingré**, celebrated for his work on comets, published an elegant edition of the *Astronomica* in Latin and French. The dean of Manilius scholars, however, is A(lfred) E(dward) Housman (1859–1936), the noted British poet, whose edition of the *Astronomica* is painstaking and pure. The best source for the modern reader, at once rigorous and readable, is edited by G. Goold.

Robert Alan Hatch

Alternate name

Mallius, Marcus

Selected References

Bentley, Richard (1739). *M. Manilii Astronomicon*. London: H. Woodfall. (Brilliant classical treatment of the text.)
Creech, Thomas (1697). *The five books of M. Manilius, containing a system of the ancient Astronomy and Astrology; together with the Philosophy of the Stoicks*. London, J. Tonson.
——— (1700). *Lucretius His Six Books of Epicurean Philosophy and Manilius His Five Books Containing a System of the Ancient Astronomy and Astrology: Together with The Philosophy of the Stoicks*. London.
Goold, George P. (ed.) (1977). *Astronomica*. Loeb Classical Library, no. 469. Cambridge, Massachusetts: Harvard University Press.
——— (1998). *M. Manilii Astronomica*. Rev. of 1985 edn. Stuttgart: Teubner.
Housman, A. E. M. (1903–1931). *Manilii Astronomicon liber primus (-quintus)*. 5 Vols. London: Grant Richards.
Pingré, A. G. (1786). *Marci Manilii Astronomicon libri quinque, accessere Marci Tulli Ciceronis Arataea, cum interpretatione Gallica et notis*. 2 Vols. Paris. (Reprint, 1970; Superb French translation.)
Regiomontanus (1472?). *M. Manilii Astronomicon primus (-quintus)*. Nuremberg. (Regiomontanus appears to have simply copied an available manuscript obtained from Italy with possible minor emendations.)
Scaliger, J. J. (1600). *Manilii Astronomicon*. Leiden. (Presented the first substantial scholarly text of the *Astronomica*.)
——— (1655). *Marci Manilii Astronomicon*. Bockenhofferi, Strasburg, 1655. (Reprinting by Johann Heinrich Boecler, with commentary by Ismaël Boulliau and Thomas Reinesius.)
Sherburne, Edward, Sir (1675). *The Sphere of M. Manilius made an English poem; with annotations and an astronomical appendix*. London: Nathanael Brook.

Mañjula

➔ **Muñjāla**

Maraldi, Giacomo Filippo

Born **Perinaldo near Imperia, (Liguria, Italy), 21 August 1665**
Died **Paris, France, 1 December 1729**

As one of the earliest members of the Paris Observatory staff, astronomer and geodesist Giacomo Maraldi, sometimes identified as Maraldi I, conducted the first systematic observations of the surface features of Mars. The son of Francesco Maraldi and Angela Cassini (sister of **Giovanni Cassini**), Maraldi studied near his hometown until 1687 when, at his uncle's request, he moved to Paris and joined the observatory staff as an observer. Maraldi's main task was the production of a new stellar catalog that remained incomplete on his death; it was never published. He appears to have been a careful observer and was certainly a mainstay in the observatory's work. Maraldi observed a wide variety of phenomena, including planets, satellites, eclipses, and variable stars (including the discovery of R Hydrae). He observed six comets, calculating several orbits. It was he who first realized that the corona belongs to the Sun, not to the Moon.

Maraldi supported his uncle in the controversy with **Ole Römer** concerning the velocity of light, which the latter believed to be finite. Römer's theory, which he used to account for discrepancies between predicted and observed times of eclipses and occultations for the Galilean satellites of Jupiter as observed from Earth at various times of the year, was later shown to be correct.

One of Cassini's programs was a geodesic survey of France. Maraldi was brought into the operation to extend the Paris meridian south, working with **Jacques Cassini** (Cassini II), J. de Chazelles, and Pierre Couplet during 1701/1702. In 1718, he assisted in a survey of the Paris–Amiens meridian to Dunkirk with Cassini II.

During 1702/1703, Maraldi resided in Rome, where he worked on producing a meridian for the Church of the Carthusians and making observations.

Maraldi is best remembered for his Mars observations. He observed Mars at every opposition from 1672, eventually determining a rotational period of 24 hours 40 m. On the surface, Maraldi identified what later became named as Syrtis Major, Mare Sirenum, and Mare Tyrrhenum. He also monitored the polar caps. Late in his life Maraldi brought his nephew, **Giovanni Maraldi**, onto the Paris staff. Maraldi was associated with the Paris Academy of Sciences from 1694.

Richard A. Jarrell

Alternate name

Maraldi I

Selected Reference

Taton, René (1974). "Maraldi, Giacomo Filippo (Maraldi I)." In *Dictionary of Scientific Biography*, edited by Charles Coulston Gillispie. Vol. 9, pp. 89–91. New York: Charles Scribner's Sons.

Maraldi I

➔ **Maraldi, Giacomo Filippo**

Maraldi, Giovanni Domenico [Jean-Dominique]

Born **Perinaldo near Imperia, (Liguria, Italy,) 17 April 1709**
Died **Perinaldo near Imperia, (Liguria, Italy), 14 November 1788**

As an observational astronomer at the Paris Observatory, Giovanni Domenico Maraldi made an accurate determination of the difference in longitude to the Greenwich Observatory and contributed important observations of comets and nebulae. He was related, though somewhat indirectly, to the famed Cassini family and the great astronomer **Giovanni Cassini** – indeed it seems he was likely named after his famous relative. But the Cassini family produced at least six astronomers and one somewhat well known botanist between the mid-17th century and the mid-19th century with no fewer than three using the name Giovanni Domenico or Jean Dominique (depending on whether the individual lived in France or Italy). Giovanni Maraldi was sometimes referred to as Maraldi II and was the nephew of **Giacomo Maraldi** (Maraldi I), who was, in turn, the nephew of Giovanni Cassini.

Maraldi II came to Paris in 1727 from his home, also the birthplace of his uncle and of Cassini. Perinaldo, though in Italy, was a mere 55 km from Nice, France. In 1731, Maraldi II was made a member of the Paris Academy of Sciences and was employed at the Paris Observatory, then home of his uncle Maraldi I and of several of the younger members of the Cassini clan including his cousin **Jacques Cassini**, with whom he performed some of his most memorable observations. Maraldi II retired in 1772 and returned to Perinaldo. There is actually mention of yet a third Maraldi (Giovanni Filippo: 1746–1797), sometimes referred to as Maraldi III, who observed planetary satellites at Perinaldo, but only one obscure source makes this reference.

When Maraldi II first arrived in Paris he was assigned the task of carrying out geodesic measurements using the eclipse times of Jupiter's satellites. Using this technique, he found a longitude difference between Paris and Greenwich of 9 min 23 s compared to the modern value of 9 min 20.93 s. Later Maraldi was to observe several comets, starting in 1742 and including the great comet of 1743/1744 (discovered by Dirk Klinkenberg and **Jean-Philippe Loys de Chéseaux**). Maraldi II also observed Halley's comet (IP/Halley) in 1759 and calculated several comet orbits. When not observing comets Maraldi observed the transits of Mercury and Venus and helped to publish 25 volumes of the *Connaissance des Temps* and **Nicolas de La Caille**'s catalog of southern stars, *Coelum Australe Stelliferum*. In September 1746, he observed comet C/1746 P1 De Chéseaux (which had been discovered by Loys de Chéseaux) along with his cousin Jacques Cassini. In the process of observing this comet, Maraldi discovered two globular clusters that would eventually be included in **Charles Messier**'s historic catalog: M2 in Aquarius and M15 in Pegasus.

Ian T. Durham

Alternate name

Maraldi II

Selected Reference

Delambre, J. B. J. (1827). *Histoire de l'astronomie au dix-huitième siècle*. Paris: Bachelier.

Maraldi II

➤ **Maraldi, Giovanni Domenico [Jean-Dominique]**

Marius

➤ **Mayr, Simon**

Markarian, Beniamin Egishevich

Born **Shulaver, (Armenia), 29 November 1913**
Died **Yerevan, (Armenia), 29 September 1985**

Armenian observational astronomer Beniamin Markarian is remembered for the discovery of many hundreds of galaxies, the Markarian galaxies, characterized by more emission in the ultraviolet [UV] part of the spectrum than found in normal spirals or ellipticals. Many of them also have compact nuclei and are related to the quasars. Mkn 421 was the first object outside the Milky Way recognized as a source of photons as energetic as teraelectron volts.

Markarian graduated with a degree in mathematics from Yerevan State University in 1938 and began postgraduate work with **Viktor Ambartsumian,** defending his thesis in 1944 and beginning work with Ambartsumian at the Byurakan Astrophysical Observatory, where he remained until his death. He was elected to the Armenian Academy of Sciences in 1971, received a state prize in 1950, and served as president of the Commission on Galaxies of the International Astronomical Union (1976–1979), in which position he was eventually succeeded by his close colleague Edward Khachikian (1991–1994).

Markarian's early work was in spectroscopy, at relatively high resolution, of white dwarfs and stars in open clusters. But in 1965 Ambartsumian asked him to take over a project that was to obtain low-resolution (objective prism) spectra of a very large numbers of galaxies to look for faint blue ones with strong UV emission. Ambartsumian expected that these would be related to Seyfert and other active galaxies and fit into his (now obsolete) model that indicated diffuse material in the Universe, including gas and star clusters, was initially expelled from much denser material at galactic centers. Markarian primarily used an objective prism with an angle of 1.5° (producing a dispersion of about 2000 Å/mm) at the prime focus of the 1-m Schmidt telescope at Byurakan. By 1967 he had compiled a list of 70 UV-excess galaxies, eventually examining more than 2,000 plates, covering 17,000 square degrees of the sky and containing about 15,000 objects per plate. The final list of objects with strong UV continua took the numbers up to about 1,500. A second spectroscopic survey was begun by Markarian and his Byurakan colleagues in 1978; he continued to participate in the discovery and characterization of new Seyfert galaxies and quasi-stellar objects until his death. Markarian never traveled outside the Soviet Union, and some of the follow-up spectroscopy was carried out at Mount Palomar by his colleague Khachikian, who has continued work on UV galaxies since Markarian's death. Many of the objective-prism images from the 1960s were recorded on Kodak spectroscopic plates imported into Armenia with considerable difficulty.

Ian T. Durham

Acknowledgment

The author gratefully acknowledges Professor Edward Khachikian for providing most of the information on Markarian.

Selected References

Khachikian, E. Y. (1968). "Two New Seyfert Galaxies from Markian's List of Galaxies with Strong UV Continua." *Astronomical Journal* 73: 891–892. (Paper abstract.)

——— Personal communication, 2002.

Markarian, B. E. (1951). *Soobshcheniya Byurakanskoj Observatorii* 9: 1–40.

——— (1967). "Galaxies with an Ultraviolet Continuum" (in Russian). *Astrofizika* 3: 55–68.

Markarian, B. E. and E. Y. Oganesian, (1961). "White Dwarfs in Praesepe (NGC 2632)." *Soobshcheniya Byurakanskoj Observatorii* 29: 71–80.

Markarian, B. E., E. Y. Oganesian, and S. N. Arakelian (1965). "A Detailed Photometric and Colorimetric Study of Galaxies in the Constellation of Virgo" (in Russian). *Astrofizika* 1: 38.

Markgraf, Georg

Born **Liebstadt near Dresden, (Germany), 30 September 1610**
Died **São Paulo de Loanda, (Angola), 1643 or 1644**

Naturalist Georg Markgraf made the first systematic astronomical observations of the southern skies. At the age of 16, Markgraf began a tour of Central European universities including Strasbourg, Basle, Ingolstadt, Altdorf, Erfurt, Wittenberg, Leipzig, Griefswald, Rostock, and Stettin, before finally matriculating at the University of Leiden in September 1636. Markgraf officially studied medicine. One of his instructors there was Jacob Gool, the noted astronomer and arabist. Another Leiden astronomer, Samuel Kechel, was also a close associate. By June 1636 Markgraf was in Brazil, where he was to remain for the next 8 years, probably in the employment of Johan Maurits van Nassau-Siegen, who was expanding Dutch interests in South America. Markgraf busied himself in compiling detailed maps of the region, collecting specimens of flora and fauna, and making astronomical observations.

Markgraf's astronomical ambition may have been to become a New World equivalent of **Tycho Brahe**. Establishing an observatory in the Vrijburg Palace at Mauritsstad, he was almost certainly the first European to pursue systematic astronomy in the Southern Hemisphere. Markgraf recorded the meridian altitudes of stars and planets as well as a solar eclipse (13 November 1640) and a lunar eclipse (4 April 1642). He planned a treatise to be entitled *Progymnastica mathematica Americana* – intriguingly similar to Brahe's *Astronomia insauratae progymnasmata* – but apparently never produced as much as a rough draft. Sometime after August 1643 Markgraf left Brazil for São Paulo de Loanda in Angola, a new Dutch possession in Africa, where he soon contracted a tropical fever and died. Johan de Laet acquired Markgraf's natural history papers; this valuable material was published in *Historia naturalis Brasiliae* (1648). Markgraf's astronomical observations and calculations, including those for some 26 horoscopes, remain unpublished in collections of the Gemeentearchief (Municipal Archives) Leiden and the Observatoire de Paris.

Keith Snedegar

Selected Reference

North, J. D. (1979). "Georg Markgraf: An Astronomer in the New World." In *Johan Maurits van Nassau-Siegen 1604–1679... Essays on the Occasion of the Tercentenary of His Death*, edited by E. van den Boogaart, pp. 394–423. The Hague: Johan Maurits van Nassau Stichting.

Markov, Andrei Andreevich

Born **Ryazan, Russia, 14 June 1856**
Died **Petrograd (Saint Petersburg, Russia), 20 May 1922**

Russian mathematician Andrei Markov had the good luck to work with Pafnuty Chebyshev (of the polynomials) at Saint Petersburg (1874–1878). Many of his contributions were in the area of probability theory, including a refinement of the central limit theorem invented by **Pierre de Laplace**. He is best known for the Markov chains. Roughly these describe systems and processes whose future can be predicted from (completely known) current conditions with no knowledge of the past history of the system. Some astronomical systems, for instance clusters of stars (treated as point masses), can be thought of as Markovian. In practice, the precise knowledge of everything about the system at one time is never available. Markov's contemporary A. Lyapunov wrote down criteria for deciding when imprecise knowledge of a system would lead to its future behavior evolving in totally unpredictable directions. Such systems are called chaotic and can be recognized by the so called Lyapunov exponent.

Virginia Trimble

Selected Reference

Anon. (1994). "Markov." In *The Biographical Dictionary of Scientists*, edited by Roy Porter, 2nd ed., p. 462. New York: Oxford University Press.

Markowitz, William

Born **Mlec, (Austria), 8 February 1907**
Died **Pompano Beach, Florida, USA, 10 October 1998**

William Markowitz devoted most of his career to improving astronomical measurements for determining time, and then to establishing new time systems based on atomic standards rather than astronomical measurements. His efforts, as director of the United States Naval Observatory [USNO] Time Service Department, resulted in greatly improved international cooperation on matters related to time.

Markowitz was the son of Hyman and Rebecca (*née* Baumstein, from Poland) Markowitz. In 1910, he immigrated with his family to Chicago, Illinois, USA. His early interest in astronomy developed at Crane Technical High School in Chicago and Crane Junior College, where Markowitz took a course in astronomy. He entered the University of Chicago and obtained his B.S. (1927), M.S. (1929), and Ph.D. in astronomy (1931). Markowitz married Rosalyn Shulemson in 1943; they had one son, Toby.

After teaching at Pennsylvania State College, Markowitz joined the USNO in 1936, working under Paul Sollenberger and with **Gerald Clemence**. He later served as director of the USNO's Time Service Department from 1953 until his retirement in 1966. Markowitz's principal research interests concerned the rotation of the Earth and the motion of its pole. The polar motion occurring at decadal time scales is named the Markowitz wobble for him.

One of Markowitz's early duties was operating the Photographic Zenith Tube [PZT]. In 1949, he and Sollenberger designed an improved version for the observatory's new station near Miami, Florida. The variation of latitude, determined with the PZT, was one of Markowitz's chief research interests. Analysis of these data led to his contributions on the study of polar motion.

Markowitz directed the Time Service Department during a period of increasing demands for more uniform and accurate time. Ephemeris time, based on the orbital motion of the Earth, was proposed in the early 1950s to provide a more uniform time scale than that based on the Earth's rotation. Markowitz devised a practical means for its determination by inventing the dual-rate Moon camera bearing his name. The first Markowitz Moon camera was placed in operation at the Naval Observatory in June 1952, and 20 such cameras were used around the world during the International Geophysical Year (1957/1958). With data from these cameras, Markowitz worked with Louis Essen at the National Physical Laboratory in England to calibrate newly developed atomic clocks in terms of the Ephemeris second. The fundamental frequency of cesium atomic clocks, 9,192,631,770 Hz, which they determined, has defined the "second" internationally since 1967. At the International Astronomical Union [IAU] meeting in Dublin in 1955, Markowitz proposed the system of UT0, UT1, and UT2, which went into effect within months and remains today.

Markowitz participated in experiments synchronizing time using artificial satellites and atomic clocks transported by airplanes. He served as president of the IAU Commission on Time from 1955 to 1961, and was active in the International Union of Geodesy and Geophysics, the American Geophysical Union, and the International Consultative Committee for the Definition of the Second.

After retiring from the Naval Observatory, Markowitz served as professor of physics at Marquette University (1966–1972), and adjunct professor at Nova University in Florida.

Steven J. Dick and *Dennis D. McCarthy*

Selected Reference

Dick, Steven J. and Dennis D. McCarthy (1999). "William Markowitz, 1907–1998." *Bulletin of the American Astronomical Society* 31: 1605.

Marrākushī: Sharaf al-Dīn Abū ʿAlī al-Ḥasan ibn ʿAlī ibn ʿUmar al-Marrākushī

Flourished **(Egypt), second half of the 13th century**

Marrākushī was one of the major astronomers in 13th-century Egypt. As his name indicates, he was originally from Maghrib, but his major astronomical activities took place in Cairo during the second half of the 13th century. It is not too surprising, given the turmoil affecting al-Andalus and Maghrib at that time, that a scholar from the westernmost part of the Islamic world would decide to emigrate to Egypt, whose capital Cairo was already established as the major cultural center of the Arab–Islamic world. Unfortunately, Marrākushī does not figure in any biographical sources, so we must rely on the scanty evidence provided by his own work in order to shed some light on his life.

Marrākushī is best known for his remarkable *summa* devoted to spherical astronomy and astronomical instrumentation, entitled *Jāmiʿ al-mabādiʾ wa-ʾl-ghāyāt fī ʿilm al-mīqāt* (Collection of the principles and objectives in the science of timekeeping), which is intended as a comprehensive encyclopedia of practical astronomy. This work is the single most important source for the history of astronomical instrumentation in Islam. It was the standard reference work for Mamluk Egyptian and Syrian, Rasūlid Yemeni, and Ottoman Turkish specialists of the subject.

This voluminous work (most complete copies cover 250 to 350 folios of text, diagrams, and tables) has occasionally been qualified as a mere compilation of older sources without original contents. While it is true that this synthetic work heavily depends upon the works of predecessors, it is definitively original and without any precedent. In fact, no single part of the work can be proven to reproduce the words of an earlier author, except for the few sections where Marrākushī clearly states from whom he is quoting. In those occasional cases where an earlier source is mentioned, Marrākushī's text always turns out to be either a major rewriting of the original or an independent paraphrase.

The *Jāmiʿal-mabādiʾ wa-ʾl-ghāyāt* is well written and logically organized, and employs a relatively literate style that is unusual for a work on technical topics. The author is clearly a very competent astronomer and also occasionally displays his knowledge of ancillary disciplines such as philosophy.

The *Jāmiʿ* is made up of four books on the following topics:

(1) On calculations, in 67 chapters. This book gives exhaustive calculatory methods (without proofs) concerning chronology, trigonometry, geography, spherical astronomy, prayer times, the solar motion, the fixed stars, gnomonics, *etc.*
(2) On the construction of instruments, in seven parts. The first part concerns graphical methods in spherical astronomy and gnomonics. The second through the seventh parts then treat the construction of portable dials, fixed sundials, trigonometric and horary quadrants, spherical instruments, instruments based upon projection, and observational and planetary instruments.
(3) On the use of selected instruments, in 14 chapters.
(4) The work ends with a "quiz" – *i.e.*, a series of questions and answers – in four chapters, whose aim is to train the mental abilities of the students.

An interesting confirmation of Marrākushī's Maghribi origin is provided by his geographical table: 44 of the 135 localities featured in the list of latitudes are written in red ink to indicate that the author visited these places personally and determined their geographical latitude *in situ* through observation. These 44 locations begin along the Atlantic coast of today's western Sahara, include numerous cities and villages in the Maghrib, two cities in al-Andalus (Seville and Cádiz), and continue along the Mediterranean coast *via* Algiers, Tunis, and Tripoli to end up in Alexandria, Cairo, Minya, and Tinnis. Marrākushī's western Islamic heritage is also apparent in the fact that his chapters on precession and solar theory depend upon the works of **Zarqālī** and **Ibn al-Kammād**.

Marrākushī appears to have written his major work in Cairo during the years 1276–1282. First, a solar table is given for the year 992 of the Coptic calendar (Diocletian era), corresponding to the years 1275/1276. Also, some examples of chronological calculations are given for the year 1281/1282, and his star table in equatorial coordinates is calculated for the end of the same year.

The arrival of Marrākushī in Cairo coincided with the establishment of the first offices of *muwaqqits* (timekeepers) in Egyptian mosques. His work can thus be seen as fulfilling a specific demand of Mamlūk Egyptian society (more specifically, the mosque administration, the muezzins and *muwaqqits*, instrument-makers, interested students, *etc.*). But the lack of any reference to the profession of the *muwaqqit* or to the milieu of the mosque would seem to indicate that Marrākushī was an independent scholar without institutional affiliation. The motive he gives for writing his *magnum opus* is the inadequate education of instrument-makers and their methodological failures. His introduction suggests that his target audience was instrument-makers, *i.e.* artisans and practitioners of applied science, who were not professional astronomers. However, this is somewhat contradicted by the technical level of the book, which certainly assumes the reader to know at least the basics of arithmetic, geometry, spherics, algebra, and trigonometry. Thus the *Jāmiʿ al-mabādiʾ wa-ʾl-ghāyāt* seems more likely to be a comprehensive reference work of intermediate to advanced level intended for active and apprentice *muwaqqits*, and for specialists of timekeeping and instrumentation who were associated with them.

Marrākushī must have died, most probably in Cairo, between the years 1281/1282 and *circa* 1320, since two early 14th-century sources refer to him as being deceased (an anonymous treatise on timekeeping entitled *Kanz al-yawāqīt*, datable to 723 H/1323 and preserved in MS Leiden Or. 468, f. 91r, and a treatise on instrumentation by **Najm al-Dīn al-Miṣrī** composed in Cairo *circa* 1330).

François Charette

Selected References

Charette, François (2003). *Mathematical Instrumentation in Fourteenth-Century Egypt and Syria: The Illustrated Treatise of Najm al-Dīn al-Miṣrī*. Leiden: E. J. Brill.

Delambre, J. B. J. (1819). *Histoire de l'astronomie du moyen âge*. Paris. (Reprint: New York: Johnson Reprint Corp., 1965). (Delambre used the unpublished manuscript of J. J. Sédillot's translation of Marrākushī for his section on Islamic astronomy.)

King, David A. (1991). "al-Marrākushī, Abū ʿAlī al-Ḥasan b. ʿAlī." In *Encyclopaedia of Islam*. 2nd ed. Vol. 6, p. 598. Leiden: E. J. Brill.

——— "The Astronomy of the Mamluks." *Isis* 74 (1983): 531–555. (Reprinted in King, *Islamic Mathematical Astronomy*, III. London: Variorum Reprints, 1986; 2nd rev. edn., Aldershot: Variorum, 1993.)

——— (1993). *Astronomy in the Service of Islam*. Aldershot: Variorum.

——— (1996). "On the Role of the Muezzin and the Muwaqqit in Medieval Islamic Society." In *Tradition, Transmission, Transformation: Proceedings of Two Conferences on Pre-modern Science Held at the University of Oklahoma*, edited by F. Jamil Ragep and Sally P. Ragep, with Steven Livesey, pp. 285–346. Leiden: E. J. Brill.

——— (2004). *In Synchrony with the Heavens: Studies in Astronomical Timekeeping and Instrumentation in Medieval Islamic Civilization*. In *The Call of the Muezzin* (Studies I–IX). Vol. 1. Leiden: E. J. Brill.

Lelewel, Joachim (1850–1857). *Géographie du moyen-âge*. 5 Vols. and an atlas. Vol. 1, pp. 134–142 and atlas, plate 22. Brussels.

Mancha, J. L. (1998). "On Ibn al-Kammād's Table for Trepidation." *Archive for History of Exact Sciences* 52: 1–11.

Mercier, Raymond (1977). "Studies in the Medieval Concept of Precession (Part II)." *Archives internationales d'histoire des sciences* 27: 33–71.

Schmalzl, Peter (1929). *Zur Geschichte des Quadranten bei den Arabern*. Munich.

Schoy, Carl (1923). *Die Gnomonik der Araber*. Munich.

Sédillot, Jean-Jacques (trans.) (1834). *Traité des instruments astronomiques des Arabes*. 2 Vols. Paris. (Reprint, edited by Fuat Sezgin, Frankfurt an Main: Institut für Geschichte der Arabisch-Islamischen Wissenschaften). (Printed in two volumes under the editorship of his son L. A. Sédillot; French translation of the first half [Book 1 and the first three parts of Book 2] of Marrākushī's book. The rest of Book 2 was summarized in a rather inadequate fashion by L. A. Sédillot [see below]. The third and fourth books have never been investigated. This work represents one of the first Islamic astronomical texts to have been translated into a European language in the modern period and was given a prize by the Académie des inscriptions et belles-lettres in 1822.)

Sédillot, Louis-Amélie (1841). *Mémoire sur les instruments astronomiques des Arabes*. Paris. (Reprint, edited by Fuat Sezgin, Frankfurt an Main: Institut für Geschichte der Arabisch-Islamischen Wissenschaften, 1989.)

——— (1842). *Mémoire sur les systèmes géographiques des Grecs et des Arabes*. Paris.

Sezgin, Fuat (2000). "Geschichte des arabischen Schrifttums." In *Mathematische Geographie und Kartographie im Islam und ihr Fortleben im Abendland: Historische Darstellung*. Vol. 10, pp. 168–172. Frankfurt an Main: Institut für Geschichte der Arabisch-Islamischen Wissenschaften.

Martin of Bohemia

➔ **Behaim, Martin**

Martinus Hortensius [Ortensius]

➔ **Van den Hove, Maarten**

Marwarrūdhī: Khālid ibn ʿAbd al-Malik al-Marwarrūdhī

Flourished **Damascus, (Syria), 832**

Along with **ʿAlī ibn ʿĪsā al-Asṭurlābī** and a party of surveyors, Khālid ibn ʿAbd al-Malik al-Marwarrūdhī traveled to the Plain of Sinjār under orders of ʿAbbāsid Caliph **Ma'mūn** to determine the size of the Earth by making accurate measurements of one degree of latitude. Marwarrūdhī designed instruments, including an armillary and an astrolabe, for observations made in Baghdad. Following the death of **Yaḥyā ibn Abī Manṣūr**, **ʿAbbās ibn Saʿīd al-Jawharī** selected Marwarrūdhī to prepare appropriate instruments for placement at the Dayr Murrān monastery on Mount Qāsiyūn near Damascus. There, he led the yearlong series of solar and lunar observations *circa* 832, though he encountered considerable difficulties with the warping and expansion of the copper and iron instruments. The first of three generations of astronomers, he also took part in the project circa 843/844 in Baghdad concerning observations for determining the length of the spring season.

Marvin Bolt

Selected References

Barani, Syed Hasan (1951). "Muslim Researches in Geodesy." In *Al-Bīrūnī Commemoration Volume, A. H. 362–A. H. 1362*, pp. 1–52. Calcutta: Iran Society. (Includes transcriptions and an analysis of Arabic primary sources, as well as translations.)

Kennedy, E. S. "A Survey of Islamic Astronomical, Tables." *Transactions of the American Philosophical Society*, n.s., 46, pt. 2 (1956): 121–177, esp. pp. 127, 136. (Reprint, Philadelphia: American Philosophical Society, 1989.) (An important list, with excellent introduction to the topic of *zījes*.)

King, D. A. (2000). "Too Many Cooks … A New Account of the Earliest Geodetic Measurements." *Suhayl* 1: 207–241. (Provides translated texts related to Marwarrūdhī's involvement with measuring the Earth.)

Langermann, Y. Tzvi (1985). "The Book of Bodies and Distances of Ḥabash al-Ḥāsib." *Centaurus* 28: 108–128. (A recently discovered Arabic manuscript includes contemporary records of astronomical projects initiated by Caliph Ma'mūn. With English translation and discussion.)

Sarton, George (1927). *Introduction to the History of Science*. Vol. 1, p. 566. Baltimore: Published for the Carnegie Institution of Washington by Williams and Wilkins.

Sayılı, Aydın (1960). *The Observatory in Islam*. Ankara: Turkish Historical Society. (See Chap. 2, "Al Mamûn's Observatory Building Activity," pp. 50–87, for a valuable discussion, beginning with a thorough analysis of early Islamic astronomical observations.)

Māshā'allāh ibn Atharī (Sāriya)

Died ***circa* 815**

Māshā'allāh (from mā shā' Allāh, *i. e.*, "that which God intends") was a Jewish astrologer from Basra. Ibn al-Nadīm says in his *Fihrist* that his name was Mīshā, meaning Yithro (Jethro). Māshā'allāh was one of the leading astrologers in 8th- and early 9th-century Baghdad under the caliphates from the time of al-Manṣūr to **Ma'mūn**, and together with al-Nawbakht worked on the horoscope for the foundation of Baghdad in 762.

Ibn al-Nadīm lists some 21 titles of works attributed to Māshā'allāh; these are mostly astrological, but some deal with astronomical topics and provide us information (directly or indirectly) about sources (*i. e.*, Persian, Syriac, and Greek) used during this period. This valuable information also comes from the Latin translations of some of Māshā'allāh's works, some of which are no longer extant in Arabic.

A selection of the works by Māshā'allāh includes *De scientia motus orbis* (On Science of the Movement of Spheres), preserved in Latin translation, containing an introduction to astronomy as well as a study of **Aristotle**'s *Physics*, both based on Syriac sources. **Ptolemy** and **Theon of Alexandria** are mentioned, but the planetary models are pre-Ptolemaic Greek and similar to those found in 5th-century Sanskrit texts, *Kitāb fī al-qirānāt wa-'l-adyān wa-'l-milal* (A book on conjunctions, Religions, and communities), an astrological history of mankind, attempts to explain major changes based on conjunctions of Jupiter and Saturn; a discussion of eclipses is preserved in a Latin translation by John of Seville and a Hebrew translation by **Abraham ibn ʿEzra**, and a commentary on the armillary sphere. (For other works, see Sezgin.)

Misattributions have sometimes occurred because of confusion between the works of Māshā'allāh, **Abū Maʿshar,** and Sahl ibn Bishr. Indeed, the authenticity of two treatises on the astrolabe attributed to Māshā'allāh and translated into Latin has been questioned by P. Kunitzsch.

Finally, according to E. Kennedy, Māshā'allāh's son was an astronomer who composed a manuscript unifying the theories of **Khwārizmī** and **Ḥabash**.

Ari Belenkiy

Alternate name

Messahala

Selected References

Carmody, Francis J. (1956). *Arabic Astronomical and Astrological Sciences in Latin Translation: A Critical Bibliography*. Berkeley: University of California Press, pp. 23 ff.

Goldstein, Bernard R. (1964). "The Book on Eclipses by Masha'allah." *Physis* 6: 205–213. (English translation of Abrahim ibn ʿEzra's Hebrew translation of Māshā'allāh's work.)

Ibn al-Nadīm (1970). *The Fihrist of al-Nadīm: A Tenth-Century Survey of Muslim Culture*, edited and translated by Bayard Dodge. 2 Vols. Vol. 2, pp. 650–651. New York: Columbia University Press.

Kennedy, E. S. (1956). "A Survey of Islamic Astronomical Tables." *Transactions of the American Philosophical Society*, n.s., 46, pt. 2: 121–177. Reprint: Philadelphia: American Philosophical Society, 1989.

Kennedy, E. S. and David Pingree (1971). *The Astrological History of Māshā'allāh*. Cambridge, Massachusetts: Harvard University Press.

Kunitzsch, Paul (1981). "On the Authenticity of the Treatise on the Composition and Use of the Astrolabe Ascribed to Messahallah." *Archives internationale d'histoire des sciences* 31: 42–62.

Pingree, David (1974). "Māshā' allāh." In *Dictionary of Scientific Biography*, edited by Charles Coulston Gillispie, Vol. 9, pp.159–162. New York: Charles Scribner's Sons.

Rosenfeld, B. A. and Ekmeleddin Ihsanoğlu (2003). *Mathematicians, Astronomers, and Other Scholars of Islamic Civilization and Their Works (7th–19th c.)*. Istanbul: IRCICA, p. 17.

Samsó, Julio (1991). "Mā<u>sh</u>ā' allāh." In *Encyclopaedia of Islam*. 2nd ed. Vol. 6, pp. 710–712. Leiden: E. J. Brill.

Sezgin, Fuat (1978). *Geschichte des arabischen Schrifttums*. Vol. 6, *Astronomie*: pp. 127–129; Vol. 7, *Astrologie – Meteorologie und Verwandtes* (1979): 102–108. Leiden: E. J. Brill.

Thorndike, Lynn (1956). "The Latin Translations of Astrological Works of Messahala." *Osiris* 12: 49–72.

Messahala

➔ **Māshā'allāh ibn Atharī (Sāriya)**

Maskelyne, Nevil

Born **London, England, 6 October 1732**
Died **Greenwich, England, 9 February 1811**

As Great Britain's fifth Astronomer Royal and founder of the *Nautical Almanac*, Nevil Maskelyne made practical the finding of longitude at sea. Maskelyne was the third son of Edmund and Elizabeth Booth Maskelyne of Purton, Wiltshire, England. His father died when he was 11 years old. Maskelyne was educated at Westminster School and admitted to Trinity College, Cambridge University. He graduated seventh wrangler in mathematics in 1754, took Holy Orders in 1755, and became a fellow of his college. Maskelyne was elected a fellow of the Royal Society of London in 1758. He was appointed the fifth Astronomer Royal of England and director of the Royal Observatory at Greenwich in 1765; he held that office for 46 years. Maskelyne was also awarded his Doctor of Divinity (1777); he was named rector of Shrawardine, Shropshire (1775) and of North Runcton, Norfolk (1782). He married Sophia Rose in 1784; their only child, Margaret, was born the following year.

At the request of the Royal Society, Maskelyne traveled to the island of Saint Helena with Robert Waddington to observe the 6 June 1761 transit of Venus, but was defeated by clouds. On the same voyage, however, he was able to make longitude calculations using the so called "lunar distances" method advocated by Sir **Isaac Newton** and **Edmond Halley**, amongst others, and made possible by the improved lunar tables calculated by **Johann Mayer**. Maskelyne published the lunar distances method in *The British Mariner's Guide* (1763). While on Saint Helena, he carefully observed the tides, and the variation of the compass, and undertook measurements on the annual parallax of Sirius.

Finding longitude at sea was a major problem for sailors in the 18th century. Many ships had foundered as a result of not being able to determine their positions with accuracy. In 1714, the British Board of Longitude established a prize of £20,000 to facilitate the discovery of a reliable method for determining longitude at sea. It fell to Maskelyne (as Astronomer Royal) to examine the various solutions and inventions proffered to this problem. In 1763, he sailed to Barbados in order to test the reliability of John Harrison's fourth chronometer, H-4, and found its accuracy superior to the lunar distances method. Maskelyne also undertook longitude determinations by observing eclipses of Jupiter's Galilean satellites, and found this method was impractical on the deck of a ship at sea.

When Maskelyne succeeded **Nathaniel Bliss** as Astronomer Royal in 1765, he at last fulfilled the public function for which the Royal Observatory was founded by King Charles II in 1675, namely, the preparation of tables for ocean navigation. Maskelyne inaugurated publication of *The Nautical Almanac and Astronomical Ephemeris*, the first volume of which appeared in 1766 for the year 1767. It contained a compendium of astronomical tables and navigational aids, such as **James Bradley**'s tables of atmospheric refraction. Maskelyne had assisted Bradley in the preparation of such tables during the latter's tenure as the third Astronomer Royal in 1755. Maskelyne supervised publication of the *Nautical Almanac* for 50 years, from 1767 to 1816. He also published the cumulative Greenwich observations for the period from 1776 to 1811 in four volumes, containing positions of the Sun, Moon, planets, and selected reference stars. Maskelyne's work on the proper motions of several bright stars was used by Sir **William Herschel** to estimate the Sun's movement toward the constellation of Hercules.

In 1774, Maskelyne experimented with a plumb-line to determine the mean density of the Earth by measuring the gravitational deflection induced by a mountain. In the summer of the previous year, astronomer **Charles Mason** (of Mason–Dixon line fame) toured the highlands of Scotland and regions in the north of England in search of a suitable mountain. He eventually selected the peak of Schiehallion in the Cairngorm mountain range in Perthshire, Scotland. This mountain was reasonably isolated from other hills, had the desired east–west orientation (with a small north–south extent that Maskelyne sought), and had a relatively regular form to facilitate the calculation of its volume. In this experiment, Maskelyne investigated the principle and the constant of universal gravitation, confirming that the force of gravity acting between bodies is proportional to the inverse square of their separation. Charles Hutton analysed Maskelyne's data and calculated a value for the mean density of the Earth between 4.56 and 4.87 g cm^{-3}, as compared with the modern value of 5.52 g cm^{-3}. For this demonstration, Maskelyne received the Copley Medal of the Royal Society in 1775.

One of Maskelyne's correspondents was the Irish astronomer, James Archibald Hamilton, who operated a private observatory at Cookstown, County Tyrone. Hamilton communicated his observations of the 1782 transit of Mercury to Maskelyne, who commented favorably upon the results. Hamilton was later appointed the first astronomer of the Armagh Observatory in 1790. Maskelyne was requested to obtain precision clocks for the Armagh Observatory, and eventually recommended chronometer maker Thomas Earnshaw who subsequently produced two astronomical clocks for the Observatory. With Maskelyne's support, Earnshaw was awarded £3,000, under the new Longitude Act of 1774, for his innovative clock designs.

Maskelyne contributed to a number of fields of study, *e. g.*, he invented the prismatic micrometer, and edited Mason's improvements to Mayer's lunar tables. Yet, his most enduring legacy was his contributions toward the longitude problem and his establishment of the *Nautical Almanac*. Several lunar craters are named for him, Maskelyne W being the crater used as a finder by the crew of Apollo 11 during the lunar module's final descent onto the surface in 1969. The Maskelyne Islands in the Pacific Ocean are also named for our subject.

John McFarland

Selected References

Andrewes, William J. H. (1996). *The Quest for Longitude: The Proceedings of the Longitude Symposium*. Cambridge, Massachusetts: Collection of Historical Scientific Instruments, Harvard University.

Forbes, Eric G. (1974). "Maskelyne, Nevil." In *Dictionary of Scientific Biography*, edited by Charles Coulston Gillispie. Vol. 9, pp. 162–164. New York: Charles Scribner's Sons.

Forbes, Eric G., A. J. Meadows, and Derek Howse (1975). *Greenwich Observatory*. 3 Vols. London: Taylor and Francis.

Howse, Derek (1989). *Nevil Maskelyne: The Seaman's Astronomer*. Cambridge: Cambridge University Press.

Lane Hall, Mrs. A. W. (1932–1933). "Nevil Maskelyne." *Journal of the British Astronomical Association* 43: 67–77.

Ronan, Colin A. (1969). *Astronomers Royal*. Garden City, New York: Doubleday.

Sobel, Dava (1995). *Longitude: The True Story of a Lone Genius Who Solved the Greatest Scientific Problem of His Time*. New York: Walker.

Mason, Charles

Born **Wherr, Gloucestershire, England, 1730**
Died **Philadelphia, Pennsylvania, USA, 25 October 1786**

Charles Mason, a surveyor and astronomer, worked with **Jeremiah Dixon** on several astronomical expeditions including surveying the Mason–Dixon line delineating the boundary between Maryland and Pennsylvania. Although his father was a miller and a baker, Mason was educated at a private school, and became a professional surveyor. His first wife, Rebekah Peach, is believed to have been responsible for Mason's introduction to Astronomer Royal **James Bradley**, who appointed him assistant observer at the Greenwich Observatory in 1756; when Rebekah died in 1759 in Greenwich, on her tombstone was carved: "Wife of Charles Mason, Junr, ARS" (assistant of the Royal Society).

With the transit of Venus of 1761 impending, Bradley chose Mason to lead an observatory expedition to Bencoolen, Sumatra. On the voyage he was accompanied by Dixon, a surveyor and astronomer with a private observatory. They departed in November 1760 aboard *HMS Seahorse* with orders to proceed to Bencoolen unless it was in the hands of the French, in which case they would divert to Batavia. While still in the English Channel the *Seahorse* was attacked by the French frigate *Le Grand*. After a violent battle, which lasted barely an hour, the captain was able to return the ship back to Plymouth. However, upon witnessing the casualties and damage to both the ship and some of the astronomical equipment, Mason and Dixon wrote of their desire of not going to Bencoolen. Instead, Mason suggested the eastern portion of the Black Sea, from where they would be able to observe first contact, but not the planet leaving the face of the Sun.

The Royal Society not only denied their request but threatened them with a law suit, so the voyage to Bencoolen was recommenced. However, by the time they were rounding the Cape of Good Hope, they received news that Bencoolen had been taken by the French. Arriving at the Cape of Good Hope in April 1761, Mason and Dixon prepared to observe the transit from there. As luck would have it, their observations at the Cape of Good Hope were the only successful ones for the South Atlantic region; everywhere else was clouded out.

Afterward, Mason and Dixon joined **Nevil Maskelyne** on the island of Saint Helena, assisting him in various measurements, such as for tides, longitude, and the gravitational constant.

In 1763, as a result of the successful collaboration with respect to the transit of Venus, Mason and Dixon were charged with the responsibility of surveying what is still referred to as the Mason–Dixon line. The language of the original land grants to William Penn (later Pennsylvania) and to Lord Baltimore (later Maryland) was sufficiently vague that, by the mid-18th century, the argument between their respective heirs required the appointment of a commission in 1760 to adjudicate the border dispute. Three years later, Mason and Dixon were hired to survey and establish the boundary. Arriving in America in November 1763, they set up their equipment: two transits, two reflecting telescopes, and a zenith sector. Within a month, the two had measured the southernmost latitude of Philadelphia: 39° 56′ 29.1″ N and began the survey proper.

During the first few months, Mason and Dixon followed the old "temporary line" surveyed in 1739 by Benjamin Eastburn. This brought them through small townships such as Darby, Providence, Thornbury, West Town, and West Bradford. From there they continued to travel westward, as they were directed, to continue along the parallel of latitude as far as the country was inhabited. The surveyors continued until September of 1767 where, at Dunkard Creek, their Indian guide informed them it was the will of the Six Nations that the survey be stopped. They returned to England a year later in September 1768.

Because of their experience and their quality observations in 1761, Mason and Dixon were again asked to participate in an expedition for the 1769 Venus transit. Mason did not wish to participate; at the last minute, he grudgingly agreed to travel to County Donegal in Ireland. Only Dixon was willing, and he observed from the island of Hammerfest, off the Norwegian coast.

After the transit, Mason returned to England and continued his professional association with Maskelyne and the Royal Society. He was charged with aiding in the solution to a problem Sir **Isaac Newton** had devised decades earlier: Whether large land masses, such as mountains, could draw a plumb line up to 2 min out of the true perpendicular. Maskelyne had been intrigued with this problem for many years, and upon receiving funding to test it, in 1773 he commissioned Mason to select a suitable hill. Traveling to Scotland, Mason chose Schehallien, in Perthshire. And then, quite suddenly, Mason returned to England, quitting what some believe could have been his greatest scientific feat. Apparently, he had remarried, and desired to remain in England, where he worked on the *Nautical Almanac*, cataloging fixed stars and determining precise positions for the Moon.

Unfortunately, Mason's health was failing; apparently, his years in America had left him in a weakened state. Also, his second marriage added six children to his original two, and he was feeling the strain of poverty. For reasons unknown, Mason and his entire family sailed to Philadelphia, where he died shortly after arrival and was buried in Christ Church Burying Ground, in an unmarked grave.

Francine Jackson

Selected References

Pynchon, Thomas (1997). *Mason and Dixon*. New York: Henry Holt and Co.

Woolf, Harry (1959). *The Transits of Venus: A Study of Eighteenth-Century Science*. Princeton, New Jersey: Princeton University Press.

Mästlin [Möstlin], Michael

Born **Göppingen, (Baden-Württemberg, Germany), 30 September 1550**

Died **Tübingen, (Baden-Württemberg, Germany), 20/30 October 1631**

Michael Mästlin was a noted observer and mathematician himself, but is perhaps best known as the teacher of **Johannes Kepler**.

Mästlin was the son of Jakob Mästlin and Dorothea Simon (died: 1565), who were pious Lutherans; he had a younger brother and an older sister. Young Michael was sent to the monastic school in Königsbronn, and eventually he enrolled at Tübingen University in 1568. There, Mästlin studied mathematics and astronomy under Philip Apian (the son of the famous astronomer **Peter Apian**), whom Mästlin eventually replaced. Mästlin received his master's degree, *summa cum laude*, from Tübingen University in 1571. He tutored and taught there until he was called to be a deacon at the Lutheran Church in Backnang in 1576. There, Mästlin married Margarete Grüninger (1551–1588) in April 1577, who bore him three sons and three daughters; she died (possibly due to childbirth complications), with their sixth child. Mästlin then married Margarete Burkhardt, a daughter of a Tübingen professor, in 1589; they had eight more children.

Mästlin's publication of his careful observations of the comet of 1577 brought him fame as an astronomer. His reputation rose across Europe, leading to his appointment as a professor of mathematics at the University of Heidelberg in 1580. In 1584 Mästlin returned to the faculty at Tübingen, where he remained until his death.

For a while in the late 1570s, Mästlin was apparently the chief scientific advisor to his patron, Duke Ludwig III of Württemberg. Ludwig's successor, Duke Friedrich I, also relied on advice and opinions from Mästlin. At Tübingen, Mästlin was elected dean of the arts faculty several times. He was well liked by both his colleagues and his students. Mästlin was very generous both to his family and to others. He was a religious man; he followed the Lutheran line in opposing the Gregorian calendar reform partly because it was initiated by the pope. Mästlin had several students who became noted mathematicians, the most famous being Kepler. Mästlin also maintained interests in Biblical chronology and geography.

Mästlin was a prolific scholar of astronomy, writing extensively and corresponding with other astronomers throughout Europe. He can be considered the first astronomer to offer an orbit of a comet (though he did not use a proper procedure), putting the comet of 1577 in a heliocentric orbit just outside the orbit of Venus; he claimed that this supported the Copernican model of heliocentrism. Mästlin was an eager mathematician, working with spherical trigonometry to convert his observations to a useful format, and followed the published works of **Johann Müller** (Regiomontanus), Region Ontarus, Peter Apian, and **Caspar Peucer** in doing this. Mästlin read scholarly books very carefully, making extensive notes in many of his own books in a neat, small handwriting. For example, he heavily annotated his personal copies of **Nicolaus Copernicus'** *De revolutionibus* (noting, among many other things, numerous typographical errors in cataloged star positions), **Tycho Brahe's** *De mundi aetherei* on the 1577 comet (carefully assessing the positional observations), and **Johann Schöner's** 1544 treatise containing observations of Müller and **Bernard Walther** (where Mästlin seemed quite interested in eclipse measurements).

Through Mästlin's course on astronomy used his own textbook that followed Ptolemaic themes, this was likely due to the fact that basic astronomy (as taught at a low level at that time) did not need the technical aspects of Copernicus's heliocentrism and Müller's spherical trigonometry. In more advanced courses, these more technical aspects were evidently taught by Mästlin, who was widely known as a heliocentrist. That reputation had its origin in Mästlin's tract on the 1577 comet in which he placed the comet in a Venus-like orbit about the Sun, as did Brahe in his grand book on the same comet a decade later. Both Mästlin and Brahe credited the idea for such an orbit to **Abū Maʿshar**. Kepler credited Mästlin with having introduced him to Copernicus' philosophy during Kepler's student years at Tübingen (1589–1594). Kepler's mentor wrote an appendix entailing discussion of Copernican astronomy in the younger astronomer's first major publication, *Mysterium Cosmographicum* (1596, Tübingen). Mästlin maintained a long, productive correspondence with Kepler on astronomical matters. Kepler probably owed much of his own development of astronomical thought over the years to the training that he received from Mästlin.

In the late 1570s, Mästlin prepared for publication his *Ephemerides novae*, which were ephemerides of the planets based on Copernican theory (following the work of **Erasmus Reinhold**). Mästlin duly noted that the ephemerides needed correcting because the observations upon which they were based lacked accuracy, and he stated that Copernicus' theory is truer than older ideas. Following Regiomontanus, Peter Apian, and others from the previous 100 years, Mästlin joined his own generation of observers (including Brahe) in working carefully to obtain the best positional measurements possible of celestial objects and thereby improve the state of knowledge in astronomy.

Mästlin was known in his lifetime as a first-rate astronomical observer —his good eyesight is indicated by his drawing in 1579 of 11 stars in the Pleiades— and as an astronomer who was willing to challenge intelligently the old way of thinking about astronomy through the use of observations obtained in a more detailed and systematic fashion. In his early years, Mästlin improvised by using a thread to determine the position of transient objects (1572 supernova; comet of 1577) by checking their alignments with various stars. He impressed Brahe by finding that the supernova (B Cas) showed no parallax and must therefore be as distant as the other stars, attacking the Aristotelian position that the stellar region is unchanging. By 1577 Mästlin was using a clock to record times of observation; his was the first generation of astronomers where timekeeping was taken to be important, and times were noted often, despite the poor quality of timepieces then. Mästlin's tracts on comets and the 1572 supernova notably parallel Brahe's own tracts on these objects in that, unlike other typical treatises on such objects in that era, they concentrated on observations and reductions of observations while keeping astrological speculation to a bare minimum. Mästlin is also credited with being the first to publish his own finding that the unlit part of the crescent Moon glows faintly due to sunlight reflected off the Earth onto the Moon.

Though he was unable to undertake a huge observational program, such as Brahe did at Hven, Mästlin was an important influence on Brahe's work through his correspondence. Mästlin challenged his contemporaries to improve observational data rather than to just accept what had been passed down from the ancients through medieval times. He was also familiar with constructing sundials, celestial globes, quadrants, cross-staffs, and maps – all knowledge that was likely passed on to a large degree from his professor Philip Apian at Tübingen. Within 4 years of **Galileo Galilei**'s first pointing a telescope skyward, Mästlin had obtained two small telescopes which, though rather poor, showed him sunspots and the satellites of Jupiter. Mästlin remained an eager astronomical observer into his late years, making notes of his observations of the comets of 1618 and of a lunar eclipse in 1628.

Much of Mästlin's library now resides at the Municipal Library in Schaffhausen, Switzerland.

Daniel W. E. Green

Alternate names

Moestlinus

Möschlin, Michael

Selected References

Gingerich, Owen (2002). *An Annotated Census of Copernicus' De Revolutionibus (Nuremberg, 1543 and Basel, 1566)*. Leiden: Brill. (Contains Mästlin's extensive annotations from his personal copy in Schaffhausen, inscribed over a period of about 50 years. One of Mästlin's notes therein states that Andreas Osiander was the anonymous author of the controversial preface to Copernicus's book.)

Green, D. W. E. (2004). Astronomy of the 1572 Supernova (B Cassiopeiae) *Astronimische Nachrichten*, Vol. 325, p. 689.
Hellman, C. Doris (1944). *The Comet of 1577: Its Place in the History of Astronomy*. New York: Columbia University Press, pp. 137ff. (Reprint, New York: AMS Press, 1971.) (A detailed discussion of Mästlin and the context of his observations on comets.)
Jarrell, Richard A. (1971). "The Life and Scientific Work of the Tübingen Astronomer, Michael Mästlin, 1550–1631." Ph.D. diss., University of Toronto. (The most extensive discussion of Mästlin's life and career.)
Mästlin, Michael (1573). *Demonstratio astronomica loci stelle novae*. In *Consideratio novae stellae, quae mense Novembri anno salutis 1572*, by N. Frischlin, pp. 27–32. Tübingen. (Mästlin's tract was first published here. The same tract on the 1572 supernova by Mästlin also appears in Brahe's 1602 *Astronomiae instauratae progymnasmata*, pp. 544–548.)
——— (1578). *Obseruatio & demonstratio cometae aetherei, qvi anno 1577. et 1578. constitutus in sphaera Veneris, apparuit*. Tübingen Förge. (Contains Mästlin's detailed account of observations and analysis of the comet of 1577.)
——— (1581). *Consideratio & obseruatio cometae aetherei astronomica, qvi anno MDLXXX*. Heidelberg: Jacob Mylius. (Contains Mästlin's detailed account of observations and analysis of the comet of 1580.)
——— (1582). *Epitome astronomiae*. Heidelberg. (This general astronomy text based on Ptolemaic and Alfonsine principles went into numerous editions/reprintings until 1624; it was evidently used by Mästlin as his course textbook at Tübingen.)
Methuen, Charlotte (1996). "Mästlin's Teaching of Copernicus: The Evidence of His University Textbook and Disputations." *Isis* 87: 230–247. (An analysis of Mästlin's embracing of heliocentrism.)
Rosen, Edward (1974). "Mästlin, Michael." In *Dictionary of Scientific Biography*, edited by Charles Coulston Gillispie. Vol. 9, p. 167. New York: Charles Scribner's Sons.
Westman, Robert S. (1972). "The Comet and the Cosmos: Kepler, Mästlin, and the Copernican Hypothesis." In *The Reception of Copernicus' Heliocentric Theory*, edited by Jerzy Dobrzyck, pp. 7–30. Dordrecht: D. Reidel. (A discussion of Mästlin's incorporation of the comet of 1577 into a heliocentric model, with its impact on Kepler.)
——— (1975). "Three Responses to the Copernican Theory: Johannes Praetorius, Tycho Brahe, and Michael Mästlin." In *The Copernican Achievement*, pp. 285–345. Berkeley: University of California Press. (A discussion of Mästlin's acceptance of heliocentrism.)
Zinner, Ernst (1925). *Verzeichnis der astronomischen Handschriften des deutschen Kulturgebietes*. Munich: C. H. Beck, 1925, pp. 217–219, 469–470. (Details the location in 1925 of many of Mästlin's unpublished papers.)

Mathurānātha Śarman

Flourished **Bengal, (India), 1609**

Mathurānātha Śarman composed the *Ravisiddhāntamañjarī* or *Sūryasiddhāntamañjarī*, an astronomical treatise consisting of four chapters and tables, in 1609. This work uses parameters belonging to the *Saurapakṣa*, one of the traditional schools of astronomy in India. The tables are for calculating the longitudes of the planets; there are also parallax tables for computing solar eclipses. He may have composed two other works, the *Pañcaṅgaratna* and the *Praśnaratnāṅkura* or *Samayāmṛta*.

Setsuro Ikeyama

Selected References

Mathurānātha Śarman (1911). *Ravisiddhāntamañjarī*. Bibliotheca Indica, no. 198, edited by Viśvambhara Jyotiṣārṇava. Calcutta: Asiatic Society.
Pingree, David (1973). *Sanskrit Astronomical Tables in England*. Madras: Kuppuswami Sastri Research Institute, pp. 128–134.
——— (1974). "Mathurānātha Śarman." In *Dictionary of Scientific Biography*, edited by Charles Coulston Gillispie, Vol. 9, p. 175. New York: Charles Scribner's Sons.
——— *Census of the Exact Sciences in Sanskrit* (1981 and 1994). Series A. Vol. 4: 349a; Vol. 5: 274b. Philadelphia: American Philosophical Society.
——— (1981). *Jyotiḥśāstra*. Wiesbaden: Otto Harrassowitz.

Maudith, John

Flourished **1309–1343**

John Maudith produced astronomical tables in the early 14th century. In 1310 he compiled tables for the rising and setting times of stars. Maudith's tables were essentially Toledan tables recomputed for the Oxford meridian. He drafted a separate catalog of bright stars, epoch 1316, with stellar positions calculated according to **Thebit ibn Qurra's** theory of trepidation.

Very little is known of Maudith's life. Between 1309 and 1319 he was a fellow of Merton College, Oxford. The Merton *Catalogus vetus* describes him as a good astronomer and physician. After leaving Oxford, Maudith joined Richard de Bury's scholarly circle at Durham; **Thomas Bradwardine** was one of his colleagues there. Maudith later served John de Warenne, Earl of Surrey and Sussex, probably as a physician. He wrote a *Tractatus de doctrina theologica* dated to 1343.

Keith Snedegar

Selected References

Emden, A. B. (1957–1959). *A Biographical Register of the University of Oxford*. Vol. 2, pp. 143–144. Oxford: Oxford University Press.
North, J. D. (1992). "Natural Philosophy in Late Medieval Oxford" and "Astronomy and Mathematics." In *The History of the University of Oxford*. Vol. 2, *Late Medieval Oxford*, edited by J. I. Catto and Ralph Evans, pp. 65–102, 103–174. Oxford: Clarendon Press.

Maunder, Annie Scott Dill Russell

Born **Strabane, Co. Tyrone, (Northern Ireland), 14 April 1868**
Died **London, England, 15 September 1947**

Solar astronomer Annie Russell joined her husband, **Edward Walter Maunder**, in supporting amateur astronomers in Britain by editing their journal and leading solar eclipse expeditions, while continuing her own solar research and popular writing on astronomy.

Russell was the first daughter of the Reverend William Andrew Russell, a minister of the Irish Presbyterian Church, and his second wife, Hester (*née* Dill). Annie had two half-brothers from her father's first marriage and two brothers and one sister from the second. For her secondary education, she attended the Ladies' Collegiate School, Belfast (renamed Victoria College in 1887), known as the premier institution for the education of girls in Ireland. Russell decided not to work for an Irish university degree but, instead, took the Girton College open entrance examination. By studying diligently, she overcame a deficit in her early training and upon graduation won the highest mathematical honor available to a woman, Senior Optime in the mathematical *tripos*. (When Russell graduated, women were allowed to sit for the Cambridge *tripos* examinations, although they were not granted a university degree.)

Upon leaving Girton College, Russell became a mathematics teacher at the Ladies' College, Jersey, but found teaching unrewarding. After learning of a possible vacancy for a "lady computer" at the Royal Greenwich Observatory, she applied for the position, even though the pay was much less than she earned as a teacher. She accepted the post and, while measuring daily sunspot photographs, met Maunder, head of the solar photography department. In 1890, Maunder had assisted a number of leading amateur astronomers in founding the British Astronomical Association [BAA] after the collapse of the Liverpool Astronomical Society. Together, Russell and Maunder worked on the association's journal. She was its first editor, from 1894 to 1896, and served an additional term, from 1917 to 1930.

In 1895, Russell married Maunder, and they worked together on numerous astronomical projects. Annie, however, was obliged to resign her position as Walter's paid assistant at Greenwich. Walter and Annie had no children of their own, but 45-year-old Walter was a widower with five children when he married the 27-year-old Annie. Although she continued with her astronomy, much time was spent in rearing her stepchildren. In another sense, her marriage to Walter proved fortunate for Annie Maunder's career. Through her husband, she was able to borrow instruments, establish contacts with other astronomers, and travel to various eclipse sites. Probably the most important factor was Walter's view that women deserved an important place in astronomy.

Maunder made many contributions to astronomy. Shortly after her marriage, she received the Pfeiffer Research Student Fellowship, established to upgrade the Girton College research potential. As the first recipient of this fellowship (1896), she used the money to undertake a photographic study of the Milky Way. At Greenwich, she was assigned to the solar department as a photographic assistant. Maunder's work involved photographing the Sun and examining the negatives with a micrometer. Recruited during the approach of a sunspot maximum, she noted the positions of the sunspots and worked on interpreting the phenomena.

Although much of her astronomical work was done in collaboration with Walter, Annie was an important contributor to astronomy in her own right. She published numerous papers and a book, *The Heavens and Their Story* (1908). Although Walter's name appeared as coauthor, he insisted that Annie had done all of the writing. Her professional level of competence was gained from formal university training, working as a paid assistant, and from informal training by her husband.

Maunder's prodigious output included theoretical work. She developed a theory that the Earth influences the numbers and areas of sunspots, and that sunspot frequency decreases from the eastern to the western edge of the Sun's disk (as viewed from Earth). With Walter and the BAA, she went on solar eclipse expeditions and became an expert eclipse observer. One of her photographs revealed a coronal streamer extending out to six solar radii – the longest then observed to that date. Maunder published numerous reports on these eclipses, and many papers on the history of astronomy, especially early accounts of the constellations.

Gender must be considered when examining Maunder's career. While she possessed all of the requisites to be a professional scientist, as a woman, she received less than full recognition of her qualifications and contributions from male professional astronomers. Fortunately, her husband recognized the importance of women to astronomy. Through a variety of channels including the BAA, she and other women astronomers were able to make their contributions.

Maunder's papers may be found in the Archives of the British Astronomical Association, the Archives of the Royal Astronomical Society, and the Archives of the Royal Greenwich Observatory.

Marilyn Bailey Ogilvie

Selected References

Brück, Mary T. (1994). "Alice Everett and Annie Russell Maunder: Torch Bearing Women Astronomers." *Irish Astronomical Journal* 21: 281–291.

Evershed, M. A. (1947). "Mrs. Walter Maunder." *Journal of the British Astronomical Association* 57: 238.

Kidwell, Peggy Aldrich (1984). "Women Astronomers in Britain, 1780–1930." *Isis* 75: 534–546.

Ogilvie, Marilyn Bailey (1996). "Patterns of Collaboration in Turn-of-the-Century Astronomy: The Campbells and the Maunders." In *Creative Couples in the Sciences*, edited by Helena M. Pycior, Nancy G. Slack, and Pnina G. Abir-Am, pp. 254–266. New Brunswick, New Jersey: Rutgers University Press.

——— (2000). "Obligatory Amateurs: Annie Maunder (1868–1947) and British Women Astronomers at the Dawn of Professional Astronomy." *British Journal for the History of Science* 33: 67–84.

Maunder, Edward Walter

Born **London, England, 12 April 1851**
Died **London, England, 21 March 1928**

Walter Maunder is chiefly remembered for his work in the field of solar studies. His plot of the latitude drift of sunspots is known as the Maunder butterfly diagram. The lapse in sunspot numbers during the interval from 1645 to 1715, which he investigated, has been termed the Maunder minimum.

The youngest of three sons of the Reverend George Maunder, a Wesleyan minister, Maunder's basic education was acquired at the school attached to University College in Gower Street, London, and supplemented with additional courses at King's College, London. He worked briefly in a City of London bank before taking the first-ever examination set by the British Civil Service Commissioners (1872), designed to fill vacancies created at the Royal Greenwich Observatory. With his appointment as photographic and spectroscopic assistant in 1873, the observatory entered the realm of astrophysics, or the New Astronomy, as it was called by **Samuel Langley**.

Maunder spent 40 years at Greenwich and became superintendent of the solar department, working chiefly under the direction of **William Christie**. Starting in 1874, Maunder operated the photoheliograph, taking daily photographs of the Sun (originally on wet plates, later on dry), then measuring and tabulating the numbers, areas, positions, and motions of sunspots. Data acquired over the next 30 years enabled Maunder to affirm what was known about the axis of solar rotation, the equatorial drift and periodicity of sunspots, and the Sun's differential rotation from the work of **Samuel Schwabe** and **Richard Carrington**. Another important finding was the observed correlation between solar activity and terrestrial magnetic disturbances, a subject discussed in a series of papers starting in 1904.

In 1887 and 1889, **Gustav Spörer** drew attention to a long continued absence of sunspot activity, from about 1645 to 1715. Maunder summarized Spörer's papers for the Royal Astronomical Society in 1890, and began his own search of historical records. Maunder published an article, entitled "A Prolonged Sunspot Minimum," which supported Spörer's conclusions. It attracted little attention. Nor did an article with a similar title, published in 1922. Only in the mid-1970s, after solar physicist John A. Eddy took a fresh look at the evidence, were Maunder's findings confirmed. This anomaly was named the Maunder minimum. Spörer's name is given to an earlier, similar epoch.

Maunder also undertook observations of solar prominences, the spectra of comets, planets, novae, and nebulae, and was a keen follower of total solar eclipses, observing those of 1886 (Carriacou, West Indies), 1896 (Vadsö, Norway), 1898 (India), 1900 (Algeria), 1901 (Mauritius), and 1905 (Canada). He was elected a fellow of the Royal Astronomical Society in 1875, served as a council member for several years, and was its secretary (1892–1897).

In 1890, Maunder and his brother, Thomas, played a central role in the formation of the British Astronomical Association [BAA], whose purpose was "to afford a means of direction and organization in the work of observation to amateur astronomers." He served as its third president (1894–1896), and was acting secretary (1914–1915). Maunder headed the Mars Section (1892–1893), the Star Colour Section (1900–1902), and the Solar Section (1910–1925). He likewise edited the association's *Journal* for a number of years, having previously edited *The Observatory* (1881–1887).

In all of these investigations, Maunder received invaluable help from his second wife, Annie, who was academically more qualified than her husband. **Annie Maunder** graduated from Girton College, Cambridge, as senior optime in the mathematical *tripos*, and joined the Royal Greenwich Observatory staff as "lady computer" in 1891. She and Walter were married in 1895 and worked closely together on the solar data, although Annie was obliged to resign her post as Maunder's paid assistant. Together, they provided leadership to a series of successful solar eclipse expeditions sponsored by the BAA.

Maunder retired from Greenwich at the end of 1913, but was recalled to maintain the sunspot record during World War I. Between 1914 and 1916, he served as secretary of the Victoria Institute, London, a society founded in 1865 to investigate questions of philosophy and science, especially those bearing upon religion.

Maunder and his wife published several books, including a history of the Royal Greenwich Observatory (1900), and many articles and papers in the *Monthly Notices of the Royal Astronomical Society*, and the *Journal of the British Astronomical Association*. They were frequent contributors to *Knowledge* and *Nature*.

Richard Baum

Selected References

Eddy, John A. (1976). "The Maunder Minimum." *Science* 192: 1189–1202.

Hollis, H. P. (1927–1928). "Edward Walter Maunder." *Journal of the British Astronomical Association* 38: 229–233. (See also pp. 165–168.)

——— (1929). "E. W. Maunder." *Monthly Notices of the Royal Astronomical Society* 89: 313–318.

Hufbauer, Karl (1991). *Exploring the Sun: Solar Science since Galileo*. Baltimore: Johns Hopkins University Press, esp. pp. 273–277.

Maunder, E. Walter (1894). "A Prolonged Sunspot Minimum." *Knowledge* 17: 173–176.

——— (1904). "Note on the Distribution of Sun-spots in Heliographic Latitude, 1874 to 1902." *Monthly Notices of the Royal Astronomical Society* 64: 747–761.

——— (1922). "The Prolonged Sunspot Minimum, 1645–1715." *Journal of the British Astronomical Association* 32: 140–145.

Warner, Deborah Jean (1974). "Maunder, Edward Walter." In *Dictionary of Scientific Biography*, edited by Charles Coulston Gillispie. Vol. 9, pp. 183–185. New York: Charles Scribner's Sons.

Maupertuis, Pierre-Louis Moreau de

Born **Saint-Malo, (Ille-et-Vilaine), France, 28 September 1698**
Died **Basle, Switzerland, 27 July 1759**

In astronomy, Pierre de Maupertuis contributed to the understanding and diffusion of **Isaac Newton**'s theory in France and in continental Europe. He arranged for, and participated in, measurements to ascertain the shape of the Earth. In physics, Maupertuis was the first to formulate the least-action principle. He also made contributions to mathematics, biology, heredity, and moral philosophy. As a prolific intellectual, Maupertuis opened new roads in science.

Maupertuis' father, René Moreau, was a layperson. Maupertuis was raised by his overcautious mother and first educated at home. For philosophical instruction, he attended the Collège de la Marche in Paris in 1714. At his mother's request, he returned to Saint-Malo in 1716 and gave up his wish to go to sea. After a visit to Holland in 1717, Maupertuis moved back to Paris where he began musical studies, but switched to mathematics.

In 1718, Maupertuis joined the *Mousquetaires Gris* and in 1720, with the rank of lieutenant, was stationed in Lille. During his army period, he devoted all his free time to geometry. In the following year, he resigned his commission and returned to Paris. There Maupertuis joined a group of scholars, three of whom were members of the Académie royale des sciences, through whose intervention he was elected to the academy on 14 December 1723 as an *adjoint-géomètre*, the lowest position, despite having no publications.

In August 1725, a short time after the publication of his first paper devoted to the influence of shape on the properties of musical instruments, Maupertuis was promoted to *associé*. From 1723 to 1733, he published various memoirs concerning geometry, mathematics, and zoology; the memoir *Sur la question des maximis et des minimis* was the first step in his formulation of the least-action principle.

Maupertuis made his first foreign journeys, to London in 1728, to Basle – he registered there as a student – in 1729, and again to Basle the next year. These journeys played a major part in his future intellectual evolution. In London, Maupertuis was in the center of the Newtonianism, of observational science, and of watch and instrument making. He was admitted to the Royal Society on 27 June 1729 (O.S.). In Basle Maupertuis met **Johann Bernoulli**, from whom he received an excellent general scientific training and an introduction to **Gottfried Leibniz**' thought. Throughout his life, Maupertuis found friendship and support from the Bernoulli family.

During the 1730s, Maupertuis published many papers. 1731 was the year of both the publication, in England, of the *De Figuris*, his first astronomical paper, and his election as a *pensionnaire-géomètre* to the academy. The publication, in the following year, of his *Discours sur la figure des astres* is considered to be the first book promoting widely the Newtonian theory in France and continental Europe. While he presented Cartesianism and Newtonianism with some symmetry, Maupertuis did in fact support the latter.

The question of the exact shape of the Earth was of central importance, particularly to the academy, because **Jacques Cassini's** and colleagues' measurements led to a prolate model of Earth, whereas Newtonians argued for an oblate Earth. During the period 1732–1735, Maupertuis studied the consequences of the law of attraction on the Earth's shape and other celestial bodies. Because of dissension, the academy ordered two expeditions to measure the length of a degree along a meridian at two very different latitudes. **Charles de la Condamine**, **Louis Godin**, and **Pierre Bouguer** led an expedition to Peru, while Maupertuis and **Alexis-Claude Clairaut**, who already worked together, led a second one to the Gulf of Bothnia. Before sailing to Lapland, both were trained in observing and measuring by Jacques Cassini. The abbé **Réginald Outhier,** a member of the Academy of Caen and an astronomer, accompanied them and chronicled the expedition in his *Journal d'un voyage au Nord, en 1736 & 1737*. This expedition may have been one wherein **John Hadley**'s octant was first used. Whereas the expedition to Peru lasted about 10 years, the Lapland team returned to Paris on 20 August 1737, just 16 months after departure. Their measurements confirmed the oblateness of Earth.

Maupertuis made two reports to the academy (1737 and 1738) but came under attack. He carried on his argument with Cassini in his *Examen désintéréssé des différents ouvrages qui ont été faits pour déterminer la figure de la Terre*, published in 1738 or 1739. Waiting for the return of the Peru mission, the Lapland astronomers reassembled in August 1739 and made a new measurement of the arc between Amiens and Paris, measured by **Jean Picard** in 1669. To support his position on the Earth's figure, Maupertuis published three works in 1740: *Éléments de géographie*, *Degré du méridien entre Amiens et Paris*, and *Lettre d'un horloger anglois à un astronome de Pékin*, this last an ironic literary piece attacking Cassini's followers in the academy. During this period, Maupertuis carried on a wide correspondence with leading European scholars. He also taught Mme du Chatelet geometry and calculus.

Maupertuis had been elected an *associé-étranger* of the Berlin Academy in 1735 and was so informed when he returned from Lapland. When Frederick II became King of Prussia in 1740, he wished to reform his academy and invited Maupertuis to come to Berlin. In September 1740, Maupertuis arrived there for the first time. Going to meet Frederick at Mollwitz, during the War of Austrian succession in the following year, Maupertuis was taken prisoner by the Austrians, but was well received by the court in Vienna. In 1745, he settled in Berlin and, in August, married Eleonor de Bork, a noblewoman he had met on an earlier visit. Maupertuis assumed the presidency of the Berlin Academy on 3 March 1746. Although he had been active in the Paris Academy (*sous-directeur* in 1735 and 1741, directeur in 1736 and 1742), he now had to resign. He was also a member of the Académie française.

In Paris, before his official installation in Berlin, Maupertuis penned his *Discours sur la parallaxe de la Lune pour perfectionner la théorie de la Lune et celle de la Terre* (1741), *Lettre sur la comète* (1742), and *Astronomie nautique* (1743). All dealt with Newtonian solutions to various questions.

In the later 1740s and 1750s, Maupertuis turned more to speculative and natural philosophy and to his routine work for the Berlin Academy. He published many of his ideas in letters. As president, he supported astronomical work, including the first precise measurement of lunar parallax thanks to observations by **Joseph-Jérome de Lalande** (another Berlin academician) at Berlin and by **Nicolas de La Caille** at the Cape of Good Hope.

Maupertuis was the first to formulate the least-action principle, which he considered as the summit of his work. He published on statics in *Loi du repos des Corps* in 1740; he applied the ideas to optics in a paper "*Accord de différentes lois de la Nature qui avoient paru incompatibles*" (1744). To extend the ideas to mechanics, Maupertuis assumed collisions of massive points. His full ideas appeared in *Essay de Cosmologie* in 1750.

Samuel König, a Berlin academician and long-time friend, claimed that Leibniz had indicated, in a letter, that he had been the first to formulate the principle. Although in poor health, Maupertuis fought to maintain his priority, and the fight drew in many from Berlin intellectual circles. As the Leibniz letter could not be found, the academy supported Maupertuis in a meeting of 13 April 1752, forcing König to resign. This resulted in much hostility towards Maupertuis, including a virulent attack by Voltaire, in his *Diatribe du docteur Akakia* (1752), which portrayed him as an arrogant fool.

In his last years Maupertuis produced works on reproduction, heredity, and pleasure, including *Dissertation physique à l'occasion du nègre blanc* (1744) and *Vénus physique* (1745). In the *Système de la nature* of 1751, Maupertuis speculated on parental heredity, anticipating some ideas of the following century. He left Berlin for the last time in June 1756. He was reinstalled in the Paris Academy on 15 June. A final journey in 1759 took him to Bernoulli's home in Basle, where Maupertuis died. At Saint-Roch in Paris, a marble funeral stele was erected by his friends in 1766.

Monique Gros

Selected References

Beeson, David (1992). *Maupertuis: An Intellectual Biography*. Oxford: Voltaire Foundation.

Brunet, Pierre (1929). *Maupertuis: Étude Biographique*. Vol. 1. Paris: Librairie scientifique Albert Blanchart.

Glass, Bentley (1974). "Maupertuis, Pierre Louis Moreau de." In *Dictionary of Scientific Biography*, edited by Charles Coulston Gillispie. Vol. 9, pp. 186–189. New York: Charles Scribner's Sons.

Martin, Jean-Pierre (1987). *La figure de la terre*. Cherbourg, France: Isoète.

Terrall, Mary (2002). *The Man Who Flattened the Earth: Maupertuis and the Sciences in the Enlightenment*. Chicago: University of Chicago Press.

Maurolico, Francesco

Born **Messina, (Italy), 16 September 1494**
Died **near Messina, (Italy), 21 or 22 July 1575**

Francesco Maurolico, in addition to doing original work, translated and commented on works by ancient authors such as Euclid, **Apollonius**, and **Archimedes**.

Maurolico spent nearly all of his life in Sicily, where he was ordained a priest in 1521, and held various ecclesiastical as well as civil posts. His astronomical writings include a criticism of **Nicolaus Copernicus**, and a treatise on the use of the principal astronomical instruments. Maurolico's observation of the supernova of 1572 in Cassiopeia (SN B Cas) appears to predate that of **Tycho Brahe** by five days.

Katherine Bracher

Selected References

Hellman, C. Doris (1960). "Maurolyco's 'Lost' Essay on the New Star of 1572." *Isis* 51: 322–336.

Masotti, Arnaldo (1974). "Maurolico, Francesco." In *Dictionary of Scientific Biography*, edited by Charles Coulston Gillispie, Vol. 9, pp. 190–194. New York: Charles Scribners' Sons.

Maury, Antonia Caetana de Paiva Pereira

Born **Cold Springs, New York, USA, 21 March 1866**
Died **Hastings-on-Hudson, New York, USA, 8 January 1952**

American spectroscopist Antonia Maury discovered a way of recognizing supergiant stars from their spectra, even when their distances could not be measured. She was the granddaughter of **John William Draper**, the first person to photograph the Moon, the niece of **Henry Draper**, another keen amateur astronomer, and the daughter of Mytton and Virginia (*née* Draper) Maury. Her father, an Episcopalian priest, edited volumes in natural history; a sister became a paleontologist and a cousin an oceanographer. Maury was taught largely by her father and uncle, going on to Vassar College, New York, where she was one of the last students of **Maria Mitchell**, and received a BA in 1887.

In 1888, Maury was employed by **Edward Pickering** at Harvard College Observatory as part of a program to determine the spectral types of stars. The program was funded by a memorial contribution from Henry Draper's wife in his honor, and eventually published as the Henry Draper Catalog. **Williamina Fleming** and **Annie Cannon** were also employed in this project. Maury was probably the most intellectually gifted of the three women. Although she had been employed to classify objective-prism spectra into the system defined by Pickering and Fleming, she instead set up her own system. It improved on the earlier system in two ways. First was a finer gradation by the temperatures of stars; Maury was the first to recognize that the temperature sequence should be O, B, A. Second, she noticed that in some cases the spectral features were unusually hazy (her type b) and in some cases unusually sharp (her type c). These and other details that Maury recorded were regarded by Pickering as a waste of time, and it was not until about 1907 that **Ejnar Hertzsprung**, who had independently discovered supergiants by another method, recognized the importance of Maury's class c. Her own catalog, with the a, b, and c characteristics and a variety of additional kinds of information including notes on composite spectra and emission lines, appeared in the *Harvard College Annals* in 1897. Many of Maury's "b" types were later recognized as rapid rotators, an interpretation she had herself suggested.

Maury was also a pioneer in the investigation of spectroscopic binaries. Pickering had discovered the first of these, Mizar, from the doubling of its calcium K line in 1889. Maury found the second, β Aurigae, the same year and was the first to measure the orbital periods of both. She was no longer formally employed by Harvard College Observatory after 1892 but continued to analyze spectra taken there until 1935, including a large number of plates of the very peculiar eclipsing binary β Lyrae. This work appeared periodically in the *Harvard Annals Bulletin* until 1935.

Maury lectured in several east-coast colleges and served for several years as curator of Draper Park Museum. She was also a recognized ornithologist and naturalist.

Maury was, somewhat ironically, the 1943 recipient of the Annie J. Cannon Prize of the American Astronomical Society, the only major recognition she ever received. Her relationship with Harvard College Observatory, though informal, became much smoother under the directorship of **Harlow Shapley**.

Virginia Trimble

Selected References

Gingerich, Owen (1974). "Maury, Antonia Caetana de Paiva Pereira." In *Dictionary of Scientific Biography*, edited by Charles Coulston Gillispie, Vol. 9, pp. 194–195. New York: Charles Scribner's Sons.

Hoffleit, Dorrit (1952). "Antonia C. Maury." *Sky & Telescope* 11, no. 5: 106.

——— (2002). "Pioneering Women in the Spectral Classification of Stars." *Physics in Perspective* 4: 370–398.

Maury, Matthew Fontaine

Born **near Fredericksburg, Virginia, USA, 14 January 1806**
Died **Lexington, Virginia, USA, 1 February 1873**

A naval officer best known for his wind and current charts and therefore considered one of the founders of oceanography, Mathew Maury was, in effect, the first superintendent of the Depot of Charts and Instruments, though **James Gillis** might more properly be considered its founder.

Maury became a midshipman in 1825 and had three periods of sea duty through 1834. Two years later he authored a widely used textbook, *A New Theoretical and Practical Treatise on Navigation*. Promoted to lieutenant in 1836, Maury suffered a leg injury in 1839 that confined him permanently to shore duty, a situation that would greatly affect the remainder of his naval career. In 1842 Maury was named officer-in-charge of the Depot of Charts and Instruments, founded in 1830 to centralize the Navy's navigational maps and technology. Maury became the first superintendent of a newly equipped depot established by Congress with a "small observatory" in 1844. That depot quickly grew into the United States Naval Observatory and Hydrographic Office, an agency Maury headed until he joined the Southern cause in the Civil War in 1861. Maury was promoted to the rank of commander in 1858, retroactive to 1853.

During his years as the superintendent, Maury struggled to balance astronomy and hydrography. His main achievements were in the latter; the wind and current charts, based upon data that sea captains submitted to the depot, greatly shortened ocean voyages. In astronomy Maury oversaw observations made with a variety of astrometric instruments, and produced widely praised star catalogs. He struggled with limited resources, but his undoubted achievement is that he turned a small depot into the first national observatory of the United States, on a par with the Greenwich Observatory in England and other national observatories around the world. Maury was often considered an outsider among the new breed of professional American scientists, and his legacy in the scientific establishment was complicated by strong feelings stemming from his role in the Civil War.

Steven J. Dick

Selected References

Dick, Steven J. (2003). *Sky and Ocean Joined: The U. S. Naval Observatory, 1830–2000.* Cambridge: Cambridge University Press.

Williams, Francis L. (1963). *Matthew Fontaine Maury: Scientist of the Sea.* New Brunswick, New Jersey: Rutgers University Press.

Maxwell, James Clerk

Born **Edinburgh, Scotland, 13 June 1831**
Died **Cambridge, England, 5 November 1879**

It was while competing for the fourth Adams Prize that James Maxwell wrote a paper on Saturn's rings, in which he proposed that they were made of small particles, and could not be solid.

Maxwell's father was John Clerk; Maxwell was added later for inheritance purposes. His mother was Frances Cay. Maxwell was brought up in the Scottish countryside on an estate at Middlebie, Galloway, in a house called Glenlair. His mother died when he was 8 years old, and after an unsuccessful spell with a private tutor, he was educated at Edinburgh Academy, from 1841. Maxwell wrote a paper on the geometry of ovals when he was 14.

Maxwell went to Edinburgh University in 1847, then to Peterhouse College, Cambridge, in 1850, transferring to Trinity College and graduating in 1854. In 1856 he took up the post of professor of natural philosophy at Marischal College, Aberdeen, to be close to his father who was ill.

Maxwell married Katherine Mary Dewar, the daughter of the principal of Marischal College, but this did not prevent him from losing his position when Marischal College and King's College in Aberdeen were merged in 1860. In 1861 he was elected a fellow of the Royal Society.

Maxwell was turned down for a chair at Edinburgh University, and was appointed chair of natural philosophy at King's College, London until 1865. He then divided his time between Glenlair and Cambridge, where he designed the Cavendish Laboratory, which opened on 16 June 1874.

In 1866, Maxwell formulated, independently of Ludwig Boltzmann, what is now known as the Maxwell–Boltzmann kinetic theory of gases. Later, Maxwell developed the famous equations describing electromagnetism that bear his name.

David Jefferies

Selected References

Harman, P. M. (1998). *The Natural Philosophy of James Clerk Maxwell.* Cambridge: Cambridge University Press.

——— (ed.) (1990–2002). *The Scientific Letters and Papers of James Clerk Maxwell.* 3 Vols. Cambridge: Cambridge University Press.

Maxwell, James Clerk (1876). *Matter and Motion.* London: Society for Promoting Christian Knowledge. Also available from Dover.

Mayall, Margaret Walton

Born **Iron Hill, Maryland, USA, 27 January 1902**
Died **Cambridge, Massachusetts, USA, 6 December 1995**

Over a period of 24 years, Margaret Walton Mayall led the American Association of Variable Star Observers [AAVSO] to a position of international leadership among variable star organizations while providing substantial support to professional variable star astronomers and for amateur and popular astronomy through her publications. Mayall studied at the University of Delaware, Swarthmore College (BA: 1924), and Radcliffe College Harward University, where she earned an MA in Astronomy in 1927. With the help of **Leslie Comrie** at Swarthmore, Mayall found employment at Harvard College Observatory [HCO]. From 1924 to 1954 she worked at HCO as a research assistant, and later as Pickering Memorial Astronomer. She spent the summers of 1925 and 1926 as first assistant to Margaret Harwood, director of the Maria Mitchell Observatory in Nantucket, Massachusetts, and it was there that she first became interested in variable stars. It was also Nantucket where she met Robert Newton Mayall. They were married in 1927, but had no children. During World War II and for a year beyond (1943–1946), Mayall served in the research staff of the Heat Research Laboratory, Special Weapons Group, Massachusetts Institute of Technology.

While at HCO, Mayall assisted **Annie Cannon** in the classification of the spectra of faint stars and the estimation of the brightness of catalogued stars. She worked with Cannon until the latter's death in 1941, and then completed Cannon's unfinished spectral work, editing the results for publication as the second volume of *The Henry Draper Extension—The Annie J. Cannon Memorial* in 1949. Mayall published many other technical monographs while working at HCO.

Early in 1949, HCO Director **Harlow Shapley** asked Mayall to consider taking over the position of AAVSO recorder from **Leon Campbell** when he retired. The AAVSO was founded in 1911 by **William Olcott** in response to HCO director **Edward Pickering**'s efforts to collect observations of variable stars. From about 1918 to 1954, the AAVSO was headquartered at and run under the auspices of HCO. In 1949, Mayall accepted the position and was named Pickering Memorial Astronomer at HCO and AAVSO recorder. (The title was later changed to director.) She remained director of the AAVSO for 24 years until her retirement in 1973.

When Shapley retired from the HCO directorship in 1952, AAVSO's position began to change. With an inadequate budget, aging telescopes and other facilities, and aspirations for a rather different type of organization, the new HCO director, **Donald Menzel**, was forced to reconsider observatory priorities. In 1954, he announced that the AAVSO would have to move out of the observatory, and the endowment that supported the Pickering Memorial Professorship would no longer be available. With a new title of director, Mayall oversaw the transition of the AAVSO to an independent, nonprofit scientific and educational organization. As the association struggled through severe financial hardship Mayall worked without salary for a number of years to ensure the future of the AAVSO.

Through her determination, persistence, and vision, and with substantial help from **Clinton Ford** and other AAVSO leaders, Mayall secured the future of the AAVSO in many important ways. During the critical early years of independence, she actively sought out new sources of funding from government, industry, and private donors. She communicated widely with the astronomical community to solicit both technical and moral support for the AAVSO from the professional community; she established an endowment fund to secure a firmer financial footing for the AAVSO; she expanded existing programs and committees and added new ones to attract a broader membership; she added more stars to the observing program and established new systems for correlating and publishing data for professional use; and in 1967 she introduced modern data processing methods at the AAVSO, with emphasis on machine computing and plotting for publication. Mayall retired as director emeritus of the AAVSO in 1973.

As a member of the International Astronomical Union [IAU], Mayall participated in the activities of two IAU commissions, Commission 27 on Variable Stars, and Commission 29 on Stellar Spectra. She was a fellow of the American Association for the Advancement of Science, and a member of Sigma Xi, the American Astronomical Society, and the Royal Astronomical Society of Canada.

Besides her professional work in astronomy, Mayall had a lifelong interest in promoting the work of amateur astronomers, and especially in encouraging popular interest in astronomy at all levels. Perhaps her most widely recognized contribution to popular astronomy was as the co-editor, with R. Newton Mayall, of the revised editions of Olcott's *Field Book of the Skies*. She was also co-author with her husband of *Skyshooting, a book on photography for amateur astronomers*, and *Sundials and Their Construction, Astronomical Contribution and Significance*. In addition, Mayall edited and revised a new edition of **Thomas Webb**'s *Celestial Objects for the Common Telescope* that has remained in print since 1962.

In all, Margaret Mayall worked with devotion and dedication to firmly establish the AAVSO as an independent research organization. It is today the largest organization of variable star observers in the world. In recognition of her efforts, the AAVSO, in 1974, established the Margaret W. Mayall Assistantship in her honor to provide variable star research opportunities for young people at AAVSO headquarters. In 1957, the Western Amateur Astronomers awarded their G. Bruce Blair Gold Medal to Mayall for her contributions to amateur astronomy. The following year, Mayall was the recipient of the American Astronomical Society's Annie Jump Cannon Award. The IAU named minor planet (3342) Fivesparks in honor of Mayall and her husband Newton.

Michael Saladyga

Selected References

Cannon, Annie J. and Margaret W. Mayall (1949). *The Annie J. Cannon Memorial Volume of the Henry Draper Extension*. Cambridge, Massachusetts: Harvard College Observatory.

Hoffleit, Dorrit (1993). *Women in the History of Variable Star Astronomy*. Cambridge, Massachusetts: American Association of Variable Star Observers.

——— (1996). "Margaret Walton Mayall 1902–1995." *Bulletin of the American Astronomical Society* 28: 1455–1456.

Mayall, Margaret (ed.) (1955). *The Leon Campbell Memorial Volume: Studies of Long Period Variables*. Cambridge, Massachusetts: American Association of Variable Star Observers.

Mayall, R. Newton and Margaret W. Mayall (1968). *Photography for Amateur Astronomers*. New York: Dover.

Olcott, William T. (1954). *Field Book of the Skies*. 4th ed, edited by R. Newton and Margaret W. Mayall. New York: G. P. Putnam's Sons.

Robinson, Leif J. (1990). "Enterprise at Harvard College Observatory." *Journal for the History of Astronomy* 21: 89–103.

Webb, Thomas William and Margaret W. Mayall (1962). *Celestial Objects for Common Telescopes*. 2 Vols. New York: Dover. (Vol. 1, *The Solar System* and Vol. 2, *The Stars*.)

Mayall, Nicholas Ulrich

Born **Moline, Illinois, USA, 9 May 1906**
Died **Tucson, Arizona, USA, 5 January 1993**

Nicholas (Nick) Mayall, a 20th-century observational astronomer, produced major advances regarding the motions and compositions of globular clusters, established the Crab Nebula as the remnant of the supernova of 1054 that was described by the Chinese, studied the internal motions of two galaxies, and contributed to the measurement of the rate of expansion of the Universe. Mayall was also an effective observatory director for the construction, commissioning, and early years of operation of the Kitt Peak National Observatory.

Mayall was the son of an engineer, Edwin L. Mayall, Sr., and his wife Olive (*née* Ulrich) Mayall. They moved to near Modesto, California, where Mayall attended elementary school, and then to Stockton, California where he completed high school. Mayall entered the University of California College of Mining in 1924 but had to shift to another field because of his extreme color blindness. He selected astronomy and graduated in 1928. Mayall lived with and supported his divorced mother during this period by working in the university library.

Mayall then worked for 2 years (1929–1931) as a computer at the Mount Wilson Observatory assisting **Edwin Hubble**, **Alfred Joy**, **Walter Adams**, and **Seth Nicholson** by measuring and reducing their observational data. After Pluto was discovered by **Clyde Tombaugh** in 1930, Mayall and Nicholson found earlier photographs of it in the archives and produced the first definitive elliptical orbit of the planet. Mayall decided, on the basis of his experience at Mount Wilson Observatory, that he would pursue a career in astrophysics specializing in nebular spectroscopy. He earned a Ph.D. from the University of California at Berkeley in 1934 for a thesis suggested by Hubble, a census of galaxies as a function of galaxy brightness and area in the sky. Mayall's thesis advisor was **William Wright** who arranged employment for his protégé as an astronomer at the Lick Observatory.

To work on diffuse galaxies and globular clusters, Mayall designed an UV-transmitting spectrograph for the 0.9-m Crossley telescope. Using that spectrograph he determined the radial expansion of the Crab Nebula and along with **Jan Oort**, established that it was a remnant of the supernova first observed by the Chinese in 1054. He obtained measures of the internal rotation of the Andromeda Galaxy (M31) and the large spiral galaxy (M33) in Triangulum. Mayall then collaborated with his graduate school colleague, theoretician **Arthur Wyse** on the analysis of that data in 1942.

During World War II Mayall worked at the Radiation Laboratory in Cambridge, Massachusetts, the Office of Scientific Research and Development project at the Mount Wilson Observatory headquarters in Pasadena, California, and the California Institute of Technology rocket project in Pasadena and Inyokern, California.

After the war Mayall was deeply involved in obtaining a 3-m reflector (now called the Shane reflector in honor of **Charles Shane**) for the Lick Observatory; the telescope was not completed until 1960. Mayall continued to use the Crossley reflector to determine the motions of 50 globular clusters and showed that they shared only partly in the rotation of the star of the Galactic disk. With **Milton Humason** and others, Mayall obtained the redshifts of 800 galaxies. Humason used the 100-in. (2.5-m) and 200-in. (5-m) telescopes for the fainter galaxies while Mayall observed the brighter ones with the Crossley reflector. Perhaps in compensation for his color blindness, Mayall was able to see fainter objects through a telescope than other astronomers, *e.g.*, $V = 16$th magnitude with the 0.9-m Crossley Reflector. The data gathered by Humason and Mayall were analyzed by Allan Rex Sandage (born: 1926) to give a Hubble constant of 180 km/s/Mpc, a significant milepost on the way from Hubble's original value of 530 km/s/Mpc to the currently accepted value near 65–70 km/s/Mpc.

In 1960 Mayall became the second director of the Kitt Peak National Observatory [KPNO]. He oversaw the development of the site and the construction of the 4-m telescope (now called the Mayall telescope in his honor). What Mayall lacked in administrative experience (at first a liability) he more than made up through his ability to attract a first-class scientific staff because of his personal reputation as an outstanding research astronomer. Importantly, that reputation had been developed on the large telescopes at Lick Observatory, Mount Wilson, and Mount Palomar. When the Association of Universities for Research in Astronomy [AURA], the managing corporation for KPNO, took over the project from the University of Chicago, Mayall assumed responsibility for the early development of the Cerro Tololo Interamerican Observatory [CTIO] in Chile and the construction of the 4-m CTIO telescope (now called the Blanco telescope in honor of the first CTIO director, Victor Blanco).

Perhaps the most important of Mayall's contributions at KPNO was his leadership of a transition in style for major observatory management. Over strong recommendations to the contrary by the directors and staff at the Mount Wilson and Palomar observatories, Mayall guided the implementation of an organization designed to facilitate use of the major facilities at KPNO as community assets. He was a gentle director, knew everyone of 300 employees by first name, and was particularly effective in relations with the university administrators and the appropriate federal and state officials who oversaw KPNO.

Mayall had long suffered from diabetes and arthritis, and retired from KPNO in 1971 at the age of 65. Throughout his research career of nearly 30 years at the Lick Observatory he was a meticulous and outstanding observer of galaxies and clusters, and proved equally capable as the administrator of a large scientific institution.

Helmut A. Abt

Selected References

Edmondson, Frank K. (1997). *AURA and Its US National Observatories*. Cambridge: Cambridge University Press.

Humason, M. L., N. U. Mayall, and A. R. Sandage (1956). "Redshifts and Magnitudes of Extragalactic Nebulae." *Astronomical Journal* 61: 97–162.

Mayall, N. U. (1946). "The Radial Velocities of Fifty Globular Star Clusters." *Astrophysical Journal* 104: 290–323.

Mayall, N. U. and J. H. Oort (1942). "Further Data Bearing on the Identification of the Crab Nebula with the Supernova of 1054 A. D.: Part II. The Astronomical Aspects." *Publications of the Astronomical Society of the Pacific* 54: 95–104.

Osterbrock, Donald E. (1996). "Nicholas Ulrich Mayall." *Biographical Memoirs, National Academy of Sciences* 69: 188–212.

Mayer, Christian

Born **Meseritsch (Velké Meziříčí, Czech Republic), 20 August 1719**
Died **Heidelberg, (Germany), 17 April 1783**

The Jesuit astronomer Christian Mayer published the first catalog of stars that were close enough together, as seen through a telescope, that they might be considered double stars, and showed by comparison of some of these stars with **John Flamsteed**'s observations from the previous century that some possible orbital motion was detectable. Mayer could thus be considered to be the originator of the new and important field of double star astronomy. It was left to **William Herschel** to show, some two decades later, that some double stars might be linked gravitationally through his more accurate measures of stellar positions. Herschel demonstrated, with greater precision, possible motions on a shorter time frame.

Little is known with certainty of Mayer's early life except that in 1745 he entered the Jesuit order as a noviate in Mannheim, then capital of the Palatinate. Mayer taught languages and mathematics at the Jesuit school in Aschaffenburg and began active observational work in astronomy. In 1752, Mayer was appointed professor of mathematics and physics at Heidelberg University, and began publishing works in those fields but was soon concentrating his efforts on astronomy. Mayer's astronomical work eventually attracted the attention of Karl Theodor, Elector of the Palatinate, who was very interested in science. Karl Theodor first arranged for the construction of a small observatory at his summer residence in Schwetzingen and appointed Mayer state astronomer. A larger observatory was constructed in Mannheim, equipped with some of the finest instruments available from London instrument makers, the leaders in this field at the time. The new instruments included a great mural quadrant equipped with a telescopic site by Bird, installed in 1775, as well as other instruments by Dollond, Troughton, and Ramsden.

Mayer's work as an astronomer included much that was routine for the period, including participation in the measurement of a degree of the meridian with **César Cassini de Thury**, observation of the two transits of Venus from Russia, and the development of a map of the Russian Empire for Catherine II. But it was in his investigations of double stars that Mayer established his claim as a historical figure.

Ptolemy first applied the name double star to the bright naked eye pair ν^1 ν^2 Sagittarii, and in later years **Giambattista Riccioli** identified two such pairs in Capricorn and Hyades, while **Robert Hooke** noticed that γ Arietis was a telescopic double in 1665. **Jean Cassini** (Cassini I), **Giovanni Biachini**, and **Charles Messier** are all credited with recording the duplicity of one or more bright stars in subsequent years. By 1767, **John Michell** had concluded, on statistical grounds, that some of the many stars that appeared very close together telescopically might actually be gravitationally connected, but it was Mayer who first systematized the study of double stars. He first sent his lists of such stars to **Neville Maskelyne** for his comments, cataloging the right ascension and declination of various close pairs that to Mayer seemed candidates for closer observation.

Mayer published his first catalog of possible double stars in 1778. His claim that he had discovered stars with satellites was at first misinterpreted by astronomers and triggered a string of rebuttals from N. Fuss, **Maximilian Hell**, **Johann Bode**, and others who disputed the discovery of a planetary satellite of these stars. Herschel issued three catalogs of possible double stars in 1782, 1785, and 1821, listing a total of 848 examples of such pairs. Herschel's first catalogue was apparently published without prior knowledge of either Michell's or Mayer's previous efforts, but there was little overlap between the two lists of stars. It was not until Herschel reexamined his original catalog of double stars in 1802, and discovered that some of the companion stars had moved in such a way as to leave little doubt that the two stars were linked, that Michell's hypothesis and Mayer's empirical claims were given credibility. By then, also, Mayer had couched his terminology in such a way as to make clear that the coupling of stars and not planetary satellites was the object of his study. The English translation of his article in the *Transactions of the American Philosophical Society*, for example, discussed the companion stars as attendants rather than as satellites. In a table of 12 double stars in that article, Mayer supported his hypothesis by including Flamsteed's observations of eight stars from the previous century that seemingly demonstrated orbital motion had occurred over the intervening period.

Thomas R. Williams

Selected References

Hoskin, Michael A. (1964). *William Herschel and the Construction of the Heavens.* New York: W. W. Norton and Co., pp. 27–40.

Houzeau, Jean Charles (1882). "Brussels (Belgium), and Observatoire royale de Belgique." In *Vade-mecum de l'astronome*, pp. 889–890. Brussels: F. Hayez.

Mayer, Christian (1778). *Gründliche Vertheidigung neuer Beobachtungen von Fixsterntrabanten*. Mannheim: American Philosophical Society.

——— (1786). "Astronomical Observations." *Transactions of the American Philosophical Society* 2: 217–225.

Mayer, Johann Tobias

Born **Marbach near Stuttgart, (Germany), 17 February 1723**
Died **Göttingen, (Germany), 20 February 1762**

Selenographer Tobias Mayer prepared the earliest quantitative map of features on the surface of the Moon as well as lunar tables used by **Neville Maskelyne** in preparing early editions of the *Nautical Almanac*. Mayer was the son of a cartwright who left his trade in 1723 to work as foreman of a well-digging crew in Esslingen, Baden-Württemberg, where his family joined him a year later. Following the death of his father in 1731, Mayer was put into the local orphanage, and he taught himself mathematics from Christian von Wolff's *Anfangs-Gründe aller mathematischen Wissenschaften*. His mother found work in Saint Katharine's Hospital, which probably explains why Mayer came to prepare architectural drawings of the hospital at barely 14. His skill in this direction attracted the notice of a noncommissioned officer in the Swabian district artillery, then garrisoned in

Esslingen, under whose direction in 1739 Mayer produced a book on military fortifications; later that year he drew a map of Esslingen and its environs.

Mayer's first book, published around 1741, deals with the application of analytical methods to the solution of geometrical problems. His second, *Mathematischer Atlas* (1745), appeared in the period during which he briefly worked for the firm of Johann Andreas Pfeffel of Augsburg, Bavaria. Its choice of subject matter is a good index of the extent of Mayer's scientific knowledge at that time. On leaving Augsburg he joined the Homann Cartographic Bureau in Nuremberg, Bavaria, where he devoted 5 years to improving the state of mapmaking. Mayer collated geographical and astronomical data, and made observations of occultations and eclipses. He drew over 30 maps of which the Mappa Critica of Germany is considered the most significant, as it set new standards for handling geographical source materials and for applying astronomical data to the determination of latitude and longitude.

In 1747 and 1748 Mayer obtained a large number of meridian transits of the Moon, and made numerous measures of its angular diameter, to facilitate the lunar-eclipse method of fixing longitude. In addition to determining the selenographic coordinates of 89 major lunar markings, in doing so, he made allowance for the irregularity in the orbital motion and the libration of the Moon.

The *Kosmographische Nachrichten und Sammlungen auf das Jahr 1748* (Nuremberg, 1750), which Mayer edited for the newly formed Cosmographical Society, contains a description of his glass micrometer, his observations on the solar eclipse of 25 July 1748 and occultations of some bright stars, his major treatise on the lunar libration, and his consideration on why the Moon cannot have an atmosphere.

In early 1751, Mayer took a professorship at the Georg-August Academy in Göttingen. This was a nominal position, based solely on his reputation as a cartographer and practical astronomer. Shortly before leaving Nuremburg, he married Maria Victoria Gnüge, by whom he had eight children, of whom only three survived.

In 1752 Mayer drew up new lunar and solar tables, accurate to 1′. Comparing his positional values to historical observations (for instance, those made at all lunar and solar eclipses described since the invention of the telescope), he found that all discrepancies were attributable to errors in star places and the inferior quality of the instruments used. On the recommendation of **James Bradley,** Mayer's lunar tables, edited by **Nevil Maskelyne**, were used to compute the lunar and solar ephemerides for the early editions of the *Nautical Almanac.*

Mayer's further researches included elimination of errors from a 6-ft.-radius mural quadrant to be installed in Göttingen, the invention of a new method of calculating the circumstances of solar eclipses, the study of the proper motion of stars, and a catalogue of zodiacal stars. His works on each of these subjects were published posthumously in Georg Christoph Lichtenberg's *Opera inedita Tobiae Mayeri* (Göttingen, 1775). Appended to the book is a copper engraving of Mayer's map of the Moon. At 8 in. in diameter, it was the most accurate map of the visible lunar surface for half a century and was reproduced by **Johann Schröter** in his *Selenotopografische Fragmenten.*

Mayer's later work included efforts to improve land measurement, a method to find geographical coordinates independent of celestial observation. In 1765 his widow received £3,000 from the British government in recognition of her husband's claim, presented a decade earlier, for one of the prizes in connection with the quest to determine longitude at sea.

Much of the manuscript material relating to Mayer is preserved in Göttingen at the Niedersächsische Staats-und Universitäts-Bibliothek.

Richard Baum

Selected References

Forbes, Eric G. (ed.) (1972). *The Unpublished Writings of Tobias Mayer.* 3 Vols. Göttingen: Vandenhoech and Ruprecht.

——— (1971). *The Euler–Mayer Correspondence (1751–1755).* London: Macmillan.

Mayer, Julius Robert

Born **Heilbronn, (Baden-Württemberg, Germany), 25 November 1814**

Died **Heilbronn, (Baden-Württemberg), Germany, 20 March 1878**

Julius Mayer independently established the principle of energy conservation, proposed a novel theory of the Sun's energy source, and considered the effects of tidal friction on the Earth–Moon system. Mayer was the youngest of three sons of the apothecary Christian Jakob Mayer and his wife, the daughter of a bookbinder. After attending Gymnasium at Heilbronn and an evangelical seminary at Schöntal, he began studies in the medical faculty at the University of Tübingen in 1832. Although he was expelled from the university in 1837 (for belonging to a secret student society), Mayer was allowed to take the state medical examinations and received his M.D. in 1838. After a short stay in Paris, he served as physician from 1840 to 1841 on a Dutch merchant ship that traveled to the East Indies. Following that experience, Mayer settled into a medical practice at Heilbronn. In 1842, he married Wilhelmine Regine Caroline Closs; the couple had seven children (only two of whom survived to adulthood).

During his oceanic voyage, Mayer undertook certain physiological observations that led him to speculate upon the conversion of food to heat in the human body, and on the resultant work that the body could perform. He concluded that heat and work are interchangeable, and this led him to reflect upon motion and heat as manifestations of a single, indestructible *Kraft* (force) in nature, which was quantitatively conserved in any conversion process. These ideas contained rudiments of the law of energy conservation. Soon afterward, Mayer extolled what later became known as the mechanical equivalent of heat. But he was not familiar with contemporary physical research and presented his ideas with a certain metaphysical style. Not surprisingly, his initial paper (1841) was rejected by the *Annalen der Physik und Chemie* (Annals of Physics and Chemistry). But after considerable reworking, Mayer's second paper, entitled "Bemerkungen über die Kräfte der unbelebten Natur" (Remarks on the forces of inanimate nature), was published in another journal, *Annalen der Chemie und Pharmazie* (Annals of Chemistry and Pharmacy), but went largely ignored. Nonetheless, Mayer anticipated to some extent the energy conservation principle later formulated by James Prescott Joule and **Herman von Helmholtz**, among others. Mayer's work was finally recognized after the former became engaged in a priority dispute. His views were later defended by Peter Guthrie Tait and some belated recognition at last came to him. In 1870, he was named a corresponding member of the Paris Academy of Sciences; in 1871, he received the Copley Medal of the Royal Society of London.

Mayer turned his theory of energy conservation upon two astrophysical problems. In 1846, he offered a novel explanation for the source of the Sun's heat. Mayer argued that its energy arose by the conversion of mechanical energy into heat, released during the Sun's continual bombardment by solid particles (meteors) attracted across interplanetary space. Though ultimately rejected, Mayer's idea may have influenced the creation of a successor theory of the Sun's heat by its gradual contraction.

Two years later, Mayer discussed the problem of tidal friction in his work, *Dynamik des Himmels* (Dynamics of the heavens, 1848). Here, he advanced a hypothesis by which the loss of energy and slowing of the Earth's rotation were roughly compensated by the planet's gradual contraction and acceleration. But like many of Mayer's speculations, this notion went largely unheeded by mainstream scientists. It is also wrong. Some of the energy heats the Earth, but the angular momentum is transferred to the orbit of the Moon, which is expanding at about 3 cm/year.

Mayer suffered bouts of depression that were exacerbated by family problems. But while he kept up his practice as a physician in Heilbronn, Mayer remained an outsider and dilettante when it came to his scientific work. He conducted few experiments and largely shunned mathematical analysis; he was primarily a conceptual thinker. Although his ideas regarding energy conservation were eventually accepted, they did not influence further developments in these subjects, which came from the works of others.

Horst Kant

Selected References

Caneva, Kenneth L. (1993). *Robert Mayer and the Conservation of Energy*. Princeton: Princeton University Press.

Münzenmayer, H. P. (1982). "Julius Robert Mayer's Ideas on a Theory of Tidal Friction." In *Tidal Friction and the Earth's Rotation II*, edited by P. Brosche and J. Sündermann, pp. 1–3. Berlin: Springer-Verlag.

Schmolz, Helmut and Hubert J. Weckbach (1964). *Robert Mayer: Sein Leben und Werk in Dokumenten*. Heilbronn: A. H. Konrad.

Turner, R. Steven (1974). "Mayer, Julius Robert." In *Dictionary of Scientific Biography*, edited by Charles Coulston Gillispie. Vol. 9, pp. 235–240. New York: Charles Scribner's Sons.

Mayr, Simon

Born **Gunzenhausen, (Bavaria, Germany), 20 January 1573**
Died **Ansbach, (Bavaria, Germany), 26 December 1624**

As court mathematician Simon Mayr was in charge of the Ansbach calendar, made the first telescopic observations of the Andromeda galaxy, and computed tables of the mean periods of the satellites of Jupiter more accurate than **Galileo Galilei**. Some sources state that his father Reichart Mayr was the *Burgermeister* (mayor) of Gunzenhausen, but most evidently Simon was from poor family, as in 1586 he went to the Margrave's school for talented poor boys. This school was established to train poor young men for the ministry. He stayed there until 1601 when he was appointed mathematician to the Margrave of Ansbach and was sent to Prague to study with **Tycho Brahe**. After Brahe's death he went to Padua to study medicine. Mayr returned to Germany in 1605 and was appointed mathematician and physician to the Margraves, Christian and Joachim Ernst, serving the rest of his life in that position.

An observatory was built for Mayr, but little is known about his instruments. According to Mayr's own account, he learned in 1609 from an artillery officer, Freiherr Hans Philip Fuchs, that a Dutchman had tried to sell him a telescope at the Frankfurt fair. Mayr grasped the concept and reproduced a telescope, which he used mainly to observe Jupiter. He claimed in a book printed in 1614, *Mundus Iovialis Anno M.DC.IX Detectus Ope Perspicilly Belgici* (The Jovian World, discovered in 1609 by means of the Dutch Telescope), that he had first observed Jupiter's moons in December 1609, a month before Galilei. Galilei fiercely accused Mayr of plagiarism. Disputes about the plagiarism case continued for centuries. In a long treatise by J. Klug, Mayr was accused as a plagiarist, while support for Mayr was presented by J. H. C. Oudemans and J. Bosscha. As a compromise, it was suggested by J. H. Johnson that Mayr probably saw the satellites of Jupiter before Galilei; however, he evidently did not comprehend their real nature until Galilei had published his account of their discovery and the explanation of their connection with the planet. Mayr discovered the variability of magnitudes of the satellites of Jupiter and gave them names – Europa, Io, Ganymede and Callisto – that are still in use.

Mayr was an independent discoverer of M31, the Andromeda nebula. *Mundus Iovialis* contains his telescopic observations of our neighbor galaxy. Mayr also published on the comet in 1596 (C/1596 N1) and was among the first to observe the "new star" in 1604.

Mihkel Joeveer

Alternate name
Marius

Selected References
Klug, Josef (1906). "Simon Marius aus Gunzenhausen und Galileo Galilei." *Abhandlungen der Königlichen Bayerischen Akademie* der *Wissenschafter, Math.-Phys. Kl.* 22: 385–526.

Johnson, J. H. (1931). "The Discovery of the First Four Satellites of Jupiter." *Journal of the British Astronomical Association* 41: 164–171.

Oudemans, J. A. C. and J. Bosscha (1903). "Galilée et Marius." *Archives néerlandaises des sciences exactes et naturelles*, 2d ser., 8: 115–189.

McClean, Frank

Born **Glasgow, Scotland, 13 November 1837**
Died **Brussels, Belgium, 8 November 1904**

Frank McClean invested heavily in both laboratory and telescopic spectroscopy, comparing the solar spectrum with numerous terrestrial elements. His spectrographic survey of all stars brighter than 3.5 magnitude in both the Northern and Southern Hemispheres was a useful resource for astrophysicists for several years. McClean was the first astrophysicist to detect oxygen in the spectra of stars.

This wealthy civil engineer and amateur astronomer was the only son of John Robinson McClean, M.P., F.R.S., a renowned civil engineer, and Anna (*née*) Newsam McClean. Educated at Westminster School and the University of Glasgow, McClean graduated as 27th wrangler in the mathematical *tripos* at Trinity College, Cambridge in 1857. After working in his father's engineering firm, and later as a shipyard and railway engineer, McClean retired in 1870 in order to concentrate on his scientific and artistic interests. He maintained a private observatory and laboratory at Ferncliffe near Tunbridge Wells. Unlike many gentleman scientists, however, McClean engaged no laboratory assistants.

In his private laboratory McClean started experimenting with electricity in coils. After the completion of an attached observatory in 1875, he could extend his work to solar observation as well as stellar spectroscopy. McClean used a direct spectroscopic eyepiece of his own invention that was later commercialized by John Browning. Intrigued with the possibilities, McClean mounted a heliostat on his roof around 1879 to guide the light of the Sun to a fixed 4-in. refractor and spectroscopic bench in his attic laboratory. Using this apparatus, he undertook studies of the composition of the Sun as well as the effect of the Earth's atmosphere on the solar spectrum.

In December 1888, McClean presented to the Royal Astronomical Society a portfolio of photographs, enlarged 8.5 times, of the solar spectrum from the Fraunhofer line D to A, corresponding in representational mode and approximate dispersion to **Anders Ångström**'s lithographed normal map. In a subsequent publication, McClean continued Ångström's numbering sequence, labeling his plates VII to XIII, counting from the section containing D towards the violet.

Ångström's instrument had been a Nobert grating; McClean used a grating by **Lewis Rutherfurd** with 17,296 lines to the inch and with a ruled surface of about 1.74-in. width. His "Photographs of the Red End of the Solar Spectrum" covered that half of the Sun's visible spectrum not included in **Henry Rowland**'s photographic map. McClean's photographs also included "subsidiary" exposures displaying "in the same sections, both the red spectrum of the second order as before, and also the overlapping green-to-violet spectrum of the third order." The purpose of these double spectra was to act as a reference for spectroscopists in the red part of the spectrum with third-order spectra, for lack of overlappings in the same order. Commendable as was this extension of Rowland's photographic map into the yellow–red region of the spectrum (5,800 to 7,700 Å), it had the serious problem of not being sufficiently enlarged. One set of portfolio enlargements, deposited in the library of the Royal Astronomical Society, was accessible to the London-based scientific community and, to some extent, to visitors from other parts of Great Britain, but it was not easily available to researchers elsewhere. The only published part of this series of photographs was a small segment of the spectrum around the line A at 7,600 Å. Technically speaking, though, this plate was not a photograph but a lithographed sketch by the assistant secretary of the society and highly skilled lithographer William Henry Wesley (1841–1922) – "taken from the photograph." The plate is misleadingly labeled "The A group of the solar spectrum. Photographed by F. McClean," thus ignoring Wesley's lithographic drafting, and mentioning only the printing company (E. Stanford in London). Wesley did a superb job in bringing out the beautifully striated structure of this line group and the nearly plastic appearance of some of the split lines. Yet this mixture of techniques clearly did not meet the purpose, which was to complete the map photographically.

This became moot with the appearance of Rowland's second series in 1889, which included the red extreme of the spectrum. In spring of that year, McClean once again appeared before the Royal Astronomical Society to present exposures of the solar spectrum, with "parallel photographs" of iron, iridium, and titanium, produced in the same manner. However, his next two spectrographic publications adopted the contemporary photomechanical reproduction technique, even though this amounted to surrendering his do-it-yourself principle, which he had rigorously followed up to that point. In late 1890 he published "Comparative Photographic Spectra of the high sun and Low sun from H to A Showing the Atmospheric Absorption Bands" (in a sense, the photographic analogue of **Louis Thollon**'s contemporary lithographic comparative map). In November 1891 McClean completed comparative photographic spectra of the Sun and 15 metals (including platinum, iridium, osmium, palladium, rhodium, ruthenium, gold, and silver, as well as manganese, cobalt, nickel, chromium, aluminium, and copper) for the wavelength region 3,800–5,750 Å. This map of metallic spectra, compiled with the aid of a plane Rowland grating acquired in 1890, was especially useful because it was the first to include the rare metals palladium, rhodium, and ruthenium.

In 1895, McClean began a systematic study of the spectra of all stars brighter than 3.5 magnitude using a 20° objective prism. Grubb and Sons produced a special telescope for his purpose based on the telescopic cameras manufactured by the firm for the Astrographic Catalogue and *Carte du Ciel* project. McClean's survey for the northern heavens was published in 1896. The following year McClean was invited to the Royal Observatory at the Cape of Good Hope by **David Gill** to complete his survey of the whole sky. For his work at the Cape, McClean took only the objective prism and mounted it on the Cape's *Carte du Ciel* telescope. The southern skies survey was published in 1898. In reviews in The *Astrophysical Journal*, McClean's spectral charts were criticized by **Edwin Frost** as "unprofessional" because McClean failed to properly document the instrument, dates, times, and conditions under which the spectra were recorded, all details deemed necessary for the interpretation of the stellar spectra. Frost also observed that the photographic quality of the southern charts was improved over those from the north. Notwithstanding Frost's criticism, however, the McClean charts were a resource to which some astrophysicists could turn for comparative data, pending availability of the Henry Draper Catalogue. Based on his photographic spectral survey, McClean proposed a stellar spectral classification scheme similar to that of **Angelo Secchi**, but with Secchi's first class subdivided into three types corresponding to the B, A, and F stars in the Henry Draper classification, a scheme that was beginning to find favor. Thus, McClean's system was never widely adopted.

The spectral survey work produced two important results. First, because McClean arranged the survey in terms of galactic coordinates rather than in the conventional sidereal format, it clearly demonstrated that the stars in McClean's first type, which he recognized contained the spectrum of neutral helium, were concentrated near the galactic plane. Also, during his work at the Cape, McClean recognized that the spectrum of β Crucis and other stars in his Type 1, which he also identified as "helium" stars, contained the spectral lines of oxygen. Gill was able to verify the oxygen identification with the large Cape spectrograph. However, McClean's discovery of stellar oxygen was received with guarded caution by other astrophysicists who recalled the major controversy over **Henry Draper**'s earlier claim to have discovered oxygen in the spectrum of the Sun. It was for that reason that when McClean received the Royal Astronomical Society Gold Medal in 1899, it was primarily for his stellar spectral survey work, though the discovery of oxygen in the stars might equally have merited such recognition.

Aside from his scientific accomplishments, McClean also furthered the cause of scientific research by donating several expensive instruments to various institutions, including the 24-in. Victoria telescope to the Royal Observatory at the Cape of Good Hope. He also endowed the Isaac Newton studentships at the University of Cambridge for the encouragement of study and research in astronomy and physical optics.

McClean was honored with an honorary LL.D. by the University of Glasgow in 1894. Elected to membership in the Royal Astronomical Society in 1877, McClean served on its council from 1891 until his death from pneumonia. His wife, Ellen (*née* Greg), bore him two daughters and three sons.

Klaus Hentschel

Selected References

Ball, Robert S. (1899). "Address Delivered by the President, Sir Robert S. Ball, LL.D., F. R. S., on presenting the Gold Medal to Mr. F. McClean." *Monthly Notices of the Royal Astronomical Society* 59: 315–324.

Hentschel, Klaus (1999). "Photographic Mapping of the Solar Spectrum 1864–1900." Parts 1 and 2. *Journal for the History of Astronomy* 30: 93–119, 201–224.

——— (2002). *Mapping the Spectrum: Techniques of Visual Representation in Research and Teaching.* Oxford: Oxford University Press, Chap. 6.

Lankford, John (1981). "Amateurs and Astrophysics: A Neglected Aspect in the Development of a Scientific Specialty." *Social Studies of Science* 11: 275–303.

McClean, Frank (1889). "Photographs of the Red End of the Solar Spectrum from the Line (D) to the Line (A) in Seven Sections." *Monthly Notices of the Royal Astronomical Society* 49: 122–124.

——— (1890). *Comparative photographic spectra of the high sun and low sun from (H) to (A).* London: Stanford.

——— (1897). "Comparative Photographic Spectra of Stars to the 3½ Magnitude." *Philosophical Transactions of the Royal Society of London* 191: 127–138.

——— (1898). *Spectra of Southern Stars.* London: Stanford.

Newall, Hugh Frank (1907). "Frank McClean." *Proceedings of the Royal Society of London* A 78: xix–xxiii.

Thollon, Louis (1890). "Nouveau dessin du spectre solaire." *Annales de l'Observatoire de Nice* 3: A7–A112 (And Atlas with 17 plates.)

Turner, Herbert Hall (1905). "Mr. Frank McClean." *Monthly Notices of the Royal Astronomical Society* 65: 338–342.

McCrea, William Hunter

Born **Dublin, Ireland, 24 December 1904**
Died **Lewes, (East Sussex), England, 25 April 1999**

Irish–English mathematical astronomer Sir William McCrea is most widely known for work in modern, general-relativistic cosmology; he also helped to establish that the Sun and stars are made primarily of hydrogen and helium. McCrea was the eldest child of schoolmaster Robert Hunter McCrea and Margaret (*née* Hutton) McCrea. He attended elementary and grammar schools in Chesterfield, Derbyshire, winning an entrance scholarship in mathematics to Trinity College, Cambridge, and receiving a *tripos* degree in mathematics in 1926. McCrea's Ph.D. dissertation was completed under **Ralph Fowler** in 1930, after a year's visit at Göttingen University in Germany. His early work was largely on the application of mathematical methods to basic problems in quantum physics and relativity.

McCrea's academic appointments were as lecturer in mathematics at and the University of Edinburgh (1930–1932), reader in mathematics at Imperial College, London (1932–1936), professor of mathematics at Queen's University, Belfast (1936–1944). The last was interrupted by operational research in the British Admiralty, which he completed with the rank of captain. (The team was led by **Patrick Blackett**) McCrea was then Professor at Royal Holloway College, London (1946–1966), and, finally, the first research professor in theoretical astronomy at the recently established University of Sussex (1966–1972, and as an active emeritus professor for 25 years thereafter). Soon after the end of World War II, McCrea had been one of the strong advocates of a national center of theoretical astronomy. After much negotiation, the Astronomy Centre at the University of Sussex and the Institute of Theoretical Astronomy (under **Fred Hoyle**), which later merged into the overall Cambridge Institute of Astronomy, were the outcome of this. The best known of McCrea's students who remained in astronomy date from the Royal Holloway years and include Derek McNally, Gillian Peach, Michael

Rowen-Robinson, and Iwan Williams, most of whose careers were spent in London.

Two areas of McCrea's work had the longest, lasting impact on subsequent astronomical research. The first of these is cosmology. In 1933 he and **Edward Milne** of Oxford University showed that most general relativistic models of an expanding universe have close Newtonian analogs, which are somewhat easier to understand and within which approximate calculations can be done. McCrea was also the journal editor directly responsible for the acceptance of the first paper on steady-state cosmology (by Hermann Bondi and Thomas Gold) and was initially quite sympathetic to this alternative to an evolutionary universe; eventually he realized that it was in strong disagreement with observations. Late in life McCrea expressed doubts about all cosmological models (a trait he shared with **Grote Reber** and a number of other astronomers of his generation).

The other major area in which McCrea rightly receives credit is the demonstration that the Sun and stars are made mostly of hydrogen and helium. The 1925 Ph.D. thesis of **Cecilia Payne-Gaposchkin** reached this conclusion for K giants, but it was not widely accepted. Acceptance of the idea required calculations using more accurate treatments of the properties of hydrogen and other atoms and of the diffusion of radiation through ionized gases. European textbooks most often credit **Carl von Weizsacher**, American texts **Henry Russell**, British texts McCrea (particularly his work on the solar corona before 1931), and a case can perhaps also be made for **Bengt Strömgren**. The final step, showing that the main obstacle to the free passage of light through the outer layers of the Sun is due to hydrogen atoms with a second electron temporarily attached, was taken by **Ruper Wildt** in 1947.

McCrea was also an early proponent for a binary star model of the origin of blue stragglers (stars which appear to be younger than the populations around them) and came to be regarded as a source of last resort for mechanisms to explain or measure effects that otherwise defied the community, along the lines of, "Well, if Bill can't think of a way to measure the cosmological constant independent of the other cosmological parameters, perhaps it can't be done." (Actually it can.) Some of the ideas McCrea put forward, for instance accounting for the most massive stars by accretion of gas onto ones like the Sun, and mechanisms for formation of hydrogen molecules, were indeed prompted by observations most of his contemporaries thought inexplicable.

Knighthood came relatively late in McCrea's life, in 1985, long after honorary degrees from universities in Ireland, England, and Argentina, and academy memberships in Belgium, Italy, Scotland, and England. His service to the Royal Astronomical Society [RAS] (whose Gold Medal he received in 1976) was remarkable. McCrea held all four of the major offices, secretary (1946–1949), president (1961–1963), foreign Secretary (1968–1971), and treasurer (1976–1979), as well as editorships of the RAS's *Monthly Notices* and *Observatory* Magazine. He married Marian Nicol Core Webster in 1933, and was a lifelong communicating member of the established Anglican Church.

George Gale

Selected References

McCrea, W. H. (1987). "Clustering of Astronomers." *Annual Review of Astronomy and Astrophysics* 25: 1–22.

Mestel, Leon (2004). "McCrea, Sir William H." In *Oxford Dictionary of National Biography*, edited by H. C. G. Matthew and Brian Harrison. Vol. 35, pp. 166–168. Oxford: Oxford University Press.

McIntosh, Ronald Alexander

Born **Auckland, New Zealand, 21 January 1904**
Died **Auckland, New Zealand, 17 May 1977**

An accountant, and then journalist by profession, Ronald McIntosh distinguished himself as an amateur observer of meteors and also as a computer of meteor orbits and radiants. He was self-trained in mathematics and celestial mechanics. An internationally recognized authority, McIntosh's published papers (numbering over 100) served as the standard reference works on Southern Hemisphere radiants and meteor rates between 1925 and 1945, and still remain, along with more recent radar studies, among the most reliable sources of such information. The importance of his contributions was frequently acknowledged by **Charles Olivier** of the American Meteor Society. Elected a fellow of the Royal Astronomical Society, and of the International Astronomical Union's Commission 22 on Meteors, McIntosh twice was awarded a Donovan Prize and Bronze Medal from Australia for his contributions to astronomy. He was also a popular lecturer at the Auckland Planetarium for a number of decades and served as secretary of the Auckland Observatory and Planetarium Trust Board. McIntosh was married and had two children.

Thomas R. Williams

Selected Reference

Bateson, Frank (1977). "Ronald Alexander McIntosh (1904–77)." *Southern Stars* 27: 82–87.

McKellar, Andrew

Born **Vancouver, British Columbia, Canada, 2 February 1910**
Died **Victoria, British Columbia, Canada, 6 May 1960**

Canadian stellar astronomer Andrew McKellar is best remembered today for his accidental discovery (not at the time understood) of the cosmic microwave background, which is perhaps the strongest evidence for a Big-Bang universe (one that had a very hot, dense state about 15 billion years ago). The discovery came in the form of the measurement of the temperature of CN molecules (really radicals) in the interstellar medium. He found this to be about 2.3 K, with the result being confirmed the next year by **Walter Adams**.

Andrew McKellar was the son of John Hamilton McKellar and Mary Littleson of Scotland. Andrew married Mary Belgrave Crouch in 1938, and they had two children, Andrew Robert and Mary Barbara.

McKellar took an honors BA in mathematics and physics at the University of British Columbia in 1930, followed by graduate work in physics at the University of California (MA: 1932, Ph.D.: 1933). During this time, he was a student assistant at the Dominion Astrophysical Observatory. With a fellowship from the United States National Research Council, he spent 2 years at the Massachusetts Institute of Technology (1933–1935). During World War II, McKellar taught physics at the University of British Columbia (1941/1942), followed by service in the Royal Canadian Navy in operational research (1944/1945), for which he was awarded an MBE in 1947. He also spent a year as visiting professor of astronomy and physics at the University of Toronto (1952/1953). Most of McKellar's career, from 1935 until his death, was at the Dominion Astrophysical Observatory.

McKellar was an associate editor of the *Astrophysical Journal*. He served as president of the Royal Astronomical Society of Canada (1959/1960), the Astronomical Society of the Pacific (1955–1957), and the International Astronomical Union Commission on Comets. He was also active in the American Astronomical Society and was a fellow of the Royal Society of Canada.

McKellar was a leading figure in the development of molecular spectroscopy in astronomy. He determined the $^{12}C/^{13}C$ ratio in carbon stars to be 5:1, very different from the terrestrial ratio; this evidence supported the idea of a CNO- cycle hydrogen fusion is stars. He detected the molecules CH, CN, and NaH in interstellar space in 1940, realizing that their spectra in space is altered from the laboratory form due to extreme conditions. His analysis of cometary spectra showed that solar radiation is the cause of excitation of cometary molecules. Before his death, McKellar studied the chromospheres of giant stars spectroscopically by observing eclipsing binaries.

At the time of McKellar's death, the importance of his measurement of the interstellar temperature had not yet been appreciated. Both he and his contemporaries would have agreed with the judgement of **Gerhard Herzberg** that it had "of course a very restricted meaning."

Richard A. Jarrell

Selected References

Beals, C. S. (1960). "Andrew McKellar, 1910–1960." *Journal of the Royal Astronomical Society of Canada* 54: 153–156.

Jarrell, Richard A. (1988). *The Cold Light of Dawn: A History of Canadian Astronomy*. Toronto: University of Toronto Press.

Herzberg, G. (1950). *Molecular Spectra and Molecular Structure*. 2nd ed. Vol. 1. New York: Van Nostrand.

McKellar, Andrew (1940). "Evidence for the Molecular Origin of Some Hitherto Unidentified Interstellar Lines." *Publications of the Astronomical Society of the Pacific* 52: 187–192.

McLaughlin, Dean Benjamin

Born **Brooklyn, New York, USA, 25 October 1901**
Died **Ann Arbor, Michigan, USA, 8 December 1965**

American stellar spectroscopist Dean McLaughlin participated in the discovery that stars in general rotate like the Sun and that members of close binary systems are often rapid rotators. He was the son of Michael Leo Benjamin and Celia Elizabeth Benjamin McLaughlin. In 1927, he married Laura Elizabeth Hill, with whom he had five children. McLaughlin received all his degrees from the University of Michigan, BA (1923), MS (1924), and Ph.D. (1927), the last for the analysis of spectra of eclipsing binaries, under the guidance of **Ralph Curtiss** and **Richard Rossiter**. They recognized that, just before the brighter star is fully eclipsed, rotation will give the uneclipsed segment of the star an additional motion away from us, leading to an asymmetrical line profile. This is still sometimes called the McLaughlin effect or the Rossiter–McLaughlin effect.

McLaughlin served as an assistant in the Michigan astronomy department from 1922 to 1924 and then became instructor of mathematics and astronomy at Swarthmore College, from 1924 to 1927. In 1927 he returned to Ann Arbor as assistant professor of astronomy, was promoted to associate professor in 1934 and professor in 1941, and remained at Michigan for the remainder of his career. He participated in the Swarthmore College solar eclipse expedition to Sumatra in 1926 and in the Michigan expedition to Fryeburg, Maine in 1932. When World War II broke out, the McLaughlins moved to Massachusetts for 3 years while he served in the Radiation Laboratory of the Massachusetts Institute of Technology, working on radar problems. In 1940, 1951, and 1958 he was a guest observer at the Mount Wilson, Mount Palomar, and Lick observatories. McLaughlin had a great interest in field geology, and from 1951 until 1965 his summers were spent as a cooperating geologist in the Pennsylvania Topographic and Geologic Survey.

McLaughlin was a fellow of the American Association for the Advancement of Science, the Geological Society of America, and the American Astronomical Society, which he served as secretary from 1940 to 1946. He also belonged to the International Astronomical Union, the Astronomical Society of the Pacific, the Michigan Academy of Science, Arts, and Letters, the Michigan Geological Society, the Pennsylvania Academy of Science, and Sigma Xi.

After the death of Curtiss in 1929, McLaughlin directed the Michigan spectrographic programs. A careful, diligent spectroscopist, McLaughlin used Michigan's 37.5-in. reflecting telescope and spectrographs for 40 years under increasingly difficult circumstances, as the university and city encroached on the observatory. His research continued Curtiss's work on the atmospheres of Be stars, bright stars with emission lines of hydrogen in their spectra, and on peculiar variable stars. McLaughlin also became an authority on the spectral characteristics of novae; he devoted much time to unraveling their problems after an unprecedented run of good weather and seeing allowed him to create an excellent photographic record of Nova Herculis 1934. McLaughlin's expertise allowed him to observe and measure many obscure lines on the plates. He also combined spectroscopic with photometric observations.

McLaughlin was known for his good humor and gifts as a teacher and colleague. He also took part in efforts to popularize astronomy and wrote many general articles for *Popular Astronomy*, beginning as an undergraduate at Michigan. In 1961 he published an introductory textbook based upon his courses, and in 1965 he coedited a standard handbook on astrophysics with his Michigan colleague **Lawrence Aller**. The textbook was the immediate precursor of those written starting about 1970 and meant specifically for nonmajor courses in astronomy, and was widely used for general astronomy courses in the 1960s.

In the 1950s McLaughlin's geological researches led him to propose a new theory to explain the appearance of the surface features of Mars, based upon the effects of wind and volcanic activity. His ideas have found more favor in recent years, thanks to observations from space-borne vehicles. His geological researches in the United States led to studies of Triassic era rocks in Pennsylvania and New Jersey and of the pre-Cambrian in Canada.

The University of Michigan observatories had established a tradition of careful observational spectroscopy, beginning with Curtiss and W. Carl Rufus. McLaughlin and his students continued that tradition, specializing in peculiar stars, variables, and novae. McLaughlin chose important problems that lay, however, within the capabilities of the Michigan equipment; he developed a large and consistent body of observations, and he wrote studies that included not only the observational data but also shrewd interpretations. With his colleagues, Curtiss at first and later with Aller, he kept the Michigan graduate program going strong. He was a productive, respected, and significant astronomer throughout his 40-year career.

A number of records are in the Michigan Historical Collections, Bentley Library, The University of Michigan.

Rudi Paul Lindner

Selected References

Anon. (1966). "Dean B. McLaughlin." *Physics Today* 19, no. 1: 153–155.

Anon. (1969). "McLaughlin, Dean Benjamin." In *National Cyclopedia of American Biography* 51: 351–352. New York: James T. White and Co., 1969.

McLaughlin, Dean B. (1922). "The Present Position of the Island Universe Theory of the Spiral Nebulae." *Popular Astronomy* 30: 286–295, 327–339. (In part a review of the Curtis–Shapley "great debate," and an early example of his more popular exposition.)

——— (1927). "Spectroscopic Studies of Eclipsing Binaries." Ph.D. diss., University of Michigan.

——— (1937). "Notes on Spectra of Class Be." *Astrophysical Journal* 85: 181–193. (An example of his work on Be stars.)

——— (1937). "The Spectrum of Nova Herculis." *Publications of the Observatory of the University of Michigan* 6, no. 12: 107–214. (His classic monograph on Nova Herculis.)

——— (1943). "On the Spectra of Novae." *Publications of the Observatory of the University of Michigan* 8, no. 12: 149–194.

——— (1950). "Problems in the Spectra of Novae." *Publications of the Astronomical Society of the Pacific* 62: 185–195.

——— (1956). "A New Theory of the Surface of Mars." *Journal of the Royal Astronomical Society of Canada* 50: 193–200.

——— (1956). "The Volcanic-Aeolian Hypothesis of Martian Features." *Publications of the Astronomical Society of the Pacific* 68: 211–218.

——— (1959). "Mesozoic Rocks." *Pennsylvania Topographic and Geological Survey*. 4th ser., C9.

——— (1961). *Introduction to Astronomy*. New York: Houghton Mifflin.

McLaughlin, Dean B. and Lawrence Aller (eds.) (1965). *Stellar Structure*. Chicago: University of Chicago Press.

Struve, Otto and Velta Zebergs (1962). *Astronomy of the 20th Century*. New York: Macmillan.

Wright, K. O. (1966). "Michigan Stellar Spectroscopist." *Sky & Telescope* 31, no. 2: 91. (A warm appreciation of McLaughlin as astronomer and colleague.)

McMath, Robert Raynolds

Born **Detroit, Michigan, USA, 11 May 1891**
Died **Bloomfield Hills, Michigan, USA, 2 January 1962**

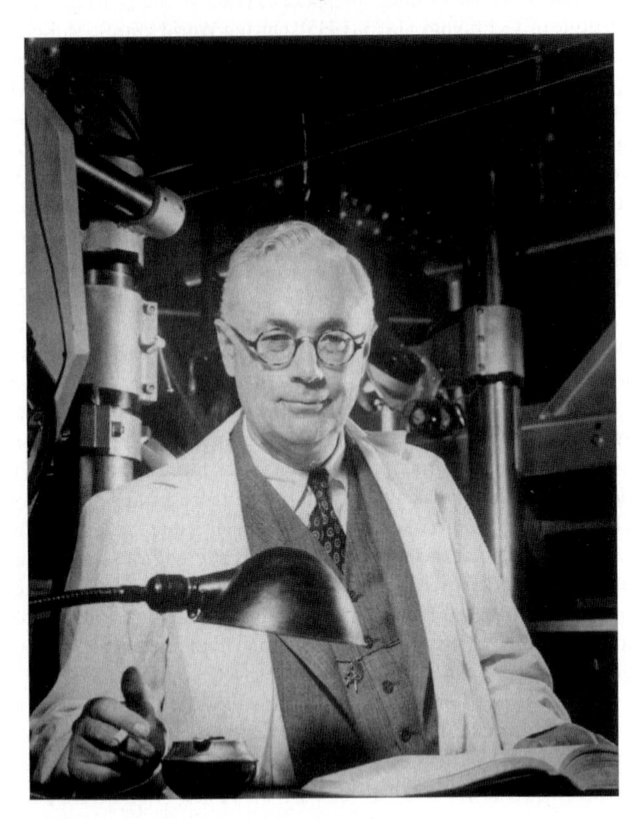

As the founder and long-time director of the McMath–Hulbert Solar Observatory [MHSO], Robert McMath enjoyed a second career in astronomy following a successful career in engineering and industrial management. Organizational skills learned in his business career served McMath and astronomy well as he led the American Astronomical Society as its president, and guided the foundation of the Association of Universities for Research in Astronomy [AURA]. McMath served as the first president and as chairman of the AURA board of directors during the early phases of the development of the Kitt Peak National Astronomical Observatory.

McMath, the son of Francis Charles and Josephine (*née* Cook) McMath, enjoyed a privileged childhood earning nearly a full year of advanced standing credits at the prestigious private Detroit University High School. He received his BCE from the University of Michigan in 1913. During and immediately after college McMath worked as a draftsman with the Canadian Bridge Company and in 1914 joined the Saint Lawrence Bridge Company as an assistant engineer; both companies were founded by his father. Anticipating the importance of aviation in World War I, already raging in Europe, McMath took flying lessons and qualified as a pilot before volunteering for service as a commissioned officer in the army. His assignment, however, was as a civil engineer building air bases for the Army Signal Corps's fledgling Air Service. Discharged with the rank of major but in poor health and on the verge of a nervous breakdown from overwork during the war, McMath recuperated for a period, before becoming general manager of Biltmore Forest Estates Company, a North Carolina real estate venture in which his father was an investor. There, during 1921 he met and married a young widow, Mary Ridgley (*née* Rodgers) Garrison, daughter of a vice president of the DuPont Company. They had a daughter, Madeline.

McMath joined Motors Metal Manufacturing Company [MMMC] the next year as assistant manager with explicit instructions from his father, a major stockholder in the firm, to assist in the liquidation of the nearly bankrupt company. McMath's preliminary assessment was that the firm could be salvaged, so he returned the company to profitability, remaining with MMMC or its successors for the remainder of his business career. McMath rose successively to vice president and general manager (1922), president (1925), and chairman of the board (1928–1954).

At an early age, McMath developed an interest in photography that coincided with a long-term family interest in astronomy. His father and grandfather both pursued astronomy as an avocational interest that spun off their training as surveyors in their careers in civil engineering. At the urging of his father, McMath took up astronomy intending to combine that interest with photography. As amateur astronomers, they participated in a solar-eclipse expedition on 24 January 1925, planning to observe the eclipse from a balloon with Detroit probate judge Henry Schoolcraft Hulbert (1869–1959). This plan had been recommended by **William Hussey**, then director of the University of Michigan's Observatory. Although the balloon ruptured during the inflation process, high winds would likely have otherwise prevented the proposed flight. This aborted attempt was only the first of an extended series of astronomical projects undertaken by the McMaths with Judge Hulbert.

In August, 1928 McMath made his first effort combining photography with astronomy. Holding a spring-driven camera between his stomach and the eyepiece of a 4-in. refracting telescope guided manually, he attempted to photograph sunrise in the Moon's craters. While the resulting frames showed decreasing shadow lengths, the quality of the images was not what McMath desired. When, at Judge Hulbert's suggestion, the University of Michigan astronomy department came to MMMC for help in fabricating the mounting for a new telescope, McMath met **Ralph Curtiss**, the new director of the University of Michigan's Detroit Observatory and chair of the astronomy department. Curtiss was impressed by the lunar sunrise film; he decided to join with the McMaths and Hulbert in creating time-lapse motion pictures of celestial objects. They designed a special 10.5-in. Newtonian reflecting telescope mounted on essentially the same mounting that MMMC fabricated for the Detroit Observatory that was completed in 1930 and housed in a dome at McMath's summer place on Lake Angelus near Pontiac, Michigan. At the suggestion of **Heber Curtis**, who was appointed head of the astronomy department following the untimely death of Curtiss, this observatory was named the McMath–Hulbert Observatory with Robert as director, a position he held until 1961.

The 10.5-in. telescope, which utilized the best mechanical and electrical engineering practices of the time, was designed to make time-lapse celestial motion pictures. It was likely the first equatorially mounted telescope that used continuously changing driving rates in both hour angle and declination. The highly refined tracking system eliminated annoying sudden motions of the image caused by manual tracking of an astronomical object during time-lapse photography. By 1931, the telescope could accurately follow the Moon or any other astronomical object smoothly without manual tracking. Curtis arranged to have McMath present films of the Moon and the satellites of Jupiter at a meeting of the National Academy of Sciences held in Ann Arbor in 1932. The films were widely acclaimed as educational tools and were distributed for many years by the *Encyclopedia Britannica.*

In December, 1931 the McMath–Hulbert Observatory was formally deeded to the University of Michigan. McMath, his father, and judge Hulbert were declared honorary curators for the new asset of the university's astronomy department. The telescope was then modified to take time-lapse photographs of solar prominences in the light of, for example, the K-line of calcium. The human eye cannot easily follow the slow changes on the Sun. But the McMath–Hulbert time-lapse technique speeded up the action by some 300 to 600 times, thus giving for the first time a dynamic presentation of solar activity. It was in these films that astronomers became aware that solar prominences originated above the surface of the Sun and flowed back into the Sun, rather than the reverse as had previously been believed.

To further pursue this work, McMath designed and constructed a 50-ft. solar tower telescope that was completed in 1936, with technical assistance from the Mount Wilson Observatory. **Edison Pettit** of the Mount Wilson staff spent three summers at the McMath–Hulbert Solar Observatory using these innovative new instruments. Later McMath designed and had built a 70-ft. tower telescope and a 24-in. Cassegrain reflector. These were not quite ready for operation when World War II broke out. Funding for these projects came from a variety of sources including family members, industry, private foundations, and the university.

Efforts at the McMath–Hulbert Observatory from 1942 until 1946 were devoted to various wartime projects. McMath worked principally with the Office of Scientific Research and Development and was awarded the President's Medal of Merit in 1948 for his accomplishments. Among the developments at the observatory was an instrument that McMath designed and built to observe in the near infrared with a Cashman lead-sulfide photocell, a product of World War II research, or a lead-telluride photocell. By connecting each to a recorder, McMath and the observatory staff were able to conduct spectral investigations in the infrared to study conditions in the atmospheres of both the Earth and Sun. In 1945, McMath was appointed professor of solar physics in the University of Michigan Astronomy Department.

After the war, research at the McMath–Hulbert Observatory continued with a vacuum spectrograph to make highly refined Doppler velocity measurements of motions within various solar features. The observatory staff included **Leo Goldberg**, **Orren Mohler**, Helen Dodson-Prince (1905–2004), and Austin Keith Pierce (1918–2005). McMath was appointed professor of astronomy in 1951, and retired as professor emeritus of astronomy, from the University of Michigan in 1961.

As president of the American Astronomical Society (1952–1954), McMath was asked to chair the National Science Foundation's [NSF] advisory panel for the National Astronomical Observatory [NAO]. It was an inspired choice on the basis of McMath's eclectic interests in business, engineering, and science, and his government service. From his wartime governmental service, McMath was acquainted with many influential individuals in Washington. Among his friends were Alan Waterman, the director of the NSF, former Michigan congressman Joseph Dodge, director of the Bureau of the Budget, and Paul Klopsteg, associate director of the NSF.

Prior to the first meeting of the NAO panel, McMath developed detailed plans for site surveys using 100-ft. towers to automatically record seeing at possible sites for the new observatory. Both daytime and nighttime seeing was considered reflecting McMath's hope that the site might include a large, modern solar telescope. By the fourth panel meeting in February 1956, site selection was well underway and a lengthy discussion of the observatory's organization took place. Additional details were discussed a year later when seven universities joined forces to form AURA. The association, selected by the National Science Board in May 1957 to construct and operate a national optical observatory, was officially incorporated on 28 October 1957 with McMath as its first President. When ill health caused him to relinquish many of his responsibilities, McMath was appointed chairman of the AURA board.

The need for a very large solar telescope was debated in the following year. At a meeting in September 1958, the AURA executive committee approved McMath's action in sending a revised solar telescope proposal to NSF. By late November, some 9 months after the decision to build the observatory on Kitt Peak, the solar telescope project, with Keith Pierce in charge, was added to the ongoing construction program.

McMath died before the solar telescope dedication in November 1962, when by resolution the AURA board, the instrument was officially named the Robert R. McMath Solar Telescope. Some 30 years later it was rededicated as the McMath–Pierce Solar Telescope.

At the 1961 annual AURA meeting, McMath suggested that plans be made for the construction of a 150-in. optical telescope. The resulting 158-in. reflector was dedicated in 1973 as the **Nicholas Mayall** telescope in honor of Mayall's long service as the second director of Kitt Peak National Observatory.

McMath served the American astronomical community ably and well during this crucial time, in which the emphasis was changing from individual projects to those of a team, and received many honors recognizing that service. He was awarded an honorary AM by his *alma mater* in 1933, and honorary DSc's in 1938 by Wayne State University, and by Pennsylvania Military College 3 years later. McMath received the John Price Witherill Medal of the Franklin Institute, Philadelphia in 1933, the Rittenhouse Medal of the Rittenhouse Astronomical Society in 1936, and the Society of Motion Picture Engineers Journal award, 1940. Two years later he was elected a fellow of the American Philosophical Society and of the National Academy of Sciences in 1958.

Many of McMath's papers are held among the Michigan Historical Collections of the University of Michigan's Bentley Historical Library. A valuable autobiography exists in the personal file at the National Science Foundation in Washington, DC.

George S. Mumford

Selected References

Bowen, Ira S. (1962). *American Philosophical Society Year Book*. Philadelphia: American Philosophical Society pp. 149–153.

Edmondson, Frank K. (1997). *AURA and Its US National Observatories*. Cambridge: Cambridge University Press, Chap. 5, pp. 26–45.

Goldberg, Leo, Orren C. Mohler, and Robert R. McMath (1949). "New Solar Lines in the Spectral Region 1.52–1.75 μ." *Astrophysical Journal* 109: 28–41.

McMath, Francis C., Henry S. Hulbert, and Robert R. McMath (1931). "Preliminary Results on the Application of the Motion Picture Camera to Celestial Photography." *Proceedings of the American Philosophical Society* 70: 371–379.

McMath, Robert R. (1938). "Recent Studies in Solar Phenomena." *Proceedings of the American Philosophical Society* 79: 475–498.

——— (1939). "Motion Pictures of Small Chromospheric Flocculi." *Monthly Notices of the Royal Astronomical Society* 99: 559–560.

——— (1953). "Tower Telescopes and Accessories." In *The Solar System*. Vol. 1, *The Sun*, edited by Gerard P. Kuiper, pp. 605–626. Chicago: University of Chicago Press.

McMath, Robert R. and Edison Pettit (1938). "Motions in the Loops of Prominences of the Sunspot Type, Class III b." *Publications of the Astronomical Society of the Pacific* 50: 56–57.

McMath, Robert R., A. Keith Pierce, Orren C. Mohler, Leo Goldberg, and Russell A. Donovan (1950). "N_2O Bands in the Solar Spectrum." *Physical Review* 78: 65.

McMath, Robert R. and Harold E. Sawyer (1939). "Location of Velocity Changes in a Class IIIb Prominence." *Publications of the Astronomical Society of the Pacific* 51: 165–166.

Mohler, O. C. and Helen Dodson-Prince (1978). "Robert Raynolds McMath." *Biographical Memoirs, National Academy of Sciences* 49: 185–202.

McVittie, George Cunliffe

Born **Smyrna (Izmir, Turkey), 5 June 1904**
Died **Canterbury, Kent, England, 8 March 1988**

British mathematician and relativist George C. McVittie made important contributions toward the problem of comparing the predictions of different general relativistic models of the Universe with observations. He was the son of a British merchant father and an Alsatian mother, born in a largely immigrant community that also included Jason Nassau (later director of the Warner and Swasey Observatory, CaseWestern Reserve University, Cleveland, Ohio, USA). With his brother, McVittie was home-schooled, developing an interest in both astronomy and Turkish archaeology. The family was on holiday in England in the summer of 1922 when a change in government in Turkey led to the sacking of Smyrna and the destruction of their home. None of them ever returned. After a spell helping his father, McVittie won a bursary and entered Edinburgh University in 1923 to study mathematics and natural philosophy. Profiting from the excellent teaching of Sir **Edmund Whittaker** and Charles Darwin, he progressed with high honors and scholarships to his master's degree.

In 1927, McVittie started research on the Maxwell–Einstein equations under the supervision of Whittaker. In 1928, he moved to Cambridge University where he received his only formal training in astronomy and, working under **Arthur Eddington,** completed a Ph.D. dissertation on unified field theories (the attempt to unify electromagnetism and general relativity, the theory of gravity, into a single set of equations) in 1930. He failed, but as did Einstein, who persisted in the attempt until near his death. In 1930, McVittie began work (part of it in collaboration with **William McCrea**) on the fate of density perturbations in a nearly homogeneous, static universe and then turned to the fate of condensations in an expanding universe, which grow more slowly than those on a static matrix – but still grow, so that galaxies indeed do form!

McVittie, held academic positions at Leeds (1933/1934), Liverpool (1934–1936), and London (a readership at King's College 1936–1948 and a professorship at Queen Mary College 1948–1952). Between 1939 and 1945 McVittie was seconded to Bletchley Park to work on deciphering of meteorological information from enemy territory and to attempt to restore accurate weather forecasting of the sort that had previously depended on international cooperation. The work of his group was enormously successful; he visited both the United States and Canada (an essential collaborator in receiving and decoding Japanese meteorological information) helping to establish their programs. McVittie was awarded the OBE for this work. Some of it was later published in meteorological journals.

During his London years, McVittie was one of the editors of *The Observatory* and of the *Quarterly Journal of Mechanics and Applied Mathematics*. He also gave some attention to **Edward Milne**'s theory of kinematical relativity, but soon concluded that the Einsteinian theory had a better chance of describing the observed universe. McVittie's 1937 book, *Cosmological Theory*, began the process of asking how valid comparisons between data and theory might be made on the scale of the whole Universe. At London, he was an inspiration to the young Arthur C. Clarke (among many other students) and served on the council of the Royal Astronomical Society.

In 1952, McVittie made big changes in both his personal and professional life, when he and his wife Mildred moved to Urbana, Illinois, USA. He became professor and head of a moribund Department of Astronomy at the University of Illinois. In the 20 years he was in Urbana, McVittie built the department into a thriving research school, with both optical and innovative radio telescopes. This suited his interest in comparing astronomical and cosmological theory with experiment; he kept in close touch with the wealth of new ideas and observational discoveries that emerged in the discipline during his time at Urbana.

McVittie's monograph *General Relativity and Cosmology* (1955, 1964) belongs to the Urbana period. Comparison of the two editions is particularly interesting. The first had described cosmological observations in terms of time and distance, using formulae that were linear approximations to the equations of general relativity and appropriate only for nearby objects. The advent of radio astronomy and discovery of much more distant galaxies and quasars prompted him to recast the entire discussion in terms of exact equations in the only directly measurable quantity, redshift, in the second edition.

In 1961, McVittie became secretary of the American Astronomical Society, holding the office until 1969. He was the last person to run the society on a part-time basis. Toward the end of his tenure, he helped oversee a number of difficult transitions, including the establishment of subdisciplinary divisions, transfer of the ownership of the *Astrophysical Journal* from the University of Chicago to the society, and the transformation of the Annie J. Cannon Prize to a research award. In curious irony given his birthplace, McVittie was by then a conservative voice counterbalancing the "young Turks." He also managed a series of publications on steady-state cosmology (which he never really believed in), stellar statistics, gravitational collapse, and the redshift-distance relationship.

On his retirement in 1972, the McVitties moved to Canterbury, England, where the University of Kent welcomed him as honorary professor of theoretical astronomy. The word "honorary" rapidly became a misnomer; he began teaching astronomy for the natural scientists and general relativity for the applied mathematics group, which he joined. McVittie supervised several mathematics doctoral students, including D. L. Wiltshire, R. P. A. C. Newman, and G. G. Swinerd, who produced important results. Also, 50 years after his 1933 paper, he published a solution of the nonlinear differential equation [NLDE] arising from a cosmological model. It is a quirk of history that his 1983 paper has eventually led to the formation of a very strong NLDE group within applied mathematics at Kent.

Some books on differential geometry offended McVittie's sense of orderliness and clarity of thought, and in the last year of his life McVittie turned to Clifford algebra as a vehicle for understanding gravitation at elementary-particle level. Up to a week before his death in March 1988, he was working on a fourth-order generalization of Einstein's equations, based on the Clifford algebraic approach.

In Canterbury, McVittie rediscovered his love for archaeology and played an active part in the Canterbury Archaeological Trust, of which he was treasurer from its foundation in 1976. He was elected to the Royal Society of Edinburgh (but not that of London). Minor planet (2417) McVittie was named for him.

J. S. R. Chisholm

Selected References

MacCallum, Malcolm A. H. (1989). "George Cunliffe McVittie (1904–1988)." *Quarterly Journal of the Royal Astronomical Society* 30: 119–122.

Runcorn, S. Keith et al. (1990). *Vistas in Astronomy* 33: 39–81. (A collection of articles on McVittie, with contributions by Roy Chisholm, W. Davidson, R. Hide, E. Knighting, W. McCrea, Donald E. Osterbrock, K. A. Pounds, and G. W. Swenson.)

Méchain, Pierre-François-André

Born **Laon, (Aisne), France, 16 August 1744**
Died **Castellón de la Plana, Spain, 20 September 1804**

Pierre Méchain discovered and computed orbits for many comets and participated in important surveys and other geodesic work in France. As the son of a humble architect, Méchain spent his formative years in his native town. His talents eventually led him to attempt a course of study at the prestigious École des Ponts et Chausées. Because of lack of funds Méchain was forced to leave the school in order to earn his living as a private tutor. A small telescope that he handed over to his father to help the latter through some financial embarrassment brought the young man to an astronomer's attention. His father sold the telescope to **Jérôme Lalande** who then took the younger Méchain under his wing. Méchain owed his career as an astronomer to Lalande. Thanks to this patronage, Méchain obtained his first scientific employment in 1772, being appointed hydrographer in the Dépot de la Marine, the French navy's cartographic service, at first in Versailles and then in Paris. During the course of this work he gained experience in drawing maps, as well as in the determination of latitudes and longitudes on land.

Méchain also specialized privately in the search for comets. This met with success with the discovery of two comets in 1781 and the suggestion that **William Herschel**'s planet, initially taken to be a comet, had a circular orbit. The following year, by calculating the orbits, he showed that two comet apparitions, those of 1532 and 1661, did not correspond to the same comet, as had been the general opinion until then. This work earned Méchain an award from the Académie des sciences and his election as a member of that institution.

Once started on the pursuit of comets, Méchain soon became the most famous comet-hunter in France, securing his reputation by the discovery of up to nine more comets in subsequent years, and also through the calculation of the orbits for all comets for which he had sufficient knowledge. Based on his reputation as an orbital calculator, Méchain took charge of the preparation of the *Connaissance des Temps*, the French yearbook of astronomical ephemerides, which he edited between the years of 1788 and 1794. His work as a cometary observer necessarily led him to discover other new nebulous objects, which were included by **Charles Messier**, also a famous comet-hunter, in his well-known catalogue.

Méchain's experience in cartography and his status as a member of the Académie des Sciences, led to his selection, with **Jean-Dominique Cassini** and a young mathematician, **Adrien-Marie Legendre**, for a major geodetic task, linking the Paris Observatory and Greenwich Observatory by a great triangulation chain. This operation resulted in a confrontation between two different methods of geodetic surveying. The British, under general Roy, used the well-known portable quadrant, whereas the French tried a new instrument, invented by another academician, **Jean Charles de Borda**, which later became known as the repeating circle. Publication of the results, which was delayed until 1791 by the bellicose tension between France and Britain, was to be Méchain's first major astronomical/geodetic work.

When the French National Assembly decided to create a new standard of measurement, based on nature, and accepted the definition of the meter proposed by the Académie des sciences, Cassini, Legendre, and Méchain were again selected to undertake a major geodetic operation, the measurement of the arc of the meridian between Dunkerque and Barcelona.

In the end, it was Méchain and **Jean Delambre** who were in charge of the operation, which was, in time, to lead to Méchain's death. Between 1792 and 1798, Méchain and Delambre carried out both geodetic and latitude measurements on both French and Spanish territory. These they presented at the International Conference that met in Paris in 1799 to establish the decimal metric system; the results were published in *Base du système métrique décimal* (Paris, 1806–1810).

In 1798, Méchain was elected a member of the Bureau des longitudes, the organization set up to oversee French astronomical observatories and operations; he went on to hold the most important institutional positions in astronomy in France. On 10 September 1798, the bureau appointed him as as *Capitaine Concierge* of the Paris Observatory. On 8 June 1801, he was entrusted with the functions of the director of the observatory, and on 5 December of that year he was nominated as president of the Bureau des longitudes. Méchain remained in this post until 31 August 1802 when the bureau decided to extend geodetic operations in Spain to link the Balearic Islands with the Spanish coast and extend the Dunkerque–Barcelona arc by 3° towards the south.

Appointed as the person responsible for this work, Méchain left Paris on 26 April 1803. He carried out various subsidiary projects and made numerous terrestrial and latitude determinations both on the mainland and in the Balearic Islands. Méchain's death from malaria interrupted the work, which was completed between 1806 and 1808 by **Jean Biot** and **François Arago**.

Antonio E. Ten
Translated by: *Storm Dunlop*

Selected References

Bigourdan, G. (1901). *Le système métrique des poids et mesures*. Paris: Gauthier-Villars.

Débarbat, Suzanne and Antonio E. Ten (eds.) (1993). *Mètre et système métrique*. Meudon: Observatoire de Paris.

Delambre, J. B. J. (1806). "Notice historique sur Méchain." *Mémoires de l'Institut* 6: 1–28.

Ten, Antonio E. (1996). *Medir el metro: La historia de la prolongación del arco de meridiano Dunkerque–Barcelona, base del Sistema Métrico Decimal*. Valencia: Instituto de Estudios Documentales e Históricos sobre la Ciencia, Universitat de València.

Mee, Arthur Butler Phillips

Born **Aberdeen, Scotland, 21 October 1860**
Died **Cardiff, Wales, 15 January 1926**

Amateur astronomer Arthur Mee, an author, journalist, and educator, combined an interest in planetary observation with organizing and coordinating astronomical activities, particularly in Wales. He was the son of George S. Mee, a pastor in the Baptist Church, and of Elizabeth (*née* Phillips) Mee. His father left the ministry to become a journalist, eventually settling in Llanelli in Wales, where Arthur spent his early years. Mee followed his father into journalism, first in Llanelli, where he married Claudia Thomas in 1888. He moved to Cardiff in 1892 to work in the *Western Mail* newspaper, for which he continued to work until his death, serving for many years as an assistant editor. He was best known professionally as a newspaper columnist.

Mee developed an early interest in astronomy and was active as an observer from the age of 17. He was noted for his visual observations with his 8-in. reflecting telescope, particularly of the Moon and Mars. He made many drawings of lunar features; these and his other observations, especially those of sunspots, meteors, and eclipses, were regularly published in amateur scientific journals. **Camille Flammarion** included some of Mee's drawings in his book *La Planète Mars*. On 11 March 1892, Mee observed Saturn with his 8-in. telescope. He was surprised to see the shadow of Titan on the disk and the satellite itself in transit a short distance away; he claimed to have been the first person to observe both simultaneously.

Shortly after moving to Cardiff, Mee founded the Astronomical Society of Wales, becoming its first president. For two decades, he edited its journal, its almanac, and later its magazine, *The Cambrian Natural Observer*. For a time, the society had over 200 members, with its publications relatively widely distributed. Through these activities, Mee was noted for his efforts to coordinate the activities of amateur astronomers and to publish their observations. He consequently had a pivotal role in Welsh amateur astronomy for the three decades preceding his death.

Mee encouraged amateur astronomers, and was prominent in popularizing astronomy as a public lecturer and as a writer. He published the book *Observational Astronomy* and the booklet "The Story of the Telescope." Mee was instrumental in arranging the gift of a 12-in. reflector as a public observatory to the city of Cardiff. He later donated his own telescope to the astronomical society in the nearby town of Barry.

Arthur Mee was known as a quiet, diffident man, short in physical stature. He was noted for his sense of humor, as is clear from his own entry in the book *Who's Who in Wales*, which he himself edited. He stated that in his early years he was intended for the medical profession, but "saved many lives by becoming a journalist." Mee developed widespread interests alongside astronomy, including history, natural history, languages, literature, and geography. He published several books, pamphlets, and articles on a diverse range of subjects, often writing under the *nom-de-plume* "Idris." He has occasionally been mistakenly confused with the English author Arthur Henry Mee, editor of the *Children's Encyclopaedia*.

Arthur B. P. Mee was suddenly taken ill and died of heart failure, survived by his wife. The crater Mee in the southern uplands of the lunar nearside commemorates him.

J. Bryn Jones

Selected References

Anon. (1921). "Mee, Arthur." In *Who's Who in Wales*, pp. 310–311. Cardiff: Western Mail Ltd.

Anon. (16 January 1926). "Death of Mr. Arthur Mee." *Western Mail*. Cardiff, p. 8.

Evans, J. S. (1923). "Seryddiaeth a Seryddwyr." Cardiff: William Lewis (Printers) Ltd.

Mee, Arthur B. P. (1892). "Note on the Transit of Titan, 1892, March 11." *Monthly Notices of the Royal Astronomical Society* 52: 423–424.

——— (1893). *Observational Astronomy*. Cardiff: Owen and Company.

——— (1909). "*The Story of the Telescope*." Cardiff.

Wilkins, H. P. (1926). "Arthur Mee." *Journal of the British Astronomical Association* 36: 123.

Megenberg, Konrad [Conrad] von

Born **probably (Germany), *circa* 1309**
Died **Regensburg, (Bavaria, Germany), 14 April 1374**

Konrad von Megenberg translated **John of Holywood's** (Sacroboscö's) *Sphere*, the standard primer on astronomy and cosmology from the early 13th to the 17th century in the Latin West, into German. His *Die deutsche Sphaera* and his *Buch der Natur*, a translation of the influential *De natura rerum*, one of the earliest works of natural philosophy to appear in German, established him as an important figure in the transmission of early scientific literature into the vernacular and an influential source in the development of scientific terminology in German. Also an observer of comets, Megenberg was a prominent scholar in German lands of the mid-14th century. He was one of the major figures at Saint Stephan's School, the predecessor to the University of Vienna, and thus he was a forerunner of such later astronomers as **Georg Peurbach** and **Johann Müller** (Regiomontanus).

Von Megenberg's exact birthplace is unknown, though he spent some of his early years in Erfurt before studying at the University of Paris, where he received his Master of Arts in 1334. At Paris he would undoubtedly have studied and then probably taught (as a lecturer in the Cistercian College of Saint Bernard) Sacrobosco's *Sphere*. While at Paris Megenberg observed the comet of 1337 (C/1337 M1), which he described briefly in his *Buch der Natur*. Megenberg is known to have been active in administration as Procurator of the English Nation, and twice a delegate to the Papal Curia at Avignon. He was listed in the Faculty of Arts at Paris until 1342, when Megenberg moved to Vienna as rector of Saint Stephan's School, where he appears to have introduced astronomy as part of the curriculum. In 1348 Megenberg took the post of Canon at the Cathedral of Regensburg, where he remained for the rest of his life.

Arnold summarizes the known biographical evidence for Megenberg's life and career. He goes on to discuss and analyze Megenberg's German *Sphaera* and also judges him the author, identified in the manuscript as Chunradus de Monte Puellarum, of a Latin commentary on Sacrobosco entitled *Super spera questiones*, although Thorndike (1949) does not draw that connection.

James M. Lattis

Alternate name

Chunradus de Monte Puellarum

Selected References

Arnold, Klaus (1976). "Konrad von Megenberg als Kommentator der 'Sphaera' des Johannes von Sacrobosco." *Deutsches Archiv für Erforschung des Mittelalters namens Monumenta Germaniae Historica* 32: 146–186.

Krüger, Sabine (1980). "Konrad von Megenberg." In *Neue deutsche Biographie*. Vol. 12, pp. 546-547. Berlin: Duncker and Humblot.

Thorndike, Lynn (1949). *The Sphere of Sacrobosco and Its Commentators*. Chicago: University of Chicago Press.

——— (1950). *Latin Treatises on Comets between 1238 and 1368 A.D.* Chicago: University of Chicago Press.

Mellish, John Edward

Born **Madison, Wisconsin, USA, 12 January 1886**
Died **Medford, Oregon, USA, 13 July 1970**

As an amateur astronomer, John Mellish discovered six comets. In his later years as an optical worker, he manufactured many telescopes and optical components for American observatories and astronomers. Raised on a farm near Cottage Grove, Wisconsin, Mellish received a limited education before taking full responsibility for his mother's farming operations. With a strong native curiosity, he invested his spare time in intense investigations of nature including the night sky. After making his own 6-in. Newtonian reflector in 1907, Mellish discovered several comets. He then published a description of the telescope making process in the *Scientific American*. The response to that article, together with his notoriety from the comet discoveries, launched Mellish on a career as an optical worker though he longed for an alternate career in observational astronomy.

Mellish languished in that indeterminate state as farmer, part-time optical worker, and amateur astronomer for 8 years before a major opportunity emerged for him. A shortage of technical staff at Yerkes Observatory prompted director **Edwin Frost** to take a closer look at Mellish on the urging of the aging **Edward Barnard**, who had himself once been an amateur discoverer of comets. When Mellish discovered his third comet in winter 1915, Frost offered him a position as an unpaid observer. A grant to Mellish from the Watson Fund of the National Academy of Sciences paid for a hired hand to work the family farm while Mellish was away. Mellish's arrival at Yerkes Observatory was delayed for over a month by his hasty courtship and marriage to 18-year-old Jessie Ruth Wood of Glencoe, Illinois.

Soon after his arrival at Yerkes, Mellish discovered what he took to be a comet in the rapidly fading dawn twilight. Although the discovery was announced to the international astronomical community by a telegram from Harvard Observatory, the object was soon shown to be a previously catalogued nebula, NGC 2261. Mellish redeemed himself within a few weeks by discovering his fourth comet. After the erroneous comet discovery was announced, **Edwin Hubble**, then a graduate student at Yerkes, studied NGC 2261 photographically. Within a year Hubble announced that, as Mellish had contended, the nebula had apparently changed in appearance. NGC 2261 has since been known as Hubble's variable nebula.

Mellish remained at Yerkes for 15 months, during which he continued to observe regularly with the Yerkes comet seeker as well as other available telescopes. Another early morning session, this time with the 40-in. Clark refractor, resulted in a second controversial Mellish observation. In the post-dawn sky in November 1915 Mellish observed what he described as craters on Mars, although Mars was many weeks beyond opposition and presented a comparatively small image. Mellish was convinced of his observation and discussed the matter with Barnard. Mellish claimed that Barnard showed him his own drawings of Mars. Those drawings displayed circular objects that Mellish took to be craters. However, Barnard's advice was that other astronomers were unlikely to believe such observations and that Mellish would be well advised not to discuss his own observations with others. In his later years, Mellish ignored Barnard's advice and discussed his observations as well as his discussion with Barnard with many other astronomers. The matter was not resolved satisfactorily until the Barnard drawings were discovered in a trunk at Yerkes Observatory in the early 1990s. The three circular markings that Mellish described were actually displayed in Barnard's drawings. However, those markings have now been identified with the Martian volcano *Olympus Mons* and two smaller nearby volcanoes known to exist in the region of Mars that Barnard observed. More recently, careful examination of modern photographs of Mars in connection with an accurate ephemeris of the planet for November 1915 indicates Mellish likely observed the giant impact crater Argyre, the mountains surrounding that crater, Neridium Montes, and the super canyon Valles Marineris, which Mellish described as a crack.

In November 1916, after the Mellish family's first child was born, Mellish accepted an appointment as the director of the Harrold Observatory in Leetonia, Ohio. He discovered his fifth comet in April 1917, but otherwise Mellish's astronomical observing from Leetonia was unproductive. Instead, during this period Mellish faced the reality that he was unlikely to succeed as an observational astronomer and shifted his interest back to optical work. With strong support from Leetonia machine tool manufacturer Elmer A. Harrold (1864–1931), Mellish perfected the telescope mountings for his telescopes. He advertised regularly to the amateur astronomy market and aggressively solicited optical work from professional observatories as well. By the time the Mellish family moved to Illinois from Leetonia in 1923, Mellish had a substantial backlog of orders for telescopes and had become well established as a supplier of optical windows, small lenses, and mirrors for use in professional observatory photometers and spectrographs. Mellish visually discovered a possible sixth comet that later was suggested as being an accidental rediscovery of comet 6P/d'Arrest.

For the remainder of his life, Mellish continued to practice recreational astronomy while maintaining a productive business in optics. His small telescopes, both refracting (up to 10 in. in aperture) and reflecting (many in apertures up to 16 in.), were inexpensive and sustained his business as a continuing workload. In addition, Mellish took on larger mirrors for professional telescopes up to 30 in. in aperture, for example a Pyrex mirror for the University of Illinois. Mellish was asked by **Otto Struve** to produce an unusual set of Schmidt system optics (20-in. × 20-in.) for the McDonald Observatory in 1939. The collapse of his second marriage (also to Jessie) and Struve's impatience with the resultant delays thwarted that effort.

During World War II, Mellish capitalized on skills he learned while making optical windows for observatory instruments by manufacturing rock-salt optics sets for infrared spectrophotometers. He was apparently responsible for 90 percent or more of the optics for these important new instruments for chemical laboratory and refinery process control analyses. He supplied rock-salt optics sets to Shell Development, Pure Oil, Dow Chemical, the University of Michigan, and many other institutions in addition to his regular optics work.

The Mellish family eventually included ten children, but both of Mellish's marriages to Jessie ended in disastrous divorces. The first of these occurred in 1933 in Saint Charles, Illinois, the culmination of a prolonged legal battle as Jessie suffered an extended period of depression. After the first divorce, Mellish moved to Escondido, California, where he reestablished his business with the help of Clarence Lewis Friend (1878–1958). The second divorce occurred in 1939 in Escondido as Jessie's struggle to reestablish herself once again failed. Mellish gave Friend two telescopes and showed him how to search for comets. Friend eventually made independent discoveries of at least four comets.

Mellish was a productive contributor to astronomy for most of his life, with six comet discoveries to his credit and as a telescope maker who supplied inexpensive telescopes for both amateur and professional applications. The International Astronomical Union named a crater on Mars in Mellish's honor.

Thomas R. Williams

Selected References

Anon. (1970). "Veteran, Amateur." *Sky & Telescope* 40, no. 3: 138.

Baker, Robert H. (1928). "The 30-inch Reflecting Telescope and Photoelectric Photometer of the University of Illinois Observatory." *Popular Astronomy* 36: 86–91.

Cross, Jr., Eugene, W. (1971). "John E. Mellish: Telescope Maker, Astronomer, and Naturalist." *Strolling Astronomer – The Journal of the Association of Lunar and Planetary Observers* 22: 215-216. (A useful obituary, details on Mellish as a naturalist.)

Gordon, Rodger W. (1975). "Mellish and Barnard–They Did See Martian Craters!" *Strolling Astronomer – The Journal of the Association of Lunar and Planetary Observers* 25: 196-199.

Gordon, Rodger W. and Sheehan, William. (2005). "Solved: The Mars Crater Mystery." Sky & Telescope 110, no. 5: 64–67.

Sheehan, William (1992). "Did Barnard and Mellish Really See Craters on Mars?" *Sky and Telescope* 84: 23-25.

Thompson, Paul (1979)." The Boy Astronomer of Cottage Grove." *Wisconsin Academy Review* 26: 34-40. (Useful details of Mellish's early life but glosses over marital problems, for which see Williams 1999.)

Williams, Thomas R. (1997)." John Edward Mellish and the Origins of the Amateur Telescope Making Movement in North America." *Journal of the Antique Telescope Society* 13: 15–19.

——— (1999). "Telescopes, Marriages, and Mars: The Life of John E. Mellish." *Sky & Telescope* 98, no. 5: 84–88.

Melotte, Philibert Jacques

Born **Camden Town, (London), England, 29 January 1880**
Died **Abinger, Surrey, England, 20 March 1961**

English astrometrist Philibert Melotte discovered the eighth satellite of Jupiter, Pasiphae, and some southern nebulae carry his name. He was the son of a (Belgian-born, hence the names) lecturer at the Royal Naval College, who enrolled him at the Roan School in Greenwich, where Melotte completed his formal education at age 15. He then joined the staff of the Royal Observatory (Greenwich) as a supernumerary computer, remaining on the staff there for his entire career. At the time, it was the custom of the Royal Observatory to recruit the bulk of its staff members for observation and data reduction from local schools. This was an ideal opportunity for Melotte, as those students who showed particular talent (and were able to pass a set of fairly challenging examinations) were allowed to take up permanent posts when they fell vacant. Melotte was one of these students who, by exceptional talent, rose within the observatory to be a department head.

In his first 10 years at Greenwich Melotte made hundreds of astrometric observations of a wide range of astronomical objects. His most celebrated discovery occurred in 1908 while he was engaged in systematic observations of Jupiter and the known Jovian moons. Melotte had set out to confirm the existence of satellites numbers six and seven (Himalia and Elara), discovered at Lick Observatory (Mount Hamilton, California) by **Charles Perrine** in 1904 and 1905. In the course of these observations he found a "moving object" that, through subsequent investigation, he demonstrated to be a satellite of Jupiter. The discovery of the eighth satellite of Jupiter brought Melotte considerable recognition and the 1909 Royal Astronomical Society's Jackson–Gwilt Medal.

Most of his first decade as an astronomer was, however, spent in measuring the photographic plates for the Greenwich zone of the *Carte du Ciel*. As a result, Melotte was the obvious person to undertake the remeasurement of the astrographic plates taken about 1905 by **John Franklin–Adams** and the compilation of a Royal Astronomical Society edition of the Franklin-Adams star charts (on a much smaller scale of 1° = 20 mm than the *Carte du Ciel* charts). In collaboration with **Sydney Chapman,** Melotte also used these plates for faint star counts, and, still later with **Knut Lundmark,** he examined the obscured regions of the plates. A 1926 catalog of southern nebulae found there includes objects still referred to by their Melotte numbers.

Melotte's early work had been supervised by H. P. Hollis, who was head of the astrographic department, and upon Hollis's retirement in 1920, Melotte succeeded him, though his appointment to appropriate rank was considerably delayed. In 1933, Sir **Harold Spencer Jones** returned from the Cape Observatory to become Astronomer Royal and director at Greenwich. He brought with him some 2,847 photographic plates, taken at a number of observatories during an observing campaign aimed at minor planet (433) Eros with the purpose of improving knowledge of distance scales within the Solar System. Most of the rest of Melotte's career was devoted to measuring and reducing these plates, some of the work being done at Abinger, where the observatory staff was moved at the start of World War II. After the war, Melotte returned to Greenwich, where he remained until retirement in 1948. He had served at the observatory for 53 years.

Melotte was well-known to support the efforts of amateur astronomers, and he willingly provided his expertise in celestial photography to many projects. He was also active in several organizations and served as the secretary of the Royal Astronomical Society's Photographic and Instruments Committee from 1913 to 1950 and secretary to the British Astronomical Association from 1913 to 1921 and again from 1926 to 1930, finally becoming its president in 1944. He was a Freemason of high rank, serving as Treasurer of the Trafalgar Chapter for 30 years, and he was a member of the Royal Naval College Lodge of Mark Master Masons. Melotte was survived by his wife and son.

Scott W. Teare

Selected Readings

Chapman, S. and P. J. Melotte (1913). "On the Application of Parallel Wire Diffraction Gratings to Photographic Photometry." *Monthly Notices of the Royal Astronomical Society* 74: 50–58.

Dyson, Sir F. W. and P. J. Melotte (1919). "The Region of the Sky between R. A. 3^h and 5^h 30^m and N. Dec. 20° to 35°." *Monthly Notices of the Royal Astronomical Society* 80: 3–6.

Hunter, A. (1962). "Philibert Jacques Melotte." *Quarterly Journal of the Royal Astronomical Society* 3: 48–50.

Melotte, P. J. (1931). "The Numbers of Stars per Square Degree, Magnitude 11 to 14, in Different Galactic Latitudes." *Monthly Notices of the Royal Astronomical Society* 91: 810–812.

Newall, H. F. (1909). "In Presenting the Jackson-Gwilt Medal to Mr. Philibert Melotte." *Monthly Notices of the Royal Astronomical Society* 69: 331–332.

Menaechmus

Flourished **(Turkey), *circa* 350 BCE**

Menaechmus, the Greek mathematician who supposedly told Alexander the Great that there is no royal road to geometry, was a pupil of **Eudoxus**, the founder of Greek mathematical astronomy. Menaechmus is said to have added more homocentric spheres to the Eudoxean planetary models, which had already been enriched by **Callippus**. But his greatest achievement was the discovery of the conic sections, which takes pride of place in **Johannes Kepler**'s *Astronomia Nova* (New astronomy) (1609).

Menaechmus defined the conic sections as the intersection of a cone of revolution with a plane perpendicular to one of the cone's generators. He distinguished three kinds, which arise, respectively, by cutting a right-angled, an acute-angled, or an obtuse-angled cone. (**Apollonius** later gave them the names parabola, ellipse, and hyperbola, by which we call them still.) The conic sections occur in Menaechmus' two solutions to the Delian problem of duplicating the cube, which can be readily explained using modern mathematical lore and notation.

Roberto Torretti

Selected References

Bulmer-Thomas, Ivor (1974). "Menaechmus." In *Dictionary of Scientific Biography*, edited by Charles Coulston Gillispie. Vol. 9, pp. 268–277. New York: Charles Scribner's Sons.

Eutocius Ascalonius (1972). *Comentarii in libros de sphaera et cylindro*, edited by Charles Mugler. Paris: Belles Lettres, pp. 58–62, 68–69.

Folkerts, Menso (1999). "Menaichmos [3]." In *Der neue Pauly: Enzyklopädie der Antike*, edited by Hubert Cancik and Helmuth Schneider, Vol. 7, col. 1213. Stuttgart: J. B. Metzler.

Heath, Sir Thomas L. (1921). *A History of Greek Mathematics*. 2 Vols. Oxford: Clarendon Press. Vol. 1, pp. 251–255; Vol. 2, pp. 110–116.

Joannes Stobaeus (1884–1912). *Anthologium*, edited by C. Wachsmuth and O. Hense. Berlin: Weidmann, 2.31.115

Kline, Morris (1972). *Mathematical Thought from Ancient to Modern Times*. New York: Oxford University Press, 1972, pp. 41–42, 47–48.

Theon Smyrnaeus (1878). *De utilitate mathematica.*, edited by E. Hiller. Leipzig: Teubner.

van der Waerden, B. L. (1988). *Science Awakening I*, (English translation by A. Dresden with additions of the author.) 5th ed. Dordrecht: Kluwer Academic Publishers.

Menelaus of Alexandria

Born **Alexandria, (Egypt), *circa* 70**
Died ***circa* 130**

Menelaus is best known for his development of spherical trigonometry.

Little is known about the life of Menelaus. It appears that he spent his early years in Alexandria, and was probably born there; after that, he seems to have moved to Rome. **Pappus** and **Proclus** both referred to him as Menelaus of Alexandria. **Ptolemy** noted astronomical observations made by Menelaus in Rome on 14 January 98. In addition, **Plutarch** related a conversation about optics involving Menelaus as an adult in Rome around the same time.

Ibn al-Nadim's *Fihrist* (a register of mathematicians, written *circa* 950) mentions six books by Menelaus, some of which were said to have been translated into Arabic at the time. They included *The Book of Spherical Propositions*, *On the Knowledge of the Weights and Distribution of Different Bodies*, three books on the *Elements of Geometry*, and *The Book on the Triangle*. The only book by Menelaus that can be found today is the first in the list above, generally referred to as the *Sphaerica*.

That book documents Menelaus' major contribution to astronomy, which was the development of spherical trigonometry. For example, the first known definition of a spherical triangle appears at the beginning: The Greek term that he used for a spherical triangle, *tripleuron*, was not commonly used by other mathematicians to refer to triangles (although it does occasionally appear in Euclid), and suggests that Menelaus was aware of the originality of his topic. In short, he appears to have been the founder of spherical trigonometry. He is best known today, within that field, for Menelaus' theorem, which has applications to astronomy.

In addition to his theoretical work, Menelaus made many astronomical observations and attempted to organize them and make estimates relating to the movement of the stars. Several 10th-century Arab astronomers (**al-Battani**, **al-Sufi**, and Hajji-Khalifa) allude to a catalog of fixed stars composed by Menelaus; this was apparently not a full catalog, and was largely based on his observations. Based on Menelaus' observations, Ptolemy suggests that Menelaus was able to estimate that the equinox was moving westward at the rate of 1° per 100 years. (A more accurate figure given today is about 1° per 72 years.) Pappus also mentions a treatise by Menelaus on the settings of the signs of the zodiac. (The calculations in this treatise would have involved the use of trigonometry.)

Kenneth Mayers

Selected References

Aintabi, M. F. (1971). "Arab Scientific Progress and Menelaus of Alexandria." In *Science et philosophie*, pp. 7–12. XIIe Congrès international d'histoire des sciences, Paris, 1968, Actes, Vol. 3A. Paris: A. Blanchard.

Dicks, D. R. (1985). *Early Greek Astronomy to Aristotle*. Ithaca, New York: Cornell University Press.
Heath, Sir Thomas L. (1960). *A History of Greek Mathematics*. 2 Vols. Oxford: Clarendon Press.
Hogendijk, Jan P. (1999–2000). "Traces of the Lost Geometrical Elements of Menelaus in Two Texts of al-Sijzī." *Zeitschrift für Geschichte der Arabisch-Islamischen Wissenschaften* 13: 129–164.
Krause, M. (1936). "De Sphärik von Menelaos aus Alexandrien." *Abhandlungen der Gesellschaft der Wissenschaften zu Göttingen* 17.
Neugebauer, Otto (1975). *A History of Ancient Mathematical Astronomy*. 3 pts. New York: Springer-Verlag.
Schmidt, Olaf (1955). "On the Theorems of Ptolemy and Menelaus" (in Danish). *Nordisk matematisk tidskrift* 3: 81–95, 127.
Tannery, Paul (1883). "Pour l'histoire des lignes et surfaces courbes dans l'antiquité." *Bulletin des sciences mathématiques* 7: 289–292.
——— (1976). *Recherches sur l'histoire de l'astronomie ancienne*. Hildesheim, New York: Georg Ohms. (Reprint of the 1893 ed. published by Gauthier-Villars, Paris. Includes bibliographical references.)
Thurston, Hugh (1994). *Early Astronomy*. New York: Springer-Verlag.
Yussupova, Gulnava (1995). "Zwei mittelalterliche arabische Ausgaben der 'Sphaerica' des Menelaos von Alexandria." *Historia Mathematica* 22: 64–66.

Menzel, Donald Howard

Born **Florence, Colorado, USA, 11 April 1901**
Died **Boston, Massachusetts, USA, 14 December 1976**

Donald Menzel combined astronomy and atomic physics to revolutionize our understanding of the physics of the Sun and gaseous nebulae. He founded three observatories and brought the Smithsonian Astrophysical Observatory to Harvard College Observatory to enrich the scientific cultures of both institutions.

Menzel, the son of Charles Theodor and Ina Grace (*née* Zint) Menzel, was raised in Leadville and Denver, Colorado. In Leadville, his father worked as a telegrapher, clerk, and ticket salesman before becoming the proprietor of the city's largest general store. By 1916, the financial success of the store allowed Charles to retire, and the family moved to Denver, where Donald completed high school. As a schoolboy Menzel pursued numerous hobbies, suggested by his insatiable scientific curiosity. These in turn provided an outlet for an almost boundless energy that was to characterize his entire professional life. One such interest, in particular, was to have important consequences later in his life. After his father taught him the Morse code at an early age, Menzel combined this skill and his interest in the emerging field of radio technology to become an amateur radio operator (W1JEX), a hobby he retained for most of his adult life.

Menzel earned BA and MA degrees from the University of Denver, where he carried out observations of eclipsing variable stars under the benign guidance of professor Herbert Alonzo Howe. Through these efforts he established contacts with Princeton University's **Raymond Dugan**, which eventually led to his enrollment for graduate study at Princeton, University in the fall of 1921. Once at Princeton, the new astrophysics being pioneered by **Henry Norris Russell** drew Menzel's attention. During the summers, Russell dispatched Menzel to the Harvard College Observatory, where he identified a few thousand star clusters near the periphery of the Large Magellanic Cloud for **Harlow Shapley**. For his Ph.D., awarded in 1924, Menzel endeavored to establish a stellar temperature scale by using the recent theory of **Alfred Fowler** and **Edward Milne** on application of **Meghnad Saha**'s equation to stellar spectra.

After graduating from Princeton, University Menzel spent a few years on the faculties of the University of Iowa and the Ohio State University before landing a solid appointment at the Lick Observatory. At Lick, he honed his talent in the application of atomic physics to the interpretation of solar and nebular spectra. Menzel's scientific apogee remains his seminal treatise on the solar chromosphere, based on his exhaustive analysis of **William Campbell**'s collection of photographic plates of the so called flash spectrum. This work, published in 1931, stands as a milestone in theoretical astrophysics. It is one of the earliest examples of quantitative astronomical spectroscopy. Menzel's findings revealed that the chromosphere was not in thermodynamic equilibrium, and that helium and hydrogen were the dominant solar chemical components. His work foreshadowed the spectacular conclusion of **Walter Grotrian** and **Bengt Edlén** that the temperature of the solar plasma increased outward into the overlying tenuous solar atmosphere. Menzel also began studies of planetary nebulae at Lick Observatory.

In the fall of 1932, Menzel accepted Shapley's invitation to join the staff of the Harvard Observatory. There, he soon organized the second outstanding example of his facility with astrophysical spectroscopy. With his students and other collaborators, Menzel authored a series of 16 papers on the physics of gaseous nebulae between 1937 and 1945. This systematic study is as valid and valuable today as when it was written. Menzel's collaborators included **James Baker**, **Leo Goldberg**, Malcolm H. Hebb, George H. Shortley, and **Lawrence Aller**. Menzel's other original contributions to the physics of gaseous nebulae included his observation that under some circumstances, light passing through a gas may be amplified through fluorescence, presaging the phenomena now known as lasers and masers. Further, Menzel's observations regarding the dependence of the visual appearance of a gaseous nebula on the three possible combinations of the opacity of the gas to Lyman continuum and Lyman α radiation served as a useful heuristic for several generations of astrophysicists.

Encouraged by the progress made in France by **Bernard Lyot** with his coronagraph, Menzel undertook to develop a similar device for observing the solar corona whenever the Sun was visible. Menzel's coronagraph, which ultimately became much like Lyot's, was installed in an observatory at Climax, Colorado in late 1940 and manned by Harvard graduate student Walter Orr Roberts (1915–1990).

When the United States entered World War II, Menzel suspended his astronomical work to teach cryptography at Radcliffe College, but was soon commissioned as a lieutenant commander in the navy. He coupled his knowledge of radio and solar physics to become an expert on radio and radar transmission. Assigned to the Office of the Chief of Naval Communications, Menzel arranged for frequent advice on solar activity from Roberts at Climax to be transmitted to the Carnegie Foundation's Department of Terrestrial Magnetism, Washington, DC, for use in radio propagation forecasts for the military services. The forecasts were made using methods of analysis Menzel helped to develop. In this effort, Menzel not only demonstrated the critical practical importance of knowledge of the

Sun and solar activity, he also established relationships with key government scientists and administrators that would later prove useful to his observatory construction programs.

Menzel remained at Harvard for the rest of his career, serving as the director from Shapley's retirement in 1952 until 1966, when he was succeeded by his own student Goldberg. In the years immediately following his assumption of responsibility for the observatory, Menzel was faced with some difficult choices. Physical facilities were deteriorating, and the staff was depleted. Menzel was eventually successful in raising funds to modernize the observatory office buildings. To make immediate room for expanding the staff, Menzel severed long-standing relationships with the American Association of Variable Star Observers [AAVSO] and *Sky & Telescope* magazine, which had previously occupied space in the observatory. In addition, he began destruction of the Harvard plate collection, seen by most astronomers as a priceless resource for research. Eventually Menzel abandoned that plan in the face of protests from many astronomers. In addition, **Bart Bok** and a few other key staff members moved on to other observatories. After this difficult period, Menzel was successful in restoring the vigor of both the academic and research programs at Harvard.

In addition to being one of the leading intellectual figures of American astrophysics, Menzel also distinguished himself as a potent scientific entrepreneur in the fine tradition of **George Hale**. During his career, Menzel established three solar observatories: The High Altitude Observatory in Colorado, the Sacramento Peak Observatory in New Mexico, and the Harvard Radio Astronomy Observatory in Texas; he also brought another, the Smithsonian Astrophysical Observatory, to Harvard.

Menzel served as a consultant to various federal agencies, notably the National Bureau of Standards, the Air Force Office of Scientific Research, and the Office of Naval Research, and as a director of the Association of Universities for Research in Astronomy that organized the National Optical Astronomical Observatories. He had a natural talent for popularizing current scientific discoveries and wrote extensively for newspapers, magazines, and even for film. He authored nearly a dozen books ranging from a practical field guide to the stars and planets and a handbook on radio transmission to scholarly monographs on quantum spectroscopy, stellar structure, and mathematical methods in physics.

Curiously, another source of Menzel's fame was his writings on Unidentified Flying Objects [UFOs]. Like Edward U. Condon who looked into this same issue, Menzel concluded that there was no solid evidence to support the claims of those who believed that extraterrestrials were routinely visiting this planet. While completely serious in his belief that there was no evidence supporting the existence of the UFOs, Menzel also exhibited his sense of humor well in this area. Colleagues came to cherish his frequent cartoons of imaginary intruders from other planets.

During the late 1960s Menzel's activity was slowed by a serious circulatory problem. He stepped down from the directorship, but continued his researches as time permitted. Menzel also continued his pilgrimages to solar eclipses often with his wife Florence whom he married in 1926. The final eclipse he observed, on 24 December 1973, was his 16th expedition.

Thomas J. Bogdan

Selected References

Anon. (2002). "Donald. H. Menzel Centennial. Symposium." *Journal for the History of Astronomy* 33. (This entire issue is devoted to papers presented at a centennial symposium in honor of Menzel in May 2001. Includes papers by Thomas J. Bogdan, David DeVorkin, Owen Gingerich, David Layzer, Ruth P. Liebowitz, Donald E. Osterbrock, and Jay Pasachoff.)

Goldberg, Leo and Lawrence H. Aller (1991). "Donald Howard Menzel." *Biographical Memoirs, National Academy of Sciences* 60: 149–167.

Hoffleit, Dorrit (2002). *Misfortunes as Blessings in Disguise: The Story of My Life.* pp. 62–69. Cambridge, Massachusetts: American Associations of Variable Star Observers.

Menzel, Donald H. (1931). "A Study of the Solar Chromosphere." *Publications of the Lick Observatory* 17, pt. 1: 1–303.

Menzel, Donald H. and Leo Goldberg (1936). "Multiplet Strengths for Transitions Involving Equivalent Electrons." *Astrophysical Journal* 84: 1–10.

Menzel, Donald H. and Chaim L. Pekeris (1935). "Absorption Coefficients and Hydrogen Line Intensities." *Monthly Notices of the Royal Astronomical Society* 96: 77–111.

Osterbrock, Donald E. (2001). "Herman Zanstra, Donald H. Menzel, and the Zanstra Method of Nebular Astrophysics." *Journal for the History of Astronomy* 32: 93–108.

Robinson, Leif J. (1990). "Enterprise at Harvard College Observatory." *Journal for the History of Astronomy* 21: 89–103.

Mercator, Nicolaus

➜ **Kauffman, Nicolaus**

Merrill, Paul Willard

Born **Minneapolis, Minnesota, USA, 15 August 1887**
Died **Pasadena, California, USA, 19 July 1961**

American stellar spectroscopist Paul Merrill discovered the presence of the short-lived radioactive element technetium in the atmospheres of a few cool, highly evolved stars, proving that these stars must have nuclear reactions going on in their interiors at the present time. He also discovered the diffuse interstellar bands.

Merrill's father was a Congregational minister, and throughout his life the son adhered to this religion. Most of his life was spent in California. Merrill attended Stanford University, earning the AB degree in physical science and mathematics in 1908, and then joined the United States Coast Survey for a short time before entering the University of California in 1909 as assistant and fellow at Lick Observatory, achieving the Ph.D. in 1913. He then taught astronomy at the University of Michigan, 1913–1916.

After leaving the University of Michigan Merrill accepted an appointment with the United States Bureau of Standards from 1916 to 1918. He experimented with treating photographic plates with dicyanin to make them sensitive to red light. Merrill arranged with professor **Edward Pickering**, director of the Harvard College

Observatory, to use Harvard's 24-in. refractor and an objective prism to photograph the spectra of stars on dicyanin-sensitized plates. Whereas ordinary photographic plates reached into the red only slightly beyond the hydrogen Hβ line at 4,861 Å, the sensitized plates reached beyond 8,000 Å. From them he was able to identify several molecules not previously known to occur in stellar spectra.

Finally, in 1919, Merrill went to the Mount Wilson Observatory where he remained not only until his retirement in 1952, but as a volunteer the rest of his life. During his lifetime he published some 260 articles and four books. While at Lick Observatory, Merrill and **Charles Olivier** jointly published two articles on the great January comet C/1910 A1. That same year Merrill published a note on the "Spectrographic Orbit of β Capricorni." Thenceforth the bulk of his publications dealt with stellar spectroscopy and variable stars, in which he did considerable original research. Merrill was also a pioneer in red and infrared photography of stellar objects.

Merrill was particularly interested in wartime advances of photographic techniques that could also be applied to astronomical work. In a 1920 article, "Progress in Photography Resulting from the War," he wrote that the government, which usually considered scientific work an unnecessary luxury, "under the stress of a great emergency acknowledged its value and supplied funds as never before." As a result, certain aspects of research were pursued with unprecedented vigor during the war. Photographs of the ground taken from aircraft had been seriously handicapped by the fact that blue light is scattered in the atmosphere, resulting in blurred photographs on blue-sensitive plates. Red wavelengths, on the other hand, are less affected by scattering. Hence Merrill's experiments with dicyanin-sensitized plates were extended for military purposes. During World War I he designed special airborne cameras and took numerous airplane flights from Langley Field to test his equipment.

At Mount Wilson Observatory Merrill's primary topics of investigation were stellar spectroscopy, infrared photometry, stars with emission lines, and variable stars, especially the spectra of long-period variables. For many years, he edited the publications of the Mount Wilson Observatory written by his colleagues as well as publishing his own work.

Besides his many research papers Merrill published four treatises. The first, in 1938, *The Nature of Variable Stars*, was intended primarily for amateurs and laymen interested in astronomy. In the preface Merrill expressed that his purpose was not only to outline current knowledge of variable stars, but also to clarify for the non-technical reader the general nature of modern astrophysics.

Merrill's second book, *The Spectra of Long Period Variable Stars* (1940), is a scholarly text giving the history of the interpretation of stellar spectra, work to which he contributed extensively. His third book, *Lines of the Chemical Elements in Astronomical Spectra* (1956), not only identified all the elements found in stellar spectra but included over 1,060 references to their discovery and identification. Merrill's final book, *Space Chemistry* was published in 1963, nearly 2 years after his death.

Among Merrill's achievements are his identification of zirconium in S-type stars as well as technetium. He was also the first to suggest that red stars form three distinct branches of the giant red stars. Progressing toward the red side of the HR giant diagram from spectral classes G or K, the most prevalent red stars are class M, showing titanium oxide in their spectra; another branch consists of R and N carbon stars; and finally the S types show zirconium oxide. The variable stars with these spectral classes also show hydrogen lines in emission. Merrill found that 88% of the long-period variables showed M-type spectra, 5% N type, 5% S type, and 2% K or R types, and clarified that the differences in spectral appearance derive from (1) temperature, (2) the ratio of carbon to oxygen, and (3) the ratio of elements around zirconium to those around titanium in the periodic table. The latter two differences are a result of nuclear reactions in the stars themselves and mixing of the products to the surfaces. Other Merrill contributions include (1) the determination of the absolute brightness of the long-period variables with **Ralph Wilson**, using the method of statistical parallax; (2) classification of stars with extended atmospheres with **Roscoe Sanford**; (3) demonstration of turbulence in interstellar gas with **Olin Wilson**; and (4) probably most importantly, the recognition that certain diffuse features, commonly found in stellar spectra (especially those centered around 4430 Å), are actually produced in the interstellar medium. Most of these features probably arise from absorptions by carbon–hydrogen bonds, but the precise substances involved are not yet certain.

Merrill received many honors during his career. He was a member of the National Academy of Sciences and received its Draper Medal in 1945. In the American Astronomical Society [AAS] he served as councilor (1931–1934), vice president (1947–1950), and president (1956–1958), and was the AAS Russell Lecturer in 1955. Merrill was elected to the Council of the Division of Physical Sciences of the National Research Council for 1941–1944, and to the Council of the Astronomical Society of the Pacific, whose Bruce Medal was conferred upon him in 1946. He was a fellow of the American Philosophical Society. Additionally he was a member of the American Association for the Advancement of Science [AAAS], and of the honorary societies, Phi Beta Kappa and Sigma Xi. In 1922 Merrill was elected a fellow, and in 1948 associate member of the Royal Astronomical Society.

In 1913 Merrill had married a school classmate, Ruth Currier. She and their son, Donald, survived him.

Dorrit Hoffleit

Selected References

Joy, Alfred H. (1962). "Paul Willard Merrill." *Quarterly Journal of the Royal Astronomical Society* 3: 45–47.

——— (1962). "Paul Willard Merrill, 1887–1961." *Publications of the Astronomical Society of the Pacific* 74: 41–43.

Merrill, Paul W. (1917). "The Application of Dicyanin to Stellar Spectroscopy." *Popular Astronomy* 25: 661. (Paper abstract.)

——— (1920). "Progress in Photography Resulting from the War." *Publications of the Astronomical Society of the Pacific* 32: 16–26.

——— (1922). "Stellar Spectra of Class S." *Astrophysical Journal* 56: 457–474.

——— (1938). *The Nature of Variable Stars*. New York: Macmillan.

——— (1940). *The Spectra of Long-Period Variable Stars*. Chicago: University of Chicago Press.

——— (1956). *Lines of the Chemical Elements in Astronomical Spectra*. Washington, DC: Carnegie Institution of Washington.

Merrill, Paul W. and C. P. Olivier (1910). "Comet α 1910." *Publications of the Astronomical Society of the Pacific* 22: 31–32.

——— (1910)." Photographs of Comet α 1910, Obtained with the Crocker Telescope." *Lick Observatory Bulletin* 5, no. 174: 182.

Mersenne, Marin

Born **Oizé, (Sarthe), France, 8 September 1588**
Died **Paris, France, 1 September 1648**

An avid astronomical correspondent, Marin Mersenne provided vital communication links between practicing scientists of his era. He made important contributions in time keeping, experimental practice, and the philosophical approach to science by religion, the latter at some personal risk. Merseene was born into a family of laborers. He spent 5 years at the Jesuit collège at La Flèche beginning in 1604, followed by 2 years of theology at the Sorbonne University in Paris. In 1611 he joined the Franciscan Order of Minims, so named because they considered themselves the least of all the religious orders. Mersenne became a priest in Paris in 1612, and from 1614 to 1619 he taught philosophy at the convent at Nevers. In 1619 Mersenne moved to the Minim convent de l'Annonciade near the Place Royal (now Place de Vosges). Other than for a few short trips, he remained there until his death.

Mersenne's greatest contribution was his continual correspondence and meetings with scientific leaders, developing an informal network for disseminating information well before the inception of scientific journals. It was said "to inform Mersenne of a discovery meant publishing it throughout the whole of Europe." After Mersenne's death, letters from nearly 100 correspondents were found in his cell. Those who visited or corresponded with Mersenne included

René Descartes, Gérard Desargues, Pierre Fermat, Thomas Hobbes, **Christiaan Huygens**, John Pell, **Galileo Galilei**, Blaise Pascal, and **Nicholas Torricelli**.

Mersenne viewed the question of Earth's motion as undecided and encouraged the search for more scientific evidence to settle the issue, yet he defended Galilei and published his work in French (*Les Méchanique de Galilée* 1634, as well parts of Galilei's *Dialogo* and *Discorsi*). Mersenne felt the church should censor some opinions, but urged moderation because he believed that "true philosophy never conflicts with the belief of the church." Mersenne was a careful experimenter who insisted on precision and repetition. "One should not rely too much only on reasoning," Mersenne wrote, as he questioned whether or not Galilei actually carried out some of the experiments on acceleration down a plane that Galilei described.

In 1636 Mersenne proposed a design for a reflecting telescope using a concave paraboloidal primary and a convex paraboloidal secondary arranged so that their focal points coincide. The electro-optics branch at the National Aeronautics and Space Administration's Marshall Space Flight Center uses an off-axis Mersenne telescope in a lidar (light detection and ranging) system.

Mersenne was the first to discover that the frequency of a pendulum is inversely proportional to the square root of its length, and it was Mersenne who proposed the use of a pendulum as a timing device to Huygens, inspiring him to invent the pendulum clock.

In other fields, Mersenne is often credited with developing the system of tuning musical instruments called equal temperament, and he experimentally developed three important principles in the acoustics of stringed instruments. Mersenne also published results on the cycloid, reported on the chemistry of tin, discussed a "sensitive plant" from the West Indies, and sought a perfect language that was natural and universal for communication of scientific ideas.

Chris K. Caldwell

Selected References

Ariotti, Piero E. (1977). "Bonaventura Cavalieri, Marin Mersenne, and the Reflecting Telescope." *Isis* 66: 303–321.

Boorstin, Daniel J. (1983). *The Discoverers*. New York: Random House.

Crombie, Alistair C. (1994). *Styles of Scientific Thinking in the European Tradition: The History of Argument and Explanation especially in the Mathematical and Biological Sciences and Arts*. 3 Vols. London: Duckworth.

——— (1975). "Marin Mersenne (1588–1648) and the Seventeenth-Century Problem of Scientific Acceptability." *Physis* 17: 186–204.

de Coste, H. (1649). *La vie du R. P. Marin Mersenne, theologien, philosophe et mathematician de l'ordre der Pères minimes*. Paris.

Dear, Peter (1988). *Mersenne and the Learning of the Schools*. Ithaca, New York Cornell University Press.

Lenoble, Robert (1943). *Mersenne: Ou la naissance du mécanisme*. 2d ed. 1971. Paris: Vrin.

MacLachlan, James (1977). "Mersenne's Solution for Galileo's Problem of the Rotating Earth." *Historia Mathematica* 4: 173–182.

Tannery, Mme Paul and Cornélis de Waard. *Correspondance du P. Marin Mersenne, religieux minime*. Vols. 1–2, Paris: Beauchesne, 1932–1933; Vols. 3–4, Paris: Presses Universitaires de France, 1945–55; Vols. 5–17, Paris: CNRS, 1959–1988.

Messier, Charles

Born **Badonviller, (Meurthe-et-Moselle), France, 26 June 1730**
Died **Paris, France, 12 April 1817**

Charles Messier, the first astronomer to search systematically for comets, published memoirs of his astronomical observations on solar-system phenomena, including his observations of 44 comets. He made independent discoveries of 20 comets. However, Messier is best known today for his catalog of the 103 brightest nebulous objects in the skies visible from the Northern Hemisphere.

Messier was the tenth of twelve children of Nicolas Messier, a catchpole, and Françoise B. Grandblaise. Six of his brothers and sisters died at young age; his father died in 1741. Messier was educated by his older brother, Hyacinthe, who was working in the administration of the princes of Salm, reigning in Badonviller at that time. When the princes gave up Badonviller in 1751, Hyacinthe left for Senones, and Charles Messier went to Paris. Messier's interest in astronomy originated when he saw the great six-tailed comet (C/1743 X1) independently discovered by **Jean-Philippe Loys de Chéseaux** and Dirk Klinkenberg (Haarlem, the Netherlands). The annular eclipse of 25 July 1748 was also visible from his hometown.

In 1751 in Paris, the astronomer of the navy **Joseph Delisle** employed Messier, because of his fine handwriting, as a clerk. In addition to a small salary, Delisle provided a room for Messier at the Hôtel de Cluny. Delisle's secretary introduced Messier to the astronomical observatory on Hôtel de Cluny and instructed him to

keep careful records of his observations. Messier's first documented observation was the Mercury transit of 6 May 1753. Delisle convinced Messier of the necessity of measuring exact positions for all observations – one of the most important preliminaries to his success as an observational astronomer. In 1754, Messier was regularly employed as a depot clerk of the navy.

In 1757 Delisle directed Messier to search for a comet expected to return in 1758. At that time, the return was no more than a prediction by **Edmond Halley**. Because of an error in Delisle's calculations, Messier looked at wrong positions for months. However, in August 1758, he discovered another comet (which had been discovered previously) and a comet-like patch in Taurus. The latter was not moving and was thus not a comet, but a nebula. Now known as the Crab Nebula, the remnant of the 1054 supernova, it became the first entry in Messier's catalog of such comet-like objects. Messier later reported that it was this pair of discoveries that prompted him to continue looking for comets with telescopes and to compile his catalog of nebulous objects that might be mistaken for comets.

Messier finally succeeded in finding comet 1P/Halley in January 1759, 4 weeks after its recovery by **Johann Palitzsch**. The secretive Delisle witheld the announcement of Messier's discovery until April; other astronomers were then skeptical about the late announcement. In 1760, Messier made his first independent comet discovery. He continued searching for and observing comets with telescopes over the decades, making original discoveries for a total of 13, and independently codiscovering another seven up to 1801.

Though Messier discovered more nebulous objects that could be mistaken for comets during his comet searching, he had little interest in these objects *per se*. In 1764, he undertook a serious scan of the skies, and compared his results to all the catalogs and lists available to him (those of Halley, **William Derham**, **Johannes Hevel**, **Nicolas de La Caille**, **Giovanni Maraldi** and **Guillaume le Gentil de la Galazière**) to compile a catalog of all such comet-like objects.

His catalog had grown to 45 objects when Messier decided to publish it in 1769, in the *Memoirs* of the French academy for 1771.

When Delisle retired in 1765, Messier continued to work at the Hôtel de Cluny, but it was not until 1771 that he was reclassified as astronomer of the navy. Because of his numerous comet discoveries, his fame spread. French King Louis XV nicknamed Messier "The Comet Ferret." He was elected to most scientific academies existing at that time, including the Academy of Haarlem (1764), the Royal Society (London, 1764), the Royal Academy of Sweden (1769), the Royal Academy of Prussia (1769), and the academies of Belgium (1772), Hungary (1772), and of Russia (1777). Although a proposal to elect Messier to the Royal Academy of Sciences in Paris had failed in 1763, on 30 June 1770 he was finally honored in his own city. He became good friends with the astronomer/mathematician **Jean Bochart de Saron**, who evaluated Messier's observations and calculated orbits of his comets to help him find them after perihelion.

At the age of 40, Messier married Marie-Françoise de Vermauchampt, who was 37; they had known each other as astronomical observers for about 15 years. On 15 March 1772, Madame Messier gave birth to a son, who was christened Antoine-Charles. Tragically, both the mother and the infant died within 11 days.

Messier started a most fruitful cooperation with a younger astronomer, **Pierre Méchain**, whom he met in 1774. Together, they produced a second version of Messier's catalog in 1780; they added 23 new objects to bring the catalogue of nebulae and star clusters to 68 entries. Their vigorous effort increased the number of entries to 103 in the third version published in 1781. Both the second and third editions of the catalog were published in the French almanac, the *Connaissance des Temps*. Twentieth-century astronomers added seven more objects from Messier's and Méchain's notes, expanding their catalog to 110 objects. Messier himself originally discovered 44 and independently codiscovered about 20 of them: 11 nebulae, 27 open star clusters, 29 globular star clusters, 44 galaxies, and 3 other objects (star cloud M24, double star M40, and M73, an asterism of 4 stars).

On 6 November 1781, Messier was severely injured when he fell about 25 ft. into an ice cellar while walking in a park with Bochart de Saron. It took Messier more than a year to recover. In November 1782, Messier resumed his assiduous observing activities, again concentrating on comets.

As a result of the French Revolution, the Royal Academy of Sciences was closed in 1793. Among the many victims of the revolution was Messier's friend Bochart de Saron, who was guillotined on 20 April 1794. In 1795, Messier entered the Bureau of Longitudes and the new National Institute of Sciences and Arts, which succeeded the academy. In 1806, he received the Cross of the Legion of Honor from Napoleon Bonaparte. Suffering from failing eyesight, Messier made his last comet observation of the great comet C/1807 R1. In 1815, Messier suffered a stroke that left him partially paralyzed. He passed away in his home in Hôtel de Cluny.

A lunar crater and the minor planet (7359) are named Messier, in his honor. The 1775 proposition of **Joseph-Jérôme de Lalande** to name a constellation for him, *Custos Messium*, was soon rejected. However, his most obvious honor is certainly the common naming of deep-sky objects for him, with their catalog numbers, for example Messier 42 or M42 for the Orion Nebula, and M31 for the Andromeda Galaxy.

Hartmut Frommert

Selected References

Broughton, Peter (1985). "The First Predicted Return of Comet Halley." *Journal for the History of Astronomy* 16: 123–133.

Gingerich, Owen (1953). "Messier and His Catalogue-I." *Sky & Telescope* 12, no. 10: 255–258, 265.

Jones, Kenneth Glyn (1991). *Messier's Nebulae and Star Clusters*. 2nd rev. ed. Cambridge: Cambridge University Press.

Mallas, John and Evered Kreimer (1978). *The Messier Album*. Cambridge, Massachusetts: Sky Publishing Corp.

Messier, Charles. "Catalogue des Nebuleuses et des amas d'Etoiles." (The third and definitive edition published in 1781 in *Connaissance des Temps* for 1784 [pp. 117–272]. Mallas's and Kreimer's contains a complete facsimile reprint. Abridged English translations are included in Jones.)

Philbert, Jean-Paul (2000). *Charles Messier: Le furet des comètes*. Sarreguemines: Edition Pierron.

Metcalf, Joel Hastings

Born **Meadville, Pennsylvania, USA, 4 January 1866**
Died **Portland, Maine, USA, 23 February 1925**

Joel Metcalf enjoyed three simultaneous and challenging careers: a pastoral career of considerable intensity; a career as a very successful amateur astronomer with the discoveries of 41 asteroids, 5 comets, and at least 10 variable stars to his credit; and a third career as an outstanding optical designer and craftsman, with four important instruments to his credit.

Metcalf was the son of Lewis Herbert and Anna (*née* Hicks) Metcalf. Lewis, a Civil War veteran, lost a leg at the first battle of Bull Run and was held at Libby Prison, Richmond, Virginia. After marrying and settling in Meadville, Lewis served as a newspaper editor and as county treasurer.

At about age 13, Joel Metcalf read **Richard Proctor**'s book, *Other Worlds Than Ours*, which triggered his lifelong interest in astronomy, an interest that was further stimulated by the close conjunctions of Jupiter and Mars in 1879 and 1881. Metcalf acquired his first telescope, a 2-in. French spyglass, in 1882. This was replaced in a few years by a 3.6-in. Fitz refractor. Metcalf graduated from Meadville Theological Seminary in 1890, continued his education for a year at Harvard Divinity School, and completed a Ph.D. at Allegheny College, Meadville, in 1892. He married Elizabeth S. Lockman, of Cambridge, Massachusetts, in September 1891. They had a son and a daughter.

Metcalf was ordained as a Unitarian minister in 1890. After organizing a church in Roslindale, Massachusetts, Metcalf accepted his first formal pastorate in Burlington, Vermont in 1893. That same year, Metcalf began to perfect his techniques for designing and polishing lenses.

In 1901, Metcalf acquired a 7-in. Alvan Clark & Sons equatorial from the estate of Elisha Arnold, a wealthy amateur astronomer of Keesville, New York. The telescope and its heavy mounting, a one-ton granite pier, and the dome were all transported across the frozen Lake Champlain, New York, in February. The dome and granite pier were nearly lost when the ice cracked and the horses bolted.

The load straddled the crack perilously, supported only by the edges of the dome, before neighbors helped Metcalf complete the transportation. After the observatory was reassembled in Burlington, Metcalf apparently limited himself to recreational observing; he published no formal scientific papers during this 10-year period of his career. The observatory was eventually donated to the University of Vermont.

After a decade of intense community service in Burlington, and exhausted to the point of a nervous breakdown, Metcalf took a leave of absence to spend a year at Oxford University in England. There, he attended an average of 24 lectures weekly on philosophy and religion. Oxford's Radcliffe Observatory director **Herbert Turner** became aware of Metcalf's intense interest in astronomy, and gave him keys to the observatory. Metcalf observed furiously in addition to his vigorous classroom participation. He published one project in the *Monthly Notices of the Royal Astronomical Society*, the astrometric positions and magnitudes of 90 stars in Cygnus. Although stressed by this dual occupation, it provided welcome relief for Metcalf; he returned to the United States exhausted but mentally refreshed.

After his return from England, Metcalf served at the First Congregational Church in Taunton, Massachusetts from 1904 to 1910. As a minister, he was very active in the social issues of the time, child labor, low wages, and social injustice, and called for slum elimination. Metcalf belonged to the Taunton Science Club, was elected president of the Ministerial Association, and in April of 1906 was elected to the Executive Committee of the Channing Conference of Unitarian churches in Rhode Island, Southern Massachusetts, and Eastern Connecticut.

By late 1905 Metcalf had reestablished himself in astronomy with a new roll-off roof observatory. In addition, he completed a 12-in. photographic telescope with which he could record a field of stars over 4° in diameter on an 8-in.-by-10-in. photographic plate. It was in Taunton, using the 12-in. camera, that Metcalf's career as a discoverer of asteroids, comets, and variable stars ignited. Following a suggestion by **Solon Bailey**, Metcalf developed a new photographic asteroid search procedure. In a reversal of the normal process, Metcalf guided his telescope so it would track the retrograde motion of an asteroid as it drifted slowly against the background of stars. The consequence of this change was that stars appeared on photographic plates as lines, but any asteroid in the field appeared as a single spot. The asteroid was not only easy to identify, but much fainter asteroids were detected because their light was all concentrated in one spot. Metcalf took two exposures one half hour apart, shifting the plate slightly between them, to avoid misidentification of plate flaws. Within 3 months Metcalf discovered three new asteroids, in late 1905 and early 1906. He described his new procedure for the information of other astronomers in the *Astrophysical Journal*. In a second *Astrophysical Journal* paper he demonstrated how the procedure could be used to study short-term light variations in asteroids. Metcalf discovered 31 asteroids from Taunton using this procedure. He also discovered six variable stars photographically, AK Her, RW Leo, SS Tau, UU Tau, UV Tau, and NSV04158 Cnc. On 15 December 1906, while searching for asteroids photographically, Metcalf discovered the short-period comet now known as 97 P/Metcalf-Brewington.

Metcalf accepted a call to a new pastoral assignment at the Unitarian Society in Winchester, Massachusetts in 1910. He made his 32nd asteroid discovery, his first from Winchester, in November 1911. In the next 3 years Metcalf discovered nine more asteroids bringing his total to 41. During his stay in Winchester, he also discovered one more variable star, NSV 04891 Leo.

Metcalf's work in optics was conducted mainly as recreation during Unitarian summer camps at South Hero, Vermont. In addition to the 12-in. photographic doublet, he developed a 7-in. folded *f*/10 refractor for use as a comet seeker. Sweeping the sky from a hill near their summer cottage with this telescope in August 1910, he discovered comet C/1910 P1 (Metcalf). His third comet, C/1913 R1 (Metcalf), was also discovered from South Hero. Other optical work computed and polished by Metcalf at South Hero included a 10-in. triplet for photographic patrol work at Harvard's Oak Ridge Observatory. The 10-in. triplet went into service at the Boyden Observatory, University of the Orange Free State, Bloemfontein, South Africa. The 12-in. doublet used for so many asteroid, comet, and variable star discoveries in Taunton, and Winchester, is now located at Oak Ridge Observatory, as is a 16-in. *f*/5.25 photographic doublet that Metcalf designed and polished under contract to Harvard.

When the United States entered World War I, Metcalf volunteered to serve the troops through the Young Men's Christian Association. Assigned to the front lines with the 3rd division, 7th infantry, he saw action at Château Thierry and the battle of the Marne, delivering letters, cigarettes, and candy to the troops, and even carrying wounded soldiers and/or their equipment during frequent 25-mile marches. He received a citation for bravery in action, and recuperated in Paris after experiencing shell shock and being exposed to mustard gas. After returning to the United States in 1919, Metcalf volunteered to go with the Unitarian Church to Transylvania (Hungary) to help with the reconstruction of 100 churches in that country.

Metcalf's last pastoral assignment was in Portland, Maine, where he served until his death. While vacationing in South Hero, he discovered two more comets, including 23P/1919 Q1 (Brorsen–Metcalf) and C/1919 Q2 (Metcalf). This was the second recorded apparition of comet 23P, a comet with an orbital period around 72 years. Metcalf also made an independent rediscovery of comet 22P/Kopff in 1919, and discovered two more variable stars, SV Hya and WZ Oph. Thus, although the rigors of Metcalf's war service had taken a severe toll on him, he was still making contributions to astronomy. At the time of his death, Metcalf was making a 13-in. photographic triplet lens of his own design. The triplet lens was later completed by C. A. Robert Lundin of Alvan Clark & Sons and used in the discovery of the planet Pluto by **Clyde Tombaugh** at Lowell Observatory.

A further appreciation of Metcalf's importance in astronomy may be gained in several quite different ways. Metcalf was chair of the Visiting Committee for Harvard College Observatory and a member of the Visiting Committee for the Ladd Observatory at Brown University, serving as chair of a search committee when Brown University needed a new observatory director. **Edward Pickering** nominated Metcalf to serve on both the Comet and Asteroid Committee of the American Astronomical Society.

Metcalf died of an aneurysm. About 900 persons attended his funeral in Portland including several justices of the Supreme Court.

Richard R. Didick

Selected References

Bailey, Solon I. (1925). "Joel Hastings Metcalf." *Popular Astronomy* 33: 493–495.

Metcalf, Joel Hastings (1906). "An Amateur's Observatory " *Popular Astronomy* 14: 211–217.

——— (1906). "A New Method for the Discovery of Asteroids." *Astrophysical Journal* 23: 306–311.

——— (1907). "A Photographic Method for the Detection of Variability in Asteroids." *Astrophysical Journal* 25: 264–266.

Stoneham, Rachel Metcalf (1939). "Joel H. Metcalf, Clergyman–Astronomer." *Popular Astronomy* 47: 22–28.

Metochites [Metoxites], Theodore [Theodoros, Theoleptos]

Born **Nicaea, (Iznik, Turkey), 1260/1261**
Died **Constantinople, (Istanbul, Turkey), 1332**

Son of George Metochites, a cleric of the Eastern Orthodox Church during the imperium of Emperor Michael VIII Paleologos, Theodore Metochites grew up in the cultural center of Constantinople. However, because his father George favored union with the Latin Church, the family was exiled. Theodore, nevertheless, received a good education and completed his *enkyklios paideia* by the time he was 20. Favored by Emperor Andronikos II, he became a close collaborator and counselor. In this capacity he made several important diplomatic missions to Cyprus, Serbia, and Thessaloniki, among others, and was appointed to several successively more important public offices. In 1304, Metochites was appointed to the highest position of the Byzantine administration, *megas logothetes*, or Grand Deputy, with duties equivalent to chancellor or prime minister, which he held until 1321.

Metochites's career ended when the emperor was deposed, and he was exiled by the new Emperor Andronikos III Paleologos. He died, as the monk Theoleptos, in 1332 at the Chora monastery in Constantinople to which he had donated his extensive library, and whose restoration work he had personally supported. Metochites's mosaic portrait in the monastery where he offers the Church of Chora to the enthroned Christ commemorates his extensive gifts to the institution.

Metochites was an exceptionally prolific writer and scholar, leaving behind works of rhetoric (royal eulogies and discourses), 20 poems, a literary testament in verse, a collection of philosophical texts, and two works on astronomy. His collection of texts, *Hypomnematismoi kai semeoses gnomikai* (Annotations and gnomic notes or Personal comments and annotations), an astonishing collection of essays and texts on history, literature, and thinking, includes material on over 70 Greek authors. It contains the most extensive commentary on Aristotelian philosophy of the late Byzantine period. Metochites's commentaries on the *Dialogues of Plato* had an important influence on the Platonic renaissance of the 15th century.

Metochites's work associated with astronomy includes his paraphrases of **Aristotle**'s works on natural philosophy and his comprehensive introduction to Ptolemaic astronomy. His *Stoicheiosis Astronomike* (Elements of astronomy) revived Ptolemaic studies in Byzantium and gives evidence of the significance of contacts with Persian and Arabic science in astronomy as practiced in the period of the early Paleologai. In this work, Metochites described earlier astronomical studies and made a clear argument for the importance of astronomy over the other branches of mathematics. He clearly distinguished between astronomy and the then popular apotelesmatics (astrology), which he condemned. In his *Semeoses gnomikai* (Annotations), Metochites provided an important critique of Aristotle.

Katherine Haramundanis

Selected References

Bydén, Börje (2002). "To Every Argument there is a Counter-Argument: Theodore Metochites' Defence of Scepticism (*Semeiosis* 61)." In *Byzantine Philosophy and Its Ancient Sources*, edited by Katerina Ierodiakonou, pp. 183-217. Oxford: Clarendon Press.

Hult, Karen with Börje Bydén (2002). *Theodore Metochites on Ancient Authors and Philosophy. Studia Graeca et Latina Gothoburgensia* 65. Göteborg: *Acta Universistatis Gothoburgensis.*

Paschos, E. A. and P. Sotiroudis (1998). *The Schemata of the Stars: Byzantine Astronomy from A.D. 1300.* Singapore: World Scientific.

Talbot, Alice-Mary (1991). "Metochites, Theodore." In *Oxford Dictionary of Byzantium*, edited by Alexander P. Kazhdan. Vol. 2, pp. 1357–1358. Oxford: Oxford University Press.

Underwood, P. A. (1966). *The Kariye Djami.* Vols. 1–4. New York: Bollingen Foundation.

Meton

Flourished **Athens, (Greece), *circa* 432 BCE**

An early careful and quantitative observer, Meton is known for his calendrical discoveries, some of which are still in use today.

Meton was the son of Pausanias, but little is known of his life apart from a few probably apocryphal stories, such as the tale, found in **Plutarch**, that he simulated madness to avoid military service. Meton was a contemporary of the famous Greek playwright Aristophanes, who characterized him in his still popular comedy *The Birds*, as a ridiculous geometer. The picture of Meton given here is, of course, a caricature, but it indicates that he was well known to contemporary Athenians, as he himself avers in the play, and it probably contains more than a grain of truth. He is shown wearing soft boots, typical of women and effeminates, and he makes a show of publicly carrying out his researches.

Like most other ancient Greek astronomers, however, Meton is known essentially for his contributions to science. He may well be the first astronomer who moved away from an approach to astronomy tied up with magic and came to conclusions based on serious, scientific investigations. He is certainly the first personality on whom those pillars of ancient astronomy, **Hipparchus** and **Ptolemy**, relied as a collector of reliable and usable data.

Meton, probably along with others (notably **Euctemon**), carried out careful observations to detect the solstices, by determining when the shadow cast by a gnomon was at its maximum and minimum. He observed that 235 lunations, or synodic months, amount approximately to 6,939 days and 16.5 hours, while 19 solar years amount to 6,939 days and 14.5 hours – a difference of only two hours. Now 19 years of lunar months amount to 228; by intercalating 235–228 = 7 months over the course of the 19-year cycle, it is possible to synchronize lunar months and solar years.

This observation has no relevance to the Gregorian calendar (which is based on the Sun) or to the Muslim calendar (which is based on the Moon, and hence runs perpetually about 11 days shorter than the solar year). Meton's observation has, however, had continuing relevance for the Jewish calendar, which determines how all Jewish festivals fall. These are observed to this day by Jews throughout the world.

The ancient Hebrew calendar was basically lunar, but the festivals such as Passover had to fall at the appropriate season of the year. Originally this calendar was *ad hoc*. When the New Moon was sighted, the Jerusalem Sanhedrin declared the new month, which might occur after 29 or 30 days (since the mean synodic month is slightly over 29.5 days long), and the message was relayed by beacons on high places to persons living at a distance. When the spring month in which Passover would be celebrated was approaching, yet there were no signs of spring, the authorities would announce the intercalation of a full month, in order to "catch up" with the Sun. Such a calendar is self-correcting, since a dubious call in respect of the sighting of the Moon or the intercalation of the month, would be corrected next time around. On account of political changes, such a system gradually became unworkable, and around the 7th century Meton's cycle was used to substitute a perpetual calculated calendar, still in use today. In it, some slight modifications known as "postponements" were introduced, in order to avoid the coincidence of certain religious celebrations on or near the Sabbath, with attendant religious problems. This calendar was more accurate than the Julian calendar, despite its relatively large swings, but will ultimately require correction (to allow for the slight error in the ancient calculation of the mean lunation period).

The Metonic cycle also determines the Christian festival of Easter. Easter was originally fully dependent on the Jewish Passover, but, beginning with the Council of Nicaea in 325, the Church attempted to fix an independent, agreed date. In the west this was eventually settled on as the first Sunday after the paschal moon occurring on or after the Vernal Equinox, reckoned as 21 March. But the determination of the paschal moon is based on Meton's cycle and does not necessarily correspond to the astronomical Full Moon, any more than the beginning of a Jewish month necessarily corresponds with the New Moon.

The Metonic cycle has its place in the Julian Period proposed in 1583 by Joseph Justus Scaliger, and named by him for his father Julius. This consists of numbered days in a period of 19 × 28 × 15 = 7,980 years, beginning on 1 January 4713 BCE. The figure was arrived at by multiplying together the numbers of years in the Metonic cycle, the solar cycle of the Julian calendar, and the ancient Roman cycle of indiction. The starting point was the nearest past year in which the cycles began together. Julian dates are still in use by astronomers.

Another place where the Metonic cycle figures is in the calendrical ciphers known as clog almanacs, found on boards or sticks, used in western Europe for hundreds of years until the 17th century. These present the so called golden number, which gives the position of a particular year within the Metonic cycle. They were first noticed in England by Richard Verstegan in his book *A Restitution of decayed intelligence* (London, 1634) p. 58:

They used to engrave upon certaine squared sticks about a foot in length the courses of the Moones of the whole yeere, whereby they could al waies certainely tell when the new Moones, full Moones, and changes should happen as also their festival daies.

Alan D. Corré

Selected References

Aristophanes (1987). *The Birds*, edited with translation and notes by Alan H. Sommerstein. Warminister: Aris and Phillips. (For Meton, see pp. 992–1020.)

Neugebauer, Otto (1975). *A History of Ancient Mathematical Astronomy*. 3 pts. New York: Springer-Verlag. (Discusses the vexed question of Meton's precise position in discovering the cycle ascribed to him. See p. 623, note 12.)

Ptolemy (1984). *Ptolemy's Almagest*, translated and annotated by G. J. Toomer. New York: Springer-Verlag. (Meton's observation of the summer solstice is mentioned in III.1.)

Schnippel, E. (1926). *Die englischen Kalenderstäbe*. Beiträge zur englischen Philologie, 5. Leipzig: B. Tauchnitz. (Deals with the wooden almanacs containing the golden number.)

Metrodorus of Chios

Flourished **Chios (Khíos, Greece), *circa* 325 BCE**

Metrodorus of Chios formulated the first theory of the origin of the Universe and made several early contributions to astronomy and cosmology. He is often confused with Metrodorus of Lampsacus, but the two have no relation to each other. Metrodorus of Chios (hereafter referred to as Metrodorus) was most likely born in the fourth century BCE, though Smith suggests the date 500 BCE is more probable. However, Freeman (1966) indicates that his father was put to death during the reign of Antigonus Gonatas between 323 and 301 BCE, putting his (Metrodorus' father's) birth date between 400 and 380 BCE. This means Metrodorus could not have been born earlier than 380 BCE.

Metrodorus was the son of Theocritus, who was a well-known statesman in Chios. Theocritus was, oddly enough, the student of yet another Metrodorus, who some have suggested is the younger Metrodorus' namesake. Theocritus was also a leader of an anti-Macedonian democratic party in Chios, but these political interests do not seem to have been inherited by his son, who was largely interested in the physical sciences and epistemology. Metrodorus was thought to have studied under **Democritus**, though some scholars have argued that Metrodorus simply learned the teachings of Democritus from Nessas. This latter point makes sense in light of the likely dates of his life, for Democritus died in 359 BCE and Metrodorus most likely was born well after 380 BCE, making him too young. Once Metrodorus had established himself as a philosopher, his pupils included Diogenes of Smyrna, who later taught Anaxarchus.

Metrodorus' views on the physical sciences are described in his book *On Nature*. Metrodorus espoused the view that the stars, like the Moon, were lit up by the Sun, and that the Milky Way marked the line of passage of the Sun. Meteorites were sparks caused by the collision of clouds drawn up by the Sun. The Sun's heat also apparently caused the winds to form. This latter fact is due to his view that the Sun was a "sediment" of the air and was constantly quenched and ignited. This occurs when air condenses, forms a cloud (of water), and descends on the Sun, which puts out the fire. However, it is reignited when the water dissipates. Metrodorus extended this concept to describe some processes of the early Universe. He said that this process of quenching and reigniting continued until the fire of the Sun dominated the water vapor around it. The stars then formed from portions of this water vapor. (Basically they were ignited clouds.) He also extended the quenching and igniting aspect to explain night and day, as well as eclipses. This concept is fascinating, as it is one of the earliest scientific (albeit completely incorrect) explanations of the formation of the Universe.

Meotrodorus' views on cosmology also contained another progressive viewpoint. It was his opinion that the world was only one of many, since "it is as unlikely that a single world should arise in the infinite as that a single ear of corn should grow on a large plain" (Guthrie, 1965, 405). In terms of the ordering of the cosmos, Metrodorus held that the Sun was the highest object in the sky, with the Moon next and beneath them the planets and stars. Metrodorus also contributed several noteworthy arithmetical puzzles, including some that dealt directly with astronomy.

Ian T. Durham

Selected References

Freeman, Kathleen (1966). The *Pre-Socratic Philosophers: A Companion to Diels.* 2nd ed. Cambridge, Massachusetts: Harvard University Press.

Guthrie, W. K. C. (1962). *A History of Greek Philosophy.* Vol. 1. Cambridge: Cambridge University Press.

——— (1965). *A History of Greek Philosophy.* Vol. 2. Cambridge: Cambridge University Press.

Smith, David Eugene (1923). *History of Mathematics.* Vol. 1. Boston: Ginn and Co. (Reprint: New York: Dover, 1958.)

——— (1925). *History of Mathematics.* Vol. 2. Boston: Ginn and Co. (Reprint: New York: Dover, 1958.)

Michell, John

Born **Nottinghamshire, England, 1724**
Died **Thornhill, (West Yorkshire), England, 21 April 1793**

Reverend John Michell demonstrated the existence of binary stars and physical star clusters, predicted the existence of black holes, and made the first realistic estimate of the distance to a star, all major intellectual achievements for an 18th-century English scientist – clergyman.

Michell is thought to have been born in Nottinghamshire, perhaps on Christmas Day 1724, but that is uncertain. There are no portraits available, and only a brief description of him as "a little short man, of a black complexion, and fat" exists in a contemporary diary. It is known that he married a Sarah Williamson in 1764 but that she died the next year.

In 1742 Michell became a student at Queens' College, Cambridge University, from which he received his MA (1752) and BD (1761) degrees. During this time he held the first of several rectorships at churches in the region.

In 1767 Michell published a paper on double stars and clusters, arguing that they were far more common than would be the case if stars were randomly scattered over the celestial sphere. For example, he estimated that there was only about one chance in 496,000 that random placement would result in a cluster like the well-known Pleiades. Consequently, he argued that most close pairs of stars must be physical binary star systems and most clusters must be physically close groups of stars, probably held together by the gravitational force. This was apparently the first use of statistical arguments in astronomy, and the result was confirmed when **William Herschel** measured the motion of the stars in a few binary pairs around each other.

Michell was also a telescope builder; he built his own reflector with a 30-in.-diameter primary mirror and a focal length of 10 ft. After his death, this telescope was purchased by Herschel, who built a newer but similar reflector.

In May 1783 Michell wrote a fascinating letter to **Henry Cavendish** (1731–1810), which Cavendish read before the Royal Society on 27 November 1783 and had published in the *Philosophical Transactions*, 1784. Starting from **Isaac Newton**'s corpuscular theory of light, published in his *Opticks*, Michell pointed out that light particles leaving an astronomical object such as a star would slow down like any other objects in a gravitational field, and that this decrease in speed might be used to provide information about the mass and distance of stars. For example, the images through a prism of different stars would be offset differently depending on the speeds of their light, and from this one could recognize differences in mass.

Michell also pointed out that if a star was massive enough, its escape speed might exceed the speed of light from the star, so that light could not escape, rendering the star invisible – the first known description of what we now call a black hole. He also tried to determine the mass of such a star on the assumption that it had the same density as the Sun, and estimated that it would have 497 times the diameter of the Sun, stating that "a body falling from an infinite height towards it, would have acquired at its surface a greater velocity than that of light, and consequently, supposing light to be attracted by the same force in proportion to its *vis inertiae*, with other bodies, all light emitted from such a body would be made to return towards it, by its own proper gravity." Such a classical Newtonian black hole star would be incredibly more massive than the Sun, of course, far more than is required by the modern theory of black holes. However, his calculation of the mass-to-radius ratio was correct.

Michell apparently did not think it likely that many such unseen stars would exist, but he pointed out that the presence could be detected by their gravitational influence on nearby objects. For example, a binary system with a black hole would seem to have a star revolving around nothing. Indeed, black holes can be detected in this fashion today.

The following year, 1784, Michell made the first realistic assessment of the distances to the stars. Reasoning by means that are analogous to modern photometric parallaxes, he used the apparent diameter and brightness of Saturn, and its known distance from the Sun, to infer that Vega, with approximately the same brightness as Saturn, must be some 460,000 astronomical units from the Sun. The actual distance is more than four times that much but Michell could not have guessed that Vega is intrinsically much brighter than the Sun. It was over 50 years before **Friedrich Bessel** measured the first stellar parallax to confirm the enormous distances that Michell had been the first to realistically estimate.

In his later years Michell addressed the problem of determining the density of the Earth. Since the radius of the Earth was well known, this was equivalent to determining the mass of the Earth and also the value of *G*, the constant in Newton's law of gravitation. For this purpose Michell devised a truly beautiful experiment to measure the force of gravitation between small lead spheres affixed to the ends of a 6-ft. wooden beam and large lead spheres placed nearby. The beam was to be suspended from a single metal fiber acting as a torsional balance, which Michell invented independently of, and perhaps before, Charles de Coulomb. Measuring the angles of rotation of the fiber with the large sphere first on one side of the beam and then the other, the force between the lead spheres could be compared with the gravitational force of the Earth on the spheres, allowing determination of the density or mass of the Earth.

Michell died before carrying out this experiment, but his apparatus was given to Cavendish, who rebuilt the apparatus and carried out the experiment. This well-known Cavendish experiment (more appropriately, the Michell–Cavendish experiment) produced a value of 5.48 for the specific gravity of the Earth, compared to the modern value of 5.52. Cavendish's 1798 paper on this experiment begins with a long description of Michell's apparatus and experimental design.

In geology, Michell carried out seminal studies of earthquakes, earning him the title "father of seismology." The great Lisbon earthquake of 1755 intrigued Michell, who developed a theory of earthquakes, read to the Royal Society of London in 1760 and published in *Philosophical Transactions*. He suggested that earthquakes were produced when two different layers of rocks rubbed against each other at a particular location many miles beneath the surface of the Earth (now referred to as the epicenter), perhaps caused by steam produced by volcanos. The earthquakes were wave motions of the solid material of the Earth, which traveled from their source to other parts of the Earth. Michell distinguished between two different types of waves of different speeds, which he expected would enable the location of the epicenter to be determined.

While at Cambridge Michell wrote a *Treatise of Artifical Magnets* (1750), describing for the first time in print how to make strong artificial magnets. He also reported experiments that showed the forces between magnets were consistent with an inverse-square law between individual poles of the magnets.

Michell's work on earthquakes and seismology led to his election in 1761 as a fellow of the Royal Society of London, a position he had coveted. It also merited him the Woodwardian Chair of Geology at Cambridge University, which he held from 1762 to 1764. In 1767, apparently in order to earn a better living, Michell became rector of the Church of Saint Michael and All Angels in Thornhill, near Leeds, Yorkshire, England, where he remained the remaining 26 years of his life, and where he was buried.

Laurent Hodges

Selected References

Cavendish, Henry (1798). "Experiments to determine the Density of the Earth." *Philosophical Transactions of the Royal Society of London* 88: 469–526.

Clerke, Agnes M. (1921–1922). "Michell, John." In *Dictionary of National Biography*, edited by Sir Leslie Stephen and Sir Sidney Lee, Vol. 13, pp. 333–334. London: Oxford University Press.

Michell, John (1760). "Conjectures concerning the cause, and Observations upon the Phaenomena of Earthquakes; particularly of that great Earthquake of the first of November, 1755, which proved so fatal to the City of Lisbon, and whose Effects were felt as far as Africa, and more or less throughout almost all Europe." *Philosophical Transactions of the Royal Society of London* 51: 566–634.

——— (1767). "An Inquiry into the probable Parallax, and Magnitude, of the fixed Stars, from the Quantity of Light which they afford us, and the particular Circumstances of their Situation." *Philosophical Transactions of the Royal Society of London* 57: 234–264.

——— (1784). "On the Means of discovering the Distance, Magnitude, & c. of the Fixed Stars, in consequence of the Diminution of the Velocity of their Light, in case such a Diminution should be found to take place in any of them, and such other Data should be procured from Observations, as would be farther necessary for that Purpose." *Philosophical Transactions of the Royal Society of London* 74: 35–57.

Michelson, Albert Abraham

Born **Strenlo (Strzelno, Poland), 19 December 1852**
Died **Pasadena, California, USA, 9 May 1931**

Albert Michelson made the first measurement of a star's diameter and many measurements of the speed of light. His longevity, extraordinary skills of observation, and his enthusiasm for research (which he often expressed as "It is such good fun!") allowed him to make significant other contributions to the fields of optics, astronomy, and the study of light.

Michelson came to the United States as a child in 1855, living with his parents first in New York and then in the western states of Nevada and California. He attended high school in San Francisco, California, where his science teacher encouraged him to continue his formal education after matriculating in 1869. Michelson was able to secure a position at the United States Naval Academy, at Annapolis, Maryland (the first Jew to be admitted) and graduated in 1873. Following this he served as a midshipman in the navy for 2 years and then was appointed to the position of instructor of physics and chemistry at the Naval Academy from 1875 to 1879. While this is an impressive start to his career, it is even more so when it is noted that there were no vacancies at the Naval Academy when Michelson was applying for admission. It was only through the favorable impression made on President U. S. Grant and high-ranking navy officials at the meeting he arranged in Washington, DC, that he was able to secure the appointment.

Michelson married Margaret McLean Heminway in 1877, and they had three children. They eventually divorced, and Michelson later married Edna Stanton in 1899; they had three children. In 1879, Michelson took up a post at the United States Naval Observatory Nautical Almanac Office in Washington to work with **Simon Newcomb**. The following year he received a leave of absence to travel to Europe, where he studied at the Collège de France and the universities of Heidelberg and Berlin. On his return to the United States in 1881 Michelson resigned from the navy.

Michelson received an appointment as professor of physics in the Case School of Applied Science in Cleveland, Ohio, in 1882, which he held until 1889 when he accepted a similar position at Clark University at Worcester, Massachusetts. Michelson was offered the position of professor and head of the Department of Physics at the University of Chicago in 1892, which he accepted, and he remained there for the rest of his career.

Throughout his career, Michelson's research focused on experimental optics, and most of his 78 papers concerned light. Some of his major accomplishments were the measurement of the velocity of light, ruling of diffraction gratings, development of one kind of interferometer, null detection of the Earth's motion through the "ether," development of the echelon spectrometer, measurement of stellar diameters, and an investigation of the metallic colorings in nature. His research influenced other eminent scientists. **Albert Einstein**, for example, probably knew about Michelson's results on the speed of light and its constancy when he formulated his Special Theory of Relativity, in which such constancy was postulated.

Michelson's lifetime of research into the nature of light began as an instructor at the Naval Academy when he was ordered to construct an experiment to measure its velocity. Using what was essentially "home built" equipment, he found he was able to make the measurements. Most surprising was that his measurements proved more accurate than any that had been previously obtained! Michelson continued to work on making more accurate measurements of the velocity of light while he was at the Case School of Applied Science, and in 1883 he reported a value of 299,853±60 km/s.

In 1887 Michelson and **Edward Morley** collaborated in investigating the motion of the Earth through hypothetical ether using the Michelson interferometer. The presence of the ether was expected to produce a 0.4 fringe difference for two heams of light travelling the same distance parallel and perpendicular to the Earth's direction of motion. However, the measured difference was less than 0.01 fringe, or in other words the effect of the ether was not detected.

Also in 1887 Michelson and Morely showed that they were able to accurately define the length of the standard meter using measurements of the wavelength of light. Using the interferometer and the red cadmium light line, they were able to determine that 1,553,163.5 wavelengths of this light made up a meter. Since the wavelength of light remains constant over a wide range of conditions, they proposed that it would be a better way to define the standard length of a meter rather than the then used "metal bar" standard for the meter. The use of light to define the meter was generally adopted in 1960.

In 1882 Michelson published the theory of his "differential-refractometer," which today is referred to as an interferometer. Much of Michelson's research time at Clark University was spent in pursuing the uses of the interferometer. The interferometer was one of Michelson's most famous inventions and can be simplistically described as a device that splits a beam of light into two parts traveling in different directions and then recombines them such that the difference in their path lengths traveled can be detected. In 1891 Michelson published his technique for using the interferometer to measure stellar diameters and also used the approach to measure the diameters of some of the Jovian satellites. It took until 1920, in collaboration with **Francis Pease** at the Mount Wilson Observatory, for Michelson's dream of measuring stellar diameters to be realized. Together they found the diameter of the star α Orionis (Betelgeuse) to be 47 milli-arcseconds.

Michelson's interest was also caught by Pieter Zeeman's discovery of how magnetic fields influence spectral lines. To explore this effect in greater detail, Michelson invented the echelon spectrometer to provide the resolving power needed to study the effect.

Michelson became a research associate of the Carnegie Institution of Washington in 1924 at the invitation of **George Hale**. At the Mount Wilson Observatory Michelson once again began his experiments in determining the velocity of light. Using projection apparatus located on Mount Wilson, light was reflected back from a receiver on Mount San Antonio some 22 miles away, to determine its velocity. Using this technique he determined the speed of light to be 299,798 km/s. The piers used for this experiment at the Mount Wilson Observatory still stand to this day on a south ridge of the mountain, and a plaque commemorating his experiments is located nearby.

Michelson was not only a man of science but also a Freemason, and he had a well-developed artistic side with many varied interests. He was an avid sketcher and often took his watercolor kit when he went on vacation. Michelson was known to play the violin, enjoy a good game of billiards, and – as a measure of the quality of his health – play tennis until he was past 70 years of age.

Michelson was able to view the beauty of art and nature through science. In a paper he wrote, "On the Metallic Coloring in Birds and Insects," he explored the nature of the metallic colors seen in hummingbirds and certain butterflies. Michelson was able to conclude that their iridescent colors were due to the same effects as the colors reflected from thin metallic films.

Michelson's final experiment was on the same topic on which he began his career, the measurement of the velocity of light. By 1926, he had made more than a thousand measurements of the speed of light using a variety of approaches. Even though the results of these experiments were the best that had ever been made, Michelson was not quite satisfied. For his final experiment he developed a completely new apparatus including a mile long, 3-ft. diameter pipe that could be evacuated of air. The results of this experiment were published in the *Astrophysical Journal* in 1935 by Pease and William Pearson, with Michelson listed as the first author. In this paper they describe the need for the new experiment, which provided a reflective light path through the pipe of eight or ten miles under a vacuum of 0.5 to 5.5 mm of mercury. The result was a measure of the speed of light in a vacuum of 299,774 ± 11 km/s.

To mention just two of Michelson's notable honors, and there are many more, he received the Nobel Prize for Physics in 1907 (the first American to do so in science) for his measurements of the speed of light; he also served as the president of the National Academy of Sciences from 1923 to 1927. Michelson's awards include honorary degrees, medals of merit in science, and membership in scientific societies. There is little doubt that he will continue to be remembered as one of the finest experimental physicists.

Scott W. Teare

Selected References

Gale, Henry G. (1931). "Albert A. Michelson." *Astrophysical Journal* 74: 1–9.

Michelson, A. A. (1918). "On the Correction of Optical Surfaces." *Astrophysical Journal* 47: 283–288.

——— (1920). "On the Application of Interference Methods to Astronomical Measurements." *Astrophysical Journal* 51: 257–262.

———(1927). *Studies in Optics*. Chicago: University of Chicago Press.

Middlehurst, Barbara Mary

Born **Penarth, Glamorgan, Wales, 15 September 1915**
Died **Houston, Texas, USA, 6 March 1995**

Welsh astronomer Barbara Middlehurst cataloged extant reports of lunar transient phenomena. The reality of these observed changes on the Moon remains controversial. With **Gerard Kuiper** she also edited two multi-volume compendia of astronomical review articles. *The Solar System* in the 1950s and *Stars and Stellar Systems* in the 1960s.

Selected Reference

Suirles-Jeffreys, Bertha. (1995) *Quarterly Journal of the Royal Astronomical Society*. "Obituary-Barbara Middlehurst". 36: 461–462.

Mikhailov, Aleksandr Aleksandrovich

Born **Morshansk near Tambov, Russia, 26 April 1888**
Died **Pulkovo near Leningrad (Saint Petersburg, Russia), 29 September 1983**

Alexsandr Mikhailov oversaw the reconstruction of the Pulkovo Observatory. For decades after World War II, Alexsandr Mikhailov was the capable and authoritative leader of the Soviet Union's astronomical community. He was the author and editor-in-chief of numerous publications and a participant in many scientific expeditions.

Mikhailov graduated in 1911 from Moscow University with a Gold Medal. He served as an astronomy professor at this university from 1918 to 1947. Concurrently, Mikhailov was a geodesy professor at what is now the Geodetic University. From 1947 to 1964 he served as director of the world-famous Pulkovo Observatory – the Central Astronomical Observatory of the Soviet Academy of Sciences. In 1964 Mikhailov was given the title academician.

Mikhailov was decorated with a plethora of Soviet and foreign awards, including the highest Soviet decoration, the Medal of a Hero of Socialist Labor (1978). Living permanently under the pressure of his high social status, especially during the Stalin regime, he – unlike many others of his rank at that time – nonetheless never volunteered to harm his colleagues, associates, and subordinates in the style of the political situation of the time.

The area of Mikhailov's personal scientific interests was very broad: astrometry (positional astronomy), stellar astronomy, gravimetry and the theory of Earth's figure, the theory of eclipses, and space research involving the Moon. He was an active solar-eclipse observer and designed several new astronomical devices for eclipse study. A great erudite, Mikhailov wrote much on the history of astronomy and was notable as a popularizer of astronomical knowledge.

An amazingly hardworking person who was fluent in several European languages, Mikailov was an illustrious communicator of Soviet science to the international astronomical community. He was elected as a member of many scientific bodies abroad. Mikailov served as a vice president of the International Astronomical Union (1946–1948), and as a vice president of the International Academy of Astronautics (1967–1977).

It was Mikhailov who oversaw the reconstruction of the glorious Pulkovo Observatory, which had been completely destroyed by the Nazis during the siege of Leningrad during World War II. The observatory was successfully restored as a great symbol of Russian science. Unfortunately, decades later Mikhailov himself became an eyewitness to the slow decline of this internationally acclaimed scientific institution.

Mikhailov married Zdeňka Kadlá in 1946. The couple had a son, Georgij Aleksandrovich Mikhailov.

Alexander A. Gurshtein

Selected References

Anon. (1984). "A. A. Mikhailov, 1888–1998. (In Memoriam)" (in Russian). *Astronomicheskii zhurnal* 61: 412–413.

Mikhailov, A. A. (1939). *A Course in Gravimetry and the Theory of the Earth's Figure* (in Russian). 2nd ed. Moscow: Gos. izdatel'stvo techniko-teoreticheskoi literatury.

——— (1945). *The Theory of Eclipses* (in Russian). Moscow: Gos.izdatel'stvo techniko-teoreticheskoi literatury.

Anon. (1986). "Mikhailov, Aleksandr Aleksandrovich." In *Astronomy: Biograficheskii spravochnik* (Astronomers: A biographical handbook), edited by I. G. Kolchinskii, A. A. Korsun', and M. G. Rodriges, pp. 220–221. 2nd ed. Kiev: Naukova dumka.

Pariysky, N. N. (1990). "A Quarter of a Century Close to A. A. Mikhailov" (in Russian). *Istorico-astronomicheskie issledovaniya* (Research in the history of astronomy) 22: 311–331.

Milankovitch [Milanković], Milutin

Born **Dalj, (Croatia), 28 May 1879**
Died **Belgrade, (Serbia), 12 December 1958**

Yugoslavian mathematician and geophysicist Milutin Milankovitch is best known for his presentation of the theory that Earth's paleoclimate resulted from interactions of three long-term astronomical cycles affecting the amount of solar energy received by Earth.

Milankovitch was awarded a doctorate in technical sciences from the Technical High School (later the Institute of Technology) in Vienna in 1904. For the next 5 years, he served as a chief engineer in the construction industry, working with reinforced concrete. In 1909, he was appointed to a professorship in physics and celestial mechanics at the University of Belgrade in Yugoslavia, where he remained until his retirement some forty-six years later. Milankovitch was a member of the Serbian Academy of Sciences and other professional organizations.

Starting around 1912, and for the next thirty years, Milankovitch convincingly demonstrated that variations in (1) the eccentricity of Earth's orbit around the Sun and (2) the obliquity of Earth's axis of rotation, along with (3) the precession of the equinoxes, all contributed to cyclical changes in climate experienced over the past 600,000 years, most notably during the Pleistocene glaciation. Because the three variations have different periods (the best-known is the 26,000 years of precession), they sometimes reinforce and sometimes nearly cancel each other, producing large swings in the amount of sunlight reaching the northern hemisphere. These long-term astronomical and climatological cycles are known as Milankovitch cycles and have gained widespread acceptance.

Jordan D. Marché, II

Selected References

Milankovitch, M. (1969). *Canon of Insolation and the Ice-Age Problem*. Jerusalem: Israel Program for Scientific Translations. (Originally published by the Royal Serbian Academy as *Kanon der Erdbestrahlung und seine Anwendung auf das Eiszeitenproblem*. Belgrade: Mihalia Curcica, 1941.)

Milankovitch, V. (1984). "The Memory of My Father." In *Milankovitch and Climate: Understanding the Response to Astronomical Forcing*, edited by A. Berger *et al.*, pp. xxiii–xxxiv. Dordrecht: D. Reidel.

Miller, John Anthony

Born **Greensburg, Indiana, USA, 16 December 1859**
Died **Wallingford, Pennsylvania, USA, 15 June 1946**

When the Lowell Observatory staff needed help in computing an orbit for the newly discovered planet Pluto, they called upon John Miller, director of the Sproul Observatory. Miller showed that Pluto could not be the "Planet X" proposed by **Percival Lowell**.

Selected Reference

Anon. (1948). "John Anthony Miller." *Monthly Notices of the Royal Astronomical Soceity*. 108: 49.

Millikan, Robert Andrews

Born **Morrison, Illinois, USA, 22 March 1868**
Died **San Marino, California, USA, 19 December 1953**

American physicist Robert Millikan's experiments confirmed that subatomic particles arrive at the Earth from extraterrestrial sources. It was he who coined the term "cosmic rays."

Millikan is best known for measuring the charge of the electron and founding the California Institute of Technology. He was awarded the 1923 Nobel Prize in Physics.

Selected Reference

Kargon, Robert H. (1982). *The Rise of Robert Millikan: Portrait of a Life in American Science*. Ithaca, New York: Cornell University Press.

Millman, Peter Mackenzie

Born **Toronto, Ontario, Canada, 10 August 1906**
Died **Ottawa, Ontario, Canada, 11 December 1990**

Peter Millman developed several pioneering programs to observe meteors and to record their spectra. He was the son of Robert M. Millman and Edith Middleton. In 1931, Millman married Margaret B. Gray with whom he had two children, Barry and Cynthia.

Millman spent most of his youth in Japan, where his parents were missionaries. His interest in astronomy developed during his secondary education at the Canadian Academy in Kobe. Millman returned to Canada to enter the University of Toronto, where he took his B.A. in astronomy with the Royal Astronomical Society of Canada Gold Medal in 1929. As an undergraduate, he worked as a summer student at the Dominion Astrophysical Observatory, Ottawa. During his graduate training at Harvard Observatory, **Harlow Shapley** suggested Millman analyze meteor spectra in the observatory's collection. After taking his M.A. (1931) and Ph.D. (1932), he remained another year at Harvard as an Agassiz Fellow.

In 1933, Millman joined the Astronomy Department at the University of Toronto. With the opening of the David Dunlap Observatory in 1935, he participated in the routine radial velocity program but also initiated visual meteor observations and developed a meteor spectra program. Millman joined the Royal Canadian Air Force in 1941; after teaching navigation to pilots, he moved to London to work in operational research. On his return in 1946, he moved to the Dominion Observatory, where he became chief of the Stellar Physics Division and developed a visual meteor program.

From the late 1940s, Millman worked with D. W. R. McKinley of the National Research Council [NRC] of Canada on visual and radar tracking of meteors by triangulation. In the early 1950s, two stations with super-Schmidt cameras were erected in Alberta in a joint United States–Canadian project to study the upper atmosphere. Millman's meteor spectrographs in Alberta and Ottawa produced a steady stream of data. In 1955, Millman joined the NRC as head of its Upper Atmosphere Research Section.

Millman was active in the Royal Astronomical Society of Canada (president: 1960–1962), president of the Meteoritical Society (1962–1966), president of Commission 22 of the International Astronomical Union [IAU] (1964–67), chair of the IAU Working Group for Planetary System Nomenclature (1973–1982), and first secretary of the Canadian Astronomical Society (1971–1977). He was also a fellow of the Royal Society of Canada and a councillor of the Smithsonian Institution. Millman was awarded the J. Lawrence Smith Medal of the United States National Academy of Sciences in

1954, awarded the Gold Medal of the Czechoslovak Academy of Science (1980), and a minor planet (2904) was named for him.

Millman was the world's leading authority on meteor spectra, being the first to undertake systematic spectroscopic studies. Beginning in his Toronto days, he organized visual meteor programs. After World War II, radar and photographic recording of meteor showers were added. During the International Geophysical Year, this became a North America-wide project, providing solid statistics on meteors. Millman was a pioneer in using aircraft for observing meteors and solar eclipses. He collaborated with **Carlyle Beals** on meteorite impact structures, and later devised a program for meteorite recovery.

Richard A. Jarrell

Selected References

Halliday, Ian (1991). "Peter Mackenzie Millman, 1906–1990." *Journal of the Royal Astronomical Society of Canada* 85: 67–78.

Jarrell, Richard A. (1988). *The Cold Light of Dawn: A History of Canadian Astronomy*. Toronto: University of Toronto Press.

Millman, Peter M. (1980). "The Herschel Dynasty–Part I: William Herschel." *Journal of the Royal Astronomical Society of Canada* 74: 134–146.

——— (1980). "The Herschel Dynasty–Part II: John Herschel." *Journal of the Royal Astronomical Society of Canada* 74: 203–215.

Millman, Peter M. and D. W. R. McKinley (1967). "Stars Fall Over Canada." *Journal of the Royal Astronomical Society of Canada* 61: 277–294.

Milne, Edward Arthur

Born **Hull, England, 14 February 1896**
Died **Dublin, Ireland, 21 September 1950**

British mathematician Edward Milne contributed many of the ideas that have made it possible to analyze the spectra of stars and determine the temperatures, densities, and chemical compositions of their atmospheres, some of those ideas carrying his name. However, he is perhaps more often remembered for a unique cosmological model in which gravitation and electromagnetism followed two different kinds of time.

Milne was the eldest of three sons (all eventually scientists) of a headmaster of a Church of England school, Sidney Milne, and a teacher, Edith Cockcroft. After completion of studies at Hymers College in Hull, Milne won a scholarship at Trinity College, apparently having achieved the highest score to date on the entrance examination. His eyesight made him ineligible for active duty in World War I, but a year and a half after beginning work (in 1914) at Cambridge, he withdrew to work on antiaircraft ballistics research for the duration of the war. Shortly after returning in 1919, Milne was elected a fellow of Trinity College (precluding the need for an advanced degree) and, in 1920, became assistant director of the Solar Physics Observatory under Hugh Frank Newall, the founder of astrophysics at Cambridge. Milne was appointed to the Beyer Professorship of Applied Mathematics at the University of Manchester in 1925 and, in 1929, became the first Rouse Ball Professor of Mathematics at Oxford University, where he and **Harry Plaskett** founded another school of astrophysics.

Milne's work of greatest lasting importance was done at Cambridge and Manchester. He showed that a star in which energy was transported by radiation could not rotate like a solid body, and that the pressure of radiation on atoms with very strong absorption lines (like the violet pair of ionized calcium) would be enough to lift atoms off the surface of a star into its chromosphere. Such "line driving" is now recognized as the cause of winds from cool, bright stars. Working with **Ralph Fowler,** Milne showed that these strong lines are produced very high in the atmospheres of the Sun and stars, and used the Saha equation to calculate the temperature at which lines would be strongest. This, in turn, led to a theoretical explanation of why some spectral features of ionized atoms look stronger in giants than in dwarfs, providing the physical underpinning of stellar luminosity criteria, including an explanation of what the light coming from the Sun ought to look. He pioneered the idea of detailed balance in stellar atmospheres (the idea that the number of transitions between a pair of levels going up and down must have a fixed ratio), leading to what were then called the Milne relations, and derived a form of the equation describing how radiation propagates through stellar gas that also carries his name. His approximation for the opacity of gas to that propagation is still used in calculations to show what the dominant effects must be in the appearance of stellar spectra. And the Milne–Eddington approximation describes absorption features and the stellar continuum as it is being formed together in all layers of the atmosphere.

At Oxford, Milne turned his attention to cosmology, as did many astronomically inclined British mathematicians, in light of **Edwin Hubble**'s discovery of the redshift–distance relation and the increasing familiarity of **Albert Einstein**'s general theory of relativity. Milne and **William McCrea** in 1934 published a set of Newtonian cosmological models that had many of the features of the relativistic ones, but were much easier for most people to understand and are still used as analogies in modern discussions of cosmology. But Milne felt that his most important contribution was what he called kinematic relativity. He started with homogeneity and isotropy (so that every fundamental observer would see not only the same physics but the same history of the Universe) as a basic assumption, not as an observation. Milne believed that he could follow this assumption to a single-model universe that would be the only self-consistent possibility, requiring certain laws for gravitation and so forth. This was not published in final form until his 1948 book, *Kinematic Relativity*. By then he had also incorporated different timescales for gravitational and electromagnetic time, the former a natural logarithm of the latter. Others have returned from time-to-time to these ideas, but so far without much impact on understanding of either the structure of the Universe or the physics in it.

Also at about the time he went to Oxford University, Milne suggested that nova explosions might be caused by the collapse of a normal star to a white dwarf. They are in fact explosions on the surfaces of white dwarfs with close companions, but the suggestion meant that the idea of collapse as a source of rapid energy release was "in the air" when **Walter Baade** and **Fritz Zwicky** put forward the (correct) idea of powering supernovae with collapses to neutron stars.

Milne was elected to the Royal Society (London) in 1926, and received medals from the Royal Astronomical Society (London), the Astronomical Society of the Pacific, and other scholarly organizations. He served as president of the London Maths Society (1937–1939) and suffered a fatal heart attack during a meeting of the Royal Astronomical Society, which he had served as president from 1943 to 1945.

Douglas Scott

Selected References

Bertotti, B. *et al.* (1990). *Modern Cosmology in Retrospect*. Cambridge: Cambridge University Press. (See G. F. R. Ellis, "Innovation, Resistance and Change," pp. 97–113 and William McCrea, "Personal Recollections," pp. 197–219.)

McCrea, W. H. (1951). "Edward Arthur Milne." *Obituary Notices of the Royal Society of London* 7: 421–443.

Milne, E. A. (1930). "*Thermodynamics of the Stars*." In *Grundlagen der Astrophysik*, pp. 62–255. Vol. 3, *Handbuch der Astrophysik*. Berlin: J. Springer.

——— (1932). *The White Dwarf Stars*. Oxford: Clarendon Press.

——— (1935). *Relativity, Gravitation and World-Structure*. Oxford: Clarendon Press.

——— (1948). *Kinematic Relativity*. Oxford: Clarendon Press.

——— (1952). *Modern Cosmology and the Christian Idea of God*. Oxford: Clarendon Press.

Milton, John

Born **London, England, 9 December 1608**
Died **London, England, 8 November 1674**

Although in no technical sense an astronomer, John Milton, the greatest epic poet of the English language, engaged the new world picture of **Nicolaus Copernicus**, **Galileo Galilei**, and **Johannes Kepler** and imaginatively stretched the boundaries of the canvas upon which it was painted.

Milton's birth coincided roughly with the invention of the telescope. The son of a well-off middle-class Puritan father, he received a rich religious and classical education at Saint Paul's School in London and at Cambridge University. After leaving university, Milton pursued a further period of personal study, and in 1638 embarked on an almost 2-year Grand Tour; this took him to Italy, where he was granted an audience with the aging Galilei. This encounter, in addition to stimulating his awareness of astronomical issues, reinforced Milton's sense of identity as a free Englishman and a Protestant: "I found and visited the famous *Galileo* grown old, a prisoner to the Inquisition, for thinking in Astronomy otherwise than the Franciscan and Dominican licencers thought" (*Areopagitica*, 1644; in Flannagan, p. 1014b).

In the 1640s and 1650s, Milton was an active defender of the antiroyalist position in the English Civil War and the Commonwealth under Oliver Cromwell, writing the official defense of the 1649 execution of King Charles I (*Defensio pro populo Anglicano*, 1651). Soon thereafter, Milton became completely blind, and some of his enemies saw the latter state as divine punishment for his regicidal propaganda. After the restoration of the monarchy in 1660, Milton narrowly escaped with his life, and the rest of his years were spent in relative seclusion. It was principally in this period, from 1660 until his death in 1674, that he produced the greatest of his poetic works, the epic *Paradise Lost* (1667), and *Paradise Regained* and *Samson Agonistes* (published together in 1671).

The narrative of *Paradise Lost* encompasses not only the fall of humankind into sin but also extensive probing of astronomical issues. In Milton's exercise of cosmological imagination, Galilei's influence can be traced in a number of ways. Galilei is the only contemporary of Milton's to be explicitly named in *Paradise Lost*, and his discoveries provided Milton with an abundant store of both poetic and scientific material. For example, Satan's shield is described as being

> like the moon, whose orb
> Through optic glass the Tuscan artist views
> At evening from the top of Fesole,
> Or in Valdarno, to descry new lands,
> Rivers or mountains in her spotty globe. (*PL* 1.287–1.291.)

Milton thus not only alludes to Galilei's telescopic examination (in *Sidereus Nuncius*) of the Moon's surface, but also uses the simile to suggest that Satan's armor, when viewed up close, may, like the Moon, appear less perfect and "celestial" than one had previously thought. Milton again references Galilei's telescopic discoveries – and again associates Satan with the word "spot," implying blemish – in calling the fiend, in his solar journey, "a spot like which perhaps / Astronomer in the sun's lucent orb/Through his glazed optic tube yet never saw" (*PL* 3.588–3.590). Milton nods in the direction of yet a further discovery of Galilei's – that the Milky Way consists of multitudinous stars – when the angel Raphael, speaking to Adam, the first man, refers to "the Galaxy, that Milky Way/Which nightly as a circling zone thou seest/Powdered with stars" (*PL* 7.579–7.581).

Because of suggestions in *Paradise Lost* that the Earth may be in the center of the world, Milton's astronomical position is often wrongly thought to be basically Ptolemaic. However, Milton undermines this premature conclusion in two main ways. One includes the Copernican references and vocabulary that he and his authoritative characters (such as Raphael) employ, even if these appear in the context of similes or surmises. Speaking to Adam, Raphael asks, "What if," in addition to the six planets, "seventh to these/The planet earth, so steadfast though she seem,/Insensibly three different motions move?" (*PL* 8.128–8.130). Earth's status as a planet and its three motions are clearly Copernican, as is Raphael's further reference to Earth "as a star/Enlight'ning her [*i. e.*, the Moon] by day, as she by night/This earth" (*PL* 8.142–8.144).

A final feature of the narrative of *Paradise Lost* that places it in the tradition of Copernicans such as Kepler and **John Wilkins** is its envisaging of the transport of living bodies across astronomical space. As Robert Burton had pointed out in 1638, the dissolution of the Aristotelian "hard and impenetrable" crystalline spheres opened up the prospect of space travel: "If the heavens then be penetrable, … it were not amiss in this aerial progress to make wings and fly up" (*The Anatomy of Melancholy*, Part 2, p. 50). In *Paradise Lost* Satan does just this, though his flight is not upwards from Earth toward the stars, but down through the stars toward the Sun and the Earth. From the rim of the Universe, downward Satan "throws"

> His flight precipitant, and winds with ease
> Through the pure marble air his oblique way
> Amongst innumerable stars, that shone
> Stars distant, but nigh hand seemed other worlds. (*PL* 3.563–3.566.)

J. Tanner has called *Paradise Lost* "perhaps the greatest description of space travel in highbrow fiction." Yet what enabled Milton to produce such a description – and so to contribute to the genre of science fiction, then in its infancy – was his immersion in the cosmological ferment of his day.

Much more cosmologically radical than these generally Copernican themes, however, was Milton's imaginative depiction of our universe of stars, Sun, and planets as itself a mere speck within an immensely larger, perhaps infinite, physical "creation" dominated by the unformed matter of chaos. In Milton's extracosmic chaos, even normal categories of time and space do not apply. It is

a dark
Illimitable Ocean without bound,
Without dimension, where length, breadth, and highth,
And time and place are lost. (*PL* 2.891–2.894.)

Moreover, these dark, chaotic materials, which are nevertheless of divine origin, may be, or may become, the stuff of new or other "worlds" (*PL* 2.916).

In any case, in the picture presented in *Paradise Lost*, the entire known stellar Universe – regardless whether its internal structural features be Ptolemaic or Copernican – is enclosed in a sphere that hangs down from the walls of heaven. And when Satan, on his journey across chaos, beholds this Universe from afar, it appears as a mere speck, "in bigness as a star/Of smallest magnitude close by the moon" (*PL* 2.1052–2.1053). Such, in the decades just before the advent of **Isaac Newton**'s cosmology, was Milton's imaginative poetic expansion of cosmography outward toward the infinite.

Dennis Danielson

Selected References

Burton, Robert (1638). *The Anatomy of Melancholy*. London. (Reprint: London: J. C. Nimmo, 1886.)
Flannagan, Roy (1998). *The Riverside Milton*. Boston: Houghton Mifflin.
Galilei, Galileo (1989). *Sidereus Nuncius, or The Sidereal Messenger*, translated by Albert van Helden. Chicago: University of Chicago Press.
Milton, John (2000). *Paradise Lost*, edited by John Leonard. London: Penguin.
Nicolson, Marjorie (1956). *Science and Imagination*. Ithaca, New York: Cornell University Press. Chap. 4, "Milton and the Telescope," was first published in *English Literary History* 2 (1935): 1–32.
Tanner, John S. (1989). "And Every Star Perhaps a World of Destined Habitation: Milton and Moonmen." *Extrapolation* 30, no. 3: 267–279.

Mineur, Henri Paul

Born **Lille, Nord, France, 7 March 1899**
Died **Paris, France, 7 May 1954**

Mathematician Henri Mineur played an important role in the development of astrophysics in France. His 1944 recalibration of the zero-point of the Cepheid period–luminosity relationship was unnoticed at the time but presaged the later work of **Walter Baade**. Mineur was the son of Paul Mineur, a mathematics teacher who taught in Lille and later at the Lycée Rollin in Paris, and Léonie Jacquet. Henri was a pupil at the Lycée Rollin along with **Jean Dufay**, who became director of the Lyon Observatory and Haute-Provence Observatory. At 18, Mineur received entry to the École Polytechnique in a good rank, and entry at first rank to the École Normale Supérieure. He chose the latter, but with France engaged in World War I, he went instead into the army. At the war's end in 1919, Mineur returned to his studies, and passed the examination (*l'agrégation*) to teach secondary-school mathematics 2 years later. After passing his doctoral thesis (on the analytic theory of continuous finite groups) in 1924, he spent the following year teaching mathematics at the Lycée Français in Düsseldorf, Germany. In 1925, he took a position as an *astronome adjoint* at the Paris Observatory, directed then by **Benjamin Baillaud**. Mineur was already familiar with the observatory, having served there in 1922/1923 as a trainee under **Paul Couderc**.

After serving various departments at the Paris Observatory, Mineur became involved in the *Carte du Ciel* project, launched in 1887 by **David Gill** and **Ernest Mouchez**. Mineur also researched stellar statistics, a topic that influenced his astronomical career and developed his skills in various domains of astronomy. He earned good reports about his research from others at Paris, including from **Ernest Esclangon**, director at that time.

In 1931, Mineur directed the work of several students, including Li Heng (1898–1989), whose work led Mineur to study stellar fields and clusters. Li Heng worked on the determination of the zero-point of the period–luminosity relationship for Cepheid variables, studied their spatial and velocity distributions, and in 1932 submitted a dissertation on his statistical researches on Cepheids. After Li Heng's departure from France, Mineur pursued and developed this line of research; in 1944, he published a paper giving corrections for the zero-point for classical Cepheids and for RR Lyrae variables. He also noted in a later publication that this result served to multiply the distances to such stars by a factor close to 1.8.

The elections of 1936 in France had given power to the Popular Front, one of whose innovations related to scientific research and led to the appointment of both Irène Joliot-Curie and Jean Baptiste Perrin as Undersecretary of State for Scientific Research. This subsequently led to the creation of a service for research in astrophysics, which comprised the Haute-Provence Observatory (near Manosque in the south of France) and an associated laboratory in Paris. The observing station and the laboratory would later be attached to the Centre National de la Recherche Scientifique [CNRS], established in October 1939. Mineur played a large role in this effort along with **Daniel Chalonge**, an *aide-astronome* at the Paris Observatory after 1933, whom Mineur had previously known as an assistant in the Sorbonne. Mineur later obtained the positions of general secretary of the service and director of the laboratory. Mineur was removed from his position in 1941 by the Vichy government, while he was engaged in the resistance; the laboratory was removed to several different locations. A part of the domain of the Paris Observatory was allotted to it. After World War II, a new laboratory building, constructed in 1952, hosted the newly named Institut d'Astrophysique de Paris. Mineur, still an assistant astronomer at the Paris Observatory, served as its director until his death. He played a large part in centralizing this hitherto widely dispersed astrophysical research, giving France an important role within it.

Mineur, an astrophysicist concerned with the Milky Way, its constituents, and stellar absorption, was also a mathematician who studied relativity, celestial mechanics, and pure mathematics. He was especially engaged with general problems relating to the treatment of data, and wrote a clear, well-written, and widely used book on the technique of least squares. He trained many students, and his courses in stellar astronomy and lectures on the expansion of the Universe enjoyed large audiences.

Mineur was a talented popularizer, helping amateur astronomers mostly through the articles he published in *L'Astronomie,* the magazine of the Société Astronomique de France, which he had joined in his youth. In declining health by 1952, Mineur attended a meeting in Rome during which Walter Baade announced and **Andrew Thackeray** confirmed, that the calculated distances in the Universe had to be multiplied by a factor of 2; on that occasion, **Clabon Allen** recalled Mineur's 1944 paper that had presented a similar result, also from consideration of Cepheid and RR Lyrae variables. Mineur's early death was a deep loss within the ranks of his generation of researchers.

Mineur married twice, to Suzanne Fromant in 1926 and to Gabrielle Cloche in 1929.

Jacques Lévy

Selected Reference

Mineur, Henri (1938). *Technique de la méthode des moindres carrés.* Paris: Gauthier-Villars.

Minkowski, Hermann

Born **Aleksotas, Russia (now Kaunas, Lithuania), 22 June 1864**
Died **Göttingen, Lower Saxony, Germany, 12 January 1909**

German mathematician Hermann Minkowski introduced the concept of "space–time." The Minkowski metric or geometry is the one appropriate to **Albert Einstein's** special theory of relativity.

Selected Reference

Rosenfeld, B. A. (1988). *A History of Non-Euclidean Geometry: Evolution of the Concept of a Geometric Space.* New York: Springer-Verlag.

Minkowski, Rudolph Leo Bernhard

Born **Strasbourg, (Bas-Rhin, France), 28 May 1895**
Died **Berkeley, California, USA, 4 January 1976**

Rudolph Minkowski made significant contributions to the understanding of gaseous nebulae, was codiscoverer of the two principal types of supernovae, and participated in the early identification of radio sources with their optical counterparts. Son of Oskar Minkowski, a professor of pathology, Minkowski attended German schools in Cologne, Greifswald, and Breslau. He began to study physics at the University of Breslau (1913) and hoped to go to Berlin the following year. These plans were disrupted by war; from 1914 to 1918, Minkowski served in the German army. After that conflict ended, he studied in Berlin, then returned to Breslau, and completed his doctoral thesis under the supervision of Rudolf Ladenburg (1921). After working at Göttingen University for a year with James Franck and Max Born, Minkowski relocated to Hamburg in 1922 and taught at University of Hamburg until 1935, when he was forced to immigrate to the United States.

In 1933, Adolf Hitler seized power in Germany; his National Socialist Party subsequently forbade persons from "non-Aryan" backgrounds to retain employment in official places like universities. In 1935, Minkowski lost his title as professor and was no longer allowed to teach. His father-in-law, judge Alfons David, had been dismissed from the court at Leipzig in 1933. But through the assistance of **Walter Baade**, who had left Hamburg for the Mount Wilson Observatory in 1931, Minkowski first secured a 1-year appointment at Baade's new institution. His position on the Mount Wilson staff was made permanent after his formal dismissal (1936) was received from the University of Hamburg. Thus, from 1935 until 1960, Minkowski worked as a research astronomer at the Mount Wilson and Palomar observatories and (after compulsory retirement) from 1961 to 1965 at the Radio Astronomy Laboratory of the University of California in Berkeley. In 1926, Minkowski married Luise Amalie David; the couple had two children.

Minkowski's research career can be divided into two phases. Prior to his emigration, he had worked chiefly on laboratory spectroscopic problems. His principal subject of investigation was the width of spectral lines, and how they were broadened by pressure and self-absorption. Minkowski published related papers on the behavior of electrons in metallic vapors and the process by which electrons pass through atoms (with Hertha Sponer). His final Hamburg paper (with **Hermann Brück**) described the atomic-beam method of determining the fine structure of spectral lines.

Yet, as early as 1933, Minkowski had turned his attention towards astrophysical problems, including features observed in the spectrum of the Orion Nebula. Upon settling in the United States, his knowledge of spectroscopy proved most useful for studying a variety of astronomical objects. Minkowski's collaboration with Baade led to a rapidly growing number of publications. These included studies on the internal motions of gaseous nebulae and the discovery of almost 200 new planetary nebulae. He and Baade also collaborated on the study of supernovae in other galaxies and supernova remnants in our Milky Way. Minkowski and Baade distinguished supernovae of two types (I and II), based upon their light curves and the presence/absence of hydrogen in their spectra. Their classification scheme also proved useful as a tool for estimating cosmological distances in space (supernovae as "standard candles"). Shortly after the Crab Nebula was recognized as a supernova remnant, Minkowski and Baade identified its small central star (later found to be a neutron star/pulsar). Minkowski also worked on the distribution of emission nebulae in our Galaxy and on the spectral features of comets.

After 1949, Minkowski became intrigued with the new field of radio astronomy. Together with Baade, he began to locate the optical counterparts of newly discovered radio sources. One of the first extragalactic counterparts identified (1954) was the radio source Cygnus A. Minkowski likewise studied the distribution of galaxies and in 1960 identified the galaxy (3C 295) having the highest known redshift (Z=0.45) at the time (prior to the discovery of quasars).

Minkowski was also responsible for the wide-field photographic sky survey conducted by the National Geographic Society and known today as the "Palomar Observatory Sky Survey" [POSS]. This work was done with the 48-in. Schmidt camera on Palomar Mountain and covered the Northern Celestial hemisphere down to a declination of −33°. The POSS has proven to be an invaluable tool for countless astronomers.

Minkowski was chosen a member of the Royal Astronomical Society and the United States National Academy of Sciences (1951). He was awarded the Bruce Gold Medal of the Astronomical Society of the Pacific in 1961 and an honorary doctorate from the University of California at Berkeley in 1968.

Ian T. Durham

Selected References

Greenstein, Jesse L. (1984). "Optical and Radio Astronomers in the Early Years." In *The Early Years of Radio Astronomy: Reflections Fifty Years after Jansky's Discovery*, edited by W. T. Sullivan, III, pp. 67–81. Cambridge: Cambridge University Press.

Nicholson, Seth B. (1961). "Award of the Bruce Gold Medal to Dr. Rudolph Minkowski." *Publications of the Astronomical Society of the Pacific* 73: 85–87.

Osterbrock, Donald E. (2001). *Walter Baade: A Life in Astrophysics*. Princeton, New Jersey: Princeton University Press.

Minnaert, Marcel Gilles Jozef

Born **Bruges, Belgium, 12 February 1893**
Died **Utrecht, the Netherlands, 26 October 1970**

Marcel Minnaert pioneered new techniques in the measurement and understanding of spectral line strengths and chemical abundances in the Sun and stars. Minnaert was the son of Jozef and Jozephina (*née* Van Overberge) Minnaert. Both of his parents were teachers at normal schools. Ever since his birth, Minnaert's father meticulously kept a diary on his only child's education, which provides a clear insight into the intellectual development of young Minnaert. Just before his father's premature death in 1903, Minnaert and his parents relocated to Ghent. In 1910, Minnaert enrolled at the University of Ghent, where he studied biology. Four years later, he was awarded his doctoral degree after completing a thesis on the effects of light on plants, entitled "Contributions à la étude de la photobiologie quantitative."

During his studies, Minnaert associated himself with Flemish students, who wished to change the language of instruction at the university from French to Dutch. This measure was introduced during the German occupation of Belgium during World War I and resulted in an urgent need for teachers at the Dutch-language university. Minnaert then went to Leiden, the Netherlands in 1915 to take up physics. He attended the lectures of leading scientists, especially **Hendrik Lorentz** and Paul Ehrenfest. After his return to Ghent in 1916, Minnaert was appointed as an associate professor of experimental physics.

In 1918, Minnaert had to flee his country, to escape prosecution by the Belgian government. He had been one of the language activists at the University of Ghent and was now accused of collaboration with the Germans. Afterwards, he was sentenced *in absentia* to 15 years of penal servitude. Minnaert moved again with his mother to Utrecht, the Netherlands. Thereafter, his main interest became photometry, or the precise measurements of light intensity. He secured a position at the Heliophysical Institute of the University of Utrecht, which, under the direction of Willem Julius, was engaged in developing new photometric techniques related to solar spectroscopy. Minnaert became an observer at the institute; by 1925 he obtained a second doctoral degree (in physics), having written a thesis on anomalous dispersion, entitled "Onregelmatige Straalkromming" (Irregular ray curvature).

Minnaert applied himself to research in solar physics, especially the formation of the Fraunhofer absorption lines in the solar spectrum. Following **Albertus Nijland**'s death in 1936, Minnaert was chosen his successor in 1937 as professor of astronomy and director of the University of Utrecht Observatory. He converted the observatory into a prominent astrophysical institute for research in solar and stellar spectra.

Minnaert's chief contributions lay in his foundation of the "curve of growth" technique in determining solar chemical abundances by spectral analysis. Along with his concept of "equivalent width" (introduced simultaneously by Harald von Kluber in 1927), this methodology established Minnaert's reputation as an outstanding astronomer. The intensities of spectral lines originate from several different broadening mechanisms, which allows them to exhibit characteristic profiles, with a residual intensity being left at the center. Minnaert introduced the measurement of line intensities in terms of the fraction of energy removed from the adjacent continuum by absorption, a quantity which he expressed as the "equivalent width" of a rectangular (*i. e.*, fictitious) absorption feature.

In 1929, Minnaert investigated how the equivalent width increases with the number of absorbing atoms. By first calibrating **Henry Rowland**'s spectral scale in terms of equivalent width, and then applying an earlier technique of **Henry Norris Russell**, he obtained an empirical result by plotting the equivalent width against the number of absorbing atoms. This relationship was called by Minnaert the "curve of growth," a term reminiscent of his earlier biological studies. Following Wilhelm Schütz's theoretical extension of the "curve of growth" in 1930, Minnaert and his students were able to analyze solar equivalent widths, which yielded explanations for the broadening of solar spectral lines and renewed theoretical understanding of line formation.

By the late 1930s, Minnaert's love of nature and his ability to popularize physics were combined into the three-volume work *De natuurkunde van 't vrije veld* (Physics of the outdoors), which appeared from 1937 to 1940. At the same time, his extensive research on solar physics, in collaboration with Jacob Houtgast and Gerard F. W. Mulders, culminated in publication of the voluminous *Photometric Atlas of the Solar Spectrum* (1940). For several decades, this proved to be a standard reference work. During World War II, Minnaert was imprisoned by the Germans for his vocal left-wing opposition to fascism.

After World War II, Minnaert investigated a variety of subjects, including the temperatures of cometary nuclei, gaseous nebulae, the atmospheric homogeneity of Venus, and lunar photometry. For over 20 years, he revised the photometric atlas of the solar spectrum, initially with a several of young astronomers, later with Houtgast, and in the final stages, with **Charlotte Moore-Sitterly**. This publication, *The Solar Spectrum*, was completed in 1966. Minnaert officially retired in 1963, but remained active as a member of the International Astronomical Union [IAU] commission that established the nomenclature of features on the Moon's farside. He also published a laboratory manual, *Practical Work in Elementary Astronomy* (1969).

During his career, Minnaert received honorary degrees from the universities of Heidelberg, Moscow, and Nice. He was a member of various academies, among which are the Koninklijke Nederlandse Akademie van Wetenschappen, the Koninklijke Academie voor Wetenschappen, and the Letteren en Schone Kunsten van België. For his research in solar and stellar photospectrometry, Minnaert received several international awards, the Gold Medal of the Royal Astronomical Society in 1947 and, in 1951, the Bruce Medal of the Astronomical Society of the Pacific. His engagement in cooperative astronomical research enabled Minnaert to hold the positions of president and vice president of several commissions within the IAU.

In 1928, Minneart married Maria Boergonje Coelingh; the couple had two sons.

Steven M. van Roode

Selected References

de Jager, Cornelius (1974). "Minnaert, Marcel Gilles Jozef." In *Dictionary of Scientific Biography*, edited by Charles Coulston Gillispie. Vol. 9, pp. 414–416. New York: Charles Scribner's Sons.

Hearnshaw, J. B. (1986). *The Analysis of Starlight: One Hundred and Fifty Years of Astronomical Spectroscopy*. Cambridge: Cambridge University Press, esp. pp. 236–240.

Minnaert, M. *De natuurkunde van 't vrije veld*. Zutphen: Thieme and Cie, 1937–1940. (The first volume was translated as *The Nature of Light and Colour in the Open Air*. New York: Dover, 1954; republished as *Light and Color in the Outdoors*. New York: Springer-Verlag, 1993.)

——— (1965). "Fourty [sic] Years of Solar Spectroscopy." In *The Solar Spectrum: Proceedings of the Symposium, Held at the University of Utrecht, August 26–31, 1963*, edited by Cornelius de Jager, pp. 3–25. Dordrecht: D. Reidel.

Minnaert, M., G. F. W. Mulders, and J. Houtgast (1940). *Photometric Atlas of the Solar Spectrum from λ 3612 to λ 8771, with an Appendix from λ 3332 to λ 3637*. Utrecht: Sterrewacht Sonneborgh.

Molenaar, Leo (2003). *Marcel Minnaert Astrofysicus*. Amsterdam: Uitgeverij Balans.

Moore, Charlotte E., M. G. J. Minnaert, and J. Houtgast (1966). *The Solar Spectrum 2935 Å to 8770 Å*. National Bureau of Standards Monograph. Washington, D.C.: U.S. Government Printing Office.

Mīram Čelebī: Maḥmūd ibn Quṭb al-Dīn Muḥammad ibn Muḥammad ibn Mūsā Qāḍīzāde

Born **Istanbul, (Turkey), 1475**
Died **Edirne, (Turkey), 1525**

Mīram Čelebī, one of the most important Ottoman mathematicians and astronomers, attempted to reconcile the mathematical (Ptolemaic) and natural philosophical (Aristotelian) traditions concerning astronomy, while writing astronomical texts that were widely used in the Ottoman Empire.

Mīram Čelebī's grandfather Muḥammad was **Qāḍīzāde**'s son; he married **ʿAlī Qūshjī**'s eldest daughter in Samarqand. His father, the scholar Quṭb al-Dīn Muḥammad, came with his grandfather ʿAlī Qūshjī to Istanbul, where Quṭb al-Dīn married Mīram Čelebī's mother, who was the daughter of Khōja-zāde, a famous scholar and philosopher of that time. His father, who had been a teacher at the Manāstır *madrasa* (school) in Bursa, died at a young age, and so Mīram Čelebī was raised by his grandfather Khōja-zāde. Mīram was educated not only by his grandfather but also by other leading scholars of the time such as Sinān Pasha. Upon his graduation, he taught at several *madrasas* (the Gelibolu, the Edirne ʿAlī Bey, and the Bursa Manāstır), becoming the most prominent figure of his time in the mathematical sciences. Indeed Sultan Bāyazīd II (died: 1512) asked him to be his teacher. Mīram Čelebī was appointed as Qāḍī ʿaskar (a high official in the Ottoman judiciary) of Anatolia during the reign of Yavuz Sultan Selīm I (reigned: 1512–1520); however, shortly thereafter he was dismissed from his post and retired. Towards the end of his life, he went on the pilgrimage to Mecca; upon his return he settled in Edirne. He was buried in the courtyard of the Qāsīm Pasha Mosque.

Mīram Čelebī, most famous for his many works in astronomy, optics, and astrology, was also well known in the fields of history and literature. (He even wrote an important work on hunting.) He wrote in Arabic and Persian (the scientific languages of his time) as well as in Turkish. Among his many students were **Muṣṭafā ibn ʿAlī al-Muwaqqit** and the famous philosopher and historian Ṭashköprülüzāde.

Mīram Čelebī inherited the scientific tradition of the Samarqand School of mathematics and astronomy represented by his great-grandfathers Qāḍīzāde and ʿAlī Qūshjī. He was also greatly influenced by **Ibn al-Haytham**'s methodology in the field of optics (*ʿilm al-manāẓir*) and tended to favor his approach of combining mathematics and natural philosophy over the more mathematical approach of both great-grandparents. In addition, Mīram Čelebī was well informed of the opinions of Kamāl al-Dīn al-Fārisī, **Ibn Sīnā**, and **Fakhr al-Dīn al-Rāzī**, among others.

One of Mīram Čelebī's most important astronomical works is his commentary on the Persian *Zīj-i Ulugh Beg*, also known as *Dustūr al-ʿamal fī taṣḥīḥ al-jadwal*, which was completed in 1499 and dedicated to Sultan Bāyazīd II. Mīram Čelebī incorporated findings from **Jamshīd al-Kāshī**'s *Zīj-i Khāqānī* and ʿAlī Qūshjī's *Sharḥ Zīj-i Ulugh Beg*. The work, written in a didactic style, provided five examples of solutions for calculating the sine of 1°. More than 30 extant copies of the *Dustūr* attest to its widespread use by Ottoman astronomers. Mīram Čelebī's mathematical bent is also indicated by a work in which he calculated the ratio of the highest mountain in the world to the diameter of the Earth, a problem going back to **Naṣīr al-Dīn al-Ṭūsī**.

The most noteworthy work written by Mīram Čelebī on the subject of theoretical astronomy is a commentary on ʿAlī Qūshjī's work *al-Fatḥiyya fī ʿilm al-hayʾa*. Unlike his great-grandfather, who sought to eliminate Aristotelian natural philosophy from astronomy, Mīram Čelebī sought to reconcile the mathematical and the natural philosophical in astronomy as he had done in optics. He completed it in 1519 following the request of many of Mīram Čelebī's students when he was teaching *al-Fatḥiyya*. The commentary was both practical and theoretical and was used as a supplementary textbook in the Ottoman *madrasas*. Mīram Čelebī stated his intention to write an appendix to his commentary in which he would analyze the problems pertaining to the models of Mercury and the Moon. Although there is no extant copy of this appendix, it is an indication of the importance of the subject as well as an example of a continuous astronomical tradition to solve difficulties related to **Ptolemy**'s planetary models.

Many of Mīram Čelebī's other astronomical works deal with instruments, including a variety of quadrants. His *Risāla dar Shakkāzī wa Zarqāla az ālāt-i raṣadiyya* (in Persian) examines two astronomical instruments invented by **Zarqālī** and their use in astronomical observations. He also wrote on the calendar, the determination of the direction to Mecca (*qibla*), and various other astronomical problems. His *Risāla fī samt al-qibla* is a comprehensive study on the determination of the *qibla* using astronomical and mathematical calculations. Moreover, in accordance with the tendencies of his time, he wrote original works in the field of astrology, such as *al-Maqāsid fī al-ikhtiyārāt* and *Masāʾil-i Mīram Čelebī* (in Turkish).

Throughout his work, Mīram Čelebī placed great importance on rational and empirical evidence for the subjects he investigated. His work in theoretical astronomy was an extension of the Samarqand tradition that his great-grandfather ʿAlī Qūshjī continued with his colleagues and students in Istanbul. Mīram Čelebī especially enriched its mathematical character. His relationships with other members of the Samarqand School who came to Istanbul (such as Sayyid Munajjim and **ʿAbd al-ʿAlī al-Birjandī**) await further research. More information is also needed on his contribution to studies on observations conducted in Istanbul at the time of Sultan Bāyazīd II.

İhsan Fazlıoğlu

Selected References

Brockelmann, Carl. (1949 and 1938) *Geschichte der arabischen Litteratur*. 2nd ed. Vol. 2: 593; Suppl. 2: 665. Leiden: E. J. Brill.

Bursalı, Mehmed Tāhir (1923). *Osmanlı Müellifleri*. Vol. 3, pp. 298–299. Istanbul, 1342 H. Matbaa-i & Mire.

Ihsanoğlu, Ekmeleddin, *et al*. (1997). *Osmanlı Astronomi Literatürü Tarihi (OALT)* (History of astronomy literature during the Ottoman period). Vol. 1, pp. 90–101 (no. 47). Istanbul: IRCICA.

Kātib Čelebī, *Kashf al-ẓunūn ʿan asāmī al-kutub wa-'l-funūn*. Vol. 1 (1941), cols. 866, 867, 870, 872, 881; Vol. 2 (1943), cols. 966, 1236. Istanbul: Milli Egition Babanlīgi Yayinlan.

Mehmed Mecdī Efendi (1989). *Ḥadā'iq al-shaqā'iq*. Istanbul, pp. 338–339. Gağri Yayinlare.

Saliba, George (1994). *A History of Arabic Astronomy: Planetary Theories during the Golden Age of Islam*. New York: New York University Press, pp. 282–284.

Salih Zeki (1913/1914). *Asar-i bakiye*. Vol. 1, pp. 199–200. Istanbul: 1329. Matbaa-i & Mire.

Storey, C. A. (1958). *Persian Literature*. Vol. 2, pt. 1. *A. Mathematics. B. Weights and Measures. C. Astronomy and Astrology. D. Geography*. London: Luzac and Co., pp. 79–80.

Suter, H. (1981). *Die Mathematiker und Astronomen der Araber und ihre Werke*. Amsterdam: APA-Oriental Press, p. 188 (no. 457).

Ṭashköprüzāde (1985). *Al-Shaqāl'q al-nuʿmāniyya fī ʿulamā' al-dawlat al-ʿuthmāniyya*, edited by Ahmed Subhi Furat. Istanbul, pp. 327–328. Istanbul Ūniversitesi, Edebiyal Falailtesi Yayinlare.

Mitchel, Ormsby MacKnight

Born **Morganfield, Kentucky, USA, 28 July 1809**
Died **Beaufort, South Carolina, USA, 30 October 1862**

Ormsby Macknight Mitchel, the founder and first director of the Cincinnati Observatory (which briefly housed the largest telescope in the United States), established the first exclusive astronomical periodical in the United States, served as second director of the Dudley Observatory, and became nationally famous as an inspirational lecturer on astronomy. He also enjoyed a distinguished United States army career during the Civil War, reaching the rank of major general.

Born in 1809 in Kentucky on what was then the American frontier, Mitchel was the youngest child of John Mitchel; both his father and his grandfather were surveyors. As a child in Lebanon, Ohio, Ormsby was tutored by his widowed mother, Elizabeth (*née* MacAlister) Mitchel, and a brother-in-law, until he entered a school opened by an elder brother, becoming an active member of the school's Thespian Society and Debating Club. In 1825, Mitchel entered the United States Military Academy at West Point, New York, studying surveying, civil engineering, and practical astronomy in addition to military subjects. After graduating in 1829, he taught mathematics at West Point, meanwhile meeting and marrying a young widowed mother, Louisa (*née* Clark) Trask, in 1831.

In 1832, Mitchel resigned his commission and went to Cincinnati, Ohio, then the fourth largest city in the United States and the most important metropolis not on the Atlantic seaboard. There, he studied law and was admitted to the Ohio bar while working as chief engineer of the Little Miami Railroad, one of the key railway lines being constructed across booming Ohio. In 1836, the newly founded Cincinnati College hired Mitchel to teach mathematics, mechanics, machinery, and astronomy; he also established a civil engineering program, one of the first in the country.

Although slightly built (5 ft. 6 in. and 130 lbs), Mitchel was a charismatic teacher and public speaker who held every audience spellbound. In 1841 and 1842, Mitchel gave a series of public lectures on the solar system to crowds of 2,000 Cincinnati residents. After the last lecture, Mitchel announced he would devote the next 5 years to founding a major astronomical observatory in Cincinnati. He also proposed a novel financing scheme: public subscription of shares of stock at $25 apiece – more than a month's wages for many laborers – for which subscribers would have the privilege beholding the heavens through the observatory's grand telescope. Within a month (on 24 May 1842), the Cincinnati Astronomical Society was formed, announcing more than $6,500 worth of stock had been subscribed. After visiting the finest telescope makers in Europe, Mitchel contracted with the firm of Merz & Mahler of Munich, Bavaria, to buy a 12-in. refractor. When it was mounted in 1845, it stood as the largest telescope in the United States and second largest in the world (surpassed only by the Pulkovo 15-in. refractor in Russia) until 1847, when another 15-in. Merz refractor was installed at Harvard College Observatory.

Having secured the telescope, Mitchel's next challenge was to construct the observatory within which to mount the telescope. He did so, meeting the deadline established in the deed for the property, but only by calling upon volunteer labor from Cincinnati citizens and with a complete dedication of his own time and energy to the project.

By the observatory's completion, however, Mitchel had big problems. As the fledgling institution had no endowment, in 1844 he had promised to serve as its director for a decade without pay beyond his college professor's salary. However, in January 1845, 4 days before the telescope arrived, Cincinnati College burned to the ground, and up in smoke went Mitchel's sole source of support for his wife and seven children as well as for the Cincinnati Observatory.

As a result, Mitchel launched himself into a career as a professional itinerant lecturer in astronomy. His timing was fortunate, as the mid-19th century was a golden age for circuit lecturers. Over the next decade, Mitchel became nationally famous for speaking without notes or visual aids aside from diagrams drawn in the air with a wand; his lectures riveted thousands of listeners in Boston, New York City, Philadelphia, New Orleans – some audiences (according to contemporary accounts) leaping to their feet and cheering as at a sporting or political event. Each year, he toured the nation from November through March when Ohio's observing conditions were poor, further augmenting his income by writing several popular books based on his lectures, among them *Planetary and Stellar Worlds* (1848), *Popular Astronomy* (1860), and *Astronomy of the Bible* (1863).

Less well-known was Mitchel's astronomical research during Ohio's clearer, warmer months. He remeasured the positions of the 2,700 double stars in **Friederich Struve**'s 1827 catalogue, especially those south of the celestial equator that never rose high above Pulkovo's horizon; he resolved many stars Struve had not marked double or triple, discovering the greenish companion to the red giant Antares (possibly his most original scientific discovery). Mitchel also determined the rotation period of Mars, monitored the position of the newly discovered planet Neptune, counted sunspots, measured the positions of planetary satellites, and observed comets and nebulae; in short, routine work typical of visual astronomers in mid-19th century observatories.

More significantly, by 1848, Mitchel was involved in early trials (conducted by the United States Coast Survey under **Sears Walker** and **Alexander Bache**) of the American method of longitudes, using the telegraph to compare local times at observatories hundreds of miles apart to determine their differences in longitude. He also proposed and experimented with a telegraph-type system within the Cincinnati Observatory itself to automate the recording of right ascension and declination for ordinary positional astronomy – the so called American method of transits. For both purposes, he invented an early chronograph, which recorded timings on a rotating flat disk, installing working prototypes at the Cincinnati and Dudley observatories.

Lastly, Mitchel founded the first exclusive astronomical periodical in the United States, the *Sidereal Messenger*. Published monthly from July 1846 through October 1848 (when publication ceased for lack of funds), it was a hybrid research journal and popular magazine, summarizing major findings by astronomers in Europe and the United States, plus detailing work at the Cincinnati Observatory.

In 1857, Mitchel's wife, who had also long been his observing assistant, was paralyzed by a series of strokes. In 1859, weighed down by her care, Mitchel reluctantly accepted an offer to become the second director of the Dudley Observatory in Albany, New York. He replaced **Benjamin Gould** after several years of controversy between the latter and the board of Albany citizens responsible for overseeing the observatory. The position included the luxuries of a house and a regular salary. But in 1861, with the start of the Civil War, Mitchel reentered active military service as a brigadier general in charge of volunteers from Ohio; the day after his departure, his wife suffered her last stroke and died.

Affectionately known in the army as "Old Stars," Mitchel served in several campaigns, the most famous being "the great locomotive chase" in Alabama, in which his men took possession of a train to cut telegraph wires and destroy bridges to disrupt the Confederate army's supply lines. His services earned him the rank of major general of volunteers. Mitchel was placed in command of the Union's Department of the South, at Hilton Head, South Carolina; however, a month later, he was stricken with yellow fever and died.

According to contemporary and later biographers, Mitchel's enduring contribution to astronomy was his oratorical eloquence that inspired hundreds of thousands of listeners and readers with wonder for the heavens – a powerful influence in inspiring wealthy philanthropists and the general populace to found scores of observatories in later 19th-century America.

Trudy E. Bell

Selected References

Anon. (1862). "Death of General O. M. Mitchel." *American Journal of Science* 84: 451–452.

Anon. (1884). "Sketch of Ormsby MacKnight Mitchel." *Popular Science Monthly* 24: 695–699.

Anon. (1944). "The Centenary of the Cincinnati Observatory, November 5, 1943." *Publications of the Historical and Philosophical Society of Ohio.* Cinncinnati: Historical and Philosophical Society of Ohio and the University of Cincinnati.

Bell, Trudy E. (Spring) (2002). "The Victorian Global Positioning System." *The Bent* 93 (2) (Spring): 14–21 (quarterly of Tau Beta Pi, the engineering honor society) (Outlines some of Mitchel's early involvement in the development of the American method of longitudes and describes his disk chronograph.)
Luther, Paul (1989). *Bibliography of Astronomers: Books and Pamphlets in English By and About Astronomers.* Vol. 1. Bernardston, Massachusetts: Astronomy Books. (Gives an extensive bibliography of Mitchel's writings.)
McCormmach, Russell (1966). "Ormsby MacKnight Mitchel's *Sidereal Messenger*, 1846–1848." *Proceedings of the American Philosophical Society* 90: 35–47.
Mitchel, F. A. (1887). *Ormsby MacKnight Mitchel, Astronomer and General: A Biographical Narrative by His Son.* Boston: Houghton and Mifflin.
Shoemaker, Philip S. (1991). "Stellar Impact: Ormsby Macknight Mitchel and Astronomy in Antebellum America." Ph.D. diss., University of Wisconsin-Madison.
Smith, Elliott (1941). "Historical Background of the Cincinnati Observatory." *Popular Astronomy* 49: 347–355.
Weddle, Kevin J. (1986). "Old Stars: Ormsby Mitchel." *Sky & Telescope* 71, no. 1 : 14–16.
Yowell, Everett (1913). "The Debt Which Astronomy Owes to Ormsby MacKnight Mitchel." *Popular Astronomy* 21: 70–74.

Mitchell, Maria

Born **Nantucket, Massachusetts, USA, 1 August 1818**
Died **Lynn, Massachusetts, USA, 28 June 1889**

Maria Mitchell, the first woman astronomer in the United States, paved the way for women in science. She trained an entire generation of women astronomers who followed her into research as well as teaching. Mitchell played a vital role in the 19th century's more enlightened attitudes about the role of women in American society and science.

One of ten children born to William and Lydia (*née* Coleman) Mitchell, Mitchell was raised in a favorable environment with the intellectual lives of her parents. As members of the Nantucket Quaker Meeting, they encouraged each of their children to read extensively and engage in thoughtful dialogue about what they were learning. At an early age, Mitchell began to help her father with his astronomical observation and computing. Her father was well-known as an astronomer who could be trusted to rate chronometers for the whaling vessels and merchant ships calling at Nantucket. Using a platform on top of their home, and later a similar structure on top of the Pacific Bank where William served as a teller, their observing was not limited to time-keeping work but included various objects in the Solar System and sweeping for comets.

Mitchell's formal education was limited to a few years in a school run by her father, followed by a few years at the Reverend Cyrus Peirce's school. She showed a special aptitude for mathematics from an early age, and learned under Peirce's guidance to the extent of his ability. For several years Mitchell assisted Peirce in the operation of his school before starting her own career as a teacher in a school she organized in 1835.The following year she accepted an additional role as librarian of the new Nantucket Athenaeum.

Mitchell experienced more frequent contact with women in the community as a librarian, but also enjoyed more time for her own studies. She taught herself to read French and German and then mastered the mathematical works of **Pierre de Laplace**, **Joseph Lagrange**, and **Karl Gauss** while continuing astronomical observations with her father when weather permitted. While sweeping the skies on the night of 1/2 October 1847, she discovered a comet near Polaris, now known as C/1847 T1 (Mitchell). This was her fourth independent discovery but the first for which the priority was properly hers. Her father immediately reported it to **William Bond**, director of the Harvard College Observatory. There were other independent discoveries of the same comet within days by others, including the director's son **George Bond** who conceded he had narrowly missed making the discovery himself on the same evening. However, William's immediate action in posting the letter to Bond ensured that the comet was credited to his daughter. The discovery by the young American woman brought her substantial notoriety in Europe as well as America. Mitchell was awarded a gold medal by King Frederich of Denmark.

The fame gave Mitchell the opportunity to increase acquaintances among leading American scientists like **Alexander Bache**, director of the United States Coast Survey [USCS], and **Charles Davis**, newly appointed superintendent of the United States Navy's Nautical Almanac office in Boston. Within 2 years, Davis offered her additional employment as a computer for the new *American Ephemeris and Nautical Almanac.* Mitchell computed ephemeris data for the planet Venus from her home in Nantucket. Bache invited Mitchell to spend one summer in Maine working with USCS observers. She continued in that dual employment as librarian and astronomical computer from 1849 until 1865. With her additional income, Mitchell traveled in the United States, learning first hand how privileged her special position on Nantucket was in comparison with women in other parts of the country. Female self-development and self-reliance was encouraged at the peak of whaling and maritime activity in Nantucket, like no where else in the United States at the time.

Mitchell had more travel opportunities when she was asked to travel as a tutor and chaperone for a Chicago banker's daughter in 1857. Mitchell accompanied the young woman to England where she was welcomed as a visitor to the Greenwich and Cambridge observatories. When the banker's daughter was forced to return to Chicago by her father's crisis during the financial panic of 1857, Mitchell stayed in Europe and visited many noteworthy astronomers including **Angelo Secchi**, **Caroline Herschel**, and **Mary Somerville**.

On her return to America, Mitchell received a 5-in. Clark refractor equipped with a micrometer as a gift "from the women of America" in recognition of her achievements as a woman astronomer. This fine instrument intensified Mitchell's desire to continue in her chosen career as an astronomer. After her mother died, Mitchell and her father moved to Lynn, Massachusetts. She was soon interviewed by a trustee for Matthew Vassar's endowment to establish the first women's college in the United States. As one of the best known women in America, Mitchell was a natural choice for the Vassar College faculty, New York, even though she lacked any formal educational credentials. Mitchell quickly accepted the position and moved with her father to the campus near Poughkeepsie, New York in 1865.

The Vassar College Observatory was equipped with a 12-in. Fitz refractor, one of the largest telescopes in the United States. Mitchell was eager to use the new instrument and valued the opportunity to have an influence in the higher education of women. Her classes in astronomy, though rigorous, were popular and well attended. Her astronomy classes and night observing sessions with the telescope created opportunities for education unlike that available to women in any other institution. With the help of her students, Mitchell conducted visual observation of double stars and planets. Thus, Mitchell's later career in astronomy was primarily as a teacher who empowered women in astronomy and the sciences rather than as a researcher.

Several of Mitchell's students were employed in the field of astronomy. The most noteworthy of these include **Antonia Maury**, **Caroline Furness**, Margaretta Palmer, and Mitchell's successor at Vassar College, Mary Whitney (1847–1921). More importantly, her students who chose to pursue careers in astronomy were important assets for the expansion of American astronomy through the first few decades of the 20th century. Working for **Edward Pickering** as "computers" at Harvard College Observatory, and later at Lick and Mount Wilson Observatories, they contributed greatly through their interpretation, classification, and measurement of spectra, as well as the more routine computations necessary for observational-data reduction.

Mitchell took a second tour in Europe in 1873, but while she visited noteworthy astronomers, she was more interested in the status and education of women. The trip had a marked influence on her subsequent life decisions, and she became more involved in improving the status and opportunities for women in the United States. She helped organize and served as president of the American Association of Women, assumed leadership roles in organizations such as the Social Sciences Association, and was the first women accepted as a member of the American Association for the Advancement of Science.

After she resigned in 1888 from Vassar College, Mitchell returned to her home, where she died. During her lifetime, Mitchell was honored by her election to the American Academy of Arts and Sciences and to the American Philosophical Society.

Thomas R. Williams

Selected References

Belserene, Emilia Pisani (1986). "Maria Mitchell: Nineteenth Century Astronomer." *Astronomy Quarterly* 5, no. 19: 133–150.

Hoffleit, Dorrit (1983). *Maria Mitchell's Famous Students*. Cambridge, Massachusetts: American Association of Variable Star Observers. (In celebration of the 75th anniversary of the Maria Mitchell Observatory.)

——— (1994). *The Education of American Women Astronomers before 1960*. Cambridge, Massachusetts: American Association of Variable Star Observers.

Keller, Dorothy J. (1974). "Maria Mitchell, an Early Woman Academician." Ph.D. diss., University of Rochester.

Kendall, Phebe Mitchell (comp. and ed.) (1896). *Maria Mitchell, Life, Letters, and Journals*. Boston: Lee and Shepard Publishers.

Kohlstedt, Sally Gregory (1978). "Maria Mitchell: The Advancement of Women in Science." *New England Quarterly* 51, no. 1: 39–63.

Lankford, John (1997). *American Astronomy: Community, Careers, and Power, 1859–1940*. Chicago: University of Chicago Press, pp. 288–308.

Merriam, Eve (ed.) (1971). "Maria Mitchell. 1818–1889." In *Growing up Female in America*, pp. 75–90. Garden City, New York: Doubleday.

Wright, Helen. (1949). *Sweeper in the Sky: The Life of Maria Mitchell*. New York: Macmillan.

Mizzī: Zayn al-Dīn [Shams al-Din] Abū ʿAbd Allāh Muḥammad ibn Aḥmad ibn ʿAbd al-Raḥīm al-Mizzī al-Ḥanafī

Born **probably al-Mizza near Damascus, (Syria), 1291**
Died **Damascus, (Syria), 1349**

Mizzī was a *muwaqqit* (*i. e.*, an astronomer appointed to a mosque who is responsible for regulating the times of prayer), an instrument maker, and the author of numerous treatises on astronomical instruments. He studied in Cairo under the well known physician and encyclopedist Ibn al-Akfānī. He was first appointed as a *muwaqqit* in al-Rabwa, a quiet locality near Damascus, and then at the Umayyad Mosque in Damascus, a position he held until his death. Mizzī authored treatises on the use of the astrolabe, the astrolabic quadrant, and the sine quadrant. In particular his treatises *al-Rawḍāt al-muzhirāt fī al-ʿamal bi-rubʿ al-muqanṭarāt* (On the astrolabic quadrant) and *Kashf al-rayb fī al-ʿamal bi-'l-jayb* (On the sine quadrant) were popular. He also wrote on the use of less common instruments, such as the *musattar* (concealed) and the *mujannaḥ* (winged) quadrants.

Although he made few original contributions to instrument making in particular or to astronomy in general, Mizzī was nevertheless an important and influential authority in the field, whose didactic treatises were appreciated by students of applied astronomy dealing with timekeeping (*ʿilm al-mīqāt*). The instruments he made were highly praised as being the best of his times and sold for considerable prices, namely 200 dirhams or more for an astrolabe, and at least 50 dirhams for a quadrant. Some five quadrants of his fabrication are extant, dated between the years 1326/1327 and 1333/1334. According to the 15th-century astronomer Ibn al-ʿAṭṭār, he also made astrolabes with mixed projections (*i. e.*, with markings obtained by a combination of stereographical projections about the North Pole and South Pole, respectively). According to his biographer al-Ṣafadī, Mizzī also excelled in oiling bows (*baraʿa fī dahn al-qisī*) and impressed his contemporaries by constructing mechanical devices such as those of **Banū Mūsā**.

François Charette

Selected References

Al-Ṣafadī. (1911). *Nakt al-himyān fī nukat al-ʿumyān*. Cairo, p. 244.

Brockelmann, Carl (1938). *Geschichte der arabischen Litteratur*. 2nd ed. Vol. 2 (1949): 155–156; Suppl. 2: 156, 1018. Leiden: E. J. Brill.

Charette, François (2003). *Mathematical Instrumentation in Fourteenth-Century Egypt and Syria: The Illustrated Treatise of Najm Al-Dīn Al-Miṣrī*. Leiden: E. J. Brill.

Combe, Étienne (1930). "Cinq cuivres musulmans datés des XIIIe, XIVe, et XVe siècles, de la Collection Benaki." *Bulletin de l'Institut français d'archéologie orientale* 30: 49–58, esp. p. 56.

Dorn, B. (1865). "Drei in der Kaiserlichen Öffentlichen Bibliothek zu St. Petersburg befindliche astronomische Instrumente mit arabischen Inschriften." *Mémoires de l'Académie impériale des sciences de St. Pétersbourg*, 7th ser., 9: 1–150, esp. pp. 16–26 and plates.

Féhérvari, Géza (1973). "An Eighth/Fourteenth-Century Quadrant of the Astrolabist al-Mizzī." *Bulletin of the School of Oriental and African Studies* 36: 115–117 and two plates.

Ibn Ḥajar (1966–67). *Durar al-kāmina fī aʿyān al-mi'a al-thāmina*. 5 Vols. Vol. 3, p. 410, no. 3392. Cairo: Dār al-Kutub al-Ḥadītha.

King, David A. (1986). *A Survey of the Scientific Manuscripts in the Egyptian National Library*. Winona Lake, Indiana: Eisenbrauns, no. C34.

——— (1993). "L'astronomie en Syrie à l'époque islamique." In *Syrie, mémoire et civilization* (exhibition catalogue), edited by Sophie Cluzan, Eric Delpont, and Jeanne Mouliérac, pp. 386–395, with descriptions of related objects on pp. 432–443. Paris: Institut du monde arabe and Flammarion. (On Mizzī see esp. pp. 391, 438.)

Suter, Heinrich (1990). "Die Mathematiker und Astronomen der Araber und ihre Werke." *Abhandlungen zur Geschichte der mathematischen Wissenschaften* 10: 165.

Moestlinus

➔ **Mästlin [Möstlin], Michael**

Mohler, Orren Cuthbert

Born **Indianapolis, Indiana, USA, 29 July 1908**
Died **Ann Arbor, Michigan, USA, 17 September 1985**

Orren Mohler served as director of the University of Michigan Observatories from 1961 until 1970 and continued as director of the McMath–Hulbert Solar Observatory [MHO] until 1979.

Mohler received his AB in 1929 from Michigan Normal College, followed by an MA (1930) and Ph.D. (1933) from the University of Michigan. From 1933 to 1940, Mohler taught astronomy at Swarthmore College, Pennsylvania, and worked as an astronomer at the private observatory of amateur astronomer Gustavus Wynne Cook (1867–1940). While working for Cook, Mohler conducted observing programs on the spectrographic binary star TX Leonis and on Nova Herculis 1934. He also attempted to improve the quantum efficiency of the Geiger–Muller counter so it could count individual protons.

In 1940, Mohler became the first full-time professional astronomer at MHO of the University of Michigan at Lake Angelus near Pontiac, Michigan. During World War II, Moher worked at MHO on military research and development, including development of the Cashman PbS infrared detector. He later exploited this experience by fitting such a detector to the MHO spectrograph. Working with **Robert McMath**, Mohler was able to extend infrared spectroscopy to well beyond previous limits, but it was apparent that local seeing was preventing full exploitation of the resolution of the excellent grating available. This prompted McMath to seek funding for the first astronomical vacuum spectrograph which, when completed, fully rewarded their effort. Mohler then published an atlas of the solar spectrum from 11,984 Å to 25,578 Å with approximately 100 times greater resolution than any previous spectrum in this region. Mohler was appointed professor of astronomy in 1956. Mohler became director of the McMath–Hulbert Observatory, and in 1962 was appointed as chairman of the Astronomy Department and director of the University of Michigan Observatories.

Mohler's primary interests were instrumentation and the history of astronomy. He was awarded the Naval Ordnance Development Award for his contributions to military research and development, and was a Fulbright Scholar.

Patricia S. Whitesell

Selected Reference

Miller, Freeman D. (1986). "Orren C. Mohler," *Physics Today* 39, no. 4: 74.

Molesworth, Percy Braybrooke

Born **Colombo, (Sri Lanka), 2 April 1867**
Died **Trincomali, (Sri Lanka), 25 December 1908**

Percy Molesworth, one of the notable British amateurs in the late 19th century, contributed much of our pre-spacecraft knowledge of Jupiter. Molesworth was the son of Sir Guildford and Mary Elizabeth (*née* Bridges) Molesworth. His father, a noted civil engineer, served as a colonial officer in Ceylon. Molesworth was educated in England at Winchester College, Winchester and received a commission in the Corps of Royal Engineers of the British army in 1886. After further education at the Royal Military Academy at Woolwich and Chatham, he served in England and Hong Kong before requesting a return to Ceylon. As captain (later major), Molesworth was stationed in Trincomali. This gave him an ideal location for observing Mars and Jupiter, using a 12.5-in. reflector, in the garden of his house high above the sea.

Molesworth was the most assiduous of all observers of Jupiter. His observations (visual transit timings, especially from 1898 to 1903) contributed greatly to establishing the patterns and permanence of the various currents. In 1900, for example, he recorded 6,758 central meridian transit timings on Jupiter! Molesworth's individual reports were published in the *Monthly Notices of the Royal Astronomical Society*. As a result of his and others' work, all the permanent slow currents that control large circulations on Jupiter had been identified by 1901. Molesworth also noted several phenomena whose importance for the atmospheric dynamics on Jupiter has become recognized in the era of spacecraft: The start of a great south tropical disturbance (discovered in February 1901), the turbulent bright spots west of the Great Red Spot, one of which he saw erupting within less than an hour, and the 90-day oscillation of the Great Red Spot itself. Molesworth also contributed substantial observations and drawings of Mars to the *Monthly Notices*; those that were not published in 1903 are still in the archives of the Royal Astronomical Society.

With his health failing, Molesworth retired in 1906 at the age of 39, intending to continue astronomical observations from his estate in Trincomali; however, he fell ill again and died of dysentery.

John Rogers

Selected References

Anon. (1909). "Major Percy B. Molesworth, R. E." *Observatory* 32: 108.

Anon. (1909). "P. B. Molesworth." *Journal of the British Astronomical Association* 19, no. 3: 143.

Anon. (1909). "Percy Braybrooke Molesworth, Major, R. E." *Monthly Notices of the Royal Astronomical Society* 69: 248–249.

Moll, Gerard

Born **Amsterdam, the Netherlands, 18 January 1785**
Died **Amsterdam, the Netherlands, 17 January 1838**

Dutch astronomer Gerard Moll directed the Utrecht Observatory for 26 years. The son of Gerard, a wealthy Amsterdam merchant, and Anna (*née* Diersen) Moll, he became a junior clerk in an Amsterdam mercantile house around 1800. It was a job he undertook with reluctance but performed satisfactorily. His spare time was devoted to mathematical and scientific studies, including astronomy. After several years, Moll's father allowed him to undertake academic studies at the University of Amsterdam. Moll was awarded a Ph.D. in 1809, and afterwards studied at the University of Paris.

In 1812, Moll was appointed director of the Observatory at the University of Utrecht. Following that institution's reorganization in 1815, Moll was also appointed professor of mathematics and natural philosophy, a post which he held until his death. The university, however, failed to provide much in the way of support or maintenance for the observatory.

In 1825, Moll was offered a chairmanship at the University of Leiden. But he declined the offer after Utrecht administrators evidently agreed to provide substantially greater support. Moll used this opportunity to acquire an improved collection of astronomical instruments and a substantial library. During his Utrecht career, he observed an annular eclipse of the Sun (7 September 1820), a transit of Mercury across the Sun's disk (5 May 1832), and an occultation of Saturn (8 May 1832).

Besides astronomical observations, Moll conducted research in several areas of physics (then called natural philosophy). He extended the work of Hans Christian Ørsted after the latter's 1820 discovery of electromagnetism. Moll repeatedly measured the speed of sound and arrived at a value very near the currently accepted figure.

Moll was given further responsibilities by the Kingdom of the Netherlands, concerning flood protection and the observation of tides along the Dutch coast. His comparative study of the British and French systems of weights and measures earned Moll a knighthood of the Order of the Belgian Lion. He was awarded honorary doctorates by the universities of Edinburgh and Dublin.

Hartmut Frommert

Selected References

Anon. (1838). "Professor Moll, of Utrecht." *Monthly Notices of the Royal Astronomical Society* 4: 110–111.

Dieke, Sally H. (1974). "Moll, Gerard." In *Dictionary of Scientific Biography*, edited by Charles Coulston Gillispie. Vol. 9, pp. 459–460. New York: Charles Scribner's Sons.

Mollweide, Karl Brandan

Born **Wolfenbüttel, (Germany), 2 February 1774**
Died **Leipzig, (Germany), 10 March 1825**

Astrometrist and mathematician Karl Mollweide is best known for his spirited defense of the Newtonian theory of color. The son of Christoph Mollweide, Karl Mollweide studied mathematics at Helmstadt University beginning in 1796, and in 1800 started teaching mathematics and physics in Halle/Saale, while continuing his studies at the University of Halle. After his graduation from the university in 1811, he became an observer at the university observatory in the old castle of Pleissenburg, and lectured in astronomy. Mollweide was elected professor of astronomy in 1812 and became a professor in mathematics at the University of Leipzig in 1814. From 1820 to 1823 he was dean of the philosophical faculty.

In 1814, Mollweide married the widow of the gatekeeper Knorr, of the hospital gate in Leipzig. Mollweide's wife was sister to the wife of another assistant at Pleissenburg Observatory. After his wife's death in 1821, his sister-in-law took care of his household. Mollweide had a stepson but no natural children by marriage.

Mollweide lived in the tower of the Pleissenburg until 1816. He worked on problems in spherical astronomy (*e. g.* stellar aberration) and astronomical position finding, as well as interpretations of various text passages from astronomers of classical antiquity.

Before his appointment in Leipzig, Mollweide was publicly engaged in defending **Isaac Newton**'s theory of colors. In that regard, he was a dedicated opponent of Johann Wolfgang von Goethe and his *Farbenlehre* (Theory of colors); Goethe never accepted Newton's theory.

Mollweide also wrote mathematical essays on the construction of magic squares, the application of **Carl Gauss**' addition and subtraction logarithms, and on the largest ellipse contained in a square. Two of his outstanding discoveries are still of great mathematical importance today. He calculated the Mollweide projection, which is often applied in the field of cartography when the whole Earth is depicted. He studied Cartesian trigonometry and developed the Mollweide formulae of plane trigonometry or sky, which are used in the calculation of the triangle.

Thomas Klöti
Translated by: *Balthasar Indermühle*

Selected References

Bruhns, Karl Christian (1879). *Die Astronomen auf der Pleissenburg*. Leipzig: A. Edelmann.

Freiesleben, H.-Christ (1974). "Mollweide, Karl Brandan." In *Dictionary of Scientific Biography*, edited by Charles Coulston Gillispie. Vol. 9, p. 463. New York: Charles Scribner's Sons.

Schmeidler, Felix (1997). "Mollweide, Karl Brandan." In *Neue deutsche Biographie*. Vol. 18, pp. 6–7. Berlin: Duncker and Humblot.

Molyneux, Samuel

Born **Chester, England, 18 July 1689**
Died **Kew, (London), England, 13 April 1728**

Samuel Molyneux, noted as an instrument maker and observational astronomer, assisted **James Bradley** in the latter's studies that led to the discovery of the aberration of light. The only son of the astronomer **William Molyneux** and Lucy Domville, Molyneux was raised by his uncle, Thomas after both parents died while he was a young boy. Educated at Trinity College, Dublin (BA: 1708, MA: 1710), Molyneux studied meteorology and was elected to the Royal Society in 1712. He became a member of parliament in 1715 (elected again in 1726 and 1727).

In 1717 Molyneux married Lady Elizabeth Capel, who inherited money and an estate at Kew, to which the couple moved. Caught by the enthusiasm of **John Hadley** for optics, Molyneux turned his scientific interests to making optical components and instruments. His design for a Newtonian reflector set the standard of construction for such instruments. He conducted experiments to find the best alloy for speculum metal by testing 150 alloys of varying compositions. Between 1723 and 1725, Molyneux worked on reflecting telescopes with **James Bradley**, then Savilian Professor at Oxford (later Astronomer Royal).

In 1725, with Bradley, Molyneux ordered a large zenith sector from instrument maker **George Graham**, in order to investigate the large parallax of γ Draconis reported by **Robert Hooke**. Molyneux and Bradley set the telescope up in Molyneux's house in Kew, looking through holes in the roof. They observed γ Draconis in December 1725 and found a large shift in position as the month progressed. However, it was in the wrong direction to be a parallactic shift; as the two continued observing through the year they saw a large annual circular motion. Bradley ordered a larger, more versatile zenith sector from Graham and erected this one at the house of Molyneux's aunt in Wanstead in August 1727. By December, Bradley had made accurate measurements by which he inferred the phenomenon of the aberration of light.

Meanwhile, Molyneux's political interests took more of his time. He became a member of the Irish Parliament in 1727 and a lord of the admiralty, ceasing work in astronomy. Molyneux died at a young age of a stroke, presumably as a result of a medical problem inherited from his mother who also died early of a brain disease.

Paul Murdin

Selected Reference

Bradley, James (1832). *Miscellaneous Works and Correspondence*, edited by S. P. Rigaud. Oxford: Oxford University Press.

Molyneux, William

Born **Dublin, Ireland, 17 April 1656**
Died **Dublin, Ireland, 11 October 1698**

William Molyneux, an influential figure in the scientific affairs of Dublin in the late 17th century, gained the respect of **Edmond Halley** and **John Flamsteed** as an astronomer and wrote *Dioptrica Nova*, the first major book in English on optics.

William was the eldest surviving son of Samuel Molyneux and Margaret, daughter of William Dowdall, of Dublin. The Molyneuxs became part of the protestant establishment that dominated Dublin social and political life. Samuel, trained as a lawyer, was a skilful artillery officer, experimenting with gunnery for many years. William received a good schooling and entered Trinity College, Dublin, at the age of 15. After graduating he went to London in 1675 to study law. However, law did not interest Molyneux greatly, and he devoted most of his time to applied mathematics and science. He returned to Dublin in 1678 and married Lucy, the youngest daughter of Sir William Domville, Attorney General of Ireland. Only 2 months later, she became ill, becoming blind and suffering from constant headaches until her death 13 years afterwards. After a vain search in England for medical relief for his wife, Molyneux settled down in Dublin and passed the time by translating **René Descartes**'s *Méditations*, published in London in 1680. He also translated **Gallileo Galilei**'s *Discorsi* from Italian for his own use.

Molyneux's wife's affliction probably led him to pose the Molyneux problem, which assumes that a man blind from birth who gains his sight is confronted by a sphere and a cube that he had previously learnt to distinguish by touch. Can he identify at first sight which is which? Molyneux believed not. John Locke, George Berkeley, and other philosophers discussed the problem.

Molyneux's first important achievement in astronomy was to record the lunar eclipse seen in Dublin on 1 August 1681. He sent his observations to a friend in London, Charles Bernard, who passed them on to Flamsteed, the Astronomer Royal, whom Molyneux had previously visited at Greenwich. This led to a correspondence between Flamsteed and Molyneux that continued for the next ten years in which problems of optics, astronomy, ballistics, and tides were discussed. Molyneux learned a great deal about optical instruments from Flamsteed's letters; the exchanges were remarkable for their cordiality, as Flamsteed was reputed to have a short temper.

In 1684 Molyneux became involved in a controversy between **Robert Hooke** and **Johannes Hevel**. In his *Machina coelestis* (1673), Hevelius had claimed that open sights were better than telescopic sights. Hooke vigorously disputed the claim and, in 1679 Halley went to Danzig to test the rival theories. Halley supported Hevelius and gave him a written testimonial to that effect. Hevelius repeated his claim in 1685, in his *Annus Climactericus* which Molyneux reviewed, showing Halley's conclusions were invalid and telescopic sights were more accurate.

In October 1683 Molyneux and some colleagues from Trinity College formed a society on the model of the Royal Society of London. It was called the Dublin Philosophical Society; it met in rooms owned by an apothecary and included a garden for plants and a laboratory. Sir William petty was the first president, and Molyneux was secretary. The society exchanged minutes with the Royal Society and the Oxford Philosophical Society. Papers read in Dublin were frequently published in the *Philosophical Transactions of the Royal Society*. The society was the forerunner of the Dublin Society (1731) that became the Royal Dublin Society (1820), promoting the utilitarian aims of the original society.

In 1685 Molyneux was elected a fellow of the Royal Society. That same year he visited his brother Thomas (afterwards Sir Thomas) who was studying medicine in Leiden, the Netherlands. The brothers visited **Christiaan Huygens** at The Hague where he showed them a telescope in his garden and a planetary clock. In Paris they visited **Jean Cassini** and saw a clockwork device for driving a telescope.

Returning through London, Molyneux visited Flamsteed, and he also ordered the construction of a combined dial and telescope of his own design, to which he gave the name *Sciothericum telescopicum*. Flamsteed later examined the device and was not impressed. The following year Molyneux published a book describing the instrument.

When James II arrived in Ireland in 1688, Molyneux fled to England, settling in Chester. There, his son **Samuel Molyneux** was born; he was a source of great interest and pride to his father and became an astronomer.

During his 2-year stay in Chester, the elder Molyneux wrote *Dioptrica Nova,* which was the first book on optics published in English. The book was received favorably and became a standard text. In a dedication to the Royal Society he made a complimentary reference to Locke, which led to a long and friendly correspondence between the two men.

After the defeat of James II in July 1690, Molyneux returned to Dublin and became involved in politics. He was elected as one of the university representatives in the new Irish parliament in October 1692 and was reelected in 1695. As a result of his concern about the effect of the English parliament's legislation on the linen and woolen industries in Ireland, early in 1698 he published the work by which he is generally best known: *The Case of Ireland's being bound by Acts of Parliament in England, stated.* It was an attempt to prove the legislative independence of the Irish parliament, and it provoked strong opposition from the English parliament.

Despite the unfavorable reaction to his book in England, Molyneux went to London in July 1698 to fulfill a long-standing promise to visit Locke. Upon his return in September, he soon suffered a recurrence of a kidney complaint and died.

Ian Elliott

Selected References

Hoppen, K. Theodore (1970). *The Common Scientist in the Seventeenth Century: A Study of the Dublin Philosophical Society, 1638–1708*. Charlottesville: University Press of Virginia.

King, Henry C. (1978). *The History of the Telescope*. New York: Dover.

Molyneux, Sir Capel, Bt. (ed.) (1803). *Anecdotes of the life of the celebrated patriot and philosopher Wm. Molyneux, author of the Case of Ireland: Published from a manuscript written by himself*. Dublin: privately printed.

Simms, J. G. (1982). *William Molyneux of Dublin*, edited by P. H. Kelly. Dublin: Irish Academic Press.

Monck, William Henry Stanley

Born **Skeirke near Borris-in-Ossory, (Co. Laois), Ireland, 21 April 1839**
Died **Dublin, Ireland, 24 June 1915**

Although trained as a lawyer and philosopher, William H.S. Monck was a highly proficient amateur astronomer who was among the first to realize the existence of dwarf and giant stars. With the assistance

of Stephen M. Dixon, Monck made the first astronomical photoelectric measurements of light in Dublin in August 1892.

Monck was the third of four sons of the Reverend Thomas Stanley Monck (1796–1858) and his wife Lydia (*née* Kennedy). His childhood was spent in a rural community 16 miles southeast of Parsonstown (now Birr), King's County, where **William Parsons**, the Third Earl of Rosse, completed the great 72-in. reflecting telescope in 1845. The sight of the Leviathian of Parsonstown may well have kindled Monck's interest in astronomy.

The Anglo–Norman family Monck that settled in Ireland descended from William Le Moyne (living in 1424) of Devonshire and his son Robert. From 1705 the Monck family was associated with the Charleville estate near Enniskerry, County Wicklow. William's grandfather, the Reverend Thomas Stanley Monck, was a younger brother of Charles Stanley, the First Viscount Monck. Charles, the Fourth Viscount Monck, was Governor General of Canada from 1861 to 1868.

William Monck was educated at home by tutors. He distinguished himself on entry to Trinity College Dublin, and in 1861 he obtained the first scholarship in science with a gold medal and a first class honors degree in logic and ethics; he also received the Wray Prize for the encouragement of metaphysical studies. He studied divinity with distinction for several years. However, instead of following his father and grandfather into the Anglican Church of Ireland, Monck turned to law and was called to the bar in 1873. He was later appointed chief registrar in bankruptcy in the High Court in Dublin. In 1878 Monck returned to academic life and was professor of moral philosophy in Trinity College (the chair formerly occupied by George Berkeley) until 1882. He wrote *An Introduction to Kant's Philosophy* (1874) and a well-received *Introduction to Logic* (1880).

In 1886 Monck became a fellow of the Royal Astronomical Society [RAS] and a member of the Liverpool Astronomical Society. On 12 July 1890 he wrote a letter to the *English Mechanic* advocating the formation of an association of amateur astronomers to cater for those who found the RAS subscription too high, or its papers too technical, or who, being women, were excluded. Moves toward setting up such a society had already been made by **Edward Maunder**, and the British Astronomical Association was established at a meeting in London on 24 October 1890.

Monck was a prolific writer, and from 1890 onwards many of his letters and articles appeared in the early volumes of the *Publications of the Astronomical Society of the Pacific*, *The Sidereal Messenger*, and *Astronomy and Astro-Physics* (the forerunner of the *Astrophysical Journal*), as well as the *Journal of the British Astronomical Association* and the *English Mechanic*. In 1899 he published *An Introduction to Stellar Astronomy*, which consisted mainly of articles that had previously appeared in *Popular Astronomy*.

In 1891 Monck purchased a 7½-in. refractor designed by **Alvan Clark** that had been owned previously by the Reverend **William Dawes**, Frederick Brodie, and Dr Wentworth Erck. Monck erected the telescope in the back garden of his residence at 16 Earlsfort Terrace, Dublin. Shortly afterward he received a request from his old college friend, George M. Minchin (1845–1914). Minchin, a professor of applied mechanics at the Royal Indian Engineering College at Cooper's Hill, London, wanted to test "on the stars" some selenium photocells that he had developed. Using a quadrant electrometer borrowed from Trinity College to record the voltage produced by the cell, Monck and his neighbor, professor Dixon, succeeded in measuring the relative brightness of Jupiter and Venus on the morning of 28 August 1892. They failed to obtain "certain" results from the stars on account of instrumental drift and other difficulties. Minchin continued to improve his cells. Working with professor **George FitzGerald** of Trinity College and William Edward Wilson (1851–1908), Minchin made measurements of stellar brightness in April 1895 and January 1896 using the 24-in. reflector operated by Wilson at Daramona Observatory in County Westmeath. These observations were reported by Minchin in *The Proceedings of the Royal Society*.

In 1894 Monck suggested that there were probably two distinct classes of yellow stars – one being dull and near, the other being bright and remote. He based this on his examination of several catalogs that displayed the early Harvard Observatory spectral classifications as well as proper motions for large numbers of stars. What Monck noticed was that those stars comparatively close to the Earth (and therefore exhibiting the largest proper motions) were not also the brightest stars. In fact, the 92 stars he identified as Capellan (modern spectral classes F and G) and the 59 Arcturian stars (modern spectral class K) had large proper motions and were about equal in brightness to the Sirian stars, of which there were only 11 of the same brightness and proper motion. Monck concluded, properly, that this meant that the Sirian stars were likely intrinsically brighter but farther away on average. This clue to the existence of dwarf and giant stars was taken up by his friend **John Gore** who estimated the size of Arcturus. If better data had been available to Monck, his discussion of proper motions and spectra might have led him to the relationship between luminosity and color later discovered by **Ejnar Hertzsprung** and **Henry Norris Russell**.

Monck may have been the first to suggest the parsec as a unit of distance. In his *An Introduction to Stellar Astronomy* (1899), he wrote:

> If we adopted as our unit of distance that of a star with a parallax of one second (in other words 206,265 times the distance of the Sun), the distance of any other star will be 1/p, where p is the parallax expressed as a fraction of a second. This would, I think, be a more convenient unit of distance than that usually adopted by astronomers; *viz.*, a year's light passage, or the distance which light travels in a year.

Monck was greatly interested in meteors; he corresponded with **William Denning** for many years. Monck wrote 39 articles or letters related to meteors in the current journals over 3 decades (1885–1914).

David DeVorkin has described Monck as a "brilliant amateur astronomer," but probably very few of his contemporaries realized the potential impact of his astronomical studies. Quiet and reserved in manner, he was an authority on politics, history, and legal matters. Monck took a great interest in educational matters and supported all schemes for improving the lot of the underprivileged. He was a formidable chess opponent but took more pleasure in solving problems than in the cut and thrust of games.

Ian Elliott

Selected References

Batt, Elisabeth (1979). *The Moncks and Charleville House – A Wicklow Family in the Nineteenth Century*. Dublin: Blackwater Press.

Bracher, Kate (1987). "Early Spectral Classification of Stars." *Mercury* 16, no. 4: 119.

Butler, C. J. and I. Elliott (1993). "Biographical and Historical Notes on the Pioneers of Photometry in Ireland." In *Stellar Photometry – Current Techniques and Future Developments: Proceedings of IAU Colloquium No. 136 Held in Dublin, Ireland, 4–7 August 1992*, pp. 3–12. Cambridge: Cambridge University Press.

De Vorkin, David H. (1984). "Stellar Evolution and the Origin of the Hertzsprung–Russell Diagram." In *Astrophysics and Twentieth-Century Astronomy to 1950: Part A*, edited by Owen Gingerich, pp. 90–108. Vol. 4A of *The General History of Astronomy*. Cambridge: Cambridge University Press, p. 94

Dixon, S. M. (1892). "The Photo-Electric Effect of Starlight." *Astronomy and Astrophysics* 11: 844.

Minchin, G. M. (1895). "The Electrical Measurement of Starlight. Observations made at the Observatory of Daramona House, co. Westmeath, in April, 1895. Preliminary Report." *Proceedings of the Royal Society of London* 58: 142–154.

——— (1896). "The Electrical Measurement of Starlight. Observations made at the Observatory of Daramona House, co. Westmeath, in January, 1896. Second Report." *Proceedings of the Royal Society of London* 59: 231–233.

Monck, W. H. S. "Letter". *English Mechanic*, 12 July 1890.

——— (1892). "The Photo-Electric Effect of Starlight." *Astronomy and Astrophysics* 11: 843.

——— (1895). "The Spectra and Colours of Stars." *Journal of the British Astronomical Association* 5: 416.

——— (1899). *An Introduction to Stellar Astronomy: with illustrations*. London: Hutchinson and Co. (Especially p. 51 for the quote on the origin of the parsec.)

Monnier, Pierre-Charles le

Born **Paris, France, 23 November 1715**
Died **Herils, Calvados, France, 3 April 1799**

Pierre Le Monnier, an outstanding observational astronomer, brought English astronomical ideas and instrumentation to France. Le Monnier's father (also named Pierre) was professor of philosophy at the Collège d'Harcourt and member of the Académie royale des sciences. His brother, Louis Guillaume Le Monnier, also an academician, was professor of botany at the Jardin du roi, physician (*premier médecin ordinaire*) to King Louis XV and King Louis XVI, and a physicist. In 1763, Pierre-Charles married a daughter of the wealthy Norman family of Cussy, with whom he had three daughters, one of whom married **Joseph de Lalande**, another who married Pierre Charles' brother Louis.

Le Monnier began his astronomical career early with the assistance of his father. As early as 1731, when Le Monnier was still quite young, he indicated his skill with observations of the opposition of Saturn. In 1732, he was allowed to observe in Paris. In 1735, Le Monnier presented an elaborate lunar map to the academy and was admitted as *adjoint géomètre* on 23 April 1736. By 1746, he rose to *pensionnaire* and became professor of physics at the Collège royal (Collège de France). He was eventually elected to the Royal Society of London, the Berlin Academy, and the Académie de la Marine.

Mainly as an observer, Le Monnier greatly advanced astronomical measurement in France. He was a favorite of Louis XV, who procured for him some of the best astronomical instruments in France. Le Monnier studied the Moon, determined the positions of many stars, conducted extensive research in terrestrial magnetism and atmospheric electricity, and wrote about astronomical navigation.

Le Monnier's career accelerated with the geodesic expedition of 1736 to Lapland. He accompanied **Pierre de Maupertuis**, **Alexis Clairaut,** and the Swedish astronomer–physicist **Anders Celsius** to measure a degree of an arc of meridian at high latitude. The equipment of the expedition included a 9-ft. zenith sector, a transit instrument, and a clock, all by **George Graham**, the leading English maker of the day. In addition to surveying the degree of meridian, Le Monnier made observations of atmospheric refraction at various latitudes and in different seasons.

In 1741, Le Monnier published his *Histoire céleste* as a review of observations made in France from the founding of the Académie royale des sciences. The transit instrument he describes there and illustrates in detail was designed by Graham. Le Monnier introduced his *Description* by relating it to that previous account of the transit instrument, to Maupertuis' account of the zenith sector, and to the general textbooks written by Nicolas Bion in France and Robert Smith in England. Le Monnier thought that the mural quadrant was capable of being a universal instrument for all fundamental measurements in astronomy. Intimately familiar with the English mural quadrants, having acquired one of 5-ft. 4 in. radius from Jonathan Sisson in 1743, Le Monnier had also examined the original Graham quadrant in detail at Greenwich in 1748, and obtained a similar instrument of 8-ft. radius from John Bird in 1753. It was clear that, as the mural quadrant was becoming the principal instrument in the observatories of Europe, so London makers were moving into a position where they dominated the market in precision instruments. Le Monnier played an important part in this process, especially in relation to France, and his *Description* is one of its most impressive monuments.

Then Le Monnier published *La théorie des comètes* (1743), which was largely a translation of **Edmond Halley**'s *Cometography* but with several additions, and constructed the first transit instrument in France at the Paris Observatory and the great meridian at the Church of Saint Sulpice, Paris, also during the year 1743.

In 1748, Le Monnier travelled to England and observed an annular eclipse of the Sun in Scotland. An indefatigable observer, he undertook a star catalog and in 1755 produced a map of the zodiacal stars. He made several observations of Uranus before it came to be recognized as a planet by **William Herschel** in 1781. In fact, the third Astronomer Royal, **James Bradley**, recorded Uranus as a star in 1748, 1750, and 1753, while **Tobias Mayer** sighted it in 1756. But Le Monnier recorded Uranus 12 times between 1764 and 1771, including six observations in January 1769. Apparently he did not recognize Uranus, which was near its stationary point in the sky at the beginning of 1769, so its movement among the stars would not have been obvious.

Le Monnier introduced two constellations, now obsolete. The Reindeer (*Le Renne* or *Tarandus Vel Rangifer*) was created in 1743 for commemorating Lapland. Some faint stars between Camelopardalis and Cassiopeia formed it. The Solitary Thrush (*Le Solitaire* or *Turdus Solitarius*) was an Indian bird created in 1776 near the tail of Hydra.

Le Monnier's work on lunar motion was the most extensive and the most important of his time. In the first edition of the *Principia*, **Isaac Newton** had shown that the principal inequalities of the Moon could be calculated from his law of universal gravitation; in the second edition Newton applied these calculations to the observations of **John Flamsteed**. His methods, however, added little to the theory that **Jeremiah Horrocks** had suggested long before. Flamsteed calculated new tables based on Horrocks' theory incorporating Newton's corrections, but he did not publish them. They appeared for the first time in *Institutions astronomiques* (1746), Le Monnier's most famous work. It was basically a translation of **John Keill**'s *Introductio ad veram astronomiam* (1721), but with important additions and with new tables of the Sun and the Moon. The textbooks of Lalande and **Nicolas de La Caille** later largely replaced this book, which was the first important general manual of astronomy in France.

Le Monnier supported the view of Halley that the irregularities of the Moon's motion could be discovered by observing the Moon regularly through an entire cycle of 223 lunations, which is the Saros cycle of approximately 18 years and 11 days, with the assumption that the irregularities would repeat themselves throughout each cycle. Le Monnier and Bradley each began such a series of observations; Le Monnier continued this work for 50 years.

During the 1740's Clairaut, **Jean d'Alembert**, and **Leonhard Euler** were racing to create a satisfactory mathematical lunar theory, which demanded an approximate solution to the difficult three-body problem. In the ensuing controversy, Le Monnier seconded d'Alembert who used Le Monnier's *Institutions astronomiques* as the basis for his tables. In 1746, Le Monnier presented a memoir to the academy describing his observations of the inequalities in the motion of Saturn caused by the gravitational attraction of Jupiter. His results were important for improving the lunar theory, because a good explanation of the perturbations of Saturn also required a treatment of the three-body problem.

The best lunar tables of the 18th century were those of Tobias Mayer, which were based on Euler's theory but with the magnitude of the predicted perturbations taken from observations. Although Le Monnier admired English methods in astronomy, he adhered to the task of correcting Flamsteed's tables even while confessing the superiority of Mayer's.

As a regular correspondent with Bradley, Le Monnier was apparently the only astronomer with a sufficiently long and sufficiently accurate enough series of stellar observations to verify Bradley's discovery of the nutation of the Earth's axis in 1748. Le Monnier was the first to apply nutation in correcting solar tables.

Le Monnier's most prominent pupil was Lalande, who attended his lectures in mathematical physics at the Collège Royal. To determine lunar parallax, Lalande, on Le Monnier's strong recommendation, went to Berlin in 1751 to take measurements of the Moon for comparison with those taken by La Caille at the Cape of Good Hope.

Christian Nitschelm

Alternate name

Lemonnier, Pierre-Charles

Selected References

Delambre, J. B. J. (1827). *Histoire de l'astronomie au dix-huitième siècle*. Paris: Bachelier.

Hankins, Thomas (1972). "Le Monnier, Pierre-Charles." In *Dictionary of Scientific Biography*, edited by Charles Coulston Gillispie. Vol. 8, pp. 178–180. New York: Charles Scribner's Sons.

Lalande, Jérôme (1803). *Bibliographie astronomique avec l'histoire de l'astronomie depuis 1781 jusqu'à 1802*. Paris.

Robida, Michel (1955). *Ces bourgeois de Paris: Trois siècles de chronique familiale, de 1675 à nos jours*. Paris: R. Julliard.

Monnig, Oscar Edward

Born **Fort Worth, Texas, USA, 4 September 1902**
Died **Fort Worth, Texas, USA, 3 May 1999**

Oscar Monnig functioned as an active amateur astronomer as well as a publisher of an astronomical newsletter that provided an important link between members of the North American community of amateur astronomers. As a collector of meteorites, he assembled one of the largest and most diversified collections in the world.

Monnig was one of three sons of William and Alma (*née* Wandry) Monnig. His father and uncle, George B. Monnig, founded Monnig Wholesale, Inc., a successful dry goods business that eventually grew to include five retail department stores in the Fort Worth area. Monnig finished his formal education at the University of Texas at Austin earning an LL.D. degree in 1925. After practicing law for 5 years, Monnig joined the family business. He succeeded his brother Otto in 1974 as chief executive officer of the company, continuing in that capacity until the business was sold in 1983, by which time his health had begun to deteriorate.

Monnig's first love was astronomy. He always insisted on calling himself an "amateur," having had no formal training in any related field. He once signed up for a course in astronomy in college, but the class was cancelled for lack of enrollment; therefore his knowledge was all from individual study. Monnig was part of a small band of dedicated observers that included Robert Brown, James Logan, Sterling Bunch, and Blakeney Sanders. Together, they founded the Las Estrellas Observatory under dark skies south of Fort Worth. Work at the observatory emphasized variable star and meteor observing, photographic observation of comets and planets, and lunar occultation timing. With the advice of Canadian astronomers **Peter Millman** and Ian Halliday, gained during a solar eclipse expedition in August, 1932, Monnig successfully captured early spectrographic images of meteors by placing a prism in front of a camera lens. The group's meteor work was so regular and of such high quality that **Charles Olivier** began identifying them as the "Texas Observers", and the title was adopted by the group.

In 1928, Sterling Bunch published two issues of a newsletter that he titled *The North Texas Astronomical Bulletin*, but his effort floundered for lack of writers. In 1931, Monnig restarted the publication, but titled it *The Texas Observers Bulletin*, as an informal newsletter that carried various observations about astronomical subjects "of interest to amateurs." Monnig continued the publication of this bulletin until 1947. It was mimeographed and mailed in standard business envelopes to more than 200 subscribers around the United States and Canada. In some issues, Monnig would include photographic prints of interest in the envelope, but would leave space on the mimeographed page for the photograph to be pasted in after it was received by the subscriber. The *Texas Observers Bulletin* was recognized favorably by such professionals as **Otto Struve** and **Bart Bok**, who encouraged similar efforts on the part of other amateurs.

Over time, Monnig's astronomical interest became strongly focused on meteoritics, a field that he chose because "it was a small area where he could find joy." His most lasting legacy is the internationally recognized meteorite collection that he amassed. The Monnig Meteorite Collection is now housed in the Department of Geology at Texas Christian University in Fort Worth. His collecting started in the early 1930s when meteorites were mere curiosities and the idea of making a collection of these objects was alien to most people. Because the family business required his visiting large areas of the United States southwest, Monnig left word that he would buy any "strange rock that turned out to be a meteorite" as he traveled. As a result he bought many "hen-house door stops" from ranchers and farmers. Also, an important opportunity to collect fresh meteorites occurred when newspapers or radio would report a "fireball." Monnig would try to establish where a possible fall would have occurred and would then go there at the first opportunity, sometimes successfully finding meteorites. Over time and his travels, he built up a network of observers and meteorite hunters around Texas, Oklahoma, and Arkansas. Monnig encouraged others to get involved by publishing a small but well-written brochure describing the appearance and physical characteristics of a meteorite together with information on how to contact him when a suspected meteorite was found. When Monnig could not visit the area of a possible fall, he frequently solicited the help of Robert Brown who traveled to the area to coordinate a search. When a new discovery was made, Monnig often teamed up with **Harvey Nininger**, one of the few other meteorite collectors in the early days. Because his duties as a businessman were primary, Monnig's impact on science exists largely in his collection of many rare meteorites that otherwise might have been lost.

Monnig was one of the charter members of the Society for Research on Meteorites (later known as the Meteoritical Society). In addition to being a fellow of the society, he served as a councilor from 1941 to 1950 and from 1958 to 1966; he also served as the society secretary from 1946 to 1950. In 1984, Monnig was honored as the first recipient of the Texas Lone Star Gazer Award.

On 2 January 1941, Monnig married Juanita Mickle whom he met while pursuing a meteorite fall in East Texas; they had no children.

Arthur J. Ehlmann

Selected References

Ehlmann, Arthur J. and Timothy J. McCoy (1999). "Memorial: Oscar E. Monnig (1902–1999)." *Meteoritics and Planetary Science* 34: 817.

Monnig, Oscar (ed.) *Texas Observers Bulletin*. (Two volumes of this work which exist in the archives of Texas Christian University are thought to be the only relatively complete run of this publication.)

West, John and David Swann. "An Interview with Oscar Monnig and Bob Brown." Parts 1 and 2. *Meteor News* 67 (Oct. 1984): 1–6; 68 (Jan. 1985): 1–5.

West, John O. (1987). "He Touched Unearthly Stones!" *StarDate* 15, no. 10: 3–5.

Williams, Thomas R. (2000). "Oscar E. Monnig, 1902–1999." *Bulletin of the American Astronomical Society* 32: 1682–1683.

Montanari, Geminiano

Born **Modena, (Italy), 1 June 1633**
Died **Padua, (Italy), 13 October 1687**

Geminiano Montanari was the first to recognize the variability of Algol. The son of Giovanni Montanari and Margherita Zanasi, he was left fatherless at age 10. Montanari began his studies in Modena. After an agitated youth, at 20 he went to Florence to study law; he remained there for 3 years. After one of the frequent and violent quarrels he had during his life (to testify a strong character), Montanari left Florence and moved to Salzburg, Austria, where he took a law degree in 1656.

During a stay in Vienna, where he carried on legal practice, Montanari befriended Paolo del Buono, **Galileo Galilei**'s pupil and Florentine diplomat at the imperial court. With his help Montanari took up mathematical and scientific studies, which always interested him.

After a long journey in Austria and the eastern Carpathians, Montanari returned to Italy in 1658, dropped law, and became astronomer to the Grand Duke of Tuscany. In 1661, he returned to Modena as astronomer to Duke Alfonso IV d'Este. There

Montanari married Elisabetta, who collaborated actively with him. Montanari also met the Marquis Cornelio Malvasia, the duke's military chief who was interested in astronomy. In 1649 Malvasia had called the young **Gian Domenico Cassini** to Bologna. Montanari observed with Cassini in Malvasia's private observatory at Panzano, near Bologna. The collaboration produced a large volume, the *Ephemerides novissimae motuum coelestium*, published by Malvasia in 1662.

Montanari became skilled in the design and construction of astronomical instruments including good objective lenses. (One, dated 1666, is in the Museo della Specola in Bologna.) He was one of the first to invent the bifilar micrometer, which he used to draw a lunar map for Malvasia's *Ephemerides*. A device for the exact positioning of telescopes was described in *La livella diottrica* (The leveling diopter) of 1674.

In 1664, with Malvasia's assistance, Montanari obtained the chair of mathematics at the University of Bologna, where Cassini had been teaching astronomy since 1650. His Bologna years were the happiest of his life; he became a promoter of Bolognese cultural development.

In 1667, Montanari discovered that the star Algol (β Persei) was fainter than usual. Between 1668 and 1677 Montanari followed the variability of Algol, sending his results to the Royal Society of London. He also gave a description of Algol's variability in his 1671 *Sopra la sparizione d'alcune stelle ed altre novitá celesti* (On the disappearance of some stars and other celestial novelties), in which he also reported his suspicion of the variability of the star now known as R Hydrae. Montanari perceived neither the regularity nor the period of variation of Algol because of the deterioration of his eyesight, which prevented regular observation. These observations by Montanari struck another blow against Aristotelian immutability, still in contention decades after Galilei's *Sidereus Nuncius*. Montanari's ideas on this phenomenon, which were contrary to popular opinion, were thus violently attacked. Montanari's observational diaries, including his Algol observations, passed to **Francesco Bianchini** and **Eustachio Manfredi**, who recorded them in Bianchini's *Astronomicae ac geographicae observationes* (Verona, 1737); unfortunately the diaries were later lost.

Montanari was also a keen observer of comets and other celestial phenomena, as demonstrated by the observations he made of the meteor that crossed the sky of central Italy in 1676 or those of the comet of 1682, the same observed by **Edmond Halley**. He believed comets to be above the Moon, *pace* the Aristotelians, because he was able to measure the parallax (with a telescope equipped with a micrometer) and the distance, confirming **Tycho Brahe**'s and Cassini's observations. He mistakenly maintained that meteors are similar to lightning and that rocks sometimes found at impact sites are terrestrial in origin.

Typical of natural philosophers of this period, Montanari was also interested in many other natural phenomena. In physics, Montanari experimented on the behavior of liquids in capillary tubes, ascribing it to the major or minus capacity of liquid to stick to the matter of a vessel, which became the topic of a decade-long polemic against Donato Rossetti. Montanari also undertook biological and medical experiments, including blood transfusion between animals in 1668 at Udine. Interested in meteorology, Montanari studied tornados and was the first to use the term "atmospheric precipitation." He noted the utility of the barometer for weather forecasting and as an altimeter. Montanari also devoted himself to ballistics, writing a short manual in which the gunners could find tabulated values for elevations corresponding to gun-ranges. He also worked in hydraulics, leaving the results of his work to his pupil Domenico Guglielmini. With the help of Bolognese and Paduan intellectuals Montanari published several tracts intended to discredit astrological prognostication.

Montanari organized and promoted the Accademia della Traccia (derived from the Florentine Accademia del Cimento), important for scientific debate and for instruction that he gave to his best pupils. It is clear from his works that he was an authentic Galilean, and he maintained a position based on a clear distinction between "metaphysics" and "natural philosophy."

In 1678 Venice created a chair of astronomy and meteors at Padua. Lured by a large salary, Montanari took the position. In addition to teaching, for which he was famous for clarity and scientific rigor, he performed other duties for the Republic, from inspection of rivers and the protection of the lagoon of Venice, to fortifications, artillery training, and organization of the mint and monetary problems. Due to all these duties, some dangerous to his health, Montanari became blind. He died suddenly of an apoplectic stroke.

Fabrizio Bònoli
Translated by: *Giancarlo Truffa*

Selected Reference

Bònoli, F. and E. Piliarvu (2001). *I lettori di astronomia presso lo Studio di Bologna dal XII al XX secolo*. Bologna: Clueb.

Moore, Joseph Haines

Born **Wilmington, Ohio, USA, 7 September 1878**
Died **Oakland, California, USA, 15 March 1949**

American stellar spectroscopist Joseph Moore has as his monuments a catalog of the radial velocities of all northern stars brighter than visual magnitude 5.51, compiled with **William Campbell**, and the third, fourth, and fifth editions of the definitive catalog of orbital elements of spectroscopic binaries. He was the only child of John Haines and Mary Ann (*née* Haines) Moore. In 1907, Moore married Fredrica Chasa, a Vassar College, New York, graduate who had come to Lick Observatory as a computer; they had two daughters, Mary Kathryn (Gates) and Margaret Elizabeth (Matthews). Moore received his postsecondary education at Wilmington College, whence he graduated AB in classics in 1897, and at Johns Hopkins University, from which he received his Ph.D. in 1903. He was elected to the United States National Academy of Sciences in 1931 and was a fellow (and in 1931 vice president) of the American Association for the Advancement of Science. Moore twice served as president of the Astronomical Society of the Pacific (1920 and 1928) and was vice president of the American Astronomical Society in 1942.

Moore came from a Quaker family and remained a lifelong member of the Society of Friends. Although he began his postsecondary education in the field of classics, Moore was influenced in his last year at Wilmington College by W. W. Bennett, who taught

astronomy there and encouraged him to use the 12-in. reflector belonging to the college. At Johns Hopkins University, Moore studied astronomy under **Simon Newcomb** and physics under **Henry Rowland**, J. S. Ames, and **Robert Wood**. His doctorate was awarded for studies of the spectroscopy of the fluorescence and absorption of sodium vapor. At graduation, Moore was offered by Ames the choice of instructorships in physics at Harvard, Yale, or the University of California (Berkeley), or the position of assistant in spectroscopy at Lick Observatory under Campbell. He chose the latter although the salary was lower, and joined the staff of Lick Observatory on 1 July 1903. Moore was promoted to assistant astronomer in 1906. From 1909 to 1913, he was astronomer-in-charge of the Lick Obervatory's southern station in Chile. On his return, he was promoted to associate astronomer in 1913, and astronomer in 1918. He held the position of assistant director from 1936 to 1942 and was director from 1942 to 1945. Although Moore relinquished the directorship, for medical reasons, in November of 1945, he continued as astronomer and also taught courses at Berkeley until 1948.

Moore's first major task at the Lick Observatory was to assist Campbell in the determination of the radial velocities of all stars brighter than visual magnitude 5.51 that were observable from the Lick Observatory. This resulted in a major catalog in Volume 16 of the *Publications of Lick Observatory*. In 1932, he published *A General Catalogue of the Radial Velocities of Stars, Nebulae, and Clusters* (the predecessor of **Ralph Wilson**'s later catalog). Moore was author or joint author of three of the five *Catalogs of Spectroscopic Binaries* published by the Lick Observatory – a task that the present writer, although he never met Moore, eventually inherited from him. The last catalogue of which Moore was an author (with F. J. Neubauer) was published shortly before Moore's death. All these catalogs remained important reference works to other astronomers for several decades.

Moore took part in five expeditions from the Lick Observatory to observe total solar eclipses. Three were within the United States, one in Mexico, and one in Australia. All but the Mexican expedition, for which the weather was poor, were successful and produced useful results. Moore was in charge of the last two expeditions, in California in 1930 and to Maine in 1932. Moore's special interest in eclipses was the study of the spectrum of the solar corona. This spectrum had been first observed by **Charles Young** during the eclipse of 1868. Until 1931, when **Bernard Lyot** succeeded in observing the corona without waiting for a total eclipse, its spectrum could be observed only during the brief moments of totality. Moore's work thus helped to lay the foundations of our knowledge of the solar corona and of the physical conditions of the matter within it. Campbell and Moore also discovered the doubling of the emission lines in spectra of some planetary nebulae, which they attributed to rotation, but which is now understood as evidence for expansion of the nebulae.

Alan H. Batten

Selected References

Aitken, Robert G. (1949). "Joseph Haines Moore: 1878–1949." *Publications of the Astronomical Society of the Pacific* 61: 125–128.

Marsden, Brian G. (1974). "Moore, Joseph Haines." In *Dictionary of Scientific Biography*, edited by Charles Coulston Gillispie Vol. 9, pp. 503–504. New York: Charles Scribner's Sons. (Moore's work with Campbell on the radial velocities of bright stars is to be found in Volume 16 of the *Publications* of the Lick Observatory 1928, and the radial-velocity catalog is in Vol. 18 of the same *Publications* (1932).)

Moore, J. H. (1924). "Third Catalogue of Spectroscopic Binary Stars." *Lick Observatory Bulletin* 11, no. 355: 141–185.

——— (1932). "The Lick Observatory-Crocker Eclipse Expedition to Freyburg, Maine, August 31, 1932." *Publications of the Astronomical Society of the Pacific* 44: 341–352. (See also accounts by W. H. Wright, Donald H. Menzel, and C. D. Shane in the same issue.)

——— (1936). "Fourth Catalogue of Spectroscopic Binary Stars." *Lick Observatory Bulletin* 18, no. 483: 1–38.

Moore, J. H. and F. J. Neubauer (1948). "Fifth Catalogue of the Orbital Elements of Spectroscopic Binary Stars." *Lick Observatory Bulletin* 20, no. 521: 1–31.

Neubauer, F. J. (1949). "J. H. Moore – A Good Neighbor." *Sky & Telescope* 8, no. 8: 197–198.

Wright, W. H. (1956). "Joseph Haines Moore." *Biographical Memoirs, National Academy of Sciences* 29: 235–251.

Moore-Sitterly, Charlotte Emma

Born **Ercildoun, Pennsylvania, USA, 24 September 1898**
Died **Washington, District of Columbia, USA, 3 March 1990**

Spectroscopist Charlotte Emma Moore-Sitterly devoted her professional career to atomic spectroscopy, providing a wealth of vitally needed basic astrophysical data. Moore obtained a BA degree in mathematics from Swarthmore College, Pennsylvania, USA, in 1920. (She would be awarded an honorary doctorate in 1962.) After graduation, Moore moved to Princeton University Observatory, where she became a computational assistant to **Henry Norris Russell**, the director. She also attended graduate courses at the university. Thus began a lifelong association with Russell, until his death in 1957, and Moore's work in two broad, largely separate, fields of fundamental astrophysics.

Russell was engaged in the determination of physical properties of binary stars and on the analysis of stellar spectra based on laboratory data. The first centered on the analysis of eclipse light curves and radial velocity variations for binary stars with the aim of determining accurate masses and radii with which to compare stellar models. The data also yielded dynamical parallaxes for many of the systems (over 1,700) that supplemented trigonometric parallax determinations, especially those compiled at Yale University. The summary monograph resulting from this extended study (1940) served for decades as the standard compilation of fundamental dynamical properties for stars.

The second, atomic spectroscopy, is the field with which Moore is most closely associated. On her arrival at Princeton, she was put to the task of identification of atomic lines in stellar spectra. Laboratory spectroscopy saw a rapid advance in the last quarter of the 19th and first quarter of the 20th centuries. The introduction of electric furnaces, improved vacuum technology, broad wavelength sensitivities of photographic plates, and availability of high-resolution diffraction gratings and long-focus cameras paralleled improvements in sample purity produced for laboratory analysis. In astrophysics, however, quantitative analysis of stellar spectra had not yet started. Progress was hampered by low resolution and poor sensitivity, and by barely developed theoretical means for handling line formation in stellar atmospheres.

This began to change around the time of Moore's arrival at Princeton. By 1920, line identifications were available for a significant number of lines in the solar spectrum and, by extension, to a broad range of stellar spectra. Although lacking a firm theoretical basis with which to explain the appearance of these spectra, the still nascent quantum theory – represented by the Bohr atom and Sommerfeld's explanation of fine structure – at least provided some organizing principles. The analysis was extended to the Stark effect and Zeeman effect, all of which showed that similar groups of lines could arise from coupled energy levels. The recognition of the spin quantum number in 1925, and rapid progress on a vector model of the atom by Alfred Lande, set the stage for Moore's most valuable contributions. (The vector mode was elaborated by Russell and **Frederick Saunders** – so called spin-orbit or L–S coupling.) These were the multiplet table of the elements and compilation of atomic energy levels.

There were two critical steps involved here. First, accurate ionization energies were required to provide a zero point to the energy levels for each element and the identification of the ground-state (resonance) transitions. By the early 1930s, extensive laboratory intensity and wavelength tables were available through Massachusetts Institute of Technology, California Institute of Technology, and other laboratories; the largest sets were the so called Harrison and Paschen tables. But these were limited by the relatively low temperatures that could be achieved in sparks and furnaces. In contrast, the Sun has been used as a standard for spectroscopic measurements, since the discovery of the absorption spectrum by **William Wollaston** and **Joseph von Fraunhofer** early in the 19th century.

Spurred by the accumulation of laboratory data, particularly by **Robert Bunsen**, **Gustav Kirchhoff**, **Norman Lockyer**, and others, **Henry Rowland**, between 1895 and 1897, published the first wavelength list of the photosphere on the Ångstrom scale with identifications for 39 elements (of which several were later shown to be spurious) between 2,975 Å and 7,350 Å. This served as the benchmark list.

The construction of the Snow telescope at Mount Wilson, the first dedicated observatory in the United States for solar spectroscopy and imaging, provided an opportunity to update the list as new spectra covering the same wavelength range became available. Moore was hired by **Charles St. John** to assist with the task. She remained at Mount Wilson from 1925–1928, holding the job title of "computer."

Laboratory standards had by then produced an absolute scale, and this was adopted for the line list. The published list contained identifications for 57% of the detectable lines and assignment of 57 elements. Moore continued her work at Mount Wilson through 1932, and obtained her Ph.D. from the University of California at Berkeley in 1931 with a thesis on sunspot spectra. This represented a major step in the analysis of the solar spectrum, since both magnetic-field data and lower-temperature species (including molecules) were observed. The list was published in 1932 and 1933.

Moore returned to Princeton in 1931 as a research assistant (1931–1936) and then as a research associate (1936–1945), working mainly with Russell and remaining until 1945 when she moved to the National Bureau of Standards [NBS] to work in the spectroscopy section that was headed by William Frederick Meggers. During her second sojourn at Princeton, she married the physicist Bancroft Sitterly and was awarded the second Cannon Prize of the American Astronomical Society for her thesis work.

The product of this period of Moore's life, *A Multiplet Table of Astrophysical Interest*, became the fundamental reference for all astrophysical line identifications. It assigned spectroscopic terms and energy levels to almost 26,000 lines for up to the second ionization state of elements through thorium. First appearing in 1933 in the *Princeton University Observatory Publications* (under the title "Spectrum Lines of Astrophysical Interest"), it was revised twice, in 1945 and again in 1972. The last was issued by the NBS and received very wide circulation. It was soon followed by *An Ultraviolet Multiplet*, which extended the line lists and analyses into the vacuum ultraviolet [UV]. Also resulting from this effort was a series on atomic energy levels, which went through several revisions after first issue by NBS in 1949, a compilation that served as a basic reference for all spectrum predictions.

The utility of these publications extends far beyond their original intention – without these lists, modern stellar atmospheres analyses would be impossible. After Moore's death, they were merged with critically evaluated atomic energy levels and quantum mechanical line strengths to become the National Institute of Science and Technology [NIST] electronic database of atomic line identifications. When the UV spectral window opened after World War II, through the use of sounding rockets and (ultimately) satellites, the UV multiplet table became an indispensable guide to the new territory. Although frequently mentioned as a critical need in her publications, no similar multiplet table was produced for the infrared during Moore's lifetime. Her tables still are widely cited by atomic physicists and chemists, as well as astronomers.

The final step in the evolution of Moore's analyses of the Sun came in 1966 with the publication of the line strengths and wavelengths in the combined solar spectrum using equivalent widths based on the intensity-calibrated tracings obtained at the University of Utrecht, the Netherlands. The work, which was sponsored by a resolution of the International Astronomical Union [IAU], formed the basis for the determination of elemental abundances in the Sun. By its completion, around 24,000 lines were measured with identifications for almost 73%.

Moore remained at NBS, now the NIST, until her retirement in 1968. Thereafter, she joined the Naval Research Laboratory [NRL] and remained affiliated with NRL until her death. She was awarded the Meggers Prize of the Optical Society of America (1972) and the Bruce Medal of the Astronomical Society of the Pacific (1990). She served as vice president of the American Association for the Advancement of Science (1952) and IAU Commission 14 (1961–1967).

Steven N. Shore

Selected References

Anon. (1988) "Atomic Spectroscopy in the Twentieth Century: A Tribute to Charlotte Moore-Sitterly on the Occasion of Her Ninetieth Birthday." *Journal of the Optical Society of America* B, 5. (A special issue dedicated to Moore; see Bengt Edlén and William C. Martin, "Introduction," p. 2042; and Karl G. Kessler, "Dr. Charlotte Moore Sitterly and the National Bureau of Standards," p. 2043.)

Condon, E. U. and G. Shortly (1951). *The Theory of Atomic Spectra*. Cambridge: Cambridge University Press.

Garton, W. P. S. and W. C. Martin (1991). "C. M. Sitterly (1898–1990)." *Quarterly Journal of the Royal Astronomical Society* 32: 209–210.

Moore, Charlotte E. (1949–) *Atomic Energy Levels as Derived from the Analyses of Optical Spectra*. Washington, DC: National Bureau of Standards.

——— (1950–1952). *An Ultraviolet Multiplet Table*. Washington, DC: National Bureau of Standards.

——— (1972). *A Multiplet Table of Astrophysical Interest*. Rev. ed. Washington, DC: National Bureau of Standards.

Moore, C. E., M. G. J. Minnaert, and J. Houtgast (1966). *The Solar Spectrum 2935 Å to 8770 Å*. Washington, DC: National Bureau of Standards.

Roman, Nancy G. (1991). "Charlotte Moore Sitterly." *Bulletin of the American Astronomical Society* 23: 1492–1494.

Russell, H. N. and C. E. Moore (1940). *The Masses of the Stars*. Chicago: University of Chicago Press.

St. John, Charles E. *et al.* (1928). *Revision of Rowland's Preliminary Table of Solar Spectrum Wave-Lengths*. Washington, DC: Carnegie Institution of Washington.

Morgan, Augustus de

Born **Madura (Madurai, Tamil Nadu, India), 27 June 1806**
Died **London, England, 18 March 1871**

Though not an astronomer, Augustus de Morgan, one of the most notable British mathematicians and logicians of the 19th century, served the Royal Astronomical Society in leadership positions for over 3 decades, including his service on the council, and as secretary and editor of the *Monthly Notices*. His influence on the organization and its members was substantial and positive.

De Morgan's father John was a lieutenant colonel in the British Army in India. Born with only one eye, de Morgan was raised in England. Though he did poorly at school, at the age of 16 de Morgan entered Cambridge University where he studied under George Peacock, professor of astronomy and geometry, and **William Whewell**, with both of whom he remained friends. Peacock, along with **John Herschel** and Charles Babbage, formed the Analytical Society famous for introducing to Cambridge advanced German and French methods of calculus and helping to develop a purely symbolic algebra. De Morgan took his BA in 1826, but because of his strong objections to the theological test required at Cambridge, he did not get a fellowship or proceed to the MA. He read for the bar in London but, in 1828, with no mathematical credentials, he was awarded the first professorship of mathematics at the new University College, London. De Morgan held the post until 1831 when he resigned on a matter of principle; he held the post a second time from 1836 to 1866, when he resigned, again on a matter of principle, once again on theological strictures, but now applied to others rather than to himself.

The publication of de Morgan's *Elements of Arithmetic* (1831) was a significant advance in providing a mathematically rigorous yet philosophically sophisticated treatment of number and magnitude useful for scientific applications. De Morgan coined the term "mathematical induction" to differentiate once and for all the purely formal technique of advancing from number n to $n+1$ (used in mathematical proofs) from the purely empirical method of hypothetical induction in science. He saw the far-reaching applications of algebraic and numerical analysis to science, and was himself fascinated by purely algebraic and numerical applications to purely empirical problems; it was he, for instance, who produced the first almanac of Full Moons (from 2000 BCE to 2000) and showed how probability theory can be used, for instance, in predicting catastrophic events in life, a technique in use today by insurance companies throughout the world. His *Trigonometry and Double Algebra*, first published in 1849, provided the first thoroughly geometric interpretation of complex numbers, which further extended their application in engineering and astronomical calculations.

De Morgan also made important contributions to symbolic logic; he saw, more than any other British luminary of the time (except, perhaps, George Boole), that logic as it had been passed down from **Aristotle** was severely handicapped in scope, due in large part to a paucity of rigorous mathematical symbolism. He showed that many more valid inferences are possible than were envisioned by Aristotle, using formulas such as the ones now known as De Morgan's Law:

$$\sim(p \vee q) = \sim p \wedge \sim q, \text{ and } \sim(p \wedge q) = \sim p \vee \sim q.$$

These laws of converses and contradictions state, in English, that the truth value of the negation, or contradictory, of the disjunction of two propositions, is the same as the conjunction of the negation of each of the propositions; likewise, the truth value of the negation, or contradictory, of the conjunction of two propositions is the same as the disjunction of the negation, or contradictory, of each of the propositions. In his *Formal Logic*, de Morgan uses the important new concept of quantification of the predicate to solve problems that were simply unsolvable in classical Aristotelian logic; when Sir William Hamilton accused him of stealing the idea from him, de Morgan replied that it was Hamilton who was the plagiarist, a charge that seems to have been settled in de Morgan's favor. His *Budget of Paradoxes*, published in 1872 and reprinted in 1954 with a new introduction by the great philosopher of science, Ernest Nagel, is a paradigm debunking book; in it, de Morgan shows step by step the fallacies by which frauds, cranks, and pseudoscientific tricksters continue to this day to titillate the public with extraordinary but ultimately false claims.

De Morgan became a fellow of the Royal Astronomical Society in 1828, joining the council in 1830. He was twice secretary of the society (1831–1838, 1848–1854). Though he was asked to become president of the society, he declined on the basis that, in his view, only practicing astronomers should assume that responsibility. In 1837 de Morgan married Sophia Elizabeth Frend, daughter of a mathematician/actuary.

Daniel Kolak

Selected Reference

Dubbey, John M. (1971). "De Morgan, Augustus." In *Dictionary of Scientific Biography*, edited by Charles Coulston Gillispie. Vol. 4, pp. 35–37. New York: Charles Scribner's Sons.

Morgan, Herbert Rollo

Born **near Medford, Minnesota, USA, 21 March 1875**
Died **Washington, District of Columbia, USA, 11 June 1957**

Yale University's Herbert Morgan prepared the N30 Catalog (1952), providing proper motions for 5,268 stars, updated since **August Kopff**'s *Fundamental Katalog des Berliner Astronomischen Jahrbuch.*

Selected Reference

Scott, F. P. (1957). "Morgan, H. R." *Science*. 126: 457.

Morgan, William Wilson

Born **Bethesda, Tennessee, USA, 3 January 1906**
Died **Williams Bay, Wisconsin, USA, 21 June 1994**

American spectroscopist William Wilson Morgan gave his name both to the Morgan–Keenan–Kellman [MKK] system of classification of spectra of stars and to the Johnson–Morgan [UBV] system of measuring stellar colors. Both are still in use.

Morgan's town of birth no longer exists. He was the son of Protestant home missionaries, and as a result of the family's itinerant life style, received his schooling at home, in Florida, Colorado, and Missouri, before finishing high school in the nation's capital. A Latin teacher in Missouri first turned his attention to astronomy with a view of the Moon through a theodolite mounted near a library window. Morgan, whose mother had been born in Virginia, started college at Washington and Lee University (Lexington, Virginia) in 1923 and had planned to complete a degree in English there. But his physics and astronomy professor, Benjamin Q. Wooten, spent the summer of 1926 at Yerkes Observatory and learned that the director, **Edwin Frost** (a spectroscopist who had been nearly blind for some years), needed an assistant to obtain daily spectroheliograms as part of the observatory's routine work.

Morgan was offered the job, and took it up, over his father's objections, in fall 1926. He remained at Yerkes Observatory the rest of his life. Morgan completed a BS in mathematics and physics (largely by correspondence) at the University of Chicago in 1927, and a Ph.D. in 1931, working with **Otto Struve** (who had succeeded Frost as director at Yerkes), with a thesis on the spectra of type A stars displaying a variety of spectral peculiarities, now called Ap stars and known to have strong magnetic fields that facilitate concentrations of rare elements like europium on their surfaces. Morgan was appointed an instructor in 1932 and progressed to full professor, becoming department chair at Chicago for 1960 to 1966 and director of both Yerkes Observatory and McDonald Observatory (1960–1967). He held a distinguished service professorship (1966–1974) and an emeritus position the rest of his life. He served as editor of the *Astrophysical Journal* from 1947 to 1952. Morgan's own Ph.D. students at Chicago included at least 16 who became professional astronomers, including J. Allen Hynek and **Armin Deutsch**.

Morgan's approach to stellar classification and measurement was an innovative, morphological, process-based one. He did not seek to calibrate any particular features in terms of stellar temperature, density, or composition. Rather, working with **Philip Keenan** and Edith Kellman, he chose specific stars to be standards of each type, and other stars were typed by comparing their spectra (obtained with similar telescopes and spectrographs) with a set of images of the standard ones. Their *Atlas of Stellar Spectra with an Outline of Spectral Classification* was published in 1943, with revisions in 1951 and 1973 by Morgan and Keenan. In the early 1950s, together with Harold L. Johnson (died: 1980), Morgan developed a three-color system for stellar photometry. The V color was nearly equivalent to the traditional visual magnitude, B was a blue color (but longward of the Balmer jump), and U (for ultraviolet) was shortward of the Balmer jump. The spectral classes and the color system together were an extraordinarily powerful tool for stellar astrophysics. The spectral type revealed what the intrinsic color and brightness of a star must be, and then the measured UBV brightness revealed the distance to the star, how much reddening and absorption of light had occurred along the way, and something about the abundances of heavy elements in the star.

With graduate students Stewart Sharpless and Donald Osterbrock, Morgan applied these techniques to measure the distances to clusters and associations of hot (OB) stars in the plane of the Milky Way. They announced in 1951 that these associations were concentrated in a few spiral arms, showing clearly for the first time that the Milky Way is a spiral galaxy, as had been suggested half a century before by **Cornelis Easton**. The discovery of a spiral pattern in the distribution of hydrogen gas in the galactic plane from its radio emission at 21 cm both confirmed and unfairly eclipsed their work.

Beginning in the 1950s, Morgan also turned his attention to galaxies. He and **Nicholas Mayall** were able to show that the integrated spectrum of a galaxy contains a good deal of information about the kinds of stars that contribute most of the light (typically populations of giants and supergiants of different ages and compositions). Morgan also worked on a more morphologically based system for the classification of the appearance of galaxies, independent of that evolved by **Edwin Hubble**. An important aspect of the system was the recognition of the special status of the very large galaxies found at the centers of some rich clusters, to which he assigned the types N (meaning nucleated) and cD (meaning supergiant, diffuse; the letter c having been the label given to supergiant stars by **Antonia Maury** some 60 years before).

Morgan received medals or other awards from the United States National Academy of Sciences, the Royal Astronomical Society, the American Astronomical Society (serving as vice president; 1968–1970), and the Astronomical Society of the Pacific. He was a member or associate of scientific academies in the United States, Denmark, Belgium, and Argentina and of the Pontifical Academy of Sciences. He married in 1928 Helen Barrett (died: 1963), the daughter of Yerkes astronomer Storrs Barrett, and, in 1966, Jean Doyle Eliot.

William Sheehan

Selected References

Garrison, Robert F. (ed.) (1984). *The MK Process and Stellar Classification.* Toronto: David Dunlap Observatory. (For the ongoing influence of the MK process.)

Osterbrock, Donald E. (1997). "William Wilson Morgan." *Biographical Memoirs, National Academy of Sciences* 72: 1–30.

——— (1997). *Yerkes Observatory, 1892–1950: The Birth, Near Death, and Resurrection of a Scientific Research Institution.* Chicago: University of Chicago Press. (For the history of Yerkes Observatory, including an excellent treatment of Morgan's work in its institutional context.)

Morin, Jean-Baptiste

Born **Villefranche-sur-Saône, (Rhône), France, February 1583**
Died **Paris, France, 6 November 1656**

Jean-Baptiste Morin, a noted astrologer and notorious controversialist, defended **Johannes Kepler**'s elliptical orbits while at the same time attacking **Nicolaus Copernicus**'s heliocentric cosmology. Baptized on 23 February 1583, his birthdate is unknown. Little is known of Morin's family origins, and he never married. After studying philosophy with Marc Antoine at Aix-en-Provance (1609–1610), Morin took degrees at Avignon (BA: 1611, MD: 1613) where he likely met **Pierre Gassendi** and began his studies with the provincial astronomer **Joseph Gautier**. Always successful in attracting patronage, Morin was first supported by Claude Dormy, Bishop of Boulogne, who sent him to Germany and Hungary to visit mines and to conduct research on metals. Thereafter, Morin received patronage from Leon d'Albert, Duke of Luxembourg, before finally being appointed professor of mathematics at the Collège royal, a post he held from 1630 until his death. It is likely Morin received this appointment through Cardinal Richelieu, no doubt due to his skill in astrology. Quarrelsome in his disputes, Morin consistently opposed Copernicanism and the New Science.

As an astronomer, Morin is chiefly remembered for his efforts to determine longitude at sea. The ensuing debate – like the whole of Morin's career – was surrounded by bitter controversy. His first public pronouncement came in 1633 in a conference at Théophraste Renaudot's Bureau d'adresse, and thereafter a string of pamphlets added fresh fuel to a debate that dominated the decade. Morin's method was sound in principle but impractical. Derived in part from the methods of **Gemma Frisius** (employing spherical triangles but rejecting the use of clocks), it differed in emphasizing the measurement of lunar distances from a fixed star. But the method required especially accurate ephemerides and precise observations of angular distances. The controversy involved the most celebrated astronomers of the day, and given the influence of Richelieu, several commissions were established that included Abbé Chambon, Claude Mydorge, Étienne Pascal, Pierre Hérigone, and later Jean de Beaugrand. Despite receiving 1,000 livres to develop a suitable quadrant, Morin failed to demonstrate his theory. For his part, Morin boasted he had discovered new methods for finding parallaxes and refractions, the equation of time, the height of the pole, the obliquity of the ecliptic, and for determining the more subtle lunar motions.

Morin's public humiliation soon spawned new controversy. Shifting the debate from longitude, Morin launched a barrage of attacks against Copernicanism and atomism, sometimes defending **Aristotle** (1624) against the mechanical philosophy. Morin railed against **Galileo Galilei**, **René Descartes**, and a dozen others, particularly Gassendi and **Ismaël Boulliau**. Significantly, Morin's arguments against Copernicanism were rooted in astrology, and here he endorsed the Tychonic model. As astrologer and champion of the anti-Copernicans, Morin nevertheless embraced **Johannes Kepler**'s elliptical orbits, further asserting he had improved Kepler's methods and the accuracy of the Rudolphine Tables. **Thomas Streete** and **Nicolaus Kauffman** (Mercator) apparently agreed.

Driven by ambition and animosity, Morin's career tells us a good deal about the troubled context of the New Science. The most important polemicist of his generation, Morin seemed to pick his battles on the basis of political advancement, not scientific merit. As a *caractère mélancolique,* he frequently called his opponents "imbeciles," "imposters," and "ignorant buffoons" as well as "plagiarists," "pygmies," and "pathetic pretenders." Vehement and vulgar, his attacks wreak with scatological references. For himself, Morin reserved the rarified title "Complete Restorer of Astronomy." Boulliau, who did his best to ignore Morin, called him the *astronome papier* (paper astronomer). But Morin's published attacks soon escalated from simple incivility to public charges of impiety and atheism. Unpublished evidence suggests Morin wrote Cardinal Mazarin to expose Gassendi's heresies, he advised that Bernier be arrested, he urged that his enemies and their books be burned, and he maliciously predicted the death of Gassendi and Boulliau. Opinion varied. Gui Patin thought Morin "touched in the head", Pierre Bayle thought him a "fake savant" touched with genius. In any case, Morin broke unspoken rules, and his opponents destroyed his reputation. His methods, theories, and judgment were publicly condemned, his astrology was ridiculed, and his Latin was lampooned. Morin was attacked, avoided, and finally ignored.

Morin's often ugly career in astronomy was supported handsomely by astrology. Good evidence suggests he maintained strong ties with the court of Louis XIII, and by tradition, he was present at the birth of Louis XIV. Morin dedicated a number of publications to Richelieu and regularly cast horoscopes for king, queen, and the royal family. Mazarin awarded him an annual pension of 2,000 livres in 1645, and he reportedly earned 4,000 livres per year as astrologer, a princely sum. Following his death, Morin's *magnum opus*, the *Astrologia gallica* finally came to press in 1661, a large folio edition of some 850 pages. The most influential work of its kind, Morin extended principles pioneered by **Johann Müller** (Regiomontanus), boasting new astrological theories of elections and astrological houses. The volume appeared posthumously with assistance from the Queen of Poland, Louise-Marie de Gonzague, who supplied 2,000 thalers toward publication. The recommendation came from her secretary, Pierre Desnoyers, a major correspondent of Gilles Personne de Roberval and Boulliau. An advocate of both astrology and the New Science, Desnoyers, like many contemporaries, used the terms "astronomer" and "astrologer" interchangeably.

Morin was buried at Saint-Étienne-du-Mont. A decade later, in 1666, J.-B. Colbert proclaimed academics could no longer publish on astrology, and in 1682, by royal proclamation, astrological almanacs were forbidden in France. In the end, if Morin was the last great astrologer, he has yet to receive the historical attention he legitimately deserves.

Robert Alan Hatch

Selected References

Bayle, Pierre. *Dictionnaire historique et critique.* New ed. Vol. 10. Rotterdam, 1697; Paris, 1820.

Delambre, J. B. J. (1821). *Histoire de l'astronomie moderne.* Vol. 2, pp. 236–274. Paris: Courcier.

Martinet, M. (1986). "Jean-Baptiste Morin (1583–1656)." In *Quelque savants et amateurs de Science au XVIIe siècle*, edited by Pierre Costabel and Monette. Paris., pp. 69–87.

Moréri, Louis (1759). *Le grande dictionnaire historique.* Vol. 6, pp. 785–788. Paris: Libraires Associés.

Pares, Jean (n.d.) "Jean-Baptiste Morin (1583–1656) et la querelle des longitudes de 1634 à 1647." Ph.D. diss., University of Paris.

Tronson, Guillaume (1660). *La vie de Maistre Jean Baptiste Morin.* Paris.

Morley, Edward Williams

Born **Newark, New Jersey, USA, 29 January 1838**
Died **West Hartford, Connecticut, USA, 24 February 1923**

American chemist Edward Williams Morley is best known for his collaborative experiments with physicist **Albert Michelson**, which failed to detect an "ether-drift" effect on the speed of light measured in different directions relative to the Earth's motion. The negative result of the Michelson–Morley experiments may have helped inspire **Albert Einstein**'s special theory of relativity.

The son of a Congregational Minister, Morley received his BA (1860) and MA (1863) from his father's *alma mater*, Williams College, Williamstown, Massachusetts. There he studied with astronomer Albert Hopkins, who, during Morley's postgraduate year, guided his mounting of a transit instrument with which Morley then measured the college's latitude from observations of stars. In 1865, Morley read a paper about his results before the American Academy of Arts and Sciences and published it in their *Proceedings*. His work on these observations and their reduction has been credited with instilling in him the careful experimental nature that later led to his greatest successes. (Williams College's Morley Science Center was named for him in 2000.)

Intending to follow in his father's professional footsteps as well, Morley studied at Andover Theological Seminary, where he earned a license as a minister of the gospel in the Congregational Church (1861–1864). Morley spent the final months of the Civil War as a relief agent in the United States Sanitary Commission at Fort Monroe, Virginia, attending convalescing Union soldiers. He then returned to the seminary, where he studied until he was appointed teacher at South Berkshire Academy, Marlboro, Massachusetts.

In 1868 Morley moved to Ohio, where he briefly served as pastor of a Congregational Church in Twinsburg. Preaching did not agree with him, however, and he accepted an appointment to the faculty of Western Reserve College (1869), where he began a program of chemistry teaching and research, and where he served as chairman of the chemistry and natural history programs (1882–1906). Morley also taught chemistry and toxicology at the Cleveland Medical School (1873–1888). His interest in chemistry dated back to his boyhood, when he devoured not only popular works but also textbooks.

In 1887, he gave a series of lectures on Charles Darwin, natural selection, and evolution, which he preferred to call "development".

Over the course of his career, Morley focused on three significant problems. In the 1870s, he refined the techniques of gas analysis to demonstrate that colder air contained less oxygen than warmer air, thus proving the theory of Yale meteorologist **Elias Loomis** that attributed winter cold snaps to cold air falling from high altitudes.

In the 1880s, in order to evaluate the hypothesis of English chemist William Prout that all the elements were based on hydrogen, Morley developed procedures to determine the precise relative atomic weights of oxygen and hydrogen that eliminated impurities in the gases he tested, as well as every likely source of experimental error. Morley's determination in 1895 that the relative atomic weight of oxygen was 15.789 led him to conclude that Prout's hypothesis was invalid. In recognition of Morley's apparent resolution of a long-lived problem, Morley's colleagues elected him to the National Academy of Sciences (1897) and to the presidencies of the American Association for the Advancement of Science (1895–1896) and of the American Chemical Society (1899).

Yet Morley is best remembered for his work in the 1880s in collaboration with physicist A. A. Michelson of the Case School of Applied Science (later the Case Institute of Technology and then part of Case Western Reserve University), which failed to detect any "ether drift," taken as a given by all the contemporary wave theories of light. One way in which Morley improved Michelson's techniques was to eliminate sources of optical distortion by mounting his interferometer on a stone platform floating in mercury. Neither scientist, however, believed the null result of their efforts, and after Michelson's departure from Case in 1888, Morley continued to search for ether drift in collaboration with Michelson's successor, Dayton C. Miller.

Miller, who did not give up easily, reported a positive result at the December/January 1924–1925 meeting of the American Association for the Advancement of Science, for which he received a prize designated for the best paper of the meeting. Corecipient was **Edwin Hubble**, for his paper on the discovery of Cepheid variables in spiral nebulae, establishing their nature as separate galaxies, well outside the Milky Way.

Morley was married to Isabel Ashley Birdsall; they had no children. After Morley's retirement to West Hartford, Connecticut, in 1906, he stayed active by analyzing, in the chemistry laboratory at his home, geological samples collected in Indonesia by a neighbor. He traveled to England to accept the Davy Medal of the Royal Society in 1907 at the hands of Lord Rayleigh.

Naomi Pasachoff and *Jay M. Pasachoff*

Selected References

Hamerla, R. R. (2003). "Edward Williams Morley and the Atomic Weight of Oxygen: The Death of Prout's Hypothesis Revisited." *Annals of Science* 60: 351–372.

Livingston, Dorothy Michelson (1973). *The Master of Light: A Biography of Albert A. Michelson*. New York: Charles Scribner's Sons. (Discusses the history and significance of the Michelson–Morley experiment.)

Williams, Howard R. (1957). *Edward Williams Morley: His Influence on Science in America*. Easton, Pennsylvania: Chemical Education Publishing Company. (Contains a full list of Morley's publications.)

Morrison, Philip

Born **Somerville, New Jersey, USA, 7 November 1915**
Died **Cambridge, Massachusetts, USA, 22 April 2005**

American theoretical physicist and astrophysicist Philip Morrison is most often mentioned in connection with the 1959 suggestion, with Guiseppe Cocconi, that the most natural means of communication across interstellar distances would be radio waves, probably at a wavelength close to the 21 cm emitted by neutral atomic hydrogen gas.

Born to Moses and Tilly (*née* Rosenbloom) Morrison, Philip attended high school in Pittsburgh, Pennsylvania, and received a BS in physics in 1936 from the Carnegie Institute of Technology (now Carnegie Mellon). He earned a Ph.D. from the University of California (Berkeley), working in theoretical physics with **Robert Oppenheimer**, in 1940 and moved quickly through teaching positions at San Francisco State College (1940–1941) and the University of Illinois (1941–1942), where he was recruited by fellow Oppenheimer student Robert F. Christy for the Metallurgical Lab (atomic bomb project) at the University of Chicago.

Morrison worked initially on the design of the reactors to be built in Hanford, Washington, for producing plutonium from uranium. By 1944, as the Manhattan Project became more intense, he and many others moved to Los Alamos, New Mexico, site of the laboratory under Oppenheimer's direction. Morrison and Marshall Holloway were responsible for the final readiness and assembly of the plutonium bombs, both the one tested at the Trinity site on 13 July 1945 and the one dropped over Nagasaki on 9 August (for which he traveled to the Tinian Island take-off point). They and colleagues **Luis Alvarez** and Robert Serber wrote a letter, also to be dropped over Japan, which they hoped would reach a Japanese colleague and make clear the United States could deploy additional bombs if necessary. Morrison was also part of the assessment team sent to Japan soon after. The experience left him a life-long opponent of nuclear war, and indeed war in any form. He was for many years active in the Federation of American Scientists (serving as its chairman: 1972–1976), a group supporting peaceful uses of science.

After another year at Los Alamos (directing a plutonium reactor that he dubbed Clementine, because of its location "in a cavern, in a canyon . . .") Morrison took up an associate professorship at Cornell University. While there, he initially remained engaged in nuclear physics, coauthoring with **Hans Bethe** a classic text in the field. Morrison moved from Cornell to the Massachusetts Institute of Technology in 1965, officially retiring in 1986, but remaining actively engaged in physics research and teaching until shortly before his death.

Morrison's interests gradually shifted to astrophysics, initially the propagation of cosmic rays in the Solar System – he rightly associated the variation of the flux reaching Earth with solar effects – and their origins (which he associated with supernova remnants like the Crab Nebula and active galaxies). He wrote the first review article on γ-ray astrophysics, before the first source had been seen, predicting in 1958 that supernova remnants and active galaxies should be sources – but so should be interstellar space if sufficiently high-energy cosmic rays hit neutral gas making pions. All indeed are, though not such bright sources as Morrison had hoped.

It is worth noting that in 1959 when Cocconi and Morrison suggested the 21-cm search wavelength, it lay in an accessible region between the short-wave emission by the Earth's atmosphere and the longer wavelength emission by the Galaxy. The 3-K microwave background radiation had not yet been discovered. The suggestion led to the first modern search for radio signals from extraterrestrials, by Frank Drake in 1960.

Morrison guided a couple of dozen future physicists to and beyond Ph.D.s. Those who remained in cosmic-ray physics or astrophysics included Howard Laster, Kenneth Brecher, James Felten, Leo Sartori, Alberto Sadun, and Minas Kafatos.

Morrison and his students had for many years a remarkable record of bringing concepts of fundamental physics to bear on new astronomical results very quickly. This included:

(1) cooling of stellar remnants by neutrino emission (with Hong-Yee Chiu);
(2) calculation of X-ray production by various mechanisms (with James Felten), which permitted a calculation of how much hot gas would be required to produce the X-rays discovered to be coming from rich clusters of galaxies even as they were finishing the calculation – it turned out to be comparable with the total amount of mass needed to hold the clusters together gravitationally – the modern number is 30% of that;
(3) a fluorescent theory of supernova light emission (which shared some of the virtues with the radioactive decay mechanism that was ultimately adjudged correct);
(4) an analogy between active galaxies and pulsars as soon as the latter were discovered;
(5) a prediction of X-ray emission from the Crab Nebula and radio galaxies (later seen, though the mechanism is probably different);
(6) the suggestion with Brecher that the emission from γ-ray bursts must be beamed into a narrow cone;
(7) the association of a subset of active galaxies (including M 82) with star formation fueled by recent infall of new gas rather than with a central black hole; and
(8) a shadowing mechanism to account for the jet found to be sticking out of the edge of the Crab Nebula in the 1980s.

In spite of all these remarkable theoretical achievements, Morrison may well have made his greatest impact as a communicator of science. For decades, thousands of students in introductory astronomy courses have begun the term by watching a film he coproduced with his wife and Charles and Ray Eames, *Powers of Ten*, which illustrates the enormous range of length scales from subatomic physics to the cosmos. (The book version of 1980 was in collaboration with Phylis Morrison.) He was the narrator and guiding spirit of several educational television programs (*e. g., Whisper from Space*, on the cosmic microwave background radiation, and *Ring of Truth*, a series dealing with how science works), and from 1968 to 1994 wrote virtually all of the book reviews that appeared in *Scientific American*, the December issue always featuring children's books, coreviewed by the Morrisons together. His first marriage, to Emily Morrison of Boston, ended in divorce in 1983, and his second wife and collaborator, Phylis Singer Morrison, died in 2002.

Morrison received honorary degrees from Case Western Reserve, Rutgers, and Denison universities, and medals and prizes from the American Physical Society, the American Chemical Society, and several other organizations. He was elected to the United States National Academy of Sciences.

It is probably useful in understanding Morrison's career to know that a childhood attack of polio left his legs considerably weakened, so that he was dependent on canes for getting around from midlife onward and later on a wheelchair.

Virginia Trimble

Selected References

Morrison, Philip (1995). *Nothing is Too Wonderful to be True*. Woodbury, New York: AIP Press.
Morrison, Philip, John Billingham, and John Wolfe (eds.) (1979). *The Search for Extraterrestrial Intelligence*. New York: Dover.
Trimble, Virginia (2005). "Philip Morrison." *Bulletin of the American Astronomical Society*. 37: 1552–1554.

Möschlin, Michael

➜ **Mästlin [Möstlin], Michael**

Mouchez, Ernest Amédée Barthélemy

Born **Madrid, Spain, 21 August 1821**
Died **Wissous near Paris, France, 25 June 1892**

Ernest Mouchez, a notable hydrographer and cartographer, directed a rejuvenation of the Paris Observatory and helped initiate the *International Carte du Ciel and Astrographic Catalogue* project. Mouchez's parents were French: His father, Jacques Barthélemy Mouchez, was a perfumer and wig maker to the spouses of the Spanish King Ferdinand VII. He was a widower when he married Mouchez's mother, Louise Cécile Bazin. Mouchez was sent to Paris for his studies, living with family friends. After attending the lycée in Paris and Versailles, he entered the École Navale.

After graduation, Mouchez served in the marine as an officer, rising to the rank of *contre-amiral* by 1878. During this period, he produced nearly 150 coastal maps of Asia, South America, and Algeria. (These last maps were the only ones available to Allied troops when they landed in Algeria during World War II.) Mouchez explored some 4,000 km along the Brazilian coast. Mouchez's work in hydrography and geographic astronomy contributed to the training of navigators, particularly for the determination of longitudes. He directed one of the voyages to observe the transit of Venus of 9 December 1874 at Saint Paul Island in the Indian Ocean, obtaining high-quality visual and photographic data.

In 1862, Mouchez married Carlota Finat (born: 1843), the second daughter of his half-sister Sophie, who was 7 years older than him. They had six children; their older daughter later married **Camille Guillaume Bigourdan**.

Mouchez succeeded **Urbain Le Verrier** as director of the Paris Observatory in 1878. In this position, he managed to liberate the astronomers from administration, opened the observatory to foreign exchanges and was a patron, along with Sir **David Gill** of the Cape of Good Hope Observatory, of the *Carte du Ciel*, inaugurated in 1887. This often-criticized enterprise of 18 observatories from all over the world found its value a century later: The accurate proper motions of almost one million stars from the *Carte du Ciel* catalog formed the first epoch for comparison with the "Brahe" catalog produced by the astrometric Hipparcos mission (1989–1992).

Under Mouchez's directorship, the observatory built the coudé equatorial from **Maurice Löwy**'s plans, with which the famous lunar map was obtained, along with **Henri Deslandre**'s spectroheliograph, which long provided standard references. In 1885, Mouchez formed a section on Astronomie physique, combining it with sections for Météorologie and Physique du Globe and placing them all under **Charles Wolf**. In 1890, he asked Deslandres to direct a section called Spectroscopie stellaire. This service, related to the new field of astrophysics, was discontinued when Deslandres moved to the Observatoire d'astronomie physique in Meudon, near Paris. Mouchez was also responsible of the first complete network of synchronous clocks in Paris, of the creation of the Paris Observatory museum, and the organization of regular visits for the public, still in existence. In 1884 he founded the first periodical devoted to astronomy in France called *Bulletin astronomique*. A few hours after presiding over a meeting of the Paris Observatory Council, Mouchez died suddenly at his cottage.

Mouchez was a remarkable organizer, having inherited a difficult situation from Le Verrier. Mouchez developed the Paris Observatory along the lines that **Friedrich Struve** had suggested to Le Verrier when the latter was thinking about reorganizing French astronomy. Mouchez was a member of the Bureau des longitudes from 1873 and of the Académie des sciences from 1875.

Solange Grillot

Selected Reference

Mouchez, Robert (1970). *Amiral Mouchez, Marin, Astronome et Soldat, 1821–1892.* Paris: Editions Cujas.

Moulton, Forest Ray

Born **Osceola County, Michigan, USA, 29 April 1872**
Died **Wilmette, Illinois, USA, 7 December 1952**

Forest Moulton is perhaps best remembered for his collaboration with **Thomas Chamberlin** on what became known as the Chamberlin–Moulton hypothesis.

Moulton was born in a log cabin built by his father Belah Moulton, a Civil War veteran, on the family's 160-acre homestead in Michigan. His mother, Mary (*née* Smith) Moulton, was impressed by rays of sunlight filtering through the surrounding forest; hence her son's name. Moulton was educated at home by his mother, next in a one-room school, and then at Albion College, from which he received his B.A. degree in 1894. He attended the University of Chicago as a graduate student in 1895, was appointed assistant in astronomy in 1896, received his Ph.D. in astronomy and mathematics in 1899, and then joined the university's faculty.

Moulton rose to full professor in the astronomy department. His field was celestial mechanics, including the three-body problem and its application to the motion of the Moon under the influence of the Earth and Sun. His 1902 text, *Introduction to Celestial Mechanics*, later revised in 1914, was widely utilized. More elementary texts were his *Introduction to Astronomy*, published in 1906, and *Descriptive Astronomy*, in 1912. With his colleagues at Chicago, Moulton developed a survey course and the accompanying text, *The Nature of the World and Man*, was published in 1926. It was reissued in 1937, and revised in 1939, as *The World and Man as Science Sees Them*. Moulton also wrote *Consider the Heavens*, a popular book, in 1935. Always interested in popularizing astronomy, Moulton acted as an informal advisor to Frederick Charles Leonard (1896–1960) and his Society for Practical Astronomy that enjoyed some success nationally in the period from 1910 to 1916. Leonard went on to found the astronomy department at the University of California at Los Angeles and established himself as an authority in meteoritics.

During World War I, Moulton was commissioned a major in the Army and placed in charge of the Ballistics Branch of the Army Ordnance Department at Aberdeen, Maryland. There, he developed new methods for calculating the trajectories of artillery projectiles. In 1926, he published *New Methods in Exterior Ballistics*. Also of a technical nature was Moulton's 1930 textbook, *Differential Equations*.

Moulton was a member of an interdisciplinary research team headed by Chamberlin, the chairman of the Geology Department at the University of Chicago. The team investigated mutual problems in geophysics and astronomy. Beginning in 1903, half of Moulton's salary was covered by Chamberlin's grant from the Carnegie Institution of Washington.

In his studies on the Earth's changing climate in the geological past, Chamberlin had begun to question the plausibility of **Pierre de Laplace**'s nebular hypothesis. Laplace had postulated that a hot fluid had cooled, condensed, and gradually shrunk to the present size of the Sun; rings of gas shed by the shrinking Sun had condensed into the planets of the Solar System. Chamberlin, however, realized that glaciation and salt deposits indicated a colder and more arid climate in the past, incompatible with the warm and moist conditions postulated by Laplace's theory.

More troubling was the fact that a majority of the Solar System's angular momentum resided in the orbits of the jovian planets, while its mass is heavily concentrated in the Sun—an unlikely occurrence within Laplace's nebular hypothesis. This unsymmetrical distribution of matter and angular momentum suggested to Chamberlin that the Solar System had been formed by the near-collision of a nebulous cloud and the proto-Sun. From his analysis of the dynamic considerations, Moulton concluded that the original solar nebula had perhaps been similar to a spiral nebula. Astronomers have since abandoned the Chamberlin–Moulton hypothesis, and have assumed an alternate version of the nebular hypothesis, despite its unsolved dynamical problems.

Moulton served as a private consultant to the directors of the Meteor Crater Exploration and Mining Company (1929/1930), organized by **Daniel Barringer**, and produced the most thorough analysis regarding the probable size, mass, and speed of the incoming projectile. His calculations cast serious doubt on the existence of a still-buried meteoric mass.

By the early 20th century, the calculation of orbits as practiced by Moulton had become somewhat obsolete. Observational astrophysics held much greater research promise than did theoretical astronomy. The booming business and stock market of the 1920s, in which Moulton's younger brothers were participating, offered an alluring alternative career, and Moulton resigned his position at the University of Chicago in 1926 to become director of the Utilities Power and Light Company in Chicago, where he remained until 1937. His company barely survived the Great Depression. Moulton also served as a trustee and director of concessions for Chicago's Century of Progress Exposition (1933/1934). He closed the concession's books in 1936 with a profit.

In the 1930s, after leaving the University of Chicago, Moulton retained his interest in popularizing astronomy. He conducted a weekly radio broadcast of interest to aspiring amateur astronomers that was heard throughout the midwestern United States. On each broadcast, Moulton offered copies of his books as a prize for the best weekly essay he received. More than a few of the recipients of these books went on to pursue careers in science and engineering, including Hugh M. Johnson who had a distinguished career as an X-ray astronomer.

Afterwards, Moulton undertook another major career move, serving as Permanent Secretary of the American Association for the Advancement of Science [AAAS] from 1937 to 1946, and as Administrative Secretary from 1946 to 1948. Under his direction, the Association doubled its membership and gained control of the Association's journal, *Science*, from the J. McKeen Cattel family. The archives of the AAAS contain records of Moulton's work there from 1937 to 1948.

Moulton was twice married and divorced: to Estelle Gillete, from 1897 to 1938, with whom he had four children; and to Alicia Pratt, from 1939 to 1951.

Norriss S. Hetherington

Selected References

Brush, Stephen G. (1978). "A Geologist among Astronomers: The Rise and Fall of the Chamberlin-Moulton Cosmogony." Parts 1 and 2. *Journal for the History of Astronomy* 9: 1–41, 77–104.

Gasteyer, Charles E. (1970). "Forest Ray Moulton." *Biographical Memoirs, National Academy of Sciences* 41: 341–355.

Hetherington, Norriss S. (1994). "Converting an Hypothesis into a Research Program: T. C. Chamberlin, His Planetesimal Hypothesis, and Its Effect on Research at the Mt. Wilson Observatory." In *The Earth, the Heavens and the Carnegie Institution of Washington: Historical Perspectives after Ninety Years*, edited by Gregory A. Good, pp. 113–123. History of Geophysics, vol. 5. Washington, D.C.: American Geophysical Union.

Hoyt, William Graves (1987). *Coon Mountain Controversies: Meteor Crater and the Development of Impact Theory*. Tucson: University of Arizona Press, esp. pp. 264–293, 306–317.

Osterbrock, Donald E. (1997). *Yerkes Observatory, 1892–1950: The Birth, Near Death, and Resurrection of a Scientific Research Institution*. Chicago: University of Chicago Press.

——— (1999). "Moulton, Forest Ray." In *American National Biography*, edited by John A. Garraty and Mark C. Carnes, Vol. 16, pp. 27–28. New York: Oxford University Press.

Tropp, Henry S. (1974). "Moulton, Forest Ray." In *Dictionary of Scientific Biography*, edited by Charles Coulston Gillispie. Vol. 9, pp. 552–553. New York: Charles Scribner's Sons.

Mouton, Gabriel

Born **Lyon, France, 1618**
Died **Lyon, France, 28 September 1694**

Gabriel Mouton is most widely known for his work on a universal standard of length based on a "geometric foot" or one-thousandth of a *mille*. Mouton lived his whole life in Lyons. After obtaining a doctorate in theology he became, in 1646, a Vicar of Saint Paul's Church in that city. In spare time from his clerical duties he studied mathematics and astronomy.

Mouton's *mille* was proposed as a minute of longitude on the Earth's surface, a distance that eventually became known as a nautical mile. In 1670 Mouton developed a system of decimal divisions for his length standard. In order for this standard to be reproducible, Mouton proposed a pendulum of the standard length and measured the number of oscillations in 30 min. His value of 3959.2 oscillations was to be an easy way to verify that the pendulum was of proper length. A similar concept, but based on a pendulum with a period of 1 s in Paris, was later proposed as the basis for a universal system of measurement. However, the metric system as originally devised was not based on the length of a pendulum, but on a meter as one ten-millionth of the distance between the Earth's pole and Equator. Several of the principles used in Mouton's system were incorporated into the International System of Units [SI].

Mouton's astronomical accomplishments included determining the apparent diameter of the Sun at apogee. As an experimentalist, he constructed an astronomical pendulum. And, as a calculator, he was able to present a practical way to compute ordered tables of numbers such as logarithmic tables of trigonometric functions.

Mouton's work includes *Observationes diametrorum soles et lunae apparentium*, published in Lyons in 1670.

Donald W. Hillger

Selected Reference

Speziali, Pierre (1974). "Mouton, Gabriel." In *Dictionary of Scientific Biography*, edited by Charles Coulston Gillespie. Vol. 9, pp. 554–555. New York: Charles Scribner's Sons.

Mrkos, Antonín

Born **Střemchoví, Moravia, (Czech Republic), 27 January 1918**
Died **Prague, Czech Republic, 29 May 1996**

Antonín Mrkos is remembered as a discoverer of comets and minor planets, and for measuring the precise astronomical coordinates of these objects. The son of farmers, Mrkos studied at several secondary schools, including an ecclesiastical gymnasium, before entering the Technical University in Brno in 1938. His studies were interrupted by the onset of World War II, and during 1939–1943, he taught at primary schools at Nové Město and Ždárec in Moravia.

After a short stay in Austria, Mrkos took a position in 1945 under director **Antonin Bečvář** at the Skalnaté Pleso Observatory in Slovakia. There, Mrkos participated in the famous visual comet-hunting program using 25 × 100 Somet–Binar binoculars, as well as the cometary astrometric program with the 0.6-m reflector. Following his successful visual discovery of three comets (including 45P/Honda–Mrkos–Pajdušáková) in 1948, he made photographic recoveries of three periodic comets during 1949/1950. Mrkos was the principal observer with the 0.6-m reflector during 1946–1956.

Noting that the observatory's location (on the eastern slope of the second highest mountain, Lomnicky' Štít) in the High Tatras made it impossible to hunt for evening comets in the western sky, Mrkos decided to observe from the 2,634-m peak, to which he would regularly climb, shunning the more convenient cable cars. He also worked as a meteorologist for the weather station there. During a 4-year interval starting in early 1952, Mrkos visually discovered six more comets, including 18D/Perrine–Mrkos, an accidental rediscovery of a comet lost since the beginning of the century. With his private 0.5-m reflector, he recovered the intermediate-period comet 13P/Olbers in 1956.

Mrkos's most famous discovery, of C/1957 P1 (Mrkos) on 2 August 1957, was of one of the century's brightest comets, whose tail he detected while measuring the night sky glow at Lomnicky' Štít. Shortly afterwards, Mrkos traveled to Antarctica for 2 years with a Soviet expedition as part of the International Geophysical Year (1957/1958).

On his return to Slovakia, Mrkos made his 11th and final visual comet discovery in late 1959. During 1961–1963, he participated in another Soviet Antarctic expedition, working at Molodeznaya and Novolazarevskaya, after which he spent a year on the staff of the Geophysical Institute of the Czechoslovak Academy of Sciences in Prague, before taking a position (which he held until his retirement) in the astronomy department of Charles University. In addition, Mrkos was made director of the observatory at České Budějovice in southern Bohemia, and this led to the establishment of a planetarium and of the new observatory he was to direct on Klět Mountain.

Mrkos made the first photographic observations of comets with the 1-m reflector and 0.4-m Maksutov at Klět Mountain in 1968, and the 1969 recovery of his comet 45P was an early success. The Klět activity was extended to cover minor planets in 1977, and for many years it was the most regular contributor of data to the International Astronomical Union [IAU] Minor Planet Center. It was Mrkos's habit to spend 20 days each month at the observatory and 10 days in Prague reducing and typing up the observations prior to mailing.

Because of the dedication of Mrkos and assistants such as Růžena Petrovičová and Zdeňka Vávrová, at the time of his death, the program ranked as the sixth most successful ever launched for the discovery of minor planets that had been numbered. At that time, Mrkos was listed as the 11th most prolific discoverer. Two photographic discoveries of comets were credited to him at Klět, bringing his lifetime total to 13. Among Mrkos's more interesting discoveries of minor planets is the near-earth object (5797) 1980 AA, later named "Bivoj." Mrkos was elected president of IAU Commission 6 during the triennium 1985–1988.

Following his retirement on 31 December 1991, Mrkos had hoped to continue his photographic astrometric work from a site closer to Prague. He located a small camera at an abandoned geodetic observatory near Ondřejov and planned to put it to use. An extended stay in the hospital in 1994/1995 delayed these plans, and on a subsequent visit to the remote site, he found that it had been vandalized. Particularly sad was the theft of his Somet–Binar binoculars, which had played a role in the visual discovery of 11 comets, and which Mrkos always had nearby when he was observing.

Brian G. Marsden

Selected Reference

Marsden, Brian G. (1996). "Antonín Mrkos (1918–1996)." *International Comet Quarterly* 18: 182–183.

Muʿādh

➋ **Ibn Muʿādh: Abū ʿAbd Allāh Muḥammad ibn Muʿādh al-Jayyānī**

Mukai, Gensho

Born **Hizen (Saga and Nagasaki Prefectures), Japan, 1609**
Died **Kyoto, Japan, 1677**

Gensho Mukai introduced Western astronomical ideas in Japan during the early Edo period; however, he retained a traditional Chinese cosmology. At age five, he moved to Nagasaki, where he learned astronomy under Kichiemon Hayashi and became a Confucian scholar as well as a physician. In 1658, Mukai moved to Kyoto, where he practiced medicine and remained until his death. Though little seems to have been written about his life, Mukai no doubt found his place among the intelligentsia of Confucian scholars in Kyoto in the early Edo Period (1603–1867). He is probably remembered as much for what was seen as his personification of Confucian ideals as he is for specific works that he wrote. Mukai's comparison of Eastern and Western concepts also seems to have had an influence not only on scholars who immediately followed him, but on Japanese society as a whole. He inspired his children, who exhibited notable intellect and creativity. Mukai's second son, Kyorai (1651–1704), was a poet and student of the famous *haiku* master, Basho Matsuo.

Mukai is probably best known in the context of Japanese science for introducing concepts of Western astronomy in the early Edo era, a time of growing isolation from the outside world. Under *Tokugawa* policies of seclusion, Nagasaki was the only link to the West, and indirect contact with European sources allowed some acquisition of western astronomical knowledge. This contact led to what has been termed *Nanban* astronomy, a scholarship that involved translation, commentary, and comparative evaluation of European and Chinese-derived epistemologies. As a part of such activity, Mukai completed a set of commentaries around

1650 titled *Kenkon Bensetsu* (Western cosmography with critical commentaries). This work was based on a translation by Christopher Ferreira, whose Japanese name was Chuan Sawano, of what no doubt were several European works, but most especially **Christoph Clavius**'s *In Sphaeram Ioannis de Sacro Bosco, Commentarius* (1607).

In Mukai's commentaries, for perhaps the first time in Japan, sphericity of the Earth was explicitly recognized. However, while Mukai was instrumental in introducing such concepts into Japan, his commentaries show a disdain if not outright antagonism at times for what he saw as contradictions to classic Chinese precepts such as the theory of five elements (wood, fire, earth, metal, and water). He never abandoned his beliefs in Neo-Confucian principles, and with his strong contrasts of European and Chinese models, was no doubt influential in making such principles central to intellectual thought in early Edo Japan.

Mukai saw the Earth as inextricably related to the sky, and the European view of nature was simply something he could not recognize. Although he did accept western astronomical knowledge as it related to enhance the classical concerns of traditional far-eastern astronomy, Mukai continually denounced physical theories based on the Aristotelian four elements. For example, whereas **Tycho Brahe** showed that trepidation was really a matter of observational error, Mukai dismissed the theory of north–south oscillation as an explanation for trepidation not because of observational discrepancies but because the idea of such oscillation did not fit within what he felt was the harmony of the five-elements theory.

Steven L. Renshaw and *Saori Ihara*

Selected References

Ho, P. K. (2000). *Li, Qi, and Shu; An Introduction to Science and Civilization in China*. New York: Dover Publications. (Originally published in 1985. Perhaps less forceful in arguing that goals of epistemology in east and west were fundamentally the same, this introduction to Needham's classic *Science and Civilization in China*, Vol. 3, *Mathematics and the Sciences of the Heavens and the Earth*, presents the reader with an approachable view of the philosophies and cosmologies that guided Chinese thinking about the sky and influenced much thought in Japan both before and during the Edo period. Readers who wish to gain a fundamental understanding of the basis for Mukai's stark contrast of western and eastern concepts of astronomy will find this volume invaluable.)

Nakayama, S. (1969). *A History of Japanese Astronomy: Chinese Background and Western Impact*. Boston: Harvard University Press. (This work remains the primary source in English for gaining a glimpse into the underlying methods and mathematics of astronomical development in Japan. Nakayama presents an extensive section on Mukai's commentaries, and with detailed explanations of both western and eastern precepts, gives a clear picture of the epistemological base from which most work in early Edo Japan was developed.)

——— (1978). "Japanese Scientific Thought." In *Dictionary of Scientific Biography*, edited by Charles Coulston Gillispie, Vol. 15 (Suppl. 1), pp. 728–758. New York: Charles Scribner's Sons. (Nakayama clarifies differences between epistemological views in Europe and Japan both before and during the era in which Mukai lived. He pays particular attention to differences in the way nature was conceived in the east and the west. In contrast to Needham, he argues that the ultimate goals of science in Europe and in China were indeed not the same. Such background helps the reader to understand the philosophical and social base from which Mukai and others of Nanban astronomy developed their comparisons.)

Sugimoto, M. and D. L. Swain (1989). *Science and Culture in Traditional Japan*. Tokyo: Charles E. Tuttle and Co. (This volume presents a comprehensive and scholarly view of scientific development in Edo Japan. Mukai's activities may be seen against the backdrop of a society whose leaders sought to increasingly isolate the country from outside influences. Sugimoto and Swain pay special attention to contrasts between traditional Chinese derived epistemologies and those seeping into Japan indirectly from the West.)

Muler, Nicolaus

Born **Brugge, (Belgium), 1564**
Died **Groningen, the Netherlands, 1630**

Nicolaus Mulerius, a professor of medicine and mathematics at the University of Groningen, edited the 1617 (Amsterdam) version of **Nicolaus Copernicus**'s *De Revolutionibus*. In it, he included extensive notes and a thesaurus of observations to supplement the text of the two earlier editions.

Alternate name

Mulerius

Selected Reference

Swerdlow, N. M. and O. Neugebauer (1984). *Mathematical Astronomy in Copernicus's De Revolutionibus*. 2 pts. New York: Springer-Verlag.

Mulerius

➜ **Muler, Nicolaus**

Müller, Edith Alice

Born **Madrid, Spain, 5 February 1918**
Died **Spain, 24 July 1995**

With **Leo Goldberg** and **Lawrence Aller**, Spanish astrophysicist Edith Müller determined the abundances of chemical elements in the Sun. She served as general secretary of the International Astronomical Union and was the first woman to do so. She was based in Switzerland for many years.

Selected Reference

Blaauw, Adriaan (1996). "Edith Alice Müller, 1918–1995." *Bulletin of the American Astronomical Society* 28: 1457–1458.

Müller, Johann

Born **Königsberg, (Bavaria, Germany), 6 June 1436**
Died **Rome, (Italy), 8 July 1476**

Johann Müller (Regiomontanus) published valuable astronomical ephemerides and mathematical texts and devised new instruments and methods of observation. He was to 15th-century astronomy as **Nicolaus Copernicus** and **Tycho Brahe** were to 16th-century astronomy.

Johann Müller himself used the scholarly name "Johannes de monte regio" or similar names. Though he evidently never used it himself, since the 16th century he has been known as Regiomontanus, from the Latin for Königsberg (King's mountain).

Young Johann was a child prodigy and was sent off to the University of Leipzig at the age of 11. Three years later, he visited Vienna and stayed on as a student under **Georg Peurbach** at the university there, receiving his bachelor's degree in 1452 (at age 15); upon receiving his master's degree in 1457, Regiomontanus was appointed to the University of Vienna's faculty. He was tremendously knowledgeable about ancient, medieval, and contemporary scholarly works, and he knew Latin and Greek.

A visit by Cardinal Bessarion to Vienna in 1460 led to Regiomontanus traveling to Rome and many other places in Europe, before settling in Nuremberg in 1471, where he set up a print shop to publish scholarly works with technical diagrams. As was typical of the astronomers of his day, Regiomontanus was very much both an astrologer and an astronomer, though he seemed to spend more time in his later years on astronomical science and mathematics than on astrology. He was also a religious man, and his personal trademark sign was a cross on a hill with a background of stars. Regiomontanus was supposedly called by the Pope to Rome in 1475 or 1476, where he died of unknown causes (though the plague has been suggested as a possibility), suddenly terminating an increasingly productive career in revolutionizing astronomy.

Johann Gutenberg had published the first printed astronomical calendar in 1448, and while at Leipzig, young Regiomontanus was drawn toward checking the accuracy of that calendar, finding errors between the predicted positions of the planets and their true locations in the sky. This led him on an intense astronomical career to build an improved astronomy that would forever hold him as the foremost astronomer of 15th-century Europe. During his lifetime, Regiomontanus created much impact by traveling to visit astronomers elsewhere, by corresponding with astronomers in technical discussions, by seeking out classical writings on astronomy and encouraging publication of new translations of them, and by writing and publishing his own widely consulted works on astronomy. Among those with whom Regiomontanus dealt was **Paolo Toscanelli** of Florence, who made perhaps the most accurate positional observations of comets in the 15th century. One of the problems that Regiomontanus eagerly attacked was that of a comet's position, size, and distance, and his "Sixteen Problems" regarding these unknown quantities became a much-cited work after **Johann Schöner**'s publication of the manuscript in 1531, upon the appearance of Halley's comet, (IP/Halley)

Regiomontanus' *Ephemerides* were heavily used in Europe because they were seen as authoritative in predicting eclipses, phases of the Moon, positions of the Sun, Moon, and planets, and church calendar information.One of the reasons that ephemerides such as these were so popular was the strong European adherence to astrology in that era. Copernicus and Christopher Columbus both annotated their own copies of *Ephemerides*, and it has been argued that both Columbus and **Amerigo Vespucci** must have used Regiomontanus' almanacs on their voyages overseas; Columbus likely used his copy for impressing the American natives in Jamaica by successfully predicting the lunar eclipse of 14 September 1494. Copernicus was heavily influenced by Regiomontanus' *Epitome* and *Ephemerides* in his writing of *De revolutionibus*.

Regiomontanus wrote an impressive book, *De triangulus omnimodis* (On triangles), a treatise on plane and spherical trigonometry that was first published in 1533. This book was much used by European astronomers in the 16th and 17th centuries to determine celestial positions. Regiomontanus wrote this organized text on geometry and trigonometry specifically for use by astronomers, realizing that astronomers needed these mathematical tools to advance their knowledge of the motions of celestial objects. Even though his treatise was built on the work of ancient and medieval mathematicians, Regiomontanus added his own theorems and employed algebraic techniques for solving geometric problems; he utilized his geometric procedures in his other astronomical work, as can be seen in his treatise on comets.

Regiomontanus was also interested in making astronomical instruments for improved astronomical observations – one of the essential ingredients for building a better astronomy. Making better clocks seems to have been a part of these endeavors, as he attempted to construct a planetary clock. Had Regiomontanus lived to old age, his impact on astronomy would doubtless have been much greater, given his talents and enthusiasm. He envisaged publishing many more books on astronomy – translations into Latin of historical works and also many of his own new works on the subject.

Numerous manuscripts and letters written by Regiomontanus are extant in various libraries, attesting to a prolific career of correspondence and writing on topics astronomical and mathematical. His instruments, books, and papers passed to **Bernard Walther** of Nuremberg at the death of Regiomontanus. Walther kept these estate items to himself, unwilling to share them, until his death in 1504, when they began to be scattered through various sales. Schöner purchased much of the estate material on astronomy belonging to Peurbach, Regiomontanus, and Walther around 1522, eventually publishing much of the unpublished materialof the three men. Even the famous artist **Albrecht Dürer** eventually purchased Regiomontanus' copy of Euclid's *Elements* at auction.

Daniel W. E. Green

Alternate name

Regiomontanus

Selected References

Regiomontanus and G. Peurbach. (1496). *Epytoma Joanis De Monte regio In almagestum ptolomei*. Venice: Johannes Hamman. (Translation from Greek to Latin of a condensed version of Ptolemy's *Almagest*, with commentary, by Peurbach and Regiomontanus. Their commentary included some mention of where Ptolemy had erred, which caught Copernicus's attention and was thus influential.)

Hughes, Barnabus (trans.) (1967). *Regiomontanus on Triangles*. Madison: University of Wisconsin Press. (Recent English translation of Regiomontanus's monumental work on geometry and trigonometry.)

Pedersen, Olaf (1978). "The Decline and Fall of the *Theorica Planetarum*: Renaissance Astronomy and The Art of Printing." In *Science and History: Studies in Honor of Edward Rosen*, edited by Erna Hilfstein *et al.*, pp. 157–185. *Studia Copernicana*, Vol. 16, 161ff. Wroclaw: Ossolineum, Polish Academy of Sciences Press. (Discussion of Regiomontanus' printing philosophy and his impressions of Peurbach's work, with some translation of Regiomontanus' passages into English.)

Regiomontanus. (1474). *Ephemerides*. Nuremberg: Johann Muller of Königsberg.

——— (1474). *Kalendar fuer 1475–1531*. Nuremberg: Johann Muller of Königsberg.

——— (1972). *Joannis Regiomontani opera collectanea*, edited with an introduction by Felix Schmeidler. Osnabrück: Otto Zeller. (Recent publication of ten of the main written works attributed to Regiomontanus.)

Schöner, J. (ed.) (1544). "Ioannis de Monteregio et Barnardi Waltheri eivs discipvli ad solem obseruationes," and "Ioannis de Monteregio, Georgii Pevrbachii, Bernardi Waltheri, ac aliorum, Eclipsium, Cometarum, Planetarum ac Fixarum obseruationes." In *Scripta clarissimi mathematici*. Nuremberg. Johannem Montanum & Ulricum Neuber. (Contains astronomical observations of Regiomontanus and his colleagues, though with some typographical errors.)

Snell, W. (1618). *Coeli & siderum in eo errantium Observationes Hassiacae Lantgravii … Quibus accesserunt Ioannis Regiomontani & Bernardi Walteri Observationes Noribergiacae*. Leiden: Justum Colsterum. (Contains astronomical observations of Regiomontanus and his colleagues, though with some typographical errors.)

Steele, John M. and F. Richard Stephenson (1998). "Eclipse Observations Made by Regiomontanus and Walther." *Journal for the History of Astronomy* 29: 331–344. (Translation of eclipse observations with technical assessment.)

Thorndike, Lynn (1934, 1941). *A History of Magic and Experimental Science*. Vol. 4 (1934): p440ff; Vol. 5 (1941): 332ff. New York: Columbia University Press. (Extensive discussions of Regiomontanus and his impact on astronomy in late-medieval and early-modern times.)

Zinner, Ernst (1964). *Geschichte und Bibliographie der astronomischen Literatur in Deutschland zur Zeit der Renaissance*. Stuttgart: Anton Hiersemann, p. 7ff. (List of all known publications and reprintings of Regiomontanus' tracts, together with some interesting text putting Regiomontanus into context for his time period.)

———(1990). *Regiomontanus: His Life and Work*, translated by Ezra Brown. Amsterdam: Elsevier Science Publications. (The most extensive biography of Regiomontanus, recently translated from German to English.)

Müller, Karl

Born **1866**
Died **22 October 1942**

With **Mary Blagg**, Viennese selenographer Karl Müller wrote the definitive reference *Named Lunar Formations* (1935), sponsored and adopted by International Astronomical Union Commission 17.

A formation on the Moon is named for him.

Selected Reference

Chong, S. M., Albert Lim, and P. S. Ang (2002). *Photographic Atlas of the Moon*. Cambridge University Press.

Müller, Karl Hermann Gustav

Born **Schweidnitz (Swidnica, Poland), 7 May 1851**
Died **probably Potsdam, Germany, 7 July 1925**

Karl Müller's career, devoted to photographic photometry, produced valuable contributions in the form of catalogs of both photometric data and of variable stars.

Educated at Leipzig University and Berlin University, Müller worked briefly for **Arthur von Auwers** and then assisted **Hermann Vogel** at the Astrophysical Observatory at Potsdam from its beginning in 1877. During 1917–1921 he served as that institution's Director.

From early in his career, Müller pursued the study of terrestrial lines in the solar spectrum. His investigation of the Sun's brightness over its disk resulted in several expeditions to climb high mountains (including Tenerife at age 59). Müller's planetary and asteroid work recorded changes in brightness as a function of phase. He began in 1886, assisted by **Paul Kempf**, the construction of the *Potsdam Durchmusterung*, giving visual magnitudes and colors of all stars in the northern sky down to magnitude 7.5. (With **Ernst Hartwig**, Müller compiled a catalog of more than 1,600 variable stars from 1916 to 1921.)

Selected Reference

Munch, Wilhelm (1926). "Gustav Muller." *Astrophysical Journal*. 63: 141.

Muñjāla

Flourished **Deccan, (India), possibly 900**

Muñjāla was the author of a remarkable work, the *Laghumānasa*, which is an abridged version of a larger work called *Bṛhanmānasa*. Very little is known about the life of Muñjāla, except that he was a *brāhmaṇa* belonging to the *Bhāradvājagotra*, and that he lived in Deccan.

The *Laghumānasa* was very popular among the astronomers from Kerala, and it is mentioned by **Bīrūnī**. **Parameśvara** wrote a commentary on it, and quotations from it are found in the works of Bhāskaracārya and Munīśvara.

The *Laghumānasa* appears to be the first siddhāntic text to treat the precession of the equinoxes. Muñjāla gives the number of "ayana" revolutions to be 199,669 in a kalpa, and the "ayanāṃśa" to be 6° 54′ in 932 and the year of zero "ayanāṃśa" as 522. Muñjāla was the first Indian astronomer to introduce corrections to the Moon's equation that account for what today is called evection. Muñjāla anticipates Bhāskaracārya in understanding that the sine and cosine are related in a way that we would express today by saying the derivative of a sine function is a cosine function.

Narahari Achar

Alternate name

Mañjula

Selected References

Bose, D. M., S. N. Sen, and B. V. Subbarayappa (1971). *A Concise History of Science in India*. New Delhi: Indian National Science Academy.

Dikshit, S. B. (1896). *Bhāratīya Jyotisha*. Poona, English translation by R. V. Vaidya. 2 pts. New Delhi: Government of India Press, Controller of Publications, 1969, 1981.

Muñjāla. *Laghumānasa*, edited with the commentary of Parameśvara by B. D. Āpaṭe. Ānandāśrama Sanskrit Series, 123. Poona; 1944; 2d ed. Poona, 1952. (There is also an English translation with notes by N. K. Majumdar, Calcutta, 1951.)

Pingree, David. *Census of the Exact Sciences in Sanskrit*. Series A. Vol. 4 (1981): pp. 435a–436a; Vol. 5 (1994): 312b. Philadelphia: American Philosophical Society.

Shukla, K. S. (1990). *A Critical Study of the Laghumānasa of Mañjula*. New Delhi: Indian National Science Academy.

Muñoz, Jerónimo

Born **Valencia, Spain, *circa* 1520**
Died **Salamanca, Spain, 1592**

Jerónimo Muñoz, a leading Spanish astronomer and geographer of the 16th century, observed the supernova of 1572 (SN B Cas) and speculated upon its cosmological significance. Muñoz began his studies at the University of Valencia, from which he received his Bachelor of Arts degree in 1537. He continued studies at other European centers and received a Master's degree. Two of his principal instructors were **Oronce Finé** and **Gemma Frisius**. Being an expert Hebraist, Muñoz for some time was professor of Hebrew at the University of Ancon. He returned to Valencia sometime before 1556; in 1563, he was appointed to the chair of the Hebrew Department in Valencian General Studies. In 1565, Muñoz was also named head of the Mathematics Department. In 1578, he accepted the chair of mathematics at the University of Salamanca. Muñoz was also awarded the chair of that institution's Hebrew Department.

In Valencia and Salamanca, Muñoz taught arithmetic, geometry, and trigonometry, geometrical optics, astronomy and its applications, astronomical instruments, cartography, geography, and astrology. Although Muñoz did not publish many works, his personal manuscripts, and students' copies of notes and texts prepared for his classes, are preserved at the libraries of Salamanca, Barcelona, Madrid, Munich, the Vatican, Naples, and Copenhagen.

In Spain, Muñoz enjoyed a wide reputation as mathematician, astronomer, geographer, Hellenist, and Hebraist. His fame at other places in Europe was due primarily to his published work on the supernova of 1572 entitled *Libro del Nuevo Cometa* (Book of the new comet) (Valencià, 1573). This account of the "comet" was prepared at the request of Phillip II. It was then translated and reprinted in French (1574). Muñoz corresponded about the "comet" with several astronomers, including the Viennese doctor and mathematician Bartholomaeus Reisacherus, and the imperial doctor of Bohemia, **Tadeá Hájek z Hájku** (Hagaecius). From the latter, but especially through the work of **Cornelius Gemma** (*De Naturae Divinis Characterismis*, 1575), which included an extensive account of Muñoz's observations, his own data and conclusions eventually reached **Tycho Brahe**. Brahe then dedicated a chapter with commentary in his *Astronomiae Instauratae Progymnasmata* (1602) to the work of Muñoz.

As an observational astronomer, Muñoz achieved the highest precision in determining stellar positions. He should likewise be counted among the astronomers who exposed the cosmological implications of the "comet," recognizing that maintenance of the Aristotelian dogma regarding an incorruptible sky (violated by the supernova's appearance) was untenable. Although Muñoz called the object a comet, he recognized that it looked more like a star than a comet. His reason for classifying it as a comet had much to do with his desire to find natural causes for the phenomenon, without calling upon divine omnipotence (*potentia dei absoluta*), which other astronomers and mathematicians such as Frisius, **Thomas Digges**, Hagaecius, or Brahe himself did invoke.

Muñoz's work on the supernova of 1572 must be placed in the context of his ambitious scheme to revise Aristotelian cosmology and Ptolemaic astronomy. This can best be seen by Muñoz's comments about the second book of **Pliny's** *Natural History* and in his translation (with numerous additions) of the *Commentary on the mathematical composition of Ptolemy* by **Theon of Alexandria**. Within his comments on Pliny, read at the University of Valencia in 1568, Muñoz presented some cosmological ideas similar to the Stoic tradition. He rejected the idea of the sphere of fire and considered the cosmos to be a continuous medium that became progressively more rarified as one moved away from the center (the Earth) and eventually trailed off into an immense vacuum. Using various arguments, Muñoz also rejected the idea of crystalline spheres and thought that the planets moved through the cosmic medium upon their own natures.

Muñoz clearly showed that the discipline of astronomy could successfully address questions from natural philosophy. On the other hand, his astronomical observations and comparisons to available planetary predictions, gleaned from **Ptolemy** to **Nicolaus Copernicus** and **Erasmus Reinhold**, led him to doubt the viability of Ptolemy's parameters, those of subsequent astronomers, and his own contemporaries. But the reform in astronomy required a radical transformation in the realm of instrumentation and a systematic program of observations – things that remained out of reach for a modest university professor. As for cosmology, Muñoz never accepted the Copernican system. Instead, he proposed a cosmology consisting of a fluid heaven and a theory of planetary movements resulting from different sources. That cosmology's qualitative features were described without the use of mathematical astronomy.

Muñoz was a successful geographer and cartographer; these were subjects to which he further applied his mathematical knowledge. He determined precise geographic coordinates (*e. g.*, latitudes) of numerous places on the Iberian Peninsula, of which he then drew a map. He applied geodetic methods of triangulation, which he learned from his teacher Frisius and which he taught in his own classes with practical examples. The oldest surviving map of the Kingdom of Valencia, appearing in Abraham Ortelio's atlas, *Theatrum Orbis Terrarum* (1570), was prepared from Muñoz's data.

Muñoz trained a large number of students including prominent military engineers and treaty writers such as Diego de Alava, university professors, and several leading cosmographers from the late 16th to the early 17th centuries who were in service to the Spanish monarchy.

Victor Navarro-Brotóns
Translated by: *David Valls-Gabard*

Selected References

Muñoz, Jerónimo (1573). *Libro del nuevo cometa*. Valencia: Pedro de Huete. (Facsimile reprints, for this and the succeeding works, along with a preliminary study by Victor Navarro, "The Astronomical Work of Jerónimo Muñoz," are available from Valencia Cultural, 1981.)

——— (1574). *Littera ad Bartholamaeum Reisacherum*. Valencia: Pedro de Huete.

——— (1578). *Summa del prognostico del cometa*. Valencia: Juan Navarro.

——— (2004). *Introducción a la astronomía y la geografía*, editied and Spanish translation by Victor Navarro, *et al*. Preliminary study by V. Navarro and V. Salavert. Valencia: Consell Valencià de Cultural.

Navarro-Brotóns, Victor and Enrique Rodríguez Galdeano (1998). *Matemáticas, cosmología y humanismo en la España del siglo XVI: Los "Cosmentarios al segundo libro de la Historia natural de Plinio" de Jerónimo Muñoz*. Valencia: Instituto de Estudios Documentales e Históricos sobre la ciencia, Universitat de Valencia.

N

Naburianu [Naburianus, Nabû-ri-man-nu]

Flourished **Probably 5th century BCE**

Naburianu was a Babylonian astronomer whose name is mentioned by **Pliny** in his first-century text, *The Natural History*. In 1923, he was identified with an astronomer by the name of Nabû-?-man-nu by the cuneiform scholar Paul Schnabel. The name occurs in the colophon of a lunar ephemeris of System A from Babylon for the year 263 of the Seleucid era, or 48/47 BCE. Schnabel reconstructed the name to Nabû-ri-man-nu and concluded that he was responsible for discovering the System A method of computation of the Moon's position, dating him to around 427 BCE. Schnabel's claim has been effectively challenged by **Otto Neugebauer**, who points out that the inclusion of Nabû-ri-man-nu's name on the ephemeris is hardly evidence that he invented the system of computation on which it is based.

Nicholas Campion

Selected References

Hunger, Hermann and David Pingree (1999). *Astral Sciences in Mesopotamia*. Leiden: Brill. (Schnabel's views on Naburianu are summarized herein.)

Neugebauer, Otto (1975). *A History of Ancient Mathematical Astronomy*. 3 pts. New York: Springer-Verlag.

Najm al-Dīn al-Miṣrī: Najm al-Dīn Abū ʿAbd Allāh Muḥammad ibn Muḥammad ibn Ibrāhīm al-Miṣrī

Flourished **Cairo, (Egypt), *circa* 1300–1350**

Little is known of the life of the Cairene applied astronomer Najm al-Dīn al-Miṣrī, who was a contemporary of **Mizzī**. Several works, though, help document his scientific activities. Following are some of them:

1. A concise treatise on spherical astronomy entitled *Treatise on the Universal Operations [of Timekeeping] by Calculation*.
2. A short treatise on approximate methods of timekeeping.
3. A huge set of tables covering 419 folios, extant in two codices, which form the first and second halves of a single copy that was later split. In the main table, the time since the rising of the Sun or a star is tabulated in terms of three arguments. With nearly 415,000 entries, this is the single largest mathematical table ever compiled before the late 19th century.
4. An anonymous treatise, which can be attributed to Najm al-Dīn al-Miṣrī, gives detailed instructions on how to use these as universal auxiliary tables for solving all problems of spherical trigonometry for any terrestrial latitude. (The tables and the commentary have been analyzed in Charette, 1998.)
5. The previous item forms the prologue of an illustrated treatise – also anonymous – on the construction of over 100 different astronomical instruments (astrolabes, quadrants, sundials, *etc.*). This work has been recently shown to be by Najm al-Dīn al-Miṣrī (Charette, 2003). The text and its accompanying illustrations represent one of the richest and most astounding medieval sources on the topic of astronomical instrumentation.

Although Najm al-Dīn's writings suggest that he was not a first-rate astronomer, especially on the theoretical level, his intuitive and practical, "hands-on" approach to timekeeping (*mīqāt*) and instrumentation did yield original results.

François Charette

Selected References

Charette, François (1998). "A Monumental Medieval Table for Solving the Problems of Spherical Astronomy for All Latitudes." *Archives internationales d'histoire des sciences* 48: 11–64.

______ (2003). *Mathematical Instrumentation in Fourteenth-Century Egypt and Syria: The Illustrated Treatise of Najm al-Dīn al-Mißrī*. Leiden: E. J. Brill.

Napier, John

Born **Merchiston Castle near Edinburgh, Scotland, 1550**
Died **Edinburgh, Scotland, 4 April 1617**

John Napier spent much time in devising methods of facilitating and shortening astronomical calculations.

Napier was educated at Saint Andrews University and traveled on the Continent in his youth. Settling down at his family seat of Merchiston, he devoted his attention to many subjects, writing on theology and politics, and investigating various branches of science. His invention of logarithms with base e was announced in 1614. The invention was of great assistance to **Johannes Kepler**, and made possible the rapid development of astronomy in the 17th century. He also created the mathematical tool known as Napier's bones. (Local legend considered Napier to be a wizard.)

Napier was married, widowed, and remarried. His descendants still hold the title of Lord Napier.

Selected Reference

Napier, John (1966). *The Construction of the Wonderful Canon of Logarithms*, translated by William Rae MacDonald. London: Dawsons of Pall Mall. (Facsimile reprint of the 1889 edition; includes a catalog of the various editions of Napier's works.)

Nasawī: Abū al-Ḥasan ʿAlī ibn Aḥmad al-Nasawī

Born **Rayy, (Iran), 1002/1003**

Nasawī was an astronomer and mathematician whose name indicates that his family was originally from Nasā, a town in ancient Khurāsān that is in present-day Turkmenistan. He spent most of his life in his birthplace. In the introduction to his book, *Bāz-nāma* (On caring for falcons), Nasawī states that he served in the army, had been in the service of the kings, and trained birds of prey for 60 years, since age eight. Bayhaqī remarks that Nasawī lived until the age of 100. However, the date of his death is unclear.

Nasawī's disciple Shahmardān Rāzī, as well as **Naṣīr al-Dīn al-Ṭūsī**, refer to Nasawī as *al-ustādh al-mukhtaṣṣ* (distinguished teacher), probably due to his expertise in mathematics and astronomy. The famous Iranian poet Nāṣir-i Khusraw (1003–1088) writes in his *Safar-nāma* that he met Nasawī in Simnān (Iran) in 1046, where the latter was teaching Euclid's *Elements*, medicine, and arithmetic. Nasawī also quoted from discussions he had had with **Ibn Sīnā**, which led Nāṣir-i Khusraw to conclude that Nasawī had been a disciple of Ibn Sīnā. It has been claimed that Nasawī was also a disciple of **Kūshyār ibn Labbān**, but Nasawī would have been too young when Kūshyār died.

Nasawī wrote several astronomical works, only one of which is extant. *Kitāb al-lāmiʿ fī amthilat al-Zīj al-jāmiʿ* (Illustrative examples of [the 85 chapters] of [Kūshyār's] *Zīj-i jāmiʿ*) is also called *Risāla fī maʿrifat al-taqwīm wa-'l-asṭurlāb* (A treatise on the almanac and the astrolabe).

Only a few of the tables from *al-Zīj al-Fākhir* (The glorious astronomical tables) have survived following the Leiden manuscript of Kūshyār's *Zīj-i jāmiʿ*. These tables indicate that the values used for the planetary mean motions are extracted from **Battānī**'s *zīj*, confirming remarks in *al-Zīj al-mumtaḥan al-ʿarabī*, a recension of **Muḥammad ibn Abī Bakr al-Farisī**'s *Zīj* preserved in Cambridge.

Ikhtiṣār ṣuwar al-kawākib (Summary of the constellations) is dedicated to al-Murtaḍā, the Shiʿite leader from Rayy. This nonextant work was a summary of **ʿAbd al-Raḥmān al-Ṣūfī**'s book on the constellations.

Nasawī was also a noted mathematician and wrote works on arithmetic, geometry, and spherics. Among his works are his *al-Muqniʿ fī al-ḥisāb al-Hindī*, a treatise on Indian arithmetic whose purpose was, among other things, to be useful for both businessmen and astronomers. Chapter 4 of *al-Muqniʿ* deals specifically with sexagesimal reckoning used in Islamic astronomy. *Al-Tajrīd fī uṣūl al-handasa* (An abstract of Euclid's *Elements*) was composed for those who wanted to learn geometry in order to be able to understand **Ptolemy**'s *Almagest*.

Nasawī also wrote works on philosophy, pharmacology, and medicine.

Hamid-Reza Giahi Yazdi

Selected References

Al-Bayhaqī, Ẓahīr al-Dīn (1350 AH). *Tatimmatṣiwān al-ḥikma*, edited by M. Shāfiʿ. Lahore, pp. 109–110.

Al-Nasawī, Alī ibn Aḥmad. *Al-Muqniʿ fī al-ḥisāb al-hindī*. Facsimile ed. in Ghorbani, *Nasawī-nāma*, Tehran, 1351 H. Sh.

Kennedy, E. S. (1956). "A Survey of Islamic Astronomical Tables." *Transactions of the American Philosophical Society*, n.s., 46, pt. 2: 121–177. (Reprint, Philadelphia: American Philosophical Society, 1989.)

Nāßir-i Khusraw. *Safar-nāme*, edited by M. Dabīr Siyāqī. Tehran, 1354 H. Sh. (repr. 1363 H. Sh.).

Rosenfeld, B. A. and Ekmeleddin Ihsanoğlu (2003). *Mathematicians, Astronomers, and Other Scholars of Islamic Civilization and Their Works (7th–19th c.)*. Istanbul: IRCICA, pp. 140–141.

Sadiqi, Gh. H. "Hakīm Nasawī." *Majalle-ye Danishkade-ye Adabiyyat-i Tehran* (Journal of the Faculty of Letters) 6, no. 1 (1337 H. Sh./1958): 17–26.

Saidan, A. S. (1974). "Al-Nasawī." In *Dictionary of Scientific Biography*, edited by Charles Coulston Gillispie. Vol. 9, pp. 614–615. New York: Charles Scribner's Sons.

Sezgin, Fuat. (1974 and 1978). *Geschichte des arabischen Schrifttums*. Vol. 5, *Mathematik*: 345–348; Vol. 6, *Astronomie*: 245–246. Leiden: E. J. Brill.

Nasmyth, James Hall

Born **Edinburgh, Scotland, 19 August 1808**
Died **South Kensington, (London), England, 7 May 1890**

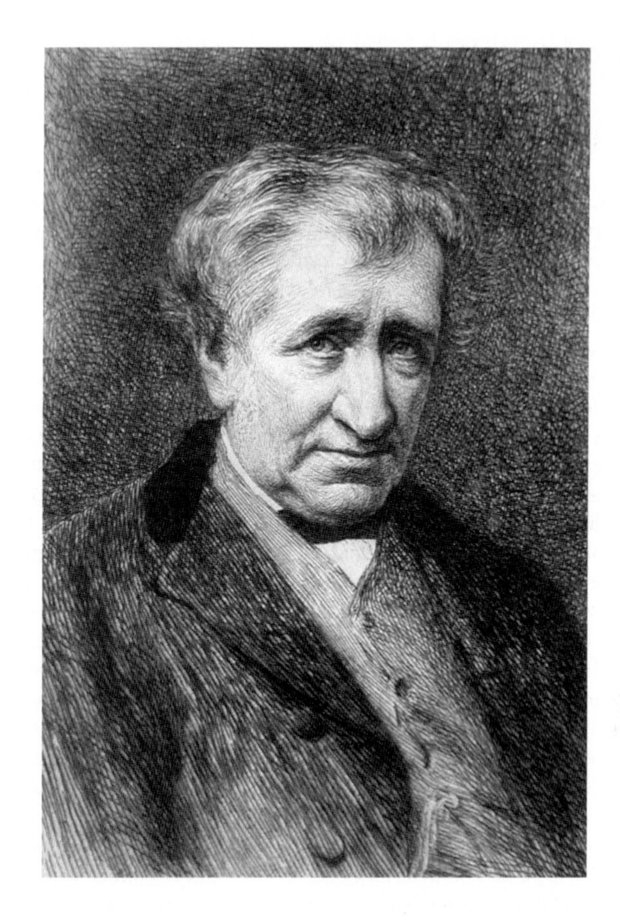

James Nasmyth advocated volcanic origin of the craters and other lunar features and devised a novel arrangement for the optical path of a Cassegrain telescope, yielding what is now identified as the Nasmyth focus in reflecting telescopes. The marriage of the painter Alexander Nasmyth and Barbara (*née* Foulis) Nasmyth yielded four sons and seven daughters, of which James was the youngest. On 16 June 1840, Nasmyth married Anne Hartop, daughter of the manager of the ironworks in Barnsley near Manchester, England. There were apparently no children from this marriage.

James Nasmyth showed exceptional mechanical aptitude and took an early interest in foundries and chemical laboratories. In his late teens, he made model steam engines and constructed, under a commission from the Scottish Society of Arts, a primitive automobile that he called a "steam road-carriage." After working as an associate of the noted machine toolmaker and engineer Henry Maudslay, Nasmyth established his own factory at Patricroft near Manchester. There he manufactured machine tools, hydraulic punches, pile drivers, and steam hammers, the latter an invention for which he bitterly disputed priority. The enterprise proved so profitable that Nasmyth was able to retire to the town of Penshurst in Kent at the age of 48.

Encouraged by his close friend, the Liverpool brewer and accomplished amateur astronomer **William Lassell**, Nasmyth became a telescope maker. At Patricroft, he used a succession of Newtonian reflectors, culminating in a 13-in. instrument. Ever inventive and attempting to improve objects that attracted his attention, Nasmyth introduced both an improved mirror grinding and polishing machine, and an improved formulation of speculum metal used to cast mirrors. After moving to Kent, Nasmyth erected a 20-in. telescope of innovative design. By introducing a planar tertiary mirror inclined at an angle of 45° between the primary and secondary mirrors of a Cassegrain reflector, the necessity of perforating the primary mirror was obviated, and a more accessible eyepiece position was provided in a hollowed-out trunnion bearing for the tube mounting. Neglected for decades, this "bent Cassegrain" or "Nasmyth" configuration became widely adopted in the 20th century on many large professional instruments featuring heavy spectrographs and large modern alt-azimuth mountings.

Always keenly interested in the Moon, Nasmyth became one of the first to contribute to **William Birt**'s committee to map the Moon. Unlike Birt, who looked for evidence of ongoing changes in the lunar surface, Nasmyth focused on the origins of the features that he saw scattered over its crowded surface. A self-taught amateur, he enlisted a more learned collaborator for his work: **John Carpenter**, a computer and later assistant astronomer at the Royal Greenwich Observatory. In their book on the Moon, an influential work about the lunar surface, Nasmyth and Carpenter accepted that lunar craters had been formed by volcanic action. But they did not attempt to gloss over the fundamental difference of form between terrestrial craters and their lunar counterparts. Terrestrial volcanic craters consist of a caldera or hollow atop a mountain, but the interior bowls of lunar craters are almost invariably depressed below the level of the surrounding surface. And yet despite the apparent differences in form, Nasmyth and Carpenter nevertheless affirmed **John Herschel**'s view that lunar craters manifest "the true volcanic character in its highest perfection."

Nasmyth and Carpenter endorsed a version of **Pierre de Laplace**'s cosmogony, in which particles of diffused primordial matter condensed and aggregated into a spherical planetary body consisting of a solid shell encompassing a molten core. Consistent with his empirical approach, Nasmyth did not take the formation of such a solid shell on faith but confirmed at an iron-works that cold slag floated on molten slag. During an 1865 visit to Vesuvius, then undergoing an eruption, he and Carpenter observed a lake of molten lava on the plateau or bottom of the crater on which vast cakes of the same lava that had solidified were floating.

Nasmyth applied these principles to the evolution of the Moon. He theorized that cooling from the outside inward, the Moon's encasing outer shell would cause pressures to build up in the still-molten interior until a portion of the molten interior burst through the shell "with more or less violence according to the circumstances." As each successive shell cooled and contracted in the attempt to accommodate itself to the core beneath it, the skin would crease into alternating ridges and depressions, "like a long-kept shriveled apple." Nasmyth and Carpenter saw everywhere on the surface of the Moon evidence of this general process.

The main problem in understanding the origin of the lunar features remained their sheer size relative to those of the Earth. The Vesuvian crater was less than 2 miles across, tiny by lunar standards. Even the largest terrestrial volcanic craters measure only some 15 miles across, yet lunar craters such as Tycho and Copernicus, with diameters of 50 and 56 miles respectively, were not even the largest specimens of their kind.

Nasmyth and Carpenter, proposed a "fountain model" solution: The reduced surface gravity of the Moon, combined with the negligible resistance of its thin atmosphere, would cause lunar volcanoes to eject a pyramid of matter around an orifice. Their theory required the ejected material to fall symmetrically in order to build up the crater walls and produce a ring structure. Later, the vent hole would fill with lava to produce the crater floor or give rise to the central peaks found in many of the lunar formations. As the lunar crust grew thicker and more rigid, and the reserves of lava dwindled, these last gasps of the Moon's dying internal activity that created the central peaks found in so many craters were, according to Nasmyth and Carpenter, evidence that such internal activity had ceased long ago.

Nasmyth's and Carpenter's theory was widely accepted and overshadowed other theories of the origin of the craters until well into the 20th century. The impression made by their book was greatly enhanced by the inclusion of 24 photographs of exquisite plaster-of-Paris models of lunar formations. Taken under oblique lighting, these plates made the book almost as much a work of art as a piece of scientific literature. The models were very deceptive, however, as they consistently depicted the Moon's surface as far more jagged than it is in reality. These rugged landscapes and sharp, craggy peaks provided evocative illustrations in popular books on the Moon for decades.

Nasmyth also made many observations of the Sun. His depiction of the granulation of the solar photosphere as a series of luminous filaments that he called "willow-leaf structure" remained in vogue for several decades but was ultimately proven to be an illusion.

Thomas A. Dobbins

Selected References

Ashbrook, Joseph (1984). "James Nasmyth's Telescopes and Observations." In *The Astronomical Scrapbook*, edited by Leif J. Robinson, pp. 141–147. Cambridge, Massachusetts: Sky Publishing Corp., esp. pp. 144–145.

Both, Ernst E. (1961). *A History of Lunar Studies*. Buffalo, New York: Buffalo Museum of Science.

Chapman, Allan (1998). *The Victorian Amateur Astronomer: Independent Astronomical Research in Britain, 1820–1920*. Chichester: John Wiley and Sons.

King, Henry C. (1955). *The History of the Telescope*. Cambridge, Massachusetts: Sky Publishing Corp.

Sheehan, William P. and Thomas A. Dobbins (2001). *Epic Moon: A History of Lunar Exploration in the Age of the Telescope*. Richmond, Virginia: Willmann-Bell.

Smiles, S. (ed.) (1883). *James Nasmyth, Engineer: An Autobiography*. London: Harper and Brothers.

Nasṭūlus: Muḥammad ibn ʿAbd Allāh

Flourished **10th century**

Nasṭūlus is credited with constructing two astrolabes. The first, dated 927/928, is considered the oldest surviving astrolabe (though not the first ever constructed). This elegant instrument is preserved in the Kuwait Museum of Islamic Art. It has a single plate (for latitudes 33° and 36°) on the back of which are four quadrant scales and a shadow scale. The throne bears the inscription, "Made by Nasṭūlus (or Basṭūlus) in the year 315." The second astrolabe, of which only the *mater* is still extant, bears no date but was probably constructed around 312 hijra (925). It is preserved in the Museum of Islamic Art in Cairo; the inscription "Made by Nasṭūlus" appears on the throne. It contains the earliest and only geographical list to appear on an instrument before *circa* 1100. The purpose of the gazetteer on the *mater* is evidently to show which plates should be used in different cities. Most of the latitudes included are derived from **Khwārizmī**'s geographical table, although the remainder may have been taken from other early sources such as **Battānī** (*circa* 910). Although no original plate has survived, the instrument has various Mamluk additions, dated 1314.

We know almost nothing about this astronomer, and even his name remains in doubt. Some historians have interpreted the manuscripts to refer to someone with a Greek name, perhaps Βατύλος/βαθύλος or Απόστολος. However, it is unclear whether he is a Muslim or Christian. King claims that he was a Muslim based on the testimony of the 10th-century astronomer **Sijzī**, who states that a certain Muḥammad ibn ʿAbd Allāh (clearly a Muslim name), known as Nasṭūlus, was the first person to design the astrolabe with a crab-shaped rete. Sijzī adds that Nasṭūlus also invented the hours drawn on the face of the alidade and the operation with the azimuth on the back of the astrolabe. This statement was later repeated by **Bīrūnī** in his *Istīʿāb*, in which he adds that Nasṭūlus was one of the people who worked on instruments for determining eclipses. On the other hand, M. Hinds suggests Nasṭūlus might refer to the Christian sect of the Nestorians, and Kunitzsch points out that the form Nasṭūrus was attested in 10th-century Egypt, and was used by Christian men. Nasṭūlus would then be just another form of Nasṭūrus.

Mònica Rius

Alternate name

Basṭūlus

Selected References

Ibn al-Nadīm (1970). *The Fihrist of al-Nadīm: A Tenth-Century Survey of Muslim Culture*, edited and translated by Bayard Dodge. 2 Vols. New York: Columbia University Press, Vol. 2, p. 671.

King, David A. (1978). "A Note on the Astrolabist Nasṭūlus/Basṭūlus." *Archives internationales d'histoire des sciences* 28: 117–120. (Reprinted in King, *Islamic Astronomical Instruments*, IV. London: Variorum Reprints, 1987.)

——— (1990). *The Earliest Islamic Astrolabes (Tenth to Eleventh Centuries)*. Preprint, pp. 16–22.

——— (1995). "Early Islamic Astronomical Instruments in Kuwaiti Collections." In *Kuwait – Arts and Architecture—A Collection of Essays*, edited by Arlene Fullerton and Géza Fehérvári, pp. 76–96. Kuwait. Oriental press, UAE.

_____ (1999). "Bringing Astronomical Instruments Back to Earth: The Geographical Data on Medieval Astrolabes (to ca. 1100)." In *Between Demonstration and Imagination: Essays in the History of Science and Philosophy presented to John D. North*, edited by Lodi Nauta and Arjo Vanderjagt, pp. 3–53, esp. pp. 9, 10, 11, 22, 29, 35, 39. Leiden: E. J. Brill.

King, David A. and Paul Kunitzsch (1983). "Nasṭūlus the Astrolabist Once Again." *Archives internationales d'histoire des sciences* 33: 342–343. (Reprinted in King, *Islamic Astronomical Instruments*, V. London: Variorum Reprints, 1987.)

Maddison, Francis and Alain Brieux (1974). "Basṭūlus or Nasṭūlus? A Note on the Name of an Early Islamic Astrolabist." *Archives internationales d'histoire des sciences* 24: 157–160.

Sezgin, Fuat (1978). *Geschichte des arabischen Schrifttums*. Vol. 6, *Astronomie*, pp. 178–179. Leiden: E. J. Brill.

Seemann, Hugo and T. Mittelberger (1925). "Das kugelförmige Astrolab nach den Mitteilungen von Alfonso X. von Kastilien und den vorhandenen arabischen Quellen." *Abhandlungen zur Geschichte der Naturwissenschaften und der Medizin* 8: 32–40. (A discussion of the spherical astrolabe and its origins in the Arabic world.)

Neison, Edmund

Nevill [Neville], Edmund Neison

Nemicus, Tadeá

Hájek z Hájku, Tadeá

Nayrīzī: Abū al-ʿAbbās al-Faḍl ibn Ḥātim al-Nayrīzī

Flourished **Baghdad, (Iraq), last half of the 9th century**

Nayrīzī is reputed to have been among the best mathematicians and astronomers of his day, though not much biographical information is known. In astronomy, his best-known work, a commentary on the *Almagest* of **Ptolemy**, is no longer extant. This must have been one of the earliest commentaries to be written in Arabic, because the *Almagest* had been first translated into Arabic only a century earlier. He is also credited with the composition of two *zījes* (astronomical tables used for predicting planetary motions). The longer was said, by the bio-bibliographer Ibn al-Qifṭī, to have been based on the *Sindhind*, an Indian classic in astronomy. The shorter was, presumably, based upon the *Almagest*. These works were cited by several astronomers from the ʿAbbāsid period, although they are no longer extant. Three shorter, more specialized treatises survive: (1) on the spherical astrolabe; (2) on finding the *qibla* direction (the direction toward Mecca, toward which pious Muslims pray five times a day); and (3) on constructing hour lines in a hemispherical sundial. **Ibn Yūnus**, in his own *zīj*, criticized some elements of Nayrīzī's astronomical work while praising him as a renowned mathematician.

Gregg DeYoung

Selected References

King, David A. (1993). *Astronomy in the Service of Islam*. Aldershot: Variorum. (Reprints of several papers dealing with mathematical astronomy, including discussions of the problem of the *qibla* direction.)

Sabra, A. I. (1974). "Al-Nayrīzī." In *Dictionary of Scientific Biography*, edited by Charles Coulston Gillispie. Vol. 10, pp. 5–7. New York: Charles Scribner's Sons. (An overview of what is known of Nayrīzī's life and scientific work, as well as the biographic and bibliographic source materials and important secondary materials.)

Schoy, Carl (1922). "Abhandlung von al-Faḍl b. Ḥātim an-Nairîzî: Über die Richtung der Qibla." *Sitzungsberichte der Bayerischen Akademie der Wissenschaft zu München*: 55–68. (A German translation of Nayrīzī's treatise on finding the direction of the *qibla*.)

Nernst, Walther Hermann

Born **Briesen, Wąbrźezno, Poland, 25 June 1864**
Died **Muskau, (Sachsen), Germany, 18 November 1941**

After elucidating the third law of thermodynamics, for which he won the 1920 Nobel Prize in Chemistry, physicist Walther Nernst worried about the consequences of the second law for the fate of the Universe. Entropy seemed to require an eventual "heat death" to everything. **Victor Hess**'s discovery of cosmic rays gave Nernst hope that perhaps there was a constant source of new matter and energy at large, so that the Universe could go on – much as it is – indefinitely. Nernst called this a steady-state cosmology.

Selected Reference

Mendelssohn, K. (1973). *The World of Walther Nernst: The Rise and Fall of German Science, 1864–1941*. Pittsburgh: University of Pittsburgh Press.

Neugebauer, Otto E.

Born **Innsbruck, (Austria), 26 May 1899**
Died **Princeton, New Jersey, USA, 19 February 1990**

Austrian–German–American mathematician and historian of mathematical astronomy Otto Neugebauer meticulously demonstrated the technical content of ancient mathematical astronomy and the ingenuity in abstract thinking of ancient mathematicians and astronomers.

Otto Neugebauer's father, Rudolph Neugebauer, was a railroad engineer. The Protestant family moved to Graz, Austria, when Otto was still young. There Otto attended the Akademisches Gymnasium, studying mathematics, mechanics, and technical drawing, in addition to the Greek and Latin required by the curriculum. In 1917 he enlisted in the Austrian army, ostensibly to avoid taking the Greek examination to receive his graduation certificate. Neugebauer became a lieutenant of artillery, spending the remainder of the war as a forward observer on the Italian front.

In 1919 after his discharge, Neugebauer entered the University of Graz to follow a course in electrical engineering and physics, transferring to the University of Munich in 1921. There he attended lectures by Arnold Sommerfeld and Arthur Rosenthal. After the death of his parents and as the result of Austrian hyperinflation, Neugebauer lost his entire inheritance and suffered a difficult winter, but in 1922 he changed the focus of his education, moving to the Mathematisches Institut at the University of Göttingen, where he studied under Richard Courant, the new director of the institute, and with Edmund Landau and Emmy Noether. By 1923 Neugebauer became an assistant at the institute, and in 1924, special assistant to Courant, and was put in charge of the library. During 1924 he spent time at the University of Copenhagen with Harald Bohr, with whom he published his only paper in pure mathematics.

During this time Neugebauer studied Egyptian mathematics, publishing a seminal document on the *Rhind Papyrus*, a late Egyptian mathematical document.

In 1927 Neugebauer received his *venia legendi* for the history of mathematics, and became a *Privatdozent* (lecturer). He soon married Grete Bruck, a fellow student, also a mathematician: Their two children, Margo and Gerry (a distinguished infrared astronomer), were born in 1929 and 1932, respectively. The following year Neugebauer traveled to Leningrad, Russia, to work with Wilhelm Struve in preparing the *Moscow Papyrus*, an important text in Egyptian mathematics, for publication. In 1929 Neugebauer founded a Springer series devoted to the history of mathematical sciences, astronomy, and physics, the *Quellen und Studien zur Geschichte der Mathematik, Astronomie und Physik*. Under his editorship, its focus would be primarily on Egyptian mathematics.

Starting in 1927, Neugebauer had learned Akkadian in an investigation of Babylonian mathematics, which eventually enabled him to establish the origin of the sexagesimal system, and collect a substantial corpus of texts later published as *Mathematische Keilschrift-Texte*, in several volumes. This corpus demonstrates the richness of Babylonian mathematics.

Neugebauer was founding editor of the review journal *Zentralblatt für Mathematik und ihre Grenzgebiete* and a Springer series of short monographs on current mathematics. But when Adolph Hitler became chancellor of Germany, Neugebauer was removed from his job at the Courant Institut for presumed political unreliability. Obtaining a professorship at the University of Copenhagen starting in 1934, he prepared a series of lectures on Egyptian and Babylonian mathematics, and planned a volume on Greek mathematics.

In his 1928 review of *The Venus Tablets of Ammizaduga* by Stephen Langdon *et al.*, Neugebauer demolished the earlier chronology of the Old Babylonian dynasty. In a paper a decade later, he similarly cast doubt on the use of the Sothic cycle for establishing the origin of the Egyptian calendar. Neugebauer became intensely interested in astronomical cuneiform texts, which were primarily ephemerides in the form of arithmetic functions for computing lunar and planetary phenomena. He developed a method using linear Diophantine equations to check these functions, with the result that many previously unrelated cuneiform fragments were joined and dated. This work showed that some functions ran continuously for hundreds of years, and provided the basis for much significant work to follow. After publishing some of his results in 1938, Neugebauer had his work interrupted by difficulties with Springer and his editorship on the *Zentralblatt*, from whose board he resigned in December. Neugebauer was immediately offered a position at Brown University in the United States, which he readily accepted, moving to Providence, Rhode Island, and founding *Mathematical Reviews* in 1939. He shortly afterward applied for American citizenship.

At Brown University, Neugebauer quickly published several papers on ancient astronomy and mathematics, later reprinted in his book *Astronomy and History* (1983). His most famous work is *The Exact Sciences in Antiquity*, a survey of Egyptian mathematics and astronomy, and their relation to Hellenistic science and its descendents. Neugebauer's treatment of the subject emphasizes the transmission of ideas as they were developed, with many cultures adding to the corpus of understanding and observations of astronomical phenomena.

With his collaborator Abraham Sachs, Neugebauer published all the known Assyrian astronomical texts as *Astronomical Cuneiform Texts* (1955), most dating from the last three centuries BCE. In his preface he commended the spirits of the ancient scribes of Enu ma-Anu-Enlil, who "by their untiring efforts ... built the foundations for the understanding of the laws of nature ... they also provided hours of peace for those who attempted to decode their lines of thought two thousand years later."

Neugebauer published several analyses of Egyptian astronomical documents, tomb ceilings, coffin lids, zodiacs, and papyri, collecting these works in the three-volume work *Egyptian Astronomical Texts* (1960–1969). His later *History of Ancient Mathematical Astronomy* (1975) indicated, however, that the Egyptian contribution to mathematical astronomy was minimal.

Along with the chief librarian at Brown University, a classicist and papyrologist, Neugebauer published *Greek Horoscopes* (1959), the standard work on the subject, containing an introduction to the methods of Greek astrology. He planned a history of mathematical astronomy from antiquity to **Johannes Kepler**, and published many ancient texts from several languages, including Greek, Latin, Indian, Arabic, and Ethiopic. Neugebauer's *History of Ancient Mathematical Astronomy* established the history of ancient astronomy on a new foundation and demonstrated the continuity of the science from ancient times to the present. The material included planetary and lunar theory, astrological sources, the works of **Ptolemy** and their derivatives, chronology, astronomy, and his own methods that had proved so useful. In later publications he dealt with astronomy in the Middle Ages, Byzantine sources, and analysis of **Nicolaus Copernicus**'s *De revolutionibus*, Ethiopic astronomy, chronology, and the calculations of the ecclesiastical calendar, with an analysis of the primitive astronomical section of the *Book of Enoch*.

Neugebauer became professor of the history of mathematics at Brown University, and was named the Florence Pirce Grant University Professor at Brown University in 1960. After his retirement, he moved to the Institute for Advanced Study at Princeton University, where he spent the remainder of his life. Some of the work Neugebauer completed during his last years was published posthumously.

Neugebauer received honorary doctorates from Saint Andrews University (Scotland), Princeton University, and Brown University; was elected to membership in academies of science and the arts of Denmark, Belgium, Austria, Britain, Ireland, France, and the United States; received the Balzan Prize (1986), the Franklin Medal (American Philosophical Society, 1987), and awards from the History of Science Society, the Mathematical Association of America, and the American Council of Learned Societies; and in 1967, became the only historian of science ever awarded the Russell Lectureship of the American Astronomical Society. His approach to the history of mathematical astronomy continues in the work of many influential scholars.

Katherine Haramundanis

Selected References

Pannekoek, A. (1961). *History of Astronomy*. London: Allen and Unwin.

Sachs, J. and G. J. Toomer (1979). "Otto Neugebauer, Bibliography, 1925–1979." *Centaurus* 22: 257–280.

Swerdlow, Noel M. (1993). "Otto E. Neugebauer." *Proceedings of the American Philosophical Society* 137: 139–165.

Neumann, Carl Gottfried

Born **Königsberg (Kaliningrad), Russia, 7 May 1832**
Died **Leipzig, Germany, 27 March 1925**

Carl Neumann developed new techniques in mathematical physics that have found diverse applications in dynamical astronomy and other research areas. Neumann was born into a noted academic family; his aunt was married to the astronomer and mathematician **Friedrich Bessel**. His father, Franz Ernst Neumann, a professor of physics at Königsberg University, contributed to advancements in the wave theory of light and the mechanical theory of heat. Neumann attended Königsberg University and completed his doctorate in 1855. After post-doctoral work at the University of Halle, Neumann became a *Privatdozent* (lecturer) there until his promotion to assistant professor in 1863, when he accepted a position at the University of Basle. After 2 years, he moved to the University of Tübingen, before settling at the University of Leipzig in 1869. One of Neumann's prominent students was **Hugo von Seeliger**. He remained there until his retirement (1911). In 1864, Neumann married Hermine Mathilde Elise Kloss.

Neumann's initial work on the Galilean–Newtonian theory of mechanics influenced mathematicians, physicists, and astronomers of the time, many of whom studied under him. His chief contributions were in mathematical physics: the application of mathematical techniques to the study of mechanics and electrodynamics, especially harmonic analysis ("potential functions") used in the solutions to partial differential equations (*e. g.*, Laplace's equation). These tools have since become standard for the mathematical modeling of physical processes involving electricity, electromagnetism, fluid dynamics, and gravitation.

Today, Neumann's methods and techniques are found in a surprising array of applications, including structural analysis, contact problems with friction (*e. g.*, the Dirichlet–Neumann algorithm), domain decomposition algorithms for the treatment of elastic body problems, spectral properties of the Neumann Laplacian involving stars and comets, and even the global potential functions used in programming the paths of robots.

Neumann was a cofounding editor (with Alfred Clebsch) of the prestigious and influential journal *Mathematische Annalen*, led subsequently by **Felix Klein** and David Hilbert.

Daniel Kolak

Selected Reference

Wussing, H. (1974). "Neumann, Carl Gottfried." In *Dictionary of Scientific Biography*, edited by Charles Coulston Gillispie. Vol. 10, p. 25. New York: Charles Scribner's Sons.

Nevill [Neville], Edmund Neison

Born **Beverley, (Humberside), England, 27 August 1849**
Died **Eastbourne, (East Sussex), England, 14 January 1940**

Edmund Nevill, regarded for a short time as the preeminent selenographer in Britain, provided a firm basis for later lunar studies. Nevill was educated at Harrow and New College, Oxford. During the Franco–Prussian war of 1870/1871, he joined the French army, serving on the staff of marshal Ney. After the war, Nevill followed a journalistic career and for a time was parliamentary reporter for *The Standard* (London), as well as a theater critic. He worked alone and without financial sponsorship, and produced a contribution of outstanding importance in its day. He wrote under the pseudonym "Edmund Neison" in the conviction that the holder of an ancient name should not make a career in science. Nevill played an excellent game of tennis and was a golf enthusiast. His wife, Mabel (*née* Grant), whom he married in 1894 and who was South Africa's tennis champion for 11 years, survived him by several years.

Following an early interest in astronomy, Nevill became a fellow of the Royal Astronomical Society in 1873 at the age of 24. Using the classic work of **Wilhelm Beer** and **Johann von Mädler**, Nevill initiated a serious study of the Moon with a 6-in. refractor and a 9.5-in. With–Browning reflector from his residence in Hampstead, London. His voluminous book on the Moon was an important text; though based largely on the work of Beer and Mädler and in places merely a translation of *Der Mond*, it skillfully integrated all contemporary data to produce one of the most useful lunar reference works available in the English language and established Nevill's place in the history of astronomy.

In 1873 and 1874, Nevill argued for the existence of an appreciable lunar atmosphere and endorsed the idea that the Moon's craters represented the results of "vast volcanic convulsions," although he felt under obligation to discuss, at some length, Beer and Mädler's view that the surface of the Moon showed an "entire dissimilarity to that of the Earth." This was the general impression furnished by the small telescope that Beer and Mädler had employed, but with resemblances more compelling following the work of **Jean Chacornac**, Nevill found that closer examination with powerful instrument revealed far greater terrestrial analogy in the structures of the Moon than otherwise appears even possible. He reported that while a general analogy is often traceable between terrestrial volcanic regions and the more disturbed portions of the lunar surface, through powerful telescopes the larger craters appeared "less and less like volcanic orifices or craters." Indeed, their enclosing walls lost their regularity of outline and appeared instead as confused masses of mountains broken by valleys, ravines, and depressions, an irregularly broken surface.

Although Nevill accepted the probability of minor changes on the Moon and was convinced that the change reported by the German selenographer **Johann Schmidt** in 1867 within the small crater Linné on the plain of the Mare Serenitatis had been real, he departed from the conventional view that the alteration signified a volcanic upheaval, suggesting that it represented the comparative whimper of a landslide. However, Nevill expressed doubt when the German astronomer **Hermann Klein** reported in 1877 that Hyginus N was of recent origin.

Nevill had earlier written about the possibility of a lunar atmosphere (1873) and in greater detail on that subject in *Popular Science Review* in October 1874. Nevill was unafraid to voice his opinions. He was a founding member of the short-lived Selenographical Society (1878–1882), and served as its secretary until his career took an unexpected turn in 1882.

Though chiefly known for his work on lunar morphology, Nevill had a very strong interest in lunar theory and in 1877 confirmed the reality of an inequality in the longitude of the Moon, produced by the action of Jupiter, an effect that had been detected by Newcomb in 1876. The value obtained by Nevill for this term is close to that currently accepted. That same year he published a memoir containing the theoretical foundations of a complete analytical development of lunar theory, essentially a simplification of earlier treatments.

As part of the preparations to observe the transit of Venus in December 1882, a long-discussed plan to set up an observatory in Durban, South Africa was finally activated through the efforts of Harry Escombe and **David Gill**. Nevill was offered and accepted the post of government astronomer at Natal, and sailed for Durban arriving on 27 November 1882, a few days before the transit. He encountered many difficulties, but the weather at least was in his favor and the phenomenon was successfully observed. It was an auspicious start; Nevill conceived many plans, including observations to perfect the tables of lunar motion, and the establishment of the observatory as a meteorological center. He also became government chemist and official assayer for Natal, acting sometimes as pathologist in instances of suspected poisoning. However, promises of official funding were unfulfilled, and things became more difficult until, in 1911, the observatory was closed.

Nevill gave up his scientific career and returned to England. On retirement he settled at Eastbourne on the South Coast where he remained agile and unimpaired in mind and body unto the last, indulging his varied interests, especially chemistry. In the 1870s he had, with C. T. Kingzett, pressed for the setting up of a body to represent the profession of chemistry. At a meeting in the rooms of the Chemical Society (of which he was a fellow) on 26 April 1876, a committee was appointed as a first step toward the foundation of the Institute of Chemistry. Nevill was an original member of its council, serving from 1877 to 1880, and for some time was honorary corresponding secretary for the Institute of Natal Province. In 1935 he was awarded the Medal of the Royal Chemical Society.

Nevill was inactive, astronomically speaking, for almost three decades after returning to the United Kingdom from Africa, and he was assumed to have long since died. Twice, he refused the invitation to become a Fellow of the Royal Society, though in 1908 he was finally persuaded to accept. Nevill also declined the presidency of the Royal Chemical Society, and though a fellow, seldom, if ever, attended meetings of the Royal Astronomical Society. He researched Babylonian history, wrote several novels (but never submitted them for publication), and in 1886 published a popular outline of astronomy.

Richard Baum and *Thomas A. Dobbins*

Alternate name

Neison, Edmund

Selected References

Ashbrook, Joseph (1958). "The Selenographical Journal." *Sky & Telescope* 17, no. 12: 623, 628.

Both, Ernst E. (1961). *A History of Lunar Studies*. Buffalo, New York: Buffalo Museum of Science.

Moore, Patrick (1965). "E. N. N. Nevill: 'Edmund Neison.'" *Journal of the British Astronomical Association* 75: 223–227, 328–330.

Neison, Edmund (1876). *The Moon and the Condition and Configuration of Its Surface*. London: Longmans, Green, and Co.

Sheehan, William P. and Thomas A. Dobbins (2001). *Epic Moon: A History of Lunar Cartography and Nomenclature*. Richmond, Virginia: Willmann-Bell.

Newcomb, Simon

Born **Wallace, Nova Scotia, (Canada), 12 March 1835**
Died **Washington, District of Columbia, USA, 11 July 1909**

The commanding figure of United States astronomy in the 19th century, Simon Newcomb systematized and brought unparalleled precision to our knowledge of the Solar System.

The oldest of seven siblings born to parents of New England extraction, Newcomb grew up in the British colony of Nova Scotia, later part of Canada. His father, John Burton Newcomb, was an itinerant village schoolmaster; his mother, Emily Prince, was the daughter of a New Brunswick magistrate.

Largely home-schooled or self-taught, Newcomb spent his childhood in various parts of Nova Scotia and Prince Edward Island. Around age 15, he grew estranged from his mother's Calvinist beliefs and never really settled on an alternate faith. Apprenticed at 16 to a quack doctor, he ran away in 1853 and joined his father

in the United States, gaining a schoolteacher's position at Massey's Cross Roads, Maryland.

Through relentless study of authors ranging from Euclid to **Isaac Newton**, Newcomb acquired an increasingly profound understanding of mathematics. He submitted his first paper to **Joseph Henry**, secretary of the Smithsonian Institution, and was encouraged to persevere. Having moved closer to Washington, DC, Newcomb started to draw on the riches of the Smithsonian library. Henry soon recommended the United States Coast Survey as a suitable outlet for his talents. Referred in turn to the Nautical Almanac Office, then in Cambridge, Massachusetts, he arrived in December 1856 and became an astronomical computer. Newcomb simultaneously enrolled in Harvard's Lawrence Scientific School, studying mathematics under **Benjamin Peirce** and obtaining his B.Sc. in 1858. He never looked back.

When the Civil War began in 1861, Newcomb was appointed to the United States Naval Observatory in Washington, replacing staff that had departed to join the Confederacy. In 1863, he wedded Mary Caroline Hassler, granddaughter of the founder of the Coast Survey, Swiss-born Ferdinand Hassler. They would have three daughters, Anita, Emily, and Anna. In 1864, Newcomb was naturalized as a United States citizen.

In 1875, Newcomb was offered the directorship of Harvard College Observatory but turned it down, just as he declined the opportunities to replace Henry at the Smithsonian Institution or to head the Coast and Geodetic Survey. Instead, he accepted the superintendency of the Nautical Almanac Office in 1877. By then, he was accumulating honors: the Gold Medal of the Royal Astronomical Society, the Huygens Gold Medal of the Dutch Academy of Science, memberships in European scientific societies, and honorary degrees.

Newcomb served as president of the American Association for the Advancement of Science [AAAS] in 1877. He shared responsibility for the 1874 and 1882 American observations of the transits of Venus. In 1876, he lectured at the newly created Johns Hopkins University in Baltimore, Maryland, turning his course into a best-selling book, *Popular Astronomy* (1878). Newcomb was formally named a professor of mathematics and astronomy at Johns Hopkins University in 1884.

In 1890, the Royal Society of London awarded Newcomb the Copley Medal, and the Paris Académie des sciences chose him in 1895 to replace **Hermann von Helmholtz** as a foreign associate. In 1898, Newcomb received the Astronomical Society of the Pacific's first Catherine Wolfe Bruce Medal, and became, in 1899, the first president of the Astronomical and Astrophysical Society of America, later the American Astronomical Society. In 1907, France made him a Commander of the Légion d'honneur. Elevated by the United States to the rank of rear admiral, he was buried with full military honors in Arlington National Cemetery.

Newcomb's first important scientific result was his demonstration that the asteroids could not have originated, as **Heinrich Olbers** had suggested in 1803, from the fragmenting of a single planet. The work's 1862 publication in the German research journal *Astronomische Nachrichten* attracted notice abroad.

By then, Newcomb was already mastering the practical side of astronomy at the Naval Observatory, standardizing procedures, and tracking down systematic errors in stellar position catalogs. He took part in eclipse expeditions of 1860, 1869, 1870, and 1878, and played a key role in procuring for the observatory the nation's then largest telescope, a 26-in. refractor, inaugurated in 1873.

On the theoretical side, Newcomb analyzed the orbits of the Solar System's then-known outermost two planets, Uranus and Neptune, and published greatly improved tables for both by 1874. But he won fame by tackling a subject that was starting to exercise astronomers, namely, the gap between the observed motion of the Moon and theoretical attempts to represent it. By 1869, **Peter Hansen**'s 1857 lunar theory, based on data from 1750 to 1855, was clearly deviating from observations.

Newcomb realized that he could recover precise positional data from the Paris Observatory's records of lunar occultations, which yielded results stretching back to 1675. His reduction of the occultations was finished by 1888, although Newcomb was unable to complete the entire analysis before his retirement. With the inclusion of eclipses reported by **Ptolemy**, the observations spanned some 2,600 years. In a summary of his results, Newcomb identified a fluctuation that could not be attributed to gravity. It was later established as arising from the variable rotation (mostly slowing) of the Earth, as Newcomb had suspected.

Ensconced at the Nautical Almanac Office, Newcomb initiated a comprehensive redetermination of the constants of dynamical astronomy, from the best data obtained since 1750 by the world's observatories, in order to prepare new tables and formulas for the construction of ephemerides. One of the investigation's fruits was an improved value for the anomalous advance of Mercury's perihelion. (He favored some slight deviation from Newtonian gravity over the hypothetical intra-Mercurian planet Vulcan as the explanation.) With the help of several skilled mathematicians, Newcomb essentially completed this monumental endeavor by 1894.

Newcomb's retirement in 1897 did not end his career. In 1899, he published new tables of Uranus and Neptune, superseding his own earlier effort. A modest congressional stipend and later Carnegie grants from 1903 onward, enabled him to remain active in the international movement to standardize astronomical constants and basic data. Newcomb produced new catalogs of reference stars and rederived the principal constant of precession. Many of his numerical values endured until the advent of satellites and electronic computers led to superior determinations.

In other work, Newcomb refined the calculation of Jupiter's mass. In 1892, he deduced the approximate rigidity of the Earth. In the early 1900s he made the first quantative estimate of the background light of the night sky (equivalent approximately to one fifth magnitude star per square degree) and showed that the zodiacal light actually extends almost to the north ecliptic pole.

Newcomb also sought to better determine the distance to the Sun. Since the transits of Venus had proven disappointing, he measured anew the speed of light in order to find the distance directly from Earth's orbital velocity. His value for the speed of light was confirmed by the 1882 experiments of **Albert Michelson**. Though uncertainty persisted, Magnus Nyrén's 1883 constant of aberration yielded a result that turned out to be substantially correct.

Among the most distinguished scientists of his time, Newcomb was a celebrated figure who wrote astronomy textbooks and popularizations, works on economics, various opinion pieces (including a notorious refutation of the possibility of flying machines), and even some fiction.

First and foremost, Newcomb raised 19th-century positional astronomy to a pitch of perfection widely admired by his peers. Upon taking charge of the *Nautical Almanac*, he proposed changes in the ephemerides that were partially adopted by 1882. Results of his work were used by United States almanac makers from 1901 to 1960. Internationally, Newcomb's constants, theories, and tables for the Sun and the inner planets were not wholly superseded until 1984, so that his influence on astronomical ephemerides spans over a century. His contributions have been overshadowed by the more glamorous field of astrophysics, but Newcomb's precise determination of Mercury's perihelion advance lent inescapable significance to **Albert Einstein**'s derivation of the same within general relativity.

Most of Newcomb's works are found in the *Astronomical Papers Prepared for the Use of the American Ephemeris and Nautical Almanac*, 1879–1913, and in his 1895 volume, *The Elements of the Four Inner Planets and the Fundamental Constants of Astronomy*. Over 40 of his books and papers are available as microforms from the Canadian Institute for Historical Microreproductions. Newcomb's personal papers are held at the United States Library of Congress. He is widely held to have been the prototype of Walt Whitman's "Learned Astronomer."

Jean-Louis Trudel

Selected References

Archibald, Raymond Clare (1924). "Simon Newcomb, 1835–1909: Bibliography of His Life and Work." *Memoirs of the National Academy of Sciences* 17: 19–69. (Vol. 10 of the *Biographical Memoirs, National Academy of Sciences.*)

Campbell, W. W. (1924). "Biographical Memoir Simon Newcomb." *Memoirs of the National Academy of Sciences* 17: 1–18. (Vol. 10 of the *Biographical Memoirs, National Academy of Sciences.*)

Dick, Steven J. (2003). *Sky and Ocean Joined: The U.S. Naval Observatory, 1830–2000*. Cambridge: Cambridge University Press.

Marsden, Brian G. (1974). "Newcomb, Simon." In *Dictionary of Scientific Biography*, edited by Charles Coulston Gillispie. Vol. 10, pp. 33–36. New York: Charles Scribner's Sons.

Moyer, Albert E. (1992). *A Scientist's Voice in American Culture: Simon Newcomb and the Rhetoric of Scientific Method*. Berkeley: University of California Press.

Newcomb, Simon (1903). *The Reminiscences of an Astronomer*. Boston: Houghton, Mifflin, and Co.

Norberg, Arthur L. (1974). "Simon Newcomb and Nineteenth-Century Positional Astronomy." Ph.D. diss., University of Wisconsin-Madison.

Newton, Hubert Anson

Born **Sherborne, New York, USA, 19 March 1830**
Died **New Haven, Connecticut, USA, 12 August 1896**

Hubert Newton, who made some of the first rigorous studies of meteor orbits and their distribution in interplanetary space, was among the first to recognize meteors, fireballs, and meteorites as the same objects differing only in mass and velocity.

Newton was the fifth son of 11 children born to educators William and Lois (*née* Butler) Newton. At age three, he witnessed the spectacular Leonid meteor storm; his later prediction concerning its maximum activity in 1866 brought him fame. Newton was schooled locally until 1846 when he entered Yale College. He showed great aptitude for mathematics and the physical sciences, and won election to the Phi Beta Kappa Society. On 14 April 1850, Newton married Anna C. Stiles, with whom he parented two daughters.

Following his graduation in 1850, Newton studied mathematics privately for several years before accepting the position of mathematics tutor at Yale College in 1853. He was elected professor of mathematics in 1855, a position he held until his death. Newton also became director of the university's observatory. Under his charge, a research program on meteor photography was developed, principally carried out by **William Elkin**.

In the early 1860s, Newton conducted a survey to find accounts of outbursts similar to the 1833 Leonid storm, then regarded as an unpredictable phenomenon. He correctly deduced that the Perseids are distributed around the Sun in an elliptically shaped ring, though his calculated orbit differed significantly from today's accepted orbit, first determined by **Giovanni Schiaparelli** in 1866.

In an important 1864 paper, Newton coined the term "meteoroid" to describe interplanetary debris and suggested that the orbits of meteoroids closely resemble those of some comets. His review of 13 historical observations of the Leonids revealed cyclical outbursts with a period of about 33 years, and that the shower had been active since at least 902. Newton successfully predicted a Leonid meteor storm for 1866. He concluded that the Earth passes through the densest part of the meteor swarm during these times, strong evidence that the meteors are not uniformly distributed throughout their orbiting ring. Newton calculated that the Leonid meteors are spread out over 40 million miles along their orbit, with a stream thickness exceeding 100,000 miles.

Newton found five orbits consistent with the data, noting that analysis of the stream's nodal regression would single out one of them. In 1867, **John Couch Adams** showed that Leonid meteoroids move along highly elliptical orbits, with periods of 33.25 years and aphelia as distant as Uranus. Calculations by Carl A. Peters, Schiaparelli, **Urbain Le Verrier**, and **Theodor von Oppolzer** showed that Leonid meteoroids follow orbits nearly identical to that of a comet observed in 1866 (now identified as comet 55P/Tempel–Tuttle). This was the second occasion for evidence linking a comet with a meteoroid stream, following Schiaparelli's 1866 connection of the Perseids with comet 109P/Swift–Tuttle.

Further evidence that meteoroids are cometary decay products came from Newton's study of the Bielid (also known as Andromedid) meteors, which had produced impressive displays in 1798, 1830, 1838, 1841, and 1847; the irregular intervals defied simple explanation. Newton observed over 1,000 Bielids per hour on 24 November 1872, inspiring him to investigate the swarm named for their connection to comet 3D/1826 (Biela). After discovering the parent comet's close encounters with Jupiter in 1772 and 1841, Newton determined the planet's perturbing effect on the meteoroids' orbits and its contribution to the comet's breakup into meteoric fragments. Newton later demonstrated how Jupiter and Saturn could "capture" passing comets, perturbing the meteoroids coming from these comets from parabolic to elliptical orbits.

Newton's studies of the Bielids led to his investigation of how meteoroids burn up in the Earth's atmosphere and on how they might produce meteorites on the ground. He developed a sizeable collection of meteorites that he later donated to Yale's Peabody Museum of Natural History. Newton noticed that many specimens showed a smoothed side, which he interpreted as the "leading side" of the meteorite that bore the brunt of heating and melting by its passage through the atmosphere, and a rougher "trailing" side, which he believed escaped the worst of the ablative process. Studying meteor trails, Newton suggested that irregularly shaped meteoroids would experience differential air resistance, leading to a trail that would straighten after leading parts melted and smoothed and would diverge when asymmetric parts lead.

Newton confirmed this theory after studying a remarkable photograph, made by amateur astronomer John Lewis, of a fireball that appeared over Ansonia, Connecticut, on 13 January 1893. Newton measured Lewis's photographic plate to determine the distances from the fireball's trail to several bright stars that appeared on the plate. From the varying brightness of the fireball's trail, Newton concluded that it rotated more rapidly during the latter portions of its flight through the atmosphere – the first mention in the scientific literature of meteoroid rotation. From his study of the Ansonia fireball, Newton urged that photography of meteor trails would allow their orbits to be determined with considerably greater accuracy.

Newton's investigations turned to the origin and distribution of comets. In an 1878 paper, he argued that the distribution of cometary aphelia and inclinations were consistent with the interstellar-capture theory proposed by **Pierre de Laplace**. While this idea for the origin of comets is no longer held to be true, Newton importantly showed for the first time that long-period comets could be captured into short-period orbits through the gravitational perturbations of the planet Jupiter.

Newton's interests in meteoric phenomena were wide-ranging. He analyzed the nebular hypotheses of **Immanuel Kant** and Laplace, each of whom had attempted to explain the Solar System's origin through the contraction of a primordial cloud. Applying these theories to the origin of comets, Newton thought that Laplace's version of the nebular hypothesis accounted for most of the long-period comets, with orbital inclinations greater than 30°, while Kant's version, which argued that the comets were closely associated with the planets and should show orbits of smaller inclinations, better explained the short-period comets. Newton also wrote several papers and gave many lectures on the history of meteors and meteorites, including the worship of meteorites by ancient cultures and American Indians. He served as an expert witness in an Iowa dispute over meteorite ownership. Newton aggressively advocated adoption of the metric system of weights and measures, and lobbied for the 1866 enactment legalizing the metric system in North America.

Upon his retirement in 1886, Newton addressed the American Association for the Advancement of Science on the relationship between meteorites and shooting stars (meteors). He also speculated on the processes responsible for the origin and formation of meteorites; this talk included the first-ever discussion of cooling rates, a tool used by modern meteoriticists to analyze the origins and evolutionary histories of primordial Solar System material. Newton died on the date of the Perseid meteors' maximum activity, having witnessed a fine display of these meteors just the night before.

Newton served as president of the American Association for the Advancement of Science during 1885 and 1886. In 1872, he was elected foreign associate of the Royal Astronomical Society, and in 1892 was elected as a foreign member of the Royal Society of London. Newton was awarded the J. Lawrence Smith Gold Medal of the National Academy of Sciences for his work on meteors in 1888. He was a founding member of the National Academy of Sciences. The University of Michigan awarded Newton an honorary Doctor of Laws degree in 1868.

Martin Beech

Selected References

Anon. (1897). "Hubert Anson Newton." *Monthly Notices of the Royal Astronomical Society* 57: 228–231.

Beech, M. (2002). "The Mazapil Meteorite: From Paradigm to Periphery." *Meteoritics and Planetary Science* 27: 649–660.

Gibbs, J. W. (1897). "Memoir of Hubert Anson Newton." *Biographical Memoirs, National Academy of Sciences* 4: 99–124.

______ (1897). "Obituary." *American Journal of Science* 3: 359–378.

Hoffleit, D. (1988). "Yale Contributions to Meteoric Astronomy." *Vistas in Astronomy* 32: 117–143.

______ (1992). *Astronomy at Yale, 1701–1968.* Memoirs of the Connecticut Academy of Arts and Sciences, Vol. 23. New Haven: Connecticut Academy of Arts and Sciences.

Hughes, D. W. (1982). "The History of Meteors and Meteor Showers." *Vistas in Astronomy* 26: 325–345.

Littmann, M. (1998). *The Heavens on Fire: The Great Leonid Meteor Storms.* Cambridge: Cambridge University Press.

Newton, Isaac

Born **Grantham, Lincolnshire, England, 25 December 1642**
Died **London, England, 20 March 1727**

Sir Isaac Newton was born as a fatherless child on Christmas day. He was then given by his mother Hannah at 3 years to be reared by his grandmother. The young Isaac did not receive undue parental nurturing. There were stories of how his youthful inventions alarmed the inhabitants of Grantham village, such as a night-flying kite that carried a lit candle. The sundial he constructed as a youth is now owned by the Royal Society of London.

After a grammar school education in Grantham, Newton entered Trinity College, Cambridge, in June 1661 and was chosen as a scholar in 1664. In 1669, the college elected him a fellow and the university, through the influence of **Isaac Barrow**, the incumbent, appointed him Lucasian Professor of Geometry.

In December 1671 Newton presented a 2-in.-diameter reflecting telescope – the first ever constructed – to the Royal Society, which led to his election as a fellow. The telescope had a short lifetime because its mirror surface clouded over in a fortnight: It took over a century for nontarnishing reflectors to be made. This was swiftly followed by Newton's 1672 "New theory of light and colour," sometimes viewed as his first scientific paper. This had the effect of promoting his new reflecting telescope design by exaggerating the chromatic aberration from which refracting telescopes suffered. This exaggeration, which disturbed **Robert Hooke** and **John Flamsteed**, was reinforced in Newton's *Opticks* of 1704, and effectively blocked achromatic lens development until 1740.

Using a prism and a chink of sunlight, Newton claimed to demonstrate that white light was composed of various colored rays that had merely been separated by the prism. Hooke disagreed, commenting that he could not see the necessity for such an inference. Then, drawing from his alchemical studies, in a 1675 letter to the Royal Society, Newton formulated his immortal concept of the seven colors of the rainbow.

Newton was taught the new physics of **René Descartes**, and accepted the Cartesian vortex theory of planetary motions, adhering to it until the early 1680s, but he modified it with his own view of a downward-flowing gravity ether: This had a "sticky and unguent" nature as it pulled objects downward, as he explained in his 1675 letter. He was, at the time, immersed in the alchemical tradition, and this theory emerged from it. Modern Newtonian scholarship has shown that Newton's early computations in the plague years concerning the *conatus recedendi* (or tendency of the huge ethers rotating round the Sun to recede) cannot be seen as an early perception of the inverse-square law of gravitational attraction, contrary to several centuries of interpretation.

Around 1679/1680, in addition to his arduous alchemical labors on such matters as preparing the elixir and fixing antimony, Newton's major interest lay in decoding the Apocalypse (Revelation) in order to analyze a presumed theological heresy of the fourth century concerning the Holy Trinity. The Platonist philosopher Henry More at Trinity College recorded the enthusiasm with which Newton participated in discussion on such issues. We should therefore hesitate before accepting the received notion that Newton then linked **Johannes Kepler**'s first two laws of planetary motion to dynamical principles, as he later claimed and as many books have repeated. But no documents of this character exist (as science historian D. T. Whiteside demonstrated) dateable prior to the autumn of 1684, when, at **Edmond Halley**'s bidding, he struggled with the great problem, and solved it.

As a student, Newton had observed the comet of 1664 (C/1664 W1), but it was too distant for any orbital parameters to be inferred. The comets C/1680 W1 and 1P/1682 Q1 (Halley) were decisive for his thinking, with characteristics that seemed to be pointing to features of the to-be-born gravity theory. That of 1680 had its perihelion a mere fraction of the solar radius, yet was well outside the plane of the Solar System and so had little implication for the solar-vortex theory. Newton scrutinized it and received data from Flamsteed, after which he declined to believe what Flamsteed was telling him, that "ye two comets," one of which faded away in the evening sky and the other of which reappeared in the morning sky a week later, were one and the same. Years later, the comet merited 17 pages of his *Principia* for its parabolic orbit – but, in 1680, discussing its motion in the context of his vortex-theory with Flamsteed, he preferred **Giovanni Cassini**'s view that it was in orbit round Sirius. Hooke's seminal words to him, written on 6 January 1680, that throughout the Universe a force of gravity worked so that "the Attraction is always in a duplicate proportion to the distance from the Center Reciprocall ..." had hitherto lain dormant in his mind. Then the only bright, periodic comet (later named after Halley) conveniently turned up in 1682, orbiting within the ecliptic plane but in the reverse direction to the planets, and this acted as a trigger.

Newton's alchemical laboratory fire then went out for a couple of years. In the summer of 1684, following a visit of Halley, a more austere, left–brain process began as he apprehended that the "two comets" of 1680 were in fact one. In November of 1684 Halley received a draft of *De Motu*, which employed the inverse-square law. Newton there demonstrated the link to Kepler's first and second laws, using a cumbersome logic based upon relative volumes. Dealing with small changes in an elliptical orbit, it was a rudimentary integration procedure. The proof thus laboriously constructed required the rest of *De Motu* as its context, because it used the concepts there developed of force, impulse, and momentum conservation.

In the spring of 1685, Newton accomplished his "moon-test" computation, justly his most famous. For his predecessors, the 27.3-day sidereal lunar orbit had carried an astral meaning, from the Moon's passage against the starry constellations, but Newton ignored that and viewed it only as resulting from a central force. He became able in the 1680s to compute acceleration by a centripetal force. "If stopped," he explained, the Moon would fall a distance in 1minute, equal to the distance an object on Earth would fall in 1 second; a 60-fold ratio was employed, related to the 60 Earth-radii lunar distance. There was no computation of acceleration, of "g," despite the many textbooks that have averred this. Newton was now able to treat uniform circular motion as accelerated, toward its center.

Then began the great synthesis of many physical ideas in the 2 1/2 years during which Newton wrote his *Principia*, from the autumn of 1684 to March of 1687. **Nicolaus Copernicus** had made the Sun stationary, which Newton transformed into the immobility of the Solar System's center of mass. Newton incorporated the work of **Galileo Galilei**, who had first discerned accelerated motion in free fall, where distance fallen in equal times goes as the sequence of odd numbers and is the same for all objects, replacing the old notion that heavy bodies fall faster. Descartes, and now Newton, affirmed that one physics should link the Earth and sky, demolishing the old duality between the "sublunary" world and the immutable heavens: Newton extended the work of **Jean Buridan**, who had developed the notion of impetus, whereby a body keeps moving, in place of **Aristotle**'s notion that a body moves so long as it is pushed. Robert Boyle had described a vacuum at the top of a mercury column, which Newton now envisaged throughout the immensity of space; Newton derived Kepler's three laws of planetary motion, which had ellipses replacing circular epicycles; Barrow, Newton's mathematics teacher at Trinity, had taught a rudimentary calculus concerning "just nascent quantities," which Newton employed to describe the motions of bodies; last but not least, from Hooke came Newton's inverse-square law of gravitational attraction. A new Universe gleamed, rational to the core.

In dealing with the three-body problem, Newton's calculations were given to five decimal places and eight-figure accuracy, generating a huge error (200%) for lunar mass. He found the Earth–Moon mass ratio to be 22:1 rather than the currently accepted 81:1. Newton thus left to posterity an ultra-dense Moon. As a result, his first computation of the Earth–Moon barycenter in 1713 (for the *Principia's* second edition) located it outside the Earth, from which derived the main error in his historic computation, linking the fall of an apple to the lunar orbit.

Newton also explained why the Earth has two tides a day, a question that had so baffled Salviati and Sagredo in Galileo's *Dialogue*, and indeed many previous natural philosophers. Newton formulated the inverse-cube law of tidal pull, whereby "the force of the moon to move the sea varies inversely as the cube of its distance from the earth." This accounted for the Moon having a larger tidal pull than the Sun, although having only a tiny fraction of its gravity. Thereby he could explain why there are two high tides a day aligned with the Moon. Newton intuited this law with little by way of explanation, so his contemporaries such as Halley and **David Gregory** attempting to explain this tidal argument could do so only in a qualitative sense.

The mighty synthesis thus accomplished had no practical use to astronomers. British ephemerides (for planetary and lunar positions) were not improved: Paris became the main center of their production over this period. After his 1693 nervous breakdown, Newton made one further scientific endeavor. He grappled with lunar theory in 1694/1695, using Flamsteed's new, high-precision data. This was the supreme scientific problem of the age, holding out the promise of finding longitude at sea. Could Newton explain the Moon's erratic path using his gravity theory, since the rest of the Universe obeyed it? He could not (in Whiteside's view). His hitherto respectful partnership with Flamsteed suffered from this, with a (successful) ploy of laying the blame for the failure upon the astronomer, as if he had demurred in sending the data. A fruit of this struggle appeared in 1702, with a lunar "theory" which was, paradoxically, not evidently based upon gravitational principles. This 1702 *opus* was the most frequently reprinted work of Newton's in the first half of the 18th century: In seven steps of "equation" it obtained a final lunar longitude, accurate to several arc minutes.

A modified version appeared in Book III of the *Principia's* 2nd edition of 1713. Thus began the idea of ancillary equations, as a means of solving the three-body problem. Newton reintroduced epicycles into astronomy, a century after Kepler had banished them: His neo-Horroxian 1702 lunar theory was laden with four of them, and as such they reappeared in his *Principia*. French sources could never believe that this model with its wheels moving upon wheels had been deduced from the gravity theory, while English histories soon managed to retell the story using the mid-18th-century theories of **Leonhard Euler** or **Johann Tobias Mayer** as being "Newtonian."

In his *Algebra* of 1685, **John Wallis** commented upon a mathematical tract of the 1660s by Newton, *De analysi*, which Newton would not allow to be published; while admiring certain conventions and nomenclature, Wallis perceived in it no germ of a new fluxions theory, nor did anyone else in the 17th century, despite wide circulation of the manuscript. Only retrospectively, during the great fluxions battle with **Gottfried Leibniz** at the beginning of the 18th century, were such claims first advanced. (The *Principia* contained integral but not differential calculus, the former having developed somewhat earlier than the latter.) Newton's *Arithmetica Universalis* published in 1707 and taken from his mathematical lecture notes of the 1680s, compiled by **William Whiston**, enjoyed a much greater popularity in its time than either the *Principia* or *Opticks*, but it contained no trace of fluxions, Newton's term for the differential calculus, and rather argued against the concept of introducing arithmetical terms into geometry.

Albert Einstein once declared that, "the solution of the differential law is one of Newton's greatest achievements," but the equations $F = ma$ and $F = m\, dv/dt$ were invented around 1750 by Euler in Berlin; no one in Newton's lifetime had heard about them. The Berlin Academy of Sciences showed no inclination to view Euler's great discoveries as having been anticipated. What the *Principia* stated was, merely, "change of motion is proportional to motive force impressed" with quantity of motion having been earlier defined as the product of mass and velocity. That was a statement about impulse as proportional to change in momentum and uses no rate-of-change concept. As I. Bernard Cohen has observed, Newton never wrote anything resembling F = kmv. The *Principia* with its geometrically structured proofs, achieved a depth of inscrutability unmatched by any other scientific text. Much of what is called "Newtonian" science is the reformulation of Newton's work using Leibnizian calculus, a task accomplished largely on the Continent in the 18th century.

The myths that surround the image of Newton tend to exaggerate the extent to which he used "fluxions," but not always. For example, it is often asserted that he developed the equation $F = GMm/r^2$, a formula which was not, in fact, published in his lifetime. In "a famous but delusive phrase" (Rupert Hall), Newton averred in 1712 that his masterwork had first been composed in fluxional terms and then, later on, recast into a geometrical format. Generations of historians have reaffirmed that Newton had first composed his *Principia* in fluxional form and then recast it into its inscrutable geometric format, but not until 1975 did. D. Whiteside disprove this notion and lay it to rest.

"Newton's method of approximation" was invented in 1845 by John Simpson, known today for his "Simpson's method" for finding the approximate area under a curve. It is an iterative technique where the same equation is reused, and employs the Leibnizian calculus. Newton's own method of approximation, described in his *De Analysi* and which he used to solve the Kepler equation for elliptical motion, was neither iterative nor fluxional. For each step of approximation it generated a new and different equation. Simpson was not eminent enough to hang onto the credit for his invention, which became attributed to Newton in the latter half of the 18th century.

Few paid Newton more golden compliments than did Leibniz: "taking mathematics from the beginning of the world to the time of Sir Isaac, what he had done was much the better half," he wrote to the Queen of Prussia in 1701. But after his mistreatment by the Royal Society in the fluxions dispute, he described Newton as "a mind neither fair nor honest." Leibniz first published papers on the differential calculus in 1684; these were seminal for the European development of the subject. Newton's first work on the subject appeared in 1704, *De Quadratura*, which gave what we would call implicit functions. It did not describe time-dependent functions or how to find the gradient of a curve, and was primarily about methods of integration.

Newton wrote over a million words on chemistry alchemy, and believed that transmutation could possibly make or unmake gold, as expressed in his one published chemical alchemical text, *De Natura Acidorum* (1710), which described that process. He read alchemical texts eagerly, but seems not to have written like an alchemist; he sought no path of redemption or perfection through such labors. Ultimately his relation to the western alchemical tradition was that of terminator. Once his *Opticks* had affirmed the atomic view ("God in the Beginning form'd Matter in solid, massy, hard, impenetrable moveable Particles ... even so very hard, as never to wear or break in pieces"), the colorful language of alchemy had to transform into particulate affinity theory during the 18th century.

Hardly was the ink dry from his *Principia* in March 1687 when Newton was elected to represent Cambridge University in parliament in an attempt to defy the king's promotion of Roman Catholic professors. This morally courageous act put his career at risk and got him sternly rebuked from the feared judge Jeffreys. Two years later the king had fled the country, Jeffreys was in the Tower, and Newton was in parliament. When in 1695 he became warden of the Mint, the nation's recoinage was successful, and the Bank of England first floated paper money, a difficult exercise in credibility. When Newton was elected president of the Royal Society in 1704, its membership and prestige climbed steadily. From being the most reclusive of scholars, where most of the tales about him concern his absent-mindedness, he became a man of public affairs: Member of parliament, Justice of the Peace, knight, president of the Royal Society, and master of the Mint. In his religious views, Newton was probably a mortalist (disbelieving in human survival after death) and an anti-Trinitarian, either of which would have utterly debarred him from holding public office.

In the last year of his life, in a Kensington garden far from the bustle and fumes of London and while having tea with **William Stukeley**, Newton first told his story of the apple. Thus had the law of gravity dawned upon him. He located it in 1666, as London burnt and the plague raged. Earlier narrations beginning in the 1690s had involved his Mother's garden at Grantham but lacked mention of this fruit. The neo-Biblical simplicity of this story proved irresistible, and it has flourished ever since.

Nicholas Kollerstrom

Selected References

Bechler, Zev (1975). "'A Less Agreeable Matter': The Disagreeable Case of Newton and Achromatic Refraction." *British Journal for History of Science* 8: 101–126.

Cohen, I. Bernard (1975). *Isaac Newton's Theory of the Moon's Motion (1702)*. Folkestone, Kent: W. Dawson.

——— (1980). *The Newtonian Revolution*. Cambridge: Cambridge University Press.

——— (1999). *The* Principia, *Mathematical Principles of Natural Philosophy: A New Translation*. Berkeley: University of California Press.

Densmore, Dana (2003). *Newton's* Principia: *The Central Argument*. Santa Fe: Green Lion.

Fauvel, John *et al.* (1988). *Let Newton Be! A New Perspective on His Life and Works*. Oxford: Oxford University Press.

Gjertsen, Derek (1986). *The Newton Handbook*. London: Routledge and Kegan Paul.

Guicciardini, Niccolò (1998). "Did Newton Use His Calculus in the *Principia*?" *Centaurus* 40: 303–344.

Hall, A. R. (1980). *Philosophers at War: The Quarrel between Newton and Leibniz*. Cambridge: Cambridge University Press.

Kollerstrom, Nicholas (2000). *Newton's Forgotten Lunar Theory: His Contribution to the Quest for Longitude*. Santa Fe: Green Lion.

Sabra, A. I. (1981). *Theories of Light from Descartes to Newton*. Cambridge: Cambridge University Press.

Westfall, Richard (1980). *Never at Rest: A Biography of Isaac Newton*. Cambridge: Cambridge University Press.

Whiteside, Derek (1970). "Before the *Principia*: The Maturing of Newton's Thoughts on Dynamical Astronomy, 1664–1684." *Journal of the History of Astronomy* 1: 5–19.

——— (1976). "Newton's Lunar Theory: From High Hope to Disenchantment." *Vistas in Astronomy* 19: 317–328.

Wilson, Curtis (1989). "The Newtonian Achievement in Astronomy." In *Planetary Astronomy from the Renaissance to the Rise of Astrophysics: Part A*, edited by René Taton and Curtis Wilson, pp. 231–274. Vol. 2A of *The General History of Astronomy*. Cambridge: Cambridge University Press.

Nicetus

➤ **Hicetus**

Nicholas Cusanus

➤ **Krebs, Nicholas**

Nicholas of Lynn [Lynne]

Flourished **1386**

Nicholas was a Carmelite friar in Oxford. His almanac (then called *Kalendarium*) acknowledged – and provided a remedial table to correct for – the error in the Julian calendar, which by Nicholas's time had led to a 12 March equinox date. The *Kalendarium* of Nicholas was used by **Geoffrey Chaucer** in his *Canterbury Tales.*

Selected Reference

Eisner, Sigmund (ed.) (1980). *The Kalendarium of Nicholas of Lynn*, translated by Gary Mac Eoin and Sigmund Eisner. Athens: University of Georgia Press.

Nicholas of Cusa

➤ **Krebs, Nicholas**

Nicholson, Seth Barnes

Born **Springfield, Illinois, USA, 12 November 1891**
Died **Los Angeles County, California, USA, 2 July 1963**

American observational astronomer Seth Nicholson is probably best known for the discoveries of four satellites of Jupiter, numbers 9–12, but his most significant contribution may have been the discovery, with **Charles St. John** that the atmosphere of Venus contains at most a vanishingly small amount of water and molecular oxygen. Nicholson was the son of a schoolteacher and principal with a master's degree in geology from Cornell University. He received a BS from Drake University in 1912 (and an honorary LLD in 1949). He and Drake University classmate Alma Stotts went on to the University of California (Berkeley) and Lick Observatory for graduate work in astronomy and married in 1913, having three children. Nicholson was very prompt in completing the traditional Lick Observatory requirement to determine the orbit of a comet accurately enough for publication with a paper on the orbit of comet C/1912 V1 coauthored by O. Lanzendorf that same year. He supported his graduate studies by serving as an instructor of astronomy, completed the Ph.D. in 1915, and accepted an appointment at Mount Wilson Observatory, where he remained formally until his retirement in 1957 and informally for the rest of his life. Nicholson served as a civilian with the Office of Scientific Research and Development in 1944 and later with the Atomic Energy Commission.

Nicholson discovered his first jovian satellite (Sinope) while still a graduate student and a volunteer assistant at Lick when most of the astronomers were on an eclipse expedition to Russia. Therefore he had extra time for observing with the 36-in. Crossley. He had been assigned the task of photographing the known, faint outer satellites in order to improve their orbits; and, on the plates taken to track the eighth moon (discovered in 1908 by **Philbert Melotte**) found Jupiter IX. In 1938, while other Mount Wilson astronomers were at the general assembly of the International Astronomical Union in Stockholm, Nicholson once again had extra telescope time, this time at the 100-in. Hooker, to survey the region around Jupiter. He set the telescope to track Jupiter, which left faint stars as elongated trails on the plates and satellites carried along by Jupiter as small circular images. This technique revealed Jupiter X and XI. Jupiter XII came in 1951.

Systematic observation of Neptune showed Nicholson that it should be possible to determine the mass of its satellite Triton from the motion of Neptune around their mutual center of mass. This was done by **Harold Alden** in 1942.

Nicholson spent many years on the solar program initiated by **George Hale**. He collected data on sunspot numbers and magnetic fields over several sunspot cycles and looked for correlations with terrestrial phenomena using a spectroheliograph sensitive to polarization. In 1922, he and St. John looked hard for features due to water vapor and molecular oxygen in the spectrum of Venus. They concluded that the amount of oxygen must be less than 0.1% of what would be found above the same ground area on Earth and that there was less water vapor than would make a 1-mm layer if it precipitated out as liquid.

The most important long-term contribution was Nicholson's development with **Edison Pettit** in the 1920s of a vacuum thermocouple as a detector for infrared radiation beyond the longest wavelengths to which photographic emissions are sensitive. This enabled them to measure the thermal infrared emission from, and therefore the surface temperatures of the Moon, and terrestrial planets, and temperatures of the visible layers in the atmospheres on the Jovian planets. They found, for instance, that the subsolar point on Mars was sometimes above freezing and that the difference between day and night temperatures on Venus was relatively small, indicating that, although its rotation is slow, it does not always keep the same face toward the Sun.

Nicholson was somewhat unfairly drawn into the controversy about the rotation and distance of the spiral nebulae. He remeasured some of the plates on which **Adriaan van Maanen** thought he had seen motion outward along the arms (corresponding to rotation with leading arms) and tentatively confirmed the motions, but later, under the guidance of **Edwin Hubble**, he measured some of the plates again and came to agree with Hubble that there was no detectable proper motion.

Nicholson was elected to the National Academy of Science in 1937 and received the Bruce Medal of the Astronomical Society of the Pacific [ASP] at their June 1963 meeting in San Diego. He was by then too ill to attend and participated in the ceremony *via* a telephone hookup. Nicholson had twice served as president of the ASP and edited its *Publications* for some years, remaining on its board of directors, publications committee, and lecture committee until his death. He was a scoutmaster from 1923 to 1938, served on a boy scout troop committee and as a commissioner, and received scouting's Silver Beaver Award.

Norriss S. Hetherington

Selected References

Anon. "Obituary". (3 July 1963). *Los Angeles Times*, pt. 2, pp. 1–2.

Anon. (1963). "Seth B. Nicholson." *Physics Today* 16, no. 9: 106.

Anon. (1963). "Seth B. Nicholson Dies." *Sky &Telescope* 26, no. 2: 63.

Hale, George E. and Seth B. Nicholson (1925). "The Law of Sun-Spot Polarity." *Astrophysical Journal* 62: 270–300.

Hale, George H., Frederick Ellerman, S. B. Nicholson, and A. H. Joy (1919). "The Magnetic Polarity of Sun-Spots." *Astrophysical Journal* 49: 153–178.

Herget, Paul (1971). "Seth Barnes Nicholson." *Biographical Memoirs, National Academy of Sciences* 42: 201–227.

Hetherington, Norriss S. (1990). *The Edwin Hubble Papers: Previously Unpublished Manuscripts on the Extragalactic Nature of Spiral Nebulae*. Tucson: Pachart. (For Nicholson's role in the controversy over the purported rotation of spiral nebulae.)

Petrie, R. M. (1963). "Award of the Bruce Gold Medal to Seth B. Nicholson." *Publications of the Astronomical Society of the Pacific* 75: 305–307.

Pettit, Edison and Seth B. Nicolson (1922). "The Application of Vacuum Thermocouples to Problems in Astrophysics." *Astrophysical Journal* 56: 295–317.

______ (1930). "Lunar Radiation and Temperatures." *Astrophysical Journal* 71: 102–135.

______ (1933). "Measurements of the Radiation from Variable Stars." *Astrophysical Journal* 78: 320–353.

Sheehan, William (2001). "The Historic Hunt for Moons." *Mercury* 30, no. 2: 23–27. (For an outline of the historical context of planetary satellite discoveries, including Nicholson's four Jovian moons.)

Warner, Deborah Jean (1974). "Nicholson, Seth Barnes." In *Dictionary of Scientific Biography*, edited by Charles Coulston Gillispie. Vol. 10, p. 107. New York: Charles Scribner's Sons.

Niesten, Jean Louis Nicholas

Born **Vise near Liège, Belgium, 4 July 1844**
Died **probably Brussels, Belgium, 27 December 1920**

Louis Niesten joined the staff of the Royal Observatory (Brussels) in 1878, and in that year discovered the Great Red Spot on Jupiter – independently of, and 2 months after, **Carr Pritchett**. He was an assiduous observer of the planets, especially Venus, Jupiter, and Mars.

Selected Reference

Poggendorff, J. C (1926). "Niesten." In *Biographisch-literarisches Handwörterbuch*. Vol. 5, p. 906. Leipzig, Verlag Chemie.

Nietzsche, Friedrich Wilhelm

Born **Röcken bei Lützen, Sachsen, (Germany), 15 October 1844**
Died **Weimar, Germany, 25 August 1900**

German philosopher Friedrich Nietzsche was influential during an era when scientists were working out the ramifications of the laws of thermodynamics. In this light, his vision of "eternal recurrence" in infinite time may be viewed as a cosmological theory or as a concept in thermodynamics, related to the more quantiative Poincaré recurrence time.

Selected Reference

Krueger, Joe (1978). "Nietzschean Recurrence as a Cosmological Hypothesis." *Journal of the History of Philosophy* 16: 435–444.

Nightingale, Peter

Flourished **Denmark, 1290–1300**

The life of calendricist Peter Nightingale is only fragmentarily known. It is here assumed that Petrus de Sancto Audomaro (Peter from Saint Omer) is also identical to Peter Nightingale.

Possibly he is identical to the unknown astronomer at Roskilde, Denmark, who in 1274 measured day by day the declination of the Sun and computed the corresponding lengths of the day, data later used by Nightingale and **William of Saint Cloud** in their calendars.

Nightingale must have left Denmark in the 1280s to study at Bologna. There, he obtained a master's degree, and lectured on astronomy, or rather astrology, at the Faculty of Medicine. His popular Tabula Lunae, a diagram to determine the zodiacal sign of the Moon for every month, most likely is from this period. He also made a similar diagram to find the reigning planet for every day.

Around 1292, Nightingale left Bologna for Paris, at this time a flourishing center of astronomy comprising names like **Campanus of Novara**, **John of Sicily**, and William of Saint Cloud. In Paris, Nightingale worked out a new edition of **Robert Grosseteste**'s calendar, which had expired in 1283 at the end of its 76-year period based on four Metonic cycles of 19 years. Nightingale made the correction needed for another 76 years starting in 1292, improved the Moon tables, and added information about the Sun's declination and length of the day. This calendar and its subsequent prolongation became a standard in much of Europe in the 14th century.

Another achievement of Nightingale in practical astronomy was his improved equatorium, found in the "Tractatus de Semissis" from 1293. An equatorium is an analog computing device, based on the Ptolemaic model of the planetary system, designed to find the longitudes of the planets. Earlier equatoria had the drawback of requiring a separate unit for each planet. In Nightingale's equatorium they were merged into a single unit, thereby reducing considerably the number of graduated circles to be drawn. He seems to have worked together with William of Saint Cloud on this project.

Apart from the highlights described earlier, mathematical tables and commentaries to astronomical works of others are found among the extant manuscripts of Nightingale. Of the commentaries, the one on the combined quadrant and astrolabe by **Jacob ben Mahir** is of particular importance because it explained and spread knowledge of this ingenious instrument.

After his years in Paris, the only known information about Nightingale's life is that in 1303 he held a position as canon at the Roskilde Cathedral.

Truls Lynne Hansen

Alternate names

Petrus (Philomena) de Dacia
Petrus Dacus [Danus]

Selected References

Pedersen, Fritz Saaby (ed.) (1983–1984). *Petri Philomenea de Dacia et Petri de S. Audomaro opera quadrivialia*. Vols. 1 and 2. Copenhagen: G. E. C Gad. (Includes the extant original works of Peter Nightingale, all in the form of manuscripts in Latin.)

Pederson, Olaf (1968). "The Life and Work of Peter Nightingale." *Vistas in Astronomy* 9: 3–10.

______ (1993). *Early Physics and Astronomy*. Cambridge: Cambridge University Press.

Zinner, E. (1932). *Nordisk astronomisk tidsskrift* 13: 136–146.

Nikolaus von Cusa

➲ **Krebs, Nicholas**

Nīlakaṇṭha Somayāji

Born **Tṛkkaṇṭiyūr, (Kerala, India), 14 June 1444**
Died **after 1501**

Nīlakaṇṭha Somayāji was one of the foremost names of the Kerala School, which produced several outstanding astronomers and mathematicians over the centuries. The end of the Kerala School (about 1600) seems to have coincided with the fall of the Hindu Vijayanagar empire. Nīlakaṇṭha was the son of Jātaveda and a performer of the Soma sacrifice. His student days were spent in the house of another Kerala astronomer, **Parameśvara**, although his own teacher was Parameśvara's son, Dāmodara.

Nīlakaṇṭha's most important text is the *Tantrasaṅgraha*, a comprehensive treatise on astronomy written in 1501. The *Tantrasaṅgraha* reveals Nīlakaṇṭha to be a follower of the *dṛggaṇita* system of astronomy founded by Parameśvara. He prepared several other important texts, including an extensive commentary on the *Āryabhaṭīya* of **Āryabhaṭa I**. He also wrote the *Golāsara*, which explains the parameters of his planetary models, the celestial sphere, and other computational principles.

Nīlakaṇṭha combined the earlier and incomplete heliocentric models of Āryabhaṭa I and Parameśvara within his *Tantrasaṅgraha* and developed a fully heliocentric system, wherein the five planets moved in eccentric orbits around the Sun. He also made significant contributions to geometrical theorems and infinite series.

A. Vagiṣwari

Selected References

Chattopadhyay, Anjana (2002). "Nilakantha." In *Biographical Dictionary of Indian Scientists: From Ancient to Contemporary*, p. 935. New Delhi: Rupa.

Pingree, David (1974). "Nīlakaṇṭha." In *Dictionary of Scientific Biography*, edited by Charles Coulston Gillispie. Vol. 10, p. 129. New York: Charles Scribner's Sons.

______ (1978). "History of Mathematical Astronomy in India." In *Dictionary of Scientific Biography*, edited by Charles Coulston Gillispie. Vol. 15 (Suppl. 1), pp. 533–633. New York: Charles Scribner's Sons.

Sarma, Ke Ve (1972). *A History of the Kerala School of Hindu Astronomy*. Hoshiarpur: Vishveshvaranand Institute.

Nininger, Harvey Harlow

Born **Conway Springs, Kansas, USA, 17 January 1887**
Died **Westminster, Colorado, USA, 1 March 1986**

Harvey Nininger has been called the father of modern meteoritics. As an exceptionally successful meteorite hunter, Nininger collected specimens of 226 meteorite falls not previously identified, of which only eight were actually seen to fall. He began collecting and studying meteorites in 1923, during a period when the leading scientists of the day thought there was nothing left to learn from them. By the time Nininger retired, the Space Age had dawned and meteorites had become a key component of research into the origin of the Earth and the Solar System.

Nininger was the son of farmers James Buchanan and Mary Ann (*née* Bower) Nininger. In their simple home there were only two books, the Bible and a mail order catalog. Science was considered the work of the devil. As a result, Nininger was 20 years old before he passed an eighth-grade equivalency test and entered Northwestern State Normal College at Alva, Oklahoma, eventually matriculating from McPherson College in Kansas. After receiving a B.S. in Natural History from McPherson College in 1914 and an A.M. from Pomona College in California in 1916, Nininger worked as a biologist, serving both as a professor of biology at several small colleges and as a field entomologist for the United States Department of Agriculture.

In 1923, while teaching at McPherson College, Nininger learned of the existence of meteorites through an article in *Scientific Monthly*. Three months later, a brilliant fireball passed over the town as he was walking home. Nininger set out to find the meteorite that produced the fireball. Although he did not find that meteorite, he did find another one, and his life changed forever. Meteorites became his passion: Nininger set out to learn everything he could about them. He also started a program to collect meteorites, although he was not able to generate much support for this endeavor. In 1928, George Merrill, head curator of the Smithsonian Institution, had told him, "Young man, if we gave you all the money your program required and you spent the rest of your life doing what you propose, you might find one meteorite." However, Nininger was not dissuaded and devised a successful program of collecting meteorites based on educating the people who work the land and offering to buy the meteorites that they found. In 1930, Nininger resigned his professorship at McPherson College and moved to Denver, Colorado, where he joined the staff of the Colorado Museum of Natural History (now the Denver Museum of Nature and Science). He spent the next 16 years searching for and collecting meteorites and investigating meteorite craters. In 1932, Nininger and Frederick Leonard (1896–1960), professor of astronomy at the University of California at Los Angeles, founded the Society for Research on Meteorites (later renamed the Meteoritical Society).

In 1946, Nininger left Colorado with his collection of eight tons of meteorites to establish the American Meteorite Museum, near Winslow, Arizona. His intent was to use the tourist income from the museum admission and sale of meteorite specimens to support his studies of the nearby Barringer Meteor Crater. In 1939, he had been granted a formal permit to continue his investigation of the crater. Although Nininger had originally accepted the idea that the meteorite lay in the bottom of the crater, he eventually was convinced by the calculations of **Forest Moulton** that the meteorite had exploded on impact. His discovery in 1948 of metallic spheroids in the area around the crater provided proof of the explosion theory. These spheroids were produced by condensation of the vaporized meteorite from the fireball that resulted from the impact. Thousands of tons of them, representing the majority of the mass of the impactor, are present in the soil surrounding the crater.

Rapid expansion of interest in meteorites inevitably led to differences of opinion and controversy. Nininger was a participant in several of these early disputes. In 1941, Lincoln LaPaz published a theory of meteorites composed of antimatter, a topic of high interest as popular knowledge of the emerging disciplines of atomic and nuclear physics began to reach the broader scientific community. LaPaz speculated that craters otherwise devoid of evidence of a meteoric origin might have been formed by antimatter or "contraterrene" meteorites that exploded as they annihilated normal terrestrial matter. Nininger properly invoked Occam's razor in pointing out that much more plausible explanations could be put forward to justify the existence of such craters. The dispute escalated in a series of rebuttals on both sides of the argument before gradually dying out during World War II as the society ceased all professional publishing activities. The debate over the contraterrene meteorites was only the first of several acrimonious disputes between Nininger and LaPaz. Their feud came to a head at the 1949 meeting of the Meteoritical Society, when Nininger and his wife Addie resigned from the society after a priority dispute erupted at the meeting. By that time, the society was polarized by the contentious debates which appear to have involved, in addition to technical issues, LaPaz's view that Nininger's effort to earn a living by selling meteorites was unprofessional and not in the best interest of science. Their fights nearly destroyed the society in the view of a number of members who witnessed the years of wrangling.

In late 1948, Nininger's permit to collect meteorites on Barringer Crater Company property was withdrawn, limiting his research on the crater to government land. Shortly thereafter, highway traffic was rerouted so that it no longer passed in front of the museum. In 1953, the American Meteorite Museum was moved to Sedona, Arizona, where it operated until Nininger retired in 1960.

The meteorites that Nininger collected are a permanent legacy to his work. About 20% of his collection was purchased by the British Museum of Natural History in 1958. The remainder of his collection was acquired by the Arizona State University [ASU] in 1960. The Nininger Meteorite Collection became the centerpiece of ASU's Center for Meteorite Studies, which currently houses the largest university-based meteorite collection in the world.

Nininger received an honorary Sc.D. degree from McPherson College in 1967 and an honorary Sc.D. from Pomona College in 1976. The Meteoritical Society honored him with its highest award, the Frederick C. Leonard Medal, in 1967. During his lifetime he wrote four books, three booklets, and more than 150 scientific papers.

On 5 June 1914, Nininger married Addie Delp, and their union resulted in three children, Robert, Doris, and Margaret. Addie was an integral part of Nininger's meteorite program, running the household (both at home and during extended field trips), acting as field assistant, helping to run the Meteorite Museum, serving as chair of the Committee on Catalog of the Society for Research on Meteorites, and coauthoring "The Nininger Collection of Meteorites." Although Addie died in 1978, Nininger continued to live a full life, attending the Meteoritical Society's 50th anniversary meeting in Mainz, Germany in 1983. Nininger's field program was carried on by his son-in-law Glenn I. Huss (husband of Margaret), who operated the American Meteorite Laboratory in Denver until his death in 1991 and who collected an additional 239 meteorites previously unknown to science.

Gary Huss and *Peggy Huss Schaller*

Selected References

Anon. (1986). "In Memory of Harvey Harlow Nininger." *Meteoritics* 21: 239.

Hoyt, William Graves (1987). *Coon Mountain Controversies: Meteor Crater and the Development of Impact Theory*. Tucson: University of Arizona Press, pp. 337–343.

Huss, Glenn I. (1986). "Remembrance of Harvey Harlow Nininger." *Meteoritics* 21: 551–552.

Marvin, Ursula B. (1993). "The Meteoritical Society: 1933 to 1993." *Meteoritics* 28: 261–314.

Monnig, Oscar E. (1967). Forward to "Meteorites and Tektites." *Geochimica et Cosmochimica Acta* 31, no. 10. (Special issue.)

Nininger, Harvey Harlow (1952). *Out of the Sky: An Introduction to Meteoritics*. Denver: University of Denver Press.

——— (1956). *Arizona's Meteorite Crater: Past – Present – Future*. Sedona, Arizona: American Meteorite Museum.

——— (1971). *The Published Papers of Harvey Harlow Nininger: Biology and Meteoritics*, edited by George A. Boyd. Tempe: Arizona State University.

——— (1972). *Find a Falling Star*. New York: Paul S. Erikson.

Nininger, Harvey Harlow and Addie Delp Nininger (1950). *The Nininger Collection of Meteorites: A Catalog and History*. Winslow, Arizona: American Meteorite Museum.

Nīsābūrī: al-Ḥasan ibn Muḥammad ibn al-Ḥusayn Niẓām al-Dīn al-Aʿraj al-Nīsābūrī

Born **Nīshāpūr, (Iran)**
Died **(Iran), 1329/1330**

Niẓām al-Dīn al-Aʿraj al-Nīsābūrī composed several widely studied astronomy texts in the 14th century, which indicate the integration of astronomy within a tradition of religious scholarship in Islamic civilization. He was born into a Shīʿa family with roots in Qum.

The sources say little about Nīsābūrī's early life and education. By mid-1303, Nīsābūrī had begun to write *Sharḥ Taḥrīr al-Majisṭī* (Commentary on the recension of the *Almagest*), a commentary on **Naṣīr al-Dīn al-Ṭūsī**'s *Taḥrīr al-Majisṭī* (Recension of the *Almagest*) of **Ptolemy**. As was true for many commentaries in Islamic science, Nīsābūrī did not simply explain the meanings of the original text but included the results of his own work as well. In the *Sharḥ*, Nīsābūrī devoted much space to observations of the obliquity of the ecliptic and to **ʿUrḍī**'s work on instrument construction. Nīsābūrī also investigated whether Venus and Mercury had been observed to transit the Sun, an observation that would determine the position of the Sun with respect to Mercury and Venus. In 1304, Nīsābūrī arrived in Azerbaijan; by 1306 he was in Tabrīz, the largest city in Azerbaijan, where he completed the *Sharḥ*. In Tabrīz, Nīsābūrī also began to study with the astronomer **Quṭb al-Dīn al-Shīrāzī**.

Nīsābūrī completed his second major text, *Kashf-i ḥaqā'iq-i Zīj-i Īlkhānī* (Uncovering of the truths of the Īlkhānid astronomical handbook), in 1308/1309. The *Kashf*, a commentary on Ṭūsī's astronomical handbook entitled *Zīj-i Īlkhānī*, refers to the *Sharḥ*. Nīsābūrī wrote the *Kashf* right after the *Sharḥ* inasmuch as the *Kashf* focused on topics that were closely connected to the *Sharḥ*, such as the observation and prediction of planetary positions.

The *Tawḍīḥ al-Tadhkira* (Elucidation of the *Tadhkira*), a commentary on Ṭūsī's *al-Tadhkira fī ʿilm al-hay'a* (Memento on astronomy), was Nīsābūrī's third and final text on astronomy. A cross-reference to a *Tadhkira* commentary in the *Sharḥ* shows that Nīsābūrī had begun to compose the *Tawḍīḥ* before he finished the *Sharḥ*.

In the *Tawḍīḥ*, Nīsābūrī investigated theoretical topics, such as non-Ptolemaic models for planetary motions, and topics that combined theory and observations, such as physical hypotheses that accounted for the observed variations in the obliquity of the ecliptic. Although the *Sharḥ* and the *Tawḍīḥ* evinced a mastery of the technical innovations of Islamic astronomy, Nīsābūrī did not make significant advances with the most difficult questions. Shīrāzī, however, did, and the weight of Shīrāzī's reputation may explain the coincidence of the date of the appearance of the *Tawḍīḥ* with the date of Shīrāzī's death in 1311.

Īlkhānid ministers patronized Nīsābūrī's scientific work. The Īlkhānids were the descendents of Hülegü Khān (died: 1265), who had patronized the construction of the famous observatory at Marāgha, Azerbaijan, where both Ṭūsī and Shīrāzī worked. Nīsābūrī dedicated the *Sharḥ* to Khwāja Saʿd al-Dīn Muḥammad ibn ʿAlī al-Sāwajī. Sāwajī was chief minister (along with Rashīd al-Dīn) under Īlkhānid Sultan Ghāzān (reigned: 1295–1304) and continued in that post until 1312 when Rashīd al-Dīn had him executed. Shīrāzī's acquaintance with Sāwajī would have provided a way for Nīsābūrī to gain Sāwajī's patronage. There is a 1309 copy of the *Kashf* dedicated to al-Sāwajī. Nīsābūrī dedicated the *Tawḍīḥ* to a certain ʿAli ibn Maḥmūd al-Yazdī.

Because the *Sharḥ* and the *Tawḍīḥ* were clearly written and intended for nonexpert astronomers, they became important components of a tradition of religious scholarship that included astronomy. Many manuscripts of the *Sharḥ* and *Tawḍīḥ* have ownership statements from the libraries of *madrasas* (colleges of religious studies). Two reports attest to how the *Tawḍīḥ* was the most important text at **Ulugh Beg**'s *madrasa* in Samarqand for the study of the *Tadhkira*. Later works on Islamic astronomy, also with *madrasa* library ownership statements, refer to Nīsābūrī as *al-shāriḥ* (the commentator).

Nīsābūrī's best-known text, his Quran commentary entitled *Gharā'ib al-Qur'ān wa-raghā'ib al-furqān* (The curiosities of the Quran and the *desiderata* of the demonstration), demonstrates the importance of science for religious scholars. Nīsābūrī in general relied heavily on Fakhr al-Dīn al-Rāzī's (died: 1209) al-*Tafsīr al-kabīr* (The great commentary), but frequently disagreed with Rāzī about the use of science and philosophy *(falsafa)* to portray nature. The *Gharā'ib* reflected Nīsābūrī's scientific education and privileged the views of the natural philosophers *(falāsifa)*, while Rāzī had favored the positions of the theologians *(mutakallimūn)*. Through subtle rewordings and emendations of scientific detail, Nīsābūrī rebutted Rāzī's critique of science and *falsafa* in his portrayal of nature. Nīsābūrī completed *Gharā'ib* in 1329/1330, a date which the bio-bibliographers consider to be the date of his death.

Robert Morrison

Selected References

Al-Nīsābūrī, al-˘asan ibn Mu˛ammad (1992). *Gharā'ib al-Qur'ān wa-raghā'ib al-furqān*. Beirut: Dār al-maʿrifa. (Reprint of Cairo: al-Ma†baʿa al-kubrā al-amīriyya, 1905. Contained in margins of al-‡abarī's *Jāmiʿ al-bayān fī tafsīr al-Qur'ān*).

Morrison, Robert Gordon (1998). "The Intellectual Development of Niẓām al-Dīn al-Nīsābūrī (d. 1329 A. D.)." Ph.D. diss., Columbia University.

______ (2002). "The Portrayal of Nature in a Medieval Qur'an Commentary." *Studia Islamica* 94: 115–137.

______ (2005). "The Role of Portrayals of Nature in Medieval Qur'ān Commentaries." *Arabica* 52: 182–203.

Ragep, F. J. (1993). *Naßīr al-Dīn al-‡ūsī's* Memoir on Astronomy (*al-Tadhkira fī ʿilm al-hay'a*). 2 Vols. New York: Springer-Verlag.

Rashīd al-Dīn, Faḍl Allāh (1959). *Jāmiʿ al-tawārīkh*, edited by Bahman Karīmī. Vol. 2. Tehran.

Rosenfeld, B. A. and Ekmeleddin Ihsanoğlu (2003). *Mathematicians, Astronomers, and Other Scholars of Islamic Civilization and Their Works (7th–19th c.)*. Istanbul: IRCICA, pp. 238–239.

Nishikawa, Joken

Born **Hizen, (Nagasaki and Saga Prefectures), Japan, 1648**
Died **Nagasaki, Japan, 1724**

Joken Nishikawa strove to identify the practical and theoretical merits and defects of both Chinese and European astronomy. He was the son of a provincial official from Nagasaki. Joken was his penname, but he also used Tadahide, among others, as a first name throughout his life. Nishikawa learned astronomy from Kentei Kobayashi and developed Confucian scholarship under the tutelage of Soju Nanbu. At the age of 50, he retired and devoted himself to astronomy and calendar studies. As his writings on astronomy, geography, and pragmatic matters received recognition, Nishikawa was invited to Edo (present-day Tokyo) by the eighth *Shogun* Yoshimune Tokugawa in 1719 to discuss his ideas about Eastern and Western knowledge.

Nishikawa's productivity was evident in a number of fields spanning the intellectual and purely philosophical as well as the pragmatic and empirical. He passed on a sense of scholarship to his children and others. Nishikawa's third son, Masayoshi (1693–1756), learned astronomy from his father and was assigned to the Tenmongata (Bureau of Astronomy) as an official astronomer in 1740.

It is easy to oversimplify the changes that occurred in scientific thinking in 17th-century Edo Japan. This period started with a strong reliance on Chinese classics and the theory of five elements as exemplified in **Gensho Mukai**'s commentaries on western astronomical concepts. As the century closed, these traditions were giving way to an amalgamation of study that might be termed "pure" astronomy in the modern sense.

Nishikawa was somewhat of a pioneer in this intellectual movement; his greatest contribution was to delineate matters of speculative philosophy relative to matters best studied by empirical means. In his 1712 work, *Tenmon Giron* (Discussions of the principles of astronomy), Nishikawa discussed the moral and physical dualism of "two heavens." One was the heaven of *Meiri,* which Nishikawa considered a realm of philosophical speculation and most similar to the more classic idea of Li. The second was that of *Keiki,* which he considered the realm of empirical investigation and most like that of the classic Qi.

While some have considered Nishikawa to be closed to western concepts and rigid in his Confucian upbringing, his work must be viewed within its context. His intellectual work helped to free his own empirical investigations as well as those of many who followed. In *Ryougi Shusetsu* (An explanation of collected materials on celestial and terrestrial globes, 1714), he concerned himself with geographical as well as astronomical phenomena from the standpoint of both empiricism and pragmatism. While not rejecting Chinese Classics, he felt that western-based methods in areas such as navigation were clearly superior to those that had been coupled with the more mystic sides of classical Chinese works. He maintained skepticism toward the traditions of ancient Chinese classics as well as European ideas that he felt did not conform to empirical verification. Nishikawa abandoned traditional portent astrology but just as strongly rejected the zodiacal astrology of the west. To those who followed, he advocated relying on empirical verification of postulates rather than blind acceptance of any dogma, whether from China or Europe.

Steven L. Renshaw and *Saori Ihara*

Alternate name

Tadahide

Selected References

Ho Peng, Yoke (2000). *Li, Qi and Shu; An Introduction to Science and Civilization in China*. Mineola, New York: Dover. (Originally published in 1985. Originally written as an introduction to Joseph Needham's classic *Science and Civilization in China*, Volume 3: *Mathematics and the Sciences of the Heavens and the Earth*. (This volume on science in China presents the reader with an excellent view of the philosophies and cosmologies that guided Chinese thinking about the sky and influenced thought in Japan both before and during the Edo period (1603–1867). As the name suggests, it is an excellent source for gaining an understanding of the Chinese concepts of Li and Qi, important in considering the delineations that Nishikawa made.)

Nakayama, Shigeru (1969). *A History of Japanese Astronomy: Chinese Background and Western Impact*. Cambridge, Massachusetts: Harvard University Press. (Nakayama presents an extensive section on Nishikawa's commentary, and with detailed explanations of both western and eastern precepts, gives a clear picture of the epistemological base from which most work in early Edo Japan was developed.)

——— (1978). "Japanese Scientific Thought." In *Dictionary of Scientific Biography*, edited by Charles Coulston Gillispie. Vol. 15 (Suppl. 1), pp. 728–758. New York, Charles Scribner's Sons. (In this extensive article, Nakayama clarifies differences between epistemological views in Europe and Japan at the time Nishikawa lived. Whereas his explanation of Chinese classics require the reader to find supplemental material in order to grasp their full impact, this article provides a necessary base to understand the development of empiricism in the early Edo era as a mode of inquiry based on aspects of Aristotelian cosmology and pragmatic social need.)

Sugimoto, Masayoshi and David L. Swain (1989). *Science and Culture in Traditional Japan*. Rutland, Vermont: Charles E. Tuttle and Co. (This volume presents a comprehensive and scholarly view of scientific development in Edo Japan. Unlike many of his predecessors in Nagasaki, Nishikawa lived and worked in a time when the study of western concepts was somewhat freed from its previous relation to Christianity and the possible negative consequences of such a link. Sugimoto and Swain provide an excellent social backdrop for placing Nishikawa's work within a historical context.)

Nordmann, Charles

Born **Saint-Imier, now Bern Canton, Switzerland, 18 May 1881**
Died **Paris, France, 28 August 1940**

Charles Nordmann was an early pioneer of radio astronomy, the creator of multicolor photometry techniques, and a noted popularizer of astronomy. After receiving his *Licencié ès sciences* in 1899, he spent the following year at the Meudon Observatory, near Paris, under the directorship of **Pierre Janssen**. There, Nordmann began a study of the Sun, its activity, and effects on the Earth's magnetic field. From 1902 to 1903, he worked in the magnetic service at the Nice Observatory.

For his doctoral research at the University of Paris, Nordmann attempted to test the prediction of **Henri Deslandres** that the Sun ought to emit Hertzian (*i. e.*, radio) waves. He constructed a horizontal antenna, 175 m long, and installed it on the glacier of the Bossons (on Mont Blanc). Despite Nordmann's best efforts, no definite signals were ever received. Radio waves were not to be detected from the Sun until 1942. Nonetheless, Nordmann was awarded his degree (1903) under the guidance of **Jules Poincaré**.

That same year, Nordmann became an *auxiliaire* at the Paris Observatory and in 1905, was given the position of *aide-astronome*. **Maurice Löwy**, the director, encouraged him to pursue astronomical photometry. In 1906, as *astronome adjoint*, Nordmann devised a three-filter visual photometer that employed colored solutions to isolate the red, green, and blue regions of the spectrum. These experiments were performed on the observatory's 23-cm coudé telescope. His preliminary research investigated the apparent time differential (as a function of wavelength) on the observed minimas of eclipsing binary stars, a phenomenon independently described by **Gavril Tikhov**. Nordmann's observations were designed to test the potential dispersive powers of the interstellar medium on the speed of light, an idea ultimately rejected. However, recognition that the color index of a star (obtained by multicolor photometry) provided an effective measure of its surface temperature (given by the Planck curve) was first accorded to other observers.

During World War I, Nordmann worked in the field of sound-ranging techniques to locate the positions of German artillery. For these efforts, he was awarded the *Croix de guerre* in 1915 and made an *Officier de la Légion d'honneur* in 1918 (a *Chevalier* after 1912).

In 1920, Nordmann was appointed *astronome titulaire* at the Paris Observatory and began teaching at the École Supérieure des Postes et Télégraphes. He continued research in multicolor photometry and was succeeded in that effort by **Daniel Chalonge**. A significant part of Nordmann's work, however, was devoted to writing popular accounts of scientific discoveries, *e. g.*, *Einstein and the Universe: A Popular Exposition of the Famous Theory* (Paris, 1921).

Jacques Lévy

Translated by: *Suzanne Débarbat*

Selected References

Esclangon, E. (1942). "Charles Nordmann." *Rapport annuel sur l'état de l'Observatoire de Paris, pour l'année 1940*, p. 2. Paris: Imprimerie Nationale.

Hearnshaw, J. B. (1996). *The Measurement of Starlight: Two Centuries of Astronomical Photometry*. Cambridge: Cambridge University Press, esp. pp. 101–103, 371–373.

Nordmann, Charles (1982). "A Search for Hertzian Waves Emanating from the Sun." In *Classics in Radio Astronomy*, edited by Woodruff Turner Sullivan, III, pp. 158–160. Dordrecht: D. Reidel.

Poggendorff, J. C. "Nordmann." In *Biographisch-literarisches Handwörterbuch*. Vol. 5 (1926): 911; Vol. 6 (1938): 1874–1875; Vol. 7b (1980): 3648. Leipzig and Berlin.

Norton, William Augustus

Born **East Bloomfield, New York, USA, 25 October 1810**
Died **New Haven, Connecticut, USA, 21 September 1883**

William Norton wrote what was likely the earliest astronomical textbook produced in the United States; he was also a significant contributor to 19th-century ideas on the structure and behavior of cometary tails, terrestrial magnetism, and solar activity. By 1870, Norton went so far as to suggest, though without describing an exact mechanism, a link not only between solar coronal activity and Earth's auroral displays, but also with the observed activity in cometary tails.

Born the son of Herman and Julia (*née* Strong) Norton, William graduated in 1831 from the United States Military Academy at West Point, at that time arguably the premier technical, scientific, and engineering school in America. From 1831 to 1833, Norton served as both an artillery officer in the field during the Black Hawk War, and as an assistant professor of Natural and Experimental Philosophy at West Point. Norton left the army to become professor of natural philosophy and astronomy at the University of the City of New York (later renamed New York University) from 1833 to 1839. He then took a post as chairman (1839–1850) of mathematics and philosophy at the then struggling Delaware College (later renamed the University of Delaware), becoming interim president for a year (1850). A brief stint followed at Brown University, Providence, Rhode Island, as professor of natural philosophy and civil engineering. Norton finally settled down as professor of civil engineering at Yale's Sheffield Scientific School from 1852 until his death. He married Elizabeth Emery Stevens in 1839.

Seemingly inspired by the Great March Comet of 1843 (C/1843 D1), Norton acquired an early and lasting fascination with the behavior of comets, terrestrial magnetism, and the Sun. In 1844, Norton published an article, "On the Mode of Formation of the Tails of Comets," in the *American Journal of Science and Arts*. There, he explained the formation of cometary tails as originating from the evaporation of matter in the nucleus that was then swept away by the "repulsive force" of the Sun, which he suggested was magnetic rather than electrical in nature. Norton further explained the complexity of structures observed in cometary tails through a combination of the Sun's "repulsive force" *and* the centrifugal forces arising from a postulated rotating cometary nucleus. In 1859 and 1861, Norton presented evidence supporting his ideas from the widespread observations made upon Donati's comet (C/1858 L1). His studies of comets, terrestrial magnetism, and the Sun reached their fullest development in an 1870 paper, "The Corona Seen in Total Eclipses of the Sun." Here, Norton attempted to link phenomena of the corona and terrestrial magnetism by suggesting that the solar atmosphere itself was in fact an intense auroral display.

Norton was a natural and very successful teacher of astronomy; his textbooks reflected his experience in making the subject more understandable to students. *An Elementary Treatise on Astronomy in Four Parts* was one of the first astronomy textbooks published in America by an American astronomer. Norton's text, concentrating on both mathematical and practical aspects of the subject, was typical of those of the period, but differed in the greater degree to which it utilized illustrations and diagrams and in its more concise language as compared to others. This volume went through four editions, the last being published in 1881 as *Treatise on Astronomy, Spherical and Physical*. While at Yale, Norton authored a second textbook, *First Book of Natural Philosophy and Astronomy* (1858).

Norton's scientific interests extended to thermodynamics and molecular physics. He was elected a member of the American Philosophical Society, the National Academy of Sciences, and a corresponding member of the National Institute for the Promotion of Science.

Gary L. Cameron

Selected References

Anon. (1883–1884). "William Augustus Norton." *Proceedings of the American Academy of Arts and Sciences* 19: 530–534.

Anon. (1979). "Norton, William Augustus." In *Biographical Dictionary of American Science: The Seventeenth through the Nineteenth Centuries*, edited by Clark A. Elliot, p. 191. Westport, Connecticut: Greenwood.

Norton, William A. (1845). *An Elementary Treatise on Astronomy in Four Parts.* New York: Wiley and Putnam.
Trowbridge, W. P. (1886). "Memoir of William A. Norton." *Biographical Memoirs, National Academy of Sciences* 2: 189–199.

Norwood, Richard

Born **Stevenage, Hertfordshire, England, 1590**
Died **Bermuda, 1675**

Richard Norwood was notable for his works on navigation and surveying. He wrote an autobiographical journal in 1639 in which he described his father (Edward?) as a gentleman who had suffered financial adversity. After Richard's formal schooling ended, he was apprenticed to a fishmonger at age 15. Contact with seafaring men started him on a series of voyages and adventures and initiated his devotion to the study of navigation, which he pursued by studying every book on the subject that came his way. Norwood married Rachel Boughton in 1622; of their four children, their second son, Matthew, followed his father's interests.

In the 17th century, much effort in astronomy and mathematics was motivated by the practical needs of navigation and surveying, areas to which Norwood made notable contributions. He carried out the first survey of Bermuda in 1614/1615, taught mathematics in London from 1627 to 1637, and, in a futile attempt to escape religious strife, returned to Bermuda where he continued teaching. In 1663, Norwood completed a second survey, which still stands as the basis of boundaries on the island.

Besides his journal and surveys, Norwood wrote at least seven books, covering aspects of fortification, mathematics, and navigation. He was arguably the first person to provide, in his *Trigonometrie* of 1631, clear explanations of great circle sailing and how to use logarithms in solving plane and spherical triangles. While this book also contained tables of the declination of the Sun based on tables by Edward Wright and of a few stars precessed from **Tycho Brahe**'s catalog, Norwood's only book dealing specifically with an astronomical topic was *A table of the suns true place ... made according to many exact observations of the Sunne, taken in severall Years last past in the Somer Islands* [Bermuda] (1656).

His very influential *Sea-mans Practice* (1637) contained Norwood's measurement of the length of a degree, or after multiplying by 360, the circumference of the Earth: 25,036 English miles, very close to the modern mean value of 24,873 miles. It stood as the best available value for a long time; **Isaac Newton** used it in the 1713 and 1726 editions of his *Principia*. To arrive at this remarkable result, Norwood measured the latitude of York and London, England, and found the distance between those cities by chaining and pacing, correcting for winding roads, ascents, and descents. The vital implication for navigators was that a nautical mile (1 min of latitude) was not the 5,000 ft. that they had traditionally used but rather 6,120 ft. as Norwood stated, very close to our present-day value of 6,080 ft. For astronomers, knowing the size of the Earth accurately is the first step in measuring extraterrestrial distances.

The fact that three of Norwood's books were reprinted as recently as the 1970s is an indication of the lasting interest he holds for scholars. These and his *Journal* (published in 1945) are accessible primary sources.

Peter Broughton

Selected References

Multhauf, Lettie S. (1974). "Norwood, Richard." In *Dictionary of Scientific Biography*, edited by Charles Coulston Gillispie. Vol. 10, pp. 151–152. New York: Charles Scribner's Sons.
Waters, D. W. (1958). *The Art of Navigation in England in Elizabethan and Early Stuart Times.* New Haven: Yale University Press.

Novara, Domenico Maria da

Born **Ferrara, (Italy), 29 July or 1 August 1454**
Died **Bologna, (Italy), 15 or 18 August 1504**

Domenico da Novara was a teacher of **Nicolaus Copernicus**.

Domenico's family name was Ploti; they originated in Novara, a city in northwestern Italy, and later moved to Ferrara in northeastern Italy, where Domenico was born. He obtained both the title of Doctor of Arts and Doctor of Medicine, but we do not know at which university he studied. A possible record of his education is contained in an astrological text ascribed to Dominicus. The author testifies he was a pupil of **Johann Müller** (Regiomontanus), who traveled in Italy between 1460 and 1467 and later between 1472 and 1475, and lived in various Italian cities for long periods, including Ferrara. In any case, da Novara taught astronomy at Bologna University from 1483 till his death.

During this period da Novara was requested to publish prognostications for every year; many of them (in Italian and Latin) have survived and represent the only works we know by him. Apart from the astrological judgments they contain, the interesting parts of these publications are the preambles, where he discusses his scientific and philosophical theories.

The most famous was the prologue for the prognostication of 1489, in which da Novara presented his theory on the shift of the terrestrial polar axis. This was based on the comparison of the latitudes of Cádiz in Spain and several places in Italy, determined in his own time, with those reported in **Ptolemy**'s *Geography*. His conclusions were wrong, because they were based on unreliable data, but they are important because they represent one of the first attempts to suppose an Earth not at rest.

From later sources we know da Novara computed the obliquity of the ecliptic in 1492, obtaining a value very close to the actual one for this year.

Between the end of 1496 and the beginning of 1497, Copernicus registered in the Public Register of the German College in Bologna University. No record of his life in Bologna can be found in Copernicus's works, but **Rheticus** later reported that Copernicus lived in da Novara's house and helped him in his astronomical observations. Three of these observations are indeed reported by Copernicus: the observation of Aldebaran eclipsed by the Moon on 9 March 1497, and the observations of the conjunction of Saturn with the Moon on 9 January 1500 and 4 March 1500.

Giancarlo Truffa

Alternate name

Ploti Ferrariensis

Selected References

Bilinski, Bronislaw (1975). *Alcune considerazioni su Niccolò Copernico e Domenico Maria Novara (Bologna 1497–1500)*. Wroclaw.

Biskup, Marian (1973). *Regesta Copernicana* (Calendar of Copernicus' Papers). *Studia Copernicana*, Vol. 8. Wroclaw: Ossolineum, pp. 39–42.

Bònoli, F. and D. Piliarvu (2001). *I lettori di astronomia presso lo Studio di Bologna: dal XII al XX secolo*. Bologna CLUEB, pp. 118–121.

Bònoli, F., A. C. Colavita, and C. Mataix Loma (1995). "L'ambiente culturale bolognese del Quattrocento attraverso Domenico Maria Novara e la sua influenza in Nicolò Copernico. " *Memorie della Società astronomica italiana* 66: 871–880.

Jacoli, Ferdinando (1877). "Intorno alla determinazione di Domenico Maria Novara, dell'obliquità dell'eclittica." *Bullettino di bibliografia e di storia delle scienze matematiche e fisiche* 10: 75–88.

Prowe, Leopold (1883). *Nicolaus Coppernicus*. Vol. 1, pp. 236–246. Berlin, Weidmann.

Rosen, Edward (1975). "Copernicus and His Relation to Italian Science." In *Convergno internazionale sul tema Copernico e la cosmologia moderna*, pp. 27–38. Rome: Accademia nazionale de Lincei.

Sighinolfi, Lino (1920). "Domenico Maria Novara e Nicolò Copernico allo Studio di Bologna." *Studi e memorie per la storia dell' Università di Bologna*, ser. 1, 5: 207–236.

Thorndike, Lynn (1941). *A History of Magic and Experimental Science*. Vol. 5, pp. 234–236. New York: Columbia University Press.

Westman, Robert S. (1993). "Copernicus and the Prognosticators: The Bologna Period, 1496–1500." *Universitas*, no. 5: 1–5.

Zinner, Ernst (1990). *Regiomontanus: His Life and Work*, translated by Ezra Brown. Amsterdam: North-Holland, pp. 240–241.

Numerov [Noumeroff], Boris Vasil'evich

Born **Novgorod, Russia, 17/29 January 1891**
Died **Orel, (Russia), 13 September 1941**

Boris Numerov was one of the principal organizers of Soviet astronomy after the 1917 Bolshevik Revolution. Both a theoretician and a practitioner, Numerov specialized in celestial mechanics, astrometry (positional astronomy), and gravimetry. He was the key Soviet figure in applied celestial mechanics before World War II. His principal legacy was the Institute of Theoretical Astronomy of the Russian Academy of Sciences, which has unfortunately closed very recently. Numerov was elected a corresponding member of the Soviet Academy of Sciences in 1929.

Numerov was a graduate (1913) of Saint Petersburg University and for a short time (1913–1915) a supernumerary astronomer at the Pulkovo Observatory. Like the renowned astronomer Sir **Arthur Eddington**, Numerov debuted as an observer with a zenith-telescope. From 1917 until his arrest in 1936, Numerov taught at Leningrad University where he was named a professor in 1924. In 1919, he founded and operated the Computing Institute, which, in 1923, was merged with the Astronomical–Geodetical Institute, also initiated by Numerov, to become the Leningrad Astronomical Institute. In 1943, the latter was transformed into the Institute of Theoretical Astronomy – the principal center for celestial mechanics research in Russia. Apart from his other duties, in 1926/1927, Numerov was the director of the Voeikov Main Geophysical Observatory and, in 1931–1933, chief of the Department of Applied Mathematics within the State Optical Institute. He fostered the development of the Abastumani Astrophysical Observatory in Georgia.

Numerov arranged computations for the *Astronomicheskii Ezhegodnik SSSR* (USSR Astronomical Almanac; first issued in 1921), which provided annual information on astronomical events of all kinds. He produced a number of astronomical tables and handbooks. He paid full attention to the data obtained by astrometrical and gravimetrical instruments and strove to improve them. Numerov performed significant applied research, including gravimetrical measurements to identify potential ore and oil deposits. He initiated the creation of an annual ephemeris of minor planet elements and positions, published to this day. Numerov also developed a method for using minor planet data to determine the Equator and equinox for star catalogs. His disciples worked in many institutions of the then USSR.

Internationally known for his scientific and organizational activities, Numerov was arrested (along with many other Leningrad astronomers, including Pulkovo's director, **Boris Gerasimovich**), at the beginning of Stalin's Great Terror. Imprisoned and tortured as a supposed saboteur and fascist spy, Numerov was executed by a firing squad in prison, before the Nazis captured the city. Many aspects of Numerov's heritage are visible even today. Minor planet (1206) Numerowia was named for him, as is a crater on the farside of the Moon.

Alexander A. Gurshtein

Selected References

Kulikovsky, P. G. (1974). "Numerov, Boris Vasilievich." In *Dictionary of Scientific Biography*, edited by Charles Coulston Gillispie. Vol. 10, pp. 158–160. New York: Charles Scribner's Sons.

McCutcheon, Robert A. (1991). "The 1936–1937 Purge of Soviet Astronomers." *Slavic Review* 50, no. 1: 100–117.

Numerova, A. B. (1983). "Boris Vasil'evich Numerov (1891–1941)" (in Russian). *Istoriko-astronomicheskie issledovaniia* (Research in the history of astronomy) 16: 193–218.

Numerova, A. B. and V. K. Abalakin (1984). *Boris Vasil'evich Numerov, 1891–1941* (in Russian). Leningrad: Nauka.

Nunes, Pedro

Born **Alcácer do Sal, Portugal, 1502**
Died **Coimbra, Portugal, 11 August 1578**

Pedro Nunes is chiefly known for his theoretical work on celestial navigation, and his translations into Portuguese of works by **John of Holywood** (Sacrobosco) and **Ptolemy**. Nunes was the son of Jewish parents converted to Christianity. During his youth, Portugal was a leader in voyages of exploration and discovery; in the year of Nunes's birth, Vasco da Gama undertook his second voyage to India, and Brazil had already been discovered 2 years earlier. In 1517, Nunes started his university studies in humanities, philosophy, and medicine at the University of Lisbon, and, by 1521/1522, he had gone to Salamanca to study at the university there. In Salamanca Nunes married Dona Guiomar Aires in 1523, and they had six children (two boys and four girls).

Nunes returned to Portugal in 1524/1525, and on 16 November he was named cosmographer of the kingdom. At that time he was Doctor of Medicine. Nunes's nomination was due to his wide knowledge of astronomy – the course of medicine included studies of mathematics and astronomy, required for the practice of astrological medicine.

At the University of Lisbon, Nunes was in charge of lecturing on the subjects of moral philosophy, logic, and metaphysics. In 1532, he went to Évora as tutor of the Infants D. Luís and D. Henrique, future cardinal and king of Portugal. Another pupil was D. João de Castro, who would become one of the greatest Portuguese navigator pilots and was the future Viceroy of India. On 16 October 1544 Nunes was named lecturer of mathematics at the University of Coimbra, where he had been transferred from Lisbon in 1537. He lectured on that subject until his 25th year jubilee in 1562.

On 22 December 1547, Nunes was named First Cosmographer of the kingdom. He became so famous that in the year before his death, he was consulted by Pope Gregory XIII about his project of calendar reform. Some studies about Pedro Nunes state that he had been **Christoph Clavius**'s professor when the latter attended classes at the College of Arts at Coimbra in 1556/1557, but such information is erroneous.

Nunes's work is vast, not just limited to mathematics and the nautical sciences; he also wrote poetry. Nunes was considered the greatest Portuguese navigator of the Renaissance, and yet he did not set foot overseas. His fame was due to his scientific work, mainly nautical. His worries about the practical workability of knowledge showed a preponderance of the utilitarian mind over the speculative one. To him, "science" had its sphere of action limited to the "certain and true" things. This practical aspect is seen in the style Nunes used to write out his texts and notes, endowed with great clarity and rigor concerning the presentation of rules and the demonstration of theorems.

Nunes's work can be divided into two major groups: The first is translations and annotations, the second, original works.

Nunes's *Tratado da Sphera com a Theorica do Sol e da Lua* and the *1° Livro da Geografia de Ptolomeu Alexandrino* include the annotated translations of three works. The *Tratado da Sphera* is the translation of the book of Sacrobosco with his own notes attached; the *Theorica do Sol e da Lua* comprises the translation of the first three subjects of **Georg Peurbach**'s *Theoricas dos Planetas*, with 17 added notes of his own. The third translated work is the first book of geography by Ptolemy, also with notes by Nunes.

After some questions posed by the Portuguese navigator Martim Afonso about the direction and course line of a ship at sea, Nunes was the first to develop the theory of loxodromics, demonstrating that the line drawn by a ship at the sea's surface, when it cuts all the meridians on the same oblique angle, is not a great circle, as they used to think, but a spheric spiral that rounds the Earth's poles an infinite number of times. He also demonstrated that the only circular course lines are the meridians and the parallels that equal the angles of the course of 0 and 90°. Consequently, he wrote *Tratado sobre certas duvidas da navegação que Martim Afonso propoz ao Author*, as well as *Tratado em defensão da Carta de marear com o regimento da altura*, where Nunes also demonstrated a new way to calculate the distance to the pole.

One of Nunes's masterpieces is *De Crepusculis Liber unus; Allacen Arabis vetustissimi Liber de Crepusculis*. In this work he solved the problem of how to find the day with the smallest twilight, and also presented a method for measuring angles with more accuracy. At that time, in order to measure the altitudes of the stars and planets, astronomers used the cross-staff, the quadrant, and the plane astrolabe, all of which devices lacked precision. In his work *De Regulis et Instrumentus*, Nunes described his invention thus:

> ... due to the smallness of the device, it is not possible to subdivide its parts, therefore it is not possible to estimate the part of the height that should be needed to add to the whole number of degrees. It would be convenient to describe, inside the device's surface area, 44 concentric circles, dividing the external quadrant in 90 equal parts, the closest in 89, the following in 88, and so on, always in the same sequence.

Later, Nunes's idea would be improved by Jacob Curtio and Clavius, but it was Pierre Vernier who made possible the practical workability of this solution to the problem of reading fractions of angles, and whose name is associated with it today.

Fernando B. Figueiredo

Selected References

Albuquerque, Luís de (1989). *A náutica e a ciência em Portugal*. Lisbon: Gradiva.

Anon. (1945). *Obras Completas de Luciano Pereira da Silva*. Vol. 3, p. 264. Lisbon: Agência Geral das Colónias.

Da Silva, Luciano Pereira (1925). *As obras de Pedro Nunes: Sua cronologia bibliográfica*. Arquivos de História e Bibliografia, separata, Vol. 2, n° 1. Coimbra: Imprensa da universidade.

Delerue, Raul Esmeriz (1992). "Pedro Nunes, Contributo para uma síntese referenciada da sua bibliográfia." *Revista da Biblioteca Nacional* 2: 129–148.

Gomes Teixeira, F. (1934). *História das matemáticas em Portugal*. Lisbon: Academia das Ciências de Lisboa.

Martyn, John R. C. (1991). "Pedro Nunes – Classical Poet." *Euphrosyne* 19: 231–270.

——— (1992). *Pedro Nunes (1502–1578): His Lost Algebra and Other Discoveries*. New York: Peter Lang.

Nunes, Pedro (1537). *Tratado da sphera*. Lisbon: Germão Galhardo.

——— (15??). Astronomici Introdutorii de Sphere Epitome.?

——— (1542). *De Crepusculis Liber Unus*. Lisbon: Luis Rodrigues.

——— (1546). *De Erratis orontii Finaei Coimbra:* João Barreiros e João Alvares.

——— (1566). *Petri Nonii Salaciensis Opera. Basileia: Exofficina Henric Petrina.*

——— (15??). De Arte Atque Ratione Navigandi.?

——— (1567). *Libro de álgebra en aritmética y geometria*. Anvers: Juan Stelsio.

——— (1940–1960). *Obras de Pedro Nunes*. 4 Vols. Lisbon: Academia das Ciências de Lisboa.

Nušl, František

Born **Jindřichův Hradec, (Czech Republic), 3 December 1867**
Died **Prague, (Czech Republic), 17 September 1951**

František Nušl was one of the founders of modern Czech astronomy. He was the son of Ignác Nušl and Františka Nušlová (*née* Novotna) and was educated at the Czech Charles-Ferdinand University in

Prague (bachelor of philosophy 1905). He and his wife Aloisie (*née* Doležalová) had two sons and a daughter. Nušl taught mathematics and physics at a high school before receiving his Ph.D. and was appointed an associate professor of mathematics at the Czech Technical University, Prague (full professor: 1911). He was director of the National Astronomical Observatory in Prague from 1918 and professor of astronomy at Charles University until his retirement in 1937. Nušl served as president of the Czechoslovak Astronomical Society (1922–1948) and vice president of the International Astronomical Union (1928–1935).

Nušl's work was largely in practical astronomy and construction of astrometric instruments. In 1899, he constructed the first model of a mirror astrolabe with an artificial mercury horizon to measure the time of star transit over the almucantar (parallel of altitude) using the method of equal altitudes. A professional instrument, later called a circumzenithal, was constructed in 1901 in cooperation with Josef Jan Frič, an amateur astronomer, engineer, and owner of a fine mechanics and optics factory in Prague. Nušl used the instrument in a photographic mode in 1907 to measure short-term variations of refraction of light in the atmosphere. He and Frič presented a portable model at the International Union of Geophysics at its 1924 general assembly in Madrid, Spain, and it was later used in international longitude measurements and to determine astronomical coordinates of the Czechoslovak fundamental triangulation network. During the same period, he designed two other less successful instruments with mirrors and mercury horizons: A diazenithal (to measure the time a star crosses a particular azimuth) and a radiozenithal (to measure the time a star crosses a great circle, inclined to the horizon). Nušl also improved the state of micrometers and regulators in use at his observatories, made use of the then new radio telegraphic signals for the time service, and applied the instruments to problems of geodesy.

Beginning in 1898, Nušl assisted Frič in building a private observatory at Ondrejov, near Prague. It was donated to the state in 1928, merged with the National Astronomical Observatory (with Nušl as the first director), and is now the home of the Astronomical Institute of the Academy of Sciences of the Czech Republic. Nušl was also active in visual observations of meteors and in popularization of astronomy and physics. His publications appeared in Czech, German, and French.

Jan Vondrák

Selected References

Nušl, F. (1907). "Über allgemeine Differenzformeln der sphärischen Aberration." *Bulletin international de l'Académie des sciences de Bohème.*

______ (1909). "Einige Bemerkungen zu der Abbeschen Theorie der optischen Abbildung." *Sitzungsberichte der Königlich-Böhmischen Gesellschaft der Wissenschaften.*

______ (1909). "Kritische Übersicht der Triangulierungen der Umgebung von Prag." *Sitzungsberichte der Königlich-Böhmischen Gesellschaft der Wissenschaften.*

______ (1929). "Micromètre impersonnel de l'appareil circumzénithal." *Rozpravy II. třídy České Akademie* 38, č.14a. Prague.

Nušl, F. and J. J. Frič (1902). "Note sur deux appareils sans niveaux pour la détermination de l'heure et de la latitude." *Bulletin astronomique* 19: 261–274.

______ (1903). "Étude sur l'appareil circumzénithal." *Bulletin international de l'Académie des sciences de Bohême.*

______ (1904). "Mitteilung über das Diazenital." *Astronomische Nachrichten* 166: 225–228.

______ (1906). "Deuxième étude sur l'appareil circumzénithal." *Bulletin international de l'Académie des sciences de Bohème.*

______ (1908). "Première étude sur les anomalies de réfraction." *Bulletin international de l'Académie des sciences de Bohème.*

______ (1925). "Troisième étude sur l'appareil circumzénithal." *Publications de l'Observatoire national de Prague*, nos. 1 and 2.

O

O'Connell, Daniel Joseph Kelly

Born **Rugby, Warwickshire, England, 25 July 1896**
Died **Rome, Italy, 14 October 1982**

The O'Connell effect was discovered by Daniel O'Connell, British–Irish director of the Vatican Observatory. It is a variation in the period of eclipsing binary stars. Tidal effects, star spots, circumstellar disks, and other models have been proposed to explain it.

Selected References

Petit, Michel (1987). *Variable Stars*, translated by W. J. Duffin. Chichester, New York: Wiley.
Wayman, P. A. (1982). "Daniel J. K. O'Connell, S. J.: In Memoriam." *Irish Astronomical Journal* 15: 349–350.

Odierna [Hodierna], Giovanbatista [Giovan Battista, Giovanni Battista]

Born **Ragusa, (Sicily, Italy), 13 April 1597**
Died **Palma di Montechiaro, (Sicily, Italy), 6 April 1660**

Giovanbatista Odierna was an early observer of Jupiter and Saturn and one of the first to discover and describe nebulae. He was the son of Art Vita Dierna, who was either a mason or a shoemaker according to different sources, and Serafina Rizo. Very probably, his family lived under inferior, more probably poor conditions. While mostly referred to as Odierna, it seems that the man himself always used"Hodierna" as his name; it is not completely clear why and how he changed the name.

Odierna was self-educated, at least in science. As a young man, he observed the three comets of 1618/1619 (C/1618 Q1, C/1618 V1, and C/1618 W1) from Ragusa, with a Galilean type telescope and fixed magnification 20. He became a Roman Catholic priest and was ordained at Syracuse, Sicily, in 1622. From 1625 to 1636, Odierna served as a priest in Ragusa, and taught mathematics and astronomy at his hometown.

Odierna was an enthusiastic follower of **Galileo Galilei**. In 1628 he wrote the *Nunzio del secolo cristallino*, an appraisal of Galilei's *Siderius Nuncius*. Odierna was particularly impressed by Galilei's resolution into stars of the Milky Way and the "nebulae" like Praesepe; this generated a lifelong interest in nebulae, although most of his astronomical work concentrated on the bodies of the Solar System.

In 1637, Odierna followed Carlo and Gulio Tomasi, the Dukes of Montechiaro, to the newly founded Palma di Montechiaro. They gave him a house and a piece of land on which to live and funded his publications; he served them first as a chaplain and then as a parish priest. In 1644, Odierna earned a doctorate in theology. In 1645, he was named archpriest, in 1655 court mathematician.

Besides his duties as priest, Odierna practiced astronomy, as well as natural philosophy, physics, botany, and other sciences. He studied light passing through a prism and formulated a vague explanation of the rainbow. He developed an early microscope and studied, for example, the eyes of insects and the fangs of vipers. He also studied meteorological phenomena.

Odierna's contributions to astronomy, though interesting and remarkable in particular if one takes his isolated life into account, have had at best little impact, because his publications had only limited circulation and were hardly known outside Sicily. Therefore, standard tracts on the history of astronomy rarely spend more than a few lines on this early priest astronomer. Also, his astronomy always tended to be mixed up with astrology. In 1646 and 1653, Odierna observed Saturn and created drawings showing the planet with its ring quite correctly; he had a short correspondence with **Christiaan Huygens** on this subject around 1656. His *Protei caelestis vertigines sev. Saturni systema*, published in 1657, is among his best-known publications.

In 1652, Odierna observed eclipses of Jupiter's satellites, as well as the passages of their shadows over the disk of the planet. In 1656, he published in the *Medicaeorum Ephemerides*, probably his best-known work, the first published ephemerides of the Galilean satellites, based on an improved theory of the motion of Jupiter's moons by the contribution of three types of periodic disturbances – analogous to contemporary planetary theory. In his 1656 *De Admirandis Phasibus in Sole et Luna visis*, Odierna gives a treatise on the appearance of the Sun and the Moon, including sunspots and eclipses.

Perhaps Odierna's most interesting work is his 1654 *De systemate orbis cometici; deque admirandis coeli characteribus*, which unfortunately was forgotten and ignored until it was recovered by G. Serio *et al.* (1985). Odierna thought there were profound differences between comets and nebulae: Because of the motion and changing appearance of comets, he thought them to be made up of a more terrestrial matter, while nebulae should be made up of stars, and thus *Lux Primogenita*. In the first part he follows Galilei's ideas on comets. In the second, more interesting part, he describes and lists 40 nebulae he had observed, with finder charts and some sketches, which Odierna classifies according to their resolvability into stars as Luminosae (star clusters to the naked eye), Nebulae (appearing nebulous to the unaided eye, but resolved in his telescope), and Occultae (not resolved in his telescope). About 25 of them can be identified with real deep-sky objects (mostly open clusters); the others are either asterisms or insufficiently described for identification.

Today, the original discovery of between ten and 14 deep-sky objects, and two independent rediscoveries (of the Andromeda and the Orion Nebula), are attributed to Odierna. The formal list includes: the Alpha Persei cluster (Melotte 20), NGC 6231, M6, NGC 6530, M36, M37, M38, M47, M41, and NGC 2362, as well as probably M33, NGC 752, M34, and NGC 2451 and perhaps NGC 2169 and NGC 2175. Odierna 's deep-sky discoveries occurred within a larger project he endeavored, compiling a sky atlas, *Il Cielo Stellato Diviso in 100 Mappe*, a work he never completed.

Hartmut Frommert

Selected References

Pavone, Mario (1986). *La vita e le opere di Giovan Battista Hodierna*. Ragusa, Sicily: Didattica Libri Eirene Editrice.

Pavone, Mario and Maurizio Torrini (eds.) (2002). *G. B. Hodierna e il "secolo cristallino": Atti del convegno di Ragusa, 22–24 ottobre 1997*. Florence: Leo S. Olschki.

Serio, G. Fodernà, L. Indorato, and P. Nastasi (1985). "G. B. Hodierna's Observations of Nebulae and His Cosmology." *Journal for the History of Astronomy* 16: 1–36.

Oenopides of Chios

Born **Chios, (Khíos, Greece), *circa* 490 BCE**
Died ***circa* 420 BCE**

Little is known about the life of mathematician and astronomer Oenopides; his place of birth in Chios is reasonably well documented, and there is circumstantial evidence indicating that he spent time in Athens as a young man.

Oenopides made a number of contributions to mathematics and astronomy. He is thought to have settled on the value of 24° as an estimate for the angle that the ecliptic makes with the celestial equator. (**Eudemus** attributed the concept of the ecliptic to Oenopides, although the Babylonians were aware that the apparent path of the Sun through the zodiacal constellations was inclined to the plane of the Equator.) This is uncertain, because there is no explicit reference to this estimate in the Greek sources; however, **Proclus**, discussing Euclid IV, 16, says that the construction of a 15-sided regular polygon within a circle was included because it is useful in astronomy. (The 360° division of the circle was not yet developed as a common usage, and this figure would generate central angles of 24°.) It is possible that Oenopides originated both the estimate (24°) and the construction of the 15-sided figure. **Plato** appears to allude to Oenopides's research on the ecliptic when he includes him in the *Erastae*.

Oenopides seems to be best known for his research on the Great Year. As knowledge of astronomy progressed in classical times, this concept came to refer to the period after which the motions of the Sun, the Moon, and all of the planets would repeat themselves. Aelian and **Aëtius** give Oenopides credit for an estimate of 59 years for the period of the Great Year; it is not clear which, if any of the planets were intended to be accounted for within this period. It seems likely that Oenopides made his estimate based on a lunar month of roughly 29 and one-half days, and a 365-day solar year. This could quickly lead to the ratio of 730 lunar months for every 59 years; since 59 is a prime number, it would then provide a possible figure for the Great Year period. It appears that Oenopides attempted to confirm this estimate based on observations throughout his life.

Oenopides's contributions to mathematics may have a wider significance as well. In his commentary on Euclid's *Elements*, Proclus cites Oenopides as the originator of two theorems (I.12 and I.23), both having to do with elementary constructions. Ivor Bulmer-Thomas in the *Dictionary of Scientific Biography* makes the interesting conjecture that "it may have been [Oenopides] who introduced into Greek geometry the limitation of the use of instruments in all plane constructions ... to the ruler and compass." This may be significant for astronomy because it suggests that Oenopides developed a serious interest in the methodology of mathematics, with a particular focus on the distinctions between theoretical and applied mathematics.

Kenneth Mayers

Selected References

Dicks, D. R. (1985). *Early Greek Astronomy to Aristotle*. Ithaca, New York: Cornell University Press.

Heath, Sir Thomas L. (1960). *A History of Greek Mathematics*. 2 Vols. Oxford: Clarendon Press.

——— (1956). *The Thirteen Books of Euclid's* Elements. New York: Dover.

Mau, Jürgen (1972). "Oinopides." In *Der kleine Pauly*. Vol. 4, cols. 263–264. Munich: Alfred Druckenmüller.

Morrow, Glenn R. (1970). *Proclus: A Commentary on the First Book of Euclid's Elements*. Princeton, New Jersey: Princeton University Press.

Neugebauer, Otto (1975). *A History of Ancient Mathematical Astronomy*. 3 pts. New York: Springer-Verlag.

Steele, A.D. (1936). "Ueber die Rolle von Zirkel und Lineal in der griechischen Mathematik." *Quellen und Studien zur Geschichte der Mathematik* 2: 3.

Tannery, Paul (1888). "La grande année d'Aristarque de Samos." *Mémoires de la Société des sciences physiques et naturelles de Bordeaux* 4: 79–96.

——— (1976). *Recherches sur l'histoire de l'astronomie ancienne*. Hildesheim, New York: Georg Ohms. (Reprint of the 1893 edition published by Gauthier-Villars, Paris. Includes bibliographical references.)

Thurston, Hugh (1994). *Early Astronomy*. New York: Springer-Verlag.

Offusius, Jofrancus

Born **Geldern, (Nordrhein-Westfalen, Germany), before 1530**
Died **possibly Paris, France, after 1557**

Astrologer Jofrancus Offusius is an almost unique example of a scholar in the French cultural area who studied the technical details of **Nicolaus Copernicus**'s *De revolutionibus* in the first two decades after its publication in 1543. Offusius's examination of Copernicus's ideas is well documented by the annotations and by appended manuscript pages in his copy of *De revolutionibus* (now preserved in the National Library of Scotland), and his influence is attested by eight further copies of his annotations made by his students or by their students. The longest annotation occurs in the most complex part of the book, where Copernicus deals with the lunar parallax. While his major criticism there is marred by an erroneous start, Offusius's series of short corrections to the later part of the paragraph demonstrate that he could handle the detailed calculations with more accuracy than Copernicus did.

Very few details are known regarding Offusius's life. **Girolamo Cardano** refers to Franciscus Offusius Geldrensis when discussing cipher codes in his *De rerum varietate* (Basel, 1557), from which we deduce his place of birth. Offusius records in his *De revolutionibus* that he had been in Seville, Spain in 1550; later he was in England for a year under the patronage of the Elizabethan *magus* **John Dee**. In 1557, Offusius published in Paris *Ephemerides* for that year based on calculations derivative from Copernicus's work. It was during his time in Paris that he must have operated an atelier for astronomy students including Jean Pierre de Mesmes, who worked with Offusius for some time between 1552 and 1557. In one of his own annotations, de Mesmes referred to his teacher as "not one of the common astronomers." Circumstantial evidence suggests that the 16th-century central European astronomers **Paul Wittich** and/or **Tadeá Hájek z Hájku** saw Offusius' own annotated copy, for Wittich's own annotations show influence from that specific copy and include the name "Jofrantius" in the margin.

Pontus de Tyard, a 16th-century astronomer and philosopher, mentions meeting Offusius in Dieppe, presumably in 1556, and remarked that his tables (ephemerides) were "different from all the rest." In the preface to his *Ephemerides*, Offusius indicated that he had worked on planetary influences for 14 years, and had become increasingly dissatisfied with the *Alfonsine Tables*, which he rejected as worthless. Apparently for January 1557, Offusius used the Copernican *Prutenic Tables*, but for the remainder of the year he employed some not very successful modification of them. In the book's preface, he announced that Mercury would twice pass in front of the Sun that year, predictions that proved quite wrong.

After 1557 Offusius disappeared from the scene, but in 1570, his widow published *De divina astrorum facultate*, a book that deals with planetary distances, a topic rarely addressed between Copernicus's *De revolutionibus* and **Johannes Kepler**'s *Mysterium cosmographicum* of 1596. The preface to the book is dated 1556. Using a geocentric framework, Offusius attempted to find an esthetically pleasing principle to establish the planetary distances to be used for his theory of astrological aspects. He worked within the traditional notions that the Sun was 19 times further than the Moon, and that Saturn was about 19 times farther than the Sun. He then had to insert two other planets within each of these two intervals, which he achieved with proportional spacing so that the ratio of the distances of each pair was the same, 8/3 (nearly the cube root of 19). His starting point must have been the solar distance of 576 terrestrial diameters, the square of 24, whose beauty pleased him greatly. In an obscure way Offusius argued that his proportions were related to numbers associated with the Platonic polyhedra, a feature that **Tycho Brahe** commented on favorably and that no doubt helped persuade him to take seriously Kepler's later use of these polyhedra to explain the spacing of the planets in the Copernican system.

Owen Gingerich

Selected References

Gingerich, Owen (2002). *An Annotated Census of Copernicus'* De Revolutionibus *(Nuremberg, 1543 and Basel, 1566)*. Leiden: Brill.
Gingerich, Owen and Jerzy Dobrzycki (1993). "The Master of the 1550 Radices: Jofrancus Offusius." *Journal for the History of Astronomy* 24: 235–253.
Stephenson, Bruce (1994). *The Music of the Heavens: Kepler's Harmonic Astronomy*. Princeton, New Jersey: Princeton University Press, pp. 47–63.

Öhman, K. Yngve

Born **Stockholm, Sweden, 27 March 1903**
Died **Lund, Sweden, 17 June 1988**

Yngve Öhman, the founder of Swedish solar astronomy, designed, built, and used the first narrow-band interference filter (the birefringent quartz polarizing monochromator) useful for monitoring solar activity. The son of Karl E. and Ida (*née*) Cassel Öhman, he saw comet 1P/Halley at age seven and built a small telescope by 17. Öhman began studies at Uppsala University in 1922, receiving a Ph.D. in 1930 for work with **Bertil Lindblad** on luminosities and distances of stars. Other important early influences at Uppsala were **Östen Bergstrand**, **Hugo von Zeipel**, and **Knut Lundmark**. The thesis work was done partly with spectrograms taken at the Mount Wilson Observatory, California, by **Milton Humason** and **Paul Merrill** and brought back to Sweden by Lindblad. Öhman recognized that the spectral absorption features due to CaH at 6365 Å and to neutral calcium at 4227 Å were much stronger in dwarf stars than in giants of the same (relatively low) temperature, and so could be used to estimate stellar intrinsic luminosities (and so stellar distances) even in quite low-dispersion spectra.

Öhman held positions at Stockholm Observatory most of his life (1930–1938 as assistant or observer, with a year away at Mount Wilson in 1933/1934; as observer at the Saltsjöbaden Observatory of Stockholm 1939–1953; and as occupant of a personal chair of astrophysics 1953 to 1968, when he retired). He also spent brief periods at Uppsala, Boulder, Colorado, USA and Mount Wilson again. As early as 1929, Öhman had predicted that the light from solar prominences and from comets should be polarized because of fluorescence, and was eventually able to verify this in both cases (comet C/1940 R2, and the solar case within days of commissioning his new filter).

Robert Wood first showed that one could separate out a single spectral feature using a sandwich with a quartz slab between two polarizing filters. It is a property of quartz that two perpendicular polarizations of light travel at different speeds, and so arrive at the other side of the slab out of phase, producing an interference pattern, in which some wavelengths come through at almost full strength and others are not transmitted at all. Öhman extended this idea to a "layer cake" with four quartz slabs (relative thickness 1:2:4:8, all of order millimeters) and five films of Polaroid around and between them. The slab thicknesses were chosen so that the 6563 Å wavelength of hydrogen came through, and the next transmission peaks were nearly 1000 Å away and so could be blocked by ordinary colored glass filters. Such a device picks out a bandwidth of about 50 Å. Thus it is not as good a monochrometer as the spectroheliograph of **George Hale** and **Henri Deslandres**, but has the enormous advantage that you can image the whole Sun at once, instead of one thin slice at a time. The filter can be tuned to somewhat different wavelengths by tilting it, and it automatically acts as a polarimeter as well. Such devices are widely used in solar observatories today. The idea had occurred somewhat earlier to **Bernard Lyot**, but he had not yet constructed a device at the time Öhman started using his.

Öhman's main field was observational solar astronomy. He developed solar patrols. At first, these observations were performed at the Satlsjöbaden Observatory. However, because of the weather conditions at Stockholm, setting up a solar observatory elsewhere was the goal. Together with **Donald Menzel**, Öhman first studied the conditions for solar research in northern Sweden. The region around Abisko research station was a better site than Stockholm, and for a while the alternative of Abisko seemed viable, but more favorable conditions were found on the island of Capri, where a station for solar research was established by Öhman and the Royal Swedish Academy of Sciences. Öhman's interest in Capri as a site for solar observations had been aroused when Axel Munthe's villa *San Michele* had been willed to the Swedish state and opened for use by scientists and artists. Öhman also developed good contacts with the foundation Centro Caprense di Vita e di Studi, from which initial support was given.

Regular observations at Capri began in the summer of 1952. Soon the station ran a continuous solar flare patrol. Flares, prominences, and other solar phenomena were studied with photometric and spectroscopic methods. Several Swedish astronomers worked at the Capri station as assistants, gaining important training in solar observation techniques. The best known of Öhman's students in solar and stellar astronomy are Jan Stenflo, Kirsten Fredga, and Dainis Dravins.

Öhman's involvement in international astronomy began as early as 1938, when he chaired the local organizing committee for the 1938 general assembly of the International Astronomical Union [IAU] in Stockholm, and he was president of an IAU solar commission in 1952–1955. Both Öhman and the Capri observatory were active in the International Geophysical Year (1956–1958), and he served on committees for the European Space Research Organization and the journal *Solar Physics*. The Royal Swedish Academy of Sciences elected him to membership and awarded one of the several medals Öhman received for his contributions to international solar astronomy.

Öhman's interests were not confined to solar astronomy: The construction of scientific payloads for use on sounding rockets, the influence of geomagnetic storms on power distribution systems, solar energy, and other alternatives to fossil fuels were only some of the fields in which he worked. He held several patents. Öhman's archive shows that he was active with scientific and technological questions up until his death.

Öhman's papers are at the Center for the History of Science, Royal Academy of Sciences, Stockholm.

Gustav Holmberg

Selected References

Bernhard, Carl Gustaf (1989). *The Research Station for Astrophysics Capri and La Palma*. Stockholm: Royal Swedish Academy of Sciences.

Elvius, Aina (1989–1990). "Yngve Öhman." *Kungliga Fysiografiska Sällskapet i Lund Arsbok*: 111–116.

Hearnshaw, John B. (1986). *The Analysis of Starlight: One Hundred and Fifty Years of Astronomical Spectroscopy*. Cambridge: Cambridge University Press.

Holmberg, Gustav (1999). *Reaching for the Stars: Studies in the History of Swedish Stellar and Nebular Astronomy, 1860–1940*. Lund: Lund University.

Hufbauer, Karl (1991). *Exploring the Sun: Solar Science since Galileo*. Baltimore: Johns Hopkins University Press.

Stenflo, Jan-Olof. Interview with author.

Wyller, Arne E. (1989). "Ohman, Yngve (1903–1988)." *Solar Physics* 119: 1–3.

Olbers, Heinrich Wilhelm Matthias

Born **Arbergen near Bremen, (Germany), 11 October 1758**
Died **Bremen, (Germany), 2 March 1840**

Credited with the independent discovery of four comets and two of the first four asteroids discovered, physician Heinrich Olbers was one of the leading astronomers of the early 19th century. Functioning as an amateur in his private observatory, Olbers was widely respected by his astronomical colleagues. Though he neither discovered nor resolved Olbers' paradox, it is invariably given his name.

Young Olbers lived with his father, Johann Jürgen Olbers, a Protestant minister, and 15 brothers and sisters. His interest in astronomy originated at the age of about 14, but due to a limited curriculum in the exact sciences at the humanistic Gymnasium in Bremen, he was forced to study mathematics on his own. Olbers's first attempts in astronomy were the calculation of the solar eclipse of 1774, the calculation of the orbit of comet C/1779 A1 (Bode) from his own observations (in 1779), and a discovery of a comet (C/1780 U1, independently discovered by Jacques Leibax Montaigne (1716–1785?) of Limoges, France. While studying medicine at Göttingen University, Olbers also attended lectures in mathematics and physics. In medicine he specialized on ophthalmology; after graduation in 1781 he moved to Vienna, where he spent one year practicing in hospitals. During that year Olbers also visited the Vienna Observatory and enjoyed life in the aristocratic society.

Toward the end of 1781 Olbers opened a private ophthalmologic practice in Bremen and soon attracted many patients. He slept only 4 hours a day, reserving his nights for astronomical observations. During the next 30 years he was able to practice medicine while working in astronomy at a level comparable to the best professional astronomers. In 1820 Olbers decided to devote all his time to astronomy, so he closed his medical practice.

Olbers converted one part of the second floor of his Bremen house into an observatory, the equipment included movable telescopes and comet seekers made by such renowned instrument producers as Dollond and Fraunhofer. By fortunate chance, one of the best-equipped private observatories in the region was in nearby Lilienthal. After 1786, Olbers collaborated with its owner, **Johann Schröter**. **János von Zach** visited Olbers in Bremen as well as the Lilienthal observatory in September 1800 and published a description of both observatories in his journal *Monatliche Correspondenz* in 1801. During that visit, on 20 September, Schröter, Zach, Olbers, **Carl Harding,** and about 20 other outstanding astronomers from Germany and the rest of Europe founded *Vereinigte Astronomische Gesellschaft*, the first astronomical society in the world.

When Olbers discovered the comet of 1796 (C/1796 F1), he elaborated a new method for calculating its parabolic orbit using only a few observed positions. The methods developed by **Joseph-Jérôme de Lalande** and **Pierre de Laplace**, common at that time, were based on subsequent approximations, and lead often to erroneous results. However, Olbers's method was much simpler, satisfactorily precise, and reliable. Zach published it in Weimar (1797) under the title "Ueber die leichteste und bequemste Methode, die Bahn eines Kometen aus einigen Beobachtungen zu berechnen." The Olbers method was widely adopted and used throughout the 19th century. Olbers computed about 20 orbits of comets, including an orbit for comet C/1798 X1 (an independent discovery by Olbers, but credited to **Alexis Bouvard**), as well as for the periodic comet 13P/1815 E1 that he discovered.

When the first-known minor planet (1) Ceres disappeared behind the Sun, astronomers were not successful in finding it when it emerged from the Sun's glare a few months later. **Carl Gauss** developed a new method of determination of an elliptic orbit, and computed the expected positions. On 1 January 1802, Olbers found Ceres near the place given by Gauss, exactly one year after the original discovery of Ceres by **Giuseppe Piazzi** in Palermo. As a result, a close friendship developed between Olbers and the younger Gauss. Olbers discovered two other minor planets – (2) Pallas in 1802 and (4) Vesta in 1807 – the third, (3) Juno, was found at Lilienthal by Harding.

In 1812, Olbers suggested that a comet's tail was composed of particles driven away from the nucleus of the comet in the anti-solar direction by repulsive forces from the Sun. Olbers supposed that the forces were electrical in nature. He also offered the insight that the direction and orbital speed of the comet as well as the gravitational influence of the Sun all had a marked influence on the morphology of the comet's tail.

The question "Why the sky is dark at night?" had earlier been raised by **Johannes Kepler**. He pointed out that in the infinite space, uniformly filled by stars, every line of sight had to end on a surface of a star, so that the sky had to appear as bright as the stellar surface, *e. g.*, as the Sun. Like **Edmund Halley** and **Jean Loys de Chéseaux**, Olbers came independently to the same conclusion. He published it in 1823 under the title "Über die Durchsichtigkeit des Weltraums" in *Berliner astronomisches Jahrbuch fuer das Jahr 1826*. Since that time the problem has been referred to as Olbers' paradox. The contemporary (not correct) solution assumed slightly absorbing matter dispersed between the stars, but the correct solution (the finite age and energy density of the Universe) was later given as a part of relativistic cosmology.

In 1837, Olbers pointed out, based on his study of the Leonid meteor showers of 1799 and 1833, that maxima in the strength of that shower might occur on intervals of 3, 6, or 34 years and predicted another great storm in 1867. However, though he studied other celestial phenomena, Olbers's main interest continued to be comets. His preoccupation with the subject caused him to assemble one of the finest libraries on the history of astronomy of its time. With a particular emphasis on cometography, the Olbers library was thought to be "essentially complete" and formed a valuable addition to the library of the Pulkovo Observatory when purchased by **Friedrich Struve** from the Olbers estate.

Olbers himself saw his major contribution to astronomy in the fact that he influenced the young **Friedrich Bessel** (later director of the observatory in Königsberg) to enter a career as a professional astronomer. Bessel's interest in astronomy was aroused after listening to Olbers's lectures. Later, after meeting Olbers on the street in 1804, Bessel showed him his calculations for the orbit of Halley's comet (IP/Halley) based on **Thomas Harriot**'s observations in 1607. Impressed by the young man's self-taught mathematical ability, Olber's suggested some additions and arranged to have the computations published.

In two marriages, Olbers was the father of a daughter and a son. His first wife, Dorthea Köhne, died while giving birth to their daughter in 1786, only a year after their marriage. Olbers remarried in 1789, and Anna Adelheid Lurssen gave birth to a son. Anna died in 1820.

Martin Solc

Selected References

Anon. (1840). "Dr. Olbers." *Proceedings of the Royal Society of London* 4: 267–269.
Bessel, Friedrich (1844). "Über Olbers." *Astronomische Nachrichten* 22: 265–270.
Erman, A. (1852). *Briefwechsel zwischen W. Olbers und F. W. Bessel*. 2 Vols. Leipzig. (For the correspondence with Bessel.)

Harrison, Edward (1987). *Darkness at Night: A Riddle of the Universe*. Cambridge, Massachusetts: Harvard University Press.
Jaki, Stanley L. (1969). *The Paradox of Olbers's Paradox*. New York: Herder and Herder.
——— (1970). "New Light on Olbers's Dependence on Chéseaux." *Journal for the History of Astronomy* 1: 53–55.
Schilling, C. (1894–1909). *Wilhelm Olbers, sein Leben und seine Werke*. 2 Vols. in 3 pts., and supplement. Berlin. (For a complete list and texts of about 200 Olbers publications; the correspondence with Gauss is also included.)
Stein, Walter (ed.) (1958). *Von Bremer Astronomen und Sternfreunden*. Bremen: A. Geist.
Struve, F. G. W. (1842). "Bericht Über die Bibliothek der Hauptsternwarte in Pulkova." *Astronomische Nachrichten* 19: 307–312. (A report on the purchase of Olbers's library.)
Struve, Otto (1963). "Some Thoughts on Olbers's Paradox." *Sky & Telescope* 25, no. 3: 140–142.
Vsekhsvyatskii, S. K. (1964). *Physical Characteristics of Comets*. Jerusalem: S. Monson.
Yeomans, Donald K. (1991). *Comets: A Chronological History of Observation, Science, Myth, and Folklore*. New York: John Wiley and Sons.

Olcott, William Tyler

Born **Chicago, Illinois, USA, 11 January 1873**
Died **George's Mills, New Hampshire, USA, 6 July 1936**

Tyler Olcott founded and nurtured the American Association of Variable Star Observers [AAVSO], served as its secretary for nearly 25 years, and wrote a number of popular books on astronomy.

Olcott, the son of William Marvin and E. Olivia (*née* Tyler) Olcott was raised in Norwich, Connecticut, in the ancestral Tyler home in which he lived for his entire life. Educated at Trinity College, Hartford, Connecticut, Olcott studied at the New York Law School. Though admitted to the state bars in New York and Connecticut, he never practiced law as a career. Olcott married Clara Hyde of Yantic, Connecticut, in 1902; they had no children.

Olcott's introduction to astronomy occurred during a 1905 summer vacation, and he was instantly captivated by the night sky. As he taught himself the stars and constellations he wrote a book on the subject of the naked-eye sky, *A Field Book of the Stars* published in 1907. This was followed by the acquisition of a small telescope with similar results: *In Starland with a 3-inch Telescope*, reflecting his deepening appreciation of the night sky, was published in 1909. After acquiring skills with the telescope, Olcott sought out an application of those skills in the service of astronomy. He found an ideal application in variable star astronomy as he chanced upon a presentation on the subject at a meeting of the American Association for the Advancement of Science. There, he heard **Edward Pickering** describe a scientific need for amateur observers. When Olcott wrote to Pickering asking for more information, the Harvard College Observatory [HCO] director dispatched one of his staff, **Leon Campbell,** to Connecticut to train Olcott and deliver the necessary charts and reporting forms to him. Olcott was so delighted with the experience that he wrote an article describing both the techniques and the need for observers for *Popular Astronomy* and volunteered to facilitate the work of other interested amateur observers. That invitation led Frederick Leonard (1896–1960) to recruit Olcott as the variable star section leader for the Society for Practical Astronomy [SPA]. In addition, with encouragement from Pickering, Olcott undertook the formation of the AAVSO in 1911 and within a year had resigned from the SPA.

From 1911 until his death, Olcott voluntarily served as the recording secretary of the AAVSO and in various other capacities, corresponding with amateur and professional astronomers and prospective new members. He traced and distributed hundreds of variable star charts, instructed observers, acted as liaison between the variable star observers and HCO, compiled AAVSO variable-star observations and notes for use by Pickering and for publication each month in *Popular Astronomy*, and gave numerous public talks on astronomy and variable star observing. In the process Olcott established contacts with noted astronomers at home and abroad.

Olcott published the AAVSO's initial report of 208 observations in the December 1911 issue of *Popular Astronomy*. By the end of its first year the AAVSO had published over 6,000 observations of variable stars in *Popular Astronomy*. From that point on, the number of observations being sent to the AAVSO each year rose dramatically: to 11,600 in 1912; 17,400 in 1917; 19,300 in 1921; and 26,900 in 1924. Under Olcott's guidance, the AAVSO grew very rapidly from a handful of amateur and professional observers in 1911 to about 480 observers in 1936. By the time Olcott died, the AAVSO was receiving nearly 55,000 observations per year.

With rapid growth from 1911 through the 1920s and 1930s, and under the guidance and encouragement of Olcott, the AAVSO membership gained a strong sense of identity and personality. Olcott organized the AAVSO's first regular annual meeting at HCO in November 1915. In October 1918, the AAVSO was incorporated under the laws of the Commonwealth of Massachusetts. HCO then provided a room for the AAVSO's headquarters, and the organization continued to grow. United by their love of astronomy, telescope making, and variable star observing, the AAVSO members' common interest was strengthened by their sharing the common purpose and scientifically important goal of providing researchers with long-term variable star measurements in a reliable and efficient way. The AAVSO is now the largest variable star organization in the world, receiving, processing, and disseminating some 400,000 variable star observations each year, and maintaining an international database that now contains over 11 million observations of variable stars.

Olcott was a life member of the American Astronomical Society, the British Astronomical Association, and the Société Astronomique de France. He was made a fellow of the Royal Astronomical Society of Great Britain, and a fellow of the American Association for the Advancement of Science. He was also appointed to two terms as a member of the Visiting Committee of the Department of Astronomy of Harvard University.

The Field Book of the Skies (first published in 1929) became widely known as the definitive handbook for amateur astronomers, and was published in revised editions after Olcott's death.

The AAVSO council presented, posthumously, the Merit Award of The American Association of Variable Star Observers – the association's highest honor – to William Olcott as "the Founder and Life Secretary of our Association whose words and writing and patient guidance have led many to know and love the stars."

Michael Saladyga

Selected References

Fortier, Edmund (1990). "The Legacy of William Tyler Olcott." *Sky & Telescope* 80, no. 5: 536–539.

Olcott, William Tyler (1907). *A Field Book of the Stars*. New York: G. P. Putnam's Sons.

——— (1909). *In Starland with a Three-Inch Telescope*. New York: G. P. Putnam's Sons.

——— (1911). *Star Lore of All Ages*. New York: G. P. Putnam's Sons.

——— (1914). *Sun Lore of All Ages*. New York: G. P. Putnam's Sons.

——— (1923). *A Book of the Stars for Young People*. New York: G. P. Putnam's Sons.

Olcott, W. T. and E. W. Putnam (1929). *Field Book of the Skies*. New York: G. P. Putnam's Sons.

Pickering, David B. (1936–1940). "William Tyler Olcott." *Variable Comments* 3, no. 6: 21–24.

Waagen, Elizabeth O. (1996). "William Tyler Olcott, 1873–1936." *Journal of the American Association of Variable Star Observers* 24: 50–58.

Olivier, Charles Pollard

Born **Charlottesville, Virginia, USA, 10 April 1884**
Died **Bryn Mawr, Pennsylvania, USA, 14 August 1975**

Charles Olivier contributed to several fields of astronomy: In meteors, he corrected an erroneous belief about meteor shower radiants, and established and guided the American Meteor Society [AMS] for over six decades. He also discovered and studied double stars and made invaluable contributions to the standardization of variable - star visual photometry.

The Olivier family lived in Charlottesville, the site of the University of Virginia. Olivier's parents, George Wythe and Katharine (*née* Pollard) Olivier, owned the University Book Store, purchased with the proceeds of her family estate, and operated a boarding house in their home. Katharine had been raised in the surrounding Albemarle County, and the couple knew most of the university faculty well.

A year before Charles' birth, the university dedicated the Leander McCormick Observatory and its 26-in. Alvan Clark refractor. **Ormond Stone**, director of the observatory and a frequent guest at the Olivier home, encouraged young Olivier to become an astronomer. While still a boy, he eagerly learned how to use the Clark refractor and smaller observatory instruments.

Olivier accompanied the Leander McCormick Observatory's staff to Scottsville, Virginia, in an attempt to photograph the 1899 Leonid meteor shower, his first scientific expedition, and later to Winnsboro, South Carolina, for the 28 May 1900 solar eclipse. During the summer of 1901, Stone employed Olivier to measure stellar magnitudes in variable star fields using a wedge photometer.

While an undergraduate at the University of Virginia (1901–1905), Olivier published variable star observations and meteor shower reports in *Popular Astronomy*. In August 1905, two months after he earned his BA degree, Olivier volunteered as a member of a United States Naval Observatory [USNO] solar eclipse expedition to Daroca, Spain. While there, he met Samuel Mitchell, who would later play an important role in his career. Back in the United States, Olivier began graduate work at the University of Virginia, supported by a Vanderbilt Fellowship, earning a master's degree in astronomy in 1908.

Conducting graduate research at the Lick Observatory from July 1909 to December 1910, Olivier studied stellar spectra and acted as **Heber Curtis**'s assistant in an apparition-long study of comet 1P/Halley. He published double star measurements, a report about the 1909 Perseid meteor shower, and a photographic report about the great January comet C/1910 A1 (with **Paul Merrill**); he also demonstrated that the η Aquarid meteors are likely debris from Halley's comet based upon their orbital similarities.

Olivier earned his Ph.D. in astronomy from the University of Virginia in June 1911. His dissertation, "175 Parabolic Orbits and other Results Deduced from Observations of 6200 Meteors," was drawn mainly from his personal observational data. One of Olivier's important conclusions in the dissertation attacked a notion prevalent in the early 20th century, advanced by **William Denning**, that meteor radiants were stationary in the sky and were intermittently active for periods of days to months. Olivier argued that due to orbital motions of the Earth and the meteor streams this was impossible. He used his own meteor plot data to demonstrate that radiants actually move against the background sky, and only exist for a few days. Over a period of 20 years beginning with his 1911 dissertation, Olivier's theoretical and observational arguments demolished the credibility of Denning's stationary radiants.

In 1911, Olivier accepted a position as an assistant professor of physics and astronomy at Agnes Scott College. While there, he volunteered to direct the meteor section of Frederick Leonard's (1896 – 1960) Society for Practical Astronomy, and simultaneously founded the American Meteor Society [AMS], an organization that continues to promote and collect meteor observations to the present day.

During the summer of 1913, Olivier volunteered at the Yerkes Observatory, and then in June 1914 accepted an adjunct professorship at the University of Virginia, where Mitchell had just been appointed director of the Leander McCormick Observatory. During his early career at McCormick Observatory, Olivier contributed to Mitchell's photographic parallax program. When the United States entered World War I, Olivier volunteered but was disqualified from service on physical grounds, and instead was appointed to the Scientific Staff of the Aberdeen Proving Grounds in Maryland. His supervisor from October 1918 to January 1919 was **Forest Moulton** who assigned Olivier the problem of adapting astronomical photographic techniques to the task of finding cannon shell burst ranges at night. After World War I, Olivier collaborated with Mitchell and Harold Lee Alden (1890–1963) on the visual photometry of variable star fields, and continued his earlier work measuring parallax plates, from which he also discovered a number of new double stars. By the end of his life, Olivier claimed discovery of 198 new double stars.

Olivier married Mary Frances Pender on 18 October 1919. They had two daughters, Alice Dorsey and Elise Pender Olivier during their 15-year marriage.

Olivier obtained leave from his faculty and observational duties from October 1923 to July 1924 to do library research at the USNO in Washington. The product of this work was his first book, *Meteors*, published in 1925. Mary Frances helped him prepare the text for publication. The book, dedicated to Olivier's father who died in 1923, included a comprehensive review of meteor observation history, techniques of meteor observation and height determination, the connection between comets' orbits and meteor streams, and meteorites. *Meteors* also continued Olivier's attack on the belief that meteor showers could have long-lasting, stationary radiants. Olivier explained how inappropriate observational and data reduction practices of the past had contributed to that erroneous belief.

Olivier was appointed secretary of the American Astronomical Society's [AAS] Committee on Meteors in 1916. As *Meteors* was going to press 8 years later, Olivier was elected president of the Meteor Commission of the International Astronomical Union [IAU], a position he held for 10 years. In 1927, Olivier was appointed to the IAU's Commission on Double Stars.

In 1928, Olivier became professor of astronomy and director of the Flower Observatory at the University of Pennsylvania. There, Olivier concentrated upon double star measurements and the photometry of variable star fields using the 18-in. Brashear refractor. Olivier supervised the consolidation of the Flower Observatory with the private observatory of Gustavus Wynne Cook (1867–1940), bequeathed to the university. The Cook Observatory included a 28.5-in. Cassegrain reflector and a 15-in. siderostat refractor.

In 1930, Olivier published his second book, dedicated to Mary Frances. Intended as a sequel to *Meteors*, *Comets* was a nonmathematical treatment of what was then known about the origin, physical composition, and visual presentations of various comets, and associated meteor showers. It was based on Olivier's observations since about 1900. He included a chapter about comet collisions with the Earth that anticipated the 1990s popularization of comet catastrophes.

Until its cessation in 1951, Olivier published monthly articles about AMS results in *Popular Astronomy*. Thereafter, he published periodic reports about the society's findings in *Meteoritics* and *Flower and Cook Observatory Publications*. These reports and Olivier's frequent correspondence with AMS observers to suggest observational goals and to praise or cajole the members' participation, were crucial to the vitality of the organization and its longtime productivity. Olivier directed the American Meteor Society from 1911 until 1973, when he selected David Dering Meisel (born: 1940), an AMS member who had become a professional astronomer, as the new executive director.

After Mary Frances Olivier died in 1934, Olivier raised his daughters by himself, although his sister, Katharine Olivier Maddux, helped him on many occasions. Olivier got married a second time, to Ninuzza Seymour, on 23 October 1936. This marriage ended in a divorce. Olivier did not marry again until 24 July 1950, when he wed Margaret Ferguson Austin who survived him.

Appointed emeritus professor of the University of Pennsylvania in 1954, Olivier continued reducing data produced by members of the American Meteor Society for another 20 years in retirement. In 1974, at age 90, he published his sixth, and final, catalog of hourly sporadic meteor rates.

Olivier was a tireless observational astronomer for 73 years. He also popularized astronomy in newspaper articles, on the radio, and by giving personal presentations to astronomy clubs and community groups. In 1979, the IAU named a lunar crater in his honor.

Richard J. Taibi

Selected References

Anon. (1975). "Meteor Expert." *Sky & Telescope* 50, no. 4: 231.

Anon. (18 August 1975). "Olivier, Charles." *New York Times*.

Beech, Martin (1991). "The Stationary Radiant Debate Revisited." *Quarterly Journal of the Royal Astronomical Society* 32: 245–264.

Olivier, Charles P. (1925). "175 Parabolic Orbits and other Results Deduced from Over 6,200 Meteors." *Transactions American Philosophical Society*, n.s., 22: 5–35.

——— (1925). *Meteors*. Baltimore: Williams and Wilkins Co.

——— (1928). "Measures of 357 Double Stars." *Astronomische Nachrichten* 233: 393–412.

——— (1930). *Comets*. Baltimore: Williams and Wilkins Co.

——— (9 March 1952). "A Report on the Department of Astronomy and the two Astronomical Observatories during the directorship of Charles P. Olivier." Typed report. History of the Observatory File. The Flower and Cook Observatory Archives, University of Pennsylvania, Philadelphia.

——— (1967). "History of the Leander McCormick Observatory *circa* 1883 to 1928." *Publications of the Leander McCormick Observatory* 11, pt. 26: 203–209.

——— (circa 1967). "Scientific Publications by Charles P. Olivier." Typed list. Charles P. Olivier Papers. Manuscript Archives, American Philosophical Society Library, Philadelphia, Pennsylvania.

——— (*circa* 1975). "Notes on life and work of C.P. Olivier, which would not be found probably in Who's Who or American Men of Science." Typed list with written annotations. History of the Observatory File. The Flower and Cook Observatory Archives, University of Pennsylvania, Philadelphia.

Olmsted, Denison

Born **East Hartford, Connecticut, USA, 18 June 1791**
Died **New Haven, Connecticut, USA, 13 May 1859**

Denison Olmsted was a prominent member of the scientific community in antebellum America. Remembered for pioneering contributions in astronomy, geology, and meteorology, he authored a number of widely used textbooks, including *An Introduction to Astronomy* (1839) and *A Compendium of Astronomy* (1841).

Son of farmers Nathaniel Olmsted and Eunice Kingsbury, Olmsted entered Yale College in 1809 and received his bachelor's degree in 1813. After teaching in New London, Connecticut for two years, he earned a master's degree (1816) from Yale. In 1817, he was appointed to a professorship of chemistry at the University of North Carolina, which he accepted after spending another year of preparation in that subject. A member of the short-lived American Geological Society (founded at Yale in 1819), Olmsted was also instrumental in creating the first state-sponsored geological survey (of North Carolina, 1823–1824), whose reports helped to stimulate other states to conduct geological surveys, starting in the 1830s. In 1825, he returned to Yale as a professor of mathematics, natural philosophy, and astronomy, where he spent the rest of his life. One of Olmsted's most notable students was **Jonathan Lane**.

Olmsted's most notable contributions to astronomy arose from his observations of the 13 November 1833 Leonid meteor storm. He noted that the radiant point, in the sickle of Leo, followed the diurnal movement of the sky. In papers published in the *American Journal of Science and Arts*, Olmsted correctly argued for the shower's origin in interplanetary space, whose meteoric particles traveled along parallel paths and were consumed by fire in the Earth's atmosphere. Along with American astronomers **Edward Herrick** and **John Locke**, he helped to establish the annual nature of these (and other) meteor showers. Olmsted likewise investigated the aurora borealis and the zodiacal light, publishing theories of their purported origins in a related "nebulous body."

Using a small telescope at Yale in 1835, Olmsted and tutor **Elias Loomis** were the first American astronomers to recover comet 1P/Halley at its predicted return.

Jordan D. Marché, II

Selected References

Goodrum, Matthew R. (1999). "Olmsted, Denison." In *American National Biography*, edited by John A. Garraty and Mark C. Carnes. Vol. 16, pp. 696–697. New York: Oxford University Press.

Littmann, Mark (1998). *The Heavens on Fire: The Great Leonid Meteor Storms*. Cambridge: Cambridge University Press, esp. pp. 1–3, 13–32.

Treadwell, Theodore R. (1946). "Denison Olmsted an Early American Astronomer." *Popular Astronomy* 54: 237–241.

Olympiodorus the Younger [the Platonist, the Neo-Platonist, the Great]

Born **Alexandria, (Egypt), 495–505**
Died **after 565**

Olympiodorus the Younger is remembered for having continued the Platonic tradition at Alexandria following the closing of the Greek Academy at Athens by Emperor Justinian of Byzantium in 529. His most influential writings include commentaries on **Plato**'s *Phaedo*, Alcibiades I, Gorgias, and Philebus. Olympiodorus's commentary on Plato's life is most widely cited. Although his authorship of other works is disputed, it is agreed that the extant remains of Olympiodorus are student notes (or copies of notes) taken from his public lectures delivered at Alexandria.

Olympiodorus's works are also valued for information they provide on the life and work of earlier thinkers, notably Damascius, Iamblichus, Syrianus, and significantly, **Ptolemy**. In his *Phaedo*, Olympiodorus offers several valuable sentences, suggesting that Ptolemy lived near Alexandria at Canobus, where he devoted himself to astronomy for 40 years. Further, it was there that he had stone tablets carved to preserve his astronomical discoveries. Details about these inscriptions made at the temple of Serapis have been disputed since the 17th century. Prompted by their interest in Ptolemy, Isaac Vossius and **Ismaël Boulliau** were the first to study Greek manuscript copies of Olympiodorus found in Florence and Paris.

Robert Alan Hatch

Selected References

Norvin, William (ed.) (1913). *Olympiodori philosophi In Platonis Phaedonem commentaria*. Leipzig: Teubner.

Westerink, Leendert G. (ed.) (1976–1977). *The Greek Commentaries on Plato's Phaedo*. 2 Vols. Amsterdam: North-Holland.

Omar Khayyām

➔ **Khayyām: Ghiyāth al-Dīn Abū al-Fatḥ ʿUmar ibn Ibrāhīm al-Khayyāmī al-Nīshāpūrī**

Oort, Jan Hendrik

Born **Franeker, The Netherlands, 28 April 1900**
Died **Wassenaar, The Netherlands, 5 November 1992**

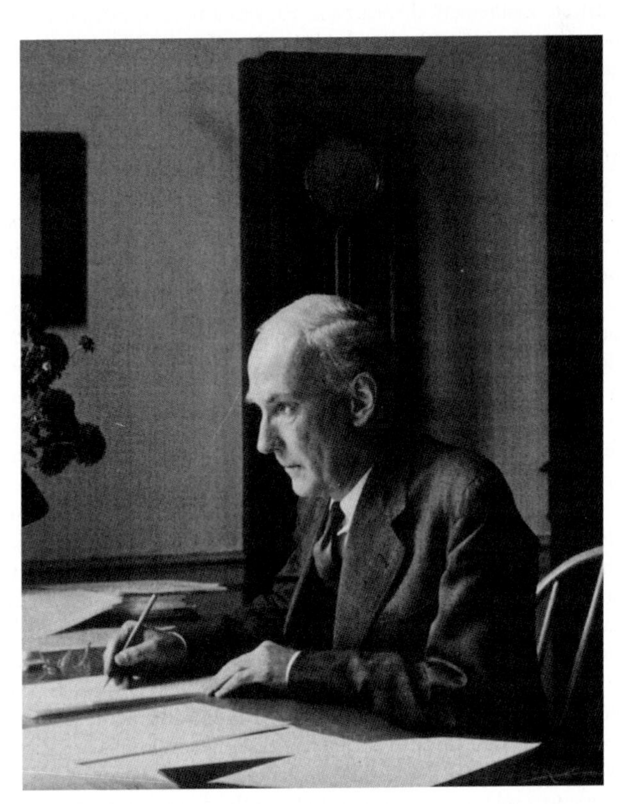

Dutch astronomer Jan Oort is eponymized in the Oort cloud (of potential future comets on the outskirts of the Solar System), in the Oort constants (of galactic rotation), and in the Oort limit to the density of mass in the disk of the Milky Way Galaxy.

Oort was the son of physician Abraham H. Oort and Hannah Faber. He and his wife, Johanna Maria Graadt van Roggen had two sons and a daughter.

Oort began studies at the University of Groningen in 1917 and was quickly inspired by **Jacobus Kapteyn** to take up astronomy. He completed his doctoral dissertation, on "The Stars of High Velocity," under **Pieter van Rhijn**, following Kapteyn's death and after 2 years at Yale Observatory (1922–1924), where he learned astrometry from **Frank Schlesinger**. Oort's thesis already contained the idea that stars moving faster than about 65 km/sec (an Oort limit) in a particular direction would probably escape from the Galaxy.

Oort spent the rest of his career at the Leiden Observatory and Leiden University as staff member (1924–1930), lecturer (1930–1935), professor extraordinary (1935–1945), full professor and director of the observatory (1945–1970), and in retirement until shortly before his death. The only exception was the period of World War II, 1940–1945, when the university was closed by the occupying forces, and Oort withdrew to a small village in the center of Holland. During his early years at Leiden, Oort was most influenced by **Ejnar Hertzsprung** and **Willem de Sitter**. He, in turn, was formal advisor or mentor to a large fraction of the next two generations of Dutch astronomers.

Jan Oort's approach to science, strongly determined by Kapteyn, was one of close confrontation between observation and interpretation, with a minimum of speculation and, if possible, avoidance of intricate mathematical treatment. He was a man of extraordinary perception, gifted with remarkable intuition for choosing promising courses of research.

Oort received numerous awards, among which were the Kyoto Prize, the Balzan Prize, and the Vetlesen Prize. He was a member of 16 foreign science academies including all the major ones and the recipient of ten honorary degrees.

Oort's scientific activities spanned more than 70 years; his first paper dates from 1922, his last one from 1992. The leading motive in nearly all of his research was the problem of the structure of the Universe at large. In the early 1920s this was still synonymous with the problem of the structure of the Milky Way; only later in the 1920s did it become apparent that the Galaxy is just one among numerous, more-or-less similar, stellar systems. By the time of Oort's last papers, research on the Universe concentrated on the large-scale features in the spatial distribution of stellar systems and on their relation to the initial stages of the Universe.

Rotation of the Galaxy. In a paper of 1927, which brought him international fame, Oort confirmed **Bertil Lindblad**'s hypothesis that the (flattened) Galaxy is a rotating stellar system. He demonstrated by means of available observational data on stellar distances, radial velocities, and proper motions that the bulk of the stars move in nearly circular orbits around a center, the direction of which coincided with the center of the system of globular clusters as proposed by **Harlow Shapley**. The proof lay in the differential motions exhibited by stars located at different distances from this center for which the distance could then be estimated at about 20,000 light years. In a subsequent paper of 1928 on the dynamics of this rotating system, Oort showed that it offered a natural explanation for a variety of phenomena observed earlier, among which were the ellipsoidal velocity distribution of stellar velocities and the peculiar distribution of the directions of the largest stellar velocities – the socalled asymmetry in high velocities. The Oort constants describe this differential rotation in the form of derivatives.

Hidden and observable matter in the Galaxy. In a paper of 1932 Oort derived the strength of the field of force in the stellar system, in particular the force perpendicular to the plane of symmetry (the galactic plane), from a comparison of the velocity distribution of the stars with their density distribution. In the region near the Sun, this force could not be satisfactorily explained by adding the contributions from matter present in the form of stars and interstellar matter [ISM], and Oort suggested the possible existence of an additional component, called "hidden matter" or "dark matter." In his George Darwin Lecture in 1946 he discussed the physics of the ISM: the formation of dust particles from the gaseous medium, the interaction with hot stars, and possible mechanisms of star formation. The upper limit to the local mass density in the galactic disk, implied by the velocities of stars perpendicular to it, is the "Oort limit."

Radio-astronomy, spiral structure, and the galactic center. Following up on the discovery of cosmic emission at radio wavelengths by **Karl Jansky** and **Grote Reber**, Oort was the first to realize its far-reaching implications for astronomy. Of two fields of application, the measurement of the continuum spectrum and that of emission lines, he pursued in particular the latter after van de Hulst's prediction, in 1943, of the 21-cm emission line due to neutral interstellar hydrogen, HI. Its measurement became the primary objective, first of the Dwingeloo Radio Observatory (operational in 1956) and, next, of the Westerbork Synthesis Radio Telescope (in 1970) in the Netherlands, both largely the result of Oort's efforts. The measurement resulted in the mapping of the spiral-like distribution of HI in the Galaxy and of the HI structural patterns in other nearby galaxies. Oort's discovery, with C. A. Muller and Ernst Raimond, of clouds of neutral hydrogen gas apparently falling at high speed into the Milky Way from its halo or from inter-galactic space, was an important product of this program. Another topic of particular interest in Oort's research was the central region of the Galaxy with its explosive nature and small-scale spiral structure. He spoke in favor of a central black hole at least from the 1980s.

The Oort and Walraven map of polarization of the light of the Crab Nebula was probably the last serious optical astronomy done from the Netherlands, as the community turned to radio astronomy and to optical observatories in more favorable locations. It laid the ground for later studies by **Jan Woltjer** and others demonstrating the need for an ongoing energy source in the supernova remnant, later found in the form of a pulsar.

Large-scale structure in the distribution of galaxies. Whereas already early in his career Oort on several occasions had expressed his deep interest in the large-scale structures on a cosmological scale, he addressed these thoroughly only in the later stages of his career. His substantial review article on superclusters published in 1983 became the standard reference for many subsequent investigations.

Comets. Oort's interest in the origins of comets arose because a student, A. J. J. van Woerkum, had begun a thesis on the topic with Woltjer and became Oort's student after his first advisor's death. Oort postulated the existence of a spherical reservoir of comets beyond the domain of the major planets, now called the Oort cloud. Some comets will occasionally be perturbed in their orbit by passing stars; as a result comets may now be classified as either "new" ones that for the first time pass through the region of the Earth's orbit or "old" (periodic) ones that as a consequence of Jupiter's gravitation remain captured in the inner regions.

European Southern Observatory. Following up on a suggestion by **Walter Baade** of Mount Wilson Observatory, Oort in 1954 called a meeting of leading European astronomers and proposed that they join financial and personnel resources for the establishment of a joint southern observatory, a proposal he pursued with utmost perseverance. After a long struggle, requiring much diplomacy and insistence on the part of Oort, **André-Louis Danjon**, **Otto Heckmann**, and **Bertil Lindblad**, the international convention was signed in 1962.

International Astronomical Union [IAU]. A strong advocate of international collaboration, Oort contributed considerably to the activities of the IAU. From 1935 until 1948 he was its general secretary, except during the years 1940–1945 (World War II) during which the secretariat was filled temporarily by **Walter Adams** of Mount Wilson Observatory. Oort's efforts contributed considerably to the resumption of cooperation within the IAU among astronomers from the formerly hostile countries. During the years 1958–1961 Oort was president of the IAU. At the time of the General Assembly [GA] of the International Union in 1976 at Grenoble, France, Oort was the only remaining member to have attended all of the GAs since the first in Rome in 1922.

Adriaan Blaauw

Selected References

Blaauw, Adriaan (1991). *ESO's Early History*. Garching, Germany: ESO. (For Oort's role in the establishment of the ESO.)

——— (1992). *ESO Historical Archives [EHA]: Inventory per December 1992.* Garching: ESO. See also the same author's *Archives of the International Astronomical Union: Inventory for the Years 1919–1970.* Paris: IAU, 1999. (For archival material.)

——— (1993). "Jan Hendrik Oort" (in Dutch). *Zenit* 20: 196–210.

——— (1994). *History of the IAU.* Dordrecht: Kluwer. Academic Publishers. (For Oort's role in IAU.)

Blaauw, Adriaan and Maarten Schmidt (1993). "Jan Hendrik Oort (1900–1992)." *Publications of the Astronomical Society of the Pacific* 105: 681–685.

Katgert-Merkelijn, J. K. (1997). *The Letters and Papers of Jan Hendrik Oort.* Dordrecht: Kluwer Academic Publishers (Describes the nearly complete collection of archival material kept in the University Library, Leiden. It includes a complete list of Oort's publications. See also "A Short Biography of Jan Hendrik Oort," pp. xv–xxx.)

Oort, J. H. (1926). "The Stars of High Velocity." *Publications of the Kapteyn Astronomical Laboratory*, no. 40. (Oort's thesis.)

——— (1927). "Observational Evidence Confirming Lindblad's Hypothesis of the Rotation of the Galactic System." *Bulletin of the Astronomical Institutes of the Netherlands* 3: 275–282. (Most of Oort's early papers appeared in this journal.)

——— (1928). "Dynamics of the Galactic System in the Vicinity of the Sun." *Bulletin of the Astronomical Institutes of the Netherlands* 4: 269–284.

——— (1932). "The Force Exerted by the Stellar System in the Direction Perpendicular to the Galactic Plane and Some Related Problems." *Bulletin of the Astronomical Institutes of the Netherlands* 6: 249–287.

——— (1950). "The Structure of the Cloud of Comets Surrounding the Solar System, and a Hypothesis Concerning Its Origin." *Bulletin of the Astronomical Institutes of the Netherlands* 11: 91–110.

——— (1965). "Stellar Dynamics." In *Galactic Structure*, edited by Adriaan Blaauw and Maarten Schmidt, pp. 455–511. Vol. 5 of *Stars and Stellar Systems*. Chicago: University of Chicago Press. (An impressive review of the status of our knowledge of the Galaxy in the early 1960s.)

——— (1977). "The Galactic Center." *Annual Review of Astronomy and Astrophysics* 15: 295–362. (A review article which summarizes work on the central regions of the Galaxy.)

——— (1981). "Some Notes on My Life as an Astronomer." *Annual Review of Astronomy and Astrophysics* 19: 1–5.

——— (1983). "Superclusters." *Annual Review of Astronomy and Astrophysics* 21: 373–428.

Pecker, Jean-Claude (1993). "La vie et l'oeuvre de Jan Hendrik Oort." *Compte rendus: La vie des sciences* 10: 535–540.

Radhakrishnan, V. (1993). "Jan Oort and Radio Astronomy." *Current Science* 65: 134–138.

Van de Hulst, H. C. (1994). "Jan Hendrik Oort (1900–1992)." *Quarterly Journal of the Royal Astronomical Society* 35: 237–242.

Van de Hulst, H. C., C. A. Muller, and J. C. Oort (1954). "The Spiral Structure of the Outer Part of the Galactic System Derived from the Hydrogen Emission at 21cm Wave Length." *Bulletin of the Astronomical Institutes of the Netherlands* 12: 117–149.

Van den Bergh, Sidney (1993). "An Astronomical Life: J. H. Oort (1900–1992)." *Journal of the Royal Astronomical Society of Canada* 87: 73–76.

Van Woerden, Hugo, Willem N. Brouw, and Henk C. van de Hulst (eds.) (1980). *Oort and the Universe.* Dordrecht: D. Reidel.

Woltjer, Lodewijk (1993). "Jan H. Oort." *Physics Today* 46, no. 11: 104–105.

Öpik, Ernst Julius

Born **Port Kundu, (Estonia), 23 October 1893**
Died **Bangor, Co. Down, Northern Ireland, 10 September 1985**

Ernst Öpik was among the first to put forward a number of ideas now regarded as part of mainstream astronomy, including a way of measuring distances to other galaxies (before it was generally agreed that there were other galaxies), a composition discontinuity as the cause of red-giant structure, and the need for a three-body process to carry nuclear reactions beyond the formation of helium to carbon and heavier elements.

Öpik was educated at Tallinn High School in Estonia and graduated with honors and a gold medal in 1911. At school, and later at university, he supported himself and helped his parents by teaching mathematics, science, and languages. He had to choose between music and science and decided on a career in astronomy.

Öpik graduated with first class honors from Moscow Imperial University in 1916 and held teaching posts there for 3 years. During the Bolshevik Revolution he volunteered for the White Russian Army. From 1919 to 1921 he headed a new astronomy department at the Turkestan University in Tashkent and then returned to his native Estonia as associate professor at Tartu University, where he obtained his doctorate in 1932 with a thesis on meteor observations.

Between 1930 and 1934 Öpik held the post of research associate and visiting lecturer at Harvard University and Harvard College Observatory. During this time he founded the meteor research group at Harvard. He then returned to Estonia and established his family home there.

After the German occupation of Estonia in 1941, and with the Soviet armies about to reoccupy the country in 1944, Öpik and his family fled to the west by horse and cart. They eventually reached Hamburg Observatory where they were given succor. Öpik became professor of astronomy in the Baltic University in Hamburg, which was set up to accommodate displaced scholars and students from the east. When his plight became known, **Eric Lindsay**, a Harvard graduate and director of Armagh Observatory from 1937, invited Öpik to Armagh, Northern Ireland, in 1947 as a research associate. He remained at Armagh until his retirement in 1981.

While he was still an undergraduate at Moscow University in 1916, Öpik wrote a paper on the densities of 40 visual binaries, which he derived from their corrected surface brightness. He excluded from his survey one binary with known orbital elements, o_2 Eridani (40 Eri B). Taking its spectrum as A0, its parallax as 0.17 arc seconds, and the mass ratio as unity, he found a density 25,000 times that of the Sun. Assuming that this was an impossible result, Öpik sought another explanation. He was not the first to think the density of a white dwarf was incredible; in 1905, **John Gore** of Dublin estimated the density of Sirius B at nearly 32,000 times the Sun's and thought it "entirely out of the question."

In 1921 Öpik published an article in the Russian astronomy magazine *Mirovedenie* that included an estimate of the distance to the Andromeda Nebula from dynamical considerations. By connecting the observed velocity of rotation of the Nebula to the centripetal acceleration, and hence its gravitational attraction and mass, he obtained a distance of 785 kpc. The following year a similar paper appeared in the *Astrophysical Journal* and, assuming a new value for the mass–luminosity ratio, his estimate was now 450 kpc. These values compare well with a modern estimate of 690 kpc considering that they predate the extragalactic scale based on Cepheid luminosities.

In 1932 Öpik set out to investigate the stability of a large cloud of meteors and comets attached gravitationally to the Solar System. He calculated the perturbation effects of stars passing through the cloud, making the reasonable assumption that objects in very large orbits about the Sun would spend most of their time at aphelion. He decided that a system of comets could persist without great losses for over 3 billion years, in orbits about the Sun, at distances up to 5 parsecs. Subsequent studies have reduced this distance by a large factor, but the basic conclusion is still valid. **Jan Oort** introduced his concept of a cloud of comets surrounding the Sun in 1950.

In 1938 Öpik tackled the problem of the origin of red giant stars, which must have sizes up to hundreds of times that of the Sun, because they are very bright, although very cool. He realized that in stars like the Sun, nuclear burning takes place only in the central 10% by mass. As the available hydrogen fuel is depleted, there is no longer enough pressure to support the central regions; the core collapses; and, as a result, the outer envelope expands to an enormous size. Öpik's hand calculations were confirmed about 10 years later by **Fred Hoyle** and **Martin Schwarzschild** using electronic calculators.

In 1950 Öpik put forward a theory to explain the occurrence of ice ages on the Earth with periods of several hundreds of millions of years. He suggested that transport of fresh hydrogen into the solar core could increase the rate of energy generation, affecting temperatures on the Earth. Other factors are now thought to be more important.

Öpik made many contributions to knowledge of the planets and the minor bodies of the Solar System. He anticipated the desert-like nature of the surface of Venus, and his prediction of craters on Mars was confirmed 15 years later by planetary probes. His statistical studies of Earth-crossing comets and asteroids led to a better understanding of the dynamics of these bodies and their influence on the Earth. In recognition of this work, Minor planet (2099) was named for Öpik.

In 1950, on the initiative of Lindsay and with the active support of **Hermann Brück**, the *Irish Astronomical Journal* began publication with Öpik as editor. Over the following three decades the journal became a channel for his views on astronomical and other matters. During this time he was also a visiting professor in the Department of Physics and Astronomy at the University of Maryland.

Öpik was an accomplished amateur musician, both as a performer and composer. He wrote some 3,000 pieces for piano as well as some choral works. An appraisal of his musical attainments is given by Mary de Vermond of the University of Maryland in a special issue (1972) of the *Irish Astronomical Journal*.

For his contributions to astronomy, Öpik received many honors. He was fellow of the Estonian Academy of Sciences from 1938, member of the Royal Irish Academy from 1954, foreign associate of the United States National Academy of Sciences [NAS] from 1975, and foreign honorary member of the American Academy of Arts and Sciences from 1977. He received honorary degrees of Doctor of Science from Queen's University, Belfast, in 1968 and from the University of Sheffield in 1977. He received the J. Lawrence Smith Medal of the NAS in 1960, the F. C. Leonard Medal of the International Meteoritical Society in 1968, one of six Gold Medals of the American Association for the Advancement of Science in 1974 awarded in connection with the fourth centenary of **Johannes Kepler**'s birth, the Gold Medal of the Royal Astronomical Society in 1975, the Catherine Wolfe Bruce Gold Medal of the Astronomical Society of the Pacific in 1976, and the *Grand Prix* of the Louis Jacot Foundation "La Pensee Universelle" in 1978. He was elected a fellow of the Royal Astronomical Society in 1949.

On his retirement at the age of 88, Öpik went to live in the seaside town of Bangor. He continued as an associate editor of the journal until his death. He was survived by his second wife Alide, and by one son and five daughters. Öpik's grandson, Lembit Öpik, has served as a Member of Parliament at Westminster in London.

Ian Elliott

Selected References

Anon. (1986). "Ernst Julius Opik, 1893–1985: The Man and the Scientist." Memorial Issue dedicated to Ernst Julius Öpik. *Irish Astronomical Journal* 17: 411–442.

Bennett, J. A. (1990). *Church, State and Astronomy in Ireland: 200 Years of Armagh Observatory*. Armagh: Armagh Observatory in association with the Institute of Irish Studies and the Queen's University of Belfast. (See also the earlier special issue dedicated to Öpik, *Irish Astronomical Journal* 10 (1972): 1–92.)

Gore, John Ellard (1905). "The Satellite of Sirius." *Observatory* 28 (1905): 55–57.

Longair, Malcolm S. (1995). "Astrophysics and Cosmology." In *Twentieth Century Physics*, edited by Laurie M. Brown, Abraham Pais, and Sir Brian Pippard. Vol. 3, pp. 1691–1821. Bristol: Institute of Physics.

Öpik, E. J. (1916). "The Densities of Visual Binary Stars." *Astrophysical Journal* 44: 292–302.

——— (1922). "An Estimate of the Distance of the Andromeda Nebula." *Astrophysical Journal* 55: 406–410.

——— (1932). "Note on Stellar Perturbations of Nearly Parabolic Orbits." *Proceedings of the American Academy of Arts and Sciences* 67: 169–183.

——— (1938). "Stellar Structure, Source of Energy, and Evolution." *Acta et Commentationes Universitatis (Dorpatensis)* 33, no. 9: 1–118.

——— (1950). "Secular Changes of Stellar Structure and the Ice Ages." *Monthly Notices of the Royal Astronomical Society* 110: 49–68.

——— (1977). "About Dogma in Science, and Other Recollections of an Astronomer." *Annual Review of Astronomy and Astrophysics* 15: 1–17.

Wayman, P. A. and D. J. Mullan (1986). "Ernst Julius Öpik." *Quarterly Journal of the Royal Astronomical Society* 27: 508–512.

Oppenheimer, J. Robert

Born **New York City, New York, USA, 22 April 1904**

Died **Princeton, New Jersey, USA, 18 February 1967**

American physicist J. Robert Oppenheimer features in world history as the leader of the Manhattan Project, which developed the first American atomic (fission) bombs, but his contributions to astronomy concern the structure of neutron stars (using the eponymous Tolman–Oppenheimer–Volkoff equation of state) and the unstoppable collapse of objects larger than some critical mass.

Oppenheimer was the elder of two sons. (The younger, Frank Oppenheimer, was also a physicist.) Robert graduated from the Ethical Culture High School in New York at age 16. He received his A.B. in physics from Harvard University in 1925 (which also awarded him an honorary D.Sc. in 1947, one of many honorary degrees) and his Ph.D. in physics in 1927 from Göttingen University in Germany for work with Max Born on what is now called the Born–Oppenheimer approximation (for quantum mechanical calculations of systems of particles with very different masses; they worked on molecules). After brief fellowships in Leiden and Zürich (1928/1929), he took up joint appointments at the California Institute of Technology and the University of California, Berkeley, progressing from assistant professor (1929–1931) to associate (1931–1936) to full professor (1936–1947).

Oppenheimer apparently had a gift for languages, and after his brief stays in Göttingen and Leiden was able to write scientific papers and converse in German and Dutch. His adult nickname, Oppie, was an Americanization of the Dutch. Oppenheimer married Katharine Harrison in 1940. Neither of their children, Peter and Katharine, evinced any interest in science.

Oppenheimer was director of the Los Alamos National Laboratory from 1943 to 1945 and professor of physics and director of the Institute for Advanced Study in Princeton from 1947 to 1966. He served as chair of the General Advisory Committee of the Atomic Energy Commission from 1946 to 1952, was president of the American Physical Society (1948), and was elected to a number of national academies of science (United States, Danish, Brazilian, and most remarkably, the Japanese).

It can be argued that his most important long-term influence on physics was in building up the large theory groups at Caltech and Berkeley, so that an American no longer had to go to Europe, as Oppenheimer had, for advanced training. Among his students who have become important astrophysicists are **Philip Morrison** and Robert Christy (professor emeritus, Caltech). The best known of his post-doctoral fellows in the area of astrophysics and gravitation was probably Leonard Schiff, who developed the theory of one of the classic astronomical tests of general relativity.

Oppenheimer's work on topics relevant to astrophysics was confined to a narrow time window, beginning with a 1937 attempt, with Robert Serber, to identify a particle newly discovered in the cosmic rays (now called the μ meson or muon) with a particle predicted by Hideki Yukawa to carry the strong or nuclear force. The Yukawa particle, now called the pion, was later discovered in the cosmic rays and, although its mass is similar to that of the muon, it is otherwise fundamentally different.

Two papers, with Serber and with **George Volkoff**, in 1937–1938 contained the first careful calculations of the structure of neutron stars, which had been predicted in 1933/1934 by **Walter Baade** and **Fritz Zwicky**. They showed in particular that there was an upper limit to the mass of a stable neutron star, similar to the Chandrasekhar limit for white dwarfs. The number they found, 70% of the mass of the Sun, was too small because they had not been able to include the effect of the nuclear force, but only those of quantum mechanics and general relativity. A 1939 paper with **Hartland Snyder** was entitled "On Continued Gravitational Contraction," and dealt with the behavior of a mass above the critical limit. It is generally considered to be a prediction of black holes. Oppenheimer did not live to see observational confirmation of either neutron stars (the 1967/1968 discovery of pulsars) or black holes (the 1972 measurement of the mass of the compact object in the X-ray binary Cygnus X-1).

In the postwar period, Oppenheimer did very little original scientific research, but did make some contributions on the role of mesons in generating cosmic-ray showers. He mentored a number of younger physicists in this period, including Tsung-Dao Lee and Chen Ning Yang, who shared the 1957 Nobel Prize in Physics for the prediction of parity nonconservation.

The last 25 years of Oppenheimer's life were inevitably dominated by the development of the atomic bomb (regrets for which he expressed the form of saying that "physicists had known sin") and its aftermath. His period of influence in the United States government and postwar development of atomic weaponry and energy ended abruptly when his security clearance was withdrawn in 1953. The subsequent trial divided the American physics community, including especially those who had worked with him closely at Los Alamos. He was given the Fermi Award in 1963 by the Atomic Energy Commission in a not uncontroversial partial apology.

President Kennedy had been scheduled to present the Fermi award on 2 December 1963. As it turned out, the presentation was made by President Lyndon B. Johnson, with the words:

> Dr. Oppenheimer, I am pleased that you are here today to receive formal recognition for your many contributions to theoretical physics and to the advancement of science in our nation. Your leadership in the development of an outstanding school of theoretical physics in the United States and your contributions to our basic knowledge make your achievements unique in the scientific world.

Oppenheimer's response typified the courtly grace, or pomposity, that had made him either loved or hated by so many:

> I think it is just possible, Mr. President, that it has taken some charity and some courage for you to make this award today. That would seem to me a good augury for all our futures . . . These words I wrote down almost a fortnight ago [that is, before Kennedy's death]. In a somber time, I gratefully and gladly speak them to you.

He developed throat cancer soon after, perhaps as a result of inveterate pipe-smoking and perhaps also from the effects of radiation exposure during the Los Alamos years, and died very shortly after being filmed in a tribute to Enrico Fermi.

Virginia Trimble

Selected References

Kipphardt, Heinar (1968). *In the Matter of J. Robert Oppenheimer*. New York: Hill and Wong.

Rabi, I. I. *et al.* (1969). *Oppenheimer*. New York: Scribner.

Rhodes, Richard (1986). *The Making of the Atomic Bomb*. New York: Simon and Schuster.

Rigden, John S. (1995). "J. Robert Oppenheimer: Before the War." *Scientific American* 273, no. 1: 76–81.

Schiff, Leonard (1967). "J. Robert Oppenheimer – Scientist, Public Servant." *Physics Today* 20, no. 4: 110–111.

Smith, Alice Kimball and Charles Weiner (eds.) (1980). *Robert Oppenheimer: Letters and Recollections*. Cambridge, Massachusetts: Harvard University Press. (Includes a complete listing of Oppenheimer's technical publications.)

Stern, Philip M. (1969). *The Oppenheimer Case: Security on Trial*. New York: Harper and Row. (There is an enormous literature concerning the security hearings, including a complete publication of the transcripts.)

Oppolzer, Egon Ritter von

Born **Vienna, (Austria), 1869**
Died **Vienna, (Austria), 15 June 1907**

Egon Oppolzer was the son of **Theodor von Oppolzer**. He was educated at the University of Vienna and the University of Munich; in 1897 he became an assistant at the Prague Observatory, where he discovered the variability of the minor planet (433) Eros. Oppolzer was appointed a professor of astronomy at the University of Innsbruck in 1901.

Selected Reference

Seeliger, H. (1907) "Todes-Anzeige: Egon von Oppolzer." *Astronomische Nachrichten* 175 : 239–240.

Oppolzer, Theodor Ritter von

Born **Prague, (Czech Republic), 26 October 1841**
Died **Vienna, (Austria), 26 December 1886**

Though trained as a physician, Theodor von Oppolzer excelled at celestial mechanics and published a monumental volume on the elements of all solar and lunar eclipses, visible from 1207 BCE to 2163. Oppolzer's father, Johann von Oppolzer, was a well-known physician, specialist on internal diseases, and professor of medicine at the Universities of Prague, Leipzig, and Vienna. Oppolzer received a private education (1851–1859) before he entered the Piaristen-Gymnasium in Vienna, from which he graduated *summa cum laude*. While his teachers encouraged his abilities in mathematics and science, Oppolzer fulfilled his father's wish that he study medicine at the University of Vienna, where he received his medical degree in 1865. Yet, it seems that he never wanted to work as a physician. In 1865, Oppolzer married Coelestine Mauthner von Markhof; the couple had six children. One of his sons, **Egon von Oppolzer**, became professor of astronomy at the University of Innsbruck.

Oppolzer's intellectual talents enabled him to pursue mathematical astronomy even through the course of his medical studies. His parents supported his interests by financing a private observatory located on the outskirts of Vienna, which was christened "Sternwarte Wien-Josefstadt." This observatory was equipped with a 7-in. refracting telescope, a meridian circle, and a wide-field comet-seeker. Before the new Vienna Observatory was erected in the late 1870s, Oppolzer's principal telescope was likely unsurpassed in the Austro–Hungarian monarchy.

Two of the principal influences on Oppolzer's astronomical interests were **Maurice Löwy** and **Edmund Weiss**, both of whom then worked at the original Vienna Observatory. Loewy and Weiss observed minor planets and comets and computed their orbital elements and ephemerides. Oppolzer's first publication concerned the orbit of comet C/1861 G1. By the time his medical studies were completed, he had published some 56 papers in astronomical journals, dealing chiefly with orbit determinations. The following year, Oppolzer became a *Privatdozent* (lecturer) at the University of Vienna, without having acquired a Ph.D.

Oppolzer's preliminary research dealt with the orbital determinations and ephemeris calculations of asteroids and comets. Yet, he soon found ways to improve upon the cumbersome methods and auxiliary tables used at that time. These techniques enabled Oppolzer to recover the lost minor planets (62) Erato, (73) Klytia, and (91) Aegina. In 1870, he published the first volume of a textbook on orbit determination, whose methods for computing an elliptical orbit from three or four positions converged toward a solution much more rapidly than the method of **Carl Gauss**. In that same year, Oppolzer was appointed an extraordinary professor of astronomy and advanced geodesy at the University of Vienna.

Oppolzer undertook a major scientific and administrative task in the supervision of his nation's geodetic studies, which included a series of longitude measurements between Austrian cities and other European capitals. Reflecting certain patriotic motivations, this large-scale undertaking also evinced international cooperation in science on an unprecedented scale. Its ambitious goals were to conduct a topographic survey and establish the precise shape of the Earth. In 1873, Oppolzer became director of the Austrian Gradmessungs-Bureau (Geodetic Survey); in the following year, he was appointed Austrian representative to the International European Geodetic Congress (and later became its vice president). Between 1873 and 1876, Oppolzer personally supervised some 40 longitude determinations. Apart from his role as manager, he contributed ideas for new instruments that were applied in other countries.

In connection with the problem of the Earth's figure, Oppolzer was interested in gravity determinations. He improved existing methods of measurement by the use of reversible pendulums and eliminated sources of error. In 1883, the first determination of the absolute intensity of gravity was made in the basement of the new Vienna Observatory. Some years before its adoption in 1875, Oppolzer had promoted establishment of the metric system in the Austro–Hungarian monarchy. For these scientific and administrative activities, he was decorated with a series of medals from various countries and elected a member of numerous scientific institutions and academies. Oppolzer was awarded an honorary doctorate from the University of Leiden in 1871.

In spite of these responsibilities, Oppolzer never gave up astronomy. He was engaged with preparations for an expedition to Jassy in Romania to observe the 1874 transit of Venus. In 1875, Oppolzer had been offered the directorship of the Gotha Observatory but declined the invitation. In that same year, he was made an ordinary professor; by 1879, he had been elevated to the rank of full professor at the University of Vienna. Those years also witnessed construction of the new Vienna Observatory. After its director, **Karl von Littrow**, was released from teaching, the bulk of that activity fell to Oppolzer and Weiss. The former also prepared the second volume of his textbook on orbit determination, involving perturbation theory, which appeared in 1880.

Oppolzer's studies in celestial mechanics culminated with his refinements to the theory of the Moon's motion and the Sun–Earth–Moon system in connection with solar and lunar eclipses. Oppolzer possibly became interested in this subject on the occasion of the total solar eclipse of 18 August 1868. As a participant in the Austrian expedition to Aden, he was responsible for determining the position of the observing station and the times of contacts during the eclipse.

To arrive at a more accurate determination of the Moon's orbit, it was necessary for Oppolzer to compare historical and contemporary eclipse observations by means of a uniform theory. He began his investigations with the best available lunar theory of **Peter Hansen**. Yet, even here he found irregularities that could not be explained by the theory. Oppolzer then modified Hansen's theory and developed appropriate tables that allowed more efficient calculations to be made. In 1881, he published his "Syzygien-Tafeln für den Mond" (Syzygy tables for the Moon), which enabled the times of new and full Moons to be calculated in the distant past or future.

Oppolzer's study of eclipses reached its climax with the 1885 publication of his monumental *Canon der Finsternisse* (Canon of Eclipses). This massive undertaking required the assistance of ten human computers, half of whom worked as volunteers, half of whom were paid privately by Oppolzer. The calculations were performed by two independent groups and their results compared for accuracy. Described in 1887 as "one of the greatest works of calculation which has ever been accomplished by man," the *Canon* contains the elements for approximately 8,000 solar and 5,200 lunar eclipses, including visibility tracks of the solar eclipses. When completed, the *Canon* was of special use for chronological research, offering precise dates of historical events that otherwise might be known only to the nearest century.

Appearing well before the advent of mechanical or digital calculations, Oppolzer's *Canon* was reprinted (and translated into English) as late as 1962. In turn, his work has stimulated the computation of other eclipse catalogs, some more extended and specialized than the original. A newer compilation (1983), based on the more refined lunar theory of **Ernest Brown**, and generated by a digital computer, provides a significant test of the accuracy of Oppolzer's *Canon*.

Oppolzer did not live long enough to finish his modified lunar theory; his widow financed the partial completion of his manuscript by enlisting those computers who had worked on the *Canon der Finsternisse*.

A crater on the Moon and several minor planets are named for the subject or his relatives: (1492) Oppolzer, (237) Coelestina (his wife), (228) Agathe (his youngest daughter), and (153) Hilda (a daughter who died as a child).

Anneliese Schnell

Selected References

Carton, Wilhelm H. C. (1989). "Oppolzer's Great Canon of Eclipses." *Sky & Telescope* 78, no. 5: 475–478.

Ferrari d'Occhieppo, Konradin (1974). "Oppolzer, Theodor Ritter von." In *Dictionary of Scientific Biography*, edited by Charles Coulston Gillispie. Vol. 10, pp. 218–220. New York: Charles Scribner's Sons.

Firneis, M. G. (1987). "Leben und Wirken des Theodor Ritter von Oppolzer." In *Geodätische Arbeiten Österreichs für die Internationale Erdmessung*, n.s., Vol. 5: 18–36.

Oppolzer, Theodor Ritter von (1887). *Canon der Finsternisse*. Denkschriften, Mathematisch-Naturwissenschaften Classe der Kaiserlichen Akademie der Wissenschaften, Vol. 52. Vienna. (Reprint, New York: Dover, 1962.) (Translated by Owen Gingerich.)

Weiss, Edmund (1887). "Nekrolog über Theodor von Oppolzer." *Astronomische Nachrichten* 116: 95–96.

Oresme, Nicole

Born **diocese of Bayeux, (Calvados), France, *circa* 1320**
Died **Lisieux, (Calvados), France, 11 July 1382**

French bishop, scholastic philosopher, economist, and mathematician Nicole Oresme is considered today as one of the principal forerunners of modern science. Oresme's contributions in mathematics and physics are considered decisive ideas for the development of modern science in the 16th and 17th centuries. Editions of Oresme's work were published well into the Renaissance.

Probably Oresme took his philosophical training at the University of Paris under **John Buridan**, whose influence in Oresme's works is evident. By 1348 he had a scholarship in theology at the college of Navarre at Paris, of which he became grand master (head) in 1356. Oresme left Navarre after his appointment as a canon at Rouen (1362). Later he was appointed Canon at the Sainte-Chapelle in Paris (1363) and Dean of the Cathedral of Rouen (1364). At the behest of King Charles V of France, from about 1370 Oresme translated and annotated many works of **Aristotle**, including *On the Heavens*, from Latin into French. In his commentaries he expressed his critique of several Aristotelian tenets by developing many original astronomical, physical, and economic ideas. As a reward for this extended and difficult work, Charles V had him appointed Bishop of Lisieux in 1377. Little is known of Oresme's last years.

Oresme is the author of more than 30 works, the majority of which are still unpublished and remain in manuscript form. As a scholastic philosopher he is famous for his critique of several Aristotelian positions. Oresme rejected two of Aristotle's main definitions, replacing them with his own – he rejected the definition of the place of a body as the inner boundary of the surrounding medium in favor of a definition of place as the space occupied by the body, and he replaced the definition of time as the measure of motion with a definition of time as the successive duration of things, independent of motion.

Oresme's main contributions to astronomy, mathematics, and kinematics are contained in several works produced throughout his life. His two major scientific works are the *Tractatus de configurationibus qualitatum et motum* (Treatise on configurations of qualities and motions), and the *De proportionibus proportionum* (On ratios of ratios). Oresme's main astronomical views are exposed in his early works *Questiones de Celo* and *Questiones de Spera*, in a work opposing astrology, *Questio contra divinatores*, and in his later work written in French, *Le livre du ciel et du monde* (Book on the heavens and the world), a translation of and commentary on Aristotle's *On the Heavens*.

A very interesting characteristic of his argumentation is that it often permitted Oresme to suggest unorthodox and radical philosophical ideas while disclaiming any commitment to them. For example, in treating the question of the plurality of worlds, he stressed the possibility that God by His omnipotence could create such a plurality, though he finally rejected this in favor of a single Aristotelian cosmos. Another famous example of Oresme's argumentation is his study of the rotation of the Earth. He brilliantly argued against the "proofs" of Aristotle for a stationary Earth, mainly in his *Le livre du ciel et du monde*. With a series of arguments based mainly on the idea of the complete relativity in the detection of motion and the demonstration that all observed phenomena can be saved equally well by the diurnal rotation of the Earth as by the rotation of the heavens, Oresme explained why we cannot exclude the possibility of a rotating Earth. Though he finally concluded, "The truth is, that the earth is not so moved but rather the heavens," he added, "However I say the conclusion [about the nonexistence of a such rotation] cannot be demonstrated but only argued by persuasion."

Another original astronomical idea of Oresme is the metaphor of the heavens as a mechanical clock, which is considered the first attempt to understand mechanically the celestial motions. He suggested the possibility that God implanted in the heavens at the time of their creation special forces and resistances – differing from those on Earth – by which the heavens move continually like a mechanical clock. Through his opposition to astrology Oresme was also able to expose other aspects of his views about the relation of celestial and terrestrial phenomena. For Oresme terrestrial phenomena arise from natural and immediate causes rather than from celestial influences, with the exception of the influences of the light of the Sun. Only ignorance, he claimed, causes men to attribute terrestrial phenomena to the heavens, to God, or to demons. Oresme composed his anti-astrology dissertation in Latin, but so strong was his desire to divert people from the false science that he produced a short tract against the practice in French.

Finally we must mention Oresme's idea to develop mathematical arguments in order to prove the probable incommensurability of the ratio of any two celestial motions. He started from a suggestion of the theologian–mathematician **Thomas Bradwardine** that an arithmetic increase in velocity corresponds to a geometric increase in the original ratio of force to resistance. Oresme went on to give an extraordinary elaboration of the problem of relating ratios exponentially by a treatment of fractional exponents conceived as ratios of ratios. His idea was the distinction between irrational ratios of which the fractional exponents are rational and those of which the exponents are themselves irrational. Based on this treatment he claimed (without any real proof) that the ratio of any two celestial motions is probably incommensurable. This excluded precise predictions of successively repeating conjunctions, oppositions, and other astronomical aspects with the methodology of astrologers. Oresme presented his original method for manipulating ratios in an independent work, the *Algorism of Ratios*.

Dimitris Dialetis

Selected References

Clagett, Marshall (1959). *The Science of Mechanics in the Middle Ages*. Madison: University of Wisconsin Press. (Reprint, 1979.)

——— (1974). "Oresme, Nicole." In *Dictionary of Scientific Biography*, edited by Charles Coulston Gillispie. Vol. 10, pp. 223–230. New York: Charles Scribner's Sons.

——— (ed.) (1968). *Nicole Oresme and the Medieval Geometry of Qualities and Motions: A Treatise on the Uniformity and Difformity of Intensities Known as* Tractatus de configurationibus qualitatum et motuum. Madison: University of Wisconsin Press.

Duhem, Pierre (1954, 1958). *Le système du monde*. Vols. 7 and 8. Paris: A. Hermann.

——— (1909). *Un précurseur français de Copernic: Nicole Oresme (1377)*. Paris.

Grant, Edward (1994). *Planets, Stars and Orbs: The Medieval Cosmos, 1200–1687*. Cambridge: Cambridge University Press.

——— (ed.) (1966). *Nicole Oresme,* De proportionibus proportionum *and* Ad pauca respicientes. Madison: University of Wisconsin Press.

——— (1971). *Nicole Oresme and the Kinematics of Circular Motion:* Tractatus de commensurabilitate vel incommensurabilitate motuum celi. Madison: University of Wisconsin Press.

Kirschner, Stefan (2000). "Oresme's Concepts of Place, Space, and Time in His Commentary on Aristotle's *Physics*." *Oriens-Occidens: Sciences, mathématiques et philosophie de l'Antiquité à l'Âge classique* 3: 145–179.

Meunier, Francis (1857). *Essai sur la vie et les ouvrages de Nicole Oresme*. Paris.

Oriani, Barnaba

Born **Garegnano near Milan, (Italy), 17 July 1752**
Died **Milan, (Italy), 12 November 1832**

Barnaba Oriani was an accomplished observer and observatory director who contributed to planetary astronomy. He was of humble origins. His father Giorgio worked as a mason; his mother was Margherita Galli. Oriani began his studies in the Certosa of Garegnano, then at the College of San Alessandro, Milan, where he was educated and supported by the Barnabites (a religious order of the Catholic Church). He later joined the Barnabites, and, after studying the humanities, physical and mathematical sciences, philosophy, and theology, was ordained as a priest at the age of 23 in 1776. Despite his youth, he was admitted to the competition for teaching mathematics at the Como Gymnasium. His rejection diverted him to pursue his love of astronomy; shortly after his ordination Oriani was appointed to the staff of the Observatory of Brera in Milan. He became assistant astronomer in 1778.

For 5 months in 1786 Oriani traveled to Brussels, Amsterdam, and London, where he met a host of scientific luminaries. In 1787 Oriani was offered the full professorship of astronomy and directorship of the observatory to be founded in Palermo. He excused himself as he was not willing to leave his observatory in Brera. In any event, it was **Giuseppe Piazzi** who became the director of Palermo. These two met only once, in Brera in 1789, when Piazzi was returning from England and France. Piazzi had in Oriani a sincere and loyal friend; their correspondence spanning several decades became ever more frank as the years went on.

When observing the comet of 1779 (C/1799 A1), which had been discovered by **Johann Bode** on 6 January and independently by **Charles Messier** on 18 January, Oriani observed three "nebulous stars," namely M49, M60, and M61. M49 had been discovered in 1771 by Messier. Oriani found M49 on 22 April. (Messier had discovered it in 1771.) M60 had been discovered first by **Johann Köhler** on 11 April, then by Oriani the next day and by Messier on 15 April. M61, however, was the original discovery of Oriani on 5 May 1779, and was seen by Messier 6 days later.

Oriani's greatest work involved the orbit calculation of Uranus after its discovery by **William Herschel** in March 1781. After **Jean Bochart de Saron**, **Anders Lexell**, and others had shown that Uranus was not on a parabolic orbit and had obtained an approximate circular orbit, Oriani calculated the planet's elliptical orbit in 1783. He improved this calculation in 1789 by taking into account the perturbations of Jupiter and Saturn.

In 1786 Oriani conducted trigonometrical operations for measuring an arc of the meridian over Lombardy. He continued the ephemeris begun by **Joseph Lagrange**, and inserted in it many able and valuable discussions on the lunar theory, on refraction, and various practical and theoretical matters. His skill in spherical trigonometry enabled him to be the first in computing the path and perturbations of the first minor planet, (1) Ceres (discovered by his friend Piazzi in 1801). Oriani became one of the leading observers of the first four asteroids.

In 1802 Oriani was made director of the Observatory of Brera, and he became one of the founding members of the National Institute of the Italian Republic. That year he was sent to Bologna by order of the Italian Republic and found "astronomy almost abandoned." Oriani tried to revive the observatory there by offering its directorship to Piazzi, but he refused.

Oriani was long known in scientific circles as abbot and professor. When Napoleon established the republic in Lombardy, Oriani refused absolutely to swear hatred toward monarchy; the new government modified the oath of allegiance in his regard, retained him in his position at the observatory, and made him president of the commission appointed to regulate the new system of weights and measures. When the republic was transformed into the Napoleonic kingdom, Oriani received the decorations of the Iron Crown and of the Legion of Honor, was made count and senator of the kingdom, and was appointed (in company with Angelo de Cesaris) to measure the arc of the meridian between the zeniths of Rimini and Rome.

In private life Oriani was of a pleasing and amiable disposition, and a great encourager of study amongst the youths of his acquaintance. A longtime acquaintance, **Giovanni Plana** of Turin, wrote of him in tribute: "His splendid talents, by which he has rendered illustrious not only the Observatory of Milan, but the whole of Italy, made him highly respected, and caused his loss to be seriously deplored."

Clifford J. Cunningham

Selected References

Phillips, Edward C. (1913). "Oriani, Barnaba." In *Catholic Encyclopedia*, edited by Charles G. Herbermann *et al.* Vol. 11, pp. 302. New York: Encyclopedia Press.

Tagliaferri, Guido and Pasquale Tucci (1989). "The Visit to the Low Countries in 1786 of the Astronomer Barnaba Oriani of Milan." In *Italian Scientists in the Low Countries in the XVIIth and XVIIIth Centuries*, edited by C. S. Maffioli and L. C. Palm, pp. 277–290. Amsterdam: Rodopi.

——— (1993). "Laplace, Oriani and the Italian Meridian Degree." *Proceedings of the 1st EPS Conference on History of Physics in Europe in the 19th and 20th Centuries, Como, September 2–3, 1992*, edited by Fabio Bevilacqua, pp. 93–100. Bologna: Italian Physical Society.

Orontius Finaeus

➲ **Finé, Oronce**

Osiander, Andreas

Born **Gunzenhausen, (Bavaria, Germany), 19 December 1498**
Died **Königsberg (Kaliningrad, Russia), 17 October 1552**

German theologian, reformer, astronomer, and mathematician Andreas Osiander's most notorious role in astronomy was the writing of an anonymous preface to the first edition of **Nicolaus Copernicus**'s *De Revolutionibus Orbium Coelestium*, suggesting that the heliocentric idea was merely a mathematical hypothesis and not the actual way the Solar System was laid out.

Osiander was admitted to the University of Ingolstadt on 9 July 1515, but left without a degree. He was ordained a priest in Nuremberg in 1520. In 1522 Osiander joined the cause of the Reformation, and became one of the leading spokesmen of Lutheranism. He frequently participated as a representative of Nuremberg's clergy at gatherings of Lutherans, among others the Marburg Conference (1529), the Diet of Augsburg (1530), and the signing of the Schmalkaldic Articles (1537). In 1548 Osiander refused to agree to the pro-Catholic Augsburg Interim, the compromise temporary settlement of the religious wars. His refusal made it necessary for him to leave Nuremberg, and he went to Königsberg. In 1549 Duke Albrecht of Prussia, who regarded Osiander as his spiritual father, appointed him as pastor of one of Königsberg's churches and professor of theology at the newly founded university of the city. Osiander retained the support of the duke until his sudden death.

Osiander was a rigid man, given to strong opinions that he did not express with moderation. His style fomented theological divisions in Königsberg that subsequently involved the whole German Evangelical Church. As a theologian Osiander developed a mystical interpretation of the Lutheran doctrine of justification by faith. His views found supporters among the strict Lutherans, who refused any compromise with Rome or Calvinism, and were opposed by the moderate wing, headed by Philip Melanchthon and the group of Wittenberg University, who strove for reconciliation.

Though Osiander was primarily engaged in theological matters, he also entertained a deep interest in mathematics and astronomy. It is known that Osiander corresponded with the mathematician **Girolamo Cardano** about horoscopes for some 5 years. But his place in the history of astronomy is mainly due to his involvement in the first edition of Copernicus's *De revolutionibus orbium coelestium*, and the famous preface that he inserted, in which he warned readers that the book was not intended to propose more than a mathematical hypothesis. Osiander was approached for the publication of the book by **Rheticus** professor at the University of Wittenberg and an admirer of Copernicus. Osiander is said to have been shocked by the implications of Rheticus's *Narratio prima*, the first published account of the Copernican system. He never accepted the possibility that the proposed system could be anything more than another attempt to "save the phenomena."

On 20 April 1541, in letters to Rheticus and Copernicus, Osiander stated his opinion, while Rheticus was waiting in Frombork (Frauenburg) for Copernicus to put the final touches on the manuscript of *De revolutionibus*. Osiander urged the inclusion in the introduction of the statement that even if the Copernican system provided a basis for better and easier astronomical computations, it might still be false. Historians agree that Copernicus firmly rejected Osiander's recommendation. At the time of the printing of the book, Copernicus was taken seriously ill in Frombork, and Rheticus supervised the printing of the manuscript in the printing shop of Johannes Petreius in Nuremberg. In the final phase of the printing Rheticus was appointed professor of mathematics in the University of Leipzig and was obliged to go there; he left Osiander to see through the final stages of the publication. Osiander surreptitiously slipped in an unsigned preface, denying that the book intended to propose more than a mathematical hypothesis, and represented as impossible a task of discovering how the Universe is laid out.

Osiander's statement was a reassertion of the traditional position regarding the astronomical method known as "save the phenomena." Copernicus appears to have believed in the reality of his system, but because Osiander's preface was unsigned, it was widely believed to represent the views of Copernicus. For this reason it was thought by most in succeeding years that Copernicus himself had not really believed that the Earth could move.

When copies of *De revolutionibus* reached Rheticus in Leipzig, he became enraged and sent to the City Council of Nuremberg a sharp protest written by Tiedemann Giese, one of the closest friends of the deceased Copernicus. Petreius replied that he had received the preface in a form undifferentiated from the rest of the material. Whereas Osiander never publicly acknowledged his authorship of the interpolated preface, he did so privately, and finally in 1609 in *Astronomia nova* **Johannes Kepler** stressed that Osiander, not Copernicus, had penned this preface.

Dimitris Dialetis

Selected References

Copernicus, Nicholas (1978). *On the Revolutions*, edited by Jerzy Dobrzycki with translation and commentary by Edward Rosen. Baltimore: Johns Hopkins University Press.

Graubard, Mark (1964). "Andreas Osiander: Lover of Science or Appeaser of Its Enemies." *Science Education* 48: 168–187.

Möller, W. (1887). "Osiander: Andreas." *Allgemeine Deutsche Biographie*. Vol. 24, pp. 473–483. Leipzig: Duncker and Humblot.

Rosen, Edward (1974). "Osiander, Andreas." In *Dictionary of Scientific Biography*, edited by Charles Coulston Gillispie. Vol. 10, pp. 245–246. New York: Charles Scribner's Sons.

Wrightsman, Bruce (1975). "Andreas Osiander's Contribution to the Copernican Achievement." In *The Copernican Achievement*, edited by Robert S. Westman, pp. 213–243. Berkeley: University of California Press.

Outhier, Réginald [Réginaud]

Born **La Marre, (Jura), France, 16 August 1694**
Died **Bayeux, (Calvados), France, 12 April 1774**

Réginald Outhier was an observational astronomer who also undertook geodetic work in Lapland and France. He studied in Poligny, Dole, and Besançon, the new capital of Franche-Comté. Outhier became a priest of the diocese of Besançon, acting as vicar at Montain near Lons-le-Saunier. It is there that he first became interested in astronomy.

In 1726 Outhier invented a moving globe that showed the apparent movement of the Sun and the Moon, and the movement of the nodes of Moon's orbit. This globe, 5 in. of diameter, was executed by J. B. Catin; it is represented in the *Machines de l'Académie* and was praised by A. Thiout in his treatise on watches and clocks (1741). As a consequence of his work, Outhier was named on 1 December 1731 a correspondent of **Jacques Cassini** (then in 1756 of **César Cassini de Thury**) at the Academy of Sciences in Paris, where he was invited to present his globe.

Paul d'Albert de Luynes, Bishop of Bayeux since 1729, who occupied his leisure in astronomy, gnomonics, and meteorology, invited Outhier to come to Bayeux to act as his secretary. Upon his arrival, Outhier sketched in the bishop's library a large meridian with lines marking the time from 5 min before and after real noon.

In 1733, Outhier joined the team of Cassini II, which measured the perpendicular to the Paris meridian; he participated in the work from Caen to Saint-Malo and (the last measures) in Bayeux. Then Outhier drew a map of the diocese of Bayeux (published in 1736). On the basis of these works, the Academy of Sciences designated him, in 1735, a member of the Lapland expedition commanded by **Pierre de Maupertuis**. It also provided him with a pension of 1,200 livres.

Outhier left Bayeux in December 1735 and, after 5 months of preparation, the Lapland expedition members proceeded to Dunkerque, where the Swedish physicist, **Anders Celsius** (coming from London with some instruments) met them. They took a ship on 2 May 1736, arriving in Tornea, near the Arctic Circle, on 22 June. Outhier wrote the "Journal du Voyage" in which he described the measures of the arc of the meridian done from 1736 till 1737 and the measure of a baseline on the ice in winter. He also drew several maps of the areas covered. This work, published in 1744, described the charming simplicity of life in Sweden and the habits of the Laplanders. It was translated into English in 1777.

Back in Bayeux, Outhier continued his geodetic work for the *Carte de France* at the request of Cassini III. In 1735 and 1736, the Paris Observatory triangulated the coasts of Picardy and Brittany and, in 1737, César Cassini de Thury and **Giovani Maraldi** followed the coast southward, leaving Abbé Outhier in charge of finishing the remaining Norman coasts near Cherbourg.

From 1749 until 1755, Outhier drew up the diocese almanacs. In 1750, he reported a light earthquake felt from Avranche to Cherbourg. He published in 1755 a map of stars of the Pleiades, the 35 main stars of which had been observed by **Pierre-Charles Le Monnier** and the others by himself. At Bayeux, Outhier made meteorological and astronomical observations: for example, the transit of Venus of 6 June 1761 with a 36-in. focal length refractor with a micrometer, and the lunar eclipse of 8 May 1762. He also traveled to Sens and drew topographical maps of the Meaux and Sens dioceses, probably called there by Monsignore de Luynes.

Outhier was named in 1748 Canon of the Bayeux cathedral, a benefice he left in 1767 to concentrate upon his scientific work. He had participated in two of the most important scientific operations of 18th-century France, the *Carte de France* and the Lapland expedition. Outhier was a correspondent of the Paris Academy of Sciences; he was also member of the Berlin Academy, and in France of the Caen and Besançon academies.

Simone Dumont

Selected References

Condorcet, M. J. A. N. (1847). "Eloge du cardinal de Luynes." In *Oeuvres*, published by M. Condorcet, A. O'Connor, and F. Arago. Vol. 3. Paris.

Hoefer, Ferdinand (1858). *Nouvelle biographie générale*. Paris.

King, Henry C. (1978). *Geared to the Stars: The Evolution of Planetariums, Orreries, and Astronomical Clocks*. Toronto: University of Toronto Press, p. 282.

Lalande, Jérôme (1970). *Bibliographie astronomique avec l'histoire de l'astronomie depuis 1781 jusqu'ā 1802*. Paris, 1803. (Facsimile edition. Amsterdam: J. C. Gieben.)

Martin, Jean-Piere (1987). *La figure de la terre*. Cherbourg, France: Isoète.

Terrall, Mary (2002). The *Man Who Flattened the Earth: Maupertuis and the Sciences in the Enlightenment*. Chicago: University of Chicago Press.

Ovid

Born **Sulmona (Sulmona, Abruzzo, Italy), 20 March 43 BCE**
Died **Tomis (Constanza, Romania), 17**

In a number of his poems the Roman poet Publius Ovidius Naso describes celestial mythology, constellations, and calendrical lore related to the sky. The chief source for his life is one of his own poems, *Tristia* 4.10. Descended from an equestrian family, he was born in

Sulmo, a mountain town some 90 miles northeast of Rome. Ovid and his brother were sent to Rome at an early age for training in rhetoric and to prepare for careers in public life. After an extended tour of the Greek East, Ovid returned to Rome to hold a few minor public offices but chose rather to devote himself to poetry under the patronage of Marcus Valerius Messalla Corvinus (64 BCE –8). The poet soon became the darling of the brilliant social life of the capital.

In 8, Emperor Augustus, for reasons not fully known, exiled Ovid to Tomis, the modern Constanza, a barely Romanized outpost on the western coast of the Black Sea. Life there was a sad and harsh change from his former life at Rome. After years of unsuccessful appeals to Augustus and to his successor Tiberius, the poet died there. Ovid was married three times and had at least two children; his last wife remained behind in Rome during his exile. Ovid has greatly influenced subsequent western literature and art, particularly with his love poetry.

Although clearly informed by literary rather than scientific considerations, much astronomical material is found in three of Ovid's works. His *Aratea*, a Latin version of **Aratus**'s *Phaenomena*, is now lost except for two brief fragments. Two other major works incorporating celestial references do survive, however, and both seem to have been composed at about the same time, before and during the poet's exile: the *Fasti* and the *Metamorphoses*. Taken together, these two works complement one another in Ovid's incorporation of celestial storytelling.

The 15 books of the *Metamorphoses* retell in dactylic hexameters almost the whole range of classical mythology based on the idea of change. The tales are mostly Greek in origin, but the last three books are concerned with Roman myth and history right up to Ovid's own day. An Ovidian cosmogony begins the *Metamorphoses* (1.5–88). Reminiscent of the beginning of **Hesiod**'s *Theogony*, the account details the evolution of the Universe out of disorder (*chaos*) and the role of the four elements in the process. Ovid also ascribes to divine intervention the eventual creation of the world and the appearance of humans. In Book Two, the story of Phaethon, with his attempt to drive the chariot of the Sun, features engaging passages with astronomical themes, including a description of the palace of the Sun and the constellations that the reckless young man encounters in his uncontrolled ride through the heavens (2.1–327). Elsewhere, astral myth in the form of catasterism appears. The most prominent of such episodes is that of Callisto and Arcas (2.410–530), which is also retold at *Fasti* 2.153–192. The tale is exemplary for Ovid's literary methodology: As in a number of other tales of terrestrial beings transported to the heavens, the poet reworks Aratus's version. The *Metamorphoses* concludes with the poet's description of the celestial portents and the comet, the *Sidus Iulium*, which appeared upon the death of Julius Caesar (15. 745–870).

The six books of the *Fasti*, Ovid's unfinished religious calendar in elegiac couplets, cover the year from January to June and relate stories associated with Roman holidays. The reader of the work is presumed to be conversant with the heavens, and the poet intersperses among the work's longer narratives calendrical references to astronomical occurrences. The *Fasti* likewise incorporates astral myths of many constellations, often presenting differing versions for a given star group and offering variants on the accounts in Aratus. This borrowing and adaptation is primarily a literary process (*variatio*), and at least some of the technical errors may also be attributed to the literary rather than the scientific bases of the work. Nevertheless, Ovid's so called "Eulogy of the Astronomers" (1.295–310) introduces the concepts of Stoicism, linking his work with a similar philosophical approach taken by Aratus.

The number of familiar constellations and stars in the *Fasti* is substantial, with 27 of the former traceable to those described by Aratus. The accounts vary in length and complexity, ranging from the involved to the simple, and some star groups (Hyades, Pleiades, Orion) are mentioned several times. As in the *Metamorphoses* Ovid also incorporates political and cultural themes into some of the tales. The concept of ruler catasterism, as promoted by Hellenistic monarchs and eventually supportive of the Augustan imperial program, is clearly evident in places. Most particularly this is true of the star Capella (5.111–28), to be linked ultimately to the constellation Capricorn, Augustus's own birth sign and a public symbol for the return of the Golden Age.

John M. McMahon

Alternate name

Ovidius Naso, Publius

Selected References

Gee, Emma (2000). *Ovid, Aratus and Augustus: Astronomy in Ovid's Fasti.* Cambridge: Cambridge University Press. (Gee offers useful appendices on Ovid's incorporation of Aratean elements into the *Fasti* and on the inaccuracies associated with his astronomy. See pp. 193–208.)

Ideler, J. (1822–1823). "Über den astronomischen Theil der *Fasti* des Ovid." *Abhandlungen der Königlichen Akademie der Wissenschaften zu Berlin aus den Jahren 1822–23*: 137–169. (For the technical aspects of Ovid's astronomy this work remains essential.)

Newlands, Carole E. (1995). *Playing with Time: Ovid and the Fasti.* Ithaca, New York: Cornell University Press. (For Ovidian use of astronomical material in literary contexts.)

Ovid. *Die Fasten*, edited by Franz Bömer. 2 Vols. Heidelberg: C. Winter, 1957–1958 (German and Latin text); *Fasti*, translated by Sir James George Frazer. 2nd ed. Loeb Classical Library, no. 253. Cambridge, Massachusetts: Harvard University Press, 1989 (Latin and English text; Frazer's earlier 1929 edition is the definitive English commentary); *Ovid's Fasti*, translated by Betty Rose Nagle, with introduction and notes. Bloomington: Indiana University Press, 1995; *Fastorum libri sex*, edited by E. H. Alton, D. E. W. Wormell, and E. Courtney. 4th ed. Stuttgart: B. G. Teubner, 1997 (The standard Latin text); *Fasti*, translated by A. J. Boyle and R. D. Woodard, with introduction, notes, and glossary. London: Penguin, 2000.

——— *Metamorphoses*, translated by Frank Justus Miller. 2nd ed. Loeb Classical Library, nos. 42 and 43. Cambridge, Massachusetts: Harvard University Press, 1984 (Latin and English text); *The* Metamorphoses *of Ovid*. A verse translation by Allen Mandelbaum. New York: Harcourt Brace, 1993; *The* Metamorphoses *of Ovid*. A verse translation by David R. Slavitt. Baltimore: Johns Hopkins University Press, 1994; *The* Metamorphoses *of Ovid*, translated by Michael Simpson. Amherst: University of Massachusetts Press, 2001; Ovid, *Metamorphoses*, translated by Z. Philip Ambrose. Newbury port, Massachusetts: Focus Publishing, 2004; Ovid, *Metamorphoses*, translated by Davis A. Racburn. London: Penguin, 2004; Ovid, *Metamorphoses*, edited by R. J. Tarrant. New York: Oxford University Press, 2004. (This supersedes the longtime standard Latin text by W. S. Anderson.)

Ramsey, John T. and A. Lewis Licht (1997). *The Comet of 44 B.C. and Caesar's Funeral Games.* Atlanta: Scholars Press, esp. Chap. 4, pp. 61–94 and Chap. 7, pp. 135–153. (For the *Sidus Iulium*.)

Ovidius Naso, Publius

➲ **Ovid**

P

Page, Thornton L.

Born **New Haven, Connecticut, USA, 13 August 1913**
Died **Houston, Texas, USA, 2 January 1996**

Spectroscopist Thornton Page, son of prominent American physicist Leigh Page, was almost a martyr to astronomy; his near-fatal fall from the 82-in. telescope platform is recounted in **David Evans** and Derral Mulholland's history of the McDonald Observatory. Page's post-world-war work on binary galaxies foreshadowed what was to become the missing mass problem. Page is known in popular circles for having participated on the "Robertson panel," convened in January 1953 by the United States of America's Central Intelligence Agency as a "Scientific Advisory Panel on Unidentified Flying Objects."

Selected References

Evans, David S. and J. Derral Mulholland (1986). *Big and Bright: A History of the McDonald Observatory*. Austin: University of Texas Press.

Osterbrock, Donald E. (1996). "Thornton L. Page, 1913–1996." *Bulletin of the American Astronomical Society* 28: 1461–1462.

Palisa, Johann

Born **Troppau, (Opava, Cech Republic), 6 December 1848**
Died **Vienna, Austria, 2 May 1925**

Johann Palisa was an excellent astronomical observer, discovering many minor planets whose orbits he helped to determine. He came from a poor family; his father dealt in salt. Palisa married twice. Two of his daughters married astronomers, Erna to Friedrich Bidschof, astronomer in Vienna and Trieste, and Hedwig to Joseph Rheden, a keen observer and collaborator of Palisa's. Palisa's mathematical abilities appeared quite early; after finishing school in Troppau, he began to study mathematics and astronomy at the University of Vienna in 1866. The astronomers of the old Vienna Observatory (**Karl von Littrow**, **Edmund Weiss**, and **Theodor Ritter von Oppolzer**) enabled him to earn some money doing observational and computational work. In 1870, after completing his military service, Palisa received a position at the observatory, located in the center of town, measuring positions of stars and planets with a meridian circle and making drawings of sunspots.

In 1871, Palisa worked for several months at the Geneva Observatory. He applied for and, with the recommendation of Weiss, received the position of the director of the Austrian Naval Observatory at Pola (now Pula, Croatia, on the Adriatic Sea). Palisa was responsible for time service observations and the Austrian Navy ship chronometers.

Taking advantage of the better observing conditions at Pola, Palisa took up Oppolzer's suggestion to observe minor planets with the 6-in. refractor there. The discovery of new ones by visual means required precise star charts including faint objects. Palisa used the *Bonner Durchmusterung* charts, as well as those of **Christian A. Peters** and **Jean Chacornac**, but he soon started to draw his own charts of small areas along the ecliptic. Furthermore, orbital determinations required precise positions of reference stars. For determining positions, Palisa used the Pola's 6-in. meridian circle by Troughton & Simms. In March 1874, Palisa discovered minor planet (136) Austria and by 1880, a total of 28 minor planets. He emphasized the importance of reobserving asteroids in order to secure good orbit determinations. In 1876, Palisa received the Lalande Prize of the Paris Academy of Sciences for the rediscovery of (66) Maja, lost since its discovery in 1861.

Meanwhile, a new observatory in Vienna featured a 27-in. Grubb refractor, then the world's largest. Weiss offered Palisa a position, which he accepted, giving up his rank of navy captain to become simply a member of an institute. By the end of 1880, Palisa moved to Vienna, where he received a Ph.D. in 1884. In May 1881, he discovered minor planet (220) Stephania; while in Vienna, he found a total of 93 asteroids using visual techniques. Palisa focused on determining their orbital elements and preparing star charts, using Vienna's 12-in. Clarke refractor. For over 40 years, he worked every clear moonless night. Palisa made careful notes about stellar positions in his copy of the *Bonner Durchmusterung*. In 1902, he published these notes under the title "Stern-Lexikon."

Palisa put forth his efforts into determining positions of objects recently discovered by **Maximilian Wolf** in Heidelberg and by

August Charlois in Nice. The Paris Academy of Sciences honored his work again in 1906, awarding him the Valz Prize, with **Maurice Loewy** declaring that Palisa had accomplished more in this field than other astronomers put together. Nearly every object he discovered was numbered and named, and could be reobserved during its next opposition. Only one, (719) Albert, was lost, finally rediscovered by the Spacewatch Project at Kitt Peak in spring 2000.

Palisa's style of work changed after 1892. Along with Wolf, he began to work on the production of a new photographic atlas along the northern region of the ecliptic: the Wolf–Palisa charts. Survey plates taken in Heidelberg were superimposed with precise coordinate grids prepared in Vienna. Their handy size made these prints especially valuable for direct work at the telescope. From 1908 to 1914, 180 sheets covering 50 square degree areas along the ecliptic were printed in Vienna; about 125 copies were sold. The price of each was set to earn enough money for the production of the next. World War I stopped the work; after 1918, it was impossible to sell enough copies, ending the production of these charts.

Palisa received an invitation to participate in a solar-eclipse expedition in 1883 at the Caroline Islands in the Pacific Ocean. Others in the expedition included **Pierre Jules Janssen** and **Étienne Trouvelot** from Paris and **Pietro Tacchini** from Rome. Astronomers thought they might find a new planet located between the Sun and Mercury, though the search was unsuccessful. Palisa was a government guest on board a French navy ship but had to find funding to return to Vienna *via* the United States, where he visited various observatories.

With the assistance of the Austrian emperor and the Academy of Sciences, as well as government and some private funds, he started naming minor planets he discovered in commemoration of those who supported his research.

In Vienna, Palisa was well known as an astronomy popularizer. He fascinated audiences with clear explanations of difficult problems. In 1910, he even rented the Vienna Musikvereinssaal, the concert hall of the best Viennese orchestras, to explain Halley's comet (IP/Halley) to the public. Vienna's immense growth around the turn of the century placed the observatory near the center of town, with deteriorating observing conditions. Palisa campaigned for a good telescope outside of the city, a wish left unfulfilled because of economic conditions.

After Weiss's retirement in 1909, Palisa hoped to be the Vienna Observatory's new director, but Josef von Hepperger (1855–1928), professor of theoretical astronomy, was appointed instead. For a short time, Palisa protested by giving up observing, but he could not stand to live without a telescope. He was forced to retire at the age of 71, but received special permission to continue his work and to use the observatory's equipment.

Palisa received many honors: the *Ritterkreuz* des Franz Josephs-Ordens (Austrian distinction: 1874); Lalande Prize of the Academy of Sciences, Paris (1876); Valz Prize of the Academy of Sciences, Paris (1906); Member of the Astronomische Gesellschaft (1875); *Bürger von Wien* (honorary citizen of Vienna, 1921); and Minor Planet (914) Palisana. A lunar crater (Palisa) is named in his honor.

Anneliese Schnell

Selected References

Albrecht, Rudolf, Hans-Michael Maitzen, and Anneliese Schnell (2001). "Early Asteroid Research in Austria." *Planetary and Space Science* 49: 777–779.

Ashbrook, Joseph (1978) "An Elusive Asteroid: 719 Albert." *Sky & Telescope* 56, no. 2: 99–100.

Oppenheim, Samuel and Joseph Rheden (1925). "Johann Palisa." *Vierteljahresschrift der Astronomischen Gesellschaft* 60: 187–197.

Palisa, Johann (1874). "Entdeckung eines Planeten." *Astronomische Nachrichten* 83: 207–208.

——— (1874). "Jahresbericht der Sternwarte Pola." *Vierteljahresschrift der Astronomischen Gesellschaft*: 227–240.

——— (1884). "Bericht über die während der totalen Sonnenfinsternis vom 6. Mai 1883 erhaltenen Beobachtungen." *Sitzungsberichte de Akademie der Wissenschaften in Wien* 88: 1018–1031.

——— (1902). "Stern-Lexikon." *Annalen der Universitäts-Sternwarte in Wien* 17.

Palisa, Johann and Friedrich Bidschof (1899). "Katalog von 1238 Sternen." *Denkschriften der Akademie der Wissenschaften in Wien* 67.

Palisa, Johann and Johann Holetschek (1908). "Katalog von 3464 Sternen." *Annalen der Universitäts-Sternwarte in Wien* 19.

Schmadel, Lutz D. (1991). *Dictionary of Minor Planet Names*. 4th ed. Berlin: Springer-Verlag.

Palitzsch, Johann

Born **Prohlis near Dresden, (Germany), 11 June 1723**
Died **Leubnitz near Dresden, (Germany), 21 February 1788**

Johann Palitzsch was the first to view Halley's comet (IP/Halley) on its 1758 return. Coming from a farm family, he studied astronomy at home and attended the Mathematischer Salon in Dresden. Thanks to his recovery of Halley's comet, Palitzsch became, in 1759, instructor in astronomy to the young Elector Friedrich August III. He is reputed to have recognized Algol's variability.

Palitzsch was instrumental in introducing both the potato and the lightning rod to Saxony. A small museum honoring him is being planned on the outskirts of Dresden.

Richard A. Jarrell

Selected Reference

Becker, G. A. (1985). "The Christmas Comet of Johann Palitzsch." *Griffith Observerer*. 49(12): 2.

Palmer, Margaretta

Born **Branford, Connecticut, USA, 1862**
Died **New Haven, Connecticut, USA, 30 January 1924**

Margaretta Palmer obtained the first doctorate in astronomy awarded to an American woman by an American institution. (Winifred Edgerton's 1886 doctorate had been awarded by the department of mathematics at Columbia University, whereas **Dorothea Klumpke Roberts** earned the equivalent degree, a *Docteur ès Science*, from the University of Paris in 1893.) Palmer was descended from a Branford, Connecticut, family that settled there in colonial days. She attended Vassar College, New York, where she was a student of America's first woman astronomer, **Maria Mitchell**. Palmer graduated in 1887 and for the next 2 years served as Mitchell's assistant. When Mitchell retired in 1888, she pleaded with the college president to find a place for Palmer, whom she described as "remarkably faithful and conscientious." Because the college did not have funding for Palmer's salary, it has been assumed that her employment under Mitchell was sustained through private funding provided by Mitchell herself.

In 1889, Palmer began work at Yale University. She was to carry out the reductions of observations of the minor planet (7) Iris that had been acquired the preceding year. In 1892, Yale opened its graduate school to women students. Palmer took advantage of this opportunity and acquired her Ph.D. in 1894. Her dissertation dealt with an improved orbit for comet C/1847 T1, which had been discovered by Mitchell and had catapulted her to lasting fame.

In 1894, Palmer undertook an enormous computational task, unfortunately left unfinished, related to the reduction of observations of the satellites of Jupiter. She set up 1,128 observational equations of condition with 13 unknowns. Illness interrupted completion of this project. If only she could have had the benefit of modern computing facilities! She also worked on the compilation of atmospheric refraction tables for correcting the apparent positions of celestial objects to their true coordinates.

The "Yale Index to Star Catalogues" was begun under **William Elkin**'s directorship of the Yale Observatory in 1897, with the assistance of J. S. Newton, daughter of the previous director. Soon, Palmer was put in charge of this monumental task. Such an undertaking had been suggested as early as 1878 by **Arthur von Auwers** of Germany, but was not begun, as the *Geschichte des Fixsternhimmels* [GFH], until 1897. The first of 24 volumes for Northern Hemisphere stars were not published until 1922, the last in 1936, and the 24 volumes for Southern Hemisphere stars appeared between 1937 and 1952. The "Yale Index" was never published, but Palmer made it known that anyone needing astrometric data for any particular stars would be furnished references to, or the actual observations from, such sources – a service indispensable for astronomers determining the proper motions of stars.

Elkin retired in 1910, and only acting directors supervised the work started under his regime. In 1918, **Ernest Brown**, a mathematics professor renowned for his analyses of the motions of the Moon, was instructed to close the Observatory until the end of World War I. Palmer was the only employee left, and her services were considered so valuable that she was given an appointment in the college's library, working half time classifying scientific and mathematical books, and allowed to continue her astronomical research the rest of the time.

When **Frank Schlesinger** was appointed Observatory director in 1920, he found Palmer to be an ideal assistant, familiar with previous projects and especially qualified to work with him on the first general catalog of trigonometric stellar parallaxes. This too was a tremendous undertaking, but one for which her previous work on the "Yale Index to Star Catalogues" proved invaluable. Sadly, Palmer did not live to see the completion of the catalog of parallaxes. She died as a result of an automobile accident.

Palmer had been a member of the American Astronomical Society from 1915 until her death. During that interval, she gave two oral reports at meetings of the society, one in 1918 on the "Yale Index" and another in 1921 on the orbit of comet 35P/1788 Y1 (Herschel–Rigollet).

Dorrit Hoffleit

Selected References

Albers, Henry (2001). *Maria Mitchell: A Life in Journals and Letters*. Clinton Corners, New York: College Avenue Press, esp. pp. 302–304, 354.

Hoffleit, Dorrit (1992). "The Quiescent Years and a Lady Astronomer." In *Astronomy at Yale, 1701–1968*, pp. 92–96. New Haven: Connecticut Academy of Arts and Sciences.

Palmer, Margaretta (1894). "Determination of the Orbit of the Comet 1847 VI." *Transactions of the Astronomical Observatory of Yale University* 1: 183–207.

——— (1923). "The GFH." *Popular Astronomy* 31: 78–82.

Pannekoek, Antonie

Born **Vaasen, the Netherlands, 2 January 1873**
Died **Wageningen, the Netherlands, 28 April 1960**

Antonie Pannekoek made valuable contributions to our early understanding of the spectra of stellar atmospheres, but is best remembered for his majesterial history of astronomy. The son of Johannes and Wilhelmina Dorothea (*née* Beins) Pannekoek, he developed his interest in astronomy at an early age. As a young amateur Pannekoek was a careful and skilled observer. In 1891 he detected independently the brightness variations of Polaris (α Ursa Minoris), and was the first to determine its approximate periodicity. Concerned about the value of his data – the amplitude of variation is only 0.11 m – Pannekoek waited until 1913, when photographic and photometric data had become available, to propose the Cepheid nature of the star. Prior to his entrance into the university, Pannekoek also developed a light curve for β Lyrae and began a lifetime series of visual and photographic observations of the Milky Way. Pannekoek studied at Leiden University, where he earned a Ph.D. defending a thesis on the eclipsing system Algol (β Persei) in 1902 and acted as a staff member between 1899 and 1906. He married Johanna Maria Nassau Noordewier in 1903.

Pannekoek was launched on a successful career in astronomy when, in 1906, he had observed, independently from **Ejnar Hertzsprung**, that apparently bright red stars included some that must be substantially more luminous than others. His discovery of what later came to be known as red giants and red dwarfs was based on the significant differences between the parallaxes and proper motions of the groups of stars. Paradoxically, however, Pannekoek left astronomy to pursue his second passion, that of social justice and reform. As a political activist, one properly described as an antiauthoritarian Marxist, he spent the years from 1906 to 1914 as a teacher in Berlin, where he became one of the main leaders of the antirevisionist movement. Pannekoek contributed extensively to leftist publications in Europe and the United States. In 1914 he was deported from Germany. He returned to the Netherlands as a high school teacher while continuing his political writing and activism.

Pannekoek resumed his astronomical work in 1915, and was one of the first (in 1920) to recognize the relevance of **Meghnad Saha**'s law for deriving stellar diagnostics from spectroscopic observations. With that flash of insight, Pannekoek dropped the work he had to take up the theoretical study of stellar atmospheres and made important contributions to the field. **Marcel Minnaert** credited him with initiating the study of astrophysics in the Netherlands after Pannekoek founded, in 1921, what is now known as the Pannekoek Institute of Astronomy at the University of Amsterdam. He understood at an early date that certain line ratios in stellar spectra, which are correlated with stellar luminosity, actually probe the gravity in the stellar atmosphere. One of the striking results of his work on stellar atmospheres was his 1946 suggestion that in some pulsating stars the sound wave that moves outward converts gradually to a shock wave and ejects some material at the time of maxium radius on each pulsation cycle, a remarkably prescient suggestion that was later confirmed.

Throughout his career Pannekoek excelled in the careful measurement of astronomical information from photographic plates. The galactic studies he carried out during his early years in Amsterdam are characterized by a thorough understanding of what the eye can see on such plates. Pannekoek's knowledge of astronomical photography enabled him to pioneer, with **John Plaskett**, the quantitative analysis of photographic stellar spectra, on which the research during his later career strongly relied. Using spectra of Deneb (α Cygni) he had taken at the Dominion Astrophysical Observatory (near Victoria, British Columbia, Canada) in 1929, and a technique he invented (using log tables and mechanical calculators) for numerically integrating the equation of radiative transfer that arises from **Arthur Eddington**'s approximation, Pannekoek was the first astrophysicist to develop an empirical stellar (nonsolar) curve of growth similar to those developed for solar spectra by Harald von Klüber (1901–1978) and Minnaert. His 1930 chapter on ionization in stellar atmospheres in the *Handbuch der Astrophysik* was an important early fundamental contribution to the field. By 1935, Pannekoek had done extensive theoretical modeling of the line widths that could be expected in stellar spectra taking into account every known factor. The line widths were systematically in error, especially in cooler stars, but he predicted positions of maximum line intensity correctly. The missing factor in Pannekoek's equations was soon recognized when **Rupert Wildt** demonstrated the previously unappreciated role of negative hydrogen ions in the production of continuous absorption in cool stars.

In 1938, Pannekock made the first theoretical calculation of the widths of the Balmer series of hydrogen spectral lines applying the Stark effect to demonstrate the broadening of the wings of those lines as a function of stellar gravity. In other work, Pannekoek contributed to the understanding of metal concentrations on stellar opacity, showing that at high effective temperatures, hydrogen concentration was the dominant influence on stellar opacity and color, while at lower temperatures the effect of metallicity dominated. His analysis showed where the two effects were nearly balanced, proposing an effective temperature of about 10,500 K for stellar class A0, very nearly the value that has since been accepted.

Pannekoek's studies of the Milky Way galaxy continued for 60 years, and involved visual drawings and photographic surveys of both the Northern and Southern Hemisphere skies. He added isophote lines to show contours of brightness that have formed the basis for many popular star atlases. Pannekoek also studied the distribution of giant stars in the Milky Way using statistical approaches to the spectral parallaxes of A and K stars and spectral parallaxes of individual B stars. In this work he identified the clusters of the early stars later called associations. Pannekoek's work on the distribution of stars in the Milky Way Galaxy had a profound effect on thinking about galactic structure; his approach was later successfully applied by **Viktor Ambartsumian**, **William Morgan**, **Albert Whitford**, and Arthur Dodd Code (born: 1923).

In 1942 Pannekoek's career as professor in Amsterdam ended when he was dismissed by the German occupation forces. Until his death he remained active as a leading socialist theoretician and pursued individual astronomical studies. During and after World War II, Pannekoek's personality as both a leading scientist and a warm humanist found a remarkable expression in the historical work. His *A History of Astronomy* appeared first in 1951 in Dutch under the title *De groei van ons wereldbeeld*, which may be better translated as "The emergence of our view of the world." His historiography reflects both of his lifetime preoccupations as he examined the history of astronomy from Babylonian to modern times in relation to the social and political systems extant in each relevant period. This remarkable book then not only offers an accurate and well-documented introduction to the history of the discipline, but also helps the modern reader to see the cosmos with the eyes of ancient people who shared with Antonie Pannekoek the delight of the beauty of the heavens. He gave full credit to Hertzsprung for distinguishing giant and dwarf stars despite his own contribution to the topic. The accuracy and insight Pannekoek offers on matters astronomical outweighs his occasional forays into political polemicism, and the book remains one of the standards in the field.

Harvard University conferred an honorary Ph.D. upon Pannekoek in 1936, while in 1951 the Royal Astronomical Society honored him for his astrophysical contributions by awarding him its Gold Medal.

Christoffel Waelkens

Selected References

Aller, Lawrence H. (1951). "Interpretation of Normal Stellar Spectra." In *Astrophysics: A Topical Symposium*, edited by J. Allen Hynek, p. 50. New York: McGraw-Hill.

Bok, Bart J. (1960). "Two Famous Dutch Astronomers." *Sky & Telescope* 20, no. 2: 74–76.

Hearnshaw, J. B. (1996). *The Measurement of Starlight: Two Centuries of Astronomical Photometry*. Cambridge: Cambridge University Press, pp. 239–243, 271, 402–403.

Pannekoek, Antonie (1930). "Die Ionisaton in den Atmosphären der Himmelskörper." In *Handbuch der Astrophysik*, edited by G. Eberhard, A. Kohlschütter, and H. Ludendorff, pp. 256–350. Berlin: Verlag von Julius Springer.

——— (1969). *A History of Astronomy*. New York: Barnes and Noble.

Struve, Otto and Velta Zebergs (1962). *Astronomy of the 20th Century*. New York: Macmillan, pp. 95, 496–498.

Papadopoulos, Christos

Born **Prussa, (Turkey), 20 February 1910**
Died **Westcliff near Johannesburg, South Africa, 8 May 1992**

Christos Papadopoulos created what is likely to be the last great photographic atlas of the whole sky taken with one camera in one consistent sequence of well-controlled photographic charts corrected to visual magnitudes. Educated as a civil engineer in Greece, Papadopoulos fought with Greek resistance forces during World War II and then immigrated to South Africa after the war. Successful as a manager of engineering construction, he retired at the peak of his capacities to devote his time to two avocational interests: cinematography and astronomy.

Papadopoulos's greatest contribution to astronomy came in his combination of these two avocational interests in the creation of a photographic all-sky star atlas. The goal of this project was to completely photograph the sky with film and filter combinations that produced, as nearly as possible, a star atlas in which the photographic representation of the sky matched, to the greatest extent possible, the telescopic visual appearance of the sky at visual magnitudes as faint as 13.5. Similar undertakings by private individuals had previously been attempted only by **John Franklin-Adams** early in the 20th century, and again by Hans Verhenberg (1910–1991) in the middle of the same century. Neither effort was intended, however, to match the photographic representation to the visual appearance of the sky in a photometric sense.

Papadopoulos completed his Herculean task with the publication, in 1979/1980, of *The True Visual Magnitude Photographic Star Atlas*. One thousand photographic exposures of the southern, equatorial, and northern regions were taken from Papadopoulos's private observatory at his home in Westcliff, Johannesburg, and at the Stamford Observatory, Stamford, Connecticut, USA (by Charles E. Scovil). The plates were carefully matched on depth of exposure and sky brightness to produce the final atlas.

Papadopoulos was honored for this achievement by the presentation of the Astronomical Society of South Africa's Gill Medal in 1981.

Thomas R. Williams

Selected References

Gray, M. A. (1982). "The Presentation of the 1981 Gill Medal." *Monthly Notes of the Astronomical Society of Southern Africa* 41, nos. 11 and 12: 94–96.

Overbeek, M. D. (1992). "Christos Papadopoulos 1910–1992." *Monthly Notes of the Astronomical Society of Southern Africa* 51, nos. 9 and 10: 81–83.

Pappus of Alexandria

Flourished **probably 4th century**

Primarily a mathematician, Pappus also wrote commentaries on **Ptolemy**, **Aristarchus**, and **Theodosius**, and summaries of other astronomers such as **Apollonius** and **Eratosthenes**.

Pappus's dates from the literary sources are uncertain, though he was apparently computing an eclipse in 320. His most recent cited source is Ptolemy; a scholiast puts Pappus in the reign of Diocletian (284–305); and the *Suda* has him a contemporary of **Theon of Alexandria**, and flourishing in the reign of Theodosius (379–395).

His surviving work provided and continues to provide points of departure for developments in the history of mathematics. He, with Diophantus, provided François Vieta with his point of departure in the development of algebra; one problem in Collection 7 provided **René Descartes** with his point of departure for his *Geometrie*; and Pappus's work continues to appear as starting points for articles in mathematical journals.

Credited with being halfway between compiler–commentator and originator, Pappus was fully capable of extending the work of his predecessors. His greatest – or at least most famous – individual achievement is the universalizing of Euclid 1.47, the Pythagorean theorem. Pappus further extended work of **Archimedes** and (apparently) originated the theorems that were rediscovered and named after Paul Guldin.

As reporter of the *status quo* of the mathematics of his time, it is owing to Pappus that we have the Greeks' threefold division of problems: (1) solvable with straightedge and compass, called plane; (2) solvable with one or more conics, called solid; and (3) solvable with a more complicated curve, called linear. Modern mathematicians still find this classification appropriate, though they would rename (3) and further split it into algebraic problems and transcendental problems.

Pappus's principal surviving work is the *Collection*, a treasury of all the mathematics of his time, including some of his own, in eight books. We have fragments from his other works: a commentary on Ptolemy's *Almagest*, a commentary on Euclid 10, and a geography.

The *Collection* contains such mathematical topics as arithmetic, geometric, and harmonic means, the universalizing of Euclid, the globularizing of the spiral of Archimedes, and epitomes of Euclid, Apollonius, Aristaeus, and Eratosthenes; there is also a summary of mechanics. The astronomy is in book 6, which gives commentary on Theodosius's *Spherics*, **Autolycus**'s *On the Moving Sphere*, Aristarchus's *On the Size and Distance of the Sun and the Moon*, plus Euclid's *Optics and Phenomena*.

Thomas Nelson Winter

Selected References

Baragar, Arthur (2002). "Constructions Using Compass and Twice-Notched Straightedge." *American Mathematical Monthly* 109: 151–164.

Cuomo, Serafina (2000). *Pappus of Alexandria and the Mathematics of Late Antiquity*. Cambridge: Cambridge University Press.

Heath, Sir Thomas L. (1921). *A History of Greek Mathematics*. Vol. 2. Oxford: Clarendon Press. (Reprint, New York: Dover, 1981.)

Hewsen, R. H. (1971). "The *Geography* of Pappus of Alexandria: A Translation of the Armenian Fragments." *Isis* 62: 186–207.

Junge, G. (1936). "Das Fragment der lateinischen Uebersetzung des Pappus-Kommentars z. 10. Buche Euclids." *Quelle u Studien zur Geschichte der Mathematik, Astronomie, und Physik*, ser. b, 3: 1–17.

Klein, Jakob, E. Brann (trans.) (1968). *Greek Mathematical Thought and the Origin of Algebra*. Cambridge, Massachusetts: MIT Press.

Netz, Reviel (1999). *The Shaping of Deduction in Greek Mathematics*. Cambridge: Cambridge University Press.

Pappus (1876–1878). *Pappi Alexandrini Collectionis quae supersunt*, edited by F. Hultsch. 3 Vols. Berlin: Weidmann (Reprint, Amsterdam: A. M. Hakkert, 1965. [Still the standard text of Pappus. Superseded in part by Hewsen, Jones, Junge and Thomson, and Rome.])

——— (1930). *The Commentary of Pappus on Book X of Euclid's* Elements, edited and translated by G. Junge and W. Thomson. Cambridge, Massachusetts: Harvard University Press.

——— (1931–1943). *Commentaires de Pappus et de Theon d'Alexandrie sur l'* Almagest, edited by A. Rome. 3 Vols. Vatican.

——— (1986). *Book 7 of the Collection*, edited with translation and commentary by Alexander Jones. 2 Vols. New York: Springer-Verlag.

Parameśvara of Vāṭaśśeri [Parmeśvara I]

Born **Ālattūr, (Kerala, India), *circa* 1360**
Died ***circa* 1455**

Parameśvara, one of the foremost astronomers of Kerala, hailed from the village of Ālattūr (Aśvatthagrāma in Sanskrit), and his house, Vāṭaśśeri, was situated on the confluence of the river Nīla with the Arabian Sea. He was a *Ṛgvedin*, of the *Aśvalāyana Sūtra*, and belonged to the *Bhṛgugotra*. He was a pupil of Rudra I. He carried out astronomical observations near his house for some 45 years. He also observed a large number of eclipses between 1393 and 1432, which are recorded in his work *Siddhāntadīpikā*. Nothing else is known about the life of Parameśvara.

Parameśvara was a prolific writer and authored some 30 works. These include original treatises and commentaries on other works of astronomy and astrology. Among his original works on astronomy might be mentioned the following: *Dṛggaṇita* (1430); a work on spherics, *Goladīpikā* (1443); and three works on the computation and rationale of eclipses, *Grahaṇāṣṭaka*, *Grahaṇamaṇḍana*, and *Grahaṇanyāyadīpikā*. He also commented on a large number of astronomical works including the *Āryabhaṭīya*, *Sūryasiddhānta*, *Laghumānasa*, and *Līlāvatī*. Many of his works are yet to be published.

Narahari Achar

Selected References

Parameśvara (1916). *Goladīpikā*, edited by Ganapati Sāstrī. Trivandrum Sanskrit Series, no. 49. Trivandrum.

——— (1957). *Goladīpikā*, edited by K. V. Sarma. Adyar Library Pamphlet Series, no. 32, Adyar, India: Adyar Library and Research Centre.

——— (1963). *Dṛggaṇita*, edited by K. V. Sarma. Hoshiarpur: Vishveshvarananda Institute.

Pingree, David *Census of the Exact Sciences in Sanskrit*. Series A. Vol. 4 (1981): 187b–192b; Vol. 5 (1994): 211b–212a. Philadelphia: American Philosophical Society.

Sarma, K. V. (1972). *A History of the Kerala School of Hindu Astronomy*. Hoshiarpur: Vishveshvarananda Institute.

Parenago, Pavel Petrovich

Born **Ekaterinodar (Krasnodar), Russia, 20 March 1906**
Died **Moscow, (Russia), 5 January 1960**

Parenago was a doyen of Russian scholars of the Galaxy and the founding father (1938) and head (until his death) of the department of stellar astronomy at Moscow State University [MSU]. Among his many accomplishments, he played a significant role in fostering astronomical observations by amateurs. His name appears in the Kukarkin–Parenago relation between the periods and amplitudes of many different kinds of variable stars.

Parenago graduated from the faculty of physics and mathematics at MSU in 1929, concurrently working at the Astronomical–Geodetic Institute, which was soon incorporated into the present-day Sternberg State Astronomical Institute of MSU (1931). In 1935, he was awarded a doctorate in the physical and mathematical sciences (without the defense of a dissertation).

Parenago's main research field was stellar astronomy, especially studies of the structure, kinematics, and dynamics of our Galaxy. He investigated the spatial distributions and kinematics of many different classes of stars, especially variable stars. These results added important parameters to the concept of stellar populations and stellar subsystems. Parenago also made an important contribution to stellar dynamics, by creating a theory of the gravitational potential of the Galaxy. His works on the stellar luminosity function, the color–luminosity diagrams for different classes of stars, and interstellar extinction in the Galaxy were widely recognized for many years. Parenago's investigations were based on statistical treatments of the large databases that he had assembled from every accessible source.

Under the auspices of the International Astronomical Union [IAU], Parenago and Boris V. Kukarkin compiled the *Obshchii katalog peremennykh zvezd* (General catalog of variable stars, 1948), whose later editions are still being edited at Moscow. This catalog remains the most extensive of its kind to this day. Another of Parenago's catalogs, of the stars in the Orion Nebula (1954), achieved lasting usage, being an important collection of data on this region of star formation. Parenago authored one of the early textbooks on stellar astronomy (1938), which was repeatedly revised.

In 1948, Parenago was awarded the first Bredikhin Prize, the top astronomical award presented by the Soviet Academy of Sciences. The Order of Lenin was bestowed on Parenago in 1951, and he was elected a corresponding member of the USSR Academy of Sciences in 1953.

Yuri N. Efremov

Selected References

Anon. (1960). "Pavel Petrovich Parenago." *Soviet Astronomy – AJ* 4: 183–184. (This sketch first appeared in Russian in *Astronomicheskii zhurnal* 37 (1960): 191–192.)

Kukarkin, B. V. and P. P. Parenago (1948). *Obshchii katalog peremennykh zvezd* (General catalog of variable stars). Moscow: Izd-vo Akademii nauk SSSR.

Kulikovsky, P. G. (1974). "Parenago, Pavel Petrovich." In *Dictionary of Scientific Biography*, edited by Charles Coulston Gillispie. Vol. 10, pp. 317–319. New York: Charles Scribner's Sons.

Parenago, P. P. (1954). "Issledovaniia zvezd v oblasti tumannosti Oriona" (Investigations of stars in the area of the Orion Nebula). *Trudy Gosudarstvennogo astronomicheskogo instituta im. P. K. Shhternberga* 25: 1–547.

——— (1954). *Kurs zvezdnoi astronomii* (Textbook on stellar astronomy). 3rd ed. Moscow: Gostechizdat.

Sharov, A. S. (1983). "P. P. Parenago and his Life in Science" (in Russian). *Istoriko-astronomicheskie issledovaniia* (Research in the history of astronomy) 16: 219–232.

Vorontsov-Veliaminov, B. A. *et al.* (1961). "Pavel Petrovich Parenago (1906–1960)" (in Russian). *Istoriko-astronomicheskie issledovaniia* (Research in the history of astronomy) 7: 335–394.

Parkhurst, Henry M.

Born **New Hampshire, USA, 6 March 1825**
Died **New York, New York, USA, 21 January 1908**

American amateur Henry Parkhurst observed variable stars and developed an accurate theoretical–observational relationship between orbital parameters and brightnesses for asteroids. He communicated with **Edward Pickering** on matters photometric, having initiated such measurements of long-period variable stars in 1883 and publishing a volume on a decade's worth of observations in 1893. Afterward, he published in the *Astronomical Journal*.

Selected Reference

Rothenberg, Marc and Thomas R. Williams (1999). "Amateurs and the Society during the Formative Years." In *The American Astronomical Society's First Century*, edited by David H. DeVorkin, pp. 40–52. Washington, DC: Published for the American Astronomical Society through the American Institute of Physics.

Parmenides of Elea

Born ***circa* 515 BCE**
Died **after 450 BCE**

Few details of cosmologist Parmenides's life are known. His dates are deduced from a story related by **Plato** that Parmenides and Zeno visited Socrates in Athens around 450 BCE, at which time Parmenides was reputedly aged about 65. An alternative chronology, ultimately deriving from **Apollodorus**, places Parmenides's birth around 540 BCE. However, this tradition is wrong and probably results from confusing Parmenides's birth with the founding of Elea.

Parmenides was the son of Pyres and appears to have originally come from Eastern Greece. He settled in Elea on the west coast of Lucania in Italy, where, according to ancient tradition, he was active in politics and renowned as a wise law-maker.

Parmenides was influenced by both **Xenophanes** and the Pythagorean Ameinias Diochaites, in whose name he erected a shrine. Indeed, in Antiquity Parmenides was considered to have been closely associated with the Pythagoreans, and in particular the Pythagorean school at Croton, southeast of Elea. Both his philosophical and astronomical ideas were similar to those of the Pythagoreans, so this connection is probably real. Parmenides founded the "Eleatic" school of philosophy, where his successors were Zeno and Melissus.

Parmenides made a greater contribution to philosophy than to astronomy. He denied that change was possible by maintaining that "what is" could not disappear, and similarly that that which does not exist could not come into existence. He denied the possibility of movement on similar grounds. He explained the apparent ubiquity of change and movement by suggesting that the senses are unreliable and do not reveal the true nature of the world. These speculations, and in particular their refutation, had profound implications on the development of Greek thought.

Parmenides's astronomical ideas are not easy to reconstruct. His surviving writings are few and fragmentary, and later sources are confused and contradictory in the views that they attribute to him. He appears to have considered the Universe to consist of a system of concentric rings or bands of fire, alternating with bands of darkness. These details come mostly from a later source, which is so heavily abridged as to be nearly incomprehensible. However, the bands are mentioned in surviving fragments of Parmenides's own writing. In the middle of the bands, guiding everything, was a divinity, personified as Justice or Necessity.

Parmenides's extant writings include the assertion that the Universe is both finite and spherical. He was not necessarily the first to hold this view, which seems to have been common among early Pythagoreans. However, his work is the first espousal of it for which the attribution is completely certain. Parmenides was speaking of the Universe as a whole, rather than just the Earth or the heavens. However, later sources misconstrue him to have said that the Earth is spherical, which his extant writings, at least, do not.

Parmenides described the Moon as an "alien light," wandering around the Earth and "always looking towards the rays of the Sun." This statement is usually taken to imply that he realized that the Moon shines by reflected sunlight. This interpretation seems reasonable, but is not undisputed. If it is correct then Parmenides is the first Greek known to hold this view. (Plato recorded that **Anaxagoras** held this opinion, but did not say that he was the first to do so.)

The extant fragments of Parmenides also mention the aether, the constellations (the signs in the aether), the Sun, the furthermost heaven, and the Milky Way, one of the first extant mentions of it in Greek.

The confused later sources attribute various ideas to Parmenides, in addition to the sphericity of the Earth, including among others, dividing the spherical Earth into five climatic zones, recognizing the Morning and Evening star as the same object, and considering the Sun, Moon, and stars as being made of compressed fire. However, these are all ideas that later sources often indiscriminately attributed to various pre-Socratics.

Unlike other early Greek philosophers, Parmenides wrote in hexameter verse rather than prose. About 150 lines of his poem *On Nature* survive, mostly through **Simplicius**, who quotes it in his commentaries on **Aristotle**. A further six lines are extant only in a Latin translation. Parmenides's ideas are described, with greater or lesser fidelity, by Plato, Aristotle, **Theophrastus,** and numerous later (and less reliable) sources.

A. Clive Davenhall

Selected References

Dicks, D. R. (1970). *Early Greek Astronomy to Aristotle*. London: Thames and Hudson. (For a discussion of Parmenides's astronomical ideas.)

Diels, Hermann and Walther Kranz (1952). *Die Fragmente der Vorsokratiker*. Vol. 1, pp. 217–246. Berlin: Weidmann. (For the surviving sections of *On Nature*, and most of the secondary sources.)

Heath, Sir Thomas L. (1913). "Parmenides." In *Aristarchus of* Samos, *the Ancient Copernicus*. Oxford: Clarendon Press, pp. 62–77. (For a discussion of Parmenides' astronomical ideas; a facsimile edition was published in 1997 by Sandpiper Books.)

Tarán, Leonardo (1965). *Parmenides: A Text with Translation, Commentary, and Critical Essays*. Princeton: Princeton University Press. (For a modern translation and discussion.)

Parsons, Laurence

Born **Birr Castle, King's county (Co. Offaly), Ireland, 17 November 1840**

Died **Birr Castle, King's county (Co. Offaly), Ireland, 29 August, 1908**

Laurence Parsons, the eldest son of **William Parsons**, the Third Earl of Rosse, shared his father's enthusiasm for astronomy, continuing the study of nebulae and star clusters at Birr Castle, undertaking pioneering work on the infrared emission of the Moon and becoming the first to obtain what is now recognized as an excellent estimate of its surface temperature.

Parsons was educated at home by tutors, including the Reverend T. T. Gray and John Purser, later professor of mathematics in Belfast. The Third Earl and the Countess of Rosse took a keen interest in their children's education. The educational regimen included open-air activities on the estate and practical work in their father's well-equipped workshops. Known in his youth by his courtesy title of Lord Oxmantown, Parsons entered Trinity College, Dublin, as a nonresident student and excelled in mathematics and physics. He graduated in 1864 and immediately started to observe and sketch nebulae with the 3-ft. and 6-ft. reflectors. As a young man, Parsons had many opportunities, both in Birr and in London, to meet distinguished scientists who were friends of his father; this undoubtedly strengthened his ambition to be an astronomer.

In 1865, **Robert Ball** was appointed assistant astronomer and tutor to Laurence's younger brothers. Ball and Parsons differed by only a few months in age; they spent many nights together observing with the Birr telescopes. Parsons's first scientific paper in the *Monthly Notices of the Royal Astronomical Society* in 1866 described a water clock that he invented to drive an 18-in. equatorial telescope. His next paper, published by the Royal Society, collated all the observations of the Orion Nebula that had been made at Birr since 1849; it included an engraving of the nebula that **John Dreyer** judged as being "always of value as a faithful representation of the appearance of the Orion nebula in the largest telescope of the nineteenth century."

The Third Earl died in October 1867, and Parsons succeeded to the title and the estates. The same year, he was elected a fellow of the Royal Society and of the Royal Astronomical Society; he was also appointed High Sheriff of the county. In 1868, Parsons became a representative peer for Ireland in the House of Lords in Westminster. In 1870, he married Frances Cassandra Harvey-Hawke, only child of the fourth Lord Hawke and his second wife Frances (*née* Fetherstonhaugh).

In 1868, the Fourth Earl commenced the work for which he is best remembered, the study of radiant heat from the Moon. He was the first to make infrared measurements of any astronomical body other than the Sun. Parsons started by using a single thermocouple connected to a galvanometer but discovered immediately that the signals were affected by changes in the temperature of the 3-ft. telescope and by the ambient air temperature. He had the idea of using two identical thermocouples, connected in opposition and placed side by side in the focal plane of the telescope. One thermocouple was exposed to the Moon and the other to the sky.

At that time, no filters were available to transmit infrared radiation so Parsons used a plate of glass to block the infrared and differenced the measurements with and without the plate. He calibrated his observations by comparing his lunar measurements with those from blackened cans containing water at various temperatures. Parsons studied how the lunar radiant heat varied with the phase of the Moon and how the atmospheric attenuation varied with the distance of the Moon from the zenith. The results were summed up in his Bakerian Lecture, which he delivered to the Royal Society in March 1873. His final estimate of the lunar surface temperature was 197° F. A reanalysis of the Earl's data by William Merz Sinton (1925) in 1958 gave a value of 158° F, in good agreement with modern estimates. It was always a source of regret to Parsons that his contemporaries did not fully appreciate this achievement.

Parsons was a prolific inventor and never happier than when busy in his own workshops. As commercial thermocouples were not sensitive enough for his lunar measurements, he made his own and continued to perfect the design until a few years before his death.

Parsons was also continually trying to improve the drives of the Birr telescopes. In 1869, he fitted a clock drive to the great 6-ft. telescope that improved its ease of use but was never entirely satisfactory as the mount was not an equatorial type. In 1874, he replaced the old wooden altazimuth mounting of the 3-ft. speculum with an equatorial mounting of metal designed by B. B. Stoney and built in Dublin. While the new mount was a considerable improvement on the wooden one, it was not good enough for celestial photography.

Parsons was assisted in his investigations by a succession of talented assistants. These included Charles E. Burton, **Ralph Copeland**, Dreyer, and **Otto Boeddicker**.

The Fourth Earl took a keen interest in the development of the steam turbine, which had been invented by his youngest brother, Charles Algernon Parsons, and which revolutionized electric power generation and marine propulsion. Laurence served as chairman and director of the companies formed to exploit the invention. Charles frequently sought his advice on technical and business matters. In 1899, Laurence was made an associate of the Institute of Naval Architects in recognition of his contributions to marine technology.

Lord Rosse was elected chancellor of the University of Dublin in 1885 and remained in office until his death. He was made a Knight of the Order of Saint Patrick in 1890 and was Lord Lieutenant of King's County from 1892. Parsons served as president of the Royal Dublin Society (1887–1892) and the Royal Irish Academy (1896–1901). He received honorary degrees of DCL from Oxford (1870) and LLDs from Dublin (1879) and Cambridge (1900). The Institution of Mechanical Engineers made him an honorary member in 1888.

Following a gradual decline in health over 2 years, the Fourth Earl died and was buried in the old churchyard of Birr. He was survived by his wife and three children.

Ian Elliott

Alternate name

Fourth Earl of Rosse

Selected References

Moore, Patrick (1971). *The Astronomy of Birr Castle*. Birr: Telescope Trust.

Parsons, Laurence [Lord Oxmantown] (1868). "An Account of the Observations on the Great Nebula in Orion, made at Birr Castle, with the 3-feet and 6-feet Telescopes, between 1848 and 1868. With a drawing of the Nebula." *Philosophical Transactions of the Royal Society of London* 158: 57–73.

Parsons, Laurence [The Earl of Rosse] (1869). "On the Radiation of Heat from the Moon." *Proceedings of the Royal Society of London* 17: 436–443.

——— (1870). "On the Radiation of Heat from the Moon–No. II." *Proceedings of the Royal Society of London* 19: 9–14.

——— (1873). "On the Radiation of Heat from the Moon, the Law of its Absorption by our Atmosphere, and of Its Variation in Amount with her Phases." *Philosophical Transactions of the Royal Society of London* 163: 587–627. (The Bakerian Lecture.)

——— (1879). "Observations of Nebulae and Clusters of Stars Made with the Six-foot and Three-foot Reflectors at Birr Castle, From the Year 1848 up to the Year 1878." *Scientific Transactions of the Royal Dublin Society*, n.s., 2: 1–178.

Scaife, W. Garrett (2000). *From Galaxies to Turbines – Science, Technology and the Parsons Family*. Bristol: Institute of Physics.

Sinton, William M. (1986). "Through the Infrared with Logbook and Lantern Slides: A History of Infrared Astronomy from 1868 to 1960." *Publications of the Astronomical Society of the Pacific* 98: 246–251.

Parsons, William

Born **York, England, 17 June 1800**
Died **Monkstown, Co. Dublin, Ireland, 31 October 1867**

William Parsons, Third Earl of Rosse, a skilled engineer, ingenious scientist, and dedicated astronomer, constructed a reflecting telescope larger than any previously made, the largest in the world for seven decades. With it he discovered the spiral nature of many nebulae.

William was the eldest son of Sir Lawrence Parsons, Second Earl of Rosse and Alice (*née* Lloyd). The Parsons family came to Ireland from England at the end of the 16th century and settled in Birr, King's County (county offaly) (Parsonstown, King's County), in 1620. The Second Earl had been a prominent member of the Irish parliament since 1782, representing the University of Dublin and then his own county. With the passing of the Act of Union in 1800, his interest in politics waned and he devoted his time to the development of the town of Birr and the education of his family. William, with his brothers and his sisters, was educated at home by a series of tutors and governesses and with the active involvement of his parents.

When the title of Earl of Rosse passed to his father in 1807, William as eldest son assumed the courtesy title of Lord Oxmantown. He entered Trinity College, Dublin, in 1818 and then, with his brother John, transferred to Magdalen College, Oxford, in 1821. The following year he graduated with first class honors in mathematics. In 1823, Parsons was elected to represent King's County in the Westminster parliament; he held his seat until 1834, when he resigned in order to concentrate on his scientific interests.

Parsons became a member of the Astronomical Society of London (soon to become the Royal Astronomical Society) in 1824, just 4 years after its establishment. Through such a connection, he could have met **John Herschel**, whose father **William Herschel** had pioneered the building of large reflecting telescopes and their use in scanning the skies; in any event, Parsons and John Herschel exchanged numerous letters on astronomical matters. Parsons resolved to make a reflecting telescope as large as existing resources would allow. He could not benefit from the entire scope of knowledge of previous telescope makers, including the Herschels, for they took pains to keep some of their methods secret.

Parsons established a workshop and foundry at Birr Castle and trained his estate workers in all the practical skills that were required. After a long series of experiments to determine the optimum mixture of copper and tin for a speculum mirror, he settled on an alloy of four atoms of copper to one of tin, which he took to be in the ratio of 2.15:1 by weight. Parsons first tried making large mirrors from small segments of speculum soldered to brass plates. He invented a special machine (since widely adopted) for grinding, polishing, and parabolizing mirrors in a systematic way. The machine was driven by a small rotary steam engine of his design and made under his direction at Birr in 1827. On the strength of these achievements Parsons was elected to membership in the Royal Society in 1831. The same year he was appointed Lord Lieutenant for King's County.

In 1834, Parsons married Mary Wilmer-Field (1813–1885), the eldest daughter of a wealthy landowner who lived near Bradford, England. Mary inherited estates that were valued at £88,000 as well as a cash settlement of £8,700. Among her many interests, she became a pioneer photographer and one of the founders of the Photographic Society of Ireland.

After building a new forge, foundry, and workshop, Parsons resumed his experiments by constructing a 36-in.-diameter segmented mirror. After many trials and tribulations, in 1839 he succeeded in casting a perfect 36-in. speculum disk in one piece; it weighed 1¼ tons. A key factor in his success was the design of the casting mold, which had a base of closely packed steel strips through which gases could escape. The performance of this mirror encouraged him to attempt the casting of a 6-ft. monolithic mirror. When his father died in February 1841, William assumed the title of Third Earl of Rosse.

After many failures, two 6-ft. mirrors were successfully cast in 1842 and 1843, each weighing more than 3 tons. To avoid distortion when it was pointed in different directions, the mirror was supported by a system of 81 "equilibrated levers" suggested by **Thomas Grubb**.

The telescope tube resembled a giant barrel, 54 ft. long and 7 ft. in diameter, bulging to 8 ft. in the middle. The tube was pivoted about a huge universal joint at its base and slung with chains from two massive masonry walls, 23 ft. apart and parallel to the meridian. Horizontal movement was limited to 10° on either side of the meridian; a vertical range of nearly 110° was possible. The cost of the entire telescope was estimated between £20,000 and £30,000.

In February 1845, **Thomas Romney Robinson** of the Armagh Observatory and **James South** of London were present for the initial observations. Despite unfavorable weather, they caught a glimpse of a magnificent double star and numerous faint stars shining in M67. In April 1845, Parsons made a pencil drawing of the M51 nebula that caused a sensation when it was displayed at a meeting of the British Association for the Advancement of Science in Cambridge in June 1845. As the spiral arms suggested some sort of motion, the nebula was called "The Whirlpool." Parsons went on to discover 15 other spiral nebulae.

In autumn 1845 observational work was brought to a halt by the failure of the Irish potato crop and the resulting Great Famine. The Earl and Countess of Rosse devoted all their time and most of their income to alleviating the terrible effects of the famine. By 1848, when observations were resumed, there were many other demands on Parsons's time, so he employed a succession of able assistants, most notably **George Stoney** and **Robert Ball**. As the fame of the Great Telescope spread, visitors came from all over the world to see it and, if the weather allowed, to view the heavens.

Parsons received many honors. He was president of the Royal Society from 1848 to 1854 and was awarded its Gold Medal in 1851. He was made a knight of Saint Patrick in 1845; Napoleon III created him a knight of the Legion of Honor in 1855. Parsons was a member of the Royal Irish Academy (1822) and a member of the Imperial Academy of Science of Saint Petersburg (1852). He received honorary degrees from Cambridge (1842) and Dublin (1863). Parsons was chancellor of the University of Dublin from 1862 until his death. From 1845, he was a representative peer for Ireland in the Westminster parliament. The International Astronomical Union named the lunar crater at 17°. 9 S and 35°. 0 E in his honor.

In 1867, the Third Earl's health declined, and he took a house by the sea, just south of Dublin. He died after an operation to remove a tumor on his knee, and was buried in the old church of Saint Brendan in Birr; some 4,000 of his tenants attended his funeral.

Of the 11 children born to the Earl and Countess of Rosse, only 4 survived to adulthood. The eldest, **Laurence Parsons**, followed his father's interest in astronomy; Randal became a canon in the Church of England; Richard Clere became a successful railway engineer; and the youngest, Charles Algernon, became world famous as the inventor of the steam turbine.

After the death of the Fourth Earl in 1908, the Leviathan was partially dismantled, and one 6-ft. speculum was sent to the Science Museum in London. The great tube lay recumbent for many years until, as a result of the efforts of the sixth and seventh Earls, funding for restoration was secured in 1994. The telescope was completely reconstructed in 1996 to form the centerpiece of a historic science museum at Birr Castle.

Ian Elliott

Alternate name

Third Earl of Rosse

Selected References

Bennett, J. A. and Michael Hoskin (1981). "The Rosse Papers and Instruments." *Journal for the History of Astronomy* 12: 216–229.

Dewhirst, David W. and Michael Hoskin (1991). "The Rosse Spirals." *Journal for the History of Astronomy* 22: 257–266.

Ellison, M. A. (1942). "The Third Earl of Rosse and his Great Telescope." *Journal of the British Astronomical Association* 52, no. 8: 267–271.

Hoskin, Michael (1982). "The First Drawing of a Spiral Nebula." *Journal for the History of Astronomy* 13: 97–101.

——— (1990). "Rosse, Robinson, and the Resolution of the Nebulae." *Journal for the History of Astronomy* 21: 331–344.

——— (2002). "The Leviathan of Parsonstown: Ambitions and Achievements." *Journal for the History of Astronomy* 33: 57–70.

Moore, Patrick (1971). *The Astronomy of Birr Castle*. Birr: Telescope Trust.

Parsons, Charles A. (ed.) (1926). *The Scientific Papers of William Parsons, Third Earl of Rosse, 1800–1867*. London: P. Lund, Humphries and Co.

Parsons, William (1828). "Account of a New Reflecting Telescope." *Edinburgh Journal of Science* 9: 25–30.

——— (1828). "Account of an Apparatus for grinding and polishing the Specula of Reflecting Telescopes." *Edinburgh Journal of Science* 9: 213–217.

——— (1850). "Observations on the Nebulae." *Philosophical Transactions of the Royal Society* 140: 499–514.

——— (1861). "On the Construction of Specula of Six-feet Aperture; and a Selection of Observations of Nebulae Made with Them." *Philosophical Transactions of the Royal Society of London* 151: 681–745.

Scaife, W. Garrett (2000). *From Galaxies to Turbines: Science, Technology and the Parsons Family*. Bristol: Institute of Physics.

Tubridy, Michael (1998). "The Re-construction of the 6-foot Rosse Telescope of Ireland." *Journal of the Antique Telescope Society*, no. 14: 18–24.

Pawsey, Joseph Lade

Born **Ararat, Victoria, Australia, 14 May 1908**
Died **Sydney, New South Wales, Australia, 30 November 1962**

Australian radar and radio astronomer Joseph Pawsey led the group that, coming out of radar work in Australia during World War II, developed into one of the world's outstanding radio astronomy groups, whose early contributions included the demonstration of the very high equivalent temperature of emission from the solar corona, the precise location of the radio source Taurus A (leading to its identification with the Crab Nebula supernova remnant), and the development of interferometric and image processing techniques adopted by radio and optical astronomers and in ionospheric research and other areas. Pawsey was born to poor farming parents, Joseph Andrew Pawsey and Margaret Pawsey, in Ararat, Victoria, Australia. He did not start school until age eight, when he attended a small local school. At age 14, he received a government scholarship to attend boarding school at Wesley College in Melbourne, and then another to attend the University of Melbourne in 1926. Pawsey earned a B.Sc. (Honors) degree in 1929 and an M.Sc. (First Class Honors in Natural Philosophy) in 1931, and was granted an 1851 Exhibition Research Scholarship to Cambridge University, where he worked in the Cavendish Laboratory with J. A. Ratcliffe.

At the Cavendish Laboratory, while working toward his Ph.D., Pawsey studied the effects of the ionosphere on radio propagation. His observations of the reflection of radio waves from the E region of the ionosphere led to the discovery of irregularities, which move rapidly due to strong winds. This proved to be of pivotal importance in later ionospheric physics research.

After receiving his Ph.D. in 1934, Pawsey worked for 5 years at EMI Electronics Ltd., near London, on the design of aerials, especially the television transmitter being designed at Alexandra Palace. It was during this period that he met and married Greta Lenore Nicoll from Battleford, Saskatchewan, Canada, a marriage that resulted in two children, Margaret and Stuart, born in England, and another, Hastings, later in Australia. The time spent at EMI established him as an expert in antenna design, a skill that was used and developed later in his radio astronomy research. During this period, Pawsey was directly associated with 29 patents for devices that remained in wide use for several decades. However, only one external publication was written, due to EMI's policies to restrict access to research results.

After World War II broke out, Pawsey returned to Australia and took a position in Sydney with the Radiophysics Division of the Council for Scientific and Industrial Research [CSIR], later to become the Commonwealth Scientific and Industrial Research Organization [CSIRO]. During the war, he built up and led a team of engineers and physicists in studying radar transmission and reception and developing radar systems for the military. Because his work was classified, few publications were made public.

Pawsey had been interested in the prewar observations that **Karl Jansky** and **Grote Reber** had made, showing the existence of radio emission from the plane of the Galaxy. Although the science of radio astronomy had not yet been christened, Pawsey saw that the group he had assembled during the war could be kept intact if they could continue to study these phenomena and, encouraged by E. G. ("Taffy") Bowen, the chief of the Radiophysics Laboratory of CSIR, he led his group of researchers into this area. In 1945, using equipment that was originally part of the military defense of Sydney, located on a high cliff overlooking the ocean, Pawsey set up a radio receiving antenna operating at a frequency of 200 MHz. He used the interference between the direct rays and those reflected off the ocean to study the radio emission from the Sun, an approach known in optics as the Lloyd's mirror technique.

Two important results were published in *Nature* in 1946. The first was that intense radiation was emitted from a sunspot group, and this was so intense as to be nonthermal in origin. In that paper, Pawsey discussed the possibility of getting one-dimensional information about the source using two-beam interferometry with Fourier synthesis. This became one of the most powerful methods for studying the radio sky. The second discovery was that, at a wavelength of 200 MHz, the Sun has a lower limit of emission corresponding to a temperature of 1 million degrees. Soon afterward, from these observations, Pawsey and Tabsley established the intensity of thermal emission from the "quiet" Sun.

Pawsey's group studied the size, position, movement, spectrum, and growth and decay of various sources of radio emissions on the Sun. The group invented the swept-lobe interferometer (for location of rapidly moving sources on the Sun), the swept-frequency receiver (providing a spectrum of disturbances), and the grating interferometer, which had a resolving power of as little as one-twentieth of a degree, a much finer resolution than had been possible before. These techniques are now in general use in radio astronomy. Pawsey and S. F. Smerd reviewed much of this work in a chapter of a book on the Solar System edited by Gerard Kuiper. This, and an article written with E. R. Hill on cosmic radio waves, remained for many years essential reading for any student in the subject.

Members of the group discovered discrete radio sources in the Milky Way and external galaxies, and by accurately locating them were able to identify them optically. To achieve this, first the Lloyd's mirror technique, then two-aerial and radio-link interferometers were developed for the first time, followed by a series of linked antennas in the shape of a cross. The first survey of neutral hydrogen in the sky by the Pawsey group gave the first clear evidence of the spiral structure of our Galaxy. Again, these methods later became normal procedures in radio astronomy observatories around the world. Among the best-known members of the group were John Bolton (who located the Crab Nebula source and later built up the radio astronomy group at the California Institute of Technology) and Bernard Y. Mills, whose association with the T-shaped interferometer led to its frequently being called the Mills cross.

In 1955, with Ronald N. Bracewell, Pawsey wrote a book on radio astronomy that became the standard text in the field for many years. This was translated into Russian in 1958 with **Iosef Shklovsky** as editor. In 1962 Pawsey accepted the position of director of the American National Radio Astronomy Observatory in Green Bank, West Virginia, USA. During a visit to Green Bank in 1962 before taking up the post, he was diagnosed with brain cancer and, returning to his home, died. During his last illness, assisted by his devoted colleagues to get to his office each morning, Pawsey wrote the introduction to and edited a special radio astronomy issue of the *Proceedings of the Institution of Radio Engineers in Australia*, published in February 1963, which became a minor landmark in the instrumental aspects of radio astronomy.

Pawsey was a Foundation Fellow of the Australian Academy of Sciences (which now awards a Pawsey Medal in his honor), a foreign fellow of the Royal Society of London (which awarded him its Hughes Medal in 1960), and president of Commission 40 (radio astronomy) of the International Astronomical Union [IAU] from 1952 to 1958, in which role he played a key part in the definition of the IAU system of galactic coordinates, enabling the location of astronomical objects to be described in relation to the plane and center of the Milky Way, rather than in relation to the rotation axis and orbit of the Earth. The IAU named a lunar crater for him in 1970.

Stuart F. Pawsey

Selected References

Blackall, Simon (ed.) (1988). *The People Who Made Australia Great*. Sydney: Collins Publishers.

Christiansen, Wilbur N. and Bernard Y. Mills. (1964) "Biographical Memoirs – J. L. Pawsey." In *Australian Academy of Science Year Book*.

Kuiper, Gerard (ed.) (1955). *The Solar System*. Chicago: University of Chicago Press. (Contains a chapter by Pawsey on radiowave techniques.)

Lovell, A. C. Bernard (1964). "Joseph Lade Pawsey." *Biographical Memoirs of Fellows of the Royal Society* 10: 229–243.

Pawsey, J. L. and R. N. Bracewell (1955). *Radio Astronomy*. Oxford: Clarendon Press.

——— (1961). "Australian Radio Astronomy." *Australian Scientist*. (Describes five important items that had been developed under his direction.)

Pawsey, J. L. and E. R. Hill (1961). "Cosmic Radio Waves and Their Interpretations." *Reports on Progress of Physics* 24: 69–115.

Payne-Gaposchkin [Payne], Cecilia Helena

Born **Wendover, Buckinghamshire, England, 10 May 1900**
Died **Cambridge, Massachusetts, USA, 5 Dec 1979**

Cecilia Payne (later Payne-Gaposchkin) demonstrated in her 1925 Ph.D. dissertation that nearly all stars have the same chemical composition, with the apparent enormous differences due largely to the wide range of stellar temperatures. She also showed that this composition was dominated largely by hydrogen and helium (which was not immediately accepted) and later became a noted expert on novae and other kinds of variable stars.

Her father was Edward John Payne, a historian, barrister, and scholar at University College, Oxford; her mother, Emma Pertz, a painter and copyist in oils, was a granddaughter of Chevalier G. H. Pertz, Hanoverian scholar and member of Parliament. Cecilia

was their oldest child, soon followed by Humfry and Leonora. When her father died, Cecilia was only four, and her mother was left with three small children whom she raised "by a miracle of courage and self-sacrifice" in the environment of Edwardian England. Cecilia Payne married **Sergei Gaposchkin** in 1934, with whom she had three children: Edward Michael, Katherine Leonora, and Peter John Arthur. All three have had some involvement in astronomy.

After attending elementary school in Wendover, Payne had the opportunity to further her education when the family moved to London. Even at an early age she had learned much science independently, fascinated, for example, by the chemical elements. With a keen interest in science or possibly classics, she attended Saint Mary's College, Paddington, London, England, from 1913 to 1917 and Saint Paul's Girls School, Brook Green, Hammersmith, from 1918 to 1919.

Payne was awarded the Mary Eward Scholarship for Natural Sciences and was thus enabled to attend Newnham College, Cambridge University, Cambridge, England (1919–1923). There she first pursued the study of natural sciences with a concentration on botany, but, inspired by a lecture by **Arthur Eddington**, she changed her course of study to include more astronomy, graduating in 1923. She wrote her first paper in astronomy on the proper motions of stars in the cluster M36 in 1923.

Impressed by a lecture given by **Harlow Shapley**, then director of the Harvard College Observatory in 1922, Payne traveled to the United States for further study and in pursuit of a research career in astronomy. Payne was the first recipient of the Ph.D. in astronomy from Harvard College Observatory, which she received in 1925, as the first of Shapley's many students.

Her thesis, published as *Stellar Atmospheres*, applied **Meghnad Saha**'s is theory of ionization to establish the temperatures of the cool giants and the relative abundances of the chemical elements in their atmospheres. The first result, that nearly all stars had essentially the same abundance ratios, much like the terrestrial ratios for elements heavier than carbon, was incorporated into mainstream astronomical thinking immediately. The second result, that hydrogen and helium were by far the most common elements, was not. Shapley and **Henry N. Russell**, who had also read the work in advance of publication, recommended that Payne modify this conclusion and speak of "anomaous excitation" and a concentration of light elements on the surfaces of the stars. Nevertheless, the initial conclusion was essentially right, and has led to the thesis being described as "the best Ph.D. thesis in astronomy ever written" and Payne being described as the greatest woman astronomer of all time. Additional observations and analysis by Russell, **William McCrea**, **Carl von Weizäcker**, and others led to the accepted fraction of hydrogen and helium in the stars and Sun gradually increasing from a percent or two in 1925, to 10% in 1930, to more than 90% by 1960.

The 1920s were probably the happiest period of Payne's life. During this time she wrote several papers discussing spectral analysis and application of the Saha equation. Payne's second monograph, *The Stars of High Luminosity* (1930), established the temperature scale and uniform composition for the hotter stars of types 0, B, and A. Her collaborators included Shapley, **Leon Campbell**, **Donald Menzel**, Frederick Wright, **Fred Whipple**, and, from 1934 onward, very often Sergei Gaposchkin. On instruction from Shapley, Payne turned her attention from spectroscopy (which was to be Menzel's bailiwick) to variable stars.

Payne-Gaposchkin wrote a textbook, *Introduction to Astronomy* (1954), a monograph on *Variable Stars* (1938) with Gaposchkin, a definitive monograph, *The Galactic Novae* (1964), and an acclaimed popular account of stellar evolution, *Stars in the Making* (1953). The latter was credited by some young astronomers as their inspiration for entering the field. Her last book was *Stars and Clusters*, summarizing much that was known on this topic. She had a deep familiarity with individual stars, and even with specific spectral lines, and discussed them and recalled their characteristics as though they were friends.

In addition to her work in astrophysics and spectroscopy, Payne-Gaposchkin spent many years working with variable stars, including those enigmatic objects – the novae – and made significant contributions to the understanding of their nature. She frequently worked with photometric observations made by her husband, and they often published together. In their study of the galaxies, the Large Magellanic Cloud and the Small Magellanic Cloud, the two made roughly a million observations of variables, from which they were able to estimate the distance to these objects.

Payne-Gaposchkin was indefatigable in her research endeavors and was highly valued by her colleagues. She received her MA and D.Sc. from Cambridge University, England, in 1952. Payne-Gaposchkin also made occasional forays into history, contributing papers to the *Journal of the History of Science* and writing obituaries of several astronomers. She wrote "The Nashoba Plan for Removing the Evils of Slavery: Letters of Frances and Camilla Wright, 1820–1829" published in the *Harvard Library Bulletin*, 1975, based on a collection of letters passed down in her family.

In addition to her scientific work, Payne-Gaposchkin was an editor of the observatory publications for over a decade, and had teaching duties. Unfortunately, she suffered from overt discrimination both at Harvard and in the astronomical community because she was a woman; she was not even considered for certain positions, in spite of the extraordinary caliber of her scientific work. For example, it was not thought possible for her to make her own observations at remote observatories, as the accommodations would not permit a single woman even to visit the site. This excluded her from positions for which less talented men could readily apply.

For many years Payne-Gaposchkin had no official position at Harvard University and received a very low salary. Eventually, after the retirement of Harvard's president Lowell, she was named Phillips Professor of Astronomy. After Shapley retired as the observatory director, Payne-Gaposchkin received a professorship at Harvard, the first woman to hold this title. She then became chairman of the Department of Astronomy, the first woman to become chair of any department at Harvard University. After her retirement from this institution, she worked for some years at the Smithsonian Astrophysical Observatory, doing research exclusively.

Payne-Gaposchkin was remembered with affection and admiration by her colleagues; she was called "an astronomer's astronomer," and considered a genius. She was an inspiration to her many students and members of the general public as a role model and articulate scientist. Her sense of humor was subtle, and she was addicted to puns.

Among the honors received in Payne-Gaposchkin's lifetime were the first Annie Jump Cannon Prize of the American Astronomical Society, in 1934; honorary degrees from Smith College and elsewhere; and prizes and lectureships of the American Philosophical Society, the Franklin Institute, and the American Astronomical Society. At the latter, she was the first woman to deliver the Henry Norris Russell Prize (being introduced by the first woman to be president of the Society, E. Margaret Burbidge) in 1976. Minor planet (2039) Payne-Gaposchkin and a feature on Venus were named for her. A number of her own Ph.D. students have made important contributions to astronomy, including **Helen Sawyer Hogg**, **Joseph Ashbrook**, Elske Smith van Panhuijs, Frank Drake, Paul Hodge, and Andrew Young.

Katherine Haramundanis

Selected References

Gingerich, Owen (1982). "Cecilia Payne-Gaposchkin." *Quarterly Journal of the Royal Astronomical Society* 23: 450–451.

Haramundanis, Katherine (ed.) (1996). *Cecilia Payne-Gaposchkin: An Autobiography and Other Recollections*. 2nd ed. Cambridge: Cambridge University Press.

Hearnshaw, J. B. (1986). *The Analysis of Starlight: One Hundred Years of Astronomical Spectroscopy*. Cambridge: Cambridge University Press.

Hoffleit, D. (1979). "The End of an Era: Cecilia Payne-Gaposchkin and Her Last Book." *Journal of the American Association of Variable Star Observers* 8, no. 2: 48–51.

Lang, Kenneth R. and Owen Gingerich (eds.) (1979). *A Source Book in Astronomy and Astrophysics, 1900–1975*. Cambridge, Massachusetts: Harvard University Press.

Leverington, David (1996). *A History of Astronomy from 1890 to the Present*. London: Springer-Verlag.

Öpik, Ernst (1979). "Cecilia Payne-Gaposchkin." *Irish Astronomical Journal* 14: 69.

Payne, C. H. (1925). *Stellar Atmospheres*. Harvard Observatory Monographs, no. 1. Cambridge, England: W. Heffer and Sons.

——— (1930). *The Stars of High Luminosity*. Harvard Observatory Monographs, no. 3. New York: McGraw-Hill.

Payne-Gaposchkin, C. H. (1964). *The Galactic Novae*. New York: Dover.

Payne-Gaposchkin, C. H. and S. I. Gaposchkin (1938). *Variable Stars*. Harvard Observatory Monographs, no. 5. Cambridge, Massachusetts: Harvard College Observatory.

Payne-Gaposchkin, C. H. and K. G. Haramundanis (1970). *Introduction to Astronomy*. 2nd ed. Englewood Cliffs, New Jersey: Prentice-Hall.

Smith, Elske van Panhuijs (1980). "Cecilia Payne-Gaposchkin." *Physics Today* 33, no. 6: 64–65.

Whitney, Charles A. (1980). "Cecilia Payne-Gaposchkin: An Astronomer's Astronomer." *Sky & Telescope* 59, no. 3: 212–214.

Payne, William Wallace

Born **Somerset, Michigan, USA, 10 May 1837**
Died **Elgin, Illinois, USA, 29 January 1928**

William Payne is remembered as the 19th-century founder of Goodsell Observatory at Carleton College and as the independent publisher of three popular astronomical journals. The son of Jesse and Rebecca (*née* Palmer) Payne, William earned bachelor's (1863) and master's (1864) degrees from Hillsdale (Michigan) College with proficiencies in mathematics and languages. While a teacher in the Cambria Township (Hillsdale County) schools, Payne studied law and received his LL.B. degree in 1866 from the Chicago Law School. He relocated to Mantorville, Minnesota, and formed a partnership with Robert Taylor but grew discontented in the practice. Payne returned to teaching and was chosen superintendent of Dodge County schools. He cut his editorial teeth by launching *The Minnesota Teacher and Journal of Education* (*circa* 1867–1871), which was later united with *The Chicago Teacher* to become *The Western Journal of Education*. In 1870, Payne married Josephine Vinecore; the couple had one daughter, Jessie.

In 1871, Carleton College president James W. Strong hired Payne as a professor of mathematics and natural philosophy at the college's Northfield campus. Remarkably, Payne undertook construction of an astronomical observatory, though the college, founded in 1866, consisted of but three buildings. By 1878, a small wooden observatory, housing a clock, a transit instrument, and an 8-in. Clark refracting telescope, was completed. Time signals derived from astronomical observations were first relayed by telegraph from the unfinished structure in 1877. This service, eventually the largest in the northwest, provided time for more than 12,000 miles of railroad lines. Payne influenced railroad officials to adopt standard time upon its inauguration in 1883. From 1887 to 1897, Charlotte R. Willard operated Carleton's time service. While it was under Payne's guidance, the United States Signal Corps (1881) and later the National Weather Service (1883) designated the observatory as an official meteorological station.

By the 1880s, astronomy was the most vital and important of the college's programs; Payne successfully advocated the need for more precise astronomical equipment and a larger observatory. A 5-in. meridian circle was installed, and in 1890, funds were secured for installation of a 16-in. refractor – then the sixth largest telescope in the nation. The new brick observatory was named after the college's founder, deacon Charles M. Goodsell. Its plan of work was to be threefold: "[u]ndergraduate instruction ...; a school for practical astronomy ...; and original research."

In 1882, Payne launched the first of three journals that were to spread Carleton's name throughout the astronomical community. *The Sidereal Messenger* aimed to bring an understanding of developments in astronomy to wider audiences, chiefly instructors, amateur astronomers, and the public. Sprinkled with Protestant natural theology, *The Messenger* reflected Payne's deep religious sentiments. Payne privately managed its expenses, and his subscriptions grew. When astronomer **George Hale** sought to create a research journal devoted to astrophysics, he found it necessary to compromise on a joint publication coedited with Payne. For three years (1892–1894), *Astronomy and Astro-Physics* was published at Northfield. In 1895, the University of Chicago acquired Hale's interest and the *Astrophysical Journal* was born (1895).

Many of *The Messenger*'s former subscribers found its transformation from a popular to a technical journal undesirable. Circulation of *Astronomy and Astro-Physics* actually declined even as new subscribers were added from a handful of astrophysics practitioners. To reclaim his general readers, in 1893 Payne launched another privately owned journal, *Popular Astronomy*. Subtitled "A Review of Astronomy and Allied Sciences," *Popular Astronomy* reiterated *The Messenger*'s forum on celestial events, news of the profession, essays, and a distinctive focus on pedagogy. For more than five decades, *Popular Astronomy* served as the principal channel of communication, or "trade" journal, within the American astronomical community.

In 1892, Payne represented astronomy on a subcommittee chaired by Johns Hopkins University chemist Ira Remsen. The subcommittee reported jointly to Harvard University president Charles W. Eliot and the National Educational Association's Committee on Secondary School Studies, which was popularly known as the Committee of Ten. One of the committee's recommendations was that astronomy courses should be reduced from college prerequisites to elective subjects. While seemingly an innocuous decision, its cumulative effect was to bring about a decline in astronomy education after 1900; that was an outcome strongly antithetical to Payne's own views.

Payne's teaching was conducted by the lecture-recitation method. He advocated the "mental discipline" model of pedagogy and favored adoption of textbooks that students might "read and reread thoroughly and exhaustively." One Carleton student, who punned, "He never knew pleasure, who never knew Payne," immortalized Payne's reputation as an instructor.

Additional faculty were hired to support the growth of Carleton's astronomy program, including alumni **Herbert Wilson** (1879), who succeeded Payne as editor of *Popular Astronomy* (1909–1926), and **Edward Fath** (1902). The college trained some of the era's leading women astronomers, including **Anne Young** (1892), director of Mount Holyoke College Observatory, and **Mary Byrd**, who earned a Carleton Ph.D. in 1904 and directed the Smith College Observatory. Payne received an honorary degree from Hillsdale College (Ph.D., 1894) and Carleton conferred a similar honor (Sc.D., 1916) at the college's golden anniversary.

In response to a controversy surrounding Carleton's second president, William H. Sallmon, who was himself forced to resign, Payne (along with several other faculty members) resigned in 1908. Although Payne had resigned, he was still awarded a Carnegie Endowment pension in recognition of his outstanding service to the college. The college purchased *Popular Astronomy* from Payne. After Wilson's retirement, **Curvin Gingrich** continued its publication as editor from 1926 to 1951. When Gingrich died in 1951, Carleton College elected to abandon the publication of *Popular Astronomy*.

A new demand for Payne's services arose from President Theodore Roosevelt's directive that the National Bureau of Standards conduct tests of accuracy on portable watches. In a replay of his Carleton appointment, the Elgin National Watch Company hired Payne in 1909 to establish an astronomical observatory and time service in Illinois. He retired as director emeritus of the Elgin Observatory in 1926.

Apart from his role in creating the Goodsell Observatory and Carleton's astronomy department, Payne's contributions lay chiefly in the realm of practical service and popularization. His formal training was completed before photographic plates or methods of spectral analysis were widely adopted, which may in part explain why he never conducted research. Payne recognized the prejudices that researchers associated with popularization, yet was never deterred by those prejudices. The astronomical journals that he founded and edited brought acclaim to the college that far outlasted his own services. Goodsell Observatory is listed on the National Register of Historic Places, as a site where important contributions were made to Minnesota astronomy education and the "scientific literary field" embraced by Payne's journals.

Carleton College Archives, Northfield, Minnesota, retains selected papers of Payne, chiefly in its President's Annual Reports (1873–1895), Series 42, Box 1, together with a biographical file. Payne's extensive correspondence was discarded after his departure from the college, a significant loss for the history of American astronomy.

Jordan D. Marché, II

Selected References

Fath, E. A. (1928). "William Wallace Payne." *Popular Astronomy* 36: 267–270.

Gingrich, Curvin H. (1943). "Popular Astronomy: The First Fifty Years." *Popular Astronomy* 51: 1–18, 63–67.

Greene, Mark (1988). *A Science Not Earthbound: A Brief History of Astronomy at Carleton College*. Northfield, Minnesota: Carleton College.

Marché II, Jordan D. (2005). "Popular Journals and Community in American Astronomy, 1882–1951." *Journal of Astronomical History and Heritage* 8, no. 1: 49–64.

Osterbrock, Donald E. (1995). "Founded in 1895 by George E. Hale and James E. Keeler: The *Astrophysical Journal* Centennial." *Astrophysical Journal* 438: 1–7.

Payne, William W. (1927). "Elgin Observatory." *Popular Astronomy* 35: 1–10.

Pearce, Joseph Algernon

Born **Brantford, Ontario, Canada, 7 February 1893**
Died **Victoria, British Columbia, Canada, 8 September 1988**

Joseph Pearce was a stellar astrophysicist who, with **John Plaskett**, confirmed the rotation and scale of the Milky Way Galaxy. Pearce was the son of Joseph William Pearce and Clarissa Augusta Rounds. He entered the University of Toronto in 1913, but interrupted his studies in 1915 to join the Canadian army as a signals officer. After suffering wounds in France in 1916, he returned to Canada, rising to the rank of major by 1919. Pearce completed his BA at Toronto in 1920, acting as a class assistant to **Clarence Chant** during term, while working as a magnetic observer for the Dominion Astrophysical Observatory [DAO] in Victoria, British Columbia, during the summers. At Chant's insistence, Pearce obtained a research fellowship at the Lick Observatory in 1922 and began his Ph.D. studies with **Robert Aitken**. In 1924, Plaskett required a replacement at the DAO and, on Aitken's recommendation, Pearce was hired, despite not having finished his degree. It was not until 1930, after much prodding by Plaskett, that he obtained the Ph.D. Pearce married Esther Mott in 1917, and they had two children, Josephine and Richard. Esther Pearce died in 1945, and he married Elizabeth Allan in 1947.

Plaskett had completed his survey of O stars when Pearce arrived and recruited him to work on the B-star program, a survey of all B stars brighter than magnitude 7.5 and north of declination −11°, altogether some 1,056 stars. The work was undertaken with the DAO's 72-in. reflector. Plaskett had keenly followed the work of **Bertil Lindblad** and **Jan Oort** on the possible rotation of the Galaxy and had conferred with Oort in Leiden in 1927. Although the B-star survey was incomplete – it would be finished in early 1929 and published in 1930 – Plaskett and Pearce had sufficient data to test the theory. With useful data for about 500 O and B stars, they were able to locate the galactic center near to where Oort found it and near the point **Harlow Shapley** indicated from his globular cluster measurements. Plaskett and Pearce published their preliminary results in the *Monthly Notices of the Royal Astronomical Society* in 1928 and their final results in the *Publications of the Dominion Astrophysical Observatory* in 1930. They calculated a galactic rotation speed of 275 km/s at the distance of 10 kpc from the galactic center.

The O- and B-star data also provided Plaskett and Pearce with the possibility of solving the problem of the location and motion of the interstellar gas. **Arthur Eddington**, in 1926, argued that interstellar calcium was spread throughout the Galaxy in clouds that were relatively stationary as stars moved through them. Ca II lines would show up only in very hot stars. By 1929, possibly with preliminary data from Plaskett and Pearce, **Otto Struve** and **Boris Gerasimovic** posited that interstellar calcium did move, but at half the rate of the stars. In 1930, Plaskett and Pearce, with their extensive data, showed that this seemed to be the case. Plaskett retired in 1935, and Pearce embarked upon an expanded B-star program with Robert M. Petrie, adding stars brighter than magnitude 9 and north of declination +20°. The final results did not appear until 1962. Pearce, like other DAO staff, computed a number of spectroscopic binary orbits, and observed stars in the Pleiades and Hyades. Few of these results were published.

Pearce became assistant director of the DAO in 1936 and director in 1940 on **William Harper**'s death. He held the position until 1951, when he turned over the reins to Petrie. Pearce remained on staff until 1958, when he retired.

Pearce was a long-time supporter of the Royal Astronomical Society of Canada, being a key figure in the Victoria Centre and a member of the national council for a decade, culminating in his tenure as president in 1940. He was also a vice president of the American Astronomical Society in 1943/1944, and a member of the American Association for the Advancement of Science, the Astronomical Society of the Pacific, and the Société Astronomique de France. He was a fellow of the Royal Society of Canada (president 1949/1950). Like other DAO staff members, Pearce was a contributor to International Astronomical Union commissions, as a member of Commissions 27, 42, and 30 (Radial Velocities, of which he was president from 1948 to 1952).

Richard A. Jarrell

Selected References

Jarrell, Richard A. (1988). *The Cold Light of Dawn: A History of Canadian Astronomy*. Toronto: University of Toronto Press.

Wright, Kenneth O. (1989). "Joseph A. Pearce, 1893–1988." *Journal of the Royal Astronomical Society of Canada* 83: 3–7.

Pearson, William

Born **Whitbeck, (Cumbria), England, 23 April 1767**
Died **South Kilworth, Leicestershire, England, 6 September 1847**

Reverend Dr. William Pearson cofounded the Astronomical Society of London (now the Royal Astronomical Society [RAS]), made a number of elaborate astronomical clocks and demonstration instruments, and published a valuable treatise on practical astronomy. The son of yeoman farmer William and Hannah Pearson, the younger William pursued his education and career vigorously in spite of his modest origins, spending the first half of his working career as a highly successful schoolmaster, and then evolving as a beneficed clergyman in his later years.

As an amateur astronomer who was active in the astronomical community at the beginning of the 19th century, Pearson's earliest astronomical interests appear to have been focused strongly on the design and construction of clocks, orreries, and planetary machines. Well acquainted with instrument maker Edward Troughton (1753–1836), Pearson utilized gears with substantially more teeth than the standard horological practice of the day and produced smoothly functioning and effective clocks, watches, and demonstration machines. Pearson's interest in observational astronomy flourished in later years, leading to the publication of his *Practical Astronomy*, which was more appreciated in the decades after his death than after its publication in 1829.

Thomas R. Williams

Selected References

Turner, J. L. E. and H. H. Dreyer (eds.) (1987). *History of the Royal Astronomical Society Volume I: 1820–1920*. Oxford, U. K.: Blackwell Scientific Publications.

Gurman, S. J. and S. R. Harrat (1994). "Revd. Dr. William Pearson (1767–1847): A founder of the Royal Astronomical Society." *Quarterly Journal of the Royal Astronomical Society*. 35: 271.

Peary, Robert Edwin

Born **Cresson, Pennsylvania, USA, 6 May 1856**
Died **Washington, District of Columbia, USA, 20 February 1920**

Robert Peary, Arctic explorer and naval officer, was the son of Charles Nutter Peary and Mary (*née* Wiley) Peary. Peary's father died when he was 2 years old, leaving him to be raised by his mother, who had a strong influence on his life. Peary attended Portland (Maine) High School and then Bowdoin College, from which he received a degree in civil engineering in 1877. For 2 years, he worked as a surveyor at Fryeburg, Maine, before joining the United States Coast and Geodetic Survey as a draughtsman. In 1888, Peary married Josephine Diebitsch of Portland, and the couple had two children.

Peary was commissioned as a civil engineer in the United States Navy from 1881 to 1891. His interest in Arctic exploration developed from a private trip he made to Greenland in 1886. On several of his later expeditions, he was accompanied by his wife. Peary decided to use northern Greenland (and its coast) as points of departure for trying to reach the North Pole. He made observations of the Sun to determine his positions and set several new records for northern latitude, including Greenland's northernmost point, Cape Morris Jesup.

Peary's final push toward the North Pole began in 1908, with support from the National Geographic Society and other private patrons. Peary's claims to have reached the North Pole in April 1909, ahead of competitor Frederick Cook, were widely accepted in his day. For his reputed accomplishment, Peary received worldwide recognition and a number of awards, including a Gold Medal from the Congress. However, these claims have not stood up to later scrutiny, and substantial discrepancies exist in Peary's account. Most polar authorities no longer accept Peary's assertion to have reached the North Pole, or many other alleged discoveries.

During an earlier Greenland expedition, Peary recovered three massive fragments of an iron meteorite from the Cape York shower. These are known from Eskimo folklore as "The Tent," "The Woman," and "The Dog." Peary's wife later sold these meteorites to the American Museum of Natural History, where they remain on display.

Peary's popular writings, which described the numerous hardships, trials, and disappointments he experienced, nonetheless served to inspire later generations. He was buried at the Arlington National Cemetery. Peary's collections are housed in the National Archives, but they were not publicly opened until the 1980s.

Raghini S. Suresh

Selected References

Green, Fitzhugh (1926). *Peary: The Man Who Refused to Fail*. New York: G. Putnam's Sons.

Heckathorn, Ted (1999). "Peary, Robert Edwin." In *American National Biography*, edited by John A. Garraty and Mark C. Carnes. Vol. 17, pp. 217–219. New York: Oxford University Press.

Lankford, Kelly L. (2003). "Home Only Long Enough: Arctic Explorer Robert E. Peary, American Science, Nationalism, and Philanthropy, 1886–1908." Ph.D. diss., University of Oklahoma.

Peary, Robert E. (1910). *The North Pole: Its Discovery in 1909 Under the Auspices of the Peary Arctic Club*. New York: Frederick A. Stokes.

Weems, John Edward (1967). *Peary, The Explorer and the Man, Based on His Personal Papers*. Boston: Houghton Mifflin.

Pease, Francis Gladhelm

Born **Cambridge, Massachusetts, USA, 14 June 1881**
Died **Pasadena, California, USA, 7 February 1938**

American optician and spectroscopist Francis Pease is most widely remembered for his contributions to the design and construction of the 60-, 100-, and 200-in. telescopes at Mount Wilson and Palomar observatories, but he also obtained the first accurate rotation curves of spiral galaxies and with **Albert Michelson** made the first direct measurements of the diameters of stars other than the Sun. Pease graduated from the Armour Institute of Technology in Chicago (now part of the Illinois Institute of Technology) in 1901 with a BS in mechanical engineering, and received honorary MA and Sc.D. degrees from the institute in 1924 and 1927. While a student, he worked evenings at the Petitdidier optical shop, and his employers recommended him to **George Ritchey**, the chief optician at Yerkes Observatory. Ritchey's father had also been one of Pease's teachers at Armour. Thus, with Ritchey, **Walter Adams**, and **Ferdinand Ellerman**, Pease moved west with **George Hale** in 1904 as one of the founding staff members of Mount Wilson Observatory. He remained there the rest of his life, apart from a year of war work in 1918 as chief draftsman in the engineering section of the National Research Council.

The history of astronomy has focused more closely on discoveries and theories than on the instruments used to make them and verify them; Pease's reputation as an astronomer consequently is less than it might otherwise have been. On the other hand, his role at Mount Wilson Observatory has been somewhat exaggerated, at the expense of Ritchey's reputation. Even before Ritchey was fired in 1919, Hale, the Mount Wilson director, had worked to make Ritchey an "unperson," preventing him from receiving outside recognition. Pease customarily is credited in whole or in large part for many of the instruments at Mount Wilson and Mount Palomar observatories, including the 60-inch 100-inch and 200-inch reflecting telescopes, the 60-foot and 150-foot tower telescopes, and the 20-foot and 50-foot interferometers. Actually, Ritchey led the work on the 60-inch reflecting telescope, and also on the 100-inch until November 1912, when Pease was placed in charge of the design for its mounting. Even then, Ritchey continued to figure the mirror.

Pease also made observations with the instruments at Mount Wilson Observatory. Not only did his observing experience contribute to his design skills, but some of Pease's observations were significant in themselves. During August, September, and October of 1917, he managed to take a 79-hour exposure of the Andromeda Nebula (M31). From the spectrograms taken with this exposure, and from an even longer one of 84 hour made a year earlier by Adams, Pease confirmed the spectroscopic rotation of the nebula.

From 1917 through 1919, working with the 60-inch reflector at Mount Wilson Observatory, Pease took some 66 plates of the spiral nebula M33. A nova appeared on four of the plates, and its observed magnitude was consistent with the distance to the nebula determined by **Edwin Hubble** from Cepheid stars also found in the nebula.

In 1928 Pease was the first to identify a planetary nebula in a globular cluster (M15). Previously cataloged as a star, the planetary nebula is now named Pease 1. It was, for many years, the only known planetary in a globular cluster, and they are still very rare.

Pease's most famous astronomical discovery was the first measurement of the diameter of a star other than our own Sun. Physicist **Albert Michelson** redesigned his stellar interferometer, and in the summer of 1920 had it mounted on the 100-inch reflecting telescope at Mount Wilson Observatory. He had to return to the University of Chicago at the end of the summer, and left Pease in charge of the measurements. In December Pease reported success, having determined an astounding diameter of 240 million miles for the star Betelgeuse.

A decade later Pease built a 50-foot interferometer, potentially capable of measuring stellar diameters half the size of what the 20-foot interferometer had measured. However, the instrument was not a complete success; thermal gradients allowed deflections in the support beam, which in turn allowed unacceptable fluctuations in the optical path lengths from the outer mirrors to the eyepiece.

Pease also worked with Michelson on a more accurate determination of the velocity of light. The measurements were begun in the summer of 1922 and continued over subsequent summers, to and beyond Michelson's death in May 1931, first between Mount Wilson and Mount San Antonio (1924–1938) and later on the Irvine Ranch in Orange County, California (1930–1934).

Although Pease was a member of most of the renowned astronomical societies, he received no major awards and held no major offices in them. In 1905, he married Carline T. Furness, who must not be confused with **Caroline Furness**, director of Vassar College Observatory for many years. Pease's most ambitious telescope design was surely the 300-inch one made in 1926.

Norriss S. Hetherington

Selected References

Adams, Walter S. (1938). "Francis G. Pease." *Publications of the Astronomical Society of the Pacific* 50: 119–121.

Anderson, J. A. (1939). "Francis G. Pease." *Journal of the Optical Society of America* 29: 306–307.

Anon. (1938). "Dr. Francis G. Pease." *Nature* 141: 542–543.

Anon. (1938). "Obituary." *Observatory* 61: 198.

Anon. "Obituary notices." *Los Angeles Times*, 8 February 1938, pt. II, pp. 1, 5; and 9 February 1938, pt. II, p. 4.

Berendzen, Richard and Richard Hart (1974). "Pease, Francis Gladhelm." In *Dictionary of Scientific Biography*, edited by Charles Coulston Gillipie. Vol. 10, p. 473. New York: Charles Scribner's Sons.

Hetherington, Norriss S. (1990). *The Edwin Hubble Papers: Previously Unpublished Manuscripts on the Extragalactic Nature of Spiral Nebulae*. Tucson: Pachart, p. 108. (Pease's discovery of a nova in M33 and the use of it by Edwin Hubble were not published, but are noted herein.)

Livingston, Dorothy Michelson (1973). *The Master of Light: A Biography of Albert A. Michelson*. Chicago: University of Chicago Press. (See pp. 274–277 for the measurement of the diameter of Betelgeuse and pp. 302–307, 318–319, 328–338 for Michelson and Pease on the speed of light.)

Michelson, A. A. and F. G. Pease (1921). "Measurement of the Diameter of α Orionis with the Interferometer." *Astrophysical Journal* 53: 249–259.

Michelson, A. A., F. G. Pease, and F. Pearson (1935). "Measurement of the Velocity of Light in a Partial Vacuum." *Astrophysical Journal* 82: 26–61.

Osterbrock, Donald E. (1993). *Pauper and Prince: Ritchey, Hale, and Big American Telescopes*. Tucson: University of Arizona Press. (A major study of Pease as a designer of astronomical instruments is yet to be written. He is mentioned, in connection with his optical mentor, George Willis Ritchey herein.)

Pease, F. G. (1915). *Publications of the Astronomical Society of the Pacific* 27: 134.

——— (1918). "The Rotation and Radial Velocity of the Central Part of the Andromeda Nebula." *Proceedings of the National Academy of Sciences* 4: 21–24.

——— (1920). "Photographs of Nebulae with the 60-Inch Reflector, 1917–1919." *Astrophysical Journal* 51: 276–308. (For his observations of M33.)

——— (1928). *Publications of the Astronomical Society of the Pacific* 40: 342. (For a report of his identification of the planetary nebula in M15.)

——— (1930). "The New Fifty-foot Stellar Interferometer." *Scientific American* 143, no. 4: 290–293.
Pendray, G. (1935). *Men, Mirrors, and Stars*. New York: Funk and Wagnalls Co., pp. 198–220.
Strömberg, Gustav (1938). "Francis G. Pease, 1881–1938." *Popular Astronomy* 46: 357–359.

Peek, Bertrand Meigh

Born **Boscombe, Dorset, England, 27 December 1891**
Died **Melbourne, Victoria, Australia, May 1965**

Bertrand Meigh Peek is best remembered as the author of many of the Jupiter *Memoirs* of the British Astronomical Association [BAA] and especially of the classic book *The Planet Jupiter*.

Peek had a traditional English upbringing through private schools and the University of Cambridge, where he was a three-time winner of the mathematics prize and a tennis champion. His highest degree at Cambridge was an MA. Peek rose to the rank of major in the British army, serving with The Hampshires Regiment in India during World War I. After the war, Peek took up a career as a teacher, and became headmaster of a school in Solihull, Birmingham, England. Peek's other avocational activities included a continued involvement in sports, chess (a member of the Anglo-Soviet match teams), and music – he composed at least one symphony. He was an early amateur radio operator.

Although Peek was interested in many fields of amateur astronomy, and contributed results to the Double Star and other BAA sections, it was his lifelong interest in planetary astronomy that produced his greatest contributions. Peek was briefly director of the BAA's Mars Section and then Saturn Section. In 1933, while director of the Saturn Section, Peek carried out an exhaustive mathematical analysis of the motions of **William Hay**'s white spot on Saturn. Then, in 1934, he swapped posts with the then director of the Jupiter Section, Reverend **Theodore Phillips**, who had done so much to direct amateur observations to form a scientifically reliable body of work. Like Phillips, Peek maintained an active correspondence with many amateur astronomers, including Hugh M. Johnson (born: 1923) and **Walter Haas**, active observers in the United States who, as Peek noted, helped sustain the work of the section during World War II.

Often Peek would visit Phillips's observatory to observe with him. Peek continued Phillips's high standards both of observation and of analysis. He insisted on a careful scientific approach to his own and others' observations, taking care to exclude subjective effects and emphasizing numerical results – particularly the Jovian wind speeds that could be deduced from visual transit measurements. Although color changes on Jupiter are perhaps an equally important phenomenon, of which Peek made careful observations, he became skeptical of the possibility of reaching reliable conclusions from these subjective impressions and therefore devoted little attention to Jovian colors in his writings.

Peek retired from the directorship of the BAA Jupiter Section in 1949, when his health declined, then embarked on writing *The Planet Jupiter*. In this book, only the second book published on Jupiter, he summarized the observed phenomena of the atmosphere, with notably lucid narrative and occasional dry wit. *The Planet Jupiter* was the definitive text for a generation until spacecraft visited the planet. Peek's health recovered somewhat, and he resumed observing to codiscover the 1955 South Tropical Disturbance.

An active participant in BAA affairs, Peek served as the BAA president from 1938 to 1940, and the association owed much to his leadership and stabilizing influence during World War II.

John Rogers

Selected References

Fox, W. F. (1965). "Bertrand Meigh Peek." *Journal of the British Astronomical Association* 76: 165.
Peek, B. M. (1958). *The Planet Jupiter*. London: Faber and Faber.

Peirce, Benjamin

Born **Salem, Massachusetts, USA, 4 April 1809**
Died **Cambridge, Massachusetts, USA, 6 October 1880**

Benjamin Peirce established an American presence in celestial mechanics, trained a number of leading astronomers, and played an important role in the development of the institutional structure of American science.

Peirce was the son of Benjamin and Lydia Ropes (*née* Nichols) Peirce. The Peirces were among the oldest families in the United States; Peirce's ancestor, John Pers of Norwich, England, came to the New World in 1637. Peirce attended the Salem Private Grammar School where he became acquainted with **Nathaniel Bowditch**, father of his classmate Henry Ingersoll Bowditch. Peirce entered Harvard University in 1825, at a time when the university was in a dire financial crisis. When Nathaniel Bowditch became one of Harvard's trustees the next year, he forced a thorough reorganization of the university, including the dismissal of a mathematics professor whose grasp of mathematics, according to Bowditch, was less than that of "Peirce of the Sophomore class." Recognizing the capability of the young Peirce, Bowditch employed him to help read proofs of his translation of **Pierre de Laplace**'s *Traité de méchanique céleste*.

After graduating in 1829, Peirce taught for 2 years at a private school before becoming a tutor in mathematics at Harvard University in 1831. Appointed temporary head of the Department of Mathematics in 1832, Peirce became permanent head of the department when his predecessor retired for medical reasons. Peirce received his MA in 1833, and was appointed a professor of mathematics and natural philosophy. In the same year, he married Sarah Hunt Mills. They would have a daughter and four sons, including two, James Mills Peirce and Charles Sanders Peirce, who would themselves become mathematicians.

Peirce was active in computing the orbits of comets and developing the mathematics of perturbation functions in celestial mechanics. In 1842, he was appointed Perkins Professor of Mathematics and Astronomy. His public lectures on the great March sungrazing comet C/1843 D1 helped stimulate public support for expansion of the Harvard College Observatory and acquisition of the 15-in. Merz and Mahler refractor, at that time one of the world's three largest refractors. Peirce was the first in the United States (1848) to give lectures in celestial and analytical mechanics.

Peirce's interest in celestial mechanics would lead him into several controversies. After **Urbain Le Verrier** and **John Adams** discovered that the irregularities in the motion of Uranus could be accounted for by assuming the existence of a hitherto undiscovered planet, and made detailed predictions of where such a planet might be found, on 23 September 1846 the German astronomer **Johann Galle** discovered Neptune in very nearly the position predicted by Le Verrier. However, Le Verrier's predicted distance was far in excess of the actual distance, which led Peirce and United States Naval Observatory astronomer **Sears Walker** to conclude that the discovery of Neptune, far from being a triumph of celestial mechanics, was in fact little more than a coincidence. This contention was bitterly disputed on both sides of the Atlantic and added to an already intense debate over Le Verrier's priority in comparison to Adams.

At about the same time, Peirce's activities expanded to include administrative affairs of the university as well as the institutional structure of science in the United States. In 1846, he was asked to draw up a plan for what became Harvard University's Lawrence Scientific School. Peirce was a member of a committee that drafted and distributed the constitution of the American Association for the Advancement of Science [AAAS] as that organization emerged from the American Association of Geologists and Naturalists in 1848. In these activities, Peirce joined with other influential figures in mid-19th-century American science such as **Joseph Henry**, first director of the Smithsonian institute; **Alexander Bache**, director of the United States Coast Survey; and others who corresponded and met frequently. In their correspondence they described themselves as the "scientific Lazzaroni." In residence primarily in Cambridge and Washington, and well connected politically and socially to elites in both centers of American culture, the Lazzaroni tended to act in concert on matters involving the institutional structure of science in America.

Peirce's connections with the Lazzaroni would lead him to another controversy. In 1856, as part of the Dudley Observatory's scientific council, Peirce found himself in the midst of a power struggle. The scientific council, consisting of Peirce, Bache, Henry, and **Benjamin Gould**, first director of the Dudley Observatory, had coordinated a plan of detailed astronomical observations with the Coast Survey. However, Gould's delays in implementing the plan and demands for further improvements in the observatory created a major feud between the scientific council and the observatory's trustees over governance of the institution. The trustees prevailed; in 1859, the scientific council was effectively dissolved.

Peirce did not abandon his teaching and scientific pursuits during these years of external involvement. In a paper presented orally at a meeting of the AAAS in 1851, Peirce showed that Saturn's rings could not be solid, but must instead be fluid. Peirce's paper credited **George Bond** of Harvard College Observatory with reaching the same conclusion observationally the previous year, but failed to mention that Bond offered a mathematical argument to support his observations, leading to yet another acrimonious dispute. It would be several years before **James Maxwell** demonstrated theoretically that the rings must be solid particles rather than a fluid. Peirce's students at Harvard University included the astronomers **George Hill**, **Percival Lowell**, and **Simon Newcomb**. Peirce was responsible for Newcomb's postgraduate commissioning as professor of mathematics at the United States Naval Observatory.

Peirce considered himself a candidate to replace **William Bond** as director of the Harvard College Observatory when the latter died in 1859. That placed him in direct competition with the younger Gould, who had been forced to resign at Dudley Observatory, as well as with Bond's son George, who was selected to fill the post. Peirce's candidacy disrupted his previously friendly relationship with Gould irreparably.

Peirce shared with other Lazzaroni members a desire for a more exclusive national venue in which leading scientists might share their research and influence national science policy. In 1863, he joined with Bache, Louis Agassiz, and Gould to work with US Senator Henry Wilson in writing the congressional act that established the National Academy of Sciences. Peirce was, of course, one of the 50 elite scientists selected for initial membership in the academy. The fact that Bond, his rival at Harvard, was not among the 50 scientists contributed to their continuing animosity.

In 1867, Henry prevailed upon Peirce to accept appointment as the director of the United States Coast Survey after Bache's death. Gould had been working part-time for the Survey as director of longitude determination since 1852. In that year, he published a statistical method for discarding discrepant observations that was widely adopted as Peirce's criterion but was later discredited. Peirce's 1870 work on linear associative algebra is considered the first major original contribution to mathematics produced in the United States and marks the beginning of the acceptance of American mathematics as a field separate from astronomy by European mathematicians. Peirce was highly effective as an administrator of the Coast Survey until he returned to full-time teaching at Harvard University in 1874.

Peirce was honored on both sides of the Atlantic with membership in scientific societies: He became a member of the American Philosophical Society and one of 50 foreign members of the Royal Society of London in 1852, a fellow of the American Academy of Arts and Sciences in 1858, and an honorary fellow of the University of Saint Vladimir in Kyiv in 1860. When Peirce died, his pallbearers included James Joseph Sylvester, J. Ingersoll Bowditch, Newcomb, and Oliver Wendell Holmes, a Harvard classmate who later wrote a tribute to Peirce.

Jeff Suzuki

Selected References

Beach, Mark (1972). "Was there a Scientific Lazzaroni? "In *Nineteenth-Century American Science*, edited by George H. Daniels, pp. 115–132. Evanston, Illinois: Northwestern University Press.

Dupree, A. Hunter (1957). "The Founding of the National Academy of Sciences – A Reinterpretation." *Proceedings of the American Philosophical Society* 101: 434–440.

Eliot, Charles W., A. Lawrence Lowell, W. E. Byerly, Arnold B. Chace, and R. C. Archibald (1925). "Benjamin Peirce." *American Mathematical Monthly* 32: 1–30.

James, Mary Ann (1987). *Elites in Conflict: The Antebellum Clash over the Dudley Observatory*. New Brunswick, New Jersey: Rutgers University Press.

Jones, Bessie Zaban and Lyle Gifford Boyd (1971). *The Harvard College Observatory: The First Four Directorships, 1839–1919*. Cambridge, Massachusetts: Harvard University Press.

Matz, F. P. (1895). "Benjamin Peirce." *American Mathematical Monthly* 2: 173–179.

Menand, Louis (2001). *The Metaphysical Club*. New York: Farrar, Straus and Giroux.

Peirce, Benjamin (1852). "Criterion for the Rejection of Doubtful Observations." *Astronomical Journal* 2: 161–163.

——— (1882). *Linear Associative Algebra*, edited by C. S. Peirce. New York: Van Nostrand. Originally published in *American Journal of Mathematics* 4 (1881): 97–229. (Revision of 1870 edition.)

Peiresc, Nicolas-Claude Fabri de

Born **Belgentier, (Var), France, 1 December 1580**
Died **Aix-en-Provence, (Bouches-du-Rhône), France, 24 June 1637**

In addition to fostering scientific correspondence, Nicolas Peiresc discovered the Orion nebula and tracked the satellites of Jupiter in order to solve the longitude problem. He was the son of Réginald Fabri, descendant of a Pisan family, and Margareta Bomparia, both of whom represented notable Provencal lineages and connections. After Peiresc attended Jesuit schools in Aix and Avignon, his father and uncle sent him on an extended trip to Italy (1599–1602) to prepare him further for the family post in the parliament of Provence. During his first year in Italy, Peiresc studied in Padua, where he met **Galileo Galilei** before settling in Montpellier to study law. After finishing his legal studies there, Peiresc attained a doctorate degree in civil law in Aix (1604). When his uncle died on 24 June 1607, leaving open the family *parlement* position, Peiresc immediately filled the seat and held it for 30 years until his own death. For much of his adult life, Peiresc was at the center of an important correspondence network as a mediator to whom others looked for diplomatic solutions. For example, when Galilei was put under house arrest, Peiresc warned Cardinal Barberini that the failure to change Galilei's verdict might yield a comparison with Socrates' trial and similar condemnation.

Among his many interests and activities, Peiresc dedicated time to astronomical observations. In November 1610, while repeating some of Galilei's observations published in *Sidereus Nuncius*, Peiresc and cleric Joseph Gaultier de la Valette (1564–1647) were apparently the first to observe a nebula in the constellation of Orion. Peiresc also observed the moons of Jupiter with the help of Gaultier and mathematician **Jean Morin**, and subsequently wrote a commentary he never published.

Peiresc's most important and practical astronomical contribution stems from his work on longitude calculations. The main problem of determining longitude involves finding an accurate timekeeper. In the early 17th century, the regular motions in the heavens provided the most accurate clock. Peiresc originally planned to use the satellites of Jupiter as that celestial clock. Between November 1610 and May 1612, Peiresc made regular observations of the Jovian moons. Near the end of this period of observation, Peiresc felt his calculations were adequate enough for testing. He sent his assistant Jean Lombard to make observations of the moons of Jupiter in locations as far away as North Africa, Malta, and the Levant. The local time difference between the appearance of a configuration of Jupiter's satellites as they appeared in Aix (according to Peiresc's tables) and the appearance of that same configuration observed in Malta (by Lombard) could be used to calculate the difference in longitude between the two locations. After Lombard's mission failed due to the difficulty of this technique, Peiresc largely abandoned work in astronomical observations for 16 years.

Between 1616 and 1623, Peiresc lived in Paris, where he met and associated closely with the circle of thinkers surrounding the librarians Pierre and Jacques Dupuy. It was through the Dupuy brothers that Peiresc met **Marin Mersenne** and others. Peiresc never married; his relationships did not extend beyond the intellectual friendships he had with such men as the Dupuy brothers and Mersenne. In 1618, Louis XIII granted Peiresc the abbacy of a monastery in Guîtres, north of Bordeaux, making Peiresc's ties to the church stronger and his distance from marriage further. Peiresc returned to Aix in the summer of 1623; in the next year, he took the tonsure in order to regularize his position as abbé of the Guîtres monastery.

By 1628, Peiresc again took up the task of establishing longitude positions from his home in Aix, but with a different plan. He determined to use observations of lunar and solar eclipses made in different cities to establish the separation of longitude between them. To begin this new project, he requested others (among them the Dupuys) to send him observations of eclipses that occurred in January and February of 1628. He later distributed the observed times of the eclipses, which could then be compared to astronomical tables. According to the observations made in Paris and Aix in 1628, Peiresc calculated that the separation in longitude between the two cities was 3° 30′ 2″ greater than the previous standard.

Because of plague and public unrest between 1629 and 1631, Peiresc temporarily abandoned Aix for his country home in Belgentier, where he was unable to continue his astronomical observations. In 1632, Peiresc returned to Aix, where he resumed his telescopic observations and his larger project of gathering observations of eclipses from diverse locations in order to make longitudinal calculations. To aid him in this project, he recruited priests and Jesuits stationed in various locations from Rome to Mount Sinai. Despite the condemnation of Galilei in 1633, Peiresc explained to his recruits that making observations would not bring harm to souls and could even encourage others to follow in their footsteps.

For making eclipse observations, Peiresc stressed the need to use a telescope, but the network of observers he assembled did not always perform as he wished. Complications included letters and instruments lost in the mail, bad weather, sick observers, and faulty clocks – a crucial problem given the importance in determining the precise time of any given observation. These and other difficulties made Peiresc's task of compiling observations all the more difficult. He was, however, able to find some success with a lunar eclipse on 28 August 1635; it resulted in correcting, and reducing by about 1,000 km, the length of the Mediterranean found on contemporary maps. To further increase the accuracy of such observations, Peiresc, along with the help of **Pierre Gassendi** and others, established the Provençal school of astronomy, where he could instruct future observers and achieve more uniformity in his project. However, after his death, the school folded, with most of the remaining students and teachers moving to Paris.

Derek Jensen

Selected References

Baumgartner, Frederic J. (1991). "The Origins of the Provençal School of Astronomy." *Physis*, n.s., 28: 291–304.

Chapin, Seymour L. (1957). "The Astronomical Activities of Nicolaus Claude Fabri de Peiresc." *Isis* 48 (1957): 13–29.

Gassendi, Pierre (1657). *The Mirrour of True Nobility and Gentility: Being the Life of the Renowned Nicolaus Claudius Fabricius, Lord of Pieresk, Senator of the Parliament at Aix*, translated by W. Rand. London: Printed by J. Streater for Humphrey Moseley. (Still the standard for biographical information on Peiresc.)

——— (1992). *Gassendi-Peiresc correspondance*. Le Chaffaut: Terradou.

Humbert, Pierre (1933). *Un amateur: Peiresc, 1580–1637*. Paris, Desclée de Brouwer.

Miller, Peter N. (2000). *Peiresc's Europe: Learning and Virtue in the Seventeenth Century*. New Haven: Yale University Press.

Pearl, Jonathan L. (1978). "Peiresc and the Search for Criteria of Scientific Knowledge in the Early 17th Century." *Proceedings of the Annual Meeting of the Western Society for French History* 6: 110–119.

——— (1984). "The Role of Personal Correspondence in the Exchange of Scientific Information in Early Modern France." *Renaissance and Reformation* 20: 106–113.

Sarasohn, Lisa T. (1993). "Nicolas-Claude Fabri de Peiresc and the Patronage of the New Science in the Seventeenth Century." *Isis* 84: 70–90.

Tolbert, Jane T. (1999). "Fabri de Peiresc's Quest for a Method to Calculate Terrestrial Longitude." *Historian: A Journal of History* 61: 801–819. (On Peiresc's large-scale project of longitudinal calculations.)

Pèlerin de Prusse

Born **Chelm (Chelmno), Poland, mid-to-late 1330s**
Died **after 1362**

Pèlerin de Prusse is known less for his original work than for disseminating knowledge of astronomy and astrology. He studied at the University of Paris under the direction of the well-known scientist and mathematician **Albert of Saxony**, and graduated as master of arts by 1359. In that year Pèlerin requested and received permission from the University of Paris to deliver "extraordinary lectures" (public lectures given outside ordinary lecture hours) on a book that may have been about either astronomy or astrology: The document referring to the lectures uses the equivocal Latin word *astrologia*, and the book to be lectured on is not specified. It is perhaps worth noting that Pèlerin's associate Robert le Normand had in the previous year been given similar permission for extraordinary lectures on **Ptolemy**'s astrological *Tetrabiblos* and on the pseudo-Ptolemaic *Centiloquium*, also astrological.

No doubt on the basis of his academic work, Pèlerin was soon appointed court scholar to Charles, Duke of Normandy (later King Charles V of France). He was but one of a number of astrologers–astronomers associated at one time or another with Charles' court, but he seems to have been especially favored. Court records call him Charles' "beloved clerk," and Pèlerin was in the early 1360s installed in his own rooms, with his manservant, in Charles' new palace, the Hôtel Saint-Pol.

Two surviving works by Pèlerin testify to the nature of his contributions to astronomy. One is the *Livret de elecions* (1361), an astrological treatise with some astronomical side benefits, such as the promotion of a relatively new and sophisticated instrument, the planetary equatorium. The other work, the *Practique de astrolabe* (1362), is on a more familiar instrument, the planispheric astrolabe. The *Practique* is a strictly astronomical work, based largely on a Latin treatise said to have been translated from the Arabic of **Māshā 'allāh ibn Atharī**. Pèlerin expressly, and accurately, disclaims any originality for his work. His role is to put knowledge of the stars into French at the command of Charles, who was at the time sponsoring similar efforts by other scholars, including the above-mentioned Le Normand as well as the famous scholar-bishop **Nicole Oresme**. The same kind of vernacularizing role was played later by the English writer **Geoffrey Chaucer**, whose *Treatise on the Astrolabe* (*circa* 1391) was also based on Māshā 'allāh. What George Sarton says about Chaucer can be said with equal justice about Pèlerin: "the study of [his] scientific knowledge is important not so much from the point of view of the history of science *stricto sensu*, but rather for the understanding of popular diffusion of scientific ideas of his time."

Pèlerin's date of death is inferred from the date of his last known written work.

Edgar Laird

Alternate names

Preussen, Pilgrim Zeleschicz von
Peregrinus de Prussia

Selected References

Berger, Harald (2000). "Zu zwei Gelehrten des 14. Jahrhunderts." *Sudhoffs Archiv* 84: 100–103. (For documentation of Pèlerin's university career.)

Laird, Edgar and Robert Fischer (1995). *Pèlerin de Prusse on the Astrolabe*. Binghamton: State University of New York. (Pèlerin's *Practique* and excerpts from his *Livret* are presented.)

Shore, Lys Ann (1989). "A Case Study in Medieval Nonliterary Translation: Scientific Texts from Latin into French." In *Medieval Translators and Their Craft*, edited by Jeanette Beer, pp. 297–327. Kalamazoo, Michigan: Medieval Institute. (Pèlerin is mentioned among other translators.)

Peltier, Leslie Copus

Born **near Delphos, Ohio, USA, 2 January 1900**
Died **Delphos, Ohio, USA, 10 May 1980**

As a prolific variable star observer as well as the independent discoverer of 12 comets and six novae in over six decades of observing, Leslie Peltier established himself as the leading amateur astronomer of his time. His autobiography, *Starlight Nights*, lead countless readers into astronomy. **Harlow Shapley** called him "the world's greatest non-professional astronomer."

Peltier was born to Stanley W. Peltier, a strawberry farmer and distributor, and Resa (*née* Copus) Peltier, a schoolteacher. Both parents were avid readers; their home was filled with books on many subjects. Peltier's love of books, reading, and his woodworking, designing, and architectural skills were absorbed from his parents.

Peltier's elementary education in a one-room schoolhouse was typical of the time. Living on a farm gave him the independence to study and observe whatever interested him as he went about his farm chores. He taught himself the geology, flora, and fauna of the Delphos area. At 5 years of age, through the kitchen window Peltier had noticed bright stars in the night sky, which his mother identified as the Seven Sisters, or Pleiades. But it was not until a dark night 10 years later that Peltier suddenly realized that he knew much about many aspects of nature but not about the stars. That night marked the beginning of his avid interest in astronomy. The librarian at the Delphos Public Library suggested that he read Martha Evans Martin's *The Friendly Stars*. Using this simple but well-written book, Peltier learned about the bright stars starting with Vega.

Peltier purchased his first telescope, a 2-in. French spyglass, with 18 dollars that he earned by picking 900 quarts of strawberries on the family farm for 2 cents per quart. The telescope had a focal length of 36 in. with eyepieces for 35× and 60× magnification. Thus began the long and successful observing career that would span more than six decades.

The next book Peltier consulted was **William Tyler Olcott**'s *A Field Book of the Stars*. Olcott invited those with small telescopes who were interested in assisting professional astronomical research to write to him. Peltier wrote immediately; Olcott's response described the systematic observations of variable stars by the American Association of Variable Star Observers [AAVSO], and included an AAVSO application. When Peltier returned the application, he received charts and instructions for observing. On 1 March 1918 Peltier made his first variable star observation of R Leonis. On 1 March every year thereafter, Peltier observed R Leonis in commemoration of that first observation. Beginning with his first report, for March 1918, Peltier sent consecutive monthly reports to the AAVSO until his death, never missing a single month. Peltier's monumental total of 132,123 observations will insure that he remains among the leading variable star observers of all time.

In November 1918, the AAVSO offered to loan Peltier a 4-in. Mogey refractor. News of the young observer was spreading among professional astronomers, and shortly thereafter Princeton's **Henry N. Russell** loaned Peltier a 6-in. refractor, a comet seeker of short focus, through the AAVSO. In 1921, Peltier's father helped him build his first full observatory; it was ready for observing in January 1922.

Around 1937, the idea for the transportable, rotating observatory came to Peltier, perhaps due to his experience and expertise in furniture design. His idea was to be seated in a comfortable chair while observing variable stars or hunting for comets, and have the whole observatory revolve with the chair and telescope. Using both new and junkyard parts, Peltier built such an observatory. The chair included an adjustable headrest for comfort during long observing sessions, and a hot plate to warm his feet. Peltier's unique design later became famous as the "Merry-Go-Round observatory."

In 1959, the Miami University of Ohio donated a 12-in. Clark refractor, and a dome to contain the telescope. With the larger telescope, Peltier was equipped to observe much fainter stars and so he modified his observing program, concentrating on stars fainter than 11th magnitude. One important aspect of the new program was Peltier's observation of cataclysmic variable stars during their quiescent phase.

Peltier made independent discoveries of 12 comets, between 13 November 1925, and 26 June 1954; 10 of these comets bear his name. He carved the year of each comet discovery into the wooden tube of his 6-in. comet seeker. Peltier also made independent discoveries of six novae or recurring novae (Nova Aurigae 1918; Nova Cygni 1920; RS Ophiuchi 1933; DQ Herculis and CT Lacertae [both 1934]; and Nova Herculis 1963).

Except with his close friends, Peltier was a shy man and was uncomfortable with strangers. However, through the years Peltier met many prominent astronomers who came to visit him in Delphos, including **George van Biesbroeck**, **William Morgan**, **Bart** and Priscilla **Bok**, **Donald Menzel**, **Polydore Swings**, **John Hall**, **William Hiltner**, Clyde Fisher, and Walter Scott Houston. In 1932, Houston persuaded Peltier to travel with him to an AAVSO meeting in Cambridge, Massachusetts – a rare occurrence for Peltier who seldom left Delphos.

Peltier met Dottie Nihiser, daughter of a local beekeeper, in 1925. They shared many interests including geology and a love of nature. She attended Ohio Wesleyan University and had a great interest in archaeology. On the other hand, Peltier never finished high school. When his brother left for World War I, Leslie dropped out of high school and took over more farm chores, but his acquisition of knowledge never stopped. He married Dottie on 25 November 1933; she bore two sons, Stanley H. and Gordon J. Peltier. In 1934, Peltier left the farm to work at the Delphos Bending Company, designing children's toys and juvenile furniture. He was still employed by the company at the time of his death.

During the 1930s and 1940s, Peltier wrote articles on comet hunting, nature, and equipment design that appeared in magazines such as *The American Photographer*, *Popular Science*, *Nature Magazine*, and *Sky & Telescope*. He was the subject of articles in many popular newspapers and magazines. In 1965 Peltier's first book, *Starlight Nights, the Adventures of a Star-Gazer* was published. *Starlight Nights* is an autobiographical ode to the joys of observing both the night sky and nature. The language of the book is poetic, humorous, and beautifully descriptive in the style of a 19th-century naturalist. In 1972, Peltier published *Guideposts to the Stars: Exploring the Skies throughout the Year*. Departing from astronomy, in 1977 he published *The Place on Jennings Creek*, a natural history of their home, Brookhaven, and its surrounding areas.

Peltier was awarded a honorary D.Sc. degree by Bowling Green State University (Ohio) in July 1947. In August 1965, at the dedication of **Clinton Ford**'s observatory near Wrightwood, California, the mountain on which the observatory stood was christened Mount Peltier. Minor Planet (3850) was named Peltier in his honor in 1989. From 1925 to 1954 Peltier received the Astronomical Society of the Pacific's Donohoe medal for most of his comet discoveries. AAVSO honored Peltier with its First Merit Award in 1934, and with the Nova Award in 1963. Peltier also received the G. Bruce Blair Award from the Western Amateur Astronomers. After his death, the Astronomical League established the Leslie C. Peltier Award "for significant contributions to observational astronomy" and made the first award posthumously to Peltier himself.

Brenda G. Corbin

Selected References

Anon. (12 May 1980). "Delphos Astronomer: Heart Attack Claims Life of Leslie Peltier." *Delphos* (Ohio) *Herald*.

Hurless, Carolyn (1980). "Leslie Peltier Remembered." *Sky & Telescope* 60, no. 2: 104–105.

——— (1980). "Our Friend, Leslie Peltier: A Personal Reminiscence." *Journal of the American Association of Variable Star Observers* 9 no. 1: 32–34.

Peltier, Leslie C. (1965). *Starlight Nights: The Adventures of a Star-Gazer*. New York: Harper and Row.

——— (1972). *Guideposts to the Stars: Exploring the Skies throughout the Year*. New York: Macmillan.

——— (1977). *The Place on Jennings Creek*. Chicago: Adams Press.

——— (1986). *Leslie Peltier's Guide to the Stars: Exploring the Sky with Binoculars*. Milwaukee: AstroMedia, Kalmbach Publishing. (The foreword by Walter Scott Houston includes personal reminiscences.)

Peregrinus de Maricourt, Petrus

Flourished **France, *circa* 1269**

Petrus Peregrinus is best known for his *Epistula de magnete*.

Of Peregrinus' life, almost nothing is known except what is revealed by his works and suggested by his name. Maricourt is almost certainly a reference to the village of Méharicourt in Picardy, and the appellation Peregrinus indicates that he was a crusader. But since his *Epistula de magnete* was written on 8 August 1269 from the siege of Lucera in Italy, the assault on which had been declared a crusade, it need not be assumed that he had visited the Holy Land.

Epistula de magnete is addressed to Peregrinus's dear friend Sigerus de Foucaucourt. Although not primarily an astronomical work, the text does describe two instruments with astronomical significance: an instrument incorporating a magnetic compass that could be used to determine the azimuth of celestial bodies and a magnetic clock, in the form of a lodestone terella (spherical magnet), which Petrus claimed would mimic the diurnal rotation of the heavens. This text was read by, and influenced, **William Gilbert**, and in the plagiarized form of the *De Natura Magnetis* (1562) of Jean Taisnier, it was studied by **Johannes Kepler** some years before Gilbert's own *De magnete* (1600) was published. It may, therefore, have contributed to Kepler's conceptualization of celestial forces.

Peregrinus also wrote, sometime after 1263, a *Nova Compositio Astrolabii Particularis*, a treatise on the construction of the astrolabe notable for its clarity, its comprehensiveness, and its unusual choice of projection. Petrus described both the standard stereographic projection from one pole and the universal projection of **Zarqālī** in which the West Hemisphere and East Hemisphere of the celestial sphere are projected onto a single plane, but opted for a projection in which the North Celestial Hemisphere and South Celestial Hemisphere were both projected onto the equatorial plane. However, this treatise does not seem to have been very popular; it survives in only four manuscripts, and does not appear to have been printed in the early modern period.

Adam Mosley

Selected References

Grant, Edward (1974). *A Source Book in Medieval Science*. Cambridge, Massachusetts: Harvard University Press. (For a reprint of Brother Arnold's translation of Peregrinus' 1269 *Epistula de magnete*. See also A. Schoedinger, *Readings in Medieval Philosophy*. Oxford: Oxford University Press, 1996.)

Kepler, Johannes (1945). *Gesammelte Werke*. Vol. 13, *Briefe 1590–1599*, edited by Max Caspar. Munich: C. H. Beck. (Kepler's reading of Taisnier is attested in his letter to Herwart von Hohenburg of 26 March 1598.)

Petrus Peregrinus de Maricourt (1995). *Opera*, edited by Loris Sturlese and Ron B. Thomson. Pisa: Centro di cultura medievale della Scuola normale superiore. (For an annotated edition of Petrus's works.)

Roller, Duane H. D. (1959). *The* De Magnete *of William Gilbert*. Amsterdam: Menno Hertzberger. (The relationship between Petrus Peregrinus's work and Gilbert's is discussed herein.)

Schlund, E. (1911). "Petrus Peregrinus von Maricourt. Sein Leiben und seine Schriften." *Archivum Franciscanum Historicum* 4: 436–455, 633–643. (For a survey of what can be concluded from the scant information about Petrus's life it is still worth consulting this source.)

Peregrinus, de Prussia

➔ **Pèlerin de Prusse**

Perepelkin, Yevgenij Yakovlevich

Born **Saint Petersburg, Russia, 4 March 1906**
Died **probably 1937**

Pulkovo Observatory's Yevgenij Perepelkin produced a nonhomogeneous model of the solar chromosphere in the 1930s. He was imprisoned and executed during a Stalin purge. Craters on both the Moon and Mars are named for him.

Selected Reference

Vitinsky, Yu. I. (1981). *The Main Astronomical Observatory of the Academy of Sciences of the USSR at Pulkovo*. Leningrad (Saint Petersburg), Russia: Nauka.

Péridier, Julien Marie

Born **Sète, Hérault, France, 3 February 1882**
Died **Le Houga, Gers, France, 19 April 1967**

An electrical engineer by profession, Julien Péridier was an accomplished amateur astronomer who devoted the later years of his life to the development and operation of a substantial private observatory and scientific library. Péridier's early interest in astronomy can be traced to 1900 when he first observed variable stars. In 1933, Péridier established an observatory near the village of Le Houga in southwest France. It was equipped with a double 8-in. refractor with optics by André Joseph Alexandre Couder, a 12-in. Newtonian reflector that was made by George Calver, and a small transit telescope from Troughton & Simms. For nearly 30 years, Péridier observed actively from this station, while hosting young French astronomers who used his facilities for their own research as well as the observatory's programs. The main subjects of the research carried out at Le Houga were planetary physics, photometry, and stellar photometry, especially studies of variable stars and flare stars. There was also some work on double stars and galaxies. Péridier not only directed and sponsored this research but also published the results in the *Annales Astrophyśique* and also a score of *Publications de l'observatoire du Houga*, copies of which were exchanged with major observatories around the world. The Le Houga Observatory was selected by **Donald Menzel** as one of the sites from which Harvard College Observatory successfully observed the occultation of Regulus by Venus in July 1959. The last major project at the Le Houga Observatory was a 5-year National Aeronautics and Space Administration-sponsored program, conducted jointly with Harvard University from 1961 to 1965, involving multicolor photoelectric photometry of the Moon and planets with the 12-in. reflector.

Gérard de Vaucouleurs, who was both a collaborator at Le Houga from 1939 to 1949 and Péridier's longtime friend, likened him to the great private sponsors of American astronomical research including **Percival Lowell** and **Robert McMath,** as well as the great French amateur **René Jarry-Desloges**. This brief biography was extracted from an obituary prepared by de Vaucouleurs.

Thomas R. Williams

Selected Reference

Vaucouleurs, G. de (1968). "Obituary." *Quarterly Journal of the Royal Astronomical Society* 9: 228–229.

Perrin, Jean-Baptiste

Born **Lille, Nord, France, 30 September 1870**
Died **New York, New York, USA, 17 April 1942**

French physico-chemist Jean Perrin was one of the early enthusiasts for nuclear (subatomic) energy sources for the Sun and stars, along the lines pursued more thoroughly by **Arthur Eddington**. He was the son of an army officer, who died soon after Jean's birth. He entered the Paris École Normale Supérieure in 1891, receiving his doctoral degree in 1897 for work on cathode rays and X-rays. Perrin showed that cathode rays are deflected in magnetic fields and so must carry negative charges, part of the evidence that led J. J. Thompson to the discovery of the electron.

Perrin began teaching at the University of Paris (Sorbonne) in 1897, and he was given a chair in physical chemistry there in 1910. Perrin remained at the Sorbonne until 1940, when he emigrated to the United States. Perrin was married in 1897 to Henriette Duportal; they had two children. Although he did not die in France, Perrin was eventually (1948) reburied in the Panthéon in Paris.

Perrin's work mainly focused on the nature of molecules. The atomic theory, which claimed that elements are made up of discrete particles called atoms and that chemical compounds are made up of molecules, was not fully appreciated at the end of the 19th century, and it had important opponents like Ernst Mach (1838–1916) and Wilhelm Ostwald. Robert Brown in 1827 had described the motion of very small particles, suspended in a fluid; in 1905 **Albert Einstein** gave some quantitative explanations for Brownian motion. Perrin studied colloidally suspended particles undergoing Brownian motion, and in 1908 started a series of experiments on this subject. Only then did he learn of Einstein's work, and finally – by using the "ultramicroscope" – he was able to confirm Einstein's predictions experimentally. Perrin was able to work out the size of the water molecule and a precise value for Avogadro's number. These results made clear that atomism was more than just a useful hypothesis. Already in 1913 Perrin had summed up the then known facts on molecules in his influential book *Les Atomes*. He was awarded the 1926 Nobel Prize in Physics.

During World War I, Perrin served in the Engineering Corps of the French army, working on remote acoustical detection of submarines and artillery fire, and inventing for the purpose a device called the telesite meter. In later years Perrin also became involved with institutional and organizational development of science in France. Thus in the late 1930s he was responsible for establishing both the Centre National de la Recherche Scientifique and the Palais de la Découverte in Paris at the 1937 International Exposition. He also was influential in the establishment of the Institut d'Astrophysique in Paris, and in the construction of the large Observatoire de Haute Provence. Perrin held honorary doctorates from several universities and in 1923 was elected a member of the French Academy of Sciences and served as its president in 1938.

Horst Kant

Selected References

Nye, Mary Jo (1972). *Molecular Reality: A Perspective on the Scientific Work of Jean Perrin*. London: Macdonald.

Perrin, Jean Baptiste (1913). *Les atomes*. Paris.
——— (1950). *Oeuvres scientifiques de Jean Perrin*. Paris.
Ranc, Albert (1945). *Jean Perrin – Un Grand Savant au Service du Socialisme*. Paris.
Townsend, J. S. (1943). "Jean Baptiste Perrin." *Obituary Notices of Fellows of the Royal Society* 4: 301–305.

Perrine, Charles Dillon

Born **Steubenville, Ohio, USA, 28 July 1867**
Died **Villa General Mitre, Argentina, 21 June 1951**

Charles Perrine, discoverer of Himalia (the sixth) and Elara (the seventh) satellites of Jupiter and nine new comets, began his astronomical career in the United States, but spent a large portion of it studying the Southern Hemisphere skies as director of the Argentine National Observatory at Córdoba, Argentina.

The son of Peter and Elizabeth Dillon (*née* McCauley) Perrine, Charles graduated from high school in Steubenville, Ohio, in 1884, and by 1886 had moved to San Francisco, California, where he worked as a business secretary until 1893. Interested in astronomy since high school, Perrine volunteered to observe the 1 January 1889 solar eclipse as part of an expedition organized by the Lick Observatory and became acquainted with **Edward Holden**. By 1893, Perrine had convinced Holden that he should be employed at Lick Observatory as the observatory secretary.

Perrine began his career in astronomy assisting Holden with nighttime celestial photography. By 1895, he had demonstrated sufficient aptitude that he was appointed assistant astronomer while continuing to serve as the observatory secretary. That same year, using the observatory's 12-in. Clark refractor, Perrine discovered the first of many comets credited to him. He was given full-time responsibility as an assistant astronomer 2 years later. As **James Keeler**'s assistant, Perrine continued to expand his observing repertoire, gaining skill on the Crossley 36-in. reflector as Keeler struggled to subdue the mechanical problems that plagued that instrument. After Keeler died and **William Campbell** was appointed to replace him, Campbell promoted Perrine to a full status as an astronomer. Mechanical upgrading of the Crossley reflector was incomplete at that time, so Campbell gave Perrine full responsibility for the instrument and its program. One of the many modifications Perrine made to the instrument was to place the photographic plate holder inside the tube at the primary focus, thus eliminating light loss from one reflection. Perrine's celestial photographs taken with the Crossley were excellent and in fact in some cases were substituted for photographs taken by Keeler before the instrument was fully functional in the *Keeler Memorial Volume* of the *Publications of the Lick Observatory*.

Perrine already had received five Astronomical Society of the Pacific's Donohoe Medals, each one for the discovery of "an unexpected" comet, when he was awarded the Paris Academy of Sciences' Lalande Prize and Gold Medal in 1897 for his comet discoveries. In total, between 1895 and 1902 Perrine discovered nine new comets and recovered four returning periodic comets. He made good use of the Crossley reflector for several other discoveries. Using the 36-in. telescope, he discovered the apparent superluminal motion of the expanding light bubble around Nova Persei (1901). Thought to be a nebula, the visual appearance was actually caused by the light from the nova event reflected from the surrounding interstellar medium as the light moved outward from the star. Perrine studied this phenomenon using photographic, spectroscopic, and polarization techniques. In 1904, he discovered Himalia, and found Elara in 1905.

Perrine had a deep interest in solar eclipses. Between 1900 and 1908, he led one Lick Observatory eclipse expedition to Sumatra (1901) and participated in several others. Campbell placed Perrine in charge of the Lick Observatory's observations of the 1901 opposition of the minor planet (433) Eros to measure the Earth–Sun distance. Perrine used Lick Observatory observations of Eros to publish an estimate of the solar parallax based on the Eros data in 1904.

Perrine left the Lick Observatory when he was appointed director of the Argentine National Observatory in Córdoba, Argentina, in 1909. He continued a program of modernization of Córdoba's facilities for conventional astrometry, the traditional program at Córdoba Observatory since it had been founded by **Benjamin Gould**. Equipment for the modernization had been ordered by Gould's successor, **John Thome**, who died before the Repsold meridian circle was completed in Hamburg, Germany. Perrine eventually completed the publication of all of the observatory's astrometric observations in 16 volumes of the *Resultados del Observatorio Nacional Argentino* along with an additional volume containing photographs from the 1910 return of Halley's comet (IP/Halley).

However, astrometry did not interest Perrine. His lasting contribution was the creation of the Astrophysical Station at Bosque Alegre (50 km southwest of Córdoba), which houses a 60-in. reflecting telescope. Almost from the day he arrived at Córdoba, Perrine was committed to establishing a leading position for Argentina in the emerging field of astrophysics. He persuaded the national government to fund the 60-in. telescope for which the mirror and mounting were to be produced in the shops at Córdoba. After upgrading the mechanical shops in anticipation of work on the 60-in. telescope, Perrine was successful in building a 30-in. reflecting telescope in the Córdoba shops and achieved noteworthy results with this instrument, the largest in the Southern Hemisphere. Unfortunately, he underestimated the difficulty of working the larger glass, and the project dragged on for many years. The delays in finishing the 60-in. mirror created a severe political problem for Perrine in the xenophobic atmosphere that dominated Argentina in the late 1920s and 1930s. He retired from the director's position of the Córdoba Observatory in 1936, but continued to live in South America until his death. Perrine was replaced by the Argentine astrophysicist Enrique Gaviola, who wisely arranged to have the large mirror finished by J. W. Fecker in the United States.

As a consequence of Perrine's farsightedness as the third director of the observatory, Córdoba became, for a time, the main astrophysical station in the Southern Hemisphere. Perrine wrote more than 200 papers between 1896 and 1947 on a variety of astronomical topics, including radial velocities of stars, comets, solar eclipses, the nature of globular clusters, nebulae, nova, and astronomical instrumentation.

In 1905, Perrine married Bell Smith of Philadelphia, Pennsylvania, USA. He was honored by Santa Clara College, California, in 1905 when that institution conferred an honorary D.Sc. upon him. Perrine was elected president of the Astronomical Society of the Pacific in 1902 and a foreign associate of the Royal Astronomical Society in 1904.

Scott W. Teare

Selected References

Anon. (1908). "Photographs of Nebulae and Clusters, Made with the Crossley Reflector." *Publications of the Lick Observatory* 8 (Keeler Memorial Volume).

Bobone, Jorge (1951). "Charles D. Perrine." *Publications of the Astronomical Society of the Pacific* 63: 259.

Hodge, John E. (1977). "Charles Dillon Perrine and the Transformation of the Argentine National Observatory." *Journal for the History of Astronomy* 8: 12–25.

Jones, H. Spencer (1952). "Charles Dillon Perrine." *Monthly Notices of the Royal Astronomical Society* 112: 273–274.

Osterbrock, Donald E. (1999). "Perrine, Charles Dillon." In *American National Biography*, edited by John A. Garraty and Mark C. Carnes. Vol. 17, pp. 357–358. New York: Oxford University Press.

Osterbrock, Donald E., John R. Gustafson, and W. J. Shiloh Unruh (1988). *Eye on the Sky: Lick Observatory's First Century*. Berkeley: University of California Press, pp. 104, 123, 141–142, 158–162.

Perrine, C. D. (1904). "Recent Spectrographic Observations of *Novae* with the Crossley Reflector." *Astrophysical Journal* 19: 80–83.

——— (1907). "Results of the Search for an Intermercurial Planet at the Total Solar Eclipse of August 30, 1905." *Publications of the Astronomical Society of the Pacific* 19: 163.

——— (1947). "The Total Solar Eclipse of May 20, 1947, in Córdoba." *Publications of the Astronomical Society of the Pacific* 59: 188–189.

Sersic, J. L. (1971). "The First Century of the Cordoba Observatory." *Sky & Telescope* 42, no. 6: 347–350.

Vsekhsvyatskii, S. K. (1964). *Physical Characteristics of Comets*. Jerusalem: Israel Program for Scientific Translations.

Perrotin, Henri-Joseph-Anastase

Born **Saint-Loup, Tarn-et-Garonne, France, 19 December 1845**
Died **Nice, Alpes-Maritimes, France, 29 February 1904**

Henri Perrotin achieved much of his recognition for his observations of planets and asteroids, as well as for his work in celestial mechanics on the determinations of their orbits. His astronomical observations were published in 50 notes in *Comptes rendus de l'Académie des sciences* from 1875 to 1903 and in *Astronomische Nachrichten* from 1875 to 1889.

Perrotin was born into a family of modest means in the southwest of France, and received scholarships for his education at the lycée in Pau. He began his astronomical career with **Félix Tisserand**, a professor of celestial mechanics at the Faculté des Sciences of Toulouse. When Tisserand became director of the Toulouse Observatory in 1873, he appointed Perrotin as astronomer. The year after, the novice astronomer discovered his first asteroid, which he called (138) Tolosa. Perrotin discovered five more, the last one in 1885.

Following the method used by **Urbain Le Verrier** for the big planets, particularly Jupiter and Saturn, Perrotin started to develop the first precise orbital theory of the minor planet (4) Vesta, the topic

of his thesis in 1879. In this work, he used a perturbation function of the eighth order in the eccentricities and inclinations, an achievement that has been applied by astronomers to verify recent theories of Vesta.

In 1880, Perrotin was directed by the Bureau des longitudes to banker Raphaël Bishoffsheim to install the private observatory that Bischoffsheim was founding in Nice on Mont Gros. After visiting the most important observatories in Europe to study their organization and development, Perrotin supervised the installation on Mont Gros the next year. As its first director, a position he held for more than 20 years until his death, he devoted himself to setting up a well-equipped observatory, and to supervise its growth. His work and research there related to astrometry (of asteroids, comets, double stars, and satellites), astrophysics (pertaining to the study of planetary surfaces and the velocity of light), and celestial mechanics (orbits of asteroids and planets).

In 1882, the Académie des sciences designated Perrotin as the leader of the expedition to observe the transit of Venus in Patagonia, at Carmen de Patagonès (Argentina) on the banks of the Rio Négro. His return to Nice marked the beginning of the most active part of Perrotin's career. He provided the impetus for a variety of work concerning the construction of instruments and scientific research. Perrotin greatly contributed to the renown of the observatory, supervising the installation of many instruments, including the 30-in. refractor (1886), one of the world's largest at the time. He used it to carry out observations of double stars, and of large and small planets.

Perrotin tried to verify two findings of **Giovanni Schiaparelli**: his discovery of the so called canals of Mars and his determination of Venus's rotational period as 225 days. On both counts, Perrotin confirmed the Milan astronomer's results, though later findings disproved the former and showed the inaccuracy of the latter. Yet his own detailed observations of the Martian surface proved of great interest, and his drawings of Venus's surface (1890) show that 70 years before their widespread recognition, he had visually noticed the now famous Y- and Ψ-shaped markings recognized today in the Venusian clouds.

Perrotin also organized meteorological and magnetic observations, and contributed to physics by determining the velocity of light. He made a series of accurate observations using the slotted-wheel technique developed by **Armand-Louis Fizeau**, and applied it to beams sent between Mont Gros and Mont Vinaigre, the highest point of the massif of Estérel 46 km away. The 299,880 km/s value obtained in 1902 was regarded as the best until the measurements taken by **Albert Michelson** in 1926.

Perrotin founded the *Annales de l'Observatoire de Nice* in 1887. He managed the publication of the first ten volumes, and wrote several of them. He was twice awarded the Lalande Prize by the Académie des sciences (1875 and 1884), and was a corresponding member of the Académie des sciences (1892) and of the Bureau des longitudes (1894).

Raymonde Barthalot

Selected References

Anon. (1904). "Todes-Anzeige: Joseph Anthanase Perrotin." *Astronomische Nachrichten* 165: 253–256.

Anon. (1904). "M. Henry Perrotin." *Nature* 69: 468.

Anon. (1904). "Obituary." *Observatory* 27: 176–177.

Perrotin, Henri (1880). "Théorie de Vesta." *Annales de l'Observatoire astronomique magnétique et météorologique de Toulouse* 1: B1–B90. This first precise theory of the asteroid also appeared in *Annales de l'Observatoire de Nice* 3 (1890): B1–B118; 4 (1895): A3–A71.

——— (1881). *Visite à divers observatoires d'Europe*. Paris: Gauthier–Villars, (A description of the 31 most important European observatories at that time.)

——— "Observations des canaux de Mars." *Comptes rendus de l'Académie des sciences* 106 (1888): 1393–1394; "Observations de la planète Mars." *Comptes rendus de l'Académie des sciences* 115 (1892): 379–381; and "Sur la planète Mars." *Comptes rendus de l'Académie des sciences* 124 (1897): 340–346.

——— "Observations de la planète Vénus à l'Observatoire de Nice." *Comptes rendus de l'Académie des sciences* 111 (1890): 587–591; and "Observations de Vénus sur le mont Mounier". *Comptes rendus de l'Académie des sciences* 122 (1896) 442–446. (Results of the period of rotation of Venus and the drawings [eight] showing the Y and Psi structures.)

——— (1903). "Parallaxe solaire déduite des observations d'Eros." *Bulletin astronomique* 20: 161–165.

Stephan, E. (1904). "J. A. Perrotin." *Annales de l'Observatoire de Nice* 8: i–iv.

Peters, Christian August Friedrich

Born **Hamburg, (Germany), 7 September 1806**
Died **Kiel, Germany, 8 May 1880**

During the second third of the 19th century, a time of widespread emphasis on astrometrical precision, Christian Peters studied from and worked with the first two observers of stellar parallax and deduced the orbit of an unseen companion of the star Sirius. Born in a family of merchants, Peters displayed a precocious ability in the mathematical sciences. He received encouragement from **Christian Schumacher**, the genial founder and first editor of the leading astronomical scholarly journal, *Astronomische Nachrichten*. Peters's first astronomical publications, completed before he was 20, appeared in that journal in 1826.

On matriculating at the University of Königsberg in East Prussia (presently Russia) and studying under **Friedrich Bessel**, Peters joined the leading exponent of the scientific value of precision measurement. Peters's scientific career developed in the direction suggested by Bessel, who advocated the application of advanced mathematical techniques to improve precision by paying close attention to the analysis of both stochastic and systematic errors. Bessel's first extended application of careful error analysis involved the seconds pendulum; Peters's 1834 Ph.D. dissertation discussed the effect of air resistance on the pendulum. Bessel successfully applied his techniques to the centuries-old search for stellar parallax in 1838, inspired by the preliminary announcement by **Friedrich Struve** in 1836 of a parallax for the star Vega. Peters, who worked from 1839 untill 1849 under Struve's direction at the Pulkovo Observatory, some 400 miles up the Baltic Sea coast from Königsberg, made the critical comparison of the emerging measures of stellar parallax his specialty.

Peters worked productively as one of Struve's four assistants, publishing in 1842 his determination of the motion of the pole star and the constant of nutation. He became an adjunct (1842) and then an extraordinary (1847) member of the Saint Petersburg Academy of Sciences. In 1852, Peters won the Royal Astronomical Society of London Gold Medal for his work at Pulkovo Observatory. He also engaged in a polemical exchange with **Johann von Mädler**, Struve's successor at the University of Dorpat observatory.

In 1844, Bessel announced that irregularities he detected in the proper motions of Sirius and Procyon implied that these stars were orbited by unseen companions. After Bessel's death in 1846, August Ludwig Busch, Bessel's successor at Königsberg, made it possible for Peters to return as full professor of astronomy. In 1851, Peters produced a 60-page paper calculating the orbit of the companion of Sirius, despite Struve's doubts about its existence. While testing what was then the world's largest objective lens, United States telescope-maker **Alvan Clark** observed the predicted companion for the first time in 1862.

In 1854, Peters accepted the prestigious directorship of the Altona Observatory, in a suburb of Hamburg, and with this position also the editorship of the *Astronomische Nachrichten*. While at the professional summit of the German-speaking astronomical community for the next quarter of a century, Peters published a more technical journal (that appeared more frequently) than before, but he was opposed by astronomers working within the Russian empire. The practice of separate publication of the most important results from Pulkovo in the *Astronomische Nachrichten* came to an abrupt end as soon as Peters became editor. German astronomers complained of a greater degree of partisanship in the journal, and its circulation suffered.

After Germany was unified in 1870, the imperial authorities agreed to Peters's 1864 suggestion to move the instruments of the Altona Observatory 50 miles to Kiel and build a larger observatory for the university there. Peters served as professor of astronomy at Kiel from 1874 until his death.

Michael Meo

Selected References

Freiesleben, Hans-Christian (1974). "Peters, Christian August Friedrich." In *Dictionary of Scientific Biography*, edited by Charles Coultston Gillispie. Vol. 10, pp. 542–543. New York: Charles Scribner's Sons. (A concise survey of Peters's career, but it unfortunately speaks of Peters becoming "assistant director" of Pulkovo Observatory, an error which appeared in the usually authoritative *Biographisch-literarisches Handwörterbuch* of J.C. Poggendorff [Leipzig, 1863], Vol. 2, col. 413, but which was corrected both in Vol. 3 [1898], p. 1026, and in the eulogy by Winnecke cited later, referred to by Freiesleben.)

Novokshanova-Sokolovskaya, Z. K. *Vasilii Yakovlevich Struve*. Moscow: Izdatel'stvo Nauka, 1964. (Provides an account of the competition with F. G. W. Struve, and Peters's collaboration in the determinations at Pulkovo of a number of stellar parallaxes. It is evident from the chronological list of Struve's works in Novokshanova-Sokolovskaya, pp. 249–271, that the practice of publishing each important contribution separately in the *Astronomische Nachrichten* stopped as soon as Peters assumed the editorship.)

Olesko, Kathryn M. (1991). *Physics as a Calling: Discipline and Practice in the Königsberg Seminar for Physics*. Ithaca, New York: Cornell University Press. pp. 1–99. (A thorough treatment of Bessel's influence within the University of Königsberg and on Peters.)

Peters, C. A. F. (1842). *Numerus constans nutationis ex ascensionibus rectis stellae polaris in* Specula Dorpatensi *annis 1822 ad 1838 observatis deductus*. St. Petersburg.

——— (1851). "Ueber die eigene Bewegung des Sirius." *Astronomische Nachrichten* 32: 1–58.

——— (1853). "Recherches sur la parallaxe des étoiles fixes." *Mémoires de l'Académie impériale des sciences de St-Pétersbourg*, 6th ser., 5.

Winnecke, [Friedrich August Theodor] (1881). "Nekrolog." *Vierteljahrsschrift der Astronomischen Gesellschaft* 16: 5–8. (States that Peters took the editorship of the *Astronomische Nachrichten* over the opposition of "the leading Russian astronomers," which phrase likely refers to Otto Wilhelm Struve, director of Pulkovo at the time.)

Peters, Christian Heinrich Friedrich

Born **Coldenbüttel, (Schleswig-Holstein, Germany), 19 September 1813**

Died **Clinton, New York, USA, 18 July 1890**

Christian Peters, who played a subsidiary role in the rise of the American astronomical community to international prominence, was one of several senior scientists who emigrated from Europe to the United States prior to the American Civil War.

Peters received a classical education at the Gymnasium in Flensburg, Germany, obtained a Ph.D. from the University of Berlin in 1836 in theoretical physics, and continued his studies under **Carl Gauss** at the Göttingen Observatory, after which he devoted himself to positional astronomy. Under Gauss's influence, German astronomers of the 1830s engaged in a substantial program of geodetic mapping. Peters worked on a survey of Mount Etna in Sicily from 1838 until 1843, and was then promoted to director of the government trigonometric survey of Sicily. When in 1848 that island was swept by an antimonarchial revolution, one of a dozen or so that year throughout Europe, Peters supported the revolutionaries. He left for neutral Turkey when the monarchial troops invaded and reestablished the royal government.

During 5 years in Constantinople with few prospects for scholarly work, Peters managed to learn Arabic and Turkish. He then joined the group of skilled scientists going to the United States. In 1854, at the urging of the American ambassador to Turkey, he emigrated to Massachusetts, where **Benjamin Gould**, who had known him in Göttingen, helped him get a position in the United States Coast and Geodetic Survey, working under Gould's supervision. Peters arrived in his new country the same year as **Franz Brünnow** came from the University of Berlin Observatory to head the new observatory at the University of Michigan.

Mercantile philanthropists in Albany, New York, had subscribed generously to the establishment of a research-grade astronomical observatory, the first American observatory devoted solely to research, and invited Gould to be director; he agreed to act as a scientific consultant during construction. Although there is little doubt that the head of the Coast Survey, **Alexander Bache**, would have allowed Gould to be present during construction, he preferred to remain in Cambridge, Massachusetts, and to send Peters to Albany. The contrast between the diffident and competent Peters and the nervous and abrasive Gould so struck the businesslike trustees that they invited Peters, who resigned from the Coast Survey, to step in as director of what was to be called the Dudley Observatory. Gould, Bache, and the head of the Smithsonian Institution, **Joseph Henry**, interpreted Peters's acceptance of the position as betrayal, and by personal intervention and material inducements persuaded the trustees to release Peters, who was then shunted to the directorship of the Hamilton College Observatory in Clinton, New York.

As director there from 1858 to 1890 under difficult circumstances, Peters conducted a program of precision positional astronomy to the standards of accuracy demanded by contemporary research. The 13.5-in. refractor on an equatorial mount at Hamilton was the largest American-made refractor in the United States when Peters arrived. Its quality had convinced Gould to engage its maker, Charles A. Spencer of Canastota, New York, to build the main research instrument of the Dudley. Noting the acclaim afforded Brünnow's discovery of several new asteroids, Peters undertook the colossal project of mapping all stars, down to 14th magnitude, within a zone of 30° on either side of the ecliptic, during which effort he discovered at least 42 new asteroids of his own. Most of his results appeared in the *Astronomische Nachrichten*, the leading scholarly astronomical journal of the day, although he made use of the *Astronomical Notices* as well, founded and edited by Brünnow while he was in the United States (1858–1862).

Two problems frustrated Peters's ambitious research program. The first was inadequacy of means. He wrote several times to his close friend **George Bond**, director at the Harvard Observatory, lamenting his arrears in pay; on 1 February 1863, he mentioned consulting a lawyer about obtaining the previous year's salary. Fortuitously, in 1867, the owner of a railroad living in nearby Delphi Falls donated enough money to Hamilton to endow its astronomer with a modest salary; the Litchfield professor of astronomy from then on directed the now-renamed Litchfield Observatory. Peters's plan of work envisioned 182 charts of carefully determined star positions, but only 20 were ever published.

A second development doomed Peters's plan: instrumental change. The 1870s and 1880s saw the increasing use of photography in positional astronomy – Peters himself was one of three American representatives participating in the International Astrophotographic Congress held in Paris in 1887 – but all work at Hamilton College was visual. Even so, the quality of his work elevated Peters to the elite National Academy of Sciences in 1876, the Royal Astronomical Society of London in 1879, and the French Legion of Honor in

1889. He was selected as chief of the 1874 American transit of Venus expedition sent to New Zealand.

The Coast and Geodetic Survey, the United States Naval Observatory, the Nautical Almanac, and the Harvard College Observatory were the leaders in the emerging American participation in contemporary astronomical research in the second half of the 19th century. The careful, numerous, and hard-won contributions of Christian Peters to positional astronomy were a minor but helpful part of that process.

Michael Meo

Selected References

Ashbrook, Joseph (1984). "The Adventures of C. H. F. Peters." In *The Astronomical Scrapbook*, edited by Leif J. Robinson, pp. 56–66. Cambridge, Massachusetts: Sky Publishing Corp. (Overemphasizes Peters' work on asteroids, but constitutes the most detailed recent discussion of his career.)

Bruce, Robert V. (1987). *The Launching of Modern American Science, 1846–1876*. New York, Alfred A. Knopf, pp. 237ff. (Bruce points out that Peters's career at Hamilton College, far from the "empty dazzle" that Gould and his backers lamented, was devoted to just the compilation of star catalogs and positional computations for which Gould had intended his never-acquired heliometer.)

Dick, Steven J., Wayne Orchiston, and Tom Love (1998). "Simon Newcomb, William Harkness and the Nineteenth-century American Transit of Venus Expeditions." *Journal for the History of Astronomy* 29: 221–255. (See pp. 224–232 for details of Peters's leadership of the 1874 American expedition to New Zealand to observe the transit of Venus.)

James, Mary Ann (1987). *Elites in Conflict. The Antebellum Clash over the Dudley Observatory*. New Brunswick, New Jersey: Rutgers University Press. (See especially, Chap. 6: "Conflict Begins: The Peters Problem." In discussing in detail the struggle over the direction of the Dudley Observatory, James uses "manic-depressive" in describing Gould, and "calculating" for Peters.)

Jones, Bessie Zaban and Lyle Gifford Boyd (1971). *The Harvard College Observatory: The First Four Directorships, 1839–1919*. Cambridge, Massachusetts: Harvard University Press. (See p. 122 for Peters's financial concerns.)

Lankford, John (1997). *American Astronomy. Community, Careers, and Power, 1859–1940*. Chicago: University of Chicago Press. pp. 165ff. (He characterizes Peters's work as a "failed career," since "the application of photography to astronomy … rendered visual mapping of the sky obsolete." Further, according to Lankford, Peters was "demanding, indeed dictatorial, in the fashion of a European institute director.")

Peters, C. H. F. (1882). *Celestial Charts Made at the Litchfield Observatory*. Clinton, New York.

Peters, C. H. F. and Edward Ball Knobel (1915). *Ptolemy's Catalogue of Stars*. Carnegie Institution of Washington Publication No. 86. Washington DC: Carnegie Institution of Washington.

Warner, Deborah Jean (1974). "Peters, Christian Heinrich Friedrich." In *Dictionary of Scientific Biography*, edited by Charles Coulson Gillespie. Vol. 10, p. 543. New York: Charles Scribner's Sons. (Contains the most balanced account of Peters' career; her count of 48 discoveries of asteroids, however, differs from the usually reliable *Poggendorff's Biographisch-literarisches Handwörterbuch*, which counts 42. Vol. 3, p. 1027. Leipzig, 1898.)

Peter of Ailli

➔ **D'Ailly, Pierre**

Petit, Pierre

Born **Montluçon, (Allier), France, 8 December 1594 (?) or 31 December 1598**

Died **Lagny-sur-Marne, (Seine-et-Marne), France, 20 August 1677**

Physicist, mathematician, astronomer, and instrument maker Pierre Petit was christened on 31 December 1598. He was one of seven children born to Pierre Petit, then *contrôleur en l'élection* in Montluçon, a town situated in the Bourbonnais region of France, and Marie (*née* Bonnelat) Petit. As a young man, Petit assumed the duties of his father's office in 1626. But being more interested in scientific matters, he gave up this position and went to Paris in 1633. During his lifetime, Petit served in many official capacities, as a *commissaire provincial d'artillerie*, as an engineer and geographer to King Louis XIV, and as *intendant général des fortifications* in France.

While at Paris, Petit became a friend of **Marin Mersenne**, and was included in meetings of Mersenne's circle. Petit also became acquainted with Etienne Pascal and his mathematician son, Blaise Pascal. It was from Mersenne that Petit learned about **Evangelista Torricelli**'s experiments with the barometric vacuum and, helped by the Pascals, he successfully repeated them. Following Mersenne's death, this group met at the residence of Henri-Louis Habert de Montmor, and became known as the Académie de Montmor. There, Petit was introduced to **Adrien Auzout**, **Jean Picard**, and **Christiaan Huygens**. In conjunction with Auzout and Picard, Petit helped to perfect an instrument known as a filar micrometer, used to measure the very small angles of celestial objects viewed through a telescope. Petit also produced the earliest surviving sketch (1662) of the "magic lantern," or lantern projector, in his correspondence with Huygens.

By 1664, however, the Académie de Montmor had exhausted its funds. An appeal for assistance was directed to Jean-Baptiste Colbert, French statesman and financier. Two years later, King Louis XIV established the Académie royale des sciences, following many of the suggestions proposed by the academicians. Petit, however, was not among the founding members of the Académie des sciences, perhaps because he held a full-time government position. In 1667, however, Petit was named a corresponding member of the Royal Society of London.

Petit's most important astronomical work was his *Dissertation sur la nature des comètes* (1665). Therein, he speculated that a bright comet seen in December 1664 was the same as another viewed in 1618. Petit surmised that the comet had an elliptical orbit with a period of 46 years. He even went so far as to predict the return of this comet in 1710. Petit, however, was wrong about the particulars of this assertion. (Comets C/1664 W1 and C/1618 W1 were neither identical nor periodic). Yet his suggestion that comets might be periodic phenomena helped to diminish lingering fears and superstitions about their mysterious nature.

Petit also published a study on magnetic declination. His astronomical equipment included graphometers, quadrants, and objective lenses fashioned by the optician **Giuseppe Campani**. Petit married and had, among his children, one daughter, "Marianne", who became a *bénédictine* at Lagny-sur-Marne. A street and place, situated in the old part of Montluçon, bear Petit's name.

Suzanne Débarbat

Selected References

Fichman, Martin (1974). "Petit, Pierre." In *Dictionary of Scientific Biography*, edited by Charles Coulston Gillispie. Vol. 10, pp. 546–547. New York: Charles Scribner's Sons.

Genuth, Sara Schechner (1997). *Comets, Popular Culture, and the Birth of Modern Cosmology*. Princeton, New Jersey: Princeton University Press.

Hankins, Thomas L. and Robert J. Silverman (1995). *Instruments and the Imagination*. Princeton, New Jersey: Princeton University Press, esp. pp. 48–49.

Petit, Pierre (1665). *Dissertation sur la nature des comètes au roy avec un discours sur les prognostiques des eclipses et autres matières curieuses*. Paris. Louis Billaine.

Yeomans, Donald K. (1991). *Comets: A Chronological History of Observation, Science, Myth, and Folklore*. New York: John Wiley and Sons, Chap. 4, "The Comet of 1664: Confusion Reins," pp. 69–94.

Petrus Apianus

➋ **Apian, Peter**

Petrus Dacus [Danus]

➋ **Nightingale, Peter**

Petrus de Alliaco

➋ **D'Ailly, Pierre**

Petrus [Philomena] de Dacia

➋ **Nightingale, Peter**

Pettit, Edison

Born **Peru, Nebraska, USA, 22 Sept 1889**
Died **Tucson, Arizona, USA, 6 May 1962**

Edison Pettit, American astronomer best known for accurate measurements of the color temperatures of planets and stars, was the son of George Knox Pettit and Martha Ann Knox. He married Elizabeth Schmauser and, later Hannah Bard Steele. Their children were Helen Bard Pettit Knaflich and Marjorie Steele Pettit Meinel.

Edison Pettit was given no middle name because his parents thought "Edison" was a unique first name. (He was born at the time of Thomas Edison's inventions relating to electrical power.) George Pettit owned a sawmill and converted its machinery from steam-belt drive to steam-generated electricity drive. Young Edison Pettit, in fact, earned money for his education by helping to his father's growing electrical power plant wire up the village.

Pettit received his B.Ed. from Nebraska Normal College, Peru, Nebraska, in 1911. Following this he taught, first, science at Minden (Nebraska) High School, and then physics and astronomy at Washburn College, Topeka, Kansas, before going on to Yerkes Observatory to work on a Ph.D.

Pettit was a Ph.D. student at Yerkes when the army drafted him into World War I. At the Induction Center, when the soldiers manning the induction desk saw he gave "astronomer" as his job on his papers, they commented, "Of what use is an astronomer?" Fortunately, the officer in charge recognized Pettit's worth and assigned him to the United States Army Signal Corps. There Pettit worked with the optical scientist, **Robert Wood** at Johns Hopkins University. He was given the task of measuring the optical properties of materials and of assisting Wood in experiments with gas-filled balloons used as reconnaissance platforms. On the first flight, they nearly brought down the municipal gas supply of Baltimore.

Following the war, Pettit finished his Ph.D. at Yerkes Observatory where he had met his second wife, Hannah, the first woman to earn a Yerkes/University of Chicago Ph.D. in astronomy (guided by **Edwin Frost,** then director). They both extensively used the Yerkes 40-in. telescope. One night, while he was rewinding the drive weights, Edison's necktie got caught in the telescope drive mechanism and threatened to slowly strangle him. From the library, Hannah heard him cry out. She got scissors, ran to the dome, and severed his necktie. After that Edison never wore a necktie, only a black bow tie. Edison Pettit received his Ph.D. in 1920 for studies in solar physics, especially the spectra of prominences. Hannah's was for astrometric work. In 1920, shortly after establishing a solar observatory on Mount Wilson in California, **George Hale** asked Edison to join him on the staff of Mount Wilson Observatory [MWO]. Pettit accepted, and asked if there also was a position available on the staff for Dr. Hannah Pettit. Hale replied that it was not possible for women to be on the staff: only men. This remained true for five decades. Much of Edison Pettit's MWO work was done with the 60-ft. solar tower, to be followed in a few years by the 150-ft. tower.

In the winter of 1925–1926, the Pettit family went to the University of Arizona, Tucson, where Edison did research at the Tucson Tuberculosis Sanitarium (now the Tucson Medical Center) on ultraviolet transmission of glasses and the possible beneficial effect of solar ultraviolet radiation on tuberculosis patients. (His first wife, Elizabeth, had died of tuberculosis.) In 1930, he again took his family to Tucson where he assisted **Andrew Douglass** in upgrading the 36-in. telescope of the university's Steward Observatory, at that time located on the campus.

In addition to his solar work, Pettit had a laboratory at the MWO Pasadena offices to support his developments in radiation measurements of astronomical sources. In the 1920s, he built an ultrasensitive thermocouple and, with **Seth Nicholson**, made the first measurements of the temperature of the Moon, both the solar-illuminated and night surfaces, which showed the fine granular nature of the lunar surface.

Pettit and Nicholson extended temperature measurement to Venus, Mars, Jupiter, and Saturn, using the MWO 100-in. telescope. The values they obtained were confirmed in the 1970s when spacecraft first visited these planets. In 1944, Pettit was asked to use his expertise to build three thermocouples and to hand-deliver them to Alamogordo, New Mexico, where they were used to monitor the output of the first nuclear explosion.

During the summers of 1930–1936, Pettit took his family to Yerkes Observatory and made solar prominence observations using Hale's spectroheliograph on the 40-in. refractor. Pettit's attention was focused on why eruptive prominences show several sudden jumps in their upward motion before blasting into space. To follow this up he went to Michigan in the summers of 1936–1938 to test and then to use the new spectroheliokinematograph at the McMath–Hulbert Observatory.

In 1939, Pettit built a home observatory equipped with an Alvan Clark 6-in. refractor. For years this old telescope had lain in a barn, covered with trash, on a farm in Michigan. It now had been offered for sale. The owner would not let a buyer inspect it, but Pettit could see that although the telescope had been neglected, it was at least intact. After the owner accepted his fair price offer, another astronomer slyly told her that Pettit had offered too little, so she raised the price.

Nonetheless, Pettit took the 6-in. home to Pasadena and refurbished it so that it looked and worked like new. Now he could do much more of his observing from his home rather than on Mount Wilson, an advantage since his wife Hannah was a semi-invalid.

Shortly after **Yngve Ohman** invented the quartz-polarizing monochromator, Pettit designed and built one in his small home machine shop, adding a time-lapse movie camera. Attaching this at the eyepiece of his telescope made possible photographing solar prominences from the convenience of home. Assisted by Marjorie, he obtained enough movies of eruptions to enable him to refine a set of "laws" describing the motion of eruptive prominences. Understanding why they erupt in this manner required the discovery of magnetohydrodynamics and model of solar flares; subtleties in their behavior still puzzle solar theorists.

The Clark refractor saw additional duty as Pettit frequently had groups of amateur astronomers and the public come to look through the 6-in. at the planets, especially during the excellent 1940 opposition of Mars. That telescope and monochromator are still in use, but now in a small observatory in Prescott, Arizona.

A backyard telescope has other advantages. One winter morning in 1941, Pettit went out before dawn to pick up the morning newspaper. He was startled to see a new bright star shining low in the south. He immediately ran to the backyard, opened the sliding roof, and turned the telescope on the new star – Nova Puppis. Reading off its coordinates and measuring its magnitude, he sent a telegram to Harvard College Observatory with the news. However, Pettit was about 2 hours too late to be the discoverer of a nova, since a telegram had just been received from South America where it had been discovered first.

As photoelectric observations took center stage for precision brightness measurements, Pettit's interests moved from solar and planetary observations to measuring the radial brightness distributions of over 500 galaxies. During his retirement years, he used his home machine shop to build spectrographs and small instruments for a number of observatories and universities. Following the death of Hannah, and a subsequent stroke, Edison moved to Tucson to live with Marjorie Meinel and her family.

Pettit received a honorary degree from Carthage College (LL.D.) in 1935, and craters both on the farside of the Moon and on Mars are named for him.

Marjorie Steele Meinel

Selected References

Anon. (1963). "Report of the President." In *Year Book 62*. Washington, DC: Carnegie Institution of Washington. (See pp. 51–52 for an obituary notice.)

Nicholson, Seth B. (1962). "Edison Petit, 1889–1962." *Publications of the Astronomical Society of the Pacific* 74: 495–498.

Oriti, Ronald A. (July 1962). "Some Recollections of Edison Petit." *Griffith Observer* 26: 90–98.

Peucer, Caspar

Born **Bautzen, (Sachsen, Germany), 1 June 1525**
Died **Dessau, (Sachsen-Anhalt, Germany), 25 September 1602**

In his textbooks Caspar Peucer argued for the importance of studying astronomy; he also wrote on the new star of 1572, and defended certain aspects of astrology.

Peucer was the child of Gregor Beucker, a *Bürger* of Bautzen, and Ottilie Simon. He was educated at the University of Wittenberg, matriculating there in March 1543, and graduating MA in September 1545. He went on to distinguish himself as a professor at Wittenberg, first in lower mathematics (1550), then higher mathematics (1554), and finally, after graduating as a doctor of medicine in 1560, in the medical faculty. His first wife, Magdalena, was the daughter of Philipp Melanchthon; she bore him three sons and seven daughters, but died in 1575. Between 1576 and 1586, Peucer was imprisoned as a crypto-Calvinist by August von Saxony, whom he had previously served as counselor and physician. His release was effected through the petition of Joachim Ernst von Anhalt, whom he then likewise served in both of these capacities. In 1587 Peucer married Christine Schild, a widow of Bautzen; this second marriage was without any issue.

Peucer's significance to astronomy lies partly in his authorship of astronomical texts, and also in the position he held within the astronomical community of the later 16th century. Peucer's professors at Wittenberg had included Melanchthon, **Rheticus**, and **Erasmus Reinhold**, and as a professor of higher mathematics (astronomy) in his own right, and an author of textbooks that had their origin in his lectures, Peucer played a large part in propagating both the Philippist view of the importance of astronomy within the arts curriculum and the so called Wittenberg interpretation of the Copernican world system. Peucer's pupils included Johannes Praetorius, Victorin Schönfeld (who became professor of mathematics at Marburg), and Jørgen Dybvad (later professor of theology, natural philosophy, and mathematics at Copenhagen). In addition, he was consulted by princes, including Landgrave **Wilhelm IV** of Hessen-Kassel, about such phenomena as the supernova of 1572 (SN B Cas), and he included among his correspondents the Danish astronomer **Tycho Brahe**. Peucer's standing within the astronomical community is suggested by the fact that Brahe attempted to use Peucer as the arbiter of his disputes with **Christoph Rothmann**, and addressed to him an important epistolary defense of his claim to priority in the development of the geoheliocentric world system.

The most astronomically significant of Peucer's own works are the *Elementa doctrinae de circulis coelestibus*, first published in 1551, and the *Hypotheses astronomicae*, first published under his name in 1571, after a previous version had been published in 1568 without his consent. His treatise on the new star of 1572 was published at Wittenberg, and later reproduced by Brahe. Peucer's *De dimensione Terrae* of 1550 was essentially a geography textbook, but one heavily indebted to the techniques of spherical astronomy. His *Commentarius de praecipuis divinationum generibus* of 1553 is worthy of mention since it included a section on astrology, the practice of which provided one of the chief motivations for astronomical study in the early modern period; in defending the legitimacy of certain forms of astrological divination Peucer was again following the lead of his father-in-law Melanchthon.

Adam Mosley

Selected References

Blair, A. (1990). "Tycho Brahe's Critique of Copernicus and the Copernican System." *Journal for the History of Ideas* 51: 355–377. (For part of Peucer's correspondence with Brahe.)

Dreyer, J. L. E. (ed.) (1913–1929). *Tychonis Brahe Dani opera omnia*. 15 Vols. Copenhagen. (See Vol. 7, pp. 127–141 for Peucer's most important letter written to him by Tycho; see Vol. 3, pp. 132–139 for the treatise on the new star of 1572.)

Gingerich, Owen (1973). "From Copernicus to Kepler: Heliocentrism as Model and as Reality." *Proceedings of the American Philosophical Society* 117: 513–522.

——— (1973). "The Role of Erasmus Reinhold and the Prutenic Tables in the Dissemination of Copernican Theory." In *Colloquia Copernicana II*, pp. 43–62, 123–125. Studia Copernicana, Vol. 6. Wroclaw: Ossolineum.

Kolb, Robert (1976). *Caspar Peucer's Library: Portrait of a Wittenberg Professor of the Mid-Sixteenth Century*. St. Louis: Center for Reformation Research.

Kühne, H. (1983). "Kaspar Peucer. Leben und Werk eines großen Gelehrten an der Wittenberger Universität im 16. Jahrhundert." *Letopis* B 30: 151–161.

Moran, Bruce T. (1981). "German Prince-Practitioners: Aspects in the Development of Courtly Science, Technology, and Procedures in the Renaissance." *Technology and Culture* 22: 253–274. (For Peucer's relationship with Landgrave Wilhelm IV of Hessen-Kassel.)

Weichenhan, Michael (1995). "Astrologie und Natürliche Mantik bei Casper pencer." In *Too Jahre Wittenberg: Stadt, Universität, Reformation*, edited by Stefan Oehmig, pp. 213-224. Weinar: Böhlarus.

Westman, Robert S. (1975). "The Melanchthon Circle, Rheticus, and the Wittenberg Interpretation of the Copernican Theory." *Isis* 66: 165–193.

——— (1975). "The Wittenberg Interpretation of the Copernican Theory." In *The Nature of Scientific Discovery*, edited by Owen Gingerich, pp. 393–429. Washington, DC: Smithsonian Institution Press.

Peurbach [Peuerbach, Purbach], Georg von

Born **Peurbach near Linz, (Austria), possibly 30 May 1423**
Died **Vienna, (Austria), 8 April 1461**

Georg von Peurbach continued a strong tradition in which the University of Vienna was considered to have the best astronomy scholars in Europe during the 15th and early 16th centuries.

Peurbach was born sometime after 1421, the date of 30 May 1423 coming from a horoscope published as late as 1550. Peurbach received a bachelor's degree in 1448 and his master's degree 5 years later, both at the University of Vienna. As a lecturer at the university there, he was one of the leaders in reviving classical Greek and Roman literature in the arts and sciences. After observing an occultation of Jupiter by the Moon in 1451, Peurbach spent the last decade of his life making observing instruments, lecturing on astronomy and the classics, collaborating with men such as **Johann Müller** (Regiomontanus) on astronomical observing and theory, and serving as imperial astrologer to the king of Hungary. Peurbach died of unknown causes.

Peurbach evidently made many astronomical observations during the last 10 years of his life with his famous student Regiomontanus, including taking measurements of lunar eclipses during 1457–1460 whereby the two astronomers carefully noted observing location and times, altitudes of the Moon and various stars, and degree to which the Moon was seen inside the Earth's umbral shadow. Such observational detail was not only highly unusual for medieval observations, but it also helped to set a precedent for 16th-century observers who tried to emulate the work of the Vienna astronomers. Peurbach observed Halley's comet (IP/Halley) in 1456, using instruments to record the ecliptic positions of the comet's head and tail on two nights and to make some assessment of the comet's distance from parallax measures. That a comet's position would be measured as seriously as a planet's indicated the beginning of moving away from **Aristotle**'s claim that comets were merely atmospheric phenomena; this was also apparently the first attempt to seriously determine a comet's distance from parallax, a procedure elaborated upon by Regiomontanus and widely discussed and refined in the 16th and 17th century by others.

Led by Peurbach and Regiomontanus, the Vienna group sought actively to reform astronomy by improving on theory through the beginning of systematic observation of the planets, Sun, Moon, and stars. Peurbach's *Nouae theoricae planetarum* (New theories of the planets), which involved revised theories of the Ptolemaic system (evidently augmented by the ideas of Arabic astronomers), was published posthumously in numerous editions from 1472 to 1596, edited by such notable scholars as Regiomontanus, **Peter Apian**, **Erasmus Reinhold**, and Philip Melanchthon. No less than 56 printings of this Latin text appear to have been made up to 1653, with additional printings in other languages. (While still working toward his master's degree, Regiomontanus apparently heard Peurbach's lectures on this text in 1454 at Vienna.)

Peurbach also prepared and issued tables predicting eclipses of the Sun and Moon (a practice continued by Regiomontanus), and

Peurbach seems also to have supervised the collecting and copying of astronomical manuscripts, leading evidently to the establishment of the scientific printing press by Regiomontanus to publish astronomical tracts (including Peurbach's *Nouae theoricae planetarum* and the ancient poet **Manilius**'s *Astronomicon*).

At the request of Cardinal Bessarion, Peurbach began in 1460 a translation, with commentary, of **Ptolemy**'s *Almagest*; this was cut short by Peurbach's death, but was continued to completion by Regiomontanus and eventually published under the title *Epitome of the Almagest* in 1496. The *Epitome* was an important reference in the following decades for **Nicolaus Copernicus** during his preparation of *De Revolutionibus*. Peurbach also influenced Regiomontanus for the development of advanced trigonometric relationships that would be used by astronomers in the century to come.

Peurbach's *Theoricae novae planetarum* was published by Regiomontanus from his printing press in Nuremberg. Though many Peurbach manuscripts seem to have circulated (particularly on astronomical theory and practice, including instrumentation), other work by Peurbach was even more delayed in terms of printing, his observations not being fully published until nearly a century after his death by **Johann Schöner**.

Daniel W. E. Green

Selected References

Aiton, E. J. (1987). "Peurbach's *Theoricae novae planetarum*: A Translation with Commentary." *Osiris* 3: 5–43. (Translation into English of Peurbach's most important treatise.)

Gassendi, P. (1655). *Tychonis Brahei, Equitis Dani, Astronomorum Coryphaei, Vita . . . Accessit Nicolai Copernici, Georgii Pevrbachii, & Ioannis Regiomontani . . .* The Hague: Adriani Vlacq, pp. 333ff. (Includes the first extensive biography of Peurbach.)

Hellman, C. Doris and Noel M. Swerdlow (1978). "Peurbach (or Peuerbach), Georg." In *Dictionary of Scientific Biography*, edited by Charles Coulston Gillispie. Vol. 15 (Suppl. 1), pp. 473–479. New York: Charles Scribner's Sons.

Jervis, Jane L. (1985). *Cometary Theory in Fifteenth-Century Europe*. Wroclaw: Ossolineum, Polish Academy of Sciences Press. *Studia Copernicana*, Vol. 26, p. 86ff. (Details the 1456 observations of comet 1P/Halley by Peurbach.)

Pedersen, Olaf (1978). "The Decline and Fall of the Theorica Planetarum: Renaissance Astronomy and the Art of Printing." In *Science and History: Studies in Honor of Edward Rosen*, edited by Erna Hilfstein *et al.*, pp. 157–185. *Studia Copernicana*, Vol. 16, p. 161ff. Wroclaw: Ossolineum, Polish Academy of Sciences Press. (Details the role of Peurbach and Regiomontanus in the downfall of Aristotelian and Ptolemaic theory.)

Peurbach, Georg von (1472). *Theoricae novae planetarum*. Nuremberg. (Peurbach's premier treatise.)

Peurbach, G. and Regiomontanus (1495). *J. de Sacro-Bosco Sphaericum opusculum, contraque cremonensia in planetarum theoricas deliramenta J. de Monteregio disputationes, necnon G. Purbachii in motum planetarum theoricae*. Venice. (Reprinted nine times up to 1513; commentary on Sacrobosco based on ideas of Regiomontanus and Peurbach.)

——— (1496). *Epytoma Joanis De Monte regio In almagestum ptolomei*. Venice: Johannes Hamman. (Books I–VI were translated by Peurbach, but left to Regiomontanus to complete upon the older man's early death.)

Schöner, J. (ed.) (1544). "Ioannis de Monteregio et Barnardi Waltheri eivs discipvli ad solem obseruationes," and "Ioannis de Monteregio, Georgii Pevrbachii, Bernardi Waltheri, ac aliorum, Eclipsium, Cometarum, Planetarum ac Fixarum obseruationes." In *Scripta clarissimi mathematici*. Nuremberg: Ioannem montanum & ulricum Neuber. (Includes observations of Peurbach.)

Steele, John M. and F. Richard Stephenson (1998). "Eclipse Observations Made by Regiomontanus and Walther." *Journal for the History of Astronomy* 29: 331–344. (English translations of Peurbach's eclipse observations.)

Zinner, Ernst (1990). *Regiomontanus: His Life and Work*, translated by Ezra Brown. Amsterdam: Elsevier Science Publications, pp. 17ff. (Includes a fair amount of information on Peurbach, particularly in his relationship with Regiomontanus.)

Pfund, August Hermann

Born **Madison, Wisconsin, USA, 28 December 1875**
Died **Baltimore, Maryland, USA, 4 January 1949**

American spectroscopist and colorimetrist August Pfund received a BS from the University of Wisconsin in 1901 and a Ph.D. from Johns Hopkins University, Baltimore in 1906. He held research positions from 1906 to 1910, an associate professorship (1910–1927), and a professorship at John Hopkins University from 1927 until his retirement. Pfund appears occasionally in astronomy books because of his discovery of the fifth series of hydrogen lines, that is, those arising from the $n = 5$ electron shell, the third in the infrared after the Paschen and Brackett series. Pfund lines have wavelengths longer than 2.28 μm. There is no limit to such series, and transitions from $n = 108$ and 109 can be studied at radio wavelengths.

Pfund received medals from the Franklin Institute and the Optical Society of America.

Virginia Trimble

Selected Reference

Anon. (1944). "Pfund." In *American Men of Science*, 7th ed. Lancaster, Pennsylvania: Science Press, p. 1387.

Silverman, Shirleigh and John Strong (1949). "A. Herman Pfund" *Journal of the Optical Society of America* 39, 325.

Phillip of Opus

Flourished **(Greece), fourth century BCE**

Ancient sources attribute a number of astronomical "firsts" to Phillip. However, these sources date from centuries after Phillip's time.

Selected Reference

Neugebauer, Otto (1975). *History of Ancient Mathematical Astronomy* New York: Springer–Verlag.

Phillips, Theodore Evelyn Reece

Born **Kibworth, Leicestershire, England, 28 March 1868**
Died **Walton-on-the-Hill near Headley, Surrey, England, 13 May 1942**

Reverend Theodore Phillips led the British Astronomical Association [BAA]'s Jupiter Section for 31 years. In the process, from both his own outstanding observations and those of other observers, Phillips compiled an unprecedented continuous record of the visual surface features of Jupiter, a record that for the most part cannot be duplicated from any other source. His analysis of the currents in various zones and belts of Jupiter is a model for all such efforts.

The son of an Anglican minister, Reverend Abel Phillips, Theodore matriculated at Oxford, where he graduated with a BA degree from Saint Edmund's Hall, Oxford, England, in 1891 and earned a MA at Oxford University in 1894. Ordained as a minister in the Church of England in 1891, Phillips served as the curate of the parish of the Holy Trinity at Taunton, and then was appointed to a similar role at Hendford, Yeoville, in 1895.

It was around 1895 that a parishioner, aware of Phillips's interest in science, made a gift of a 3-in. Grubb refractor to Phillips. When Phillips chanced upon Saturn during his first observing with this telescope, the beauty of the object created an enthusiasm for astronomy that would influence him for the rest of his life. In 1896 Phillips joined the BAA and began systematically observing Jupiter and Mars with a 9¼-in. reflector on an alt-azimuth mount. With respect to Jupiter, Phillips soon recognized the importance of the pioneering work of **Arthur Williams**, which had in the main been largely ignored by the directors of the Jupiter Section. Soon after Phillips joined the section in 1896, the apparition memoirs began to reflect this new awareness. In 1901 Phillips was appointed director of the BAA Jupiter Section. He changed little at first, but by 1914 had completely revised the reporting process for Jupiter. Instead of preparing drift charts for each individual observer, the practice that emerged from his initial emphasis on the work of Williams, Phillips consolidated the results of all observers for each zone or belt, in the process not only simplifying the reporting but also making the analysis much more rigorous and likely more accurate. Takeshi Sato, a planetary observer, called Phillips the "Father of the Jupiter Observation." William Edwin Fox (1898–1988), a later BAA Jupiter Section director, characterized Phillips by stating "as a planetary draughtsman, he was unsurpassed."

Phillips sustained his energetic role in astronomy in spite of increasingly demanding professional assignments. In 1901, the same year he successfully assumed responsibility for the Jupiter Section, he was reassigned to a curacy in the parish of Saint Saviour, Croyden, followed by another assignment as curate in Ashstead, Surrey. While assigned at Croyden, Phillips met and married Millicent Kynaston; they had one son.

Phillips's involvement in astronomy continued to grow during this period. Having been elected a fellow of the Royal Astronomical Society [RAS] in 1899, Phillips was elected to the RAS Council in 1911. He served nearly continuously in that body until his death, including terms as secretary (1919–1925) and president (1927–1929). Phillips is one of only two amateur astronomers to serve as the RAS president after 1907, the other being **William Steavenson**.

Reverend Phillips assumed even greater church responsibility when, in 1916, he became the rector of Headley. This was, however, seen as a longer-term assignment, so he built an observatory near the rectory that eventually housed two telescopes, an 8-in. Cook refractor, and a 12-in. Calver equatorial reflector that was later replaced by an 18-in. With reflector. Phillips observed double stars with the refractor but employed the reflectors for his planetary studies. Phillips was fond of using the filar micrometer, which had reached its engineering peak in the 19th century, and contributed a long series of double star measures in annual reports to the *Monthly Notices of the Royal Astronomical Soceity*.

Phillips's planetary observing was not limited to Jupiter. For a time Phillips was known as one of the world's more prolific Mars' canalists along with BAA observers **Percy Molesworth** and **Eugéne Antoniadi**. When he observed Mars for the first time (with the 9¼-in. reflector), Phillips is reported to have declared "My experience of Martian observation this winter has led me to believe that Mars is not nearly so difficult an object as is commonly supposed, and that many of the canals are easy." Phillips observed at every opportunity during 43 oppositions of Mars. His drawings, which eventually displayed no canals, reflect the care and skill he evidenced in his Jupiter observations.

Phillips was known as an inspiring person and a gifted speaker. It would have been interesting to listen to a sermon on the heavens delivered by the amateur astronomer–minister. Headley Observatory was known as the "Mecca" of British amateur astronomy at that time. Every year in June, Phillips and Millicent would entertain guests at Headley Observatory, including young noteworthies like Steavenson, Frederick James Hargreaves (1891–1970), and Reginald Lawson Waterfield (1900–1986). In the warm summer afternoons, Theodore, Millicent, and guests would stroll along the wildflower covered countryside area of Nottingham.

It is characteristic of Phillips's broad interests in astronomy that in his second presidential address to the BAA, he attacked a theoretical problem rather than an observational topic. At the time, professional astronomer **Herbert Turner** was actively promoting amateur involvement in what is now known as data mining. (See also **Mary Blagg**.) At Turner's suggestion, Phillips undertook the harmonic analysis of the light curves of about 80 long-period variable stars. He demonstrated that those long-period variables could be classified into two groups based on the constancy of the third harmonic in one case but a simple linear relationship between the second and third harmonics in the other. Phillips's study became a classic in the literature of variable-star analysis.

In 1923 Phillips and Steavenson edited a book entitled *Splendour of the Heavens* published by Hutchinson & Company. It first appeared as two volumes containing 979 pages and was very readable, discussing solar and stellar spectroscopy in great detail. It featured almost 1,000 illustrations and numerous fine colored plates that contributed to the book's popularity. Top British astronomers and BAA section directors collaborated as authors of *Splendour*, which was widely used as a text for teaching nonastronomers at the time. Chapters were included on ancient constellations and Chinese astronomy in addition to more conventional astronomical topics. Astronomer Freeman J. Dyson claimed that he learned science more from books than from teachers: "My favorite book was *The Splendour of the Heavens*, a huge and lavishly illustrated compendium of popular astronomy ..."

Phillips was honored by the RAS with their presentation of the Jackson-Gwilt Medal and Prize in 1918, and was the first recipient of the BAA's Walter Goodacre Medal in 1930. Just weeks before his death, Oxford University conferred the degree D. Sc. *honoris causa* on Phillips, which fortunately he was able to receive in person.

Robert McGown

Selected References

Davidson, M. (1942). "Honour for Rev. T. E. R. Phillips." *Observatory* 64: 228–231.
Peek, B. M. (1942). "Theodore Evelyn Reece Phillips." *Journal of the British Astronomical Association* 52: 203–208.
Phillips, T. E. R. and W. H. Steavenson (eds.) (1923). *Splendour of the Heavens*. London: Hutchinson and Co.
Sheehan, William P. (1996). *The Planet Mars: A History of Observation and Discovery*. Tucson, Arizona: University of Arizona Press, chap. 9.
Steavenson, W. H. (1943). "Theodore Evelyn Reece Phillips." *Monthly Notices of the Royal Astronomical Society* 103: 70–72.
Tayler, R. J. (ed.) (1987). *History of the Royal Astronomical Society*. Vol. 2, *1920–1980*. Oxford: Blackwell Scientific, pp. 12–15.

Philolaus of Croton

Born **Croton, (Crotone, Calabria, Italy),** ***circa*** **460 BCE**

Philolaus held that the Universe and everything in it are constituted by the harmonious combination of unlimited and limiting principles.

Philolaus was born in southern Italy, but immigrated to Greece proper, presumably for political reasons. He was active in Thebes until shortly before Socrates' death in 399, and perhaps also in Phlio, a small city near Corinth. The ancient story that **Plato**'s *Timaeus* was plagiarized from Philolaus' only book is surely false, but may be indicative of the latter's interests and outlook. He stated the main Pythagorean tenet as follows: "Everything that is known has number, for without this nothing can be thought of or known" (fr. 5).

Most scholars agree today that the system of the world with a moving Earth, attributed by **Nicolaus Copernicus** and **Galileo Galilei** to the Pythagoreans, was due in fact to Philolaus. In this system the Earth, however, circumambulates not the Sun, as in the systems of **Aristarchus** and Copernicus, but a distinct cosmic source of heat and light, which Philolaus called "hearth of the universe," "house of Zeus," "mother of the gods," and "altar, bond and measure of nature." We earthians cannot see it because it is permanently eclipsed by the "counter-earth," which in turn we cannot see because it constantly shows us its unilluminated face. Ten divine bodies "dance" around the hearth, *viz.*, the firmament of the fixed stars, the five planets, after these the Sun, under it the Moon, then the Earth, and then the counter-earth. The number 10 = 4 + 3 + 2 + 1 was accorded a privileged status by the Pythagoreans. Thus, the counter-earth is not simply postulated *ad hoc* to ensure the occultation of the hearth, but derived so to speak from formal requirements of a well-built world. According to Philolaus, the Sun is like glass, receiving the reflection of the cosmic fire and filtering its light and warmth to us, so that in a sense there are two suns, the primary fire and its mirror image.

Aristotle reports that in this Pythagorean system, the quick motion of all the said bodies was supposed to produce an incredibly strong sound. By assuming that their speeds are in the same ratio as musical concords, it was concluded that the sound caused by the circular movement of the heavenly bodies is a harmony. Because this beautiful "music of the spheres" continually strikes our ears since our birth, we do not hear it.

Roberto Torretti

Selected References

Burkert, Walter (1972). *Lore and Science in Ancient Pythagoreaism*, translated by Edwin L. Minar, Jr. Cambridge, Massachusetts: Harvard University Press.
Diels, Hermann and Walther Kranz (1954). *Die Fragmente der Vorsokratiker*. 7th ed. Berlin: Weidmann, sect. 44.
Guthrie, W. K. C. (1962). "Pythagoras and the Pythagoreans." In *A History of Greek Philosophy*. Vol. 1, *The Earlier Presocratics and Pythagoreans*, pp. 146–340. Cambridge: Cambridge University Press.
Kirk, G. S., J. E. Raven, and M. Schofield (1983). "Philolaus of Croton and Fifth-Century Pythagoreanism." In *The Presocratic Philosophers: A Critical History with a Selection of Texts*, pp. 322–350. 2nd ed. Cambridge: Cambridge University Press.
Riedweg, Christoph (2000). "Philolaos [2]." In *Der neue Pauly: Encyclopädie der Antike*, edited by Hubert Cancik and Helmuth Schneider. Vol. 9, cols. 834–836. Stuttgart: J. B. Metzler.

Philoponus, John

Born **(Egypt),** ***circa*** **490**
Died **(Egypt),** ***circa*** **570**

John Philoponus (literally "Lover of work"), a Christian philosopher, scientist, and theologian, is one of the most important and certainly most original natural philosophers in Late Antiquity. His life and work are closely connected to the city of Alexandria and the Alexandrian Neoplatonic school. Although the geocentric Aristotelian–Neoplatonic tradition formed his intellectual roots and concerns, he was also an original thinker who eventually broke with that tradition in many respects, helping to lead eventually to the demise of the predominance of Aristotelianism in the natural sciences.

Philoponus' *œuvre* falls into three different parts. First, there are the commentaries on **Aristotle**, which belong to the early period (until about the mid-530s); some of these depend heavily on the lectures of Philoponus' teacher **Ammonius**. Then, in a string of polemical treatises, Philoponus turns to attacking squarely fundamental doctrines of Neoplatonic–Aristotelian natural science. One aim of these treatises may have been to integrate fully the still largely pagan school at Alexandria into a Christian social and intellectual context. In this he failed; leadership of the school remained in pagan hands well into the 6th century. Philoponus seems to have dissociated himself from the school, because from the 540s onward he was writing theological–exegetical works, most importantly a commentary on the biblical creation myth (*De opificio mundi*).

From the point of view of astronomy, Philoponus' two most important contributions to the development of science are his theory of matter and the theory of the impetus, which he seems to have embedded into a whole new view of explaining natural and forced motion of bodies. According to this new theory, the continuation of motion (*e. g.*, of an arrow) is no longer explained by the surrounding medium (air) acting as moved mover (as Aristotle thought), but by invoking the notion of a "force" that is imparted to the moved object by the original mover and that resides in it for the duration of the movement. Philoponus suggests (*De opificio mundi*, I 12) that one might well understand celestial motion in this way, *viz.*, that the celestial bodies have received a powerful impetus at the time of their creation, by which they continue to move in accordance with the will of God.

Philoponus' theory of matter is significant as a landmark in the history of cosmology. He vehemently opposed the Aristotelian dichotomy of the cosmos into a divine, eternal superlunary region and a sublunary region that is subject to generation and corruption. According to him, matter is uniform throughout, and the material basis of the heavens is in fact no different from the material basis of the terrestrial elements and the more complex structures they give rise to. Philoponus' ideas were viciously opposed by the contemporary Neoplatonist **Simplicius**, whose influential writings assured that Philoponus never received the kind of acknowledgment and attention he deserved.

Although this does not play a central role in his writings on natural philosophy, there is enough evidence to suggest that Philoponus was also an expert astronomer. Our oldest description of the construction and use of an astrolabe, written in the first half of the 6th century, is attributed to him; **Theon**'s earlier work on it is now lost. Moreover, we have fairly accurate knowledge of the date of one of his treatises (*De aeternitate mundi contra Proclum*) because he mentions that in the 245th year after Diocletian (529) there occurred a conjunction of all planets in Taurus; since the Sun counted as one of the planets, this conjunction was unobservable and could be inferred only on the basis of long-time observation and theory.

Christian Wildberg

Alternate names

John the Grammarian
John of Alexandria

Selected References

Fladerer, Ludwig (1999). *Johannes Philoponos:* De opificio mundi: *Spätantikes Sprachdenken und christliche Exegese*. Stuttgart, Leipzig: Teubner.

Haas, F. A. J. de (1997). *John Philoponus' New Definition of Prime Matter: Aspects of Its Background in Neoplatonism and the Ancient Commentary Tradition*. Leiden: E.J. Brill.

Hainthaler, T. (1990). "Johannes Philoponus, Philosoph und Theologe in Alexandria." In *Jesus der Christus im Glauben der Kirche*. Vol. 2/4, *Die Kirche von Alexandrien mit Nubien und Äthiopien nach 451*, edited by A. Grillmeier, pp. 109–149. Freiburg: Herder.

Krafft, F. (1988). "Aristoteles aus christlicher Sicht. Umformungen aristoteliescher Bewegungslehren durch Johannes Philoponos." In *Zwischen Wahn, Glaube und Wissenschaft*, edited by J.-F. Bergier. Zürich.

Lee, Tae-Soo (1984). *Die griechische Tradition der aristotelischen Syllogistik in der Spätantike*. Göttingen: Vandenhoeck and Ruprecht.

Philoponus, John (1839). *On the Use and Construction of the Astrolabe* (in Latin), edited by H. Hase. Bonn: E. Weber. See also *Rheinisches Museum für Philologie* 6 (1839): 127–171. (Reprinted and translated into French in *Traité de l'astrolabe*, edited by A. P. Segonds. Paris: Librairie Alain Brieux, 1981; translated by H. W. Green in *The Astrolabes of the World*, edited by R. T. Gunther, Vol. 1/2, pp. 61–81. Oxford, 1932. (Reprint, London: Holland Press, 1976.)

——— (1887–1888). *Commentary on Aristotle's* Physics (in Latin and Greek), edited by H. Vitelli. Vol. 16 and 17 of *Commentaria in Aristotelem Graeca*. Berlin: Reimer. For partial translations, see: *On Aristotle's* Physics *2*, translated by A. R. Lacey. London: Duckworth, 1993; *On Aristotle's* Physics 3, translated by M. J. Edwards. London: Duckworth, 1994; *On Aristotle's* Physics 5–8, translated by Paul Lettinck. London: Duckworth 1994; *Corollaries on Place and Void*, translated by David Furley. London: Duckworth, 1991. (Philoponus's most important commentary in which he challenges Aristotle's tenets on time, space, void, matter, and dynamics; there are clear signs of revision.)

——— (1897). *De opificio mundi* (On the Creation of the World), edited by W. Reichardt. Leipzig: Teubner. (A theological–philosophical commentary on the Creation story in the book of Genesis written probably 557–560.)

——— (1899). *De aeternitate mundi contra Proclum* (On the eternity of the world against Proclus), edited by Hugo Rabe. Leipzig: B. G. Teubner. (Reprint, Hildesheim: Olms, 1984. A detailed criticism of Proclus's 18 arguments in favor of the eternity of the world.)

——— (1901). *Commentary on Aristotle's Meteorology* (in Latin and Greek), edited by M. Hayduck. Vol. 14, pt. 1 of *Commentaria in Aristotelem Graeca*. Berlin: Reimer.

——— (1987). *Against Aristotle, on the Eternity of the World.* Fragments reconstructed and translated by Christian Wildberg. London: Duckworth. (A refutation of Aristotle's doctrines of the fifth element and the eternity of motion and time consisting of at least eight books.)

Sambursky, S. (1962). *The Physical World of Late Antiquity*. London: Routledge and Kegan Paul, pp. 154–175.

Scholten, C. (1996). *Antike Naturphilosophie und christliche Kosmologie in der Schrift "De opificio mundi" des Johannes Philoponos*. Berlin: De Gruyter.

Sorabji, Richard (1983). *Time, Creation and the Continuum*. London: Duckworth, pp. 193–231.

——— (1988). *Matter, Space, and Motion*. London: Duckworth, pp. 227–248.

——— (ed.) (1987). *Philoponus and the Rejection of Aristotelian Science*. London: Duckworth.

Verbeke, G. (1985). "Levels of Human Thinking in Philoponus." In *After Chalcedon: Studies in Theology and Church History*, edited by C. Laga, J. A. Munitz, and L. van Rompay, pp. 451–470. Leuven: Peeters.

Verrycken, Koenraad (1990). "The Development of Philoponus's Thought and Its Chronology." In *Aristotle Transformed*, edited by Richard Sorabji, pp. 233–274. London: Duckworth.

Wildberg, Christian (1988). *John Philoponus' Criticism of Aristotle's Theory of Aether*. Berlin: De Gruyter.

——— (1999). "Impetus Theory and the Hermeneutics of Science in Simplicius and Philoponus." *Hyperboreus* 5: 107–124.

Wolff, Michael (1971). *Fallgesetz und Massebegriff: Zwei wissenschaftshistorische Untersuchungen zur Kosmologie des Johannes Philoponus*. Berlin: De Gruyter.

——— (1978). *Geschichte der Impetustheorie: Untersuchungen zum Ursprung der klasischen Mechanik*. Frankfurt am Main: Suhrkamp.

Piazzi, Giuseppe

Born **Ponte in Valtellina, (Graubunden, Switzerland), 16 July 1746**
Died **Naples, (Italy), 22 July 1826**

Giuseppe Piazzi was the discoverer of the first minor planet, (1) Ceres, and published two important star catalogs. Piazzi was the ninth of ten children born to Bernardo Piazzi and his wife, Antonia Francesca Artaria. He came from a family of nobles; his father worked in a land registry office in the Villa di Tirano.

In an age of limited career choices, Piazzi became a Theatine monk in Milan, which gave him the opportunity of broadening his knowledge in the classics and mathematics. He finished his novitiate at the abbey of San Antonio. Studying at Theatine colleges in Milan, Turin, Rome, and Genoa, Piazzi acquired a taste for mathematics and astronomy, earning a doctorate in these subjects. He then taught philosophy at Genoa, mathematics at the new University of Malta (1770–1773), and mathematics at Ravenna (1773–1779). In 1779, Piazzi was appointed professor of dogmatic theology in Rome.

It was not until 1780 that Piazzi found a permanent home when he accepted the chair of mathematics at the Academy of Palermo in Sicily. In 1787, Piazzi became a professor of astronomy and resolved to build one of Europe's finest observatories. He soon obtained a grant from Prince Tomaso d'Aquino Caramanico, Viceroy of Sicily. To equip his new institution, and to facilitate communication with other European astronomers, Piazzi traveled to France and Great Britain. There, he accompanied a group of scientists who were determining the longitude difference between the Paris and Greenwich observatories. While in England, Piazzi became acquainted with Sir **William Herschel** and observed the solar eclipse of 3 June 1788 with Astronomer Royal **Nevil Maskelyne**. But the most substantive result of Piazzi's visit was the 5-ft. vertical circle, a masterpiece of 18th-century technology, which was completed for him by instrument-maker Jesse Ramsden.

In 1789, Piazzi set up his observatory in the Santa Ninfa tower of the royal palace, for the purpose of creating a new catalog of fixed stars, which he considered as the basis of fundamental astronomy. The observatory at Palermo was the southernmost observatory located in Europe and offered unequaled access to the southern skies. Piazzi estimated the possible systematic errors of the Ramsden circle to be between 1 and 3 arc seconds. After a decade of laborious observations, he published his first catalog of 6,748 stars in 1803. With it, Piazzi was able to show that the proper motions of stars are not the exception but the rule. He also made famous the star 61 Cygni, whose abnormally large proper motion led him to call it the "flying star." The Institut de France awarded Piazzi's catalog the Lalande Prize in 1803. Baron **János von Zach**, editor of the *Monatliche Correspondenz* (the world's first astronomical journal), pronounced Piazzi's catalog as "epochal."

Piazzi once described his own personality in a letter to a friend, **Barnaba Oriani**: "My temper is fiery, and even though I am older I cannot suppress it. I have formed many wrong judgments because of it, but even if I am wrong my heart is not bad." Piazzi remained youth-oriented, flamboyant, and inclined to quick judgment. His health was delicate and illness interrupted his observations many times.

Even before publication of his star catalog, Piazzi had become famous. On 1 January 1801, he discovered the "missing planet" between Mars and Jupiter, which he named Ceres Ferdinandea, after the patron goddess of Sicily and its King Ferdinand. While looking for a seventh-magnitude star in Taurus with Ramsden's circle, Piazzi spotted a somewhat fainter object not previously cataloged. He continued to observe the moving point of light until 12 February, after which it became lost in twilight. He was criticized by astronomers all over Europe for not sharing news of his discovery sooner. Thereafter, the Göttingen mathematician **Carl Gauss** calculated the object's orbital elements by a new method that enabled it to be recovered at the end of the same year. Piazzi was subsequently offered the directorship of the Bologna Observatory but declined the invitation.

Piazzi was anxious to create another star catalog, but for all his enthusiasm, his eyesight had begun to fail. By 1807, he had to entrust work on the new catalog to his assistant, **Niccolò Cacciatore**. In 1813, Piazzi published the second catalog of 7,646 stars; it too received a prize from the Institut de France.

During this period, Piazzi was charged by the government of Naples with renovating the system of weights and measures used in Sicily. In 1808, he published an essay on the subject and was given another annual pension by King Ferdinand for establishment of the metric system. In turn, the Great Comet C/1811 F1 prompted him to publish his views on its nature. Piazzi supposed that comets were not formed along with the planets, but were produced from time to time in deep space, where they eventually dissipated.

In 1817, Piazzi was invited to Naples to examine the Capodimonte Observatory being constructed there. Though he at first declined the invitation to become its director, Piazzi finally agreed to become director-general of both the Naples Observatory and the Palermo Observatory. King Ferdinand gave him 93,500 lire in value of land, which he could sell or administer as he wished. Piazzi was also appointed president of the Royal Academy of Sciences in Naples; Cacciatore succeeded him as director of the Palermo Observatory. While at Naples, Piazzi introduced many innovations but sorely missed Palermo.

Piazzi received an unusual honor from admiral **William Smyth**, the son of an American-born loyalist who came to England after the Revolution. Smyth was also an astronomer and became acquainted with Piazzi. When the admiral's son was born in 1819, he was named **Charles Piazzi Smyth** and later became the Astronomer Royal for Scotland.

In 1825, Piazzi wrote, "It is a month since I have been in Naples, having left Palermo with regrets. Perhaps I will never see it again." The following year, he succumbed to cholera. In 1871, his native village remembered him by erecting a statue in Piazza Luini. Minor planet (1000) is named Piazzia in his honor.

Clifford J. Cunningham

Selected References

Abetti, Giorgio (1974). "Piazzi, Giuseppe." In *Dictionary of Scientific Biography*, edited by Charles Coulston Gillispie. Vol. 10, pp. 591–593. New York: Charles Scribner's Sons.

Cunningham, Clifford J. (1988). "The Baron and His Celestial Police." *Sky & Telescope* 75, no. 3: 271–272.

——— (1992). "Giuseppe Piazzi and the 'Missing Planet.'" *Sky & Telescope* 84, no. 3: 274–275.

——— (2002). *The First Asteroid: Ceres, 1801–2001*. Surfside, Florida: Star Lab Press.

——— (2004). *Jousting for Celestial Glory: The Discovery and Study of Ceres and Pallas*. Surfside, Florida: Star Lab Press.

Foderà Serio, Giorgia (1990). "Giuseppe Piazzi and the Discovery of the Proper Motion of 61 Cygni." *Journal for the History of Astronomy* 21: 275–282.

Foderà Serio, Giorgia and Pietro Nastasi (1985). "Giuseppe Piazzi's Survey of Sicily: The Chronicle of a Dream." *Vistas in Astronomy* 28: 269–276.

King, Henry C. (1979). *The History of the Telescope*. New York: Dover, esp. pp. 166–168. (On Ramsden's 5-ft. vertical circle.)

Picard, Jean

Born **La Flèche, (Sarthe), France, 21 July 1620**
Died **Paris, France, 12 October 1682**

Jean Picard made notable contributions to early precision astronomy, geodesy, cartography, and hydraulics. Picard, the son of a bookseller, attended lectures at the Jesuit College of La Flèche, where **René Descartes** had been a student. He studied Greek and Latin literature and theology, and was initiated to astronomy during a course on Aristotelian philosophy. Picard's earliest known astronomical work occurred on 21 August 1645, when he assisted **Pierre Gassendi** in the observation of a solar eclipse. He also attended Gassendi's lectures on astronomy at the Collège de France in Paris. Like his mentor, Picard became an ordained priest (1650) and then traveled throughout Europe, learning the Italian and German languages. He held several ecclesiastical positions, as *abbé* and *prieur* at Rillé and Brion, and was also a schoolmaster. By and large, his astronomical studies were conducted privately, although Picard became an informal member of the Académie de Montmor.

Together with **Adrien Auzout** and **Pierre Petit**, Picard devised a movable-wire (filar) micrometer and used it to measure the angular diameters of the Sun, Moon, and planets. In 1666, he was named a member of the Académie royale des sciences, an appointment that brought Picard's astronomical and geodetic skills to the fore. Thereafter, he applied both telescopic sights and cross hairs to other scientific instruments used for angular measurements, chiefly quadrants and sectors, and proposed that meridian observations be conducted by the method of corresponding heights. His assistant, **Philippe de la Hire**, was the first to implement Picard's suggestion of establishing a mural quadrant in the meridian plane (1683).

Armed with these newer techniques, Picard undertook his principal investigation, namely the measurement of an arc of the meridian. Supported by the Académie des sciences, this operation sought to determine a more precise value for the radius of the Earth. Picard employed the method of skeleton triangulation along the meridian between Paris and Amiens. His results, published as *La Mesure de la Terre* (1671), attained a precision some 30 to 40 times greater than previously achieved. Picard's meridian line eventually led to the first accurate trigonometric survey of France. He subsequently applied these methods to the creation of a precision map of the Paris area (*Carte des Environs de Paris*, 1678), which superseded all former cartographic ventures.

In 1671, Picard traveled to **Tycho Brahe**'s former observatory on the island of Hven to accurately determine its location, so that Brahe's observations could be compared directly with those at Paris. Picard (at Hven) and **Jean Cassini** (at Paris) used observations of the eclipses of Jupiter's satellites to determine the longitude difference of the two observatories. It was the first attempt to employ this method simultaneously, which was made possible by the ephemerides of Jupiter's satellites prepared by Cassini. During this project, Picard observed an annual displacement of the pole star (Polaris), which was not explained until 1728 by **James Bradley** as due to the combined effects of nutation and the aberration of starlight.

Picard brought back to Paris his Danish assistant, **Olaus Römer**, and a copy of Brahe's registers of observations. In 1673, he moved into the newly constructed Observatoire de Paris, where Cassini was installed. Picard participated on several expeditions to determine precise coordinates of various cities and harbors, for the purpose of creating a new map of France. This map, drawn and published by de la Hire (1693), afforded corrections as great as 150 km in longitude and 50 km in latitude over previous cartographic methods.

Picard also played a role in hydraulics, helping to solve the problem of supplying water to the fountains at Versailles. He oversaw the survey when the Grand Canal was dug in order to create artificial ponds or tanks needed to supply the fountains. Out of this survey arose Picard's correction of the apparent level, due to the curvature of the Earth. His posthumous treatise on the subject, *Traité du Nivellement* (1684), became a standard reference for more than a century.

In 1982, an international conference was held at Paris to commemorate the tercentenary of Picard's death and to highlight his role in the institutionalization of science in France during the 17th century.

Raymonde Barthalot

Selected References

Armitage, Angus (1954). "Jean Picard and His Circle." *Endeavour* 13: 17–21.

Picard, Jean (1671). *La Mesure de la Terre*. Paris: Imprimérie Royale.

——— (1680). *Voyage d'Uranibourg ou observations astronomiques faites en Dannemarck*. Paris: Imprimérie royale.

——— (1684). *Traité du Nivellement, avec une Relation de quelques Nivellements faits par ordre du Roy*. Paris: Etienne Michallet.

——— (1741). "Journal autographe des observations, 1666–1682." In *Histoire Céleste, ou Recueil de toutes les observations astronomiques faites par ordre du Roy*, edited by P. C. Le Monnier, pp. 1–267. Paris: Briasson.

Picolet, Guy (ed.) (1987). *Jean Picard et les débuts de l'astronomie de précision au XVIIe siècle: Actes du colloque du tricentenaire*. Paris: CNRS.

Taton, Juliette and René Taton (1974). "Picard, Jean." In *Dictionary of Scientific Biography*, edited by Charles Coulston Gillispie. Vol. 10, pp. 595–597. New York: Charles Scribner's Sons.

Piccolomini, Alessandro

Born **Siena, (Italy), 13 June 1508**
Died **probably Siena, (Italy), 12 March 1579**

Alessandro Piccolomini is best known for producing the first star atlas, in 1540, with 47 star maps designed for the general reader.

Piccolomini was born into an illustrious family whose members included Pope Pius II and Pope Pius III. As a young man in his hometown of Siena, he joined the Accademia degli Intronati, an organization that fostered the development of the use of Italian as a literary language on par with Latin. Around 1538 Piccolomini moved to Padua, where he eventually became a professor of philosophy at the university and joined the Accademia degli Infiammati, a group that promoted Italian works as well as had an interest in the fields of astronomy and mathematics. Piccolomini rose in the ranks of the church and in 1574 became the Archbishop of Patras, in Greece. He never went to Greece, however, and remained in Italy as coadjutor to the Archbishop of Siena until his death. During his lifetime Piccolomini was an important literary and philosophical figure, writing numerous comedies, sonnets, and philosophical treatises, and translating several classical works into Italian.

Although not an astronomer, Piccolomini did produce a few treatises on mathematical astronomy. These works were written in Italian because of his strong interest in promoting the literary use of the language. Piccolomini's first astronomical work was *De la sfera del mondo* (On the sphere of the Universe, Venice, 1540), a cosmographical work in which he defends the Ptolemaic universe. Though there are a few inferences in later editions that hint at a sympathy with the Copernican system, Piccolomini constantly supported the Ptolemaic viewpoint throughout his life. In his *Della filosofia naturale* (On natural philosophy, Rome, 1551) he mounted a staunch defense of a central and immovable Earth. His most original astronomical work was his *La prima parte dele theorique ovvero speculationi dei pianeti* (Part one of the theories or speculations of the planets, Venice, 1558), in which he endeavored to save the appearances of the motions of the planets in terms of the Ptolemaic theory. Piccolomini particularly noted that the Ptolemaic theory does not represent reality but is merely a useful tool for the astronomer.

Piccolomini's best-known astronomical work is undoubtedly *De le stele fisse* (On the fixed stars, Venice, 1540), his companion volume to *De la sfera del mondo*. *De le stele fisse* is famous as being the very first star atlas and contains a series of woodcut star maps that seem particularly suited to the casual stargazer. The two books were dedicated to Laudomia Forteguerri, and designed so that she, and any other reader, would be able to recognize the different constellations and find them in the sky. The atlas proved to be quite popular with readers and went through at least 12 editions from 1540 to 1595. Piccolomini produced 47 star maps for the book, one for each of the Ptolemaic constellations except for Equuleus, which he considered too insignificant for inclusion. Unlike previous depictions of the constellations, he did not draw in the mythological images associated with them, and so they appear as simple star patterns. Piccolomini included stars from **Ptolemy**'s first through fourth magnitudes, omitting the fainter fifth magnitude stars, and used different star symbols for each of the magnitude classes. The stars were also labeled a, b, c, d, *etc.*, usually in order of their brightness. The accompanying text helps to explain the particulars about each constellation, including the legends associated with them. In the individual maps the constellations are drawn to fill the page and, therefore, the different maps are not to the same scale. They are also drawn so that the traditional constellation pattern is most recognizable and so the direction of north varies from one map to the next. There is also a lack of a coordinate grid on the maps; no coordinates are listed on the subsequent star tables, and little attention is paid to the accuracy of star positions with respect to each other. As a result, *De le stele fisse* was not of much use to professional astronomers, but this is not surprising as Piccolomini was writing for a more popular audience.

Some sources incorrectly list 1578 as the year of Piccolomini's death.

Ronald Brashear

Selected References

Cerreta, Florindo (1960). *Alessandro Piccolomini letterato e filosofo senese del Cinquecento*. Monografie di storia e letteratura senese, Vol. 4. Siena: Accademia Senese degli Intronati. (There is no good biography or monograph on Piccolomini in English. This biography is the best work about Piccolomini.)

Gingerich, Owen (1981). "Piccolomini's Star Atlas." *Sky & Telescope* 62, no. 6: 532–534. (A very informative article on *De le stele fisse* and its printing history.)

Suter, Rufus (1969). "The Scientific Work of Alessandro Piccolomini." *Isis* 60: 210–222. (While many articles have been written about Piccolomini's literary contributions, there is very little available on his astronomical writings. This is the best source for the latter.)

Pickering, Edward Charles

Born **Boston, Massachusetts, USA, 19 July 1846**
Died **Cambridge, Massachusetts, USA, 3 February 1919**

Edward Pickering created the world's largest programs in photometric, photographic, and spectroscopic stellar research. These programs helped American astronomy to attain a dominant role within the discipline by the early 20th century.

Pickering was the son of Edward and Charlotte (*née* Hammond) Pickering. The Pickering family had deep roots in Massachusetts; Edward's ancestor John Pickering had emigrated from England to Salem in 1636. Edward was born into a family well connected with people of influence in Boston. He attended Boston Latin School, but his interest in science developed during studies at Harvard College's Lawrence Scientific School. On his graduation at age 19, Pickering taught mathematics at the Lawrence School for 2 years before becoming an instructor in physics at the Massachusetts Institute of Technology [MIT]. At the age of 22, he was appointed Thayer Professor of Physics at MIT, and in 1874, he married Elizabeth Wadsworth Sparks, daughter of former Harvard president Jared Sparks. The couple had no children. Upon the death of **Joseph Winlock**, Pickering was named the fourth director of the Harvard College Observatory, an office he assumed on 1 February 1877. He would remain the observatory's director until his own death. Initially named Phillips Professor, Pickering would be renamed Paine Professor of Astronomy when that position was created in 1887.

Pickering is remembered as a pioneer in the new science of astrophysics. On becoming director of the observatory, he decided not to focus on the astronomy of stellar positions and motions, a research field that occupied many of the large observatories of the day. Instead, he began by emphasizing a still undeveloped field, the measurement of the brightnesses of stars. In 1884, Pickering published *Harvard Photometry*, a catalog of the visual magnitudes of 4,260 stars. In this and later works, Pickering adopted British astronomer **Norman Pogson**'s proposal that a difference of five magnitudes should correspond to a brightness ratio of 100 times. The Pogson scale was soon universally adopted, in part because of Pickering's advocacy. *Harvard Revised Photometry*, a catalog of the visual magnitudes of 9,110 stars brighter than magnitude 6.5, followed in 1908. A supplement to this catalog included measurements of the magnitudes of 36,682 additional stars.

In order to achieve worldwide coverage of the sky, it was necessary for Pickering to establish an observing station in the Southern Hemisphere. Pickering's brother, **William Pickering**, helped found Harvard's Boyden Station at Arequipa, Peru, in 1891. The bulk of the Peruvian photometric observations, however, were acquired and reduced by another member of the observatory staff, **Solon Bailey**. Rather than employing the technique of an artificial comparison star, Pickering's early photometric measurements were chiefly conducted by the polarization method, wherein the magnitudes of two stars were visually compared.

The magnitudes in *Harvard Revised Photometry* are visual magnitudes. His brother, however, convinced Pickering of the advantages of astronomical photography, after sensitive dry emulsions for photographic plates were developed. Many stars could be recorded on a single exposure, and the photograph became a permanent record that could be examined at leisure. Starting in the 1880s, the Harvard College Observatory also pioneered the development of photographic photometry. With data-collection facilities located in both hemispheres, the observatory accumulated an unmatched archive of photographic plates of the heavens. In 1903, Pickering published the first *Photographic Map of the Entire Sky*, a set of 55 photographs that showed stars as faint as the 12th magnitude.

Variable stars were another focus of research at the Harvard College Observatory under Pickering's directorship. In 1881, he proposed a system for classifying the types of variables that were known. Later, Pickering's assistant, **Henrietta Leavitt**, discovered the period–luminosity relationship among Cepheid variable stars, by examining photographs of the Small and Large Magellanic Clouds obtained at Harvard's southern station. Bailey likewise employed photography to discover more than 500 variable stars in globular clusters. When Pickering assumed the directorship, only some 200 variable stars were known. At the time of his death, however, more than 3,000 variable stars had been discovered at the observatory, almost all by photographic methods.

Pickering provided technical support and encouragement to **William Olcott** after the latter founded the American Association of Variable Star Observers [AAVSO] in 1911. The early AAVSO had a membership that included professional astronomers like **Anne Young** as well as amateur astronomers.

Pickering initiated routine patrol photography of the sky with wide-field cameras. These patrol photographs provide an important record of the appearance of the sky in times past and are another of Pickering's legacies.

Under Pickering's directorship, the observatory carried out spectroscopic investigations that were perhaps even more important than its photometric programs. In 1885, Pickering initiated a program of objective prism spectroscopy. A thin prism was placed in front of the objective lens of a wide-field telescope. The images of many stellar spectra could then be recorded on a single photographic plate. In 1889, he reported the discovery of the first spectroscopic binary star, ζ Ursae Majoris.

When Pickering began this spectroscopic work, several systems had already been proposed for classifying stellar spectra. These systems, based primarily on visual observations, had stars grouped into only a handful of spectral types. Yet, all of them proved inadequate to the wealth of details revealed by the new photographic surveys. At Harvard, new classification schemes were developed for the thousands of stellar spectra acquired by the objective prism surveys.

The first Harvard spectral catalog, the *Draper Memorial Catalogue*, was published in 1890, and classified 10,351 stars into 15 spectral types. The second Harvard catalog, initiated by Pickering's assistant **Antonia Maury**, abandoned the original Draper scheme and substituted 22 groups represented by Roman numerals. Maury's system incorporated further subdivisions based on the widths of certain spectral lines. However, the next Harvard catalog, produced by **Annie Cannon**, returned to an elaboration of the Draper scheme. It was this catalog that introduced the now familiar sequence of spectral types, OBAFGKM, along with their decimal subdivisions. Pickering lobbied for acceptance of this classification system at the 1910 meeting of the International Union for Cooperation in Solar Research. A committee on the classification of stellar spectra was formed, which gave support to the Harvard system. That system gained worldwide acceptance, and was the forerunner of the MKK spectral classification system widely used today.

In 1911, Pickering began a more ambitious program in the spectral classification of stars, which culminated in publication of the *Henry Draper Catalogue*. Therein, Cannon classified the objective prism spectra of some 225,300 stars of the ninth magnitude and brighter. Although the bulk of her classification work was completed by 1915, publication of the *Henry Draper Catalogue* was completed only after Pickering's death by his successor, **Harlow Shapley**.

Abundant recognition came to Pickering during his lifetime. Among his awards were two Gold Medals from the Royal Astronomical Society, a knighthood of the Prussian Order *Pour la Mérite*, and the Bruce Medal of the Astronomical Society of the Pacific. He received honorary degrees from six American and two international universities. At the age of 27, he was elected to the National Academy of Sciences. Pickering played a role in the formation of the American Astronomical Society [AAS] (founded in 1899) and in 1905 was elected its second president, a post he likewise held until his death.

When Pickering was chosen president of what was then called the Astronomical and Astrophysical Society of America [AASA], the organization had met for only 6 years. During his long presidency, the society changed its name to the AAS and became a more effective instrument for fostering the development of an increasingly professional science. Pickering presided over the establishment of numerous research committees that, with varying degrees of success, attempted to encourage cooperation among astronomers and to establish professional standards within the discipline. From its inception, the AAS was an organization that served chiefly the interests of professional astronomers. Pickering was sympathetic to amateurs, but during his presidency, the society's membership and direction were firmly consolidated in professional hands.

While Pickering himself performed the bulk of measurements for his visual photometric surveys, numerous women assistants helped to carry out the photographic and spectroscopic programs he supervised. The Harvard College Observatory under Pickering's tenure has been likened to a factory system of mass production, with Pickering as the observatory's chief executive officer. Pickering saw astronomy as a field in which women could make important contributions to science. Yet, most of the women hired under Pickering's directorship were consigned to routine work in the reduction of astronomical data. Even so, his assistants Maury, Leavitt, Cannon, and **Williamina Fleming** were among the most important women astronomers of their time.

Pickering never abandoned his belief in the importance of securing large collections of astronomical data. Toward the end of his life, this emphasis caused friction with some younger astronomers who emphasized research programs driven by astrophysical questions over data gathering. Even these critics, however, appreciated the significance of Pickering's data and its ready availability to the astronomical community.

Pickering's papers (comprising 68 linear feet) are in the Harvard University Archives.

Horace A. Smith

Selected References

Bailey, Solon I. (1931). *The History and Work of Harvard Observatory, 1839 to 1927*. New York: McGraw-Hill.

——— (1934). "Edward Charles Pickering." *Biographical Memoirs, National Academy of Sciences* 15: 169–189.

DeVorkin, David H. (1981). "Community and Spectral Classification in Astrophysics: The Acceptance of E. C. Pickering's System in 1910." *Isis* 72: 29–49.

——— (1999). "The Pickering Years." In *The American Astronomical Society's First Century*, edited by David H. DeVorkin, pp. 20–36. Washington, DC: Published for the American Astronomical Society through the American Institute of Physics.

Hearnshaw, J. B. (1986). *The Analysis of Starlight: One Hundred and Fifty Years of Astronomical Spectroscopy*. Cambridge: Cambridge University Press.

——— (1996). *The Measurement of Starlight: Two Centuries of Astronomical Photometry*. Cambridge: Cambridge University Press.

Jones, Bessie Zaban and Lyle Gifford Boyd (1971). *The Harvard College Observatory: The First Four Directorships, 1839–1919*. Cambridge, Massachusetts: Harvard University Press.

Lankford, John (1997). *American Astronomy: Community, Careers, and Power, 1859–1940*. Chicago: University of Chicago Press.

Plotkin, Howard (1990). "Edward Charles Pickering." *Journal for the History of Astronomy* 21: 47–58.

——— (1999). "Pickering, Edward Charles." In *American National Biography*, edited by John A. Garraty and Mark C. Carnes. Vol. 17, pp. 476–478. New York: Oxford University Press.

Pickering, William Henry

Born **Boston, Massachusetts, USA, 15 February 1858**
Died **Mandeville, Jamaica, 16 January 1938**

William Pickering influenced the selection of mountaintop sites for astronomical observatories in the Western Hemisphere, pioneered the application of photography to astronomy, and was a noted popularizer of the discipline. His later works concerned visual observations of the Moon and planets, but in this regard, he strayed progressively further from mainstream science.

Pickering was descended from a notable New England family; his parents were Edward and Charlotte (*née* Hammond) Pickering. His older brother, **Edward Pickering**, was appointed director of the Harvard College Observatory, [HCO] in 1877. William graduated from the Massachusetts Institute of Technology in 1879 and taught physics there until 1883. In 1887, he was named an assistant professor at HCO. But while retaining this title until his retirement in 1924, Pickering fashioned for himself an eclectic and peripatetic career.

Pickering and his brother were the first to recognize the favorable seeing conditions that existed on California's Mount Wilson in 1889. He subsequently mounted Harvard's Boyden refractor at the site and conducted photographic studies. Roughly a decade elapsed before **George Hale** performed similar tests at Mount Wilson, prior to establishing his solar and astronomical telescopes on the mountain's peak. In 1891, meanwhile, Pickering reerected the Boyden telescope at Arequipa, Peru, to further his brother's photometric and photographic program on Southern Hemisphere stars. There, he devised a standard scale (from 1 to 10) for rating the atmospheric seeing conditions.

But the routine acquisition of data proved unsuitable for Pickering, whose attention was directed instead toward visual studies of the planets, particularly Mars. After being recalled to Massachusetts, Pickering became acquainted with **Percival Lowell**, and was soon commissioned by him to establish an observatory near Flagstaff, Arizona. Pickering thus played a role in the founding of three American observatories, and demonstrated the advantages of arid, high-altitude conditions for the optimization of astronomical research.

Pickering's early success with dry-plate emulsions convinced his brother of the enormous potential of photographic methods for astronomical data collection. This opportunity was soon exploited by Edward as one of three great programs that he established at Harvard. On behalf of the observatory, Pickering then undertook an expedition to Jamaica in 1899. Finding its atmospheric conditions suitable, he returned the following year with a 12-in. refractor of 135-ft. focus. With it, Pickering secured some of the finest images then obtained of the Moon's surface, which were published in his photographic atlas (1903).

From his lunar studies, Pickering argued for the volcanic origins of most lunar surface features. He concluded that the rival impact theory could not explain the uniform circularity of craters. In support of this hypothesis, Pickering amassed geological evidence from places as widely separated as the Hawaiian Islands and the Azores, and drew from it what he believed to be convincing analogies for the volcanic nature of the Moon's features. On the basis of albedo changes that he observed during the course of each lunar "day," Pickering advanced suggestions that either hoarfrost, or some type of vegetation, could explain the apparent phenomena, despite the absence of any measurable lunar atmosphere.

Much of Pickering's later life was devoted to the search for a trans-Neptunian planet. Starting in 1907, he predicted the existence of no less than seven unseen worlds over the next 24 years. Names for Pickering's hypothetical planets bore letters of the alphabet, starting with "Planet O." He employed a simplified graphical process to analyze the measured residuals of Uranus and Neptune. While none of Pickering's hypothetical planets were found to exist, his search nonetheless catalyzed a similar investigation at Lowell Observatory, from which the eventual discovery of Pluto was announced in 1930.

Pickering's most notable discovery, on photographic plates taken at Peru, concerned Saturn's ninth satellite (Phoebe) and his demonstration of its retrograde orbit. Phoebe was the first planetary satellite found to possess this property. For his investigations of the Saturnian system, Pickering was awarded the Lalande Prize of the Paris Academy of Sciences in 1905 and the Janssen Medal in 1909. Nonetheless, his scientific methods remained largely those of the 19th century. Pickering's visual observations and highly speculative theories were greeted with increased skepticism and ostracism by 20th-century astronomers.

Pickering was married to Anne Atwood; the couple had two children. A crater on the Moon bears his name.

Jordan D. Marché, II

Selected References

Hoyt, William Graves (1976). *Lowell and Mars*. Tucson: University of Arizona Press.

——— (1980). *Planets X and Pluto*. Tucson: University of Arizona Press.

——— (1987). *Coon Mountain Controversies: Meteor Crater and the Development of Impact Theory*. Tucson: University of Arizona Press.

Marsden, Brian G. (1974). "Pickering, William Henry." In *Dictionary of Scientific Biography*, edited by Charles Coulston Gillispie. Vol. 10, pp. 601–602. New York: Charles Scribner's Sons.

Pickering, William H. (1903). *The Moon: A Summary of the Existing Knowledge of Our Satellite, with a Complete Photographic Atlas*. New York: Doubleday, Page and Co.

Plotkin, Howard (1993). "William H. Pickering in Jamaica: The Founding of Woodlawn and Studies of Mars." *Journal for the History of Astronomy* 24: 101–122.

Sadler, Philip M. (1990). "William Pickering's Search for a Planet beyond Neptune." *Journal for the History of Astronomy* 21: 59–64.
Strauss, David (2001). *Percival Lowell: The Culture and Science of a Boston Brahmin.* Cambridge, Massachusetts: Harvard University Press, esp. pp. 173–179.

Pigott, Edward

Born **England or France, 1753**
Died **Bath, England, 27 June, 1825**

Edward Pigott's chief claim to fame is as one of the founders of the study of variable stars.

Pigott was the son (probably the oldest) of Nathaniel Pigott and Anna Mathurina de Beriol. Nathaniel was the grandson of William, Eighth Viscount Fairfax; Edward's full name is sometimes given as Edward Fairfax Pigott. Little is known of Edward's early life and formal education, although he appears to have acquired a good knowledge of French. A portrait in the Junior School of Ampleforth College, Yorkshire, England, long believed to be of Edward Pigott (and published as such) may, in fact, be that of a relative, Gregory.

Nathaniel Pigott was a surveyor and astronomer. Edward's interests in these subjects were probably stimulated by helping his father. They observed the 3 June 1769 transit of Venus together from Caen, France (where they then lived), and many years later, the 3 May 1786 transit of Mercury from Louvain, Belgium. In 1771, the family moved to Wales.

Edward Pigott surveyed the region of the Severn estuary in 1778/1779, discovered a nebula in Coma Berenices in 1779, and the short-period comet D/1783 W1. He also made observations of the satellites of Jupiter and studied the method of longitude determination by lunar transits and is believed to have observed the great comet of 1807.

In 1781 the family moved to Bootham, Yorkshire, where father and son constructed an observatory. There, Edward began a partnership with **John Goodricke**, discoverer of the periodicity of the light variation of Algol (β Persei). Pigott himself discovered the variability of η Aquilae on 10 September 1784. By early December of the same year, he had determined its period to be 7 days, 4 hours, 38 minutes, about 24 minutes longer than the accepted modern period. Piggott tried to set up a photometric system so that he could estimate the variations in stellar brightness more precisely.

It was not until the 17th century, after the application of the telescope to celestial observation, that astronomers began to accept that the brightness of some stars varied. Pigott and Goodricke were the first to show that stellar variability is often periodic. Pigott has some claim to be regarded as the founder of the study of stellar variability. In 1786, he published a catalog of 50 variable stars writing, "these discoveries may, at some future period, throw fresh light on astronomy." Later he announced the variability of the stars now known as R Scuti and R Coronae Borealis. During a lull in the Napoleonic wars, Pigott returned to the continent but was caught by a renewal of hostilities and detained for a while at Fontainebleau, where, nevertheless, he managed to write a new study of the period of R Scuti and tried to explain stellar variability in terms of a model of rotating spotted stars. He also inferred the existence of "dark stars" and surmised that aggregations of such objects might account for phenomena like the "Coal Sack." Pigott was apparently released from detention through the good offices of Sir Joseph Banks. Although Pigott knew Banks and both the **Herschels** (**William** and **John**), and published some 20 papers in the *Philosophical Transactions of the Royal Society of London*, he does not seem to have received much recognition for his work during his lifetime. He is known to have retired to Bath and appears to have withdrawn from astronomical work by the end of his life.

Alan H. Batten

Selected References

Clerke, Agnes M. (1921–1922). "Pigott, Edward." In *Dictionary of National Biography*, edited by Sir Leslie Stephen and Sir Sidney Lee. Vol. 15, pp. 1169. London: Oxford University Press.
Hoskin, Michael (1979). "Goodricke, Pigott and the Quest for Variable Stars." *Journal for the History of Astronomy* 10: 23–41.
Kopal, Zdeněk (1974). "Piggott, Edward." In *Dictionary of Scientific Biography*, edited by Charles Coulston Gillispie. Vol. 10, pp. 607–608. New York: Charles Scribner's Sons.
Pigott, Edward (1785). "Observations of a New Variable Star." *Philosophical Transactions of the Royal Society of London* 75: 127–136.
——— (1786). "Observations and Remarks on Those Stars which the Astronomers of the last Century suspected to be changeable." *Philosophical Transactions of the Royal Society of London* 76: 189–219.
——— (1805). "An Investigation of all the Changes of the variable Star in Sobieski's Shield, from Five Year's Observations, exhibiting its proportional illuminated Parts, and its Irregularities of Rotation; with Conjectures respecting unenlightened heavenly Bodies." *Philosophical Transactions of the Royal Society of London* 95: 131–154.

Pingré, Alexandre-Guy

Born **Paris, France, 4 September 1711**
Died **Paris, France, 1 May 1796**

Alexandre Pingré was a diligent observational astronomer, calculator, and noted historian of astronomy. Pingré's parents sent him to the college of Senlis directed by the regular canons of the Congregation of France (the Génovéfains). A good student, he entered the Congregation at age 16. By 1735 he was a teacher of theology in the college. Because of the persecutions against the Jansenistes, Pingré was dismissed from his chair and sent away from Paris to obscure colleges to teach grammar. Fortunately, in Rouen, Pingré met the surgeon Claude-Nicolas Le Cat who, in 1749, asked him to join—as an astronomer—the academy he had founded there. Thus, at the age of 38, Pingré undertook his first scientific work by calculating the lunar eclipse of 23 December 1749; 4 years later he observed the transit of Mercury on 6 May 1753. The Académie royale des sciences in Paris then named Pingré a correspondent of **Pierre-Charles Le Monnier**, and his congregation called him to Paris, to the abbey of Sainte-Geneviéve, where the abbot installed a small observatory and offered him some instruments. In Paris Pingré met Le Monnier, who assigned him the calculations for the four-volume *Etats du Ciel à l'usage de la marine*, which appeared from 1754 to 1757. Later, **Joseph-Jérôme Lalande** integrated this nautical almanac into the *Connaissance des temps*. In 1756, Pingré became a member of the academy.

For the second edition (1770) of the *Art de vérifier les dates* of **Nicolas La Caille**, Pingré calculated the solar eclipses up to 1900 and later (1787) the eclipses in the Northern Hemisphere for 10 centuries before our era. Pingré designed several sundials in Paris. At the request of the Provost of the Guilds, he created in 1764, on the tower-column of the former Hôtel de Soissons, a very original sundial, cylindrical with several horizontal stilettos.

In 1760, the academy designated Pingré to observe the 1761 transit of Venus from Rodrigue Island in the Indian Ocean. He also made three voyages to examine the marine chronometers of Berthoud and Le Roy. During one of these voyages, he observed the 1769 transit of Venus from Cap François in Santo Domingo, Haiti. From the numerous observations of the transit received in Paris, Pingré deduced a solar parallax of 8.8″.

In 1769, Pingré was named astronomer–geographer in the place of the deceased **Joseph Delisle**. At the same time he was chancellor of the University, librarian of the Sainte-Geneviéve Library, and a member of the Académie de marine.

Besides his regular observations, Pingré wrote lengthy works, interrupted from time to time by his travels. Beginning in 1757, he undertook research on comets from Antiquity. This large historical and theoretical work, *Cométographie*, was published in two volumes in 1783/1784. From 1756, he worked on a history of 17th-century astronomy at the request of Le Monnier, who passed to him the manuscripts of **Ismaël Boulliau**. Pingré completed this work at the age of 80. Jerôme de Lalande reported on it to the academy in 1791 and then, with Le Monnier, undertook its publication. Upheaval during the Revolution halted printing. The manuscript was located by **Camille Bigourdan** in the archives of the Paris Observatory; Bigourdan published it in 1901.

At Lalande's request Pingré translated the *Astronomiques* of **Manilius**, a Latin poet of the Augustan century, and a poem by **Aratus**. During the Revolution, with the suppression of the academies, Pingré remained the librarian of Sainte-Geneviéve but with few resources. With the establishment of the Institut National in 1795, he was elected to the astronomy section, attending meetings almost until his death.

A regular canon of Sainte-Geneviéve, Pingré was also a dignitary of freemasonry, being an active correspondent with the lodges of Bourbon Island (now Réunion). His leisure activities included reading (in particular Horace in Latin), music, and, at the end of his life, botany. Esteemed as a scientist, he was also a man of great goodness, modesty, and piety.

Simone Dumont

Selected References

Delambre, J. B. J. (1827). *Histoire de l'astronomie du dix-huitième siècle*. Paris: Bachelier.

Dumont, S. and S. Débarbat (1999). "Les académiciens astronomes, voyageurs au XVIIIe siècle." *Comptes rendus de l'Académie des sciences de Paris*, série 2 b: 415. (With abridged English version.)

Lacroix, A. (1938). "Pingré." *Figures de savants*. Vol. 3. Paris: Gautier-Villars.

Lalande, J. (1970). *Bibliographie astronomique avec l'histoire de l'astronomie depuis 1781 jusqu'à 1802*. Paris, 1803. (Facsimile edition. Amsterdam: J. C. Gibben.)

——— (1789). "Cadran." In *Dictionnaire encyclopédique de mathématiques*, edited by J. d'Alembert. Vol. 1, p. 281. Paris.

Prony, G. C. F. M. (1798). "Eloge de Pingré." *Mémoires de l'Institut national des sciences* 1: 26.

Pişmiş, Paris Marie

Born **Istanbul, (Turkey), 1911**
Died **Mexico City, Mexico, 1 August 1999**

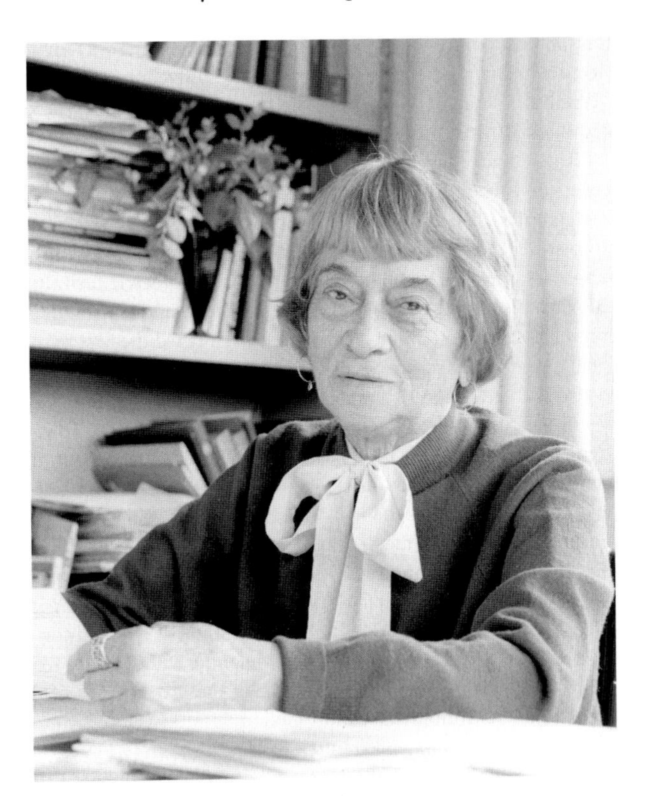

Paris Pişmiş was the first formally trained astronomer in Mexico. She was born into an Armenian family living in Istanbul, Turkey. In defiance of her parents' wishes and native tradition, she became one of the first women in Turkey to attend Istanbul University, earning her Ph.D. in mathematics in 1937, under the supervision of professors **Erwin Freundlich**, who was then in Istanbul as a refugee from the Nazi persecution, and Richard von Mises. Pişmiş worked as a research assistant at the Istanbul University Observatory from 1935 to 1937.

Before World War II, Pişmiş traveled to the United States to pursue postdoctoral studies. She worked as an assistant astronomer at the Harvard College Observatory (1938–1942). There, she met astronomers **Harlow Shapley**, **Bart Bok**, **Sergei Gaposhkin**, **Cecilia Payne Gaposhkin**, and **Donald Menzel**. Pişmiş also met Félix Recillas, a Mexican mathematician and an astronomy student who became her husband in 1942. In that year, she moved with him to Mexico and joined the staff of the newly created Tonantzintla Observatory in Puebla, where she worked alongside **Guillermo Haro**, who became the observatory's director. Pişmiş started teaching astronomy to students of physics and mathematics. Her two children, Elsa and Sevín Recillas, who also became scientists, were born during that time.

After visiting appointments at Princeton University and Yerkes Observatory, Pişmiş moved to Mexico City in 1948 and joined the Observatorio Astronómico Nacional de Tacubaya, part of the Universidad Nacional Autónoma de Mexico [UNAM], which is now the Instituto de Astronomía. She spent most of her time on research, teaching, and advising students. At UNAM, Pişmiş taught formal courses in astronomy, and transmitted her passion for doing scientific research to many students (*e. g.*, Arcadio Poveda, Eugenio Mendoza, Enrique Chavira, and others).

Pişmiş was also a restless traveler and was engaged with different groups doing astronomical research all over the world. She was always interested in new ideas and new techniques. She was invited to lecture at many scientific institutions and universities, traveling to Istanbul, Heidelberg, Ankara, Byurakan, the Canary Islands, Paris, Buenos Aires, and so forth.

Pişmiş made significant contributions in observational and theoretical astronomy. She discovered 20 open stellar clusters and three globular clusters that are named for her. She was among the first to study the kinematic behavior of the gas associated with hot young stars, introducing the Fabry–Perot interferometric technique to Mexico. Pişmiş studied the effects of interstellar absorption in the observed distribution of star clusters. She performed the first photometric observations of young stellar clusters carried out in Mexico. She also worked theoretically to explain the origins of spiral structures in galaxies and the observed galactic velocity fields. Toward the end of her life, Pişmiş became interested in the morphology and kinematics of the so called mildly active galactic nuclei. Her scientific output totaled more than 100 research articles.

Another of Pişmiş's major contributions was to foster the publication of Mexican astronomical journals. She was a member of the board of editors of the *Revista Mexicana de Astronomía y Astrofísica*, from its foundation in 1974 untill her death. She was editor of the *Boletín de los Observatorios de Tonantzintla y Tacubaya*, from 1966 to 1973. Pişmiş was likewise editor of the proceedings of the International Astronomical Union Colloquium 33 (*Observational Parameters and Dynamical Evolution of Multiple Stars*) in 1975. She supervised the edition of three volumes of the *Astrophotometric Catalogue of Tacubaya* (1966).

Pişmiş was a member of the International Astronomical Union, and was appointed Mexico's representative to that organization. She was a member of the Royal Astronomical Society, the American Astronomical Society, and the Academia Mexicana de Ciencias. She received a science teaching award from UNAM, plus honorary doctorates from the same institution and from the Instituto de Astrofísica, Óptica y Electrónica.

Paris's former student, Deborah Dultzin, wrote:

> Listening to her lectures, learning to observe with her, and later on, being initiated into the wonderful world of scientific research work by her, was an inspiring experience. She spoke about a scientist's life as something wonderful for a woman, and one could see that she really enjoyed it.

Pişmiş was interested not only in science, but also in all aspects of culture. Fluent in several languages, she loved literature, painted, played the flute and piano, and was also a good singer and dancer. She inspired admiration in all those who knew her, and was a role model for women in science all over Mexico and beyond. In collaboration with her grandson, Gabriel Cruz-González, Pişmiş prepared an autobiography, *Reminiscences in the Life of Paris Pişmiş: A Woman Astronomer* (1998).

Nidia Irene Morrell

Selected References

Pişmiş, Paris and Gabriel Cruz-González (1998). *Reminiscences in the Life of Paris Pismis: A Woman Astronomer*. Universidad Nacional Autónoma de México, Instituto de Astronomía, Instituto Nacional de Astrofísica, Optica y Electrónica.

Torres-Peimbert, Silvia (1999). "Paris Marie Pismis, 1911–1999." *Bulletin of the American Astronomical Society* 31: 1607–1608.

Plana, Giovanni Antonio Amedeo

Born **Voghera, (Lombardy, Italy), 6 November 1781**
Died **Turin, Italy, 20 January 1864**

Mathematical physicist and astronomer Giovanni Plana wrote more than 100 memoirs dealing with mathematical analysis, geodesy, astronomy, celestial mechanics, and heat theory, including his most important work, the theory of the lunar movements. Plana was the son of Antonio Maria Plana and Giovanna Giacoboni. From 1796, Plana studied in Grenoble, France, where he met and befriended Stendhal (Henry Beyle). In 1800, Plana entered the École Polytechnique where **Joseph Lagrange** taught analysis and mechanics, Gaspard Monge taught geometry, and **Pierre de Laplace** taught astronomy. He also became friends with **Siméon-Denis Poisson**. In 1803, Plana became professor of mathematics at the Turin Artillery School, located in Alessandria, and became professor of astronomy at Turin University in 1811. Before assuming the responsibilities of his chair, Plana practiced astronomy at the Brera Astronomical Observatory in Milan under the scientific direction of **Barnaba Oriani**. In 1816, he became director of the Turin Astronomical Observatory. In 1817, Plana married Alessandra Maria Lagrange, a niece of the great mathematician.

In Milan in 1811, Oriani proposed to the young astronomer **Francesco Carlini** some research in the theory of the Moon, probably on Laplace's suggestion. Plana was involved with this project and collaborated closely with Carlini, at least until 1820, when they won a prize of the Paris Académie des sciences. The prize, promoted by Laplace in 1818, was to be assigned to whoever succeeded in the construction of lunar tables based solely on the law of universal gravity. (The prize was shared with **Charles Damoiseau.**) Despite the prize, Laplace criticized the memoir of Carlini and Plana. A bitter dispute ensued. Plana determined the tone of the dispute, perhaps remembering the discussions in Grenoble when the young students sided excitedly with Lagrange, *homme à principes*, or with Laplace, *homme à théorèmes*. In the end, Laplace admitted that the two Italian astronomers were more accurate than him. The Académie des sciences published Damoiseau's memoir but not that of Carlini and Plana, who decided to publish a complete theory of the Moon. But the collaboration was complicated and, in the end, Plana decided to continue his work alone, which was published in 1832 in three big volumes.

Plana's work is considered a milestone in celestial mechanics. The theory was derived solely from the principle of universal gravitation and borrowed only the essential data from the observations. Plana's lunar theory had completed Laplace's program, along the road drawn by **Isaac Newton**, for one of the most complex celestial phenomena.

In 1825 and 1827, Plana and Carlini published the data of their observations obtained by the measurement of the mean parallel (45°) linking the French geodesic net to the northern Italian one, from Bordeaux to Fiume (today Rijeka), crossing the Alps. They were also able to explain the anomalies of Giovanni Battista Beccaria's measurements of the meridian arc between Mondovì and Andrate, as they noted the deviation of the plumb line due to the presence of high mountains.

In 1822, Plana inaugurated the new Turin Observatory and the new instrument, a meridian circle of Reichenbach. He published the results of his 1822, 1823, 1824, and 1825 observations and his theoretical considerations on astronomical refraction. Other astronomical memoirs dealt with Foucault's pendulum, the movement of a body launched from the Moon to the Earth, and comets.

Plana won the Copley Medal of the Royal Astronomical Society. He was president of the Turin Accademia delle Scienze, foreign member of the Académie des sciences, and corresponding member of the most important European academies. Plana's archives are in Turin in the Accademia delle Scienze. Plana's letters to Carlini about the theory of the Moon and other subjects are in the Archives of the Brera Astronomical Observatory.

Pasquale Tucci

Selected References

Agostinelli, Cataldo (1964–1965). "Della vita e delle opere di Giovanni Plana." *Atti dell'Accademia delle scienze di Torino* 99: 1177–1199.

De Béaumont, Élie (1873). "Éloge historique de Jean Plana." *Mémoires de l'Académie des sciences de l'Institut de France* 38: CVII–CLXXV.

Leschiutta, Sigfrido and Maria Rolando Leschiutta (1992). "Giovanni Antonio Amedeo Plana, astronomo reale." *Giornale di fisica* 33: 111–125.

Maquet, Albert (1965). "L'astronome royal de Turin Giovanni Plana (1781–1864)." *Mémoires de l'Académie des sciences* 36: 2–253.

Realis, Savino (1886). "Plana." *Bullettino di bibliografia e di storia delle scienze, matematiche e fisiche* 19 : 121–128.

Sclopis, Federico (1842). "Della vita di Giovanni Plana." *Memorie della Reale accademia delle scienze di Torino* 22: 51–63.

Tagliaferri, Guido and Pasquale Tucci (1999). "Carlini and Plana on the Theory of the Moon and Their Dispute with Laplace." *Annals of Science* 56: 221–269.

Tricomi, Francesco Giacomo (1964–1965). "Giovanni Plana (1781–1864)." *Atti dell'Accademia delle scienze di Torino* 99: 267–279.

Plancius, Petrus

Born **Dranoutre, (Belgium), 1552**
Died **probably Amsterdam, The Netherlands, 1622**

Petrus Plancius is known as a cartographer and globe maker, and also as a Calvinist theologian and Bible scholar. Plancius's education took him to England and France; he subsequently moved north from Flanders during the Dutch Wars of Independence and settled in the Netherlands. His interest in missionary work arose from his religious convictions, and gave impetus to his cartographic activities, both celestial and terrestrial.

Plancius's 1589 celestial globe, published with **Michael van Langren**, included Southern Hemispheric stars that had not been previously depicted, namely *Crux*, *Triangulum Antarcticus*, and the *Nubeculae Magellani*. He encouraged **Pieter Keyser** and Frederick de Houtman to make further observations of the southern skies during the voyage of the *Hollandia* to the East Indies in 1595–1597. These observations supplied the data for the formation of 12 further constellations, which appeared on Plancius's globe of 1598 published by Hondius and which were subsequently adopted by **Johann Bayer**. On a globe published in 1612 with Pieter van den Keere, Plancius added another eight constellations, in between old ones. Some of these, of biblical reference, have not survived; others such as Monoceros, Camelopardalis, and Columba have endured.

Peter Nockolds

Alternate name

Platevoet, Petrus

Selected References

Keuning, Johannes (1946). *Petrus Plancius, theoloog en geograaf*. Amsterdam: Van Kempen en Zoon.

Stevenson, Edward Luther (1921). *Terrestrial and Celestial Globes*. 2 Vols. New Haven: Published for the Hispanic Society of America by the Yale University Press.

Warner, Deborah J. (1979). *The Sky Explored: Celestial Cartography, 1500–1800*. New York: Alan R. Liss.

Zinner, Ernst (1967). *Deutsche und niederländische astronomische Instrumente des 11–18. Jahrhunderts*. Munich: Beck, 2nd ed. 1979.

Planman, Anders

Born **Hattula, (Finland), 1724**
Died **Pemar, (Finland), 25 April 1803**

Anders Planman observed the transits of Venus in 1761 and 1769 for the Swedish Academy of Sciences. Planman, the son of lieutenant Pehr Planman and Ingeborg Leufstadius, was born in a Swedish family in Finland (then part of Sweden). He studied in Åbo (Turku) and Uppsala, where he became associate professor (*docent*) in astronomy in 1758. In 1763, Planman became a professor of physics in Åbo, where he remained until his retirement in 1801. He was a member of the Science Society at Uppsala (*Vetenskapssocieteten*) and from 1767 of the Royal Swedish Academy of Sciences.

Planman's most important contribution in the field of astronomy came with the transits of Venus in 1761 and 1769, which he observed in the far north of Finland. Planman's expeditions were financed by the Academy of Sciences in Stockholm, and they were part of a larger effort by Swedish astronomers to contribute to the measurement of the solar parallax, which they did to an extent second only to Britain and France. Though his own 1761 measurements were somewhat uncertain, Planman was given the task of calculating a value for the parallax based on all the data received by the academy. This work resulted in values between 8.2″ and 8.7″. As a preparation for the transit in 1769, Planman developed a new method to calculate the parallax, and his own observations in 1769 were considered among the best in Europe. He now found the parallax to be 8.5″. Planman participated in discussions concerning the possible existence of an atmosphere on Venus and also on the much-debated issue of the reliability of the observations carried out by **Miska Höll** in northern Norway (Vardöhus).

By making an observational effort great enough to command the attention of all of Europe, and by giving Planman the main responsibility of handling its observational data, the Academy of Sciences created a platform for Planman that for a few years made him an international authority on the important question of the solar parallax. Otherwise his scientific life was uneventful.

Sven Widmalm

Selected References

Amelin, Olle (1995–1997). "Anders Planman." *Svenskt biografiskt lexikon*. Vol. 29. Stockholm: Svenskt biografiskt lexicon.

Lindroth, Sten (1967). *Kungl. Svenska vetenskapsakademiens historia.* 2 Vols. Stockholm: Almqvist and Wiksell.

Woolf, Harry S. (1959). *The Transits of Venus: A Study of Eighteenth-Century Science*. Princeton, New Jersey: Princeton University Press.

Plaskett, Harry Hemley

Born **Toronto, Ontario, Canada, 5 July 1893**
Died **Oxford, England, 26 January 1980**

Harry Plaskett made observational and theoretical contributions to the understanding of stellar atmospheres and solar physics, especially through studies of spectral line formation, solar granulation, and the recognition of large-scale mass movements on the Sun's surface. Son of **John Plaskett** and Rebecca Hemley, Plaskett graduated from Ottawa Collegiate School before enrolling at the University of Toronto, where he was awarded his B.A. degree in 1916. During World War I, Plaskett served in the Canadian Field Artillery and rose to first lieutenant among his brigade in France. Afterward, he spent a year at Imperial College, London, in the laboratory of spectroscopist **Alfred Fowler**.

Returning to his native land, Plaskett spent eight years (1919–1927) as a research astronomer at the newly established Dominion Astrophysical Observatory in Victoria, British Columbia, where his father was director. At the time, its 72-in. reflector was the second largest telescope in the world. In 1921, Plaskett married Edith Alice Smith; the couple had two children.

Plaskett was then called to Harvard University (1928–1932), where he served as lecturer, associate professor, and professor of astrophysics. In 1932, Plaskett was chosen as the Savilian Professor of Astronomy at Oxford University, a post that he retained until his retirement in 1960. At Oxford University, he built up a "school" of solar physics that produced a long line of research students. Plaskett had two solar telescopes with spectrographs constructed there, and campaigned (starting in 1946) for a large, modern reflecting telescope in the British Isles. This ambition was realized (after his retirement) with the completion of the Isaac Newton telescope at Herstmonceux in 1967.

Plaskett's research involved analysis of the Sun and stars by means of spectroscopy and spectrophotometry. He was the first to identify four lines of ionized helium in the spectra of O-type stars and deduced their corresponding temperatures of excitation, according to the ionization theory of **Meghnad Saha**. This work led to a new understanding and classification of these very hot stars. Plaskett likewise measured the absorption profiles of the so called magnesium "b" lines in the solar spectrum, as a function of their position on the Sun's disk. These measurements constituted the first practical tests of the relative strengths of scattering versus absorption in the Sun's atmosphere and significantly improved the theory of spectral line formation. Plaskett's detailed study of the phenomenon of solar granulation provided convincing evidence that these small-scale structures were manifestations of Bénard convection cells arising in the photosphere's outer layers. Finally, Plaskett's studies of the Sun's rotation led to his announcement of large-scale velocity fields and especially meridional currents on its surface. He recognized the importance of hydrodynamic considerations to the explanation of such motions, which served as a precursor to later work in solar oscillations and helioseismology.

Plaskett was elected a fellow of the Royal Society of London, and served as secretary (1937–1940) and president (1945–1947) of the Royal Astronomical Society [RAS]. Like his father, Plaskett was awarded the Gold Medal of the RAS (1963) and delivered the Halley Lecture at Oxford University (1965). From 1932 to 1935, he served as president of the International Astronomical Union's [IAU] commission on spectrophotometry. Saint Andrews University conferred an honorary doctorate upon him (1961).

Jordan D. Marché, II

Selected References

Adam, Madge G. (1986). "Plaskett, Harry Hemley." In *Dictionary of National Biography, 1971–1980*, edited by Lord Blake and C. S. Nicholls, pp. 673–674. Oxford: Oxford University Press.

McCrea, W. H. (1981). "Harry Hemley Plaskett." *Biographical Memoirs of Fellows of the Royal Society* 27: 445–478.

Plaskett, Harry H. (1933). *The Place of Observation in Astronomy*. Oxford: Clarendon Press. (Inaugural lecture.)

Plaskett, John Stanley

Born **Hickson, (Ontario, Canada), 17 November 1865**
Died **Esquimalt, British Columbia, Canada, 17 October 1941**

Canadian astronomer John Plaskett is probably best remembered for his role in the establishment of the Dominion Astrophysical Observatory and the building of its 72-in. telescope, briefly the largest in the world. But he also showed (together with **Otto Struve**) that the material that produces stationary absorption lines in the spectra of spectroscopic binaries is truly interstellar and not just circumstellar.

Plaskett was one of a number of children of Ontario farmers Joseph and Annie P. Plaskett, and his father's death when he was only 16 forced him to defer his own education until some of the younger children had been taken care of. He married Rebecca Hope Hemley in 1892, and the elder of their two sons, **Harry Plaskett**, followed in his father's footsteps, eventually becoming Savillian Professor of Astronomy at Oxford University. Trained as a mechanic, John Plaskett acquired some engineering experience in Ontario and Schenectady, New York, USA, and entered the University of Toronto as an assistant to the professor of physics and as a mature student, earning his BA in physics and mathematics in 1899 at the age of 33. He was hired initially at the Dominion Observatory in Ottawa primarily as Mechanical Superintendent, but immediately set to work to establish an astrophysics department, install a number of new instruments, and initiate several new research programs, including solar rotation and radial–velocity measurements of stars.

Plaskett and **William King**, by then director of the Dominion Observatory at Ottawa, soon became dissatisfied with the 15-in. refractor available to them and set out to persuade the Canadian government to fund a 72-in. reflector, somewhat modeled on the Mount Wilson 60-in. telescope, but with innovations of Plaskett's own (like the central hole in the primary mirror for the Cassegrain focus) adopted in later large reflectors. They succeeded by 1913, and a 72-in. reflector was ordered from Warner and Swasey. Plaskett became the founding superintendent (later director) of the Dominion Astrophysical Observatory in Victoria, British Columbia, retiring in 1935, but continuing to serve as a consultant on astronomical instrumentation, particularly for the McDonald Observatory 82-in. telescope, until his death.

Plaskett's leadership had a major role in the development of Canadian excellence in the measurement of stellar velocities and their interpretation. His own work included collaboration with Joseph Pearce on radial velocities of hot, bright stars, which confirmed the work of **Bertil Lindblad** and **Jan Oort** on the rotation of the Milky Way. The investigations with Struve showed, first, that there is gas that is truly interstellar and, second, that it shares the rotation of the galactic stellar system. He was part of the International Astronomical Union commissions on variable stars, stellar classification, and radial velocities, reflecting the unusual breadth of his accomplishments. Plaskett's 1939 drawing of the Milky Way, making clear the relationships among the disk, bulge, halo, and globular clusters is widely reproduced even now.

Plaskett received the Gold Medal of the Royal Astronomical Society, of which he was a fellow, the Rumford Medal of the American Academy of Arts and Science, the Henry Draper Medal of the National Academy of Science, and the Bruce Medal of the Astronomical Society of the Pacific. He was active in the Royal Astronomical Society of Canada and the American Astronomical Society and was a fellow of the Royal Society of Canada and a recipient of its Flavelle Medal. Plaskett received honorary degrees from Toronto, Pittsburgh, British Columbia, McGill, and Queen's universities.

Richard A. Jarrell

Selected References

Beals, C. S. (1941). "John Stanley Plaskett." *Journal of the Royal Astronomical Society of Canada* 35: 401–407.

Jarrell, Richard A. (1988). *The Cold Light of Dawn: A History of Canadian Astronomy*. Toronto: University of Toronto Press.

Redman, R. O. (1942). "J. S. Plaskett." *Observatory* 64: 207–211.

Platevoet, Petrus

➜ **Plancius, Petrus**

Plato

Born **Athens, (Greece), *circa* 428 BCE**
Died **Athens, (Greece), 348/347 BCE**

Plato's astronomy, though never systematized like that of **Eudoxus** or **Aristotle**, continued to influence readers for millennia, through the beauty and coherence of his images and myths.

Plato was born as Aristocles to Ariston and Perictionê, in one of the wealthiest Athenian families, descended from Solon's brother through his mother. He had two older brothers, Glaucôn and Adeimantos, a sister Potonê, and a younger half-brother Antiphôn. Well-educated, Plato in his early 20s came under the influence of Socrates, upon whose execution he left Athens, traveling especially to south Italy and Syracuse. He returned to Athens around 390 or 385 BCE and founded his school in the grove sacred to the hero Akademos. In the 360s BCE Plato again visited Syracuse twice, vainly attempting to teach philosophy to its tyrant Dionysios II. He was unmarried but well loved by his students.

The pervasive dramatic irony of his dialogs, his own express preference for probable accounts, plus millennia-deep scholarship continue to challenge Plato's readers. Certain ideas recur, and may rather confidently be attributed to Plato (not just to characters in the dialogs), but many details may simply be decorative.

Plato discusses astronomy in six dialogs – *Phaedo*, *Republic* (books 7 and 10), *Politicus*, *Timaeus*, *Laws* (books 7 and 12), and *Epinomis* – composed in about that order over a generation. The first three weave astronomy into myth, and all six show a steadily growing appreciation of astronomy. Plato always sought to reveal cosmic design, penetrate phenomena to hidden mathematical reality, and inspire students to contemplate higher truths. Thus, and also because contemporary celestial observations eluded any coherent account, Plato preferred theory over observation.

The 6th-century commentator **Simplicius** reported that **Sosigenes** in the 2nd century quoted **Eudemus** (in the generation after Plato) as saying that Plato's astronomical program was to find regular circular motions of the planets to explain their apparent irregular motions. **Geminus** attributed that program to the

Pythagoreans, who said that even in human affairs, noble men do not alter speed or course, so one must hypothesize celestial uniformity, while **Plutarch** claimed that Plato saved astronomy from reproach by subordinating natural laws to divine principles.

In the *Phaedo*, equilibrium holds the spherical Earth centered in the spherical heaven. Earth is very large, with dimples filled by water, mist, and air, in which people gather like ants around frog ponds; atop the highest peaks lies Earth's true surface exposed to the surrounding bright, clear *aithêr*, through which the stars move.

In the *Republic*, Plato first insists on the utility of astronomy, then its importance in education: The proper goal of astronomical study is to direct the mind *via* theoretical exercise away from the mutable world to the eternal world of truth, just as number and geometry do, and music should. A mythic celestial mechanics completes the *Republic*, in which the whole *kosmos* is a spindle (hung from a rainbow-hued pillar of light), its whorl formed of eight nested whorls, representing the fixed stars and the seven "wanderers" (*planetai*). The outermost whorl, broad and "spangled" with stars, rotates the same as the spindle itself; the seven inner whorls revolve gently in the opposite direction. The eighth (innermost) lunar whorl moves the fastest, illuminated by the next (seventh) and brightest solar whorl. That and the next two (the sixth Venus and fifth Mercury) all move together, with the next greatest swiftness, and so on in descending order out to the second (Saturn). The third (Jupiter) is the whitest – Venus is almost as white, and the fourth (Mars) is ruddy and "recycling" (probably because although Jupiter and Saturn retrograde about once per year, and Venus and Mercury follow the Sun, Mars traverses more than a full circuit of the zodiac between retrogradations). This planetary order – Moon, Sun, Venus, Mercury, Mars, Jupiter, and Saturn outermost – persists in later works. Each whorl has upon it a Seirên singing a single tone, all eight in harmony. The spindle lies in the lap of the goddess Necessity, around whom sit the three Fates: Clotho turning the outer whorl, Atropos turning the seven inner whorls, and Lachesis playing with both. (Plato omits planetary names, and the well-known zodiacal inclination; no one has convincingly explained the whorl-widths – outermost widest, then Venus, Mars, Moon, Sun, Mercury, Jupiter, and Saturn narrowest.)

The myth in the *Politicus* tells that the revolution of the whole Universe affects the course of earthly events, and alters its direction at great intervals. Plato proposed that time is created by the rotation of the *kosmos*, and that when the *kosmos* reverses, so too time reverses in its course.

The *Timaeus* gives a different model, not physical but fundamentally mathematical and deeply religious. The whole divinely ordered *kosmos* is a living spinning spherical creature pervaded by soul. The world-soul contributes to a band, half the length of which, called the "Same," forms the celestial equator moving all the fixed stars with the same motion, while the other half, called the "Other," being divided lengthwise into seven parts, corresponds in a way never completely described to a pair of interleaved geometric series 1, 2, 3, 4, 8, 9, 27, and to the seven planets. Venus and Mercury remain close to the Sun, due to an otherwise unspecified "contrary power." Each planet has its own soul, the cause of its unique motion, and is thereby an "instrument of time." Mere mortals do not understand their motions and periods, but all planets together completing their cycles marks a "perfect" year of the *kosmos*, whose value scholars have spilt much ink vainly computing. Each fixed star is spherical, with a soul whose motion is axial rotation. The Earth herself is wound round the *kosmic* pole, and by resisting the rotation of the whole (unlike the planets) remains the unmoved "guardian of day and night." The *Timaeus*, with its theory of matter founded on the Platonic solids, provided a geometric cosmological model that carried an enormous influence through the time of the Renaissance.

In the *Laws*, astronomy is considered to be useful (for better regulation of the civic calendar), just as arithmetic and surveying are otherwise useful, as well as for mental exercise and training. But the members of the all-powerful Nocturnal Council must study astronomy to learn the primacy in time of soul and the order and divinity of the stars, for only such men are fit for such high rank. True astronomical education will inculcate deep faith, rather than the disbelief caused by the soulless astronomy of mere bodies of stone. (Plato here alluded to **Anaxagoras** and **Democritus**.) Indeed, it is only by asserting that the planets do not truly wander that the wise and devout man may validly study astronomy without committing blasphemy, because the Sun and all stars are ensouled and self-moved.

All ancient philosophers accepted the deeply Platonic *Epinomis*, but some modern scholars have denied his authorship. It argues that astronomy provides the best education for statesmen, and that the celestial bodies are worship-worthy beings through whom we learned number. The stars are fiery-bodied living beings, proven very large, and the Sun "leads" Mercury and Venus.

Plato's world-soul in harmony with all parts of the Universe offered a benign and humanocentric *kosmos*, whose appeal only Epicureans sought to resist before the modern period. Plato's approval of the musical harmony of the stars (which he may have intended as decoration, not definition), and his belief that the planets were divinities worthy of worship, ensured that astrology would be hard to resist, but his insistence that astronomy, with mathematics and music, were the highest studies of which the human mind is capable ensured their survival across centuries when little else did.

Paul Keyser

Selected References

Gregory, Andrew (2000). *Plato's Philosophy of Science*. London: Duckworth.

Kalfas, Vasilis (1996). "Plato's 'Real Astronomy' and the Myth of Er." *Elenchos* 17: 5–20.

Plaut, Lukas

Born **Kumamoto, Japan, 5 June 1910**
Died **Haren, the Netherlands, 4 October 1984**

Lukas Plaut investigated the intrinsic properties and space distribution of variable stars and eclipsing binaries. He was one of the identical twins born to Joseph Plaut and Katharina Lewy. Plaut received his primary education in Japan and at the Real-Gymnasium of München, Germany

(1925–1929), and attended the Friedrich Wilhelm University in Berlin from 1929 until spring 1933, where he studied variable star astronomy with **Paul Guthnick** at Babelsberg Observatory. Because of his Jewish parentage, Plaut was expelled from the observatory in April 1933, and so he moved to Leiden Observatory in the Netherlands, then under the directorship of **Willem de Sitter**. Here he earned a bachelor's degree (1936), a master's degree (1938), and a doctorate in 1939, with thesis on photographic photometry of two variable stars, guided by **Ejnar Herzsprung**. Occupation of the Netherlands by the German army in 1940 forced Plaut to move first to Groningen (the Kapteyn Laboratory, followed by a modest teaching job), then to a labor camp, and finally to the concentration camp in Fürstenau, Germany.

After liberation of the Netherlands, Plaut returned to Groningen, as a member of the staff of the Kapteyn Laboratory (1945–1964) and as professor of astronomy (1964–1975). There he carried out the work that led to two impressive publications. One, in 1950, was a critical compilation of the orbital elements of 117 eclipsing binaries. The next one, in 1953, was based on the compilation. It presented a thorough discussion of the frequency distribution of the orbital elements (*Groningen Publications* Numbers 54 and 55).

In 1953, at a conference on coordination of galactic research (published as International Astronomical Union Symposium Number 1, 1955), **Walter Baade** proposed a large, systematic survey of faint variable stars of the RR Lyrae type in order to explore the stellar population in the interior regions of the Galaxy, and to arrive at an improved estimate of the distance of the galactic center. Plaut undertook the execution of the program at the Kapteyn Laboratory, which came to be known as the Groningen–Palomar survey of faint variable stars. For four areas at low and intermediate latitudes, carefully chosen in consultation with Baade, Plaut collected long series of photographic plates with the Palomar Schmidt telescope from 1956 to 1959. The variables were detected with an instrument specially designed at the Kapteyn Laboratory by J. Borgman, based on the principle of eliminating all nonvariable stars in the field by combining a negative and a positive photographic image. Observational data and provisional analyses of the about 2,500 variables detected were published by Plaut in the years 1966–1973. The final outcome of this impressive project was the joint paper by Plaut and **Jan Oort**, dealing with the determination of the distance of the galactic center – 8.7 kiloparsecs – and the space distribution of the various types of variable stars in the region within 5 kiloparsecs from the galactic center.

After retirement, Plaut attempted to continue research, but the effects of the persecution and humiliation he had undergone between 1933 and 1945, led to mental disturbances and gradual lack of contact with the world.

A nearly complete set of Plaut's publications, as well as many of his notebooks used for lectures and research, is kept in the archives of the Kapteyn Institute of Groningen University.

Adriaan Blaauw

Selected References

Blaauw, A. (1985). "Leven en werk van Lukas Plaut" (in Dutch). *Zenit* 12: 152–153.

Henkes, Barbara (1998). "Het Vuil, de Sterren en de Dood; Lukas Plaut en Stien Witte: Portret van een 'gemengd' huwelijk"(in Dutch). In *Parallelle levens,* edited by Corrie van Eijl, pp. 91–116. Amsterdam: Stichting beheer IISG.

Pliny the Elder

Born **Novum Comum, Gallia Cisalpina, (Como, Italy), 22 or 23**
Died **Stabiae, (Campania, Italy), August 79**

Pliny's *Natural History* described the known state of astronomy and astrology, and was influential for 1,000 years or more.

Pliny's *Natural History*, in 37 volumes, contained, he wrote, "twenty thousand noteworthy facts obtained from one hundred authors … with a great number of other facts in addition that were either ignored by our predecessors or have been discovered by subsequent experience." This composition was Pliny's last, for the natural spectacles that he so admired led him to his death: As commander of the Imperial Fleet at Misenum, he had set sail partly for a closer look at the eruption of Mount Vesuvius, and partly to rescue those stranded by the same eruption. Landing at Stabiae, he judged the situation safe enough to bathe, sup, and snore through the night. But just before dawn, Pliny was awakened for immediate evacuation. The exertion and excitement must have proven too much for the corpulent, unfit asthmatic: Arriving at the beach, he fell down and died, apparently from a heart attack.

Pliny was born in an equestrian family – whose status gave access to a public service career. His sister married a wealthy landowner and bore a son – Pliny the Younger – whom the Elder adopted as his own. Apart from the Elder's own writings (only his *Natural History* is extant), the Younger's letters to the historian Tacitus supply our best observations of Pliny's life, death, and person.

Before the *Natural History*, Pliny had composed at least six other works on several topics – grammar, oratory, history, cavalry technique, and a biography – and had followed a distinguished military and government career. He was well traveled and a close observer of all that went on around him.

Pliny shunned most of the material luxuries common to his time, preferring instead to study and write, and to quietly enjoy natural phenomena. His effort was to become educated in everything, as he saw the Greeks to be, but he noted that the Greek mastery of individual subjects had produced no single work summarizing the whole. So Pliny took this task – the unification of the whole *enkuklios paideia*, "all-encompassing learning" – upon himself.

Pliny's survey became an encyclopedia, the *Natural History*. Its coverage is necessarily superficial and in many places wrong, but it is neither sterile nor wholly uncritical: Pliny stated and judged his sources, often expressing wonder, surprise, or doubt. Setting aside the introductory and bibliographical Book I, the *Natural History* opens with astronomy and cosmology in Book II. Pliny began with a generally Stoic overview of the cosmos, earlier investigations into its nature and size, and its arrangement. Although he gave some relative dimensions of its parts, he dismissed attempts upon its overall dimensions as "madness, downright madness." Eliminating theology from the study of nature, Pliny rejected inquiry into God's existence, shape, and form as "a mark of human weakness." Divinity, he retorted, belongs to that which *obviously* governs the rest of the cosmos: the Sun.

Having set theology apart from astronomy, Pliny went on to describe planetary motion and eclipses quantitatively, giving periodicities and ranges of motion in coordinate systems that are unfortunately inconsistent and unspecified. He stated the order of the geocentric planetary orbits, and the interplanetary spacings according to Pythagorean harmonic theory, which he immediately rejected as "a refinement more entertaining than convincing." But the planetary order was justified by its ability to explain, at least qualitatively, the planetary elongations and speed variations. Planetary motions are governed by orbital curvature and solar rays, the Sun clearly being, Pliny observed, the governing power of the cosmos, controlling the planets from their midst and the Earth from above.

Pliny's treatment of astrology is noteworthy for its detail and balance: While Pliny certainly believed in astrological influence, he was unhesitant in denouncing astrologers. He enumerated biological evidence for celestial influence: Certain plants bloom or move according to the Sun. "Persistent research" showed that the Moon influences shellfish growth, ant activity, and eye diseases in cattle. Storms can be caused by the stars, and winds by the Sun, and the various winds arise on dates measured by stellar risings and settings. Despite this, Pliny considered astrologers mistaken in assigning individual stars to individual people, and he thought it implausible to attribute chance to stellar influence: Because stellar motions are predetermined, such "chance" would not be chance at all.

Why all this attention to astronomy and astrology, in a culture that valued practical engineering and government so much more than theoretical science? Pliny gave two reasons: to combat superstitious fears of gods and nature, and because astronomy and astrology genuinely do influence life, especially through agriculture. Pliny devoted a book to astronomical farming calendars, giving content from the Greeks and Babylonians, noting their disagreements on astronomy and also on what farmers should do at different times of the year. Despite these disagreements he considered them useful, given the absence of any simpler, nonastronomical system.

The shallowness of Pliny's *Natural History* – hardly avoidable in any single-author encyclopedia – did not prevent its long-lived influence on the astronomy of several important medieval writers, particularly **Martianus Capella** and **Bede**, and hence on all who studied their works in subsequent centuries. This tradition perpetuated Pliny's planetary data and, perhaps more importantly, his doctrine that solar rays govern planetary movement. Overall, Pliny's *Natural History* became widely used for general education. Around the 12th century, his solar-ray doctrine ceded ground to element-based theories of Aristotelian and Platonist origin. The absence of epicycles in Pliny's planetary system inspired further rejection of his astronomy in the Renaissance, compounded by growing awareness of his unreliability more generally. In recent decades he is read as representative of knowledge not in our day, but in his own.

The rest of Pliny's works are lost. They include a work *De iaculatione equestri* (On horseback javelin-throwing); the 20-volume *Bella Gemaniae* (German Wars); the 6-volume *Studiosi* on oratorial training; the 8-volume grammar, *Dubius sermo*; a two-volume biography of his patron, *De vita Pomponi Secundo*; and 30 volumes on the *History of Rome*.

Alistair Kwan

Alternate name

Plinius secundus

Selected References

Beagon, Mary (1992). *Roman Nature: The Thought of Pliny the Elder*. Oxford: Clarendon Press.

Eastwood, Bruce S. (1989). *Astronomy and Optics from Pliny to Descartes: Texts, Diagrams and Conceptual Structures*. London: Variorum Reprints.

French, Roger and Frank Greenaway (eds.) (1986). *Science in the Early Roman Empire: Pliny the Elder, His Sources and Influence*. London: Croom Helm.

Gaius Suetonius Tranquillus (1997). *On Famous Men* (De viris illustris), translated by J. C. Rolfe. In Vol. 2 of *The Lives of the Caesars*. Loeb Classical Library, no. 38. Cambridge, Massachusetts: Harvard University Press.

Irby-Massie, Georgia L. and Paul T. Keyser (2002.) *Greek Science of the Hellenistic Era: A Sourcebook*. London: Routledge.

Pliny the Elder (1938–1963). *Natural History*, translated by H. Rackham. 10 Vols. Loeb Classical Library. Cambridge, Massachusetts: Harvard University Press.

Pliny the Younger (1969). *Letters and Panegyricus*. 2 Vols. Translation by Betty Radice. Vol. 1. Loeb Classical Library, no. 55. Cambridge, Massachusetts: Harvard University Press. (Letters III.5, VI.16, and VI.20.)

Zirkle, Conway (1967). "The Death of Gaius Plinius Secundus (23–79 AD)." *Isis* 58: 553–559.

Plinius Secundus

➔ **Pliny the Elder**

Ploti Ferrariensis

➲ **Novara, Domenico Maria da**

Plummer, Henry Crozier Keating

Born **Oxford, England, 24 October 1875**
Died **Oxford, England, 30 September 1946**

Henry Plummer, who served as Andrews Professor of Astronomy and Astronomer Royal of Ireland, conducted valuable research on variable stars and stellar motions. He is commemorated in Plummer models for the distribution of stars in globular clusters and other spheroidal systems. Plummer was the eldest son of William Edward Plummer, an astronomer and then senior assistant at the Oxford University Observatory, and Sara Crozier. The elder Plummer later became director of the Observatory of the Mersey Docks and Harbour Board and reader in astronomy at the University of Liverpool. Henry Plummer's education began at Saint Edmund's School, Oxford, and continued at Hertford College under a scholarship. He took first class honors in Mathematical Moderations and Finals, as well as the Open Mathematical Scholarship and a second class in Final Honours School of Natural Science (physics). Plummer then spent a year as a lecturer in mathematics at Owen's College, Manchester, before returning to Oxford in 1900 to spend a year as assistant demonstrator at the Clarendon Laboratory.

In 1901, Plummer accepted the position of second assistant in the University Observatory under the directorship of **Herbert Turner**. By this time, he had published several papers on astronomy and been elected a fellow of the Royal Astronomical Society [RAS]. Despite its low salary, the post offered him his first opportunity to work professionally in astronomy. With the exception of a year spent as a fellow at the Lick Observatory (1907), where he worked on spectroscopy, Plummer remained at the University Observatory until 1912. There, he coordinated its participation in the International *Carte du Ciel* Astrographic Chart and Catalogue, and investigated systematic errors in the positions of images on photographic plates. During this time, Plummer also produced a number of papers on other topics.

In 1912, Plummer succeeded **Edmund Whittaker** as Andrews Professor of Astronomy at the University of Dublin and Astronomer Royal of Ireland. He was placed in charge of Dunsink Observatory, which, owing to poor funding, could make only one serious contribution to astronomy, that being the observations of variable stars. But the observatory's location in the secluded countryside was not conducive to Plummer's lifestyle; he was a bachelor and was described as a "townsman." Added to that were the political troubles that arose in Ireland after the end of World War I. As a result, he faced the very difficult decision of giving up his career in astronomy in 1921 in favor of returning to England as chair of mathematics at the Artillery College, Woolwich (later the Military College of Science), a post he held until his retirement in 1940.

Plummer's responsibilities at the military college left him little time for astronomy. Nonetheless, he remained tied to the astronomical community and while at Woolwich managed to publish about a dozen astronomical papers. Plummer regularly attended meetings of the RAS and was elected the Society's vice president (1936–1937) and president (1939–1941). He had been elected a fellow of the Royal Society in 1920, just prior to his departure from Ireland. In 1924, Plummer broke his lengthy solitude by marrying his longtime friend Beatrice Howard. The couple did not have any children. In 1940, upon his retirement, he and Beatrice returned to Oxford where, in 1942, he gave the Halley Lecture. Beatrice died in the spring of 1946, from which Plummer never fully recovered.

Plummer's contributions to astronomy were many and varied. His first published work describes a graphical method of solving Kepler's equation. A scan of Plummer's early papers reveals topics as varied as using projective geometry to solve binary star orbits, measuring occultations of stars by the Moon, compensating for errors in a siderostat, and studying images formed by parabolic mirrors. Two of his papers on comet 2P/Encke discredited the then-current model of solar-radiation pressure as the cause of its anomalous acceleration. Plummer's work on the *Astrographic Catalogue* not only flexed his mathematical muscle but also displayed his skill with astronomical equipment. His most remembered text, *An Introductory Treatise on Dynamical Astronomy* (1918), was a significant contribution to celestial mechanics and planetary theory. Plummer returned to this subject again in 1932 after learning of numerical investigations carried out at the Copenhagen Observatory.

By far Plummer's most prolific areas of research were studies of variable stars and of stellar motions. At Dunsink Observatory, Plummer and his assistant, Charles Martin, conducted photometric observations of variable stars, in conjunction with measurements of their radial velocities. Plummer showed that eight of those stars, including ζ Geminorum, could not be spectroscopic binaries, as had been assumed, because their atmospheres displayed radial pulsations. This was among the first evidence gathered for the theory of pulsations in classical Cepheid variables, a theory developed more fully by **Arthur Eddington**.

In 1911, Plummer derived an expression for the spatial distribution of stars in globular clusters and their apparent two-dimensional representations on photographic plates. From this relationship, he concluded that most clusters had condensed from primitive spherical nebulae while maintaining convective equilibrium. Plummer also began a lengthy study of the parallaxes of B- and A-type stars in the Milky Way. His results showed a relatively simple way to estimate the parallax of a star from known values of its proper motion and radial velocity. More importantly, he emphasized a distinction between the two principal velocity fields of stars in the Milky Way system: (1) those that are moving parallel to the plane of the galactic disk, and (2) those that exhibit a more spherical (or random) distribution. After Plummer's death, these two dynamical systems were recognized as related to the two stellar populations defined by **Walter Baade**, for which any model of galaxy formation must account.

Plummer was also very interested in the history of science. In 1939, he was asked by the Royal Society to edit the complete works of **Isaac Newton** before the tercentenary of Newton's birth (1942). Unfortunately, World War II put most of the project on hold, and it remained incomplete on Plummer's death. (It was eventually completed by the Society.) At the Society's triple centenary celebration of Newton, **Galileo Galilei**, and **Edmond Halley** on 9 October 1942, Plummer delivered the address on Newton. His 1942 Halley Lecture was entitled, "Halley's Comet and Its Importance."

Ian T. Durham

Selected References

Danby, J. M. A. (1975). "Plummer, Henry Crozier." In *Dictionary of Scientific Biography*, edited by Charles Coulston Gillispie. Vol. 11, p. 49. New York: Charles Scribner's Sons.

Greaves, W. M. H. (1948). "Henry Crozier Plummer." *Obituary Notices of Fellows of the Royal Society* 5: 779–789.

——— (1959). "Plummer, Henry Crozier Keating." In *Dictionary of National Biography, 1941–1950*, edited by L. G. Wickham Legg and E. T. Williams, pp. 677–678. London: Oxford University Press.

Plummer, H. C. (1918). *An Introductory Treatise on Dynamical Astronomy*. Cambridge: Cambridge University Press.

Smart, W. M. (1947). "Henry Crozier Plummer." *Monthly Notices of the Royal Astronomical Society* 107: 56–59.

Whittaker, Edmund (1946). "Henry Crozier Plummer." *Observatory* 66: 394–397.

Plutarch

Born **Chaironeia, Bœotia, (Greece), *circa* 45**
Died ***circa* 125**

Though most famous for his biographies, Plutarch also wrote a dialog on the Moon, in which the participants discuss the Moon's appearance and possible habitability, and how it is able to remain in orbit. Plutarch is best known as a biographer, but none of his contemporaries seems to have written a biography about him – nearly all that we know about Plutarch comes from clues in his own writings. Plutarch lived in a time of peace and contemplation, and seems to have spent most of his life in or near his beloved hometown. He did, though, manage to travel widely: He studied at Athens under the Peripatetic philosopher Ammonius of Lamptrae. He served his town as building commissioner and archon, and participated in a diplomatic mission to the Roman proconsul. Plutarch's travels took him throughout Greece, and he ventured as far as Alexandria, Asia Minor, north Italy, and Rome. In Rome, Plutarch spent about 15 years attending to matters of state, but also lectured on philosophy, probably in Greek, for he never found time to learn Latin. His social circle embraced Rome's most prominent, among whom Mestrius Florus gave him Roman citizenship with an official Roman name, Mestrius Plutarchus. On returning from Rome, Plutarch became a priest at Delphi (near Chaironeia), where he apparently remained until the end of his life. Two inscriptions there testify to his presence: One under a statue of Hadrian, and one under a statue of Plutarch himself. Though both statues have been destroyed, the inscriptions, plus a third inscription at Chaironeia, indicate the esteem in which Plutarch was held.

Plutarch's fame originates in his abundant and eloquent writings. Most famous is the *Lives of Famous Men*, a collection of parallel biographies that contrasts the virtues and vices of Greece's and Rome's prominent public figures. Less famous are 78 essays collected as the *Moralia*, of which about 12 are thought to be written by other authors. The *Moralia* are, as their name suggests, mostly of moral tone. Some of them, such as the dialog *De facie in orbe lunae apparet* (on the Moon's face) present scientific topics for nontechnical readers.

In *De facie*, six interlocutors discuss what might lie behind the Moon's uneven appearance. The discussion presents viewpoints from several Greek schools – including the Academics, Epicureans, Peripatetics, and Stoics – and considers astronomy, cosmology, geography, and the optics of reflection. The interlocutors weigh the several philosophical viewpoints against each other for physical plausibility, after which one of the interlocutors summarily concludes that the face is due to topographical features in its Earth-like terrain. The conversation then turns to whether the Moon might be so Earth-like as to be habitable, what kinds of plants and animals might live there, and why they would not fall off. There is also some discussion of whether the lunar beings are terrestrial souls who, by being either deceased or not yet born, lack bodies. Understandably, in debates on extraterrestrial life and the plurality of worlds, *De facie* was routinely cited well into early modernity.

Other highlights of *De facie* include discussion on how the heavy Moon can remain aloft. One proposed explanation seems tantalizingly close to modern thought: Comparing the Moon's revolutions to a stone whirled in a sling, Plutarch writes, "the moon is saved from falling by its very motion and the rapidity of its revolution … for everything is governed by its natural motion unless it be diverted by something else." There is a description of a total solar eclipse, one of only a few from classical times that mentions stars appearing in the darkness, and perhaps the only classical description of the temperature drop and the solar corona.

Along with the rest of his writings, *De facie* remained influential for centuries after Plutarch wrote it. It has been copied, edited, and emulated by many with scientific motives, the last of whom included **Johannes Kepler**. Kepler was strongly attracted to Plutarch's discussion on lunar habitation: *De facie* is an important influence behind Kepler's *Somnium* on lunar astronomy. Some of Plutarch's other writings include astronomical material, too, in particular his expansions on **Aratus**'s *Phænomena* and a commentary on **Hesiod**'s *Works and Days*. These have received less attention than *De facie*, which, since Kepler's day, has received little attention indeed.

Alistair Kwan

Selected References

Adler, Ada (ed.) (1928–1938). *Suidae lexicon*. 5 Vols. Stuttgart: B. G. Teubner. (Reprint, 1967–1971.) (This 10th-century Byzantine encyclopedia provides the closest that we have to a substantial biography of Plutarch. Under his own entry, Lamprias is described as Plutarch's son who compiled the incomplete catalog of 227 of Plutarch's works, known as the Lamprias Catalogue. Plutarch, in his own writings, mentions four sons, but none is called Lamprias.)

Cherniss, Harold (1951). "Notes on Plutarch's *De facie in orbe lunae*." *Classical Philology* 46: 137–158. (Commentary prepared while translating *De facie* for the Loeb edition of 1957, discussing translation and textual difficulties as well as astronomical aspects of the work.)

Irby-Massie, Georgia L. and Paul T. Keyser (2002). *Greek Science of the Hellenistic Era: A Sourcebook*. London: Routledge.

Mossman, Judith (ed.) (1997). *Plutarch and His Intellectual World*. London: Duckworth.

Plutarch (1957). *Concerning the Face Which Appears in the Orb of the Moon* (De facie in orbe lunae apparet), translated by Harold Cherniss. In Vol. 12 of Plutarch's *Moralia*. Loeb Classical Library, no. 406. Cambridge, Massachusetts: Harvard University Press. (The beginning of the dialog is lost, though this seems to be only a small part. Also of interest are Vol. 1 [no. 197], which contains general details about Plutarch and the sources for his works, and Vol. 15 [no. 429] which contains the Lamprias Catalogue of Plutarch's works and many fragments, including some from Plutarch's *Explanations of Aratus' Phænomena* and *Commentary on Hesiod's Works and Days*.)

Poczobut, Marcin [Martin Poczobutt]

Born **Slomiank, (Lithuania), 30 October 1728**
Died **Daugavpils, (Latvia), 17 November 1810**

Lithuanian astronomer Marcin Poczobut directed the astronomical observatory at Vilnius University and contributed to the refinement of cartography in Eastern Europe. He was born in the Gardinas region of Lithuania. His father, Kazimier Odlanicki Poczobut, came from a noble family once promoted to *bojar* by King Sigismund I in 1536. His mother was Helena Hlebowicz.

Poczobut became a Jesuit in 1745 and followed that society's tenets until it was banned in 1773. After his graduation from Vilnius University in 1751, he continued studies of Greek, Latin, and mathematics at Prague University. He studied next in France and Italy from 1761 to 1763, where he acquired much of his knowledge of mathematics and astronomy. Upon returning to his homeland, Poczobut was appointed professor of mathematics and astronomy (1764) and later rector of Vilnius University (1780–1799). As third director of the university's astronomical observatory (1764–1808), he added new instruments, including a sextant of 6-ft. radius (1765) and a mural quadrant of 8-ft. radius (1777). To obtain this equipment, Poczobut traveled to England and France, visiting the Greenwich and Paris observatories.

With the large sextant, Poczobut measured the precise length and width of Vilnius for the first time. In recognition of this work, King Stanislaus II bestowed upon him the title of Royal Astronomer (1767), and likewise honored the observatory. Poczobut made other contributions to geodesy and cartography. He determined the geographic coordinates of some 20 points in Lithuania and Latvia, and together with Jan Sniadecki, submitted a plan of dividing Lithuania, Poland, and White Russia into 400 quadrangles. He directed the work of A. Rostanas, J. Bistrickas, and others. Poczobut also participated in K. Perthes's cartographic work in western Lithuania and Courland (now part of Latvia).

Poczobut conducted regular astronomical observations, the most famous of which involved some 60 precise positions of the planet Mercury, from which **Joseph de Lalande** calculated an improved orbit of the planet. In 1766, Poczobut published a book about eclipses of the Moon, which compared data from the observatories of Paris, Vilnius, Warsaw, Krakow, and London.

In 1777, Poczobut asked the French Academy of Sciences to honor the last Polish monarch, King Stanislaus (Poniatowski) II, with a constellation. For this, he selected a V-shaped group of stars mentioned in **Ptolemy**'s *Almagest* that lay just outside of Ophiuchus. The designation did not last, but Poniatowski's bull may still be found on some older star charts; the bull's head is formed by the stars 67, 68, and 70 Ophiuchi.

For a brief time, Poczobut was a member of the provisional government of Lithuania formed in Vilnius during the 1794 rebellion against the Russians. Along with 18 others, he signed the Proclamation of the Lithuanian Supreme Council on 30 April 1794.

Poczobut exhibited extraordinary interest in the first minor planet, (1) Ceres, discovered in 1801. During his 73rd year, Poczobut reportedly "continued to search for and tirelessly observe this body with such persistence and effort that he lost consciousness several times." After Ceres was designated by the sign of a sickle, Poczobut sang its praises with a Latin distich, translated as: "Whoever taught the sickle to cut the stalks of standing corn/The toothed sickle shall become for you the garland of Ceres."

Ancient Egypt was another subject of Poczobut's fascination. He published two papers on the probable age of the Dendera zodiac, recovered from the temple of Hathor by Napoleon Bonaparte's armies and transported to the Louvre.

In 1808, Poczobut resigned as head of the Vilnius Observatory and moved to a monastery. The Polish poet Adam Mickiewicz mentioned him in his epos, *Pan Tadeusz*. He wrote that Poczobut was a priest and astronomer who had finished his life in peace and silence.

Poczobut was named a Knight of the White Eagle and was a recipient of the Order of Saint Stanislaus. He was elected a member of the Royal Society of London (1769) and a corresponding member of the French Academy of Sciences (1781).

Clifford J. Cunningham

Selected References

Anon. (1975). "Poczobutt, Martin." In *Encyclopedia Lituanica*, edited by Simas Suziedis and Juozas Jakstas. Vol. 4, pp. 305–306. Boston: Juozas Kapocius.

Cunningham, Clifford J. (2002). *The First Asteroid: Ceres, 1801–2001*. Surfside, Florida: Star Lab Press.

Girnius, A. (1987). *Martynas Poczobutas: Zymusis astronomas ir geodezininkas*. Rome.

Rabowicz, Edmund (1983). "Poczobut (Poczobut Odlanicki) Marcin." In *Polski Slownik Biograficzny*, edited by Wladyslaw Konopczynski *et al.* Vol. 27, pp. 52–62. Krakow: Gebethnera i Wolffa.

Slavenas, P. (1955). "Vilnius senoji astronomine observatorija." *Astronominis Kalendorius*. Vilnius.

Tchenakal, V. L. (1961). "Martin Poczobutt i Peterburgskaia Akademia Nauk." *Istoriko-astronomicheskie issledovaniia* 7: 297–305.

Poe, Edgar Allan

Born **Boston, Massachusetts, USA, 19 January 1809**
Died **Baltimore, Maryland, USA, 7 October 1849**

Edgar Allan Poe's *Eureka* has gained attention because of its potential affinity with what in the late 20th century emerged as the standard "Big-Bang" cosmology.

Poe was the second child of actors David Poe and Elizabeth Arnold Poe, both of whom died before their son turned three. Poe then became the foster son of Fanny Allan and her husband, John Allan, a businessman in Richmond, Virginia. Although best known as a writer of haunting poems and exquisitely plotted fantastic tales, Poe has gained attention in the history of astronomy for his speculative but seemingly prescient cosmogony, which opened a cosmic narrative unfolding from a singular "primordial particle."

In 1848, a year before his premature death, Poe published *Eureka: A Prose Poem*, dedicated to **Alexander von Humboldt**, German author of the multivolume treatise *Kosmos*. In *Eureka*, Poe sets out "to speak of the physical, metaphysical and mathematical – of the material and spiritual universe – of its essence, its origin, its creation, its present condition and its destiny." While assuming (Newtonian) absolute space, Poe distinguishes "the universe of stars" from "the universe proper," and propounds an intriguing evolutionary account of the former, the sidereal Universe. The plot begins with "a particle absolutely unique, individual, undivided," and it evolves accordingly from that singularity: "From that one particle" are "irradiated spherically – in all directions – to immeasurable but still to definite distances in the previously vacant space – a certain inexpressibly great yet limited number of unimaginably yet not infinitely minute atoms."

Some have seen Poe as pointing toward a solution to Olbers's paradox (Harrison), or even toward the anthropic cosmological principle as an explanation for the large size of the Universe (Cappi). Strictly speaking, Poe's "primordial particle" was speculative, nonscientific, and perhaps merely a curiosity. On a broader historical canvas, however, it stands as a noteworthy mid-19th-century example of the impulse to offer an evolutionary account of cosmic history, and as a perennial example of the longing of the human imagination to tackle questions concerning the physical origin of the Universe.

Dennis Danielson

Selected References

Benton, Richard P. (ed.) (1975). *Poe as Literary Cosmologer: Studies on Eureka*. Hartford, Connecticut: Transcendental Books. (This collection includes a facsimile of the 1848 *Eureka*.)

Cappi, Alberto (1994). "Edgar Allan Poe's Physical Cosmology." *Quarterly Journal of the Royal Astronomical Society* 35: 177–192.

Danielson, Dennis (ed.) (2000). *The Book of the Cosmos: Imagining the Universe from Heraclitus to Hawking*. Cambridge, Massachusetts: Perseus, esp. "The Primordial Particle," pp. 307–311.

Harrison, Edward Robert (1987). *Darkness at Night: A Riddle of the Universe*. Cambridge, Massachusetts: Harvard University Press.

Poe, Edgar Allan (1848). *Eureka: A Prose Poem*. New York: Putnam.

——— (1902). *The Complete Works of Edgar Allan Poe*. Vol. 16, edited by James A. Harrison. New York: Crowell.

Quinn, A. H. (1998). *Edgar Allan Poe: A Critical Biography*. Baltimore: Johns Hopkins University Press. (Originally published 1941.)

Pogson, Norman Robert

Born **Nottingham, England, 23 March 1829**
Died **Madras, (India), 23 June 1891**

Norman Robert Pogson discovered eight asteroids and 20 variable stars but is usually remembered for his recommendation leading to the adoption of a standard value of the photometric constant, which standardizes the old Greek magnitude system of stellar brightnesses.

Although his father, George Pogson, an established hosiery manufacturer, intended a career in business for his son, Norman developed an early interest in science. It is likely that his mother Sarah's intuitive understanding of this saved him from a Dickensian existence as an apprentice in the textile business. Pogson was twice married, first to Elizabeth Jane Ambrose. They had 11 children. Elizabeth died of cholera in 1869. He then married Edith Louisa Stopford in 1883. She bore him three children. A daughter of his first wife, Elizabeth Isis Pogson, served her father as assistant at the Madras Observatory from 1873 to 1881. She later became Meteorological reporter for Madras. Elizabeth was first proposed for fellowship in the Royal Astronomical Society in 1886, being admitted at last in 1920.

A family friend, **John Hind**, an assistant at George Bishop's South Villa Observatory in London, recommended Pogson's appointment as assistant at South Villa in 1851. Less than a year later he accepted an assistantship at the Radcliffe Observatory, Oxford, under the directorship of **Manuel Johnson**. Pogson's main duty was to make routine observations of stellar positions with the meridian circle, but his enthusiasm was such that he made additional observations even outside the required working hours. Johnson lent Pogson's services to **George Airy** for the Astronomer Royal's famous Harton Colliery experiments to measure the density of the Earth in 1854. Pogson's painstaking attention to detail made a positive impression on the Astronomer Royal. During his Oxford years, Pogson began actively searching for asteroids and monitoring variable stars. The French Academy of Sciences awarded him the Lalande Medal for his discovery of the minor planet (42) Isis in 1856.

That same year Pogson proposed the adoption of a light ratio of 2.512 (the fifth root of 100) for two stars that differ in brightness by one magnitude. The resulting magnitude scale eventually became an international standard after both Harvard and Potsdam observatories adopted it for their photometric programs in the 1870s.

Pogson's desire for more freedom in his astronomical activities brought him into conflict with Johnson. In the mid-1850s he gained the attention of gentleman–scientist John Fiott Lee, who maintained a private observatory at Hartwell House near Aylesbury, Buckinghamshire. Lee offered Pogson the directorship of the Hartwell Observatory, and Pogson accepted. The position provided more freedom of action but was not considered to be of professional standing, a status Pogson ardently desired. On Johnson's death, Pogson applied unsuccessfully for the Radcliffe Observership; a year later, with the support of Airy and Sir **John Herschel**, he was appointed government astronomer at the Madras Observatory in South India.

What first appeared to secure professional recognition for Pogson eventually led to the total eclipse of his career. In a remote location, the Madras Observatory was in disrepair when Pogson arrived, and the astronomer's salary proved inadequate to support his family in India. The government denied him any additional assistance in running the facility, and contemplated its closure more than once during his tenure. Pogson's oversensitive personality did not aid his cause. A quarrel with the Royal Astronomical Society over a planned survey of southern stars soured his relations with the British scientific community. His observations of the solar eclipse of 1868 – he had been one of the first to observe the bright line spectrum in the corona – did not receive due credit because the Indian government authorized the printing of only three copies of his report.

Pogson labored on, embittered by his relations with officialdom and mourning his first wife and eldest son, both of whom died in Madras. He withheld publication of meridian circle observations and telegraphic longitude determinations. Finally, in 1884 a sympathetic governor of Madras, Sir Mountstuart Elphinstone, encouraged Pogson to publish his results. Elphinstone also proposed him for fellowship in the Royal Society, but the application failed. Thereafter, Pogson ceased regular observation, although he remained government astronomer until his death from liver cancer. At the time of his death, he was widely, if mistakenly, considered to have been a scientific nonentity.

Pogson was in fact one of the great visual observers of his time. A pioneer of variable star astronomy, he toiled for many years over an atlas of variables comprising charts of some 60,000 stars. **Edward Pickering** arranged for a grant from the Bruce Fund to support publication of the atlas, but Pogson died before he was able to complete the project. Fragmentary results were finally published in 1908 in the *Memoirs of the Royal Astronomical Society*.

Keith Snedegar

Selected References

Brook, C. L. (ed.) (1908). "Observations of Thirty-One Variable Stars by the Late N. R. Pogson." *Memoirs of the Royal Astronomical Society* 58.

Hearnshaw, J. B. (1996). *The Measurement of Starlight: Two Centuries of Astronomical Photometry*. Cambridge: Cambridge University Press, pp. 74–78.

Pogson, Norman Robert (1882). *Discoveries, observations, calculations, &c., made successively at London, Oxford, Hartwell, and Madras, 1847–82*. Madras.

Shearman, T. S. H. (1913). "Norman Robert Pogson." *Popular Astronomy* 21: 479–484.

Poincaré, Jules-Henri

Born **Nancy, Meurthe-et-Moselle, France, 29 April 1854**
Died **Paris, France, 9 July 1912**

The French mathematician and philosopher Henri Poincaré made contributions to a wide range of scientific problems, including many in celestial mechanics. Poincaré came from an eminent French intellectual family. His father was professor of medicine at the University of Nancy, and an uncle inspector-general of roads and bridges in France. His cousin, Raymond Poincaré, was a lawyer, statesman, and president of the French republic during World War I, and his sister married the philosopher Emile Boutroux. Henri Poincaré was happily married to the great granddaughter of the zoologist Etienne Geoffroy Saint-Hilaire.

From infancy, Poincaré experienced bad health, including diphtheria at age five. His motor coordination was poor, he had great difficulty learning to write, and he found drawing impossible. Poincaré was not able to read from a blackboard, but his memory was such that he could memorize mathematical formulae and theorems upon hearing them, and remember the exact contents of books after a single reading.

Poincaré studied at the École Polytechnique and the École Supérieure des Mines (School of Mines), both in Paris. His inability to draw was a problem for a potential engineer, but he revealed himself as a brilliant student and genius in mathematics. After graduating as a mining engineer, Poincaré earned a doctorate in mathematical sciences at the Paris Faculty of Sciences, with a thesis on differential equations. Appointed professor of mathematics at the University of Caen in 1879, he moved to the University of

Paris 2 years later, where he held the chair of experimental physics until his death. Considered the last universalist in mathematics, Poincaré published more than 500 papers and 30 books on nearly all branches of pure and applied mathematics, including theoretical physics and mathematical astronomy.

In 1889, Poincaré received a prize offered by King Oscar II of Sweden and Norway for work on the stability of orbits in the n-body problem (in which more than two gravitating point masses mutually interact). This was on the advice of German mathematician Karl Weierstrass, who pointed out that, although Poincaré had not solved the problem in full generality, his results "inaugurated a new era in the history of celestial mechanics." Poincaré developed new mathematical techniques to approach the n-body problem and showed that there is no analytical solution already for the case of three bodies: The orbits can have very irregular and chaotic form. Thus Poincaré's work was a predecessor of modern chaos theory as applied, for instance, to the mutual perturbations of the orbits of the outer planets.

Poincaré summarized his research on celestial mechanics in his fundamental work *Les méthodes nouvelles de la méchanique céleste* (three volumes, published 1892–1899) and his more practical *Leçons de méchanique céleste* (1905–1910). He applied the results of his study of the equilibrium of rotating fluid bodies (*Figures d'équilibre d'une masse fluide*) to the problem of the origin of the Solar System (*Leçons dur les hypotheses cosmogonique*). He concluded that the planets could not have formed from rotating bodies that contracted as they cooled (as implied in the hypothesis of **Pierre de Laplace**) because the rotating mass would separate into two distinct, unequal masses. The process may be relavant to the formation of binary stars. Poincaré also applied the method to the rings of Saturn. In 1906, his mathematical investigations of electromagnetism yielded results analogous to those put forward some months earlier by **Albert Einstein** in his special theory of relativity.

Besides his purely scientific work, Poincaré wrote some very accessible books on the philosophy of science and mathematics. These books became very popular and were translated into many languages. For his literary talent, he was elected member of the Académie française in 1908, while he was already member of the Académie des sciences, the Royal Society, and many other academies and scientific societies.

Tim Trachet

Selected References

Bell, E. T. (1937). *Men of Mathematics*. New York: Simon and Schuster.

Lebon, Ernest (1913). *Notice sur Henri Poincaré*. Paris: Hermann.

Poincaré, Henri. (1916–1956). *Œuvres de Henri Poincaré*. 11 Vols. Paris: Gauthier-Villars.

Poisson, Siméon-Denis

Born **Pithiviers, (Loiret), France, 21 June 1781**
Died **Paris, France, 25 April 1840**

Siméon-Denis Poisson's greatest contribution to astronomical and physical theory, the Poisson bracket, was generated by the mathematical development of perturbation calculation for the Solar System. He is also remembered for poisson statistics, appropriate for samples with small numbers of members, as often happens in astronomy.

Poisson came from a modest family background. His father, a former soldier, had purchased a low-ranking administrative post in Pithiviers. In 1817, he married Nancy de Bardi, an orphan born in England to *émigré* parents.

The French Revolution, which Poisson supported enthusiastically, made it possible for him to advance to the presidency of the district. Poisson was guided by his father toward those professions to which access had been made easier by First Republican social legislation. Thus, enrolled in the École Centrale of Fontainebleau, he took advantage of his instruction to obtain first place in the national competitive examination to enter the new École Polytechnique, to which he was admitted in 1798.

At the école, Poisson impressed the eminent **Pierre de Laplace** and his colleague, **Louis Lagrange**, with his intelligence, industriousness, and perspicacity. With formidable mathematical talent and enjoying the steady support of the highly placed Laplace, Poisson advanced rapidly to positions of increasing responsibility and eminence: instructor at the École Polytechnique upon his graduation in 1800; deputy professor in 1802; professor of analysis and mechanics in 1806; astronomer at the Bureau des longitudes in 1808; professor of mathematics at the newly established Faculty of Sciences at the Sorbonne; and culminating in election as member of the physics section of the elite Institut de France in 1812.

J. Heilbron terms the mathematical worldview within which Poisson worked as the first modern "standard model." It interpreted physical law as the operation of weightless, "imponderable" fluids, two each for electricity and magnetism, one for heat, one for light, and one for the newly discovered infrared radiation. These imponderables were believed to operate in a manner analogous to gravity, as inspired by the delicate experiments of Charles-Auguste de Coulomb demonstrating how the attraction between static electrical charges varied, as did gravitational attraction, with the inverse square of their separation. Laplace was the leader of this enterprise and Poisson his outstanding disciple.

In the analysis of perturbations, one may begin with the positions of a number of celestial bodies, all mutually attracting one another gravitationally, along radius vectors. If we call the number of vectors r, their attraction will satisfy r second-order differential equations, which can be written in the generalized coordinates introduced by Lagrange. Whether or not these equations are analytically soluble, the integrals of the set of differential equations depend upon $2r$ arbitrary constants. In a supplement to Book VIII of Laplace's *Mécanique céleste* that appeared in 1808, Lagrange developed the implications of Poisson's observation, reported to the Academy of Sciences early that year, that the $2r$ arbitrary constants must satisfy r physical constraints: Specifically, Lagrange proved that in the presence of a perturbation that forces the arbitrary constants to be treated as functions of time, the derivatives of the desired functions with respect to time are the solutions of a linear system in which the coefficients of the unknowns are independent of time.

Using a mathematical transformation, Poisson, in 1809, then extended and generalized Lagrange's result, modifying the variables so that they retain the same form – the Poisson bracket – even upon the introduction of the perturbation function. In the following decades William Rowan Hamilton and Carl Jacobi used the Poisson bracket as an essential element in their geometric reformulation of the dynamical equations of physics, and in the 1920s, Paul Dirac identified the Poisson bracket as crucial in the mathematical structure of Werner Heisenberg's novel quantum mechanics. It is of interest to note that Poisson also anticipated the δ-function made famous by Dirac's employment of it.

Poisson also attempted to complete a didactic presentation of the "standard model" in a series of clearly written and widely read textbooks. He assumed the positions of examiner of graduating students at the École Militaire in 1815 and that of the École Polytechnique the following year. He was named to the Royal Council of the university, the highest educational consultative body in the restored monarchy in 1820.

Another work of Poisson that is of great significance to astronomy is his *Recherches sur le mouvement des projectiles dans l'air* (1839), which first discussed in print the importance of a term, discovered by his doctoral student Gustave Gaspard de Coriolis, to correct for the deviations from the law of motion arising from a rotating frame of reference. A decade later this work inspired the striking experiment of the Foucault pendulum, which demonstrates in a dramatic way the Earth's rotation on its axis. Poisson, however, chose not to mention Coriolis's name. By the time of his death, Poisson had published some 300 papers and books.

Poisson was elected foreign associate of the Royal Society of London in 1818 (and received its Copley Medal in 1832); he was a member of virtually all the academies of the day from Boston to Saint Petersburg. That this adherent of the First Republic was honored by the empire and made a Peer of France by the Orleanist monarchy in 1837 is evidence of his political discretion.

Michael Meo

Selected References

Costable, Pierre (1978). "Poisson, Siméon-Denis." In *Dictionary of Scientific Biography*, edited by Charles Coulston Gillispie. Vol. 15 (Suppl. 1), p. 480–490. New York: Charles Scribner's Sons.

Grattan-Guinness, Ivor (1970). *The Development of Mathematical Analysis from Euler to Riemann*. Cambridge, Massachusetts: MIT Press.

Heilbron, J. L. (1993). *Weighing Imponderables and Other Quantitative Science around 1800*. Berkeley: University of California Press.

Pollio, Marcus

➤ **Vitruvius, Marcus**

Pond, John

Born **London, England, 1767**
Died **Blackheath, (London), England, 7 September 1836**

John Pond, England's sixth Astronomer Royal, raised the observing program at the Royal Greenwich Observatory to a new standard of excellence. The son of a well-to-do retired businessman, Pond developed an interest in astronomical observations while being tutored at home by **William Wales**, who had served as astronomer and conavigator to Cook on the second voyage of the *HMS Resolution*. As a teenager Pond detected errors in Greenwich observations. After enrolling at age 16 as a chemistry student at Trinity College, Cambridge, he withdrew before completing his degree because of ill health. Sent abroad to warmer climates to recuperate, Pond made astronomical observations during his travels.

Upon his return to England in 1798, Pond began making astronomical observations from his private observatory near Bristol, at Westbury in Somerset, with an altazimuth refractor equipped with 2½-ft.-diameter circles designed by Edward Troughton, one of England's leading makers of scientific and astronomical instruments. Pond's work proved that the Greenwich Observatory quadrant had been deformed by age. This work, published in the *Philosophical Transactions of the Royal Society* in 1806, led to his election the following year as a fellow of the society as well as the institution of a program by **Neville Maskelyne**, the fifth Astronomer Royal, to upgrade Greenwich's instrumentation.

Following marriage in 1807, Pond established his home in London, where he continued working in astronomy using excellent instruments whose construction was supervised by Troughton. After his appointment in 1811 as Astronomer Royal when Maskelyne retired, Pond devoted most of the next quarter-century to upgrading the equipment, staff, and procedures of the Greenwich Observatory. Troughton had designed a new mural circle, ordered by Maskelyne, which enabled Pond to obtain the data for his 1813 catalog of the north-polar distances of 84 stars. The observations made with the circle in 1813/1814 also enabled Pond to challenge the accuracy of claims by **John Brinkley**, the first director of the Dublin Observatory, to have detected stellar parallax of a number of fixed stars. Not only was Pond's assertion that effects of parallax could not be detected with contemporary instruments later proved right, but also his challenge helped stimulate **Friedrich Bessel**'s subsequent successful work in that field. During Pond's tenure, Troughton designed for the Greenwich Observatory a new transit instrument and a 25-ft. zenith telescope, which were erected in 1816 and 1833, respectively. Before ill health forced Pond to retire in 1835, he completed his crowning achievement, a catalog of 1,113 stars, observed more precisely than ever before, which was published in 1833. He is best remembered, however, for beginning the fundamental modernization of the national observatory, a task carried on with zeal by his successor **George Airy**, who regarded Pond as the "principal improver of modern practical astronomy."

Naomi Pasachoff and *Jay M. Pasachoff*

Selected References

Clerke, Agnes M. (1921–1922). "Pond, John." In *Dictionary of National Biography*, edited by Sir Leslie Stephen and Sir Sidney Lee. Vol. 16, pp. 76–78. London: Oxford University Press.

Forbes, Eric G. (1975). *Greenwich Observatory*. Vol. 1, *Origins and Early History (1675–1835)*. London: Taylor and Francis.

Ronan, Colin A. (1969). *Astronomers Royal*. Garden City, New York: Doubleday and Co., pp. 128–129.

Pons, Jean-Louis

Born **Peyre, (Hautes-Alpes), France, 24 December 1761**
Died **Florence, (Italy), 14 October 1831**

Jean-Louis Pons was the world's most successful visual discoverer of comets. Born into a poor family, Pons did not leave his mark on the astronomical community until 1801, when he logged his first discovery of a comet on 11 July. Pons's discovery (c/1801 N1), which he shared with **Charles Messier**, proved to be Messier's last. During his lifetime, Pons discovered or codiscovered a total of 37 comets. Of these, 26 are credited to his name.

In 1789, Pons gained a post as concierge at the Observatory of Marseilles and received instruction on the telescope from the astronomers. He was a fast learner and soon was allowed to observe with all the instruments. His favorite telescope was one with a 3° field of view. (This suggests a magnification around 15×.) Pons was known to possess an extraordinary ability to remember the star fields he had observed and thus to recognize changes within them.

In 1813, Pons was promoted to assistant astronomer and in 1818, assistant director of the Marseilles Observatory. In that year, he received the first of three Lalande Prizes for his discovery of three comets. **Joseph de Lalande**, director of the Paris Observatory, established the Lalande Prize in 1802, awarded each year for the outstanding achievement in astronomy. Pons's receipt of three Lalande Prizes may itself be unequaled.

On 26 November 1818, Pons discovered a comet that was later shown by German mathematician and astronomer **Johann Encke** to follow an elliptical orbit. Moreover, Encke demonstrated that Pons's comet (now designated 2P/1818 W1) was identical to that observed in 1786 by **Pierre Méchain** (2P/1786 B1) and in 1795 by **Caroline Herschel** (2P/1795 V1). The comet's orbital period is a scant 3.3 years, and Encke correctly predicted its return in 1822. For this reason, the comet is known today as 2P/Encke, although Encke himself always referred to it as Pons's comet. On account of gases that are vented from its rotating nucleus, the comet's orbital period decreases by about 2.5 hours per revolution. With its short period and one of the smallest perihelion distances (0.3 AU), comet 2P/Encke has perhaps evolved more rapidly than other periodic comets. It is associated with the annual Taurid meteor shower.

In 1819, Pons was appointed director of the observatory at the Royal Park La Marlia, near Lucca, Italy. He was called to that post by the widow of the former King Louis of Etruria, Duchess Maria Luisa of Bourbon, on the advice of Baron **János von Zach**, also of the Marseilles Observatory. In 1821, Pons received a second Lalande Prize, this time shared with Joseph Nicholas Nicollet, for discovering additional comets from his new observatory. After the Duchess died in 1824, Pons was invited to become director of the Florence Observatory by the Grand Duke of Tuscany, Leopold II. There, he found a total of seven more comets, and was awarded his third Lalande Prize in 1827, shared with **Jean Gambart**. Pons also received a Silver Medal from the Royal Astronomical Society for his discovery of two comets in 1822, at the same time that Encke was awarded the society's Gold Medal.

After 1827, Pons's eyesight began to fail, and he was forced to stop observing in 1831. Although he had no formal training in astronomy, Pons was a remarkable observer and shared the astronomical stage with many well-known scientists. Although possessing enormous patience and perseverance, Pons did not record the positions of his comets with much precision, making it more difficult for others to confirm his discoveries. While some comet discoveries are still made by visual means, most new comets are found today using photographic and electronic (charge-coupled device) imaging techniques. He is commemorated with a crater on the Moon's surface.

Robert D. McGown

Selected References

Chapin, Seymour L. (1975). "Pons, Jean-Louis." In *Dictionary of Scientific Biography*, edited by Charles Coulston Gillispie. Vol. 11, pp. 82–83. New York: Charles Scribner's Sons.

Roemer, Elizabeth (1960)."Jean Louis Pons, Discoverer of Comets." *Astronomical Society of the Pacific Leaflet*, no. 371.

Pope Sylvester II

➔ **D'Aurillac, Gerbert**

Popper, Daniel Magnes

Born **Oakland, California, USA, 11 August 1913**
Died **Los Angeles, California, USA, 9 September 1999**

Daniel Popper excelled in determining the properties of stars in binary systems. Much of modern stellar astrophysics is built on the foundation of stellar luminosities, radii and spectral types as a function of mass as determined by Popper.

The Popper family of Oakland, California, had already produced several community leaders by the time Daniel Popper received his AB and Ph.D. degrees at the University of California at Berkeley in 1934 and 1938, respectively. During this period, Popper's first published article, with **Lawrence Aller** and Alfred Mikesell, was on the spectrum of Nova Herculis 1934. His dissertation under **Arthur Wyse** was on the spectrophotometry of Nova Lacertae 1936. In this study, Popper correctly identified the greatly broadened lines of hydrogen and other common elements. In a third early paper on exploding stars, this time on supernova 1937C, Popper correctly differentiated the greatly broadened spectral lines in that object from those observed in Nova Herculis and Nova Lacertae, a difference that was not understood for an additional 40 years.

Although he began his career by working on exploding stars (novae and supernovae), Popper soon moved onto **Henry Norris Russell**'s "Royal Road to the Stars" – the study of eclipsing binary stars – and this is what turned out to be his life's work. Popper's early career featured short assignments at a number of locations beginning with a Martin Kellogg Fellowship at Eastman Kodak,

in Rochester, New York. There, he met his future wife, Catherine Salo; they were married in 1940 and had one son. On the recommendation of Lick Observatory director **William Wright**, Popper landed a job as resident astronomer at the newly dedicated McDonald Observatory in West Texas in 1939, remaining there for 3 years. After a year at Yerkes Observatory (1942–1943) Popper returned to Berkeley to join the war effort at the Radiation Laboratory, returning to Yerkes/University of Chicago as an instructor and then assistant professor at the end of World War II.

Popper was hired by the University of California at Los Angeles [UCLA], when that institution formed a new astronomy department in 1947, and remained there for the rest of his career. As a guest investigator at the Mount Wilson and Palomar observatories, one of his first results was the direct measurement of the gravitational redshift of a white dwarf, 40 Eridani B; his results agreed with the prediction of the general theory of relativity within the uncertainties inherent in the theory and the measurements.

Early in Popper's career, there was no real understanding of what today is thought of as stellar evolution. That understanding was gained in the 1940s, 1950s, and 1960s beginning with **Hans Bethe**'s idea that stars are powered by nuclear fusion reactions in their cores. Popper devoted his career to providing very accurate masses, radii, and luminosities of many pairs of stars (eclipsing binaries). Such systems are critical to understanding stellar evolution because both stars have the same age; determination of their properties provides perhaps the most critical tests of the theory of stellar evolution.

When Popper began his work, he soon came to realize that many of the previous spectroscopic workers had not recognized important subtleties inherent in the interpretation and measurement of eclipsing binary spectrograms and in the analysis of their light curves. He came to mistrust the accuracy of most of the earlier work – with good reason. Popper conveyed his misgivings forcefully to the astronomical community in a series of 18 articles entitled "Rediscussion of Eclipsing Binaries." In his published work, he established the guiding principles and detailed procedures needed to achieve accurate results. Following Popper's lead, this field of study advanced to routinely produce accuracies of better than 1% in both masses and radii.

One of Popper's few extended stays away from UCLA came in 1964 when, on a National Science Foundation Senior Research Fellowship, he worked on the inauguration of the stellar intensity interferometer program in Narrabri, New South Wales, Australia. While there, Popper helped the program focus on establishing accurate stellar temperature scales.

Popper was honored as the 1984 Karl Schwarzchild Lecturer of the Astronomische Gesellschaft.

Claud H. Lacy

Selected References

Popper, Daniel M. (1940). "A Spectrophotometric Study of Nova Lacertae, 1936." *Astrophysical Journal* 92: 262–282.

——— (1954). "Red Shift in the Spectrum of 40 Eridani B." *Astrophysical Journal* 120: 316–321.

——— (1967). "Determination of Masses of Eclipsing Binary Stars." *Annual Review of Astronomy and Astrophysics* 5: 85–104.

——— (1980). "Stellar Masses." *Annual Review of Astronomy and Astrophysics* 18: 115–164.

——— (1991). "The McDonald Observatory and I Way Back Then." In *Frontiers of Stellar Evolution*, edited by David L. Lambert, pp. 19–22. San Francisco: Astronomical Society of the Pacific.

——— (1998). "*Hipparcos* Parallaxes of Eclipsing Binaries and the Radiative Flux Scale." *Publications of the Astronomical Society of the Pacific* 110: 919–922.

Trimble, Virginia (1999). "Daniel M. Popper, 1913–1999." *Bulletin of the American Astronomical Society* 31: 1608–1610.

Poretsky, Platon Sergeevich

Born **Elizavetgrad (Kirovograd, Russia), 15 October 1846**

Died **Zhoved, (Chernigov Guberniya), Russia, 22 August 1907**

Platon Poretsky was a Russian mathematician and astronomer who developed a strong interest in mathematical logic. Poretsky, son of an army doctor, graduated from the department of physics and mathematics at Kharkov University. In 1870, he was awarded a fellowship by the department of astronomy and worked as an astronomer at the Kharkov Observatory. In 1876, Poretsky moved to Kazan University, where he was appointed its chief astronomical observer. There, he conducted meridian observations of stars in a zone assigned to the Kazan Observatory and published two volumes on the results.

In 1886, Poretsky defended a master's thesis that addressed the nature of errors associated with the observatory's meridian circle. Its theoretical part dealt with reducing the number of equations and unknowns in systems of cyclical equations relating to practical astronomy. But by a decision of the Department of Physics and Mathematics, Poretsky was awarded a doctorate in astronomy instead, based on the extraordinary quality of his work. He then became a *Privatdozent* (lecturer) in spherical trigonometry at Kazan University. Poretsky was also appointed secretary and treasurer of the physical–mathematical section of the Kazan Society of Natural Scientists (1882–1888) and supervised the publication of its *Proceedings*.

In succeeding years, Poretsky developed a deep interest in the emerging discipline of mathematical logic and became the first Russian scientist to lecture on the subject in 1888. His publications on mathematical logic (from 1884 until his death) were extensive and focused on the elaboration of Boolean algebras, on the study of the logics of classes and propositions, and the application of logical methods to the theory of probability. His influence on the development of mathematical logic was substantial, as is evidenced, for example, in the works of Archie Blake and Louis Couturat. Poretsky's poor health, however, forced him to take early retirement in 1889.

Poretsky's mathematical papers are preserved in the Kazan University Archives.

Yuri V. Balashov

Selected References

Bazhanov, V. À. (1992). "New Archival Material Concerning P. S. Poretskii." *Modern Logic* 3, no. 1: 93–94.

Dubyago, D. I. (1908). "P. S. Poretskiy." *Izvestiya fizichesko matematicheskogo obshchestva pri Kazanskom universitete*, 2nd ser., 16, no. 1: 6.

Poretsky, P. (1886). "On the Question of Solution of Normal Systems Encountered in Practical Astronomy, with Application to Determination of Errors in the Division of the Meridian Circle at Kazan Observatory." Ph.D. diss., Kazan University.

Styazhkin, N. I. (1969). *History of Mathematical Logic from Leibniz to Peano*. Cambridge, Massachusetts: MIT Press, esp. pp. 216–247.

Youschkevitch, A. P. (1975). "Poretsky, Platon Sergeevich." In *Dictionary of Scientific Biography*, edited by Charles Coulston Gillispie. Vol. 11, pp. 94–95. New York: Charles Scribner's Sons.

Porter, John Guy

Born **Battersea, (London), England, 5 November 1900**
Died **Hailsham, (East Sussex), England, 13 September 1981**

John Porter was a sophisticated astronomical computer as well as a dedicated astronomical educator. While a schoolteacher of mathematics and chemistry, he devoted himself as an amateur astronomer to computational astronomy, becoming the director of the British Astronomical Association [BAA] Computing Section in 1937. Porter became an expert on the computation of the orbits of meteors. By the end of World War II, Porter had refined his skills in this area so extensively that he was awarded a Ph.D. for his research on the orbits of meteors and comets. After the war, Porter and **John Prentice** were invited to join (Sir) **Alfred Lovell** at Jodrell Bank Observatory where they assisted in the analysis of radar observations of meteor streams. They showed from velocity profiles that meteors were members of the Solar System and not interstellar particles. Porter's book, *Comets and Meteor Streams*, spelled out these techniques for general use.

By 1949, Porter's reputation in computation was well established. He joined the staff of Her Majesty's Nautical Almanac Office at Herstmonceux, where he was responsible for computing the *Nautical Almanac and Astronomical Ephemeris*. Porter continued to lead the BAA Computing Section from this new venue, though many of his efforts on behalf of the *BAA Handbook* and other projects were by then nearly anonymous.

As a member of the International Astronomical Union, Porter served as president of Subcommission 20A on Orbits of Comets, and as chairman of the Working Group on Orbits and Ephemerides of Comets from 1955 to 1964. After a series of short-term appointments at Yale University in New Haven, Connecticut, USA, and then at the Jet Propulsion Laboratory in Pasadena, California, USA, Porter and his wife retired to Hailsham in about 1962. After his retirement, Porter published a comprehensive catalog of the most reliable orbits of 583 periodic comets observed between 240 BCE and 1965.

Quite apart from the successful career outlined earlier, Porter had a second equally successful career as a popularizer of astronomy. His first radio broadcast, delivered in 1946 on the discovery of Neptune, led to a series of popular weekly broadcasts that extended 15 years. Porter also developed a short series of television programs on astronomy for children, but he preferred the relative anonymity conferred by radio.

Porter was the recipient of the Walter Goodacre Medal and Gift of the BAA in 1965 and of the Jackson Gwilt Medal and Gift of the Royal Astronomical Society in 1968.

Thomas R. Williams

Selected Reference

Wilkins, G. A. (1983). "John Guy Porter." *Quarterly Journal of the Royal Astronomical Society* 24: 364–367.

Porter, Russell Williams

Born **Springfield, Vermont, USA, 13 December 1871**
Died **Pasadena, California, USA, 22 February 1949**

Arctic explorer, artist, and telescope maker Russell Porter was the cofounder of the amateur telescope making movement in the United States and architectural draftsman of the 200-in. Hale Telescope at Palomar Mountain.

Porter studied civil engineering at the University of Vermont and architecture at the Massachusetts Institute of Technology [MIT] but abandoned his coursework to pursue arctic exploration under the influence of admiral **Robert Peary**. Porter served as surveyor, astronomer, and artist on the ill-fated Fiala–Ziegler Expedition that failed to reach the North Pole (1903–1905). He also surveyed and mapped Mount McKinley in Alaska, but did not reach its summit.

Following these adventures, Porter temporarily settled at Port Clyde, Maine, where he married Alice Belle (1907); the couple had two children. There, he constructed his first telescopes and observatories, employing new optical and mechanical designs for instruments having fixed eyepieces. During World War I, he performed optical work at the United States National Bureau of Standards and was an instructor in design at MIT.

Porter published a quartet of drawings (1916) in *Popular Astronomy* that he labeled "Moonscapes." These depicted what an observer would see, if transported to the lunar surface in the vicinity of the crater Gassendi. Porter noted that "having himself spent many years above the Arctic Circle," he "was struck by a strange likeness of the moon's general aspect to our own polar regions." In turn, he completed "A New Projection of the Moon," depicting its appearance as if each point were simultaneously experiencing sunrise – a clever aid to helping the novice observer.

In 1919, Porter returned to his birthplace as an optical associate at the Jones and Lamson Machine Company. The following year, he founded the Amateur Telescope Makers of Springfield, whose clubhouse (Stellafane) on nearby Breezy Hill became the focus of the group's astronomical activities. The "Springfield" mounting was another of that group's innovations. A 1921 article written by Porter, "The Poor Man's Telescope," caught the attention of *Scientific American* assistant editor **Albert Ingalls**. A three-part series of articles, prepared and illustrated by Porter, launched the amateur telescope making movement, whose early phase culminated in publication of the three-volume classic, *Amateur Telescope Making, Book One, Two, and Three.*

In 1928, Porter was hired by **George Hale** to assist with the design of the 200-in. reflector on Palomar Mountain. Porter's earlier invention of the split-ring equatorial mounting was eventually incorporated into the Hale telescope (completed in 1948). Working only from blueprints, Porter produced the exquisite series of "cutaway drawings" of the giant reflector that revealed its intricate construction. He also designed the site-survey telescopes used in testing the Palomar seeing conditions, the new astrophysics laboratory at the California Institute of Technology, plus the mounting and dome for its Schmidt telescope. Additional freelance work included conceptual drawings of the Griffith Observatory, Los Angeles, and the star projector constructed by the Morrison Planetarium, San Francisco.

Porter was awarded honorary doctorates by two Vermont institutions: Norwich University (1946) and Middlebury College (1949). A crater on the Moon has been named for him.

Jordan D. Marché, II

Selected References

Florence, Ronald (1994). *The Perfect Machine: Building the Palomar Telescope.* New York: Harper Collins.

Ingalls, Albert G. (ed.) (1967). *Amateur Telescope Making: Book One.* 4th ed. New York: Scientific American. (Books Two and Three complete the trilogy.)

Willard, Berton C. (1976). *Russell W. Porter: Artic Explorer, Artist, Telescope Maker.* Freeport, Maine: Bond Wheelwright.

Williams, Thomas R. (1991). "Albert Ingalls and the ATM Movement." *Sky & Telescope* 81, no. 2: 140–143.

——— (2000). "Getting Organized: A History of Amateur Astronomy in the United States." Ph.D. diss., Rice University, esp. Chap. 6, "The Amateur Telescope Making Movement," pp. 140–169.

Posidonius

Born **Apameia, (Syria), 135 BCE**
Died **Rhodes, (Greece), 51 BCE**

Posidonius is responsible for an early measurement of the circumference of the Earth.

Posidonius was from a Greek family though he was born in Syria. He was raised in the Greek tradition and completed his education in Athens under the great Stoic philosopher Panaetius of Rhodes. His name is sometimes listed as Posidonius of Apameia, while at other times it is listed as Posidonius of Rhodes. The former obviously refers to the place of his birth, while the latter refers to the place where he ultimately taught.

Presumably, the influence of Panaetius is what brought Posidonius to Rhodes. Sometime after 100 BCE he is known to have become the head of the Stoic school at Rhodes, where he taught both **Cicero** and Pompey (the Great). In 86 BCE, Posidonius was sent as an envoy to Rome, where he met Gaius Marius, the Roman politician and general. It is also probable that it was around this time that he first met Pompey, who later visited him in Rhodes and became his pupil for a time. It is a remarkable testament to his abilities that we know as much about him as we do, as only fragments of his own writings have survived.

Posidonius is chiefly remembered for providing a value for the circumference of the Earth. There were, in fact, four attempts to measure the circumference of the Earth between the time of **Aristotle** and that of **Ptolemy**. Posidonius estimated that the distance from Rhodes to Alexandria was 5,000 stadia. (According to some estimates, a stade is roughly 185 m, though the one used by Posidonius may have been shorter.) He then observed that if the star Canopus was exactly on the horizon at Rhodes, then it was 1/4 of a sign (1/4 of 30°) about the horizon at Alexandria. This meant that the circumference of the Earth had to be 240,000 stadia, the accuracy of which depends on the value of the unknown stade. His result, whether based on observation or, more likely serving as an illustration, was likely a bit too large and not as accurate as that of **Eratosthenes**, but Posidonius's methods were a precursor to understanding the concept of latitude. It is interesting to note that later scholars indicate that Posidonius actually used different values both for the distance from Rhodes to Alexandria and for the height at which the star Canopus rose above the horizon. Both numbers as given earlier are actually incorrect for that approximate date. (Obviously, the distance never changes, but precession has changed Canopus's position from Posidonius's time.) Posidonius also made an estimate of the size of the Sun, but for its calculation used a value for the size of the Earth different from the one he had himself computed, thereby revealing something of his own convictions about their accuracy.

Most of our knowledge about Posidonius's astronomy comes to us from **Cleomedes**. Posidonius also showed a great interest in the earth sciences, having developed theories of clouds, mist, wind, rain, lightning, earthquakes, frost, hail, and rainbows in a work on *Meteorology*.

Ian T. Durham

Selected References

Heath, Sir Thomas L. (1931). *A Manual of Greek Mathematics*. Oxford: Clarendon Press. (Reprint, New York: Dover, 1963.)

Neugebauer, Otto (1975). *A History of Ancient Mathematical Astronomy*. 3 pts. New York: Springer-Verlag.

Smith, David Eugene (1923). *History of Mathematics*. Vol. 1. Boston: Ginn and Co. (Reprint, New York: Dover, 1958.)

——— (1925). *History of Mathematics*. Vol. 2. Boston: Ginn and Co. (Reprint, New York: Dover, 1958.)

Swetz, Frank J. (1994). *From Five Fingers to Infinity*. Chicago: Open Court.

Pouillet, Claude-Servais-Mathias-Marie-Roland

Born **Cusance, (Doubs), France, 16 February 1790**
Died **Paris, France, 13 June 1868**

Mathias Pouillet (the name by which he was known) provided the first precise measurements of solar radiation, which he performed with instruments of his own design. Pouillet was the second of ten children of Ignace Denis Pouillet, a papermaker, and of Marie Françoise Rolland. He is said to have attended the lycée of Besançon, before he went to the collège of Tonnerre (Yonne), where he spent 2 years as a teacher of mathematics and where he received his baccalauréatit in 1811. Pouillet then became a student at the École Normale Supérieure, Paris from 1811 to 1813, where he earned his *licence ès sciences*, and he stayed on there as a tutor until 1815, subsequently becoming a *maître de conférences* in physics in the same establishment until 1822. In the meantime, he had earned the *agrégation pour les sciences* (in 1819). From 1817 to 1826, Pouillet also acted as a deputy teacher for the astronomer **Jean-Baptiste Biot** at the Faculté des sciences of Paris. In 1820, Pouillet was appointed professor at the Collège Bourbon (now the Lycée Condorcet), a post he held until 1829, and from 1826 to 1838 also assistant professor at the Faculté des sciences of Paris. He married Henriette Pichon in 1827, and in the same year he was invited to become the tutor of several children of the future king of France, Louis-Philippe.

Pouillet held very important positions at the Conservatoire (then royal) des Arts et Métiers: He was professor and assistant director from 1829 to 1832 (and also teacher of physics at the École Polytechnique for a year in 1831) and administrator (which in fact meant director) from 1832 until 1849, when he was dismissed because of the events of 13 June: an attempted revolt organized by A. A. Ledru-Rollin on the premises of the Conservatoire that Pouillet had not been able to control. In 1838, after the death of the incumbent, P. -L. Dulong, Pouillet held the chair of physics at the Faculté des sciences. He was compulsorily retired in 1852 because he refused to swear an oath of allegiance to the imperial government that took power after the coup d' état of 2 December 1851. He had lost both his children a few years earlier. Deeply affected by these personal and professional tragedies, he devoted his last years to the Académie des sciences and to experimental research, which he mainly carried out in his cottage at Épinay-sur-Seine.

Throughout his scientific career, Pouillet was a much-appreciated teacher of physics, lucid in his theoretical lectures and adept in the art of demonstrating experiments at whatever level he was teaching. A number of his lecture texts were printed in 1828, and he published his famous book *Éléments de physique expérimentale et de météorologie* during the years 1827–1830. A popular version of it was first published in 1850. As a member of the Société d'Encouragement pour l'Industrie Nationale, Pouillet also took an important part in reports on industrial exhibitions and many civil engineering works.

Pouillet carried out researches in various fields that were published as about 40 memoirs from 1816 onward, especially in the *Comptes rendus* of the Académie des Sciences. These dealt with optics, electricity, magnetism, meteorology, photography, photometry, thermal phenomena, and even with studies on the laws of population. In optics he investigated diffraction phenomena. In electricity, he improved the measurement of weak currents by means of his tangent and sine galvanometers. This allowed him in 1837 to verify very accurately Ohm's laws, which had been originally expounded in 1827.

Pouillet's astronomical research was mainly concerned with the measurement of solar radiation and the determination of the atmospheric effects on the recorded values. In 1824, he submitted a letter on the subject to the Académie des sciences through Dulong, and the same year he read a paper before the academicians. In 1838, Pouillet presented a substantial memoir in which he described the pyrheliometers he had designed for the purpose and reported the results of his observations. The data he obtained led to the first successful determination of a "solar constant," which he defined as the flux of total solar radiation received by a surface perpendicular to the Sun rays, for a mean Sun–Earth distance, at the upper limit of the atmosphere. He obtained a value of 1.7633 cal min^{-1} cm^{-2} (*i. e.*, 1,228 W m^{-2}), which turns out to be only 10% lower than the modern value, and hence better than the much overestimated values later given by Jules Violle in 1875 (2.54) and by Alexei Hansky in 1897 (3.4). In 1856, Pouillet designed an instrument he called an *actinographe*, which allowed him to record on a strip of photographic paper the effect produced by the intensity of the image of the Sun as it moved along the strip during the day. J. Stefan successfully used these actinometric observations together with Violle's ones to derive a value for the surface temperature of the Sun.

Pouillet was made a *Chevalier* of the Légion d'honneur in 1828, and an *Officier* in 1845. Meanwhile he had been elected a member of the Académie des sciences in 1837. He served several terms as a deputy for the French department of Jura and sat in the National Assembly from 1837 to 1848 on the government side.

Françoise Launay

Selected References

Gosse, J. (1983). "Claude Pouillet (1790–1868), thermicien oublié." *Revue générale de thermique* 257: 385–388.

Pluvinage, P. (1984). "Quelques épisodes de la carrière d'un grand physicien franc-comtois: Claude Servais Mathias Pouillet (1790–1868)." *Mémoires de la Société d'emulation du Doubs* 26: 59–77.

Pound, James

Born **Bishop's Canning, Wiltshire, England, 1669**
Died **Wanstead, (London), England, 16 November 1724**

James Pound made useful observations of Jupiter, Saturn, and the solar parallax, though he is remembered chiefly for the effects of the encouragement and astronomical training he provided to his nephew, **James Bradley**. Pound was educated at Oxford University from 1687 to 1694, when he received both his BA and his MA. By 1697, he had also obtained a medical degree and took orders in the Anglican Church. From 1699, he served as a chaplain in India, where he survived a massacre in 1705. In 1706, Pound returned to England, and was appointed as rector of Wansted in Essex the following year. He was elected to the Royal Society on 30 November 1699. On 14 February 1710, he married Sarah Farmer, a widow; they had a daughter, also named Sarah, born in 1713. After Pound's wife passed away in 1715, Pound was married in October 1722 to Elizabeth Wymondesold, who had a considerable fortune.

Pound made several astronomical observations in 1715, reporting his findings in *Philosophical Transactions*. In 1717, he mounted a 123-ft. focal length objective made by **Christiaan Huygens**; he used this "aerial" telescope to observe Saturn, its moons and rings, and also Jupiter with its satellites. Pound was one of the few observers who made significant contributions using this difficult form of telescope. **Isaac Newton** incorporated Pound's observations into the third edition of his *Principia*; **Pierre de Laplace** used his data to calculate Jupiter's mass.

Pound's financial circumstances enabled him to fund Bradley's education and astronomical interests. He made many observations with Bradley, and furthered his career by introducing him to **Edmond Halley** and **Samuel Molyneux**. Pound visited the latter regularly at his observatory in Kew. In 1717, Pound and Bradley attempted to measure the solar parallax, concluding that the Sun's distance lies between 93 and 125 million miles. The following year, Pound observed the components of the double star γ Virginis in order to determine a stellar parallax, a task Bradley continued in his observations of γ Draconis, which led to Bradley's discovery instead of the aberration of starlight in 1729. Together, Pound and Bradley also observed the opposition of Mars (1719) and the transit of Mercury on 29 October 1723.

Selected References

Clerke, Agnes M. (1921–1922). "Pound, James." In *Dictionary of National Biography*, edited by Sir Leslie Stephen and Sir Sidney Lee. Vol. 16, pp. 232–233. London: Oxford University Press.

Pound, James (1719). "New and Accurate Tables for the Ready Computing of the Eclipses of the First Satellite of Jupiter, by Addition only." *Philosophical Transactions* 30: 1021–1034.

Poynting, John Henry

Born **Monton, (Greater Manchester), England, 9 September 1852**
Died **Birmingham, England, 30 March 1914**

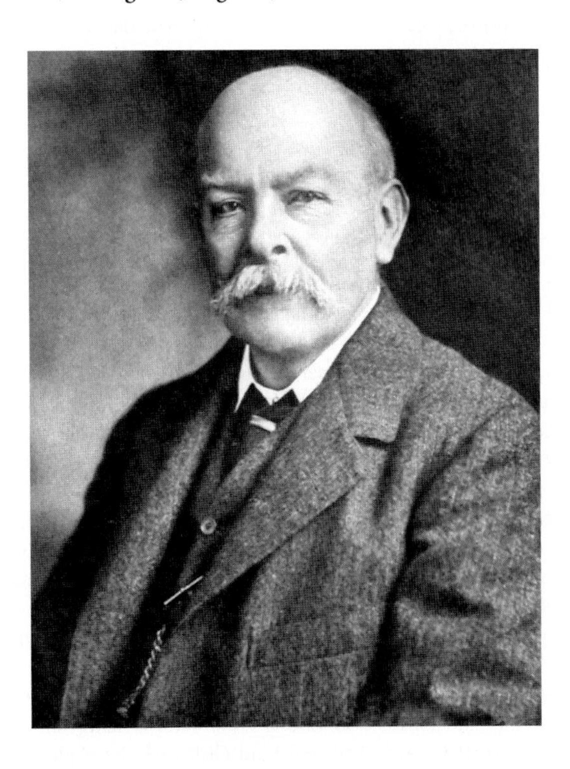

John Poynting was a physicist, mathematician, and inventor. He was the youngest son of Reverend T. Elford Poynting and Elizabeth Long. He attended Owens College, Manchester, and Trinity College, Cambridge (1872–1876), and was elected fellow of Trinity College in 1878. He was appointed professor of physics at the Mason College Birmingham (later University of Birmingham) in 1880 and remained there until his death. He married Maria Adney in 1880 and had a son and two daughters.

Poynting is well known for the Poynting vector, which describes the rate at which energy is carried by electromagnetic radiation. It was first introduced in his paper "On the Transfer of Energy in the Electromagnetic Field" (1884). He was also known for his discovery that infrared radiation causes small particles that orbit the Sun to spiral toward and plunge into the Sun. This idea was later developed by the American physicist **Howard Robertson** and is known as the Poynting–Robertson effect. Poynting also invented a method for finding the absolute temperature of celestial objects. He calculated the mean density of the Earth and made a determination of the gravitational constant. The results were published in *The Mean Density of The Earth* (1894) and in *The Earth: Its Shape, Size, Weight and Spin* (1913).

The Cambridge University Press published Poynting's *Collected Scientific Papers* in 1920. His other works include *The Earth* (1913) and *The Pressure of Light* (1910). Poynting was awarded the Adams Prize of Cambridge for his essay on *The Mean Density of The Earth* in 1894. He received the Hopkins Prize of the Cambridge Philosophical Society (1903) and the Royal Medal from the Royal Society (1905).

Suhasini Kumar

Selected Reference

Lightman, Bernard (ed.) (1997). *Victorian Science in Context*. University of Chicago Press.

Prager, Richard

Born **Hanover, Germany, 30 November 1883**
Died **Boston, Massachusetts, USA, 20 July 1945**

German variable-star astronomer Richard Prager compiled a three-volume catalog of variable stars and their literature, which was the primary source of information on these between 1925 and 1949. He was educated at Hamburg (1901) and Göttingen (1903) universities, receiving a Ph.D. from Berlin in 1908 and holding a position as an assistant at the Berlin Academy of Sciences for the next year. In 1909, Prager accepted a position as a section chief at the Observatorio Astronomico Nacional in Santiago de Chile, where he worked with **Friedrich Ristenpart** during the latter's abortive attempt to rejuvenate the facility. There Prager made a number of measurements of positions of Solar System objects including 139 of comet 1P/Halley during its 1910 appearance.

Returning to Germany in 1913, Prager was first assistant and then observer at the Berlin–Babelsbürg Observatory, being appointed professor in 1916. Together with **Paul Guthnick**, he exploited the then revolutionary potassium iodide cell to perform precision photoelectric photometry of variable stars. Of Jewish ancestry, Prager was imprisoned by the Nazis at Potsdam in 1938, but released to immigrate to London when the British Home Office accepted his application for asylum at the urging of the British astronomical community. He lectured to the Royal Astronomical Society in January 1939 before going on to Harvard College Observatory under a program for displaced scholars (in which **Harlow Shapley** played a significant part). Prager carried out some mathematical work for the navy during World War II, but he never entirely recovered from the physical hardships of imprisonment and separation from his family, and died after a succession of illnesses.

Prager's three volumes of *Geschichte und Literature des Lichtwechsels der Weranderlichen Sterne*, the last published in English from Harvard in 1941 as *History and Bibliography of the Light Variations of Variable Stars*, were superceded only after 1949, when, at the request of the Commission on Variable Stars of the International Astronomical Union, Boris V. Kukarkin and **Pavel Parenago** took up the task of compiling and evaluating the literature on this important astronomical subject. Prager served as secretary of the Astronomische Gesellschaft (the German astronomical society) from 1930 to 1936, but was debarred from membership in the International Astronomical Union because Germany was not permitted to join until after World War II.

Leif J. Robinson

Selected References

Ashbrook, Joseph (1984). "A South American Tragedy." In *The Astronomical Scrapbook*, edited by Leif J. Robinson, pp. 99–103, esp. p. 101. Cambridge, Massachusetts: Sky Publishing Corp.
Beer, Arthur (1945). *Observatory* 66: 186.
Gaposchkin, Sergei (1945). "Richard Prager." *Sky & Telescope* 4, no. 11: 2.

Prentice, John Philip Manning

Born **Stowmarket, Suffolk, England, 14 March 1903**
Died **Stowmarket, Suffolk, England, 6 October 1981**

John Prentice discovered the Giacobinid (Draconid) meteor shower and Nova DQ Herculis 1934, but his main contributions to astronomy came through his careful supervision of the British Astronomical Association [BAA] Meteor Section for over 30 years. Prentice played an important role in establishing radar observation as a primary technique for studying meteors.

A lawyer by profession, Manning Prentice (as he was known to his friends) acquired an interest in astronomy as a schoolboy. He first began observing the Moon and planets with a small refractor, and later meteors with his naked eye. After joining the BAA in 1919, Prentice continued his meteor observations and was appointed leader of the BAA Meteor Section in 1923, holding that position until 1954.

In 1915, Reverend Martin Davidson (1880–1968) pointed out that the orbit of the short-period comet 21P/Giacobini – Zinner passed close enough to that of the Earth that it might result in a meteor shower on about 10 October each year. No such activity was seen until **Andrew Crommelin** calculated that on 10 October 1926 the orbits of the comet and the Earth would intersect. Acting on Crommelin's prediction, Prentice took up a routine watch the previous evening and was rewarded with the observation of a strong meteor shower – Prentice estimated meteors appeared at a rate of 17 per hour for one observer – with a radiant very near that projected by Davidson. **William Denning** published a similar conclusion about this shower at an earlier date than Prentice, but priority for the discovery clearly belongs to Prentice as the earliest observer. A spectacular return of the Giacobinid meteor shower in 1933, when the Earth crossed the comet's orbit 80 days after the passage of the comet (as opposed to 19 days before the comet's passage in 1926), confirmed not only the relationship of the shower with the comet but also revealed that the duration of the shower was sharply limited to only 4½ hours. The zenith hourly rate for the

1933 Giacobinid shower was estimated at between 4,000 and 6,000 meteors per hour. Unfortunately, the skies were cloudy in England, and the information on this shower was provided by Reverend William Frederick Archdall Ellison (1834–1936) at the Armagh Observatory and by observers on the European Continent.

Prentice's familiarity with the night sky was an important prerequisite for his program of meteor observation and also contributed to his discovery, early on the morning of 13 December 1934, of a nova in the constellation of Hercules. While taking a break from his tiring routine of meteor counting, Prentice noticed something wrong in the appearance of the head of Draco. The problem was quickly traced to an interloping star that was promptly reported to the Royal Greenwich Observatory. Prentice's expeditious reporting facilitated valuable premaximum spectroscopic observation of Nova DQ Herculis. An independent discovery of this nova was made within hours at Delphos, Ohio, by another amateur astronomer, **Leslie Peltier**.

In June 1937, Prentice married Elizabeth Mason Harwood; their union resulted in the birth of four children. The resulting obligations, as well as those associated with his leadership in the congregation of the Stowmarket Congregational Church, must have increased the burden of his avocational interests, but Prentice's zeal for observation remained comparatively undiminished. A further complication arose as a result of the bombing of the church in 1941. For 14 years thereafter, Prentice shouldered a heavy burden as he led the membership's effort to rebuild the structure and preserve the integrity of the congregation. Prentice was active as a leader of youth activities in the church and served as a lay minister and church secretary as well.

After World War II, (Sir) **Bernard Lovell** contacted Prentice and his BAA meteor section for assistance in tracing the relationship between meteor activity and apparently spurious radar signals, interference that could not be traced to cosmic-ray activity. During the Perseid shower in August 1946, Prentice traveled to Jodrell Bank to assist with the correlation of radar signal reception with the appearance of specific meteors; the correlation was immediately evident. Prentice and **John Porter** worked with Lovell and others at Jodrell Bank to apply this technique to good advantage in investigations of meteor streams during daylight hours, and streams of meteors too faint to be seen with the naked eye at night.

In 1935, the BAA and American Association of Variable Star Observers honored Prentice for his discovery of Nova DQ Herculis by awarding him their Walter Goodacre Medal and D. B. Pickering Medal, respectively. In 1953, the Royal Astronomical Society honored his work on meteors with the award of its Jackson-Gwilt Gift and Medal while the University of Manchester conferred an honorary master of arts degree on Prentice in that same year.

Thomas R. Williams

Selected References

Lovell, Sir Bernard (1982). "J. P. M. Prentice." *Quarterly Journal of the Royal Astronomical Society* 23: 452–460.

Prentice, J. P. M. (1939). "Report on the Giacobinids." *Journal of the British Astronomical Association* 50: 27.

Prentice, J. P. M., A. C. B. Lovell, and C. J. Banwell (1947). "Radio Echo Observations of Meteors." *Monthly Notices of the Royal Astronomical Society* 107: 155–163.

Preussen, Pilgrim Zeleschicz von

➲ **Pèlerin de Prusse**

Pritchard, Charles

Born **Alberbury, Shropshire, England, 29 February 1808**
Died **Oxford, England, 28 May 1893**

Reverend Charles Pritchard pursuaded astronomers to take accurate positional measurements from photographic plates. Pritchard was the fourth son of William Pritchard. He married Emily Newton in 1834 and Rosalind Campbell in 1858.

Pritchard was educated at Merchant Taylors' and Christ's Hospital schools, then Saint John's College, Cambridge. He graduated first in 1830 and then received an MA from the same institution in 1833. Pritchard became a fellow of Saint John's College in 1832 and an honorary fellow in 1886. He earned another MA from Oxford in 1870 and a DD in 1880. Pritchard was named a fellow of New College in 1883.

Pritchard had a successful career as headmaster of Clapham Grammar School, London (1834–1862). He was particularly noted for his teaching of mathematics and sciences. Pritchard also equipped the school with a small observatory. Upon his retirement to the Isle of Wight he became involved in the current controversies between science and religion. Pritchard was opposed to Darwinism and used his scientific accomplishments in defense of the Christian religion. He was subsequently appointed as Hulsean Lecturer (Cambridge) in 1867 and then as Savilian Professor (Oxford; at age 62) from 1870 to 1893.

At Oxford, Pritchard oversaw a complete renovation of the observatory, obtaining new instruments and appointing competent assistants. He engaged the observatory in a number of research projects of specific duration. Pritchard started work on a study of lunar libration using photography; however, this work was not successful and was never published. He compiled a catalog of star magnitudes using a wedge photometer; this was published as the *Uranometria nova Oxoniensis* (1885). Pritchard embarked on a program to measure stellar parallax using photography. He also committed the Oxford Observatory to the *Carte du Ciel* project, obtaining a suitable telescope and initiating the survey.

Pritchard was awarded the Gold Medal of the Royal Astronomical Society in 1886 and a medal of the Royal Society in 1892. He was a fellow of the Royal Astronomical Society from 1849 (its president from 1866 to 1868), fellow of the Royal Society from 1840, and also a fellow of the Geological Society and Cambridge Philosophical Society.

Letters and some other papers by Pritchard are kept in the Library of the Royal Astronomical Society.

Mark Hurn

Selected References

Anon. (1893). "Obituary." *Journal of the British Astronomical Association* 3: 434–436.

Anon. (1893). "Obituary." *Observatory* 16: 256–259.

Anon. (30 May 1893). "Obituary." *Times* (London).
Anon. (1894). "Charles Pritchard." *Monthly Notices of the Royal Astronomical Society* 54: 198–204.
Clerke, Agnes M. (1921–1922). "Pritchard, Charles." In *Dictionary of National Biography*, edited by Sir Leslie Stephen and Sir Sidney Lee. Vol. 16, pp. 403–406. London: Oxford University Press. See also revised biography of Anita McConnell in the *Oxford Dictionary of National Biography*, O.U.P., 2004.
E. D. (1893). "Rev. Charles Pritchard, D. D." *Proceedings of the Royal Society of London* 54: iii–xii.
Hennessey, R. A. S. (July 1998). *Astronomy Now*: 19–20.
Pritchard, Ada (comp.) (1897). *The Life and Work of Charles Pritchard*. London. Seely & co. (This contains biographical memoirs by friends and family.)

Pritchett, Carr Waller

Born **Henry County, Virginia, USA, 4 September 1823**
Died **Independence, Missouri, USA, 19 March 1910**

Starting in 1878, and in every apparition since, Jupiter's Great Red Spot has been continuously visible. In that year, it was first spotted by Reverend Carr Pritchett of Glascow, Missouri, USA. (Sporadic sightings of the Great Red Spot appear to go back to the time of **Giovanni Cassini**.)

Selected Reference

See, T. J. J. (1910). "Pritchertt, Carr Waller." *Popular Astronomy*. 18: 348.

Proclus

Born **Byzantium (Istanbul, Turkey), *circa* 411**
Died **Athens, (Greece), 17 April 485**

Head of the Athens Academy in the 5th century, Proclus promoted astronomy in mathematics, cosmology, physics; in empirical observation and instrumentation; and in higher education, proposing that celestial objects have their own self-movement in free space, that our system can be heliocentric, and that cosmic space consists of pure light. He was the last major thinker of Antiquity, and also the one who systematized Greek knowledge in the form it was transmitted to Islam and western Europe.

Proclus' Greek–speaking parents, Patricius and Marcella, moved from Byzantium to Xanthus, a district of Lycia in Asia Minor, probably by 415. Proclus studied rhetoric, Roman law, and Latin at Alexandria. He visited Byzantium at the time of the revival of advanced schools inspired by the Athenian-born Empress Eudocia (425). There he experienced a momentous conversion to Athenian philosophy. On his return to Alexandria, Proclus studied **Aristotle** and mathematics, which included astronomy. He then traveled to Athens (430), where he was quickly embraced by the leaders of the Neoplatonic "academy," and studied the philosophy of Aristotle, **Plato**, and other Greek thinkers. Proclus rose to become head of the premier center of higher learning of the Roman Empire at the age of 25. Around 450, he traveled for a year's sabbatical to the region of Lydia in Asia Minor, to avoid persecution by the Christian authorities. His chief surviving book on astronomy was written just after this trip. In Athens, Proclus pursued a career as a teacher, author, administrator, and influential figure within the empire, which left his followers in awe, as recorded by his biographer, Marinus. The biography concludes with a valuable reference to two solar eclipses at Athens, one precisely observed (14 January 484) and one predicted (19 May 486).

Proclus' integration of astronomy and physics with his metaphysics freed scientific thinking from constraints owed to Aristotle and **Ptolemy**. But he was not just a theoretician. He promoted practical and empirical knowledge, for they combine critical reasoning with direct perception of the appearances of reality. Proclus' chief astronomical treatise, the outline *(Hypotyposis) of the Astronomical Hypotheses*, contains a detailed description of how to construct and calibrate the spherical armillary astrolabe, and how to make observational measurements of the Moon and the stars; he noted the existence of optical binary stars. He also refers to the use of Heron's water clock to measure the Sun's diameter, to the meteoroscope instrument, and to the actual construction of ephemerides tables. He made some of the last reliable astronomical observations of Antiquity (475).

Proclus' views on astronomy are part of his responses to the core questions: How much can the appearances of things tell us about their deep nature? What is the nature of reality? For Proclus the celestial objects are "self-substantiated" agents with the power to move freely of their own accord. They orbit according to their own natural, unimpeded motion, and move in the three dimensions of free space without the need of celestial spheres. Proclus further proposed that every fixed star and planet must have its own spin around its axis. He went so far as to suggest that the Earth itself is like a star, and if it were not for the inertness characteristic of its predominant physical property, it should move circularly.

Proclus advanced the heliocentric view. He accepted that the celestial bodies revolve around the Earth as center but only in their capacity as earth-like bodies. Since it is their self-power that really matters, they should be arranged around the center of most power: the Sun. The Sun is also in the middle of a system consisting of the five observed planets, the Moon, and the Earth's four elements. Further, Proclus speculated that every planet has its own group of attendant "satellites."

Proclus poured scorn on astronomers such as Ptolemy, who believed that inventing epicyclic and eccentric spheres could explain away the apparent irregular movement of the planets and the Sun and the Moon. Proclus took these irregularities seriously, as problems challenging us to reconcile our current level of understanding with that proper to deeper reality. He also rejected **Hipparchus**' discovery of the "precession," as described by Ptolemy, and Ptolemy's interpretation that the precession involves a backward movement of all the stars. For Proclus the stars are fixed in their constellations and do not precess. Proclus' *Outline of the Astronomical Hypotheses* contained the only critical evaluation of Ptolemy in antiquity, and rejected his speculations on many counts.

Proclus also rejected Aristotle's fifth element for the heavens. For Proclus the heavens have the same four constituents (fire, air, water, and earth) as the Earth, but in a different state of matter. This contains the "summits" of all the elements, where the properties of fire prevail. He concluded that the celestial bodies must have some Earth properties also, to be opaque (as in eclipses) and have gravity. Above all, he asserted that there is one science for both the heavens and Earth, not separate ones.

In cosmology, Proclus was the first to propose that space must be a three-dimensional body of a special kind that allows normal bodies to coexist with it. He speculated that there is a cosmic space, a body of pure, invisible light, in which the entire Universe is immersed.

Proclus instilled astronomical interest in his students, including Marinus, his successor at the Athenian School; **Ammonius**; and Ammonius' brother **Heliodorus**. The latter records Proclus' observation of an occultation of Venus by the Moon (475) from Athens. Ammonius' students **Simplicius** and the Christian **John Philoponus** wrote the major commentaries on Aristotle, but followed Proclus on the rejection of Aristotelian physics, and accepted most of his views on the celestial bodies and the elements.

Through the Byzantine emigrés Gemistos Plethon and John Bessarion, Proclus and Ptolemy spread to Renaissance Europe. By the 16th to 17th centuries Proclus' mathematical and astronomical achievements gained wide recognition. Proclus' *Commentary on Euclid* was highly regarded and discussed in **Galileo Galilei**'s circle. **Nicolaus Copernicus** cited it in the *Revolutions of the Heavenly Spheres*, and **Johannes Kepler** did likewise in the *Harmonices Mundi* (1619). Kepler quoted Proclus repeatedly and praised him as a true precursor of the heliocentric theory.

Proclus' name has been given to a lunar crater near the Sea of Tranquillity.

Lucas Siorvanes

Selected References

Allatius, L. (ed.) (1635). *Paraphrase of Ptolemy's* Tetrabiblos (in Greek and Latin). Leiden. (English translation by J. Ashmand. London, 1822. Based on Ptolemy's chief astrological work; unconfirmed attribution to Proclus.)

Anon. (1559). *Anonymous Commentary on the* Tetrabiblos *of Ptolemy* (in Greek and Latin). Basel. (Astrological commentary on Ptolemy; unconfirmed attribution to Proclus.)

Marinus (1896). *Le Commentaire de Marinus aux Data d'Euclide*, edited and translated into French by M. Michaux. Louvain, 1947; Greek text in *Euclidis opera omnis*, edited by J. L. Heiberg and H. Menge. Vol. 6, p. 234. Leipzig: Teubner.

——— (2001). *Proclus ou sur bonheur*, edited by H. D. Saffrey and A. Ph. Segonds, with Concetta Luna. Paris: Les Belles Lettres. (Greek text and French translation.)

O'Meara, Dominic J. (1989). *Pythagoras Revived: Mathematics and Philosophy in Late Antiquity*. Oxford: Clarendon Press.

Proclus (1899). *Eighteen Arguments on the Eternity of the World against the Christians* (in Greek). In *De aeternitate mundi contra Proclum*, by John Philoponus, edited by Hugo Rabe. Leipzig: Teubner. (Reprint, Hildesheim: Olms, 1984.) (On the celestial objects and cosmology).

——— (1903–1906). *In Platonis* Timaeum *commentaria* (Commentary on Plato's Timaeus). 3 Vols. Edited by E. Diehl. Leipzig: Teubner. (Reprint, Amsterdam: Hakkert, 1965. [Greek text]; French translation: *Commentaire sur le* Timée, edited by A. J. Festugière. 5 Vols. Paris: Vrin, 1966–1968; English translation by Thomas Taylor. 1810. Reprint, with corrections and Greek text pagination, Frome, Somerset: Prometheus Trust, 1998. [Source material on cosmology.])

——— (1909). *Hypotyposis astronomicarum positionum* (Outline of the astronomical hypotheses), edited and translated by C. Manitius. Leipzig: Teubner. (Reprint, Stuttgart, 1974.) (Parallel Greek text and German translation; critical examination of Ptolemy's *Planetary Hypotheses*, and source for astronomical references.)

——— (1912). *Institutio physica* (The elements of physics), edited by A. Ritzenfeld. Leipzig: Teubner. (Parallel Greek text and German translation); (English translation by Thomas Taylor. 1831. Teaching manual consolidating Aristotle's *Physics* books 6 and 7 and *On the Heavens* book 1 in a demonstrative method.)

——— (1970). *A Commentary on the First Book of Euclid's* Elements, translated by Glenn R. Morrow. Princeton, New Jersey: Princeton University Press. (Reprint, with foreword by I. Mueller, 1992; Greek text: *In primum Euclidis Elementorum librum commentarii*, edited by G. Friedlein. Leipzig: Teubner, 1873. [On the value of mathematical sciences, including astronomy.])

Segonds, A. Ph. (1987). "Proclus: Astronomie et philosophie." In *Proclus, lecteur et interprète des anciens*, edited by J. Pépin and H. D. Saffrey, pp. 319–334. Paris: CNRS.

Simplicius (1882–1895). *In Aristotelis* Physicorum *commentaria* (Commentary on Aristotle's Physics), edited by H. Diels. Vols. 9–10 of *Commentaria in Aristotelem Graeca*. Berlin: Reimer. (Greek text; fragments and reports on Proclus); English translations in Ancient Commentators on Aristotle Series: *Corollaries on Time and Place*, edited by J. O. Urmson and L. Siorvanes. Ithaca, New York: Cornell University Press, 1992 and *Against Philoponus on the Eternity of the World*, edited by C. Wildberg. London: Duckworth, 1991.

——— (1894). *In Aristotelis* De caelo *commentaria* (Commentary on Aristotle's On the heavens), edited by J. L. Heiberg. Vol. 7 of *Commentaria in Aristotelem Graeca*. Berlin: Reimer. (Greek text; fragments and reports on Proclus.)

Siorvanes, Lucas (1996). *Proclus: Neo-Platonic Philosophy and Science*. New Haven: Yale University Press. (Comprehensive account and new research; Chap. 5 on astronomy, Chap. 4 on matter, space, motion, *etc.*)

Tihon, Anne (1976). "Notes sur l'astronomie grecque au Ve siècle de notre ère (Marinus de Naplouse – Un commentaire au *Petit commentaire* de Théon)." *Janus* 63: 167–184.

Proctor, Mary

Born **Dublin, Ireland, 1862**
Died **probably London, England, 1957**

Mary Proctor never became a professional astronomer, but was widely known for numerous articles and books on the subject.

Proctor was the daughter of **Richard Proctor** and his first wife, Mary (*née* Mills) Proctor. Mary's father was a well-known astronomer, lecturer, and writer, born in London and a graduate of Saint John's College, Cambridge. It was from him that Mary acquired her astronomical knowledge and her love for writing. From an early age, she took pride in the care of his library; as a young woman, she arranged his letters and corrected the galley proofs of his books. She graduated from the College of Preceptors at London in 1898.

Proctor's mother died in 1879. When her father remarried in 1881, the Proctor family immigrated to America and settled in Saint Joseph, Missouri. Proctor studied writing and assisted her father in the production of a new journal, *Knowledge*, which he founded and edited that same year. Among her earliest published writings was a series of articles on comparative mythology. She also began a secondary career as a lecturer on astronomy, following a very successful appearance at the World's Columbian Exposition at Chicago in 1893. Proctor's ambition to write a book was first realized with the 1898 publication of *Stories of Starland*, which was adopted by the New York City Board of Education. She attended classes at Columbia University and taught astronomy in private schools.

Many of Proctor's articles and books were aimed at younger audiences, and Proctor became known as the "children's astronomer." Her literary works included *Wonders of the Sky*, *Everyman's Astronomy*, *The Romance of Comets*, *Legends of the Stars*, and *Half-Hours with the Summer Stars*. Her books were easy to read, accurate, informative, and well illustrated. Proctor was widely respected by professional astronomers, and in 1898 was elected to membership in the American Association for the Advancement of Science. She was elected a fellow of the Royal Astronomical Society in 1916.

In *The Book of the Heavens*, published in 1924, Proctor wrote: "Some of my readers may become great astronomers; for that reason I have [included] ... accounts of ... some leading observatories of the world [so] that boys and girls may see astronomers of today at work." She never married. A 12-mile diameter lunar crater has been named for her.

Patrick Moore

Selected References

Bailey, Martha J. (1994). "Proctor, Mary." In *American Women in Science: A Biographical Dictionary*, pp. 309–310. Denver: ABC-CLIO

Creese, Mary R. S. (1998). "Mary Proctor." In *Ladies in the Laboratory? American and British Women in Science, 1800–1900*. pp. 237–238. Lanham, Maryland: Scarecrow Press.

Petrusso, Annette (1999). "Mary Proctor." In *Notable Women Scientists*, edited by Pamela Proffitt, p. 470. Detroit: Gale Group.

Proctor, Richard Anthony

Born **Chelsea, (London), England, 23 March 1837**
Died **New York, New York, USA, 12 September 1888**

Richard Proctor was principally an expositor of science, especially astronomy. He wrote prodigiously on the latter topic, and gained considerable fame through his articles and books. In 1854, Proctor worked as a clerk in the London Joint Stock Bank. A year later, he enrolled at King's College, London, where he studied theology and mathematics. Proctor graduated in 1860 from St. John's College, Cambridge, as 23rd Wrangler. That year he also married his first wife, Mary Mills.

"A few months after leaving Cambridge," Proctor tells us, "in my quiet home at Ayr, free of all anxieties about maintenance (for I had inherited ample means), I began in a very modest and quiet way the study of some astronomical matters, to which my attention had been attracted by two books picked up at a book stall in Glasgow." His first published article took him 6 weeks to prepare. "Often I would not complete more than four or five lines in a day with which I was satisfied, so that ... my work went on very slowly indeed." The piece, entitled "Colours of Double Stars," appeared in *Cornhill Magazine* (1865).

As a form of therapy, following the death of his eldest son, Proctor undertook a "work which would occupy me," he wrote, "for at least a year." In due course, his first astronomical treatise, *Saturn and Its System*, was completed (1865). This contained the first popular account of **James Maxwell**'s theory of the discrete nature of the planet's ring system.

Although commercially unsuccessful at first, the book established Proctor's reputation among those best equipped to judge its merits, and in 1866 he was elected a fellow of the Royal Astronomical Society. That same year, however, Proctor suffered a serious financial crisis when the New Zealand bank, in which the bulk of his assets were invested, collapsed.

Out of necessity, Proctor taught mathematics at the Military School, Woolwich. But by the autumn of 1867, he had definitely settled on a literary career. Passing "through experiences enough to age a man ten years in as many weeks," Proctor eventually succeeded, and in 1872 became one of the secretaries of the Royal Astronomical Society, and editor of its journal, the *Monthly Notices*. From that time onward, his output was prodigious and diverse.

It was Proctor's intention to produce original works that formed a standard embodiment of the astronomical knowledge of the last quarter of the 19th century. His exhaustive account of the Venus transits of 1874 and 1882 was characterized by one biographer as "the best popular exposition of the nature and use of that phenomenon that has yet appeared." Proctor wrote, in all, around 57 books, his last being *Old and New Astronomy* (1892), a large treatise on which he reported, "I intend ... putting into final form the results of my studies, now scattered through my essays, lectures and magazine articles." This was left unfinished at the time of Proctor's death, but enough materials had been assembled that his friend, **Arthur Ranyard**, could complete the task.

In 1867, Proctor prepared a map of Mars from the 1864–1865 drawings by Reverend **William Dawes**. Employing considerable cartographic skills, he applied names to supposed continents, seas, bays, and straits, thus showcasing current beliefs about physical conditions on the planet. A small forked bay-like marking, named after Dawes (now Sinus Meridiani), was adopted as the Martian prime meridian. Proctor also deduced an extremely accurate rotation period of Mars, from historic drawings of the red planet. In 1870, Proctor plotted the positions and proper motions of roughly 1,600 stars. As a result, he found that certain clusters shared a common direction of travel through space, a phenomenon he dubbed "star drift." Around 1873, Proctor conjectured that the craters of the Moon were impact features and not the result of volcanic eruptions. It would be decades before this idea gained wider acceptance.

Proctor's first wife, Mary, died in 1879. He then went on a 2-year lecture tour of the United States. In 1881, Proctor met and married the widow of Robert J. Crawley; the couple settled in Saint Joseph, Missouri (her hometown). That same year, he founded the popular science magazine *Knowledge*, and continued as its editor until his death from yellow fever. One year earlier, Proctor and his wife had moved to Orange Lake, Florida. His daughter, **Mary Proctor**, followed in her father's footsteps and published many popular books on astronomy. A crater on Mars has been named in Proctor's honor.

Richard Baum

Selected References

Hoyt, William Graves (1987). *Coon Mountain Controversies: Meteor Crater and the Development of Impact Theory*. Tucson: University of Arizona Press, esp. pp. 18–19, 21–26.

Luther, Paul (1989). *Bibliography of Astronomers: Books and Pamphlets in English by and about Astronomers*. Vol. 1. Bernardston, Massachusetts: Astronomy Books. (See pp. 175–202 for Proctor's major works.)

Noble, William (1888). "Richard A. Proctor." *Observatory* 11: 366–368.

North, J. D. (1975). "Proctor, Richard Anthony." In *Dictionary of Scientific Biography*, edited by Charles Coulston Gillispie. Vol. 11, pp. 162–163. New York: Charles Scribner's Sons.

Proctor, Richard A. (1895). "Autobiographical Notes." *New Science Review* 1: 393–397. (Quoted above.)

Ranyard, Arthur Cowper (1889). "Richard Anthony Proctor." *Monthly Notices of the Royal Astronomical Society* 49: 164–168.

Profatius

➲ **Jacob ben Makhir ibn Tibbon**

Prosperin, Erik

Born **Närlinge, Upland, Sweden, 25 July 1739**
Died **Upsala (Uppsala), Sweden, 4 April 1803**

Uppsala University's Erik Prosperin calculated an orbit for the new, seventh planet spotted by **William Herschel**. Prosperin wanted to call the discovery Neptune; instead, this name eventually went to the *eighth* planet from the Sun.

Selected Reference

Alexander, A. F. O'D. (1965). *The Planet Uranus: A History of Observation, Theory and Discovery*. New York: American Elsevier.

Przybylski, Antoni

Born **Rogozno, Poland, 1913**
Died **Queanbeyan, New South Wales, Australia, 21 September 1984**

Polish–Australian astrophysicist Antoni Przybylski discovered (1960) one of the most unusual stellar spectra: Przybylski's Star (HD 101065) is the only astronomical object to exhibit the presence of holmium. Przybylski was a protégé of **Richard Van der Riet Woolley**.

Selected Reference

Gascoigne, S. C. B. (1985). "Dr Antoni Przybylski." *Proceedings of the Astronomical Society of Australia* 6, no. 2: 275.

Ptolemy

Flourished **Alexandria, (Egypt), second century**

Hundreds of years of Greek geometrical astronomy were systematized, with rigorous demonstrations and proofs, by Claudius Ptolemaius. His intent was to do for applied mathematics what Euclid had done for pure mathematics (geometry). Ptolemy produced handbooks containing all that was known about astronomy, optics, geography, astrology, musical theory, geometrical constructions using spherical projection, the structure and size of the Universe, and mechanics. Although his primary goal was to summarize what was already known, Ptolemy also advanced astronomical knowledge so far as to earn for himself a reputation as the greatest astronomer of the ancient world. He showed how, based upon observation and empirical data, geometrical models could be constructed that simulated nature. His astronomical textbook surpassed all that had gone before and dominated future astronomy for well over a millennium. In replacing much of previous astronomy, however, Ptolemy helped cause the loss of a vast body of earlier data.

Of the man himself almost nothing is known. His recorded observations purportedly were made – some assert they were fabricated – between the 9th year of Hadrian's regime (125) and the 4th year of Antoninus Pius (141) "in the parallel of Alexandria." Most writers assume that Ptolemy worked in Alexandria, Egypt, but according to **Olympiodorus the Younger**, a philosopher teaching in Alexandria in the 6th century, Ptolemy worked for 40 years at Canopus, a town 15 miles to the east of Alexandria (and hence in the same parallel). One tradition even has it that Ptolemy in 147 erected in the temple of Seraphis at Canopus a pillar to commemorate

his discoveries. Wherever he actually resided, Ptolemy rightly is associated with Alexandria, whose library provided him with the observations of his predecessors, upon which he constructed his great synthesis.

Ptolemy's mathematical systematic treatise of astronomy, *The Mathematical Syntaxis*, soon attracted the appellation *megiste*, the Greek adjective for "greatest," which was transliterated into Arabic. With the addition of the definite article *al*, Ptolemy's complete exposition of mathematical astronomy became, upon passing from Arabic into Medieval Latin, the *Almagest*. Ptolemy's treatise was lost in the west soon after its completion, but was copied and studied in the Byzantine Empire. Manuscript versions in the original Greek and dating from the 9th century are extant, as are Arabic translations dating from the 11th century and later. So is a Latin translation from Arabic made in 1175. This was rendered, in 1515, into the first printed version of the *Almagest*.

The *Almagest* begins with a brief introduction to the nature of astronomy and a presentation of the necessary trigonometric theory and spherical astronomy. Then come theories of the Sun and the Moon, an account of eclipses, discussion of the fixed stars, and finally a discussion of the planets. The motivation underlying Ptolemy's study is found in an epigram:

> I know that I am mortal and the creature of a day; but when I search out the massed wheeling circles of the stars, my feet no longer touch the Earth, but, side by side with Zeus himself, I take my fill of ambrosia, the food of the gods.

Ptolemy proposed to begin with reliable observations and attach to this foundation a structure of ideas using geometrical proofs. Had he completely replaced the Greek deductive geometrical science he inherited with an inductive observational procedure, the result would have been a scientific revolution. However, determining the reliability of observations other than from their agreement with the very theory that were to be used to confirm proved a major problem for Ptolemy.

Ptolemy sought to explain the apparent irregularity of the Sun's motion as a combination of regular circular motions (defined as motions that cut off equal angles in equal times at the center of the circle). In the eccentric hypothesis, the circle carrying the Sun was not centered on the Earth, and thus regular motion as viewed from the center of the Sun's orbit appeared irregular when viewed from the Earth. In the epicycle hypothesis, a small circle (the epicycle) had its center fixed on a large circle (the deferent), and the combination of their regular motions was irregular. In the case of the Sun, either hypothesis could produce the observed motion. The hypotheses were interchangeable in a mathematical sense, though not in a physical sense.

Ptolemy next presented a table of the motions of the Moon and showed that the eccentric circle and the epicycle hypotheses could produce the same appearances. He reported lunar observations of greater accuracy than ever made before, using an astrolabe, which he described in detail.

To reproduce the more complex movements of the planets, Ptolemy found it necessary to employ both hypotheses simultaneously. Furthermore, difficulties in matching theory to observation eventually forced Ptolemy to violate his own definition of uniform circular motion. The violation would become one of the major causes of dissatisfaction with Ptolemy's system, leading to **Nicolaus Copernicus**' revision and revolution 14 centuries later.

In several instances in the *Almagest*, reported observations match corresponding theory more accurately than could be expected of random observations subject to probable errors, and Ptolemy is thus suspected of having fabricated the purported observations. There exists, however, a close agreement between Ptolemy's numerical parameters and modern observational values, so Ptolemy must have had a large, if unreported, body of real observations from which he derived his accurate parameters. Once a theory and its quantitative parameters were determined from a large body of real observations, Ptolemy next might have selected from among the observations a few in best quantitative agreement with the theory and then presented these examples to illustrate – not necessarily to determine, or to prove – the theory. Furthermore, Ptolemy was working within the tradition of Greek geometrical astronomy originally concerned almost totally with geometrical procedure and very little, if at all, with specific numerical results. The objective of the *Almagest* could have been didactic. Ptolemy may not have intended to deceive his readers, but he was less than candid concerning the manner in which he arrived at his results and was most remiss if his conduct is judged against the ethics of modern science.

Another question involving Ptolemy and his astronomy is whether the many circular motions compounded to determine the trajectory of a planet had a physical reality. Did Ptolemy envision actual physical structures in the heavens carrying around the planets? Or were his planetary theories merely means of calculating the apparent places of the planets without pretending to represent the true system of the world? His lunar theory predicted the Moon's positions in longitude and latitude accurately but greatly exaggerated the monthly variation in the Moon's distance from the Earth. Hence, the argument goes, Ptolemy could not have intended that the theory be interpreted realistically. In one of his other books, however, Ptolemy showed a concern with the physical world. In the *Planetary Hypotheses* he nested the mechanism of circles for each planet inside a spherical shell between adjoining planets. And a passage in the *Almagest* is susceptible to the interpretation that a construction in the heavens made not of wood, nor of metal, nor of other earthly material, but of some divine celestial material offering no obstruction to the passage of one part of the construction through another, controlled the motions of the planets.

After all is said – the charges of fraud leveled, the scientific shortcomings revealed, and the unanswerable questions exhausted – the historical influence and significance remain. Ptolemy's *Almagest* was the culmination of Greek astronomy, unrivaled in Antiquity, surpassing all that had gone before, and not itself surpassed for some 1400 years, until the time of Copernicus.

Norriss S. Hetherington

Alternate name

Claudius Ptolemaius

Selected References

Britton, John Phillips (1992). *Models and Precision: The Quality of Ptolemy's Observations and Parameters*. New York: Garland.

Evans, James (1993). "Ptolemaic Planetary Theory," and "Ptolemy's Cosmology." In *Encyclopedia of Cosmology: Historical, Philosophical, and Scientific Foundations of Modern Cosmology*, edited by Norriss S. Hetherington, pp. 513–526 and 528–544. New York: Garland.

Graßhoff, Gerd (1990). *The History of Ptolemy's Star Catalogue*. New York: Springer-Verlag.

Hetherington, Norriss S. (1996). "Plato and Eudoxus: Instrumentalists, Realists, or Prisoners of Themata?" *Studies in History and Philosophy of Science 27*: 271–289.

——— (1997). "Ptolemy on Trial for Fraud: A Historiographical Review." *Astronomy and Geophysics* 38, no. 2: 24–27.

——— (1999). "Plato's Place in the History of Greek Astronomy: Restoring *both* History *and* Science to the History of Science." *Journal of Astronomical History and Heritage* 2, no. 2: 87–110.

Neugebauer, Otto (1975). *A History of Ancient Mathematical Astronomy*. Pt 1. New York: Springer-Verlag.

Newton, Robert R. (1977). *The Crime of Claudius Ptolemy*. Baltimore: Johns Hopkins University Press.

Pedersen, Olaf (1974). *A Survey of the* Almagest. *Acta Historica Scientiarum Naturalium et Medicinalium, Edidit Bibliotheca Universitatis Haunienis* 30. Denmark: Odense University Press.

Ptolemy (1952). *The Almagest*, translated by R. Catesby Taliaferro. In *Ptolemy, Copernicus, Kepler*. Vol. 16 of *Great Books of the Western World*. Chicago: Encyclopaedia Britannica.

——— (1984). *Ptolemy's Almagest*, translated and annotated by G. J. Toomer. New York: Springer-Verlag.

Taub, Liba Chaia (1993). *Ptolemy's Universe: The Natural Philosophical and Ethical Foundations of Ptolemy's Astronomy*. Chicago: Open Court.

Puiseux, Pierre-Henri

Born **Paris, France, 20 July 1855**
Died **Fontenay, Jura, France, 28 September 1928**

Pierre Puiseux produced a series of studies on the Moon, addressing its motion, its surface, and its internal structure. He was the son of Viktor Puiseux (1820–1883), an astronomer and mathematician as well as a member of the Academy of Sciences, and of Laure Louise Jeannet (died: 1858). In 1865, Puiseux began to attend the high school in Saint-Louis; in 1875, he became a university student at the École Normale Supérieure, where he received his Ph.D. in mathematical sciences in 1879. On 20 June 1883, he married Laurence Elisa Marie Béatrice Bouvet, with whom he had seven children. Puiseux was well-known as a mountain climber. He succeeded in several first ascents and contributed to the exploration of several mountains in France, Switzerland, Sweden, and Italy. He frequently published mountaineering articles in *Annuaire du Club alpin français* as well as in the journal *Montagne*.

Puiseux worked his way up the career ladder at the Observatory of Paris, starting as a student astronomer (1879) before taking the positions of assistant astronomer (1881), associate astronomer (1885), astronomer (1904), and, after his retirement in 1917, honorary astronomer. From 1880, he lectured at the Faculty of Science in Paris, since 1897 as a honorary professor. In 1904, Puiseux was appointed to a lectureship in celestial mechanics.

In his dissertation of 1879, Puiseux analyzed the secular acceleration of the Moon. As an astronomer at the observatory, he devoted his studies to the photography of celestial bodies, to the realization of an international celestial map, and to the analysis of the Moon's topography and libration. He was sent abroad for scientific missions several times, for example, to Fort-de-France (Martinique) in 1882 for the observation of the transit of Venus, and to Cistierna (Spain) in 1905 for the observation of a solar eclipse. In addition to his books on the internal structures of the Earth and the Moon, and his photographic atlas of the Moon, Puiseux wrote numerous treatises published in *Annales de l'Observatoire de Paris*, in *Comptes-rendus de l'Academie des sciences*, and in the *Bulletin de la Société Astronomique de France*.

From 1906 to 1917, Puiseux supervised the photographic mapping of the sky. In 1907 and 1910, he served as secretary at the congresses of the International Union for Solar Research, which took place in 1910 at Mount Wilson Observatory in California. In 1909, he participated as the secretary at an international conference for the edition of celestial maps.

Puiseux was awarded the Valz Prize of the Academy of Sciences in 1892, the Lalande Prize of the Astronomical Society of France in 1896, and the Janssen Prize of the Academy of Sciences in 1908. He became knight of the Legion of Honor in 1900, president of the Astronomical Society of France in 1911, member of the Astronomy Section of the Academy of Sciences in 1912, associate member of the Royal Astronomical Society in 1917, and honorary president of the French Alpine Club. In 1935, a crater on the Moon was named for Puiseux to credit his scientific achievements (latitude 27°.8 S, longitude 39°.0 W, diameter 24 km).

Thomas Klöti
Translated by: *Andreas Verdun*

Selected References

Charle, Christophe and Eva Telkes (1989). *Les professeurs de la faculté des sciences de Paris: Dictionnaire biographique, 1901–1939*. Paris.

Collard, Auguste (1931). "Un astronome français, Pierre Puiseux (1855–1928)." *Bulletin de la Société belge d'astronome, de météorologie et de physique du globe* 1: 14–24; 2 (1931): 60–68.

Curinier, C. E. (1901–1918). *Dictionnaire national des contemporains*. 6 Vols. Paris: Office Général dÉdition.

Puiseux, P. (1908). *La Terre et la Lune, forme extérieure et structure interne*. Paris: Gauthier-Villars

Puiseux, P. and M. Löwy (1896–1910). *Atlas photographique de la Lune*. Paris: Imprimerie nationale.

Rivière, C. (1931). "Puiseux, Pierre." *Annuaire École normale supérieure* pp. 32–36.

Purcell, Edward Mills

Born **Taylorville, Illinois, USA, 30 August 1912**
Died **Cambridge, Massachusetts, USA, 7 March 1997**

American experimental physicist Edward Purcell is honored within astronomy for the 1951 discovery, with his student Harold C. Ewen, of the 21-cm line of neutral hydrogen, which had been predicted in 1944 by Henk C. van de Hulst, but it was for the 1945 discovery of nuclear magnetic resonance [NMR] that he shared the 1952 Nobel Prize in Physics with Felix Bloch, who had demonstrated the same phenomenon at nearly the same time, using a different technique.

Edward Purcell was the son of Elizabeth Mills, a high school Latin teacher, and Edward A. Purcell, the manager of the Taylorville telephone exchange (and later of the Illinois Southeastern Telephone Company). He probably owed some of his laboratory skills to very early familiarity with electrical equipment, including the magnetos of old crank telephones. Purcell earned a BS in electrical engineering at Purdue University in 1933, spent a year as a foreign exchange student at the Institute of Technology in Karlsruhe, Germany, and entered the Physics Department at Harvard University in 1934. He married Elizabeth Busser, who had also been an exchange student in Germany. Purcell received his Ph.D. in 1938 with a thesis on a spherical condenser mass spectrograph, carried out under Kenneth Bainbridge, but was also introduced to the quantum theory of magnetism by John H. van Vleck.

Purcell was appointed an instructor in the Harvard Department of Physics in 1938, rising through the ranks to assistant professor in 1941, associate professor in 1946, full professor in 1949, and the endowed Gerhard Gade University Professorship in 1961. The first, and probably most distinguished, of his Harvard Ph.D. students was Nicolaus Bloembergen (Nobel Prize in Physics, 1981; Gerhard Gade Professor Emeritus and professor of optical sciences at the University of Arizona). He continued to make significant contributions in the area of NMR until about 1955. The Purcells raised two sons, Frank and Dennis.

During World War II, Purcell worked full-time at the Massachusetts Institute of Technology's [MIT] Radiation Laboratory. He headed the Fundamental Development Group responsible for X-band and K-band radar techniques. Purcell contributed to two volumes of the well-known *MIT Radiation Laboratory Series*.

The associate director, I. I. Rabi, arranged numerous discussions and lectures at the end of the war, to assist the scientists at the Radiation Laboratory in their return to physics research and academic life. Purcell, Henry C. Torrey, and Robert V. Pound discussed the transitions of water molecules in the microwave K-band, as well as the nuclear spin magnetic resonance transitions, observed by Rabi and coworkers with molecular beam techniques. They conceived of the idea of looking for NMR in condensed matter. A literature search revealed that C. J. Gorter in the Netherlands had attempted to detect this phenomenon earlier without success. In December 1945, the three MIT scientists demonstrated the proton magnetic resonance at about 30 MHz in a magnetic field of about 7,000 oersted. They had borrowed the large electromagnet used by J. C. Street for cosmic ray research, in a shed adjacent to the Lyman Laboratory of Physics at Harvard.

Somewhat by accident, Purcell had heard of the van de Hulst prediction of a spectral feature produced by neutral hydrogen when the spin of the electron flips from parallel to antiparallel with the spin of the proton, and suggested a search for it to Harold C. Ewen, who was looking for a thesis topic where he could apply his background in microwave engineering. Ewen built a state-of-the-art microwave radiometer, and they put it at the focus of a small horn antenna, pointed at the sky from the top floor of the Lyman Laboratory of Physics. When the Milky Way passed overhead, the effective radiation temperature at 21 cm increased, leading to recognition of the predicted feature on 25 March 1951. Ewen and Purcell shared the 1988 Tinsley Prize of the American Astronomical Society for their discovery. The line is of enormous continuing importance in astronomy, because it enables us to trace out the amounts and motions of the most abundant element in the Universe, even when there are no stars nearby to light it up.

Van de Hulst was visiting Harvard at that time. Together with Professor **Jan Oort** and a microwave engineer C. H. Mueller, he had been working for some time on the same problem. They used a radio telescope antenna in the Netherlands that had belonged to the Germans during World War II. The Dutch group had experienced an experimental setback caused by a fire. When van de Hulst notified the group of the discovery at Harvard by Ewen and Purcell, they were quickly able to confirm it. Purcell asked the editor of the magazine *Nature* to postpone the publication of the letter he and Ewen had submitted, so that the results of the Dutch group could appear simultaneously. Purcell also had correspondence with Australian radio astronomer **Joseph Pawsey**, so three short announcements appeared in *Nature* **168**, 356 (1951) at the same time.

Purcell remained interested in astronomy to the end of his career, working on the properties of interstellar dust particles in the 1960s and 1970s and on the orientation of rotating dust grains in the magnetic field of interstellar space with **Lyman Spitzer, Jr**.

His wide-ranging intellectual curiosity is evident from two other scientific explorations outside the field of astronomy. He spent considerable time in the search for the magnetic monopole postulated by **Paul Dirac**. Purcell told the author that his unpublished notes on this subject would read like a fable "Hunt for the Unicorn." No evidence for the existence of the magnetic monopole has been found to date.

Purcell's interest in bacterial locomotion met with much more success. It was the result of collaboration with Howard C. Berg, who had been a junior fellow in the Society of Fellows at Harvard, while Purcell was a senior fellow. Berg demonstrated that *E. coli* bacteria move not by a reciprocal motion of their flagella (tails), but rather by a helical twisting motion. Purcell realized that viscous forces completely dominate inertial effects. It is the hydrodynamic regime of low Reynolds number. His paper and popular talk "Life at Low Reynolds Number" bear witness to his superb teaching ability and physical insight. Berg and Purcell received the biological Physics Prize of the American Physical Society in 1984.

In addition to undertaking wide-ranging research, Purcell was an outstanding teacher, receiving the 1968 Oersted Medal of the American Association of Physics Teachers and writing an outstanding textbook on electricity and magnetism and a series of short articles illustrating order-of-magnitude (envelope back) calculations of many different physical phenomena. He was a member of the Science Advisory Board for the US Air Force and of the Presidential Science Advisory Committee [PSAC] under presidents Eisenhower, Kennedy, and Johnson. Purcell received the National Medal of Science in 1969, and was elected to the United States National Academy of Sciences, the American Philosophical Society, and the Royal Society (London) as a foreign honorary member. He served as president of the American Physical soceity in 1969.

Nicolaas Bloembergen

Selected References

Berg, H. C. and E. M. Purcell (1977). "Physics of Chemoreception." *Biophysical Journal* 20: 193–219.

Bloembergen, N., E. M. Purcell, and R. V. Pound (1948). "Relaxation Effects in Nuclear Magnetic Resonance Absorption." *Physical Review* 73: 679–712. (Often referred to as BPP.)

Ewen, H. I. and E. M. Purcell (1951). "Observation of a Line in the Galactic Radio Spectrum: Radiation from Galactic Hydrogen at 1,420 Mc./sec." *Nature* 168: 356–358.

Mattson, James and Merrill Simon (1996). "Edward M. Purcell." In *The Pioneers of NMR and Magnetic Resonance in Medicine*. Ramat Gan, Israel: Bar-Ilan University, pp. 171–260. (Purcell said he was pleased with this account and that it accurately reflected his scientific endeavors.)

Pound, Robert V. (2000). "Edward Mills Purcell." *Biographical Memoirs, National Academy of Sciences* 78: 183–204.

Purcell, E. M. (1969). "On the Alignment of Interstellar Dust." *Physica* 41: 100–127.

——— (1977). "Life at Low Reynolds Number." *American Journal of Physics* 45: 3–11.

Purcell, E. M. and Paul R. Shapiro (1977). "A Model for the Optical Behavior of Grains with Resonant Impurities." *Astrophysical Journal* 214: 92–105.

Purcell, E. M. and L. Spitzer, Jr. (1971). "Orientation of Rotating Grains." *Astrophysical Journal* 167: 31–62.

Purcell, E. M., H. C. Torrey, and R. V. Pound (1946). "Resonance Absorption by Nuclear Magnetic Moments in a Solid." *Physical Review* 69: 37–38.

Pythagoras

Born **Samos, (Greece), *circa* 570 BCE**
Died **Metapontum or Croton (Crotone, Calabria, Italy), *circa* 480 BCE**

Pythagoras was a curious combination of a charismatic guru and mathematical genius, who founded an influential movement characterized by belief in reincarnation, moral and religious purity, and a predilection for numerical explanation. The Pythagorean doctrine that the nature of things consists in mathematical structure led both to **Plato**'s theory of forms and to Greek astronomers' lasting preference for simple, constant motion.

Born the son of Mnesarchus, a gem engraver, the young Pythagoras traveled widely for many years, acquiring both scientific information and religious lore. He is reported to have heard **Thales** and **Anaximander** and to have studied with Egyptian priests and with the Chaldeans. These undoubtedly encouraged his mathematical and astronomical interests. On the religious side, Pythagoras associated with Zoroasterism or the Magi, was initiated into numerous mysteries, and went into the cave on Mount Ida with Epimenides, a famous Cretan miracle-worker. **Heraclitus**, a champion of the empirical search for natural regularities and despiser of omnivorous polymathy, accused Pythagoras of patching together an idiosyncratic pseudo-wisdom from these eclectic sources.

Returning to Samos when he was about 40 and finding the rule of the tyrant Polycrates disagreeable, Pythagoras immigrated to Croton. An eloquent speaker, he quickly achieved prominence, preaching virtue, self-control, and a simple lifestyle to various audiences, including – contrary to custom – married women apart from their husbands and boys apart from their parents. He established a quasi-religious, quasi-philosophical organization, under whose leadership the city prospered so conspicuously that Pythagoreans rose to power in several neighboring cities. Pythagoreans were divided into a larger body of *acusmatici* (hearers) and a smaller inner circle of *mathematici* (learners) who adopted a moderately ascetic life characterized by mathematical and astronomical research, secret doctrines, abstinence from animal food, communal living, and various purificatory rituals connected with belief in reincarnation. Around 500 BCE Kylon, a rejected applicant to the order, led a coup in which many prominent Pythagoreans were assassinated. Pythagoras escaped, finding refuge in Metapontum. He is variously said to have lived to the age of 80, 90, or 100, and may have returned to Croton.

Although **Diogenes** disputed it, most likely Pythagoras, like Socrates and Jesus, wrote nothing. Because Pythagoreans kept their more innovative doctrines secret and honorifically attributed later discoveries to the founder, identifying a doctrine as Pythagoras's own is highly conjectural. Clearly he instituted a way of life based on a core of distinctive cosmological and anthropological beliefs. Pythagoras is reported to have been the first to call the Universe *kosmos* (although **Anaximander** certainly used the term), meaning that it is an ordered whole. The Universe is not mindless matter but a living god, breathing in the surrounding emptiness. Human souls are alienated portions of the divine world-soul, immortal but repetitively embodied in various forms, including nonhuman species. Pythagoras taught the kinship of all life; accordingly, he regarded the animal sacrifices pervasive in Greek religion as parricidal. Pythagoreans even avoided eating anything that seemed to contain soul, such as fava beans.

Pythagoras, using the monochord, discovered that the musical scale instantiated certain whole-number ratios: the octave (2:1), the fifth (3:2), and the fourth (4:3). Because the integers in these ratios add up to ten, ten was regarded as the perfect number and the key to all mathematical truth. He evidently generalized the idea that music consists of numbers to everything else. (**Aristotle** wrote that the Pythagoreans claimed that all things – perceptible objects, souls, and even moral qualities – *are* numbers.) There are two fundamental principles: limit and unlimited. The cosmos and its contents are products of the imposition of limit on the unlimited, various kinds of thing being distinguished by different numerical formulae. Various qualities, including moral virtues, are just special types of mathematical harmony. Living an orderly, nonviolent, scholarly life insures better rebirths and eventual reunion with the cosmic soul. Pythagoras called this manner of living "philosophy" and himself a "philosopher," originating a sense of those terms subsequently adopted by Plato. Pythagoreans allegedly suppressed the proof that the hypotenuse of a right triangle whose other sides were numbers (*i. e.*, integers) could not be a number, since this anomaly contradicted their thesis that all things, including spatial magnitudes, are numbers. Plato subsequently avoided the problem by constructing the cosmos from geometrical, rather than arithmetical, units.

In applying his seminal discovery to astronomy, Pythagoras imputed musical form, as well as mathematical order, to the heavens. The idea behind the doctrine of the "harmony of the spheres" seems to be that the movements of such huge objects must make sounds; and, as the pitch of a vibrating string varies with its length, the pitch of a celestial object varies with the radius of its orbit. Pythagoras may have assigned the same three ratios he identified for strings to the stars, the Sun, and the Moon. Later Pythagoreans assigned other musical ratios to each of the known planets. We do not hear this celestial music because it is omnipresent and invariant, and our auditory apparatus only detects vibratory change. Obviously, music is temporal, and time itself seems to be a serial order of events. For Pythagoreans, time consists in the repetitive orbitings of the celestial bodies, themselves embodiments of harmonious numerical arrangements. In later Antiquity the "harmony of the spheres" was mostly a curiosity; but Pythagoras's belief that heavenly bodies must move in a constant and mathematically elegant way, despite observational evidence to the contrary, continued to dominate astronomical theory. The centuries-long project of "saving the appearances" of planetary retrograde motion by hypothesizing uniformly rotating nested spheres (**Eudoxus**, Aristotle) or a combination of uniformly rotating deferents and epicycles (**Apollonius**, **Hipparchus**, **Ptolemy**, even **Nicolaus Copernicus**) testifies to the persistence of Pythagoras's preference for uniform motion.

James Dye

Selected References

Burkert, Walter (1972). *Lore and Science in Ancient Pythagoreanism*, translated by Edwin L. Minar, Jr. Cambridge, Massachusetts: Harvard University Press, chap. 4.

Diogenes Laertius (1925). *Lives of Eminent Philosophers*, translated by R. D. Hicks. Vol. 2. Loeb Classical Library, no. 185. Cambridge, Massachusetts: Harvard University Press. Book 8, Chap. 1.

Guthrie, W. K. C. (1962). "Pythagoras and the Pythagoreans." In *A History of Greek Philosophy*. Vol. 1, *The Earlier Presocratics and the Pythagoreans*, pp. 146–340. Cambridge, Massachusetts: Cambridge University Press.

Kirk, G. S., J. E. Raven, and M. Schofield (1983). "Pythagoras of Samos." In *The Presocratic Philosophers: A Critical History with a Selection of Texts*, pp. 214–238. 2nd ed. Cambridge: Cambridge University Press.

Q

Qabīṣī: Abū al-Ṣaqr ʿAbd al-ʿAzīz ibn ʿUthmān ibn ʿAlī al-Qabīṣī

Flourished **(Iraq), second half of the 10th century**

Qabīṣī, an astronomer and astrologer, came from one of two villages called Qabīṣa in Iraq. He studied **Ptolemy**'s *Almagest* under ʿAlī ibn Aḥmad al-ʿImrānī of Mosul, a mathematician and teacher, and dedicated several works (nos. 2, 3, 4, and 6, as given below) to Sayf al-Dawla, the Ḥamdānid Emir of Aleppo between 945 and 967. Otherwise, details of Qabīṣī's life are little known.

Qabīṣī's extant works are the following:

(1) A commentary on **Farghānī**'s *Kitāb al-fuṣūl* (also referred to as *Kitāb fī jawāmiʿ ʿilm al-nujūm*).

(2) A treatise on the distances and volumes of the planets (*Risāla fī 'l-abʿād wa-'l-ajrām*). This treatise provides distances and volumes for the planets other than those of the Sun and the Moon, which had already been given in the *Almagest*. Qabīṣī's account of Mercury was quoted twice by **Bīrūnī** in his *al-Qānūn al-masʿūdī* (Vol. X, Chap. 6).

(3) Book on the introduction to astrology (*Kitāb al-mudkhal ilā ṣināʿat aḥkām al-nujūm*), comprising five chapters. Qabīṣī's most famous work, this book is preserved in several Arabic manuscripts and in a Latin translation of which there are more than 200 manuscripts as well as 12 editions printed between 1473 and 1521. His text was the main book used in universities in the medieval Latin world where astrology was taught as part of the curriculum in medicine.

(4) A treatise for the examination of astrologers (*Risāla fī imtiḥān al-munajjimīn*). This treatise contains 30 astronomical or astrological questions and answers. Qabīṣī divides astrologers into four categories according to their intellectual level: The complete astrologer; the one who knows facts such as the shape of the celestial sphere but can not prove them; the astrologer who accepts things uncritically, like a blind man – the majority of astrologers fall into this category; and one who does not know anything about astronomy and astrology, relying only upon the operations of instruments.

(5) A work on the conjunction of the planets in the zodiacal signs and their prognostications for the revolutions of the years is attributed to Qabīṣī in Latin (*De coniunctionibus planetarum in duodecim signis et earum pronosticis in revolutionibus annorum*).

(6) A mathematical work in Arabic on numbers.

Qabīṣī wrote several other works that are not extant. We know of them because he refers to them in his surviving works. These include a treatise on the size of the Earth, referred to in (2) and (6) as *Risāla fī masāfat al-arḍ*, part of which is quoted at the end of (6); a book on the explanations of astronomical tables, referred to in (2) as *Kitāb fī ʿilal al-zījāt*; a book on affirming the validity of astrology, referred to in the preface of (3) as *Kitāb fī ithbāt ṣināʿat aḥkām al-nujūm*, which was a response to the criticism of astrology by ʿAlī ibn ʿĪsā, an astronomical instrumentmaker of the 9th century; *Kitāb fī al-namūdārāt, i. e.*, a book on the *namūdārs*, the method to fix a person's ascendant when the time of birth is unknown, referred to in the fourth chapter of (3); and a book referred to in the introduction of (4) as *Shukūk al-Majisṭī* (Doubts on the *Almagest*).

Keiji Yamamoto

Alternate name

Alcabitius

Selected References

Al-Qabīṣī (2004). *The Introduction to Astrology*, edited and translated by Charles Burnett, Keiji Yamamoto, and Michio Yano. London: Warburg Institute.

Ibn al-Nadīm (1970). *The Fihrist of al-Nadīm: A Tenth-Century Survey of Muslim Culture*, edited and translated by Bayard Dodge. 2 Vols. New York: Columbia University Press.

Pingree, David (1975). "Al-Qabīṣī." In *Dictionary of Scientific Biography*, edited by Charles Coulston Gillispie. Vol. 11, p. 226. New York: Charles Scribner's Sons.

——— (1978). "Al-Ḳabīṣī." In *Encyclopaedia of Islam*. 2nd ed., Vol. 4, pp. 340–341. Leiden: E. J. Brill.

Sesiano, Jacques (1987). "A Treatise by al-Qabīṣī (Alchabitius) on Arithmetical Series." In *From Deferent to Equant: A Volume of Studies in the History of Science in the Ancient and Medieval Near East in Honor of E. S. Kennedy*, edited by David A. King and George Saliba, pp. 483–500. Annals of the New York Academy of Sciences, Vol. 500. New York: New York Academy of Sciences.

Sezgin, Fuat. *Geschichte des arabischen Schrifttums*. Vol. 5, *Mathematik* (1974): 311–312; Vol. 6, *Astronomie* (1978): 208–210; Vol. 7, *Astrologie – Meteorologie und Verwandtes* (1979): 170–171. Leiden: E. J. Brill.

Yāqūt, Shihāb al-Dīn Abū ʿAbd Allāh al-Rūmī (n.d.). *Mujʿam al-buldān*. Vol. 4, pp. 308–309. Beirut.

Qāḍīzāde al-Rūmī: Ṣalāḥ al-Dīn Mūsā ibn Muḥammad ibn Maḥmūd al-Rūmī

Born **Bursa, (Turkey), *circa* 1359**
Died **Samarqand, (Uzbekistan), after 1440**

Qāḍīzāde al-Rūmī was known for his works in mathematics and astronomy, which were used extensively as teaching texts. He left his native Bursa, where his grandfather had been a prominent judge and his father an eminent scholar, and traveled to Persia in order to gain a higher level of proficiency in the philosophical and mathematical sciences. His nickname indicates his family's standing (Qāḍīzāde = son of the judge) and his origins (Rūmī = from what had been part of the eastern Roman Empire). He studied with many learned scholars in Khurāsān and Transoxiana, among whom was the famous theologian **al-Sayyid al-Sharīf al-Jurjānī** at the court of Tīmūr in Samarqand. Qāḍīzāde, however, felt that Jurjānī was deficient in the mathematical sciences. After Tīmūr's death, Qāḍīzāde found both a student and a patron in Tīmūr's grandson **Ulugh Beg**, also in Samarqand.

Qāḍīzāde joined a group of scholars in the circle of Ulugh Beg that taught mathematics and astronomy, as well as other sciences. He became the head of the madrasa (school) of Samarqand, and Ulugh Beg often attended his lectures. Qāḍīzāde also became one of the directors of the Samarqand Observatory after the death of **Jamshīd al-Kāshī** in 1429, and he undertook its observational programs assisted by **ʿAlī al-Qūshjī**, who continued the program after Qāḍīzāde's death.

Qāḍīzāde was not known for his innovations or creativity. He was most famous for his commentaries on **Maḥmūd al-Jaghmīnī**'s astronomical compendium entitled *al-Mulakhkhaṣ fī ʿilm al-hay'a al-basīṭa* (1412) and **Shams al-Dīn al-Samarqandī**'s geometrical tract *Ashkāl al-ta'sīs* (completed: 1412); the large number of extant manuscripts of both commentaries indicates their enduring popularity as teaching texts. Therefore, it is not surprising that one also finds supercommentaries on Qāḍīzāde's commentaries written by many scholars–teachers including Sinān Pāshā (died: 1486), **ʿAbd al-ʿAlī al-Bīrjandī** (died: 1525/1526), **Bahā' al-Dīn al-ʿĀmilī** (died: 1621), and Qāḍīzāde's student **Fatḥallāh al-Shirwānī** (died: 1486). All of these individuals continued the tradition established at Samarqand, thereby disseminating the mathematical sciences throughout Ottoman and Persian lands. Also noteworthy is that the marriage of Qāḍīzāde's son to Qūshjī's daughter would eventually sire the famous Ottoman astronomer–mathematician **Mīram Čelebī** (died: 1525).

A number of other astronomical works are sometimes attributed to Qāḍīzāde, including a supercommentary on Ṭūsī's commentary (*taḥrīr*) of the *Almagest* and a treatise on the sine quadrant, but it is not clear which of these are authentic. The ascription of a commentary (*Sharḥ*) on **Naṣīr al-Dīn al-Ṭūsī**'s major astronomical work *al-Tadhkira fī ʿilm al-hay'a* (Biblioteca Medicea Laurenziana or. MS 271) to Qāḍīzāde is certainly not correct; this manuscript is actually an incomplete copy of the commentary by Jurjānī.

Among Qāḍīzāde's mathematical works is a treatise on determining the value of sin 1°, for which he seems to have relied heavily on the work of Kāshī. Qāḍīzāde's only philosophical or theological work is a supercommentary on **Athīr al-Dīn al-Abharī**'s *Hidāyat al-ḥikma*, although he intended to write a refutation of parts of Jurjānī's famous commentary on the Persian ʿAḍud al-Dīn al-Ījī's (*circa*: 1281–1355) *Mawāqif.*

F. Jamil Ragep

Selected References

Bagheri, Mohammad (1997). "A Newly Found Letter of Al-Kāshī on Scientific Life in Samarkand." *Historia Mathematica* 24: 241–256.
Brockelmann, Carl. *Geschichte der arabischen Litteratur.* 2nd ed. Vol. 1 (1943): 616–617, 624, 674–675; Vol. 2 (1949): 275, 276; Suppl. 1 (1937): 850, 865. Leiden: E. J. Brill.
Fazlıoğlu, İhsan (2003). "Osmanlı felsefe-biliminin arkaplanı: Semerkand matematik-astronomi okulu." *Dîvân İlmî Araştırmalar* 1: 1–66.
Kennedy, E. S. (1960). "A Letter of Jamshīd al-Kāshī to His Father: Scientific Research and Personalities at a Fifteenth Century Court." *Orientalia* 29: 191–213. (Reprinted in E. S. Kennedy, *et al.* (1983). *Studies in the Islamic Exact Sciences,* edited by David A. King and Mary Helen Kennedy, pp. 722–744. Beirut: American University of Beirut.)
Rosenfeld, B. A. and Ekmeleddin Ihsanoğlu (2003). *Mathematicians, Astronomers, and Other Scholars of Islamic Civilization and Their Works (7th–19th c.).* Istanbul: IRCICA, pp. 272–274.
Sayılı, Aydın(1960). *The Observatory in Islam.* Ankara: Turkish Historical Society.
Ṭashköprüzāde (1985). *Al-Shaqāi'q al-nuʿmāniyya fī ʿulamā' al-dawlat al-ʿuthmāniyya,* edited by Ahmed Subhi Furat. Istanbul: Istanbul University. pp. 14–17.

Qāsim ibn Muṭarrif al-Qaṭṭān: Abū Muḥammad Qāsim ibn Muṭarrif ibn ʿAbd al-Raḥmān al-Qaṭṭān al-Ṭulayṭulī al-Qurṭubī al-Andalusī

Flourished **Cordova, (Spain), 10th century**

Qāsim ibn Muṭarrif al-Qaṭṭān may well represent the earliest astronomers in Islamic Spain (al-Andalus) of whom we have knowledge. Though known as a reciter of the Quran (*muqri'*) and traditionalist with the sobriquet *al-shaykh al-ra'īs* (Principal Shaykh), only one of his works is extant, a study of cosmological and astronomical subjects. However, in the biographical dictionaries there is no reference to Qaṭṭān's interest in cosmology or astronomy. From what we know of the lives of his teachers, we can deduce that he was born at the end of the 9th or the beginning of the 10th century. An analysis of Qaṭṭān's work offers only two chronological details: a quotation from **Maslama al-Majrīṭī** and the following statement in the title of the star table: "We found its longitude in the ecliptic in the year 300 of the Hijra" (912–913).

If the attribution is correct, Qaṭṭān's work, entitled *Kitāb al-hay'a* (Book on cosmology), would be the first extant Andalusī treatise on astronomy. The only known manuscript is preserved in the Süleymaniye Library in Istanbul (Carullah Efendi 1279, folios 315r–321v). The work is a compendium of all the Andalusī cosmological and astronomical knowledge of the time and draws upon a variety of traditions. The most prominent is that of eastern Islam, which flourished in the 10th century after successfully combining the old cosmology and astronomy of Greece and India. There are also echoes of an old Latin astrological tradition, still in use in the Iberian Peninsula.

The text consists of 30 numbered and five unnumbered chapters. The unnumbered chapters differ from the others in several aspects and do not seem to belong to the work. The chapters are as follows: 1–8: signs of the Zodiac and lunar mansions; 9–11: planets and cosmographical subjects; 12: stars; 13–16: Moon and Sun; 17–27: subjects related to the calendar, *i. e.*, years, months, days, hours (22 and 23 are devoted to clocks); and 28–30: description of the cosmos, both the superlunary and sublunary world. The other five chapters – the ones without numbering – deal with the Sun, the Moon, and terrestrial latitudes. Some of the chapters that purport to explain the physical structure of the cosmos show a clear dependence on **Aristotle**, while others draw upon **Ptolemy**, in particular on his *Planetary Hypothesis*, which very probably reached the author through the *Kitāb al-aʿlāq al-nafīsa* of the eastern geographer Abū ʿAlī Aḥmad ibn ʿUmar ibn Rustah.

The clocks described are a sundial, the description of which coincides almost word for word with the one found in the *Kitāb al-asrār fī natā'ij al-afkār* (Biblioteca Medicea Laurenziana, MS Misc. Or 152, folio 47r), explicitly attributed to **Ibn al-Ṣaffār**, one of Maslama's disciples. This clock is unlike extant Islamic clocks, although we know of at least two texts that describe a similar instrument (the *balāṭa* described in the *Zīj* by **Ibn Isḥāq al-Tūnisī** and the one in the commentary to the *Mišná* by **Maimonides**). These clocks seem to date back to biblical times. The second one, called *thurayya*, is a "fire clock" because the hours are indicated by the burning of oil. A description of a similar clock is found in a work by a certain Yūnus al-Miṣrī. Qaṭṭān's clock derives from a clock calculated for Baghdad, which probably reached al-Andalus through Tunis, perhaps thanks to the well-known epistolary relationship between Ḥasdāy ibn Shaprut of Cordova, and the Tunisian **Dunash ibn Tamīm**.

The star table contains 16 stars. It is a standard table of the kind that accompanies a treatise for constructing an astrolabe, although, in view of the errors found, it was probably derived from a reading of the coordinates of an instrument calculated for Cordova, namely ecliptic coordinates (longitude and latitude) and the degree of the zodiac that rises with a star and diurnal arc. It is the first star table documented in al-Andalus and is clearly influenced by **Battānī** and Maslama.

In a number of chapters there are signs that the author does not have a thorough understanding of the field. However, the work is important because it demonstrates the emergence of astronomical and cosmological knowledge from a range of traditions in 10th-century al-Andalus, the period in which science was beginning to develop in this area. Although the author is Andalusī, the manuscript is eastern, suggesting that it reached a fairly wide readership. The text is largely nonspecialist and was probably used in the non-scientific circles in which the author undoubtedly moved.

Mercè Comes

Selected References

Casulleras, J. (1993). "Descripciones de un cuadrante solar atípico en el occidente musulmán." *Al-Qanṭara* 14: 63–69.

——— (1994). "El contenido del *Kitāb al-Hay'a* de Qāsim b. Muṭarrif al-Qaṭṭān." In *Actes de les I Trobades d'Història de la Ciència i de la Tècnica*, edited by J. M. Camarasa, H. Mielgo, and A. Roca, pp. 75–93. Barcelona: Institut d'Estudis Catalans. Societat catalana d'Historia de la Ciència i de le Tècnica.

Comes, M. (1993–1994). "Un procedimiento para determinar la hora durante la noche en la Córdoba del siglo X." *Revista del Instituto Egipcio de Estudios Islámicos* 26: 263–272.

——— (1994)."La primera tabla de estrellas documentada en al-Andalus." In *Actes de les I Trobades d'Història de la Ciència i de la Tècnica,* edited by J. M. Camarasa, H. Mielgo, and A. Roca, 95–109. Barcelona: Institut d'Estudis Catalans. Societat catalana d'Història de la Ciència i de le Tēcnica.

Comes, M. (2006). "Ibn Muṭarrif al-Qaṭṭān, Qāsim." In *Biblioteca de al-Andalus. Enciclopedia de la Cultura andalusí. Fundación Ibn Tufaye. Amería* . Vol. 1, [893], 304–306.

Rosenthal, Franz (1955). "From Arabic Books and Manuscripts V: A One-Volume Library of Arabic Philosophical and Scientific Texts in Istanbul." *Journal of the American Oriental Society* 75: 14–23, esp. 21.

Samsó, Julio (1992). *Las ciencias de los antiguos en al-Andalus*. Madrid: Mapfre.

Sezgin, Fuat (1978). *Geschichte des arabischen Schrifttums*. Vol. 6, *Astronomie*, 197–198. Leiden: E. J. Brill.

Qaṭṭān al-Marwazī: ʿAyn al-Zamān Abū ʿAlī Ḥasan ibn ʿAlī Qaṭṭān al-Marwazī

Born **Marw (Merv, Turkmenistan), 1072/1073**
Died **Marw (Merv, Turkmenistan), October 1153**

Qaṭṭān al-Marwazī was a prominent scholar of the 11th and 12th centuries, whose only extant work, a treatise on astronomy, entitles him to be ranked among the leading observational astronomers of his age. He was born in Marw, an ancient city in Persia, which had become by then one of the most prosperous cities of Great Khurāsān, a vast and flourishing province on the eastern borders of the Islamic world and home to many outstanding scientists, philosophers, religious scholars, saints, and mystics. At the time Marw had ten large public libraries, one of them housing 12,000 books.

Living in a city with a rich cultural milieu, Qaṭṭān al-Marwazī grew up to become an expert in many fields of science and wisdom. Like other erudite and encyclopedic savants of the Islamic Middle Ages, he wrote books in most areas of knowledge including astronomy, medicine, prosody, engineering, and literature. His writings were regarded highly among the learned circles of Marw. Though well versed in different disciplines, Qaṭṭān al-Marwazī's main occupation was medicine.

Sources describe Qaṭṭān al-Marwazī as a master of Greek sciences and an ardent exponent of Greek philosophy. Being a student of Lawkarī, who himself was a student of Bahmanyār, the most distinguished disciple of **Ibn Sīnā**, Qaṭṭān al-Marwazī belongs to the third generation of scholars who have fully benefited from the Avicennian tradition.

None of Qaṭṭān al-Marwazī's numerous writings, however, have survived save a book on astronomy written in Persian and entitled *Gayhānshenākht* (Knowledge of the Cosmos). According to the author, the book was so titled because "he who understands this book will have a coherent knowledge of the configuration of the cosmos, and its system will be clear to him." The book, however, is not confined to cosmology in the proper sense of the term, but, as is usual for the works of its genre in the Islamic tradition, covers a wider range of subjects such as the configuration of the Earth and certain topics in geography. Therefore, it falls within the context of cosmographical works. Furthermore, the treatise also includes what we usually find in the works dedicated to the calendar and issues related to the "passage of time." The book, therefore, comprises a range of topics from the celestial movements, eccentrics and epicycles, apogees, planetary sectors, the ecliptic, the fixed stars, lunar and solar eclipses, the meridian, and the azimuth to the sizes of the Earth and other planets, chronology, and even some minor hints regarding astrology.

Gayhānshenākht may thus be placed within the corpus of what was known as *hay'a basīṭa, i. e.*, plain or simplified astronomy. These works were simplified forms and summaries of astronomy that gave a coherent and unified account of the discipline. The main audience for such works were ordinary, educated people for whom astronomy had a greater appeal than other sciences, in part because of its applications in religious matters, and in part because it dealt with the realm of the unknown. Therefore, despite the fact that Arabic was the prime language of science and letters throughout Islamdom, Qaṭṭān al- arwazī, out of an inner obligation, chose to write a simple and easy-to-understand book on astronomy in Persian for the educated public and for beginners who wished to have a share of the art.

Qaṭṭān al-Marwazī seems to have been involved in other aspects of astronomy. His status as an observational astronomer is well established by the fact that he mentions in several places his engagement with astronomical measurements. Furthermore, Qaṭṭān al-Marwazī claims to have written other books on astronomy, including a *zīj* or astronomical handbook, which requires direct participation of the observer. Nevertheless, his interest was not limited to pure astronomy, a science that in his view "is based on certitude and demonstration" and "into which no discrepancy shall find a way." He shows interest in astrology as well, which for him is "a science of analogy and conjecture." By this, however, Qaṭṭān al-Marwazī does not mean to belittle astrology but rather to place each within its own proper domain, since he promises to write a book on that subject, too.

Despite the very little information available to us about the man and his works, we may conclude that Qaṭṭān al-Marwazī was one of the most prominent scientific figures of his time. In a series of correspondences between him and Rashīd al-Dīn Waṭwāṭ, himself a great literary figure of his age, Rashīd al-Dīn Waṭwāṭ does not fail to acknowledge him as "a scholar for whom not even a minute replica can be found across either east or west" even though the author is being accused by Qaṭṭān al-Marwazī of plundering his library. Furthermore, his stature as a great astronomer may be substantiated by the fact that two centuries later Ibn Taymiyya, a renowned religious scholar in Damascus, singles out Qaṭṭān al-Marwazī's name as someone very skillful in astronomy, while discussing the question of lunar crescent visibility.

A clan of the Turkish Ghuzz (Oghuz) tribe from eastern Asia invaded Marw. Being taken captive, Qaṭṭān al-Marwazī is said to have shouted words of insult at his captors, which led to his tragic death. They tortured him to death by filling his mouth with soil.

Behnaz Hashemipour

Selected References

al-Bayhaqī, ʿAlī ibn Zaid (1935). *Tatimmat ṣiwān al-ḥikma*, edited by M. Shafīʿ. Lahore: University of Panjab.

al- Dhahabī, Shams al-Dīn, Muḥammad ibn Aḥmad (1995). *Tārīkh al-Islām wa-wafayāt al-mashāhīr wa-'l-aʿlām*. Beirut: Dār al-Kitāb al-ʿArabī.

al-Ṣafadī, Ṣalāḥ al-Dīn Khalīl ibn Aybak (1991). *Al-Wāfī bi-'l-wafayāt*. Stuttgart: Orient-Institut der deutschen morgenländischen Gesellschaft.

al-Suyūṭī, Jalāl al-Dīn ʿAbd al-Raḥmān (n.d.). *Bughyat al-wuʿāt fī ṭabaqāt al-lughawiyyīn wa-'l-nuḥāt*. Vol. 1, edited by Muḥammad Abū al-Faḍl Ibrāhīm. Beirut: Maktabat al-ʿAḥṣriyya.

al-Waṭwāṭ, Rashīd al-Dīn (1897). *Rasā'il*. Vol. 2. Egypt: Maṭbaʿat al-Maʿārif.

Hashemipour, Behnaz (in press). "Gayhānshenākht: A Cosmological Treatise" (in Persian). In *Proceedings of the 2nd International Conference on the Science in the Iranian World*, edited by N. Poorjavadi and Ziva Vesel. Tehran: Iran University Press.

Ibn, Taymiyya (1981). *Majmūʿ fatāwā*. Vol. 25, edited by al-Raḥmān al-ʿĀṣimī. Rabat: Maktabat al-Maʿārif.

Michot, Yahya M. (January 2001). "Pages spirituelles d'Ibn Taymiyya: Contre l'astrologie." *Action – Le Billet d'Oxford* 41: 10–11, 26.

Ragep, F. Jamil (1993). *Naṣīr al-Dīn al-Ṭūsī's Memoir on Astronomy (al-Tadhkira fī ʿilm al-hay'a)*. 2 Vols. New York: Springer-Verlag.

Storey, C. A. (1958). *Persian Literature*. Vol. 2, pt. 1. A. *Mathematics*. B. *Weights and Measures*. C. *Astronomy and Astrology*. D. *Geography*. London: Luzac and Co.

Yāqūt al-Ḥamawī al-Rūmī (1993). *Muʿjam al-udabā': Irshād al-arīb ilā maʿrifat al-adīb*. Vol. 3, edited by Iḥsān ʿAbbās. Beirut: Dār al-Gharb al-Islāmī.

Qian Lezhi

Flourished **China, 5th century**

Qian Lezhi, an astronomer of the Liu-Song dynasty (420–479) of the Southern Dynasties (420–589), made detailed armillary spheres that replicated the motion of the sky. About his life we know nothing except that he was the *Taishiling*, the highest official of the Imperial Bureau of Astronomy and Calendrics.

In 436, by command of the emperor, Qian Lezhi made a copper armillary sphere. Historical materials about the device show that the "armillary" was not a real armillary sphere, but a celestial globe demonstrating the apparent motions of celestial bodies. The device was constructed mainly of rings, with a small sphere in the center representing the Earth. Both diameters of the equatorial and ecliptic rings were 6.08 *chi*. (One *chi* equaled about 24 cm during this period.) The circumference of each ring was 18.26 *chi*. One *du* in the sky – ancient Chinese divided the circle into 365 *du* – corresponded to 0.05 *chi* along the circumference of the rings. There was a polar axis paralleling the rotation axis of the Earth. Semicircular rings representing the longitudes of the lunar mansions connected vertically to the equatorial ring. In the area of the North Celestial Pole, the Big Dipper and pole star were shown, but sources do not mention how they were fixed on the device. It is possible that dots representing the Big Dipper and the pole star appeared at the corresponding places. The Sun, the Moon, and five planets (Venus, Jupiter, Mercury, Mars, and Saturn) were fixed on the ecliptic ring.

Besides this device, there was a clepsydra or waterclock connected to the celestial globe, which was also driven by water from the water clock so that the celestial part of the globe moved synchronously with the rotation of the Earth. Therefore, the device could demonstrate the apparent motion of celestial bodies.

Qian Lezhi's thoughts on the design of the instrument were apparently influenced by astronomers **Zhang Heng**, and especially, Ge Heng (3rd century). Ge Heng made an "armillary sphere," which was actually also a moving celestial globe, with an unmoving sphere representing the Earth installed inside the globe itself.

In 440, Qian Lezhi made another, smaller "armillary sphere" that was 2.2 *chi* in diameter. On the surface of the sphere, the lunar mansions and other asterisms were indicated, and the Sun, Moon, and five planets were fixed on the ecliptic. In ancient atlases such as those of **Gan De**, **Shi Shen**, and Wu Xian, the stars were indicated by black dots. Adapting **Chen Zhuo**'s method, Qian Lezhi used white, black, and yellow pear shapes. Some sources say that other color combinations, such as red, black, and white, were used.

Li Di

Selected References

Shen Yue (441–513). *Song shu* (History of the [Liu-] Song Dynasty). Chap. 23 in *Tianwen zhi* (Monograph on astronomy and astrology). Part 1.

Wei Zhen (*circa* 629). *Sui Shu* (History of the Sui Dynasty). Chap. 23 in *Tianwen zhi* (Monograph on astronomy and astrology). Part 1.

Quetelet, Lambert Adolphe Jacques

Born **Ghent, (Belgium), 22 February 1796**
Died **Brussels, Belgium, 17 February 1874**

Lambert Quetelet became one of the most influential social statisticians of the 19th century, but is also remembered for his studies of meteor showers and their apparent radiants.

Quetelet was the son of François-Augustin-Jacques-Henri Quetelet and Anne Françoise Vandervelde. He was educated at the lyceum of his native town. In 1815, when that school was converted to the College of Ghent, he was appointed a professor of mathematics. In 1819 Quetelet was awarded the institution's first Ph.D., for a dissertation on the theory of conic sections. That same year, he was appointed professor of mathematics at the Athenaeum of Brussels, and was soon elected to the Royal Academy of Sciences and Arts. Quetelet was married in 1825; he and his wife had two children.

During the 1820s, Quetelet began a campaign to found an observatory in Belgium. Upon this suggestion, he was commissioned to go to Paris and study the practice of astronomy under **Dominique Arago**, director of the Paris Observatory. While there in December 1823, Quetelet observed a subdivision in the A-ring of Saturn, using the observatory's 10-in. achromatic refractor. He also learned probability theory from Jean Baptiste Joseph Fourier and **Pierre de Laplace**. Belgium's Royal Observatory was not completed until 1833, although Quetelet served as its director after 1828. There, he gave special attention to meteorological and geophysical observations.

Following the very intense Leonid meteor shower of 12/13 November 1833, Quetelet's attention was directed toward the occurrence and annual periodicity of meteor showers. In March 1837, he predicted the return of the Perseid meteors during the coming August. That same year, Quetelet produced the first catalog of historical sightings of this shower. Over time, he amassed more than 300 records of the appearance of this and other suspected meteor showers. Quetelet, however, was not alone in his recognition of the Perseid meteors. Two Americans, Connecticut bookseller **Edward Herrick** and Cincinnati physician **John Locke**, had independently documented the shower's annual nature. Quetelet is also regarded as the codiscoverer of the Orionid and Quadrantid meteor showers.

Quetelet's other astronomical (and meteorological) observations included numerous solar and lunar eclipses, planetary and stellar occultations, the aurora borealis (and its magnetic anomalies), comets, asteroids, and bolides. He determined the longitude difference between Brussels and other European observatories by means of telegraph signals. A catalog of more than 10,000 stars, observed by Quetelet and his associates between 1857 and 1878, was published by his son Ernest in 1887.

Quetelet's most famous work, *On Man and the Development of His Faculties: A Treatise on Social Physics* (1835), laid the foundations of sociology and introduced his concept of the "average man." In 1853, he organized the first international statistics conference and was appointed its president. Toward the end of his life, Quetelet published several histories of the physical and mathematical sciences in Belgium. At the time of his death, he was regarded as Belgium's most revered scholar.

Richard Baum

Selected References

Anon. (1875). "Lambert Adolphe Jacques Quetelet." *Monthly Notices of the Royal Astronomical Society* 35: 176–180.

Freudenthal, Hans (1975). "Quetelet, Lambert-Adolphe-Jacques." In *Dictionary of Scientific Biography*, edited by Charles Coulston Gillispie. Vol. 11, pp. 236–238. New York: Charles Scribner's Sons.

Hankins, Frank Hamilton (1908). *Adolphe Quetelet as Statistician*. New York: Columbia University Press.

Littmann, Mark (1998). *The Heavens on Fire: The Great Leonid Meteor Storms*. Cambridge: Cambridge University Press, esp. pp. 86–89.

Sauval, J. (1997). "Quetelet and the Discovery of the First Meteor Showers." *WGN: Journal of the International Meteor Organization* 25: 21–23.

Qunawī: Muḥammad ibn al-Kātib Sīnān al-Qunawī

Born **Probably Istanbul, (Turkey)**
Died **Istanbul, (Turkey), *circa* 1524**

Muḥammad al-Qunawī, astronomer and *muwaqqit* (timekeeper), lived in Istanbul and pioneered the Turcification movement of the Greco–Hellenic and classical Islamic astronomical literature. Very little is known about his life. However, Qunawī's name indicates that he came from Qunya (Konya, Turkey). Sinān, his father, served in the Ottoman State Chambers as a scribe, and so he became known as Ibn Kātib Sīnān, *the son of Sinān the Scribe*.

In his work entitled *Kitāb al-Aṣl al-muʿaddil,* Qunawī states that "he had met all the important astronomers of the time" (Istanbul Archeology Museum, MS 1255/4, 156b). These would have been from among his Ottoman intellectual circle of friends and students who had studied both the astronomical works of **ʿAlī al-Qūshjī**, thus connecting them with the mathematical–astronomical tradition of Samarqand, and the achievements of *ʿilm al-mīqāt* (astronomical timekeeping) of classical Islam, which had reached its apex with the works of **Khalīlī** and **Ibn al-Shāṭir** in 14th-century Damascus.

After completing his education, Qunawī worked for some time as the official *muwaqqit* in several religious institutions including the New (Yeni) Mosque in Edirne. In this capacity, he offered several works in the service of various Sultans: his *Hadiyyat al-mulūk* to Sultan Bāyazīd II, his *Faḍl al-dā'ir* to Sultan Selīm I, and his *Mīzān al-Kawākib* to Sultan Süleymān I (the Magnificent).

Qunawī wrote 11 books on astronomy: seven in Arabic and four in Turkish. Thus his works were not confined to the Turkish-speaking areas of Istanbul, the Balkans, and Anatolia, but could be used in Arabic-speaking areas, such as Cairo, Egypt, as well. Qunawī's works in Turkish provide us with insight into the growing needs of the Ottoman state bureaucracy. In fact, the word *al-Ihkwān* (usually meaning "brothers"), mentioned in the title of his Turkish book *Hadiyyat al-ihkwān,* actually refers to the *muwaqqit*s, who were part of this bureaucracy. Qunawī's Turkish writings helped inculcate an attitude among Ottoman astronomers that contributed to the translation of the Hellenic and Islamic astronomical heritage from Arabic and Persian into Turkish from the beginning of the 16th century onward and paved the way for the Turcification of the language of astronomy.

Most of Qunawī's works were devoted to timekeeping and astronomical instruments. He was thus following one particular tradition of Islamic astronomy whereby it was "in service" to religious, administrative, and social needs of Islamic civilization that placed a high value on precise calculations (dependent upon the mathematical sciences, especially astronomy) and instruments for attaining them. These were used for regulating the prayer times, determining the *qibla* or local direction to Mecca, and ascertaining the beginning and the end of important national and religious days and months (*e. g.*, the month of Ramadan). Each locality needed its own set of tables and calculations, and Qunawī's were for the capital city of Istanbul. Among his achievements, he simplified the standard usage of astronomical instruments, especially quadrants (*al-rubʿ al-mujayyab, rubʿ al-muqanṭarāt,* and *rubʿ al-dā'ira*), and he invented a new method for astronomical calculations in his *al-Aṣl al-muʿaddil.* Qunawī also translated the introductory part of Khalīlī's *mīqāt* tables (which provided solutions to all the standard problems of spherical astronomy for all latitudes) under the title *Tarjamah-i jadāwil-i āfāqī* or *Tarjamah-i risāla fī al-awqāt al-khamsa wa-jadāwil al-raṣad.* To the group of tables that Khalīlī prepared for each degree of latitude, he added a special table for an unknown location at latitude 40° 30′ N.

In the preface to his *Tarjamah-i jadāwil-i āfāqī*, Qunawī says "some of our sons wanted, from this poor man, to learn about sine tables; and so we translated this work into Turkish ..." (Süleymaniye Library, Ayasofya MS 2594, 1b). This is an indication that he was teaching astronomy courses in the *muwaqqithāne*s (timekeeping institutions attached to mosques) and that the language for learning and education was Turkish.

Qunawī's Arabic work entitled *Mīzān al-kawākib* contains time calculation tables by means of stars; the tables have over 500 pages, and include nearly 250 million registers. The main tables show the time from sunset (evening) to sunrise, dawn, and midday for a degree of solar longitude and full vertical rise. One can simply observe a star reaching the last point instantly and also read its rise from a different table prepared by the author; one can enter solar longitude through the rise on the main table and determine the nighttime. According to D. King (1986, p. 248), these tables represent an original Ottoman contribution in determining the astronomical time *via* tables.

After his death, Qunawī's works were developed further by **Muṣṭfā ibn ʿAlī al-Muwaqqit**, the chief astronomer to Sultan Süleymān the Magnificent.

İhsan Fazlıoğlu

Selected References

Bağdadlı, İsmail Paşa (1955). *Hadiyyat al-ʿārifīn*. Vol. 2, p. 225. Istanbul: Milli Eğ itim Bakanlíge Yayinlari.

Brockelmann, Carl. *Geschichte der arabischen Litteratur*. 2nd ed. Vol. 2 (1949): 302–303; Suppl. 2 (1938): 327; Suppl. 3 (1942): 1275. Leiden: E. J. Brill.

Bursalı, Mehmed Tahir (1342 H [1923]). *Osmanlı Müellifleri*. Vol. 3, p. 301. Istanbul: Matbaa-i Amire.

İhsanoğlu, Ekmeleddin *et al.* (1997). *Osmanlı Astronomi Literatürü Tarihi (OALT)* (History of astronomy literature during the Ottoman period). Vol. 1, pp. 84–90 (no. 46). Istanbul: IRCICA.

Kātib Čelebī. *Kashf al-ẓunūn ʿan asāmī al-kutub wa-'l-funūn*. Vol. 1 (1941), col. 317; Vol. 2 (1943), cols. 1904, 2042. Istanbul.

King, David A. (1986). "Astronomical Timekeeping in Ottoman Turkey." In *Islamic Mathematical Astronomy*, XII. London: Variorum Reprints, pp. 247–248.

——— (1993). *Astronomy in the Service of Islam*. Aldershot: Variorum.

——— (1996). "On the Role of the Muezzin and the *Muwaqqit* in Medieval Islamic Society." In *Tradition, Transmission, Transformation: Proceedings of Two Conferences on Pre-modern Science Held at the University of Oklahoma*, edited by F. Jamil Ragep and Sally P. Ragep, with Steven Livesey, pp. 285–346. Leiden: E. J. Brill.

Suter, H. (1981). *Die Mathematiker und Astronomen der Araber und ihre Werke*. Amsterdam: APA-Oriental Press, p. 187 (no. 455).

Uzunçarsılı, İsmail Hakkı (1982). *Osmanlı Tarihi*. Vol. 2, pp. 631, 633. Ankara: Tūrh Tarih Kurumu.

Qūshjī: Abū al-Qāsim ʿAlāʾ al-Dīn ʿAlī ibn Muḥammad Qushči-zāde

Born **probably Samarqand, (Uzbekistan)**
Died **Istanbul, (Turkey), 1474**

ʿAlī al-Qūshjī was a philosopher–theologian, mathematician, astronomer, and linguist who produced original studies in both observational and theoretical astronomy within 15th-century Islamic and Ottoman astronomy. He contributed to the preparation of **Ulugh Beg**'s *Zīj* at the Samarqand Observatory, insisted on the possibility of the Earth's motion, and asserted the need for the purification of all the scientific disciplines from the principles of Aristotelian physics and metaphysics.

Qūshjī was the son of Ulugh Beg's falconer, whence his Turkish name Qushči-zāde. He took courses in the linguistic sciences, mathematics, and astronomy as well as other sciences taught by scholars in the circle of Ulugh Beg. These included **Jamshīd al-Kāshī**, **Qāḍīzāde al-Rūmī**, and Ulugh Beg himself. It has been claimed that he was also taught by **al-Sayyid al-Sharīf al-Jurjānī**; if so, Qūshjī would have been quite young.

In 1420, Qūshjī secretly moved to Kirmān where he studied astronomy (*circa* 1423-1427) with Mollā Jāmī as well as the mathematical sciences. Upon his return to Samarqand *circa* 1428, Qūshjī presented Ulugh Beg with a monograph (*Ḥall ishkāl al-muʿaddil li-l-masīr*) in which he solved the problems related to Mercury; Ulugh Beg was reported to have been quite pleased. Sources say that Ulugh Beg referred to Qūshjī as "my virtuous son" (= "ferzend-i ercümend" [Nuruosmaniye MS 2932, f. 2b]). Indeed, after the death of Qādīzāde, it was Qūshjī whom Ulugh Beg commissioned to administer the observational work at the Samarqand Observatory that was required for his *Zīj* (astronomical handbook). Qūshjī, often referred to as "ṣāḥib-i raṣad" (head of observation), contributed to the preparation and correction of the *Zīj*, but it is unclear to what extent and at what stage. This question becomes especially problematic in view of Qūshjī's criticisms of it, and his pointing out of mistakes, in his *Sharḥ-i Zīj Ulugh Beg* (Commentary on Ulugh Beg's *Zīj*).

Upon Ulugh Beg's death in 1449, Qūshjī, together with his family and students, spent a considerable time in Herat where he wrote his theological work, *Sharḥ al-Tajrīd*, a commentary to **Naṣīr al-Dīn al-Ṭūsī**'s work *al-Tajrīd fī ʿilm al-kalām,* which he presented to the Timurid Sultan Abū Saʿīd. After Abū Saʿīd's defeat by Uzun Ḥasan in 1469, Qūshjī moved to Tabrīz where he was welcomed by the latter. It is said that Qūshjī was sent to Istanbul to settle a dispute between Uzan Ḥasan and Mehmed the Conqueror; after accomplishing the mission, he returned to Tabrīz. However, around 1472, Qūshī, together with his family and students, left permanently for Istanbul either on his own or because of an invitation from Sultan Mehmed.

When Qūshjī and his entourage approached Istanbul, Sultan Mehmed sent a group of scholars to welcome them. Sources say that in crossing the Bosporus to Istanbul, a discussion ensued about the causes of its ebb and flow. Upon arrival in Istanbul, Qūshjī presented his mathematical work entitled *al-Muḥammadiyya fī al-ḥisāb* to the Sultan, which was named in his honor.

Qūshjī spent the remaining two to three years of his life in Istanbul. He first taught in the Ṣaḥn-i Thamān Madrasa (founded by Sultan Mehmed); then he was made head of the Ayasofya Madrasa. In this brief period, Qūshjī educated and influenced a large number of students, who, along with his writings were to have an enormous impact on future generations. He was buried in the cemetery of the Eyyūb mosque.

Qūshjī, especially when compared with his contemporaries such as Kāshī and Qāḍīzāde, was a remarkable polymath who excelled in a variety of disciplines including language and literature, philosophy, theology, mathematics, and astronomy. He wrote works in all these fields, producing books, textbooks, and short monographs dealing with specific problems. His commentaries often became more popular than the original texts, and themselves became the subject of numerous commentaries. Thousands of copies of Qūshjī's works are extant, many of which were taught in the madrasas.

Qūshjī's philosophy of science, which had important repercussions for the history of astronomy, is contained in his commentary to **Ṭūsī**'s *Sharḥ al-Tajrīd.* Besides being one of the most important theological works in Islam, Qūshjī lays down the philosophical principles of his conception of existence, existents, nature, knowledge, and language. As for the mathematical sciences, Qūshjī in general tried to free them from hermetic–Pythagorean mysticism and to provide an alternative to Aristotelian physics as the basis for astronomy and optics. He sought to define body (*jism*) as being predominantly mathematical in character. Qūshjī claimed that the essence of a body is composed of discontinuous (atomic) quantity while its form consists of continuous (geometrical) quantity. When a body is a subject of the senses, it then gains its natural properties (qualifications).

One consequence of Qūshjī's anti-Aristotelian views was his striking assertion that it might well be possible that the Earth is in motion. Here Qūshjī followed a long line of Islamic astronomers who rejected **Ptolemy**'s observational proofs for geostasis; Qūshji, though, refused to follow them in depending on **Aristotle**'s philosophical proofs, thus opening up the possibility for a new physics in which the Earth was in motion. Qūshjī's views were debated for centuries after his death, and he exerted a profound influence on Ottoman–Turkish thought and scientific inquiry, in particular through the *madrasa* and its curriculum. His influence also extended to Central Asia and Iran, and it has been argued that he may well have had an influence, either directly or indirectly, upon early modern European science to which his ideas bear a striking resemblance.

Qūshjī wrote five mathematics books, one in Persian and four in Arabic. His *Risāla dar ʿilm al-ḥisāb* (Persian), written during his stay in Central Asia (along with his enlarged Arabic version of this work, *al-Risāla al-Muḥammadiyya fī al-ḥisāb*), were taught as a mid-level textbook in Ottoman *madrasas.* In these works, in accordance with the principles he laid down in the *Sharḥ al-Tajrīd*, he tried to free mathematics from hermetic–Pythagorean mysticism. As a result, Ottoman mathematics took on a practical character, which hindered traditional studies such as the theory of numbers.

In the field of astronomy, one of Qūshjī's most important contributions is in the observational program for the *Zīj-i Ulugh Beg* and in his corrections to the work, both before and after publication. In addition, he has nine works on astronomy, two in Persian and seven in Arabic. Some of them are original contributions while others are pedagogical. In his theoretical monograph entitled *Ḥall ishkāl al-muʿaddil li-l-masīr*, Qūshjī criticizes and corrects opinions and ideas pertaining to Mercury's motions mentioned in Ptolemy's *Almagest*. Another work is his *Risāla fī anna aṣl al-khārij yumkinu fī al-sufliyayn* that deals with the possibility of using an eccentric model for Mercury and Venus, which, as he says, goes against both Ptolemy and **Quṭb al-Dīn al-Shīrāzī**.

Qūshjī's *Risāla dar ʿīlm al-hayʾa* (Persian), written in Samarqand in 1458, was commonly used as a teaching text; there exist over eighty manuscript copies of it in libraries throughout the world. It was also translated into Turkish. Two commentaries were written on it, one by **Muṣliḥ al-Dīn al-Lārī**, the other by an anonymous author. Lārī's commentary was widely taught in Ottoman *madrasas.* Qūshjī's *Risāla* was also translated into Sanskrit and thus represents the transmission of Islamic astronomy to the Indian subcontinent. Qūshjī wrote an enlarged version of the work in Arabic under the name *al-Fatḥiyya fī ʿilm al-hayʾa,* which was presented to Sultan Mehmed in 1473. This work was taught as a middle-level textbook, and was commented on by Gulām Sinān (died: 1506) and Qūshjī's famous mathematician-astronomer great-grandson **Mīram Čelebī**. It was also translated into Persian by Muʿīn al-Dīn al-Ḥusaynī and into Turkish by Seydî Ali Reîs. In the *Risāla* and the *Fatḥiyya*, Qūshjī followed the principles he had laid down in his *Sharḥ al-Tajrīd* and excluded an introductory section on Aristotelian physics that had customarily introduced almost all previous works of this kind.

İhsan Fazlıoğlu

Selected References

Brockelmann, Carl. *Geschichte der arabischen Litteratur.* 2nd ed. Vol. 2 (1949): 305; Suppl. 2 (1938): 329–330. Leiden: E. J. Brill.

Fazlıoğlu, İhsan. "Ali Kusçu." In *Yasamları ve Yapıtlarıyla Osmanlılar Ansiklopedisi*, edited by Ekrem Çakıroğlu. Vol. 1, pp. 216–219. Istanbul: YKY, 1999.

İhsanoğlu, Ekmeleddin *et al.* (1997). *Osmanlı Astronomi Literatürü Tarihi* (*OALT*) (History of astronomy literature during the Ottoman period). Vol. 1, pp. 27–38 (no. 11). Istanbul: IRCICA.

——— (1999). *Osmanlı Matematik Literatürü Tarihi* (*OMLT*). (History of mathematical literature during the Ottoman period). Vol. 1, pp. 20–27 (no. 3). Istanbul: IRCICA.

Pingree, David (1996). "Indian Reception of Muslim Versions of Ptolemaic Astronomy." In *Tradition, Transmission, Transformation: Proceedings of Two Conferences on Pre-modern Science Held at the University of Oklahoma*, edited by F. Jamil Ragep and Sally P. Ragep, with Steven Livesey, pp. 471–485, p. 474. Leiden: E. J. Brill.

Ragep, F. Jamil. (2001). "Ṭūsī and Copernicus: The Earth's Motion in Context." *Science in Context* 14: 145–163.

——— (2001). "Freeing Astronomy from Philosophy: An Aspect of Islamic Influence on Science." *Osiris* 16: 49–71.

Saliba, George (1993). "Al-Qūshjī's Reform of the Ptolemaic Model for Mercury." *Arabic Sciences and Philosophy* 3: 161–203.

Salih, Zeki (1329). *Asar-i bakiye.* Vol. 1, pp. 195–199. Istanbul. (1913/1914): Matbaa-i Amire.

Ṭashköprüzāde (1985). *Al-Shaqāi'q al-nuʿmāniyya fī ʿulamā' al-dawlat al-ʿuthmāniyya*, edited by Ahmed Subhi Furat. Istanbul, pp. 159–162. Istanbul üniversitesi, Edebiyat Fakültesi Yayinlave.

Ünver, A. Süheyl (1948). *Astronom Ali Kusçu, Hayatı ve Eserleri.* Istanbul: Kenan Matbaası.

Qusṭā ibn Lūqā al-Baʿlabakkī

Born **Baʿlabakk, (Lebanon), probably *circa* 820**
Died **(Armenia), probably *circa* 912–913**

Qusṭā ibn Lūqā (Constantine, son of Loukas), a scholar of Greek Christian origin working in Islamic lands in the 9th century, did work in astronomy that included translations of Greek astronomical works and original compositions. In addition, he composed and translated mathematical, medical, and philosophical works. Qusṭā's scholarly reputation extended far and wide, and he was noted for his scientific achievements (especially in medicine, where his authority surpassed **Ḥunayn ibn Isḥāq** according to the bibliographer Ibn al-Nadīm [died: *circa* 990]). He reportedly collected Greek scientific manuscripts from Byzantine lands; his translations and revisions of these formed an important part of his scholarly activities. Qusṭā was fluent in Greek (as well as Syriac), as demanded by his scientific translations, and he also mastered Arabic, a language in which he produced many original scientific compositions. Qusṭā's scholarly career, which was centered in Baghdad, is notable for his association with numerous patrons, who are particularly important for establishing his biography as well as the chronology of his work. These include various members of the ʿAbbāsid caliphal family, government officials, and a Christian patriarch; the most likely interpretation of the evidence places the bulk of his work in the second half of the 9th century.

The scientific works of Qusṭā include several astronomical compositions, which cover both the theoretical and the practical aspects of astronomy. The best known are:

(1) *Kitāb fī al-ʿamal bi-'l-kura al-nujūmiyya* (On the use of the celestial globe; with some variations as to title), which contains 65 chapters and was widely disseminated through at least two Arabic recensions as well as Latin, Hebrew, Spanish, and Italian translations;
(2) the extant astronomical work, *Hay'at al-aflāk* (On the configuration of celestial bodies; Bodleian Library MS Arabic 879, Uri, p. 190), which is one of the earliest compositions in theoretical (*hay'a*) astronomy;
(3) *Kitāb al-Madkhal ilā ʿilm al-nujūm* (Introduction to the science of astronomy – astrology);
(4) *Kitāb al-Madkhal ilā al-hay'a wa-ḥarakāt al-aflāk wa-'l-kawākib* (Introduction to the configuration and movements of celestial bodies and stars);
(5) *Kitāb Fī al-ʿamal bi-'l-asṭurlāb al-kurī* (On the use of the spherical astrolabe; Leiden University Library MS Or. 51.2: *Handlist*, p. 12); and
(6) *Kitāb Fī al-ʿamal bi-'l-kura dhāt al-kursī* (On the use of the mounted celestial sphere).

The two introductory astronomical titles (3 and 4), reported in the lists of Ibn al-Nadīm's *Fihrist* and Ibn Qifṭī (died: 1248), respectively, are not extant, unless the latter is the same as the theoretical work mentioned in (2). F. Sezgin suggests that these two works are the same; however, they are listed as two distinct titles by Ibn Abī Uṣaybiʿa (died: 1269). Work (5) is sometimes questioned as a work by Qusṭā but seems to represent a variation in title of (1). Although E. Wiedemann (1913) treats (6) as an independent work, it also seems to be a variation in title of (1). This leaves Qusṭā with at least four distinct astronomical compositions, two of which (1 and 2) are extant.

Qusta's works also include translations of the so called Little Astronomy or "Intermediate Books" (*Kutub al-mutawassiṭāt*), texts studied after Euclidean geometry in preparation for Ptolemaic astronomy. Extant among these are the Arabic versions of **Theodosius**'s *Spherics* (*Kitāb al-Ukar*) and **Autolycus**'s *Rising and Setting* [*of Fixed Stars*] (*Kitāb al-Ṭulūʿ wa-'l-ghurūb*). In addition to other extant translations, such as Hero of Alexandria's "On the Raising of Heavy Objects" (*Fī rafʿ al-ashyā' al-thaqīla*), Qusṭā is associated with Arabic versions of **Aristotle**'s *Physics* as well as the later commentaries of Alexander of Aphrodisias and **Philoponus** on certain of their books. This dual translation program fits well with his statements about the "cooperation" of natural philosophy and geometry in optics as a mixed mathematical science, a genre to which astronomy and mechanics also belong.

Elaheh Kheirandish

Alternate name

Costa ben Luca

Selected References

Brockelmann, Carl (1943). *Geschichte der arabischen Litteratur*. 2nd ed. Vol. 1, pp. 222–223. Leiden: E. J. Brill. (Contains lists of manuscripts for Qusṭā's works including five astronomical titles. Entries i, k, g, and f in Section I correspond to nos. 1, 2, 5, and 6 above.)

Browne, E. G. (1902). *A Literary History of Persia*. 2 Vols, Vol. 1, p. 278. London: T. F. Unwin. (Contains an 11th-century Persian poem referring to Qusṭā ibn Lūqā.)

Gabrieli, G. (1912). "Nota bibliographica su Qusṭā ibn Lūqā." *Atti della R. Accademia dei Lincei: Rendiconti, classe di scienze morali, storiche e filologiche* 21: fasc. 5–6 : 341–382. (Contains a list of 69 of Qusṭā's compositions and 17 translations, including six astronomical titles [nos. 1–6 above, numbered respectively as nos. 40, 37, 54, 37, 67, 40], with references to historical and modern sources and manuscript copies and titles [pp. 348–350; p. 348: no. 40: Q. N. is problematic].)

Gutas, Dimitri (1998). *Greek Thought, Arabic Culture: The Graeco-Arabic Translation Movement in Baghdad and Early ʿAbbāsid Society (2nd–4th/8th–10th centuries)*. London: Routledge. (Contains a section on the problematic question of Qusṭā's early patrons.)

Harvey, E. Ruth (1975). "Qusṭā ibn Lūqā al-Baʿlabakkī." In *Dictionary of Scientific Biography*, edited by Charles Coulston Gillispie, Vol. 11, pp. 244–246. New York: Charles Scribner's Sons. (Contains a list of Qusṭā's works including four of the astronomical titles listed above [nos. 1, 2, 3, and 5] with reference to relevant manuscripts, reference works, and secondary sources up to 1975, with the exception of the important article of Wiedemann in the first edition of the *Encyclopaedia of Islam*.)

Hill, D. (1986). "Ḳusṭā b. Lūḳā al-Baʿlabakkī." In *Encyclopaedia of Islam*. 2nd ed. Vol. 5, pp. 529–530. Leiden: E. J. Brill. (Contains references to some of Qusṭā's works, including two on astronomy [with English titles, apparently nos. 1 and 3 above], and a short bibliography.)

Ibn Abī Uṣaybiʿa (1882–1884). *ʿUyūn al-anbā' fī ṭabaqāt al-atibbā'*, edited by August Müller. 2 Vols, Vol. 1, pp. 144–245. Cairo-Königsberg. (Contains a list of over 60 of Qusṭā's works including three astronomical titles [nos. 1, 3, and 4 above].)

Ibn al-Nadīm (1970). *The Fihrist of al-Nadīm: A Tenth-Century Survey of Muslim Culture*, edited and translated by Bayard Dodge. 2 vols, Vol. 1, p. 295; Vol. 2, pp. 694–695. New York: Columbia University Press. (Contains a list of over 30 of Qusṭā's compositions including 2 astronomical titles [nos. 1 and 3 above].)

Kheirandish, Elaheh (1999). *The Arabic Version of Euclid's* Optics (Kitāb Uqlīdis fī Ikhtilāf al-manāẓir). 2 Vols. New York: Springer-Verlag. (Contains discussions of Qusṭā's optical work, with reference to relevant sources and discussions ["Intermediate Books", mixed mathematical sciences, *etc.*].)

al-Qifṭī, Jamāl al-Dīn (1903). *Ta'rīkh al-ḥukamā'*, edited by J. Lippert, Vol. 1, pp. 262–263. Leipzig: Theodor Weicher. (Contains a list of over 20 of Qusṭā's works including two astronomical titles [nos. 1 and 4 above].)

Ragep, F. J. (1993). *Naṣīr al-Dīn al-Ṭūsī's* Memoir on Astronomy *(al-Tadhkira fī ʿilm al-hay'a)*. 2 Vols. New York: Springer-Verlag. (Contains as part of its exhaustive treatment of Ṭūsī and his important astronomical work, *al-Tadhkira*, discussions on several aspects of *ʿilm al-hay'a* ["cosmography, configuration"].)

Rashed, Roshdi (1997). *Oeuvres philosophiques et scientifique d'Al-Kindī*. Vol. 1, *L'optique et la catoptrique*. Leiden: E. J. Brill. (Contains the Arabic text and French translation of Qusṭā's *Kitāb Fī ʿilal mā yaʿriḍu fī al-marāyā min ikhtilāf al-manāẓir*.)

Saliba, George (August 2001). "The Social Context of Islamic Astronomy." In *Proceedings of the Conference: Islam and Science*. Amman: Royal Institute for Inter-Faith Studies, forthcoming. (Contains a discussion of Qusṭā's *Hay'at al-aflāk* [composition date given as 860].)

Sezgin, Fuat *Geschichte des arabischen Schrifttums*. Vol. 3, *Medizin* (1970): 270–274; Vol. 5, *Mathematik* (1974): 285–286; Vol. 6, *Astronomie* (1978): 180–182. Leiden: E. J. Brill. (Contains a list of the manuscripts of Qusṭā's works including four astronomical titles [nos. 1, 2, 3, and 4 above], the last three suggested as possibly referring to the same work.)

Wiedemann, E. (1913). "Ḳosṭā b. Lūḳā, al-Baʿalbakkī." In *Encyclopaedia of Islam*. 1st ed., Vol. 4, pp. 1081–1083. Leiden: E. J. Brill. (Contains, in addition to a biography, references to his works including four astronomical titles [nos. 1, 3, 5, and 6 above, listed as separate works], with reference to the problems involved, including the attribution of no. 6 to Qusṭā [pp. 1082–1083], with a short bibliography.)

Wilcox, Judith (1985). "The Transmission and Influence of Qusta ibn Luqa's 'On the Difference between Spirit and the Soul.'" Ph.D. diss., City University of New York.

——— (1987). "Our Continuing Discovery of the Greek Science of the Arabs: The Example of Qusṭā ibn Lūqā." *Annals of Scholarship* 4, no. 3: 57–74. (Contains a more recent account of Qusṭā's scientific and philosophical works and relevant sources.)

Worrell, W. H. (1944). "Qusta ibn Luqa on the Use of the Celestial Globe." *Isis* 35: 285–293. (Contains, in addition to relevant references, a useful list of six variant Arabic titles often assumed as representing different works [including nos. 1 and 6 above], an English summary and discussion based on the manuscript copy in Michigan.)

R

Rāghavānanda Śarman

Flourished **Rāḍha, Bengal, (India), 1591–1599**

Rāghavānanda Śarman was a Brāhmaṇa who followed the Saurapakṣa school of astronomy. He composed several astronomical works in the late 16th century, including the *Viśvahita*, a set of astronomical tables whose epoch is 1591; the *Dinacandrikā*, another set of astronomical tables whose epoch is 1599; and the *Sūryasiddhāntarahasya*, a commentary on the *Sūryasiddhānta*, written in 1591. Little is known of these three works though they have been published.

Setsuro Ikeyama

Selected References

Pingree, David (1975). "Rāghavānanda Śarman." In *Dictionary of Scientific Biography*, edited by Charles Coulston Gillispie. Vol. 11, p. 264. New York: Charles Scribner's Sons.

——— (1994). *Census of the Exact Sciences in Sanskrit*. Series A. Vol. 5, pp. 411a–412b. Philadelphia: American Philosophical Society.

Rāghavānanda Śarman (1913). *Dinacandrikā*, edited by Bhagavatīcaraṇa Smṛtitīrtha. Calcutta. (With Bengali commentary and translation.)

——— (1913). *Viśvahitam*, edited by Viśvambhara Jyotiṣārṇava and Śrīśacandra Jyotīratna. Bibliotheca Indica, Vol. 222. Calcutta: Asiatic Society. (Incorrectly attributed to Mathurānātha Śarman.)

——— (1915). *Sūryasiddhāntarahasya*, edited by Rajanīkānta Vidyāvinoda. Calcutta.

Raimarus Ursus

➤ **Bär, Nicholaus Reymers**

Rainaldi, Carlo Pellegrino

➤ **Danti, Egnatio**

Ramée, Pierre de la

➤ **Ramus, Peter**

Ramus, Peter [Petrus]

Born **Cuts, (Oise), France, 1515**
Died **Paris, France, 26 August 1572**

Peter Ramus rejected scholasticism and outlined new ways to view and to teach knowledge, influencing astronomy's transition from medieval to modern form. He was born in an impoverished noble family originally from Liège. There appears to have been little that was unusual about Ramus's early education. At around 12, after two unsuccessful attempts to enter the University of Paris, he enrolled at the Collège de Navarre, where he earned money working as a servant to wealthier peers. Having to attend school as a day student, however, meant that Ramus did not complete his master of arts until he was 21. Of special benefit was the friendship that Ramus then forged with his later patron, Charles de Guise, who was to become Cardinal of Lorraine and, eventually, of Guise. The support of Guise disappeared after Ramus's conversion to Protestantism around 1562, but the long and active intervention of the cardinal on Ramus's behalf was responsible for much of the fame he gained during his life.

Ramus's career began sedately enough as an instructor at the Collège du Mans, but he soon moved, with his longtime collaborator in rhetoric, Omer Talon, to the Collège de l'Ave Maria, also in Paris. There, in 1543, Ramus published the first of many attacks on Aristotelian and scholastic education that would make him notorious and that led directly or indirectly to his death. Ramus's assertion that the medieval approach to learning, with its long tradition of minutely analyzing and carefully commenting on selected classical authors, was bankrupt proved scandalous to the faculty of Paris, and in short order they secured a ban on his teaching of dialectic (the rhetorical practice of logical principles).

This edict simply led Ramus to concentrate his efforts on the instruction of eloquence and mathematics. He was forced to teach himself the latter subject, since it had been minimal in his university curriculum. But taken together, the two apparently disparate disciplines reveal exactly how much of Ramus's work remains entrenched in premodern modes of thought. To Ramus, the study of numbers and the study of literature were but two aspects of one unified and coherent body of knowledge, which in all its facets was directed only at "disputing well," as he put it in his 1543 *Structure of Dialectic*.

The prospects of Ramus began to improve in 1545 when he was invited to become an instructor at the more prestigious Collège de Presles, where Talon again joined him. Ramus was shortly thereafter promoted to principal of the school and in 1547, following the succession of Henry II to the throne and the subsequent elevation of Charles de Guise in the estimation of the court, all restrictions on Ramus's teaching were lifted; he was left entirely free to comment on the whole range of university subjects. In 1551, he was made a *regius* professor of eloquence and philosophy – the only time these two fields were combined under one title. In 1565, he was chosen dean of the *regius* professors, who came collectively to be called the Collège de France and who, as a body, had been intended by Francis I to represent a humanistic alternative to the resolutely scholastic education of the University of Paris. After this, his fortunes began to decline, falling slowly at first and then precipitously. Suspected even in the late 1550s of being a secret Protestant, Ramus had to flee Paris for Fontainebleau under royal protection during the religious troubles of 1562. He returned in 1563, but having made his conversion public in the intervening year, Ramus entered the city this time without the critical support of Guise. Harried by a growing number of antagonists, Ramus left Paris again in 1567 and again took refuge with a sympathetic member of the royal family.

From 1568 to 1570, Ramus toured the prominent Protestant centers of learning in Germany and Switzerland. During these years, too, he began to broaden his vision of pedagogical reform and wrote on subjects ranging from theology to astronomy. He found the academic atmospheres of Heidelberg and Geneva, though, uncertain at best and returned to Paris for the last time in 1570. The royal family, especially the queen mother, continued to defend Ramus publicly, but he was murdered during the Saint Bartholomew's massacre.

Ramus does not very directly affect the history of astronomy, but his work was influential throughout the 17th and 18th centuries in Europe and America in a surprising number of ways. The topical organization of many modern encyclopedias can be traced to his interest in systematic thought, and **Francis Bacon**'s search for an inductive method is in part attributable to his own youthful interest in Ramus's writing. Primarily, Ramus advocated a regular approach to the study of any given subject. He believed that all realms of knowledge could be construed schematically, and that any discipline of learning would conform to three "laws" of universality, homogeneity, and generality. In other words, a "natural method" of learning would be observably true in every posited application of the art or science, specific in its description, and arranged in such a way as to exhibit the general truths on which it was based.

Ramus considered astronomy a branch of physics and, as such, part of the quadrivium of the university curriculum. Characteristically, he did develop a chart of physics, in which the stars (still very much after the manner of **Aristotle**) were classed as elements of constant simple matter, opposed to the immaterial essences of God and intelligence. Beyond this, Ramus seems merely to have appreciated astronomy as an application of mathematics, though he was critical of both **Nicolaus Copernicus** and **Ptolemy** for propounding their respective theories of the Solar System without observation. Copernicus thus had violated the first law of method. It is, in the end, one of the most telling marks of Ramus's influence that **Johannes Kepler** later claimed, with evident pride, that his own work had at last satisfied the demands of Ramus on astronomy.

H. Clark Maddux

Alternate name

Ramée, Pierre de la

Selected References

Graves, Frank Pierrepont (1912). *Peter Ramus and the Educational Reform of the Sixteenth Century*. New York: Macmillan.

Howell, Wilbur Samuel (1961). *Logic and Rhetoric in England, 1500–1700*. New York: Russell and Russell.

Miller, Perry (1982). *The New England Mind: The Seventeenth Century*. Cambridge: Harvard University Press.

Ong, Walter J. (1983). *Ramus, Method, and the Decay of Dialogue: From the Art of Discourse to the Art of Reason*. Cambridge: Harvard University Press.

Sloane, Thomas O. (1985). *Donne, Milton, and the End of Humanist Rhetoric*. Berkeley: University of California Press.

Raṅganātha I

Flourished **Kāśī, (Vārānasi, Uttar Pradesh, India), 1603**

Raṅganātha I was born into a family of astronomers; his father was Ballāla. Raṅganātha I wrote a popular Sanskrit commentary, the *Gūḍhārthaprakāśikā* (1603), on the anonymous *Sūryasiddhānta* (10th or 11th century), one of the most popular astronomical works in Sanskrit. Raṅganātha I's son, Munīśvara, was also an astronomer and composed a Sanskrit astronomical treatise, the *Siddhāntasārvabhauma* (1646), along with a commentary on the *Siddhāntaśiromani* of **Bhāskara II**.

The *Sūryasiddhānta* was the fundamental text of the Saura School, one of four principal schools of astronomy active during the Hindu classical period (late 5th to 12th centuries). Those editions of the *Sūryasiddhānta* that have come down to us, including the first English translations, are mostly based on Raṅganātha I's text, and so have acquired considerable influence.

Selected References

Dikshita, Sankara Balakrshna (1969). *Bhāratīya Jyotish Śāstra* (History of Indian Astronomy), translated by Raghunath Vinayak Vaidya. Delhi: Manager of Publications (Government of India).

——— (1981). *Bhāratīya Jyotish Śāstra* (History of Indian Astronomy), translated by Raghunath Vinayak Vaidya. Vol. 2. New Delhi: India Meteorological Department.

Pingree, David (1975). "Raṅganātha." In *Dictionary of Scientific Biography*, edited by Charles Coulston Gillispie. Vol. 11, p. 290. New York: Charles Scribner's Sons.

——— (1978). "History of Mathematical Astronomy in India." In *Dictionary of Scientific Biography*, edited by Charles Coulston Gillispie. Vol. 15 (Suppl. 1), pp. 533–633. New York: Charles Scribner's Sons.

Raṅganātha II

Flourished **Kāśī, (Vārānasi, Uttar Pradesh, India), 1640 or 1643**

Raṅganātha II was born into a rival family of astronomers sharing the same name as **Raṅganātha I**; his father was Nṛsiṃha. He wrote a Sanskrit astronomical work, the *Siddhāntacūdāmaṇi* (1640 or 1643), along with a number of other astronomical and mathematical works. Raṅganātha II's father had written a commentary on the anonymous *Sūryasiddhānta* (10th or 11th century). Raṅganātha II's brother, **Kamalākara**, likewise composed a Sanskrit astronomical treatise, the *Siddhāntatattvaviveka* (1658).

The *Siddhāntacūdāmaṇi* consists of 12 chapters and covers many of the standard topics discussed in Hindu astronomy of the classical period (late 5th to 12th centuries). It is also based on the *Sūryasiddhānta*.

Raṅganātha II also composed the *Bhaṅgīvibhaṅgīkaraṇa*, a detailed work on the theory of planetary motions that criticized an earlier work of Munīśvara, son of Raṅganātha I. Raṅganātha II's astronomical writings included the *Lohagolakhaṇḍana*, a work on the celestial sphere, and the *Palabhākhaṇḍana*, a guide to determining terrestrial latitudes from observations of the stars.

Selected References

Dikshita, Sankara Balakrshna (1969). *Bhāratīya Jyotish Śāstra* (History of Indian Astronomy), translated by Raghunath Vinayak Vaidya. Delhi: Manager of Publications (Government of India).

——— (1981). *Bhāratīya Jyotish Śāstra* (History of Indian Astronomy), translated by Raghunath Vinayak Vaidya. Vol. 2. New Delhi: India Meteorological Department.

Rankine, William John Macquorn

Born **Edinburgh, Scotland, 5 July 1820**
Died **Glasgow, Scotland, 24 December 1872**

Scottish railroad engineer William Rankine, among many other activities ranging over science and engineering, wrote down in 1869 a set of equations connecting the density, pressure, and temperature of gases on the two sides of a shock wave. These were generalized in 1887 by Pierre Henri Hugoniot (1851–1897) and, in the form of the Rankine–Hugoniot relations, can be used to describe, for instance, propagation of a supernova remnant moving into the interstellar medium at a speed faster than that of sound.

Virginia Trimble

Selected Reference

Duab, Edward E. (1970). "Waterston, Rankine, and Clausius on the Kinetic Theory of Gases." *Isis*. 61: 105.

Ranyard, Arthur Cowper

Born **Swanscombe, Kent, England, 21 June 1845**
Died **London, England, 14 December 1894**

Arthur Ranyard not only observed the features of the Sun but also interpreted other solar observations in order to understand the sun's physical characteristics. He moved to London at an early age, and studied mathematics at University College, London, and Pembroke College, Cambridge. As a student, Ranyard and George de Morgan (son of **Augustus de Morgan**, the well-known mathematician) played a leading role in the founding of the London Mathematical Society. After taking a Cambridge law degree, Ranyard was called to the Bar at Lincoln's Inn, and practiced as a barrister for the rest of his life. He became a member of the Royal Astronomical Society in 1863, and served as its secretary in 1873.

Despite his success at law, astronomy was always Ranyard's avocation and true passion. He was especially interested in the Sun, and his diligence and high intelligence helped him rise quickly above the level of a *dilettante*. He traveled, at his own expense, to view several solar eclipses, including one in 1870 viewed from Sicily, and another in August 1878 from Cherry Creek (near Denver), Colorado, USA, which Ranyard observed with **Charles Young**'s party. The next year, he published his most important work on solar eclipse observations.

On the basis of his comprehensive review of historical eclipse observations, Ranyard discovered that the form of the extended and much-attenuated outer atmosphere of the Sun, the corona, varied with the sunspot cycle. As **Agnes Clerke** put it:

> "When sun-spots are numerous, the corona appears to be most fully developed above the spot-zones, thus offering to our eyes a rudely quadrilateral contour. The four great luminous sheaves forming the corners of the square are made up of rays curving together from each side into 'synclinal' or ogival groups, each of which may be compared to the petal of a flower."

At sunspot minima, the corona has a more shapeless, roughly circular, and amorphous appearance.

Ranyard mounted another expedition to Sohag, in Upper Egypt, for the eclipse of 17 May 1882. Totality lasted only 74 s, but was memorable because of the unexpected appearance of the solar-eclipse comet X/1882 K1, seen and photographed during the eclipse and never seen again.

Yet Ranyard was not primarily an observer; rather, he was a keen student of historical records made by others. His analysis of these observations was frequently highly perceptive and produced new results or cleared up long-standing enigmas. For instance, he and **John Hind** independently debunked the notion, based on a single drawing by the German amateur Johann Pastorff, that the great comet C/1819 N1 had been visible in transit in front of the Sun.

His work on periodicities related to the sunspot cycle led him to take an interest in Jupiter. Its cloud markings were subject to occasional outbreaks; were these related, possibly, to the sunspot cycle? Ranyard was a member of a group organized in March 1876 by the Royal Astronomical Society for the study of Jupiter. (Other members included **William Huggins**, William Noble, Alexander Lindsay, **Laurence Parsons**, and **Thomas Webb**.) Ranyard concluded from examining all the observations then available that deeper tinges of color and the eruption of equatorial "porthole" markings seemed more likely to appear during years of sunspot minima. This was a pioneering effort, based on insufficient data; unfortunately, it has not been borne out by modern studies.

After the death of the prolific astronomy popularizer **Richard Proctor** in 1888, Ranyard produced revised editions of some of Proctor's works, such as his *Old and New Astronomy*, and became editor of *Knowledge*, a London illustrated scientific magazine that had been founded by Proctor. In this role, he introduced large-scale photogravure reproductions of astronomical photographs, including reproductions of many of **Edward Barnard**'s pioneering wide-angle photographs of the Milky Way obtained at Lick Observatory, California, USA, with the Willard portrait lens. In commenting on the wild and mysterious dark markings revealed in these photographs, Raynard first perceived that they were best explained by assuming the existence of masses of dark absorbing matter rather than as holes or gaps in the star clouds. In making this suggestion, he departed from the view that had been hitherto maintained by **William Herschel** and **John Herschel**. It was a major breakthrough, but its time had not yet come. Indeed, Barnard himself continued to struggle with the nature of the dark markings for many years before finally convincing himself that Ranyard's explanation had to be the correct one.

In 1893, Ranyard accompanied Barnard on the latter's triumphant Grand Tour of the Continent following his discovery of the fifth satellite of Jupiter. Barnard found that even when suffering from hay fever, as he did on the train from Boulougne to Paris, Ranyard never lost his politeness. When the two men visited **Camille Flammarion** at Juvisy, Ranyard became so much a part of the occasion that Flammarion remarked that he was "a perfect Frenchman." He was as much appreciated for his personal charm as for his perceptive knowledge of astronomy. Soon after his return from Europe, Ranyard was taken ill. He died of cancer.

William Sheehan

Selected References

Clerke, Agnes M. (1893). *History of Astronomy during the Nineteenth Century*. 3rd ed. London: Adam Charles and Black.

Hale, George Ellery (1895). "Arthur Cowper Raynard" *Astrophysical Journal* 1: 168–169.

Hockey, Thomas (1999). *Galileo's Planet: Observing Jupiter before Photography*. Bristol: Institute of Physics.

Poggendorff, J. C. (1898). "Ranyard." In *Biographisch-literarisches Handwörterbuch*. Vol. 3, pp. 1088–1089. Leipzig: Barth.

Ranyard, Arthur Cowper (1879). "Observations Made during Total Solar Eclipses." *Memoirs of the Royal Astronomical Society* 41: 1–792.

Sheehan, William (1995). *The Immortal Fire Within: The Life and Work of Edward Emerson Barnard*. Cambridge: Cambridge University Press.

Wesley, W. H. (1895). *Knowledge*: 25–27.

Rauchfuss, Konrad

Born **Frauenfeld, Thurgau, Switzerland, *circa* 1530**
Died **Strasbourg, (Bas-Rhin), France, 26 April 1600**

Konrad Rauchfuss is best known as a mathematician, and his name is connected with the astronomical clock in the cathedral of Strasbourg.

Petrus Dasypodius, the father of Konrad, was one of the Swiss humanists; he taught at the school of Johannes Sturm in Strasbourg, wrote humanistic treatises, and held lower church positions. His son studied at this academy with Christian Herlinus and became professor there in 1558.

Rauchfuss devoted himself mainly to the teaching of mathematics. He recognized early the low level of knowledge and teaching of mathematics in his time, and he decided to translate into Latin the basic Greek mathematical texts. He edited and commented on works by Euclid, Heron of Alexandria (his favorite author), **Ptolemy**, and then propositions of the works by **Theodosius of Bythinia**, **Autolycus**, and Barlaamo. A remarkable textbook *Analyseis geometricae sex librorum Euclidis* (1566) contains the proofs of the first six books of Euclid's *Elementa*; Rauchfuss wrote it together with his teacher Herlinus. The numerous textbooks of Rauchfuss were in use for many years at many European universities.

Rauchfuss's astronomical clock in the cathedral of Strasbourg replaced the original one from 1352 to 1354. That clock had been taken down about 1550, and nothing was left from it. After several reconstructions and improvements in the following centuries, today's clock still follows the principal design by Rauchfuss, and contains some original parts. The Rauchfuss clock was installed in the cathedral between 1571 and 1574, and Rauchfuss described it in detail in the book *Heron mechanicus* (Basel, 1580). A remarkable portrait of **Nicolaus Copernicus**, painted on the clock by Tobias Stimmer (1539–1582), confirms that Rauchfuss appreciated Copernicus's work, but he never became an adherent of the Copernican cosmological system.

Martin Solc

Alternate name

Cunradus Dasypodius

Selected References

Blumhoff, J. G. L. (1796). *Vom alten mathematiker Dasypodius*. Göttingen. (Dasypodius's own books, editions, and translations come to more than 20 titles that were printed mainly in Basel. They are listed herein.)

Dasypodius (1578). *Brevis doctrina de cometis*.

——— (1578). *Scholia in libros apotelesmaticos Cl. Ptolemaei*. Basel.

——— (1580). *Heron mechanicus*.

——— (1580). *Horologii astronomici description*.

——— (1580). *Warhafftige Ausslegung und Beschreybung des astronomischen Uhrwercks zu Straszburg*.

Lehni, R. (1997). *Strasbourg Cathedral's Astronomical Clock*. Paris: Éditions La Goélette.

Oestmann, Günther (2000). *Die astronomische Uhr des Straßburger Münsters*. 2nd ed. Diepholz: GNT Verlag Diepholz.

Schmidt, W. (1898). "Heron von Alexandria, Konrad Dasypodius und die Strassburger astronomischen Münsteruhr." *Abhandlungen zur Geschichte der Mathematik* 8: 75–194.
Zinner, Ernst (1943). *Entstehung und Ausbreitung der Coppernicanischen Lehre.* Erlangen: Mencke, p. 273.

Rayet, Georges-Antoine-Pons

Born **near Bordeaux, Gironde, France, 12 December 1839**
Died **Floirac near Bordeaux, Gironde, France, 14 June 1906**

Georges Rayet was a French astronomer who, with **Charles Wolf** at the Paris Observatory in 1867, detected a class of rare, exceptionally hot stars whose spectra show strong broad emission lines of helium, carbon, and nitrogen. Wolf–Rayet stars, as they are known after their discoverers, are about twice the size of the Sun, and have a rapidly expanding outer shell – the source, it is thought, of the emission lines. The residual hydrogen envelopes of these stars have been stripped away (by stellar winds or mass transfer), revealing their deeper layers. Many central stars of planetary nebulae are of this type.

Rayet had no formal schooling until the age of 14, when his family moved to Paris. He was admitted to the École Normale Supérieure (1859); he graduated with a physics degree in 1862 and taught for a year before obtaining a post in the new weather forecasting service set up by **Urbain Le Verrier** at the Paris Observatory. In 1873, its operation was entrusted to him, but within the year the two men disagreed over storm forecasts, and Rayet was dismissed. He then lectured on physics at the Faculty of Sciences of Marseilles, and in 1876 was appointed professor of astronomy at Bordeaux.

Following a government initiative to build several new observatories, Rayet was asked to undertake a survey of the history and instrumentation of the world's observatories. He was subsequently offered the directorship of the observatory to be built at Floirac. From 1879, he held this position in tandem with his Bordeaux appointment. Having installed modern astrometric equipment at Floirac, Rayet organized a program of positional measures of stars, nebulae, and the components of binary systems.

Rayet's first collaboration with Wolf occurred in 1865 when they photographed an eclipse of the Moon. On 20 May of the next year, the pair noted bright emission lines, widened into bands, in the continuous spectrum of a nova (T Coronae Borealis, the first nova to be examined spectroscopically). In the following year, they made the observations that would immortalize their names in the Wolf–Rayet stars, when they discovered a similar appearance in the spectra of three eighth magnitude stars in Cygnus.

In 1868, Rayet undertook responsibility for the spectroscopic work on an expedition to the Malay Peninsula to observe a total eclipse of the Sun. With Wolf, Rayet also made significant observations of comet C/1874 H1 (Coggia), and observed the total solar eclipse of 1905 from Spain. He was an enthusiastic supporter of the International *Carte du Ciel* Astrographic Chart and Catalogue, and in the year before his death published the first volume of the *Catalogue photographique de l'Observatoire de Bordeaux.*

Richard Baum

Selected References

Anon. (1984). "Rayet, George Antoine Pons." In *The Biographical Dictionary of Scientists: Astronomers*, edited by David Abbott, pp. 131–132. London: Frederick Muller.
Esclangon, Ernest (1906). "Nécrologie: M. Georges Rayet." *Astronomische Nachrichten* 172: 111–112.
Wolf, C. and G. Rayet (1866). "Note sur deux étoiles." *Comptes rendus de l'Académie des sciences* 62: 1108–1109.
——— (1867). "Spectroscopie stellaire." *Comptes rendus de l'Académie des sciences* 65: 292–296.

Raymond of Marseilles

Flourished **(France), 1141**

Raymond was among the earliest Latin scholars to adapt Arabic science to the needs and requirements of a well-educated Latin-reading public. While not making translations himself, he was well acquainted with Arabic texts translated in Toledo, in the fields of astronomy, astrology, and probably alchemy.

The works that can be securely attributed to Raymond are three substantial texts that refer to each other. The first is *Liber cursuum planetarum*, an adaptation to the meridian of Marseilles of the astronomical tables originally drawn up for the meridian of Toledo and translated from, or based on, Arabic sources. Raymond's adaptation was made in 1141 and is accompanied not only by his own explanation on how to use the tables (*regula*), but also by a long essay justifying the study of astronomy and, in particular, astrology. This essay begins with a substantial poem, and the essay itself quotes liberally from classical sources and the church fathers. The second work is *Liber iudiciorum*, a handbook of judicial astrology, based on 12th-century translations from Arabic of astrological works by **Abu Ma'shar, al-Qabisi**, and Sahl ibn Bishr, and on earlier Latin material attributed to "Argaphalau" and **Ptolemy**. The third is *Liber de astrolabio*, a text on the construction and use of the astrolabe.

Raymond was not an innovator, and his astronomical and mathematical competence is not outstanding. Nevertheless, his work had considerable influence in the 12th century, especially in England, where a version of the *Liber iudiciorum* was prepared for Robert, Duke of Leicester, and where **Roger of Hereford** adapted the tables of Marseilles for his local meridian.

Charles Burnett

Selected References

D'Alverny, M.-T. (1967). "Astrologues et théologiens au XIIe siécles." In *Mélanges offerts à M.-D. Chenu*, pp. 31–50. Bibliothèque thomiste 37. Paris: Librairie philosophique J. Vrin. (For his defense of astrology.)

Poulle, E. (1964). "Le traité d'astrolabe de Raymond de Marseille." *Studi Medievali*, 3a ser., 5: 866–900.

Reber, Grote

Born **Chicago, Illinois, USA, 22 December 1911**
Died **Tasmania, Australia, 20 December 2002**

Grote Reber was the first person who knowingly built a radio telescope. With it, Reber pioneered the exploration of the sky at radio wavelengths, locating the first known discrete radio objects, Cygnus A and Cassiopeia A.

Reber's father, Schuyler Colfax Reber was a lawyer and part owner of a canning factory, while his mother, Harriet (*née* Grote) Reber taught elementary school where Grote grew up in Wheaton, Illinois, a suburb of Chicago. Among Harriet's pupils in the seventh and eighth grade at the Longfellow school in Wheaton was **Edwin Hubble**, with whom Reber later corresponded. At the time of his graduation with a BA in electrical engineering from the Armour (now Illinois) Institute of Technology in 1933, Reber was already interested in **Karl Jansky**'s 1932 discovery of radio emission from our Galaxy. Reber sought employment at the Bell Laboratories, Holmdell, New Jersey, in order to work with Jansky. Unsuccessful in that strategy, Reber accepted employment with various manufacturers of electronic equipment in the Chicago area.

After failing to interest others in his plans to follow up on Jansky's work, Reber, an amateur radio enthusiast (call letters W9GFZ), decided to build his own dish antenna in order to achieve a higher angular resolution than that of Jansky's rather broad-beamed

rotating antenna. Jansky made observations at a wavelength of 14.6 m (a frequency of 20.53 MHz). Reber realized that observing at a higher frequency would also facilitate a higher angular resolution and, if the radiation was of thermal origin, stronger signals.

Reber financed and designed the telescope himself and, with little assistance, constructed it in the 4 months from June to September 1937 at his mother's home. The telescope, a meridian transit instrument consisting of a wooden frame with a galvanized iron reflecting surface 9.6 m in diameter with a focal length of 6.1 m, was able to observe declinations between −32.5° and +90°. Reber also designed and built the antenna feeds and receivers; his work as an engineer for radio manufacturing firms in Chicago gave him access to state-of-the-art technology.

Reber first observed in 1938 at a wavelength of 9 cm (3,300 MHz), close to the shortest wavelength possible at the time, and then with a more sensitive receiver at 33 cm (910 MHz) in late 1938 and early 1939. Observations of various parts of the Milky Way, the Sun and other bright stars, and planets were unsuccessful, which, as Reber characteristically put it, was "rather dampening to the enthusiasm." His lack of success did, however, indicate that the radio emission was not thermal. It is now known that radio emission at centimeter and meter wavelengths is predominantly produced by synchrotron radiation.

Reber designed a new receiver to operate at 187 cm (162 MHz). Early observations with this receiver suffered from human electrical interference, primarily automotive ignitions. To minimize these problems, Reber worked during the day, slept in the early evening, and observed after midnight. By early April 1939, it was clear that Jansky's cosmic static was coming from the Galaxy. The trend of stronger emission near the galactic center seen by Jansky was confirmed, but Reber's efforts to detect individual celestial objects including the Sun were not successful.

Reber's results, like Jansky's before him, were readily accepted for publication in the *Proceedings of the Institute of Radio Engineers*. Reber also submitted his results to the *Astrophysical Journal*; however the editor, **Otto Struve,** accepted Reber's paper only after astronomers from the University of Chicago visited Wheaton to inspect his equipment. Even then, when Reber's paper was published in 1940, the section considering theoretical aspects of his results was removed!

After Reber improved the sensitivity of his receiver, his survey of the sky undertaken between early 1943 and mid-1944 confirmed that diffuse radio emission came from the galactic plane, but that there were also several peaks of emission: one in Cassiopeia, which is now known to be the galactic supernova remnant Cassiopeia A, and a broader peak in Cygnus, later resolved into the radio galaxy Cygnus A and the galactic Cygnus X complex. Observations of the Sun in September 1943 found it to be an intense source of radio emission. Reber's 1944 papers in the *Proceedings of the Institute of Radio Engineers* were the first published reports of radio emission from the Sun; he was unaware of classified earlier detections of the Sun during the war of which that by **James Hey** is best known. Reber found that near the maximum of the 11-year solar cycle, the Sun was not only very bright at radio wavelengths but, as a series of observations at 62.5 cm (480 MHz) from mid-1946 onward revealed, it was also the source of intense, short-lived bursts of radio emission.

News of Reber's discoveries reached the Dutch astronomer **Jan Oort**, who was one of the few who fully appreciated their significance. Oort realized that radio observations, unlike optical observations, did not suffer from obscuration caused by dust and gas, and would provide a powerful probe if a spectral line at radio wavelengths could be found. Oort assigned the problem to Hendrik van de Hulst, who discovered that neutral hydrogen had a potentially detectable hyperfine transition at 21 cm (1400 MHz). Van de Hulst met Reber in 1945, and while it was not clear that the line would be detectable, Reber started preparing his telescope to observe at that frequency.

In 1947, Reber accepted a position with the National Bureau of Standards [NBS] in Washington; his work in Wheaton came to an end. Reber moved his telescope to Sterling, Virginia, and then in 1959 moved it again to the National Radio Astronomy Observatory at Green Bank, West Virginia, where it remains to this day. The telescope now sits on an azimuth rail that had been designed and built but never installed in Wheaton. Reber attempted to generate interest in his design for a 220-ft. (67-m) radio telescope, but was again ahead of his time. The United States' interest in radio astronomy lagged behind Australia's, the Netherland's, and the United Kingdom's.

In 1951, Reber left the NBS and moved to Hawaii. This move marked the beginning of his long association with Research Corporation, a private foundation for the advancement of science established in New York in 1912. Reber installed antennas on the 10,020-ft. summit of the Haleakala volcano on the island of Maui. There he used the Lloyd's mirror interferometric technique in which a reflection off water acts as the second mirror pioneered by John Gatenby Bolton (1922–1993) and colleagues in Australia. Observations were attempted at frequencies near 20, 30, 50, and 100 MHz. However, anomalies introduced by the ionosphere rendered the first three frequencies unusable. At 100 MHz, the bright radio sources Cas A and Cyg A were detected and considerable asymmetry inferred for both sources from the differences in their rising and setting interferometric patterns.

Having discovered the important role of the ionosphere in low-frequency observations, Reber moved in 1954 to Tasmania, near the south magnetic pole, where ionospheric effects were known to be much smaller. He collaborated with Graeme Reade Anthony "Bill" Ellis (born: 1921) to determine the lowest frequency at which they could detect extraterrestrial radio emission. They were able to observe several times at 0.91 MHz, and on three occasions for half an hour or so at 0.52 MHz during the 1954/1955 solar minimum. Reber then built an array of 192 dipoles covering 223 acres and carried out a survey of the sky between February 1963 and May 1967 at 144 m (2.1 MHz). Reber found that the radio appearance of the sky at these longer wavelengths is the inverse of that at shorter wavelengths, with the sky being brighter at the galactic poles, and darker at the galactic center. Reber's very low frequency observations were confirmed (albeit with lower angular resolution) by observations with the Radio Explorer Satellites in the late 1960s and 1970s. The radio darkening toward the galactic plane is thought to arise from absorption by plasma in our galaxy.

Reber returned to Tasmania in the 1970s after a 4-year stay at the Ohio State University, which had awarded him an honorary Doctor of Science degree in 1962. Reber was also honored with the Henry Norris Russell Lectureship from the American Astronomical Society and the Catherine Wolfe Bruce Award from the Astronomical Society of the Pacific in 1962, the Elliot Cresson Gold Medal from the Franklin Institute in 1963, the Karl G. Jansky Lectureship of the National Radio Observatory in 1975, and the Jackson-Gwilt Medal and Gift of the Royal Astronomical Society in 1983.

The self-professed stubbornness with which Reber persevered through many months of unsuccessful observing in Wheaton also attended his refusal to accept the Big-Bang model of the expanding Universe. He favored an interpretation of redshift in terms of energy loss due to multiple Compton scatterings, although this theory can be readily falsified. However, it was Reber's determination to succeed in spite of initial setbacks that ensured that radio astronomy would develop into a thriving, fundamental field of research, and that opened the door to the possibility of astronomical observation at many other nonvisual wavelengths.

Philip Edwards

Selected References

Jansky, Karl G. (1932). "Directional Studies of Atmospherics at High Frequencies." *Proceedings of the Institute of Radio Engineers* 20: 1920–1932.

——— (1933). "Electrical Disturbances Apparently of Extraterrestrial Origin." *Proceedings of the Institute of Radio Engineers* 21: 1387–1398.

Kellermann, K. I. (1999). "Grote Reber's Observations on Cosmic Static." *Astrophysical Journal* 525: 371–372.

——— (2003). "Grote Reber, 1911–2002." *Bulletin of the American Astronomical Society* 35: 1472–1473.

——— (2003). "Grote Reber (1911–2002)." *Nature* 421: 596.

Reber, Grote (1940). "Cosmic Static." *Proceedings of the I.R.E.* 28: 68–70.

——— (1958). "Early Radio Astronomy at Wheaton, Illinois." *Proceedings of the I.R.E.* 46: 15–23.

——— (1983). "Radio Astronomy between Jansky and Reber." In *Serendipitous Discoveries in Radio Astronomy*, edited by K. Kellermann and B. Sheets. Green Bank, West Virginia: National Radio Astronomy Observatory, p. 43.

Sullivan III, W. T. (ed.) (1984). *The Early Years of Radio Astronomy: Reflections Fifty Years after Jansky's Discovery*. Cambridge: Cambridge University Press.

Recorde, Robert

Born **Tenby, (Dyfed), Wales, *circa* 1510**
Died **Southwark, (London), England, 1558**

Robert Recorde was the first writer in English on arithmetic, geometry, and astronomy.

Recorde was the son of Thomas Recorde and Rose Johns (or Jones). He received his BA and was elected fellow of All Souls College, Oxford University, in 1531. Recorde then moved to Cambridge, where he studied mathematics and medicine, and graduated with an MD at the University of Cambridge in 1545. He may have returned to Oxford University to teach mathematics, rhetoric, anatomy, music, astrology, and cosmography, but details are lacking. In fact, it seems that his reputation as a teacher rests on the views expressed in his treatises about how best to teach mathematics. Recorde moved to London where he practiced as a physician from 1547. In 1549 he was comptroller of the Mint at Bristol. The same year he became entangled in political intrigues and was accused of treason by William Herbert; this marked the beginning of a permanent quarrel with Herbert that had serious consequences for Recorde's later career. In May 1551 he was appointed general surveyor of the mines and money by the king. Recorde died in the King's Bench prison for failure to pay a penalty in a libel suit that Herbert had brought against him, yet left a little money to relatives in a will that was admitted to probate on 18 June 1558.

Recorde seems to have been a polymath who collected historical and other ancient manuscripts. His major claim to fame rests on his precociousness in mathematics, and he was the first to introduce algebra into England. Recorde's *Pathway to Knowledge* (1551) contains the first use of the term "sine" in English, and it also has a woodcut portrait of Recorde. He introduced a number of arithmetical symbols into English, and his works in mathematics remained standard authorities until the end of the 16th century.

The *Pathway*, subtitled "First Principles of Geometry," explains solar and lunar eclipses and contains a list of astronomical instruments. The *Castle of Knowledge* (1556) is the first comprehensive treatise on astronomy and the sphere to be printed in English, containing Ptolemaic astronomy in an elementary form with a brief, favorable reference to the Copernican theory. On the basis of this reference, some have considered him the first in England to adopt the Copernican system. Probably because this is an elementary text, Recorde cautioned students against rejecting the theory until they are more advanced in the study of astronomy. Still, he implied that the new theory could save the phenomena as well as the older system. The *Castle* is based on **Ptolemy**, **Proclus**, **John of Holywood**, and **Oronce Finé**, but Recorde examined the standard authorities, correcting their textual errors. There are also a number of other works that are no longer extant but that he referred to in the *Castle*. Among these is *The Treasure of Knowledge* (1556), probably on the higher part of astronomy.

Recorde is important primarily as an educational theorist. He insisted on a definite order in the study of the various branches of mathematics. He emphasized simple explanation of fundamental ideas, deferring demonstration to a later stage. Recorde used the dialog form along with visual aids and applications to practical problems. From a historiographical point of view, he is also important for his critical use of authorities and sources. When one considers the Renaissance reverence for ancient authorities, Recorde's caution is exemplary. He is in this regard representative of the more cautious acceptance of **Aristotle** that one finds, especially among English authors in the 16th century. After praising Ptolemy for his diligence, Recorde continues:

> yet muste you and all men take heed, that both in him and in al mennes workes, you be not abused by their autoritye, but evermore attend to their reasons, and examine them well, ever regarding more what is saide, and how it is proved, then who saieth it: for autoritie often times deceaveth many menne."

It seens as if several generations of English students were introduced to mathematics by his writings, thus his reputation as the founder of the English school of mathematical writers.

André Goddu

Selected References

Clarke, Frances Marguerite (1926). "New Light on Robert Recorde." *Isis* 8: 50–70.
Easton, Joy B. (1966). "A Tudor Euclid." *Scripta Mathematica* 27: 339–355.
——— (1967). "The Early Editions of Robert Recorde's *Ground of Artes*." *Isis* 58: 515–532.
Johnson, Francis R. (1937). *Astronomical Thought in Renaissance England: A Study of the English Scientific Writings from 1500 to 1645*. Baltimore: Johns Hopkins University Press.
——— (1953). "Astronomical Text-books in the Sixteenth Century." In *Science, Medicine and History*, edited by E. A. Underwood, pp. 285–302. London: Oxford University Press.
Johnson, Francis R. and S. V. Larkey (1935). "Robert Recorde's Mathematical Teaching and the Anti-Aristotelian Movement." *Huntington Library Bulletin*, no. 7: 59–87.

Karpinski, Louis C. (1913). "The Whetstone of Witte (1557)." *Bibliotheca Mathematica*, 3rd ser., 13: 223–228.
Recorde, Robert (1543). *Ground of Artes*. London: Reynold Wolfe.
——— (1547). *Urinal of Physick*. London: Reynold Wolfe.
——— (1551). *Pathway to Knowledge*. London: Reynold Wolfe.
——— (1556). *Castle of Knowledge*. London: Reynold Wolfe. (References to an edition of 1551 are mistaken.)
——— (1557). *Whetstone of Witte*. London: John Kingston.
Sedgwick, William F. (1921–1922). "Recorde, Robert." In *Dictionary of National Biography*, edited by Sir Leslie Stephen and Sir Sidney Lee. Vol. 16, pp. 810–812. London: Oxford University Press.

Rede, William

Born **England, *circa* 1320**
Died **England, *circa* 1385**

William Rede's chief astronomical work was a recomputation of the Alfonsine Tables for the Oxford meridian, epoch 1340. These Oxford tables enjoyed wide circulation during the 14th and 15th centuries. He also lectured on the *theoria planetarum* and calculated horoscopes for his Merton colleague John Ashenden.

Rede was raised from childhood by Nicholas of Sandwich, a wealthy landholder in the south of England. Rede came up to Oxford in the mid-1330s; he was a fellow of Merton College between 1344 and 1357, where he was a protégé of the astronomer **Simon Bredon**. From the late 1350s onward, Rede held a number of church offices, and was elected Bishop of Chichester in 1368. A significant benefactor of Merton College, he gave it over 100 books, a collection of mathematical instruments, and funds for construction of the Mob Quad library.

Keith Snedegar

Selected References

Emden, A. B. (1957–1959). *A Biographical Register of the University of Oxford*. Vol. 3, pp. 1556–1560. Oxford: Oxford University Press.
North, J. D. (1992). "Natural Philosophy in Late Medieval Oxford" and "Astronomy and Mathematics." In *The History of the University of Oxford*. Vol. 2, *Late Medieval Oxford*, edited by J. I. Catto and Ralph Evans, pp. 65–102, 103–174. Oxford: Clarendon Press.

Redman, Roderick Oliver

Born **Rodborough near Stroud, Gloucestershire, England, 17 July 1905**
Died **Cambridge, England, 6 March 1975**

Roderick Redman designed and built a variety of astronomical instruments, applied them to problems in solar and stellar spectroscopy, and revitalized optical astronomy at Cambridge University after World War II. He was the only son and the eldest of four children of Roderick George and Elizabeth Miriam (*née* Stone) Redman. His father, who ran an outfitter's shop inherited from his own father, had left school at age 11, but felt strongly that his son, at least, should receive a full education. Redman senior was an active member, choirmaster, and organist of the Stroud Baptist Church, and Redman junior acquired a lifelong interest in music, being himself a fine organist.

Educated at Marling School, Stroud, Redman won an Open Exhibition in mathematics and physics to Saint John's College, Cambridge, and began his studies there in 1923 at the age of 18. He took the mathematical *tripos* with distinction in 1926, gaining a number of valuable studentships including the Isaac Newton Studentship. He was taken on as a research student by Sir **Arthur Eddington** to work on a theoretical problem in dynamical astronomy, obtaining his Ph.D. in 1930.

Following a traveling studentship at the Dominion Astrophysical Observatory in Victoria, British Columbia, Canada, Redman was appointed assistant director of the Solar Physics Observatory, Cambridge, under **Frederick Stratton** in 1931. Redman remained in Cambridge until 1939, occupied mainly with solar spectroscopy and spectrophotometry. He then moved to South Africa, as chief assistant at the Radcliffe Observatory, Pretoria. Unfortunately, with the outbreak of World War II, the observatory's 74-in. telescope under construction in England, which he had hoped to use, was left incomplete. Redman, however, made the best of what was available to him – the finder on the new telescope's mounting – to perform a valuable photometric program in collaboration with the Royal Observatory at the Cape.

In 1946 Redman was elected a fellow of the Royal Society, and in 1947 was invited to return to Cambridge University as professor of astrophysics and director of the combined University Observatory and Solar Physics Observatory. These two institutions, though on the same site, had different histories; they were now merged to form the Cambridge Observatories. Redman thus inheriting the posts of both Eddington and Stratton.

Redman's first task was a radical modernization. Old instruments were replaced by two new telescopes – a 16-in. Schmidt camera and a 36-in. reflector – and a new solar outfit with a Babcock grating spectrograph. Under Redman's directorship, Cambridge University was transformed into a leading center for astrophysical research, and he himself, a perfectionist in everything he touched, became a highly valued adviser on various national and international bodies. His own fields of activity included successful observations of four total solar eclipses in the course of his scientific life, the last being those of 1952 in Khartoum, where under ideal weather conditions he obtained exceptionally fine high-resolution spectra of the chromosphere, and of 1954 in Sweden.

Among the instruments Redman designed and built were a solar monochrometer, a slit spectrograph that could take a rapid series of coronal images during solar eclipse, a Fabry photometer (at Radcliffe), and, before leaving Cambridge for Pretoria, the spectrograph for the 74-in reflector. In the postwar period, he contributed to the design and construction of a photon-counting photometer and an assortment of narrow-band spectrometers used by his students, who included a large fraction of those who passed through the Cambridge Observatories during his directorship. His research results with these included the demonstration that the residual flux at the centers of strong solar absorption lines was much less than had previously been thought; several determinations of the temperature of the chromosphere; and a database of velocities for single and binary stars later useful as velocity standards and for studies of galactic dynamics.

An early campaigner for British telescopes at better sites than Cambridge, Redman participated in the site testing that led to the location of the Anglo–Australian Observatory and served as a consultant to the Anglo–Australian Telescope Board. He served the Royal Astronomical Society in every capacity except treasurer, and including president (1959–1961). He was president of several commissions and working groups of the International Astronomical Union, was a member of the council of the Royal Society (1953–1954), and held a variety of positions of responsibility in his college (Saint John's) and the astronomical community. He was director of the observatories and, very briefly, of the integrated Institute of Astronomy until his retirement in 1972.

Redman was survived by his Canadian wife, Kathleen, (*née* Bancroft), whom he married in 1935, and their four children.

Marry T. Brück

Selected References

Blackwell, D. E. and D. W. Dewhirst (1976). "Professor R. O. Redman." *Quarterly Journal of the Royal Astronomical Society* 17: 80–86.

Griffin, R. F. and Sir Richard Woolley (1976). "Roderick Oliver Redman." *Biographical Memoirs of Fellows of the Royal Society* 22: 335–357.

Redman, R. O. (1955). "The Widths of Narrow Lines in the Spectrum of the Low Chromosphere, Measured at the Total Eclipse of February 25, 1952." *Vistas in Astronomy* 1: 713–725.

——— (1960). "The Work of the Cambridge Observatories." *Quarterly Journal of the Royal Astronomical Society* 1: 10–25. (Royal Astronomical Society Presidential Address, 12 February 1960.)

——— (1961). "Photometry in Astronomy." *Quarterly Journal of the Royal Astronomical Society* 2: 96–106. (Royal Astronomical Society Presidential Address, 10 February 1961.)

Regener, Erich Rudolph Alexander

Born **Schleussenau near Posen (Poznań, Poland), 12 November 1881**

Died **Stuttgart, (Germany), 27 February 1955**

Physicist Erich Regener conducted research on cosmic rays and was the first to outfit a V-2 rocket for taking measurements in the upper stratosphere. Regener, son of a Royal land surveyor, attended Gymnasia in Bromberg, Marienburg (West Prussia), Stargard (Pommern), and yet again in Bromberg, and concluded his school years by sitting for his *Abitur* (final exam) in 1900. He then studied chemistry and physics at the Friedrich Wilhelm University in Berlin. Working under the supervision of Emil Warburg, he was awarded a Ph.D. in 1905 for a dissertation on the influence of ultraviolet radiation on the oxygen–ozone equilibrium. Regener was twice married, first to Victoria Mintschin and then to Gertrud Heiter. He had two children by his first wife.

From 1906 to 1913, Regener worked as an assistant at the Military Technical Academy in Berlin-Charlottenburg. In 1909, he qualified as a lecturer at the University of Berlin by completing a *Habilitationsschrift* (thesis) in nuclear physics, using α particles to determine the elementary electric charge, a subject that Heinrich Rubens had encouraged him to pursue. In 1912, he was appointed honorary professor and, in 1914, succeeded Richard Börnstein as full professor of physics and meteorology and head of the Physics Institute at the Agriculture College in Berlin. Concurrently, Regener taught physics at the Berlin Veterinarian College. From 1915 to 1918, he worked in the battlefields, first as an X-ray field-technician, and after 1917, as an assistant to Fritz Haber, where he was conducting research on gas warfare at the Kaiser Wilhelm Institute. In 1920, Regener became full professor of physics and director of the Physical Institute at the Technische Hochschule in Stuttgart, where he remained active until 1937 and, after the war, from 1945 until his retirement in 1951.

In 1937, Regener was removed from his civil service position because his first wife, Victoria, was a Russian-born Jew. She and his two children were forced to emigrate; the couple was later reunited. (After Victoria's death in 1949, Regener remarried.)

To continue his work, Regener obtained support from the Kaiser Wilhelm Gesellschaft and established the organization's research facility for studies of the stratosphere at Friedrichshafen, on the shores of Lake Constance. Here, he successfully led the institute during the war years by obtaining contracts and money from the National Ministry of Aviation and the German Research Institute for Aviation. After the war, Regener's facility became associated with the new Max Planck Gesellschaft and in 1952 became known as the Max Planck Institute for the Physics of the Stratosphere. The institute was moved to Lindau after Regener's death and was one of the two founding organizations underlying the contemporary Max Planck Institute for Aeronomy.

Regener continued experimental practices that his teacher, Rubens, had been using. His *Habilitationsschrift* led him to undertake research on radioactive compounds and to devise ways of making the trajectories of such radiation visible. Regener's interest in radiation was expanded in the 1920s to involve cosmic radiation. The significance of cosmic rays for astronomy was already apparent, although its corpuscular nature was not yet established. Regener studied the absorptivity of this radiation and performed measurements both at great depths underwater (in Lakes Constance and Alpsee), and at great heights in the atmosphere (up to 30 km). He developed ballooning techniques and automatic meteorological equipment by which it became possible to study other characteristics of the upper atmosphere (composition, moisture, temperature), which were of interest to the emerging aviation industry. With measurements of the ozone concentration in the stratosphere, Regener returned to one of the themes in his dissertation.

Regener endeavored to have his automatic meteorological equipment reach ever-greater heights. He explicitly stressed the advantage and superiority of unmanned meteorological balloons over more expensive manned balloons. In making this argument, he anticipated the contemporary discussion between supporters and opponents of manned space flights. Plans for launching measurement equipment from a balloon located 30 km above the ground were laboratory-tested in 1938, but never successfully realized.

In 1942, Regener and his colleagues Erwin Wilhelm Schopper and Hans Karl Paetzold were given the assignment of developing a scientific application for the V-2 rockets being constructed at the military research facility in Peenemünde, a village on the Baltic coast. In January 1945, instruments were installed in the nose of a rocket (the Regener container), but it was never launched. Portions of the instruments, however, were subsequently taken to the United States, where the first high-altitude research rockets were flown in 1946 from White Sands, New Mexico. In the final years of his life, Regener planned to send measurement equipment into the upper atmosphere using the French Véronique rocket, but he did not live to see the launch.

Bernd Wöbke

Selected References

De Vorkin, David H. (1992). *Science with a Vengeance: How the Military Created the US Space Sciences after World War II*. New York: Springer-Verlag, esp. Chap. 3, "Erich Regener and the V-2," pp. 23–44.

Fraunberger, F. (1975). "Regener, Erich Rudolph Alexander." In *Dictionary of Scientific Biography*, edited by Charles Coulston Gillispie. Vol. 11, pp. 347–348. New York: Charles Scribner's Sons.

Paetzold, H. K., G. Pfotzer, and E. Schopper (1974). "Erich Regener als Wegbereiter der extraterrestrischen Physik." In *Zur Geschichte der Geophysik: Festschrift zur 50jährigen Wiederkehr der Gründung der Deutschen Geophysikalischen Gesellschaft*, edited by Herbert Birett *et al.*, pp. 167–188. Berlin: Springer-Verlag.

Pfotzer, Georg (1985). "On Erich Regener's Cosmic Ray Work in Stuttgart and Related Subjects." In *Early History of Cosmic Ray Studies: Personal Reminiscences with Old Photographs*, edited by Yataro Sekido and Harry Elliot, pp. 75–89. Dordrecht: D. Reidel.

Régis, Pierre-Sylvain

Born **La Salvetat de Blanquefort, (Lot-et-Garonne), France, 1632**
Died **Paris, France, 1707**

Pierre Régis was a proponent of Cartesian philosophy and astronomy. Educated at the Jesuit College at Cahors, and at the University of Paris in theology, he taught at Toulouse and Montpellier, until he succeeded **Jacques Rohault** in Paris in 1680. He was admitted to the newly reformed Académie royale des sciences in 1699.

Régis was a student and follower of Rohault, a very popular lecturer, and one of the principal defenders of an experimentally grounded version of **René Descartes**'s natural philosophy. Régis's *System of Philosophy* of 1690, while based broadly on Descartes, is an eclectic mix of Descartes, **Pierre Gassendi**, and Rohault's probabilistic version of Cartesianism; after Rohault's 1671 *Treatise* it was the most important systematic popularization of Cartesian natural philosophy. The second and third books of Part II of the *System* are devoted to cosmogony and astronomy, and take as their starting point Descartes's theory of vortices.

Régis's one contribution to Cartesian cosmology is in the area of terrestrial gravity, which was a key element in the defense of vortex theory. Descartes had accounted for weight in terms of the complex dynamics of fluid matter around the Earth, which had the effect of pushing bodies toward the center. The theory had been further developed by **Christiaan Huygens** and Rohault. Régis questioned whether Rohault's account of the circulation of fluid matter would actually explain the pushing of bodies toward the center, arguing that it would instead push the body toward some point on the axis between the center of the Earth and the center of the parallel on which the bodies were situated.

Stephen Gaukroger

Selected Reference

Mouy, Paul (1934). *Le développement de la physique cartésienne, 1646–1712*. Paris: Vrin. (Reprint, New York: Arno Press, 1981.)

Regiomontanus

➔ **Müller, Johann**

Regius, Hendrick

Born **Utrecht, The Netherlands, 29 July 1598**
Died **Utrecht, The Netherlands, 19 February 1679**

Henricus Regius was one of **René Descartes'** first disciples, and one of the principal representatives of Cartesianism in the Netherlands, but his polemical style and intransigent approach soon brought him into conflict with the authorities. He studied in Franeker, Groningen, Leiden, Paris, Montpellier, Valence, and Padua. He taught medicine at the University of Utrecht from 1638.

Descartes showed Regius the material he had intended to publish as *The World*, and Regius had probably based his lectures on this material. In 1646, Regius published his *Fundamenta physices*, which offered a form of Cartesian natural philosophy stripped of the metaphysical foundations that Descartes had provided to legitimate his approach in his *Principles of Philosophy* (1644). Descartes had provided a detailed physical defense of a cosmological system in which the cosmos was indefinitely extended, and comprised an indefinite number of Solar Systems, each with its own planetary system. Although Regius' principal concern was with physiology, he did offer a complete system, so included Cartesian cosmology in Part II of the *Fundamenta*. Even though his account of cosmology here hardly strays at all from that of Descartes, he presented the system as a whole in a way that Descartes considered incautious and superficial, opening up Cartesianism to objections that Descartes' own carefully presented formulations had been designed to avoid. However, in his public dispute with Regius, Descartes was forced to clarify a number of elements of his natural philosophy, although none of these bore directly on cosmological issues.

Stephen Gaukroger

Alternate names
Henricus Regius
Roy, Hendrick de

Selected Reference
Mouy, Paul (1934). *Le développement de la physique cartésienne, 1646–1712.* Paris: Vrin. (Reprint. New York: Arno Press, 1981.)

Regnerus

➲ **Frisius, Gemma Reinerus**

Reichenau, Hermann von

➲ **Hermann the Lame**

Reinhold, Erasmus

Born **Saalfeld, (Thuringia, Germany), 22 October 1511**
Died **Saalfeld, (Thuringia, Germany), 19 February 1553**

Erasmus Reinhold is best known for his preparation of the Prutenic Tables, calculated from **Nicolaus Copernicus**'s heliocentric theory.

Reinhold's father Johann (1479–1558) studied in Leipzig and Cologne. He served in the Chancellory of George, Duke of Saxony, and as secretary to the Abbot of the Benedictine monastery at Saalfeld, and held various civic offices in Saalfeld, including that of Mayor. Johann Reinhold prospered sufficiently to enable his son to go from the *Stadtschule* (municipal school) in Saalfeld to Wittenberg University, where he dedicated himself primarily to the study of mathematics under Jacob Milichius. After graduation, Erasmus Reinhold was made professor of higher mathematics (*i. e.*, astronomy). He was elected dean of the philosophical faculty on repeated occasions and, in 1549–1550, rector of the university. Reinhold enjoyed a certain advancement through his close relationship to Philipp Melanchthon.

The most significant of Reinhold's publications were his commentary on the planetary theory of **Georg Peurbach** (1542, and subsequent editions); a textbook for the more advanced study of astronomy that he certainly based on his own lectures; a small work, *De Horizonte*, that appeared as supplement in a number of impressions of **John of Holywood**'s *Libellus de sphaera* (as printed in Wittenberg); and above all the *Tabulae Prutenicae* (Prutenic Tables). Reinhold was the most influential astronomer in Protestant Germany.

Through his fellow professor **Rheticus** Reinhold in Wittenberg had a very early opportunity to acquaint himself with the heliocentric system of Copernicus. Immediately upon its appearance, Reinhold worked his way attentively through Copernicus's *De revolutionibus orbium coelestium*. Following a partial recalculation of observational data (where Copernicus had made several mistakes) and on the basis of Copernicus's improved elements for planetary orbits, the constant of precession, the length of the year, and the obliquity of the ecliptic, as well as his own theory of lunar motion and approach to triangulation (among other factors), Reinhold produced new planetary tables. After many years of work, they appeared first in Tübingen in 1551, commissioned by Melanchthon and Duke Albert of Prussia, and named *Tabulae Prutenicae* in honor of the latter.

Reinhold did not endorse the heliocentric system, but recognized that the mathematical theory of this system represented a significant advance in relation to new observational data. He made no statement on the question of the physical reality of the heliocentric system; as a strongly "classically" minded astronomer, adhering to the conceptual model of a division between physics and astronomy, he did not accept that such a theory raised any problem. For Reinhold the question was irrelevant as he subscribed to the traditional scientific concept of astronomy "saving the appearances."

The Prutenic Tables rendered the heliocentric system operable for practical astronomy, in particular for the calculation of calendars and horoscopes. From the 1570s, Reinhold's tables were one of the most important astronomical tables, and they were instrumental in Copernicus being generally recognized, from the end of the 16th century, as one of the most important astronomers. The accuracy of the planetary positions as calculated proved, with time, to be a significant factor in the acceptance of the totality of the heliocentric system, rather than just its mathematical parameters. (Subsequently, larger errors became evident once again, as **Johannes Kepler** and others showed.) The tables were significant for the furtherance of the heliocentric world-system, even though Reinhold himself always recognized the geocentric.

Reinhold's brother Johannes became professor of mathematics in Greifswald in 1549, but died in 1552. Reinhold's son, Erasmus Jr. (1538–1592) studied mathematics and medicine in Wittenberg under the care of Melanchthon, and then in Jena, and became doctor of medicine and municipal doctor in Amberg and Saalfeld. Later he became *Bergvogt* (mountain steward) to the Elector of Saxony, and wrote works on land surveying as well as calendars, which appeared regularly for many years in Erfurt.

Jürgen Hamel
Translated by: *Peter Nockolds*

Selected References
Gingerich, Owen (1993). "Erasmus Reinhold and the Dissemination of the Copernican Theory." In *The Eye of Heaven: Ptolemy, Copernicus, Kepler.* New York: American Institute of Physics, pp. 221–251.

Hamel, Jürgen (1994). *Nicolaus Copernicus: Leben, Werk und Wirkung*. Heidelberg: Spektrum.

——— (1994). "Die Rezeption des mathematisch-astronomischen Teils des Werkes von Nicolaus Copernicus in der astronomisch-astrologischen Kleinliteratur um 1600." In *Cosmographica et Geographica: Festschrift für Heribert M. Nobis zum 70. Geburtstag*, edited by Bernhard Fritscher und Gerhard Brey. Vol. 1, pp. 315–335. Munich: Institut für Geschichte der Naturwissenschaften.

Henderson, Janice Adrienne (1991). *On the Distances between Sun, Moon and Earth according to Ptolemy, Copernicus and Reinhold. Studia Copernicana*, Brill's Series, Vol. 1. Leiden: E. J. Brill.
Koch, Ernst (1908). *Magister Erasmus Reinhold aus Saalfeld.* Saalfelder Weihnachtsbüchlein, 55. Saalfeld, pp. 3–16.
——— (1909). *Dr. Erasmus Reinhold [jr.].* Saalfelder Weihnachtsbüchlein, 56. Saalfeld, pp. 3–28.
Reinhold, Erasmus (2002). "Der Kommentar zu 'De revolutionibus' des Nicolaus Copernicus." In Nicolaus Copernicus, *Gesamtausgabe*. Vol. 8/1. Berlin.
——— *Commentarius in opus revolutionum Copernici.* MS lat. fol. 391. Staatsbibliothek Berlin, Preussischer Kulturbesitz.

Reinmuth, Karl Wilhelm

Born **Heidelberg, (Germany), 4 April 1892**
Died **Heidelberg, (Germany), 6 May 1979**

Before the advent of automated search techniques, Karl Reinmuth became the world's most successful asteroid hunter. He discovered a total of 389 minor planets, including some of the first known to exist outside of the main asteroid belt.

A lifelong resident of Heidelberg, Reinmuth studied at the Ruprich-Karls University. His doctoral thesis (1916), entitled "Photographische Positionsbestimmung von 356 Schultzschen Nebelflecken," reported on the position determinations of 356 nebulae, originally cataloged in 1875 by Herman Schultz of Uppsala.

Reinmuth joined the staff of the Königstuhl Observatory near Heidelberg as a volunteer in 1912, working under the supervision of director **Maximillian Wolf**, the first astronomer to apply photographic techniques to the discovery of asteroids. Reinmuth located his first new minor planet, (796) Sarita, on 15 October 1914. Upon Wolf's death, Reinmuth was appointed director of the Königstuhl Observatory in 1932 and served in that capacity until his retirement in 1957.

After World War I, Reinmuth was given responsibility for resuming a sky survey begun by Adam Massinger, another of Wolf's assistants. Massinger had been urged to compile a photographic record of more than 4,000 objects included in **John Herschel**'s *A General Catalogue of Nebulae and Clusters of Stars* (1864). Photographic plates from the Heidelberg Observatory, some dating back to 1900, were used in this collection. The project was halted by the outbreak of war; Massinger was killed in battle at Ypres. Reinmuth continued working on the project until it was completed in 1924.

In 1926, Reinmuth published the catalog entitled *Die Herschel-Nebel nach Aufnahmen der Königstuhl-Sternwarte.* Despite its German-language title, the volume listed the approximate dimensions, position angles (where relevant), and partial descriptions in English of the objects (chiefly galaxies) originally given by Herschel. At that time, Reinmuth's catalog was the only such listing that contained position angles of galaxies in the northern heavens.

Between 1919 and 1937, Reinmuth discovered all of the Trojan asteroids (seven) recognized during that period. Trojan asteroids have nearly circular orbits and are located at the leading or trailing Lagrangian points (L4 and L5) of Jupiter's orbit. During a routine asteroid patrol, Reinmuth discovered a faint comet on 22 February 1928 now known as 30P/Reinmuth.

During the 1930s, a competition of sorts developed between Reinmuth and **Eugène Delporte**, of the Belgian National Observatory at Uccle, in the discovery of Earth-approaching asteroids. Following Delporte's detection of minor planet (1221) Amor in 1932, Reinmuth recorded the trail of a nearby, rapidly moving asteroid whose orbit was found to cross that of the Earth. It was given the name (1862) Apollo, after the Greek Sun god. Reinmuth's discovery proved to be the first object in a class of asteroids sharing similar orbital properties. Apollos are one of three classes of Earth-approaching asteroids, along with the Amors and Atens, whose perihelion distances lie inside the orbit of Mars. Apollo-type asteroids have semimajor axes (a) that are greater than or equal to 1 AU and perihelion distances (q) that are less than or equal to 1.017 AU. Reinmuth discovered another member of the Apollo class of minor planets, (69230) Hermes, which remained "lost" by astronomers until its accidental rediscovery in 2003. For over 50 years, Hermes held the close-approach record of only 800,000 km from Earth.

In 1980, evidence was presented by physicist **Luis Alvarez** and colleagues that an asteroid's collision with the Earth was responsible for the Cretaceous–Tertiary extinction of the dinosaurs. Consequently, the potential danger of near-earth asteroids [NEAs] has been taken far more seriously by astronomers than in Reinmuth's day. Automated search techniques enable the discovery of NEAs to be made at a remarkable pace – a feat that surely would be the envy of Reinmuth.

Reinmuth is commemorated with the naming of minor planet (1111) Reinmuthia.

Robert D. McGown

Selected References

Cunningham, Clifford J. (1988). *Introduction to Asteroids: The Next Frontier.* Richmond, Virginia: Willmann-Bell, esp. pp. 93–111.
Peebles, Curtis (2000). *Asteroids: A History.* Washington, DC: Smithsonian Institution Press, esp. pp. 56–80.
Reinmuth, K. W. (1926). *Die Herschel-Nebel nach Aufnahmen der Königstuhl-Sternwarte.* Berlin: de Gruyter.
Schubart, J. (1980). "Karl Reinmuth." *Mitteilungen der Astronomischen Gesellschaft*, no. 50: 7–8.

Renieri, Vincenzio

Born **Genoa, (Italy), 30 March 1606**
Died **Pisa, (Italy), 5 November 1647**

A friend and disciple of **Galileo Galilei**, Vincenzio Renieri is best remembered for his work on the satellites of Jupiter. Although little is known about his early life, Renieri joined the Olivetan Order in 1623, which initially took him to Rome and a decade later to Siena, where he met Galilei (1633). Thereafter, Renieri made frequent visits to Arcetri and soon became an intimate friend of Galilei. Working closely with Vincenzio Viviani, Renieri was given the task of continuing Galilei's observations of the satellites of Jupiter and improving the computational tables. To that end, Galilei entrusted Renieri with all of his papers dealing with Jupiter's satellites, which were to be perfected and sent to the States-General of Holland as the basis for determining longitude at sea. Having accepted the chair in mathematics at Pisa (previously held by Dino Peri and earlier still by Galilei), Renieri continued to supplement Galilei's observations while improving his methods.

Unfortunately, Renieri published only one work, the *Tabulae mediceae secundorum mobilium universales* (Florence, 1639). Although the title is tantalizing, Renieri's widely discussed tables for Jupiter's satellites were not published during his lifetime. **Jean-Baptiste Delambre**, among others, garbled the story about Renieri's sole publication, the *Tabulae mediceae*, not only indicating it was published before his fourth birthday but also questioning whether Renieri's satellite observations ever existed. The *Medicean Tables* say nothing about the Medicean satellites; instead, they represent a typical effort to improve **Johannes Kepler**'s *Rudolphine Tables* for the planets. Deploring the gross errors in earlier ephemerides (Ptolemaic, Alphonsine, Prutenic), Renieri claimed a simpler method for calculating longitudes by means of a two-step procedure, which he applied first to the superior planets, then to the inferior. His treatment of planetary latitudes is similar but more complicated. Renieri concluded by comparing his results (for the middle terrestrial latitudes), arguing that his tables were superior to the Rudolphine, Tychonic, Danish, and Landsbergian.

In the end, Renieri's observations of Jupiter's satellites were not printed until the mid-19th century. Although the Grand Duke of Tuscany ordered the ephemerides published (later echoed by his brother, Prince Leopold), Renieri died prematurely, and his manuscript was lost. (A questionable report suggests it was stolen by one Giuseppe Agostini.) The loss was unfortunate, as the synodic periods Renieri supplied were remarkably accurate, clearly superior to values given by **Simon Mayr** and other contemporaries. **Alexandre Pingré** indicates Renieri had an excellent telescope but made many observations with a mediocre quadrant. Several of Renieri's observations are cited in **Giovanni Riccioli**'s *Almagestum novum* (pt. I) .

Robert Alan Hatch

Selected Reference

Delambre, J. B. J. (1821). *Histoire de l'astronomie moderne. 2 Vols.* Paris: Courcier.

Respighi, Lorenzo

Born **Cortemaggiore, (Emilia-Romagna, Italy), 7 October 1824**
Died **Rome, Italy, 10 December 1889**

Alongside **Angelo Secchi** and **Giovanni Schiaparelli**, Lorenzo Respighi can be considered the most important Italian astronomer of the second half of the 19th century. He was a pioneer in solar spectroscopy.

Respighi completed his early studies in Parma and his higher studies at the University of Bologna, where he obtained his degree *ad honorem* in philosophy and mathematics in 1847. At age 18, he was inducted into the Accademia delle Scienze of Bologna, and four years later he became a permanent member of the Accademia Benedettina. In 1849, Respighi was appointed as substitute to the chair of mechanics and hydraulics at Bologna. In 1851, he was appointed professor of optics and astronomy, a chair vacant for 3 years, and in 1855 he was also named director of the Astronomical Observatory. Respighi held these positions until 1864, when he was dismissed for refusing to take an oath of loyalty to the government of the Kingdom of Italy, which succeeded the papal government in Bologna.

Respighi moved to Rome in 1865, and in August of that year Pope Pius IX appointed him professor of optics and astronomy at the University of Rome La Sapienza and director of the Roman Observatory at the Capitol, a post he held until his death.

Respighi's scientific work began with a mathematical opus entitled *Principi del calcolo differenziale*, which **Augustin Cauchy** presented at the Académie des sciences in Paris. He subsequently became interested in positional astronomy, geodetic astronomy, and instrumental astronomy, but his main focus was physical astronomy (spectroscopy) and he is considered one of the pioneers in this field of study in Italy. In positional astronomy, Respighi compiled three catalogs, published between 1877 and 1884, of the mean declinations of over 2,500 stars at the 1875.0 epoch. Each star was observed several times using Ertel's meridian circle at the Bologna Observatory, both directly and applying the new method for observing zenithal stars by reflection in a basin of mercury. During these observations, he did extensive research into the aberration of light, conducting experiments using a water-filled telescope. He also conducted experiments on the variation of the speed of light in a vacuum, again with Ertel's meridian circle, using a thick piece of glass made specifically for this purpose with numerous air bubbles inside. Respighi also observed planets and comets, discovering three comets (C/1862 W1, C/1863 G2, and C/1861 Y1). In geodetic astronomy, Respighi determined the latitude of the Bologna Observatory and the absolute declination value for the city, as well as the latitude of the Capitol Observatory and of the prime meridian of Monte Mario in Rome. He also studied meteorology, writing several dissertations on the climate of Bologna, in which he reduced and discussed the meteorological and magnetic data accumulated by the Bologna Observatory in 45 years of observations (1814–1859).

Respighi's most important astrophysical researches involved stellar scintillation and the study of solar physics. Between 1869 and 1885, he conducted systematic observations of the border of the Sun, using the method invented by **William Huggins** in 1869. Known as "the widened slit method," it entailed placing the slit of the spectroscope tangential to the border of the Sun, widening the slit to include the entire height of the solar prominences, and placing a piece of red glass in front of the eyepiece. Respighi also studied the relationship between sunspots and solar prominences, deducing that at the points of the border of the Sun where a sunspot occurs or disappears, the chromosphere looks thinner. In 1870, he participated in the Italian expedition to observe the solar eclipse in Sicily, and in 1873 he began a series of measurements of the diameter of the Sun to study its variations in relation to the sunspot cycle. Lastly, Respighi noticed that the spectral lines over the sunspots look widened and deformed, a phenomenon that was later studied by

George Hale and is due to the Zeeman effect. He also studied the spectrum of the aurora borealis and compared this spectrum to the one of zodiacal light.

When head of the Capitol Observatory, Respighi engaged in a long dispute with Secchi about who had invented the use of the objective prism to study the stellar spectra and the nature of sunspots. He discovered that on the instrument invented by **Joseph von Fraunhofer** in 1815, the angle of refraction was too large, so it could not be used for spectra. Thus, Respighi had a prism made with an angle of refraction of just 12° and mounted it on the equatorial telescope at the Capitol. With this instrument, on 15 February 1869, he was able to show French physicist **Marie Cornu** the excellent stellar spectra he had obtained. The instrument used by Respighi spread rapidly and became one of the chief aids in astrophysics. With Secchi, **Pietro Tacchini**, **Giuseppe Lorenzoni**, and A. Nobile, in 1871 Respighi helped establish the Italian Society of Spectroscopists, although he rarely participated actively due to his disagreement with Secchi.

Respighi was a member of the European Commission for the Degree and of the Royal Superior Commission of Weights and Measures. Likewise, he was a member of numerous academies, such as the Accademia delle Scienze of Bologna, the Accademia dei Lincei, the Italian Society of Science known as the Società dei XL, the Geneva Society of Physics and Natural History, and the Royal Society of London. He was conferred with the Order of Saints Mauritius and Lazarus, the Order of the Crown of Italy, and the Military Order of the Portuguese Crown.

Respighi's manuscripts are in the Historical Archives of the Department of Astronomy of the Bologna University and in the Historical Archives of the Rome Astronomical Observatory in Monte Porzio.

Fabrizio Bònoli

Selected References

Ferrari, G. S. (1891). *Lorenzo Respighi: Suo elogio nell'anniversario della sua morte.* Rome.

Tacchini, P. (1889). "Lorenzo Respighi." *Memorie della Società degli spettroscopisti italiani* 18: 200.

Rheita, Antonius Maria Schyrleus de Schyrle [Schierl, Schürle] Johann Burchard

Born **Reutte, Tyrol, (Austria), 1604**
Died **Ravenna, (Italy), 1659 or 1660**

Antonius de Rheita was a telescope maker, observer, and supporter of the Tychonic system. He described the bands of Jupiter and developed the compound eyepiece.

Rheita was born in a noble family; "de Rheita" was derived from his birthplace. Rheita was educated in the Augustinian abbey of Indersdorf, Bavaria. On 14 October 1623, he enrolled at the University of Ingolstadt, where optics and astronomy were taught in the tradition of **Christoph Scheiner** and **Johann Cysat**. Three years later, Rheita left without a degree to become a Capuchin monk in the monastery of Passau, taking the name Antonius Maria. In 1636, he left Passau to become reader in philosophy at the Capuchin monastery of Linz. There, he met Philipp Cristoph von Sötern, the elector of Trier, for whom he worked for several years. In 1642, Rheita was in the Capuchin monastery of Cologne, where he conducted astronomical research and constructed telescopes that would result in his first publication, *Novem stellae circa Jovem visae, circa Saturnum sex, circa Martem nonnulla* (Nine stars seen [or observed] around Jupiter, six around Saturn, several around Mars), in 1643. That year, he met the instrument and telescope maker Johann Wiesel in Augsburg, where Rheita presumably was to give support to the nearby Bridgettine monastery in Altomünster of his brother Elias. One year later, Rheita was in Antwerp to prepare the publication of his main work *Oculus Enoch et Eliae*, before returning, in 1645, to Trier, again in the service of Sötern until the latter's death in 1652.

In those years, Rheita ran a workshop that produced telescopes for, among others, the Archbishop-Elector of Mainz. In Belgium to prepare a new edition of his *Oculus Enoch et Eliae* in 1653, he was informed of accusations against him by the Inquisition. Imprisoned in Bologna and later in Ravenna, Rheita's plans to build an observatory in Mainz would never materialize. On 27 November 1659 or 14 November 1660 (depending on the source used) he died, in unclear circumstances, in confinement.

In his *Oculus Enoch et Eliae*, Rheita defended the Tychonic world system, criticizing the Copernican system as established by **Philip Lansbergen**. In this he followed Scheiner and Cysat, and his friend and supporter Eryceus Puteanus. Defending the physical reality of the geoheliocentric world system was part of Rheita's contribution to the Counter-Reformation.

Rheita also published a map of the Moon in his *Oculus Enoch et Eliae*. With the exception of **Francesco Fontana**'s unpublished lunar drawings of 1629 and 1630, this was the first map to present the Moon south up, as seen through a telescope consisting of two convex lenses. However, Rheita's map was uninfluential.

Rheita's main contribution to astronomy was in the field of telescope design. In his 1643 work *Novem stellae circa Iovem, circa Saturnum sex, circa Martem nonnullae* Rheita claimed to have discovered a number of new satellites of the superior planets. In particular, on 29 December 1642, he said he discovered five satellites of Jupiter, above the four satellites already discovered by **Galileo Galilei** in 1610. The letter spread very rapidly to Paris. **Pierre Gassendi** published his answer together with Rheita's letter in April 1643. Gassendi correctly claimed that the new satellites were fixed stars. Rheita argued that those satellites could only be revealed by his recently invented binocular telescope. *Oculus Enoch et Eliae* referred to the binocular telescope, with Henoch and Elias each symbolizing one eye. With the same instrument, Rheita observed the bands on Jupiter, as he announced in a letter of 18 June 1651 to the Elector of Mainz.

More important for telescope design was the compound eyepiece, developed in collaboration with Wiesel around 1644. In 1611, **Johannes Kepler** had already introduced a third convex lens, that is, an erector lens, to reinvert the image seen through his telescope; it consisted of two convex lenses. Rheita's telescope consisted of four convex lenses, one objective lens and three ocular lenses. (Rheita introduced a field lens beside an erector lens.) The focal point of the objective lens coincided with the focal point of the third ocular lens. The introduction of the third ocular lens allowed overcoming the main problem of the telescopes of the first half of the 17th century: the limitation placed on magnifications by the progressive restriction of the field of view. Moreover, Rheita discovered that the angle of view was a function of the diameter of the ocular lenses. As a consequence, the ocular lenses were often made larger than the objective lens, resulting in an inversion of the trumpet shape of the tube with respect to earlier telescopes. The compound eyepiece also limited chromatic aberration.

Rheita's design of telescopes became known throughout Europe, mostly through telescopes produced by Wiesel in Augsburg. The same design was soon used in England by Richard Reeve (of Hartlib's circle), in Holland by **Christiaan Huygens**, in Italy by **Giuseppi Campani,** and in France by another Capuchin monk, Chérubin d'Orléans. Rheita in his book suggested that polishing on paper glued into a spherical mould would allow for precise polishing of spherical lenses of long focal lengths. This new polishing technique, together with the compound eyepiece, allowed for longer and longer telescopes, and the astronomical discoveries that went along with it, during the second half of the 17th century.

Sven Dupré

Selected References

Goercke, Ernst (1991). "Der Schliff asphärischer Linsen nach Angaben des Rheita." *Technikgeschichte* 58: 179–187.

Keil, Inge (2000). *Augustanus Opticus: Johann Wiesel (1583–1662) und 200 Jahre optisches Handwerk in Augsburg*. Berlin: Akademie Verlag.

——— (ed.) (2003). *Von Ocularien, Perspicillen und Mikroskopen, von Hungersnöten und Friedensfreuden, Optikern, Kauflauten und Fürsten: Materialien zur Geschichte der optischen Werkstatt von Johann Wiesel (1583–1662) und seiner Nachfolger in Augsburg*. Augsburg: Wissner-Verlag.

Thewes, Alfons (1983). *Oculus Enoch … Ein Beitrag zur Entdeckunsgeschichte des Fernrohrs*. Oldenburg: Verlag Isensee.

Van Helden, Albert (1977). "The Development of Compound Eyepieces, 1640–1670." *Journal for the History of Astronomy* 8: 26–37.

Whitaker, Ewen A. (1999). *Mapping and Naming the Moon: A History of Lunar Cartography and Nomenclature*. Cambridge: Cambridge University Press.

Willach, Rolf (2001). "The Development of Telescope Optics in the Middle of the Seventeenth Century." *Annals of Science* 58: 381–398.

Rheticus

Born **Feldkirch,Vorarlberg, (Austria), 16 February 1514**
Died **Kassa (Košice, Slovakia), 4 December 1574**

Georg Rheticus was among the first to adopt and spread the heliocentric theory of **Nicolaus Copernicus**.

Georg von Lauchen, later known as Rheticus, was born in an Austrian town near the Swiss border. His father, Georg Iserin, was the town doctor and a government official; he taught his son until 1528, when he was tried on a charge of sorcery, convicted, and beheaded. One of the consequences of this execution was that his name could no longer be used; therefore, Georg's mother, an Italian noblewoman named Tommasina de Porris, reverted to her maiden name. Since "de Porris" means "of leeks" in Italian, Rheticus preferred to translate it into German "von Lauchen." Later, he took the additional name of Rheticus, after the ancient Roman province of Rhaetia in which he had been born.

Rheticus studied first at the Latin school in Feldkirch, then at the Frauenmünsterschule in Zürich until 1531. In 1533 he matriculated from the University of Wittenberg, where he received the title of *magister artium* in 1536. Soon afterward, thanks to Philipp Melanchthon's support, Rheticus was appointed to teach arithmetic, geometry, and astronomy at the University of Wittenberg.

In October 1538, Rheticus went on leave to visit leading astronomers and mathematicians: **Johannes Schöner** in Nüremberg, **Peter Apian** in Ingolstadt, Joachim Camerarius in Tübingen, and Copernicus in Frombork (Frauenburg). In September 1541, Rheticus went back to Wittenberg, where he was elected dean of the arts faculty. A few months later he was offered a post as professor of higher mathematics at the University of Leipzig, where he began teaching in October 1542.

In 1545, Rheticus left Leipzig to study abroad. After making a short stay in Feldkirch, he spent some time in Italy. Toward the end of 1546, he suffered a severe mental disorder in Lindau, a town on Lake Constance; this aroused some unfounded rumors about his death. But his health recovered to allow Rheticus to teach mathematics and astronomy at Constance for 3 months in late 1547. Then he studied medicine in Zürich with Konrad von Gesner.

In February 1548, Rheticus went back to Leipzig, where, with Melanchthon's influence, he was made a member of the theological faculty. In this period he was deeply engrossed in his university duties and sent many books to the press, among them a Latin translation of Euclid's *Elements* (1549), a calendar and ephemeris (1550), and the *Canon doctrinae triangulorum* (1551), which was the first publication to contain all the six trigonometrical functions. In April 1551, Rheticus was accused of having a homosexual affair with one of his students, and the consequent scandal compelled him to escape from Leipzig. His friends, such as Melanchthon, stopped supporting him, and he was tried in his absence by a town court. Rheticus was condemned to 101 years of exile, and all his possessions in Leipzig were impounded.

After leaving Leipzig, Rheticus spent some time at Chemnitz, before establishing himself in Prague. In 1551/1552 he studied medicine at the University of Prague. In 1553 Rheticus made a trip to Vienna, and the following year he moved to Kraków, Poland, where he remained for 20 years as a practicing doctor, though he continued to devote himself to mathematics and astronomy. In this period he worked on the trigonometric tables, projected and constructed astronomical instruments, and carried out astronomical observations and alchemy experiments.

In 1574 Rheticus left Kraków and went to Kassa (Kosice), by request of the local magnate Johannes Ruben. Here he was visited by Valentine Otho, who was at that time a student of mathematics at the University of Wittenberg. Rheticus died shortly afterward. He left an unfinished manuscript, which was then completed by Otho and published with the title of *Opus palatinum de triangulis* (1596); another book, the *Thesaurus mathematicus*, was edited by Bartholomeo Pitiscus (1613). Thanks to these posthumous works, Rheticus can be considered one of the most important authors of trigonometrical tables.

Rheticus was among the first to adopt and spread the heliocentric theory of Copernicus. Rumors of this hypothesis had reached Rheticus at Wittenberg. In May 1539 he traveled to Poland to visit Copernicus, and ended up staying at Frombork for 2 years, during which time he was able to persuade Copernicus to let him study the virtually completed *De revolutionibus*. Rheticus became enthusiastic over the heliocentric theory and tried to convince Copernicus to publish it. Since his efforts were rewarded with no success, Rheticus wrote a brief summary of the theory, which was published in Danzig at the beginning of 1540, explicitly authorized by Copernicus, under the title of *Narratio prima de Libris Revolutionum eruditissimi viri et mathematici excellentissimi, reverendi Domini Doctoris Nicolai Copernici Torunnaei Canonici Varmiensis* (First report on the Books of the Revolutions of the learned gentleman and distinguished mathematician, the Reverend Doctor Nicolaus Copernicus of Torun, Canon of Warmja).

Rheticus's booklet, written in the form of a letter to Schöner, was the first illustrated account of Copernicus's heliocentric theory. The *Narratio prima* is not, however, a pure summary of *De revolutionibus*; it has a different structure. First of all, Rheticus explained the questions about the motion of the fixed stars and the precession of the equinoxes: Curiously, therefore, at the beginning of his booklet he did not speak of the three motions of the Earth, but introduced them at the end. Rheticus's expository method was contrary to that used by Copernicus. While the latter started from the statement that the Earth moves and then tried to demonstrate it by analyzing the apparent motions of the stars, Rheticus expounded these motions of the stars in order to be able to assert that the motions of the Earth are the unique way of explaining them. This expository order was a consequence of the pedagogical aims that Rheticus wanted to achieve, but perhaps there is also another reason: Rheticus seemed to underline that the apparent motions of the stars and those of the Sun, which the astronomical tradition had considered as separated matters, are strictly correlated and can be coherently explained only by assuming a moving Earth. For many decades the *Narratio prima* remained the best popularization of the heliocentric theory. The first edition of 1540 was enthusiastically received, and a second edition was published in Basel less than a year later. The *Narratio* was reprinted as an appendix of Copernicus's *De revolutionibus* in 1566 and of **Johannes Kepler**'s *Mysterium cosmographicum* in 1596.

Probably as a result of the *Narratio's* success, when Rheticus left Frombork in 1541, Copernicus allowed him to take a complete copy of *De revolutionibus* to arrange for its publication. Rheticus entrusted publication of the manuscript to Johann Petrius in Nüremberg, but he could not supervise the entire work and left oversight of the printing to a Lutheran theologian, **Andreas Osiander**, who made some unauthorized additions to the manuscript. When Rheticus received the first copies of the printed book in April 1543, he saw that the title had been changed: Instead of *De revolutionibus*, the printed version read *De revolutionibus orbium coelestium*. The worst change, however, was the insertion of an anonymous preface, which affirmed that the book contained a mere mathematical hypothesis, not a description of the real universe. Rheticus suspected that Osiander had made the changes and probably did not approve; however, he did not take an official position against the preface, since perhaps he considered it a way of making the heliocentric theory more acceptable for ecclesiastics and theologians.

Marco Murara

Alternate name

Lauchen, Georg Joachim von

Selected References

Burmeister, H. (1966–1968). *Georg Joachim Rhetikus, 1514–1574: Eine Bio-Bibliographie*. 3 Vols. Wiesbaden: Pressler-Verlag.

Hooykaas, R. (1984). *G. J. Rheticus' Treatise on Holy Scripture and the Motion of the Earth: With Translation, Annotations, Commentary, and Additional Chapters on Ramus-Rheticus and the Development of the Problem before 1650*. Amsterdam: North-Holland.

Rosen, Edward (ed.) (1971). *Three Copernican Treatises: The* Commentariolus *of Copernicus, the* Letter against Werner, *the* Narratio prima *of Rheticus*. 3rd rev. ed. New York: Octagon Books.

Westman, Robert S. (1975). "The Melanchthon Circle, Rheticus, and the Wittenberg Interpretation of the Copernican Theory." *Isis* 66: 165–193.

Rho, Giacomo

Born **Milan, (Italy), 1593**
Died **Beijing, China, 27 April 1638**

Italian Jesuits Giacomo Rho and **Johann Schall von Bell** were appointed by the emperor in Beijing to reform the Chinese calendar. Rho brought the Tychonic cosmological model to China.

Selected Reference

Sivin, Nathan (1995). *Science in Ancient China: Researches and Reflections*. Aldershot: Variorum.

Ricci, Matteo

Born **Macerata, (Marche, Italy), 6 October 1552**
Died **Beijing, China, 11 May 1610**

Matteo Ricci carried out an astonishingly successful mission to Chinese scholars during which he greatly improved the maps of China and introduced Euclidean geometry and trigonometry there.

Ricci's parents were Giovanni Battista Ricci and Giovanna Angiolelli. Matteo entered the Jesuit order in 1571, and after his courses in rhetoric, philosophy, and theology he studied mathematics at the Roman College under **Christoph Clavius**.

In 1577 Ricci departed Rome to join the Jesuits working in China and arrived there in 1583, where he worked for 27 years. He first opened a residence in Nanking for himself, his fellow Jesuits, and his scientific instruments. (For a time suspicious landlords would drive Ricci and his companions from their dwellings, until they hit on the plan of renting haunted houses; then no one bothered them.) Gradually Ricci was welcomed to the academies and gained many influential friendships. Later he became the court mathematician in Beijing, and there he stayed for the rest of his life.

Jesuits were practically the sole source of Chinese knowledge about western astronomy, geometry, and trigonometry. Appointments in the Astronomical Bureau provided the Jesuits with access to the ruling elite, whose conversion was their main object. Mathematical and astronomical treatises demonstrated high learning and proved that the missionaries were civilized and socially acceptable.

Matteo Ricci brought trigonometry to China, where it had remained in primitive form until the Jesuits came. Ricci's successors, **Ferdinand Verbiest** and **Johann Schall von Bell**, used geometric and trigonometric concepts to bring about a revolution in the sciences of astronomy, the design of astronomical instruments, mapmaking, and the intricate art of making accurate calendars.

The Jesuits were inveterate mapmakers and were continually traveling around the empire of China, even though travel conditions were quite inconvenient. The *Philosophical Transactions of the Royal Society* recounts 52 such journeys by Ricci and Verbiest alone.

Ricci designed and displayed a great world map, which brought about a revolution in traditional Chinese cosmography. For the first time the Chinese had an idea of the distribution of oceans and landmasses. This was the beginning of his major contribution to the diffusion of knowledge through his more than 20 works in Chinese on such topics as mathematics and astronomy.

For 20 years Ricci tried to reach the emperor in person, but the emperor was a recluse not accustomed to seeing even his own people. Then, unexpectedly, the emperor summoned Ricci and his companions to inquire about a ringing clock brought to him by the Jesuits. His own scientists had failed to fix it when it stopped. Since the emperor could not receive these foreigners in person, he had an artist draw full-length portraits of them, so that they could have a vicarious interview.

Another opportunity was occasioned by an eclipse of the Sun: The prediction of the expected time and duration made by the emperor's own Chinese astronomers differed considerably from the Jesuit prediction. When the latter prediction proved correct, the place of the Jesuit mathematicians was secure. It is interesting that the Jesuits taught the Chinese the heliocentric theory, unaware that **Galileo Galilei**'s trial had taken place, and it had been forbidden by Rome to be taught as a proven fact. There was a more than 5-year lag in communications.

Ricci's books *Geometrica Practica* and *Trigonometrica* were translations of Clavius's works into Chinese. In 1584 and 1600 he published the first maps of China ever available to the west. As author, the Chinese geometrical works for which Ricci is remembered were books on the astrolabe, the sphere, measures, and isoperimetrics. But especially important was his Chinese version of the first six books of Euclid's *Elements*, which was written in collaboration with one of his pupils and entitled *A First Textbook of Geometry*.

The prestige Ricci gained in the highest cultural spheres by his wisdom, scientific knowledge, and capacity for philosophical speculation won him a hearing when he spoke of the gospel message. Without any trace of superiority in his manner, he used a process of dialog that was characterized by an esteem and respect for everyone. Ricci's success was due to his personal qualities, his complete adaptation to Chinese customs (choosing the attire of a Chinese scholar), and his authentic knowledge of mathematics, physics, and astronomy. (His works included Chinese texts on religious and moral topics, as well as writings on scientific topics.) It is still possible to visit Ricci's 8-ft.-high tomb in the Beijing suburbs.

Joseph F. MacDonnell

Selected References

Baddeley, John F. (1916). *Russia, Mongolia, China, 1602–1676*. New York: Burt Franklin.

Gallagher, Louis J. (trans.) (1953). *China in the Sixteenth Century: The Journals of Matthew Ricci*. New York: Random House.

Institutum Historicum (1932–). *Archivum Historicum Societatis Iesu*. Rome: Institutum Historicum.

Masotti, Arnaldo (1975). "Ricci, Matteo." In *Dictionary of Scientific Biography*, edited by Charles Coulston Gillispie. Vol. 11, pp. 402–403. New York: Charles Scribner's Sons. (Reference to Ricci is also found in Vol. 3, p. 311; Vol. 4, p. 457; Vol. 7, p. 19; Vol. 10, p. 103; and Vol. 14, pp. 162–164.)

Needham, Joseph (1959). *Science and Civilization in China*. Vol. 3, p. 173. Cambridge: Cambridge University Press.

Ricci, Matteo (1595). *Ki ho youen pen* (Geometrica practica). Beijing.

——— (1607). *Hoen kai tong hien tou cho* (Explanation of the celestial sphere). Beijing.

Sommervogel, Carlos (1860–1960). *Bibliothèque de la Compagnie de Jésus*. 12 Vols. Brussels: Société Belge de Libraire. (Thirty-six entries are found in Sommervogel; two examples are *Ki ho youen pen* and *Hoen kai tong hien tou cho.)*

Wallace, William A. (1977). *Galileo's Early Notebooks*. Notre Dame: Notre Dame University Press.

Riccioli, Giovanni Battista

Born **Ferrara, (Italy), 17 April 1598**
Died **Bologna, (Italy), 25 June 1671**

Giovanni Riccioli was a pioneer in lunar astronomy who first named craters and mountains on the Moon for scientists. He also perfected the use of the pendulum for time measurement.

Riccioli entered the Jesuit order in 1614, and studied rhetoric, philosophy, and theology in Parma and Bologna. During this period he began his studies in astronomy.

Riccioli published one of the earliest books on astronomy: *Almagestum Novum* (Bologna, 1651). His chapter concerning the Moon contains two large maps (28 cm in diameter), one of which shows – for the first time – the effects of librations and introduces new lunar nomenclature. Almost all of his names for lunar objects are still in use today. It is the first map to name craters and mountains for scientists and prominent people instead of abstract concepts. A copy of Riccioli's lunar map stands at the entrance to the lunar exhibit at the Smithsonian Institute, and was described in detail in the *Philosophical Transactions* of the Royal Society. It was meticulously drawn by another Jesuit, **Francesco Grimaldi**.

Jesuit astronomers in Rome such as Riccioli had a great advantage over others because they were able to gather information from their former pupils spread out across the globe, in places as disparate as South America, Africa, China, Japan, and India. There were many Jesuit astronomical observatories throughout the world that efficiently gathered data concerning lunar and solar eclipses as well as transits of Venus. By the time of the (temporary) suppression of the Jesuit Society in 1773, no fewer than 30 of the world's 130 astronomical observatories were run by Jesuits. This information enabled Riccioli to compose a table of 2,700 selenographical objects, incomparably more accurate than anything previously known.

Riccioli also made many important measurements in order to refine his existing astronomical data, such as the radius of the Earth and the ratio of water to land on its surface. He compiled star catalogs and described sunspots, the movement of a double star, and the colored bands of Jupiter.

In physics, Riccioli went beyond the preliminary work of **Galileo Galilei** and succeeded in perfecting the pendulum as an instrument to measure time, thereby laying the groundwork for a number of important later applications. Riccioli once persuaded nine of his fellow teachers to count 87,000 oscillations over the course of a day, enabling him to identify an error of three parts in a thousand. In his book *God and Nature*, David Lindberg notes that it was Riccioli, not Galilei, who first accurately determined the rate of acceleration of a falling body. Noting the collaborative efforts of Jesuits, he argues that Jesuit scientists, rather than the Academia del Cimento or the Royal Society, formed the first true scientific society. **Athanasius Kircher**, for example, in his ability to collect observations from a worldwide network of informants was more than a match for **Marin Mersenne** in Paris or Henry Oldenburg in London, and he published this information in massive encyclopedias, which were indispensable in disseminating scientific data and theories.

Riccioli contributed also to geography, publishing tables of latitude and longitude for many different locations. This facilitated later developments in cartography.

Joseph F. MacDonnell

Selected References

Campedelli, Luigi (1975). "Riccioli, Gianbattista." In *Dictionary of Scientific Biography*, edited by Charles Coulston Gillispie. Vol. 11, pp. 411–412. New York: Charles Scribner's Sons.

Oldenburg, Henry (ed.) (1665–1715). *Philosophical Transactions of the Royal Society*. Vols. 1–30. London.

Sommervogel, Carlos (1890–1960). *Bibliothèque de la Compagnie de Jésus*. 12 Vols. Brussels: Société Belge de Libraire.

Riccò, Annibale

Born **Modena, (Italy), 15 September 1844**
Died **Rome, Italy, 23 September 1919**

Annibale Riccò carried out an exhaustive study of solar prominences, of which he amassed over 20,000 observations, and deduced the existence of the solar wind.

Riccò was educated at the University of Modena, where he received his bachelor's degree in mathematics (1866) and doctorate in natural sciences (1868). He also earned a degree in civil engineering from the Polytechnic Institute of Milan (1868) and began his career as an architect.

Riccò's involvement with astronomy began as an observatory assistant at Modena. In 1872, he cofounded the Society of Italian Spectroscopists, whose efforts were devoted chiefly toward an improved understanding of the Sun. Temporarily an instructor of physics at Naples, Riccò was subsequently appointed an astronomer at the Royal Observatory in Palermo. In 1885, he founded an observatory at Catania, and in 1890 became director of the Catania Observatory and Mount Etna Observatories and acquired the chairmanship of astrophysics at the University of Catania. Riccò held these posts for the remainder of his life.

Riccò's data revealed the solar prominences to be of two principal types, either quiescent or active. He was among the first to argue that dark "filaments," which appeared superimposed on the solar disk, were themselves prominences. In 1892, Riccò demonstrated that delays of roughly forty to forty-five hours occurred between the crossings of sunspots on the solar meridian and resulting terrestrial magnetic disturbances. Particles emitted by the Sun, and whose existence Riccò inferred, are now recognized as the "solar wind."

Riccò oversaw the Catania Observatory's involvement with the International *Carte du Ciel* Astrographic Chart and Catalogue project, and attempted, albeit unsuccessfully, to photograph the Sun's corona without an eclipse from the top of Mount Etna with American astronomer **George Hale** (1894). Perhaps in relation to his solar studies, Riccò observed numerous bright comets, including C/1908 R1 (Morehouse) and 1P/1909 (Halley). In 1899, he became coeditor, and in 1905, editor, of the *Memoire della Società degli Spettroscopisti Italiani*, founded by **Angelo Secchi** and **Pietro Tacchini**. Riccò traveled abroad to witness several total solar eclipses.

Riccò's solar research netted him some prestigious awards, including the Royal Prize of the Accademia dei Lincei, and a Knight of the Crown of Italy. Riccò was thrice elected an executive officer of the International Union for Cooperation in Solar Research [IUCSR] and, in 1919, a vice president of its successor, the International Astronomical Union [IAU]. Both appointments reflected his strong commitments toward fostering solar research and international scientific cooperation.

Jordan D. Marché, II

Selected References

Abetti, Giorgio (1920). "Annibale Riccò, 1844–1919." *Astrophysical Journal* 51: 65–72.

——— (1975). "Riccò, Annibale." In *Dictionary of Scientific Biography*, edited by Charles Coulston Gillispie. Vol. 11, p. 412. New York: Charles Scribner's Sons.

Newall, H. F. (1920). "Annibale Riccò." *Monthly Notices of the Royal Astronomical Society* 80: 365–367.

Richard of Wallingford

Born **England, *circa* 1291**
Died **England, *circa* 1335**

Richard of Wallingford is probably best remembered for the astronomical clock he designed for the abbey of Saint Albans. He also wrote books about astronomical instruments and about trigonometry.

Richard was born in either 1291 or 1292 to William, a blacksmith, and Isabella. William died when Richard was 10, whereupon Richard was adopted by William de Kirkeby, Prior of Wallingford. He attended Oxford for 6 years, graduating with an AB degree, then assumed the monastic habit at Saint Albans. He was ordained deacon in 1316 and priest in 1317. His abbot sent him back to Oxford, where he studied for 9 years and received a Bachelor of Divinity degree in 1326.

In the same year the presiding abbot died, and Richard was elected the new abbot of Saint Albans. Despite contracting leprosy at about that time, which led to blindness in one eye and later robbed him of his power of speech, Richard was nonetheless a strong abbot. He brought Saint Albans out of debt, quelled riot among the townspeople, and survived internal struggles that led to a papal inquisition.

Richard's greatest achievement was the clock that he designed and had built for the abbey. It was the first purely mechanical clock of which there is a complete record. It not only struck the hours of the day, but also indicated the positions of the Sun, Moon, and planets, as well as lunar phases and solar eclipses; it may also have had a tidal dial. It was a clock of unprecedented complexity and accuracy for its time and was unsurpassed in quality for the next 200 years. When King Edward III chided him for building a clock instead of more churches, Richard replied that many abbots could build churches, but after he was dead, no one could complete the work of the clock. The clock itself is no longer in existence, a victim of the dissolution of the monasteries in the 16th century.

While studying at Oxford, Richard produced several works of mathematics and astronomy that are also remarkable for their time. The *Quadripartitum* and *de Sectore* comprised the most comprehensive compendium of trigonometry in Europe before the end of the 1400s. This also was the first instance where trigonometry was separated from its applications (*e. g.*, astronomy) and dealt with in an abstract manner. Richard wrote the *Exafrenon*, a treatise on the prediction of weather and natural events that were believed to be governed by planetary influence. He also wrote two manuscripts devoted to astronomical instruments he invented. The *Albion* described his version of an *equatorium* to calculate planetary positions and eclipses. It was a completely new design with few moving parts, and capable of very accurate predictions. Versions of the albion were created by Simon Tunsted, **John of Gmunden**, and **Johann Müller** (Regiomontanus), and it was influential in the design of the later instruments of **Johannes Schöner** and **Peter Apian**. The *Rectangulus*, made from seven straight rods, rather than the disks and rings of typical armillaries, was an observational instrument that also allowed one to directly transform coordinate systems. Since it was based on straight lines, it could be constructed more easily and accurately, but it was more difficult to read than the armillary or torquetum.

Michael Fosmire

Selected References

Bond, John David (1922). "Richard Wallingford (1292?-1335)." *Isis* 4: 459–465.

North, John D. (1976). *Richard of Wallingford: An Edition of His Writings*. 3 Vols. Oxford: Clarendon Press.

Richaud, Jean

Born **Bordeaux, Gironde, France, 1 October 1633**
Died **Pondicherry, (India), 2 April 1693**

French Jesuit missionary Jean Richaud made the first astronomical discoveries from India using a telescope. In 1689, he observed that the stars α Centauri and α Crucis are in fact double. (The binary nature of α Crucis had been independently noted by fellow Jesuit Jean de Fontenay, at the Cape of Good Hope, 4 years earlier.)

Selected Reference

Rao, N. Kameswara, A. Vagiswari, and Christina Louis (1984). "Father J. Richaud and Early Telescope Observations in India." *Bulletin of the Astronomical Society of India* 12: 81.

Richer, Jean

Born **France, 1630**
Died **Paris, France, 1696**

Giovanni Cassini sent Jean Richer to South America in order to witness the 1672 opposition of Mars. Richer's observations, and those made simultaneously at the Paris Observatory, were used to refine the length of the Astronomical Unit. While in present-day French Guyana, Richer measured that a pendulum swung more slowly than at Paris. He attributed this to weaker gravitational acceleration near the Equator, in accordance with the Newtonian theory of an oblate Earth.

Selected Reference

Lynn, W. T. (1910). "Jean Richer." *Observatory* 33: 102.

Riḍwān al-Falakī: Riḍwān Efendi ibn ʿAbdallāh al-Razzāz al-Falakī

Born **Cairo, (Egypt)**
Died **Cairo, (Egypt), 7 August 1711**

Riḍwān Efendi al-Falakī was an Egyptian–Ottoman astronomer known for his production of astronomical tables as well as various instruments and globes. He was also noted for the many students that he trained. There is little information on his birth, youth, and education. However, we know that Riḍwān al-Falakī studied in Cairo and received his astronomical education from distinguished scholars. Indeed, he never left Cairo except in 1680, when he visited Mecca for the *ḥajj* (pilgrimage). Besides writing on astronomy, Riḍwān al-Falakī wrote a number of books on mathematics and geometry. According to the sources on Ottoman astronomy, his works were so abundant that the drafts of his books were considered a camel's load. At the request of the timekeeper Ḥasan Efendi, in 1700 and 1701 he prepared spheres and astronomical devices upon which he marked the Arabic names of stars that he located through observation. Among Riḍwān al-Falakī's many students in astronomy, only Yūsuf al-Jamāli (the servant of Ḥasan Efendi) is known.

The titles of 17 of Riḍwān al-Falakī's astronomical works are known, most of which are extant. All were written in Arabic. Several works are adaptations of the work done at the Samarqand

Observatory under **Ulugh Beg**. His *Zīj al-mufīd ʿalā uṣūl al-raṣad al-jadīd al-Samarqāndī*, or *al-Zīj al-Riḍwānī*, is an astronomical handbook with tables based on *Zīj-i Ulugh Beg* but adapted for Cairo's latitude. It consists of four parts in addition to an introduction and various tables. Riḍwān al-Falakī's *al-Durr al-farīd ʿalā al-raṣad al-jadīd* is possibly a commentary written on Ulugh Beg's *Zīj*; it contains an introduction, 12 sections, and a conclusion. *Asnā al-mawāhib fī taqwīm al-kawākib* is another work he adapted from *Zīj-i Ulugh Beg* for Cairo's latitude.

Riḍwān al-Falakī is also known for his works on timekeeping. Of these, probably the most extensive is *Dustūr uṣūl ʿilm al-mīqāt wa-natījat al-naẓr fī taḥrīr al-awqāt*. Other treatises treat eclipses, lunar-crescent visibility, sundials, and Jupiter–Saturn conjunctions. For a listing of his works, see Ihsanoğlu *et al.* (1997), and Rosenfeld and Ihsanoğlu (2003).

Salim Aydüz

Selected References

Al-Ziriklī, Khayr al-Dīn (1980). *Al-Aʿlām*. Vol. 3, p. 27. Beirut.
Bağdadlı, İsmail Paşa. *Īḍāḥ al-Maknūn*. Vol. 1 (1945): 82, 447, 621; Vol. 2 (1947): 81. Istanbul.
——— (1951). *Hadiyyat al-ʿĀrifīn*. Vol. 1, p. 369. Istanbul.
Brockelmann Carl. *Geschichte der arabischen Literatur*. 2nd ed. Vol. 2 (1949): 471; Suppl. 2 (1938): 487. Leiden: E. J. Brill.
Çakıroğlu, Ekrem (ed.) (1999). *Yaşamları ve Yapıtlarıyla Osmanlılar Ansiklopedisi*. Vol. 2, pp. 459–460. Istanbul: Yapi Kredi Yayinlari.
İhsanoğlu, Ekmeleddin *et al.* (1997). *Osmanlı Astronomi Literatürü Tarihi (OALT)* (History of astronomy literature during the Ottoman period). Vol. 1, pp. 377–384. Istanbul: IRCICA.
Jabartī, ʿAbd al-Raḥmān (1978). *ʿAjā'ib al-āthār fī al-tarājim wa-'l-akhbār*. Vol. 1, pp. 130–131. Beirut.
Kaḥḥālah, ʿUmar Riḍā (1985). *Muʿjam al-mu'allifīn*. Vol. 4, p. 165. Beirut.
King, David A. (1986). *A Survey of the Scientific Manuscripts in the Egyptian National Library*. Winona Lake, Indiana: Eisenbrauns, pp. 107–108.
——— (2004). *In Synchrony with the Heavens: Studies in Astronomical Timekeeping and Instrumentation in Medieval Islamic Civilization*. Vol. 1, *The Call of the Muezzin* (Studies I–IX). Leiden: E. J. Brill.
Rosenfeld, B. A. and Ekmeleddin Ihsanoğlu (2003). *Mathematicians, Astronomers, and Other Scholars of Islamic Civilization and Their Works (7th–19th c.)*. Istanbul: IRCICA, pp. 386–387.

Ristenpart, Frederich Wilhelm

Born **Frankfurt, (Germany), 8 June 1868**
Died **Santiago, Chile, 9 April 1913**

Frederich Ristenpart prepared the master star catalog *Geschichte des Fixsternhimmel* in Germany. However, his failed attempt to modernize the Chilean National Observatory led to suicide. The sad story is recounted by Ashbrook (1984).

Selected Reference

Ashbrook, Joseph (1984). "A South American Tragedy." In *The Astronomical Scrapbook*, edited by Leif J. Robinson, pp. 99–103. Cambridge, Massachusetts: Sky Publishing Corp.

Ritchey, George Willis

Born **Tuppers Plains, Ohio, USA, 31 December 1864**
Died **Azusa, California, USA, 4 November 1945**

George Ritchey may arguably be called the most visionary and yet least appreciated designer and user of large telescopes in early-20th-century America. Although Ritchey has become well known for the Ritchey–Chrétien telescope design, developed along with French astronomer Henri Chrétien, roughly half a century was to elapse before its innovative traits gained widespread acceptance. At the same time, however, Ritchey was among the first astrophotographers to demonstrate the true potential of the modern reflecting telescope. While employed at the Mount Wilson Observatory, California, Ritchey produced remarkable photographs of astronomical objects, especially of the "spiral nebulae." It was in these spiral nebulae, then considered to be merely clouds of gas or dust, that Ritchey discovered faint novae, suggesting that they were in fact galaxies external to the Milky Way. A perfectionist in much of his work, Ritchey was perhaps the foremost optician and instrument maker of his day. Yet, it was also Ritchey's curse that he could be a difficult and sometimes temperamental person to work with. He tended to be secretive and possessive of his techniques, over optimistic in some of his claims, and dismissive of others' ideas. As a result, Ritchey's personal disputes with **George Hale** and others cost him much of his well-deserved reputation.

Ritchey was born into a small farming community. His father was a skilled and, at times, fairly prosperous furniture maker. Although the family had its high and low points, Ritchey succeeded in gaining a reasonably good education for the times. He graduated from Hughes High School in Cincinnati, Ohio (1881), and entered the School of Design at the University of Cincinnati the following year. It was during this time that Ritchey became interested in astronomy, and he was doubtless lucky to be living so near to one of the oldest observatories in America, the Cincinnati Observatory. Between 1882 and 1887, Ritchey was in and out of school as his ambitions to become an astronomer competed with economic reality. He married Lillie May Gray in 1885; the couple was to have two children. While attending classes, Ritchey worked part-time at the observatory, in a small home shop where he built his first telescope, and at the family furniture business. It was for economic reasons that Ritchey left college and became an industrial-arts instructor at the Chicago Manual Training School in 1888. However, the move to Chicago turned out to be Ritchey's great opportunity to become an astronomer.

Ritchey met Hale in 1890, and both realized that they could be of use to one another. Ritchey, who had also taken up photography as a hobby, was an occasional visitor and contract employee at Hale's Kenwood Physical Observatory. The subsequent establishment of the University of Chicago's Yerkes Observatory resulted in Ritchey being hired by Hale as a full-time optician and observer. Both Ritchey and Hale had come to the conclusion that the days of the unchallenged superiority of the refracting telescope were over, and that large reflectors, cheaper to build for their aperture and not susceptible to chromatic aberration, represented the future of large astronomical telescopes.

While still employed at the Chicago Manual Training School, Ritchey worked on a number of projects for Hale, including a 24-in. mirror for a special solar telescope, and perfected his mirror-making techniques on smaller instruments for himself. Ritchey's first major project at the Yerkes Observatory was the construction of a 24-in. $f/4$ Newtonian reflector. This short focal length telescope, expressly built for astrophotography, was considerably "faster" than any normal refracting telescope of the same aperture and much larger than special photographic lenses then available. It had a wide field of view, and could record faint nebulosity with much shorter exposures than competing instruments.

Ritchey proved to be more than just an optician, and with the 24-in. and other telescopes at Yerkes, began his career as an astrophotographer. For a time, he challenged **Edward Barnard** and others with his published photographs of star clusters, nebulae, and the Moon. Many journals and astronomy texts of the time were graced by his photographs. It was during this time that Ritchey began publishing a number of scholarly articles, *e. g.*, on his design for an improved mounting system for large telescope mirrors, and reports of his observations, including those changes observed in the nebulosity around Nova Persei (1901). These changes are now considered to be the first observation of a light echo.

Ritchey's next project was a 60-in. reflecting telescope, originally intended for a "western station" of Yerkes Observatory that instead became the Mount Wilson Observatory. It was with the 60-in. that he began his series of photographic observations of the Moon, planets, nebulae, star clusters, and most importantly, the spiral nebulae. In 1910, Ritchey reported that his photographs of spirals (like M81) revealed "soft star-like condensations" that he called nebulous stars. Adopting this condensation model, he concluded that his photographs "strongly oppose, if they do not effectually preclude, the theory that the spiral nebulae are distant systems of stars like our Milky Way." Starting in 1917, however, Ritchey began to discover novae associated with the spirals, and later identified them on plates taken back in 1909. This evidence seemingly contradicted his earlier stance against the spirals as independent stellar systems. As a result, it was largely left to Lick Observatory astronomer **Heber Curtis**, and Mount Wilson Observatory astronomer **Edwin Hubble**, to correctly conclude that the spiral nebulae were in fact external galaxies.

Ritchey's greatest achievement, yet also the project that would see his downfall, was the mirror of the 100-in. Hooker telescope at Mount Wilson Observatory. Construction of this telescope (the largest attempted up to that time) was begun in 1906 but only completed in 1919. The project was plagued with problems and controversies from the beginning. There was considerable difficulty in procuring a high-enough quality glass disk to suit Ritchey, who advocated, and conducted extensive experiments on, the concept of a lightweight, cellular mirror. Ritchey also objected to using a domed observatory, preferring a lightweight, roll-off structure that would eliminate the detrimental effects to atmospheric steadiness caused by the release of heat absorbed by a conventional masonry building during the day. His seemingly endless testing, experimenting, and arguments over design concepts cost him the support of Hale and others. In the end, his personal disputes with Hale also cost him his job at Mount Wilson Observatory and tarnished his reputation in the astronomical community. With the completion of the 100-in. telescope, Ritchey had produced another superb mirror. Yet, he was effectively shut out of Hale's next great telescope project – the 200-in. Mount Palomar reflector. Ritchey was to spend the next several years either working on projects in France, or in semiretirement.

As early as 1910, Ritchey met French astronomer and optician Henri Chrétien. Ritchey and Chrétien recognized the inherent flaws of Newtonian and Cassegrainian optical designs, namely the increase of the aberration called coma toward the edge of the field of view. It causes stars to appear as elongated, comet-shaped blurs. Together, they developed the Ritchey–Chrétien telescope (employing hyperbolic primary and secondary mirrors) that corrected this flaw and potentially produced near-perfect star images in photographs. But Ritchey's loss of reputation, caused by his disputes with Hale, cast serious doubts on the new type of telescope. Given a laboratory at the Paris Observatory, Ritchey completed a prototype of the new design, a 0.5-m (20-in.) reflector, in 1927. Yet, American astronomers virtually ignored the potential benefits offered by the Ritchey–Chrétien design.

In 1930, Ritchey returned to the United States and eventually obtained support to construct a 1-m (40-in.) Ritchey–Chrétien reflector for the United States Naval Observatory. But plagued by poor weather conditions at its original site in Washington, the instrument languished until the telescope was reerected at Flagstaff, Arizona, where its design was thoroughly vindicated. Starting in the 1960s, many of the largest optical telescopes constructed have employed the Ritchey–Chretien design.

Excluded from all aspects of the 200-in. Palomar reflector, Ritchey nonetheless continued to produce designs of giant reflecting telescopes, up to 8-m aperture, none of which was ever built. Yet, he had contributed toward many important developments, both observationally and instrumentally, in early-20th-century astronomy. His principal legacies remain the 60- and 100-in. reflectors at Mount Wilson Observatory. While some of Ritchey's ideas concerning large telescopes seemed radical during his lifetime, many have since come into standard practice in modern astronomical instrumentation. The Ritchey–Chrétien telescope, lightweight cellular mirror, the stress-reducing mirror cell, and even the concept of the "observatory environment," have since been adopted by the astronomical community that, under Hale's influence, had once excluded Ritchey.

Gary L. Cameron

Selected References

Berendzen, Richard (1975). "Ritchey, George Willis." In *Dictionary of Scientific Biography*, edited by Charles Coulston Gillispie. Vol. 11, p. 470. New York: Charles Scribner's Sons.

Osterbrock, Donald E. (1993). *Pauper and Prince: Ritchey, Hale, and Big American Telescopes*. Tucson: University of Arizona Press.

——— (1999). "Ritchey, George Willis." In *American National Biography*, edited by John A. Garraty and Mark C. Carnes. Vol. 18, pp. 546–548. New York: Oxford University Press.

Rittenhouse, David

Born **Paper Mill Run near Germantown, Pennsylvania, (USA), 8 April 1732**

Died **Philadelphia, Pennsylvania, USA, 26 June 1796**

David Rittenhouse, a noted self-educated man of many dimensions, rose from obscure beginnings as a clock- and scientific instrument-maker in Norriton, Pennsylvania, to prominence in the world of science as an effective observational astronomer and experimental scientist.

The origin of Rittenhouse's clockmaking knowledge is unclear, though it is certain that he inherited a chest of tools from a maternal uncle, and that his father purchased the additional tools he originally needed to enter the trade of clockmaking. The tall clocks Rittenhouse made in the roadside workshop he opened in Norriton around 1749, if not unusual in their mechanism, nevertheless were masterpieces of craftsmanship. In three he included small orreries, and in the period 1767–1771 he designed and built two large vertical orreries. One of these was purchased by the Pennsylvania General Assembly for the College of Philadelphia (later the University of Pennsylvania), the other by the College of New Jersey (now Princeton University).

Indeed, it was his clockwork orreries and telescopes that led Rittenhouse to astronomy and ultimately brought him to the attention of the scientific community in Philadelphia. Four years after his marriage to Eleanor Coulston on 20 February 1766, he and his family took up residence in Philadelphia. Eleanor died giving birth to the second of two children; David remarried in 1772, this time to Hannah Jacobs.

However, it was the transit of Venus in 1769 that turned Rittenhouse to serious observational astronomy, and earned him a place among the world's astronomers. He had previously taught himself mathematics and physical sciences by reading, mostly from **Isaac Newton**'s *Principia*. For the transit, Rittenhouse first prepared a proposal to the Philosophical Society in which he recommended that the society establish two stations to observe the transit. He volunteered to equip a station at Norriton as one of the two sites. After the society approved his plan, Rittenhouse constructed a transit telescope (possibly the first telescope made in America), in addition to an equal-altitude instrument and an 8-day clock, all for use at Norriton.

The observational techniques Rittenhouse reported were of greater significance than the data he obtained, more for their inventiveness than for their innovation. For instance, around 1785 he resolved the difficulty in lining up his meridian telescope on a distant mark by installing a collimating lens system enabling him to use a much closer mark. Such a lens system was not new, so he cannot be credited with its origin, but his inventiveness was showcased. The same can be said of his use of spider web in his telescope, for he did not know that it had been previously used by **Francesco Fontana**.

In Philadelphia, Rittenhouse established an astronomical observatory and made many astronomical observations, provided data for almanacs, and lectured on astronomy. In 1786, he published a paper describing his invention and study of a plane transmission grating, in this case a series of closely spaced fine wires wrapped on frames. Using one of these frames he observed up to six orders of diffracted spectra, measured the angular displacement of each, and from the data developed a workable theory of diffraction to account for his observations, but he took the experiments no further. It would be left for **Joseph von Fraunhofer** and Augustin Fresnel to carry them forward.

Rittenhouse also took part in the Mason–Dixon survey of the boundary between Pennsylvania and Maryland (1763), and carried out surveys of other state and colonial boundaries, as well as canals and rivers, usually with instruments of his own construction. He served in various capacities during the Revolution, and in 1792 became the first director of the United States Mint. He was one of the earliest members elected to the American Philosophical Society and succeeded Benjamin Franklin as president of the society (1791). Rittenhouse was elected a foreign member of the Royal Society of London (1795).

Richard Baum

Selected References

Ford, Edward (1946). *David Rittenhouse, Astronomer-Patriot, 1723–1796*. Philadelphia: University of Pennsylvania Press.

Hindle, Brooke (1964). *David Rittenhouse*. Princeton, New Jersey: Princeton University Press. (Reprint, New York, Arno Press, 1980.)

——— (ed.) (1980). *The Scientific Writings of David Rittenhouse*. New York: Arno Press.

Ritter, Georg August Dietrich

Born **Lüneburg, (Niedersachsen, Germany), 11 December 1826**
Died **Lüneburg, (Niedersachsen, Germany), 26 February 1908**

August Ritter, a pioneer in the theory of stellar structure, obtained fundamental results by applying the relatively recent laws of thermodynamics, as enunciated by Rudolf Clausius and **William Thomson** (Lord Kelvin), to a gaseous model of the Sun. He took up the problem where **Jonathan Lane** had left it and carried the mathematical development forward to where **Robert Emden** could turn it into a fully developed theory of a perfect gaseous sphere.

The entire subject of solar physics in the period from 1870 to 1890 was in a state of rapid flux, invigorated by the prospect of investigating the physical constitution of the stars by means of spectral analysis. Ritter's highly mathematical contributions, coming from a relatively unknown faculty member at a small polytechnic school, found few appreciative readers at the time.

Ritter began his studies in 1843 at the polytechnic school in Hanover, and continued them at Göttingen University, where he received his Ph.D. in 1853. He then returned to the Hanover polytechnic school as an instructor in 1856. Ritter was subsequently appointed a professor of mechanics at the Aachen Technische Hochschule when it opened in 1870. Some idea of his circumstances there is suggested by the testimony of Otto Lehmann, who came to Aachen in 1883 from the larger secondary school at Mühlhausen. The space for research at Aachen was so small that Lehmann could not find room for the limited apparatus he brought with him. It took him far longer to do research at Aachen, he claimed, but he felt compensated by the increased time he had there for research.

In a series of 18 papers published between 1878 and 1883 in the leading German physics journal, *Annalen der Physik*, and in a separate textbook of 1879, Ritter employed a meteorological model to derive the fundamental differential equations regulating both a thermodynamically stable gaseous sphere heated by gravitational contraction and one subject to cyclical pulsations. He showed that stable convective currents acting throughout the sphere produced a temperature gradient inversely proportional to its radius.

A 1939 assessment by **Subrahmanyan Chandrasekhar** characterized Ritter's investigations as "a classic [,] the value of which has never been adequately recognized," in which "almost the entire foundation for the mathematical theory of stellar structure was laid by him" (1939, pp. 178, 179). Astronomers **Otto Struve** and Velta Zebergs pointed out that Ritter's 1879 paper on the theory of radial pulsations in stars was advanced long "before the [observed] variation in radial velocity [among classical Cepheids] was known." More importantly, Ritter had derived a "relation between density and period of pulsation" that "remarkably closely approximates the observations" (1962, p. 314).

Michael Meo

Selected References

Chandrasekhar, S. (1939). *An Introduction to the Study of Stellar Structure*. Chicago: University of Chicago Press, esp. "Bibliographical Notes," pp. 176–180.

Jungnickel, Christa and Russell McCormmach (1986). *Intellectual Mastery of Nature: Theoretical Physics from Ohm to Einstein*. Vol. 2, p. 55. Chicago: University of Chicago Press.

Poggendorff, J. C. "Ritter." In *Biographisch-literarisches Handwörterbuch*. Vol. 3 (1898): 1125; Vol. 4 (1904): 1256; Vol. 5 (1926): 1056. Leipzig and Berlin.

Ritter, A. (1898). "On the Constitution of Gaseous Celestial Bodies." *Astrophysical Journal* 8: 293–315. (An English translation and reprint of Ritter's 16th paper from the *Annalen der Physik*. The "Editorial Introduction," pp. 293–295, contains a complete bibliography of Ritter's 18 papers.)

Struve, Otto and Velta Zebergs (1962). *Astronomy of the 20th Century*. New York: Macmillan.

Ritter, Johann Wilhelm

Born **Samitz, (Chojńow, Poland), 16 December 1776**
Died **Munich, (Germany), 23 January 1810**

German chemist, physicist, and physiologist Johann Ritter came to work at Jena with **Alexander von Humboldt** in 1795. His subsequent career and life were brief and chaotic, partially owing to issues of philosophy of science then under bitter discussion in Germany. In 1801, while at the court of the Duke of Gotha and Altenberg, Ritter discovered that paper soaked in a solution of silver iodide was blackened most by the radiation coming from the Sun just slightly off the end of the spectrum beyond the last visible violet light. Ritter was, therefore, the discoverer of ultraviolet radiation, the year after **William Herschel** discovered infrared from its heating power.

Virginia Trimble

Selected Reference

Porter, Roy (ed.) (1994). 'Ritter.' In *The Biographical Dictionary of Scientists*, 2nd ed., p. 583. New York: Oxford University Press.

Roach, Franklin Evans

Born **Jamestown, Michigan, USA, 23 September 1905**
Died **Tucson, Arizona, USA, 21 September 1993**

American spectroscopist and photometrist Frank Roach was for many years the world expert on how bright the sky is at night and the various sources that contribute to the brightness. He was the oldest of four children of optometrist Richard F. Roach and Norwegian immigrant Ingeborg (*née* Torgerson) Roach and was educated in the public schools of Wheaton, Illinois (including the high school also attended by **Edwin Hubble** and **Grote Reber**), and Highland Park, California. Roach started on a premedical curriculum at Wheaton College, and won a scholarship to the University of Michigan, but decided to use it to finish a bachelor's degree immediately rather than a medical degree involving a longer period of education that his family could ill afford. A succession of temporary, nonscientific jobs continued to interrupt Roach's education, but he succeeded in completing, at the University of Chicago, an MS in 1930, with a thesis on absorption lines of ionized sulfur in stellar spectra and a Ph.D. in 1934 with a thesis on red and near-infrared stellar spectra, both under the direction of **Otto Struve** who had, in 1932, succeeded **Edwin Frost** as director of Yerkes Observatory. In 1932, Roach was sent to Perkins Observatory, Ohio, where he used their 69-in. telescope to develop and use a spectrograph intended for the new McDonald Observatory, a Chicago–Texas collaboration.

Roach was appointed as the first astronomer stationed at McDonald Observatory and, while there, collaborated with **Christian Elvey** on photographic photometry of reflection nebulae, his first venture into accurate measurements of faint, extended sources. During his time as associate professor of astronomy at the University of Arizona, beginning in 1936, Roach continued both stellar spectroscopy (mastering the details of atomic physics essential for understanding diffuse skylight) and photometry with Elvey. During World War II, Roach was first a member of the California Institute of Technology rocket program, working at Eaton Canyon near Pasadena and then at what became the United States Naval Ordnance Test Station [NOTS], China Lake. Though he was reassigned to work on high-explosives research in connection with the Manhattan Project, Roach returned to NOTS after the war, where he established with Elvey a pioneering night-sky research group. After a year in Paris (1951/1952) collaborating with **Daniel Barbier**, the French expert on night-sky emission, Roach moved to the National Bureau of Standards [NBS], in Boulder, Colorado. He established, in 1961, another night-sky observatory on Mount Haleakala, in Hawaii, and moved there after his 1966 retirement from the NBS.

Roach provided the standard photometer against which all the others were calibrated during the International Geophysical Year (1958), acted as an advisor to astronauts on what they could reasonably expect to see in space and how to interpret it, and was part of the scientific team for the "Scientific Study of Unidentified Flying Objects" conducted by the University of Colorado in 1967, under contract to the United States Air Force.

Roach's greatest achievement, however, remained the identification of the various sources of diffuse skylight. These include the glow of ionized atoms in the Earth's atmosphere (in layers that, as he showed, moved up and down through the day and night and the seasons), aurorae (atomic glows driven by particles coming from solar flares), faint stars, and sunlight reflected by dust in the plane of the Solar System (zodiacal light). His 1973 book, *The Light of the Night Sky*, written with Janet L. Gordon (who had been part of the University of Arizona astronomy department during his time there), continues to be cited on the subject. Unfortunately, of course, the skies we see now are very much brighter than those Roach measured because of the increased input of artificial light, reflected back by high-lying dust (some of which is also artificial). In the course of his career, Roach published more than 100 papers.

Gordon and Roach were married in 1977, following the death of his first wife, Eloise Blakslee, who had been the daughter of a Yerkes Observatory photographer. Roach received a Department of Commerce Gold Medal in recognition of his outstanding work on upper-atmosphere physics.

Nadia Robotti and *Muttco Leone*

Selected References

Chamberlain, J. W. (1961). *Physics of the Aurora and Airglow*. New York and London: Academic Press. (For detailed references on his many papers on photometric studies of aurorae and airglow, see the bibliography herein.)

Condon, Edward U. and Daniel S. Gillmor (1969). "Visual Observations Made by U.S. Astronauts." In *Final Report of the Scientific Study of the Unidentified Flying Objects*. New York: Bantam Books, pp. 176–209. (For his activity on visual observations by astronauts, as a principal investigator for the US Air Force's sponsored research at the University of Colorado.)

Elvey, C. T. and Franklin E. Roach (1937). "A Photoelectric Study of the Light from the Night Sky." *Astrophysical Journal* 85: 213–241. (His first paper on the light in the night sky.)

Osterbrock, Donald E. (1994). "Franklin Evans Roach, 1905–1993." *Bulletin of the American Astronomical Society* 26: 1608–1610.

Roach, Franklin E. and Janet L. Gordon (1973). *The Light of the Night Sky*. Dordrecht: D. Reidel. (Summarizing his earlier work on upper-atmosphere research.)

Roberts, Alexander William

Born **Farr, (Highland), Scotland, 4 December 1857**
Died **Alice, (Eastern Cape), South Africa, 27 January 1938**

Using light curves he derived from his own very precise visual photometry, Alexander Roberts pioneered the computation of eclipsing binary-star orbits, shapes, and densities. Roberts demonstrated, simultaneously with **Henry Norris Russell**, that the components of some of these stellar systems are tenuous, low-density stars with gigantic diameters.

Roberts's father apparently died while Alexander was an infant as only his mother's maiden name, Ann Campbell, is known. The family moved from Farr to Leith, where Roberts received his education at the Saint James Schools, and further preparation for a career in teaching at Moray House and the Watt Institution and School of Arts. After completing his education, in 1878 Roberts accepted a teaching post at Wick, Caithness, where he apparently met his future wife, Elizabeth Dunnett. In 1881, he matriculated at the University of Edinburgh for further education.

At an early age, Roberts read a copy of **James Ferguson**'s *Astronomy Explained on Sir Isaac Newton's Principles* and thereafter contemplated an astronomical career. In 1882, he applied for a position as an assistant at the Edinburgh Observatory, but in response received a discouraging letter from **Charles Smyth**, the Astronomer Royal for Scotland. Instead, after graduating from the university, Roberts accepted a call to missionary teaching at the Free Church Mission College in Lovedale, Cape Colony, South Africa. Roberts arrived at Lovedale in July 1883 and was followed a year later by Elizabeth; they were married in 1884 and had three children.

After settling in Lovedale, Roberts took up his avocational interest in astronomy. After some preliminary recreational observing, he made an accurate determination of the latitude of his observing station. The next step of his orientation was to map the night skies, which he did carefully with an old theodolite and binoculars. Roberts noted not only the position of each visible star but also numbered each star in a sequence that represented their relative brightness. In this familiarization process, Roberts revisited each chart a day, a week, or a month later, each time reranking the stars according to their brightness. Using this process he discovered a number of new variable stars. By 1891, when construction of his observatory was completed, Roberts's orientation to the southern night skies was also complete, and he began to pursue serious astronomical research. With the encouragement of **David Gill**, director of the Royal Observatory, Cape of Good Hope, Roberts began to observe southern variable stars. By 1894, he had discovered a number of additional variable stars so that the total of his discoveries had grown to 20. As the number of known variable stars in the southern skies grew, Gill recommended that Roberts pay particular attention to eclipsing binary stars, of which Algol was the best-known example.

As his observing program matured, Roberts considered the conditions that might influence the accuracy of his observations, for example, the effect of position angle of the variable star and comparison stars in the field of view, and also the influence of relative proximity to the variable star and the comparison star in the field of view to his nose. After these effects were noticed, Gill arranged for the donation of a special photometer by Sir John Usher. Roberts conducted an exhaustive series of observations with this special photometer in which the field of view could be rotated to six different fixed positions. Using the Usher photometer, then a wedge photometer on loan from Oxford University, and later a 4-in. meridian photometer on loan from Harvard College Observatory, Roberts conducted his variable star photometry program with unprecedented levels of precision for visual observations.

For nearly 30 years, Roberts was among the most prolific and exacting variable star observers in either hemisphere. Roberts determined the orbital elements and absolute dimensions of eclipsing binaries from an analysis of their light curves. Because of his confidence in the precision of his light curves, Roberts took the extra steps necessary in computing the orbits of binary stars to demonstrate that their shape was that of oblate spheroids as had been suggested theoretically by **George Darwin** and **Jules Poincaré**. Roberts took the process a step further, and computed the relative diameters and masses of the individual stars in some of these systems. From these data, he concluded that the stars were distended and tenuous objects. In contrast, Russell reached similar conclusions based only on the combined mass of the binary pairs and did not venture the computation of the masses of individual stars.

Roberts's style of reporting for his results was to prepare accurate light curves reflecting his observations, and to draw theoretical conclusions based on his mathematical analysis of the light-curve data. Thus his reporting, in the leading astronomical journals of the times, was fairly compact and did not include his reduced observations. Eventually, Roberts had accumulated over a quarter of a million observations; professional colleagues urged him to publish his reduced data. Around 1915 or so, the reality of his dilemma began to sink in. In order to accommodate requests, from the likes of **Edward Pickering** and **Ejnar Hertzsprung**, Roberts would have to give up observing and spend his full effort at reduction and tabulation of his results. The data reduction effort was as monumental as the effort that had been invested in making the observations. At Pickering's request, Roberts prepared his observations for publication, but unfortunately the format in which he presented his data was not acceptable to Harvard College Observatory for publication in their *Annals*. Though he appealed for assistance from Pickering, Hertzsprung, and others, Roberts insisted that the work was his alone; he was unwilling to train natives to help with this effort, likely for lack of the financial resources with which to pay them for their work. Tragically, Roberts's massive accumulation of data remained unpublished at the time of his death. The data are now held by the American Association of Variable Star Observers and will eventually be reduced and added to their archives, a fitting home for this valuable resource. Well known in the field of binary-star astronomy in his era, Roberts's published work is well cited by authorities in and after his time including **William Campbell** and **Zdněk Kopal**.

In 1893, Roberts was placed in charge of a 3-year program for training South African born teachers in a new normal school at Lovedale. Over the next 27 years, he trained over 4,000 natives, and became well known throughout the colony as a result of the efforts of this large corps of teachers. Through these extensive contacts, Roberts became increasingly involved in South African race relations. In 1920, South African Prime Minister Jan Smuts appointed Roberts to represent the interests of native Africans in the all-white Senate, which Roberts did with distinction. Roberts also served on a number of commissions investigating various problems including a racially charged riot in Port Elizabeth and the Bondelzwarts rebellion in southwest Africa. While the 1920s saw the end of Roberts's own astronomical research, he used his position in government to support the establishment of the University of Michigan's Lamont-Hussey Observatory at Bloemfontein. He also worked to promote astronomy among South Africans. He lectured widely and corresponded with young enthusiasts. **Alan Cousins** was one of the future South African astronomers inspired by Roberts's personal influence.

The first recipient of an honorary Doctor of Science degree from University of the Cape of Good Hope (now the University of South Africa), Roberts was also elected fellow of the Royal Astronomical Society, the Royal Society of Edinburgh, and the Royal Society of South Africa. He served as president of the South African Association for the Advancement of Science in 1913, as president of the Astronomical Society of South Africa in 1927 and 1928, and as a South African delegate to the 1925 International Astronomical Union General Assembly held at Cambridge, England.

Roberts's personal papers are deposited in the Cory Historical Library of Rhodes University.

Keith Snedegar

Selected References

Campbell, W. W. (1913). *Stellar Motions*. New Haven: Yale University Press.

Evans, David S. (1988). *Under Capricorn: A History of Southern Hemisphere Astronomy*. Bristol: Adam Hilger.

Kopal, Zdeněk (1946). *An Introduction to the Study of Eclipsing Variables*. Harvard Observatory Monographs, no. 6. Cambridge, Massachusetts: Harvard University Press.

——— (1959). *Close Binary Systems*. London: Chapman and Hall, pp. 425, 444, 445, 530, 545.

McIntyre, Donald C. (1938). "Alexander William Roberts (1857–1938)." *Journal of the Astronomical Society of South Africa* 4, no. 3: 116–124.

Roberts, Alexander W. (1899). "Density of Close Double Stars." *Astrophysical Journal* 10: 308–314.

——— (1903). "On the Relation Existing Between the Light Changes and the Orbital Elements of a Close Binary Star System, with Special Reference to the Figure and Density of the Variable Star RR Centauri." *Monthly Notices of the Royal Astronomical Society* 63: 527–549.

——— (1905). "Further Note on the Density and Prolateness of Close Binary Stars." *Monthly Notices of the Royal Astronomical Society* 65: 706–710.

Smuts, the Rt. Hon. General J. C. (1938). "Dr. A. W. Roberts." *Journal of the Astronomical Society of South Africa* 4, no. 3: 93–94.

Warner, Brian (1987). "Roberts, Alexander William." In *Dictionary of South African Biography*, edited by W. J. de Kock. Vol. 5, pp. 644–645. Pretoria: National Council for Social Research.

Williams, Thomas R. (1988). "Roberts of Lovedale and Eclipsing Binary Stars." In *Stargazers: The Contributions of Amateurs to Astronomy*, edited by S. Dunlop and M. Gerbaldi, pp. 48–49. Berlin: Springer-Verlag.

Roberts, Isaac

Born **Groes, (Clwyd), Wales, 27 January 1829**
Died **Crowborough, (East Sussex), England, 17 July 1904**

Isaac Roberts, a pioneer astrophotographer, demonstrated that long exposures in large, well-mounted reflecting telescopes could record details of nebulae not visible to the naked eye. His photographs of the Andromeda, Orion, and many other nebulae surpassed all prior efforts.

The son of a farmer, Roberts moved with his family to Liverpool, England, where his father, William Roberts, took a position as bookkeeper in 1835. Roberts received only an elementary education. The remainder of his great store of knowledge was self-acquired.

In 1844, Roberts was apprenticed to a local building firm for a period of 7 years. After his apprenticeship, he remained with the firm, eventually becoming its manager. When the owners of the firm died, Roberts was retained by the families to supervise the closing of the company. Roberts then opened his own building firm, enjoying considerable success as a hard-working and diligent businessman. He took every possible opportunity to increase his knowledge, principally by means of study at the Mechanics Institute, which emphasized natural philosophy and experimental methods of investigation. In 1875, he married Ellen Anne Cartmell; their marriage was childless and ended by her death. By 1888 Roberts had amassed sufficient funds to retire. He then began to take part in the scientific work which had long fascinated him.

Interested in geology during his working days, Roberts became a fellow of the Geological Society, and wrote several papers that are still of interest today. But after a few years, he turned to astronomy and worked in that field for the remainder of his life.

Astrophotography was not new when Roberts came upon the scene. But, in 1878, when dry photographic plates were introduced, Roberts was among the first to realize the implications of the new emulsion for astronomy. An astronomical photograph need not end when all of the stars and nebulous details visible to the eye are recorded. The exposure could be continued, with more light accumulated on the plate to register fainter objects.

In 1879 Roberts purchased a 7-in. Cooke refractor, which he mounted in his private observatory at Maghull, near Liverpool. He began his experiments in astrophotography with this telescope. At first he intended to produce a photographic star atlas, and toward that end attended the 1887 organizational meeting for the *Carte du Ciel* project in Paris. The magnitude of the effort impressed Roberts, but he eventually decided not to participate in it.

At this time, **Andrew Common** had produced a magnificent photograph of the Orion nebula that won a Gold Medal from the Royal Astronomical Society [RAS] in 1884. Unfortunately, no one seemed to be following up in this exciting new field. Since Roberts had achieved some success in photographing nebulae and clusters, he decided that this avenue was the one through which he could make the most effective contribution.

Roberts ordered a 20-in. silver-on-glass reflector from the firm of **Howard Grubb**. The 100-in.-focal-length instrument was mounted equatorially with a very accurate clock drive. Plates were exposed at the mirror's prime focus, in order to preserve as much light as possible. Roberts added a novel touch to the mounting. In the place of a counterweight, he mounted the 7-in. refractor. It was an odd-looking arrangement, but it worked, and Grubb sold several other instrument-pairs in this style.

After a lengthy period of adjustment, the photographic results began to flow. The first exceptional images were of the Orion nebula and the Pleiades. The Orion exposure revealed more detail and extended the image to six times the area of Common's prize-winning picture. The image of the Pleiades revealed photographically, for the first time, the nebulosity that envelopes this brilliant cluster. Roberts exhibited these images at the January 1886 meeting of the RAS. He reported having taken over 200 images of objects outside the Solar System during 1885. For 15 years, Roberts regularly exhibited his latest photographs at the RAS meetings.

Roberts was active in the organization of astronomers during this period. He played a role in the formation of the Liverpool Astronomical Society [LAS], but objected to the rapid expansion of that organization into one of worldwide scope. He withdrew from active participation in the LAS, but was later active in the formation of the British Astronomical Association.

Roberts's greatest photographic achievement was a 4-hour exposure of the Andromeda nebula (known today as the Andromeda Galaxy). This image was presented to the RAS in 1888. It recorded photographically, for the first time, the spiral nature of the nebula. The image caused quite a sensation. It compares well with modern images of this galaxy. Though he did not know it, he had resolved a number of individual stars in the gigantic spiral.

In 1890 Roberts moved to Crowborough, Sussex, seeking a location with more favorable observing conditions. The new observatory was 780 ft. above sea level. He found the atmosphere to be much steadier at Crowborough. At the new location, Roberts employed a professional astronomer, **William Franks**, as his observer; and most of the observational work at Crowborough was carried out by Franks.

In 1896, Roberts went on an eclipse expedition to Norway. His party did not see the eclipse, but while on the trip he did meet **Dorothea Klumpke**, a San Francisco native and a professionally trained astronomer working on the *Carte du Ciel* project at the Paris Observatory. Roberts and Klumpke were married in 1901. The second Mrs. Roberts moved her professional astronomical activities to Roberts's observatory at Crowborough. The two collaborated on astronomical projects, mainly the analysis and publication of his plates, until Roberts's death.

Roberts died suddenly. On the morning of his death he had been working on his plates. Four years later his ashes were entombed in a stone monument in Birkenhead cemetery near Liverpool. The monument is fascinating; it includes engravings of some of his most important photographs. After settling his estate, Dorothea returned to Paris, where she continued to be active in astronomy. She maintained control of Roberts's photographic plates, making them available to other astronomers for research purposes. In 1928, to commemorate the centenary of Roberts's birth, she published a two-volume photographic atlas based on his work.

The failure of the Great Melbourne Telescope in the 1860s had cast a pall over the suitability of the reflecting telescope for astronomical work. However, over the years, Roberts had the opportunity to experiment with many cameras and telescopes. He saw that the future of observational astronomy lay with big reflectors, and took every opportunity to describe and explain this fact to other astronomers. By the time of his death, his assertions had borne fruit not only through his own work, but also that of **George Ritchey** with his 24-in. Newtonian reflector at Yerkes Observatory, and **James Keeler** with the 36-in. Crossley Reflector at Lick Observatory. The giant instruments erected by Ritchey and **George Hale** at Mount Wilson Observatory bore witness to the value of Roberts's early assessment of the reflector's potential, as did the welter of smaller reflectors installed in most contemporary observatories in the first half of the 20th century. Roberts's work played no small part in this revolution.

Roberts was an avid musician. He sang with the Liverpool Philharmonic Choral Society. He was fluent in the Welsh language. Toward the end of his life he became an agnostic, expressing the view that revealed religion had no place in the Universe that he had explored. Politically, Roberts was a liberal, and he was active in education reform.

During his career as a scientist, Roberts received numerous medals and honors. Among these were election to the Royal Society (1890), an honorary Doctor of Science from Trinity College, Dublin (1892), and the Gold Medal of the Royal Astronomical Society (1895). A crater on the Moon has been named Roberts to honor both Isaac Roberts and **Alexander Roberts**, a South African amateur variable star astronomer.

Leonard B. Abbey

Selected References

Ball, Robert S. (1905). "Isaac Roberts." *Proceedings of the Royal Society* 75: 356–363.

Hollis, Henry Park (1920). "Roberts, Isaac." In *Dictionary of National Biography, Supplement, January 1901–December 1911*, edited by Sir Sidney Lee, pp. 209–211. London: Oxford University Press.

James, Stephen H. G. (1993). "Dr Isaac Roberts (1829–1904) and His Observatories." *Journal of the British Astronomical Association* 103: 120–122.

Roberts, Isaac (1893–1899). *Photographs of Stars, Star-Clusters, and Nebulae, Together with Records of Results Obtained in the Pursuit of Celestial Photography*. Vols. 1 and 2. London: *Knowledge* Office.

Roberts, Dorothea Klumpke (1928?–1932). *Isaac Roberts' Atlas of 52 Regions: A Guide to William Herschel's Fields of Nebulosity*. Vols. 1 and 2. Paris: Bonheur.

Vaucouleurs, Gérard de (1987). "Discovering M31's Spiral Shape." *Sky & Telescope* 74, no. 4: 595–598.

Robertson, Howard Percy

Born **Hoquiam, Washington, USA, 27 January 1903**
Died **Pasadena, California, USA, 26 August 1961**

American mathematical physicist Howard ("H.P.") Robertson is honored by the names of the Poynting–Robertson effect (of light on small dust particles) and the Robertson–Walker metric, which describes the curvature of space–time within the framework of general relativity. He was the oldest of five children of a family of modest means, whose father died when he was 15. Nonetheless, all the children attended the University of Washington, where Robertson earned a bachelor's (1922) and a master's (1923) degree.

At the University of Washington, Robertson came under the tutelage of the famous mathematician and relativist, Eric Temple Bell. They had a tempestuous and engaging intellectual relationship, which they both valued highly over the years.

Two years later, Robertson obtained his Ph.D. from the California Institute of Technology (Caltech). His thesis was entitled "On Dynamical Space-times which contain a Conformal Euclidean 3-space," and ran a mere 38 pages in length. While at Caltech, he interacted with the mathematical physicists, Paul S. Epstein and Harry F. Bateman.

The next 2 years were spent studying in Göttingen and Munich, Germany as a National Research Fellow. Robertson returned to Caltech as an assistant professor in 1927. From 1929 to 1947 he was at Princeton University, again returning to Caltech as a professor in 1947. From 1940 to 1943, Robertson served with the National Defense Research Council, and was with the London Mission of the Office of Scientific Research and Development from 1943 to 1946. From 1944 to 1947, he was an expert consultant to the Office of the Secretary of War. He was awarded the United States Medal of Merit in recognition of his wartime services to his country.

After World War II, Robertson was a much-sought-after scientific advisor to numerous branches of government and industry, and he was very effective in applying science to military strategy and tactics. In this capacity he was the chief scientific advisor to general M. Gruenther, then the Supreme Allied Commander in Europe (1954–1956). From 1956 to 1960, Robertson was a member of the Defense Science Board. He was also the chief science advisor to the second and third directors of the Central Intelligence Agency, admiral Sidney W. Souers and general Hoyt S. Vandenberg. Robertson was a trustee of the Systems Development Corporation, the Institute for Defense Analysis, and the Carnegie Endowment for International Peace. He was also a director of the Northrop Aviation Corporation.

Robertson was endowed with exceptional mathematical powers coupled with a deep insight into physical processes. His early scientific efforts were concerned with the study of differential geometry, in which he was strongly influenced by the work of Bell, Luther Eisenhart, Oswald Veblen, and **Herman Weyl**. In 1931, for example, he translated Weyl's classic tome, *The Theory of Groups and Quantum Mechanics*, into English. He also published works on quantum mechanics, notably a short paper in *The Physical Review* pointing out the connection between uncertainty in the simultaneous measurement of two noncanonical variables and the commutation properties of their associated operators (1929). He also made a seminal contribution (1940) to T. von Karman and L. Howarth's theory of isotropic turbulence by applying invariant theory to the categorization of the velocity correlation tensors that feature prominently in their equations.

Robertson's best-known contributions were in the theory of relativity and its applications to cosmology. He developed the theory of uniform cosmological spaces, *i. e.*, spaces with spatial isotropy and homogeneity, and deduced the form of the line element common to all these spaces (1929). These have subsequently been called Robertson–Walker spaces. Robertson was deeply intrigued by the consequences of the observed red shift–distance relationship, which led to an extensive and long-lasting working association with **Edwin Hubble**, **Milton Humason**, and **Richard Tolman**.

Robertson is also well known for his work on the absorption and re-emission of light by a particle revolving around the Sun (1937). His fully relativistic treatment of the problem superseded **John Poynting**'s (1903) classical formulation, and in fact led to a significant quantitative correction to the classical result. Robertson's calculations indicated the presence of a tangential drag that acts to reduce the angular momentum of the body, causing it to spiral in toward the Sun. The effect is important for small dust particles, and implies that the immediate neighborhood of the Sun should be cleared of these particles on astrophysically interesting timescales. This process is commonly referred to as the Poynting–Robertson effect.

Robertson had a long-standing fascination with sports cars, and was well known for his fast driving on the Caltech campus and in the surrounding area. In early August of 1961 he was involved in a high-speed automobile accident. He died from a pulmonary embolism brought on by the injuries sustained in the accident.

Thomas J. Bogdan

Selected References

Fowler, William A. (1962). "Howard Percy Robertson." *Quarterly Journal of the Royal Astronomical Society* 3: 132–135.

Kragh, Helge (1996). *Cosmology and Controversy: The Historical Development of Two Theories of the Universe.* Princeton, New Jersey: Princeton University Press, pp. 12–17, 21, 56, 280.

Robertson, H. P. (1933). "Relativistic Cosmology." *Reviews of Modern Physics* 5: 62–90.

——— "Kinematics and World-Structure." Parts 1–3. *Astrophysical Journal* 82 (1935): 284–301; 83 (1936); 187–201, 257–271.

Taub, A. H. (1961). "Prof. H. P. Robertson." *Nature* 192: 797–798.

Robinson, Thomas Romney

Born **probably Lawrencetown near Bainbridge, Co. Down, (Northern Ireland), 23 April 1792**
Died **Armagh, Ireland, (Northern Ireland), 28 February 1892**

Thomas Romney Robinson (sometimes John Thomas Romney Robinson) was director of Armagh Observatory for more than 58 years. A child prodigy, he became one of the most respected practical astronomers of his time. He is remembered for the anemometer design that bears his name.

Romney Robinson (as he is commonly referred to) was the eldest son of Thomas Robinson, an English portrait painter, and his wife Ruth Buck. Thomas named his son after his mentor, George Romney, and set up business in Dublin about 1790. Later the family moved to County Down, and then in 1801 settled in Belfast where Romney attended Belfast Academy.

Robinson was precocious, being able to read poetry by three and to write verse by five. By the age of 12 a volume of his poems was published in Belfast with a list of 1,600 subscribers. He showed an interest in the machinery used in the linen and shipbuilding industries and experimented with chemistry and electricity. Robinson entered Trinity College, Dublin, in January 1806 and graduated in 1810, the year of his father's death. He gained fellowship in 1814. Robinson was elected a member of the Royal Irish Academy in 1816 and read papers on high-temperature furnaces and electricity. He lectured in Trinity College as deputy to Bartholomew Lloyd, the professor of natural and experimental philosophy, and provided his students with a useful textbook in his *System of Mechanics* (1820).

However, Robinson found that teaching left him with little time for research. He was a close friend of **John Brinkley**, the Andrews' Professor of Astronomy and director of Dunsink Observatory. In 1821 he relinquished his fellowship, married, took holy orders, and became the rector of the parish of Enniskillen, County, Fermanagh. In 1823 Robinson was appointed director of Armagh Observatory, and the following year he was appointed vicar of the parish of Carrickmacross, which was closer to Armagh.

Armagh Observatory had been established by the (Anglican) Church of Ireland in 1790 but had not attained much distinction under its first two directors, James Archibald Hamilton and William Davenport. Robinson set the observatory on a productive course. Thanks to the generosity of the primate, Lord John George Beresford, Robinson was able to commission badly needed new instruments. He ordered a transit instrument and a mural circle from Thomas Jones, a leading London instrumentmaker. The new transit was in place by 1827, and observations began on a program of positional measurements. The publication of the transit observations for the years 1828, 1829, and 1830 was funded by Beresford. The mural circle was delivered in 1831 but did not come into operation until late in 1834.

The third instrument that Robinson commissioned was a 15-in. reflecting telescope of an innovative design, which was highly significant for the future development of telescope technology. It was a Cassegrain, rather than a Newtonian, telescope, then the standard. It was the first large reflecting telescope to be mounted equatorially with a clock drive. The primary mirror of speculum was supported by a novel lever support system to avoid distortion. The maker was **Thomas Grubb** who, as engineer to the Bank of Ireland in Dublin, was well known for his machines for printing banknotes. Early in 1833, Robinson approached Grubb to see if he would make an equatorial mount for the 13.3-in. objective by Chauchoix of Paris, which had been purchased by **Edward Cooper** for his private observatory at Markree Castle, County Sligo. Cooper's telescope was installed in April 1834, and the 15-in. telescope was installed at Armagh Observatory the following year. With the staunch support of Robinson,

Grubb and his son **Howard Grubb** went on to establish a telescope-making firm of international renown.

Robinson was closely associated with **William Parsons**'s ambitions to build large reflectors at Parsonstown (Birr). In November 1842, accompanied by his friend Sir **James South**, Robinson witnessed the casting of the speculum mirror for the great 6-ft. telescope and described the scene in graphic detail in the *Proceedings* of the Royal Irish Academy. He advised Rosse to adopt an equatorial mount but the latter opted for a universal joint arrangement that was less demanding mechanically.

By February 1845 the great telescope was nearly finished, and Robinson returned to Birr, accompanied once again by South. After waiting for fine weather, the first observations were made on 10 March, and 40 nebulae from **John Herschel**'s catalog were examined. The following month, Robinson reported in glowing terms on the performance of the telescope to the Royal Irish Academy.

Robinson's experience of observing with large telescopes at Birr convinced him that there was a need for a large reflector in the Southern Hemisphere. In 1849, when Robinson was president of the British Association, the association proposed that the government should provide the Cape Observatory with a large reflector. A Southern Telescope Committee was formed, and a proposal for a 4-ft. Cassegrain reflector by Grubb on a German equatorial mount was made in 1853, but was rejected by the government. The project was revived in 1862 when the State of Victoria in Australia decided to erect a large telescope in Melbourne. After prolonged negotiations, which also involved the offer of a gift of a telescope from **William Lassell**, the State of Victoria ordered the telescope from Grubb in February 1866. The telescope was built in Dublin and delivered to Melbourne in 1869. Unfortunately, the resources of the Melbourne Observatory were not sufficient to maintain the speculum mirror in proper condition, and this, combined with other factors, led to disappointing results.

Meanwhile, Robinson, with the help of assistants, had maintained a strenuous program of meridian observations. This culminated in 1859 in the publication of *Places of 5,345 Stars Observed from 1828 to 1854 at the Armagh Observatory* printed at government expense. For this work, Robinson was awarded the Royal Medal of the Royal Society in 1862.

One early investigation pursued by Robinson was the determination of the exact longitude of the Armagh Observatory. In 1838 he organized a comparison of the longitude of the Greenwich, Dunsink, and Armagh observatories by transporting 15 chronometers between the 3 locations. The following year, he compared the longitudes of Armagh and Dunsink by firing rockets from the summit of an intervening mountain to synchronize observations at the two observatories.

In 1845, the meridian program at Armagh was threatened by a proposal to build a railway link between Armagh and a neighboring town. As the railway line would pass within 160 yards of the transit instrument; Robinson was concerned that vibration would interfere with the accuracy of the observations. Robinson strenuously opposed the proposal, and the railway was prohibited from approaching within 700 yards of the observatory. Robinson was one of the first of observatory directors to have to deal with this sort of problem, and his expertise was later called upon in similar situations.

Regular weather observations were part of the observing routine at Armagh Observatory from its foundation, and the erection of a wind gauge was proposed in 1839. Robinson reviewed previous designs and, acting on an idea suggested to him many years earlier by Richard Lovell Edgeworth, he constructed a horizontal windmill with four hemispherical cups, and described it to the British Association in 1846. He continued to refine his anemometer, and an improved version was described to the Royal Irish Academy in 1855.

In his later years at Armagh, Robinson had to cope with declining financial resources as a result of the disestablishment of the Church of Ireland and the reduction in rents from lands owned by the observatory. This situation reinforced his conservative political views and led to his vigorous opposition to Irish political reform.

Robinson was an eloquent and forceful speaker and soon came to prominence as a public man of science. He regularly attended meetings of the British Association for the Advancement of Science and played a central role in bringing the association to Dublin in 1835. He served as president of the Royal Irish Academy from 1851 to 1856 and played an important part in securing new premises for the Academy in 1851. Robinson was elected a fellow of the Royal Society in 1856 and awarded honorary degrees by the Universities of Dublin, Oxford, and Cambridge. The 24-km-diameter lunar crater at 59°.0 N and 45°.9 W is named in his honor.

Robinson married twice. His first wife was Eliza Isabelle Rambaut of an Irish Huguenot family, and they had three children. Eliza died in 1839. In 1843 he married Lucy, youngest child of Richard Edgeworth and half-sister of Maria, the novelist. Robinson's daughter by his first marriage, Mary Susanna, married the Irish mathematical physicist **George Stokes**, with whom Robinson corresponded regularly on scientific matters.

Ian Elliott

Selected References

Bennett, J. A. (1990). *Church, State and Astronomy in Ireland: 200 Years of Armagh Observatory.* Armagh: Armagh Observatory, in association with the Institute of Irish Studies and the Queen's University of Belfast.

Glass, Ian S. (1997). *Victorian Telescope Makers: The Lives and Letters of Thomas and Howard Grubb.* Bristol: Institute of Physics.

Hoskin, Michael (1989). "Astronomers at War: South v. Sheepshanks." *Journal of the History of Astronomy* 20: 175–212.

King, Henry C. (1979). *The History of the Telescope.* New York: Dover.

Roche, Édouard Albert

Born **Montpellier, Hérault, France, 17 October 1820**
Died **Montpellier, Hérault, France, 18 April 1883**

Édouard Roche is remembered for the study of equipotential surfaces, called Roche lobes, and calculation of the distance from a planet at which satellites will be torn into rings, called the Roche limit.

Roche continued a tradition in his family, whereupon several members became professors at the University of Montpellier. There, he earned his *docteur ès sciences* degree in 1844 but spent the next 3 years at the Observatoire de Paris, working under **Dominique Arago**. While at Paris, Roche was introduced to **Urbain Le Verrier** and **Augustin Cauchy**.

In 1849, Roche accepted the position of *chargé de cours* at Montpellier and in 1852 was appointed professor of pure mathematics. The relative isolation of Montpellier likely helped foster the emergence of Roche's original ideas. He was elected a corresponding member of the Académie des sciences in 1873, but was later denied full membership in the organization. Roche's life was afflicted by poor health; he was forced to take a leave of absence in 1881 and succumbed to a lung inflammation.

Roche devoted himself to topics that lay generally outside of mainstream astronomical research in his time. He studied the equilibrium figure of a rotating fluid body that was subjected to an external gravitational force caused by another body. This condition is extremely important for determining the equipotential surfaces around a pair of point masses like a binary star. Where one (or both) of the components assumes a tear-drop shape, the space(s) so filled is called the Roche lobe. Related to this investigation was Roche's calculation of the minimum distance at which a satellite (of equal density) may revolve above its parent planet. The boundary, inside of which the tidal forces are strong enough to disrupt the satellite into smaller pieces, is called the Roche limit. This condition plays a crucial role in the formation of planetary-ring systems.

Roche likewise explained the streamlined shapes of cometary envelopes, under the assumption of a repulsive force originating in the Sun that diminished with the square of its heliocentric distance. His theoretical explanation was offered decades before such radiation pressure was discovered in the form of the "solar wind." In turn, Roche presented a thorough analysis of the "nebular hypothesis" of **Immanuel Kant** and **Pierre de Laplace**. His investigation brought a decided coherence to the notion before it was challenged by rival cosmogonic theories. Roche's treatment of the internal density distribution of the Earth, with appropriate modifications, is still of value today.

Zdeněk Kopal (1989, p. 2) has written that Roche's principal accomplishments "were too far ahead of their time to be appreciated fully."

Martin Solc

Selected References

Kopal, Zdeněk (1989). *The Roche Problem and Its Significance for Double-Star Astronomy*. Dordrecht: Kluwer Academic Publishers. esp. "Introduction," pp. 1–5.

Lévy, Jacques R. (1975). "Roche, Édouard Albert." In *Dictionary of Scientific Biography*, edited by Charles Coulston Gillispie. Vol. 11, p. 498. New York: Charles Scribner's Sons.

Roche, E. (1850). "Mémoires divers sur l'equilibre d'une mass fluide." *Mémoires de l'Académie des sciences et lettres de Montpellier* 2: 21–32.

——— (1873). "Essai sur la constitution et l'origine du systeme solaire." *Mémoires de l'Académie des sciences et lettres de Montpellier* 8: 235–327.

Roeslin, Helisaeus

Born **Plieningen, (Baden-Württemberg, Germany), 17 January 1545**

Died **probably Haguenau, (Bas-Rhin, France), 14 August 1616**

Helisaeus Roeslin was one of the first to recognize that comets are astronomical bodies rather than atmospheric bodies and, with **Nicholas Bär** (Raimarus Ursus) and **Tycho Brahe**, formulated a geo-heliocentric model of the planets. Roeslin studied astronomy and medicine at the University of Tübingen from 1561. Samuel Eisenmenger (Siderokrates) became his teacher in astronomy, astrology, and alchemy. Roeslin finished his studies in 1569 with the degree of a doctor of medicine. He joined Eisenmenger in the employ of the Markgraf von Baden. Influenced by Eisenmenger, Roeslin devoted himself to the spiritualistic theory of Kaspar von Schwenkfeld, but showed religious tolerance during his whole life. In 1569 he became physician in Pforzheim, married, and became an employee of the Pfalzgraf von Pfalz Veldenz. There Roeslin became a well-known doctor. From 1582 onward he was physician in the city of Haguenau.

Roeslin's astronomical interests rose with two events, the supernova of 1572 (SN B Cas) and the great comet of 1577. He developed a theory of the comet that placed it first in the Earth's atmosphere. Later he withdrew this idea and agreed with the theory of **Christoph Rothmann** and Brahe, in which comets belong to the sphere of the planets. In his publication in 1597 he explained the nature of comets as heavenly bodies in the sphere of the planets. Roeslin claimed that comets are not atmospheric phenomena and that this is in clear contradiction to **Aristotle**'s ideas. With the observation of comets there is no way to maintain the idea of the unchangeable heavenly regions and their perfection and divinity. The tail of a comet he explained as light focused by the comet itself.

Therefore, for Roeslin, comets are "secondary stars," which have not been created during the "first 6 days of the world," but were "born" later. Also, they are not permanent bodies and after some time of visibility they disappear again. Moreover, comets move in circular orbits around the Sun, but not only in the zodiacal belt. With this work, Roeslin gained major significance in the history of cometary science. He is one of the first astronomers who spoke clearly of regular orbits of comets. Roeslin's opinion was contrary to the widely held views of Rothmann, Brahe, and **Johannes Kepler**. His works from 1597 and 1609 are quite significant, because the theory of comets as heavenly bodies was written in German for the first time, and he also clearly pointed out the contradiction to the physics of Aristotle.

Among Brahe, Bär, and others, Roeslin developed the idea of a geo-heliocentric system (*De opere Die creationis*, 1597), which only differs from the others in details. Roeslin still considered the planets to be fixed on spheres with no space in between. The example of Roeslin shows how widespread the theory of geo-heliocentricity was at that time. Roeslin and Kepler had several public debates, but Kepler did not succeed in convincing Roeslin of the heliocentric system. They reconstructed different birth dates of Jesus Christ (Roeslin 1.25 years BCE, Kepler 5 years BCE), and in astrology Kepler could not follow Roeslin's ideas and conclusions concerning comets and planetary motion.

Under the pen name "Lambert Floridus Plieninger," Roeslin published a paper concerning the calendar reform of Pope Gregory XIII. Like many other authors of his time (*e. g.*, **Wilhem IV** and **Michael Mäestlin**), Roeslin carried on the controversy against the new calendar, mainly for theological reasons, and claimed the pope wished to use the calendar to regain power over the Protestant church.

Jürgen Hamel
Translated by: *Peter Habison*

Selected References

Diesner, P. (1935). "Leben und Streben des elsaessischen Arztes Helisaeus Roeslin (1544–1616)." *Elsass-Lothringisches Jahrbuch* 14: 115–141.

Granada, Miguel A. (1996). *El debate cosmológico en 1588: Bruno, Brahe, Rothmann, Ursus, Röslin.* Naples: Istituto Italiano per gli Studi Filosofici.

Hamel, Jürgen (1994). "Die Vorstellung von den Kometen seit der Antike bis ins 17. Jahrhundert-Tradition und Innovation." In *Georg Samuel Dörffel (1643–1688): Theologe und Astronom*, pp. 97–122. Plauen: Vogtland-Verlag.

——— (1998). *Geschichte der Astronomie von den Anfängen bis zur Gegenwart.* Lecce: Birkhäuser.

List, Martha (1948). "Helisaeus Roeslin: Arzt und Astrologe." *Schwaebische Lebensbilder* 3: 468–480.

Roeslin, Helisaeus (2000): *De opere Dei Creations.* Ristampa anastatica dell' edizione Francoforte 1597 a cura di miguel Angel Granada. Lecce: Conte Editore.

Schofield, Christine Jones (1981). *Tychonic and Semi-Tychonic World Systems.* New York: Arno Press.

Roger of Hereford

Flourished **England, 1176–1178**

Roger of Hereford was an astronomer working in the region of Hereford in the late 12th century. His alternative names perhaps reflect an English name "Young," "Lénfant," or "Childe."

Roger adapted the astronomical tables of Toledo for the meridian of Hereford in 1178, using as his basis the version composed for the meridian of Marseilles by **Raymond of Marseilles**. Two years earlier (1176) he wrote a work on the ecclesiastical *computus* (for calculating the church calendar), in which he compared unfavorably the traditional Latin learning on the *computus* with the new learning from Hebrew and Arabic sources.

Other writings attributed to Roger are several works concerning astrology, which may all derive from a *Liber de quattuor partibus iudiciorum astronomie*; this draws on the corresponding work of Raymond of Marseilles, as well as on translations of Arabic astrological texts made by **John of Seville** and Hermann of Carinthia. Roger may too have written on alchemy, if the *De rebus metallicis* once existing in Peterhouse, Cambridge, is authentic.

Roger was held in high esteem by contemporary English scholars, which is indicated by the fact that Alfred of Shareshill dedicated his translation of the Aristotelian text on botany, *De vegetabilibus*, to him. To Roger may be attributed the invention of a new way of calculating horoscopes mathematically. He did much to further the study of the mathematical sciences in England, and provides a link between the pioneers in this study, **Petrus Alfonsi** and **Adelard of Bath**, and **Robert Grosseteste**, who joined the bishop's household in Hereford while Roger was still active there.

Charles Burnett

Alternate names

Rogerus Infans
Rogerus Puer

Selected References

French, Roger (1996). "Foretelling the Future: Arabic Astrology and English Medicine in the Late Twelfth Century." *Isis* 87: 453–480.

Haskins, Charles H. (1927). *Studies in the History of Mediaeval Science.* 2nd ed. Cambridge, Massachusetts: Harvard University Press, pp. 123–126, 128.

North, John D. (1986). *Horoscopes and History.* London: Warburg Institute, pp. 39–42.

Rogerus Infans

➔ **Roger of Hereford**

Rogerus Puer

➔ **Roger of Hereford**

Rohault, Jacques

Born **Amiens, (Somme), France, 1620**
Died **Paris, France, 1665**

Jacques Rohault was a self-taught mathematician and experimentalist, and a popularizer of natural philosophy. He was the son of Ambroise Rohault and Antoinette de Ponthieu. In the mid-1650s, he began a series of extremely popular lectures on natural philosophy. These lectures popularized **René Descartes**'s natural philosophy, but laid great emphasis on experiment and observation, and omitted the metaphysical foundations that Descartes had gone to great trouble to provide. Rohault became the leading defender of Cartesianism in France, and his *Traité de physique* (Treatise on physics) (1671) became the leading textbook of the age. It offered a probabilistic reading of natural philosophy, but avoided any detailed mathematics, and completely ignored **Johannes Kepler**'s work. It presented natural philosophy as an observational discipline rather than a mathematical one. Nevertheless, the *Traité* was exceptional in the scope and clarity of its treatment. As far as cosmology is concerned, the keystone of Cartesian cosmology is the vortex theory, which explains the formation of stars and planets, and the stability of planetary orbits, in terms of the rotation of matter around central points, and Rohault offered a detailed defense of this theory, which was extended to gravity, magnetism, and other phenomena.

In 1697, Samuel Clarke brought out an English edition of the *Traité*, establishing it as the major natural philosophy textbook in England. Clarke's edition does nothing to challenge the vortex theory in its earlier versions, and makes no mention of gravitation, but in later editions the notes took on a strongly Newtonian flavor, so that the *Traité* became, in its later incarnations, a curious hybrid of Cartesianism and Newtonianism.

Stephen Gaukroger

Selected Reference

Rohault, Jacques (1723). *A System of Natural Philosophy, Illustrated with Dr Samuel Clarkes Notes ... Done into English by John Clarke*. 2 Vols. London. (Reprint, New York: Johnson Reprint Corp., 1969.)

Römer [Roemer], Ole [Olaus]

Born **Aarhus, Denmark, 25 September 1644**
Died **Copenhagen, Denmark, 19 September 1710**

Ole Römer was a multifaceted Danish scientist and public servant, most noted for his discovery and determination of the finite velocity of light. He also constructed the first meridian transit circle incorporating a telescopic sight.

At the University of Copenhagen, Römer studied medicine under the brothers Thomas and **Erasmus Bartholin**. The latter was also a physical scientist who discovered the phenomenon of double refraction in crystals of Iceland spar, and tutored Römer in astronomy and mathematics.

In 1671, Römer's life was changed by the arrival of **Jean Picard**, who came to Denmark to determine the exact location of Uraniborg, **Tycho Brahe**'s old observatory. Römer assisted Picard in this task, during which they determined longitude by timing eclipses of Jupiter's satellite I (Io), after which they both traveled to Paris in 1672. There Louis XIV had established the most magnificent observatory in Europe, and appointed Römer as tutor to the dauphin.

During his 9 years in France, Römer turned his hand to a variety of tasks, as astronomers in those days were often general scientists and practical engineers. He devised improved instruments, such as clocks and micrometers, as well as supervising hydraulic works near Paris and in the provinces. What brought him fame, however, were his observations and interpretation of the times at which eclipses of Io occurred. This was a subject of important commercial and military importance at the time, for these phenomena offered the possibility of solving the intractable problem of determining one's longitude at sea. Because of the difficulty of making the necessary observations from the heaving deck of a ship, this method never fulfilled its promise.

Improvements in timing the beginnings of such eclipses, due to the improved accuracy and precision of contemporary clocks, had disclosed worrisome and unexplained problems in reconciling observations made at different times. These discrepancies might be explained if light had a finite velocity, but the prevailing opinion among scientists was that propagation was instantaneous.

In 1675, Römer predicted that the onset of an eclipse that was to begin on 9 November of that year would be 10 min later than otherwise expected, based on the speed of light being some 140,000 miles per second. This did indeed happen, and his estimate was not far off the modern determination of just over 186,000 miles per second. This demonstration was not universally accepted, and it was not until **James Bradley**'s 1729 discovery of the aberration of starlight, which demanded a finite velocity for light, that Römer was fully vindicated.

In 1679, Römer visited England where he met such contemporaries as **Isaac Newton**, **Edmond Halley**, and **John Flamsteed**. In 1681, he returned to Denmark at the request of Christian V, where he became in modern terms the royal scientific advisor, as well as director of the Copenhagen Observatory. Until his death, Römer held a bewildering variety of public positions, such as master of the Mint (similar to the position that Newton held), various appointments of a military and engineering nature, privy councilor, and even the effective mayor of Copenhagen.

In spite of these duties, Römer continued his scientific work, including building his own observatory, Tusculaneum, outside Copenhagen, designing new instruments for astronomical observations, and inventing a new type of thermometer, which later bore the name of Fahrenheit. During these last years, Römer also accumulated a vast store of astronomical observations, almost all of which perished in the 1728 Copenhagen fire. His first wife, Anna Maria Bartholin (daughter of Erasmus), whom he married in 1681, died in 1694; they had no children.

Much of what we know about Römer's astronomical work comes from **Peter Horrebow**'s 1735 book *Basis astronomiae sive astronomiae pars mechanica*.

Ronald A. Schorn

Selected Reference

Cohen, I. Bernard (1940). "Roemer and the First Determination of the Velocity of Light (1676)." *Isis* 31: 327–379.

Rooke, Lawrence

Born **Deptford, (London), England, 13 March 1622**
Died **Deptford, (London), England, 27 June 1662**

Lawrence Rooke held the astronomy chair at Gresham College in London and was a founding member of the group associated with the college that became the Royal Society of London soon after his death. He was educated at King's College, Cambridge (BA: 1643; MA: 1647). From 1650 to 1652, Rooke was a fellow of Wadham College, Oxford, before taking up the chair of astronomy at Gresham College in 1652. In 1657, he changed to the chair in geometry, which he held until his death. He was married and had four daughters and five sons.

Rooke was widely known as a learned, industrious scholar by contemporaries. Rooke's primary astronomical contributions are of observational, practical importance. His careful observations of cometary paths and his research on determining longitude at sea by observing lunar eclipses and the satellites of Jupiter exemplify Rooke's support for the "new philosophy" of science in 17th-century England. **Seth Ward** published Rooke's observations comet C/1652 R1. Essays on eclipses of the Moon and of the satellites of Jupiter appeared posthumously in the *Philosophical Transactions* of the Royal Society and in Thomas Sprat's *History of the Royal Society*.

Robinson M. Yost

Selected References

Keller, A. G. (2004). "Rooke, Lawrence (1619/20–1662)." In *Oxford Dictionary of National Biography*, edited by H. C. G. Matthew and Brian Harrison. Vol. 47, pp. 695–696. Oxford: Oxford University Press.

Ronan, C. A. (1960). "Laurence Rooke (1622–1662)." *Notes and Records of the Royal Society of London* 15: 113–118.

Rosenberg, Hans

Born **Berlin, Germany, 18 May 1879**
Died **Istanbul, Turkey, 26 July 1940**

German observational astronomer Hans Rosenberg made the first plot of stellar brightness versus spectral type, now called a Hertzsprung–Russell diagram, under the guidance of **Karl Schwarzchild**.

Rosenberg earned a Ph.D. from the University of Strasbourg for work under E. Becker on the variability of χ Cygni, a Mira-type variable. He received his *Habilitation* degree from Tübingen in 1910 for a thesis entitled "The relation between Brightness and Spectral Type in the Pleiades." This included a graph of apparent magnitude versus spectral class for the Pleiades, made at the suggestion of Schwarzschild, which was apparently the first of what we now call Hertzsprung–Russell diagram (for plots made by **Ejnar Hertzsprung** and **Henry Norris Russell** a few years later).

Rosenberg became head of the university observatory at Tübingen in 1912 and professor of astronomy there in 1916, in the midst of service in the German army (1914–1918). He was appointed professor of astronomy and director of the observatory at the University of Kiel in 1926, from which positions he was removed in 1935 under the *Reichsbürgergesetz* (roughly, laws pertaining to the citizens of the country) because he was Jewish. Fortunately, Rosenberg had taken a position as a guest lecturer at Yerkes Observatory (University of Chicago) in 1934 for a 3-year period. He went on to be professor of astronomy and director of the observatory in Istanbul, Turkey, from 1938 until his death.

Rosenberg was a pioneer of photoelectric photometry, applying the technique to comets, variable stars, and spectroscopic binaries. He led solar eclipse expeditions to Finland in 1927 and Thailand in 1929.

Christian Theis

Selected References

Gleissberg, W. (1940). "Hans Rosenberg." *Publications of the Istanbul Observatory* 13: 2.

Schmidt-Schönbeck, C. (1965). *300 Jahre Physik und Astronomie an der Kieler Universität*. Keil: F. Hirt.

Theis, Ch. *et al.* (1999). "Hans Rosenberg und Carl Wirtz – Zwei Kieler Astronomen in der NS-Zeit." *Sterne und Weltraum* 2: 126.

Rosenberger, Otto

Born **Tukkim, (Latvia), 10 August 1800**
Died **probably Halle, Germany, 23 January 1890**

Of German descent, Otto Rosenberger was assistant at Königsberg Observatory and later professor at Halle. He predicted the perihelion passage of Halley's comet (IP/Halley) in 1835 with great accuracy.

Selected Reference

Anon. (1891). "Otto August Rosenberger." *Monthly Notices of the Royal Astronomical Society*. 51: 202.

Ross, Frank Elmore

Born **San Francisco, California, USA, 2 April 1874**
Died **Pasadena, California, USA, 21 September 1960**

Frank Ross is known for the Ross correcting lens and his lists of new proper-motion stars. His father, Daniel Ross, a building contractor, lost a fortune during the California gold-mining boom. In 1882, the family moved to San Rafael, California, where Ross attended grammar school and cultivated an interest in mathematics. He entered the University of California at Berkeley and received his BS degree in 1896. He got his Ph.D. from the same institution in 1901.

Ross was appointed as an assistant at the Nautical Almanac Office, Washington, District of Columbia (1902), research assistant in the Carnegie Institution (1903–1905), and director of the International Latitude Station at Gaithersburg, Maryland (1905–1915). He joined the Eastman Kodak Company (1915), and carried out important investigations on the physics of the photographic process until 1924 when he joined the Yerkes Observatory of the University of Chicago as Associate Professor. Promoted to professor in 1928, Ross retired to Pasadena, California, in 1939.

To increase the size of the usable field of large reflectors, Ross invented a correcting lens system still in use. At the telescope himself, he discovered many stars with large proper motions and numerous variable stars. Ross also built upon **William Wright**'s pioneering work at the Lick Observatory by photographing Mars in the light of five different colors, during the opposition of 1926.

In 1927, Ross imaged Venus in the ultraviolet with the 60- and 100-in. reflectors of the Mount Wilson Observatory in order to register dusky markings that he interpreted as atmospheric disturbances. No detail was visible in the red and infrared, and he concluded the upper atmosphere of Venus is composed of thin cirrus-like cloud, while the lower part is exceedingly dense and yellowish. Thirty years elapsed before astronomers finally took note of this important revelation and understood what it signified.

Ross photographed large-scale structures not previously recognized in the Milky Way. This study culminated in 1934 with the publication of the *Atlas of the Northern Milky Way* with Mary Calvert, **Edward Barnard**'s niece, and Kenneth Newman.

Richard Baum

Selected References

Morgan, W. W. (1967). "Frank Elmore Ross." *Biographical Memoirs, National Academy of Sciences* 39: 391–402.

Ross, Frank E. (1926). "Photographs of Mars, 1926." *Astrophysical Journal* 64: 243–249.

——— (1928). "Photographs of Venus." *Astrophysical Journal* 68: 57–92.

Ross, Frank E., Mary R. Calvert, and Kenneth Newman (1934). *Atlas of the Northern Milky Way*. Chicago: University of Chicago Press.

Rossi, Bruno Benedetto

Born **Venice, Italy, 13 April 1905**
Died **Cambridge, Massachusetts, USA, 21 November 1993**

Italian–American cosmic-ray and X-ray physicist Bruno Rossi is honored within astrophysics as the guiding spirit of the first rocket-borne detectors intended to look for X-ray sources outside the Solar System. The first of these, flown in 1962, saw one strong source and the X-ray background, and opened a whole new window on the cosmos.

Bruno Rossi, the son of Rino Rossi and Lina Minerbi, received his BA degree from the University of Padua and his Ph.D. from the University of Bologna in 1927. On completion of his university studies, Rossi was appointed assistant to Antonio Garbasso at the University of Florence in 1928. There his lifelong interest in the nature and origins of cosmic radiations was inspired by a paper of Walther Bothe and **Werner Kolhörster** describing their discovery of charged cosmic-ray particles that penetrated 4.1 cm of gold. They concluded that most of the local cosmic rays are not gamma rays as was generally believed at the time, but energetic charged particles.

Within a few weeks of reading the paper, Rossi invented an electronic coincidence circuit with which nearly simultaneous pulses from two or more Geiger counters could be recorded with a time resolution better than 1 millisecond. It was the first electronic AND circuit, a basic logic element of future electronic computers. Its applications by Rossi in a series of pioneering experiments carried out in the period from 1930 to 1933 marked the effective beginning of electronic methods in nuclear and particle physics. Rossi demonstrated the presence in cosmic rays of penetrating charged particles, now called muons, capable of traversing more than 1 m of lead. He identified the "soft" component that interacts in thin layers of lead to produce secondary showers of particles. In an experiment carried out in Eritrea, Rossi measured the east–west asymmetry in the intensity of cosmic rays that he had predicted in 1930 in a theoretical analysis of the deflection of charged particles by the Earth's magnetic field. The direction of the asymmetry proved that the charges of the primary particles are predominantly positive. In the course of the same experiment, Rossi discovered the nearly simultaneous arrival at ground level of many particles generated by a single primary of very high energy, the phenomenon now called extensive air showers.

In 1932, Rossi was called to the University of Padua to establish a new physics institute. He married Nora Lombroso in April of 1938. In September of the same year, he was dismissed from his university position in accordance with the racial laws of the facist state. Recognizing the looming danger, the Rossis left Italy. After brief visits to the Bohr Institute in Copenhagen, Denmark, and the laboratory of **Patrick Blackett** at the University of Manchester, England, they traveled on to Chicago, Illinois, USA, where **Arthur Compton** had invited Rossi to participate in a cosmic-ray symposium at the University of Chicago during the summer of 1939.

The muon and its possible instability were major topics at the symposium. Afterward, in Compton's laboratory, Rossi constructed an apparatus with which he carried out the first of a series of experiments on Mount Evans in Colorado and at Cornell University, Ithaca, New York, that proved the radioactive decay of muons, demonstrated the relativistic dilation of the mean life of rapidly moving muons, and ultimately determined the precise value of the mean life of muons at rest. The latter was achieved at Cornell University, where he was appointed associate professor in 1940.

In 1943 Rossi was called to Los Alamos, New Mexico, to participate in the development of the atomic bomb. There he headed the group that developed the special electronic instrumentation required for the urgent experiments in nuclear physics. Among the new instruments were fast ionization chambers that Rossi used in the dangerous implosion experiments and, ultimately, in measuring the exponential rise of the chain reaction of the plutonium bomb detonated on 16 July 1945 at Alamogordo.

In 1946 Rossi was appointed professor of physics at the Massachusetts Institute of Technology [MIT]. Here, he formed the Cosmic Ray Group with several of his young colleagues from Los Alamos, other scientists returning to academic work from wartime laboratories, visitors from Asia and Europe, and numerous students. From 1946 to 1960, various members of the group, inspired and guided by Rossi, carried out a wide variety of studies on the properties of the primary cosmic rays, on the propagation of the primary and secondary cosmic rays in the atmosphere, and on new unstable particles produced in the interactions of cosmic rays with matter. Among the studies with special significance in the developing field of high-energy astronomy were a series of experiments on extensive air showers that determined the arrival directions and energy spectrum of the primary cosmic rays with energies in the range up to 10^{20} electron volts.

In the late 1950s, Rossi seized new opportunities for cosmic exploration offered by the advent of space vehicles and computers. His group developed the MIT "plasma cup" for measuring the properties of ionized gas in interplanetary space. Together with a magnetometer prepared at the Goddard Space Flight Center, it was launched aboard the space probe Explorer 10 on 25 March 1961. The measurements revealed the boundary of the geomagnetic cavity and determined the density, supersonic speed, and direction of the solar plasma flowing just outside the cavity. Ever more sophisticated plasma detectors developed by the MIT group were sent in the following years around the planets and the boundary of the Solar System.

Rossi was also eager to explore the sky for possible X-ray sources with detectors carried above the atmosphere. At Rossi's suggestion, an effort to accomplish that purpose was undertaken at American Science and Engineering, Inc., a local company, which was founded in 1958 by a former student of Rossi, and for which Rossi was cheif scientific consultant. With support from the United States Air Force, an experiment, developed under the direction of Riccardo Giacconi, was launched on 18 June 1962. It discovered the first X-ray star, Sco X-1, and an unresolved X-ray background, thereby inaugurating the field of extrasolar X-ray astronomy.

In his later years, Rossi wrote and spoke extensively about space physics and X-ray astronomy, and published his scientific autobiography. He received honorary doctorates from the universities of Palermo, Durham, and Chicago and prizes from Italy, the United States, Bolivia, Germany, and Israel (the Wolf Prize). He was elected to the major academies of science in Italy, England, and the United States and is remembered by the Rossi Prize in High Energy Astrophysics of the American Astronomical Society.

George W. Clark

Selected References

Bridge, H. S., H. Courant, H. DeStaebler, Jr., and B. Rossi (1954). "Possible Example of the Annihilation of a Heavy Particle. *Physical Review* 95: 1101–1103.

Bridge, H. S., C. Peyrou, B. Rossi, and R. Safford (1953). "Study of Neutral V Particles in a Multiplate Cloud Chamber." *Physical Review* 91: 362–372.

Clark, G. *et al.* (1957). "An Experiment on Air Showers Produced by High-Energy Cosmic Rays." *Nature* 180: 353–356.

Gregory, B. P., B. Rossi, and J. H. Tinlot (1950). "Production of Gamma-Rays in Nuclear Interactions of Cosmic Rays." *Physical Review* 77: 299–300.

Hulsizer, R. I. and B. Rossi (1948). "Search for Electrons in the Primary Cosmic Radiation." *Physical Review* 73: 1402–1403.

Linsley, J., L. Scarsi, and B. Rossi (1961). "Extremely Energetic Cosmic-Ray Event." *Physical Review Letters* 6: 485–486.

Giacconi, R., F. Paolini, and B. Rossi (1962). "Evidence for X Rays from Sources Outside the Solar System." *Physical Review Letters* 9: 439–443.

Rossi, B. (1930). "Method of Registering Multiple Simultaneous Impulses of Several Geiger's Counters." *Nature* 125: 636.

——— (1930). "On the Magnetic Deflection of Cosmic Rays." *Physical Review* 36: 606.

——— (1931). "Magnetic experiments on the cosmic rays." *Nature* 128: 300–302.

——— (1932). "Absorptions mels ungen der durchdringenden Korpuscularstrahung in einem Meter Blei," *Die Natururissenschaften* 20: 65.

——— (1932). "Nachweis einer Sekundärstrahlung der durchdringenden Korpuscularstrahlung," *Physikalische zeitschrift* 33: 304

——— (1933). "Über die Eigenschaften der durschdringenden Korpuscularstrahlung in Meeresniveau", *Zeitschrift für physik* 82: 151–178.

——— (1934). "Directional Measurement on the Cosmic Rays near the Geomagnetic Equator." *Physical Review* 45: 212–214.

——— (1934). "Misure sulla distribuzione angolare di intensita della radiazione penetrante all'Asmara." *Ricerca scientifica* 5, no. 1: 579–589.

——— (1935). *Rayons cosmiques*. Paris: Herman and Co.

——— (1952). *High-Energy Particles*. New York: Prentice-Hall.

——— (1957). *Optics*. Reading, Massachusetts: Addison-Wesley.

——— (1964). *Cosmic Rays*. New York: McGraw-Hill.

——— (1990). *Moments in the Life of a Scientist*. Cambridge: Cambridge University Press.

Rossi, B. and N. Nereson (1942). "Experimental Determination of the Disintegration Curve of Mesotrons." *Physical Review* 62: 417–422.

Rossi, B. and Olbert (1970). *Introduction to the Physics of Space*. New York: McGraw-Hill.

Rossi, B. and L. Pinsherle (1936). *Lezioni di fiscia sperimentale*. Padova.

Rossi, B. and H. Staub (1949). *Ionization Chambers and Counters*. New York: McGraw-Hill.

Rossi, B. *et al.* (1939). "The Disintegration of Mesotrons." *Physical Review* 56: 837–838.

Rossiter, Richard Alfred

Born **Oswego, New York, USA, 19 December 1886**
Died **Bloemfontein, South Africa, 26 January 1977**

Sharp eyesight allowed Richard Rossiter to discover the largest number of double stars observed by anyone up to his time, while his dedication kept his observatory functional during tough economic times.

Rossiter earned his B.A. degree at Wesleyan University in 1914, his M.A. from the University of Michigan in 1920, and his Ph.D. from the University of Michigan in 1923. He had married Jane van Dusen in 1915. The Rossiters had two children, Laura Rossiter (Kohlberg) and Alfred Rossiter.

After graduating from Wesleyan, Rossiter taught mathematics at the Wesleyan Seminary at Genesee, New York, until 1919. Upon completion of his doctorate, Rossiter remained at Michigan as an assistant professor of astronomy. In 1926, Rossiter left Michigan for South Africa, where he was appointed director of the Lamont–Hussey Observatory in Bloemfontein until his retirement in 1952. He was a member of the Astronomical Society of Southern Africa and served as its president in 1940.

Rossiter's doctoral dissertation, directed by **Ralph Curtiss**, was an intensive study of the eclipsing binary star, β Lyrae. Rossiter amassed over 400 spectrograms of the star during his research. He measured shifts in opposite directions of the brighter spectral lines and proved from these observations that the star was rotating rapidly. While **Frank Schlesinger** had provided evidence for suspected stellar rotations before this time, Rossiter's observations, along with those of **Dean McLaughlin** for Algol, offered the first convincing proof for what is now known as the Rossiter–McLaughlin rotation effect. In recent years, the effect has been used to deduce the existence of planets orbiting some stars.

William Hussey, head of Michigan's astronomy department, had long intended to establish a southern station in order to prosecute his survey of double or binary stars, begun at the Lick Observatory under director **Robert Aitken**. Hussey's friend Robert P. Lamont, by then a wealthy industrialist, offered to finance such an expedition. By the mid-1920s, the MacDowell firm had completed a 27-in. refractor for that purpose. Hussey began examining observing sites in the neighborhood of Bloemfontein. He sought out Rossiter to take part in the expedition, originally planned to take just a few years. But in October 1926, while *en route* to South Africa, Hussey died in London, and Rossiter took over the research agenda.

Once in Bloemfontein, Rossiter chose an excellent site, set up the observatory, and, with two younger colleagues, began the double-star survey. Morris K. Jessup returned to the United States in 1930 and Henry F. Donner followed 3 years later, although Donner returned for a single season in 1948. The number of double stars discovered at the Lamont–Hussey Observatory (7,368) is a record. Rossiter's personal harvest of new pairs (5,534) is itself a world record.

During the Depression, Lamont ceased to fund the observatory and Rossiter obtained local funds during the 7 years that Michigan could not support the program. Rossiter became a staunch member of the Bloemfontein community and never returned to the United

States. He also hosted **Earl Slipher** of the Lowell Observatory during two close oppositions of Mars. There, Slipher used a special camera to take an excellent series of photographs of the red planet. During the 1950s, Karl Henize used the site for important work on emission nebulae, while Frank Holden continued the double-star survey work.

Records of the Lamont–Hussey Observatory, including correspondence from Rossiter, are found in the Michigan Historical Collections, Bentley Library, University of Michigan.

Rudi Paul Lindner

Selected References

Holden, Frank (1977). "R. A. Rossiter: Obituary Notice." *Monthly Notes of the Astronomical Society of Southern Africa* 36: 60–62.

Rossiter, Richard Alfred (1955). "Catalogue of Southern Double Stars." *Publications of the Observatory of the University of Michigan* 11.

Struve, Otto and Velta Zebergs (1962). *Astronomy of the 20th Century*. New York: Macmillan, esp. pp. 225–227.

Rothmann, Christoph

Born **Bernburg, (Sachsen-Anhalt, Germany),** ***circa*** **1560**
Died **possibly in Bernburg, (Sachsen-Anhalt, Germany),** ***circa*** **1600**

Christoph Rothmann constructed the first modern star catalog based entirely on his own observations.

He was one of the most outstanding astronomers of the 16th century, but we have scant information on Rothmann's life. His enrollment at the University of Wittenberg on 1 August 1575 is authenticated. There is no reliable information about the course of his studies or which academic degree he received.

According to all we know, Rothmann concerned himself thoroughly with mathematics and astronomy in Wittenberg. He expressed that to **Tycho Brahe** as well as in his extant manuscript works. In November 1584 he started to work with Landgrave **Wilhelm IV** of Hessen-Kassel as an astronomer at the local observatory. (Dates usually referred to in the literature turn out to be wrong.) Rothmann stayed in Kassel until mid-1590. In the summer of that year he went on a journey to Brahe, studied his instruments, and left Brahe's island on September 1. His further life is still a mystery. In spite of all commitments and arrangements with the count, Rothmann never returned to Kassel, but took up residence in Bernburg, where he later wrote a treatise on the sacraments, in particular on the sacrament of baptism.

In connection with his work on the star catalog initiated by Wilhelm IV, Rothmann became an excellent observer. Rothmann's catalog elaborated for the epoch of 1586 shows an extremely small standard deviation (related to the fundamental star Aldebaran) of $\pm 1.2'$ in right ascension and $\pm 1.5'$ in declination. (For comparison, Brahe's respective values are: $\pm 2.3'$ for right ascension and $\pm 2.4'$ for declination.)

The Kassel stellar catalog comprises 383 stars; the corresponding observations date from the years 1585–1587. It represents a decisive breakthrough in modern astronomical observation, and it was the first one since the time of **Hipparchus** and **Ptolemy** to be completely based on observations by the compiler. Rothmann did not work with instruments of large dimensions, but used smaller metal instruments (*e. g.*, an azimuthal quadrant) with particular precisely manufactured sighting devices as well as precise clocks constructed by **Jost Bürgi**. Furthermore, he paid great attention to a precise calibration of the instruments, considered refraction to correct his observational results, and carried out numerous single measurements for each stellar location. The Kassel star catalog was included as an exemplary one by **John Flamsteed** in his *Historia coelestis Britannicae volumen tertium*.

After observing the comet of 1585 (C/1585 T1), Rothmann was one of the first astronomers to conclude that comets are cosmic objects. Rothmann had observed the comet from 8 October to 8 November, did not find a parallax, and determined the distance between the comet and the Earth to be 500,000 German miles; therefore it must have been far beyond the sphere of the Moon. In connection with previous observations by Brahe and other scholars, he followed the anti-Aristotelian conclusion that the comet's substance is by no means different from the elemental region below the Moon, and that therefore the doctrines of the ether as well as that of a particular celestial region of fire are completely unfounded from a scientific point of view. Rothmann considered comets as vapors (fumes) arising from the Earth, which shone in sunlight; this appears as a conservative element at first glance, but it led him to the insight of the material unity of the Earth with the comets as celestial bodies and therefore of the material unity of the whole cosmos up to the planetary spheres.

Peter Apian had observed that comet tails are always turned away from the Sun. Rothmann seems to have noticed this common fact, and used it to clarify the nature of comet tails. As he discovered from his own observations and those by Wilhelm IV of the comet of 1558 (C/1558 P1), comet tails are directed exactly neither to the Sun nor to the planets. Therefore, they must represent an independent body of specific matter by themselves.

Rothmann's unorthodox comet doctrine shows him as an independent thinker who tried to get to conclusions from his own observations and reflections. But it is always evident that for Rothmann, precise observations were the starting point and the basis for theoretical conclusions. This applies also to his investigation of refraction. Rothmann had noticed differences in star distances depending on their height over the horizon, and united it in a table. (Brahe did the same simultaneously, and Rothmann conducted long written discussions with him about that.) He concluded from the specific progression of the refraction and of the equality of refraction for planets and fixed stars that there are neither rigid spheres for the movement of the planets nor a separate sphere of fire or a crystal sky. Therefore, there is nothing else than air and the planets moving in it between the sphere of the fixed stars and the Earth. Hence, refraction has no other cause than a diversion of light at the transition from the air of the sky to the air close to the body of the Earth, being mixed with earthly vapors up to a height of approximately 102 kilometers. Contrary to that, Brahe clung to the opinion that refraction originates at the transition of light from the celestial ether to the atmosphere of the Earth. In this case, Rothmann argued, refraction had to continue until the zenith, which neither he nor Brahe had ever measured. So Rothmann gave up a general division into two world regions below and above the Moon.

Because at this point Rothmann had already given up essential elements of Aristotelian physics, he was able to take the last step in accepting the heliocentric system of **Nicolaus Copernicus** as the true construction of the world. In Kassel he became one of the first convinced disciples of the heliocentric system, as is documented in numerous places in his letters to Brahe and in his treatise on the comet of 1585.

Rothmann's correspondence with Wilhelm IV is in the Hessian State Archive, Marburg. Rothmann's works, as original manuscripts, are in the Murhard University and University Library in Kassel.

Jürgen Hamel and *Eckehard Rothenberg*
Translated by: *Günther Görz*

Selected References

Barker, Peter (1993). "The Optical Theory of Comets from Apian to Kepler." *Physis*, n.s., 30: 1–25.

Brahe, Tycho. *Epistolarum astronomicarum libri*. 1596. Vol. 6 of *Tychonis Brahe opera omnia*, edited by J. L. E. Dreyer. Copenhagen, 1919. Subsequent editions, Nuremberg, 1601 and Frankfurt am Main, 1610. (For Rothmann's correspondence with Tycho Brahe.)

Flamsteed, John (1725). *Historia coelestis Britannicae volumen tertium*. London.

Goldstein, Bernard R. and Peter Barker (1995). "The Role of Rothmann in the Dissolution of the Celestial Spheres." *British Journal for the History of Science* 28: 385–403.

Granada, Miguel A. (1996). *El debate cosmológico en 1588: Bruno, Brahe, Rothmann, Ursus, Röslin*. Naples: Istituto Italiana per gli Studi Filosofici.

Hamel, Jürgen (1998). *Die astronomischen Forschungen in Kassel unter Wilhelm IV: Mit einer Teiledition der deutschen Übersetzung des Hauptwerkes von Copernicus um 1586*. (The astronomical research in Kassel under Wilhelm IV: With a partial edition of the German translation of the main work of Copernicus c. 1586.) *Acta Historica Astronomiae*, Vol. 2. Thun: Harri Deutsch.

Rothmann, Christoph. *Catalogus stellarum fixarum ex observationibus Hassiacis, ad annum 1586*. 36 fol., 2° MS astron. 7.

——— *Observationum stellarum fixarum liber primus*. Kassel 1589, 88 Bl., 2° MS astron. 5/7. (An edition by M. A. Granada, J. Hamel and L. von. Mackensen is under preparation.)

——— *Astronomia: In qua hypotheses Ptolemaicae ex hypothesibus Copernici corriguntur et supplentur: et inprimis intellectus et usus tabularum Prutenicarum declaratur et demonstrator*. 183 fol., 4° MS astron. 11.

——— *Organon mathematicum, contens logistica sexagenaria, doctrina sinuum, et doctrina triangulorum. Quo congnito quilibet suo marte Ptolemaeum, Copernicum, Regiomontanum, reliquosque artificos facilime intelligere*. 255 fol., 4° MS math. 29.

——— (2003). *Christoph Rothmanns Handbuch der Astronomie von 1589*, edited by Miguel A. Granada, Jürgen Hamel, and Ludolf van Mackensen. *Acta Historica Astronomiae*, Vol. 19. Frankfurt am Main: Harri Deutsch, 2003.

Rowland, Henry Augustus

Born **Honesdale, Pennsylvania, USA, 27 November 1848**
Died **Baltimore, Maryland, USA, 16 April 1901**

Henry Rowland is chiefly remembered for his invention and manufacture of the concave diffraction grating, the extraordinary precision that it brought to the science of solar astrophysics, and his

Photographic Map of the Normal Solar Spectrum (1888). Rowland was the son of Henry Augustus Rowland, Sr., a Presbyterian clergyman, and Harriette Heyer. In the spring of 1865, he enrolled at the Phillips Academy in Andover, Massachusetts, to study Latin and Greek, with the intention of preparing for the ministry. But in the fall of that year, Rowland transferred to the Rensselaer Polytechnic Institute in Troy, New York, from which he graduated in 1870 with a degree in civil engineering. After employment as a railroad surveyor for a year and as a teacher at the College of Wooster in Ohio, he returned to Rensselaer as an instructor of physics and became an assistant professor 2 years later. His demonstration that the magnetic permeability of iron, steel, and nickel varied with the applied magnetizing force convinced **James Maxwell** that Rowland was a promising experimentalist.

In 1875, Daniel Coit Gilman, founding president of the Johns Hopkins University, Baltimore, appointed Rowland as the institution's first professor of physics. Rowland held this chair until his premature death. In that capacity, he advised 165 graduate students, including 45 doctoral students, 30 of whom find mention in James McKeen Cattell's *American Men of Science*. Rowland was instrumental in establishing physics as a research discipline in America.

In 1890, Rowland married Henrietta Troup Harrison; the couple later had three children. In the same year, however, Rowland learned that he had diabetes, which was then untreatable. The more commercially oriented activities he conducted toward the end of his life, such as developing and marketing a multiplex telegraph, his role as chief design consultant to a construction company for installing electric generators at Niagara Falls, and his filing of at least 19 patent applications, were aimed at assuring a future livelihood for his family. At his own request, Rowland's cremated ashes were masoned into the wall of his laboratory close to the ruling engine he had devised.

While working in the Berlin laboratory of **Hermann von Helmholtz**, Rowland demonstrated for the first time that moving charges on a rotating disk create magnetic effects resembling those from an electric current. He purchased scientific instruments in Europe totaling more than $6,000 to equip his Baltimore laboratory and its associated workshop. Rowland's lab at Johns Hopkins became the best equipped of his generation in the United States and attracted many research students. His early work included precision measurements of the value of the unit of electrical resistance, the ohm, and his determination of the mechanical equivalent of heat, for which he was awarded the Rumford Medal in 1884.

In the early 1880s, Rowland improved the design of the "ruling engine," which guided a carefully chosen, sharp, diamond point across a speculum metal surface to produce parallel ruled lines at a fixed separation of as little as one thousandth of a millimeter. The straightness of the lines was achieved by guiding the point along two parallel rails of hardened metal. More difficult was the point's repositioning after each ruling for up to 110,000 lines in a single grating. Rowland's success lay in his using a well-machined ruling screw made of special flawless steel in a painstaking grinding process that could take up to 14 days without interruption. Rowland once claimed that "there was not an error of half a wave-length, although the screw was nine inches long."

Rowland's gratings were much larger and more regular than any constructed by his American predecessors, **Lewis Rutherfurd** and William August Rogers, and were free of the periodic errors that caused the appearance of pseudo-lines (termed ghosts) in the diffraction spectra of other gratings. Throughout his career, Rowland built three such ruling engines, the first in 1881, which could rule up to 14,400 lines per inch. Afterward, he constructed two others (in 1889 and 1894) that ruled 20,000 and 15,000 lines per inch, onto surfaces up to 25 square inches.

In 1882, Rowland devised the concave diffraction grating. With a radius of curvature between 3 and 6 m, the ruled metal surface acted not only as a diffraction grating, but also as a focusing lens, thus obviating the use of glass lenses with their unwanted light absorption from the ultraviolet spectrum. During the late 19th and early 20th centuries, all major spectroscopists acquired at least one of Rowland's gratings. They allowed spectral work to be performed on a broader range of frequencies and with much higher efficiency and precision (by roughly a factor of 10) as compared with other gratings. They were manufactured in Baltimore by Rowland's chief mechanician, Theodore Schneider, and passed a rigorous examination by his assistant, Lewis E. Jewell. Rowland's distributor, the Pittsburgh instrument maker **John Brashear**, also supplied the polished curved surfaces. By January 1901, sales from the gratings totaled more than $13,000; between 250 and 300 gratings were sold at cost to physical and chemical laboratories as well as to astronomical observatories the world over.

While many of his students recorded the wavelengths of various emission spectra, Rowland and Jewell concentrated on photographic mapping of the entire visible solar spectrum at an unprecedented resolution. Rowland's tabular inventory of roughly 20,000 solar spectral lines was published in the *Astrophysical Journal* (1895–1897), and his spectral-atlas was distributed in two series of large-scale prints. Every noteworthy spectroscopic laboratory and many astronomical observatories obtained these publications, which formed the standard of solar spectroscopy for several decades.

Rowland's tables provided the basis on which the Balmer (and other) wavelength series were derived. The precision of his instruments also permitted detection of the Zeeman effect and proof of the magnetic polarities of sunspots and related solar phenomena discovered by **George Hale** in 1908. Around 1890, the first indications of a solar redshift were discovered in these data by Jewell, who attempted to interpret them as Doppler effects induced by solar convection currents. Rowland himself dismissed the effect as some type of artifact. It was not until around 1960 that interpretation of this effect as a gravitational redshift (predicted by **Albert Einstein** in 1907) was finally confirmed.

Rowland was a founding member and first president of the American Physical Society. In recognition of his importance in the institutionalization of American physics, he was awarded LL.D. degrees from Yale University in 1895 and from Princeton University in 1896. His diffraction gratings and photographic maps of the solar spectrum won him the Gold Medal of the French Académie des sciences and a grand prize at the Paris Exhibition of 1890, along with the Draper Medal of the National Academy of Sciences. Rowland served as a delegate for the United States at various international scientific congresses. He was elected a foreign member of the Royal Society of London and about a dozen other learned societies and academies.

Rowland's papers are preserved at the Milton S. Eisenhower Library, Johns Hopkins University, Baltimore.

Klaus Hentschel

Selected References

Hentschel, Klaus (1999). "Photographic Mapping of the Solar Spectrum, 1864–1900." Parts 1 and 2. *Journal for the History of Astronomy* 30: 93–119, 201–224.

——— (1999). "Rowland, Henry Augustus." In *American National Biography*, edited by John A. Garraty and Mark C. Carnes. Vol. 19, pp. 9–11. New York: Oxford University Press.

Hufbauer, Karl (1991). *Exploring the Sun: Solar Science since Galileo*. Baltimore: Johns Hopkins University Press, esp. pp. 67–72.

Kevles, Daniel J. (1975). "Rowland, Henry Augustus." In *Dictionary of Scientific Biography*, edited by Charles Coulston Gillispie. Vol. 11, pp. 577–579. New York: Charles Scribner's Sons.

——— (1987). *The Physicists: The History of a Scientific Community in Modern America*. Cambridge, Massachusetts: Harvard University Press, esp. Chap. 3, pp. 25–44.

Mendenhall, Thomas C. (1905). "Biographical Memoir of Henry Augustus Rowland." *Biographical Memoirs, National Academy of Sciences* 5: 115–140.

Rowland, Henry A. (1902). *Physical Papers of Henry Augustus Rowland*. Baltimore: Johns Hopkins Press.

Sweetnam, George Kean (2000). *The Command of Light: Rowland's School of Physics and the Spectrum*. Philadelphia: American Philosophical Society.

Roy, Hendrick de

➔ **Regius, Hendrick**

Rudānī: Abū ʿAbdallāh Muḥammad ibn Sulaymān (Muḥammad) al-Fāsī ibn Ṭāhir al-Rudānī al-Sūsī al-Mālikī [al-Maghribī]

Born **Tārūdānt, (Morocco), *circa* 1627**
Died **Damascus, (Syria), 1683**

Rudānī, also known as *al-Maghribī*, was a 17th-century scholar who lived in the Ottoman territories and was known for his work on astronomical instruments. In addition to astronomy, he was a poet and also wrote on mathematics, *hadith* (traditions of the Prophet), Qur'ān interpretation, and grammar. There is no information about Rudānī's elementary education or about his family background. He received his education in the *madrasas* (schools) of Morocco and Algeria. Then he traveled to the east, visiting Egypt, Damascus, and Istanbul and receiving education from eminent scholars along the way. Eventually, Rudānī moved to the Ḥijāz in Arabia, where he became one of the most respected scholars in the area, and was appointed governor. But due to a conflict, he was exiled to Damascus.

In the field of astronomy, Rudānī wrote works on instruments, timekeeping, and the *qibla* (direction to Mecca). He sought practical solutions and ways to simplify the calculations. With these purposes in mind, Rudānī invented a sphere, called *al-jayb al-jāmiʿa*, which was a spherical device in which another sphere (painted blue) with a different axis was attached to it. This second sphere was divided into two parts in which the zodiacal signs with their sections and regions were drawn. The purpose of this device was to facilitate timekeeping with the use of this one instrument. The device, easily constructed, was a universal instrument (*i. e.*, it could be used for different longitudes and latitudes). Unfortunately, there is no existing sample of this device, but Rudānī wrote a book describing it, in Arabic, entitled *al-Nāfiʿa fī ʿamal al-jāmiʿa*. It was written in Medina in 1662 and contains 45 parts and a conclusion. Rudānī's best-known work in the field of astronomy is *Bahja al-ṭullāb fī al-ʿamal bi-'l-asṭurlāb*, a book written in Arabic on how to make and use an astrolabe. There are 13 extant copies of this particular work. Interestingly, Rudānī also wrote three other works on the same subject. Other astronomical works by Rudānī include one on prayer times and another on the calendar in rhyme.

Salim Ayduz

Selected References

Al-ʿAyyāshī, Abū Sālim ʿAbdallāh ibn Muḥammad (1899). *Riḥlat al-Shaykh al-Imām Abi Sālim al-ʿAyyāshī*. Vol. 2, p. 30. Fez.

Al-Kattānī, ʿAbd al-Ḥayy ibn ʿAbd al-Kabīr (1982). *Fihris al-fahāris*. Beirut, p. 317.

Al-Muḥibbī, Muḥammad (1966). *Khulāṣat al-athar fī aʿyān al-qarn al-ḥādī ʿashar*. Vol. 4, pp. 204–208. Beirut.

Al- Ziriklī, Khayr al-Dīn (1980). *al-Aʿlām*. Vol. 6, pp. 151–152. Beirut.

Brockelmann, Carl. *Geschichte der arabischen Litteratur*. 2nd ed. Vol. 2 (1949): 610–611; Suppl. 2 (1938): 691. Leiden: E. J. Brill.

Daḥlān, Aḥmad ibn Zaynī (1887). *Khulāṣat al-kalām fī bayān umarā' al-balad al-ḥarām*. Egypt, pp. 102–104.

Ibn Sūdah, ʿAbd al-Salām ibn ʿAbd al-Qādir (1950). *Dalil mu'arrikh al-Maghrib*. Tetouan, p. 340.

İhsanoğlu, Ekmeleddin, *et al.* (1997). *Osmanlı Astronomi Literatürü Tarihi (OALT)* (History of astronomy literature during the Ottoman period). Vol. 1, pp. 317–321. Istanbul: IRCICA.

İzgi, Cevat (1997). *Osmanlı Medreselerinde İlim*. Vol. 1, pp. 118–119. Istanbul.

Kaḥḥālah, ʿUmar Riḍā (1985). *Muʿjam al-mu'allifīn*. Vol. 11, p. 221. Beirut.

Suter, Heinrich (1981). *Die Mathematiker und Astronomen der Araber und ihre Werke*. Amsterdam: APA-Oriental Press, p. 203 (no. 527).

Rümker, Christian Karl Ludwig

Born **Stargard, (Mecklenburg-Vorpommern, Germany), 18 May 1788**
Died **Lisbon, Portugal, 21 December 1862**

In 1822 Karl Rümker made the second ever successful recovery of a periodic comet (2P/Encke).

The son of Justus Friedrich Rümker, Court-Councillor of the duchy, young Karl was educated at the Grey convent in Berlin. With a talent for mathematics, Rümker was then sent to the Builders' Academy in Berlin because his family hoped he would become a master-builder. Instead, he moved to Hamburg in 1807 where he taught mathematics privately. Napoleon's continental blockade made the economic situation increasingly difficult, and around 1809, Rümker traveled to England in search of maritime employment. He served for several years in the merchant navy but in July 1813 was press-ganged for service on the man-of-war, *HMS Benbow*. What for many would have been a misfortune proved a stroke of luck for Rümker. The captain of the *Benbow* learned that he was no ordinary seaman but a teacher of mathematics. Rümker then served on a succession of naval vessels as a "teacher of sea cadets" with an officer's rank.

Rümker's naval service took him to the Mediterranean where he met the Austrian astronomer **János von Zach**, editor of the *Correspondence Astronomique*. Zach encouraged Rümker's astronomical skills and published his first scientific observations. Publication of Rümker's observations at Malta in the *Edinburgh Philosophical Journal* (1819) brought his name to wider scientific attention. He was discharged from the navy in that year and returned to Hamburg, where he taught at the School of Navigation.

The appointment of Sir **Thomas Brisbane** as the prospective governor of the British colony of New South Wales (Australia) saw Rümker abandon his Hamburg post and return to England in 1821. Brisbane had sought the appointment in part so that he could establish a Southern Hemisphere observatory. Rümker learned of Brisbane's desire for an astronomical assistant, applied for the position, and was accepted. Rümker's salary was to be £200.

Brisbane's party, including Rümker and a second assistant, James Dunlop, arrived in Sydney toward the end of 1821. The modest observatory Brisbane erected near Government House at Parramatta was completed in April 1822; observations began almost immediately. The main task of the observatory was to determine accurate positions of stars located between the zenith and the South Celestial Pole. Rümker served as principal astronomer, the relatively unskilled Dunlop as his assistant, and Brisbane participated as his duties allowed. Rümker and Dunlop observed comet 2P/Encke on 2 June 1822. This was only the second occasion on which the predicted return of a comet had been fulfilled. Brisbane praised Rümker's "zeal, assiduity [and] intelligence" for the discovery.

Nonetheless, strains developed between the punctilious astronomer and the private patron. In June 1823, Rümker suddenly left Parramatta to farm the land he had previously been granted at Picton, to the southwest of Sydney. Rümker devoted himself to the development of Stargard, as he called the property, and proved himself an able farmer. After a year, he resumed astronomical observations with a small telescope, discovering three comets in 1824/1825. Brisbane's attempts at reconciliation were rebuffed, and it was not until after he departed from the colony (1825) that Rümker was reinstated at Parramatta. The local government purchased Brisbane's instruments and books so that the observatory could be operated once again.

Rümker resumed work at Parramatta (but without Dunlop's assistance) in 1826, now as a government servant. The new governor, Sir Ralph Darling, suggested to authorities in London that Rümker be made "Government Astronomer." Without receiving official confirmation, Darling appointed Rümker to this position at a salary of £300. No small part of Darling's motivation for the renewed patronage was the prospect of measuring an arc of the meridian, as urged by the Royal Society of London. Rümker, however, felt this could not be achieved with the instruments at hand and returned to London to expedite the procurement of suitable equipment and to see through the publication of his observations.

Rümker arrived in London toward the middle of 1829. At first all went well; his Parramatta observations were to be published by the Royal Society (at government expense) at the end of that year. But then he was caught up in the dispute between the president of the Royal Astronomical Society, Sir **James South**, and the Royal Society itself. South had offered his Troughton transit circle for Parramatta, but when he sought to profit on its sale, a less expensive one was procured from Thomas Jones. Compounding this snub to South were the past grievances of Brisbane aired toward Rümker. Brisbane sought control of the Parramatta observation books from Rümker, whom he thought was withholding them. As a result of this vitriolic campaign against him, Rümker was formally dismissed from government service on 18 June 1830. Dunlop returned to Parramatta.

Rümker returned to Hamburg and was appointed director of the School of Navigation where he had previously taught. While there, Rümker published a *Preliminary Catalogue of Fixed Stars* (1832) from his Parramatta observations and dedicated the work to Brisbane. In 1832, he had an illegitimate son, George Friedrich Wilhelm Rümker, with his housekeeper, Maria Louise Bernadine Melcher, whom he did not marry. Years later (1848), Rümker married an astronomically minded English spinster, Mary Ann Crockford, who had reportedly independently discovered comet C/1847 T1 (Mitchell).

In 1833, Rümker was appointed director of Hamburg Observatory, in addition to his Navigation School position. An assiduous dedication to his tasks is reflected in the numerous papers he contributed to various scientific journals, along with his publication of a *Handbuch der Schiffarts-Kunde* (Manual of the theory of navigation). He was elected to memberships in both the Royal Society and the Royal Astronomical Society in London. In 1850, the King of Hanover presented Rümker with a Gold Medal for Arts and Science. Four years later, he was awarded the Gold Medal of the Royal Astronomical Society.

Yet, this heavy load took its toll on Rümker's health. Too ill to travel to London to receive his 1854 medal, Rümker was permanently crippled by a severe fall in 1857. He resigned from his posts but was given an indefinite leave of absence. His son George succeeded him as director of the Hamburg Observatory. Hoping to relieve a chronic lung complaint, Rümker traveled with his wife to the warmer climate of Southern Portugal where he died. He was buried in the English Cemetery at Estrella.

Besides the recovery of comet 2P/1822 L1 (Encke) and the publications already mentioned, Rümker's work contributed substantially to *A Catalogue of 7385 Stars, Chiefly in the Southern Hemisphere: Prepared from Observations Made in the Years 1822, 1823, 1824, 1825, and 1826, at the Observatory at Paramatta, New South Wales*, prepared by William Richardson and published in 1835. Toward the end of his life, Rümker was again reducing his Parramatta observations but did not live to complete the task. These observations were eventually reduced and published by **Friedrich Ristenpart** (1909) and J. E. Baron de Vos van Steenwijk (1923). Rümker's manuscripts appear to have been destroyed in World War II.

Rümker made significant contributions to Southern Hemisphere astronomy in the years before the Cape Town Observatory was operational. His career illustrates the difficulties of professional scientists emerging from a context dominated by amateurs and the perils of patronage.

Julian Holland

Selected References

Bergman, George F. J. (1960). "Christian Carl Ludwig Rümker (1788–1862): Australia's First Government Astronomer." *Journal of the Royal Australian Historical Society* 46: 247–289.

——— (1967). "Rumker, Christian Carl Ludwig (1788–1862)." In *Australian Dictionary of Biography*. Vol. 2, *1788–1850*, pp. 403–404. Melbourne: Melbourne University Press.

Saunders, Shirley (1990). "Astronomy in Colonial New South Wales: 1788 to 1858." Ph.D. diss., University of Sydney.

Rumovsky, Stepan Yakovlevich

Born **Stary Pogost (near Vladimir), Russia, 29 October/9 November 1734**

Died **Saint Petersburg, Russia, 6/18 July 1812**

Stepan Rumovsky was a Russian astronomer, mathematician, geodesist, and humanist scholar, who determined the most accurate value of the solar parallax during the 18th century. One of the best students at the Alexander Nevsky seminary, Rumovsky was chosen in 1748 to study at the university of the Saint Petersburg Academy of Science. There, he attended lectures by **Mikhail Lomonosov**, Georg Wilhelm Richmann, and other prominent Russian academicians.

Having majored in mathematics, Rumovsky became an adjunct of the academy in 1753 and was sent to Berlin to continue his mathematical training under **Leonhard Euler**.

After returning to Saint Petersburg, Rumovsky taught mathematics and astronomy at the university (1756–1812) and held various positions at the Saint Petersburg Academy – director of the geographical department (1766–1786), after Lomonosov's death; director of the academy's astronomical observatory and professor of astronomy (1763–1812); full member of the academy (1767); and academy vice president (1800–1803). He was also elected an honorary member of the Swedish Academy of Sciences (1763).

In 1761, Rumovsky took part in an expedition to Selenginsk (near Lake Baikal) to observe the transit of Venus. In 1769, he supervised a more comprehensive project to observe that century's final transit from several locations. Rumovsky himself observed it from the Kola Peninsula. Using data from all such observations conducted in 1761 and 1769, he calculated the value of the solar parallax at 8.67 arc-seconds, which was closer to its presently accepted value (8.79 arc-seconds,) than that derived by any of his contemporaries.

In 1762, Rumovsky compiled and published the first catalog of astronomically determined geographic coordinates for 62 sites in Russia. His determinations were remarkably precise (as noted by **Friedrich Struve** and others) and were incorporated into the 1790 edition of the *Berliner Astronomisches Jahrbuch*.

Rumovsky's more than 50 scientific works cover astronomy, geodesy, mathematics, and physics. He knew several modern languages as well as Latin, and translated literary and scientific works of Cornelius Tacitus, **Georges Leclerc**, and Euler into Russian.

In 1803, Rumovsky became a member of the Russian School Administration Board, which was charged with preparing educational reforms, and also superintendent of the Kazan department of education. In this capacity, he put much effort into the founding of the Kazan University. Rumovsky recruited, among others, Austrian astronomer **Johann von Littrow** and the German mathematician Johann Martin Bartels, who later became a teacher of **Nikolai Lobachevsky**.

Yuri Balashov

Selected References

Kulikovsky, P. G. (1975). "Rumovsky, Stepan Yakovlevich." In *Dictionary of Scientific Biography*, edited by Charles Coulston Gillispie. Vol. 11, pp. 609–610. New York: Charles Scribner's Sons.

Pavlova, Galina Evgen'evna (1979). *Stepan Iakovlevich Rumovskii, 1734–1812*. Moscow: Nauka.

Runge, Carl [Carle] David Tolme

Born **Bremen, (Germany), 30 August 1856**
Died **Göttingen, Germany, 3 January 1927**

Mathematician and physicist Carl Runge is most likely to be recognized by modern astronomers for the Runge–Kutta method of integrating complex differential equations numerically, which is of

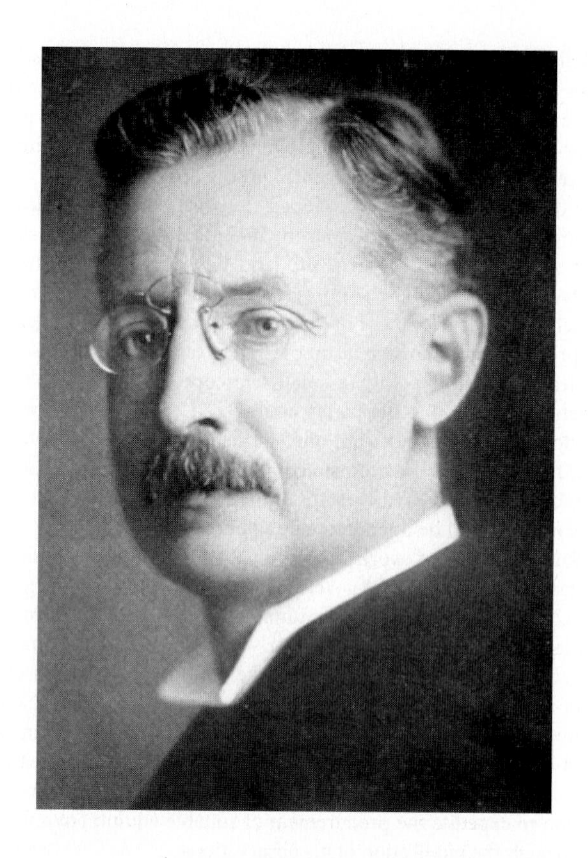

importance in calculating stellar structure and evolution. His most extensive contributions were, however, in laboratory and astrophysical spectroscopy.

Runge was the third son of Julius Runge and his wife Fanny Runge (*née* Tolme). The first years of his life were spent in Havana, Cuba, where his father, a successful merchant, was Danish consul. After the family's return to Bremen, where his father died unexpectedly, Carl attended the Lyceum and entered the University of Munich, in 1876, to pursue literature and philosophy. Soon, however, he turned to mathematics and left the following year for the University of Berlin where he studied with Karl Weierstrass and Leopold Kronecker and received his doctorate in mathematics in 1880. Runge was appointed a lecturer in mathematics at the University of Berlin 3 years later. During this time he worked on problems in algebra and number theory. Runge caught the attention of physiologist Emil du Bois-Reymond's family and married their daughter Aimée in 1887. They had six children, two sons and four daughters, one of whom, Iris, would be his biographer. A year prior to the marriage, Runge's father-in-law-to-be had helped him to obtain a professorship at the Technical High School of Hanover, a position he would hold for 18 years.

By a year after his arrival at Hannover, Runge had undergone a complete reorientation in his research interests, which now included mathematics, spectroscopy, astrophysics, and geodesy. His most important contributions are probably those in applied mathematics. With M. W. Kutta, Runge's name is associated with the Runge–Kutta methods to integrate differential equations numerically.

In a publication of 1885, **Johann Balmer** presented a formula that well represented the visible series of lines in the spectrum of hydrogen. Runge decided to find a comparable formula giving the spectral lines for each of the elements. He began his studies using published data for the spectra of lithium, potassium, calcium, and zinc. Runge found a number of formulae, but inaccuracies in the data made the results suspect. He discussed the matter with his colleague spectroscopist Heinreich Kayser, who agreed to help. In the interim, Rowland gratings and photographic techniques had become available so that more accurate measures of wavelengths could be made. For the next 7 years, until Kayser was appointed to the University of Bonn, they worked together. Runge calculated the series and helped with the experimental work.

Essentially, Runge had found that spectral series could be represented by adding an additional term to Balmer's formula. As the work progressed, emphasis was placed on the precision of the data, methods of data reduction, and evaluation of constants. Meanwhile, the Swedish physicist **Johann Rydberg** had worked out an empirical equation from which he was able to deduce Balmer's relation. Rydberg's formula connected directly with the theory of atomic structure formulated by **Niels Bohr** in which Runge took enthusiastic interest. Theoretical justification for spectral series of more complex atoms than hydrogen had, however, to await the development of quantum mechanics in the 1920s and 1930s. The precise data of Kayser and Runge were valuable in testing these more elaborate pictures of atomic structure.

After Kayser left, Runge continued on alone for 6 months until, following William Ramsay's discovery of terrestrial helium, Runge persuaded Friedrich Paschen (who had come to Hanover in 1891 as Kayser's teaching assistant and who had yet to discover the series of hydrogen lines bearing his name) to join him in an investigation of the helium spectrum. Shortly thereafter, they identified all of helium's principal lines and were able to arrange them into not one but two series. This was taken to mean that helium was a mixture of two elements until in 1897 the two scientists showed that oxygen also had more than one system or series.

By 1900, Runge and Paschen had turned to an investigation of the Zeeman effect – the splitting of spectral lines in a magnetic field. Together, they established the main Zeeman types and the series character of several groups of spark-spectra lines of the alkali earth elements.

In his spectroscopy, Runge was always on the lookout for astrophysical applications. New spectra were compared with **Henry Rowland**'s solar spectrum obtained with a finely ruled grating. Runge doubted the discovery of helium in the Sun by **Norman Lockyer**, until a particular yellow line was seen to be double, like the one produced by terrestrial helium. His work with Kayser and Paschen caught the attention of several British and American spectroscopists and astrophysicists. He visited England in 1895 and America in 1897, to which he returned as a visiting professor at Columbia University in 1907.

Paschen departed for the University of Tübingen in 1901; Runge remained behind. Among his peers, he was like a man without a country. Mathematicians considered him a physicist, and physicists thought of him as a mathematician. By 1904, however, **Hermann Klein** had managed to prevail upon colleagues to bring Runge to Göttingen as the first full professor of applied mathematics in Germany. He held this position until retirement in 1923. His research came to a virtual halt as he became involved with symposia, colloquia, and interaction with peers and students. On his death, Runge was succeeded on the collaborating editor board of the *Astrophysical Journal* (where he had served since 1903) by his former colleague Kayser.

George S. Mumford

Selected References

Forman, Paul (1975). "Runge, Carl David Tolme." In *Dictionary of Scientific Biography*, edited by Charles Coulston Gillispie. Vol. 11, pp. 610–615. New York: Charles Scribner's Sons.

Paschen, Friedrich (1929). "Carl Runge." *Astrophysical Journal* 69: 317–321.

Runge, Iris (1949). *Carl Runge und sein wissenschaftliches Werk*. Göttingen: Vandenhoeck and Ruprecht.

Rushd

➲ **Ibn Rushd: Abū l-Walīd Muḥammad ibn Aḥmad ibn Muḥammad ibn Rushd al-Ḥafīd**

Russell, Henry Chamberlain

Born **West Maitland, New South Wales, (Australia), 17 March 1836**

Died **Sydney, New South Wales, Australia, 22 February 1907**

Henry C. Russell made important contributions to Australian astronomy and meteorology, and played a leading role in the development of Australian scientific societies. The son of the Honorable Bourn Russell, a New South Wales legislator, and Jane (*née* Mackreth) Russell, Henry was educated privately and attended the University of Sydney, graduating with a BA in 1859. One year later, he married Emily Jane Foss; at the time he died, he was survived by her, four daughters, and a son.

In 1859, Russell was appointed computer at the Sydney Observatory; in 1870, he became the observatory director, a post he would occupy until his retirement in 1905. From the start, Russell was equally committed to astronomy and meteorology. In astronomy, Russell began by building up the instrumentation at the Sydney Observatory. The existing 18.4-cm Merz refractor and small transit telescope were joined by a 29.2-cm Schroeder refractor in 1874, a 15.2-cm Troughton and Sims transit telescope in 1877, and a 33-cm astrograph with a 26-cm guidescope in 1890. A two-story wing, complete with new dome, was added to the observatory in 1877. Russell and others at the Sydney Observatory observed comets, eclipses, Jovian features, transits of Mercury and Venus, the η Carinae region, the open cluster κ Crucis, and double stars. In 1871, he and Melbourne Observatory director, **Robert Ellery**, organized an expedition to far north coastal Queensland to observe a total solar eclipse, but inclement weather dashed their hopes on the vital day.

One of the comets that attracted Russell's attention was the great comet C/1881 K1. Russell was the first astronomer to subject comet C/1881 K1 to spectroscopic analysis, his sole foray into the exciting new field of spectroscopic astronomy. Because the account of his observations was published in a local journal, it did not reach a wide international audience.

Australia was well placed to observe the 1874 and 1882 transits of Venus. Russell organized ambitious observing programs for both events. He used the 1874 transit as leverage to gain government funding for new telescopes, and in 1874 equipped and manned four observing stations. All Sydney Observatory stations recorded the transit of 1874; as a result Sydney Observatory data played a key role in **George Airy**'s calculation of the solar parallax. In 1892, the government belatedly published an attractive and well-illustrated book about the 1874 transit. In stark contrast to the successes of 1874, cloudy weather in New South Wales prevented successful observation of the 1882 transit.

Double stars held a special fascination for Russell. Starting in 1870, he began by systematically reobserving those listed in **John Herschel**'s *Cape* monograph. Russell and his assistants then went in search of new double stars, and discovered about 500 of them.

One scientific area that Russell pioneered in Australia was astronomical photography. During the second half of the 1880s, he obtained a series of exquisite images of the Magellanic Clouds and various regions of the Milky Way, some of which were published in *Monthly Notices of the Royal Astronomical Society*.

An ardent supporter of the International Astrographic or *Carte du Ciel* Project, Russell attended the inaugural meeting in Paris in 1887 and pledged the involvement of both Sydney and Melbourne Observatories. But unlike other participating observatories, Russell only purchased the Grubb optics for the astrographic telescope. He then supervised construction of the instrument itself in Sydney. This ambitious project took much of Russell's time in the last decade of his career; it also prevented the Sydney Observatory from effective involvement in astrophysical investigations. Conversely, the commitment to the Astrographic Project ensured the survival of the observatory when the government sought to close it in the late 1920s. Stellar data supplied by Sydney and other participating observatories are now being used in conjunction with Hipparcos space-probe measures to derive useful proper motions.

In a nation where agricultural prosperity was a vital economic ingredient, knowledge of the weather was paramount. Russell dramatically increased the number of weather stations operating in New South Wales – from 55 in 1870 to more than 1,600 by 1898. After 1877, he published daily weather maps in the *Sydney Morning Herald* newspaper, and authored two books on the climate of New South Wales and many research papers on Australian meteorology.

In addition to using scientific instruments, Russell liked to experiment with their design and manufacture. He invented a tide gauge and a number of different meteorological instruments. Between 1878 and 1880, he designed and constructed a horseshoe-style equatorial mounting that foreshadowed the mounting of the 5-m Hale telescope. Russell made the 38.1-cm primary mirror used with the horseshoe mounting.

Apart from astronomy and meteorology, Russell continued the geomagnetic program initiated by his predecessor, George Roberts Smalley (1822–1870), and expanded the observatory's tidal studies to include readings at a number of other ports in New South Wales. He also arranged for the observatory's time service to be extended to the colony's second-largest city, Newcastle, where a time ball was installed.

Russell played a leading role in the Royal Society of New South Wales. For a number of years he served as its president, and in addition he founded its short-lived Astronomy Section. Russell cofounded the Australasian Association for the Advancement of Science and served as its inaugural president. A fellow of the Royal Astronomical Society from 1871, he was elected a fellow of the Royal Society of London in 1886. Russell's contribution to science was further recognized in 1890 when Queen Victoria made him Companion of the Order of Saint Michael and Saint George.

A vigorous and exceedingly blunt person, Russell accumulated some enemies along the way. In 1877, he received a parcel bomb, and in 1889 he was attacked by an observatory worker. His strained relations extended to many in the local amateur astronomical community. During the 1890s, he feuded openly with **John Tebbutt**, arguably Australia's foremost astronomer, and with the Lands Department's leading astronomer, Joseph Brooks. However, after Tebbutt, Russell had greater international visibility than any other Australian astronomer.

Wayne Orchiston

Selected References

Bhathal, R. (1991). "Henry Chamberlain Russell – Astronomer, Meteorologist and Scientific Entrepreneur." *Journal and Proceedings of the Royal Society of New South Wales* 124: 1–21.

Evans, David S. (1988). *Under Capricorn: A History of Southern Hemisphere Astronomy*. Bristol: Adam Hilger.

Orchiston, W. (1988). "From Research to Recreation: The Rise and Fall of Sydney Observatory." *Vistas in Astronomy* 32: 49–63.

——— (2000). "Illuminating Incidents in Antipodean Astronomy: H. C. Russell and the Origin of the Palomar-type Mounting." *Journal of the Antique Telescope Society* 19: 13–15.

——— (2002). "Tebbutt vs Russell: Passion, Power and Politics in Nineteenth Century Australian Astronomy." In *History of Oriental Astronomy*, edited by S. M. Razaullah Ansari, pp. 169–201. Dordrecht: Kluwer: Academic Publishers.

Russell, H. C. (1881). "Spectrum and Appearance of the Recent Comet." *Journal and Proceedings of the Royal Society of New South Wales* 15: 81–86.

——— (1890). *Photographs of the Milky-Way and Nubeculae Taken at Sydney Observatory*. Sydney: Government Printer.

——— (1892). *Observations of the Transit of Venus, 9 December, 1874; Made at Stations in New South Wales*. Sydney: Government Printer.

Walsh, G. P. (1976). "Russell, Henry Chamberlain (1836–1907)." In *Australian Dictionary of Biography*. Vol. 6, *1851–1890*, pp. 74–75. Melbourne: Melbourne University Press.

Wood, H. W. (1959). *Sydney Observatory 1858 to 1958*. Sydney Observatory Papers No. 31. Sydney: Government Printer.

Russell, Henry Norris

Born **Oyster Bay, New York, USA, 25 October 1877**
Died **Princeton, New Jersey, USA, 18 February 1957**

American astronomer Henry N. Russell demonstrated how a star's brightness is related to its spectral type in the so called Hertzsprung–Russell diagram, invented a method to compute the densities of binary stars, and shaped the development of contemporary astronomy by merging astronomy with astrophysics.

The son of a Presbyterian minister, Russell received his early schooling at an Oyster Bay Dames' School. His interest in astronomy dated from early childhood, when, as a 5-year-old, he viewed with his parents the transit of Venus across the Sun. As a teenager he lived with an aunt in Princeton, New Jersey, where he was educated at Princeton Preparatory School and then Princeton University. His intense focus on his undergraduate studies at the latter earned him not only highest honors from his professors (1897) but also two distinctions from his classmates, who described him both as the least socially adept amongst them and also as "Our Star." Russell also did his graduate work at Princeton University, where he figured out a new way to calculate the orbits of binary stars around each other. His dissertation, "The General Perturbations of the Major Axis of Eros Caused by the Action of Mars" (1900), combined his interests in astronomy and mathematics.

Soon after receipt of his Ph.D., Russell spent 2 years in postdoctoral studies at the Astrophysical Observatory of Cambridge University. There he engaged in a program of measurement of stellar parallaxes (distances), using photographic methods. Working with **Arthur Hinks**, he used the Sheepshanks polar reflector fed by a siderostat, a telescope combination said to combine all the disadvantages of a reflector with all the disadvantages of a refractor. Thus, many of the stellar distances used first in his Hertzsprung–Russell diagram were Russell's own work.

In Cambridge, Russell also studied orbit theory and dynamics with **George Darwin**, who, like Russell's Princeton mentor, **Charles Young**, believed that data should not be collected indiscriminately, as was the practice of most astronomers of the day, but rather with a focus on a specific problem. Some years later Russell would summarize this point of view, first to the membership of the National Academy of Sciences and then to the readers of *Popular Astronomy* (1919):

> The main object of astronomy, as of all science, is not the collection of facts, but the development, on the basis of collected facts, of satisfactory theories regarding the nature, mutual relations, and probable history and evolution of the objects of study.

Russell also used the columns he wrote for nearly a half-century (1900–1943) for *Scientific American* as platforms to promote growth in astrophysics. Likewise, the two-volume textbook, *Astronomy* (1926, 1927), which he coauthored with two colleagues (**Raymond Dugan** and **John Stewart**), changed the focus of the teaching of astronomy through its extensive coverage of astrophysics and stellar evolution.

In 1905 Russell returned to Princeton University, where he spent virtually his entire career until his retirement in 1947. He rose through the academic ranks from instructor (1905), to assistant professor (1908), to full professor (1911) and director of the university observatory (1912). Additional appointments, to research associate of the Mount Wilson Observatory (1921) and to a named Princeton professorship (the C. A. Young Research Professorship) endowed by his undergraduate classmates (1927), followed.

During his first years on the Princeton faculty, Russell's analysis of his trigonometric parallax data led him to discover, contemporaneously with, but independently of, Danish astronomer **Ejnar Hertzsprung**, the correlation between a star's intrinsic brightness and its spectral type. Hertzsprung was first to work out a diagram showing the relationship between temperature and luminosity for a group of stars (1911/1912), but few astronomers came across its publication in a photographic journal. When Russell plotted his diagram in late 1913, leading to the publication of the Hertzsprung–Russell diagram the following year, it had an enormous effect on the scientific community. The diagram remains one of the most important in astrophysics. He also popularized the terms "giant" and "dwarf" in stellar evolutionary vocabulary, though he probably did not coin them.

Before presenting the Hertzsprung–Russell diagram, Russell earlier (1909–1913) used the correlation it displays to revive the late-19th-century theory of stellar evolution developed by **Norman Lockyer** and **August Ritter**. According to this theory as updated by Russell, stars begin life as huge, cool, red bodies. As gravitational contraction leads them to become hotter and denser, their color changes from red, to yellow, white, and blue. Eventually stars contract so much that the perfect gas laws no longer apply, and in their final stages they shrink until they end their lives as small, cool, red bodies. In his 1929 article on stellar evolution in the *Encyclopaedia Britannica*, Russell summarized his understanding of the topic, which became widely accepted. It was replaced by another theory, in which stars evolve from the main sequence to become red giants after exhausting their central hydrogen fuel, in the years around World War II.

Russell went on to apply the new quantum mechanics to astronomical problems, defining modern astrophysics. His work with **Frederick Saunders** led to a theory of atomic spectra, for atoms in which more than one electron contributes to the formation of spectral features, that is still known as Russell–Saunders coupling.

Russell's influence has been much discussed in recent years in connection with the question of whether **Cecilia Payne-Gaposchkin** received enough credit for realizing that, as we currently know, hydrogen is the major constituent of the stars. The then Miss Payne concluded as much for the atmospheres of stars in her 1925 Radcliffe College thesis, but Russell's doubts led her to soften her conclusion. Subsequent work, especially by **Donald Menzel**, also at Harvard College Observatory, by **William McCrea**, studying the solar corona, and by Russell himself with the new quantum mechanics, established the supremacy of hydrogen. Only with Russell's thorough analysis did the result become generally accepted.

In addition to being interested in stellar evolution, Russell was also interested in the broader philosophical issue of cosmic evolution, and his *Solar System and Its Origin* (1935) helped pave the way for subsequent research.

Toward the end of World War I, and in the months following it, Russell served as a consulting and experimental engineer in the Army Aviation Service's Bureau of Aircraft Production. There his studies of aircraft navigation problems required him to make observations in open aircraft at altitudes as high as 16,000 ft. Aside from this hiatus, his professional life focused exclusively on astronomy and astrophysics. One field of lifelong interest was binary stars. Russell worked out a method to calculate their masses by studying their orbits, and another to compute distances from Earth by using both orbits and masses. He did trailblazing work on eclipsing variables – binary star pairs in which one member periodically hides the other from the viewpoint of Earth, causing variations in brightness. Russell's last published paper was on binary stars in the Magellanic Clouds.

Russell's connection with Mount Wilson Observatory, which began 3 years after World War I ended, led him to spend several months a year there for the next two decades, until the United

States entered World War II in 1941. During his time in southern California, at the request of the observatory's director, **George Hale**, Russell assisted the staff members in sorting through the spectroscopic data gathered there over the years in order to teach them how to apply cutting-edge physical theory to astronomy. Russell set an example by analyzing line spectra to uncover atomic structure with a view to extending a theory of Indian astronomer **Meghnad Saha**, which demonstrated the role temperature and pressure play in stellar spectra. Russell's quantitative analysis at Mount Wilson Observatory of the abundance of the elements in the solar atmosphere also helped transform theoretical astrophysics into a recognized field.

Although Russell turned down offers to direct the observatories at Harvard and Yale, which were much larger than Princeton's, he nonetheless played a role at Harvard Observatory that was similar to the one he played at Mount Wilson Observatory. In 1921, the same year Russell's association with Mount Wilson Observatory began, his first prominent graduate student, **Harlow Shapley**, was appointed director of the Harvard College Observatory. Shapley immediately put Russell on the visiting committee, where he would exert significant influence in replacing the entrenched habits of the past with newer research perspectives. Other students of Russell's at Princeton University also went on to careers in the major astronomical institutions in the United States, helping to spread Russell's conviction that astronomy and astrophysics should be indistinguishable and that astronomical research should be grounded in theory as well as empiricism.

Unlike his grandfather, father, and brothers, Russell did not choose the Presbyterian ministry as a profession, and he rejected the idea of life after death. During the 1920s at Princeton University, he advocated abolishing compulsory undergraduate chapel attendance, arguing the pointlessness of preaching to a captive but uninterested audience. Nonetheless, he lectured frequently on the ways in which science supported religion and morality, particularly in uncovering the unified design of nature. His lecture series on this subject, published as *Fate and Freedom* (1927), also explores the conflict between the idea of a deterministic universe and the belief in free will. He argued by analogy that free will was as real as the pressure of a gas or other statistical physical phenomena.

Russell and his wife, born Lucy May Cole, had four children, Lucy May, Elizabeth Hoxie, Henry Norris, and Emma Margaret, who married astronomer Frank Edmondson and remained a presence in the astronomical community until her death in 1999.

Russell's transformational leadership in the astronomical community earned him the sobriquets "the general" and "the dean of American astronomers." More than any other astronomer of his generation, he changed astronomy from a discipline based on data collection with telescopes to one driven by physical theory. His own Ph.D. students included Menzel, **Lyman Spitzer**, **Charlotte Moore**, and Louis C. Green (of Swarthmore College and source of the anecdote that Russell was capable of falling asleep during his own lectures).

The Henry Norris Russell Lectureship, given for lifetime achievement, is the American Astronomical Society's highest prize.

Naomi Pasachoff and *Jay M. Pasachoff*

Selected References

DeVorkin, David H. (2000). *Henry Norris Russell: Dean of American Astronomers*. Princeton, New Jersey: Princeton University Press. (This first book-length biography of Russell, by the curator of the Smithsonian's National Air and Space Museum, is likely to remain the definitive one.)

Russell, Henry Norris (1912). "On the Determination of the Orbital Elements of Eclipsing Variable Stars." Parts 1 and 2. *Astrophysical Journal* 35: 315–340; 36: 54–74.

——— (1914). "Relations between the Spectra and Other Characteristics of the Stars." *Proceedings of the American Philosophical Society* 51: 569–579.

——— (1947). "America's Role in the Development of Astronomy." *Proceedings of the American Philosophical Society* 91: 10–16.

Russell, Henry Norris, Raymond Smith Dugan, and John Quincy Stewart (1926–1927). *Astronomy: A Revision of Young's Manual of Astronomy*. 2 Vols. Boston: Ginn and Co.

Shapley, Harlow (1958). "Henry Norris Russell." *Biographical Memoirs, National Academy of Sciences* 32: 354–378.

Russell, John

Born **1745**
Died **1806**

In 1866, **Johann Schmidt** claimed that the telescopic appearance of lunar crater Linné had changed since its depiction in the work of **Gotthelf Lohrmann** and **Johann von Mädler** mere decades before. Evidence against such a historical event on the Moon was provided by English artist John Russell. His 1788 drawing was discovered to show Linné much as Schmidt himself had rendered it.

Selected Reference

Olson, Roberta J. M. and Jay M. Pasachoff (1999). "Moon-Struck: Artists Rediscover Nature and Observe." *Earth, Moon, and Planets* 85–86: 303–341.

Rutherford, Ernest

Born **Spring Grove near Nelson, New Zealand, 30 August 1871**
Died **Cambridge, England, 19 October 1937**

British Empire chemist and physicist Ernest Rutherford received the 1908 Nobel Prize in Chemistry for work on disintegration of the elements and chemistry of radioactive substances, but he is probably best known within astronomy for Rutherford scattering (of one nucleus by another) and his model of the atom that put all the positive charge, and most of the mass, at the center with a cloud of negative charge (electrons) around it. This concept led to the Bohr atom and so to the possibility of understanding conditions under which various spectral features can be produced.

Rutherford was the fourth of 11 children of a British immigrant family. His father was a technician and his mother a teacher. After attending local schools and high schools, and receiving his B.A. in 1892 and B.S. in 1894 at Canterbury College in Christchurch, New Zealand, Rutherford won a scholarship to Cambridge University in

England. There his teacher was the famous physicist J. J. Thomson. Thomson asked Rutherford in 1896 to investigate with him the effect of X-rays, just discovered by Wilhelm C. Röntgen (1845–1923), on the discharge of electricity in gases. This collaboration led Thomson to the discovery of the electron, and Rutherford to the development of an improved quantitative measuring method for ionization processes. As a consequence of this work, Rutherford turned to the study of the atomic structure, and the use of the new rays discovered by Antonie H. Becquerel (1852–1908) in 1896 and soon after named "radioactivity" by Marie Curie (1867–1934).

In 1898, Rutherford accepted a professorship at McGill University in Montreal, Canada. There he began an important series of experiments on the radiation of radioactive elements, beginning with uranium. Soon Rutherford discovered two types of radiation: The so called α radiation with a short range, and the β radiation with a longer range. (A third type, γ radiation, was found by Paul Villard [1860–1934] in 1900.) He also found that thorium and other radioactive elements emitted a gaseous radioactive product, named by him as emanation. In 1901–1902 Rutherford, together with his colleague Frederick Soddy (1877–1956), developed their disintegration theory of radioactivity. In 1903, Rutherford demonstrated that α particles carry positive charge. In 1907, Rutherford became professor of physics at Manchester University, England. At Manchester he developed a school for research into radioactivity.

Together with coworkers and assistants like Hans Geiger (1882–1945) and Ernest Marsden (1889–1970), Rutherford intensively investigated α particles (helium nuclei). As a consequence, in 1911 Rutherford presented a new model of the atom. (Former models had atoms of roughly uniform density, *e. g.*, Thomson's plum-pudding model, while in Rutherford's the positively charged particles are concentrated.) According to this theory, positively charged particles are concentrated in the massive center of an atom, while negatively charged particles orbit this nucleus at a relatively great distance. Two years later **Niels Bohr**, at that time a guest at Manchester University, combined this with the new quantum theory and introduced his new model, which became standard.

Rutherford made his next fundamental discovery in 1918/1919: He observed that the nuclei of certain light elements could be "disintegrated" by the impact of energetic α particles, and that protons were emitted during this process. This was the realization of artificial transmutation of one element into another. In 1925, his pupil **Patrick Blackett** confirmed this result with the help of a cloud chamber.

In 1919, Rutherford succeeded Thomson in the chair of physics at Cambridge University and became the director of the Cavendish Laboratory. Rutherford attracted a lot of young physicists to the Cavendish Laboratory, and he was the motivating central figure of this circle of talented researchers, among them the later Nobel Prize Laureates Blackett, John Cockcroft (1897–1967), James Chadwick (1891–1974), Petr Kapitsa (1894–1984), Cecil F. Powell (1903–1969), and Ernest Walton (1903–1995). Virtually it was Rutherford who created nuclear physics as a special discipline within physics.

Rutherford was elected fellow of the Royal Society in 1903. (He was its president from 1925 to 1930.) He was awarded its Rumford Medal (1904) and Copley Medal (1922).

Rutherford held honorary doctorates from several universities, among them Pennsylvania, Wisconsin, Birmingham, Giessen, Copenhagen, Oxford, Toronto, Cape Town, and London. In 1931, he was created First Baron Rutherford of Nelson. Rutherford was married to Mary G. Newton of New Zealand, and they had one daughter who married **Ralph Fowler**.

Horst Kant

Selected References

Andrade, E. N. C. (1964). *Rutherford and the Nature of the Atom*. Garden City, New York: Doubleday.

Badash, Lawrence (ed.) (1969). *Rutherford and Boltwood: Letters on Radioactivity*. New Haven: Yale University Press.

Birks, J. B. (ed.) (1962). *Rutherford at Manchester*. London: Heywood.

Eve, Arthur S. (1939). *Rutherford*. New York: Macmillan.

Eve, Arthur S. and J. Chadwick (1938). "Lord Rutherford." *Obituary Notices of Fellows of the Royal Society* 2: 395–423.

Oliphant, Sir Mark (1972). *Rutherford: Recollections of the Cambridge Days*. Amsterdam: Elsevier.

Rutherford, Ernest (1904). *Radio-activity*. Cambridge: University Press.

——— (1906). *Radioactive Transformations*. London: Archibald Constable.

——— (1913). *Radioactive Substances and Their Radiations*. Cambridge: University Press.

——— (1962–1965). *The Collected Papers of Lord Rutherford of Nelson*, edited by James Chadwick. 3 Vols. London: George Allen and Unwin.

Rutherford, Ernest, James Chadwick and C. D. Ellis (1930). *Radiations from Radioactive Substances*. Cambridge: University Press.

Wilson, David (1983). *Rutherford: Simple Genius*. Cambridge, Massachusetts: MIT Press.

Rutherfurd, Lewis Morris

Born **Morrisania, New York, USA, 25 November 1816**
Died **Tranquility, New Jersey, USA, 30 May 1892**

Amateur astronomer Lewis Rutherfurd pioneered the use of photography and spectroscopy.

During his student days, Rutherfurd showed a distinct aptitude for science and became an assistant to the Professor of physics and chemistry at Williams College. After graduating at age 18, however, he went on to study law and from 1837 he conducted a successful practice in New York City. Even so, he maintained an interest in science, especially astronomy, and counted among his friends the famous telescope maker Henry Fitz, whose optical methods he learned, and the astronomer **Benjamin Gould**. Rutherfurd was also an active long-term correspondent with the chemist Josiah Willard Gibbs and the physicist Ogden Nicholas Reed, who suggested and encouraged his work in spectroscopy.

Rutherfurd married Margaret Stuyvesant Chanler in 1841, and it was chiefly concern for her health that led him to give up his law practice in 1849 and to embark upon 7 years of foreign travel. While in Europe, he visited observatories and acquired a good background in astronomy and optics that would later prove of value. On his return to the United States in 1856, he built his own observatory, equipped it with an 11.25-in. Fitz refractor, and devoted the rest of his active life to astronomy. Rutherfurd was a capable visual observer; in 1862 and 1863 he reported useful measurements of Sirius B, which had only recently been discovered by **Alvan Clark** while he was testing the

18.5-in. objective being made for the University of Mississippi (and subsequently sold to the Chicago Astronomical Society for use in the Dearborn Observatory). Gould credited Rutherfurd's seven measures of the position of Sirius B as among the most important early confirmations of the discovery. They were the earliest usable observations of Sirius B with any telescope with an aperture less than 15 in. and met the high standards of **Sherburne Burnham** when the latter prepared his catalog of double stars some four decades later.

Two years later, Rutherfurd took up celestial photography in earnest. As photographic emulsions of that time were sensitive only to blue light, and refractors constructed for visual use at long wavelengths did not give sharp photographic images, his first such results were generally disappointing – with the exception of some images of the Sun and the Moon taken at reduced aperture. Accordingly, he undertook a series of important demonstrations to distinguish between the visual and blue focus. He also conducted a number of experiments to determine the actinic or photographic focus of his refractor. The care exercised in this respect, first showcased at the solar eclipse of 1860 in Labrador, resulted in excellent photographs of that event taken with a 4.25-in. refractor. (Rutherfurd increased the spacing of the lens elements to move the instrument's focus to a compromise position between the visual and actinic focal points.) That success led Rutherfurd to construct the first refracting telescope designed solely for photographic work. Working with Henry Fitz and his son Harry, Rutherfurd figured the lens using a spectroscope, a technique that was later applied to advantage by Alvan Clark and his sons. They completed that telescope in 1864; it had an aperture of 11.25 in., and was a total success, yielding excellent images of the Moon that were 1.7 in. in diameter on the photographic plate, and giving good images of stars to the ninth magnitude.

But such instruments are useless for visual work. Hence in 1868, Rutherfurd, again working with Harry Fitz, completed a 13-in. visual refractor that could be converted to a photovisual telescope by the attachment of a corrector lens to the front of the objective. (The design was later used by the Clarks as the basis for the 36-in. Lick refractor.) By 1877, Rutherfurd had accumulated over 1,400 photographs of celestial objects, chiefly the Moon and star clusters.

After 1861, Rutherfurd became more interested in the spectroscopic work of **Robert Bunsen** and **Gustav Kirchhof**, and from 1862 he began spectroscopic studies of the Sun, Moon, Jupiter, Mars, and some fixed stars. From his early spectroscopic work, Rutherfurd devised a stellar classification scheme based on gross differences in bright star spectra, which Gould and **Julius Scheiner** credited as preceding a very similar though more refined scheme proposed by **Angelo Secchi**.

With an ingenious spectrograph employing six prisms filled with carbon disulfide, Rutherfurd produced a solar spectrum that exceeded all previous results. In 1864 at a meeting of the National Academy of Science, Rutherfurd displayed an unpublished photographic solar spectrum that had three times the number of lines noted by Kirchhof and Bunsen in their spectrum of the Sun.

To assist his spectroscopic work, Rutherfurd then began to produce gratings from a screw-driven ruling machine rather than the lever-driven machines common to that era. His best efforts produced gratings with up to 17,000 lines per inch. Rutherfurd's gratings, unsurpassed until the advent of **Henry Rowland** and his concave spectral gratings, were generously distributed to other scientists who pressed them into various types of spectroscopic service.

Rutherfurd was convinced that photographic plates offered a solution to problems in astrometry. He pursued that proposition both through his experimentation with the photographic recording of stellar clusters, and through development of machines for making very precise measurements of star image positions on photographic plates. Though his efforts to demonstrate the efficacy of that process were clouded by the concerns of other astronomers about the stability of photographic films of the period, his vast collection of plates of stellar clusters has been useful in efforts to determine proper motions of stars in clusters over a century after their first exposure. The soundness of Rutherfurd's original concept has since been well demonstrated.

Rutherfurd made his last observations in 1878. He donated the 13-in. refractor to Columbia College Observatory in 1883, and 7 years later presented his entire plate collection, together with 20 folio volumes of unreduced plate measures, to the same institution.

When the National Academy of Science was formed by an act of the United States Congress in March 1863, Rutherfurd was honored by his selection for membership in the first group of 50 American scientists named in that congressional act. He was a foreign associate of the Royal Astronomical Society and received the Rumford medal of the Royal Society of London. Columbia University conferred its Doctor of Laws *honoris causa* upon Rutherfurd in 1887.

Richard Baum

Selected References

Anon. (1893). "Lewis Morris Rutherfurd." *Monthly Notices of the Royal Astronomical Society* 53: 229–231.

Ashbrook, Joseph (1984). "An Episode in Early Astrophotography." In *The Astronomical Scrapbook*, edited by Leif J. Robinson, pp. 160–165. Cambridge, Massachusetts: Sky Publishing Corp.

Gould, Benjamin Apthorp (1895). "Memoir of Lewis Morris Rutherfurd." *Biographical Memoirs, National Academy of Sciences* 3: 415–441.

Rees, John K. (1892). "Lewis Morris Rutherfurd." *Astronomy and Astro-Physics* 11: 689–697.

Rutherfurd, Lewis Morris (1862). "Companion to Sirius." *American Journal of Science* 34: 294–295.

——— (1863). "Letter on Companion to Sirius, Stellar Spectra and the Spectroscope." *American Journal of Science* 35: 407–409.

Scheiner, Julius (1898). *Astronomical Spectroscopy,* translated by Edwin Brant Frost, pp. 235–236. Boston: Ginn and Co.

Warner, Deborah Jean (1971). "Lewis M. Rutherfurd: Pioneer Astronomical Photographer and Spectroscopist." *Technology and Culture* 12: 190–216.

Rydberg, Johannes [Janee] Robert

Born **Halmstad, Sweden, 8 November 1854**
Died **Lund, Sweden, 28 December 1919**

Swedish physicist Johannes Rydberg calculated the amount of energy required to unbind the single electron of hydrogen, and this amount (13.6 eV) is often given his name as the Rydberg constant (alternatively 109,678 cm^{-1}). More recently, the phrase Rydberg matter has been used to describe neutral gas in which the electrons are located in states of very high excitation far from their nuclei.

Rydberg was the son of Maria Beata Andersson and Sven R. Rydberg, a local tradesman and boatyard operator, who died when his son was four. He married Lydia E. M. Carlsson in 1886, and they had a son and two daughters.

Rydberg studied and worked all his life at Lund University. He first went there in 1873 after completing his gymnasium studies in Halmstad; he was awarded the Ph.D. degree in mathematics in 1879, becoming a *docent* (lecturer) in mathematics in the following year. His interests progressively turned toward mathematical physics and in 1882 Rydberg was made *docent* in physics until he was appointed in 1901 to an extraordinary professorship in physics as well as to the directorship of Lund's department of physics. From 1876 to 1897, he was also assistant at the university's Physics Institute. When in 1908 a new law eliminated the rank of extraordinary professor, Rydberg automatically became an ordinary professor, a position he retained until his death, although from 1914 he was sick and was often absent from the university. Rydberg was a member of the Royal Society of London (1919), and a leading figure of the Physics Society of Lund.

While contemporary Swedish spectroscopists of renown, **Anders Ångström**, Robert Thalen, and Barnhard Hasselberg, carried out mainly experimental programs of charting the spectra of different elements that are found in the sun, Rydberg largely relied on other scientists' measurements to study the structure of spectra, in particular to find arithmetic formulae describing the wavelengths of lines and to compare them with the physical and chemical properties of elements. A mathematician by training, Rydberg carried over a mathematical approach to spectroscopy. He most notably concerned himself with the numerical analysis of regularities in spectra, producing what became known as Rydberg's formula. From the 1860s, spectroscopists had searched for patterns or regularities in the positions of spectral lines, often hoping to find harmonic ratios. **Johann Balmer** notably put forward in 1885 a formula accounting for the hydrogen spectrum. In 1889, Rydberg proposed a more general formula describing all series of all atomic line spectra, which contributed to organize the mass of available spectroscopic measurements, but which failed to lead him to his stated goal, the understanding of the nature and properties of the atom. This work, together with contemporary researches into spectral regularities by Walther Ritz, as well as Heinrich Kayser and **Carl Runge**, subsequently proved central in the elaboration of atomic theories from the 1910s onward. **Niels Bohr**'s theory of atomic structure (1913), combining **Ernest Rutherford**'s nucleus with Max Planck's quantum, for the first time gave an interpretation of Rydberg's formula and confirmed Rydberg's belief that spectral characteristics were useful in the investigation of atomic structure and properties. The numerical value of the Rydberg constant depends upon the charge and mass of the electron (not yet discovered when he put forward his formula) and Planck's constant. Though it was to Rydberg an empirical result from laboratory experiment, the constant can be derived using Bohr's theory of atomic structure.

Charlotte Bigg

Selected References

Bohr, N. (1954). "Rydberg's Discovery of the Spectral Laws." In *Proceedings of the Rydberg Centennial Conference on Atomic Spectroscopy*, edited by Bengt Edlén, pp. 15–21. *Acta Universitatis Lundensis*, 50. Lund: Gleerup. (Niels Bohr assessed the use of Rydberg's work for the construction of quantum theories of matter.)

Dörries, Matthias (1995). "Heinrich Kayser as Philologist of Physics." *Historical Studies in the Physical Sciences* 26: 1–33. (For an evaluation of Heinrich Kayser and Carl Runge.)

Hamilton, Paul C. (1993). "Reaching Out: Janne Rydberg's Struggle for Recognition." In *Center on the Periphery: Historical Aspects of 20th-Century Swedish Physics*, edited by Svante Lindqvist, pp. 269–292. Canton, Massachusetts: Science History Publications. (For a discussion of the difficulties faced by Rydberg to obtain official recognition of his work.)

Rydberg, J. R. (1890). "Recherches sur la constitution des spectres d'émission des éléments chimiques." *Kungliga Svenska vetenskapsakademiens handlingar*, n.s., 23. (Rydberg's major work.)

——— (1890). "On the Structure of the Line-Spectra of the Chemical Elements." *Philosophical Magazine*, 5th ser., 29: 331–337.

Siegbahn, Manne (1952). "Janne Rydberg, 1854–1919." In *Swedish Men of Science, 1650–1950*, edited by Sten Lindroth, pp. 214–218. Stockholm: Swedish Institute, Almqvist and Wiksell.

S

Sabine, Edward

Born **Dublin, Ireland, 14 October 1788**
Died **Richmond, (London), England, 26 June 1883**

Irish polymath Edward Sabine was the first to point out the correlation between the 11-year solar activity (sunspot) cycle and a similar cycle in the behavior of the Earth's magnetic field.

Sabine's education was completed at the Royal Military Colleges of Marlow and Woolwich, England. On 22 December 1803, he was commissioned as a second lieutenant in the Royal Artillery, becoming a captain one decade later. He would ultimately achieve the rank of major general (1859).

After his services in the war of 1812 against the United States, Sabine was chosen as astronomer on the first expedition (1818) sponsored by the Royal Society of London to search for the Northwest Passage. He served in the same capacity on the subsequent expedition from 1819 to 1820. On these voyages, he made his first observations of terrestrial magnetism that were later described before the Royal Society. Sabine reached the important conclusion that, through extensive travel, he might be able to study the Earth as an astronomical body. He argued for a worldwide study of magnetism, to be conducted through a network of widely separated stations. This idea was furthered by **Alexander von Humboldt**, who promoted the concept and aided the establishment of magnetic observatories from Germany to Beijing.

Sabine spent the years 1821 and 1822 on another mission for the Royal Society to determine the true figure of the Earth from timings of the period of a pendulum's swing. His observations were carried out at several sites on or near the Equator along the coasts of Africa and South America. He then extended those observations to Greenland and Norway. For this work, he was awarded the Royal Society's Copley Medal.

During 1825, Sabine worked with Sir **John Herschel** on a joint commission of the British and French governments charged with determining the longitude difference (by means of rocket signals) between the observatories located at Paris and Greenwich, England. Two years later, Sabine resumed his pendulum observations. In the interim, Sabine had married Elizabeth Juliana Leeves who assisted him in his work. She died in 1879; the couple had no children.

But Sabine did not lose sight of his goal of assembling worldwide magnetic data. He joined the British Association for the Advancement of Science (created by Charles Babbage as an alternative to the Royal Society). Years later, Sabine promoted reforms in the election of fellows of the Royal Society to answer Babbage's complaints.

By 1835, the British Association (at Sabine's urging) passed a resolution calling for the government to open magnetic and meteorological observatories in the colonies. When this prod proved ineffective, Sabine exerted his influence upon the Royal Society, where he had been secretary since 1827. Again, nothing happened until Herschel's return from South Africa in 1838, whereupon he gave his support, and the project was launched. An early result was the construction of a magnetic observatory at the University of Toronto for studies of the aurora borealis.

Sabine's political adroitness was further demonstrated regarding the King's Observatory at Kew. This facility had been constructed so that King George III, an amateur astronomer, could observe the 1769 transit of Venus. Thereafter, it had been used primarily as an educational center for the Royal family and to house George III's collection of scientific instruments. In the early 1840s, Sabine had the idea of converting the structure into a magnetic observatory and training personnel in its use. He suggested that the Royal Society take over the facility. Herschel objected on grounds that this was too limited a venture; the Royal Society declined to act. In turn, Sabine took his proposal to the British Association, of which he was then the general secretary. The association acquired the site in 1842 and managed the observatory until 1871, when it was transferred to the Royal Society, where Sabine was completing a 10-year term as president.

Sabine's correlation of magnetic data from the Toronto and Hobarton (Tasmania) observatories with measurements of sunspot activity compiled by German astronomer **Heinrich Schwabe** led to his most important discovery. In 1852, Sabine announced to the Royal Society that the 11-year sunspot cycle was directly correlated with the newly discovered geomagnetic cycle. This provided the first evidence of solar–terrestrial relationships (beyond gravitational attractions) and ushered in further studies of these phenomena.

Among other honors, Sabine was awarded a Royal Medal in 1849 and the Lalande Prize of the French Academy of Sciences. He was named an honorary or associate member of numerous foreign academies and scientific institutes. Sabine was knighted in 1869.

Sabine's brother Joseph, a naturalist who accompanied him to the Arctic, named a seagull he first sighted in 1819 after Edward; a lunar crater was likewise named for him in 1935.

George S. Mumford

Selected References

Anon. (1883). "Sir Edward Sabine." *Observatory* 6: 232–233.

Glaisher, J. W. L. (1884). "Sir Edward Sabine." *Monthly Notices of the Royal Astronomical Society* 44: 136–138.

Reingold, Nathan (1975). "Sabine, Edward." In *Dictionary of Scientific Biography*, edited by Charles Coulston Gillispie. Vol. 12, pp. 49–53. New York: Charles Scribner's Sons.

Sacrobosco

➔ **John of Holywood**

Ṣadr al-Sharīʿa al-Thānī: ʿUbaydallāh ibn Masʿūd al-Maḥbūbī al-Bukhārī al-Ḥanafī

Died **Bukhara, (Uzbekistan), 1346/1347**

Ṣadr al-Sharīʿa (al-thānī, *i. e.*, "the Second") was a theoretical astronomer and religious scholar who created original and sophisticated astronomical theories of time and place, and under circumstances that have long been considered devoid of original scientific research. Ṣadr was famous for his commentaries on Islamic jurisprudence (*sharīʿa*, hence his nickname Ṣadr al-Sharīʿa, "preeminent [scholar] of the *sharīʿa*"). He was called "the Second," after his great-great-grandfather, Ṣadr al-Sharīʿa al-Awwal ("the First"). Ṣadr also wrote on Arabic grammar, *kalām* (theology), rhetoric, legal contracts, and *ḥadīth* (prophetic traditions).

Ṣadr's astronomical writings are found in the third volume of his three-volume encyclopedia of the sciences, the *Taʿdīl al-ʿulūm* (The adjustment of the sciences). The first two volumes dealt with logic and *kalām*. The third volume was called *Kitāb Taʿdīl hay'at al-aflāk* (The adjustment of the configuration of the celestial spheres).

Ṣadr al-Sharīʿa represents one of several theorists who worked within the astronomical tradition of theoretical astronomy (*hay'a*). This tradition had its roots within the early Islamic period, especially with **Ibn al-Haytham**, but it began to flourish among the group of astronomers who were assembled at the Marāgha Observatory in northwestern Iran by the polymath **Naṣīr al-Dīn al-Ṭūsī**. One of the major issues that was of concern to these theorists was the irregular motion produced in several of **Ptolemy**'s models, such as that brought about by the equant, and they sought to substitute models that would adhere to the physical principle of uniformity of motion in the heavens. Ṣadr frequently cites two works from this tradition – Ṭūsī's *al-Tadhkira fī ʿilm al-hay'a* (Memoir on astronomy), and *al-Tuḥfa al-shāhiyya* (The imperial gift) of **Quṭb al-Dīn al-Shīrāzī**. He does this in order to correct their work, and to present solutions to problems they missed.

In the *Kitāb Taʿdīl hay'at al-aflāk*, Ṣadr critically reviews the planetary models of his predecessors, especially Ptolemy, and points out their weaknesses. He then describes his own models that are meant to rectify them. The most significant problems Ṣadr addresses are: the lunar prosneusis point, the equant; planetary latitude theory, and the motion of Mercury.

In the case of the Moon, Ptolemy proposed that one orb rotate uniformly around the center of the Universe while maintaining a constant distance around another point, the deferent center; Ṣadr objects to this since it produces irregular motion in the celestial realm. Furthermore, rather than measure the motion in anomaly from the visible apogee of the lunar epicycle, Ptolemy measured it from the mean epicyclic apogee aligned with a point, the prosneusis, introduced into the model solely for this purpose. In offering a physically consistent model, Ṣadr employed both a rectilinear and a curvilinear "Ṭūsī couple." Both of these devices combined circular motions in such a way as to produce a compound motion that oscillates along a line. In the rectilinear case, a smaller circle, internally tangent with a larger circle, rotates in such a manner as to produce linear motion; and in the curvilinear case, concentric spheres are made to rotate in such a way as to produce an approximate curvilinear motion along the surface of the epicycle sphere.

In the case of the upper planets (Mars, Jupiter, and Saturn), for which Ptolemy was compelled to introduce the equant point, Ṣadr followed **Mu'ayyad al-Dīn al-ʿUrḍī** and Shīrāzī, without acknowledgment, and employed an epicyclet (an epicycle on an epicycle).

The Ptolemaic theory of planetary latitude and the revisions to it made by Islamic successors attempted to provide models for the planets' deviations from the ecliptic and involved complex, nonuniform spherical motions. Ṣadr summarized the work of his three predecessors and offered his own observations. As of this date, however, this problem has been insufficiently studied, so the significance of Ṣadr's work on the theory of planetary latitude remains obscure.

The case of Mercury involved several equant-like problems and thus was particularly complicated. Ṣadr employed two geometrical tools invented by his predecessors – the "ʿUrḍī lemma" and the spherical "Ṭūsī couple" to arrive at his solution. Late medieval Islamic astronomy has as yet been insufficiently studied to assess fully the possible influence of Ṣadr on subsequent astronomers, such as **Khafrī** and others.

Ṣadr's work is also significant in that it provides a counterexample to two long-standing paradigms of Islamic intellectual history. First, Ṣadr, who was a prominent religious scholar, contradicts the conclusions of traditional Orientalist scholarship, according to which the Islamic religious establishment was virtually completely opposed to science, and this opposition was supposedly a major factor in the decline of science in Islam. Second, Ṣadr stands as a major counterexample to the prevalent view of Islamic historiography whereby Islamic culture enjoyed a brilliant flourishing from the 9th century until the 11th century, but then suffered unmitigated decline in large part due to the critiques of rational science and philosophy by such religious scholars as Ghāzālī (died: 1111). Ṣadr clearly represents a very high level of mathematical and scientific sophistication within a tradition that falls well within the period of supposed decline.

Glen M. Cooper

Selected References

Al-Shīrāẓī, Quṭb al-Dīn. *al-Tuḥfa al-shāhiyya* (The imperial gift). (There is as yet, unfortunately, no published edition or translation of this important treatise.)

Dallal, Ahmad S. (ed.) (1995). *An Islamic Response to Greek Astronomy: Kitāb Taʿdīl hay'at al-aflāk of Ṣadr al-Sharīʿa*. Leiden: E. J. Brill. (Edition of the *Kitāb Taʿdīl hay'at al-aflāk* together with extensive notes and diagrams. This book is an extraction of the main portion of Ṣadr's text from his own commentary, a somewhat dubious methodology. The commentary portion has not yet been published. If it is ever published, it will cast greater light on how Ṣadr understood his own work. This edition was the primary source for the present article.)

Ragep, F. J. (1993). *Naṣīr al-Dīn al-Ṭūsī's* Memoir on Astronomy *(al-Tadhkira fī ʿilm al-hay'a)*. 2 Vols. New York: Springer-Verlag. (Perhaps the most significant study to emerge thus far in the historiography of astronomy in Islam, in which al-Ṭūsī's treatise was pivotal.)

Saliba, George (1979). "The Original Source of Quṭb al-Dīn al-Shīrāzī's Planetary Model." *Journal for the History of Arabic Science* 3: 3–18. (Describes the motivation behind the "ʿUrḍī lemma".)

——— (1987). "The Role of the *Almagest* Commentaries in Medieval Arabic Astronomy: A Preliminary Survey of Ṭūsī's Redaction of Ptolemy's *Almagest*." *Archives internationales d'histoire des sciences* 37: 3–20. (Contains a brief survey of Ptolemaic latitude theory and Ṭūsī's attempts to rectify it.)

——— (1987). "Theory and Observation in Islamic Astronomy: The Work of Ibn al-Shāṭir of Damascus." *Journal for the History of Astronomy* 18: 35–43. (Contains a description of the "ʿUrḍī lemma.")

——— (1993). "Al-Qushjī's Reform of the Ptolemaic Model for Mercury." *Arabic Sciences and Philosophy* 3: 161–203. (Description of the innovative work of a late Islamic astronomer.)

——— (1994). "A Sixteenth-Century Arabic Critique of Ptolemaic Astronomy: The Work of Shams al-Dīn al-Khafrī." *Journal for the History of Astronomy* 25: 15–38. (Survey of the work of another late Islamic astronomer.)

——— (1996). "Arabic Planetary Theories after the Eleventh Century AD." In *Encyclopedia of the History of Arabic Science*, edited by Roshdi Rashed, pp. 58–127. London: Routledge. (Excellent survey of the development of planetary models during the so called period of decline of Islamic science. Plentiful and useful diagrams help to illustrate the complexities of this intricate subject.)

Saliba, George and E. S. Kennedy (1991). "The Spherical Case of the Ṭūsī Couple." *Arabic Sciences and Philosophy* 1: 285–291. (Presents diagrams helpful in visualizing this three-dimensional device.)

Safford, Truman Henry

Born **Royalton, Vermont, USA, 6 January 1836**
Died **Newark, New Jersey, USA, 13 June 1901**

American astronomer Truman Safford was the first director of the Dearborn Observatory (now part of Northwestern University), and later the director of Williams College's Hopkins Observatory. His 1888 catalog of north polar stars showed groups of stars with common proper motions. Born with a photographic memory, while at the telescope, Safford had no need for stellar coordinates from the *Nautical Almanac*; he had memorized all of them.

A sickly child from birth, Safford's education was mainly from his family's home library. His prodigious mental computational ability was evidenced by his preparation, by age ten, of almanac's for various cities including Boston, Cincinnati, and Philadelphia. This brought him to the attention of **Benjamin Peirce**, who arranged for Safford's further preparatory education and enrollment in Harvard. After graduating from Harvard University, Safford was employed at the Harvard College Observatory. His computation of the irregularities in the proper motion of Sirius in declination confirmed prior predictions of an unseen companion by **Wilhelm Bessel** and **Christian A, Peters** based on its variations in right ascension, and preceded the accidental discovery of the first observed white dwarf star, Sirius B, by **Alvan Graham Clark**. When the observatory director, **George Bond**, died prematurely, Safford completed and edited Bond's highly praised observations of the Orion Nebula for publication in the *Harvard Annals*.

Safford's directorship of the Dearborn Observatory was frustrated by an inadequate dome for what was then one of the largest refracting telescopes in the world, the 18.5-in. Clark refractor with which Sirius B was discovered. Because the dome could not be easily moved, Safford was forced to use the telescope in a transit mode. His search for new nebulae in the manner of **William Herschel**'s sky sweeps was essentially the most appropriate use of the telescope under the circumstances. Safford is credited with the discovery of 49 nebulae.

When the Great Chicago Fire reduced the ability of the Chicago Astronomical Society to financially support the Dearborn Observatory, Safford resigned, seeking other salaried employment to support his family. From 1872 to 1875, he provided astronomical support to the United States Army Corps of Engineers in their preparation of topographical maps of the western United States. In 1876, Safford was appointed director of the Hopkins Observatory and professor of astronomy at Williams College, where he served for the remainder of his life.

Thomas R. Williams

Selected References

Hollis, H. P. (1901). "Truman Henry Safford." *Observatory* 24: 307–309.

Knobel, E. B. (1901). "The Late Professor Safford." *Observatory* 24: 349–351.

Parshall, Karen Hunger (1999). "Safford, Truman Henry." In *American National Biography*, edited by John A. Garraty and Mark C. Carnes, Vol. 19, pp. 190–191. New York: Oxford University Press.

Safford, Truman Henry (1862). "On the Proper Motion of Sirius in Declination." *Monthly Notices of the Royal Astronomical Society* 22: 145–148.

——— (1878). "Right Ascensions of 505 Stars Determined with the East Transit Circle at the Observatory of Harvard College." *Annals of the Astronomical Observatory of Harvard College* 4, pt. 2.

——— (1884). "Mean Right Ascensions of 133 Stars near the North Pole, Observed in 1882 and 1883 at the Field Memorial Observatory of Williams College." *Proceedings of the American Academy of Arts and Sciences* 19: 324–352.

Safronov, Viktor Sergeyevich

Born **Velikiye Luki, Russia, 11 October 1917**
Died **1999**

Russian solar-system cosmogonist Viktor Safronov was a protégé of **Otto Schmidt**. In his 1968 *Evolution of the Protoplanetary Cloud* and the *Formation of the Earth and Planets*, Safronov quantified the theory for planetary formation from the accumulation of planetesimals, an idea generally attributed to **Thomas Chamberlin** and **Forest Moulton** around 1900. The Safronov theory includes the evolution of the planet's mass, obliquity, and temperature as a function of the rate of accretion.

Selected Reference

Burns, Joseph A., Jack J. Lissauer and Andrei Makalkin (2000). "In Memoriam: Victor Sergeyevich Safronov (1917–1999)." *Icarus* 145: 1–3.

Ṣāghānī: Abū Ḥāmid Aḥmad ibn Muḥammad al-Ṣāghānī [al-Ṣaghānī] al-Asṭurlābī

Flourished **Ṣāghān (near Merv, Turkmenistan)**
Died **Baghdad, (Iraq), 990**

Ṣāghānī was a mathematician, astronomer, and astrolabe maker. The 13th-century biographer al-Qifṭī reports that Ṣāghānī was an expert in geometry and cosmology (*ʿilm al-hay'a*) and was the inventor and maker of instruments of observation. He had a number of students in Baghdad. He was also one of the outstanding astronomers at the observatory (*bayt al-raṣd*) built by order of the Būyid ruler Sharaf al-Dawla (982–989) at the extremity of the garden of the royal palace.

The Sharaf al-Dawla Observatory was the first in the history of Islam to have official status of some kind. According to al-Qifṭī, its program included the observation of the seven planets. This task was entrusted by Sharaf al-Dawla to **Wījan ibn Rustam al-Kūhī**, the director (*ṣāḥib*) of the observatory and the leader of the astronomers working at the institution in 988. One of the project's achievements was the observation of the Sun's entrance into two signs (the sign of Cancer and about three months later the sign of Libra). Two official documents were drawn up to testify to the accuracy of the procedures, and Ṣāghānī was one of the signatories.

According to **Bīrūnī**, Ṣāghānī used a ring with subdivisions into 5 min and diameter of 6 *shibr*, *i. e.*, about 145 cm, for the determination of the obliquity of the ecliptic and also for measuring the latitude of Baghdad. The date of the observation is given as 984/985, and the site is specified as "Birka Zalal" in western Baghdad. Bīrūnī also mentions that Ṣāghānī determined the lengths of the seasons using similar methods.

Ṣāghānī is frequently associated with a determination of the obliquity of the ecliptic by an observation using a 21-ft. quadrant in the year 995. However, this observation with a quadrant of a very similar size has also been attributed to Ṣāghānī's contemporary, the great astronomer and mathematician **Abū al-Wafā' al-Būzjānī**, who died in 997 or 998. As Ṣāghānī died in 990, the latter attribution must be the correct one.

Ṣāghānī's work on the astrolabe, entitled *Kitāb fī kayfiyyat tasṭīḥ al-kura ʿalā saṭḥ al-asṭurlāb*, was dedicated to ʿAḍūd al-Dawla (977–983). In this treatise in 12 sections, Ṣāghānī describes his own method, which he claims to be new, of projecting the sphere onto the plane of the astrolabe. With this technique, conic sections (ellipse, parabola, and hyperbola), in addition to points, straight lines, and circles, are formed by taking as the "pole of projection" not one of the poles but some other point on the line joining them. In his book *Kitāb fī istīʿāb al-wujūh al-mumkina fīṣanʿat al-asṭurlāb*, Bīrūnī states that no one can deny that Ṣāghānī is the inventor of this projection. Ṣāghānī seems to have encouraged Bīrūnī to develop a special type of projection, the orthographic or cylindrical.

Ṣāghānī's treatise, *Risāla fī al-sāʿāt al-maʿmūla ʿalā ṣafā'iḥ al-asṭurlāb*, of which only the first chapter is extant, deals with the circular arcs that represent the hour lines on an astrolabe plate. Ṣāghānī states that many people in his time believed that these arcs pass through the projections of the north and south points. With a very clear and practically oriented explanation, he then proves that on astrolabe plates for the temperate latitudes the circular arcs for the ends of the first, second, and third seasonal hour cannot all pass through the projections of the north and south points.

Ṣāghānī also wrote a work in three parts on planetary sizes and distances.

Roser Puig

Selected References

Al-Qifṭī, Jamāl al-Dīn (1903). *Ta'rīkh al-ḥukamā'*, edited by J. Lippert, p. 79. Leipzig: Theodor Weicher.

Hogendijk, Jan P. (2001). "The Contributions by Abū Naṣr ibn ʿIrāq and al-Ṣāghānī to the Theory of Seasonal Hour Lines on Astrolabes and Sundials." *Zeitschrift für Geschichte der Arabisch-Islamischen Wissenschaften* 14: 1–30. (Hogendijk gives an edition, translation, and commentary of Ṣāghānī's only extant chapter from his *Risāla fī al-sāʿāt al-maʿmūla ʿalāṣafā'iḥ al-asṭurlāb*.)

Lorch, Richard (1987). "Al-Ṣaghānī's Treatise on Projecting the Sphere." In *From Deferent to Equant: A Volume of Studies in the History of Science in the Ancient and Medieval Near East in Honor of E. S. Kennedy*, edited by David A. King and George Saliba, pp. 237–252. *Annals of the New York Academy of Sciences*, Vol. 500. New York: New York Academy of Sciences. (Reprinted in Lorch, *Arabic Mathematical Sciences*, XVII. Aldershot: Ashgate, 1995.) (Study of the *Kitāb fī kayfiyyat tasṭīḥ al-kura ʿalā saṭḥ al-asṭurlāb*.)

Puig, Roser (1996). "On the Eastern Sources of Ibn al-Zarqālluh's Orthographic Projection." In *From Baghdad to Barcelona: Studies in the Islamic Exact Sciences in Honour of Prof. Juan Vernet*, edited by Josep Casulleras and Julio Samsó. Vol. 2, pp. 737–753. Barcelona: Instituto "Millás Valicrosa"de Historia de la Ciencia Árabe.

Sayılı, Aydın (1960). *The Observatory in Islam*. Ankara: Turkish Historical Society.

Sezgin, Fuat. *Geschichte des arabischen Schrifttums*. Vol. 5, *Mathematik* (1974): 311; Vol. 6, *Astronomie* (1978): 217–218. Leiden: E. J. Brill.

Saha, Meghnad N.

Born **Seoratali near Dacca, (Bangladesh), 6 October 1893**
Died **near New Delhi, India, 16 February 1956**

Indian theoretical physicist and astrophysicist Meghnad Saha is eponymized in the Saha equation, which permits calculation of the degree of ionization in a gas that is at a well-defined temperature and density. It is of enormous importance in analyzing the spectra of stars and nebulae and permitted **Cecilia Payne-Gaposchkin** to show that the stars consist primarily of hydrogen and helium.

Meghnad Saha was educated in local schools in Dacca – a private one after participation in a nationalist demonstration caused loss of his scholarship at a government school. He enrolled at Presidency College in 1911, with a small scholarship, awarded after Satyen Bose, a lifelong collaborator, and he applied in person. Saha received an M.Sc. in 1915, having always been an outstanding student, and was appointed as a lecturer, first in mathematics and then in physics, at the University of Calcutta in 1916. Several papers over the next 2 years on the spectrum of the Sun and the quantum theory of light earned him a D.Sc. from Calcutta in 1918, where he and Bose also prepared English translations of papers on special and general relativity by **Albert Einstein** and **Hermann Minkowski**.

Before leaving Calcutta, Saha had already begun the work for which he is remembered. This was completed, under scholarships and fellowships awarded from India, at Imperial College, London, and in the laboratory of **Walter Nernst** in Berlin. It resulted in a series of six papers, published in 1920/1921, in which he laid out a theory of ionization of gases and applied it to the spectra of the Sun and stars. Saha's thinking had been guided by a 1916 discussion of the dissociation of molecules by J. Eggert, and he drew an analogy to conclude that the ratio of the number of ionized atoms to neutral atoms of some particular element would depend both on the number of electrons present (more electrons favoring neutral atoms) and, exponentially, on the ratio of the amount of energy required to ionize the atoms to the temperature of the gas (higher temperature favoring ionized atoms).

Saha returned to Calcutta University as Khaira Professor of Physics in 1921, moving to the University of Allahabad as professor and head of the physics department in 1923, and returning once again to Calcutta as Palit Professor of Physics in 1938, from which position he retired in 1953. Saha continued to work on a variety of topics in physics and astronomy, including radiation pressure, spectroscopy, molecular dissociation, radioactivity, ionospheric physics, and the solar corona, but the remainder of his life's work really focused on teaching, service to the scientific community of India, and, finally, public service. He was instrumental in the founding of the organizations now known as the National Academy of Sciences, the Indian National Science Academy, the Indian Physical Society, and the Saha Institute of Nuclear Physics. He was elected a member of the Indian Parliament as an independent candidate in 1952. Saha's career in many ways was a mirror of the growth of scientific research and progress in India, and he died *en route* to the Office of the Planning Commission. Saha was elected a fellow of the Royal Society (London) in 1927 and has a lunar crater named for him. He married Radha Rani in 1918; they had three sons and three daughters. His equation is probably as famous outside of astronomy as in, because it is also applicable to fusion plasma, flames, explosions, and partly ionized gases in many other contexts.

The Meghnad Saha Archive is at the Saha Institute of Nuclear Physics, Calcutta.

Yatendra P. Varshni

Selected References

Anon. (1921). "Ionisation in Stellar Atmospheres." *Nature* 108: 131.

Anon. (1921). "Rubidium in the Sun." *Nature* 108: 291.

Basu, Jayanta (ed.) (1994). *The Glittering Spectrum of Meghnad Saha*. Calcutta: Saha Institute of Nuclear Physics.

Council of Scientific Industrial Research (1969). *Collected Scientific Papers of Meghnad Saha*. New Delhi: Council of Scientific Industrial Research.

Karmohapatro, S. B. *Meghnad Saha*. New Delhi: Publications Division, Government of India.

Kothari, D. S. (1959). "Meghnad Saha." *Biographical Memoirs of Fellows of the Royal Society* 5: 217–236.

Raman, V. V. (1975). "Saha, Meghnad." In *Dictionary of Scientific Biography*, edited by Charles Coulston Gillispie. Vol. 12, pp. 70–71. New York: Charles Scribner's Sons.

Saha, Meghnad N. (1920). "Elements in the Sun." *Philosophical Magazine* 40: 809–824.

——— (1920). "Ionization in the Solar Chromosphere." *Philosophical Magazine* 40: 472–488.

——— (1921). "On a Physical Theory of Stellar Spectra." *Proceedings of the Royal Society of London* A 99: 135–153.

——— (1921). "On the Problems of Temperature Radiation of Gases." *Philosophical Magazine* 41: 267–278.

Sen, S. N. (ed.) (1954). *Professor Meghnad Saha: His Life, Work and Philosophy*. Calcutta: Meghnad Saha Sixtieth Birthday Committee.

Venkataraman, G. (1995). *Saha and His Formula*. Hyderabad: Universities Press.

Sahl

➔ **Ibn Sahl: Abū Saʿd al-ʿAlāʾ ibn Sahl**

Ṣāʿid al-Andalusī: Abū al-Qāsim Ṣāʿid ibn abī al-Walīd Aḥmad ibn ʿAbd al-Raḥmān ibn Muḥammad ibn Ṣāʿid al-Taghlibī al-Qurṭubī

Born **Almería, (Spain), 1029**
Died **Toledo, (Spain), July or August 1070**

Ṣāʿid al-Andalusī was a Muslim historian, historian of science and thought, and mathematical scientist with an especial interest in astronomy. Given the near-total loss of his astronomical writings, his claim to recognition in science largely rests on his encouragement and possibly patronage – in his capacity as a well-placed functionary at the Toledan court – of a group of young, precision instrument makers and scientists, the most renowned of whom was Azarquiel (*i. e.*, **Zarqālī**). The precise extent of his involvement in the compilation of the *Toledan Tables* – widely disseminated in Latin Europe during subsequent centuries – remains uncertain, owing to the *Tables*' deficient manuscript tradition and to the fragmentariness of biobibliographic data.

Following in the footsteps of his paternal family, Ṣāʿid pursued the career of a legal official, having received a solid education in the Islamic religious disciplines; in 1068, the Dhannūnid Berber amīr of Toledo, al-Ma'mūn Yaḥyā (reigned: 1043–1075), appointed Ṣāʿid chief religious judge (*qāḍī*) of Toledo, an office his father had held earlier and that he himself was to fill until his death. His civil life thus did not stand out from among many of his contemporaries of similar background. What set him apart was his interest in history, history of science, and science itself, especially astronomy; here it may be recalled that in the present context "science" refers to what in premodern Islam often was termed "the ancient disciplines," *viz.* the syllabus of Aristotelian philosophy, logic, medicine, the mathematical sciences (including astronomy), and the occult disciplines, *i. e.*, alchemy, astrology, and magic.

The only work of Ṣāʿid's to survive intact is what has often been called his "history of science": *Al-taʿrīf bi-ṭabaqāt al-umam* (Exposition of the generations of nations) of 1068. The "nations" here intended are those said to have had a disposition toward the cultivation of learning, such as, Indians, Persians, Chaldeans, Egyptians, Greeks, al-Rūm ("Byzantines" and other Christians), Arabs, and Jews (in contrast to the others not so disposed, *i. e.*, Chinese, Turks, and Berbers). Of his other three nonextant works, he cites two there: *Jawāmiʿ akhbār al-umam min al-ʿArab wa-'l-ʿAjam* (Compendious history of nations – Arab and non-Arab) and *Maqālāt ahl al-milal wa-'l-niḥal* (Doctrines of the adherents of sects and schools). These appear to have treated historical subjects, whereas the third one, *Iṣlāḥ ḥarakāt al-kawākib wa-'l-taʿrīf bi-khaṭa' al-rāṣidīn* (Rectification of planetary motions and exposition of observers' errors) adumbrated the astronomical activity of the remaining 2 years of his life, after completion of *Generations*. In *Generations*, Ṣāʿid's view of history and of the progress of scholarship and science from their earliest appearance among (or revelation to?) humankind up to his own country of al-Andalus (Muslim Iberia) and generation has drawn considerable scholarly attention during the last decade-and-a-half, without the issue of his actual beliefs having been convincingly settled. In particular, Ṣāʿid's seeming "pessimism" concerning the cultivation of learning and science among his fellow countrymen has called for comment, given the fact that by that time he and Azarquiel must have been engaged in observations for a number of years and the apparent quickening of astronomical activities in his very hometown of Toledo immediately after the completion of *Generations*, for which the name Azarquiel has taken on nearly emblematic status.

As indicated earlier, extant sources provide but disappointingly fragmentary testimony on astronomical activity in Toledo between 1068, the date of Ṣāʿid's *Generations*, and Azarquiel's less than voluntary move to Cordova *circa* 1080 because of unsettled conditions under al-Ma'mūn's dissolute grandson Yaḥyā al-Qādir. Thus Ṣāʿid's personal contribution to the observations and research as represented by sections of the *Toledan Tables* cannot be determined exactly except in the cases of planetary motions (including the length of the solar year) and the theory of trepidation; one may not stray far from reality in assuming that the title of his treatise *Rectification of Planetary Motions and Exposition of Observers' Errors* suggests the focus of his astronomical interests and of his contribution to the *Toledan Tables*. Relative ignorance of current relevant scholarship in the Islamic East was a shared Andalusī feature in Ṣāʿid's lifetime, as evidenced not merely in *Generations* but as demonstrated far more graphically by the *Toledan Tables* themselves.

Lutz Richter-Bernburg

Selected References

Llavero Ruiz, Eloísa (1987). "Panorama cultural de Al'Andalus según Abū l-Qāsim Ṣāʿid b. Ahmad, cadí de Toledo." *Boletín de la Asociación Española de Orientalistas* 23: 79–100.

——— (trans.) (2000). *Historia de la filosofía y de las ciencias o libro de las categorías de las naciones (Kitāb ṭabaqāt al-umam)*. Madrid: Trotta.

Martinez-Gros, Gabriel (1985). "La clôture du temps chez le cadi Sāʿid, une conception implicite de l'histoire." *Revue de l'Occident musulman et de la Méditerranée* 40: 147–153.

——— (1995). "Ṣāʿid al-Andalusī." In *Encyclopaedia of Islam*. 2nd ed. Vol. 8, pp. 867–868. Leiden: E. J. Brill.

Pedersen, Fritz Saaby (2002). *The Toledan Tables: A Review of the Manuscripts and the Textual Versions*. Copenhagen: Kongelige Danske Videnskabernes Selskab.

Richter-Bernburg, Lutz (1987). "Ṣāʿid, the *Toledan Tables*, and Andalusī Science." In *From Deferent to Equant: A Volume of Studies in the History of Science in the Ancient and Medieval Near East in Honor of E. S. Kennedy*, edited by David A. King and George Saliba, pp. 373–401. *Annals of the New York Academy of Sciences*, Vol. 500. New York: New York Academy of Sciences.

Ṣāʿid al-Andalusī (1912). *Kitāb Ṭabaqāt al-umam*, edited by P. Louis Cheikho. Beirut: Imprimerie Catholique. French translation with notes by Régis Blachère as *Livre des catégories des nations*. Paris: Larose, 1935.

——— (1985). *Kitāb Ṭabaqāt al-umam*, edited by Ḥayāt Bū ʿAlwān. Beirut.

Salem, Semaʿan I. and Alok Kumar (trans. and eds.) (1991). *Science in the Medieval World: "Book of the Categories of Nations,"* by Ṣāʿid al-Andalusī. Austin: University of Texas Press. Pb. ed. 1996. (English translation, to be used with caution.)

Samsó, Julio (1992). *Las ciencias de los antiguos en al-Andalus*. Madrid: Mapfre, pp. 148–150.

——— (1994). *Islamic Astronomy and Medieval Spain*. Aldershot: Variorum.

St. John, Charles Edward

Born **Allen, Michigan, USA, 15 March 1857**
Died **Pasadena, California, USA, 26 April 1935**

American solar physicist Charles St. John made the first, not entirely successful, attempt to measure the gravitational redshift of light coming from the Sun, and compiled a definitive table of wavelengths of solar spectral features. St. John earned a BS from Michigan State College (1887), and an MA (1893) and Ph.D. (1896) from Harvard University, the latter in physics, with a thesis on electric spark spectra. He also studied at the universities of Michigan (1890–1892) and Berlin (1894–1895). St. John initially held teaching positions at the Michigan Normal College (1886–1892) and University of Michigan (1896–1897) before being appointed to an associate professorship in physics and astronomy at Oberlin College, Ohio, in 1897. He became full professor in 1899 and dean of the College of Arts and Sciences in 1906. Both Michigan Normal College and Oberlin College eventually awarded St. John honorary degrees. Among the Oberlin College students who followed him into astronomy was **Alfred Joy**. In the winter of 1908, **George Hale** set his sights on St. John to fill the position of solar physicist at his new Mount Wilson Observatory in California.

St. John had some difficulty with the decision to leave his administrative and teaching responsibilities, but within the year had succumbed to the attractions of working with Hale and

Walter Adams. He held a staff position at Mount Wilson from 1908 to 1930 and a research associateship from retirement until the time of his death. St. John was the first Mount Wilson staff member to die.

The most substantial product of St. John's years at Mount Wilson was the 1930 revision of **Henry Rowland**'s Table of Solar Lines, which increased the number of chemical elements identified in the Sun from 36 to 51. He also made extensive observations of solar rotation, *via* Doppler shifts, and showed that the solar photosphere, reversing layer, and lower chromosphere rotate at different speeds. Adams and St. John attempted an analysis of the atmosphere of Mars in 1925, reporting more water vapor and molecular oxygen than were revealed by later work.

St. John's spectrograms of the Sun, with high resolution in both wavelength and position, readily confirmed the effect associated with the name of **John Evershed**, in which gas flows outward from sunspot centers. He also found that the direction of flow reverses above the spots, confirming a 19th-century model for their structure due to **Angelo Secchi**.

Most delicate of all was the observational search for a slight redshift of solar absorption lines expected because the light would lose energy in climbing out of the solar gravitational field. The amount of shift predicted by **Albert Einstein**'s theory of general relativity (and indeed by Newtonian gravity) is only 0.02 Å for a line at 5,000 Å, or the equivalent of motion at 1.2 km/s. The rising and falling convective currents throughout the solar atmosphere, as well as the Evershed flows, are of comparable size, and St. John concluded in 1917 that the gravitational shift was not there. Later, looking at a larger number of lines, he believed that it could be separated out from other line shifts and was of about the expected size. This is now known to be the case, but the definitive result is generally attributed to much later work.

St. John was elected to the United States National Academy of Sciences and several other honorary positions as well as serving as president of the commissions of the International Astronomical Union on standard wavelengths and on solar physics for several 3-year terms each. He was active in the Congregational Church, in community affairs (the Oberlin Water Works Board, for instance), and was an enthusiastic amateur of tennis, golf, and billiards.

Katherine Haramundanis

Selected References

Christianson, Gale E. (1995). *Edwin Hubble: Mariner of the Nebulae*. New York: Farrar, Straus and Giroux.

Hearnshaw, J. B. (1986). *The Analysis of Starlight: One Hundred and Fifty Years of Astronomical Spectroscopy*. Cambridge: Cambridge University Press.

Henstchel, Klaus (1993). "The Conversion of St. John: A Case Study on the Interplay of Theory and Experiment." *Science in Context* 6, no. 1: 137–194.

Joy, Alfred H. (1935). "Charles Edward St. John." *Popular Astronomy* 43: 611–617.

Lang, Kenneth and Owen Gingerich (eds.) (1979). *Source Book in Astronomy and Astrophysics, 1900–1973*. Cambridge, Massachusetts: Harvard University Press.

Leverington, David (1995). *A History of Astronomy from 1890 to the Present*. London: Springer.

Waterfield, Reginald L. (1938). *A Hundred Years of Astronomy*. New York: Macmillan.

Salih Zeki

Born **Istanbul, (Turkey), 1864**
Died **Istanbul, (Turkey), 1921**

Salih Zeki was one of the most important mathematicians of the late Ottoman period. He was the founder of the mathematics, physics, and astronomy departments of Istanbul University and was also one of the first modern Turkish scholars to undertake research on the history of science in Turkey. After the death of his parents, his grandmother sent him to Dārüssafaka (school for orphans) when he was ten. After graduating first in his class in 1882, Salih Zeki was assigned to the Post and Telegraph Ministry (Administration). In 1884, the ministry decided to train expert cable engineers and physicists in Europe, and so he, along with several of his friends, was sent to Paris. After studying electrical engineering at the École Polytechnique in Paris, Salih Zeki returned to Istanbul in 1887 and started working at his former workplace as an electrical engineer and inspector. At the same time, he taught physics and chemistry at the Faculty of Political Sciences (1889–1900). He also served as the director of the observatory (1895) and as a member of the board of the Ministry of Education (1908). After the declaration of the Second Constitutional Government, Salih Zeki was appointed in 1910 as the principal of the Galatasaray High School. In 1912, he became Under Secretary of the Ministry of Education and in 1913 the president of Istanbul University. In 1917, he resigned as the president but continued to be a professor at the University in the Faculty of Sciences until his death.

Salih Zeki played an important role in the construction and administration of the new State Observatory (Rasadhane-i Amire), this approximately 300 years after the establishment of an observatory in Istanbul in 1575 by **Taqī al-Dīn.** With the support of the French government, an observatory was opened in Istanbul in 1868, whose purpose was to disseminate weather forecasts to other meteorological centers *via* cable. Aristide Coumbary (Coumbary Efendi), who had come to Turkey to develop the telegraph cable network, was appointed as the director. This observatory, which is the forerunner of today's Kandilli Observatory, sent Coumbary Efendi as the Ottoman delegate to the International Meteorological and Astronomical Congress that was held in Vienna in 1873; in accordance with decisions taken at the congress, official ties were established with other observatories in Europe. Every year, weather forecast summaries and reports on earthquakes that occurred in Ottoman territories were published based on the observations made at this observatory. Approximately ten meteorological stations were affiliated with this observatory when it was first established, and these stations reported their daily observations *via* cable to the observatory. The central office in Istanbul forwarded these observations, also *via* cable, to observatories in Paris, Berlin, Vienna, Saint Petersburg, and Hungary and received their reports in the same manner. At the same time, these data were entered on synoptic maps on a daily basis. The observatory council, comprising three persons, also undertook to determine time, longitudes and latitudes, and magnetic declination.

After Coumbary, Salih Zeki was appointed as the director of the observatory. After Salih Zeki's appointment as the president of Istanbul University, the observatory moved to Maçka, to the building facing the Artillery School. On 12 March 1909, during the Young Turk revolution, the observational equipment and seismographs at

Maçka were mostly destroyed. What was salvaged was later given to Kabatas High School. Now Under Secretary to the Ministry of Education, Salih Zeki recommended **Mehmed Fatin Gökmen**, one of the leading scientists at the time, to be director of the observatory. Assuming his duties in 1910, Gökmen was charged with establishing a new observatory; this was accomplished in 1911 with the building of the Kandilli Observatory, which is still in operation today.

Among Salih Zeki's main works in astronomy are a *New Cosmography* (Istanbul, 1915) and an *Abridged Cosmography* (Istanbul, 1916). He also wrote a basic physics textbook, *Hikmet-i Tabiiyye* (Istanbul, 1896), that explained the concepts of general and applied physics and was used as one of the basic textbooks in physics education in Turkey for many years. In history of science, he composed the *Asar-ı Bakiye,* which was written to extol the successes of Muslim scientists, particularly in the fields of mathematics and astronomy. It contains accounts of the historical development of mathematics, algebra, geometry, and astronomy. Salih Zeki wrote this five-volume book by using the works of Western historians of science such as J. E. Montucla, P. Tannery, and M. Cantor as well as original texts in the libraries of Istanbul. The first volume, which deals with plane and spherical geometry, and the second volume, which takes up algebra, were published in 1913/1914; however, his third, fourth, and fifth volumes, which deal with astronomy, were not published. His *Kamus-i Riyaziyat* (Dictionary of Mathematics), whose ostensible purpose was to provide a dictionary of terms for mathematics and astronomy, was also meant to introduce the biographies and works of mathematicians and astronomers. The first two volumes out of the 12 volumes of this work were published, but the other ten volumes remain in draft form. Finally, it is worth mentioning that Salih Zeki also wrote articles for a number of newspapers and magazines that introduced readers to scientific and history of science topics.

Hüseyin Gazi Topdemir

Selected References

Adıvar, A. (1982). Adnan. *Osmanlı Türklerinde İlim*. Istanbul.

——— (1992). "Salih Zeki ve Asar-ı Bakiye." *Bilim Tarihi*, no. 11: 3–8.

Bursalı, Mehmed Tahir (1923). *Osmanlı Müellifleri*. Vol. 3, pp. 279–281. Istanbul, 1342 H.

İhsanoğlu, Ekmeleddin *et al.* (1997). *Osmavnlı Astronomi Literatürü Tarihi (OALT)* (History of astronomy literature during the Ottoman period). Vol. 2, pp. 707–709 (no. 546). Istanbul: IRCICA.

Saraç, Celal (1992). "Salih Zeki Bey'in Bazı Makaleleri." *Bilim Tarihi*, no. 7: 3–9.

——— (2001). *Salih Zeki Bey Hayatı ve Eserleri*, edited by Yesim Isıl Ülman. Istanbul.

Tekeli, S., E. Kahya, M. Dosay, R. Demir, H. G. Topdemir, and Y. Unat (2001). *Bilim Tarihine Giris*. Ankara.

Samarqandī: Shams al-Dīn Muḥammad ibn Ashraf al-Ḥusaynī al-Samarqandī

Born **Samarqand, (Uzbekistan)**
Died **1302**

Shams al-Dīn al-Samarqandī, who lived in the 13th century, wrote books on *kalām* (theology), logic, mathematics, and astronomy; his works were taught for many centuries in the *madrasas* (schools) throughout the Islamic world.

Little is known about his life. After studying the standard curriculum in the basic religious sciences, Samarqandī mastered *kalām*, (logic, and geometry). His works in these fields cover the standard material of Hellenistic and Islamic knowledge, but they also contain contributions that are original both in content and method. One of the most striking features of his works is that they set forth the idea of a universe based upon geometrical forms. In this sense, he can be regarded as the founder of the movement that might be named "geometrical" *kalām* in the Islamic world.

In the field of theoretical astronomy, Samarqandī wrote a (commentaroy) *sharḥ* on **Naṣīr al-Dīn al-Ṭūsī**'s *taḥrīr* (Recension) of **Ptolemy**'s *Almagest.* He also wrote a general astronomy book, no longer extant, reportedly entitled *al-Tadhkira fī ʿilm al-hayʾa*. Finally, he prepared the *ʿAmāl al-taqwīm li-ʾl-kawakīb al-thābita*, which was a star calendar for the year 1276–1277. Unfortunately, most of Samarqandī's astronomical works have not been studied yet.

Samarqandī was most influential for his various textbooks, which provided a wealth of information about the content and methods of past scholars and greatly influenced future generations, who studied these books in various *madrasas*. His geometrical work entitled *Ashkāl al-taʾsīs* contains 35 propositions from Euclid's *Elements*; the first 30 propositions are strictly geometrical, while the last five deal with what has been called "geometrical algebra." Regarding the problem of the fifth ("parallels") postulate, he supported Euclid and considered the criticisms of earlier Islamic mathematicians to have been misplaced. The most important aspect of the book was Samarqandī's view that a study of geometry was a propaedeutic to the study of the forms of Platonic philosophy. It was used as a "middle-level" textbook for Muslim scholars in the *madrasas*, later most often with **Qāḍīzāde**'s commentary. Samarqandī also wrote widely used textbooks in the fields of *kalām*, logic, rhetoric, and philosophy.

İhsan Fazlıoğlu

Selected References

Al-Samarqandī, Shams al-Dīn (1985). *al-Ṣaḥaʾif al-ilāhiyya*, edited by Aḥmad ʿAbd al-Raḥmān al-Sharīf. Kuwait: Moktabat al-Fatāh. pp. 13–28.

Bağdadlı İsmail Paşa (1955). *Hadiyyat al-ʿārifīn*. Vol. 2, p. 106. Istanbul: Milli Egition Bahaanlīge Yayinlare.

Bingöl, Abdulkuddüs (1991). "Shams al-Din Muhammad b. Ashraf al-Samarqandi ve Qistas al-Afkar'ı." *Edebiyat Bilimleri Arastırma Dergisi* 19: 173–182.

Brockelmann, Carl. *Geschichte der arabischen Litteratur*. 2nd ed. Vol. 1 (1943): 615–617; Suppl. 1 (1937): 849–850. Leiden: E. J. Brill.

Dilgan, H. (1960). "Démonstration du Ve postulat d'Euclide par Shams-ed-Din Samarqandi." *Revue d'histoire des sciences et de leurs applications* 13: 191–196.

——— (1975). "Al-Samarqandī." In *Dictionary of Scientific Biography*, edited by Charles Coulston Gillispie. Vol. 12, p. 91. New York: Charles Scribner's Sons.

Kātib Čelebī. *Kashf al-ẓunūn ʿan asāmī al-kutub wa-ʾl-funūn*. Vol. 1 (1941), cols. 39–40, 105; Vol. 2 (1943), cols. 1074, 1075, 1326. Istanbul: Milli Egition Bahaanlīge Yayinlare.

Qurbānī, Abū al-Qāsim (1986/1987). *Zindagī-nāmah-i riyāḍīʾdānān dawrah-i Islāmī*. Tehran: Markaz-i Nasr-i Danişgah, pp. 275–288.

Suwaysī, Muḥammad (ed.) (1984). *Ashkāl al-taʾsīs li-ʾl-Samarqandī (with the Sharḥ of Qāḍīzāde al-Rūmī)*. Tunis: al-Dār al-Tunîsiyya, pp. 23–26.

Samaw'al: Abū Naṣr Samaw'al ibn Yaḥyā ibnʿAbbās al-Maghribī al-Andalusī

Flourished **(Iraq), 12th century**
Died **Marāgha, (Iran), 1174/1175**

Samaw'al was an eminent mathematician, physician, and astronomer, who composed some 85 treatises, all in Arabic. He was from a cultivated Jewish family that was originally from the Maghrib or, according to some sources, from al-Andalus. His father migrated to Baghdad and settled there. The young Samaw'al studied Hebrew, mathematics, and medicine. He traveled in the Muslim east, eventually settling in Marāgha in northwestern Iran, which was then a major city. He spent the rest of his life there as a physician in service of Jahān Pahlawān (died: 1186) of a semi-independent minor dynasty, the Atābakān. There he converted to Islam and wrote a book against Judaism, which became very controversial.

His main astronomical work is *Kashf ʿawār al-munajjimīn wa-ghalaṭihim fī akthar al-aʿmāl wa-'l-aḥkām* (Exposure of the deficiencies of the astronomers and their errors in most of [their] operations and judgments), written in 1165/1166. This treatise is divided into 25 (chapters) *bābs*, each consisting of several (sections) *faṣls*, in which he indicates the errors that he has found in the astronomical works of Greek scientists, such as Euclid, **Archimedes**, and **Apollonius**, of Islamic scientists such as **Ibrāhīm ibn Sinān**, **Abū Jaʿfar al-Khāzin**, **Bīrūnī**, **Abū Maʿshar**, **Ḥabash**, **Ṣūfī**, and **Ibn al-Haytham**, and of Indian scientists such as **Brahmagupta**. The titles of the chapters are as follows:

(1) The reason for composing this book;
(2) On finding altitudes by astrolabe;
(3) On finding altitudes by shadow;
(4) On sines;
(5) On observations;
(6) On calendars;
(7) On interpolation;
(8) On finding hour-angles from equal hours;
(9) On equation of time;
(10) On daily hours;
(11) On ascensions;
(12) On projection of rays;
(13) On latitudes of planets;
(14) On aphesis;
(15) On true horizons;
(16) On finding heights of mountains and other high objects;
(17) On positions of fixed stars;
(18) On the nature of planets;
(19) On animodars;
(20) On elections (of proper times);
(21) On oblique ascensions;
(22) On the times of conjunctions, syzygies, and transfers;
(23) On properties of inscribed polygons and their effects on the sublunar world;
(24) On syzygies of epicycles; and
(25) Types of indications.

In the last chapters (20–25), Samaw'al uses a type of philosophical argument based upon his previous chapters to explain his view regarding the effects of stars on terrestrial events. He concludes that because the stars are innumerable and the relations and effects among them are virtually incalculable, an astrologer would need to take into consideration 6,817 variables for each person, therefore making it impossible to predict the future in any meaningful way.

Samaw'al was perhaps best known for his work in mathematics, especially algebra and arithmetic. He also wrote on medicine.

Negar Naderi

Selected References

Berggren, J. L. (1986). *Episodes in the Mathematics of Medieval Islam*. New York: Springer Verlag, pp. 112–118.

Rosenfeld, B. A. and Ekmeleddin Ihsanoğlu (2003). *Mathematicians, Astronomers, and Other Scholars of Islamic Civilization and Their Works (7th–19th c.)*. Istanbul: IRCICA, pp. 184–186.

Sampson, Ralph Allen

Born **Schull, Co.Cork, Ireland, 25 June 1866**
Died **Bath, England, 7 November 1939**

British astronomer Ralph Sampson made his mark with an analysis of the dynamics of the interactions of the four large (Galilean) satellites of Jupiter. As Astronomer Royal for Scotland he also encouraged major instrumental innovations, including the development of the Shortt Free Pendulum Clock and the use of microphotometers.

Sampson was the fourth of five children of James Sampson from Cornwall and Sarah Anne (*née* Macdermott) Sampson, an Irishwoman of Huguenot descent. When he was five, the family moved to Liverpool, England, and suffered from deprivation when the father became ill and his investments in the Cornish tin mines failed. As a result, Sampson had little education until the age of 14, when he entered the Liverpool Institute. He won a scholarship to Saint John's College, Cambridge, where his tutor was **John Adams**, and he graduated as third wrangler in the mathematical *tripos* of 1888. Sampson then took up a lectureship in mathematics in King's College, London and in 1889 was awarded the first Smith's Prize and Fellowship of his college in Cambridge. He returned to Cambridge in 1890 and became the first holder of the newly established Isaac Newton Studentship in Astronomy and Physical Optics. Sampson worked for 2 years on astronomical spectroscopy with H. F. Newall and in 1893 published a paper "On the Rotation and Mechanical State of the Sun." This was a highly significant publication as it showed for the first time the importance of radiation compared to convection in the outward transport of heat generated in the Sun's interior.

In 1893, Sampson was appointed professor of mathematics in the Durham College of Science at Newcastle upon Tyne. Two years later, he moved to the chair of mathematics in Durham itself and became director of Durham Observatory. Sampson's interest in this observatory led to the installation of the Durham almucantar, an instrument in which transits of stars were observed across

a horizontal circle instead of a vertical wire in the meridian. The instrument attracted much interest, and it was used for some years for observations of the variation of latitude.

It was in Durham that Sampson undertook his greatest work, the dynamical theory of the four largest satellites of Jupiter. At that time, there were serious discrepancies between the theoretical predictions and actual observations of the four satellites. Sampson used a series of accurate observations from Harvard College Observatory to amend the existing theory of the satellite orbits, but the disagreement between theory and observation persisted. He worked out a new dynamical theory and published in 1910 *Tables of the Four Great Satellites of Jupiter*, giving the positions of the satellites from 1850 to 2000. His *Theory of the Four Great Satellites of Jupiter* appeared in 1921 and earned him the Gold Medal of the Royal Astronomical Society in 1928.

Quite a different task that Sampson worked on at Durham was the editing of the unpublished manuscripts of his old tutor, Adams, for Cambridge University Press. These were published as *The Scientific Papers of John Couch Adams*. Sampson's varied achievements were recognized by his election to the Royal Society in 1903.

In 1910, Sampson was appointed Astronomer Royal for Scotland and professor of astronomy in the University of Edinburgh. During his tenure of 27 years in Edinburgh, he made notable contributions in three main areas: the determination of time, the optical performance of telescopes, and objective methods for photometry and spectrophotometry of stars.

Sampson recognized that an observatory's clock was one of its most important instruments and deserved proper attention. At the Royal Observatory, he introduced a system for monitoring the performance of the clocks to an accuracy of one thousandth of a second. Sampson improved the temperature control of the clock chamber, and he installed radio equipment for comparing time signals from clocks in other institutions. His interest in clocks led to several substantial papers on the subject in the publications of the Royal Society of Edinburgh.

Among the clocks in the observatory was one better than all the others. It had been designed by a civil engineer, W. H. Shortt in association with the Synchronome Company. This Shortt free pendulum clock was so accurate that it could detect for the first time small irregularities in the rotation of the Earth. Shortt clocks were adopted as the standard timekeepers in many observatories until they were replaced by quartz clocks. Sampson's fundamental contributions to precise time determination were recognized by his election as the first president of the Commission de L'Heure, the international organization founded to study the problems of astronomical timekeeping.

When Sampson tried to bring into use an old 24-in. reflector at the Royal Observatory, an old interest in theoretical optics was revived. His studies of optical aberrations resulted in two papers in the *Philosophical Transactions of the Royal Society* (1913, 1914). In the latter of these, he suggested that the optical aberrations of a Cassegrain reflector could be reduced by inserting a pair of suitable lenses in the outgoing beam. Sampson suggested a similar approach for correcting the field of a Newtonian reflector. These innovative ideas were later developed by others to good effect.

In an effort to make some use of the 24-in. reflector where its poor image quality would not matter, Sampson decided in 1915 to use it for photoelectric photometry of stars using alkali metal detectors that had recently been developed in Germany. Most of the laboratory work to support this project was carried out by E. A. Baker. In 1920, the program was modified by replacing direct measurement of each star at the telescope by microphotometry of the densities of star images on photographic plates. This method was extended to scanning the spectra of stars, and the recording microphotometer became a standard instrument for stellar photometry.

Sampson applied the forgoing techniques to the analysis of objective prism spectra with a view to determining the spectral distribution of intensity of various types of stars. This led to estimates of stellar temperatures in a range of spectral types from B0 to M0. These results were published by the Royal Society of Edinburgh in 1925 and 1928.

Sampson's desire to renew the equipment of the Royal Observatory was frustrated by World War I and its aftermath. It was only in 1936 that a 36-in. reflector made by Grubb Parsons was installed in the East Dome, and a versatile Hilger spectrograph was added the following year. This new equipment allowed the spectrophotometric program to be extended to much fainter stars.

In 1937, failing health compelled Sampson to retire at the age of 71. He and his wife Ida (*née* Binney), whom he had married in 1894, settled in Bath. Sampson was survived by his wife, a son, and four daughters.

Sampson was deeply involved in the affairs of the Royal Society of Edinburgh, being a member of council for 20 years including some years as general secretary. He served as president of the Royal Astronomical Society of London from 1915 to 1917. Sampson was awarded the honorary degrees of Sc.D. from Durham and LLD from Glasgow. The International Astronomical Union named the lunar crater at 29°.7 N and 16°.5 W in his honor.

Ian Elliott

Selected References

Brück, Hermann A. (1983). *The Story of Astronomy in Edinburgh from Its Beginnings until 1975*. Edinburgh: Edinburgh University Press.

Greaves, W. M. H. (1940). "Ralph Allen Sampson." *Monthly Notices of the Royal Astronomical Society* 100: 258–263.

Sampson, Ralph A. (1895). "On the Rotation and Mechanical State of the Sun." *Memoirs of the Royal Astronomical Society* 51: 123–183.

——— (ed.) (1896–1900). *Scientific Papers of John Couch Adams*. Vol. 2, pt. 1. Cambridge: University Press.

——— (1909). "A Discussion of the Eclipses of Jupiter's Satellites 1878–1903." *Annals of the Astronomical Observatory of Harvard College* 52: 149–343.

——— (1910). "On the Old Observations of the Eclipses of Jupiter's Satellites." *Memoirs of the Royal Astronomical Society* 59: 199–256.

——— (1910). *Tables of the Four Great Satellites of Jupiter*. London: Published by the University of Durham.

——— (1921). "Theory of the Four Great Satellites of Jupiter." *Memoirs of the Royal Astronomical Society* 63.

Sanad ibn ʿAlī: Abū al-Ṭayyib Sanad ibn ʿAlī al-Yahūdī

Flourished **Baghdad, (Iraq), 9th century**

Sanad ibn ʿAlī was an active mathematician and astronomer in Baghdad during the 9th century and worked as an astrologer for Caliph **Ma'mūn**. Sanad was the son of a Jewish astrologer who worked in Baghdad and counted among his clients people from the ʿAbbāsid court. Sanad converted to Islam responding to the lure exercised by the caliph.

In his youth, Sanad studied by himself several scientific books, among them the *Almagest*. He tried to gain access to the illustrious circle of scholars around **ʿAbbās ibn Saʿīd al-Jawharī** (first half of the 9th century), who regularly met in his house to discuss the latest scholarly and social news. But being merely 20 years old at this time proved to be an obstacle. According to a story told by Aḥmad ibn Yūsuf ibn al-Dāya (died: *circa* 952) on the authority of Abū Kāmil Shujāʿibn Aslam (*circa* 850–*circa* 930), Sanad convinced Jawharī of his superior knowledge of the *Almagest*. As a result, Sanad was not only permitted to stay and take part in the talks of the illustrious circle, but Jawharī, who was a companion of the caliph, also introduced him to Ma'mūn and recommended him as a new, promising servant.

Sanad wrote four mathematical texts on algebra, Indian arithmetic, mental calculation, and Euclidean irrational quantities, the latter being one of the earliest commentaries on Book X of Euclid's *Elements*. He composed a *zīj* (astronomical handbook) and explained a method for determining the circumference of the Earth by observations of the Sun. There is also a report by **Bīrūnī** in his *The Determination of the Coordinates of Cities* (Ali, 1967, pp. 185–186) that Sanad had found the size of the Earth by measuring the dip of the horizon from the summit of a high mountain, a method later used to good effect by Bīrūnī himself; this had been done "in the company of Ma'mūn when he made his campaign against the Byzantines." His *zīj* is presumably lost, and thus it is unclear how it was related to the famous so called *al-Zīj al-mumtaḥan* (The verified *zīj*) produced by a group of astronomers from Ma'mūn's court.

Sanad built and headed an observatory behind the Bāb Shammāsiyya in Baghdad, collaborating there with a group of observers. According to an account of the Egyptian astronomer **Ibn Yūnus** of the astronomical excursions carried out by the court astronomers in Ma'mūn's lifetime, Sanad had himself written such an account in which he claimed to have participated in one of these expeditions. However, R. Mercier, and following him D. King, doubt the authenticity of both these claims.

Sonja Brentjes

Selected References

Aḥmad ibn Yūsuf al-Kātib (1975). *Kitāb al-Mukāfa'a*. Beirut: Dār al-waḥda.

Ali, Jamil (trans.) (1967). *The Determination of the Coordinates of Cities: Al-Bīrūnī's* Taḥdīd al-Amākin. Beirut: American University of Beirut.

Ibn al-Nadīm (1970). *The Fihrist of al-Nadīm: A Tenth-Century Survey of Muslim Culture*, edited and translated by Bayard Dodge. 2 Vols. New York: Columbia University Press.

King, David (2000). "Too Many Cooks ... A New Account of the Earliest Muslim Geodetic Measurements." *Suhayl* 1: 207–241.

Mercier, Raymond P. (1992). "Geodesy." In *The History of Cartography*. Vol. 2, bk. 1, *Cartography in the Traditional Islamic and South Asian Societies*, edited by J. B. Harley and David Woodward, pp. 175–188. Chicago: University of Chicago Press.

Rosenfeld, B. A. and Ekmeleddin Ihsanoğlu (2003). *Mathematicians, Astronomers, and Other Scholars of Islamic Civilization and Their Works (7th – 19th c.)*. Istanbul: IRCICA, pp. 28–29.

Sanford, Roscoe Frank

Born **Faribault, Minnesota, USA, 6 October 1883**
Died **Pasadena, California, USA, 4 April 1958**

American observational astronomer Roscoe Sanford was a firm, early supporter of the idea of "island universes" or spiral nebulae as independent stellar systems outside the Milky Way, based on data he had collected concerning their apparent sizes, brightnesses, and locations relative to the galactic plane.

Sanford received an A. B. from the University of Minnesota in 1905 and, after a year of high school teaching in Minnesota, came to Lick Observatory as an assistant to Richard H. Tucker. From 1908 to 1915, he was with Lick expeditions at San Luis, Argentina, determining positions of southern stars, and at Santiago, Chile, measuring radial velocities. Sanford returned to Lick in 1915 and completed a Ph.D. dissertation in 1917, working with **Heber Curtis**. The thesis used images of spiral nebulae, plus the idea (established in the earlier thesis of **Edward Fath**) that their spectra resemble spectra of star clusters, to conclude that they are separate galaxies, well outside the Milky Way. He attributed the absence of spirals near the galactic plane to absorption there. Much of the data used by Curtis in the 1920 Curtis–Shapley debate came from Sanford's thesis.

Sanford spent about a year at Dudley Observatory before joining the scientific staff at Mount Wilson Observatory in 1918, where he remained until his 1950 retirement. While at Mount Wilson, Sanford worked primarily on the composition and motion of carbon stars, making use of the high-resolution spectra that were possible only with the world's largest telescope. He was the first to notice the great strength of the 6707 Å line of lithium in T Tauri stars, which is a signature of their youth, and of older stars having fused all their lithium to other elements.

Virginia Trimble

Selected References

Anon. (1958). "R. F. Sanford Dies." *Sky & Telescope 17, no.* 8: 406.

Osterbrock, Donald E., John R. Gustafson, and W. J. Shiloh Unruh (1988). *Eye on the Sky: Lick Observatory's First Century*. Berkeley: University of California Press, pp. 189–190.

Santini, Giovanni-Sante-Gaspero

Born **Caprese Michelangelo, (Tuscany, Italy), 30 January 1787**
Died **Noventa Padovana near Padua, Italy, 26 June 1877**

Giovanni-Sante-Gaspero Santini was a professor, observatory director, and specialist in comet orbits. He was the third of 11 children of Gerolamo and Caterina Brizzi. His uncle, the priest Giovambattista Santini, taught him Latin, grammar, philosophy, and mathematics. In 1801, Santini entered the seminary of Prato to complete his education, and in 1802 enrolled in law at the University of Pisa. At the same time, he attended free courses in mathematics and physics. The university rector, Lorenzo Pignotti, and the politician Vittorino Fossombroni (both from Arezzo), to whom Santini's uncle had recommended him, employed him at the observatory of the Museum in Florence. To learn astronomy, he was sent to the Observatory of Brera in Milan, directed by **Barnaba Oriani**. There Santini learned how to observe and calculate orbits of minor planets from the astronomer **Francesco Carlini** while studying French, English, and German. In 1806, as a consequence of the changing political situation in Tuscany, Santini was appointed assistant astronomer under abbé Vincenzo Chiminello at the Observatory of Padua. The new Napoleonic government of the Veneto region had received Santini's excellent references supplied by professors Angelo Cesaris and Oriani. Santini found a tired and ill director in Chiminello, who had to work hard to save the *specola* during the events following the fall of the Republic of Venice. Santini started by updating the old and obsolete instruments. Then, in 1810, he purchased a transit instrument. In 1815, he calculated the precise latitude of Padua with his new Reichenbach repeating circle, and in 1822 he bought an equatorially mounted telescope.

In 1810, Santini had married Teresa Pastrovich, who died in 1843. The following year, he married Adriana Conforti, who outlived him, but, like Teresa Pastrovich, left him childless. Despite his two marriages, some biographical dictionaries call Santini "abbé," perhaps owing to the cassock he used to wear when in Pisa, or because he was confused with his uncle Giambattista.

In January 1813, Santini was appointed full professor of astronomy at the University of Padua. In 1817, after Chiminello's death, his chair was confirmed by the Austrian government, which had ruled Venice since November 1813. Santini was also appointed director of the observatory. In 1837, he installed a meridian circle, and in 1838 he started the observations that would lead him to make a star catalog of the Bessel zones between declinations +10° and −10°. Santini was able to carry out this long and laborious work with the aid of astronomer Virgilio Trettenero. After 10 years, the work was published as the *Cataloghi Padovani*, and was valued by the astronomical community for the precision of its stellar coordinates. Santini had undertaken such heavy work because he needed many comparison stars for orbit calculations. In the 1864 Encke–Galle catalog of cometary orbits, he was attributed the calculation of 17 orbits, among the many others he had published in other scientific journals. In particular, Santini calculated the orbit of the short-period comet 3D/Biela very precisely. This comet had been discovered in 1826, but was not recovered at its 1839 perihelion transit because its orbit had been greatly disturbed by the major planets, especially Jupiter. Its return in 1846 was seen thanks to Santini's exact ephemerides, calculated and published in 1842. Santini's works were well known throughout Europe, and the Observatory of Padua became famous among European ones for its research in theoretical and practical astronomy.

As astronomy professor, Santini published the two-volume treatise *Elementi di Astronomia* (1819/1820 and a 2nd edition in 1830). This work was extensively used by the famous astronomers Baron **János von Zach** and **Johann von Littrow**, and became a fundamental textbook for Italian students in the 19th century. In 1828, he published an optics treatise in two volumes (*Teorica degli Stromenti Ottici*), the only such book in Italy, which also became a milestone for students, and optical instrument makers, too. Santini taught astronomy and, as substitute professor, algebra and geometry, as well as infinitesimal calculus for nine and seven years respectively.

Santini was rector of the university in 1824/1825 and 1856/1857, and dean of the Faculty of Mathematics from 1845 to 1872. From 1866 to 1875, he was mayor of Noventa Padovana, a village near Padua where he spent the last years of his life.

Santini corresponded with many Italian and European astronomers, among whom are **George Airy, Giambattista Amici, Friedrich Argelander, Wilhelm von Biela, Francesco Carlini, John Herschel, Joseph** and **Karl von Littrow, Barnaba Oriani, Heinrich Schumacher**, Zach, **Friedrich Struve, Otto Struve**, and many others. He personally knew many of the astronomers who visited him in Padua such as John Herschel, Karl von Littrow, Otto and Wilhelm Struve, and Baron von Zach. Others he met in the autumn of 1843, during his journey across Germany with Roberto De Visiani, professor of botany and director of the Botanical Garden at the University of Padua.

Santini was a member of 21 Italian and foreign scientific societies, among which are the Academy of Padua; Royal Astronomical Society; Institut de France; Kaiserliche Akademie der Wissenschaften of Vienna; Istituto Veneto di Scienze, Lettere. ed Arti, *etc.* Nine orders of knighthood were bestowed upon him. Santini's works were published in many journals of the time.

Luisa Pigatto

Selected References

Lorenzoni, Giuseppe (1877). *Giovanni Santini, la sua vita e le sue opere.* Padua: Tipografia del Seminario.

Pigatto, Luisa (1996). "Giovanni Santini." In *Professori di materie scientifiche all'Università di Padova nell'Ottocento*, a cura di Sandra Casellato e Luisa Pigatto, pp. 35–40. Trieste: Edizioni Lint.
Poggendorff, J. C. (1863). "Santini." In *Biographisch-literarisches Handwörterbuch*. Vol. 2, cols. 749–750. Leipzig.
Stein, J. (1913). "Santini, Giovanni Sante Gaspero." In *Catholic Encyclopedia*, edited by Charles G. Herbermann. Vol. 13, pp. 462–463. New York: Encyclopedia Press.

Śatānanda

Flourished **Ujjayinī, (Ujjain, Madhya Pradesh, India), 1099**

Śatānanda was an acclaimed Indian astronomer of the Hindu classical period (late 5th to 12th centuries). He was the son of Saṅkara and Sarasvatī. Śatānanda was the author of an extremely popular astronomical manual, the *Bhāsvatī*, composed in 1099. In its eight short chapters, this work sets out the methodologies for preparing the daily almanac. But the author presents several variations from other texts of the same genre. He names his work after the word for the Sun, *bhāsvān*. In tune with his own name Śatānanda, meaning "delighting in hundreds," he uses the centesimal system for commencing the epochal position and specifies several multipliers and divisors in terms of hundreds. Again, unlike most manuals that are based on the more modern yet anonymous *Sūryasiddhānta* (10th or 11th century), he specifies that his work is drawn from the older *Sūryasiddhānta*, which had been condensed by astronomer **Varāhamihira** in his work, the *Pañcasiddhāntikā*.

Śatānanda also introduces several innovations to make the results computed by the *Bhāsvatī* more correct. For the computation of longitudes of the mean positions of planets, he uses not the *ahargaṇa* (number of elapsed days) but the *varṣagaṇa* (number of elapsed years). In specifying the beginning of the year, he uses the true *Meṣādi* (first point of Aries) and not the mean *Meṣādi* as in other texts. The positions of the Sun, Moon, and other planets are not stated in terms of *rāśis* (signs) but in terms of the *nakṣatras* (asterisms). He adopts the year 528 as the reference for precession measurements and the rate of precession as 1′ per year.

Śatānanda's work has given rise to a number of expository commentaries.

Ke Ve Sarma

Selected References

Chattopadhyay, Anjana (2002). "Satananda." In *Biographical Dictionary of Indian Scientists: From Ancient to Contemporary*, p. 1228. New Delhi: Rupa.
Misra, Ramajanma (ed.) (1985). *Bhāsvatī*. With Sanskrit and Hindi commentaries by Matruprasad Pandeya. Varanasi, Bharata: Caukhambha Samskrta Sansthana.
Pingree, David (1975). "Śatānanda." In *Dictionary of Scientific Biography*, edited by Charles Coulston Gillispie. Vol. 12, pp. 115–116. New York: Charles Scribner's Sons.
——— (1978). "History of Mathematical Astronomy in India." *Dictionary of Scientific Biography*, edited by Charles Coulston Gillispie. Vol. 15 (Suppl. 1), pp. 533–633. New York: Charles Scribner's Sons.

Saunder, Samuel Arthur

Born **London, England, 18 May 1852**
Died **Oxford, England, 8 December 1912**

English mathematician and amateur astronomer Samuel Saunder was a leading selenographer at the beginning of the 20th century and helped create both a standard system of lunar nomenclature and an accurate system of lunar coordinates. Educated at Cambridge University as a mathematician, Saunder spent his entire career as professor of mathematics at Wellington College, Berkshire.

However, Saunder's great passion was astronomy, especially the study of the Moon. Attracted to the problem of measuring the exact locations of the lunar features, Saunder used both a micrometer and photographic plates to determine the position of Möstig A, the Moon's fundamental point. His measurements – to within 0.1″ – were fifty times more accurate than those any previous observer had obtained. Saunder then measured positions for over 3,000 other central lunar formations relative to Möstig A.

After the invention of the telescope, the naming of lunar features became a source of confusion and great discord among astronomers. Saunder became acutely aware of this problem while measuring the 3,000 reference points and advocated an international committee to blend the various naming conventions that had been introduced over a three-year period. With the strong recommendations of the Royal Astronomical Society and the Royal Society, such a committee was formed in 1907 under the auspices of the International Association of Academies. Saunder was appointed to the committee and, along with **Julius Franz**, was given the task of constructing an accurate map, using the measurements that the both of them had made.

Unfortunately the deaths of both Saunder and Franz, along with the advent of World War I, prevented the successful conclusion of the project, and the committee collapsed. A successful resolution to the lunar nomenclature problem was not achieved until after the formation of the International Astronomical Union in 1919.

Leonard B. Abbey

Selected References

Blagg, M. A. and S. A. Saunder (1913). *Collated List of Lunar Formations Named or Lettered in the Maps of Neison, Schmidt, and Mädler*. Edinburgh: Neill and Co.
Saunder, S. A. (1911). "The Determination of Selenographic Positions and the Measurement of Lunar Photographs." *Memoirs of the Royal Astronomical Society* 60, pt. 1.
Turner, H. H. (1907). "Lunar Nomenclature." *Monthly Notices of the Royal Astronomical Society* 68: 134–135.

Saunders, Frederick Albert

Born **London, Ontario, Canada, 18 August 1875**
Died **South Hadley, Massachusetts, USA, 9 June 1963**

Canadian-American spectroscopist Frederick Saunders gave his name to a method, devised with **Henry Norris Russell**, called Russell–Saunders coupling, used to calculate how the electrons in atoms with more than one in the outer, active shell interact to produce the emission and absorption features. It is also called L–S coupling, where L and S are symbols for the orbital and spin angular momenta of the electron ensemble.

Saunders was the son of the eminent Canadian agriculturist, William Saunders, and Sarah Robinson, both of whom emigrated from England. He married Grace Elder in 1900; they had two children, Anthony E. and Margery (Middleton). He married Margaret Tucker in 1925.

Saunders was a physics student at the University of Toronto, taking his BA in 1895. A student of **Henry Rowland** at Johns Hopkins University, Saunders took his Ph.D. there in 1899. After a brief stay at Haverford College as physics instructor (1899–1901), he moved to Syracuse University (1901), becoming an associate professor (1903) and professor (1905).

In 1913/1914, Saunders ended his time at Syracuse with a year in Europe, working in Cambridge and in Tübingen, in the latter with Louis Paschen. Paschen was the 1908 discoverer of the first series of infrared lines of hydrogen (the Paschen series). With Ernst Back (1881–1959), Saunders studied the effect (now generally called the Paschen-Bach effect) on spectral features of magnetic fields stronger than those studied by **Pieter Zeeman**. From Paschen, Saunders learned the methods of laboratory spectroscopy, and he identified large numbers of lines with the atoms and ions responsible, especially among the alkali metals (lithium, sodium, potassium, *etc.*). While in Cambridge, he worked with **Alfred Fowler**.

From 1914 to 1918, Saunders served as professor of physics at Vassar College. He spent several months in Washington during 1918, in a group directed by **Robert Millikan**, for the National Research Council.

In 1919, Saunders joined the physics department at Harvard University at the invitation of Theodore Lyman, becoming associate professor in 1921 and professor in 1923. He retired as professor emeritus in 1941. During his retirement years, Saunders lectured at Mount Holyoke College.

Saunders's pioneering work with Russell concerned the spectra of the divalent alkaline metals (beryllium, magnesium, calcium, *etc.*). Each of these has two electrons in its outer, active shell, and Russell–Saunders coupling, which fits their spectra well, assumes that the orbital angular momenta and the spin angular momenta of the two electrons interact dominantly, with smaller interaction between the total spin and total orbital angular momenta. The opposite case, in which each electron sees primarily itself, is called jj coupling and is appropriate for elements like iron where the electrons are further from the nucleus. Saunders was elected to the National Academy of Sciences in 1925.

In later years, Saunders turned to acoustical research, particularly the acoustics of the violin. In 1930, he published a basic college textbook in physics that was widely used for about a decade.

Richard A. Jarrell

Selected Reference

Olson, Harry F. (1967). "Frederick Albert Saunders." *Biographical Memoirs, National Academy of Sciences* 39: 403–416.

Savary, Felix

Born **Paris, France, 4 October 1797**
Died **Estagel, Pyrénées-Orientales, France, 15 July 1841**

A student and then faculty member at the École Polytechnique, Felix Savary was the first to compute the orbit of a binary star in 1827, showing that the orbit is elliptical and, therefore, that **Isaac Newton**'s laws of gravity apply outside the Solar System.

The double star ξ Ursa Majoris is made up of bright components with a 60-year period; it was an ideal candidate for applying Newton's law of gravitation to the stars. **William Herschel** had discovered the pair (1780), and **Friedrich Struve** had more recently measured it, but it was Savary who just beat **John Herschel** in making the calculation (published in *Connaisance des Temps pour l'an 1830*).

Savary is better known to physicists as the colleague of André-Marie Ampère and to mathematicians for describing the geometrical figure known as the *roulette*.

Selected Reference

Aitken, Robert Grant (1935). *The Binary Stars*. New York: McGraw-Hill Book Company, Inc.

Savile, Henry

Born **Bradley, Yorkshire, England, 30 November 1549**
Died **Eton, Berkshire, England, 19 February 1622**

Henry Savile is known today primarily for his endowment of the Savilian Chair of Geometry and Chair of Astronomy at Oxford; in his day he was also noted for an Oxford series of lectures on the *Almagest*.

Savile, son of Henry and Elizabeth (*née* Ramsden) Savile, matriculated at Brasenose College, Oxford, *circa* 1561, and in 1565 he became a fellow of Merton College. Savile established his scholarly reputation in the 1570s with a brilliant series of lectures on the *Almagest*. The lectures are impressive in their use of ancient, medieval, and Renaissance sources – notably including **Nicolaus Copernicus** – to elucidate the text of **Ptolemy**'s classic work. Savile was ambivalent about the controversy over the Copernican theory. "Is it not all one," he famously replied to a colleague who asked about the movement of the Earth, "sitting at dinner whether my table be brought to me, or I goe [*sic*] to my table, so I eat my meat?" Nonetheless, his lectures did much to revitalize the teaching of mathematics in Oxford.

From 1578 to 1582, Savile toured the Continent, visiting a number of European astronomers and scholars. After returning from his travels, he was appointed Greek tutor to Elizabeth I. Handsome and eloquent, Savile proved to be a masterful courtier. His qualities won him academic preferments. In 1585, he was elected warden of Merton College, and 10 years later, despite considerable obstacles, he became provost of Eton while retaining his Merton post. Savile was an autocratic if effective administrator; under his leadership both Merton and Eton enjoyed an academic resurgence. Savile's own academic pursuits were as much historical and philological as they were astronomical. His later scholarship centered on ancient texts – The *Histories of Tacitus*, the works of Saint Chrysostom, and portions of the authorized version of the Bible. Yet he maintained a strong belief in the value of mathematical science, generously endowing the Savilian Chairs of Geometry and Chair of Astronomy at Oxford University in 1619. These professorships have been historically significant, having been held by **Christopher Wren**, **David Gregory**, **James Bradley**, **Charles Pritchard**, **Herbert Turner** (astronomy), and **Edmond Halley** (geometry) among others.

Keith Snedegar

Selected References

Carr, William (1921–1922). "Savile, Sir Henry." In *Dictionary of National Biography*, edited by Sir Leslie Stephen and Sir Sidney Lee. Vol. 17, pp. 856–859. London: Oxford University Press.

Fauvel, John, Raymond Flodd, and Robin Wilson (eds.) (2000). *Oxford Figures: 800 Years of the Mathematical Sciences*. Oxford: Oxford University Press.

Goulding, Robert (1995). "Henry Savile and the Tychonic World-System." *Journal of the Warburg and Courtauld Institutes* 58: 152–179.

Savile, Henry. "Notes for lectures on the *Almagest*." Savile MSS 29, 31, and 32. Bodelian Library, Oxford.

Sawyer Hogg, Helen Battles

Born **Lowell, Massachusetts, USA, 1 August 1905**
Died **Richmond Hill, Ontario, Canada, 28 January 1993**

American-Canadian astronomer Helen Sawyer Hogg for many years maintained the definitive catalog of variable stars in globular clusters, a task of considerable importance because these stars are keys to measuring astronomical distances and ages. Helen Sawyer was the daughter of Edward Everett Sawyer and Carrie Myra Sprague. She married **Frank Hogg** in 1930 and F. E.L. Priestly in 1985. The first marriage produced three children, Sally, David E. (a noted radio astronomer), and James.

Sawyer Hogg received an AB *magna cum laude* from Mount Holyoke College in 1926. She earned an AM in 1928 and a Ph.D. in 1931 from Radcliffe College, though her thesis work had been done with **Cecilia Payne-Gaposchkin** and **Harlow Shapley** at Harvard College Observatory, as Harvard did not at that time award science degrees to women. She, Frank Hogg, and Payne earned three of the first five astronomy Ph.D.s based on work at Harvard. Her early work with Shapley helped to establish both the direction and distance from the Sun to the center of the Milky Way Galaxy, both of which had been in dispute earlier in the century.

During this time Sawyer Hogg was an instructor at Smith College (1927) and at Mount Holyoke (1930/1931). When Frank Hogg obtained a post at the Dominion Astrophysical Observatory in 1931, the director, **John Plaskett** allowed Helen Hogg to observe with the 72-in. reflector. In 1935, the Hoggs moved to Toronto in anticipation of the opening of the David Dunlap Observatory [DDO]. At Toronto, Sawyer-Hogg was a research assistant until World War II. In 1940/1941, she was acting chair of astronomy at Mount Holyoke. With most of the male staff of the DDO in the military, Hogg began teaching in the University of Toronto Astronomy department (1941); she was later appointed assistant professor (1951), associate professor (1955), and professor (1957). Sawyer-Hogg was program director for astronomy for the National Science Foundation (1955/1956). She retired as professor emeritus in 1976.

Sawyer Hogg received the Annie J. Cannon Medal of the American Astronomical Society in 1950, the Rittenhouse Medal in 1967, the Royal Astronomical Society of Canada Service Award in 1967, the Dorothea Klumpke-Roberts Award from the Astronomical Society of the Pacific in 1983, and the Sandford Fleming Medal of the Royal Canadian Institute in 1985. A companion in the Order of Canada, she was also a Fellow of the Royal Society of Canada. Long active in the Royal Astronomical Society of Canada (president, 1957–1959; honorary president, 1977–1981), the American Association of Variable Star Observers (president, 1939–1941), the American Astronomical Society (councillor 1965–1968), and the International Astronomical Union, she also became the first president of the Canadian Astronomical Society (1971) and was a president of the Royal Canadian Institute (1964/1965). Sawyer Hogg held honorary degrees from Mount Holyoke and Waterloo, McMaster, Toronto, Saint Mary's, and Lethbridge universities.

Sawyer-Hogg was an international authority on variable stars in globular clusters. She published three catalogs of these objects in 1939, 1955, and 1973 and was the author of more than 200 papers. Sawyer-Hogg was also well known in Canada as a science popularizer – her syndicated column for the *Toronto Daily Star* was one of the most noteworthy Canadian science columns for 30 years. She later wrote a popular work, *The Stars belong to Everyone* (1976) and appeared on television programs. Sawyer Hogg continued to participate actively in astronomical conferences and share her knowledge with colleagues almost until the end of her life.

Richard A. Jarrell

Alternate name

Hogg, Helen Battles

Selected References

Clement, Christine and Peter Broughton (1993). "Helen Sawyer Hogg, 1905–1993." *Journal of the Royal Astronomical Society of Canada* 87: 351–356.

Jarrell, Richard A. (1988). *The Cold Light of Dawn: A History of Canadian Astronomy*. Toronto: University of Toronto Press.

Schaeberle [Schäberle] John [Johann] Martin

Born **Oeschelbronn, (Baden-Württemberg, Germany), 10 January 1853**
Died **Ann Arbor, Michigan, USA, 17 September 1924**

German-American instrument-designer and observational astronomer John Schaeberle was one of the first astronomers at Lick Observatory and designer of the famed Schaeberle Camera. He was

the first to see the white-dwarf companion of Procyon in 1896, predicted by **Friedrich Bessel** in 1844.

Schaeberle's father, Anton Schäberle, a master saddle maker, and mother Christina Katherina Vögele, immigrated to Michigan in 1854 with their infant son. There his name changed to John Martin Schaeberle, and he was usually called Martin. Following his early education in Ann Arbor, he moved at age 15 to Chicago to serve an apprenticeship in a machine shop. There Schaeberle became interested in astronomy, and his mechanical skills enabled him to make mirrors for reflecting telescopes. The Great Chicago Fire in 1871 ended his apprenticeship. Returning to Ann Arbor, he completed high school in a few months and enrolled at the University of Michigan, where he studied engineering and mathematics, graduating in civil engineering in 1876.

Schäberle used telescopes of his own construction to make observations, first from the rooftop of his Chicago hotel, and after 1872, at a private observatory in Ann Arbor. Using a home-built 8-in. reflecting telescope, Schaeberle discovered comet C/1880 G1 from his private observatory. The following year, using the Henry Fitz comet-seeker at the University of Michigan's Detroit Observatory, he discovered comet C/1881 N1. As a student, Schaeberle's mathematical, mechanical, and observational skills caught the attention of **James Watson**, the observatory's director, and he soon became Watson's favorite pupil. After graduation, Schaeberle became Watson's assistant, and 2 years later, was promoted to instructor in astronomy, a position he held until 1888. The German astronomical methods, introduced at Michigan by Watson's teacher and predecessor **Franz Brünnow**, were well-suited to Schaeberle's interests and abilities. When Watson moved to the Washburn Observatory in 1878, Mark W. Harrington took his place. Harrington's primary interest was in meteorology, and so Schaeberle had responsibility for the bulk of the astronomical work of the Detroit Observatory during Harrington's tenure.

In 1888, Schaeberle became one of the inaugural astronomers at the new Lick Observatory, where he remained for 10 years. There he had responsibility for observations with the Repsold meridian circle.

The total solar eclipse of January 1889, which crossed northern California, captured Schaeberle's attention and prompted him to venture to Cayenne, French Guiana, to observe the next eclipse in December 1889. His desire to formulate a mechanical theory of the solar corona (as opposed to the prevailing magnetic theories) prompted him to devise a long-focus camera that could take photographs of the Sun. He designed a photographic telescope of 40-ft. focal length, driven by clockwork, which was portable, so could easily be taken on expeditions. The Schaeberle camera produced the best photographs of the solar corona ever produced, one of which revealed a comet in close proximity to the Sun that would not otherwise have been detected. Schaeberle took his camera to Mina Bronces, Chile, in 1893 and to Akkeshi, Japan, in 1896 to observe eclipses. He organized the expeditions on his own and recruited and trained civilians on location to assist him. The Schaeberle camera continued to be used by astronomers on expeditions at locations around the world, until 1932 when **Heber Curtis** took the last coronal plates at Fryeburg, Maine, USA.

Schaeberle's persistent visual observations led to his discovery in 1896 of the 13th-magnitude companion of Procyon, using the 36-in. Lick refractor. It was only the second white dwarf to be observed (after Sirius B by **Alvan Clark**).

When **Edward Holden** resigned as director of Lick Observatory in 1897, Schaeberle was the natural choice to be acting director, a position he held until 1898 when the Lick Trustees made a political move and appointed **James Keeler** as director. Schaeberle could not accept this perceived injustice, so he returned to Ann Arbor.

Although he held no formal appointment, Schaeberle carried on his astronomical work from an observatory he built at his residence. He constructed a 24-in. telescope with a 3-ft. focal length, mounted equatorially, that he planned to equip with a modified bolometer to detect far infrared radiation from the Sun and stars. Unfortunately, the mirror broke while he was drilling a hole to modify the telescope. Half of the discarded mirror was later retrieved in the 1930s by astronomers at the University of Michigan and used as an off-axis parabola for infrared spectrometry.

Schaeberle never married. He was a founding member and first secretary of the Astronomical Society of the Pacific. He was awarded honorary degrees by the universities of Michigan (1893) and California (1898). The Astronomical Society of the Pacific awarded him a medal for discovery of his third comet on 16 April 1893. Over the course of his career, Schaeberle published more than 100 articles in scientific journals, some of which contained ingenious methods for determining instrumental constants. A lunar crater is named for Schaeberle.

Patricia S. Whitesell

Selected References

Eddy, John A. (1971). "The Schaeberle 40-ft. Eclipse Camera of Lick Observatory." *Journal for the History of Astronomy* 2: 1–22.

Hussey, William J. (1924). "John Martin Schaeberle, 1853–1924." *Publications of the Astronomical Society of the Pacific* 36: 308–313.

Whitesell, Patricia S. (Dec. 2003). "Detroit Observatory: Nineteenth-Century Training Ground for Astronomers." *Journal of Astronomical History and Heritage* 6: 90–92.

Schalén, Carl Adam Wilhelm

Born **Sweden, 11 January 1902**
Died **Sweden, 11 December 1993**

Carl Schalén discovered, independently of **Robert Trumpler**, the general interstellar absorption, but his contribution to this field has been nearly forgotten. He was the son of Claës Adam Schalén and Vivica Ebpa Charlotta Strokirk and, in 1940, married Agnes Carmen Elisabeth Rosenblad.

Schalén received his Ph.D. in astronomy at Uppsala University in 1929 as one of the students of **Östen Bergstrand** and a contemporary of **Bertil Lindblad** and **Knut Lundmark**. Schalén held the position of observator at Uppsala from 1941 to 1955 and was professor of astronomy at Lund University from 1955 until his retirement in 1968. He was elected to the Royal Academy of Sciences, Stockholm, in 1949 and was a member of the council of the European Southern Observatory very early in its history.

Schalén was a member of the group of astronomers at Uppsala University, centered around Bergstrand and Lindblad, which used stellar spectra to determine the distances to large numbers of stars. Luminosity criteria had been found that made it possible to determine the approximate distance to a star from knowledge of certain spectral details. In a series of early works, Schalén determined distances and other data for stars with the spectroscopic techniques developed by Lindblad.

Schalén early on took up an interest in the question of interstellar absorption. Is there general absorption of light as it travels through space, or is interstellar space totally transparent? The existence of localized clouds of obscuring matter had been known earlier; Schalén wanted to look for general absorption. In a paper published in 1929, he announced his finding that there is a general absorption in space that amounts to 0.5 magnitudes per kiloparsec.

These results came at about the same time as other astronomers found similar results. Trumpler at the Lick Observatory published a study in 1930 with a result that was almost identical to Schalén's. (The Schalén archive shows that he pointed out his earlier result to Trumpler, who replied that he did not know of the result when he wrote the 1930 paper.) Schalén also studied the properties of interstellar matter theoretically. He started using the theory of Gustav Mie in the early 1930s for studies of how light diffused as it passed through interstellar matter.

Carl Schalén's papers are at the Lund University library.

Gustav Holmberg

Selected References

Ardeberg, Arne. "Schalén, Carl Adam Wilhelm." In *Svenskt biografiskt lexicon.*

Holmberg, Gustav (1999). *Reaching for the Stars: Studies in the History of Swedish Stellar and Nebular Astronomy, 1860-1940.* Lund: Lund University.

Schall von Bell, Johann Adam

Born **Cologne, (Germany), 1591**
Died **China, 15 August 1666**

Johann Schall von Bell was a Jesuit missionary and astronomer in China who oversaw significant improvements in the Chinese calendar. He entered the Jesuits in 1511 and wxas determined to serve in the China mission. Schall von Bell arrived in Macao in 1619 where he studied Chinese for a few years awaiting permission to enter China, which at that time strongly opposed all foreigners, especially foreign teachers. Permission was granted, however, a few years later in an unusual way after a military attack by the Dutch Calvinists on the Portuguese Catholic settlement. He and other Jesuits helped the Portuguese quell the invasion and, when the story of their victory reached the emperor, he asked the Portuguese to help him fend off the Tartars from the north. In particular he wanted more Jesuits. Thus was Schall von Bell admitted into China; and he then made his way to Beijing, arriving when a minister hostile to Christians was being dismissed. He then took the Chinese name of Tang-Jo-Wang.

Schall von Bell was very energetic and a man of charm and self-confidence. He soon became an intimate friend of some important Chinese scientists who were quite impressed by his learning and his familiarity with astronomy. It was due to these gifts that Schall von Bell and his brother Jesuits were successful in conversing with educated Chinese about religion.

Schall von Bell was given responsibility for the calendar, which was especially important in Chinese culture – in fact, the prestige of the emperor was connected to the authenticity of the court calendar. On one occasion, Schall von Bell and his Jesuit companions were able to predict a solar eclipse that took place on 21 June 1629 more accurately than their Chinese rivals. That success opened the way for them to devote themselves with full energy to the task of calendar reform. At the same time, they were able to produce maps, astrolabes, and other scientific instruments with such effectiveness that these Europeans were eventually invited to establish an observatory within the royal palace. In 1639, the emperor expressed his esteem for Schall von Bell and his Jesuit coworkers when a procession of palace royalty arrived at the Jesuit residence.

Upon the death of the emperor in 1644, a successor was named who appointed Schall von Bell director of the national "Board of Astronomers." He later made him a mandarin, and showered the Jesuit with many favors. In 1661, the latter emperor fell seriously ill and died; he was succeeded by his son, Kang-h'si. In the palace, however, Schall von Bell's position was steadily being undermined because of the jealousy of some royal scientists. Their leader, Yang-Kuan, succeeded in having him and the Jesuits accused of high treason, and of teaching a superstitious religion. This was followed by imprisonment and a trial that resulted in the Jesuits being sentenced to a slow death. But on the day of sentencing, an earthquake intervened, followed by a great fire in the palace, which alarmed the superstitious judges and resulted in the Jesuits being set free.

Schall von Bell died after spending 47 years in China. Soon after his death, the record was righted. The emperor dismissed Yang Kuan and appointed another Jesuit, Father **Ferdinand Verbiest**, as his successor. The emperor restored all Father Schall's honors posthumously, and erected a monument at his grave that read "You leave us your undying fame and the glory of your name." Schall von Bell's tomb as well as those of fellow Jesuits, **Matteo Ricci** and Verbiest, were restored after the Cultural Revolution of this past century and were relocated on the grounds of a Communist training school. These tombs still can be visited today. Another memorial of Schall von Bell is a student hostel, named in his honor, on the campus of the Chinese University of Hong Kong.

Schall von Bell's *Trigonometria* and many other works were written and published in China. He constructed a double stellar hemisphere to illustrate planetary movement and wrote 150 treatises in Chinese on the calendar.

Joseph F. MacDonnell

Alternate name

Tang-Jo-Wang

Selected References

Institutum Historicum (1958). *Archivum Historicum Societatis Iesu*. Rome: Institutum Historicum.

Needham, Joseph (1959). *Science and Civilization in China*. Vol. 3. Cambridge: Cambridge University Press.

Sommervogel, Carlos (1890–1960). *Bibliothèque de la Compagnie de Jésus*. 12 Vols. Brussels: Société Belge de Libraire.

Scheiner, Christoph

Born **Wald (Markt Wald near Mindelheim, Bavaria, Germany), 25 July 1573**

Died **Neisse (Nysa, Poland), 18 June 1650**

Christoph Scheiner was a German mathematician, physicist, and astronomer, who was one of the first to observe sunspots.

After attending the Jesuit Latin school in Augsburg and the Jesuit college at Landsberg, Scheiner entered the Jesuit order in 1593. (In 1617, he was ordained a priest.) From 1598 to 1601, he studied mathematics and metaphysics at the university at Ingolstadt; then he worked (1602–1605) as a teacher of Latin at the Jesuit college in Dillingen. From 1605 to 1609, Scheiner studied theology in Ingolstadt. During 1610–1617, he was professor of mathematics (astronomy) and Hebrew at the university at Ingolstadt, from 1619 to 1620 professor in Innsbruck, and during 1620–1621 professor in Freiburg. In 1621, Scheiner became father confessor of Archduke Karl of Austria and Bishop of Neisse, and in 1622 he founded a Jesuit college in Neisse, and became its superior. From 1624 to 1633, Scheiner was in Rome on behalf of the college. (No details are known about this stay – perhaps there was diplomatic business.) Later he was in Vienna, and in 1636 he returned to Neisse, without resuming the post of principal of the college.

Scheiner's time was the beginning of modern scientific thinking, using experiment and observation, and the period when astronomy was influenced by the ideas of **Nicolaus Copernicus**. Scheiner, first of all, is famous for his discovery of sunspots in 1611. The telescope had been invented, and Scheiner was among the first to use it for astronomical observations. (He produced a telescope specifically for solar observations – the helioscope.) A "stained sun" was in conflict with conservative Christian doctrine, and therefore the Jesuit Scheiner had to be cautious. Thus in 1612, he communicated his observations in three letters written under the pseudonym "Apelles." As a result, a priority dispute with **Galileo Galilei** arose. (Nowadays we know that there were sunspot observations already before Scheiner and Galilei, but all seem to be independent of each other.) During his time in Rome, Scheiner wrote his main work, *Rosa Ursina sive Sol*, where he summed up all his knowledge on sunspots and other solar phenomena. He showed that Galilei made errors of observation, but although he came near to a modern understanding of the nature of sunspots, he followed the Christian doctrine in his book. If Scheiner had an influence on the Galilei prosecution, as sometimes is said, it is not proved.

Another memorable contribution of Scheiner is the invention of the panthograph (around 1603/1605, but published only in 1631), an instrument for copying plans on any scale. He also dealt with the physiological optics of the eye, and he published his results in his book *Oculus* ... in 1619 (and further results also in *Rosa Ursina*). He stated that the retina is the crucial part for the sense of viewing, and he described the function of other parts including the pupil and iris. During his last years, Scheiner wrote a refutation of the Copernican theory, which was published posthumously, but had no influence at all.

Horst Kant

Selected References

Anon. (2000). *Sonne entdecken: Christoph Scheiner 1575–1650*. Begleitbuch zur Ausstellung. Inglostadt: Stadtmuseum Ingolstadt.

Scheiner, Christoph (1619). *Oculus, hoc est, fundamentum opticum, in quo ex accurata oculi anatome ... radius visualis eruitur ...* Oeniponti: Agricola.

——— (1630). *Rosa Ursina sive Sol*. Braccianum.

——— (1631). *Pantographice seu ars delineandi*. Rome.

——— (1995). *Briefe des Naturwissenschaftlers Christoph Scheiner SJ an Erzherzog Leopold V. von Österreich-Tirol 1620–1632*, edited by Franz Daxecker. Innsbruck: Universität Innsbruck.

Braunmühl, Anton von (1891). *Christoph Scheiner als Mathematiker, Physiker und Astronom*. Bayerische Bibliothek. Vol. 24. Bamberg: Brunmih–Buchnersche Verlagsbuchhandiung

Daxecker, Franz (1996). *Das Hauptwerk des Astronomen P. Christoph Scheiner SJ: "Rosa Ursina sive Sol": Eine Zusammenfassung*. Innsbruck: Universitätsverlag Wagner.

——— (2001). "'Über das Fernrohr' und weitere Mitschriften von Vorlesungen Christoph Scheiners." In Vol. 4 of *Beiträge zur Astronomiegeschichte*, edited by Wolfgang R. Dick and Jürgen Hamel, pp. 19–23. Frankfurt am Main: Harri Deutsch.

Moss, Jean Dietz (1993). "The Significance of the Sunspot Quarrel." In *Novelties in the Heavens: Rhetoric and Science in the Copernican Controversy*, pp. 97–126. Chicago: University of Chicago Press.

Shea, William R. (1970). "Galileo, Scheiner, and the Interpretation of Sunspots." *Isis* 61: 498–519.

Scheiner, Julius

Born **Cologne, (Germany), 25 November 1858**
Died **Potsdam, Germany, 20 December, 1913**

Julius Scheiner, along with **Hermann Vogel**, made the first determination of the orbit of an eclipsing binary star (Algol) from photographic observations of its radial velocities (1889), thus confirming the eclipsing hypothesis of that star's light variations. Scheiner's career spanned the late-19th-century transformation of astronomy from its emphasis on positional data and orbital motions to the newly emerging science of observational astrophysics. He made significant contributions to both areas of specialization and trained notable practitioners in the latter.

Scheiner was the son of Jacob Scheiner, a painter of architectural subjects and landscapes. His secondary education was completed at the Realgymnasium in Cologne. He was admitted to the University of Bonn in 1878 and was drawn toward astronomy by visits to its observatory, then directed by **Eduard Schönfeld**. Appointed an assistant at the observatory, Scheiner received his Ph.D. in 1882, for a study of the brightness variations of Algol. He acquired strong experimental and technical skills that were to serve him throughout his scientific career. Scheiner was married and the father of three children.

In 1887, Scheiner was appointed an assistant at the Royal Astrophysical Observatory in Potsdam; he was later named observer (1894) and senior observer (1900). There, he began a program of research on stellar radial velocities, under the direction of Vogel. Concurrently, Scheiner was made extraordinary professor of astrophysics at the University of Berlin and trained a number of researchers, including future Yerkes Observatory director **Edwin Frost**, in the newer spectroscopic methods.

By measuring the intensity of starlight across various wavelengths of its spectrum, the principle of spectrophotometry was born. As early as 1890, Scheiner was among the first to recognize that the so called color index of a star yielded an approximate measure of its surface temperature. In collaboration with **Johannes Wilsing**, Scheiner derived temperature estimates for over 100 stars by the spectrophotometric technique, thereby aiding Vogel's system of stellar spectral classification. Wilsing and Scheiner also made one of the earliest, though unsuccessful, attempts to detect radio waves from the Sun (1896). Other astrophysical investigations were Scheiner's measurements of the effective temperature of the Sun's surface (conducted with a pyrheliometer), his visual study of the intensities of three emission lines in the spectra of gaseous nebulae, and his record of the first absorption lines visible in a spectrogram of the Andromeda Galaxy (1899), which offered important clues to the true nature of this object.

As Vogel's health declined, Scheiner assumed more of the scientific and administrative work of the Potsdam Observatory. He represented that institution at three Astrographic Congresses (1891, 1896, and 1900) convened at Paris in conjunction with the international *Carte du Ciel* project. Between 1889 and 1912, Scheiner compiled six volumes of stellar positions, embracing more than 123,000 stars, in the Potsdam zone between +31° and +40° declination. Other projects that reflected the older style of positional astronomy were Scheiner's triangulation of more than 300 stars and 100 definable points within the Orion Nebula (from which future observers could derive the motions of these components). He also published a catalog of more than 1,500 double stars and tabulated data on their frequencies of occurrence.

In addition to teaching and research, Scheiner was a noted textbook author and popularizer, whose works strongly reflected his own research contributions. Two scholarly works, *Die Spectralanalyse der Gestirne* (The spectral analysis of the stars, 1890) and *Die Photographie der Gestirne* (The photography of the stars, 1897), were complemented by his *Populäre Astrophysik* (Popular astrophysics, 1908; 2nd ed., 1912).

Jordan D. Marché, II

Selected References

Anon. (1914). "Julius Scheiner." *Monthly Notices of the Royal Astronomical Society* 74: 282–284.

Frost, Edwin B. (1915). "Julius Scheiner." *Astrophysical Journal* 41: 1–9.

Hearnshaw, J. B. (1986). *The Analysis of Starlight: One Hundred and Fifty Years of Astronomical Spectroscopy*. Cambridge: Cambridge University Press, esp. pp. 87–89, 151–153, 219–220.

——— (1996). *The Measurement of Starlight: Two Centuries of Astronomical Photometry*. Cambridge: Cambridge University Press, esp. pp. 103–104, 173–174.

McGucken, William (1975). "Scheiner, Julius." *Dictionary of Scientific Biography*, edited by Charles Coulston Gillispie. Vol. 12, pp. 152–153. New York: Charles Scribner's Sons.

Wilsing, J. and J. Scheiner (1982). "On an Attempt to Detect Electrodynamic Solar Radiation and on the Change in Contact Resistance when Illuminating Two Conductors by Electric Radiation." In *Classics in Radio Astronomy*, edited by Woodruff Turner Sullivan, III, pp. 147–157. Dordrecht: D. Reidel.

Scheuchzer, Johann Jakob

Born **Zürich, Switzerland, 2 August 1672**
Died **Zürich, Switzerland, 23 June 1733**

Johann Scheuchzer was a Swiss physician and natural philosopher, who provided one of the first descriptions of the Perseid meteor shower. He was the son of Johann Scheuchzer, the senior town

physician and calendar maker of Zürich. At first Scheuchzer attended the German, and then the Latin school, and subsequently the Zürich (Carolinum) Gymnasium. In 1692, Scheuchzer began studying natural philosophy at the University at Altdorf, near Nuremberg, Germany. A year later, he made his way to Utrecht the Netherlands and gained his medical doctorate there in 1694. During a subsequent stay at Altdorf, Scheuchzer studied mathematics and astronomy under **Georg Eimmart**, occupied himself with botany and anatomy, and collected fossils. After the death of Johann Jakob Wagner (14 December 1695), he was recalled as junior town doctor (or *Poliater*) to Zürich, and as candidate for professor of mathematics at the Carolinum. In 1697, Scheuchzer was elected to the Leopoldina, the Deutsche Akademie der Naturforscher (the German Academy of Naturalists). In the same year he married Susanna, daughter of Kaspar Vogel, a councillor and innkeeper. Further memberships followed (London, Berlin, and Bologna). In 1710, Scheuchzer became professor of mathematics at the Carolinium as well as a canon.

Influenced by John Woodward, Scheuchzer was an advocate of the Neptunist theory. He maintained that the giant salamanders and the vertebrae of saurians that he discovered were the remains of predeluge humans. Between 1702 and 1711, Scheuchzer undertook nine major journeys, during which he thoroughly investigated the natural history of Switzerland. His major works include the classic *Natural History of Switzerland* in Latin with Volumes 1–3 appearing in German in the period 1706–1708. By publishing his *Herbarium diluvianum* (in 1709), one of the first books to contain illustrations of fossil plants, he became one of the founders of palaeobotany.

In 1704, following election to the Royal Society of London, Scheuchzer sent his work for publication in the *Philosophical Transactions* – in 1706, his observation of the total solar eclipse of 12 May and in 1707, his observation of the lunar eclipse of 1706. In a drawing of the extremely high number of shooting stars that he observed on 8 August 1709, Scheuchzer depicted one of the earliest-known accounts of the Perseid meteor shower. His large map of Switzerland that appeared in 1712 was partially based on his own astronomical observations.

In the case of the book *Jobi physica sacra, oder Hiobs Natur-Wissenschaft verglichen mit der heutigen Ideen* (*Jobi physica sacra,* or The natural sciences of the Book of Job, compared with modern-day ideas.) (Zürich, 1721), the censor refused permission to print, unless Scheuchzer removed Copernican teachings and other objectionable material. Scheuchzer had to fall into line. Despite strong resistance by Zürich orthodoxy, he was eventually freely able to advocate the Copernican worldview in his *Kupferbibel*, the *Physica sacra* (1731–1735). In it, Scheuchzer discusses the *eclipsis passionalis*, the eclipse at Christ's Crucifixion, which is described by the Evangelists Matthew, Mark, and Luke, but which is an eclipse that cannot be explained as a natural event. This eclipse cannot have occured as a result of the natural laws of motion, but could only have taken place through God causing them to be violated. Scheuchzer invites comparison between a modern, mechanistic view of the Universe, based on Cartesian ideas, and the revealed truth of the Bible. For him, science serves to clarify when something must be a miracle.

Thomas Klöti
Translated by: *Storm Dunlop*

Selected References

Dürst, Arthur (1978). *Johann Jakob Scheuchzer und die Natur-Histori des Schweitzerlandes*. Zürich: Orell Füssli.

Fischer, Hans (1973). *Johann Jakob Scheuchzer, 2.August 1672–23. Juni 1733: Naturforscher und Arzt*. Veröffentlichungen der Naturforschenden Gesellschaft in Zürich, 175. Zürich: Kommissionsverlag Leemann AG.

Höhener, Hans-Peter (1986). "Johann Jakob Scheuchzer." In *Lexikon zur Geschichte der Kartographie*. Vienna: Franz Deuticke.

Leu, Urs B. (1999). "Geschichte der Paläontologie in Zürich." In *Paläontologie in Zürich – Fossilien und ihre Erforschung in Geschichte und Gegenwart*. Zürich: Zoologisches Museum der Universität Zürich.

Michel, Paul (9 Aug. 1999). "Johann Jakob Scheuchzer und die Sonnenfinsternis von 1706." *Neue Zürcher Zeitung*.

Steiger, Rudolf (1933). "Verzeichnis des wissenschaftlichen Nachlasses von Johann Jakob Scheuchzer (1672–1733)." *Vierteljahresschrift der Naturforschenden Gesellschaft in Zürich* 78.

Schiaparelli, Giovanni Virginio

Born **Savigliano, (Piedmont, Italy), 14 March 1835**
Died **Milan, Italy, 4 July 1910**

Giovanni Schiaparelli was one of the most widely known astronomical observers of the middle to late 19th century, in no small part due to his observations of Mars and their reputed canals. Born to wealthy parents, he was enrolled at age seven in the Gymnasium Lycée of Savigliano. After graduating from the gymnasium in 1850, Schiaparelli entered the University of Turin, where he excelled in applied mathematics. He graduated with honors in August 1854 with a degree in hydraulic engineering and civil architecture. Schiaparelli married Maria Comotti in 1865, and together they parented five children.

Upon leaving the University at Turin, Schiaparelli began teaching mathematics as a private tutor. In 1856, he moved to Berlin to

study astronomy under the guidance of **Johann Encke**. In 1859, Schiaparelli moved yet again, this time to the Pulkovo Observatory, where he studied with **Otto Wilhelm Struve** and **Friedrich Winnecke**. In July 1860, Schiaparelli was appointed to the position of second astronomer under **Francesco Carlini** at Brera Observatory in Milan. Upon Carlini's death, in 1862, Schiaparelli was promoted to director, a post he held until his retirement in 1900. From 1863 to 1872, Schiaparelli also held a professorship at the Royal Technical Institute at Milan, where he taught classes on astronomy, geodesy, and celestial mechanics.

During his tenure at the Milan Observatory, Schiaparelli initiated several productive observational programs. In 1861, he discovered the minor planet (69) Hesperia. With the appearance of a bright comet in 1862 (now recognized as comet 109P/Swift–Tuttle), his attention was turned toward these transitory objects, leading in 1866 to his most important enduring contribution to astronomy. In a series of letters to Father **Angelo Secchi**, Schiaparelli revealed a direct orbital coincidence between comet 109P/1862 O1 and the meteoroids encountered during the annual August meteor shower (the Perseids). His announcement was the first clear demonstration that meteors derive from comets. Schiaparelli outlined his ideas on the origin of comets and meteoroid streams in *Entwurf einer astronomischen Theorie der Sternschnuppen*. While correctly identifying meteoroids as the decay products of comets, Schiaparelli argued, wrongly as it turned out, that all comets and meteoroid streams were captured by the Sun from interstellar space.

In 1877, Schiaparelli began a series of studies of Mars, then at opposition. From these studies, he produced surface albedo maps and suggested that certain features were indicative of the planet having "seas" and "continents." Schiaparelli also believed that he saw linear features or *canali* on the planet's surface. The ambiguity of *canali*, meaning either channels or canals, left open the possibilities of their being either natural features on the Martian landscape or artificial constructions. In the event, Schiaparelli continued to observe Mars at each favorable opposition until 1890.

The apparent observation of "canals" on Mars remained a topic of great interest and controversy into the 20th century. American **Percival Lowell** championed the idea that the "canals" were, in fact, signs of intelligent life existing on Mars, but by 1915, when observations with larger telescopes than Lowell's failed to record the canals, their reality began to be doubted. His defense was that only a visual observer, taking advantage of brief moments of excellent seeing, can record the finest detail present on the martian surface. Modern observations, confirm on the one hand, that visual observers under very good skies will see more or less what Lowell and Schiaparelli saw, but, on the other hand, that there are no truly contiguous canal-like features.

While at the Milan Observatory, Schiaparelli also conducted a series of observations of Saturn, Venus, and Mercury. Between 1877 and 1878, Schiaparelli observed Venus with the aim of deducing its spin period. In contrast to all other observers at that time, Schiaparelli concluded that Venus was a very slow rotator, arguing that its spin period was somewhere between 6 and 9 months, with synchronous rotation being the most likely value (corresponding to a rotation rate of 224.7 days). Schiaparelli also concluded that Venus's spin axis was orientated perpendicularly to its orbit. Modern-day radar telescope studies of Venus have revealed that the planet in fact spins even more slowly than the synchronous rate (its spin period being 243.02 days), and that the planet spins in a retrograde sense, with the spin axis being inclined by 177° to its orbit. From observations of Mercury from 1881 to 1889, Schiaparelli concluded that it was in synchronous rotation around the Sun. This observation was generally accepted until radar measurements revealed in the mid-1950s that the spin-to-orbital-period ratio is not unity, as proposed by Schiaparelli, but actually 3/2. Between 1875 and 1900, Schiaparelli also made a series of observations of double stars.

Upon his retirement from the Milan Observatory in 1900, Schiaparelli devoted himself to the study of Babylonian and Biblical astronomy. In preparation for his theological and historical works, Schiaparelli read original texts in Hebrew, Assyrian, Greek, and Latin. His studies yielded a book on the astronomy of the Old Testament, along with a paper on the astronomical allusions contained in the book of Job. At the time of his death, Schiaparelli was working on a comprehensive review of the history of ancient astronomy; this monumental work was eventually prepared for publication, in three volumes, by his pupil Luigi Gabba in 1925.

Schiaparelli received many prestigious awards in his lifetime. The Royal Astronomical Society granted him its Gold Medal in 1872 for his work on cometary and meteoroid stream orbits. The Astronomical Society of the Pacific bestowed its highest honor, the Bruce Medal, upon him in 1902. He was also elected a fellow of the Royal Academy of Science in Turin, the Royal Astronomical Society, the Royal Society, and both the French and Viennese academies of science. In 1889, Schiaparelli became a senator of the Kingdom of Italy. Lunar, mercurian, and martian geographic features have been named in his honor, as was a minor planet (4062).

Martin Beech

Selected References

Anon. (1910). "Giovanni Virginio Schiaparelli." *Astrophysical Journal* 32: 313–319.

Anon. (1960). *Atti del Convegno per le celebrazioni del cinquantenario della morte di G. V. Schiaparelli, 1-3 ottobre 1960*. Milan.

E. B. K. (1911). "Giovanni Virginio Schiaparelli." *Monthly Notices of the Royal Astronomical Society* 71: 282–287.

Lassell, William (1872). "Address Delivered by the President, William Lassell, Esq., on presenting the Gold Medal of the Society to Signor Schiaparelli." *Monthly Notices of the Royal Astronomical Society* 32 (1872): 194–199. (A review of Schiaparelli's contributions to meteor astronomy.)

Schiaparelli, G. V. (1871). *Entwurf einer astronomischen Theorie der Sternschnuppen*, edited by George von Boguslawski. Stettin: Th. von der Nahmer.

——— (1878–). *Osservazioni astronomiche e fisiche sull'asse di rotazione e sulla topografia del pianeta Marte fatte nella Reale specola di Brera in Milano … Memoria [1-] del socio G. V. Schiaparelli*. Rome: Coi Tipi del Salviucci.

——— (1905). *Astronomy in the Old Testament*. Oxford: Clarendon Press.

——— (1909). *Osservazioni sulle stelle doppie serie seconda comprendente le misure di 636 sistemi eseguite col refrattore equatoriale Merz-Repsold negli anni 1886–1909*. Milan: U. Hoepli.

——— (1925–1927). *Scritti sulla storia della astronomia antica*. Bologna: N. Zanichelli.

——— (1968). *Le opere di G.V. Schiaparelli, pubblicate per cura della Reale specola di Brera*. New York: Johnson Reprint Corp.

Schiaparelli, G. V. and Luigi Gabba (1925). *Le più belle pagine di astronomia popolare*. Milan: U. Hoepli.

Sheehan, William (1988). *Planets and Perception: Telescopic Views and Interpretations, 1609–1909*. Tucson: University of Arizona Press. (The debate concerning the apparent observation of Martian canals has been thoroughly reviewed herein.)

Schickard, Wilhelm

Born **Herrenberg, (Baden-Württemberg, Germany), 22 April 1592**

Died **Tübingen, (Baden-Württemberg, Germany), 23 October 1635**

Wilhelm Schickard invented the first mechanical computer in 1623 to solve problems that arose in predicting planetary positions. His research included mathematics, cartography, and geodesy as well as astronomy.

Son of Lukas Schickard, he was born in a family of master joiners, builders, and vicars. Schickard was educated at the well-known Tübinger Stift and the University of Tübingen. After receiving his BA in 1609, and MA in 1611, he continued to study primarily theology and oriental languages until 1613. He surely received his education in mathematics, physics, and astronomy from **Michael Mästlin**, professor of mathematics and astronomy in Tübingen from 1584 to 1631. In 1613, Schickard became a Lutheran minister at several towns around Tübingen, and in 1619, he was appointed professor of Hebrew at Tübingen University, teaching biblical languages such as Hebrew and Aramaic. His textbook in Hebrew, *Horologium Hebraeum* of 1623, went into some 45 editions, it being his most popular book.

In 1617, Schickard first met **Johannes Kepler**, who also had studied theology in Tübingen and astronomy under Mästlin. Kepler commissioned Schickard to engrave the woodcuts and copper plates for the second part of his *Epitome* and the *Harmonice mundi* of 1618–1619. They remained friends – there exist 20 letters of Schickard addressed to Kepler, and 14 from Kepler. Upon the death of Mästlin in 1631, Schickard was appointed as professor of astronomy (in addition to his Hebrew appointment). In fact, he assisted Mästlin in his lectures from 1620, and also taught mathematics and geodesy from 1631. Schickard corresponded with many scientists including Matthias Bernegger, **Pierre Gassendi**, Daniel Mögling, **Ismaël Boulliau**, and **Maarten van den Hove**.

Schickard's first astronomical work was his paper of 1619 on his observations of the three spectacular comets of 1618 (C/1618 Q1, C/1618 V1, and C/1618 W1). There followed in 1624 his fundamental, 320-page monograph on the meteor of November 1623. He showed that meteoric studies can be as scientific as those of comets by **Tycho Brahe** and Mästlin. He was also a skilled mechanic and engraver in wood and copper plate.

Schickard's work of 1632/1633 on the transits of Mercury from 1627 on, his observational instruments having a mean error of 1′ 21″, are indeed remarkable. When he took over Mästlin's astronomical lectures in 1632, he gave out his own lectures in two parts – the theoretical part two was based on his *Picta Mathesis*, which is a remarkable attempt to present the full Copernican theory on the motion of the planets in a purely graphical way, using ruler and compass. The strength of it lies in Schickard's deep knowledge of spherical trigonometry (working out the necessary formulae in a didactic, clear way) and its graphical representation by means of descriptive geometry and stereographic projection – in fact, he tested all projection methods. Its secret was his methodical and systematic approach, the astronomy and not the mathematical theory being its goal. However, it forced him to use a purely Copernican approach. Schickard could not make use of the new astronomical laws introduced by his friend Kepler for elliptical orbits.

Schickard's brilliant achievements in the demanding area of the theory of the Moon – his prints and drawings are dated between 1624 and 1632 – reveal his full knowledge of the earlier work of **Ptolemy**, **Al-Battani**, **Al-Fargani**, **Nicolaus Copernicus**, Brahe, and **Christian Severin** (Longomontanus). His outstanding work concerning the theory of the Moon remained unfinished when he died of pestilence brought in by the Thirty Years War. The marginalia of Schickard's annotated copy of Copernicus' *De revolutionibus* are remarkable. They again reveal his skill, wide-ranging knowledge about this celestial science, and his standing as an astronomer.

Schickard is now best known for the invention in 1623 of the first mechanical computer capable of carrying out the four arithmetic operations; Pascal's arithmetic machine came later in 1642. Schickard's machine is known from his letters to Kepler, suggesting a mechanical means to help him with his logarithmic calculations of ephemerides. Unfortunately, no original copies of this calculator exist, but a working model was constructed by B. von Freytag Löringhoff from written documents in Tübingen in 1960. Schickard's calculator is a curious and striking conception (similar to **Leonardo da Vinci**'s imaginative inventions). Its capabilities, rediscovered after its reconstruction, have shown that it was indeed of practical use, though with the flaws inherent in its design.

Schickard, an expert mechanic, constructed additional scientific instruments. His *tellurium*, the first portable Copernican planetarium (reconstructed in 1977), could be used for the demonstration of the geocentric as well as the heliocentric system. His *rota hebraea* of 1621 was a device for the automation of Hebrew verb inflection. He was the first to apply the 1617 triangulation method of **Willebrord Snel** in 1624–1629 to geodesy, in particular in his surveying of Württemberg. Since systematic research concerning Schickard's *oeuvre* began only in 1957, no critical edition of any of Schickard's works has as yet appeared.

Paul L. Butzer

Selected References

Drake, Stillman and C. D. O'Malley (eds.) (1960). *The Controversy of the Comets of 1618*. Philadelphia: University of Pennsylvania Press.

Freytag Löringhoff, Bruno Baron von, and Matthias Schramm (1989). *Computus: Die astronomischen Rechenstäbchen von Wilhelm Schickard*. Tübingen: Attempto Verlag.

Kistermann, Friedrich-Wilhelm (2001). "How to Use the Schickard Calculator." *IEEE Annals of the History of Computing* 23, no. 1: 80–86.

Schramm, Matthias (1978), "Det Astronom", In Seck (1978), pp. 129–287.

Seck, Friedrich (ed.) (1978). *Wilhelm Schickard, 1592–1635: Astronom, Geograph, Orientalist, Erfinder der Rechenmaschine*. Contubernium: Beiträge zur Geschichte der Eberhard-Karls-Universität Tübingen, Vol. 25. Tübingen: J. C. B Mohr.

——— (1981). *Wissenschaftsgeschichte um Wilhelm Schickard: Vorträge bei dem Symposion der Universität Tübingen im 500. Jahr ihres Bestehens am 24. und 25. Juni 1977*. Contubernium: Beiträge zur Geschichte der Eberhard-Karls-Universität Tübingen, Vol. 26. Tübingen: J. C. B. Mohr.

——— (1995). *Zum 400. Geburtstag von Wilhelm Schickard: Zweites Tübinger Schickard-Symposion, 25. bis 27. Juni 1992*. Contubernium: Beiträge zur Geschichte der Eberhard-Karls-Universität Tübingen, Vol. 41. Sigmaringen: Thorbecke.

Schiller, Julius

Born **Augsburg, (Bavaria, Germany)**
Died **Augsburg, (Bavaria, Germany), 1627**

Little is known about Julius Schiller; he is famous for his contribution to celestial cartography, thanks to his atlas entitled *Coelum Stellatum Christianum* (Augsburg, 1627). In this work, he improved **Johann Bayer**'s *Uranometria*, on the basis of **Johannes Kepler**'s *Tabulae Rudolphinae*, both correcting stars' positions and adding stars that Bayer omitted in his atlas.

The most interesting peculiarity of Schiller's atlas, from the point of view of the history of celestial cartography, was the attempt to substitute for the constellations deriving from the ancient tradition, new Christian asterisms inspired by the Old Testament (in the Southern Celestial Hemisphere) and the New Testament (in the Northern Celestial Hemisphere). The 12 zodiacal constellations were replaced with the figures of the 12 apostles. However, Schiller's proposal was not followed by other cartographers, and his Christian constellations became a historical curiosity.

As regards stellar nomenclature, Schiller chose to use Arabic numbers rather than the Greek letters introduced by Bayer.

Davide Neri

Selected References

Anon. (1986). *Deutscher Biographischer Index*. Vol. 4, p. 1789. Munich: K. G. Saur.

Tooley, R. V. (1979). *Tooley's Dictionary of Mapmakers*. Tring, England: Map Collector Publications, pp. 565.

Schjellerup, Hans Karl Frederik Christian

Born **Odense, Denmark, 8 February 1827**
Died **Copenhagen, Denmark, 13 November 1887**

An astronomer who made a specialty of compiling reference data useful to others, Hans Schjellerup was a watchmaker in early life. In 1851, he became an assistant at the Copenhagen Observatory. There he computed planetary and cometary orbits and compiled a star catalog. In 1866, Schjellerup published a well-known catalog of red stars.

Schjellerup rediscovered, translated, and edited for publication an important work by **Al-Sufi**, which he saw as a bridge in time between the uranometry of **Ptolemy** and the work of **Friedrich Argelander**.

Schjellerup was an associate of the Royal Astronomical Society. A crater on the Moon is named for him.

Selected Reference

Anon. (1888). "Hans Carl Frederik Christian Schjellerup." *Monthly Notices of the Royal Astronomical Society*. 48: 171.

Schlesinger, Frank

Born **New York, New York, USA, 11 May 1871**
Died **Lyme, Connecticut, USA, 10 July 1943**

Frank Schlesinger is best known for his contributions to the photographic determination of stellar distances, motions, and positions. He was the youngest of the seven children of Joseph William and Mary (*née* Wagner) Schlesinger, German immigrants to the United States. In 1890, he received a B.S. degree from the College of the City of New York. For 5 years, Schlesinger worked as a surveyor before receiving a fellowship that enabled him to become a full-time graduate student at Columbia University.

Schlesinger received his Ph.D. degree from Columbia University in 1898. From 1899 to 1903, he was in charge of the station of the International Latitude Service in Ukiah, California. In 1903, Schlesinger became a research associate at the Yerkes Observatory, holding that position until he assumed the directorship of the Allegheny Observatory of the University of Pittsburgh in 1905. Following the entry of the United States into World War I, Schlesinger briefly served as an aeronautical engineer for the United States Signal Corps. In 1920, he left Allegheny to become director of the Yale University Observatory, where he remained until his retirement in 1941.

Schlesinger married Eva Hirsch of Ukiah, California, in 1900. Following her death in 1928, he married the former Mrs. Philip W. Wilcox in 1929. He had one son by his first marriage, Frank Wagner Schlesinger, who would himself become a well-known planetarium director.

Schlesinger's doctoral dissertation dealt with the measurement of star positions on photographic plates taken by **Lewis Rutherfurd**, a pioneer in astronomical photography. Schlesinger became interested in the possibility of measuring distances to stars by accurately determining their annual trigonometric parallaxes using photographic methods. The first measurements of stellar trigonometric parallaxes had been made in the 1830s, when careful observations revealed small shifts in the positions of three stars, 61 Cygni (by **Friedrich Bessel**), Vega (by **Wilhelm Struve**), and α Centauri (by **Thomas Henderson**), as the Earth orbited the Sun. Because the size of the shift in position due to parallax is inversely related to distance, these measurements allowed the distances of the three stars to be determined. However, progress throughout the remainder of the 19th century was slow. Visual observations made with instruments such as the heliometer had by the 1890s produced reliable parallaxes for only 30 stars. Early applications of photography to the determination of parallaxes, while promising, had not yet led to large improvements. Schlesinger believed that significant advances were possible using photographic methods, but he did not have an opportunity to attempt such determinations until he arrived at Yerkes Observatory.

The main instrument of the Yerkes Observatory was its 1-m refractor. That telescope's long focal length provided a photographic plate scale adequate for the accurate measurement of stellar positions, a necessary condition for parallax observations. At Yerkes, Schlesinger began to develop techniques for determining stellar parallaxes, including prescriptions for the taking of the photographic plates, for the measurement of star positions on those plates, and for the reductions needed to turn those measurements into actual parallaxes. This work would not come to full fruition until after Schlesinger left Yerkes for the Allegheny Observatory, but the classic papers on the subject that he published in 1910 and 1911 would guide not only his own work, but also that of several observatories that undertook the photographic determination of stellar parallaxes. In 1914, Schlesinger began parallax observations using the newly completed Allegheny Observatory's long-focus photographic Thaw telescope. When he assumed the directorship of the Yale University Observatory, Schlesinger oversaw the construction of a refracting telescope of 66-cm aperture in Johannesburg, South Africa, specifically designed to extend his work on the determination of stellar parallaxes to the hitherto neglected Southern Celestial Hemisphere. The methods that Schlesinger developed allowed stellar parallaxes to be determined with greater accuracy than ever before. Distances accurate to 15% or better could be determined for stars within 10 parsecs (33 light years) of the Sun, and useful results could be obtained for distances approaching 100 parsecs.

When in 1924 Schlesinger published the first *General Catalogue of Stellar Parallaxes,* he was able to list 1,870 trigonometric parallax determinations, the great majority of which were determined photographically. The second edition of the *Catalogue*, published in 1935, listed data for 7,534 stars, including trigonometric parallaxes for about 4,000 stars. Photographic observations made with long-focus refracting telescopes would remain the chief source of trigonometric parallax determinations throughout much of the 20th century.

At Yale, Schlesinger began a second large astrometric program. In the 19th century, the German Astronomische Gesellschaft had organized the measurement of the positions of stars down to the ninth magnitude using the meridian circle techniques of the day. Schlesinger began to reobserve the stars of the Astronomische Gesellschaft catalogs using wide-field photographic cameras. His goal was to remeasure the positions of the stars and, by seeing how much they had moved in the intervening years, determine their proper motions. The work on these "zone catalogues" was mainly carried out while Schlesinger was at Yale, with the assistance of Ida M. Barney. At the time of Schlesinger's death, 14 volumes of the zone catalogs had been published, including data on 92,329 stars. After Schlesinger's death, the zone catalog project would be continued at Yale under the direction of Barney, **Dorrit Hoffleit**, and **Dirk Brouwer**, eventually yielding data for more than 227,000 stars. In the production of the zone catalogs, as in his other research, Schlesinger developed reduction methods that saved time without sacrificing accuracy, an important consideration in the days before automated measuring engines and electronic computers.

In 1930, Schlesinger published the first edition of the *Catalogue of Bright Stars*. This was a compilation of data on the stars brighter than visual magnitude 6.5 that had been included in the *Harvard Revised Photometry* catalog. A second edition of this useful compilation, coauthored with **Louise Jenkins**, appeared in 1940. The *Catalogue of Bright Stars*, too, would have continued life after Schlesinger's death, with the compilation of later editions by Hoffleit and her collaborators.

Schlesinger was first elected to the council of the American Astronomical Society in 1908 and, after serving as a vice president of the society, became its president from 1919 to 1922. He was a vice president of the International Astronomical Union from 1925 to 1932, and served as president from 1932 to 1935. Schlesinger received many honors, including honorary degrees from the University of Pittsburgh and Cambridge University, the Gold Medal of the Royal Astronomical Society, the Valz Medal of the French Academy of Sciences, the Bruce Medal of the Astronomical Society of the Pacific, and the Townshend

Medal of the College of the City of New York. He was elected a member of the National Academy of Sciences in 1916 and was an honorary member of several foreign societies. Schlesinger was the organizer of the Neighbors, an informal but influential group of (male) astronomers in the northeastern United States.

Horace A. Smith

Selected References

Brouwer, Dirk (1947). "Frank Schlesinger." *Biographical Memoirs, National Academy of Sciences* 24: 105–144. (Schlesinger's life and work are described by Dirk Brouwer, his successor as director of the Yale University Observatory.)

Hoffleit, Dorrit (1992). *Astronomy at Yale, 1701–1968. Memoirs of the Connecticut Academy of Arts and Sciences*, Vol. 23. New Haven: Connecticut Academy of Arts and Sciences.

Schlesinger, Frank. "Photographic, Determinations of Stellar Parallax, Made with the Yerkes, Refractor." Parts 1–7. *Astrophysical, Journal* 32 (1910): 372–387; 33(1911): 8–27, 161–184, 234–259, 353–374, 418–430; 34 (1911): 26–36. (His groundbreaking papers on the methods for measuring trigonometric parallaxes.)

——— (1924). *General Catalogue of Stellar Parallaxes*. 3rd ed., coauthored with Louise F. Jenkins, New Haven, Connecticut: Yale University Observatory.

——— (1930). *Catalogue of Bright Stars*. New Haven, Connecticut: Tuttle Morehouse and Taylor. (2nd ed., coauthored with Louise F. Jenkins, 1940.)

Schmidt, Bernhard Voldemar

Born **Island of Naissaar near Tallinn, (Estonia), 30 March 1879**

Died **Hamburg, Germany, 1 December 1935**

German optician Bernhard Schmidt gave his name to a telescope type that permitted obtaining sharp images over a very wide field quickly. It involves a spherical primary mirror and a transparent corrector plate that largely removes the focus errors called coma and spherical aberration.

Schmidt was the first of five children of Karl Konstantin Schmidt and his wife, Maria Helene. His father was a writer, farmer, and fisherman on the island of Naissaar, in the Baltic Sea. Swedish was the language spoken on the island and in school, but at home the family spoke German. At the age of 15, Bernhard experimented with gunpowder and lost his right hand and forearm in an accident. This did not prove too much of a handicap, for later that year he built his own camera, photographed local people, and sold the pictures.

In 1895, Schmidt left Naissaar for Tallinn where he found work as a telegraph operator with a rescue team. Between 1895 and 1901, he also worked as a photographer and in the "Volta" electromotor factory. Around 1900, he made his first 5-in. diameter object glass; it was not perfect, but he improved it as he made observations of Nova Persei 1901. Schmidt moved to Göteborg, Sweden, to attend a technical school, the Chalmers Institute, but after a few months moved on to Mittweida in southeastern Germany. There, he improved his knowledge of optics with Dr. Strehl as his teacher at the Technikum Mittweida. That technical college was practically oriented; Schmidt favored hands-on practice to theoretical work. In the summer of 1903, he fashioned a mirror for the Altenburg Observatory, probably his first work geared toward professional use. Most of the mirrors Schmidt had made previously were sold to amateurs, and provided income to live on, since he got little financial support from his parents.

In 1904, Schmidt opened his own optical workshop in a small house in Mittweida, and later moved to more spacious quarters. He offered his skills to observatories to improve their optics, lenses, and mirrors, and in 1905 received a commission for a reflecting telescope from the Potsdam Astrophysical Observatory. Eventually, Schmidt was well known to astronomers all over Germany. In 1913, he was asked to rework a 50-cm telescope lens originally made by Steinheil & Sons. After Schmidt's reworking of this lens, **Ejnar Hertzsprung** was able to make some very delicate observations of double stars with this telescope at Potsdam, and it remained in use until 1967. Schmidt also sold two mirrors to the University of Prague, one of 60-cm diameter and another of 30 cm. Around 1926, he was offered work at the Zeiss optical shop in Jena, Germany, but although his own business was slowing down Schmidt had been independent all his life, and wanted to work only at his own pace, so he refused the offer.

In 1927, Schmidt sold his shop and moved to Hamburg to work at the nearby observatory in Bergedorf as a freelance optician. The director then was Richard Schorr, who knew about Schmidt's abilities. He also knew that Schmidt liked French brandy and paid him only small sums of money at a time. "The optician," as Schmidt was called, also made observations with various instruments; in 1928 he took pictures of Jupiter, Saturn, and the Moon with his own telescope.

During a journey back to Hamburg after observing the solar eclipse of 1929 in the Pacific Ocean, Schmidt discussed the possibility of a special camera for wide-angle sky photography with **Walter Baade**, an astronomer on the Bergedorf staff. With Baade's encouragement, after returning to his workshop at the observatory, Schmidt developed his now famous wide-field telescope. He completed his first version of this design in 1930; it included a spherical main mirror with a diameter of 44 cm and a corrector plate, placed at the radius of curvature of the main mirror, with a diameter of 36 cm. The corrector plate, shaped in a complex figure of a circular torus, compensated for the spherical aberration introduced by the primary mirror. The overall focal ratio was *f*/1.75, the field of view 7.5°. The very first photograph taken at night with this new instrument clearly and legibly showed a tombstone in a distant graveyard.

Schmidt himself made only this one instrument. However, when Baade joined the Mount Wilson Observatory staff in 1931 and told his new colleagues about the great success of the Schmidt design, there was an immediate and enthusiastic rush to implement this new technology in the Mount Wilson optical shops and elsewhere. Their success in this pursuit owed much to Schorr's direct intervention to ensure publication of Schmidt's only technical publication on this design. Schmidt's concept for a wide-field telescopic camera for stars and other celestial objects has been widely applied in other fields. In X-ray technology, for example, the urgent need to improve photographic recording of images prompted the employment of **Jesse Greenstein** and **Louis Henyey** to supervise the design and construction of a prototype 70-mm X-ray camera in the optical shops of Yerkes Observatory.

Schmidt's original camera is now in the museum of the Hamburg Observatory in Germany. The 48-inch Schmidt telescopes at Palomar, and in Australia and Chile, were used for the most extensive sky survey ever made. Data from these provided the initial guide star catalog for the Hubble Space Telescope, and digitized versions of the surveys still are in frequent use.

Schmidt died of pneumonia in a mental hospital, shortly after returning from a trip to the Netherlands. He was buried in a cemetery very close to the observatory. Minor planet (1743) Schmidt was named in honor of Bernhard Schmidt.

Christof A. Plicht

Selected References

Hodges, P. C. (1948). "Bernhard Schmidt and His Reflector Camera: An Astronomical Contribution to Radiology. Paper I." *American Journal of Roentgenology and Radium Therapy* 59: 122–131.

Marx, Sigfried and Werner Pfau (1992). *Astrophotography with the Schmidt Telescope*, translated by P. Lamble. Cambridge: Cambridge University Press, pp. 54–65.

Schmidt, Bernhard (1932). "A Fast, Coma-free Reflecting System." *Zentral-Zeitung für Optik un Mechanik* 1, no. 2: 62–63. (A translation of this article, and of Richard Schorr's letter transmitting it to the journal, appears in Marx and Pfau, pp. 156–157.)

Schramm, Jochen (1996). *Sterne über Hamburg: Die Geschichte der Astronomie in Hamburg.* Hamburg: Kultur- und Geschichtskonto, pp. 209–220.

Wachmann, A. A. (1995). "From the Life of Bernhard Schmidt." *Sky & Telescope* 15, no. 1: 4–9.

Williams, Thomas R. (1997). "Schmidt Telescopes." In *History of Astronomy: An Encyclopedia*, edited by John Lankford, pp. 445–447. New York: Garland.

Schmidt, Johann Friedrich Julius

Born **Eutin, (Schleswig-Holstein, Germany), 26 October 1825**
Died **Athens, Greece, 7 February 1884**

German observer Julius Schmidt compiled the most complete maps of the Moon of his generation and reported changes in the appearance of one crater that were widely accepted at the time. In an era when study of the Moon had become increasingly specialized and knowledge of its topography so comprehensive that it led to the formation of a committee of British observers to further its mapping, Schmidt worked unaided and alone. It was plausibly suggested by Harvard College Observatory astronomer **William Pickering** that Schmidt "perhaps devoted more of his life than any other man to the study of the Moon."

Schmidt was the son of Carl Friedrich Schmidt, a glazier by profession, and Maria Elisabeth Quirling. At the age of 14, young Schmidt chanced upon a copy of **Johann Schröter**'s *Selenotopographische Fragmente.* He was so fascinated by its pictures of mountains and craters that the future direction of his life was determined then and there. Schmidt immediately began to study the Moon himself, using a small telescope made with lenses ground by his father.

Schmidt's first view of the Moon through a good telescope came in July 1841, when A. C. Petersen, director of the Altona Observatory near Hamburg, showed him the imposing craters Bullialdus and Gassendi. He also saw for the first time a copy of the great 1837 map of the Moon prepared by **Wilhelm Beer** and **Johann von Mädler**. Soon afterward, Schmidt moved to Hamburg and for several years made frequent observations with the telescopes of the Altona Observatory.

A strange interlude followed in 1845, when Schmidt accepted a position in the private observatory of **Johann Benzenberg** at Bilk, near Düsseldorf. Benzenberg was preoccupied with the search for a possible intra-Mercurial planet and did not allow Schmidt to use his large refractor, apparently for no better reason than that "its outward good looks and polish might not suffer by handling." Instead, he gave Schmidt access only to a "wretched instrument." After a few months, Schmidt left Bilk in disgust and took a position under **Friedrich Argelander** at the Bonn Observatory. Although most of his time was taken up with entering meridian circle observations of stars for Argelander's great catalog, the *Bonner Durchmusterung* (Bonn Survey), he made as many lunar observations as he could.

In 1853, Schmidt left Bonn for E. von Unkrechtsberg's observatory at Olmütz (now Olomouc), in Moravia, where he made some 3,000 measurements of the heights of lunar mountains with a filar micrometer. This work was published in an 1856 treatise, entitled *Der Mond* (The Moon), in which Schmidt attempted to provide a quantitative comparison of lunar and terrestrial features. He prudently warned against taking the apparent similarities between the Moon and Earth too seriously.

On 2 December 1858, Schmidt assumed the directorship of the Athens Observatory in Greece, where he would remain for the rest of his life. When he set foot on Greek soil at Piraeus, Schmidt was still a comparatively young man, full of energy. Arriving at the observatory, he found it in a state of disrepair and neglect. Within only a year, however, Schmidt was able to restore to working order a fine 6.2-in. refractor by the Viennese optician Georg S. Plössl, which served as the main instrument for his lunar work over the next quarter of a century.

By 1865, Schmidt had assembled so many lunar observations that he began laying down his surveys of selected regions on a 6-ft. diameter map. The next year, he began to construct a 1-m map based on **Wilhelm Lohrmann**'s observations, which had been entrusted to Schmidt by Lohrmann's publisher.

At first, Schmidt planned to enter details from his own observations onto Lohrmann's map, but he soon abandoned this approach in favor of something far more ambitious – nothing less than a fresh topographic map of the Moon roughly 2-m diameter, which, like Lohrmann's original design, was to be divided into 25 sections. Schmidt hoped to record all of the details of the lunar surface visible through his 6.2-in. refractor, but gradually came to the realization that such a feat would require "more powers of endurance and a longer lifetime than are allotted to mortals."

It was while Schmidt was involved in this lengthy series of observations that he came across something startling but not entirely unexpected. In 1866, he announced that the tiny crater Linné in Mare Serenitatis had undergone a profound change. He maintained that, prior to 1866, Linné had always been recorded as a crater about 6 miles in diameter and "very deep," but had been suddenly reduced to a diffuse white patch. As a recent eyewitness of volcanic eruptions on the Aegean island of Santorini, Schmidt proposed that Linné had been filled in by a similar "eruption of fluid or powdery material."

Given the daunting complexity of lunar detail, the variable effects of shadow, foreshortening, and libration as well as the inevitable deficiencies of the selenographic record, the surprising fact is that Schmidt's claim of a definite change was widely and uncritically accepted. Despite the fact that the evidence of change was always weak, the alleged alteration of Linné would not be thoroughly discredited for more than a century.

In July 1874, Schmidt presented his lunar map to the Berlin Observatory, where it excited admiration as a performance highly creditable to "Teutonic intellect and perseverance." Before long, it was being touted as a uniquely Prussian achievement. Its 25 sections were photographed at the General Staff Office under the direction of Count von Moltke; its publication as the *Charte der Gebirge des Mondes* (Map of the Mountains on the Moon) was sponsored by the Crown Prince of Prussia himself.

After the first copies of the map appeared in 1878, the English astronomer **John Birmingham** commented:

In even a cursory examination of Schmidt's map its completion by a single observer must seem almost incomprehensible ...; but it requires protracted study to well realize the extent of the work. Any person who tries with the aid of a 6-inch telescope to give a closely detailed delineation of even a small area of the Moon, will soon conclude that the period of thirty-three years was comparatively a very short one for the accomplishment of Dr. Schmidt's great task.

In his popular book entitled *The Story of the Heavens* (1886), Sir **Robert Ball** marveled:

To give some idea of Schmidt's amazing industry in lunar researches, it may be mentioned that in six years he made nearly 57,000 individual settings of his micrometer in the measurement of lunar altitudes. His great chart of the mountains in the Moon is based on no less than 2,731 drawings.

According to Schmidt's own rather compulsive analysis, Lohrmann had charted 7,177 craters and Mädler 7,735; his own map recorded no less than 32,856. The superiority of Schmidt's map was also apparent in his record of rilles – the 71 on Mädler's map paled in comparison with his own 348.

Schmidt also had a keen interest in seismology. At the age of 20, he began to collect materials for a global earthquake catalog, and he contributed to Johann J. Noeggerath's study of the 1846 Rhineland earthquake by calculating the propagation speed of the seismic wave. In 1874, he published a study of four volcanoes: Santorini, Etna, Vesuvius, and Stromboli. Schmidt's *Studien über Erdbeben* (Studies on earthquakes), a comprehensive catalog of earthquakes recorded in southeastern Europe since ancient times, was issued the following year.

Schmidt reorganized the meteorological service of the Athens Observatory. He made meteorological observations from locations throughout Greece and regularly submitted data to the Paris Observatory. These results were presented in his 1864 work, *Beiträge zur Physikalischen von Griechenland*. Schmidt also dabbled in archeology and made a concerted effort to find the site of ancient Troy.

Schmidt was awarded an honorary doctorate from the University of Bonn in 1868. Fittingly, a crater on the Moon is named for him.

Thomas A. Dobbins and *William Sheehan*

Selected References

Ashbrook, Joseph (1984). "Julius Schmidt: An Incredible Visual Observer." In *The Astronomical Scrapbook*, edited by Leif J. Robinson, pp. 251–258. Cambridge, Massachusetts: Sky Publishing Corp.

Ball, Robert S. (1886). *The Story of the Heavens*. New York: Funk and Wagnalls.

Birmingham, John. "Schmidt's Lunar Map." *Observatory* 2 (1879): 413–415; 3 (1879): 10–17.

Freiseleben, H.-Christ. (1975). "Schmidt, Johann Friedrich Julius." In *Dictionary of Scientific Biography*, edited by Charles Coulston Gillispie. Vol. 12, p. 192. New York: Charles Scribner's Sons.

Sheehan, William and Thomas Dobbins (2001). *Epic Moon: A History of Lunar Exploration in the Age of the Telescope*. Richmond, Virginia: Willmann-Bell.

Whitaker, Ewen A. (1999). *Mapping and Naming the Moon: A History of Lunar Cartography and Nomenclature*. Cambridge: Cambridge University Press.

Schmidt, Otto Iulevich

Born **Mogilyov, (Ukraine), 30 September 1891**
Died **7 September 1956**

Russian geophysicist Otto Schmidt led the 1937 Soviet air expedition to the North Pole. Thus, he may have been the first person actually to reach latitude 90° N. His subsequent fame brought him a directorship within the Academy of Sciences. However, Joseph Stalin relieved Schmidt of this position during World War II. (Whether it was simply because of his German surname is unknown.)

Freed of administrative responsibilities, Schmidt turned – for the first time – to a research field in which he had long-standing interest: solar-system cosmogony. Schmidt hypothesized that the planets were formed from "meteoritic" material that had been gravitationally captured by the Sun as it passed through an interstellar cloud.

Selected Reference

Levin, Aleksey E. and Stephen G. Brush (1995). *The Origin of the Solar System: Soviet Research 1925–1991*. New York: American Institute of Physics.

Schöner, Johannes

Born **Karlstadt near Nuremberg, (Germany), 16 January 1477**
Died **Nuremberg, (Germany), 16 January 1547**

The *Narratio Prima* of Johannes Schöner was the first publicized account of the Copernican theory.

Little is known about Schöner's youth. He matriculated at the university in Erfurt in 1494, but apparently did not complete a degree there. After being ordained as a priest in 1500, Schöner settled in Nuremberg in 1504, where he immediately devoted time to making celestial observations. In Nuremberg, he was also able to study briefly under **Bernard Walther**, until the latter's death in 1504. On 8 January and on 18 March 1504, Schöner made observations of the planet Venus, which he sent to **Nicolaus Copernicus**. Copernicus later used these and other observations in his theory of Mercury. Schöner took priestly orders in 1515 and was appointed to a position in Bamberg. In the same year, his first terrestrial globe was printed. Due to neglect of his clerical duties in Bamberg, Schöner was relocated to the small village of Kirchehrenbach, shortly after 1520. Catholic leaders were concerned not only about Schöner's negligence; Cardinal Campeggio called Schöner's orthodoxy into question, claiming that Schöner was a "Lutheran" because he was married. By 1526, it was clear that Schöner had Lutheran affiliations, for he had accepted the chair of mathematics at the new Lutheran Gymnasium in Nuremberg upon the request of Martin Luther's right-hand man Philip Melanchthon and upon the urging of the reformer Joachim Camerarius. It is still unclear when Schöner first married or had children, but in 1527 he married Anna Zelerin, with whom he had at least three sons. One of the sons, Andreas Schöner (born: 1528), followed in his father's footsteps as a mathematician and editor. As the professor of mathematics in Nuremberg, Johannes Schöner issued regular yearly prognostications in German between 1529 and 1547 for the city of Nuremberg. During these years, he also edited and had printed many of the works of **Johann Müller** (Regiomontanus) and **Johannes Werner**.

Among Schöner's contacts was **Rheticus**, who convinced Copernicus to let him write and have printed the *Narratio Prima*, which Rheticus then dedicated to Johannes Schöner. Later, Schöner was among those who encouraged Copernicus to publish *De revolutionibus* (On the revolutions) that was published in Nuremberg in 1543. Schöner defended the legitimacy of making astrological predictions, and he held that the Copernican system was not unfavorable toward astrology.

Schöner showed an interest in acquiring the works of Regiomontanus, many of which he later edited for publication. While still working as a cleric in Bamberg in 1509, Schöner bought the almanacs of Regiomontanus for the years 1464–1484. Schöner later obtained many of Regiomontanus's unpublished manuscripts that he then edited and printed. In 1531, Schöner began with Regiomontanus's treatise on the problems of determining the magnitude and location of comets, entitled *De cometae magnitudine longitudine ac de loco eius vero problemata XVI*. Among the other works of Regiomontanus that Schöner edited for publication were *De triangulis omnibus* (1533), which was printed at the press of Johannes Petreus, and *Opusculum geographicum* (also in 1533), which contained Regiomontanus's arguments against the rotation of the Earth on an axis.

In addition to the recovery of Regiomontanus's work, Schöner printed his own work and the works of others. One of Schöner's best-received publications was his own *Tabulae resolutae*, which saw its first publication in 1536. Melanchthon praised Schöner's tables for showing "the position of all the stars and not for one year only but many centuries." Andreas Schöner edited and printed his father's collected works in 1551, which contained the *Tabulae resolutae*. The tables were printed again separately in 1587/1588. In addition to casting horoscopes and issuing annual prognostications, Schöner compiled the *Opusculum astrologicum*, which contained among other things Eberhard Schleussinger's *Assertio contra calumniatores astrologiae* (Assertion against the calumniators of astrology).

In 1544, toward the end of his life, Schöner published the observations of both Regiomontanus and Walther, including their observations of eclipses, comets, and positions of the planets and fixed stars. These observations proved to be valuable to later astronomers such as **Tycho Brahe** and **Johannes Kepler**. In 1546, Schöner's final publication was Werner's work on weather predictions.

Derek Jensen

Selected References

Klemm, Hans Gunther (1992). *Der Fränkische Mathematicus Johann Schöner (1477–1547) und seine Kirchehrenbacher Briefe an den Nürnberger Patrizier Willibald Pirckheimer*. Forchheim: Ehrenbürg-Gymnasium. (Klemm concentrates on the period of Schöner's stay in Kirchehrenbach and publishes at the end of this work transcriptions of letters between the Nuremberger theologian Willibald Pirckheimer and Schöner.)

Schöner, Johannes (1967). *Regiomontanus on Triangles*, translated by Barnabas Hughes. Madison: University of Wisconsin Press. (Schöner's publication of Regiomontanus's *De triangulis omnimodis*.)

Stevens, Henry N. (1888). *Johann Schöner: Professor of Mathematics at Nuremberg*, edited with an introduction by C. H. Coote. London: Henry Stevens and Son. (Still a valuable source on Schöner's life and works. In the "Historical Introduction" to this work, there is a short discussion of Schöner's life on pp. xxxix–xlv. The book also contains facsimile reproductions and translations of Schöner's Globe of 1523 and the introductory letter to the Globe which was written to Reymer von Streytpergk. In the back of the volume, there is an excellent bibliography of 46 of Schöner's works. However, the bibliography does not list the works of Regiomontanus that Schöner edited for publication.)

Thorndike, Lynn (1941). *A History of Magic and Experimental Science*. Vol. 5, pp. 354–371. New York: Columbia University Press. (For a valuable investigation of Schöner's life and works.)

Zinner, Ernst (1990). *Regiomontanus: His Life and Work*, translated by Ezra Brown. Amsterdam: North-Holland. (For more on Schöner's acquisition and publication of Regiomontanus's works.)

Schönfeld, Eduard

Born **Hildburghausen, (Thüringia, Germany), 22 December 1828**

Died **Bonn, Germany, 1 May 1891**

Eduard Schönfeld directed two astronomical observatories, single-handedly compiled the first southern extension of the *Bonner Durchmusterung* [BD] star catalog, and was a cofounder of the Astronomische Gesellschaft [AG]. Schönfeld was the son of merchant Joseph Schönfeld and his wife Louise (*née* Fauß) Schönfeld. His mother taught young Eduard the basics of reading and mathematics before he started school. He attended a gymnasium in Hildburghausen and studied architecture at the Hanover Technische Hochschule until he was expelled in 1849 for his participation in political events of the previous year.

Schönfeld continued his studies at Kassel and later at the University of Marburg, where he attended lessons of Christian Ludwig Gerling, who had built a small observatory. Together with Gerling's assistant, **Ernst Klinkerfues**, he planned to observe an occultation of the star γ Arietis. When Klinkerfues did not get to the observatory in time, Schönfeld made the observations himself, calculated the results, and presented them to his professor on the following morning. As a reward for this work, he received a key to the observatory and encouraging words to continue his work. His first observation, published in the *Astronomische Nachrichten*, was followed by many others.

In 1851, Schönfeld visited **Friedrich Argelander** in Bonn. Discussing Schönfeld's wish to become an astronomer, Argelander tried to discourage the student, but finally accepted him in 1852. The following year, Schönfeld became Argelander's paid assistant, when **Johann Schmidt** left Bonn for a new observatory at Olmütz (now Olomouc) in Moravia. In 1854, Schönfeld received his Ph.D. with a thesis entitled, "Nova Elementa Thetidis." Along with two other assistants, Wilhelm Julius Foerster and **Karl Krüger**, Schönfeld kept up work on the *Bonner Durchmusterung*. During this time, however, he also developed secondary interests in minor planets and variable stars. Whenever conditions allowed, he observed the stars β Persei (Algol) and S Cancri to record complete cycles in their brightness variations.

In 1859, Schönfeld left Bonn for Mannheim, invited by the Grand Duke of Baden to direct the new observatory there. In the previous year, he had declined an invitation from **Friedrich Struve** to come to the Pulkovo Observatory, indicating that work on the *Durchmusterung* was far from complete. The Mannheim Observatory, which included living quarters for the director, consisted of a single building: a 33-m (100-ft.) tower with 196 steps leading to its upper platform. For his work at Mannheim, Schönfeld ordered a 16.5-cm (6.6-in.) diameter telescope whose manufacture he supervised at the workshop of Karl Steinheil, in Munich.

In 1860, Schönfeld married Helene Noeggerath, the daughter of geology professor Johann J. Noeggerath. The couple later had three children.

Schönfeld's principal scientific work continued to be the *Durchmusterung*, but he observed minor planets, comets, variable stars, and nebulae. Of the latter he published two catalogs, in 1862 and 1875, totaling 489 objects. Two more catalogs were published in 1866 and 1874, presenting data on 119 and 143 variable stars.

In 1862, Schönfeld received another invitation from Pulkovo, this time from the younger **Otto Wilhelm Struve**, who had succeeded his father as observatory director. The Mannheim astronomer was torn between his desire to operate better equipment and his well-established career in Germany, but he again declined the invitation.

A change in his life came when he was ordered to take charge of the duties of the Eichamt (Office of Weights and Measures). This opportunity allowed Schönfeld to travel at government expense. His professional demeanor led toward his election to the Normal-Eichungskommission (National Commission of Standards). During these business trips, he had the opportunity to meet with numerous colleagues and to discuss astronomical matters. Thus, Schönfeld and Foerster invited the astronomical community to establish the AG. The society's first meeting took place at Heidelberg in 1863. Schönfeld served as its secretary from 1875 until his death.

The volumes of the *Bonner Durchmusterung* that comprised the northern skies had been published between 1857 and 1863. Upon Argelander's death in 1875, Schönfeld was appointed to his now-vacant post at Bonn. Even Argelander's son-in-law, Krüger (who then worked in Helsinki), supported Schönfeld in a letter to Foerster. In one of the towers, Schönfeld found a 15.9-cm (6.25-in.) telescope made by Hugo Schröder in Hamburg, acquired the year before by Argelander. With this more exact and larger instrument, Schönfeld continued work on the *Durchmusterung*, extending its coverage from −2° to −23° declination.

Schönfeld's observations were completed by 1881, but he continued to rework the positional data for several more years. In 1886, the *Bonner Durchmusterung des südlichen Himmels* [BDS], was printed. It reproduced 133,659 stars on 24 charts. Ironically, in his introduction to the BDS, Schönfeld indicated that future star atlases would likely be prepared by the newer methods of astrophotography; a prediction that was demonstrated in years ahead.

In 1887, Schönfeld declined an invitation to direct the Strasbourg Observatory constructed by **Friedrich Winnecke**. The University of Bonn elected Schönfeld its headmaster as a reward for his choice to stay. Other honors followed – he was elected a member of the Prussian Academy of Sciences, became a recipient of the Watson Medal of the United States National Academy of Sciences, and received another medal from Russia. In 1890, Schönfeld fell ill and went to Baden-Baden to find a cure, but continued to observe the brighter variable stars.

Minor planet (5926) Schönfeld and a lunar crater are named in his honor.

Christof A. Plicht

Selected References

Ashbrook, Joseph (1984). "How the BD Was Made." In *The Astronomical Scrapbook*, edited by Leif J. Robinson, pp. 427–436. Cambridge, Massachusetts: Sky Publishing Corp.

Schmeidler, F. (1975). "Schönfeld, Eduard." In *Dictionary of Scientific Biography*, edited by Charles Coulston Gillispie. Vol. 12, p. 200. New York: Charles Scribner's Sons.

Schönfeld, Eduard (1886). *Bonner Durchmusterung des südlichen Himmels*. Bonn: Marcus and Weber's Verlag.

Steiner, Gerhard (1990). *Eduard Schönfeld: Lebensbild eines hervorragenden Astronomen aus Hildburghausen*. Hildburghausen: Verlag Frankenschwelle.

Schreck, Johann

Born ***circa* 1576**
Died **Beijing, China, 13 March 1630**

Swiss Jesuit Johann Schreck was a friend of **Galileo Galilei**. He was among the first foreign missionaries hired by the emperor to improve the Chinese calendar. After Schreck's death, this task passed to fellow Jesuits **Johann Schall von Bell** and **Giacomo Rho**.

Alternate names

Terrentius
Terrenz, Jean

Selected Reference

D'Elia, Pasquale M. (1960). *Galileo in China: Relations through the Roman College between Galileo and the Jesuit Scientist-Missionaries (1610–1640)*, translated by Rufus Suter and Matthew Sciascia. Cambridge, Massachusetts: Harvard University Press.

Schrödinger, Erwin

Born **Vienna, (Austria), 12 August 1887**
Died **Vienna, Austria, 4 January 1961**

Austrian physicist Erwin Schrödinger is chiefly known for the development of wave mechanics, as expressed in a fundamental equation that bears his name. He was educated in Vienna by a private tutor before attending that city's academic gymnasium. At the University of Vienna, he became a protégé of Ludwig Boltzmann's successor, Fritz Hasenöhrl. Schrödinger completed his Ph.D. in theoretical physics in 1910 before accepting a research appointment in experimental physics under Franz Exner and Friedrich Wilhelm Georg Kohlrausch. After he completed his *Habilitation* (the post-doctoral requirement in Germany for teaching at a university) in 1914, World War I broke out. Schrödinger became an artillery officer but nevertheless managed to publish several important papers while serving, with distinction, on the Italian front. A tour of duty in Hungary included a battle victory and another physics paper. Upon his return to the Italian front, Schrödinger received a medal for outstanding service as commander of his battalion.

After the war, Schrödinger became a research assistant to Max Wien. In 1920, he was appointed a professor at Stuttgart University and that same year married Annemarie Bertel. There, his close association with philosopher Hans Reichenbach had a lasting influence on his subsequent work, the most important of which he did in Switzerland, at the University of Zürich. At Zürich, Schrödinger became a close colleague of **Hermann Weyl** (one of David Hilbert's students) and Peter Debye (winner of the 1936 Nobel Prize in Chemistry). All of these remarkable influences culminated in Schrödinger's crowning achievement, his development in 1926 of wave mechanics – what is now known as Schrödinger's equation – severely modifying the classical laws of mechanics on small scales. One year later, he was awarded the Max Planck Chair in Physics at the University of Berlin. In 1933, Schrödinger won the Nobel Prize for Physics (along with **Paul Dirac**).

Schrödinger's groundbreaking discovery, namely, that the corpuscular conception of matter could be explained purely in terms of waves, grew out of his deep skepticism of **Niels Bohr**'s hypothesis regarding the discontinuous nature of electron orbitals, along with his deep mathematical intuition that atomic spectra could be represented by eigenvalues. Extending Louis Victor de Broglie's revolutionary conception of matter waves, in which the behavior of atomic particles is governed by the laws of wave propagation, Schrödinger provided a theoretically satisfying and logically consistent picture of the quantum universe in which the problematical, discrete nature of matter is replaced entirely by waves. Individual atoms, in Schrödinger's wave theory, are conceived as having no determinate size, and are but vibrations in space-time extending to infinity, themselves limited to a sequence of discrete patterns governed by Schrödinger's equation. Thus, instead of dealing with fixed positions and velocities of "real" particles, Schrödinger's wave function, ψ, expresses the magnitude of the matter waves that vary across space from point to point and through time from moment to moment.

In this scenario, the probability of finding an "individual" particle at a "particular" position is determined by the absolute square of the wave function $\psi(\chi)$, giving the probability distribution for all the coordinates of the system in the state represented by the wave function. Moreover, as Schrödinger himself went on to show, his wave mechanics and Werner Heisenberg's matrix mechanics were equivalent and both accounted naturally, and in a logically consistent way, for the empirically verified quantization of energy. Thus, what is generally known as quantum mechanics is in large part a synthesis of Schrödinger's and Heisenberg's conceptually distinct yet empirically complementary theories.

In 1934, Schrödinger was offered a position at Princeton University but instead accepted a position in his native Austria at the University of Graz. Four years later, after the German *Anschluss*, the university was renamed Adolf Hitler University and Schrödinger was abruptly dismissed from his position. He fled to Rome, then Oxford, England, and taught for one year at the University of Ghent. Schrödinger then accepted an offer to join the Institute for Advanced Studies in Dublin; he remained there until his retirement in 1956, whereupon he returned to his Vienna.

Daniel Kolak

Selected References

Bitbol, Michel and Oliver Darrigol (1992). *Erwin Schrödinger: Philosophy and the Birth of Quantum Mechanics*. Gif-sur-Yvette, France: Editiones Frontières.

Hermann, Armin (1975). "Schrödinger, Erwin." In *Dictionary of Scientific Biography*, edited by Charles Coulston Gillispie. Vol. 12, pp. 217–223. New York: Charles Scribner's Sons.

Schröter, Johann Hieronymus

Born **Erfurt, (Thüringia, Germany), 30 August 1745**
Died **Lilienthal, (Niedersachsen, Germany), 29 August 1816**

Observational astronomer, telescope builder, and noted selenographer, Johann Schröter provided the first extensive description of lunar rilles and solar granulation (convection cells). He was also the first to establish the presence of an atmosphere around Venus.

Schröter's father, Paul Christoph Schröter, was a lawyer who married Regina Sophia Streckroth in 1729. In 1764, Johann went to Göttingen to study law. But the observatory director, Abraham Gotthelf Kästner, awakened a love of astronomy in the young jurist. In 1777, Schröter was appointed secretary of the Royal Chamber (of King George III) in Hanover.

Being musically inclined, Schröter got to know the family of regimental bandmaster Isaak Herschel, who had nine children. The eldest of these, **William Herschel**, was a self-taught astronomer. Returning from a visit to England, William's brother Dietrich brought Schröter his first telescope, a small achromatic Dolland refractor.

There were many parallels between the lives of William Herschel and Johann Schröter. Both were German-born, and knew from their childhoods the meaning of penury. Both had a passionate fondness for music, and each enjoyed the tender care of a devoted sister. Each had command of the greatest telescopes of his own country. Both were experts at mechanical contrivances; each was supremely energetic, patient, industrious, and conscientious.

The decisive event of Schröter's life took place in 1781, with Herschel's discovery of Uranus. In a spirit of emulation, Schröter resolved to dedicate himself toward astronomy. He resigned his court position in Hanover to assume the less-demanding post of chief magistrate of Lilienthal, a village on the moor near Bremen. He took up residence in the Amthaus (city hall) in 1782.

As a government official, Schröter made use of his first income to place an order with Herschel for a 4-ft. telescope and to erect a two-story observatory. His enthusiasm was infectious; two of his servants, coachman Arnd Harjes and gardener Harm Gefken, became keen coworkers and obviated the need for a trained mechanic. From 1800 onward, Harjes also did most of the illustrations for Schröter's books.

Larger and larger instruments were installed, eventually numbering more than a dozen. Schröter obtained two mirrors, 4.75 in. and 6.5 in. in diameter, made by Herschel himself. The larger of the two mirrors he assembled into a 7-ft. reflector that was in every respect identical with the one Herschel had used to discover Uranus. When Schröter began to use it in 1786, it was the largest telescope in Germany. Ultimately, Schröter's mirrors were also ground in Lilienthal. Gefken taught himself this art, where, adjoining the observatory, the first workshop in Germany for making reflecting telescopes was erected.

The impetus to construct telescopes in Lilienthal was given by Johann Gottlieb Friedrich Schrader, a professor of physics in Kiel, who stayed in Lilienthal in 1792/1793. With Schröter, he constructed a telescope of 25-ft. focal length. Immediately after his return to Kiel, Schrader erected a 26-ft. telescope. Schröter responded to this news by fashioning another mirror (18.5-in. diameter) with a 27-ft. focal length. It remained the largest in Germany for years to come.

Schröter was no mathematician; his strength lay in making visual observations. He chiefly studied objects in the Solar System, publishing 86 memoranda over 30 years. However, very few of his conclusions have stood the test of time. Perhaps Schröter's best-known work was his book on lunar topography, published as two volumes in 1791 and 1797 (reissued 1802). He made hundreds of drawings of lunar surface features, and described and named the "rilles."

Schröter carried out the first extensive investigation into the physical nature of the planet Mercury. From the blunted appearance of its southern cusp, which seemed unchanged from night to night, he concluded that the planet's rotation period must be nearly 24 hours (the real figure being 59 days).

Schröter was also an active solar observer from 1785 to 1795. He was the first, in 1787, to notice and comment upon the surface feature now known as "granulation." He also gave detailed descriptions of light bridges seen over the umbrae of sunspots.

Schröter carried out many observations of Venus and estimated its rotation period at 23 hours 21 minutes (the real figure being 243 days, retrograde). He found that Venus appears to be at half phase not when theoretically expected (*i. e.*, at quadrature) but a few days before or afterward. This has become known as the Schröter effect. In 1790, Schröter definitely established the presence of an atmosphere around Venus by the observed extension of its cusp (viewed in the crescent phase) beyond a semicircle.

In 1785/1786, Schröter recorded multiple transient dark spots on Jupiter, which have been interpreted as activity in the southern equatorial belt of Jupiter. His observations of Saturn led him to believe that its rings were a solid body, another erroneous conclusion.

The Bremen astronomer **Wilhelm Olbers** often stayed at Lilienthal, and loved to observe with Schröter's instruments. The inaugural meeting of the "Celestial Police" (Vereinigte Astronomische Gesellschaft), on 20 September 1800, was held at Schröter's observatory, and he was elected its president. The role of the "Police" was to search for a supposed "missing planet," located between Mars and Jupiter. After discovery of the first minor planet (1) Ceres, Olbers and Schröter regularly observed it from Lilienthal. In 1805, Schröter published the official report of the "Celestial Police," which included studies of the minor planets (1) Ceres, (2) Pallas, and (3) Juno. Discovery of the third asteroid was made by Schröter's assistant, **Karl Harding**.

In 1815, Schröter transferred his instruments to the University of Göttingen, with the stipulation that he could use them as long as he lived. When the kingdom of Westphalia annexed Lilienthal, Schröter wished to void the sale and ship the instruments to France. The French wanted them, but realized that a second sale would be illegal. Göttingen astronomer **Carl Gauss** tried to enlist the aid of French astronomer **Pierre de Laplace** in the matter, but it was too late.

Schröter was alone with only his servants when Lilienthal was engulfed in war. In April 1813, French troops "broke into the observatory … and with a fury the most unprovoked and irrational[,] destroyed or carried off the most valuable clocks, telescopes, and other astronomical and mathematical instruments," Schröter wrote. Just days before, the only copies of nearly his entire works, deposited in a government office building, were completely burned.

Soon afterward, the French troops were expelled from Germany and Schröter, reinstated as chief magistrate, attempted to rebuild Lilienthal. He fought off despair by writing up his observations of the Great Comet C/1811 F1, and then turned to his observations of Mars. Miraculously, most of those records had escaped the fire. His engraver, Tischbein of Bremen, began to make copper plates of the drawings, but Schröter's eyesight was failing. The project was unfinished at the time of his death. Schröter's work on Mars was posthumously published in 1881.

A crater on Mars is named in Schröter's honor, minor planet (4983) was named Schröteria, and a prominent lunar feature is denoted Schröter's valley. He had been elected a fellow of the Royal Societies of Göttingen, London, and Stockholm, and a member of the Russian Academy of Sciences.

Clifford J. Cunningham

Selected References

Baum, Richard (1991). "The Lilienthal Tragedy." *Journal of the British Astronomical Association* 101: 369–371.

Denning, W. F. (1904). "Schroeter and the Burning of Lilienthal in 1813." *Observatory* 27 (1904): 313.

Gerdes, Dieter (1986). *Das Leben und das Werk des Astronomen Dr. Johann Hieronymus Schroeter, geboren am 30. August 1745 zu Erfurt*. Lilienthal: Gerdes.

Oestmann, Günther (2002). "Astronomical Dilettante or Misunderstood Genius? On Johann Hieronymus Schroeter's Image in the History of Science." In *Astronomie von Olbers bis Schwarzschild*, edited by Wolfgang R. Dick, pp. 9–24. *Acta Historica Astronomiae*, Vol. 14. Frankfurt am Main: Harri Deutsch.

Sheehan, William and Richard Baum (1995). "Observation and Inference: Johann Hieronymus Schroeter, 1745–1816." *Journal of the British Astronomical Association* 105: 171.

Schüler, Wolfgang

Flourished **16th century**

While it is universally known as Tycho's supernova, the "new star" of 1572 (B Cas) was first observed 6 November by Wolfgang Schüler in Germany.

Selected References

Clark, David H. and F. Richard Stephenson (1977). *The Historical Supernovae*. Oxford: Pergamon Press.

F. Richard Stephenson and David A. Green (2002). *Historical Supernovae, and their Remnants*. Oxford University Press.

Schumacher, Heinrich Christian

Born **Bramstedt, (Schleswig-Holstein, Germany), 3 September 1780**
Died **Altona, (Hamburg, Germany), 28 December 1850**

Heinrich Schumacher was a reformer of Danish science and the founding editor of the *Astronomische Nachrichten*, the most import astronomical journal of the 19th century. Schumacher, the son of the Danish senior civil servant and chamberlain Andreas Schumacher, was initially taught by the Reverend Johann Friedrich August Doerfer, who was noted for his topographical work in Schleswig and Holstein, as well as at the Altona Gymnasium, where the headmaster was Jakob Struve, the ancestor of the subsequently renowned family of astronomers. Schumacher studied jurisprudence at Kiel, was a private tutor in Livonia, and lectured in law at Dorpat (and simultaneously was active at the observatory there). In 1808/1809, Schumacher went to Göttingen on a royal Danish scholarship to pursue his studies in astronomy under **Carl Gauss**. In 1810, he was extraordinary professor of astronomy at Copenhagen, but, because of differences with the full professor, Thomas Bugge, he remained in Altona and observed from Johann Georg Repsold's observatory in Hamburg. In 1813, Schumacher held a post for a short time at Mannheim, but was not able to carry out observations because of the poor state of the instruments there.

Following the death of Bugge, Schumacher returned to Denmark in 1815. Because the observational instruments at the Copenhagen Observatory on the Round Tower were also inadequate, he devised a project for a Danish land survey, which he carried out in close collaboration with Gauss. Given leave of absence as professor by the king, Schumacher took up residence in Danish Altona. The Danish survey that was implemented from there from 1816 onward was linked to the Hanoverian triangulation, and covered from Skagen in the north to Lauenburg in the south. It was later employed by **Friedrich Bessel** to derive the rotational flattening of the Earth. Together with Bessel and the Danish physicist H. C. Oersted, Schumacher worked to reorganize the Danish system of weights and measures (including measuring the length of a seconds pendulum at Gueldenstein Castle in Holstein), and linked the Danish units to the Prussian system. He established a small, efficient observatory in Altona.

The most important of Schumacher's spheres of activity was the publication of Danish calendars, astronomical tables, and ephemerides, specifically for seamen (from 1820 onward), as well as (from 1821) founding and publishing the astronomical journal *Astronomische Nachrichten*. He did not feel any call to carry out decade-long, nightly observational activities, which may itself have been the result of his uncertain health.

Schumacher's expertise, his careful approach, his diplomatic skills, and his acquaintance with the leading astronomers of the time, soon turned the *Astronomische Nachrichten* into the international center for astronomical communication. Of particular importance was the regular publication schedule, thanks to the direct support for the undertaking given by the Danish kings Frederik VI and the later Christian VIII, as well as by the scholarly Finance Minister Johann Sigismund von Moesting.

By the time of his death, Schumacher had edited 30 volumes of the journal. During this time, there were hardly any astronomers of significance who did not publish their work in it. Throughout this period, the world was informed of the latest discoveries either through the *Astronomische Nachrichten* or through the separately issued *Zirculare* (Circulars). The significance of the *Astronomische Nachrichten*, and Schumacher's work in making it the medium of communication for the international community of astronomers cannot be over-estimated; it was fundamental. For three decades, Schumacher undertook an incredible amount of work, which involved dealing with manuscripts and in carrying on, single-handedly, an extensive correspondence. The letters to Schumacher that have been preserved must, on their own, exceed 15,000.

Schumacher's *Astronomische Nachrichten* reached Volume 327 in 2006 and is the oldest astronomical journal in continuous publication. Its editorial offices are now at the Potsdam Astrophysical Institute.

Jürgen Hamel
Translated by: *Storm Dunlop*

Selected References

Hamel, Jürgen (2001). "Heinrich Christian Schumacher – Mediator between Denmark and Germany; Centre of Scientific Communication in Astronomy." In *Around Caspar Wessel and the Geometric Representation of Complex Numbers*, edited by Jesper Luetzen, pp. 99–117. Copenhagen: C. A. Reitzels Forlag.

Hamel, Jürgen (2002). "H. C. Schumacher – Zentrum der internationalen Kommunikation in der Astronomie und mittler Zuischen Dänemark und Deutschland In *Astronomie von Olbers bis Schwarzschild: Nationale Entwr cklungen und internationale Beziehungen in 19. Johrhundert*. Edited by Wolfgand R. Dick und Jürgen Hanel. Frankfurt: Deutsch (Acta Historica Astronomiae; 14), 89–120.

Olufsen, Christian Friis Rottboell (1853). "Biographische Notizen über den verstorbenen Conferenzrath Schumacher." *Astronomische Nachrichten* 36: 393–404.

Repsold, J. A. (1918). "H. C. Schumacher." *Astronomische Nachrichten* 208: 17–34.

Schuster, Arthur

Born **Frankurt am main, (Germany), 12 September 1852**
Died **Berkshire, England, 14 October 1934**

German–English theoretical astronomer Arthur Schuster is commemorated in the Schuster–Schwarzschild (or reversing layer) approximation for analyzing the spectra of stars to learn their chemical composition. The idea is that you can treat the situation as if there were a hot layer, the photosphere, emitting a blackbody continuum, and a cooler layer above which imposes the absorption lines. The opposite approximation, that the continuum source and absorbing atoms are uniformly mixed, is called the Milne–Eddington approximation, and real stars come somewhere in between. **Karl Schwarzschild**, **Edward Milne**, and **Arthur Eddington** appear elsewhere in this book.

Schuster was the son of a Frankfurt textile merchant and banker. In the wake of the 1866 "seven weeks war" when Frankfurt was annexed by Prussia, the family moved to Manchester, England. Schuster was educated privately and at the Frankfurt Gymnasium. He attended the Geneva Academy from 1868 until he joined his parents at Manchester in 1870. Schuster studied physics at the Owens College, Manchester, and the University of Heidelberg, where he obtained his doctorate in 1873.

After a few years at the Cavendish Laboratory in Cambridge (1875–1881), Schuster returned to Manchester to become professor of applied mathematics (1881–1888) and later professor of physics (1889–1907). After an early retirement at the age of 56, he spent his time with his own research and on the formation of the International Research Council. With his retirement, Schuster made way for **Ernest Rutherford**.

Schuster worked in many areas, many of them related to astronomy:

Spectroscopy. In 1881, Schuster refuted the speculation of **George Stoney** that spectral lines could be regarded as harmonics of a fundamental vibration. He did this using a statistical analysis of spectral lines of five elements. Schuster concluded: "Most probably some law hitherto undiscovered exists which in special cases resolves itself into the law of harmonic ratios." In 1888, **Johann Balmer** took a fairly large step forward when he delivered a lecture to the Naturforschende Gesellschaft in Basel. He represented the wavelengths λ of the spectral lines as $\lambda = h{\cdot}m^2/(m^2 - n^2)$, where m and n are integers. For the hydrogen atom, where n = 2, it would lead to wavelengths $h{\cdot}9/5$, $h{\cdot}16/12$, $h{\cdot}25/21$ …, the Balmer Series seen in the visible. While Schuster had not yet seen this, his statistical analysis had refuted the speculation of a law $\lambda = h\check{c}m$, which Stoney had proposed. The Balmer law $1/\lambda = R{\cdot}(1/m^2 - 1/n^2)$ would later be derived by quantum mechanics.

Schuster's most notable paper, on the analysis of stellar absorption features, was not published until 1905.

Electricity in gases. Schuster was the first to show that an electric current is conducted by ions (charged particles). He also showed that the current could be maintained by a small potential once ions were present. He was the first to indicate a path toward determining the charge–mass ratio e/m for cathode rays by using a magnetic field. This method would ultimately lead to the discovery of the electron.

Terrestrial magnetism. Schuster's study of terrestrial magnetism showed that there are two kinds of daily variations in the magnetic field of the Earth – atmospheric variations caused by electric currents in the upper atmosphere as well as internal variations due to induction currents in the Earth. The Schuster–Smith magnetometer is the standard instrument for measuring the Earth's magnetic field. Schuster's numerous articles examined and rejected many proposed theories of geomagnetism, usually because of shortcomings in their mathematics or physics.

X-rays. In 1896, Wilhelm Röntgen had sent copies of his manuscript to a small group of fellow scientists – Schuster in Manchester, Friedrich Kohlrauch in Göttingen, Lord Kelvin (**William Thomson**) in Glasgow, **Jules Poincaré** in Paris, and Franz Exner in Wien. In the same year, Schuster proposed that the new X-rays of Röngten were, in fact, transverse vibration of the ether of very small wavelength, that is, a short-wavelength extension of the radiation (light) implied by Maxwell's equations.

Antimatter. Schuster published two letters on antimatter in *Nature* in 1898. In them, he surmised "if there is negative electricity, why not negative gold, as yellow as our own?" For 30 years, Schuster's conjecture gathered dust. Only in 1927, did an equation by Paul Dirac predict an oppositely changed counterpart to the electron.

Expeditions. Schuster, having been invited by **Norman Lockyer** to join an expedition to Siam in 1875, to observe a total eclipse, was then asked by **George Stokes** to take charge of the whole expedition on behalf of the Royal Society. In the 19th century, some, if not all, of the world's astronomers believed in a planet inside the orbit of Mercury. This speculative intra-Mercurian planet was called Vulcan. Only a total solar eclipse would make possible seeing it.

The planet Vulcan had been a theoretical construct to solve a problem in planetary dynamics – the mystery of Mercury's orbit. This problem was only resolved in 1915 with **Albert Einstein**'s general theory of relativity, in which the orbital deviations could be explained due to relativistic effects of the Sun's huge mass bending space-time. Vulcan does not exist, and never did; the hunt for it was finally abandoned after the total solar eclipse of 1929.

No eclipse yielded an intra-Mercury planet. But Schuster photographed a comet during the total solar eclipse of 1882.

Laboratory. Schuster raised funds to construct a new laboratory in 1897 and created new departments, including a department of meteorology in 1905.

Schuster was the first secretary of the International Research Council, established under the Treaty of Versailles (which abolished all pre-World-War-I international scientific collaborations) from 1919 to 1928, and was knighted in 1920.

Oliver Knill

Selected References

Anon. (1906). "Professor Arthur Schuster: Biographical and Bibliographical Notes." In *The Physical Laboratories of the University of Manchester*, pp. 39–60. Physical Series, no. 1. Manchester: University Press.

Anon. (1997). *The Grolier Library of Science Biographies*. Vol. 9. Danbury, Connecticut: Grolier Educational.

Kargon, Robert H. (1975). "Schuster, Arthur." In *Dictionary of Scientific Biography*, edited by Charles Coulston Gillispie. Vol. 12, pp. 237–239. New York: Charles Scribner's Sons.

Muir, Hazel (ed.) (1994). "Schuster, Sir Arthur." In *Larousse Dictionary of Scientists*, pp. 460–461. New York: Larousse.

Schwabe, Samuel Heinrich

Born **Dessau, (Sachsen-Anhalt, Germany), 25 October 1789**
Died **Dessau, (Sachsen-Anhalt), Germany, 11 April 1875**

As an amateur lunar, planetary, and especially solar observer, Samuel Schwabe is best known for his discovery of the 11-year sunspot cycle. Schwabe was raised in a scientifically oriented home; his father was a prominent physician, while his maternal grandfather was a pharmacist named Haeseler. Apparently influenced by his grandfather, in 1806, Schwabe began an apprenticeship in a pharmacy in his hometown. During his later pharmaceutical studies in Berlin, he developed a lifelong interest in botany and took his first courses in astronomy.

After his grandfather's death in 1812, Schwabe took over the family pharmacy and became wealthy. He acquired a telescope from a lottery in 1825, recording his first observation of the Sun on 30 October that same year. Acting on the suggestion of **Karl Harding**, Schwabe at first scoured the solar disk in search of an intra-Mercurial planet in transit across the Sun. However, his interest gradually changed to keeping records of sunspots, which he observed every day, whenever the weather at Dessau permitted. Schwabe soon outgrew his first telescope; in 1826 he acquired the 4.8-in. Fraunhofer refractor used by **Wilhelm Lohrmann** to map the Moon until his eyesight failed.

By 1829, Schwabe's interest in astronomical research so absorbed him that he no longer had time for or interest in the pharmacy. He sold it and devoted the rest of his life to research. By 1843, he had still not discovered an intra-Mercurial planet – this is hardly surprising since none exists. However, Schwabe's sunspot data, which had been accumulated through two sunspot maxima and two minima, led him to suspect the existence of a cyclical pattern, with a period of about 10 years. Schwabe was slow to publish; his first announcement, a letter to **Heinrich Schumacher**, printed in the *Astronomical Nachrichten* (no. 495), was almost completely ignored. Clearly, as solar historian Karl Hufbauer suggests, "he was too far outside the astronomical mainstream to receive the attention that, in retrospect, he deserved."

Schwabe was basically a compulsive observer and collector, rather like his great contemporary **Johann Schmidt**. During the period when he was most active as an amateur astronomer, Schwabe was also involved as founding member and president of a local society for natural history, to which he contributed many specimens of plants and minerals. He published a two-volume work, *Flora Anhaltina*, in 1838 in which he described more than 2,000 plants.

Meanwhile, the great German scientist, **Alexander von Humboldt**, presented Schwabe at court. The discovery of the sunspot cycle at last began to take hold after Humboldt publicized Schwabe's work in his celebrated *Kosmos* (1851). Acceptance of the sunspot cycle's reality by astronomers was assured after **Edward Sabine** of the Royal Society (London) and **Johann Wolf** of the Bern Observatory independently noted the correlation between Schwabe's records of sunspot numbers and variations in terrestrial magnetism. Although studied since the 1830s, magnetism was just then being subjected to systematic analysis for the first time. Wolf's reduction of sunspot observations since **Galileo Galilei**'s time indicated a periodicity of 11 years, rather than Schwabe's suggested 10 years.

In recognition of his sunspot work, Schwabe received the Gold Medal of the Royal Astronomical Society [RAS] in 1857. It was presented to him by the well-known British discoverer of solar flares, **Richard Carrington**, in Dessau. This honor made Schwabe something of an anglophile, and accounts for his decision to bequeath his 31 volumes of drawings and observational notes dating from 1825 to 1867 to the RAS. Schwabe was elected to the membership in the Royal Society of London in 1868.

In addition to the Sun, Schwabe was always an avid observer of the Moon and planets. On 5 September 1831, he made the first drawing to indicate clearly the presence of the Great Red Spot Hollow since the observations of **Giovanni Cassini** and his nephew **Giacomo Maraldi** in the late 17th and early 18th century. Interestingly, Schwabe equated the Jovian spots with sunspots, even claiming to see sunspot-like penumbrae surrounding the Jovian spots. However, in contrast to his sunspots, Schwabe never succeeded in reconciling the Jovian spots to any kind of regularly recurring cycle.

Schwabe's observing notebooks and sketches are maintained in the RAS archives at Burlington House, London.

William Sheehan

Selected References

Erfurth, H. (1989). *Samuel Heinrich Schwabe: Apotheker, Astronom, Botaniker.* Dessau: Museum für Naturkunde und Vorgeschichte.

Hufbauer, Karl (1991). *Exploring the Sun: Solar Science since Galileo.* Baltimore: Johns Hopkins University Press.

Johnson, M. J. (1857). "Address delivered by the President, M. J. Johnson, Esq., on presenting the Medal of the Society to M. Schwabe." *Monthly Notices of the Royal Astronomical Society* 17: 126–132.

Schwarzschild, Karl

Born **Frankfurt am Main, Germany, 9 October 1873**
Died **Potsdam, Germany, 11 May 1916**

German theoretical astrophysicist Karl Schwarzschild is eponymized in the Schwarzschild solution to the equations of general relativity, the Schwarzschild horizon around black holes implied by that solution, and a number of other concepts in astrophysics. He was the eldest of seven children of Moses Martin Schwarzschild, a successful member of the Frankfurt business community, whose ancestors can be traced back in the city to the 16th century, and Henrietta Sabel. One of his sisters married **Robert Emden**. Karl Schwarzschild married Else Rosenbach in 1909. They had three children, Agatha (later Thornton, a classicist whose later career was spent in New Zealand), **Martin Schwarzschild**, and Alfred (who remained in Germany into World War II).

Schwarzschild first attended the Jewish community school, and completed the *Arbitur* degree in 1891 at the municipal gymnasium in Frankfurt. He began studying astronomy at the Kaiser Wilhelm University in Strassburg, and, after a year of military service (1893/1894) in Munich, completed his Ph.D. in 1896 with **Hugo von Seeliger** at the Ludwig Maximilian University. Schwarzschild's first job was as an assistant to Leo de Ball at Kuffner Observatory in Vienna (1896–1899), and his second at Munich as a university lecturer.

In 1901, Wilhelm Schur, the director of the Göttingen Observatory, died. After Seeliger and **Maximilian Wolf**, Karl Schwarzschild was third on the recommendation list. Neither Seeliger nor Wolf wanted to go to Göttingen. Schwarzschild was next, but in the beginning the ministry in Berlin did not want to have him as director and full professor at this post. Two more candidates were asked, who also declined. On 10 October 1901, Schwarzschild sent

his parents a telegram, which read, "Extraordinarius and Director. Arrive Monday – Karl." By 24 May 1902, he was appointed to a full professorship.

Much of Schwarzschild's work of lasting importance to astrophysics dates from the Göttingen period. Papers published in 1906 established how stars could exist stably with energy carried entirely by radiation (as first suggested by **Ralph Sampson** in 1894), established the concept of local thermodynamic equilibrium (meaning that the same temperature described the gas and the radiation in a given volume, but that radiation could flow systematically in one direction), and developed the Schwarzschild criterion for deciding when radiation could no longer carry all the energy so that convection would set in, giving rise, for instance, to the observed granulation of the solar surface. He also analyzed the aberrations in various kinds of telescopes.

Schwarzschild also considered the question of how to determine the distances of stars too far away to have measurable parallaxes, arriving at theoretical justification for a method that used only apparent brightnesses and motion across the plane of the sky. He then asked how one might best describe the motions of large numbers of stars through space, arriving (1907/1908) at an alternative to the star streams of **Jacobus Kapteyn**. Schwarzschild's velocity ellipsoid recorded the fact that the dispersion of velocities seemed to be largest in two opposite directions in the sky, next largest in a direction perpendicular to that, and smallest in the third perpendicular direction. These directions are now understood as projections of the rotation of the Milky Way Galaxy and the motions of the stars perpendicular to the galactic plane. Of his students at Göttingen, the one whose work connects most directly with modern astronomy was **Hans Rosenberg**, who, on Schwarzschild's advice, plotted the luminosity of members of the Hyades as a function of their spectral type, thereby publishing in 1910 what later became known as the Hertzsprung–Russell diagram.

In 1909, Schwarzschild was appointed successor of **Hermann Vogel** at the Astrophysical Observatory at Potsdam. Although this was the most prominent position that an astronomer could hope to hold in Germany at that time, Schwarzschild was not at all enthusiastic. His wife's family lived in Göttingen, he had a wide circle of friends, and, above all, Göttingen was a mathematical stronghold. Nevertheless he finally agreed, with the condition that his assistant **Ejnar Hertzsprung** should move to Potsdam with him. Schwarzschild quickly familiarized himself with the various fields of work being carried out at Potsdam. In 1910, he traveled to the United States to attend a meeting of the American Astronomical Society and used the opportunity to visit many of the large American observatories. He returned convinced that Germany definitely needed an observatory in the Southern Hemisphere. Schwarzschild proposed Windhoek, in German South-West Africa.

Schwarzschild's research at Potsdam included additional calculations of stellar atmospheres, including the reversing-layer (or Schuster–Schwarzschild) approximation for how absorption lines are produced in stellar atmospheres, permitting calculation of how much of each element must be present. Other papers reported a study of how dark absorption lines and continuous radiation should appear as a function of position on the solar disk and the fraction of energy carried by convection. He also analyzed observations of the tails of the two great comets of 1910 (C/1910 A1 and 1P/Halley), showing that the tails contain material extraordinarily tenuous even compared to thin air. And Schwarzschild began applying the atomic model of **Niels Bohr** to the analysis of spectra of atoms and simple molecules.

In 1914, World War I broke out and affected the work of the institute more and more. Schwarzschild immediately volunteered for service in the army. In September 1914, he was sent, as acting officer, to Namur in Belgium as head of a field weather station. The whole of 1915 he spent in the field, first in Belgium, later as a member of the artillery staff partly in France and Russia. During the Russian campaign, Schwarzschild already showed symptoms of pemphigus, a painful and then incurable skin disease (now recognized as having an autoimmune component). He was invalided home, hospitalized, and died soon after.

The year of Schwarzschild's death saw the publication of three significant papers – one on ballistics (part of his war work), one explaining the broadening of atomic lines in the presence of an external electric field (the Stark effect, discovered in 1913), and the classic description of the structure of space-time outside a spherically symmetric distribution of mass (or point mass) in the framework of the general theory of relativity, which introduced the concepts of the Schwarzschild radius and Schwarzschild horizon.

Schwarzschild received many outstanding honors and awards. Some of the most important are ordinary member of the Königliche Gesellschaft der Wissenschaften at Göttingen (1905), associate of the Royal Astronomical Society, London (1909), member of the Kaiserlich Leopoldinisch-Carolinsche Deutsche Akademie der Naturforscher (1910), and member of the königlich-preussischen Akademie der Wissenschaften in Berlin (1912).

Peter Habison

Selected References

Dieke, Sally H. (1975). "Schwarzschild, Karl." In *Dictionary of Scientific Biography*, edited by Charles Coulston Gillispie., Vol. 12, pp. 247–253. New York: Charles Scribner's Sons.
Eddington, A. S. (1917). "Karl Schwarzschild." *Monthly Notices of the Royal Astronomical Society* 77: 314–319.
Einstein, A. (1916). "Memorial Lecture on Karl Schwarzschild." *Sitzungberichte der Königlich Preussischen Akademie der Wissenschaften zu Berlin* 1: 768–770.
Hertzsprung, Ejnar (1917). "Karl Schwarzschild." *Astrophysical Journal* 45: 285–292.
Kritzinger, H. H. (1916). "Karl Schwarzschild." *Sirius* 49: 129–130.
Oppenheim, S. (1923). "Karl Schwarzschild. Zur 50. Wiederkehr seines Geburtstags." *Vierteljahresschrift der Astronomischen Gesellschaft* 58: 191–209.
Parkhust, J. A. (1916). "Karl Schwarzschild." *Science*, n.s., 44: 232–234.
Runge, C. (1916). "Karl Schwarzschild." *Physikalische Zeitschrift* 17: 545–547.
Schwarzschild, Karl (1992). *Gesammelte Werke*, edited by H. H. Voigt. 3 Vols. Berlin: Springer-Verlag. (For his publications and for additional biographical references.)
Sommerfeld, A. (1916). "Karl Schwarzschild." *Die Naturwissenschaft* 4: 453–457.
Voigt, H. H. (1989). "Von Karl Schwarzschild bis Hans Kienle – Der Weg von der klassischen Astronomie zur Astrophysik in Göttingen." *Sterne und Weltraum* 28: 12–17.

Schwarzschild, Martin

Born **Potsdam, Germany, 31 May 1912**
Died **Langhorne, Pennsylvania, USA, 10 April 1997**

Martin Schwarzschild put 20th-century understanding of stellar structure and evolution on a firm, quantitative footing by calculating the solutions to the differential equations that describe stellar physics for a range of star masses and compositions. He did so using realistic descriptions for nuclear reactions and energy transport, and evolving those solutions forward in time to reveal the effects of gradual composition changes due to the nuclear reactions. Schwarzschild also made significant contributions to the definition of stellar populations and to observations and theory of solar and stellar convection. Late in his career, Schwarzschild attempted to put the dynamical structure and evolution of elliptical galaxies on a similarly firm numerical footing.

Schwarzschild was the second child and elder son of **Karl Schwarzschild** and Else Rosenbach, the gentile daughter of a local surgeon. Martin was born in Potsdam while his father was director of the Astrophysical Observatory there. **Robert Emden**'s wife was his paternal aunt. After the death of Martin Schwarzschild's father in 1916, the family returned to Göttingen, where Martin was educated in the Gymnasium. Family friends **Carl Runge**, the numerical analyst, and Ludwig Prandtl, a pioneer in aerodynamics, guided his early studies.

Schwarzchild began work at the Göttingen University, initially in mathematics under Richard Courant. That early background provided a foundation for his later grasp of the problems of numerical solution of the nonlinear, coupled differential equations that describe stellar structure, including what is called the Courant condition (that one must not try to take time steps in an evolutionary calculation that are longer than the time required for a sound wave to cross the narrowest zone across which interior conditions change significantly). After one semester in Berlin, Schwarzschild turned to astronomy, completing a doctoral thesis in December 1935, under **Hans Kienle**. His work had initially dealt with observations of Polaris, but he switched to theory of pulsating stars when it became clear that it would be wise for him to leave Germany as quickly as possible.

After a brief stop at Leiden Observatory to meet his father's former friend and colleague **Ejnar Hertzsprung**, Schwarzchild gratefully accepted a 1-year (1936/1937) position as a Nansen Fellow, working with **Svein Rosseland** in Oslo, Norway. There, he completed the publication of his thesis and wrote on the then-puzzling problem of the source of stellar energy.

With the support of both director **Harlow Shapley** and **Cecilia Payne-Gaposchkin**, Schwarzschild received a 3-year Littauer Fellowship at Harvard College Observatory. While there, he worked on the light curves of Cepheid variables and other variable stars and met graduate student Barbara Cherry (BA, Radcliffe College) who became his wife in 1945.

Schwarzschild's first academic appointment was at Columbia University and Rutherfurd Observatory (lecturer in astronomy: 1940–1944, and assistant professor 1944–1947) under director Jan Schilt. There he published papers touching on stellar pulsation, convection, and rotation. He also worked on a new photometer and, foreshadowing the work for which he would be best known, on the use of punch-card machines to integrate differential equations.

Schwarzschild's term at Columbia University was interrupted by service in the United States Army, where he qualified for Officers' Candidate School after becoming a United States' citizen in 1942. He served in an army intelligence unit in Italy where his German accent (conspicuous to the last and much imitated affectionately by his students and colleagues) occasionally caused some confusion.

In 1947, Schwarzschild accepted a position at Princeton University, where stellar astronomy had been badly damaged by the retirement of **Henry Norris Russell**. The university wisely offered positions simultaneously to **Lyman Spitzer** and Schwarzschild; each regarded the other's presence as a major incentive for accepting the appointments. They remained friends and colleagues for 50 years, dying within less than 2 weeks of each other.

Schwarzschild made significant advances in three areas of astronomy. First was stellar populations, the recognition that kinematic and photometric classes identified by **Walter Baade** have associated differences in age and composition. A 1950 paper with Barbara Cherry Schwarzschild demonstrated that stars that move rapidly relative to the Sun (population II) are also deficient in heavy elements. Papers published the same year by Nancy Grace Roman and by Wilhelmina Iwanowska made the same point. Moreover, work with Spitzer, published the next year, was on the forefront in pointing out that the formation of population I stars must be an ongoing process.

Second was the numerical calculation of stellar structure and evolution, including the demonstration that red giants must be chemically inhomogeneous and that population II stars must be older than what was then supposed to be the age of the Universe (joint work with **Fred Hoyle** published in 1955). Much of the numerical work tracing the lives of solar-type stars on beyond hydrogen fusion to helium burning and the ejection of planetary nebulae was in collaboration with Richard Härm. Schwarzschild's 1958 monograph, *The Structure and Evolution of the Stars*, served as primary introduction to the field for astronomers for the next 15 or more years.

Third, Schwarzschild was among the first to recognize the importance of convection in stellar structure and evolution and its relationship to solar granulation. In due course, he assumed primary responsibility for a project, called Stratoscope I, in which a balloon carried a 12-in. telescope to 30 km in order to image the granulation with sufficiently good angular resolution to demonstrate that the granules were truly convection cells. Stratoscope II, with a 36-in. mirror, brought back near-infrared spectra and images of Mars, cool stars, and several galactic nuclei. Schwarzschild predicted in 1975 that the convection cells on red giants would be very large, which was eventually shown to be true in Hubble Space Telescope images of Betelgeuse.

From 1976 onward, most of Schwarzschild's work focused on the structure of elliptical galaxies. His approach was to require that the integrated gravitational contribution of all the stars in their orbits add back up to correspond to the gravitational potential in which the orbits were calculated. This work continues, in the hands of more than a dozen younger astronomers, his students and postdoctoral fellows. Schwarzschild advised a total of 23 Ph.D. students at Princeton University and several at Columbia University.

Schwarzschild was elected to the academies of science of Belgium, Norway, and Denmark, as well as the United States; held honorary D.Sc.s from Swarthmore, Columbia, and Princeton universities; and was the recipient of the United States National Medal of Science (posthumously), the Bruce Medal of the Astronomical Society of the Pacific, the 1994 Balzan Prize (shared with Hoyle), the Karl Schwarzschild Lectureship of the Astronomische Gesellschaft (those lectures being his only publications in a German journal after his 1935 departure), the George Darwin Lectureship of the Royal Astronomical Society, the Russell Lectureship of the American Astronomical Society, and membership in the Royal Society, London.

Schwarzschild was vice president of the International Astronomical Union. His partially overlapping terms as vice president and president of the American Astronomical Society [AAS] (1968–1972) were of particular importance in the history of American astronomy because he oversaw the transfer of ownership of the most prestigious publication in the field, the *Astrophysical Journal*, from the University of Chicago to the AAS. Schwarzschild was also instrumental in preventing the breakup of the Society when workers in solar physics, high-energy astrophysics, and planetary science came to feel that the AAS was no longer serving their needs. His solution, semi-autonomous divisions, has continued to serve to the present.

Schwarzschild was keenly aware of the responsibility of the scientific community to communicate the excitement and importance of science to the rest of society and took the position that projects like Apollo, even if their scientific yields were modest, were nevertheless justified as a source of inspiration for science education and technology.

Schwarzschild's papers are largely archived at Princeton University, but a couple of interesting items concerning his departure from Germany are in the Swarthmore College Library. References to many of his most important publications are found in the obituaries by Mestel (1999) and Trimble (1998).

Virginia Trimble

Selected References

Mestel, Leon (1997). "Martin Schwarzschild." *Bulletin of Astronomical Society of India* 25 (1997): 285–287.

——— (1999). "Martin Schwarzschild." *Biographical Memoirs of the Fellows of the Royal Society* 45: 469–484.

Schwarzschild, Martin (1958). *The Structure and Evolution of the Stars*. Princeton, New Jersey: Princeton University Press. (Reprint, New York: Dover, 1965.)

Trimble, Virginia (1998). "Martin Schwarzschild (1912–1997)." *Publications of the Astronomical Society of the Pacific* 109: 1289–1297.

Weart, Spencer (1977). "Transcript of interview with Martin Schwarzschild." College Park, Maryland: Niels Bohr Library, American Institute of Physics.

Schwassmann, Friedrich Karl Arnold

Born **Hamburg, (Germany), 25 March 1870**
Died **Hamburg, (Germany), 19 January 1964**

German observational astronomer Friedrich Schwassmann, eponymized in several comets, graduated in 1891 after studies at Leipzig, Berlin, and Göttingen universities. He initially held short-term appointments at the observatories in Potsdam (1893–1895), Göttingen (1896/1897), and Heidelberg (1897–1901), where he worked under the supervision of **Maximilian Wolf**. Schwassmann spent the next 2 years at the institute for testing of chronometers of the German Maritime Observatory, and the rest of his life connected with the observatory in Hamburg–Bergedorf. He was appointed as an observer in 1902 and retired in 1934, but continued work as a volunteer for the next 25 years and frequently attended seminars and lectures at the observatory.

Schwassmann is remembered largely for the comets he discovered together with his younger assistant **Arno Wachmann** from 1927 onward, including three short-period comets – 29P/1927 V1, 31P/1929 B1, and 73P/1930 J1—that each carry the name Schwassmann–Wachmann. He also discovered a couple of minor planets (at Heidelberg, with Wolf). Schwassmann's most extensive work at Hamburg–Bergedorf consisted, first, of observations of nebulae and star clusters for the second index catalogue of **John Dreyer** and, later of observations of accurate positions of stars from the selected areas of **Jacobus Kapteyn** using a double astrograph.

Martin Solc

Selected Reference

Wachmann, A. A. (1964). "Arnold Schwassmann." *Mitteilungen Astronomische Gesellschaft* 17: 42–46.

Scot, Michael

Born **possibly the Borders of Scotland, *circa* 1175**
Died **possibly the Borders of Scotland, *circa* 1234**

Michael Scot's main contribution to astronomy was through his translations of works from Arabic into Latin, including those by **Al-Bitruji** and **Aristotle**, and significantly **Ibn Rushd**'s *De motibus* *coelorum*. Scot was thus instrumental in reintroducing Aristotelian ideas to the Western world.

Scot traveled widely around Europe, serving as court astronomer and physician (or astrologer and alchemist) to Holy Roman Emperor Frederick II. Legends of his supernatural powers abound (popularized by **Dante Alighieri**, Giovanni Boccaccio, and Sir Walter Scott), and few facts are known about his life.

Douglas Scott

Selected References

Carmody, Francis J. (ed.) (1952). De motibus celorum: *Critical Edition of the Latin Translation of Michael Scot*. Berkeley: University of California Press.

Thorndike, Lynn (1965). *Michael Scot*. London: Nelson.

Scottus [Scotus] Eriugena, Johannes [John]

Flourished **(France), 9th century**

Johannes Scottus Eriugena was a scholar of the Carolingian Renaissance. His most famous astronomical work is a commentary on that of **Martianus Capella**. The claim that Eriugena's model of the Sun, stars, and planets anticipates the Tychonic system is controversial.

Selected Reference

Eastwood, Bruce S. (2001). "Johannes Scottus Eriugena, Sun-centred Planets, and Carolingian Astronomy." *Journal for the History of Astronomy* 32: 281–324.

Seares, Frederick Hanley

Born **near Cassopolis, Michigan, USA, 17 May 1873**
Died **Honolulu, Hawaii, USA, 20 July 1964**

American photometrist Frederick Seares was responsible for a large fraction of the work of measuring accurate apparent brightnesses of stars as part of a multi-observatory project to understand the distribution of stars in the Milky Way. He was the son of Isaac Newton Seares and Ella Ardelia (*née* Swartwout) Seares, and, after two family moves to Iowa and California, received his BS in 1895 from the University of California (Berkeley). The university later awarded Seares an LLD, and he also had an honorary degree from the University of Missouri.

Seares obtained a position of instructor at the University of California and married Mabel Urmy. Soon, though, he decided to continue his education in Europe, studying first at the University of Berlin for a year and then at the Sorbonne in Paris for another year. Seares then returned to the United States with his family to take a position of professor of astronomy at the University of Missouri, in Columbia, where **Harlow Shapley** was among his undergraduate students. Seares spent 8 years as the director of the university's Laws Observatory. He managed to make a number of improvements at the observatory, both in the quality of equipment and in the quality of the research efforts, and was able to influence a number of students including Shapley.

In 1909, **George Hale** brought Seares to the Mount Wilson Observatory as the new head of the Computing Division. More significant for Seares was that he was given editorial control over the observatory's publications along with his astronomical duties. These tasks put Seares in regular contact with many of the astronomers of his time, through correspondence and cooperation in joint research and publication, and he served as one of the editors of the *Astrophysical Journal* from 1927 to 1941. Many colleagues gratefully allowed the meticulous Seares to edit their work to the point of complete rewriting.

Seares was part of the United States delegation to the 1919 and 1922 conferences in Brussels and Rome that established the International Research Council (later International Council of Scientific Unions) and the International Astronomical Union [IAU].

An early research interest for Seares, which would stay with him throughout his career, was astronomical photometry. With Seares's international experience he began an involvement with the Dutch astronomer **Jacobus Kapteyn**, and his statistical efforts to determine the shape and size of the galaxy (then widely thought to constitute the whole Universe) by using detailed studies of 252 "Selected Areas" of the sky. Seares's interest was in establishing photographic–photometric standards that would provide accurate magnitude estimates for the stars in the Selected Areas. He would play a major role in establishing the magnitudes of the stars in Selected Areas 1–139. In participating in this effort, Seares would work with a number of the world's leading astronomers and lead the primary effort for constantly refined accuracy in this work.

In 1922, Seares was elected the first president of the Commission on Stellar Photometry of the IAU. For the next several decades, his work would provide the standard in stellar photometry. He also would rise to the position of assistant director at the Mount Wilson Observatory by 1925. A 1931 paper by Seares was one of the first extensive efforts to incorporate the effects of absorption by interstellar dust (discovered in 1930 by **Robert Trumpler**) in the analysis of stellar statistics.

Seares received the Bruce Medal of the Astronomical Society of the Pacific in 1940 for his work "in determining fundamental standards" in astronomy, having previously served the society for two terms on their board of directors and one as president.

Following his official retirement, Seares was appointed a research associate at Mount Wilson from 1940 to 1946 and became largely inactive in astronomy after that period. Much of his work on photometric standards has been replaced over the years by photodectric methods developed by astronomers that followed him. This, of course, would have been most pleasing to the man who always strove to obtain the highest quality in the information he published.

Richard P. Wilds

Selected References

Hale, George E., Frederick H. Seares, Adriaan van Maanen, and Frederick Ellerman (1918). "The General Magnetic Field of the Sun: Apparent Variation of Field-Strength with Level in the Solar Atmosphere." *Astrophysical Journal* 47: 206–254.

Seares, Frederick H. (1913). "The Photographic Magnitude Scale of the North Polar Sequence." *Astrophysical Journal* 38: 241–267.
——— (1914). "Photographic Photometry with the 60-Inch Reflector of the Mount Wilson Solar Observatory." *Astrophysical Journal* 39: 307–340.
——— (1915). "Photographic and Photo-visual Magnitudes of Stars Near the North Pole." *Astrophysical Journal* 41: 206–236.
——— (1915). "Color-Indices in the Cluster NGC 1657." *Astrophysical Journal* 42: 120–132.
——— (1931). "Effect of Space Absorption on the Calculated Distribution of Stars." *Astrophysical Journal* 74: 91–100.
——— (1931). "A Numerical Method of Determining the Space Density of Stars." *Astrophysical Journal* 74: 268–287.
——— (1938). "Photoeletric Magnitudes and the International Standards." *Astrophysical Journal* 87: 257–279.
——— (1940). "The Dust of Space." *Publications of the Astronomical Society of the Pacific* 52: 80–115.
Seares, Frederick H. and Edwin P. Hubble (1920). "The Color of the Nebulous Stars." *Astrophysical Journal* 52: 8–22.
Seares, Frederick H. and Mary C. Joyner (1943). "Effective Wave Lengths of Standard Magnitudes; Color Temperature and Spectral Type." *Astrophysical Journal* 98: 302–330.
——— (1952). "Photovisual Magnitudes and Color Indices in 42 Kapteyn Selected Areas." *Publications of the Astronomical Society of the Pacific* 64: 202–204.
Seares, Frederick H., Frank E. Ross, and Mary C. Joyner (1941). *Magnitudes and Colors of Stars North of +80°.* Carnegie Institution of Washington Publication No. 532. Washington, D.C.: Carnegie Institution of Washington.
Seares, Frederick H., P. J. van Rhijn, Mary C. Joyner, and Myrtle L. Richmond (1925). "Mean Distribution of Stars According to Apparent Magnitude and Galactic Latitude." *Astrophysical Journal* 62: 320–374.
Seares, Frederick H., J. C. Kapteyn, P. J. van Rhijn, Mary C. Joyner, and Myrtle L. Richmond (1930). *Mount Wilson Catalogue of Photographic Magnitudes in Selected Areas 1–139.* Carnegie Institution of Washington Publication No. 402. Washington, D.C.: Carnegie Institution of Washington.

Secchi, (Pietro) Angelo

Born **Reggio nell'Emilia, (Emilia Romagna, Italy), 18 June 1818**
Died **Rome, Italy, 26 February 1878**

A pioneer in the study of the physical characteristics of celestial bodies, Pietro Angelo Secchi, S. J., observed spectra of the stars, classified more than 4,000 of them according to a scheme he devised, and made important studies on the physical constitution of the Sun. His achievements, during the 1860s and 1870s, contributed to the rapid growth of astrophysics as a new way of studying the heavens.

Secchi's parents, who aimed to give their son an education fit for his quick mind, had him attend the Jesuit Gymnasium in his hometown. He was only 15 when he became a Jesuit novice in Rome. There, he devoted himself to the study of classical literature, philosophy, and the exact sciences. While deeply interested in all fields of knowledge, Secchi soon showed a greater concern for science. From 1841, he taught physics at Loreto College. His scientific interests came to encompass mathematics, astronomy, magnetism, chemistry, optics, and so forth.

In 1845, he was called to Rome to study theology and was ordained in 1847. The following year, as a result of the expulsion of the Jesuits from Rome, Father Secchi took refuge at the Jesuits' Stonyhurst College in England, and in Georgetown, near Washington, District of Columbia, USA. He became an assistant at the latter observatory, where he improved his astronomical knowledge. In 1849, upon the readmission of the Jesuits to Rome, Secchi returned there, and, in 1850, Pope Pius IX appointed him as director of the Collegio Romano Observatory, replacing **Francesco de Vico** who had died while in exile at London.

One of Secchi's first tasks was to upgrade the equipment and oversee the relocation of the college's observatory. He first acquired a 9.5-in. Merz refractor, identical in size to that used at Dorpat by **Friedrich Struve**, for the continued observation of binary stars. Over the next 7 years, he completed nearly 1,300 measurements of double stars. Secchi's devotion to the newer physical astronomy was a choice dictated in part by the structural limitations of the observatory. Its location in the tower of the Jesuit Church of San Ignacio made it difficult to conduct precise positional measurements of stars necessary for the preparation of an astrometric catalog. Despite the modest equipment at his disposal, Secchi used it to the utmost in coming years.

By employing the analytical techniques of **Robert Bunsen** and **Gustav Kirchhoff**, Secchi began to investigate the stars with a spectroscope in 1863, and in that year, proposed a scheme of spectral classification. He initially subdivided stars into two classes, consisting of yellow or red (colored) stars, and white stars. Although the discriminating criterion was nominally the star's color, Secchi came to associate colors with more definite spectral characteristics and founded his classification scheme upon the latter. By 1866, he had identified a third class and by 1868, announced four principal Secchi types of stellar spectra. These were characterized as follows.

Type I: White or blue-white stars (*e. g.*, Sirius, Vega), whose main spectral features were a few strong absorption lines, attributed by Secchi to hydrogen.

Type II: Yellow stars (*e. g.*, Sun, Arcturus) with more numerous, narrow absorption lines. While still visible, the hydrogen lines were less intense.

Type III: Orange or red stars (*e. g.*, Betelgeuse, α Her), with spectra having wide, dark bands and a maximum intensity on the red side. The majority of stars observed belonged to these three types, which he ordered according to the criterion of increasing complexity.

Type IV: Very red stars (*e. g.*, 19 Piscium), whose spectra showed dark bands but with conspicuous differences from Type IIIs. Secchi observed that these spectra were similar to the inverted (*i. e.*, flame) spectrum of carbon; he had discovered the carbon stars.

Type V: A rare type (*e. g.*, γ Cassiopeia), with bright *emission* lines, was announced in 1877.

This scheme was adopted in *Le Stelle* (The Stars), published by Secchi in 1877, representing a culmination of his studies in stellar spectroscopy. His classification of the spectra of some 4,000 stars represents his major contribution to stellar astrophysics. The system was employed by astronomers for roughly 50 years until superseded by the Harvard classification system. Central to Secchi's achievement was his notion that the enormous diversity of stellar spectra could be reduced to a classification scheme employing just a few basic types.

Secchi offered a qualitative interpretation of the features he observed in stellar spectra. He guessed that stellar temperatures could be the physical parameter most responsible for the differences seen in their spectra. By comparing stellar and laboratory spectra, Secchi identified hydrogen as the element whose strong lines were found in the Type I stars. He supposed that their widths could be related to the pressures existing in the stars' outer layers.

The other field of research in which Secchi made important contributions was solar physics. Secchi studied numerous phenomena on or above the Sun's surface. During the total solar eclipse of 18 July 1860, he observed and photographed the Sun's corona and prominences. By comparing his results with those obtained by **Warren de la Rue**, some 250 miles away, Secchi convincingly demonstrated that those features physically belonged to the Sun and were neither optical effects nor due to an atmosphere of the Moon, as some astronomers had argued.

In subsequent years, Secchi regularly observed solar prominences outside of the times of eclipse, by employing the spectroscopic technique originally developed by **Pierre Janssen** and independently by **Norman Lockyer**. From these observations, Secchi proposed a classification of prominences, which he distinguished as "quiescent" and "eruptive", terms still employed today. Secchi likewise studied sunspots, their distribution, and their relationship with prominences. By observing sunspots at various solar latitudes, Secchi determined that the Sun has a differential rotation and behaves more like a liquid than a solid body. He named the bright areas around sunspots "faculae", deduced (correctly) that solar granulation was attributed to the action of convection cells, and measured the effect of limb darkening.

Secchi's solar studies were summarized in *Le Soleil* (The Sun), published in 1875–1877, in which Secchi related the observed surface phenomena to an overall model of the Sun's structure. He took the Sun to be composed mainly of gas and subject to complex circulation, with surface eruptions driven by an unrecognized force (later found to be magnetic fields).

Secchi turned his spectroscope onto the planets, comets, meteors, and nebulae. He provided an early classification of the latter as "planetary," "elliptical," and "irregular." Secchi argued for the existence of an interstellar medium as the cause of those dark lanes extending the length of some nebulae especially the Andromeda Nebula.

To coordinate and communicate those spectroscopic observations made by astronomers, Secchi, **Lorenzo Respighi**, and **Pietro Tacchini** founded, in 1871, the Società degli Spettroscopisti Italiani, the first scientific society expressly devoted to astrophysics.

Secchi was concerned with other scientific subjects, including geodesy and meteorology. He invented a "meteorograph," a device able to record the time variations of atmospheric temperature and pressure on a moving sheet of paper. It was shown at the 1867 Paris Universal Exhibition and was awarded the Grand Prize, conferred upon Secchi by Napoleon III.

After the Franco–Prussian War, when Italian nationalist troops occupied Rome and the Vatican, Secchi remained loyal to the Pope, spending the next 8 years with him as a voluntary prisoner. His own death occurred only three weeks after that of the Pope.

Davide Cenadelli

Selected References

Abetti, Giorgio (1928). *Padre Angelo Secchi, il pioniere dell'astrofisica*. Milan: Giacomo Agnelli.

——— (1975). "Secchi, (Pietro) Angelo." In *Dictionary of Scientific Biography*, edited by Charles Coulston Gillispie. Vol. 12, pp. 266–270. New York: Charles Scribner's Sons.

Accademia Nazionale delle Scienze (1979). *Padre Angelo Secchi nel centenario della morte (1878–1978)*. Rome: Accademia Nazionale delle Scienze detta dei XL.

Bricarelli, Carlo (1888). "Della vita e delle opere del P. Angelo Secchi." *Pontificia Accademia dei Nuovi Lincei* 4: 41–105.

Brück, H. A. (1979). "P. Angelo Secchi, S.J., 1818–1878." *Irish Astronomical Journal* 14: 9–13.

Hearnshaw, J. B. (1986). *The Analysis of Starlight: One Hundred and Fifty Years of Astronomical Spectroscopy*. Cambridge: Cambridge University Press.

McCarthy, Martin F., S. J. (1950). "Fr. Secchi and Stellar Spectra." *Popular Astronomy* 58: 153–169.

Respighi, Lorenzo (1879). "Elogio del P. Angelo Secchi." *Voce della verità*: 5–32.

Rigge, William F. (1918). "Father Angelo Secchi." *Popular Astronomy* 26: 589–598.

See, Thomas Jefferson Jackson

Born **near Montgomery City, Missouri, USA, 19 February 1866**

Died **Oakland, California, USA, 4 July 1962**

American astronomer T. J. J. See is remembered, if at all, for erroneous, perhaps even fraudulent, claims for the detection of planets orbiting other stars, though others of his once wild-sounding ideas sound superficially like our modern understanding of, for instance, solar-system formation. See earned his undergraduate degree from the University of Missouri at Columbia in 1889 and his doctorate from the University of Berlin in 1892 with a thesis on the orbits and origins of visual binary stars. Upon returning to the United States, he spent 3 years at the University of Chicago. While there, he ran afoul of **George Hale**, the driving force behind the establishment of the Yerkes Observatory (and later those at Mount Wilson and Palomar), a circumstance that did little to advance See's career. See spent the next 2 years in the employ of **Percival Lowell**, during which time the former directed a survey of southern double stars as observed from Mexico. In this case, as in most cases during his career, See was a cause of contention, later being accused of falsifying observations.

In 1899, See was appointed a United States Navy professor of mathematics, a rather senior post for so young a man, probably due to the influence of fellow Missourian Champ Clark, a powerful member of the United States House of Representatives. See spent the next 3 years at the United States Naval Observatory, followed by a year teaching mathematics at the Naval Academy. Finally, See was transferred to the Naval Observatory at Mare Island, California, in 1903, and he stayed there until retirement in 1930. The Mare Island Observatory was a dead end, for there were no instruments capable of serious astronomical research. See was the only astronomer there, and his primary duty was related to chronometers. Moreover, he had little interaction with other astronomers, whether in the Bay Area or elsewhere.

See's early work on double stars was generally regarded as good, but he soon began to make dubious and outlandishly inflated claims for his researches on a wide variety of subjects. See's report of visual detection of planetary companions to nearby stars was published only in *Atlantic Monthly* and must have arisen entirely from his imagination. But his orbit for a planet around the secondary star of the visual binary 70 Ophiuchi appeared in the *Astronomical Journal* in 1896 and looks very much like it might have come from real data. Curiously the same star was one of those for which there were erroneous reports of planetary detections in the 1940s. In 1899, **Forest Moulton** demonstrated that See's invisible companion in the 70 Ophiuchi system did not exist; the latter's intemperate response to this refutation caused him to be banned from the pages of the *Astronomical Journal*. Later, in 1912, Moulton showed that parts of See's "capture theory" of planetary formation had, in fact, been "captured" from previously published work of Moulton's. Such peccadilloes ended See's career as a professional astronomer, and he was ostracized from most professional publications (a most unusual circumstance). Although conscientious in his own way and incredibly industrious, he toiled on alone, secure in his belief that he was one of the greatest (and least appreciated) scientists of all time –"The American Herschel" and "The Newton of Cosmogony."

Because of See's outlandish behavior, his speculations were universally scoffed at, though some of them turned out to be correct (generally not for the reasons he suggested). The most startling example was his assertion that lunar craters were due to impacts and not due to volcanism, which was the accepted theory at the time (though both ideas can be traced back very far). In that connection, See had experiments performed with projectiles fired from naval guns, resulting in miniature craters complete with central peaks. Moreover, he stated that all bodies in the Solar System that have solid surfaces carry similar scars, a prediction that has been amply confirmed.

In other areas, See believed that the orbits of the major planets are so nearly circular because of the effects of a resisting medium during their formation, but his circularizing medium was the "luminiferous ether," which he was still claiming as one of his scientific interests into the 1930s, long after the Michelson–Morley experiment had shown that no such medium exists. He also said that mountain ranges are not elevated because the Earth is cooling and shrinking. Both suggestions were considered outlandish at the time made, but not now.

In recent decades, See has become somewhat of an icon for a segment of planetary astronomers. See was an incredibly enthusiastic joiner of scientific societies (arguably to bolster his fading reputation) and belonged to at least 25 societies in astronomy, mathematics, seismology, and physics in at least five countries.

Ronald A. Schorn

Selected References

Ashbrook, Joseph (1962). "The Sage of Mare Island." *Sky & Telescope* 24, no. 4: 193, 202.

Moulton, F. R. (1899). "The Limits of Temporary Stability of Satellite Motion, with an Application to the Question of the Existence of an Unseen Body in the Binary System F. 70 *Ophiuchi*." *Astronomical Journal* 20: 33–37.

——— (1912). "Capture Theory and Capture Practice." *Popular Astronomy* 20: 67–82, esp. 76–82.

See, T. J. J. (1896–1910). *Researches on the Evolution of the Stellar Systems*. 2 Vols. Lynn, Massachusetts: Nichols Press.

Webb, William Larkin (1913). *Brief Biography and Popular Account of the Unparalleled Discoveries of T. J. J. See*. Lynn, Massachusetts: T.P. Nichols and Son.

Seeliger, Hugo von

Born **Biala (Bielsko-Biala, Poland), 23 September 1849**
Died **Munich, Germany, 2 December 1924**

Hugo von Seeliger pioneered the use of stellar statistics to derive an improved understanding of the distribution of stars in our Milky Way Galaxy. Son of the mayor of Biala, Seeliger studied physics, mathematics, and astronomy at the universities of Heidelberg and Leipzig. After writing his Ph.D. thesis under **Karl Bruhns** in 1871, he gained his first practical experience as assistant in Leipzig and, from 1873 to 1878, as observer at the Bonn Observatory, directed by **Friedrich Argelander**. Hugo von Seeliger took part in their observations for the great zone catalog of the Astronomische Gesellschaft and joined an expedition to the Auckland Islands in 1874 to observe the transit of Venus. From 1878 to 1881, he was a *Privatdozent* (lecturer) at the University of Leipzig, before he became director of the Grand-Ducal Observatory in Gotha, succeeding **Peter Hansen**.

From 1882 on, Seeliger was full professor of astronomy and director of the Munich Observatory, where he succeeded **Johann von Lamont**. Working conditions at Munich were so satisfactory that he refused later calls to Prague, Strasbourg, Vienna, and Potsdam. During his tenure in that post, Seeliger turned Munich into a major center of astronomical training. **Karl Schwarzschild** was his most prominent student. From 1896 to 1921, Seeliger was president of the Astronomische Gesellschaft, and from 1918 to 1923 he presided over the Munich Academy of Sciences. He was nominated as a corresponding member of the Prussian Academy of Sciences. In 1885, he married Sophie Stoeltzel; the couple had two sons.

Hugo von Seeliger's main achievement was in the field of stellar statistics. He studied the spatial distribution of stars having the same apparent magnitude in a given part of the sky. Numbers of stars were expected to increase by a factor of 2.5 raised to the 3/2 power (or roughly by a factor of four) per magnitude, if they were uniformly distributed. With increasing distance, r, the volume of space increases with the cube of r, while the luminosity falls with the square of r (a reflection of the inverse-square law). But his statistical studies – derived from the two Bonn sky surveys – only yielded a factor between 2.8 and 3.4, which led him to infer a diminishing number of stars in space far from the Sun. As in **Jacobus Kapteyn**'s independent research on the same topic, also written around 1900, Seeliger devised a lenticular model of the Milky Way, having the Sun close to its center, and a maximum extension in the galactic plane of some 30,000 light years. This model of the Galaxy was widely adopted until significant revisions accompanied **Harlow Shapley**'s determination of the distances to the globular clusters.

When physicist **Erwin Freundlich** began exploring gravitational redshifts as one of three experimentally testable consequences of **Albert Einstein**'s general theory of relativity, Seeliger reacted strongly. In 1916, he documented mathematical errors in one of Freundlich's publications, although his polemics reached far deeper and were meant to demolish Einstein's theory of gravitation. He also published alternative interpretations of the other two effects predicted by Einstein's theory – the precession of Mercury's perihelion, and the apparent slight deflection of star positions visible near the Sun's limb during solar eclipses. Among the predominantly conservative German astronomers, Seeliger, who had already acquired a reputation as a merciless critic, became one of the most outspoken antirelativists.

Classical cosmology is another field in which Seeliger's contributions are remembered. He had noticed that a combination of the Euclidean structure of space, a nonvanishing mean density of matter, and **Isaac Newton**'s law of gravity led to an inherent instability of the strictly Newtonian cosmos. Two decades before the advent of a dynamic (and relativistic) cosmology, this situation appeared inconceivable to him. Thus, he tried to remedy the situation by modifying Newton's law. His reflections on the status of absolute motion in Newtonian mechanics were inspired by his Leipzig mathematics teacher, **Carl Neumann**, who had introduced an alternative concept to Newton's absolute space for the definition of inertial motion.

Other interests of Hugo von Seeliger included theories of the motion of double stars and of star systems containing three or four stars (*e. g.*, ζ Cancri), the spectra of novae (such as τ Aurigae) and their interpretation, physiological optics, the photometry of Saturn's rings and of cosmic dust clouds, the zodiacal light, and anomalous refraction in the terrestrial atmosphere. His colleagues valued his combination of deep theoretical insight, mathematical proficiency, and remarkable skill in practical astronomy. A *Festschrift* honoring his services to astronomy was organized on the occasion of his 75th birthday, which occurred less than 3 months before his death.

Klaus Hentschel

Selected References

Eddington, Arthur S. (1925). "Hugo von Seeliger." *Monthly Notices of the Royal Astronomical Society* 85: 316–318.

Grossmann, E. (1925). "Hugo von Seeliger." *Astronomische Nachrichten* 223: 297–304.

Keinle, Hans (1924). *Probleme der Astronomie: Festschrift für Hugo von Seeliger, dem Forscher und Lehrer, zum 75. Geburtstage*. Berlin: J. Springer.

Paul, Erich Robert (1977). "Seeliger, Kapteyn, and the Rise of Statistical Astronomy." Ph.D. diss., Indiana University. (Revised as *The Milky Way Galaxy and Statistical Cosmology, 1890–1924*. Cambridge: Cambridge University Press, 1993, esp. Chap. 3.)

Schmeidler, F. (1975). "Seeliger, Hugo von." In *Dictionary of Scientific Biography*, edited by Charles Coulston Gillispie. Vol. 12, pp. 282–283. New York: Charles Scribner's Sons.

Wilkens, Alexander (1927). *Hugo von Seeligers wissenschaftliches Werk*. Munich: Bayerische Akademie der Wissenschaften.

Seleukus of Seleukeia

Flourished **Seleukia, (Iraq), 150 BCE**

Seleukus appears to have argued for an infinite heliocentric *kosmos*, and was the first to hypothesize a mechanism for the long-known lunar influence on the tides. The precise nature of his theory is hard to determine, because his own writings have not survived.

Seleukus was from the city of Seleukeia, on the Persian Gulf near the mouth of the Tigris and Euphrates rivers, and studied with Mesopotamian astronomers and astrologers. His apparent date is determined from the fact that he responded to Krates of Mallos (who was himself active in the decades around 165 BCE), and he must have preceded **Hipparchus**' *Geography*, composed in the decades around 140 BCE, which responds to his ideas.

Plutarch, in the *Platonic Questions* 8.1, records that Seleukos proclaimed what **Aristarchus** had only hypothesized, that the Earth rotated; since Aristarchus also hypothesized that the Earth orbited the Sun, it is usually assumed that Seleukus did so as well. One ancient objection to a heliocentric theory was the apparent absence of stellar parallax (not in fact observed until 1836 by **Friedrich Bessel**, **Friedrich Struve**, and **Thomas Henderson**), which Aristarchus answered by hypothesizing that the sphere of the fixed stars was large enough to make their parallax imperceptible. Seleukus is known to have argued for an infinite universe on philosophical grounds similar to those earlier advanced by the Pythagorean **Archytas** "if the *kosmos* had a boundary, what would happen if you penetrated it?"

A second major ancient objection to a heliocentric theory was the absence of evidence that the Earth rotated. Seleukus' tidal model held that the rotation of the Earth and the orbital motion of the Moon disturbed the *pneuma* (vital spirit) filling the intervening space, which swelled the ocean (*i. e.*, the tides provide the unambiguous evidence of the Earth's rotation). He argued that when the Moon is over the Earth's Equator, the tides are regular and their irregularity increases in proportion as the Moon is distant from the Earth's equatorial plane. He also noted that the tides differed from sea to sea, and he divided the monthly tidal cycle into seven phases.

Whether or not Seleukus advocated a fully heliocentric system is unclear, but he may be the astronomer responsible for the partly heliocentric epicyclical system recorded by **Theon of Smyrna** and by **Vitruvius**. According to that model, the hollow orbital sphere of Venus encloses that of Mercury, which in turn encloses the solid sphere of the Sun, and those three together orbit the Earth carried on a common hollow sphere.

Paul T. Keyser

Selected References

Dicks, D. R. (1960). *The Geographical Fragments of Hipparchus*. London: Athlone Press, pp. 114–115.

Heath, Sir Thomas L. (1959). *Aristarchus of Samos, the Ancient Copernicus*. Oxford: Clarendon Press, pp. 305–307.

Neugebauer, Otto (1975). *A History of Ancient Mathematical Astronomy*. 3 pts. New York: Springer-Verlag, pt. 2, pp. 610–611, 697–698.

Russo, Lucio (1994). "The Astronomy of Hipparchus and His Time: A Study Based on Pre-Ptolemaic Sources." *Vistas in Astronomy* 38: 207–248.

Seneca

Flourished **Rome, (Italy), 1st century**

Roman author Seneca asserted that comets follow fixed paths, thereby anticipating **Edmund Halley**.

Selected Reference

Sørensen, Villy (1984). *Seneca*. Chicago: University of Chicago Press.

Seng Yixing

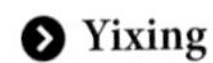

Yixing

Serviss, Garrett Putnam

Born **Sharon Springs, New York, USA, 24 March 1851**
Died **Englewood, New Jersey, USA, 25 May 1929**

Garrett Serviss's astronomy popularization spanned more than a half-century, sparked when he was a child by viewing the night sky from his father's farm. Clyde Fisher, who would become director of New York City's Hayden Planetarium, wrote that Serviss did "more to popularize astronomy than any one in America, and perhaps in the entire world."

Serviss's parents were Garrett Putnam and Catherine (*née* Shelp) Serviss, whose ancestry can be traced to pre-Revolutionary settlers in the Mohawk Valley. Serviss graduated from Cornell University in 1872 with a science degree, obtained an L.L.B degree from Columbia University in 1874, and was admitted to the New York State bar that year. Instead of practicing law, he chose journalism as a career, becoming a newspaper writer and editor, particularly with the *New York Sun*, for which he wrote a long series of science articles. Serviss resigned from the paper in 1892, took a very successful 2-year "Urania Lectures" tour of the United States, and then continued writing. He joined the American Association for the Advancement of Science and the American Astronomical Society, among others. Serviss was married twice; his first wife, Eleanore Betts died in 1906. His second wife, Henriette Gros Gatier, survived him.

Serviss is perhaps better remembered by science-fiction buffs than by the astronomical community. In retrospect, his entry into that genre was bold indeed, a sequel to H. G. Wells's *War of the Worlds*. The first installment of *Edison's Conquest of Mars* (1898) debuted in the *New York Evening Journal* 6 weeks after the last installment of the serialized Wells classic appeared in *Cosmopolitan* magazine. Nevertheless, Serviss, at best, was a pedestrian writer of science fiction.

In contrast, the several books Serviss wrote for the general public about observational astronomy were extremely well executed and remain worthwhile reading today. The first, *Astronomy with an Opera Glass* (1888; 2nd ed., 1896), was perhaps the best of the lot. The second, Serviss's *Pleasures of the Telescope* (1901), is another gem. The first is a romp through the Northern Hemisphere sky for observers abetted by only minimal optical aid. (Opera glasses were the forerunners of modern binoculars.) The second addresses the exploration of the Universe through what would be regarded at present as very small amateur telescopes. Serviss viewed from Brooklyn, New York, with a 3⅜-in. refractor. The final chapter of *Pleasures* is entitled, "Are there Planets among the Stars?" It seems that Serviss could not resist letting a little speculation spice his otherwise straightforward text.

Leif J. Robinson

Selected References

Barritt, Leon (August 1929). "Garrett Putnam Serviss: Lecturer, Writer and Scientist Dies in his 79th Year." *Monthly Evening Sky Map*, p. 2.

Clute, John and Peter Nicholls (eds.) (1993). *The Encyclopedia of Science Fiction*, pp. 1087–1088. New York: St. Martin's Press.

Fisher, Clyde (1929). "Garrett P. Serviss: One Who Loved the Stars." *Popular Astronomy* 37: 365–369.

Severin, Christian

Born **Lomborg (Longberg), Jutland, Denmark, 4 October 1562**
Died **Copenhagen, Denmark, 8 October 1647**

Longomontanus, a professor of astronomy and observer, was one of **Tycho Brahe**'s most important students and publicized the Tychonic system in the early 17th century. Christian Severin was born in a poor farmer's family – that of Soeren Poulsen and Maren Christensdatter. After the early death of his father in 1570, he had to work on the farm. During the winter months, Severin received his first tuition from a pastor in Lomborg. Beginning in 1577, he attended the cathedral school of Viborg. At the age of 26, Severin matriculated from the University of Copenhagen, inscribing his name in the Latin form of Longomontanus. Within a year, he began working for Brahe and became his only long-term student, making astronomy his life's work.

After Brahe had left his island of Hven, Longomontanus bade him farewell with a very well-meant letter of resignation dated 1 June 1597. He then continued his studies in Breslau and Leipzig universities. He finally received his MA from Rostock University. In January 1600, Longomontanus again stayed with Brahe in Benatek (close to Prague), where, according to Brahe Special wish, he was supposed to work on the Mars observations. But when **Johannes Kepler** joined Brahe as an assistant, he took over the Mars data, leaving lunar work to Longomontanus. The latter passed his lunar research to Brahe in the summer of 1600 and returned to Denmark.

Back home, in1603, Longomontanus was appointed principal of the Viborg cathedral school, which he had attended as a youth, but retained his interest in astronomy. He corresponded with Kepler on lunar theory. In 1605, he received the professorship for mathematics in Copenhagen, followed by appointment to the chair of "higher mathematics" (*i. e.*, astronomy), a post he held until his death. On the order of King Christian IV, an observatory was erected at the university in 1637. The impressive round tower still stands. There, Longomontanus inaugurated an important observational program to find precise positions for the 777 stars in Brahe's catalog. Although Longomontanus's precision was good, he could not compete with **Christoph Rothmann** working at Kassel Observatory.

Longomontanus has been described as a very easygoing, warmhearted person. In 1607, he married Dorthe Bartholin, sister to the prominent scientist Caspar Bartholin. As a mathematician, he worked on the quadrature of the circle, giving a value for π of $78\sqrt{3}/43$.

Longomontanus's textbook, *Astronomia Danica*, was first published in 1622 in Amsterdam by Willem Janszoon Blaeu. It was based upon the Tychonic geo–heliocentric system of the world. But in contrast to Brahe, Longomontanus, like **Nicholas Bär** (Raimarus Ursus) before him, incorporated daily rotation of the Earth on its axis to explain the rotation of the fixed stars. The book represented the real inheritance of Brahe – in the first half of the 17th century the Tychonic system, next to that of **Nicolaus Copernicus**, was one of the two main respected systems of the world. As the only Tychonic-based textbook, *Astronomia Danica* reached a wide audience, with two further editions (1640 and 1663) being published. The work was a complete presentation of astronomy including trigonometry (with many calculation examples), celestial circles, the obliquity of the ecliptic, terrestrial climate zones, rising and setting of objects, and a complete catalog of the Ptolemaic stars. It covered the movements of the planets according to the Tychonic, Ptolemaic, and Copernican theories as well as the production and handling of astronomic instruments, such as armillary spheres, armillae, the torquetum, the quadrant, the sextant, and the Jacob's staff – with reference to the printed instruction manuals by Brahe. An appendix is dedicated to the comets and their nature according to Rothmann, Brahe, and Kepler, with details on the novae and comets of 1572, 1577, 1607, and 1618.

Jürgen Hamel

Alternate name

Longomontanus

Selected References

Anon. (1929). *Forfatter-Lexicon for Kongeriget Danmark med tilhorende Bilande 1841 til efter 1858.* Vol. 5, pp. 181–185. Copenhagen.

Christianson, John Robert (2000). *On Tycho's Island: Tycho Brahe and His Assistants, 1570–1601.* Cambridge: Cambridge University Press.

Dreyer, J. L. E. (1890). *Tycho Brahe: A Picture of Scientific Life and Work in the Sixteenth Century.* (Reprint, New York: Dover, 1963.)

Thoren, Victor E. (1990). *The Lord of Uraniborg: A Biography of Tycho Brahe.* Cambridge: Cambridge University Press.

Severus Sebokht [Sebokt, Sebukht, Seboht]

Born **Nisibis, (Syria), *circa* 575**
Died **Kennesrin (also called Qinnesrin or Qenneshrê), (Syria), 666/667**

Severus Sebokht was one of the leading figures of ecclesiastical, philosophical, and scientific culture of late antique Syria, although little definitive information is known about his life. Born in Persian territory at Nisibis, he left his teaching post in its famous school in 612 after a doctrinal dispute among the Nestorians. Later consecrated a bishop, he pursued a career in the Syrian Monophysite church within Byzantine jurisdiction, residing as a monk at the monastery at Kennesrin on the west bank of the Euphrates, one of the chief seats of Greek learning in western Syria. He continued to write until at least 665.

Like many of his contemporaries, Severus was bicultural, partaking of the Byzantine Greek influence on western Syrian intellectual circles while fully immersed in his own Syrian cultural milieu. He does, however, criticize the contemporary Greek tendency to assume intellectual superiority and asserts his own capabilities as a native Syrian, raising a strong polemical voice against the cultural hegemony of the Greek-speaking world over that of provincials. A leading figure in the teaching and commentary tradition of Aristotelian philosophy, especially in logic and syllogisms, Severus produced a *Discourse on Syllogisms in Prior Analytics* (638) and wrote commentaries on other philosophical texts. He translated Paul the Persian's commentary on **Aristotle**'s *De interpretatione* into Syriac. Severus also played an important role in the transmission of Indian intellectual concepts into Syria and ultimately into the Islamic world. In one famous passage, he praises the Hindu decimal concept and mentions for the first time in the Greek east the nine numerical symbols used in India.

It was in astronomical matters, however, that Severus was preeminent. Syrian astronomy was predominantly Ptolemaic, and Severus himself stands as an important figure in passing on Greek astronomical knowledge to Syrian scholars and thence to Islamic civilization. He was familiar with **Ptolemy**'s *Handy Tables,* and there is some indication that he translated the *Almagest* into Syriac; in any case, he most certainly taught it in the school of Nisibis and then later in western Syria. Similarly, Severus was an important link in the transmission of the Greek tradition of the astrolabe to the east. In several passages in his astronomical works, he positions himself firmly on the side of scientific methodology and opposes speculative astrology.

Severus made two major contributions to astronomy. The first, a *Treatise on the Astrolabe,* is based on a lost work by **Theon of Alexandria**, the contents of which Severus preserved in his own work. Written in 660, it is in two parts. The first is a general description, including information about the following basic elements of the instrument – the disks, the spider, the diopter, the zones, and related aspects of the physical and mechanical parts. Instructions on its actual use comprise the second part of the work, divided into 25 chapters, of which two (12, 20) are missing. These chapters cover all the applications of the instrument – determining the hour of the day and night (1–3), finding the longitude of the Sun, Moon, and planets and the latitude of the Moon (4–6), checking the instrument (7–8), ascertaining the rising and setting times of various signs (9–10, 25) and the length of daylight during the course of the year (11), locating the geographical longitude and latitude of cities and establishing the differences of local noons (13–15), fixing the ascensions on the right sphere (16), finding latitudes of the observer and of each climate (17–18), estimating the longitude and latitude of stars and their first and last visibility (19, 21), observing the ecliptic and the declination of the Sun (22–23), and recognizing the five zones on the celestial and terrestrial spheres (24).

Severus's other astronomical work (generally entitled *Treatise on the Constellations*) was written in 660, subsequent to that on the astrolabe. Eighteen original chapters are extant. The work begins with five chapters forming a scientific critique of astrological and poetic claims about the origins and significance of the constellations. In them, Severus shows that the figures of the constellations are not arranged in the heavens through natural means but rather are a result of human imagination. Importantly, Chapter 4 features extracts from the *Phaenomena* of **Aratus** concerning many of the constellations. The remaining 13 chapters (6–18) are devoted to a scientific analysis of the heavens and the Earth. Here Severus enumerates the 46 constellations and their noteworthy stars and explains their various motions and their rising and settings. He also discusses the celestial geography of the Milky Way and the ten "circles" of the heavens, including the tropics, the equator, the meridian, the horizon, and the ecliptic. Three chapters (14–16) examine extensively the seven climatic zones, their location and extent, their relationship to the Sun, and the length of the days and nights in each, the latter in accordance with Ptolemy's *Handy Tables*. In the final two chapters, Severus treats the extent of the Earth and the sky and considers the populated and uninhabited regions of the Earth. In 665, Severus appended to this work nine additional chapters, designed to answer a variety of astronomical, cosmological, and mathematical questions posed by Basil of Cyprus, a visiting cleric. Included are treatments of the conjunctions of planets and of various points about climatic zones, the astrolabe, the determination of the date of Easter in April 665, and the date of the birth of Christ. In other passages extant in the manuscripts, Severus also writes on the phases of the Moon and on eclipses, in one case explaining lunar eclipses scientifically to dispel the popular idea that a dragon (*Ataliâ*) was responsible for such events.

John M. McMahon

Selected References

Brock, Sebastian P. (1984). "From Antagonism to Assimilation: Syriac Attitudes to Greek Learning." In *Syriac Perspectives on Late Antiquity*. Vol. 5, pp. 17–34, esp. 23–24, 28. London: Variorum Reprints. (For specialized treatment of Severus and his contemporaries.)

Gunther, Robert T. (1932). *The Astrolabes of the World*. Vol. 1, *The Eastern Astrolabes*. Oxford: University Press, pp. 82–103. (For an English version of *Treatise on the Astrolabe*, from Nau's French.)

Moosa, Matti (ed. and trans.) (2000). *The History of Syriac Literature and Sciences*. Pueblo, Colorado: Passeggiata Press, pp. 65, 108. (Originally published as I. Aphram Barsoum, *Kitāb al-Lu'lu' al-manthūr fī ta'rīkh al-ʿulūm wa-' l-ādāb al-Suryāniyya*. Hims, Syria, 1943). (Earlier treatments of Severus are now incorporated into this work, which conveniently lists and briefly discusses all of Severus's works.)

Nau, F. N. (1899). "Le traité sur l'astrolabe plan de Sévère Sabokt." *Journal asiatique*, 9th ser., 13: 56–101, 238–303. (For the *Treatise on the Astrolabe*.)

———. (1910). "La cosmographie au VIIe siècle chez les Syriens." *Revue de l'Orient chrétien* 5, no. 18: 225–254. (Assesses Severus's contributions and surveys the contents of Paris MS Syr. 346, three quarters of which is made up of his works.)

———. (1910). "Notes d'astronomie syrienne." *Journal asiatique*, 10th ser., 16: 209–228, esp. 219–224. (For Severus's explanation of lunar eclipses.)

———. "Le traité sur les 'Constellations' écrit, en 661 [*sic*], par Sévère Sébokt, évêque de Qennesrin." *Revue de l'Orient chrétien* 7, no. 27 (1929): 327–410; 8, no. 28 (1932): 85–100.

Neugebauer, Otto (1949). "The Early History of the Astrolabe." *Isis* 40: 240–256, esp. 242–245, 251–253. (For Severus's treatise on the astrolabe; discusses Severus's sources and critiques earlier discussions.)

———. (1975). *A History of Ancient Mathematical Astronomy*. 3 pts. New York: Springer-Verlag, pt. 1, pp. 7–8; pt. 2, pp. 877–878, 1041–1042. (For a brief treatment of the relationship between Greek and Syrian astronomy and its transmission to the Islamic World and Severus's treatise on the astrolabe.)

Pingree, David (1993). "The Greek Influence on Early Islamic Mathematical Astronomy." *Journal of the American Oriental Society* 93: 32–43, esp. 34–35.

——— (1994). "The Teaching of the Almagest in Late Antiquity." In *The Sciences in Greco-Roman Society*, edited by Timothy D. Barnes, pp. 73–98, esp. 94–95. Edmonton: Academic Print and Publishing.

Stautz, Burkhard (1997). *Untersuchungen von mathematisch-astronomischen Darstellungen auf mittelalterlichen Astrolabien islamischer und europäischer Herkunft*. Bassum: Verlag für Geschichte der Naturwissenschaften und der Technik, pp. 38–39. (Gives a concise description of the physical appearance of the instrument.)

Wright, W. (1966). *A Short History of Syriac Literature*. Amsterdam: Philo Press, pp. 137–139.

Seyfert, Carl Keenan

Born **Cleveland, Ohio, USA, 11 February 1911**
Died **Nashville, Tennessee, USA, 13 June 1960**

American observational astronomer Carl Seyfert is remembered in the name of the small group of galaxies, Seyfert's sextet, and, more particularly, in Seyfert galaxies, a class of spirals distinguished by bright, wide emission lines coming from gas near their centers and now believed to have nuclear black holes of millions to hundreds of millions of solar masses.

The son of a pharmacist, Seyfert was educated in the Cleveland schools, and at Harvard University, starting in 1929 in medicine, but turning to astronomy under the inspiration of **Bart Bok**. He received a BS (1933), an MA (1933), and a Ph.D. (1936), for work on colors and magnitudes of galaxies. His dissertation work was guided by **Harlow Shapley**. Seyfert married Muriel E. Mussells in 1935; their children are a daughter Gail Carol and a son Carl Keenan Seyfert, Jr., a well-known geophysicist and textbook author.

In 1936, Seyfert was appointed to the staff of the new McDonald Observatory (initially as part of the Yerkes Observatory staff), where he worked on spectra and light curves of hot stars and variable stars, including a project with **Daniel Popper** on faint B-type stars. Seyfert also worked on emission nebulae and clusters of stars in other galaxies. During his service at McDonald Observatory, Seyfert rode his horse, *Silver*, in a local cattle roundup each year.

Seyfert held a National Research Council Fellowship at Mount Wilson Observatory for the period 1940–1942, where he recognized the first half dozen examples of what are now called Seyfert galaxies, characterized by emission lines emitted by gas that is moving at very high speed at the center of the Galaxy. The first of these galaxies was actually discovered and published more than 30 years before Seyfert's key 1943 paper by **Edward Fath** but was not recognized for what it was. The best known of the original Seyfert galaxies is probably M77 (NGC 1068), and they are now regarded as a subtype (with spiral hosts) of the more general class of galaxies with active nuclei [AGNs].

In 1942, Seyfert returned to Cleveland to teach navigation to the armed forces at Case Institute, and do some war-related work in optics. He also used the facilities of nearby Warner and Swasey Observatory in collaborations with S. W. McCuskey and J. J. Nassau, on stars and planetary nebulae in the Milky Way and in the Andromeda Galaxy, M31. Seyfert and Nassau obtained the first good color photographs of nebulae and of stellar spectra during this period, using a new Schmidt telescope as an objective prism spectrograph.

In 1946, Seyfert joined the faculty of Vanderbilt University in Nashville, Tennessee. At that time, the university had only the small Barnard Observatory, equipped with a 6-in. refractor (which had once been used by **Edward Barnard**) and a modest teaching program in astronomy. With considerable vigor, Seyfert started a new series of courses, and set out to build a new observatory. Within a few years, while busy with full-time duties in teaching and research, Seyfert managed to get public support from the Nashville community. During the work-intensive planning and construction of the new observatory, he still found time to give astronomy lectures outside the university, and even appeared on television as a daily weather forecaster.

The new Arthur J. Dyer Observatory, named after one of Seyfert's strongest supporters and equipped with a 24-in. reflecting telescope, was finally completed in December 1953. Carl Seyfert became director of Dyer Observatory, a post he held for the rest of his life. Research at the new observatory included stellar and galactic astronomy, as well as new instrumental techniques. During the time at Vanderbilt, Seyfert's research included first the photometric investigation of photographic plates from Barnard Observatory and Shapley's Harvard plates, as well as studies performed with the privately owned 12-in. J. H. DeWitt telescope. In 1951, Seyfert observed and described a group of galaxies around NGC 6027, now known as Seyfert's sextet.

He was involved in instrumental innovations including the use of photomultiplier tubes and television techniques in astronomy, and electronically controlled telescope drives. Scientific results were obtained on variable stars, emission B stars in stellar associations, and the structure of the Milky Way. Seyfert was a member of several professional societies, including the American Astronomical Society, where he served on the council from 1955 to 1958, and the Royal Astronomical Society. He also served in the Associated Universities Incorporated, as a member of the board of directors of the Association of Universities for Research in Astronomy, and on the astronomy Advisory Panel of the National Science Foundation.

Carl Seyfert died in an automobile accident. He was honored by the astronomical community by the naming of Moon crater Seyfert in 1970. The 24-in. telescope at Dyer Observatory, for which he had worked so hard, now carries his name, the Seyfert telescope.

Hartmut Frommert

Selected References

Anon. (1960). "Carl K. Seyfert." *Publications of the Astronomical Society of the Pacific* 72: 434.

Hardie, Robert (1961). "Carl Keenan Seyfert." *Quarterly Journal of the Royal Astronomical Society* 2: 123–125.

Seyfert, Carl K. (1943). "Nuclear Emission in Spiral Nebulae." *Astrophysical Journal* 97: 28–40. (Seyfert's original paper on the active galaxies now called Seyfert galaxies.)

Shain [Shayn, Shajn], Grigory Abramovich

Born **Odessa, (Ukraine), 13 April 1892**
Died **Moscow, (Russia), 27 August 1956**

Astrophysicist and observatory director Grigory Shain helped to rebuild the infrastructure of Soviet astronomy in the wake of the Bolshevik Revolution and provided early evidence for nonsolar abundances in carbon stars and for a galactic magnetic field.

Shain was born into a poor joiner's family; much of his broad erudition was due to his perseverance and determination to educate himself. In 1912, he enrolled in the Faculty of Physics and Mathematics at Yurev (Dorpat) University, but his education was not completed until 1919 at Perm (following his service in World War I). Shain earned a magister's degree from the provincial Tomsk University (1920) and, in the following year, joined the staff of Pulkovo Observatory. He married astronomer Pelageya Fyodorovna Sannikova.

In 1925, Shain and his wife were dispatched to Pulkovo's astrophysics branch at Simeiz, Crimea. He remained there for the rest of his life. It is likely that Shain's peripheral location with respect to the main scientific centers of the country not only saved his life during the terrible Pulkovo purges of the 1930s, but also permitted him to maintain his high moral standards. In 1939, he was elected an academician of the Soviet Academy of Sciences, a very high rank for an astronomer, especially during the Stalinist era. On many occasions, Shain risked his own safety by signing his name to the defense of innocent victims of the regime, and he exerted great efforts to aid the families of imprisoned astronomers. In 1946/1947, he was the leader of a party of ten Soviet astronomers who studied in the United States during the 6 months prior to the beginning of the Cold War.

At the Simeiz Observatory, Shain was given charge of its 1-m reflector, constructed in the United Kingdom after World War I. He began his research in celestial mechanics, but then turned his attention to the evolution of binary stars, correctly deducing that the more massive star of a pair evolved more rapidly than its less-massive companion. In collaboration with Yerkes Observatory director **Otto Struve**, Shain pioneered studies of the rapid rotations of hot stars from measurements of their spectral line broadening.

Shain's principal accomplishments were in the fields of stellar spectrophotometry and the physics of gaseous nebulae. He offered new interpretations regarding the atmospheres of long-period variable stars, and gathered evidence of significantly higher isotopic abundances of ^{13}C to ^{12}C (compared to the solar abundance) in some other stars. This research netted Shain a Stalin Prize of the first class (1950), the supreme scientific award of that time. Using two especially fast optical systems he had developed, Shain and colleague Vera F. Gaze discovered roughly 150 new galactic emission nebulae, by recording their light in the red H-α emission. Much of this work was summarized in Shain's 1952 *Atlas diffuznykh gazovykh tumannostey* (Atlas of diffuse gas nebulae). The filamentary shapes of many of the nebulae were oriented parallel to the galactic Equator, which led Shain to postulate the existence of powerful galactic magnetic fields.

Following destruction of the original Simeiz Observatory during World War II, Shain spearheaded establishment of the newer and larger Crimean Astrophysical Observatory in 1945. He directed that institution until 1952, when he voluntarily stepped down for health reasons. Throughout his life, Shain was surrounded by a handful of disciples, such as Solomon Pikelner, and other prominent scientists, including **Iosif Shklovsky**. Additional astronomers of his circle included his wife and son-in-law, **Victor Ambartsumian**.

Shain had significant influence on contemporary Soviet and world astrophysics. As Shklovsky has written, Shain was truly "a good astronomer and a remarkable man." The Crimean Astrophysical Observatory's largest (2.6-m) telescope, along with a crater on the Moon's farside, are named for him. Shain is buried on the observatory grounds in Ukraine.

Alexander A. Gurshtein

Selected References

Kulikovsky, P. G. (1975). "Shayn, Grigory Abramovich." In *Dictionary of Scientific Biography*, edited by Charles Coulston Gillispie. Vol. 12, pp. 367–369. New York: Charles Scribner's Sons.

Pikelner, S. B. (1957). "G. A. Shain (1892–1956)." *Istoriko-astronomicheskie issledovaniia* 3: 551–607.

Shklovskii, I. S. (1991). *Five Billion Vodka Bottles to the Moon: Tales of a Soviet Scientist*. New York: W. W. Norton, esp. pp. 148–156.

Struve, Otto (1958). "G. A. Shajn and Russian Astronomy." *Sky & Telescope* 17, no. 6: 272–274.

Shakerley, Jeremy

Born **Halifax, (West Yorkshire), England, November 1626**
Died **possibly (India), *circa* 1655**

Jeremy Shakerley was an English astronomical writer who first championed **Jeremiah Horrocks**'s views on the motions of the Moon and Venus and who observed the 1651 transit of Mercury. Shakerley spent his youth in Yorkshire and by January 1648 was living in Pendle Forest, Lancashire. Around this time, he became acquainted with the works of **Johannes Kepler** and **Ismaël Boulliau**. Shakerley subsequently expanded his knowledge of astronomy with the assistance of the astrologer William Lilly, with whom he began a 3-year correspondence in 1648, and of Christopher Towneley, in whose household at Carre Hall, Burnley, Lancashire, he began living in 1649. Towneley had acquired the manuscripts of the Liverpool astronomer Horrocks after the latter's death in 1641, and Shakerley was the first person to appreciate the significance of Horrocks's works, especially those on the theory of the Moon's motion. Shakerley made references to Horrocks's lunar theory in *The Anatomy of Urania Practica* (1649), a critique of the work of the almanac maker **Vincent Wing**.

Shakerley also cited the work of Horrocks, who had predicted and observed the transit of Venus across the Sun in 1639, in making his own prediction (in his 1651 almanac *Synopsis Compendia*), that a transit of Mercury would occur on 24 October 1651. Having lost the support of Lilly in 1650 after attacking Wing, Shakerley soon thereafter immigrated to India, probably as an employee of the East India Company. The move enabled him to observe, from Surat, the 1651 transit. His observation of this transit, which was not visible in England, marked only the second occasion when observations of the phenomenon were recorded. (The French scientist **Pierre Gassendi** and two others had observed the Mercury transit that occurred on 7 November 1631.)

Virtually nothing is known about Shakerley's life after his *Tabulae Brittannicae* was published in London in 1653.

Craig B. Waff

Selected References

Chapman, Allan (1975). "Shakerley, Jeremy." In *Dictionary of Scientific Biography*, edited by Charles Coulston Gillispie. Vol. 12, pp. 342–343. New York: Charles Scribner's Sons.

Kelly, John T. (1991). *Practical Astronomy during the Seventeenth Century: Almanac-Makers in America and England.* New York: Garland.

Shams al-Dīn al-Bukhārī

Flourished **Middle to late 13th century**

Shams al-Dīn al-Bukhārī is cited in various Greek versions of Arabic and Persian astronomical handbooks (*zījes*), versions that were made in the last decade of the 13th century in Marāgha and Tabrīz. These *zījes* include *al-Zīj al-Sanjarī*, composed in Arabic in the mid-12th century by **ʿAbd al-Raḥmān al-Khāzinī** and dedicated to the Saljūq Sultan Sanjar (reigned: 1118–1157); *al-Zīj al-ʿAlāʾī*, composed in Arabic by ʿAbd al-Karīm al-Shirwānī al-Fahhād (mid-12th century), but no longer extant in Arabic; and, *al-Zīj-I Īlkhānī*, composed *circa* 1270 in Persian by **Naṣīr al-Dīn al-Ṭūsī**. The Persian text survives in many copies, and there is also an Arabic version. The Greek versions of all three are found in the following manuscripts: Florence Laur. gr. 28/17, Vat. gr. 211, and Vat. gr. 1058. The Greek version of the *Īlkhānī zīj* is much more widespread, being found in manuscripts in many collections. The Arabic version of the *Sanjarī* is found in manuscripts Vat. ar. 761, Br. Lib. Or. 6669, and Istanbul Hamidiye MS 859; one is in private possession.

A tract on the astrolabe is also attributed to Shams al-Dīn al-Bukhārī as well as a "Short syntaxis"; both are in Greek. There is nothing known of him in Persian or Arabic sources, nor is there any known reference to him outside the Greek work just mentioned. According to these sources, his *floruit* may be firmly placed at the end of the 13th century, and D. Pingree (1985) has argued for his date of birth as 11 June 1254.

These translations were made, no doubt, within the community centered at the famous observatory of Marāgha, which was under the direction of Ṭūsī and under the patronage of the Īlkhānid rulers. It is clear that Shams al-Dīn al-Bukhārī was instrumental in enabling the Byzantine scholar **Gregory Chioniades** both to obtain these translations of the tables and to learn how to use them. Shams al-Dīn's oral instruction (ἀπὸ φωνῆς τοίνυν τοῦ Σάμψ Πουχαρῆς ἀνδρὸς τὸ γένος Πέρσου) is acknowledged in the prefaces to the "Persian syntaxis" of Chioniades, *circa* 1295, and in the later "Persian syntaxis" of George Chrysococces, *circa* 1347, where we are told that the Persians were reluctant to allow a written translation of the Persian canons of the tables to be passed into Greek hands. One notes that the term "Persian syntaxis" is used somewhat loosely in the Greek texts, so that, for Chioniades, it refers to the *Zīj al-ʿAlāʾī*, while for Chrysococces it means the *Zīj-i Īlkhānī*.

Apart from Chioniades's canons for *Zīj al-ʿAlāʾī*, one finds a further work of his in 22 chapters, in which all three *zījes* are mentioned. In one of these, Chioniades relates how Shams al-Dīn al-Bukhārī calculated a lunar eclipse according to some tables he had devised on the basis of the *Zīj-i Īlkhānī*, using as an example the total lunar eclipse of 30 May 1295. These eclipse tables were presumably part of the "Short syntaxis" elsewhere attributed to him.

The last mention of Shams al-Dīn al-Bukhārī in the Byzantine sources is in the *Tribiblos*, a very prolix treatise written *circa* 1350 by Theodore Meliteniotes, covering both Ptolemaic and Persian material. This includes in its Book III a long recapitulation of the Persian material, including the Greek version of the *Zīj-i Īlkhānī*, as already given by Chrysococces. In the preface to the text, Meliteniotes mentions Σάμψ Μπουχαρῆ along with other Islamic authors (Vat. gr. MS 792, fol. 246).

Raymond Mercier

Selected References

Doyen, Anne-Marie (1979). *Le traité sur l'astrolabe le Siamps le Persan.* Mémoire, Faculté de Philosophie et Lettres de Université Catholique de Louvain. (Unpublished transcription of the text on the astrolabe, with a translation.)

Kennedy, E. S. (1956). "A Survey of Islamic Astronomical Tables." *Transactions of the American Philosophical Society*, n.s., 46, pt. 2: 121–177. (Reprint, Philadelphia: American Philosophical Society, 1989.)

Mercier, Raymond (1984). "The Greek 'Persian Syntaxis' and the *Zīj-i Īlkhānī*." *Archives internationales d'histoire des sciences* 34: 35–60. (The identification of the *Īlkhānī zīj* as the source of the tables of Chrysococces.)

Neugebauer, O. (1960). "Studies in Byzantine Astronomical Terminology." *Transactions of the American Philosophical Society*, n.s., 50, pt. 2: 3–45. (A seminal

study of Vat. Gr. 1058, a manuscript that includes the Greek versions of all three *zījes*, and much besides.)
Pingree, David (1985). *The Astronomical Works of Gregory Chioniades*. Vol. 1, *The Zīj al-ʿAlāʾī*. *Corpus des Astronomes Byzantins*, 2. Amsterdam: J. C. Gieben. (Part 1 is an edition and translation of the Greek text of the *ʿAlāʾī zīj* and related material, with an important note on Shams.)
Suter, Heinrich (1900). "Die Mathematiker und Astronomen der Araber und ihre Werke." *Abhandlungen zur Geschichte der mathematischen Wissenschaften* 10: 161, 219–220. (The older literature contains references to Shams that serve to illustrate the confusion surrounding his identity and role.)
Tihon, Anne. "Les tables astronomiques persane à Constantinople dans la premiere moitié du XIVe siècle." *Byzantion* 57 (1987). (Reprinted in Tihon, *Études d'astronomie Byzantine*. Aldershot: Variorum, 1994.) (Important overview of the manuscript sources.)
Usener, Hermann (1912–1914). *Kleine Schriften*. 4 Vols. Leipzig: B. G. Teubner, Vol. 3, pp. 288–377.

Shane, Charles Donald

Born **Auburn, California, USA, 6 September 1895**
Died **Santa Cruz, California, USA, 19 March 1983**

American observational astronomer C. Donald Shane is eponymized in the 120-in. (3-m) Shane telescope at Lick Observatory, whose design and construction he oversaw. He made his most lasting impact on astronomy through the Shane–Wirtanen counts of galaxies, completed in 1954, which helped to establish the existence of structure in the Universe on the scale of superclusters.

Shane received an AB from the University of California [UC] (now UC Berkeley) in 1915 and took up a Lick Fellowship on Mount Hamilton the next year to work toward a Ph.D. in astronomy. After brief service (1917–1918) in World War I, he returned to complete his dissertation under **William Campbell** and **Joseph Moore** on the spectra of carbon stars.

Shortly before completing his degree, Shane married fellow graduate student Mary Lea Heger, who received her Ph.D. in astronomy in 1925, working on absorption features in stellar spectra caused by cool interstellar gas. They had two children, and she devoted many years to bringing order out of the chaos of the Lick Observatory archives, named the Mary Lea Shane Archives in 1982. She also died in 1983.

Donald Shane was appointed to an instructorship in mathematics at the University of California in 1920 (mathematics and astronomy 1922–1924) and to an assistant professorship in 1924. He moved up the ranks to full professor, and in 1945 officially became a full astronomer at Lick Observatory and then its director (1945–1958), retiring to an emeritus position in 1963. During World War II, Shane served first as assistant director of the Radiation Laboratory at Berkeley and later as personnel director at Los Alamos Laboratory.

Lick flowered under Shane's directorship, with the staff increasing from 50 to about 105, the building of the 120-in. telescope (which was the second largest in the world at the time of its commission), and the renewal of both facilities and research programs. He found time to serve as president of the Association of Universities for Research in Astronomy from 1958 to 1962, the critical years during which Kitt Peak National Observatory was being established and the Cerro Tololo Inter-American Observatory in Chile was being planned. Some of his own work concerned the detailed spectra of the Sun and of stars like o Ceti (Mira).

Shane's most important contribution came, however, when he and **Carl Wirtanen** decided to photograph the entire northern sky using the 20-in. Carnegie astrograph (a wide-field telescope generally used for astrometry) to take first-epoch plates for a major Northern Hemisphere proper-motion survey. This was eventually completed as the Lick Sky Atlas. But, in the meantime, Shane and Wirtanen decided to count all the galaxies they could see on the plates, recording what they saw in a new way. Instead of representing each galaxy by a position on the sky, they divided the sky into small boxes and recorded the number of galaxies in each box. The 1954 analysis of these counts by statisticians J. Neyman, Elizabeth Scott, and Shane himself persuaded most of the astronomical community, first, that virtually all galaxies are parts of groups and clusters and, second, that there is a good deal of higher-order clustering of these into superclusters. Nearly three decades later, these counts were the best available data of their sort when cosmologist P. James E. Peebles used them to establish quantitative measures of the length scale and amplitude of that clustering behavior. The Shane–Wirtanen counts have only recently been rendered obsolete by very large surveys that include measured redshifts.

Shane was a member of the United States National Academy of Sciences (1961) and a foreign associate of the Royal Astronomical Society (London) and of the Royal Astronomical Society of New Zealand, for whose members he had provided guidance in starting an extension of the Lick Atlas to the Southern Celestial Hemisphere. The Shanes were survived by two sons.

Virginia Trimble

Selected References

Bateson, F. M. (1984). "Shane, C. Donald (1895–1983)." *Southern Stars* 30: 427.
Kron, Gerald E. (1984). "Charles Donald Shane." *Physics Today* 37, no. 2: 80.
Shane, C. D. (1958). "A New Sky Atlas." *Publications of the Astronomical Society of the Pacific* 70: 609–610.
Shane, C. D. and C. A. Wirtanen (1954). "The Distribution of Extragalactic Nebulae." *Astronomical Journal* 59: 285–304.
Vasilevskis, S. (1984). "Charles Donald Shane." *Quarterly Journal of the Royal Astronomical Society* 25: 532–533.

Shapley, Harlow

Born **near Nashville, Missouri, USA, 2 November 1885**
Died **Boulder, Colorado, USA, 20 October 1972**

American observational astronomer Harlow Shapley obtained the data that showed incontrovertibly that the Solar System is not near the center of the Milky Way Galaxy, as virtually all astronomers had thought since the time of **William Herschel**. His name is remembered in the Shapley concentration of galaxies (a very extensive supercluster) and in the Shapley–Ames catalog of nearby galaxies. Shapley was the son of Willis and Sarah (*née* Stowell) Shapley. After completing elementary school and a short business course, and before graduating from high school in 1907 (first in

a class of three from Carthage Collegiate Institution in Carthage, Missouri), he spent several years as a newspaper reporter – first in Chanute, Kansas, and then in Joplin, Missouri. Shapley then entered the University of Missouri. Finding that the new School of Journalism, which had been his goal, was not yet open, he quickly gravitated toward astronomy, under the influence of **Frederick Seares**, with whom he worked on light curves of variable stars, mostly eclipsing binaries. Shapley received a BA in 1910 and an AM in 1911, going on that year to Princeton University with a Thaw Fellowship.

Working officially with **Henry Norris Russell** and also mentored by **Raymond Dugan**, who was more observationally inclined, Shapley received a Ph.D. in 1913 for the analysis of the light curves of a large number of eclipsing binaries. He was able to measure sufficiently accurate masses and radii for the component stars to conclude that the range of stellar densities is more than 1,000, but that density is not correlated with surface temperature in the way that Russell (with his giant-and-dwarf theory of stellar evolution) was expecting. Seares had by then moved to Mount Wilson Observatory and arranged for Shapley to meet its director, **George Hale**. After a year in Europe, during which he met many astronomers, Shapley accepted a position at Mount Wilson Observatory.

In 1914, Shapley married fellow Missourian Martha Betz. Trained as a mathematician, **Martha Shapley** quickly acquired skill in analyzing variable-star light curves to extract binary properties, and published a number of papers alone and with Shapley between 1915 and 1929. Of their five children, Mildred Shapley Matthews became a planetary astronomer, Lloyd a mathematician, Alan a geophysicist, and the others (and some of the grandchildren) scientists, science administrators, and teachers of other sorts.

Before moving west, Shapley had visited Harvard astronomer **Solon Bailey** who urged him to use the 60-in. telescope on Mount Wilson (then the largest) to observe variable stars in globular clusters. This he did as soon as the opportunity presented itself. Contemporary astronomers were generally of the opinion that the Cepheid variables (for which a correlation between absolute luminosity and length of period had been discovered earlier by **Henrietta Leavitt** for stars in the Small Magellanic Cloud) were eclipsing binary pairs. Shapley carried out an analysis demonstrating that this could not possibly be true, for the separation of the stars would have to be smaller than their sizes, putting one star inside the other. In 1914, he advanced an alternative idea that the stars were pulsating radially in size, with corresponding changes in surface temperature and brightness. This proved to be correct. Shapley also tacked on to the faint end of the period–luminosity relation a class of more rapid variables, then called cluster-type (though they also occur outside globular star clusters) and now called RR Lyrae variables. This eventually caused problems.

The pulsation mechanism, however, suggested that the Cepheids could be reliable distance indicators, if only one could calibrate them with a few stars whose real luminosities (or distances) were known in some other way. **Ejnar Hertzsprung** had attempted such a calibration, using the method of statistical parallaxes invented by **Jacobus Kapteyn**, and Shapley adapted and improved this. Shapley's calibration was an important advance, but three effects combined to introduce errors into his period–luminosity relation. These were (1) neglect of galactic rotation (not discovered until a decade later by **Bertil Lindblad**), (2) his own work, which seemed to show that there was no general absorption of starlight in interstellar space (proven wrong in 1930 by **Robert Trumpler**), and (3) slightly bad luck in the statistics of very small motions on the sky for a small number of stars. He was, moreover, mistaken in thinking that the nearby Cepheids (on which the calibration was done), and those in globular clusters, were physically similar. In fact, they differ in mass by a factor of 5–10 and in brightness by factors of 2–10, so that Shapley put his globular clusters too far away when he started using Cepheids as rulers.

Shapley had come to Mount Wilson as a believer in the "island universe," or many galaxies hypothesis, under which the Milky Way and the spiral nebulae were similar kinds of systems. But the map Shapley gradually drew of the locations of the globular clusters eventually persuaded him, first, that the Solar System was very far from the center. (His center was actually within the grouping of clusters toward Sagittarius, an idea advanced earlier by Swedish astronomer **Karl Bohlin.**) Second, Shapley was persuaded that the Milky Way extended to at least 50,000 parsecs from that center, making the idea of other galaxies of comparable size most unlikely. He was further misled by his own discoveries of some novae in the Andromeda Nebula that he decided were not as bright as galactic novae (leaving only the 1885 event, now known to have been supernova 1885A, as a distance indicator) and by apparent measurements of the rotation in the plane of the sky of several spiral nebulae, carried out on Mount Wilson plates by **Adriaan van Maanen**. **Knut Lundmark** later showed that Van Maanen's measures were completely erroneous, but Shapley, who regarded van Maanen as a friend, was not convinced.

Against this background, Shapley and **Heber Curtis** engaged in a discussion of "the distance scale of the universe" before the National Academy of Sciences in Washington, in April 1920. Shapley advocated a very large Milky Way with the Sun far from the center and the spiral nebulae and globular clusters as members of this one, universal system. Curtis advocated a much smaller Milky Way, with the Solar System close to the center, and the spirals as independent similar systems of stars. The event is frequently called the Great Debate or the Curtis–Shapley debate, though it was not actually organized as a debate. In retrospect, Curtis was right about other galaxies existing (shown by **Edwin Hubble** a few years later), and Shapley was right about the noncentral position of the Solar System, which was quickly adopted by the entire community, relegating the smaller Kapteyn universe to the status of a relatively local feature in the galactic disk. The modern distance scale within the Milky Way is about half way between those of Curtis and Shapley.

The death of **Edward Pickering** in 1919 had left Harvard Observatory directorless. Those charged with naming Pickering's successor attended the debate to evaluate Shapley for the position. It was offered to him, first on a visiting basis, and then permanently in 1921. At Harvard, Shapley carried out investigations in many areas, including studies of the Magellanic Clouds, star clusters, and variable stars. The Shapley–Ames catalog of galaxies, published with Adelaide Ames in 1932, was an important survey of galaxies brighter than the 13th magnitude. That and follow-up surveys, extending to fainter magnitudes, provided early indications that galaxy clustering might be important on large scales. Indeed, Shapley eventually came to the opinion that most galaxies are clustered (the modern view) and himself recognized several of the large clusters and concentrations.

Shapley's view of the clustering of galaxies was another source of disagreement with Hubble, who saw most galaxies as part of a general field. The antipathy felt by both astronomers may have dated from their overlap at Mount Wilson Observatory. Shapley remained a civilian during World War I, arriving in 1914, while Hubble volunteered for army service in Europe, arriving at Mount Wilson in 1919 to find that Shapley had initiated some studies that he himself had intended to pursue. That antipathy gradually emerged as full-blown antagonism as their careers diverged, with Hubble garnering fame through the scientific fruits of his work with the most powerful telescope in the world while Shapley administered at the Harvard College Observatory.

In 1938, Shapley reported the discovery of the Fornax and Sculptor systems, the first of the dwarf spheroidal galaxies that are satellites of the Milky Way. He completed publication of the Henry Draper Catalogue of Spectral Classifications, a project begun under Pickering, and organized the Henry Draper Extension. Together, these surveys provided spectral classifications for 359,000 stars. Shapley also continued Pickering's support for the American Association of Variable Star Observers, whose members were mainly amateur astronomers. His support for amateurs was evident in other ways, including his founding of the Bond Astronomy Club and support for the Amateur Telescope Makers of Boston.

Harvard Observatory, despite its location next to the oldest university in the country, had been a purely research institution. Shapley built it into one of the country's strongest education programs in astronomy. Many of his early students (including **Cecilia Payne-Gaposchkin**, **Helen Sawyer-Hogg**, and **Dorrit Hoffleit**) and his early appointments to the Harvard College Observatory staff (**Henrietta Swope**, **Bart Bok**, **Donald Menzel**, and **Fred Whipple**) also appear in these pages. He moved the Harvard Southern Station from Arequipa, Peru, to Bloemfontein, South Africa, and hosted at Harvard the headquarters of the American Association of Variable Star Observers and the editorial offices of *Sky & Telescope*, the most important popular astronomy magazine in the United States.

Shapley retired from the directorship in 1952 (to be succeeded by Menzel) and from his Harvard professorship in 1956, although he maintained an active interest in astronomical innovation until very late in life, visiting, for instance, the first observatory designed to look for gravitational radiation in 1969.

A brilliant public speaker, Shapley enjoyed popularizing science, especially astronomy. He arranged an extended series of radio talks on astronomy when that medium was still comparatively young. Later, with Bok, he initiated the Harvard Books on Astronomy, popular volumes that filled a growing demand for informative books on what was happening in astronomical research. These were written by some of the leading astronomers of the time. Some books in the Harvard Books series went through four editions before the series was cancelled.

Shapley had always been an outward-looking, publicly oriented scientist, serving, for instance, as one of the first presidents of the Commission on Galaxies of the International Astronomical Union in the 1920s. In the late 1930s, he became increasingly concerned about what was happening to German, and later European, scientists and spent a great deal of time helping to resettle refugees before, during, and after the World War II. Shapley was actively involved in the processes that led to the establishment of the National Science Foundation, Science Service Inc. (the publishers of *Science News* and coordinators of the science talent search, of which he was president), and UNESCO, where he was a strong advocate for the inclusion of the "S" (science) component. He served terms as president of the American Astronomical Society, the American Association for the Advancement of Science, and Sigma Xi (the scientific research society), as well as gave very large numbers of public lectures and served as a board member of the Belgian–American Education Foundation, the Worcester Foundation for Experimental Biology, and many others. Shapley was a strong opponent of "fringe science," helping to coordinate opposition to the ideas of Immanuel Velikovksy and the flying saucerites. At some point, this pattern was perceived as threatening to United States security, and he was called before the House Un-American Activities Committee for alleged (and completely untrue) Communist connections and sympathies.

His own scientific colleagues recognized and rewarded Shapley's work in many ways. He held 16 honorary degrees (ten from within the USA and six others), was elected to the United States National Academy of Sciences and to science academies in nine other countries, and received medals and awards from the American Astronomical Society, the Royal Astronomy Society, the French Astronomical Society, and others.

Horace A. Smith and *Virginia Trimble*

Selected References

Belkora, Leila (2003). "Harlow Shapley: Champion of the Big Galaxy." In *Minding the Heavens*, pp. 245–292. Bristol: Institute of Physics.

Bok, Bart J. (1978). "Harlow Shapley." *Biographical Memoirs, National Academy of Sciences* 49: 241–291.

Bryant, Katherine L. (1992). "The Great Communicator: Harlow Shapley and the Media, 1920–1940." AB thesis, Harvard University.

De Vorkin, David H. (2000). *Henry Norris Russell: Dean of American Astronomers*. Princeton, New Jersey: Princeton University Press.

Fernie, J. D. (1969). "The Period–Luminosity Relation: A Historical Review." *Publications of the Astronomical Society of the Pacific*, 81: 707–731.
Gingerich, Owen (1988). "How Shapley Came to Harvard or, Snatching the Prize from the Jaws of Debate." *Journal for the History of Astronomy* 19: 201–207.
——— (1990). "Through Rugged Ways to the Galaxies." *Journal for the History of Astronomy* 21: 77–88.
Palmeri, Joann (2000). "An Astronomer beyond the Observatory: Harlow Shapley as Prophet of Science." Ph.D. diss., University of Oklahoma.
Shapley, Harlow (1914). "On the Nature and Cause of Cepheid Variation." *Astrophysical Journal* 40: 448–465.
——— (1915). "A Study of Orbits of Eclipsing Binaries." *Contributions from the Princeton University Observatory*, no. 3.
——— "On the Determination of the Distances of Globular Clusters." *Astrophysical, Journal* 48 (1918): 89–124. (Sixth paper in the series "Studies Based on Colors and Magnitudes in Stellar Clusters." Additional papers include "The Distances, Distribution in Space, and Dimensions of 69 Globular Clusters." *Astrophysical Journal* 48 [1918]: 154–181 and "Remarks on the Arrangement of the Sidereal Universe." *Astrophysical Journal* 49 [1919]: 311–336.)
——— (1969). *Through Rugged Ways to the Stars*. New York: Charles Scribner's Sons, 1969.
Shapley, Harlow and Adelaide Ames (1932). "A Survey of the External Galaxies Brighter than the Thirteenth Magnitude." *Annals of the Astronomical Observatory of Harvard College* 88, no. 2: 41–75.
Smith, Horace (2000). "Bailey, Shapley, and Variable Stars in Globular Clusters." *Journal for the History of Astronomy* 31: 185–201.
Smith, Robert W. (1982). *The Expanding Universe: Astronomy's 'Great Debate', 1900–1931*. Cambridge: Cambridge University Press.

Shapley, Martha

Born **Kansas City, Missouri, USA, 3 August 1890**
Died **Tucson, Arizona, USA, 24 January 1981**

Besides doing calculations for her husband **Harlow Shapley**, American Martha Betz Shapley was an authority on eclipsing binary stars in her own right.

Alternate name

Betz, Martha

Selected Reference

Kopal, Zdeněk (1981). "In Memoriam: Martha Betz Shapley (1890–1981)." *Astrophysics and Space Science*. 79: 261.

Sharaf al-Dīn al-Ṭūsī

Born **Ṭūs, (Iran), *circa* 1135**
Died **(Iran), 1213**

Although Sharaf al-Dīn al-Ṭūsī is known especially for his mathematics (in particular his novel work on the solutions of cubic equations), he was also the inventor of the linear astrolabe, a tool that derives from the planispheric astrolabe but is more easily constructed. From his name we may infer that Sharaf al-Dīn was born in the region of Ṭūs, in northeastern Iran. He spent a major part of his early career as a teacher of the sciences, including astronomy and astrology, in Damascus and Aleppo; he also taught in Mosul. Among his students was Kamāl al-Dīn ibn Yūnus, who would eventually teach Sharaf's namesake, the great **Naṣīr al-Dīn al-Ṭūsī**.

Sharaf al-Dīn al-Ṭūsī devoted several treatises to the linear astrolabe, sometimes called the staff of al-Ṭūsī. Its principle is simple – many of the important circles on the planispheric astrolabe, especially the almucantars (altitude circles) and the circles of declination, are centered on the meridian line. The main rod of the linear astrolabe is equivalent to the meridian line and contains markings to indicate the centers of these circles and their intersections with the meridian. The ecliptic (which appears on the movable rete of a standard astrolabe) is represented by the intersections of the beginnings of the zodiacal signs with the meridian when the rete is rotated. Many typical operations on a traditional astrolabe require the locations of points of intersection of these various circles. By attaching ropes to the appropriate points on the staff to act as radii, the circles and their intersections can be reconstructed and the astronomical problem solved. A scale giving chord lengths in the meridian circle extended the linear astrolabe's range of applications. Attached to a plumb line, it was also used to take observations of solar altitude. Additional markings allowed the determination of the *qibla* (the direction of Mecca) and solutions of astrological problems.

The simplicity of the linear astrolabe made it easy to construct, but its less than artful appearance rendered it unattractive to collectors. It was neither as durable nor as accurate as a planispheric astrolabe, and its operations were less intuitive. None have survived.

Glen van Brummelen

Selected References

Al-Ṭūsī, Sharaf al-Dīn (1986). *Oeuvres mathématiques: Algèbre et géométrie au XIIe siècle*, edited and translated by Roshdi Rashed. 2 Vols. Paris: Les Belles Lettres.
Carra de Vaux, R. (1895). "L'astrolabe linéaire ou bâton d'al-Tousi." *Journal asiatique*, 9th ser., 5: 464–516.
Michel, Henri (1943). "L'astrolabe linéaire d'al-Tusi." *Ciel et terre* 59: 101–107.
——— (1947). *Traité de l'astrolabe*. Paris: Gauthier-Villars.

Sharonov, Vsevolod Vasilievich

Born **Saint Petersburg, Russia, 25 February/10 March 1901**
Died **Leningrad (Saint Petersburg, Russia), 27 November 1964**

Soviet astronomer Vsevolod Sharonov was one of the leading proponents of lunar and planetary exploration before the advent of robotic spacecraft. Sharonov enrolled in the Faculty of Physics and Mathematics at Petrograd University (renamed Leningrad State University) in 1918; he did not complete his degree until 1926, having served in the Red Army (1919–1924). Most of Sharonov's professional career was spent at Leningrad State University.

In 1929, Sharonov defended his dissertation on the theory and application of the wedge photometer. He worked on problems of aerophotometry at the Institute of Air Surveys (1930–1936) and organized a photometric laboratory for the task. From 1941 to 1944, he directed the university's astrophysics laboratory, which was evacuated to Yelabuga, Tatarstan, during World War II. Sharonov was appointed professor (1944) and later director (1951) of the university's astronomical observatory. Portions of his observational work were conducted at the Pulkovo, Simeiz, and Tashkent observatories.

Sharonov worked out new methods of absolute photometry and colorimetry, which have been applied to studies of the solar corona. (He observed seven total solar eclipses between 1936 and 1963.) His photometric studies of the lunar surface compared its composition to that of terrestrial volcanic rocks, with a view toward understanding whether layers of microscopic dust covered the Moon's surface. Sharonov modeled the atmosphere of Mars and argued for the presence of the mineral limonite on the Martian surface.

In the course of his research, Sharonov devised a number of new instruments for astronomical and geophysical observations, which included a haze gauge, a diaphanometer (an instrument that measures transparency of the atmosphere), and a visual colorimeter. He was married to astronomer N. N. Sytinskaya. A crater on the Moon's surface has been named for Sharonov.

Yuri Balashov

Selected References

Kulikovsky, P. G. (1975). "Sharonov, Vsevolod Vasilievich." In *Dictionary of Scientific Biography*, edited by Charles Coulston Gillispie. Vol. 12, pp. 352–354. New York: Charles Scribner's Sons.

Sharonov, V. V. (1947). *Mars*. Moscow: USSR Academy of Sciences Press.

——— (1958). *Priroda Planet*. Moscow: Fizmatgiz.

——— (1964). *The Nature of Planets*. Jerusalem: Israel Program for Scientific Translation.

——— (1965). *Planeta Venera* (Planet Venus). Moscow: Nauka.

Sharp, Abraham

Born **Bradford, (West Yorkshire), England, 1 June 1653**
Died **Bradford, (West Yorkshire), England, 15 August 1742**

Abraham Sharp was a highly skilled instrument maker contemporary with **Isaac Newton** and **John Flamsteed**, and several such eminent people seem to have relied on the instruments that he constructed for precise measurements of the positions of the stars. Sharp was born and died at Little Horton Hall in Bradford. (His birth year was given as 1651 by the *Dictionary of National Biography*, but W. Cudworth argued that 1653 is the correct date). Like so much of Bradford's heritage, his birthplace has regrettably been demolished – the site is now occupied by a car park for Saint Luke's Hospital and the Unity Building of the University of Bradford.

Sharp attended Bradford Grammar School and worked for a period in the textile trade in York and Manchester, before moving to Liverpool where he taught and studied mathematics. Here, he met the eminent astronomer, Flamsteed, who used his influence to get employment for Sharp at the Royal Naval dockyard in Chatham. From about 1684, Sharp was associated with Flamsteed at the Greenwich Observatory, and in 1688 he was employed to make a "mural arc," a large astronomical elevation-measuring device. Sharp developed a considerable reputation for the accuracy of his graduations on such devices, and this mural arc, which had a graduated scale with a radius of 2 m, was used by Flamsteed to make the observations from which the entries in the "British Catalogue" were deduced.

Sharp left the observatory in 1690 in order to teach mathematics in London, but in 1691 he moved to another naval dockyard, Portsmouth. Unfortunately, there appears to have been a problem with his health, and he retired to his family home in Bradford in 1694. His health recovered, but he inherited the family estate on the death of his nephew, and Sharp decided to remain in Bradford, setting up a workshop for the construction of instruments. In these 49 years in Bradford, he made astronomical instruments and also undertook many extensive mathematical calculations, particularly for the generation of tables of astronomical events. Sharp also calculated π to 72 decimal places and logarithms to 61 decimal places. He appears to have obtained many contracts from Flamsteed, and several biographies are laden with quotations from Flamsteed's letters to Sharp, who seems to have been something of a sounding board for the former. A large proportion of the letters from Flamsteed are highly critical of Newton, but it is not clear whether Sharp shared these views. It may be noteworthy that there is a story that Newton visited Sharp at Little Horton Hall, and so it could be that the letters are primarily a reflection of Flamsteed's views. It is probably a fair assessment to say that Sharp made a major contribution to the refinement of astronomical measurement techniques, enabling Newton and others to test and refine their theories of planetary movements.

In the letters between Sharp and Flamsteed, there are several interesting references to striking observations of the aurora borealis from Bradford in the early 18th century. Sunspot and geomagnetic field conditions must have been very unusual indeed for this effect to have been seen so clearly so far south.

Although Sharp may appear to be "just" an instrument maker and mathematician, a measure of his significance in the astronomical community may be gained from the fact that there is a crater on the nearside of the Moon that is named for him.

Philip Edwards

Selected Reference

Cudworth, William (1889). *Life and Correspondence of Abraham Sharp*. London: Sampson Low.

Shi Shen

Flourished **China, 4th century BCE**

Uranographer Shi Shen was a court astronomer during the late Waring States Period. He shares with Kan Te and Wu Xian credit for the earliest extant Chinese celestial chart. It shows 800 stars, comparable with the number in early Greek catalogs.

Selected Reference

Teresi, Dick (2002). *Lost Discoveries: The Ancient Roots of Modern Science – From the Babylonians to the Maya*. New York: Simon and Schuster.

Shibukawa, Harumi

Born **Kyoto, Japan, 1639**
Died **Edo (Tokyo), Japan, 1715**

Harumi Shibukawa inaugurated his country's first calendar reforms in many centuries and belonged to the first generation of Japanese scholars who assimilated knowledge of western astronomical ideas and practices. Shibukawa was born into the Yasui family; his father was a professional *go* (board game) player. As a child, he was called Rokuzo. After his father's death in 1652, Shibukawa took up *go* as a profession and adopted his father's first name, Santetsu. From early childhood, he had the reputation of being a prodigy and showed a remarkable understanding of astronomy and calendar study. Shibukawa received his education from many of the leading scholars of the day. These included Ansai Yamazaki, with whom he studied Confucianism and the *Shinto* doctrine, and both Jyunsho Matsuda and Gentei Okanoi, from whom he received training in calendrical methods.

Shibukawa spent most of his life in Edo (present Tokyo) but passed time in Kyoto when he was not playing *go*. He acquired a strong reputation, not only for his excellent skill at this game but also for numerous other subjects. In his writing, Shibukawa used the pen names of Harumi and Shunkai. He attracted many passionate and eager students along with wealthy patrons who invited him to lecture. Among the latter were Mitsukuni Tokugawa (*daimyo* of Mito Province, present Ibaragi and Tochigi prefectures) and Masayuki Hoshina of Aizu Province (now a part of Fukushima prefecture); these were two of the most influential political figures in Edo Japan (1603–1867). Hoshina later recommended Shibukawa to the *shogun* as the person most qualified to carry out calendar reforms.

In many ways, Shibukawa's attempts to gain recognition for his work show not only intelligence but also a great deal of courage and perseverance. At that time, calendrical practices lay in the hands of the bureaucratic Tsuchimikado family of Kyoto. Decisions about calendar reform seem to have been based far more on whether or not this family's political prestige would be maintained and enhanced rather than whether a calendar was accurate or not. Besides, few nobles would take a *go* player seriously. Shibukawa had to pay homage to the Tsuchimikado family while trying to develop his own calendrical methods, such often being in opposition to the outdated systems then in use. In the end, empiricism, coupled with a healthy dose of pragmatism, won the day. Shibukawa's calendar was accepted, and for the first time in 800 years, calendar reform became a reality.

In developing his own mathematical astronomy, Shibukawa insisted upon a strong positivistic base. This empirical orientation influenced generations of astronomers and calendrical scholars who followed. At the same time, he concluded that ancient writers of the Chinese classics must have had reasons for seeing aspects of the world in the way they did. Shibukawa felt that no one schema could explain everything; he wrote that astronomers should be versed in both portent astrology and calendrical science. The stimulus for his interest in calendar reformation came from discrepancies between his observations and ancient records. He felt such discrepancies could be explained by variations in the length of the solar year, a perceived phenomenon that was to influence later calendar scholars as well. Within Shibukawa's inclusive approach, he felt that irregular motions of the heavenly bodies were admissible; thus, it was unnecessary to be overly concerned with spatial relationships of the Sun, Moon, and planets – issues that were of major concern in contemporary Europe.

Shibukawa is best known for his reform of Japan's lunar calendar. By the late 17th century, the *Senmyo Reki*, a lunar calendar originating in the Chinese Tang dynasty, had been in use since 862 and was 2 days behind the solar year. Shibukawa became an outspoken proponent of calendar reform. The new calendar, officially put in place in 1685, was named the *Jyokyo Reki* and was the first calendar to be constructed by a Japanese citizen. It remained in use until 1755.

In formulating his own system, Shibukawa preserved the structure and theory behind the 13th-century *Shou-shih* calendrical treatise written by **Guo Shuoujing** of the Yuan dynasty in China. However, he incorporated into the *Jyokyo Reki* the difference in longitude between Japan and China. His calendar predicted apparent solar and planetary movement more accurately than did any produced by the official Yin-Yang board headed by Kyoto's Tsuchimikado family.

Precise astronomical observation was necessary for Shibukawa's work in calendar reform, and he enthusiastically used a number of instruments including the armillary sphere. With his penchant for observation, he compiled a number of star maps including the *Tensho Retsuji no Zu* (1670), *Tenmon Bunya no Zu* (1677), and *Tensho Seisho Zu* (1699), the last being published under his son's name, Hisatada.

Following his achievements, Shibukawa was officially appointed to the Tenmongata (Bureau of astronomy) in 1684. He moved his residence to Edo in 1686, and set up an astronomical observatory. In 1692, he was promoted to the *Samurai* class, and in 1702 once again changed his surname to Shibukawa, the original surname of the Yasui family and the name by which he is perhaps best known.

Shibukawa's tomb is located in the compound of Tokai Zen temple in Shinagawa, Tokyo.

Steven L. Renshaw and *Saori Ihara*

Selected References

Miyajima, Kazuhiko (1994). "Japanese Celestial Cartography before the Meiji Period." In *The History of Cartography*. Vol. 2, bk. 2, *Cartography in the Traditional East and Southeast Asian Societies*, edited by J. B. Harley and David Woodward, pp. 579–604. Chicago: University of Chicago Press.

Nakayama, Shigeru (1969). *A History of Japanese Astronomy: Chinese Background and Western Impact*. Cambridge, Massachusetts: Harvard University Press.

——— (1975). "Shibukawa, Harumi." In *Dictionary of Scientific Biography*, edited by Charles Coulston Gillispie. Vol. 12, pp. 403–404. New York: Charles Scribner's Sons.

Sugimoto, Masayoshi and David L. Swain (1978). *Science and Culture in Traditional Japan*. Cambridge, Massachusetts: MIT Press.

Watanabe, Toshio (1986–1987). *Kinsei Nihon Tenmongakushi* (A Modern History of Astronomy in Japan). 2 Vols. Tokyo: Koseisha Koseikaku.

Shīrāzī: Quṭb al-Dīn Maḥmūd ibn Masʿūd Muṣliḥ al-Shīrāzī

Born **Shīrāz, (Iran), October/November 1236**
Died **Tabrīz, (Iran), 1311**

Quṭb al-Dīn al-Shīrāzī was one of the most prominent theoretical astronomers of the 13th century. Born into a family of physicians, he studied with his father Ḍiyā' al-Dīn al-Kāzarūnī at the then new Muẓaffarī Hospital. When Shīrāzī was 14, his father died but even at that young age he was able to assume his father's position at the hospital. He continued his studies, at first with an uncle, also a physician, and later with two prominent teachers, Shams al-Dīn Kīshī and Sharaf al-Dīn Būshkānī. These studies included most prominently the "general principles" of **Ibn Sīnā**'s *Canon of Medicine* as well as Sufi mysticism, which had been another important part of his father's life. Uncharacteristic for someone of his talents and searching intellect, Shīrāzī remained in Shīrāz until the age of 24, most likely because of the turmoil in Iran brought on by the Mongol invasions.

But the Mongols also provided Shīrāzī with a unique opportunity, that of studying at the Marāgha Observatory with its director **Naṣīr al-Dīn al-Ṭūsī**. Though Shīrāzī was probably seeking to further his medical education, he soon turned to serious studies of philosophy and the mathematical sciences, especially astronomy, and would become Ṭūsī's most prominent student. In Marāgha, he also studied with the philosopher Najm al-Dīn al-Kātibī and with the renowned astronomer **Mu'ayyad al-Dīn al-ʿUrḍī**. Even though based in Marāgha, Shīrāzī seems to have traveled a great deal for both teaching and learning. Sometime in his mid-30s, before the death of Ṭūsī in 1274, he may have become estranged from his teacher and left Marāgha. Accounts vary, but this may have had to do with the secondary role he was assigned at the observatory, or to not being named by Ṭūsī in the *Ilkhānī Zīj*, the handbook with tables that was produced at Marāgha. **Wābkanawī** states that Shīrāzī, though asked by Ṭūsī's son Aṣīl al-Dīn to help revise the *Zīj*, did so only in a perfunctory way because of his sense of having been slighted.

Sometime after leaving Marāgha, Shīrāzī traveled to Anatolia and studied for a time in Konya, perhaps meeting the famous Sufi poet Jalāl al-Dīn al-Rūmī. He was appointed chief judge in Malaṭya and Siwās and began to take an active role in political affairs, including acting as an emissary from the Mongol court to the Mamluks in 1282. Sometime around 1290, Shīrāzī retired to the city of Tabrīz in Azerbaijan where the Mongol court was located. But because of a falling out with the chief minister, he seems to have retired from government service and devoted himself to writing and teaching. It is of some interest that Shīrāzī dedicated his major philosophical encyclopedia, the *Durrat al-tāj*, to the ruler of an independent principality in western Gīlān in 1306; but later that year, the principality was brought under the control of the Mongols, and Shīrāzī was probably back in Tabrīz shortly thereafter.

Shīrāzī wrote three major works in theoretical astronomy – the *Nihāyat al-idrāk fī dirāyat al-aflāk* (The highest attainment in comprehending the orbs), dedicated to the Vizier Shams al-Dīn al-Juwaynī (who may have been responsible for his judgeship) and completed in November 1281; *al-Tuḥfa al-shāhiyya* (The imperial gift), dedicated to the Vizier Amīr Shāh ibn Tāj al-Dīn Muʿtazz ibn Ṭāhir in Siwās in July or August 1285; and *Faʿalta fa-lā talum* (You've done it so don't blame [me]), a supercommentary on the *Tadhkira fī ʿilm al-hay'a* by Ṭūsī. All have the characteristic four-part division of a *hay'a* (theoretical astronomy) work: an introduction, a section on the structure of the celestial region, a section on the structure of the terrestrial region, and a section on the sizes and distances of the celestial and terrestrial bodies. The *Nihāya* is the longest of the works, at some 300 or more pages in manuscript. It tends to present more of the work of Shīrāzī's predecessors than does the *Tuḥfa*. *Faʿalta* is a peculiar work in that Shīrāzī is ostensibly commenting on the commentary of the *Tadhkira* by a certain al-Ḥimādhī; in reality, it is a harshly worded attack on this author who, according to Shīrāzī, has plagiarized his *Tuḥfa*. This makes it an interesting work for the history of the notion of intellectual property. In addition to these straightforward astronomical works, there are also large sections related to astronomy in two of Shīrāzī's Persian works – the *Durrat al-tāj* and his *Ikhtiyārāt-i Muẓaffarī*, which was dedicated to the local ruler of a small emirate in Kastamonu. Large parts of the latter seem to be translations from the *Nihāya*.

Shīrāzī's works have not received the study they deserve, which is unfortunate since they promise to shed much light on the so called Marāgha school. Kennedy (1966) noted a number of innovative astronomical models in the *Nihāya* and the *Tuḥfa*, but Saliba showed that many of these models were due to Mu'ayyad al-Dīn al-ʿUrḍī. Shīrāzī should still be credited with new models for the Moon and Mercury (both in the *Tuḥfa*). He creatively uses what are now known as the ʿUrḍī lemma and the Ṭūsī couple to achieve combinations of uniform, circular motions (as required by ancient physics for motions in the heavens) that resolve the irregular motions resulting from **Ptolemy**'s equant for Mercury and from his choice of the center of the universe as the reference point of motion for the Moon's eccentric orb.

Shīrāzī' gives high praise to astronomy in his introduction to the *Nihāya* and echoes Ptolemy who, in his introduction to the *Almagest*, referred to physics and theology as guesswork as opposed to the true knowledge offered by the mathematical sciences. Indeed, it would seem that Shīrāzī somewhat disagreed with his mentor Ṭūsī on this point. This manifested itself in the question of the Earth's motion – Ṭūsī had held that the matter had to be left to the natural philosophers since there was no decisive observational or mathematical proof, whereas Shīrāzī, not wishing to leave such an important matter to "guesswork," insisted that there could be devised an observational test. This test took the form of two rocks of different weights thrown straight up in the air; Ṭūsī had said that in such a case a rotating Earth could carry the air and whatever was in it at the same speed, but Shīrāzī thought that objects of different weights would be carried with different speeds. Since we do not observe such an effect, the Earth must be at rest.

Shīrāzī's influence in astronomy was widespread. His words were copied and studied for several centuries. Often referred to simply as *ʿAllāma* (supremely learned), one finds citations to him by almost all later Islamic theoretical astronomers. In medicine, he was known for his extensive commentary on the first book of Ibn Sīnā's *Canon*, and he was to have a major influence on optics by recommending that his student Kamāl al-Dīn al-Fārisī undertake a study of **Ibn al-Haytham**'s *Kitāb al-manāẓir*.

F. Jamil Ragep

Selected References

Kennedy, E. S. (1966). "Late Medieval Planetary Theory." *Isis* 57: 365–378. (Reprinted in E. S. Kennedy, et al, *Studies in the Islamic Exact Sciences*, edited by David A. King and Mary Helen Kennedy, pp. 84–97. Beirut: American University of Beirut, 1983.)

Morrison, Robert (2005). "Quṭb al-Dīn al-Shīrāzī's Hypotheses for Celestial Motions." *Journal for the History of Arabic Science* 13: 21–140.

Ragep, F. J. (1993). *Naṣīr al-Dīn al-Ṭūsī's* Memoir on Astronomy *(al-Tadhkira fī ʿilm al-hayʾa)*. 2 Vols. New York: Springer-Verlag.

——— (2001). "Freeing Astronomy from Philosophy: An Aspect of Islamic Influence on Science." *Osiris* 16: 49–71.

——— (2001). "Ṭūsī and Copernicus: The Earth's Motion in Context." *Science in Context* 14: 145–163.

Saliba, George (1979). "The Original Source of Quṭb al-Dīn al-Shīrāzī's Planetary Model." *Journal for the History of Arabic Science* 3: 3–18.

——— (1996). "Arabic Planetary Theories after the Eleventh Century AD." In *Encyclopedia of the History of Arabic Science*, edited by Roshdi Rashed, pp. 58–127. London: Routledge.

Walbridge, John (1992). *The Science of Mystic Lights: Quṭb al-Dīn al-Shīrāzī and the Illuminationist Tradition in Islamic Philosophy*. Cambridge, Massachusetts: Harvard University Press. (Excellent source for Shīrāzī's life and works.)

Wiedemann, E. (1986). "Ḳuṭb al-Dīn Shīrāzī." In *Encyclopaedia of Islam*. 2nd ed. Vol. 5, pp. 547–548. Leiden: E. J. Brill. (Important for a listing of Wiedemann's articles on Shīrāzī.)

Shirwānī: Fatḥallāh ibn Abū Yazīd ibn ʿAbd al-ʿAzīz ibn Ibrāhīm al-Shābarānī al-Shirwānī al-Shamāhī

Born **Shirwān, Shamāh, (Azerbaijan), 1417**
Died **Shirwān, Shamāh, (Azerbaijan), February 1486**

The astronomer, mathematician, and teacher Fatḥallāh al-Shirwānī was part of the Samarqand school of mathematics and astronomy, which was composed of scholars who pursued the mathematical sciences including astronomy. Through his works many students were educated in the sciences, thus disseminating them in the Ottoman lands, especially in Anatolia.

Shirwānī received his primary education from his father and subsequently continued his education in Serakhs and Ṭūs. In Ṭūs, Shirwānī studied **al-Sayyid al-Sharīf al-Jurjānī**'s *Sharḥ al-Tadhkira fī ʿilm al-hayʾa*, a commentary on **Naṣīr al-Dīn al-Ṭūsī**'s seminal work on astronomy, under the Shīʿī scholar al-Sayyid Abū Ṭālib. In mid-1435 he left for Samarqand and studied mathematics, astronomy, Islamic theology (*kalām*), and the linguistic sciences under **Qāḍīzāde** at the *madrasa* (school) of Samarqand. Among the works he studied was **Niẓām al-Dīn al-Nīsābūrī**'s *Sharḥ al-Tadhkira fī ʿilm al-hayʾa*, yet another commentary on Ṭūsī's work. Clearly the *Tadhkira* occupied an important place in the school of Samarqand as well as in Shirwānī's education. Shirwānī received his diploma on 13 September 1440. During his education in the *madrasa*, he no doubt participated in astronomical activities, primarily the astronomical observations at the Samarqand Observatory. During his stay in Samarqand, he also wrote a commentary on a work of Islamic law, which he presented to **Ulugh Beg**.

In 1440, after his 5-year long education in Samarqand, Shirwānī returned to Shirwān where he lectured for some time at the *madrasas* there. On the advice of his former teacher Qādīzāde, he left for Anatolia (toward the end of the reign of Sultan Murād II [reigned: 1421–1451]) and was warmly received by Çandaroğlu Ismail Bey in Kastamonu. Subsequently, he started teaching in the *madrasas* there. Shirwānī lectured on mathematical and astronomical works, especially those of his teacher Qāḍīzāde, and on *al-Tadhkira*. Muḥyī al-Dīn Muḥammad ibn Ibrāhīm al-Nīksārī (died: 1495) and Kamāl al-Dīn Masʿūd al-Shirwānī (died: 1500) were among his prominent students.

In 1453, Shirwānī dedicated a commentary (*tafṣīr*) on the Qur'ān to the Ottoman Grand Vizier Çandarlı Khalīl Pasha in Bursa. That same year, he presented a work on music (a subdivision of the mathematical sciences) to Sultan Mehmed II. However, later in the year after the conquest of Istanbul, Khalīl Pasha was executed; having lost his patron, Shirwānī returned to Kastamonu. After these events, Shirwānī wrote a work on theoretical astronomy, which was a supercommentary on Qāḍīzāde's *Sharḥ al-Mulakhkhaṣ*. This he presented to Sultan Mehmed II in the hopes of establishing closer ties with the Ottoman court, but he was unsuccessful.

In 1465, Shirwānī set off on a pilgrimage for Mecca; *en route* he continued pursuing scientific activities, first stopping in Iraq and teaching at the *madrasas* in the region. He remained in Mecca for a time, continuing to give lectures. Shirwānī returned to Istanbul, *via* Cairo. Not receiving the attention he thought his due, he returned to his hometown of Shirwān in 1478.

Shirwānī wrote works on literature and linguistics, *kalām*, music, Islamic law, Qur'ānic exegesis, optics, and logic as well as the rational sciences. In the field of geometry, he wrote a gloss (*ḥāshiya*) to Qāḍīzāde's commentary (*sharḥ*) on **Shams al-Dīn al-Samarqandī**'s *Ashkāl al-taʾsīs*. Unfortunately this work is not extant.

In the field of astronomy, *al-Farāʾiḍ wa-'l-fawāʾid fī tawḍīḥ sharḥ al-Mulakhkhaṣ* was Shirwānī's first important work on theoretical astronomy (*hayʾa*), which was a gloss (*ḥāshiya*) on Qāḍīzāde's commentary (*sharḥ*) to **Maḥmūd al-Jaghmīnī**'s *al-Mulakhkhaṣ fīʿilm al-hayʾa al-basīṭa*. In order to explain the difficult parts, Shirwānī made use of other commentaries and class notes he took during Qāḍīzāde's lectures at the Samarqand *madrasa*; he completed the work after many rough drafts.

Shirwānī's most noteworthy work on theoretical astronomy is undoubtedly his commentary (*Sharḥ*) to Naṣīr al-Dīn al-Ṭūsī's *al-Tadhkira fīʿilm al-hayʾa*, which he completed on 11 January 1475. He emphasized that he wrote his commentary for advanced-level students to whom he lectured in the field of astronomy. His sources were other commentaries, the lecture notes of his teacher Qāḍīzāde, and his own insights.

The *Sharḥ* contains a great deal of information that often has little to do with Ṭūsī's *Tadhkira*. For example, Shirwānī provides comprehensive information about the Turkish calendar as well as other calendar systems. He also discusses Euclid's *Elements* based upon discussions he had with Qāḍīzāde, Ulugh Beg, and students at the Samarqand *madrasa*. Shirwānī also includes a registered copy of his license to teach (*ijāza*) that he obtained from Qāḍīzāde. He has a lengthy discussion on optics (*ʿilm al-manāẓir*), which was considered an ancillary branch of astronomy. He cites numerous works and authors throughout, pointing out his own views when appropriate. Although a thorough analysis of Shirwānī's text has not been yet been made, his style indicates that he was aware of the attempts by **Ibn al-Haytham** and his follower Kamāl al-Dīn Fārisī to combine physical and geometrical approaches within optics, and that this was the subject of ongoing debates in the Samarqand school.

In his *Sharḥ*, Shirwānī discusses Ṭūsī's innovative cosmology in detail. He agrees with Ibn al-Haytham in combining mathematical and natural philosophical approaches; he disagrees with his Samarqand contemporary ʿ**Alī Qūshjī**, who attempted to purge the science of astronomy of Aristotelian principles of physics and metaphysics. Further research into Shirwānī's work promises to provide important information on the history of late medieval Islamic astronomy.

İhsan Fazlıoğlu

Selected References

Akpınar, Cemil (1995). "Fethullah es-Sirvani." In *Türkiye Diyanet Vakfı İslâm Ansiklopedisi*. Vol. 12, pp. 463–466. Istanbul: Türkiye Diyanet Vakfi Yayinlare.

Bağdadlı, İsmail Pasa (1951). *Hadiyyat al-ʿārifīn*. Vol. 1, p. 815. Istanbul: Milli Egition Bahanligh Yayinlare.

Brockelmann, Carl. *Geschichte der arabischen Litteratur*. 2nd ed. Vol. 2 (1949): 269, 279; Suppl. 2 (1938): 290. Leiden: E. J. Brill.

Bursalı, Mehmed Tahir (1923). *Osmanlı Müellifleri*. Vol. 3, p. 392. Istanbul: Maṭbaa-i Amire.

Fazlıoğlu, İhsan (2003). "Osmanlı felsefe-biliminin arkaplanı: Semerkand matematik-astronomi okulu." *Dîvân İlmî Arastırmalar* 1: 1–66.

İhsanoğlu, Ekmeleddin *et al.* (1997). *Osmanlı Astronomi Literatürü Tarihi (OALT)* (History of astronomy literature during the Ottoman period). Vol. 1, pp. 42–45 (no. 16). Istanbul: IRCICA.

Kātib Čelebī. *Kashf al-ẓunūn ʿan asāmī al-kutub wa-'l-funūn*. Vol. 1 (1941), cols. 36, 67, 443; Vol. 2 (1943), cols. 1819, 1893. Istanbul.

Neubauer, Eckhard (1984). "Neuerscheinungen zur arabischen Musik." *Zeitschrift für Geschichte der Arabisch–Islamischen Wissenschaften* 1: 288–311, esp. 290–296.

Ragep, F. J. (1993). *Naṣīr al-Dīn al-Ṭūsī's* Memoir on Astronomy (*al-Tadhkira fī ʿilm al-hay'a*). 2 Vols. New York: Springer-Verlag, Vol. 1, pp. 62–63.

Ṣakhāwī, Muḥammad ibn ʿAbd al-Raḥmān. *al-Ḍaw' al-lāmi fī ahl al-qarn al-tāsi*. Vol. 4, p. 340; Vol. 6, pp. 166–167. Cairo: 1353–1355 [1935–1937].

Ṭashköprüzāde (1985). *Al-Shaqāi'q al-numāniyya fī ulamā' al-dawlat al-uthmāniyya*, edited by Ahmed Subhi Furat. Istanbul: pp. 15–16, 107–108, 273. Istanbul üniversitesi, Edebiyaṭ Falutteyi Yayinlare.

Shizuki, Tadao

Born **Nagasaki, Japan, 1760**
Died **Nagasaki, Japan, 22 August 1806**

Tadao Shizuki was a translator and commentator of works on natural philosophy; he introduced western (Newtonian) science into Japanese culture and attempted to reconcile its principles with Confucian notions. Shizuki was born into the Nakano family but was later adopted and became an eighth-generation son of the Shizuki family. His was a family of professional translators and interpreters (known as *Tsuji*) who concentrated primarily on Dutch and Japanese sources. Shizuki began practicing this profession in 1776. In the following year, however, he resigned his official position on the grounds of ill health, changed his family name back to Nakano, and began work on his own translations and commentaries. Later, he often used the pen name Ryuen Nakano.

Shizuki was the disciple of Ryoei Motoki who had begun translating western works on astronomy including explanations of the heliocentric system of **Nicolaus Copernicus**. In the Edo era (1603–1867), Nagasaki was one of few ports in Japan through which information about western knowledge could be obtained; Shizuki was one of several scholars who acquired such information. More than a translator, he wrote commentaries about the scientific materials that came into his possession. While his full understanding of the underlying principles remains open to question, Shizuki was instrumental in introducing the concepts of Newtonian mechanics into Japan. He spent some 20 years on this work and devoted the rest of his life to translations of similar materials and linguistic research on the Dutch language.

With his teacher Motoki, Shizuki struggled for most of his life to introduce western concepts of science (derived from Dutch works) into the closed society of Edo Japan. He is perhaps best known for his Japanese translation and commentary on Newtonian principles, *Rekisho Shinsho* (New Treatise on Calendrical Phenomena), completed in 1802. This work was drawn from the writings of **John Keill** but includes many of Shizuki's own ideas. Japanese science of the mid-Edo period was not advanced enough to recognize the need for an understanding of Newtonian mechanics, while an improved calendar was not easily derived from Newtonian dynamics. Shizuki's efforts, while significant for those who came afterward, were not well understood by his contemporaries.

Although he remained somewhat ensconced in classical Chinese modes of inquiry (*e. g.*, Confucian notions of Yin–Yang polarities), Shizuki's commentaries show remarkable originality. Where Keill had started with basic aspects of common experience and moved to Newtonian principles, Shizuki began with cosmological principles and worked toward fundamental mechanical laws. His treatment of **Johannes Kepler**'s third law of planetary motion was fairly accurate. Although certainly not as elaborate or complete, Shizuki's hypotheses regarding the origin of the Solar System were similar to those of **Pierre de Laplace** and **Immanuel Kant**. Japanese terms that Shizuki coined for Newtonian concepts such as gravity and centripetal/centrifugal force became standards that are still in use today.

Steven L. Renshaw and *Saori Ihara*

Selected References

Nakayama, Shigeru (1969). *A History of Japanese Astronomy: Chinese Background and Western Impact*. Cambridge, Massachusetts: Harvard University Press.

——— (1975). "Shizuki, Tadao." In *Dictionary of Scientific Biography*, edited by Charles Coulston Gillispie. Vol. 12, pp. 406–409. New York: Charles Scribner's Sons.

——— (1978). "Japanese Scientific Thought." In *Dictionary of Scientific Biography*, edited by Charles Coulston Gillispie. Vol. 15 (Suppl. 1), pp. 728–758. New York: Charles Scribner's Sons.

Watanabe, Toshio (1986–1987). *Kinsei Nihon Tenmongakushi* (A Modern History of Astronomy in Japan). 2 vols. Tokyo: Koseisha Koseikaku.

Shklovsky [Shklovskii, Shklovskij], Iosif Samuilovich

Born **Glukhov, (Ukraine), 1 July 1916**
Died **Moscow, (Russia), 3 March 1985**

Theoretical astrophysicist Iosif S. Shklovsky was one of the most remarkable personalities and scholars among the Soviet astronomy community of his time. He chose as his own most important contributions the derivation of a reliable distance scale for planetary nebulae and the recognition that the emission from the Crab Nebula is synchrotron radiation (from which he predicted optical polarization, soon after found). Many would also mention his 1964 suggestion that the energy source for quasars is accretion of gas onto giant black holes at the centers of galaxies.

Following high school graduation, Shklovsky spent 2 years working in railroad construction, before beginning the study of physics and chemistry at the Far Eastern State University in Vladivostok, Russia, in 1933. He transferred to Moscow State University [MSU], completing a first degree in 1938, a candidate degree in 1944, and a doctorate in 1949. Shklovsky was excused from active service in World War II because of extreme nearsightedness, and was evacuated with other civilian members of the MSU physics department in 1941, returning to Moscow in 1943 from Ashkhabad (Turkmenistan). He remained associated with MSU for the remainder of his career, founding the radio astronomy department at the Shternberg Institute of Astronomy [MSU] in 1953, and Department of Astrophysics at the new Institute for Space Research in 1969.

Shklovsky's candidate ("The Concept of Electron Temperature in Astrophysics") and doctoral theses dealt with the solar corona. He correctly accepted that the corona was very hot, predicted ultraviolet and X-ray emission (seen in rocket flights in 1947–1949, organized by **Richard Tousey** and **Herbert Friedman**), suggested that it might be heated by hydrodynamic waves, emphasized the coexistence of thermal and nonthermal processes, and suggested that solar radio bursts might be produced by plasma oscillations and the scattering of Langmuir waves to transverse waves.

One of the first topics in radio astronomy beyond the Sun that Shklovsky considered was the detectability of interstellar gas. He concluded in 1952 (independently of H.C. van de Hulst) that neutral hydrogen should have an observable feature. He also calculated the expected millimeter wavelengths from OH and CH molecules accurately enough that they were quickly found when technology was equal to the task 15 to 20 years later.

Shklovsky's 1953 consideration of the newly detected radio emission from the supernova remnant called the Crab Nebula led him to propose that the optical emission might also be synchrotron and so should be polarized. Two Soviet astronomers independently detected this polarization the next year. Shklovsky was also among the first to emphasize that the synchrotron spectrum should continue to X-ray energies, and the emission be extended in space (rather than compact, as in the case of a hot neutron star). This also was found to be the case.

Shklovsky received a 1960 Lenin Prize (then the highest honor in science in the USSR) for the creation of an "artificial comet." He had been thinking about fluorescence in the Earth's upper atmosphere (indeed publishing a paper from the Institute for Atmospheric Physics in 1958) and suggested that sodium vapor shot out from a probe or satellite soon after launch should also fluoresce, making possible sharp enough photographs to permit accurate measurements of trajectories.

Shklovsky went on, in the year of the prize, to formulate a model for synchrotron and other kinds of emission from expanding clouds of relativistic plasma. He applied this to radio galaxies, like Cygnus A and Centaurus A, concluding that they were nonthermal emissions at different evolutionary phases (and not the result of colliding galaxies as had been suggested by **Walter Baade** and **Rudolph Minkowski** when they first saw the optical counterparts), and that their lifetimes must be 10^7–10^8 years, essentially the modern value. The model could also be applied to radio emission from supernova remnants, and Shklovsky predicted that Cassiopea A (remnant of an explosion in about 1680) should be fading at a bit more than 1% per year. This was seen soon after. The interpretation of quasars as accretion on massive black holes came in 1964, soon after the discovery of these sources, and Shklovsky then also pointed out (1965) that quasars and Seyfert galaxies formed a continuum of source types.

The model of planetary nebulae arose as part of another dispute, over whether the gas clouds with stars at their centers came at the beginning or the end of the life of a star. Shklovsky opted (correctly) for "at the end" and, in 1956 pointed out that, if the amount of material ejected as stars like the Sun died was always about the same, then one could find the distances to the observed planetaries, and so the total number that must exist in the Milky Way and thus their birthrates. Again he was essentially right. His other contributions include:

(1) 1963 discovering of the optical variability of the first quasar, 3C273 from archival Russian plates (independent of the discovery made by Harlan Smith and **Dorrit Hoffleit** on Harvard Observatory plates);
(2) coining of the name "relic radiation" for the microwave background left from the Big Bang;
(3) drawing attention to the mix of ionization states represented in quasar absorption spectra as evidence that the absorption clouds in galaxies were distant, from both the quasars and from us, and redshifts therefore are a good distance indicator;
(4) a deep and abiding interest in the possibility of extraterrestrial life and the possibility of contacting it (which he handed on to Nicolay Kardashev, his successor as one of the leader of Soviet radio astronomy);
(5) an assortment of charming, but wrong ideas, like the possibility that one of the satellites of Mars, which seemed to have very low mass for its size, might really be a hollow spacecraft; and
(6) the education and inspiration of two generations of students.

Shklovsky himself was elected to associate membership in the Soviet Academy only in 1966 and never to full membership, though he appeared on the ballot many times. He was, however, honored with memberships, foreign associateships, or medals by the Royal Astronomical Society of Canada, the United States National Academy of Sciences, the Royal Astronomical Society, the Astronomical Society of the Pacific (Bruce Medal, 1972), and the American Academy of Arts and Sciences. He was permitted to travel outside the USSR only sporadically, and so was not able to accept all of these honors in person.

Shklovsky penned the worldwide bestseller, *The Universe. The Life. The Intelligence.* (1962), concerning humanity's place in the Universe. He was a driving force behind a number of landmark domestic and international meetings dedicated to the subject of extraterrestrial intelligence, helping to establish the topics as a legitimate one for scientific discussion.

In total, shklovsky wrote nine books and over three-hundred articles. Most were translated from Russian into foreign languages. Works such as Shklovsky's collection of essays (entitled, in English, *Five Billion Vodka Bottles to the Moon: Tales of a Soviet Scientist*; 1991), include criticisms of the Soviet regime. This book appeared postthumously in Russian only in the Mikhail Gorbachev era.

Alexander A. Gurshtein

Selected References

Friedman, Herbert (1985). "Josif Samuilovich Shklovsky." *Physics Today* 38, no. 5: 104–106.

Kardashev, N. S. (1992). *Astrophysics on the Threshold of the 21st Century*. Philadelphia: Gordon & Breach.

Shklovsky (1951). *Solar Corona*. Moscow: Gostechizdat (In Russian).

Shklovsky (1962). *The Universe. The Life. The Intellegence*. Moscow: USSR Academy of Sciences (In Russian).

Shklovsky (1991). *Convoy (Non-Fictinal Stories)*. Moscow: Novosti (In Russian).

Shklovsky (1996). *Intelligence. Life. Universe*. Moscow: Yanus (In Russian).

Shklovskii, I. S. (1960). *Cosmic Radio Waves*. Cambridge, Massachusetts: Harvard University Press.

——— (1968). *Supernovae*. Interscience Monographs and Texts in Physics and Astronomy, Vol. 21. London: Wiley and Sons.

——— (1978). *Stars, Their Birth, Life and Death*. San Francisco: W. H. Freeman.

——— (1991). *Five Billion Vodka Bottles to the Moon: Tales of a Soviet Scientist*. New York: W. W. Norton.

Shklovskii, I. S. and Carl Sagan (1966). *Intelligent Life in the Universe*. San Francisco: Holden-Day.

Sibṭ al-Māridīnī: Muḥammad ibn Muḥammad ibn Aḥmad Abū ʿAbd Allāh Badr [Shams] al-Dīn al-Miṣrī al-Dimashqī

Born **possibly Damascus, (Syria), 1423**
Died **possibly Cairo, (Egypt), *circa* 1495**

Sibṭ al-Māridīnī was a prolific author of astronomical texts, which were still being used and studied into the 19th century. Little is known with certainty about his life. It is thought that he grew up in Damascus, where his maternal grandfather, ʿAbd Allāh ibn Khalīl ibn Yūsuf Jamāl al-Dīn al-Māridīnī (died: 1406), was the *muwaqqit* (timekeeper in charge of regulating the daily rituals of the Islamic community) of the Umayyad Mosque. Later he traveled to Cairo, where tradition places him as the student of **Ibn al-Majdī**.

Sibṭ al-Māridīnī wrote extensively on mathematics and mathematical astronomy. Like his grandfather, he was especially interested in astronomical instruments. The bio-bibliographical sources list some 25 treatises, many of which exist today in multiple copies. According to the historian al-Jabartī (died: 1822), Sibṭ al-Māridīnī's works on *mīqāt* (ritual timekeeping) and on astronomical instruments were still being studied in the curriculum of Cairo's al-Azhar, one of the preeminent educational institutions in the Islamic world, at about the beginning of the 19th century.

Among Sibṭ al-Māridīnī's works related to astronomy and instruments are:

(1) *Risāla fī al-ʿAmal bi-'l-rubʿ al-mujayyab* (on using the sine quadrant);
(2) *Raqā'iq al-ḥaqā'iq* (on calculating with degrees and minutes);
(3) *Zubd al-raqā'iq* (this may be an extract from the previous treatise);
(4) *Muqaddima* (introduction) to sine problems and spherical relations;
(5) *al-Ṭuruq al-saniyya* (on sexagesimal calculations);
(6) *al-Nujūm al-ẓāhirāt* (on the *muqanṭarāt* quadrant);
(7) *Qatf al-ẓāhirāt* (apparently an extract from the previous treatise);
(8) *Hāwī al-mukhtaṣarāt* (another discussion of the *muqanṭarāt* quadrant);
(9) *Iẓhār al-sirr al-mawḍūʿ* (use of a specialized quadrant);
(10) *Hidāyat al-ʿāmil* (on another kind of specialized quadrant);
(11) *Hidāyat al-sā'il* (on the quadrant mentioned in the previous entry);
(12) *al-Maṭlab* (on the sine quadrant);
(13) *al-Tuḥfa al-manṣūriyya* (on quadrants);
(14) *Muqaddima* (introduction to construction of sundials);
(15) a treatise on the equatorial circle; and
(16) a treatise on the quadrant, astrolabe, and calendar.

Gregg DeYoung

Selected References

King, David A. (1975). "Al-Khalīlī's *Qibla* Table." *Journal of Near Eastern Studies* 34: 81–122. (Reprinted in King, *Islamic Mathematical Astronomy*, XIII. London: Variorum Reprints, 1986.) (A discussion of Māridīnī's method for finding the *qibla* direction and translation of a crucial passage appears on pp. 111–115.)

——— (1983). "The Astronomy of the Mamluks." *Isis* 74: 531–555. (Reprinted in King, *Islamic Mathematical Astronomy*, III. London: Variorum Reprints, 1986.) (A general survey of Islamic astronomical activities at the time of Māridīnī.)

——— (1986). "Ḳibla: Astronomical Aspects." In *Encyclopaedia of Islam*. 2nd ed. Vol. 5, pp. 83–88. Leiden: E. J. Brill. (Reprinted in King, *Astronomy in the Service of Islam*, IX. London: Variorum, 1993.)

Schmalzl, Peter (1929). *Zur Geschichte des Quadranten bei den Arabern*. Munich: Druck der Salesianischen Offizin. (Outdated, but still the most comprehensive general study of quadrants in Islamic culture.)

Schoy, Karl (1924). "Sonnenuhren der spätarabischen Astronomie." *Isis* 6: 332–360. (A discussion of Arabic sources for the mathematical aspects of sundial construction, including tables computed for the latitude of Cairo.)

Sid

➜ **Ibn Sid: Isaac ibn Sid**

Siguenza y Góngora, Carlos (de)

Born **Mexico City, (Mexico), possibly 14 August 1645**
Died **Mexico City, (Mexico), 22 August 1700**

Native-born Mexican scholar Carlos de Siguenza was honored with the title of Royal Cosmographer. His *Libra Astronomica* (1690) includes a rational discussion of comets, the purpose of which was to allay fears inspired by the great comet C/1680 V1.

Selected Reference

Leonard, Irving (1929). *Don Carlos de Siguenza y Góngora: A Mexican Savant of the Seventeenth Century.* Berkeley: University of California Press.

Sijzī: Abū Saʿīd Aḥmad ibn Muḥammad ibn ʿAbd al-Jalīl al-Sijzī

Born **Sijistān, (Iran),** ***circa*** **945**
Died ***circa*** **1020**

Sijzī, well known for his contributions to geometry, was also a prolific astrologer and astronomer. We possess few details of his life; his name suggests that he was born in Sijistān. His father, Abū al-Ḥūsayn Muḥammad ibn ʿAbd al-Jalīl, was also a mathematician and astronomer. Parts of Sijzī's life were spent in Sijistān and Khurāsān. In Shīrāz in 969/970, he was present (with **Kūhī**, **Būzjānī**, and others) for the famous observations of meridian transits of the Sun conducted by **ʿAbd al-Raḥmān al-Ṣūfī**. Later in life he became a friend of **Bīrūnī**, who often quoted Sijzī's results in his own works.

Of approximately 20 astrological and astronomical treatises composed by Sijzī, many were compilations and summaries of the works of others, enhanced and systematized by the addition of tables and commentary. His *Jāmiʿ al-Shāhī* contains 13 astrological works, three of which are summaries of treatises by **Abū Maʿshar**. One of these, the *Muntakhab Kitāb al-ulūf*, is an important source of information on Abū Maʿshar's *Book of Thousands*. Another of Sijzī's works, the *Kitāb al-qirānāt* (Book of Conjunctions), may be thought of as a supplement to the *Kitāb al-ulūf*. This material likely originated in Sasanian sources and deals with various topics, including astrological world history. Other astrological contributions include the *Kitāb Zarādusht ṣuwar darajāt al-falak* (The book of Zoroaster on the pictures of the degrees of the zodiac) and *Zā'irjāt li-istikhrāj al-haylāj wa-'l-kadkhudāh*, a book of horoscopes with tables based on Hermes, **Ptolemy**, Dorotheus, and "the moderns."

Sijzī seems to have had more than a passing interest in astronomical instruments. He wrote a treatise on the astrolabe that contains the geometric "method of the artisans" for drawing azimuth circles on an astrolabe, as well as descriptions of variations in the retes on astrolabes known to him. Bīrūnī describes three astrolabe variants invented by Sijzī, and in the *Exhaustive Treatise on Shadows* he discusses several of Sijzī's contributions to the theory and use of a gnomon. Sijzī's treatise *On [the Fact that] All Figures are Derived from the Circle* contains a geometric description of an instrument that could be used to find the direction of Mecca (the *qibla*). Finally, in his *Introduction to Geometry* he says:

> I made in Sijistān a great and important instrument, a model of the whole world, composed of the celestial spheres, the celestial bodies, the orbs of their motions with their sizes, their distances and their bodies, and the form of the earth, the places, towns, mountains, seas and deserts, inside a hollow sphere provided with a grid. I called it "the configuration of the universe."

Most of Sijzī's 40 mathematical works, including a unique medieval treatise on problem-solving strategies, focus on geometry in the Euclidean style. One of these treatises contains a systematic mathematical approach to establishing the 12 relations that emerge from the transversal figure in spherical trigonometry (the theorem of Menelaus). Although the work is strictly mathematical, Sijzī is explicitly aware of its fundamental importance to mathematical astronomy.

Glen van Brummelen

Selected References

Berggren, J. L. (1981). "Al-Sijzī on the Transversal Figure." *Journal for the History of Arabic Science* 5: 23–36.
——— (1991). "Medieval Islamic Methods for Drawing Azimuth Circles on the Astrolabe." *Centaurus* 34: 309–344.
Dold-Samplonius, Yvonne (1997). "Al-Sijzī." In *Encyclopaedia of the History of Science, Technology, and Medicine in Non-Western Cultures*, edited by Helaine Selin, pp. 898–900. Dordrecht: Kluwer Academic Publishers.
Frank, Josef (1918/1919). "Zur Geschichte des Astrolabs." *Sitzungsberichte der Physikalisch-Medizinischen Sozietät in Erlangen* 50–51: 275–305.
Hogendijk, Jan P. (English trans. and annot.) and Mohammad Bagheri (Arabic ed. and Persian trans.) (1996). *Al-Sijzī's Treatise on Geometrical Problem Solving*. Tehran: Fatemi.
Kennedy, E. S. (1976). *The Exhaustive Treatise on Shadows by Abū al-Rayḥān Muḥammad b. Aḥmad al-Bīrūnī.* Translation and commentary. 2 Vols. Aleppo: Institute for the History of Arabic Science.
Kennedy, E. S. and B. L. van der Waerden (1963). "The World-Year of the Persians." *Journal of the American Oriental Society* 83: 315–327.
Pingree, David (1968). *The Thousands of Abu Maʿshar.* London: Warburg Institute.

Silberstein, Ludwik

Born **Warsaw, (Poland), 17 May 1872**
Died **Rochester, New York, USA, 17 January 1948**

Ludwik Silberstein is chiefly remembered for contributions to relativity theory and for numerous textbooks in theoretical physics, mathematics, and the philosophy of science. Ludwik was the son of Samuel and Emily (*née* Steinkalk) Silberstein. He graduated from Cracow Gymnasium in 1890. Silberstein attended the Cracow, Heidelberg, and Berlin universities, receiving his Ph.D. in mathematical physics from the Berlin University in 1894. He married Rose Eisenman in 1895; the couple had three children.

Silberstein's career began at Lemberg, Poland (1895–1897), where he served as an assistant in physics. He was then appointed a lecturer in physics at the University of Bologna (1899–1904) and the University of Rome (1904–1920). In 1920, Silberstein accepted a position as research physicist at the Eastman Kodak Company in Rochester, New York, USA; he became a naturalized US citizen in 1935. Silberstein's principal efforts, however, were devoted to research and explication of **Albert Einstein**'s special and general theories of relativity, along with their philosophical implications, *e. g., Discrete Spacetime* (1936).

During the 1920s, Silberstein explored the observational consequences of the de Sitter model of the Universe (proposed by **Willem de Sitter** as a static solution to Einstein's field equations). Along with others, Silberstein argued that a proportional decrease should be observed in the frequencies of light emitted from increasingly distant sources, giving rise to an apparent velocity–distance relationship (akin to the de Sitter effect). In 1924, Silberstein calculated the expected radius of curvature of the de Sitter model of the Universe (about 40 million parsecs), which he further argued was compatible with globular cluster data (including some blueshifts) supplied by Harvard University astronomer **Harlow Shapley**. But within a few years, **Edwin Hubble**'s demonstration of the velocity–distance relationship for "spiral nebulae" (*i. e.*, galaxies) discredited not only Silberstein's calculations, but also the static de Sitter model itself. Nonetheless, historian J. D. North has written that "Silberstein's were probably the most important contributions to this subject," although they were somewhat "obscured by his polemical style" (on p. 102).

Jordan D. Marché, II

Selected References

Anon. (18 January 1948). "Dr. Silberstein, 75, a Physicist, is Dead." *New York Times*, p. 60.

Anon. (1950). "Silberstein, Ludwik." In *Who Was Who in America*. Vol. 2, p. 487. Chicago: A. N. Marquis.

North, J. D. (1990). *The Measure of the Universe: A History of Modern Cosmology*. New York: Dover.

Silberstein, L. (1924). *The Theory of Relativity*. 2nd ed. London: Macmillan.

Silvester, Bernard

Flourished **France, *circa* 1150**

Bernardus Silvestris was one of a number of well-rounded scholars who cultivated a revival of learning during the 12th century. Few personal details are available about Bernard, apart from the fact that he was a master in the schools of Tours, and was associated closely with the school of Chartres, and a friend and literary collaborator of its master Thierry. His major contribution to astronomical literature is the *Cosmographia*, a concise summary of the high medieval understanding of the creation of the cosmos and the geocentric model of the Universe.

The noted medievalist, Charles Homer Haskins, lists Bernardus among the era's great writers. Haskins points out Bernardus's debt to **Macrobius**'s *Commentary on the Dream of Scipio*, and to Thierry of Chartres, to whom the *Cosmographia* is dedicated. A translation of **Ptolemy**'s *Planisphere*, by Peter the Dalmatian, was also addressed to Thierry, providing additional evidence of interest in astronomy there. Bernard was also a master of poetry and prose.

Believing the Universe to be intelligible, the "physici" or natural philosophers of Bernardus's day sought to reconcile observation with revealed religion and classical sources. In the universities of the era, increasing attention was being paid to the *quadrivium* (arithmetic, geometry, music, and astronomy), over the foundation courses known as the *trivium* (grammar, rhetoric, and logic). Bernardus, who wrote at midcentury, stands as a representative scholar of his day, associated with a revival of Greco–Roman natural science, and as a master and advocate of the language arts.

The *Cosmographia* is divided into two sections, Megacosmos and Microcosmos. The former deals with the Universe at large, and the latter with our place in it.

In Megacosmos, we are told that the Universe, stars, and planets, are spherical and rotate upon an axis. Although it is eternal, this Universe was not always orderly and harmonious. Primary matter (*Hyle*) was in a state of chaos or conflict, subject to "random eddies." The divine intellect, *Noys*, brings order and harmony to the chaotic primal Universe. Natural philosophy (*Physis*), whose daughters are theory and practice, contemplates the Universe and generates knowledge. When *Noys* brought organization and perfection of the Universe, the simple and undivided became complex and differentiated. As part of the ordering process, the elements became distinguishable from primal matter and from each other, and eventually, the genera of plants and animals could be subdivided into species.

Bernardus considered the Sun to be one of seven planets. The Moon reflects the light of the Sun, and causes the tides. Bernardus noted that Ptolemy referred to the Moon as a planet of the Sun, and credited the Persians with first charting the heavenly bodies. Additional evidence of observational astronomy in the *Cosmographia* can be found in its mention of the major constellations and their locations. This catalog is presented in a mnemonic fashion, rather like a litany. The Milky Way is said to be a region "whose radiance is produced by clustering stars." The ecliptic is slanted with respect to the Equator; we can delimit it by noting the solstices. This accounts for changes of season.

The Earth stands at the center of a spherical Universe, within which exist four basic elements: earth, water, air, and fire. Each tends toward a place, with fire seeking the heights, earth the depths, and so forth. Energy is imparted from the outermost sphere to the inferior ones, causing them to rotate. It is noteworthy that the energy that is being imparted is energy of motion. The sublunary sphere, comprising the atmosphere and the Earth with its seas and landmasses, is the least orderly. This conforms to the sorts of observations possible in Antiquity and the Middle Ages. The courses and configurations of the stars and constellations would have appeared more regular than those of the planets.

As did **Plato**, Bernard referred to the Universe as a creature or animal, giving it both matter and a world-soul or spirit. In the Microcosmos, we see how the human race mirrors the cosmos, having a nature both spiritual and material. Astronomy can therefore be a key to self-understanding, and conversely, the understanding of the human condition can point to eternal cosmic truths. Indeed, the future can be predicted by "starry cyphers." Consequently, parts of the Microcosmos are astrological, being devoted to the influences of the various planets – Mars is associated with war, Venus with love, and so forth. In his closing sections, Bernardus discussed various human faculties.

Bernardus's model of the Universe was geocentric, and all conjectures and observations had to be reconciled with that basic premise. However, the notion that energy filters down from the stars, and the idea that the elements developed from primal, undifferentiated matter, strike more modern chords. Perhaps most important, there is a strong sense in the *Cosmographia* that physical laws exist uniformly throughout the Universe, and that we can indeed know them. *Physis*, or knowledge of that universe, depends upon both of her daughters, theory and practice.

C. Brown-Syed

Alternate name

Bernardus Silvestris

Selected References

Haskins, Charles Homer (1971). *The Renaissance of the Twelfth Century*. Cambridge, Massachusetts: Harvard University Press.

Plato (1977). *Timaeus*, translated by Desmond Lee. New York: Penguin.

Stock, Brian (1972). *Myth and Science in the Twelfth Century: A Study of Bernard Silvester*. Princeton, New Jersey: Princeton University Press.

Wetherbee, Winthrop (1990). *The Cosmographia of Bernardus Silvestris*. New York: Columbia University Press.

Sima Qian

Born **Longmen (Hancheng, Shaanxi), China, *circa* 145–135 BCE**

Died **China, *circa* 90 BCE**

Sima Qian was a Chinese historian and astronomer in the western (former) Han dynasty who helped devise a calendar in which a 13-lunar month occurred as needed, rather than always at the end of the solar year. His public name was Zichang. Sima Qian's father, Sima Tan, was also a historian and astronomer. After the death of his father in 110 BCE, Sima Qian succeeded to his father's position as *Taishiling* (historiographer and astrologer royal) in 108 BCE. In 104 BCE, Sima Qian took part in the major calendar reform of that year and also started to write a history. In 99 BCE, Sima Qian defended at court general Li Ling, who had surrendered to a vastly superior foreign force. Emperor Wu was offended by Sima Qian's outspokenness, and as a result in 98 BCE, Sima Qian was arrested and castrated for attempting to deceive the emperor. After his release, Sima Qian continued to write history, and finally completed the *Shiji* (Grand Scribe's records), a monumental work of Chinese history, in about 91 BCE.

Classical Chinese calendars are luni-solar calendars, which can be traced to the Shang (Yin) dynasty (mid-16th century to 1046 BCE). This is inferred from the oracle bone inscriptions dating from the 13th to 11th centuries BCE. The development of calendrical science in the western Zhou dynasty (1046 BCE–771 BCE) and the Chunqiu (Spring and Autumn) period (770–476 BCE) is still awaiting further research. By the end of the Zhanguo (Warring States) period (475–221 BCE), the 19-year cycle of intercalation, in which 7 intercalary months are inserted at intervals, was already in use, and the length of a year was considered to be 365 days. This type of calendar was called a *Sifen* calendar (quarter-remainder), named after the fraction of the length of a year.

At the beginning of the western Han dynasty (208–206 BCE), the Zhuanxu calendar, a kind of *Sifen* calendar of the previous Qín dynasty (221–206 BCE) was still used. In this calendar, an intercalary month was put at the end of the relevant year. A fragment of the calendar for the year 134 BCE was excavated in 1972 in Shandong province proving the calendar's actual use at the time. As an exact calendar was considered to be a symbol of the dynasty's authority, a calendar reform was proposed in 104 BCE under the reign of Wudi (Emperor Wu; reigned: 141–87 BCE). The emperor ordered that this year should be the first year of the new era *Taichu*, or grand inception.

Intellectuals, including Sima Qian, discussed the required calendar reform. After the proposal of several calendars, the one made by Deng Ping, the same as one devised by one Luoxia Hong, was finally adopted. It was used from the fifth month of the first year of the *Taichu* reign period (104 BCE) as the *Taichu* calendar. At that time, Sima Qian was the director of the Bureau of Astronomy and the Calendar, and Deng Ping was appointed deputy director. Another contributor, Luoxia Hong, is credited with the invention of the armillary sphere.

In the *Taichu* calendar, the 19-year cycle of intercalation was used as before, but the length of a year was changed to 365 385/1539 days, and that of a synodic month to 29 43/81 days. Here, the denominator 81 was selected because it was the same as the tone of the fundamental pitch pipe. In the correlative cosmology of the time, the tonal system, the calendar, and even measures of volume were all interrelated. The accuracy of the length of the year and the month in the *Taichu* calendar is almost the same as that of the *Sifen* calendar. One merit of the *Taichu* calendar was the new method of intercalation. By the beginning of the western Han dynasty, one year from the winter solstice to the next winter solstice was divided into 24 equal periods, and 24 points of time called *jieqi* ("*qi*-nodes") were established. It may be noted here that the *jieqi* in the modern East Asian classical calendars are the points in time when the Sun passes through those points whose longitude is a multiple of 15°. In the *Taichu* calendar, alternative 12 points called *zhongqi* ("central *qi*-nodes") were selected from the 24 *jieqi*, and the name (serial number) of a month was determined by the *zhongqi* that was included in that month. As the length of a synodic month is a little shorter than the interval of the *zhongqi*, sometimes a month without a *zhongqi* is produced, and such a month becomes an intercalary month. This method of intercalation was followed by later Chinese calendars.

At the end of the western Han dynasty, Liu Xin (died: 23) added to the calendar a method for the prediction of lunar eclipses, a method to calculate the position of the five planets, and the concept of the grand origin epoch. This enlarged calendar is known as the *Santong* calendar, and is recorded in the monograph on the calendar in the *Han shu* (History of the former [western] Han dynasty) (*circa* 78) by Ban Gu (32–92).

There is one curious matter in Sima Qian's *Shiji*. At the end of the monograph on the calendar, Sima Qian describes a calendrical system called *Lishu jiazi pian*. Oddly, it is not the *Taichu* calendar, but a kind of *Sifen* calendar. It may be that it is one of the rejected calendars proposed at the time of the calendar reform. Yukio Ôhashi suspects that Sima Qian tried to oppose the farfetched denominator used in the *Taichu* calendar. Comparing the length of a year and a month, converted into decimal fractions in these calendars (which are almost identical and equally inaccurate) in the *Sifen* calendar, 1 year = 365.25 days, and 1 month = 29.53085 days; in the *Taichu* calendar, 1 year = 365.2502 days, and 1 month = 29.53086 days.

From the above comparison, it is clear that the fraction used in the *Taichu* calendar was artificially selected, for metaphysical reasons, using the value of the *Sifen* calendar without any attempt to adjust it by observation. It is this artificial fraction that might have been opposed by Sima Qian. As far as the lengths of a year and a month are concerned, those of the *Sifen* calendar were recognized even by the compilers of the *Taichu* calendar. Although the *Han shu* relates that several astronomical observations were made at the time of the calendar reform, they were for the determination of a better epoch for a calendar of the same accuracy, and not for the revision of the length of a year and a month. The inaccuracy of the *Sifen* calendar was noticed already by the eastern (later) Han dynasty (25–220), and some astronomers attempted to revise the value at that time.

Alternate name

Ssu-Ma Ch'ien

Selected References

Ban Gu (*circa* BCE 78). *Han shu* (History of the former [Western] Han dynasty). (See Vol. 62 for the official biography, *Sima Qian zhuan*.)

Chavannes, Édouard (trans.) (1899). *Les Mémoires historiques de Se-ma Ts'ien*. Vol. 3, pt. 2. Paris: Ernest Leroux. (Translation of the *Shiji* that includes the astronomical and calendrical treatises.)

Chen Zungu (1980–1989). *Zhongguo tianwenxue shi* (History of Chinese astronomy.) 4 Vols. Shanghai: Shanghai renmin chubanshe (People's Publishing House of Shanghai). (A standard history of Chinese astronomy.)

Cullen, Christopher (1993). "Motivations for Scientific Change in Ancient China: Emperor Wu and the Grand Inception Astronomical Reforms of 104 B. C." *Journal for the History of Astronomy* 24: 185–203. (For the intellectual context of the calendar reformation of the *Taichu* era.)

Editorial Committee (headed by Bo Shuren) (1981). *Zhongguo tianwenxue shi* (History of Chinese astronomy). Beijing: Kexue chubanshe (Science Publishing House). (A standard history of Chinese astronomy.)

Institute of the History of Natural Sciences, The Chinese Academy of Sciences (1983). *Ancient China's Technology and Science*. Beijing: Foreign Languages Press. (Most of the works of Chinese scholars are written in Chinese, but this work by renowned Chinese scholars is written in English, and may be a convenient introduction for Western readers.)

Needham, Joseph, with the collaboration of Wang Ling (1959). *Science and Civilisation in China*. Vol. 3, *Mathematics and the Sciences of the Heavens and the Earth*. Cambridge: Cambridge University Press. (For the early history of Chinese astronomy including the contribution of Sima Qian.)

Nienhauser, William H. (ed. and trans.) *The Grand Scribe's Records*. Vol. 1, *The Basic Annals of Pre-Han China*. Bloomington: Indiana University Press, 1994; Vol. 2, *The Basic Annals of Han China*. Bloomington: Indiana University Press, 2002; Vol. 7, *The Memoirs of Pre-Han China*. Taipei: SMC Publishing, 1994. (For partial translations of the *Shiji* that do not include the astronomical and calendrical treatises.)

Ruan Yuan (1799). *Chouren zhuan* (Biographies of astronomers). Vol. 2. (Reprint, Taipei: Shijie shuju, 1962.) (For a collection of classical accounts of Sima Qian.)

Sivin, N. (1969). "Cosmos and Computation in Early Chinese Mathematical Astronomy." *T'oung Pao* 55: 1–73. (This paper was also published as an independent monograph by E. J. Brill, Leiden.)

Sun Xiaochun and Jacob Kistemaker (1997). *The Chinese Sky during the Han: Constellating Stars and Society*. *Sinica Leidensia*, Vol. 38. Leiden: E. J. Brill. (For the early history of Chinese constellations, including the contribution of Sima Qian.)

Watson, Burton (1958). *Ssu-ma Ch'ien: Grand Historian of China*. New York: Columbia University Press. (A good English biography of Sima Qian.)

——— (trans.) (1961). *Records of the Grand Historian of China*. Translated from the *Shih chi* of Ssu-ma Ch'ien. 2 Vols. New York: Columbia University Press. (Partial translation of the *Shiji*, which does not include the astronomical and calendrical treatises.)

Yabuuti Kiyosi (= Yabuuchi Kiyoshi) (1990). *Zōho-kaitei Chūgoku no tenmon rekihō* (Enlarged and revised edition of the History of astronomical calendars in China, in Japanese). Tokyo: Heibonsha. (A standard history of Chinese calendars.)

Simplicius of Cilicia

Born **Cilicia, (Turkey), *circa* 490**
Died **probably in Athens, (Greece), *circa* 560**

A mathematician primarily, Simplicius wrote one of the most detailed accounts of **Eudoxus**'s theory describing the motions of the planets.

It is often the case that Simplicius is confused with a pope and saint by the same name who died in 483, but the two were in no way related. The astronomer and mathematician Simplicius was born in Anatolia (now part of Turkey), which at the time was a Roman province and had been since the first century BCE. The first information we have on Simplicius is that he studied philosophy in Alexandria, at the school of Ammonius Hermiae. Ammonius himself was a student of **Proclus** and wrote extensive commentaries on **Aristotle**, which presumably influenced Simplicius to do the same. He later traveled to Athens to study under a Neoplatonist, Damascius, who also taught the works of Proclus.

In 529, the Christian Emperor Justinian closed all pagan schools in the Roman Empire. Simplicius then accompanied Damascius and others from the school to Persia to serve the Persian king Khosrow I, who held to traditional religion and was fighting the Roman legions on the Euphrates River and had been since before Justinian had become emperor. However, in 532 Justinian and Khosrow signed a peace accord, and this allowed Simplicius to return to Athens. In fact, the treaty reportedly was explicit about the fate of the philosophers, and allowed them complete freedom in their work and lives upon their return to the empire, though this point has been challenged historically. It is thought that Simplicius spent the rest of his life in Athens; however, his writing style changed at this point, suggesting that either of his own free will or due to political pressure, he no longer lectured.

Simplicius' contributions to mathematics were extensive and tend to overshadow his contributions to astronomy. In addition, many of his writings were actually commentaries on the writings of other mathematicians, philosophers, and astronomers, most notably Aristotle. Simplicius' commentary on Aristotle's *Physics* is of interest in that it contains considerable extracts from **Eudemus**' *History of Geometry*, which included **Hippocrates**' quadratures of "lunes" or crescent-shaped figures and an account of Antiphon's attempt to square the circle. This is an important historical link to Hippocrates' work.

In his commentary on Aristotle's *De caelo*, Simplicius gave the most detailed account that has survived of Eudoxus' famous theory of concentric spheres, a theory that was used to describe the motions of the Sun, Moon, and planets. Simplicius actually quoted largely from **Sosigenes**, the Peripatetic who, himself, drew from Eudemus' *History of Astronomy*. This Simplicius extract also contains modifications made to the model by **Callippus** and Aristotle. The theory suggested that the motion of each planet was produced by the rotation of four concentric spheres, where the inner spheres revolve around a line that is fixed in the next sphere enclosing it. The outermost sphere represented a daily rotation, while the one next to it represented motion along the Zodiac. There were two other spheres, and this set of four spheres was used to represent the motion of just a single planet. So each planet had four concentric spheres, while the Sun and the Moon only had three.

Simplicius also wrote a commentary on Euclid's *Elements, Book I*, which was later quoted by **Nayrīzī**. Simplicius referred to problems relating to gravity and expressly mentioned the work of **Archimedes** on centers of gravity. Simplicius added his own explanatory comments to this regarding the definition of the center of gravity.

Ian T. Durham

Selected References

Grimmelshausen, Hans Jakob C. von (1993). *The Adventures of Simplicius Simplicissimus*, translated by George Shulz-Behrend. Columbia, South Carolina: Camden House.

Heath, Sir Thomas L. (1931). *A Manual of Greek Mathematics*. Oxford: Clarendon Press. (Reprint, New York: Dover, 1963.)

Kretschmann, Philip Miller (1931). "The Problem of Gravitation in Aristotle and the New Physics." *Journal of Philosophy* 28: 260–267.

Simplicius (2002). *On Aristotle's "On the Heavens 1.1-4"*, translated by R. J. Hankinson. Ithaca, New York: Cornell University Press. (Since 1994 Cornell University Press has been assembling and republishing all of Simplicius's works. Some have not yet come out and are in press. In addition, Duckworth has released some of Simplicius's writings that Cornell has not produced. The complete publication of all of Simplicius's writings should be complete within a few years.)

Sīnā

➔ **Ibn Sīnā: Abu ʿAli al-Ḥusayn ibn ʿAbdallah ibn Sīnā**

Sitter, Willem de

Born **Sneek, the Netherlands, 6 May 1872**
Died **Leiden, the Netherlands, 20 November 1934**

Dutch mathematical astronomer Willem de Sitter gave his name to one of the first solutions to **Albert Einstein**'s equations of general relativity, which showed that a universe containing very little matter would, in some sense, expand, and so prepared the way for **Edwin Hubble**'s discovery of that expansion (though a different solution in fact applies). De Sitter was the son of a judge, Lamoraal U. de Sitter and Catharine Th. W. Bertling. He received his early education at Arnhem, the Netherlands, where his father was President of the Court. De Sitter studied at the University of Groningen, primarily in mathematics, under **Jacobus Kapteyn**, with whom he maintained a lifelong friendship and scientific collaboration, receiving a Ph.D. in 1901 for work involving observations of the satellites of Jupiter, made in Cape Town, South Africa (1897–1899). In Cape Town, de Sitter also met and married Eleonore Suermondt. They had three sons and two daughters.

From his position as a staff member at Groningen, de Sitter was appointed to a professorship of astronomy at Leiden University in 1908, where he became director of the observatory in 1919, holding both positions until his death. He reorganized the Department of Astronomy at Leiden, adding a department for astrophysics, and observational facilities at the Union Observatory in Johannesburg, South Africa.

De Sitter has become best known for his work on cosmology, but the earliest part of his career and much of his later life were devoted to celestial mechanics and astrometry. In his thesis of 1901, he discussed heliometer observations of Jupiter's inner moons made at the Cape Observatory, leading to an improved determination of the masses of these satellites. In a subsequent work (*New Mathematical Theory of Jupiter's Satellites*) de Sitter presented a comprehensive analysis of the observations of the satellites made since 1668. His wide knowledge of celestial mechanics also allowed him to present comprehensive discussions of the complicated interrelations among the phenomena from which the major astronomical constants are derived, in particular the combination of results from geodetic and gravity measurements with those from astronomical observations. In 1927, he published *The Most Probable Values of Some Astronomical Constants*. A second paper *On the System of Astronomical Constants*, edited by de Sitter's pupil **Dirk Brouwer**, was posthumously published in 1938. These constants include both the shape of the Earth and numbers for the length of the astronomical unit and the masses of the planets.

In order to arrive at a system of fundamental declinations, free from the systematic errors due to atmospheric refraction and flexure of the telescope that always have plagued meridian observations, de Sitter initiated in 1930 a pilot expedition to Kenya, where, right on the Equator, declinations were made by measuring the distance along the horizon between the rising and setting points of a star, which gives an angle that is two times the complement of the declination.

De Sitter was among the very first to realize the importance of Einstein's work on relativity for astronomy, and he contributed much to the introduction into the English-speaking countries of Einstein's work during World War I. He first discussed, in 1911, the small deviations in the motions of the Moon and the planets still left in the context of classical dynamics. In 1916 and 1917, de Sitter presented to the Royal Astronomical Society a series of three papers on "Einstein's Theory of Gravitation and its Astronomical Consequences." Because there was almost no communication between Germany and England during World War I, these papers were instrumental in introducing general relativity to the English scientific community, and they played an important part in the decision made by **Arthur Eddington** and others to send expeditions to observe the solar eclipse of 1919 to look for the small shifts in the positions of stars predicted by Einstein's theory. In the context of his cosmological work, de Sitter introduced a solution to the fundamental equations that define the properties of the Universe;

it soon became known as the de Sitter universe (or de Sitter space), an alternative to Einstein's solution, provided the density of matter in the Universe could be considered negligible and the Universe be allowed to expand. De Sitter's solution predicted systematic redshifts in the spectra of distant objects (though a quadratic relationship rather than a linear one, which was sought by several colleagues).

De Sitter was elected president of the International Astronomical Union for 1925–1928. One of his major concerns (shared by Eddington and others) was to reintegrate the international community, and he succeeded in extending invitations to the 1928 General Assembly in Leiden to astronomers from Germany and others of the "Central Powers," though some of these nations were not admitted to the union until after World War II. De Sitter was the author of a booklet on the history of the Leiden Observatory (1633–1933) and a founder (in 1921) of the journal *Bulletin of the Astronomical Institutes of the Netherlands*. It, in turn, was merged in 1969 with journals from France and Germany to form a single European journal, *Astronomy and Astrophysics*.

De Sitter received medals from the Royal Astronomical Society (London) and the Astronomical Society of the Pacific, as well as a number of honorary degrees.

The archives of Leiden Observatory have a collection of de Sitter's notes and letters.

Adriaan Blaauw

Selected References

Blaauw, A. (1975). "Sitter, Willem de." In *Dictionary of Scientific Biography*, edited by Charles Coulston Gillispie. Vol. 12, p. 448–450. New York: Charles Scribner's Sons.

De Sitter, W. (1925). "New Mathematical Theory of Jupiter's Satellites." *Annals Leiden Observatory* 12: 1–83.

De Sitter, W. (1931). "Jupiter's Galilean Satellites." *Monthly Notices of the Royal Astronomical Society* 91: 706–738. (George Darwin Lecture, 8 May 1931.)

——— (1932). *Kosmos: A Course of Six Lectures on the Development of Our Insight into the Structure of the Universe*. Cambridge, Massachusetts: Harvard University Press. (A historical review of research on the structure of the Universe is in his monograph.)

——— (1933). *The Astronomical Aspect of the Theory of Relativity*. Berkeley: university of California Press. (Published shortly before his death.)

Jones, H. Spencer (1935). "Willem de Sitter." *Monthly of the Notices Royal Astronomical Society* 95: 343–347.

Macpherson, H. (1933). *Makers of Astronomy*. Oxford: Clarendon Press, Chap. 8.

Oort, J. H. (1935). "Obituary". *Observatory* 58: 22–27.

Van der Kruit, P. C. and K. van Berkel (2000). *The Legacy of Kapteyn; Studies on Kapteyn and the Development of Modern Astronomy*. Dordrecht: Kluwer: Academic Publishers. (His grandson W. R. de Sitter contributed "Kapteyn and de Sitter; a Rare and Special Teacher–Student and Coach–Player Relationship," pp. 79–108.)

Sizzi, Francesco

Flourished **Italy, 1611**

Florentine Francesco Sizzi wrote *Dianoia Astronomica*, a pedantic response to **Galileo Galilei**'s *Sidereus Nuncius*. Sizzi argued that the Galilean satellites could not exist, because they would cause the quantity of "planets" to exceed the favored number seven.

Selected Reference

Drake, Stillman. (1970). "Sunspots, Sizzi, and Scheiner." In *Galileo Studies: Personality, Tradition, and Revolution*, pp. 177–199. Ann Arbor: University of Michigan Press.

Skjellerup, John Francis

Born **Cobden, Victoria, (Australia), 16 May 1875**
Died **Melbourne, Victoria, Australia, 6 February 1952**

John Skjellerup independently discovered or recovered eight comets, of which five bear his name. He was also a dedicated variable-star observer from both South Africa and Australia.

Frank Skjellerup, as he liked to be known, the son of a Danish immigrant farmer, Peder Jensen Skjellerup and his British-born wife Margaret (*née* Williamson) Skjellerup, was the tenth of 13 children. When he was still a small boy his father died. As a result, the day before his 14th birthday, Frank left school and began working as a messenger for the Victorian Post Office. He trained as a telegraph operator, and when the South African Government began recruiting telegraphers in Australia, Skjellerup was one of those selected, arriving in Cape Town in early 1900. Although he spent the rest of his working life in South Africa, when he retired in 1923 he and his wife Mary returned to Australia, settling in Melbourne.

Inspired by the Great Daylight Comet of January 1910 (C/1910 A1) and comet 1P/Halley, Skjellerup became interested in astronomy in 1910, and purchased a 7.6-cm alt azimuth-mounted refractor. In 1922, he replaced this with a Cooke refractor of identical aperture and style of mounting. These telescopes and Zeiss 8 × binoculars were his main observing aids throughout an observational career spanning more than three decades.

In 1911, Skjellerup began systematically searching for new comets. At first he used the Zeiss binoculars, but toward the end of the decade substituted the 7.6-cm refractor. Skjellerup made his first comet discovery on 11 September 1912, only to learn that Walter Frederick Gale (1865–1945) of Sydney had detected this comet (C/1912 R1) several days earlier. His second discovery would also bring disappointment. On 31 October 1915 he found a new comet and immediately notified the Cape and Union observatories, only to learn later that it was none other than the return of periodic comet 7P/Pons–Winnecke. On the morning of 19 December 1919, Skjellerup discovered his third comet (C/1919 Y1) while searching for the variable star RS Librae. No prior discovery claims emerged, and he finally secured a comet bearing his name.

Nearly a year later, on 11 December 1920, Skjellerup made his next discovery (C/1920 X1). In fact, another Cape Town amateur, Charles Clement Jennings Taylor (1861–1922), had detected this comet on 8 December, but Taylor was ill and recorded the wrong position for it. Taylor's incorrect announcement prevented confirmation by others, and Skjellerup received credit for the discovery. Skjellerup's next comet came along on 16 May 1922, but was later shown to be the same comet that **John Grigg** discovered in 1902, comet 26P/Grigg–Skjellerup. Comet 26P has one of the shortest periods of any known comet (5.1 years) and was extensively studied by the Giotto space probe. This was followed by Skjellerup's discovery of comet C/1922 W1 on 26 November 1922.

Although Skjellerup returned to Australia in 1923, it was not until 4 December 1927 that he discovered his next comet. Awakened by an unusual sound (made by his cat), he noticed that it was a beautiful clear night, could not resist the temptation to do a little comet-seeking, and discovered C/1927 X1. An observer at La Plata, Argentina, independently discovered the same comet and was granted status as a codiscoverer. Skjellerup's last independent discovery, C/1941 B2 on 21 January 1941, had been discovered 6 days earlier by South African amateur Reginald Purdon DeKock (1902–1980). In recognition of his various discoveries, Skjellerup was awarded four Donohoe Medals and two Donovan Medals. In addition to the six comets that he independently discovered, and his 7P/Pons–Winnecke recovery, Skjellerup observed at least 18 other known comets.

Skjellerup's first casual observations of variable stars were conducted in 1910, but when he and fellow-amateur Arthur William Long (1874–1939) were granted permission to use the Cape Observatory's 15.2-cm and 17.8-cm refractors, his annual tallies increased rapidly. After returning to Australia, he continued a more restricted program. During the 12-year period, from 1916 to 1927, Skjellerup made 6,773 estimates of 121 different variable stars, mainly of Mira-type variables, and also of four semi-regular stars and two R Corona Borealis stars. In addition, he recorded Nova Aquila 1918.

In addition to his observational work, Skjellerup served as secretary–treasurer, and eventually for several years as vice president, of the Cape Astronomical Association. He also served as president of the Astronomical Society of Victoria, for 3 years.

Skjellerup was survived by his wife, Mary. There were no children from their marriage.

Skjellerup's observing logs and scrapbooks are in the possession of the author of this entry.

Wayne Orchiston

Selected References

Hughes, David W. (1991). "J. Grigg, J. F. Skjellerup and Their Comet." *Vistas in Astronomy* 34: 1–10.

Orchiston, W. (1999). "Of Comets and Variable Stars: The Afro-Australian Astronomical Activities of J. F. Skjellerup." *Journal of the British Astronomical Association* 109: 328–338.

Skjellerup, J. F. (1913). "Observations of Gale's Comet (1912a)." *Journal of the British Astronomical Association* 23: 210–211.

Skellerup, P. (1961). *The Skellerup Family: A Short History 1820–1960*. Christchurch: Pegasus Press, pp. 32–34.

Slipher, Earl Carl

Born **near Mulberry, Indiana, USA, 25 March 1883**
Died **Flagstaff, Arizona, USA, 7 August 1967**

American optical astronomer Earl Slipher obtained large numbers of photographs of the planets that were used to demonstrate changes in the patterns and colors of Martian clouds and changes in the rotation period of the Great Red Spot on Jupiter (meaning that it could not be anchored to a solid surface below).

Slipher was the son of David Clark and Hannah App Slipher and the younger brother of astronomer **Vesto Slipher**. He earned B.A. (1906) and M.A. (1908) degrees from the University of Indiana and received honorary degrees from the University of Arizona and Northern Arizona State University. Slipher married Elizabeth Tidwell in June 1919; they had one daughter, Capella, and one son, Earl, Jr.

Slipher began his astronomical career by accompanying his astronomy professor at Indiana, **John Miller**, on a solar eclipse expedition to Spain in 1905. Slipher participated in four other eclipse expeditions. From 1906 to 1908 he was the Lawrence Fellow at Indiana, whereby an astronomy graduate could intern at Lowell Observatory for a year or two and receive a master's degree. **Percival Lowell** supported an expedition headed by professor **David Todd** of Amherst in 1907 and sent Slipher as Todd's assistant to observe Mars from the Andes in Peru. This expedition launched Slipher's career as a lifelong planetary observer.

Slipher was employed permanently at Lowell Observatory upon his return from Peru and remained there for his entire career, becoming a well-recognized expert on the photography of planets. He pioneered the technique of printing planetary photographs through multiple short-exposure negatives to record enhanced surface or atmospheric details seen during moments of exceptional seeing. Under Lowell's direction, Slipher participated in the search for planet X, but came to consider the project a futile effort.

Slipher eventually headed three Mars expeditions to the Southern Hemisphere at the Lamont–Hussey Observatory at Bloemfontein, South Africa, in the years 1939, 1954, and 1956. In 1954, when Mars approached unusually near to the Earth, he was instrumental in organizing and served as cochair of the International Mars Committee that coordinated observations of Mars from observatories all over the world. The committee included such well-known experts as **Harold Urey**, **Gerard de Vaucouleurs**, and amateur visual observer Thomas R. Cave, Jr. (1923–2003).

In 1960, Slipher headed a US Air Force project at Lowell to update all ground-based observations of Mars. The results were summarized in two books, *The Photographic History of Mars (1905–1961)* and *A Photographic Study of the Brighter Planets*. In 1911, he received a medal from the Astronomical Society of Mexico for his early work on eclipses.

Slipher was exceptionally active in civic affairs. He served two terms as mayor of Flagstaff, as chair of the Arizona Good Road Committee, and as representative of Coconino County in both houses of the Arizona legislature. From 1935 to 1939, Slipher was a member of the board of Flagstaff State Teachers College, now Northern Arizona University. During World War II, he served as chair of the Coconino County Selective Service Commission.

Henry L. Giclas

Selected Reference

Richardson, Robert S. (1964). "Reminiscences of E. C. Slipher." *Sky & Telescope* 29, no. 4: 208–209.

Slipher, Vesto Melvin

Born **near Mulberry, Indiana, USA, 11 November 1875**
Died **Flagstaff, Arizona, USA, 8 November 1969**

American spectroscopist Vesto Slipher is now remembered primarily as the person who obtained the spectra and measured the first radial velocities of the spiral nebulae showing that most were receding from the Earth. **Milton Humason** extended Slipher's measurements to more-and-more distant galaxies; this led to **Edwin** Hubble's discovery of the velocity–distance relation and, therefore, the expansion of the Universe.

Slipher was the son of David Clark and Hannah App Slipher; his brother, **Earl Slipher**, also was an astronomer who also spent most of his career at Lowell Observatory, being associated primarily with photography of the giant planets and their satellites. Vesto received his degrees from the University of Indiana (A.B.: 1901; A.M.: 1903; Ph.D.: 1909; honorary Ph.D.: 1929), and honorary degrees also from the University of Arizona, University of Toronto, and Arizona State University. Slipher was hired by **Percival Lowell** for his observatory partly to study the spiral nebulae, which Lowell then believed to be solar systems in the process of formation. Slipher originally shared this belief, but was later led by his own work to regard them as independent galaxies, as proved to be the case.

Slipher used a spectrograph built by **John Brashear** on the Lowell 24-in. telescope. Among his discoveries were the rotation period of Uranus (with Lowell: 1912), the details of the rotation of the rings of Saturn, the strong absorption bands in the spectra of the giant planets (later shown by **Rupert Wildt** to be due to methane and ammonia), the large velocities of the stars in globular clusters, and the fact that the spectrum of the diffuse material around the Pleiades is identical to that of the stars (also 1912). Soon after, he identified several other members of the class we now call reflection nebulae, recognized at about the same time by **Edwin Hubble**. By studying aurorae close to twilight, Slipher was able to correlate their intensity with solar activity. He was also interested in the background light of the night sky, discovered shortly before by **Simon Newcomb**.

The work for which Slipher is best known began in December 1912, when (in an exposure stretching across two nights) he obtained the first spectrogram of the Andromeda Nebula (M31) in which absorption lines (which he recognized as very much like those in solar-type stars) could be seen and have their velocities measured. He found M31 to be approaching the Solar System at about 300 km/s, the largest velocity measured up to that time. By 1925, Slipher had pushed his telescope–spectrograph combination to its limits, obtaining a radial velocity of +1800 km/s for NGC 584.

Slipher became assistant director of Lowell Observatory in 1915 (when Lowell could no longer be on site most of the time), acting director in 1916 (upon Lowell's death), and director in 1926, after which his own astronomical activity declined a good deal. He retired in 1953 and held the title director emeritus until his death. Slipher married Emma R. Munder at Frankfort, Indiana, on 1 January 1904, and they had two children, Marcia and David.

Among the honors Slipher received were the Lalande Prize of the Paris Academy of Sciences (1919), the Gold Medal of the Royal Astronomical Society (London, 1932), and the Draper Gold Medal of the United States National Academy of Sciences (1932). He was elected to Phi Beta Kappa from Indiana, and was a member of the scientific research honorary Sigma Xi and of most of the astronomical and scientific societies in the United States. At the time of Slipher's death, many had forgotten his most important contributions to astronomy; a short death notice in *Physics Today* mentions only that he had supervised the work by **Clyde Tombaugh** that led to the discovery of Pluto.

Henry L. Giclas

Selected References

Anon. (1933). "Report of Gold Medal Award". *Monthly Notices of the Royal Astronomical Society* 93: 476.

Hoyt, William Graves (1980). "Vesto Melvin Slipher." *Biographical Memoirs, National Academy of Sciences* 52: 411–449.

Smith, Robert W. (1994). "Red Shifts and Gold Medals: 1901–1954." In *The Explorers of Mars Hill*, by William Lowell Powell and others, pp. 43–65. West Kennebunk, Maine: Phoenix.

Slocum, Frederick

Born **Fairhaven, Massachusetts, USA, 6 February 1873**
Died **Middletown, Connecticut, USA, 4 December 1944**

American astrometrist Frederick Slocum contributed to the early 20th-century effort to measure parallaxes for large numbers of stars. The son of Frederick and Lydia (*née* Jones) Slocum, Frederick earned bachelor's (1895), master's (1896), and doctoral (1898) degrees from Brown University. Slocum married Carrie E. Tripp in 1899; the couple had no children.

Slocum was appointed assistant professor of astronomy at Brown University (1900–1909). He received additional training in astrophysics at Potsdam Observatory and returned to teach that subject at Yerkes Observatory (1909–1914). In 1914, Slocum was appointed professor of astronomy and director of Wesleyan University's Van Vleck Observatory, and apart from two years spent as a visiting instructor in nautical science at Brown University (1918–1920), remained in those positions until his retirement.

Slocum participated in efforts coordinated by Allegheny Observatory director **Frank Schlesinger** to determine stellar parallaxes from photographs made with Van Vleck's 20-in. refracting telescope; this work was published in 1938. Active in professional organizations, Slocum was elected vice president of Section D of the American Association for the Advancement of Science and vice president of the American Astronomical Society. He was also the recipient of an honorary degree (Sc.D., 1938) from Brown University.

Jordan D. Marché, II

Selected References

Anon. (1950). "Slocum, Frederick." In *Who Was Who in America*. Vol. 2, p. 492. Chicago: A. N. Marquis.

Slocum, Frederick (1898). "The Harmonic Analysis of the Tides and a Discussion of the Tides of Narragansett Bay." Ph.D. diss., Brown University.
Slocum, Frederick, C. L. Stearns, and B. W. Sitterly (1938). *Stellar Parallaxes Derived from Photographs Made with the 20-inch Refractor of the Van Vleck Observatory*. Middletown, Connecticut: Edwards Brothers.

Smart, William Marshall

Born **Doune, Perthshire, Scotland, 9 March 1889**
Died **Lancaster, England, 17 September 1975**

William Smart was an expert on spherical astronomy.

Smart, son of Peter Fernie Smart and Isabella Marshall Harrower, acquired his interest in science and mathematics while attending McLaren High School in Callender, Scotland. Smart's high academic abilities won him a scholarship to Glasgow University (1906–1911). He graduated with first class honors and was awarded the Cunninghame Medal for Mathematics.

Smart pursued postgraduate studies at Trinity College, Cambridge, and received his doctorate in pure and applied mathematics (1914). He was awarded several other prizes, including a First in Part I of the mathematical *tripos* examination. During World War I, Smart became an instructor–lieutenant at the Royal Naval College, Greenwich, where he developed a lifelong interest in navigation. In 1919, he returned to Cambridge as John Adams Astronomer and chief assistant at the university observatory, under the supervision of **Arthur Eddington**. Smart began to investigate stellar motions and the structure of our galaxy, which culminated in a series of papers (and more than one textbook). His most widely utilized volume, *Textbook on Spherical Astronomy*, was first published in 1931. In 1919, Smart married Isabel Macquarie Carswell; the couple had three sons.

In 1937, Smart returned to his native land upon appointment as Regius Professor of Astronomy at the University of Glasgow, succeeding Ludwig Becker. Conditions at the urbanized Dowanhill Observatory had deteriorated, however, and under Smart's directive, a smaller students' observatory was erected at Gilmore Hill (1939). But his Scottish career was interrupted by World War II; he taught celestial navigation to Royal Air Force cadets. Afterward, the postwar transformation of astronomical research effectively put an end to Smart's line of investigation. Increasingly, he devoted himself to the writing of both advanced textbooks – his *Celestial Mechanics* appeared in 1953 – and a series of popular works, including an account of the discovery of Neptune. Smart lectured widely on astronomy, and acquired many professional distinctions, notably as secretary (1931–1937), vice president (1937–1938), and later president (1949–1951) of the Royal Astronomical Society. He retired from his position in 1959.

Smart's adherence to the mathematical rigors of fundamental astronomy was not an idle pursuit. Although no longer a subject at the forefront of research, its principles nonetheless establish the basis on which virtually all other types of astronomical observation must eventually rest.

Jordan D. Marché, II

Selected Reference

Ovenden, Michael W. (1977). "William Marshall Smart." *Quarterly Journal of the Royal Astronomical Society* 18: 140–146.

Smiley, Charles Hugh

Born **Camden, Missouri, USA, 6 September 1903**
Died **Providence, Rhode Island, USA, 26 July 1977**

After receiving his Ph.D. in astronomy at the University of California in Berkeley, American mathematician Charles Smiley traveled and worked extensively in Europe, including on assignments at the Royal Greenwich Observatory, where he worked with **Leslie Comrie** and **Andrew Crommelin**, and at Cracow, Poland, where he did orbital calculations with **Thaddeus Banachiewicz** for the then recently discovered Pluto. On his return to the United States, Smiley accepted a professorship at Brown University, where his primary research involved observing 15 solar eclipses. In 1937, Smiley successfully photographed the zodiacal light during a total solar eclipse. He used a Schmidt camera of his own design and construction, a very early application of such cameras.

Also interested in ancient astronomies, Smiley in 1960 published a correlation of the Mayan and Christian calendars based exclusively on astronomical evidence. He extended this knowledge to a description of the astronomical dates identified on the Mayan codices located in Dresden, Paris, and Madrid, demonstrating their use in predicting solar eclipses anywhere on the Earth over an 8-century period.

Smiley was an active supporter of amateur telescope makers and amateur astronomers; he served as president of the American Association of Variable Star Observers.

Thomas R. Williams

Selected References

Kelley, David H. (1978). "Charles Hugh Smiley, 1903–1977." *Journal of the Royal Astronomical Society of Canada* 72: 46–47.
Reed, Donald S. (1978). "Charles Hugh Smiley." *Quarterly Journal of the Royal Astronomical Society* 19: 510–511.

Smith, Sinclair

Born **Chicago, Illinois, USA, 1899**
Died **Pasadena, California, USA, 1938**

Sinclair Smith is best known to astronomers for his measurement of the gravitating mass of the Virgo cluster of galaxies in 1937, which confirmed the very large mass-to-light ratio that had been found for the Coma cluster by **Fritz Zwicky** in 1933. Smith is a less familiar name in 20th-century cosmology for two reasons: First, his work was largely in the area we would now call instrumental physics, rather than observational astronomy, and, second, he died tragically early, of cancer.

Smith received his bachelor's degree from the California Institute of Technology (Caltech) in 1921 and his Ph.D. in 1924, also from Caltech, for work with **John Anderson** on electrically exploded wires as a method of obtaining laboratory spectra at high excitation and ionization energies. He remained in the physics laboratory of Mount Wilson Observatory the rest of his life, apart from a year (1924/1925) at the Cavendish Laboratory in Cambridge, England. A true Californian, if not quite native, Smith was an enthusiastic owner and sailor of small boats, in company with Thomas Lauritsen and other Caltech colleagues.

Smith was clearly a widgeteer par excellence. His outstanding contribution to exploding wire research was a spectrograph with microsecond temporal resolution. It focused the spectrograph slit on a rotating mirror, which, in turn, shined the light sequentially across photographic film, producing a record of the evolution of the spectrum of the wire as it vaporized. Rotating mirrors recur in several of his later devices.

Next came a vertical seismometer and an optical oscillograph. This latter device provided a permanent record of rapidly changing electric current by passing the current through a solenoid wrapped around a C_2S cell between crossed polarizers. A collimated beam of white light was shown through the device, through a prism, onto a rotating mirror, and so on to the film. As the current varied, changing the cell's rotation of the plane of polarization versus wavelength, a varying pattern of light and dark fringes was recorded on the film.

Radiometers were Smith's next major love. He studied their sensitivity as a function of temperature, using liquid air as a coolant, and adapted them for use as detectors in stellar photometry and for a laboratory registering microphotometer, also with a rotating mirror. The latter, a coworker gently recalled, was "not competitive" with photoelectric cells and other technologies.

Smith also developed a more conventional stellar photometer, with photoelectric cell and Hoffman electrometer. Attached to the 100-in. telescope, it could produce a current of 500 electrons per second for a 14th magnitude star and determine its flux to 1% in 21 s.

Smith's expertise in measurement of light intensity led to a collaboration with **Richard Tolman** to consider experimental tests of the various possible interpretations of the wave particle dualism. The device they considered appears never to have been built, perhaps because they concluded that it could not distinguish an early version of the absorber theory of radiation from more conventional interpretations. This appears to have been Smith's most nearly theoretical paper.

Perhaps the most productive widget was an *f*/l quartz spectrograph, one of the very first to use a Schmidt camera. He turned this spectrograph toward M32, along with a Wollaston prism (presumably also his own device), and the 100-in. telescope was made into an interferometer with a dark strip across its tube. He concluded that the Galaxy was unpolarized, had a slightly resolved (0.8″) nucleus, and showed no gradient of spectral type across its surface. These observations led him to rule out several now-forgotten models of elliptical galaxies in favor of the giant star clouds we all now accept.

The spectrograph was fast enough to record on the 60-in. telescope, galaxies in Virgo fainter than the ones **Milton Humason** was then studying with the 100-in. nebular spectrograph. Smith's analysis of his, Humason's, and **Vesto Slipher**'s radial velocities of cluster members is the work that brings him into the history of dark matter. He concluded (using nine of his own velocities and about two dozen others) that Virgo had a roughly isotropic distribution of velocities, no significant equipartition of energy, and a mass of 5×10^{14} h^{-1} solar masses (10^{14} at his adopted distance of 2 Mpc). He noted the similarity to Zwicky's result for Coma, and concluded that there must be either large quantities of internebular material or enormous faint extensions beyond the visible galaxies (as just found for M31 bv **Joel Stebbins** and **Albert Whitford** doing photoelectric photometry down to 27 magnitudes per square arc second).

Smith's last years were spent largely on engineering design for the 200-in. telescope, especially its control system.

Virginia Trimble

Selected References

Anderson, J. A. (1938). "Sinclair Smith." *Publications of the Astronomical Society of the Pacific* 50: 232–233.

Trimble, Virginia (1990). "Sinclair Smith (1899–1938)." In *Modern Cosmology in Retrospect*, edited by B. Bertotti *et al.*, pp. 411–413. Cambridge: Cambridge University Press.

Smyth, Charles Piazzi

Born **Naples, (Italy), 3 January 1819**
Died **Ripon, (North Yorkshire), England, 21 February 1900**

Charles Smyth, Astronomer Royal for Scotland between 1845 and 1888, was the first astronomer to argue for the importance of high-altitude observing sites and did pioneering work in solar spectroscopy.

The son of an amateur astronomer and Royal Navy officer, captain **William Smyth** and Annarella (*née* Warrington) Symth, Charles Piazzi's second name honored his godfather, the distinguished Italian astronomer **Giuseppe Piazzi**. After retiring in 1824, W. H. Smyth joined the recently formed Astronomical Society of London (later the Royal Astronomical Society [RAS]), and settled in Bedford, England. There he built the well-equipped Bedford Observatory, where the young Charles learned practical astronomy. After attending the Bedford Grammar School, Charles traveled, at the age of 16, to the Cape Observatory, South Africa, to become assistant to **Thomas Maclear**.

Smyth stayed 10 years at the Cape doing astrometry and participating in an arduous survey of **Nicolas La Caille**'s arc of the meridian, which passed through Cape Town. While in Cape Town, Smyth developed skills in the fledgling process of photography, which fascinated him throughout his life.

In 1844, **Thomas Henderson**, the first Astronomer Royal for Scotland, died in Edinburgh. Smyth's father, then president of the RAS, applied for the position on his son's behalf. When the application was accepted, the younger Smyth arrived in Edinburgh in January 1846, only 27 years old.

Located on Calton Hill, the Royal Observatory, under treasury control, suffered badly from underfunding throughout Smyth's tenure. This did not prevent him from making significant contributions both to astronomy and to Edinburgh. Smyth first reduced and published Henderson's observations. In 1852, he installed a time ball on Calton Hill to signal time for the city and the ships docked at nearby Leith. By 1861, a one o'clock gun fired from Edinburgh Castle, still to be heard daily in the city, augmented Smyth's time ball. In 1855, Smyth married Jessica Duncan, who was scientifically inclined and 4 years older than he.

Smyth was one of the first astronomers to realize the importance of high-altitude observatory sites. The idea that observing could be improved by removing the effects of the lower atmosphere had first occurred to him during his time at the Cape. In 1856, with support from Astronomer Royal **George Airy** as well as the Admiralty, he traveled to Tenerife in the Canary Islands to test his theory. Observing at an altitude of 10,000 ft. Smyth concluded that the high altitude resulted in a gain of about four magnitudes in limiting magnitude compared to that at sea level. By observing double stars, Smyth demonstrated the great improvement in atmospheric seeing at high altitudes. He estimated the heat radiated by the Moon, a pioneering step toward infrared astronomy. It was on the strength of this expedition that Smyth was elected a fellow of the Royal Society in 1857.

Smyth was only one of many distinguished British scientists who debated standardization of units (metrology) during the 19th century. Other leading protagonists included **John Herschel** and **James Maxwell**, both of whom also supported the Imperial System against metrification. Regrettably, Smyth is also frequently remembered for another legacy, his metrology of the Great Pyramid of Giza.

Interest in Egyptology was widespread during the mid-19th century. Smyth's interest ripened into the radical belief that the Great Pyramid was the repository of ancient scientific knowledge. In 1864 and 1865, Smyth went to Egypt, at his own expense, and performed the first meticulous measurements of the Giza Pyramid. The Royal Society of Edinburgh awarded him their biennial Keith Prize and Medal for this work. However, Smyth's developing theories would later garner the distrust of many scientists. The resulting criticism eventually led Smyth to resign from the Royal Society of London, the only person ever to have done so, following the society's refusal to publish papers putting forth his ideas.

Based on astronomical calculations, Smyth concluded that the Great Pyramid was built in 2173 BCE, ignoring contemporary archaeological evidence. He claimed that one of the casing stones of the Great Pyramid revealed an ancient measuring system that resembled the Imperial System, a badly concealed attempt to gain support for the retention of the Imperial System over the metric. Further, Smyth believed that the mathematical structure of the pyramid encoded the events of the Old Testament. This mystical aspect made Smyth the unwitting leader of a worldwide cult movement. The pyramid controversy consumed at least 5 years during what should have been the peak of Smyth's scientific productivity.

Exhausted by the pyramid controversies, Smyth turned his attention again to spectroscopy. With spectrographs of his own design that used the best features of contemporary instruments, Smyth investigated the main features of the solar spectrum during expeditions to Lisbon and Madeira. He used prisms at Lisbon, but a glass diffraction grating ruled by **Lewis Rutherfurd** at Madeira. Using an acetylene flame to calibrate his spectroscope, Smyth measured wavelengths of spectral lines as well as had been done up to his time. For example, his wavelength for the green auroral line was 5579 Å; the modern value is 5577 Å.

An early advocate of the importance of isolating the actual solar spectrum from the combined spectrum of the Sun and terrestrial atmosphere (the telluric spectrum), Smyth observed the Sun close to the morning horizon and again near the zenith at Lisbon. He repeated this work at Madeira, and though conditions were considerably less favorable, quantified all three factors: the actual solar spectrum, the telluric spectrum of dry air, and the effect of added water in a wet atmosphere.

When not on expeditions with his spectroscope, Smyth used his northern latitude in Edinburgh to advantage by studying the light of the aurorae. During the Great Aurora of 4 February 1872, he measured five discrete lines in the auroral spectrum, more lines at more accurate wavelengths than had previously been achieved. On the basis of this work, and work at Madeira and other southern latitudes, Smyth showed that aurorae and the zodiacal light displayed fundamentally different spectra.

An expert in laboratory spectroscopy, Smyth was the first to show that **Joseph von Fraunhofer**'s A and B lines had their origin in the same element, though he guessed wrong in declaring that element was molecular nitrogen. (It is molecular oxygen.) Because of the superiority of his spectroscope and his technique, Smyth was the first to resolve both of these lines and demonstrate the triplet nature of many lines in the oxygen spectrum, a characteristic that eventually led to the correct identification of oxygen in the spectrum of the Sun. Using Smyth's spectral data, **Alexander Herschel** discovered the "harmonic law" in which both strong and weak lines in the spectrum of a molecule follow the same arithmetic progression as their wavelengths grow progressively shorter, a tool of great value to molecular spectroscopists in following generations.

The continued underfunding of the Calton Hill Observatory and Jessica's failing health led Piazzi Smyth to retire in 1888. Disillusioned but resigned to departure from Edinburgh, they moved to Ripon, Yorkshire. In retirement Smyth continued to work, recording the ultraviolet spectrum of the Sun and producing more than 500 photographs of cloud formations, a technique for classifying clouds still widely used by meteorologists.

Smyth lived long enough to see two mountaintop observatories founded, the Lick Observatory on Mount Hamilton, California, and the Arequipa, Peru station of Harvard College Observatory, developments in which he took great satisfaction. In 1890, the University of Edinburgh conferred the degree Doctor of Laws (*honoris causa*) upon Smyth, a great honor for one who succeeded so well without a trace of collegiate education.

Alastair G. Gunn

Selected References

Brück, Hermann A. and Mary T. Brück (1988). *The Peripatetic Astronomer: The Life of Charles Piazzi Smyth*. Bristol: Adam Hilger.

Eggen, Olin J. (1955). "Charles Piazzi Smyth." *Astronomical Society of the Pacific Leaflet*, no. 313.

Hentschel, Klaus (2002). *Mapping the Spectrum: Techniques of Visual Representation in Research and Teaching*. Oxford: Oxford University Press.

Schaffer, Simon (1997). "Metrology, Metrication, and Victorian Values." In *Victorian Science in Context*, edited by Bernard Lightman, pp. 438–474. Chicago: University of Chicago Press.

Smyth, C. P. (1858). *Teneriffe; An Astronomer's Experiment*. London: Lovell Reeve.
——— (1867). *Life and Work at the Great Pyramid*. Edinburgh: Edmonston and Douglas.
——— (1882). *Madeira Spectroscopic*. Edinburgh: W. and A. K. Johnston.
——— (1890). *Our Inheritance in the Great Pyramid*. 5th ed. London: Charles Burnet and Co.
Warner, Brian (1980). "Charles Piazzi Smyth at the Cape of Good Hope." *Sky & Telescope* 59, no. 1: 4–5.
——— (1983). *Charles Piazzi Smyth: Astronomer, Artist, His Cape Years 1835–1845*. Cape Town: A. A. Balkema.

Smyth, William Henry

Born **Westminster, (London), England, 21 January 1788**
Died **Aylesbury, Buckinghamshire, England, 9 September 1865**

William Smyth, a leading figure in Britain's golden age of amateur astronomers, published a catalog of celestial objects from his own observing records that served as a stimulus to a generation of new amateur astronomers. He was the son of Joseph Brewer Smyth, an American by birth, and his English wife Caroline (*née* Pilkington). The family's romantic claim that Joseph was descended from the legendary John Smith of Virginia proved to be fiction; Joseph's own story, that as a royalist he lost large estates in New Jersey as a consequence of the American Revolution, was also highly dubious. William ran away to sea at the age of 14 as a cabin boy to avoid poverty at home. His subsequent successful career was therefore remarkable, and entirely the result of his own talents, determination, and cheerful disposition.

After some adventurous years with the navy in the Far East, Smyth saw active service at the siege of Cadiz in 1810 and was promoted for his bravery. Stationed in Palermo after the Napoleonic wars, with the rank of captain in the Mediterranean fleet, Smyth carried out a much-acclaimed hydrographic survey of Sicily, published under the auspices of the Royal Navy. His surveying activities, extending over many years, gave him experience with astronomical instruments. According to Smyth, his interest in astronomy "received its sharpest spur" in 1813 when he met **Giuseppe Piazzi**, director of the observatory of Palermo, and famous as the discoverer of the first minor planet, (1) Ceres, on 1 January 1801. Smyth helped Piazzi with proofreading the sheets of the Palermo star catalog. In 1814, Smyth married Annarella Warington, daughter of the English Consul to the Kingdom of the Two Sicilies, a cultured and artistic woman who shared all his interests and with whom he reared and educated a large family.

In 1821, Smyth was elected to the Astronomical Society of London, which had been founded only the previous year (later the Royal Astronomical Society). In 1824, he retired on half pay from the navy and resolved to devote himself seriously to astronomy. Having lived for some years in London, the family moved to Bedford, about 50 miles from the city. In 1830, Smyth set up his Temple of Urania, a beautifully equipped observatory with an excellent 6-in. Tully refractor, one of the first to be equatorially mounted and driven by clockwork. With this instrument, Smyth embarked on a program of micrometric measurements of double stars drawn mainly from Piazzi's catalog, and observations of nebulae and star clusters. On completion of his survey in 1839, Smyth dismantled the telescope and transferred it to nearby Hartwell House, the mansion of his wealthy neighbor and friend Dr. John Lee, a patron of the arts and sciences, who, under Smyth's guidance, had built his own private observatory.

From 1839 to 1844, Smyth, eventually, elevated to the rank of admiral, was again engaged in naval work, supervising the construction of floating docks in Cardiff. During this time he prepared his catalog of 850 objects, known as the *Bedford Catalogue*, which constituted Part 2 of his popular treatise, *The Cycle of Celestial Objects*, published in 1844. The catalog gained for Smyth the Gold Medal of the Royal Astronomical Society [RAS] in 1845, its highest accolade. The Smyths' home from this time onward was at Saint John's Lodge, Aylesbury, Buckinghamshire, not far from Hartwell House, where Smyth, usually accompanied by his wife, had unlimited access to his beloved equatorial for the rest of his life.

The *Bedford Catalogue*, in which the astronomical data are interspersed with a charming mixture of useful information, historical anecdotes, and classical lore, became one of the most popular works on astronomy in the English language. It gave fresh purpose to amateurs who previously believed that the only way to exercise their hobby was through meridian observations. A true classic, the book has never lost its attraction; a facsimile edition was published in 1986.

One shadow on Smyth's scientific reputation was an allegation made in 1878, long after Smyth's death, by assiduous double star observer Herbert Sadler that Smyth's observations were not original but were copied from earlier compilations. Sadler cited evidence that certain errors in earlier work were repeated in Smyth's catalog. Sadler's accusation, which caused deep offense among Smyth's admirers, had its origin in sarcastic comments published by American double-star observer **Sherburne Burnham**. The handling of the Sadler–Smyth scandal in the RAS Council was inept, leading to adverse comment in scientific journals of the period when Astronomer Royal Sir **George Airy** resigned in protest. The matter was investigated by a respected fellow of the RAS, **Edward Knobel**, who consulted Smyth's original notes, examined his micrometer, and was fortunately able to vindicate him.

Smyth, a devoted member of the RAS, was the society's foreign secretary from 1829 to 1840 and from 1843 to 1845, and president from 1845 to 1847. He was a genial member of the society's dining club, hardly ever missing a dinner and usually taking the chair, while the Smyth home was a hospitable center of scientific social life. Smyth was elected a Fellow of the Royal Society in 1826 and also belonged to the Geographical and Antiquarian Societies.

Lee was not Smyth's only astronomical disciple. Another neighbor, the physician **Thomas Maclear** (later Sir Thomas), under Smyth's inspiration abandoned the medical profession for astronomy, becoming so proficient as to be appointed His Majesty's Astronomer at the Cape, South Africa, in 1831.

Smyth had three sons who all attained eminence in their respective spheres. Sir Warington Wilkinson Smyth was a distinguished geologist. **Charles Piazzi Smyth** became Astronomer Royal for Scotland, while the youngest, Sir Henry Augustus Smyth, was an army general. One of his daughters was the wife of Baden Powell, Oxford mathematician and liberal theologian, and mother of Lord Baden Powell, Boer War hero and founder of the Boy Scout movement. Other sons-in-law were the biologist Sir William Flowers and captain Henry Toynbee, a government meteorologist.

Mary T. Brück

Selected References

Brück, Hermann A. and Mary T. Brück (1988). *The Peripatetic Astronomer: The Life of Charles Piazzi Smyth*. Bristol: Adam Hilger.

Dreyer, J. L. E. and H. H. Turner (eds.) (1923). *History of the Royal Astronomical Society, 1820–1920*. London: Royal Astronomical Society. (Reprint, Oxford: Blackwell Scientific Publications.)

Fletcher, Isaac (1866). "Admiral William Henry Smyth." *Monthly Notices of the Royal Astronomical Society* 26: 121–129.

Jeal, Tim (1989). *Baden-Powell*. London: Hutchison. (Appendix 1 of this work provides a full account of Smyth's ancestry and another of Smyth's grandchildren with interesting ancestral ties to Nevil Maskelyne.)

Knobel, E. B. (1880). "Notes on a Paper entitled 'An Examination of the Double-Star Measures of the Bedford Catalogue,' by S. W. Burnham, Esq." *Monthly Notices of the Royal Astronomical Society* 40: 532–557.

Smyth, William Henry (1844). *A Cycle of Celestial Objects*. Vol. 1, *Prolegomea*; Vol. 2, *The Bedford Catalogue*. London: privately printed. 2nd ed., enlarged by G. F. Chambers, London, 1881.

——— (1986). *The Bedford Catalogue, from A Cycle of Celestial Objects*. With Foreword by George Lovi. Richmond, Virginia: Willmann-Bell.

Snel [Snell], Willebrord

Born **Leiden, (the Netherlands), *circa* 1580**
Died **Leiden, the Netherlands, 30 October 1626**

Willebrord Snel is chiefly remembered for his discovery of the law of refraction that bears his name, and for his demonstration of the first accurate measurement of an arc of the meridian. Snel's father, Rudolph Snellius, was a professor of mathematics at the University of Leiden. There, Snel studied law but remained chiefly interested in science and mathematics. After 1600, he traveled widely in Europe and at Prague met **Tycho Brahe** and **Johannes Kepler**. Snel returned to Leiden in 1604 and began to translate and restore the mathematical works of **Apollonius**. In 1608, he was awarded a master's degree; that same year, he married Maria de Lange. The couple had eighteen children, only three of whom survived to adulthood.

After his father died in 1613, Snel assumed his position at the University of Leiden as teacher and professor of mathematics. He then applied the method of triangulation proposed by **Gemma Frisius** to measure the distance between the towns of Alkmaar and Bergen op Zoom, which lay nearly on the same meridian. For his measurements, Snel used a large quadrant of 2.1-m radius. He presented his results in the booklet, *Eratosthenes batavus* (1617). The method perfected by Snel to discover the Earth's true dimensions was later utilized by French astronomer **Jean Picard** upon a larger meridian arc. Snel, however, apparently remained a follower of the Ptolemaic (geocentric) theory of the Universe.

Snel investigated the refraction of light and succeeded where others (including Kepler) had failed in the derivation of a general law. As it is usually expressed today, the ratio of the sine of the angle of incidence to the sine of the angle of refraction is a constant for a given refractive medium, such as water or glass. Snel arrived at his law of refraction around 1621 but did not publish the finding before his death. Thus, priority for its publication rests with the *Dioptrique* (1637) of **René Descartes**, who had visited with Snel in Leiden. In the words of two later optical scientists, Snel's law "swung open the door to modern applied optics" (Hecht and Zajac, 1974, p. 2).

Jordan D. Marché, II

Alternate name

Snellius

Selected References

Hecht, Eugene and Alfred Zajac (1974). *Optics*. Reading, Massachusetts: Addison-Wesley.

Herzberger, Max (1966). "Optics from Euclid to Huygens." *Applied Optics* 5: 1383–1393.

Struik, Dirk J. (1975). "Snel (Snellius or Snel van Royen), Willebrord." In *Dictionary of Scientific Biography*, edited by Charles Coulston Gillispie. Vol. 12, pp. 499–502. New York: Charles Scribner's Sons.

Snellius

➔ **Snel, [Snell] Willebrord**

Snyder, Hartland

Born **1913**
Died **1962**

With his teacher, **J. Robert Oppenheimer**, American physicist Hartland Snyder showed (1939), using quantum mechanics and general relativity, that a collapsing massive star will continue to collapse until it forms what is now called a black hole.

Selected Reference

Michelmore, Peter (1969). *The Swift Years: The Robert Oppenheimer Story*. New York: Dodd, Mead and Co.

Soldner, Johann Georg

Born **near Feuchtwangen, (Bavaria, Germany), 16 July 1776**
Died **13 May 1833**

At Munich, Johann Soldner calculated the deflection, due to gravity, of starlight as it grazes the Sun's limb (1801). Thus he is sometimes said to have presaged **Albert Einstein**; however, because he used Newtonian gravity and a particle theory of light, Soldner's deflection is half of Einstein's 1916 value.

Selected Reference

Jaki, Stanley L. (1978). "A Forgotten Bicentenary: Johann Georg Soldner." *Sky & Telescope* 55, no. 6: 460–461.

Somerville, Mary Fairfax Greig

Born **Jedburgh, Scotland, 26 December 1780**
Died **Naples, Italy, 29 November 1872**

Mary Somerville was the first woman scientist to win an international reputation entirely in her own right rather than by working in association with a father, husband, or brother. Self-educated in mathematics and astronomy, she wrote many textbooks dealing with celestial mechanics, geography, and the sciences in general. She was the author of the first paper by a woman ever published in the *Proceedings of the Royal Society* (London).

Mary Somerville was born Mary Fairfax, the daughter of lieutenant, later vice admiral, Sir William George Fairfax. Having seen her husband off on a voyage, her mother, Margaret (*née* Charters) Fairfax gave birth to Mary while traveling from London back to the family home in Fife, Scotland. Given no systematic formal education, Mary was educated by some of her more liberal family members and by her own efforts.

At age 13, Mary was taught painting in Edinburgh by Alexander Nasmyth, father of the engineer, astronomer, and telescope maker, **James Nasmyth**. A chance remark by Alexander Nasmyth that geometry was the basis for understanding perspective as well as the foundation of astronomy set her to the study of mathematics. She studied geometry from Euclid's *Elements*, with the aid of her younger brother's tutor. Her interests broadened to algebra as a result of finding mysterious symbols in the puzzles of a women's magazine, and her brother's tutor provided algebra texts. Her father worried that the strain of abstract thought would injure the tender female frame.

In 1804, when 24 years old, Mary married her distant cousin, captain Samuel Greig. His father was a nephew of Mary's maternal grandfather. A member of the Russian navy, Greig took a post in London in order to marry Mary. Within 2 years the couple had two sons, but he died in 1807. According to Mary, her husband "had a very low opinion of the capacity of my sex, and had neither knowledge of, nor interest in, science of any kind."

With the death of her husband, Mary returned to Scotland as a widow of independent means. She took up mathematics, astronomy, and dynamics, encouraged by the circle of friends whom she had chosen. These included John Playfair (1748–1819), then professor of natural philosophy at Edinburgh, and William Wallace (1768–1843), then professor of mathematics at the Royal Military College. They guided her studies much as a doctoral student would be guided by a professor today.

In 1812, Mary married her second husband, William Somerville, also a distant cousin with naval connections. He was the son of her aunt Martha and her uncle Thomas Somerville. A doctor, William was interested in science and supportive of his wife's interests. Mary and William Somerville moved to London when he was appointed as Inspector to the Army Medical Board in 1816. He was later a physician at the Royal Hospital in Chelsea. When William was elected to the Royal Society, Mary Somerville gained access to a wide circle of prominent scientific acquaintances, including **George Airy**, Humphry Davy (1778–1829), **John Herschel**, **William Herschel**, **Henry Kater**, George Peacock (1791–1858), Thomas Young (1773–1829), and Charles Babbage (1792–1871). She frequently visited Babbage while he was designing his calculating machines. During a visit to Paris in 1817, Somerville met **Jean Biot**, **Dominique Arago**, **Pierre de Laplace**, **Simon Poisson**, Louis Poinsot (1777–1859), Emile Mathieu (1835–1890), and others.

Somerville began experiments on magnetism in 1825. She published her first paper "The Magnetic Properties of the Violet Rays of the Solar Spectrum" in the *Proceedings of the Royal Society* in 1826. Aside from **Caroline Herschel**'s astronomical observations, this was the first paper by a woman to be read at a meeting of and published by the Royal Society. She also wrote about the action of short wavelength radiation on vegetable juices, and about comets.

Somerville began translating Laplace's *Mécanique céleste* in 1827. When the book was published in 1831 under the title *The Mechanism of the Heavens*, it was more than a translation, containing a commentary on the mathematics used, and filling in the gaps in the mathematical development. According to **Nathaniel Bowditch** in a remark since echoed by many a student about many a textbook, "I never come across one of Laplace's 'Thus it plainly appears,' without feeling sure that I have got hours of hard study before me, to fill up the chasm and show how it plainly appears." When Somerville dined with Laplace in Paris in the early 1830s, he paid her a compliment during the conversation. Confused by her name from her earlier marriage, Laplace observed that only two women had ever read the *Mécanique céleste*; both being Scottish women— "Mrs. Greig and yourself."

During her visit to Paris, Somerville wrote her second book, *The Connection of the Physical Sciences*, published in 1834, which treated celestial mechanics and other sciences. The book was published in several editions. In the 1836 edition, she discussed the problematic accuracy of the orbits of the outer planets, suggesting "... [T]he discrepancies may reveal the existence, nay, even the mass and orbit, of a body placed forever beyond the sphere of vision." This passage led **John Adams**, by his own admission, to begin calculations in 1843 that led to the discovery of Neptune.

Mary Somerville was elected an honorary member of the Royal Astronomical Society – at the same time as **Caroline Herschel**, the first two women members – in 1835. She was elected to honorary membership of and offered medals by many societies, and awarded a significant civil pension. In 1838, William Somerville's health deteriorated, and the family went to the warmer climate of Italy. There she wrote *Physical Geography*, which was published in 1848 and remained in print for 50 years. Another mark of distinction for that work was that it was admonished from the pulpit in York Cathedral. She published *Molecular and Microscopic Science*, an account of chemistry and physics, in 1869 at the age of 89. William died in 1860. Her daughter Martha published Mary's autobiography in 1873.

Mary Somerville served as an inspired teacher and as a role model for aspiring women scientists. She supported women's education and women's suffrage – hers was the first signature on John Stuart Mill's 1867 petition to parliament for the right of women to vote. Somerville College in Oxford was named in her honor.

Paul Murdin

Selected References

Chapman, Allan (2001). *Mary Somerville and the World of Science*. London: Canopus.

Patterson, Elizabeth Chambers (1969). "Mary Somerville." *British Journal for the History of Science* 4: 311–339.

——— (1974). "The Case of Mary Somerville: An Aspect of Nineteenth-century Science." *Proceedings of the American Philosophical Society* 118: 269–275.

——— (1979). *Mary Somerville, 1780–1872*. Oxford: Oxford University Press for Somerville College.

Somerville, Mary (2004). *Collected Works of Mary Somerville*, edited by James A. Secord. 9 Vols. Bristol: Thoemmes.

Somerville, Mary and Martha Somerville (1873). *Personal Recollections from Early Life to Old Age, with Selections from Her Correspondence*. London: Murray.

Sorby, Henry Clifton

Born **Sheffield, England, 10 May 1826**
Died **Sheffield, England, 9 March 1908**

Henry Sorby was not a traditional astronomer or a scientist, but a productive amateur scientist particularly interested in meteorites and how they provided evidence concerning the early Solar System. He was the only child of a moderately prosperous family who had

a long association with manufacturing edge tools. In 1847, after his father died, Sorby was left with enough income to devote himself full-time to his nearly all-consuming passion, science. Why he did this is not completely clear, but in a speech to the Royal Society in 1874 he noted that

> as a young man, I had to make the choice of either wisdom or riches, and resolved to be content with moderation and devote myself to science, instead of to the accumulation of wealth and trying to rival my richer neighbours (Higham, 1963, p. 8).

Sorby used part of his inheritance to set up a laboratory and workshop at his home. He never married and continued his scientific research until 11 days before his death.

Sorby's link to astronomy comes, ironically enough, from his passion for microscopy. In 1848, Sorby began making thin sections of rocks. In this technique, a thin chip of rock is ground down to a thickness until it can be examined through a microscope by transmitted light. Although Sorby was not the first person to use this technique, he made outstanding contributions to improving it and founded the science of microscopical petrography with a paper in 1851. In this and later papers, he studied the structures of rocks and attempted to understand, usually by experiment, the structural relationships between the different constituents within the rocks and thus to determine how the rocks had formed.

In 1861, the astronomer and meteoriticist Robert Phillips Greg introduced Sorby to the subject of meteors. Greg encouraged Sorby to turn his new investigative methods to the subject of meteorites, which he duly did in 1863. Sorby noted that the thin black crust that surrounded most meteorites was quite often a black glass filled with small bubbles and that there was a sharp boundary between the crust and the main mass of meteorite. This, and other related observations, persuaded Sorby that the crust was igneous in origin and was formed when the meteorite entered the Earth's atmosphere at great speed and was thereby subject to rapid heating, thus confirming a popular theory of the time. Regarding the interiors of meteorites, Sorby noted that some meteorites are similar to brecciated rock; fragments of rock within such meteorites had subsequently been cemented together and consolidated, again confirming a then recently proposed theory.

Sorby was the first to provide an explanation for the formation of chondrules, 0.1 mm silicaceous spheres found in most stony meteorites that are observed landing on the Earth, which may have some connection to the formation of the terrestrial planets. Sorby found a laboratory analog to explain how chondrules were formed. He examined their internal structure and inferred that they were "devitrified globules of glass, exactly similar to some artificial blow-pipe beads" (Sorby, 1877).

Sorby also examined iron meteorites and was particularly interested in understanding the formation of Widmannstätten patterns. He concluded that meteoritic iron formed in a low gravity environment in which the iron was kept at temperatures just below fusion for long periods of time. Such an idea is consistent with the modern understanding for how such metal patterns form.

Sorby concluded, on the basis of the igneous nature of many of the components found in meteorites, that all meteorites were formed near the surface of the Sun and were ejected to the outer regions of the Solar System. Modern observations indicate that most meteorites originate from the asteroid belt, so Sorby's ideas would appear to be completely wrong. Recent work, however, suggests that chondrules and other related igneous objects within meteorites may, indeed, have been formed near the early Sun and ejected to the outer parts of the Solar System by bipolar jet flows, where they aggregate with other material to form the grains in the meteorites that now reach the Earth.

Sorby was one of the first planetary scientists. He used a microscope to study processes that occurred in the distant past during the formation of the Solar System. His methods and conclusions prompted discussions that continue today as we try to understand the processes in young stellar systems and the early history of our Solar System.

Kurt Liffman

Selected References

Higham, N. (1963). *A Very Scientific Gentleman: The Major Achievements of Henry Clifton Sorby*. Oxford: Pergamon Press.

Liffman, K. and M. J. I. Brown (1996). "The Protostellar Jet Model of Chondrule Formation." In *Chondrules and the Protoplanetary Disk*, edited by R. H. Hewins, R. H. Jones, and E. R. D. Scott. Cambridge: Cambridge University Press, pp. 285–302.

Shu, Frank H., Hsein Shang, and Typhoon Lee (1996). "Toward an Astrophysical Theory of Chondrites." *Science* 271: 1545–1552.

Sorby, Henry Clifton (1851). "On the Microscopical Structure of the Calcareous Grit of the Yorkshire Coast." *Quarterly Journal of the Geological Society of London* 7: 1–6.

——— (1877). "On the Structure and Origin of Meteorites." *Nature* 15: 495–498.

Sosigenes of Alexandria

Flourished **Rome, (Italy), middle of 1st century BCE**

Sosigenes was a Greek or Egyptian astronomer and mathematician of the Alexandrian School, about whom little is known. He is known as the main astronomer who helped Julius Caesar with his reform of the Roman lunar calendar, although his role in this reform is not very clear. **Plutarch** simply states, without mentioning any names, that Caesar consulted the best philosophers and mathematicians before making an improved calendar of his own. And all that **Pliny** says is that during Caesar's dictatorship Sosigenes helped him to bring the years back into conformity with the Sun. He adds that Sosigenes wrote three treatises, including corrections of his own statements. It is, in any case, not certain that Sosigenes was in Alexandria during Caesar's stay in Egypt after the battle of Pharsalos.

Caesar had a genuine interest in astronomy and composed a treatise, *De Astris*, a kind of farmer's almanac with a remarkable popularity, based on Hellenistic data that were made available to him by Sosigenes. Caesar adopted (in 45 BCE) the solar year with an average length of 365¼ days. (The year was to be independent of the Moon's motion; the ordinary year was to consist of 365 days, an extra day being added to February every fourth year.) This may have been one result of Sosigenes' advice, and the statesman's seasonal calendar another. The 365¼-day year even could have been borrowed directly from **Callippus** at the suggestion of Sosigenes. Lucan (*Pharsalia* 10.187) implies that Caesar tried to improve upon the seasonal calendar of **Eudoxus** "and my year shall not be found inferior to the calendar of Eudoxus." T. Mommsen (1887) maintains that Caesar "... with the help of Greek mathematician Sosigenes introduced the Italian farmer's year regulated according to the Egyptian calendar of Eudoxus, as well as a rational system of intercalation, into religious and official use." It is possible, but far from certain, that Sosigenes made use of Babylonian astronomical knowledge.

Little is known about Sosigenes' treatises. It is certain that one of them was based on **Eudemus**' (about 330 BCE) *History of Astronomy.* This work was an intermediary between Eudoxus's and Callippus's systems of concentric spheres, in the account which **Simplicius** gives in his commentary on **Aristotle**'s *De Caelo.* Simplicius's quotation from Sosigenes on the impossibility of reconciling the theory of concentric spheres with the observed differences in the brightness of planets at different times and the apparent difference in the relative sizes of the Sun and Moon is particularly interesting for the history of astronomy. Sosigenes showed that the apparent diameters of the Sun and Moon are not always equal, by describing the phenomenon of annular eclipses of the Sun (on *De caelo*, p. 505. 7–9, Heiberg), and doubtless **Hipparchus** had observed the differences. We also know that Sosigenes agreed with Cidenas in giving the greatest elongation of Mercury from the Sun as 22° (Pliny, *Naturalis historia* 2.39).

Dimitris Dialetis

Selected References

Huxley, G. L. (1975). "Sosigenes." In *Dictionary of Scientific Biography*, edited by Charles Coulston Gillispie. Vol. 12, p. 547. New York: Charles Scribner's Sons.

Mommsen, Theodor (1887). *The History of Rome*. Vol. 4. London.
Sarton George (1993). *Hellenistic Science and Culture in the Last Three Centuries B.C.* New York: Dover.
Simplicius (1894). In *Aristotelis* De caelo commentaria, edited by J. L. Heiberg. Vol. 7 of *Commentaria in Aristotelem Graeca*. Berlin: Reimer.

South, James

Born **Southwark, (London), England, 21 October 1785**
Died **London, England, 19 October 1867**

James South was a noted amateur astronomer who specialized in binary stars. He is usually remembered for his intemperate disposition, and his infamous quarrel with the instrument-maker Edward Troughton, facts that tarnish the mid-19th-century history of British observational astronomy, and effectively hide the truth of his contributions to scientific knowledge.

South studied surgery, became a member of the Royal College of Surgeons, and built up a thriving practice, but following his marriage to a wealthy heiress in 1816, abandoned medicine for his hobby of astronomy, an interest awakened by his friendship with the hydrographer and surveyor Joseph Huddart. He then proceeded to equip his observatory with the best telescopes and instruments then available. As a result of his friendship with **John Herschel**, with whom he now collaborated, South embarked upon a program of double star observations that lasted from 1821 to 1823, resulting in a catalog of about 380 such stars, which earned Herschel and South the Gold Medal of the recently founded Astronomical Society of London.

South temporarily moved his 5-ft. equatorial to Passy, near Paris, in 1825, where he obtained a series of multiple star measurements of such high quality that the Royal Society, to which he was elected in 1821, awarded him the prestigious Copley Medal (1826). Such was his standing as an astronomer at that time that the British and French governments openly competed to have him and his observatory in their country. He considered imigrating to France, but changed his mind when Britain gave him a knighthood plus £300 per annum to further his researches.

That same year the Royal Astronomical Society, which he had helped to found and had served in a variety of executive positions, gained a royal charter. Unfortunately, a technicality barred him from serving as its first president, a circumstance that prompted his immediate resignation from the society. South was highly critical of the Royal Society, attributing the decline of the sciences in Britain to the actions of some of its members. His attack on the *Nautical Almanac*, which he thought inferior to its Continental counterparts, was equally harsh. Such actions did not endear him to his contemporaries.

South's concern about declining standards culminated in the infamous quarrel with Troughton about the quality of the latter's workmanship. This led to an expensive lawsuit lasting from 1834 to 1838, which South lost. Bitter with rage, he destroyed the offending equipment, and auctioned off the fragments. However, he preserved the lens and toward the close of his life presented it to Dublin University. South was awarded an honorary LL.D. by Cambridge University (1833), and belonged to a number of scientific organizations in Belgium, France, Ireland, Italy, and Scotland.

Richard Baum

Selected References

Dreyer, J. L. E. and H. H. Turner (eds.) (1923). *History of the Royal Astronomical Society, 1820–1920*. London: Royal Astronomical Society. (Reprint, Oxford: Blackwell Scientific Publications, 1987.)
Hoskin, Michael (1989). "Astronomers at War: South v. Sheepshanks." *Journal for the History of Astronomy* 20: 175–212.
J. C. (1868). "Sir James South." *Monthly Notices of the Royal Astronomical Society* 28: 69–72.
Robinson, Thomas Romney (1868). "Sir James South." *Proceedings of the Royal Society of London* 16: xliv–xlvii.

Spencer Jones, Harold

Born **Kensington, (London), England, 29 March 1890**
Died **Herstmonceux, (East Sussex), England, 3 November 1960**

English positional astronomer Sir Harold Spencer Jones compiled the definitive set of data showing that certain apparent irregularities in the motions of the Moon, Sun, and planets actually arise from small variations in the rate of the Earth's rotation; but he made his firmest mark in astronomy as a science administrator. Jones was the third child of Henry Charles Jones, an accountant, and was educated at Latymer Upper School, Hammersmith and Jesus College, Cambridge, where he received first class honors degrees in mathematics (1911) and natural science (1913). Spencer Jones was elected to a fellowship at Jesus College in 1913, but the next year was appointed as one of the chief assistants at the Royal Greenwhich Observatory [RGO] by its director, Astronomer Royal **Frank Dyson**, replacing **Arthur Eddington**, who had just been elected to the Plumian Professorship at Cambridge. Spencer Jones's first task there was participation in an expedition to observe the solar eclipse of 21 August 1914 from Minsk, Russia. During World War I, he served as an inspector of optical supplies for the Ministry of Munitions. Spencer Jones married Gladys Mary Owens in 1918, and they had two children.

In 1923, Spencer Jones was appointed His Majesty's Astronomer at the Cape of Good Hope, South Africa. On his arrival, he found an inefficient staff and numerous observational programs dating back decades whose results were still awaiting reduction. Spencer Jones proceeded to reenergize the observatory, superintending a new edition of the Cape Catalog of southern stars, a reobservation of the Cape zone of the Astrographic Catalogue (*Carte du Ciel*) for measurements of proper motions, and, most important, analysis of measurements of lunar occultations from 1800 to 1922 and positional observations of Sun, Mercury, Venus, and Mars dating from 1836 to 1924. These data provided firm evidence that the rotation rate of the Earth varies on time scales from days to decades, at least, and that these variations must be taken into account when determining precise orbital motions for the Moon and planets.

The International Astronomical Union elected Spencer Jones president of its Commission on Solar Parallax in 1928, and he was, therefore, the coordinator of a 24-observatory campaign during 1930/1931 to track the opposition of Eros, which passed unusually close to Earth that winter, in the hopes of improving knowledge of the distance scale within the Solar System *via* the phenomenon of geocentric or Earth-rotation parallax. The published result (π = 8.7904 arc sec) yielded Gold Medals for Spencer Jones from both the Royal Astronomical Society and the Royal Society (London), though later work showed it was not so accurate as he had supposed, despite the 2,847 plates collected for the program (many of which were measured by **Philibert Melotte**).

Spencer Jones was recalled to England in 1933 to become the 10th Astronomer Royal and director of the Royal Observatory at Greenwich, just outside London. Again he implemented notable upgrades, bringing into operation the 36-in. Yapp telescope and the Cooke reversible transit circle (which replaced an instrument commissioned in 1850 by **George Airy**) and cooperating with the post office to provide "speaking clock" time service for England. Spencer Jones soon also recognized the decreasing suitability of the Greenwich site, close to the lights of London and embedded in Thames River fogs. Early considerations of relocation were interrupted by World War II, during which most of the Greenwich instruments were dismantled for safe storage, so that rather little astronomy, even of the conventional, positional sort, could be done there during the war or afterward.

The postwar English astronomical community was badly divided over what to do next; many (especially those engaged in imaging of distant objects) favored development of a site outside England, where the air would be clearer and drier much of the year. Several factors, however, militated in favor of Spencer Jones's preference for remaining in England. One was the unexpected availability of many country locations, as suddenly impoverished aristocrats sought to sell their homes or donate them to the government. Another was the immediate availability of a 98-in. Pyrex mirror blank (originally a trial cast for the 200-in.), acquired by the University of Michigan but never incorporated into a telescope. Herstmonceux Castle was adjudged the most suitable of the domestic sites, and Spencer Jones moved his office there in 1948. Even the existing operations were not fully transferred until after his 1955 retirement, and the 98-in. Isaac Newton telescope became operational only in the late 1960s. The observatory retained the Greenwich name, though its administrative offices were relocated again (to Cambridge) and eventually closed, while portions of the 98-in. were moved to the Canary Islands.

Spencer Jones was very much a public man. He was the only individual to hold all four of the major elective offices in the Royal Astronomical Society until Sir **William McCrea**. He was appointed president of the International Astronomical Union in 1944 by the subset of its officers who were in communication at that time, when Eddington, who had been elected in 1938, died; served as secretary general of the International Council of Scientific Unions (1955–1958); participated in the deliberations that made science part of United Nations Educational, Scientific, and Cultural Organization [UNESCO]; and was knighted for his services to the country. Spencer Jones received some 11 honorary doctorates and was elected to memberships in a comparable number of academies of science. Part of his influence appears to have been attributable to a commanding physical presence and the ability to draft coherent, persuasive arguments orally or on paper very quickly. Curiously, just as one of Dyson's early actions as director of the Royal Observatory had been to appoint Spencer Jones, who became his successor, one of Spencer Jones's first actions in 1933 was the appointment, as a chief assistant, of **Richard van der Riet Wooley** who, in turn, succeeded him as Astronomer Royal and director of the RGO (the last person for whom the two jobs automatically went together).

The official papers of Spencer Jones are in the archives of the RGO at the Cambridge University Library.

Keith Snedegar

Selected References

Anon. (1947). *John Couch Adams and the Discovery of Neptune*. Cambridge: Cambridge University Press. (As a loyal Cambridge-man, Spencer Jones contributed to this collection.)

Spencer Jones, H. (1914). "The Absorption of Light in Space." *Monthly Notices of the Royal Astronomical Society* 75: 4–16.

——— (1922). *General Astronomy*. London: E. Arnold and Co.

——— (1928). *Catalogue of 4569 Stars from Observations with the Reversible Transit Circle Made at the Royal Observatory*. London: H. M. Stationery Office.

——— (1932). "The Spectrum of Nova Pictoris (1925), 1931–32." *Monthly Notices of the Royal Astronomical Society* 92: 728–730.

——— (1935). *Worlds without End*. London: English Universities Press.

——— (1939). "The Rotation of the Earth, and the Secular Accelerations of the Sun, Moon and Planets." *Monthly Notices of the Royal Astronomical Society* 99: 541–558.

——— (1940). *Life on Other Worlds*. London: Hodder and Stoughton.

——— (1941). "The Solar Parallax and the Mass of the Moon from Observations of Eros at the Opposition of 1931." *Memoirs of the Royal Astronomical Society* 76, no. 2: 11–66.

Sphujidhvaja

Flourished **(India), third century**

Sphujidhvaja's *Yavanajātaka*, a poetic version of **Yavaneśvara**'s *circa* 150 work translated from the Greek, is the best surviving evidence for transmission of Hellenistic astronomy to India.

Selected Reference

Pingree, David (1978). *The Yavanajātaka of Sphujidhvaja*. Cambridge, Massachusetts: Harvard University Press.

Spitz, Armand Neustadter

Born **Philadelphia, Pennsylvania, USA, 7 July 1904**
Died **Fairfax, Virginia, USA, 14 April 1971**

Planetarium entrepreneur Armand Spitz contributed to the design of planetariums that could bring the sky to the public at low cost. The son of Louis and Rose (*née* Neustadter) Spitz, he was raised in the Quaker faith and maintained an affiliation with the Newtown Square, Pennsylvania, Friends Meeting throughout his life. Spitz graduated from West Philadelphia High School in 1922. He first matriculated at the University of Pennsylvania; two years later, he transferred to the University of Cincinnati, but left to pursue a career in journalism without receiving his degree. In 1928, Spitz became editor and later publisher of the Haverford *Township News*. The great economic depression, however, forced the venture into bankruptcy. Spitz's former colleague on the newspaper, Vera Golden, edited the revitalized *Township News*. Spitz and Golden later married and had two children, a daughter Verne and son Armand.

Spitz volunteered for work at Philadelphia's Franklin Institute and was eventually employed there. He prepared newspaper publicity and edited the institute's newsletter (1936–1943). By successively capitalizing on his resources as he gained experience and knowledge, Spitz acquired a host of duties at the institute, eventually becoming head of museum education. Spitz organized the institute's department of meteorology and taught courses in that subject. He was cofounder of the Amateur Weathermen of America (1946) and of the journal, *Weatherwise* (1948), later published by the American Meteorological Society. Spitz authored two texts in meteorology, a detailed history of American meteorology, and served as Philadelphia's first television weatherman.

However, Spitz's interest in astronomy was even greater than his interest in meteorology. He was excited by the presentations of **James Stokley** and other lecturers at the Fels Planetarium of the Franklin Institute. These performances fueled Spitz's desire to become a lecturer himself, but for years he was denied this opportunity because he lacked a college degree. Spitz's growing passion for astronomy was first channeled into the construction of a Springfield-mounted reflecting telescope and 3-ft. diameter replica of the Moon. He worked out an affiliation with the astronomy department of Haverford College, and gained the use of the 10-in. Clark refracting telescope in their Strawbridge Observatory to observe double stars.

Spitz's first book, *The Pinpoint Planetarium* (1940), was an important stepping-stone toward the creation of an inexpensive pinhole-style planetarium projector. Consultation with mathematical authorities, including **Albert Einstein** of Princeton, New Jersey, convinced Spitz that a twelve-sided dodecahedron could be used to project realistic images of the sky onto an artificial dome. With financial support and mechanical expertise provided by several of his meteorology students, Spitz unveiled his prototype projector in 1945. The sky portrayed by the Spitz Model A projector included roughly 1,000 stars. Although unable to match the sophistication of a Zeiss instrument costing several tens of thousands of dollars more, Spitz's Model A projector nonetheless offered a realistic display of the stars and planets visible to the unaided eye. The initial cost of the device was only $500.

Spitz joined with David M. Ludlum (born: 1910), who sold meteorological equipment, to found Science Associates in 1946. The firm marketed, among other products, the Spitz Model A planetarium projector. An important demonstration of the Model A projector occurred in 1947 when *Sky & Telescope* editor **Charles Federer** arranged for Spitz to present a planetarium program to a joint meeting of the Bond Astronomy Club and the American Association of Variable Star Observers [AAVSO], using a production version of the Model A. That meeting, which took place in Cambridge, Massachusetts, provided the first substantial publicity for the venture.

In the following years, the portability of the Model A allowed it to be taken to many similar meetings where demonstrations established the credibility of the pinhole-style projector. Spitz Laboratories was incorporated in 1949, although its products were marketed by Science Associates through 1951. Thereafter, Spitz resigned his duties at the Franklin Institute to devote full attention to the sale and installation of planetarium projectors. By the end of 1953, Spitz had sold his 100th projector; many went to institutions outside the United States. He relocated his company several times before its last facility was established at Chadds Ford, Pennsylvania, in 1969. By then, as noted by Brent P. Abbatantuono, Spitz's pinhole-style projectors had "revolutionized the availability of artificial skies."

In 1956, Spitz was chosen as national coordinator of visual satellite observations for Project Moonwatch, a program developed by the Smithsonian Astrophysical Observatory. Project Moonwatch organized a corps of amateur astronomers and others to track the first artificial satellites launched during the International Geophysical Year [IGY]. As a measure of Spitz's success, at least seventy nine stations in the United States, staffed by more than 1,200 individuals, were officially registered by the start of the IGY. Dozens of other Moonwatch groups were established around the world, especially in Japan.

Spitz was awarded an honorary D.Sc. degree in 1956 by Otterbein College at Westerville, Ohio. He gained further recognition that same year when he was named a special consultant to the National Science Foundation, serving in that capacity until 1960. He utilized the prestige associated with that position to help professionalize the American planetarium community, organizing nationwide symposia held at Bloomfield Hills, Michigan (1958), and Cleveland, Ohio (1960). Out of these symposia, the first monographs on planetarium education were produced. The IGY and post-Sputnik era caused substantial turmoil in Spitz's personal life, however. In 1957, he and Vera were divorced; in 1958 he married medical statistician Grace

C. Scholz (born: 1912). Scholz had previously served as executive secretary and president of the Astronomical League, the nation's largest association of amateur astronomers.

Federal assistance arising from passage of the National Defense Education Act of 1958 triggered a third and larger phase of American planetarium growth. But Spitz himself no longer played a significant role in this development. He retired from the company in 1961, though remaining active in the planetarium community for several more years. In 1951, Spitz had organized an association of planetarium directors, under the auspices of the American Association of Museums [AAM]. He attempted to improve the association's communications by creating a newsletter, named *The Pointer*, which was edited by Hayden Planetarium chairman Robert R. Coles (1952/1953). *The Pointer*'s significance lay in its being the first regular publication to originate within the American planetarium community. The AAM's "planetariums section" met on a yearly basis through the 1960s. Continued growth of planetariums made possible the formation of newer regional associations, and in 1970 the larger International Society of Planetarium Educators was founded. Were it not for Spitz's invention of the pinhole-style planetarium projector, such an association might never have come into existence.

In 1966, Spitz suffered a stroke that left him partially paralyzed. An annual lecture series, named in Spitz's honor, was established in 1967 by the Great Lakes Planetarium Association.

Jordan D. Marché, II

Selected References

Abbatantuono, Brent P. (1994). "Armand Neustadter Spitz and His Planetaria, with Historical Notes of the Model A at the University of Florida." Master's thesis, University of Florida.

Anon. (1975). "Spitz, Armand Neustadter." In *National Cyclopedia of American Biography*. Vol. 56, pp. 421–422. Clifton, New Jersey: James T. White and Co.

Federer, Charles A. (1971). "Armand N. Spitz – Planetarium Inventor." *Sky & Telescope* 41, no. 6: 354–355.

Marché II, Jordan D. (2005). *Theaters of Time and Space: American Planetaria, 1930–1970*. New Brunswick, New Jersy: Rutgers University Press.

Spencer, Steven M. (24 April 1954). "The Stars Are His Playthings." *Saturday Evening Post*, pp. 42–43, 97–98, 100, 102–103.

Spitz, Armand N. (1944). "Meteorology in the Franklin Institute." *Journal of the Franklin Institute* 237: 271–287, 331–357.

——— (June 1959). "Planetarium: An Analysis of Opportunities and Obligations." *Griffith Observer*: 78–82, 87.

Spitzer, Lyman, Jr.

Born **Toledo, Ohio, USA, 26 June 1914**
Died **Princeton, New Jersey, USA, 31 March 1997**

American astrophysicist Lyman Spitzer, Jr. made major contributions to our understanding of diffuse gases, especially the interstellar medium, and was among the first to strongly urge the construction of a large optical telescope in space. He was the son of Blanche C. and Lyman Spitzer, and married Doreen Canaday in 1940.

Lyman Spitzer was educated at Phillips Academy (Andover), Yale University (BA: 1935), Cambridge University (where he attended informal lectures by **Subrahmanyan Chandrasekhar** at Trinity College), and Princeton University, where he earned a Ph.D. in 1938 for work with **Henry Norris Russell** on the analysis of spectra of cool supergiant stars. Spitzer held a National Research Council Fellowship at Harvard (1938/1939), where **Harlow Shapley**, **Donald Menzel**, **Bart Bok**, and **Martin Schwarzschild** were particularly influential. He was an instructor at Yale (1939–1942), before moving to war work (1942–1946) at Columbia University, where he became the director of the sonar analysis group before returning to Yale (1946–1947). In 1947, Spitzer became director of the observatory and chair of the astronomy department at Princeton (positions he held until 1979), as successor to Russell, where one of his first actions was to persuade Martin Schwarzschild to join the Princeton group. They remained close colleagues for 50 years thereafter.

Spitzer's scientific contributions fall into a number of fairly district areas – physics of the interstellar medium; stellar dynamics; laboratory plasma physics and controlled thermonuclear fusion; and space astronomy and astrophysics.

Spitzer's career spanned the period from before the recognition of a general interstellar medium to the time when half a dozen different phases of interstellar material had been characterized, and his monograph *Physical Processes in the Interstellar Medium* served as the standard for two decades. He computed the mean free paths of electrons, ions, atoms, and dust grains, showing that the various phases tended toward pressure equilibrium, and thereby predicted a hot, "coronal" medium outside the galactic plane, later found. He was among the first to conclude that star formation must be an ongoing process, in a paper written before World War II. (It was trimmed of the star-formation section for later publication, but restored to the original text in the reprint volume of his papers.) He also pointed out the importance of magnetic fields and dust in star formation.

In the realm of stellar dynamics, Spitzer, concurrently with **Viktor Ambartsumian**, calculated the rate at which stellar encounters in clusters eject stars, introduced several new ideas into the study of dense star clusters, urged a Princeton student (Haldan Cohn) to develop numerical methods for simulating cluster evolution, and wrote another definitive monograph, *Dynamical Evolution of Globular Clusters*. Seven additional Spitzer students at Princeton were eventually also involved in cluster work. Spitzer and Martin Schwarzschild together suggested that the gravitational influence of giant gas clouds was responsible for the gradually increasing velocities of stars in the galactic plane as they age, and Spitzer went on to show that gravitational impulses from such clouds were also responsible for the dissolution of most star clusters before they reach an age of 100 million years. He and **Walter Baade** also introduced the idea of gravitational interaction between galaxies.

Spitzer was involved in the Princeton-controlled thermonuclear fusion program from its inception as Project Matterhorn (1953–1961) and through the early years of the Princeton Applied Physics Laboratory (1961–1967). His design, called the "stellerator," for a magnetically confined plasma, had obvious astronomical roots and a formal basis in lectures given at Princeton by visitor **Thomas Cowling**.

Some kinds of astronomy can be done only from above the Earth's atmosphere, and some phenomena are best studied *in situ*. Even before the war, while at Yale, Spitzer had tried to organize a program in solar ultraviolet spectroscopy and to recruit **Leo Goldberg** into it. The ultraviolet Copernicus satellite, launched in 1972, was the eventual fruit of this interest, and the spectrometer designed by his group discovered interstellar molecular hydrogen, measured the ratio of deuterium to normal hydrogen in interstellar gas, and found highly ionized atoms as evidence of the million-degree coronal component he had predicted long ago. Spitzer began urging the construction of a 3-m class telescope in space as early as 1947. He shepherded through National Aeronautics and Space Administration planning, and congressional scrutiny of what is now known as the Hubble space telescope [HST]. For many years, he chaired the Space Telescope Institute Council, the oversight group for the institute that selects observing programs and processes data for the HST. He had also urged Martin Schwarzschild to develop the Stratoscope Balloon program for high-resolution astronomy, with input as well from **James Van Allen**, whom Spitzer temporarily attracted to Princeton for the Plasma Lab. Spitzer was awarded six honorary doctorates and medals and awards from the Royal Society (London), the Royal Swedish Academy of Science, the American Astronomical Society (which he served as president during 1960–1962), the Royal Astronomical Society, and several others. He was elected to membership in the United States National Academy of Sciences, the Royal Society of Science of Liège, the American Academy of Arts and Sciences, and other both honorary and scientific service organizations.

In addition to writing many scientific papers in various astrophysical journals, Spitzer wrote important textbooks, which are useful to researchers and graduate students because they include new results from his studies.

Satoru Ikeuchi

Selected References

King, Ivan R. (1997). "Lyman Spitzer, Jr. , 1914–1997." *Bulletin of the American Astronomical Society* 29: 1489–1491.

Smith, Robert W. (1989). *The Space Telescope: A Study of NASA, Science, Technology, and Politics*. Cambridge: Cambridge University Press.

Spitzer Jr., Lyman (1952). *Interstellar Matter*. Princeton, New Jersey: Princeton University Observatory. (A set of notes prepared by John B. Rogerson from lectures by Spitzer.)

——— (1962). *Physics of Fully Ionized Gases*. 2nd rev. ed. New York: Interscience.

——— (1968). *Diffuse Matter in Space*. New York: Interscience.

——— (1978). *Physical Processes in the Interstellar Medium*. New York: Wiley-Interscience.

——— (1982). *Searching between the Stars*. New Haven: Yale University Press.

——— (1987). *Dynamical Evolution of Globular Clusters*. Princeton, New Jersey: Princeton University Press.

——— (1989). "Dreams, Stars, and Electrons." *Annual Review of Astronomy and Astrophysics*. 27: 1–17.

Spitzer Jr., Lyman and J. P. Ostriker (eds.) (1997). *Dreams, Stars, and Electrons: Selected Writings of Lyman Spitzer, Jr.* Princeton, New Jersey: Princeton University Press. (Ostriker was Spitzer's successor as Princeton Observatory director and department chair.)

Spörer, Friedrich Wilhelm Gustav

Born **Berlin, (Germany), 23 October 1822**
Died **Giessen, (Hessen), Germany, 7 July 1895**

Friedrich Spörer is best known for his refinement of our understanding of sunspots. He is commemorated in the name of the Spörer minimum, an absence of sunspots *circa* 1450.

Spörer was the son of a German merchant. He studied at the University of Berlin, where his most influential teachers were Heinrich Wilhelm Dove (1803–1873) and **Johann Encke**. His thesis on the 1723 comet (C/1723 T1) earned him a doctorate in 1843, after which Spörer worked as a "computer" for Encke, responsible for the calculation of cometary orbits.

Spörer left the Berlin Observatory in 1846 to pursue a career as a teacher. He taught mathematics and natural sciences in various Gymnasium and Grammar Schools, first in Bromberg, then in Prenzlau, and finally in Anclam, where he remained for 25 years and eventually became prorector.

Spörer's main contribution to astronomy remains his extensive studies of sunspots and their cycles, which he carried out as an avocational pursuit for many years while teaching. Starting in 1860 and working with a small and inferior telescope, he embarked on a sustained sunspot observing program aimed at determining the rotational elements of the Sun with better accuracy. His first task was to recompute the inclination of the Sun's axis with respect to the ecliptic, and the longitude of its ascending node. Spörer then studied the apparent rotational motion of sunspots, rediscovering the solar latitudinal differential rotation found by British astronomer **Richard Carrington**. He also investigated in great detail what he dubbed the "law of zones" At any given time in the cycle, sunspots are confined to two relatively narrow latitudinal bands on either side of the solar equator, and these bands gradually migrate equatorward in the course of the cycle. Although this behavior had been noted earlier by Carrington, Spörer studied it so assiduously that the phenomenon came to bear his name. Spörer also observed that successive cycles overlap slightly, in that the high-latitude sunspot bands associated with a new cycle appear while sunspots from the preceding cycle are still seen close to the Equator. Finally, he noted the common lack of symmetry between the number of sunspots observed in the Northern and Southern Solar Hemispheres.

The high quality of the observations Spörer carried out during his free time throughout those years attracted the attention of the tutor to Crown Prince Frederick Wilhelm (later Kaiser Frederick II). An immediate practical consequence for Spörer was the 1868 gift by the Crown Prince of a larger, high-quality telescope with which to pursue his sunspot research. Another apparent consequence of the recognition Spörer finally achieved also occurred in 1868, when he was invited to participate in the German eclipse expedition to the East Indies, which was unfortunately plagued by bad weather on eclipse day.

Recognition of Spörer's scientific effort eventually led, in 1874, to his appointment at Potsdam Observatory, then in the planning stage, where he became chief observer in 1882. Spörer subsequently engaged in historical researches aimed at examining whether his "law of zones" held in prior sunspot cycles. It is in the course of these investigations that he noted the striking dearth of sunspots between 1645 and 1715, as well as the pronounced North–South asymmetry in the few sunspots that were observed at the time. Surprisingly, these findings attracted relatively little attention at the time, and this curious break in the sunspot cycle is now known as the Maunder minimum, even though the British astronomer **Edward Maunder** clearly and vigorously publicized the historical sunspot researches as Spörer's work. An earlier period of reduced activity is called the Spörer minimum.

In 1885, Spörer was awarded the Valz Prize of the Institut de France. He was also elected a foreign associate of the Royal Astronomical Society in 1886, and a corresponding associate of the Società degli Spettroscopisti in Italy in 1889. Spörer retired from Potsdam Observatory on 1 October 1894. Having enjoyed perfect health throughout his life, he died suddenly of a heart attack, while on a trip to visit his children.

Paul Charbonneau

Selected References

Anon. (1896). "Professor Dr. Friedrich Wilhelm Gustav Spörer." *Monthly Notices of the Royal Astronomical Society* 56: 210–213.

Maunder, E. W. (1890). "Professor Spoerer's Researches on Sun-spots." *Monthly Notices of the Royal Astronomical Society* 50: 251–252. (Contains a short summary, in English, of Spörer's two 1889 articles.)

Spörer, G. (1889). "Sur les différences que présentent l'hémisphère nord et l'hémisphère sud du Soleil." *Bulletin astronomique* 6: 60–63.

——— (1889). "Ueber die Periodicität der Sonnenflecken seit dem Jahre 1618." *Nova Acta der Kaiserlischer Leopold-Carolinia Deutschen Akademie der Naturforscher* 53: 280–324.

Śrīpati

Flourished **Rohinīkhanda, (Mahārāṣtra, India), 1039–1056**

Śrīpati was an Indian (Hindu) astronomer. His father was Nāgadeva, and his grandfather was **Keśava** of the Kāśyapagotra. Śrīpati composed three astronomical works; the *Siddhāntaśekhara*, the *Dhīkoṭidakaraṇa* (1039), and the *Dhruvamānasa* (1056); a mathematical work, the *Gaṇitatilaka*; and two astrological works, the *Jātakapaddhati* (or *Śrīpatipaddhati*) and the *Jyotiṣaratnamālā*. Śrīpati is the only other astronomer, after **Varāhamihira** and **Lalla**, whose works spanned both astronomy and astrology. All other astronomers wrote exclusively on mathematical astronomy as far as extant texts are concerned.

Śrīpati's *Siddhāntaśekhara* followed the Brāhma School of **Brahmagupta**, one of four principal schools of astronomy active in the classical period (late 5th to 12th centuries). The *Śiddhāntaśekhara* is a treatise on mathematical astronomy and consists of 19 chapters.

At the same time, Śrīpati was much influenced by the *Śiṣyadhīvṛddhidatantra* of Lalla, a follower of the Ārya School of **Āryabhaṭa I**. Śrīpati's *Jyotiṣaratnamālā* followed the *Jyotiṣaratnakośa* of Lalla.

The *Dhīkoṭidakaraṇa* is a work of 20 verses that gives methods of calculation for lunar and solar eclipses. The *Dhruvamānasa* is, according to D. Pingree, a work of some 105 verses used for calculating a variety of lunar and planetary phenomena.

Among Śrīpati's two astrological works, the *Jātakapaddhati* is a textbook on horoscopy, while the *Jyotiṣaratnamālā* is an influential work on catarchic astrology. It contains Śrīpati's own commentary (written in Marāthī).

Selected References

Chattopadhyay, Anjana (2002). "Sripati." In *Biographical Dictionary of Indian Scientists: From Ancient to Contemporary*, p. 1389. New Delhi: Rupa.

Dikshita, Sankara Balakrshna (1969). *Bhāratīya Jyotish Śāstra* (History of Indian Astronomy), translated by Raghunath Vinayak Vaidya. Delhi: Manager of Publications (Government of India).

——— (1981). *Bhāratīya Jyotish Śāstra* (History of Indian Astronomy), translated by Raghunath Vinayak Vaidya. Vol. 2. New Delhi: India Meteorological Department.

Panse, Murlidhar Gajanan (1957). *Jyotisa-ratna-mala of Śrīpati Bhatta: A Marathi Tika on His Own Sanskrit Work*. Poona, India: Deccan College Monograph Series, no. 20.

Pingree, David (1975). "Śrīpati." In *Dictionary of Scientific Biography*, edited by Charles Coulston Gillispie. Vol. 12, pp. 598–599. New York: Charles Scribner's Sons.

——— (1978). "History of Mathematical Astronomy in India." In *Dictionary of Scientific Biography*, edited by Charles Coulston Gillispie. Vol. 15 (Suppl. 1), pp. 533–633. New York: Charles Scribner's Sons.

Ssu-ma ch'ien

➔ **Sima, Qian**

Stabius, Johann

Flourished ***circa* 1500**

Johann Stabius was court astronomer to Maximilian I in Vienna. **Albrecht Dürer**'s famous celestial charts illustrate Stabius's 1515 atlas. The star positions were provided by Conrad Heinfogel.

Selected Reference

Barton, Samuel G. "Dürer and Early Star Maps." Parts 1 and 2. *Sky & Telescope* 6, no. 11 (1947): 6–8; no. 12 (1947): 12–13.

Stark, Johannes

Born **Schichenhof, Bavaria, Germany, 15 April 1874**
Died **Traunstein, Bavaria, (Germany), 21 June 1957**

German experimental physicist Johannes Stark is important within astronomy for the Stark effect, the broadening or splitting of atomic emission and absorption lines when the atoms producing them are in an ambient electric field, such as that due to the surrounding ions and electrons in the atmosphere of a star.

Stark was born to a landed proprietor and his wife. Stark's early life included an education at the Gymnasium in Bayreuth and later in Regensburg. Upon graduation, he enrolled at Munich University in 1894 to study physics, mathematics, chemistry, and crystallography. Stark completed his doctorate in 1897 with a dissertation on **Isaac Newton**'s electrochronic rings in dim media. Upon completion of his doctorate, he worked as an assistant at the Physics Institute at Munich University, a post Stark held until 1900, when he became an unsalaried lecturer in physics at the University of Göttingen. In 1906, he moved on to a professorship at the Technische Hochschule in Hanover until 1909, when he switched to the Technische Hochschule in Aachen.

During this period, Stark studied the behavior of "canal rays" (rapidly moving positively charged particles, so called because they come out of an opening or canal in a cathode) in hydrogen gas, some of the results being published in journals of astronomy. In 1910, he was awarded the Baumgartner Prize of the Vienna Academy of Science, and he received the Rome Academy's Matteucci Medal in 1914.

In 1913, Stark showed that, when a strong electric field is applied to hot, glowing hydrogen gas, the Balmer lines split into a number of components. In the stellar context, one can think of such a field as being produced, on average, by the large number of ions and electrons around the atoms that are emitting or absorbing the lines. The effect is to broaden the Balmer and other lines, making them look stronger, rather than to split them into several separate lines. Stark's 1919 Nobel Prize in Physics was awarded both for the line-splitting discovery and for his work on canal rays.

In 1917, Stark moved to a professorship at the University of Griefswald and on to the University of Wurzburg in 1920. Beginning in the early 1920s, Stark, together with Philipp Lenard and Ernst Gehrcke, attempted to oppose the spread of nonclassical physics, both general relativity and quantum mechanics, within Germany. By the 1930s, part of the reason for their opposition was that much of the nonclassical work had been done by Jewish scientists, and the Nazi party came to support their campaign for "Aryan physics" (including, *e. g.*, the republication of a **Johann Soldner** paper from the early 1800s that derived gravitational bending of light from Newtonian mechanics and a particle theory of light). This support probably contributed to Stark's 1933 election as president of the Physikalisch-Technische Reichsanstalt (Physico-Technical Institute). The German physical community came to oppose the strange scientific ideas of Stark and Lenard by about 1940, and in 1939 Stark retired to a private laboratory, set up with money from his Nobel Prize, where he died. He was married to Luise Uepler, and they had five children.

Modern calculations of Stark broadening distinguish a linear Stark effect (due to other atoms of the same element) and a quadratic Stark effect (due to atoms of other elements and electrons). All of atomic physics rests on a foundation of quantum mechanics, ironical in the light of Stark's opposition to it, though in the case of his effect, the quantum mechanical aspect can be described by a single parameter, and the calculations of the average electric fields at a given location can be done classically.

Ian T. Durham

Selected References

Kragh, Helge (1999). *Quantum Generations: A History of Physics in the Twentieth Century*. Princeton, New Jersey: Princeton University Press.

Stark, J. (1907). "On the Radiation of Canal Rays in Hydrogen." Parts 1 and 2. *Astrophysical Journal* 25: 23–44, 170–194.

——— (1907). "Remarks on Hull's Observations of the Doppler Effect in Canal Rays." *Astrophysical Journal* 25: 230–234.

——— (1907). "Photographs of the Doppler Effect in the Spectrum of Hydrogen and of Mercury. Rejoinder to Mr. Hull's Reply." *Astrophysical Journal* 26: 63.

Whittaker, Sir Edmund (1953). *A History of the Theories of Aether and Electricity*. Vol. 2, *The Modern Theories*. New York: Harper.

Steavenson, William Herbert

Born **Quenington, Gloucestershire, England, 26 April 1894**
Died **South Marston, Wiltshire, England, 23 September 1975**

William Steavenson was renowned for his observational skills and knowledge of instruments. Steavenson was the youngest child of Reverend Frederick Robert Steavenson, rector of Quenington, Gloucestershire.

Steavenson came to astronomy early, and on 25 December 1917, he noted in one of his observation diaries that it was "the tenth anniversary of my astronomical birthday." This signified the occasion in 1907 when, having viewed the Moon through a 1.75-in. aperture telescope, Steavenson suddenly appreciated the potential of even a modest telescope. Soon after he was given an equatorially mounted 3-in. refractor on a wooden tripod, and began serious observation of the Moon, the planets (especially Jupiter, Saturn, and their satellites), and whatever comets and novae he could access.

By 1910, Steavenson had entered into published correspondence with long-standing contributors to *The English Mechanic* on such wide-ranging topics as the possible light variation of Hyperion, a suspected tiny crater on the wall of the lunar crater Aristarchus, and the detail visible on Saturn in good seeing. In March 1912, Steavenson was writing about Nova Geminorum, and reporting on the trail of a meteor he had registered while photographing the region of the nova.

Steavenson first made headlines in September 1911 when he took a series of comet photographs. This activity by a schoolboy excited wide interest and brought Steavenson to the attention of the Astronomer Royal, **Frank Dyson**, then president of the Royal Astronomical Society. He immediately proposed the youngster for a fellowship of the Society, and on 12 January 1912 Steavenson was duly elected. This in the words of his obituarist brought him more publicity and marked him "as a very active and persevering amateur astronomer." By December 1914, Steavenson had also joined the Society for Practical Astronomy [SPA] in the United States and less than a year later was appointed director of the SPA Comet Section.

Following the severe illness of his father, the family took up residence at Cheltenham, Gloucestershire, and William Steavenson entered Cheltenham College, starting in the preparatory school. Here he won a classical scholarship. He received his medical training at Guy's hospital, London.

In 1916, while still a medical student, Steavenson conducted an experiment to determine, photographically, the diameter of the fully dark-adapted pupil of a normal eye. His finding of more than 8-mm was in sharp contrast to the 5-mm standard assumption on which optical designs were then based, and was adopted for use in optical design thereafter.

Steavenson served as a civil surgeon in a military hospital at Millbank, London, during 1918 and 1919. In 1919, he was appointed a captain in the Royal Army Medical Corps, and spent 6 months in Egypt, but resigned his commission in 1920. With his mother and sister, Steavenson eventually settled in West Norwood, a suburb of London, and set up an observatory there.

In 1923, in company with Reverend **Theodore Phillips**, he edited *The Splendour of the Heavens*, a large multiauthor compendium of astronomical knowledge that proved spectacularly popular with amateurs and professionals alike, and is now a classic of its kind. In 1927, Steavenson visited observatories in North America for his own edification, and in mid-1929 he was asked to travel to South Africa, where he spent 6 months investigating seeing conditions at the possible site for the new Radcliffe Observatory.

Soon after Steavenson returned from South Africa in 1930, a 20-in. reflector replaced the 6-in. Wray refractor set up in 1922 at West Norwood. Few fine nights were wasted, and his log abounds with observations of novae, comets, planets, and the satellites of Uranus.

In 1939, the growing menace of light pollution in the metropolis caused Steavenson to move to Cambridge. He felt a larger instrument would be wasted near London. **Arthur Eddington**, then director of the University Observatory at Cambridge, gave him permission to install a 30-in. Hindle reflector and dome on the observatory grounds. By the summer of 1939, the dome was built and the new telescope installed. However, with the outbreak of hostilities in September, Steavenson went to help medical friends with their practice in Cheltenham, and the observatory remained closed until the summer of 1945 when he returned to The Hermitage, Newnham, Cambridge (later to become Darwin College).

Steavenson resumed work on 10 August 1945 and continued with unabated vigor for the next 10 years. With the greater light grasp available, Steavenson renewed his photometric studies of the satellites of Uranus, now including position measures with a homemade position-micrometer.

In 1956, Steavenson decided to abandon observing. No one knows why. Perhaps the process was becoming too strenuous; more likely it was putting his surviving eye under strain. (He had lost the right eye in a boyhood accident.) Whatever the reason, his telescope was given to the Cape Observatory and, at the age of 62, Steavenson returned to Cheltenham in the Cotswold country of his origin. Finally in 1971 he went to live with a niece at South Marston, Wiltshire, who looked after him for the last 4 years of his life.

Steavenson was elected a member of the British Astronomical Association on 28 May 1913. He served as acting director of the Saturn Section from 1917 to 1919, director of the Mars Section from 1922 to 1930, and director of the Instruments and Observing Methods Section from 1932 to 1961, a position for which he was ideally suited. Many astronomers, professional and amateur, had cause to remember his inspiring advice with gratitude. Recognition of his outstanding abilities as an observer first came when Steavenson was elected president of the British Astronomical Association for the period 1926–1928. In 1928, he was awarded the Jackson-Gwilt Medal of the Royal Astronomical Society, and in 1961 the prestigious Walter Goodacre Medal of the British Astronomical Association. His greatest honor however, was to be elected president of the Royal Astronomical Society for the period 1957–1959. Steavenson was also Gresham Professor of Astronomy from 1946 to 1964, and astronomical correspondent of *The Times* from 1938 to 1964.

Richard Baum

Selected References

Dewhirst, David W. (1977). "William Herbert Steavenson." *Quarterly Journal of the Royal Astronomical Society* 18: 147–154.

Fry, R. M. (1976). "William Herbert Steavenson, 1894–1975." *Journal of the British Astronomical Association* 86: 386–390.

Phillips, Theodore Evelyn Reece and William Herbert Steavenson (eds.) (1923). *Hutchinson's Splendour of the Heavens, A Popular Authoritative Astronomy.* London: Hutchison and Co.

Steavenson, William Herbert (1915–1916). "Note on Low Powers." *Journal of the British Astronomical Association* 26: 302–303.

——— (1964). "The Satellites of Uranus." *Journal of the British Astronomical Association* 74: 54–59.

Tayler, R. J. (ed.) (1987). *History of the Royal Astronomical Society*. Vol. 2, *1920–1980*. Oxford: Blackwell Scientific Publications, pp. 154, 216.

Stebbins, Joel

Born **Omaha, Nebraska, USA, 20 July 1878**
Died **Palo Alto, California, USA, 16 March 1966**

American photometrist Joel Stebbins was an innovator in the use of photoelectric photometry, and in 1915 applied these techniques to measure the first light curve of an eclipsing binary from which the distance to the system could be determined. He also used photoelectric photometry to make the first quantitative measurement of night-sky brightness caused by urban light pollution and to look for evidence that galaxies had changed their colors with the evolution of the Universe.

The son of Charles Sumner and Sara Ann (*née* Stubbs) Stebbins, Joel developed an early interest in astronomy. His first jobs (apart from newspaper delivery boy) included part-time surveying work for the Union Pacific Railroad, which employed his father. His marriage in 1905 produced two children, but the family member to whom he was closest seems to have been his sister Millicent, who sometimes made Wisconsin–California summer round trips with him.

Stebbins obtained a B.S. degree from the University of Nebraska in 1899 – he was awarded an honorary LL.D. in 1940 – and continued as a student at University of Wisconsin (1900/1901), before completing a Ph.D. in astronomy (spectroscopy) at the University of California, Berkeley (1903), the third such degree awarded by the department. (Wisconsin awarded him an honorary D.Sc. in 1920.) Stebbins joined the faculty at the University of Illinois first as an instructor in astronomy (1903/1904) and subsequently as assistant professor (1904–1913). After a sabbatical at the University of Munich (1912/1913), he returned as professor and director of the Illinois observatory (1913–1922). Stebbins moved to the Illinois of Wisconsin as professor and director of the Washburn Observatory in 1922, holding that position until his retirement in 1948. From 1922 to 1948, Stebbins also was a research associate at Mount Wilson. Following his retirement, on termination of the Mount Wilson research associate program, Stebbins remained affiliated with the Lick Observatory until his death.

Stebbins' work was in the precision photometric measurement of cosmic objects, and he was a pioneer of, and propagandist for, photoelectric photometry throughout the first half of the 20th century. At the time the basic limitations of accuracy and dynamic range precluded observations of faint stars and rapidly varying sources using visual and photographic photometry.

A selenium cell, which functioned as a variable resistor in a Wheatstone bridge with a galvanometer as a detector, was limited by the stability of the power supply and the sensitivity of the galvanometer. Stebbins provided the first astronomical calibrations on a selenium cell in 1908, showing that the peak response was at about 7,000 Å with a range of about 2,000 Å (to a sensitivity of about 25% of peak). The detector was, though, nearly blue and ultraviolet blind, rendering it a poor choice for photometry of nearly all stars hotter than the Sun. For this work, a new detector was required and again Stebbins was at the forefront of the search.

Stebbins attributed his inspiration for using photoelectric detectors for astronomical photometry to a 1906 lecture by his colleague F. C. Brown. The technique was similar to that of Philip Lenard (codiscoverer of the photoelectric effect); the method of detection was indirect and unamplified. A gas photocell developed by Stebbins' Illinois colleague Jacob Kunz, was employed. The blue-sensitive potassium hydride photocathode Kunz cell, equipped with filters centered at 4,300 Å and 4,800 Å, achieved much higher sensitivities than the selenium cell. It used an inert gas in a partially evacuated cell (1 mm Hg pressure) to provide modest amplification, about a factor of 10, through secondary ionizations.

Stebbins concentrated on measurement of variable stars, β Persei, β Aurigae, and δ Orionis, as demonstrations of the new technique. Fundamental stellar parameters are most easily determined using eclipsing binary light curves. Using only geometry, relative radii and luminosities of the stars can be found independent of any details of stellar structure. Colors, if they can be obtained, provide temperatures and, when supplemented by radial velocity measurements, yield bolometric luminosities and masses. Measurement of eclipse light curves was limited to visual estimates, which had both systematic and large random uncertainties, or photographic observations that could not handle some of the shorter period systems. Stebbins's introduction of photoelectric photometry eliminated both problems. Observing Algol in 1910, he was able for the first time to detect the secondary eclipse, which he measured at about 0.06 magnitudes, and also a small but distinct variation in the brightness of the binary outside eclipse that he attributed to reflection effects. These and subsequent observations served as the basis for **Henry Norris Russell**, **Harlow Shapley**, and **Joseph Moore**'s work on rectification of eclipsing-binary light curves.

Under Stebbins's leadership, Wisconsin became a center for photometry. With his students Charles Morse Huffer and **Albert Whitford** (and later **Gerald Kron** and Olin J. Eggen), Stebbins made major contributions to both binary star and variable-star photometry, obtaining the first accurate light curves for many Cepheid variable stars and compiling standard star magnitudes.

As early as 1915, Stebbins suggested that observations of reflected sunlight from solar-system bodies could be used to study possible variations in the solar constant. **Cleveland Abbe** and **Samuel Langley** initiated long-term monitoring of the solar irradiance from the ground, but uncertainties in atmospheric corrections and the problem of dealing with an extended solar disk plagued the analysis. Stebbins argued that measurements of Jovian satellites and the outer planets, which could be treated as independent point sources, could provide an alternative means for assessing the constancy of the solar luminosity. He conducted the first photoelectric observations of the Jovian moons, in 1926, and found that they have complex, phase-dependent albedos. Although not especially useful for direct study of the Sun, these measurements provided early indications of complex surface structure on small bodies in the Solar System.

During the 1920s and 1930s, Stebbins concentrated on measuring colors for B stars. Following **Robert Trumpler**'s 1930 discovery of interstellar reddening using photographic techniques, Stebbins and Whitford, in 1932, obtained photoelectric measurements that both determined the color dependence of the extinction and revised the globular cluster distance scale. Their measurement of the wavelength-dependent extinction law was a basis for **Jesse Greenstein** and **Louis Henyey**'s and Hendrich van de Hulst's theoretical work on dust grain optics. Stebbins and Whitford extended the sample to early-type stars and mapped the extinction zone for the Galaxy, showing that it corresponds to the Zone of Avoidance found for galaxies. Subsequently, during the war, Stebbins introduced the six-color system (3,530–10,300 Å) with which he and Whitford refined the reddening law and which he used to determine the phase-dependent temperature variations for δ Cephei and later, with Kron, for many other sources. Further improvements came after 1950 with the introduction of the 1P21 photomultiplier and the development by Harold Johnson and **William Morgan** of the UBV filter system, but Stebbins' work was the foundation on which all subsequent photometry was based and serves as the only precision record of the behavior of many stars over a century baseline.

Stebbins also pioneered the photoelectric observations of the solar corona and nebulae and extragalactic systems. Some of the work on nebulae was done with Whitford; they developed a new technique for photometry of extended sources, which they later applied to measuring colors of distant galaxies, looking for changes with time and finding what was called the Stebbins–Whitford effect. The Stebbins–Whitford effect, if it had proven correct, would have been the first evidence that galaxies had been significantly different in the past from the present. They believed that they had shown galaxies at a redshift of 0.1–0.2 to be much redder than zero-redshift galaxies, but the result was an artifact of the particular color bands used and the non-blackbody shape of galactic spectra moving through them (K correction). Such galaxies are, in fact, slightly bluer than contemporary ones, because they contain more young stars.

Stebbins was also an innovator in the study of light pollution and the modern effort to preserve dark astronomical sites. His photoelectric measurements of ambient light from Los Angeles and Pasadena, California, of Mount Wilson and Palomar Mountain, during 1931/1932 are the first quantitative measurements of night-sky brightness caused by urban sources.

Stebbins had a broad natural curiosity. This is evident from some of his minor papers. Examples include a paper with **Edward Fath** on using astronomical telescopes and parallax to study the altitudes of migrating birds, a report of auroral observations during the spectacular display in 1918, and an observation of the green flash from Mount Hamilton.

Stebbins's work received broad, early recognition. He was awarded the Rumford Prize of the American Academy of Arts and Sciences (1913) and the National Academy of Sciences' Henry Draper Medal (1915) for his work on development of photometric detectors, the Bruce Medal of the Astronomical Society of the Pacific (1941), the Gold Medal of the Royal Astronomical Society (1950), the Russell Prize of the American Astronomical Society (1956), and additional honorary degrees from the Universities of Chicago and California for his broad contributions to astrophysical photometry. He was elected to the United States National Academy of Sciences in 1920. Stebbins also served the American Astronomical Society in several offices, eventually as president (1940–1943).

Steven N. Shore

Selected References

Code, A. D. (1993). *Massive Stars: Their Lives in the Interstellar Medium*, edited by Joseph P. Cassinelli and Edward B. Churchwell. San Francisco: Astronomical Society of the Pacific, p. 3.

Garstang, R. H. (2002). "Light Pollution at Mount Wilson and at Palomar in 1931–32." *Observatory* 122: 154.

Hearnshaw, J. B. (1996). *The Measurement of Starlight: Two Centuries of Astronomical Photometry*. Cambridge: Cambridge University Press.

Kron, Gerald E. (1966). "Joel Stebbins, 1878–1966." *Publications of the Astronomical Society of the Pacific* 78: 214–222.

Kunz, Jakob and Joel Stebbins (1916). "On the Construction of Sensitive Photoelectric Cells." *Physical Review* 7: 62–65.

Osterbrock, Donald E. (1976). "The California–Wisconsin Axis in American Astronomy – II." *Sky & Telescope* 51, no. 2: 91–97.

Stebbins, Joel (1908). "The Color Sensitivity of Selenium Cells." *Astrophysical Journal* 27: 183–187.

——— (1915). "The Electrical Photometry of Stars." *Science* 41: 809–813.

——— (1915). "Some Problems in Stellar Photometry." *Proceedings of the National Academy of Sciences* 1: 259–262.

——— (1926). "The Light Variations of the Satellites of Jupiter and their Application to Measures of the Solar Constant." *Publications of the Astronomical Society of the Pacific* 38: 321.

——— (1945). "Six-Color Photometry of Stars. II. Light-Curves of δ Cephei." *Astrophysical Journal* 101: 47–55.

——— (1950). "The Electrical Photometry of Stars and Nebulae." *Monthly Notices of the Royal Astronomical Society* 110: 416–428. (George Darwin Lecture, 13 October 1950.)

Stebbins, Joel and A. E. Whitford (1936). "Absorption and Space Reddening in the Galaxy from the Colors of Globular Clusters." *Astrophysical Journal* 84: 132–157.

——— (1943). "Six-Color Photometry of Stars. I. The Law of Space Reddening from the Colors of O and B Stars." *Astrophysical Journal* 98: 20–32.

Stebbins, Joel, C. M. Huffer, and A. E. Whitford (1939). "Space Reddening in the Galaxy." *Astrophysical Journal* 90: 209–229.

Walsh, J. W. T. (1953). *Photometry*. 2nd rev. ed. London: Constable and Co.

Whitford, A. E. (1978). "Joel Stebbins." *Biographical Memoirs, National Academy of Sciences* 49: 293–316.

Stephan, Jean-Marie-Édouard

Born **Sainte-Pezenne, Deux-Sèvres, France, 31 August 1837**
Died **Marseilles, France, 31 December 1923**

Édouard Stephan is chiefly remembered as the director of the Observatory of Marseilles (1873–1907), where he discovered many new nebulae at a time when astronomers were vacillating about whether these were gas clouds in the Milky Way or separate large stellar systems; in fact there are some of each. He also made pioneering studies of the angular diameters of stars.

Stephan graduated at the top of his class from the École Normale Supérieure in 1862 and was promptly recruited by **Urbain Le Verrier** of the Paris Observatory. Three years later, he completed his *docteur ès sciences* degree. In 1866, Stephan was assigned to complete a transfer of the Observatory of Marseilles from the Montée des Accoules to its new site on the Plateau Longchamp. In 1873, he was appointed the observatory's official director. Stephan was also named professor of astronomy in the university of Marseilles in 1879. He held both appointments until his retirement in 1907.

At the Observatory of Marseilles Stephan's principal achievement was to catalog several hundred new "nebulae" (most of which are distant galaxies) and to measure their positions using **Léon Foucault**'s 80-cm silvered-glass reflecting telescope. Because he believed nebulae to be very distant objects, Stephan attempted to use them as fixed reference points against which to measure stellar proper motions within the Milky Way Galaxy. This strategy, however, would not be successfully accomplished before the mid-20th century (with the advent of the Lick Observatory proper motion survey, inaugurated by director **Charles Shane**).

Stephan recognized that many of his (and most other) nebulae were often found in clusters – a fact that might indicate whether gravitational attraction existed far from the Galaxy. One of these clusters, centered upon NGC 7318 in the constellation of Pegasus, is still known as Stephan's quintet. More recent research has shown that one of the five galaxies comprising it exhibits a very different redshift from the other four, despite the appearance of "bridge-like" connections linking it to the rest.

In the early 1870s, Stephan placed two parallel slits in front of the 80-cm reflector, in an attempt to measure the angular diameters of stars by means of their interference fringes. This technique had been suggested by French physicist **Armand-Hippolyte Fizeau**. Stephan found that stellar angular diameters were in all cases less than 0.16 arc seconds. It would be almost 50 years before Americans **Albert Michelson** and **Francis Pease** succeeded in measuring the angular diameter of Betelgeuse with a stellar interferometer.

Stephan's other contributions include transit timings, eclipse and cometary observations, the discovery of minor planet (89) Julia, and work on the longitude difference between France and Algeria. Appointed *Chevalier* (1868) and *Officier* (1879) of the *Légion d'honneur*, Stephan was also elected a correspondent of the Paris Académie des sciences (1879).

William Tobin

Selected References

Bigourdan, Guillaume (1924). "Jean-Marie-Éduoard Stephan." *Comptes rendus de l'Académie des sciences* 178: 21–24.

Bosler, Jean (1924). "Édouard Stephan (1837–1923)." *Journal des observateurs* 7, no. 2: 9–10.

Burnham, Jr. Robert (1978). *Burnham's Celestial Handbook: An Observer's Guide to the Universe Beyond the Solar System*. Vol. 3, *Pavo to Vulpecula*. New York: Dover, pp. 1389–1390. (On Stephan's quintet.)

Lévy, Jacques R. (1976). "Stephan, Édouard Jean Marie." In *Dictionary of Scientific Biography*, edited by Charles Coulston Gillispie. Vol. 13, pp. 36–37. New York: Charles Scribner's Sons.

Tobin, W. (1987). "Foucault's Invention of the Silvered-glass Reflecting Telescope and the History of His 80-cm Reflector at the Observatoire de Marseille." *Vistas in Astronomy* 30: 153–184.

various (1925). "Édouard Stephan." *Annales de la Faculté des Sciences de Marseille* 26: 1–13.

Stern, Otto

Born **Sohrau (Zory, Poland), 17 February 1888**
Died **Berkeley, California, USA, August 1969**

German–American experimental physicist Otto Stern is remembered for the Stern–Gerlach experiment (1922), which established the reality of space and angular momentum quantization, though his 1943 Nobel Prize in Physics was awarded for later work in developing the molecular beam technique and discovering the anomalous magnetic moment of the proton (directly responsible for the wavelength of the 21-cm transition of neutral hydrogen in the interstellar medium). Stern received a Ph.D. in physical chemistry from the University of Breslau in 1912, went as a postdoctoral associate to **Albert Einstein** in Prague, and moved with Einstein to Zürich in 1913, where he became an unsalaried *Privatdozent* (lecturer) at the Federal Institute of Technology [ETH]. In 1914, Stern became a *Privatdozent* at Frankfurt and returned there after service in World War I to Max Born's Institute for Theoretical Physics, soon switching to experimental projects.

Born had difficulty finding money for the Stern–Gerlach experiment, which was partially paid for by American financier Henry Goldman (1857–1937) of Goldman Sachs & Company. Stern was appointed to a professorship at Rostock (1921/1922) and then at Hamburg in 1923, in physical chemistry. In 1933, it became advisable for both him and Born to leave Germany, which they were able to do with financial assistance, again from Goldman. It was at Hamburg that Born did the work on molecular beam spectroscopy for which he received the Nobel Prize. His experiments also directly demonstrated the wave-like nature of whole atoms and molecules. (The wave–particle dualism for electrons had been shown by the earlier Davisson–Germer experiment.) The best known of Stern's younger associates there was Isidor I. Rabi (Nobel Prize: 1944), who brought molecular beam techniques to the United States. Stern became research professor of physics at the Carnegie Institute of Technology in Pittsburgh in 1933 (and an American citizen soon after), but it was never an entirely happy relationship, and he retired in 1945 to Berkeley, California, where he had friends in the physics community, but no opportunity to work in the laboratory or with students. Stern received honorary degrees from ETH and from the University of California, Berkeley. He never married.

Virginia Trimble

Selected References

Bederson, Benjamin (2003). "The Physical Tourist." *Physics in Perspective* 5: 87–121.

Rabi, Isidor I. (1969). "Otto Stern, Co-discoverer of Space Quantization, Dies at 81." *Physics Today* 22, no. 10: 103–105.

Sternberg [Shternberg], Pavel Karlovich

Born **Orel, Russia, 2 April 1865**
Died **Moscow, (Russia), 1 February 1920**

Pavel Sternberg's career featured diverse scientific, educational, and political accomplishments. The son of a tradesman, Sternberg graduated from the Orel Gymnasium in 1883 and then entered Moscow University in the Faculty of Physics and Mathematics. He assisted **Fedor Bredikhin** in cometary research at the university's observatory. For his investigation of the longevity of Jupiter's Great Red Spot, Sternberg was awarded a gold medal in the year of his graduation (1887).

Sternberg subsequently prepared for an academic career. He earned a master's degree (1903) with a study on polar motion and its influence upon the observatory's latitude. Sternberg was also awarded a doctoral degree (1913) with a thesis on the theory and practice of photographic astrometry, derived from work with the observatory's 15-in. double astrograph. His other principal interest concerned gravimetry, or the precise measurement of the Earth's gravitational field, and its practical applications for subsurface prospecting. Sternberg taught physics at the Kreimer Gymnasium (1887–) and became a *Privatdozent* (lecturer) at the University of Moscow (1890–). He supported the education of women students at a time when tsarist law forbade the practice. For a brief time (1916/1917), Sternberg succeeded **Vitol'd Tserasky** as director of the Moscow University Observatory.

Sternberg's prerevolutionary activities in the Bolshevik's favor permitted his name to become widely known in political circles. After the October Revolution, Sternberg held some influential positions in

Moscow, one of which was within the People's Commissariat for Education. He was further drawn into the Civil War and was appointed as the Red Army's Commissar for the east front in Siberia. While there, Sternberg fell through the ice of the Irtysh River, caught a heavy cold, and died soon after returning to Moscow.

Although his scientific accomplishments were modest, Sternberg's memory was kept alive through later unification of three small Moscow astronomical bodies into a single organization under the aegis of Moscow University (1931). This new research establishment was named the Sternberg State Astronomical Institute [GAISh]. Many revolutionary names were erased in Russia after the 1991 collapse of the former Soviet Union, but the Astronomical Institute of Moscow University continues to exist as the Sternberg Institute. In accordance with a Soviet proposal, Sternberg's name was applied to a feature on the Moon's farside.

Alexander A. Gurshtein

Selected References

Kulikovsky, P. G. (1965). *Pavel Karlovich Sternberg*. Moscow: Nauka.

——— (1976). "Sternberg, Pavel Karlovich." In *Dictionary of Scientific Biography*, edited by Charles Coulston Gillispie. Vol. 13, pp. 45–46. New York: Charles Scribner's Sons.

Stetson, Harlan True

Born **Haverhill, Massachusetts, USA, 28 June 1885**
Died **Fort Lauderdale, Florida, USA, 14 October 1964**

American observer and popularizer of astronomy Harlan Stetson was particularly interested in tracking the effects of solar activity on propagation of radio waves in the Earth's upper atmosphere. The son of Henry Allen Stetson and Jennie Sarah Rowe, he studied at Brown University (B.A.: 1908), Dartmouth College (Sc.M.: 1910), and the University of Chicago (Ph.D. in astrophysics: 1915). Stetson married Florence May Brigham on 4 September 1912; they had three children, Helen, Florence, and Harold. During his graduate training, Stetson was an assistant at several schools, as well as an instructor at Northwestern University and Dearborn Observatory in 1913/1914 and in 1916. His observational training took place at Yerkes Observatory.

In 1916, Stetson moved to Harvard University as an instructor in astronomy, being promoted to assistant professor in 1920. During his stay there, he published *A Manual of Laboratory Astronomy* (1928). Stetson left Harvard in 1929 to become professor of astronomy at Ohio Wesleyan University and director of the new Perkins Observatory. The following year, Stetson was also appointed an instructor at Ohio State University. His formal academic career ended in 1934 when he resigned, though Stetson retained a link to universities as a research associate (Harvard: 1933—1936; and Massachusetts Institute of Technology: 1936–1949).

In 1940, Stetson established his own Laboratory for Cosmic-Terrestrial Research in Needham, Massachusetts, and operated it for the next decade. He was a fellow of the American Association for the Advancement of Science and a member of the American Academy of Arts and Sciences and the Royal Astronomical Society.

As an instrumentalist, Stetson designed a photometer that could measure stellar magnitudes from photographic plates. He was responsible for the initial operation and testing of the 69-in. reflector, then one of the world's largest telescopes, at the Perkins Observatory. Stetson's most sustained interest was solar astronomy, and he participated in six solar eclipse expeditions. Solar-terrestrial connections became a specialty, leading to the publication of *Sunspots and Their Effect* (1937) and *Sunspots in Action* (1947). His interest in radio led to research into solar effects upon radio reception, resulting in *Earth, Radio and the Stars* (1934).

Stetson could be a quick, original thinker. For instance, his suggestion (quoted in the *Los Angeles Times* dated 1 November 1930) that ice ages might be caused by the Solar System passing into interstellar dust clouds dense enough to block some sunlight was made very shortly after the 1930 discovery of interstellar absorption by **Robert Trumpler**.

Stetson was a popular writer and speaker. His *Man and the Stars* (1930) was widely read. Stetson inaugurated the quarterly magazine *The Telescope* at Perkins Observatory in 1931, transferring it to Harvard in 1934 when he left Ohio. The magazine was combined with the younger *The Sky* by **Charles Federer** in 1941 to found *Sky & Telescope*.

Richard A. Jarrell

Selected References

Anon. (1964). "Harlan True Stetson." *Sky & Telescope* 28, no. 6: 340.

Anon. (1968). "Stetson, Harlan True." In *Who Was Who in America*. Vol. 4, p. 903. Chicago: Marquis Who's Who.

Stevin, Simon

Born **Bruges, (Belgium), 1548**
Died **The Hague, The Netherlands, March–April 1620**

Textbook author Simon Stevin was born in Bruges (in what is now Belgium) in 1548, the illegitimate child of Anthuenis Stevin and Catelyne vander Poort. Very little is known of Stevin's youth and education. His first job was in Antwerp as a bookkeeper and cashier in one of the city's trading houses, where he became acquainted with business practice and methods. In 1577, Stevin accepted a post with the financial administration of the Brugse Vrije, the region around the city of Brugge. A few years later we find him registered in Leiden, in the present-day Netherlands. Exactly why he immigrated to the North is not known; perhaps he disliked the Spanish oppression of the southern part of the Low Countries, or he may have had Protestant sympathies. In 1583, Stevin's name appears on the roll of the newly founded University of Leiden, where the young Prince Maurits of Orange was attending courses. From 1590 onward, Stevin worked mainly in the service of Prince Maurits.

In about 1614, at the age of 66, Stevin married the much younger Catharina Cray. They had four children: Frederic, Hendrik,

Susanna, and Levina. The second son, Hendrik, published some of his father's works posthumously.

Stevin was the first to produce a complete description of decimal fractions and the operations that can be carried out with them in a pamphlet entitled *De Thiende* (The disme, 1585), in which he also dealt with their practical applications in surveying, the measurement of weights, and the subdivision of money. The Scottish mathematician and theological writer **John Napier** also drew on Stevin's work in his invention of logarithms.

In his works on physics, Stevin was again a fount of new and innovative ideas. In *De Beghinselen des Waterwichts* (The elements of hydrostatics, 1586) Stevin gave an improved demonstration of **Archimedes**' law about the upward force acting on a body immersed in a liquid. He also succeeded in calculating the force exerted by a fluid on the bottom and walls of the vessel in which it is contained. And this led him to formulate the so called hydrostatic paradox many years before Blaise Pascal, to whom it is usually attributed.

In 1586 Stevin published his experiment in which two spheres of lead, one 10 times as heavy as the other, were dropped from a tower in Delft, fell 30 ft. and reached the ground at the same time. Stevin's report preceded **Galileo Galilei**'s first treatise on gravity by 3 years and his theoretical work on falling bodies by 18 years.

Between 1605 and 1608, the textbooks he had produced for Prince Maurits in numerous sciences (algebra, geography, astronomy, bookkeeping, statics and hydrostatics, perspective, *etc.*) were collected and published under the title *Wisconstighe Gedachtenissen* (Mathematical memoirs). Stevin supported **Nicolaus Copernicus**' heliocentric theory in *De Hemeloop* (1608), in which he showed planetary motions in both the Ptolemaic and Copernican systems. He also described how to determine the location of a place on the Earth's surface by knowing its geographical latitude and the magnetic variation of the compass needle. This method proved extremely valuable to the ships of the Dutch East India Company.

Jozef T. Devreese and *Guido Vanden Berghe*

Selected References

Dijksterhuis, E. J. (1970). *Simon Steven: Science in the Netherlands around 1600.* The Hague: Martinus Nijhoff.

Dijksterhuis, E. J. *et al.* (eds.) (1955–1966). *The Principal Works of Simon Stevin.* Amsterdam: C. V. Swets and Zeitlinger.

Halleux, Robert (ed.) (1998). *Histoire des sciences en Belgique de l'Antiquité à 1815.* Brussels: Crédit Communal.

Struik, Dirk J. (1981). *The Land of Stevin and Huygens: A Sketch of Science and Technology in the Dutch Republic during the Golden Century.* Dordrecht: D. Reidel.

Stewart, Balfour

Born **Edinburgh, Scotland, 1 November 1828**
Died **Ballymagarvey, Co. Meath, Ireland, 18 December 1887**

Balfour Stewart made pioneering contributions to the study of radiant heat, investigated solar-terrestrial relationships, and constructed a physical model of the solar cycle. The son of a tea merchant, Stewart was first educated in Dundee, and at the early age of 13, entered Saint Andrew's University. He subsequently transferred to the University of Edinburgh, where he studied natural philosophy under James David Forbes. Under parental pressure, he left the university at age 18 to undertake a business apprenticeship in Leith, which was followed by a commercial venture in Australia.

Having finally decided to opt for physics over business, Stewart returned to Britain, first for a brief stay at Kew Observatory, and, starting in 1853, as an assistant to Forbes at the University of Edinburgh. In 1859, he left Edinburgh to take on the directorship of Kew Observatory. It was during his Kew years that Stewart married Katharine Stevens, daughter of a London lawyer. He was elected to the Royal Society of London in 1862, elected to the Royal Astronomical Society in 1867, and was awarded the Royal Society's Rumford Medal in 1868. By 1870, growing tensions with the Royal Society regarding research priorities at Kew led to his resignation and acceptance of the physics professorship at Owens College in Manchester. That same year, Stewart was caught in a railway accident, suffering severe injuries from which he never fully recovered. He died while spending the Christmas holiday.

Stewart was an experimental physicist, and his most noteworthy scientific contributions were in the areas of thermodynamics. His work on this subject culminated in 1858, when Stewart read to the Royal Society of Edinburgh a groundbreaking paper putting forth a "theory of exchanges," according to which bodies radiating heat (*i. e.*, infrared radiation) at certain wavelengths also tend to absorb preferentially at those same wavelengths. Under the assumption that radiation is more than a mere surface phenomenon, but in fact pervades the interior of bodies, Stewart consistently explained his experimental data on the emissive and absorptive powers of thin plates. This was a striking anticipation of the radiation laws independently discovered a few years afterward by German physicist **Gustav Kirchhoff**. However, the greater generality and mathematical rigor of Kirchhoff's formulation largely eclipsed Stewart's earlier contribution.

Stewart was fascinated by the possibility of causal relationships between periodic phenomena in the Sun and Earth. Throughout his life, he carried out a number of investigations that sought to link meteorological phenomena to terrestrial magnetism. It was in this overall context that he made his most important contribution to astronomy, namely the development of a sunspot cycle model based on planetary influences. Stewart's approach was largely empirical; the underlying physical hypothesis was that planetary gravitational perturbations to the solar photosphere, however minute, could perturb the presumably delicate dynamical equilibrium of the solar atmosphere and trigger the formation of sunspots. Between 1864 and 1873, using data from the newly commissioned Kew photoheliograph and working in collaboration with the instrument's designer, **Warren de la Rue**, Stewart discovered a number of apparent correlations between planetary longitudes and the occurrences of sunspot groups. While the whole idea was eventually refuted as more data accumulated and the inferred correlations failed to persist, Stewart's model represented the first quantitative explanation of the sunspot cycle presented during the second half of the 19th century. His ideas were carried further by the British solar astronomer **Annie Maunder**. They have been revived from time to time down to the present, generally without credit to him.

Later in life, Stewart became interested in the scientific study of psychic phenomena, presiding for a time over the Society for Psychical Research. Presuming self-proclaimed "psychics" honest until proven otherwise, Stewart is said to have been repeatedly fooled by assorted illusionists and charlatans. His contemporary attempts at demonstrating, on physical grounds, the immortality of the soul, and the lack of any fundamental incompatibility between science and religion, were met with mixed reviews. But as professor of physics in Manchester between 1870 and 1887, Stewart educated, and with obvious success, a generation of scientific luminaries including Sir Joseph Thomson, **John Poynting**, and Sir **Arthur Schuster**.

Stewart wrote many textbooks on elementary physics that remained popular for many years. His article on terrestrial magnetism in the ninth edition of the *Encyclopædia Britannica* was extremely influential. His efforts at reconciling scientific and religious ideas were laid out in his 1875 book, *The Unseen Universe*, coauthored with Peter Guthrie Tait and first published anonymously.

Paul Charbonneau

Selected References

Charbonneau, Paul (2002). "The Rise and Fall of the First Solar Cycle Model." *Journal for the History of Astronomy* 33: 351–372.

Schuster, Arthur (1888). "Balfour Stewart." *Monthly Notices of the Royal Astronomical Society* 48: 166–168.

------ (1888). "Memoir of the Late Professor Balfour Stewart, LL. D., F.R.S." *Memoirs and Proceedings of the Manchester Literary and Philosophical Society*, 4th ser., 1: 253–272.

Siegel, Daniel M. (1976). "Balfour Stewart and Gustav Kirchhoff: Two Independent Approaches to 'Kirchhoff's Radiation Law.'" *Isis* 67: 565–600.

------ (1976). "Stewart, Balfour." In *Dictionary of Scientific Biography*, edited by Charles Coulston Gillispie. Vol. 13, pp. 51–53. New York: Charles Scribner's Sons.

Stewart, John Quincy

Born **Harrisburg, Pennsylvania, USA, 10 September 1894**
Died **Sedona, Arizona, USA, 19 March 1972**

John Stewart was in succession an accomplished engineer, astronomer, textbook author, and advocate of "social physics." He was the son of John Quincy and Mary Caroline (*née* Liebendorfer) Stewart. A graduate of Princeton University (1915), Stewart earned a doctorate in physics from his *alma mater* (1919), after a wartime interruption during which he served with the 29th Engineers in France. Upon completing his studies, he was employed from 1919 to 1921 as an engineer in the department of research and development at the American Telephone and Telegraph Company, New York. His investigations were in the area of speech and hearing, and Stewart is credited with designing the first electronically synthesized "voice."

From 1921 to 1963, Stewart was associated with Princeton University's astronomy department, where he attained the rank of associate professor in 1927. He is probably best remembered among the astronomical community as a coauthor, with **Henry Norris Russell** and **Raymond Dugan**, of the widely used two-volume textbook, *Astronomy*, itself a revision of former Princeton University astronomer **Charles Young**'s *Manual of Astronomy*. Stewart married Lillian Westcott on 17 June 1925. Their son, John Westcott Stewart, followed his father into academia, spending his career as a physicist and administrator at the University of Virginia, Charlottesville.

A highlight of the senior Stewart's astronomical career was the successful observation of a total solar eclipse from the deck of the *S. S. Steelmaker* in the Pacific Ocean during more than 7 minutes of totality on 8 June 1937. This was one of five solar eclipses at which he was present. Stewart's other interests extended to radar observations of meteors, navigation, astrophysics, meteorology, and social physics. Stewart's belief that the laws of physics should have applicability to the social sciences was incorporated into his textbook, *Demographic Gravitation: Evidence and Applications* (1948), which introduced the concept of "potentials of population." He was also the author of two works on navigation.

On Stewart's retirement from Princeton University in 1963, he moved to Sedona, Arizona. Three years later, he was appointed professor of the metaphysics of science at Prescott (Arizona) College, a post he held until just before his death.

Stewart was a fellow of the American Physical Society and the American Association for the Advancement of Science, as well as an honorary fellow of the American Geographical Society. As a member of the American Association of University Professors, he served as its national vice president (1940/1941). Stewart also belonged to the American Astronomical Society, Phi Beta Kappa, and Sigma Xi.

George S. Mumford

Selected References

Anon. (1973). "Stewart, John Quincy." In *Who Was Who in America*. Vol. 5, p. 695. Chicago: Marquis Who's Who.

Russell, Henry Norris, Raymond Smith Dugan, and John Q. Stewart (1926, 1927). *Astronomy*. Vols. 1 and 2. Boston: Ginn.

Stewart, John W. (1972). "John Q. Stewart." *Physics Today* 25, no. 6: 75.

Stewart, Matthew

Born **Rothesay, (Strathclyde), Scotland, January 1717**
Died **Catrine, (Strathclyde), Scotland, 23 January 1785**

Matthew Stewart is remembered primarily for an attempt to deduce the Sun's distance by purely geometrical means. Son of Reverend Dugald Stewart, minister of the parish of Rothesay, and Janet Bannatyne, Stewart received his early education on the Scottish Isle of Bute, then entered the University of Glasgow in 1734, intending to follow his father's wishes by pursuing an ecclesiastical career.

At Glasgow, Stewart turned to mathematics while studying with Robert Simson, with whom he developed a lifelong friendship. Simson's field of study was ancient geometry, specifically an attempt to reconstruct both **Apollonius**'s *Loci Plani* and Euclid's lost three-volume work on porisms. (A porism is essentially a geometrical proposition intermediate between a theorem and a problem; such a proposition, depending on the starting point, is either impossible or possible in an infinite number of ways.) In 1741, Stewart left Glasgow to continue his mathematical education at the University of Edinburgh under one of Simson's former students, **Colin Maclaurin**. Here Stewart deepened his expertise in more modern realms of mathematics, such as calculus and analytic geometry, including astronomical applications. At the same time, Simson periodically communicated his own progress in ancient mathematics to Stewart, who refined these studies on his own.

In 1746, while serving as minister of the parish at Roseneath, Dunbartonshire, Stewart published his breakthrough work, *Some General Theorems of Considerable Use in the Higher Parts of Mathematics*. Some of the work undoubtedly arose from Simson's continuing correspondence with Stewart, but was published with Simson's approval. *General Theorems* established Stewart's reputation in the mathematical community and led to his appointment as professor of mathematics at Edinburgh in September 1747, following Maclaurin's death.

In 1756, Stewart published an essay on a geometric analysis of **Johannes Kepler**'s second law (equal areas). His second book, *Tracts, Physical and Mathematical*, appeared in 1761. Here Stewart analyzed by purely geometrical means the motions of planets, including the perturbations of one planet on another; he further established a geometrical technique to approximate the Sun's distance by considering the observed mean angular motion of the apogee of the Moon's orbit.

The year 1763 brought two further publications – another volume of geometrical propositions plus Stewart's result for the solar distance. That result – 29,875 Earth-radii, or about 119 million miles (191 million km – was much larger than previous determinations and proved controversial. An anonymous pamphlet entitled *Four Propositions* appeared, disputing Stewart's solar distance largely on the basis of the simplifying assumptions he had made. (The pamphlet's author, John Dawson, a surgeon from Sudbury, Yorkshire, England, came forward after Stewart's death.) A harsher attack by John Landen followed in 1771.

His health in decline, Stewart retreated to his estate at Catrine in 1772. Stewart's son, Dugald, carried out his father's duties at the university and, in 1775, was elected to a joint professorship with him.

Stewart was elected a fellow of the Royal Society in 1764. Correspondence between Robert Simson and Matthew Stewart is archived at the University of Glasgow and was published in *The Proceedings of the Edinburgh Mathematical Society* XXI (1902–1903): 1–38.

Alan W. Hirshfeld

Selected References

Playfair, John (1788). "Account of Matthew Stewart, D. D." *Transactions of the Royal Society of Edinburgh* 1, pt. 1: 57–76.

Sneddon, Ian N. (1976). "Stewart, Matthew." In *Dictionary of Scientific Biography*, edited by Charles Coulston Gillispie. Vol. 13, pp. 54–55. New York: Charles Scribner's Sons.

Stöffler, Johannes

Born **Justingen, (Baden-Württemberg, Germany), 10 December 1452**

Died **Blaubeuren, (Baden-Württemberg, Germany), 16 February 1531**

Johannes Stöffler was a German mathematician, geographer, and astronomer. He was professor of mathematics at the University of Tübingen, was canon of the cathedral there, was a teacher

(1512–1514) of Philipp Melanchthon, and played a role in the effort to reconcile the Church and astronomical calendars.

Stöffler wrote numerous works, many of which were printed. His *Almanach nova plurimis annis venturis inservientia*, a calendaric work written with the astronomer Jacob Pflaum, was printed at Ulm in 1499 (and reprinted frequently, notably in Venice in 1502 and 1522). Previously Stöffler published *Johannis de Monteregio commentum in Ephemerides*. The *Almanach nova* began the debate and panic over predictions of a universal flood caused by the Great Conjunctions of all the then known planets as well as the Sun in February 1524 in the astrological sign of Pisces. This debate produced over 160 pamphlets in the 5 years before the "fated" 1524. He wrote against such catastrophic predictions, which were often attributed to him after the publication of the *Almanach nova*, in his *Expurgatio adversus divinationum XXIIII suspitiones* printed at Tübingen by U. Morhard in 1523. Stöffler pointed out in his *Expurgatio* that he had always criticized in his astronomy classes the vain and frivolous predictions, which had no scientific foundation, made by some of his contemporary astronomers. He denied that the Great Conjunction in Pisces of 1524 would signify the end of the world or a Universal Flood, neither of which is claimed in the *Almanach nova*.

Invited by Pope Leo X in June 1515 to take part in a project to reform the calendar, Stöffler published *Calendarium romanum magnum caesareae maiestati dicatum*, printed by Jacob Koebel in 1518 at Oppenheim. This was later issued in German as *Der newe gross Roemische Calendar* (Oppenheim, Jacob Koebel, 1531). As a necessary preparation for this calendar reform, he published a number of ephemerides. Stöffler maintained, as did Paul of Middenburg, that the spring equinox should not be a fixed date, as was advocated by George Tanstetter and Andreas Stiborius – with an adjustment of 1 day every 134 years. Instead, using the meridian of Tübingen and the *Alphonsine Tables*, and following the practice of the Church fathers, he suggested a variable equinoctial date for which up-to-date ephemerides were indispensable. Hence, Stöffler published *Opus ephemeridum a capite anni Christi 1532 in alios viginti proximos sequentes ad veterum imitationem accuratissimo calculo elaboratum et excussum* (Tübingen, 1513), and *Tabulae astronomicae impressae Tubinge apud Thomam Anshelmum, anno domini 1514.*

Stöffler also published an important tract on the use of the astrolabe, which was successful enough to merit two different translations into Italian in the 16th century – one by Giustiniano Veneto, *Giovanni Stoflerino regola e modi di usare l'astrolabio* (Florence, Biblioteca Nazionale, MS fondo Magliabechi 130), the second by M. A. Gandino in 1563 (Pommersfelden, Staatsbibliothek MS 153 (2713)), published as *Elucidato Fabricae ususque Astrolabii... iam denua ab eodem vix aestimandis sudoribus recognita*, Oppenheim, Iacobo Koebel, 1524.

Stöffler was renowned as a geographer, editing a commentary on the *Sphaera*, which at that time was wrongly attributed to **Proclus**. It was published, according to Gesner, in Tübingen, in 1534.

Stöffler further edited various commentaries on the *Cosmographia* of **Ptolemy**, one published at Tübingen during 1512–1514 – *Commentaria in geographiae Ptolomei libros duos.* The other, entitled *De terrarum orbe habitato in sphaerico artificiato describendo, proiectio prima terrae habitabilis in planum, descriptio orbis habitati per meridianos rectos, tabula revolutionum planetarum de radicum extractione, tabulae astronomicae,* is known only in manuscript (Heidelberg Universitaetsbibliothek MS 234, saec. xvi). Also in manuscript are other ephemerides, *Calendario De Inventione sex solemnitatum Hebreorum* (Tübingen Universitaetsbibliothek, MS 65), and he prepared a *Tractatus trium stellarum* (Munich Universitaetsbibliothek MS n.588, saec xvi).

Graziella Vescovini
Translated by: *Lorenzo Smerillo*

Alternate name

Stoeflerus

Selected References

Baldi, Bernardino. *Vite di matematici.* Biblioteca Vaticana, codex Albani 619. (Boncompagni I, cart 156, folio 244r–246v.)

Brind'Amour, Pierre (1993). *Nostradamus astrophile.* Ottawa: Presses de l'Université, p. 203. (French translation of the *Expurgatio.*)

Gesner, Conradus (1543). *Biblioteca universalis.* Zürich, p. 456.

Grafton, Anthony (1999). *Cardano's Cosmos.* Cambridge, Massachusetts: Harvard University Press. (For a translation of the *Expurgatio*; Fig. 7, p. 28 [reproduction of the ephemerides table], and Fig. 10, pp. 32–33 [astrological aspects of the planets and the Sun and the "prediction" which Grafton translates on p. 53 with bibliography n. 54.])

Hellmann, G. (1914). "Aus der Blützeit der Astrometeorologie: Johannes Stöffler, Prognose für der Jahre 1524." In *Beiträge zur Geschichte der Meteorologie* (Contributions to the history of meteorology). Vol. 1, pp. 25–67. Berlin: Behrend.

North, J. (1983). "The Western Calendar: 'Intolerabilis, horribilis et derisibilis': Four Centuries of Discontent." In *Gregorian Reform of the Calendar,* edited by G .V. Coyne, M. A. Hoskin, and O. Pedersen, pp. 75–111. Vatican City: Pontificia Academia Scientiarum, Specola Vaticana.

Oestmann, G. (1993). *Schicksalsdeutung und Astronomie: Der Himmelsglobus des Johannes Stöffler von 1493* (Meaning of destiny and astronomy: The celestial globe of Johannes Stöffler of 1493). With contributions by Elly Dekker and Peter Schiller. Stuttgart: Wuerttembergisches Landesmuseum. (Exhibit catalog.)

Oestmann, G. and Thomas Grunerz (1995). "Johannes Stöffler's Celestial Globe." *Der Globusfreund,* no. 43–44: 59–76. (Internationale Coronelli-Gesellschaft für Globen- und Instrumentenkunde Symposium, 8th, held in Prague in 1994; in English and German).

Thorndike, Lynn (1923–1958). *A History of Magic and Experimental Science.* 8 Vols. New York: Columbia University Press.

Zambelli, P. (1982). "Fine del mondo o inizio della propaganda." In *Scienze, credenze occulte, livelli di cultura.* Florence: L. S. Olschki, pp. 293–294, 310, 311.

Stoeflerus

➔ **Stöffler, Johannes**

Stoiko-Radilenko, Nicolas

➔ **Stoyko, Nicolas**

Stokes, George Gabriel

Born **Skreen, Co. Sligo, Ireland, 13 August 1819**
Died **Cambridge, England, 1 February 1903**

George Stokes was one of the leading figures of 19th-century physics and is chiefly remembered for his theoretical work, especially in hydrodynamics. His name is attached to several physical laws – the

Navier–Stokes equation governing fluid motion; Stokes's law of viscosity, relating the resistance experienced by a body moving in a fluid to viscosity; Stokes's law of fluorescence, which states that the wavelength of the light absorbed by fluorescent materials is always shorter than that emitted; Stokes's (curl) theorem, which applies to fluid dynamics and electromagnetic theory; and (better known to astronomers) the Stokes parameters of polarization for radiation.

Born to an Anglo–Irish family that included many academics and ministers of religion, Stokes obtained his early education at the Reverend R. H. Wall's school in Dublin. He moved to Bristol College, England, at the age of 16. Stokes then entered Pembroke College, Cambridge, in 1837 and graduated first in his class (senior wrangler) in mathematics (1841). Afterward, he became a fellow of the college and was appointed to the Lucasian Chair of Mathematics (1849), the post once occupied by Sir **Isaac Newton**. From his Cambridge days, Stokes became a close scientific colleague of **William Thomson** (Lord Kelvin).

Stokes was made a fellow of the Royal Society of London (1851) and was awarded its Rumford Medal (1854) for his explanation of fluorescence. He was chosen secretary of the society from 1854 until he became its president in 1885. In 1857, Stokes married Mary Susanna Robinson, daughter of the astronomer **Romney Robinson** of Armagh Observatory. The couple had three children. From 1887 to 1891, Stokes served as a Member of Parliament for Cambridge University. He received many medals and academic honors during his career and was made a baronet in 1889.

Stokes sought to explain many natural phenomena involving light waves, was deeply interested in contemporary astrophysical discoveries, and furthered the development of astronomical instrumentation. He discussed the optimization of achromatic lenses and arrived at an explanation of the criterion that **Joseph von Fraunhofer** had employed when designing telescope objectives. In 1852, Stokes employed fluorescence, one of his particular interests, as a means for detecting the ultraviolet spectrum of the Sun. He focused the solar spectrum onto a solution of quinine sulphate, which emitted a blue fluorescence except at the positions of particular absorption lines. Stokes anticipated to some extent **Gustav Kirchhoff**'s discoveries concerning the Fraunhofer lines in the solar spectrum. Yet, because he never published his conclusions on that subject, he refused to claim any credit. In 1852, he introduced a set of four quantities now termed the Stokes parameters, which are widely used to characterize the polarization state of a light wave.

Stokes's prominence in the British physics community led to invitations to serve on many committees. Those of relevance to astronomy included the Committee on Solar Physics and the Board of Visitors of the Royal Greenwich Observatory. He acted as an advisor to the telescope maker Sir **Howard Grubb** and was involved with the unsuccessful British efforts under W. V. Vernon Harcourt to improve the manufacturing of optical glass. As secretary of the Royal Society, he was closely involved in procuring a large telescope for the pioneer astrophysicist Sir **William Huggins**. Stokes was a member of the local committee set up to supervise the manufacture by Grubb of a 27-in. telescope for the Imperial Observatory of Vienna and specified the curvatures for its objective lens.

Ian S. Glass

Selected References

Kelvin, Lord (1903). "The Scientific Work of Sir George Stokes." *Nature* 67: 337–338.

Larmor, Joseph (1907). *Memoir and Scientific Correspondence of the Late Sir George Gabriel Stokes*. 2 Vols. Cambridge: Cambridge University Press.

——— (1920). "Stokes, Sir George Gabriel." In *Dictionary of National Biography, Supplement, January 1901–December 1911*, edited by Sir Sidney Lee, pp. 421–424. London: Oxford University Press.

Parkinson, E. M. (1976). "Stokes, George Gabriel." In *Dictionary of Scientific Biography*, edited by Charles Coulston Gillispie. Vol. 14, pp. 74–79. New York: Charles Scribner's Sons.

Rayleigh, Lord (1905). "Sir George Gabriel Stokes, Bart. 1819–1903." *Proceedings of the Royal Society of London* 75: 199–216. (Reprinted in Rayleigh, *Scientific Papers*. New York: Dover, Vol. 5, pp. 173–189.)

Stokley, James

Born **Philadelphia, Pennsylvania, USA, 19 May 1900**
Died **La Jolla, California, USA, 29 December 1989**

American planetarium pioneer James Stokley, son of James and Irene (*née* Stulb) Stokley, received a bachelor's degree (1922) in education and a master's degree (1924) in psychology from the University of Pennsylvania. After graduating, he taught biology and physics at Philadelphia's Central High School and wrote articles for local newspapers. An opportunity to cover the Centenary Dinner of the Franklin Institute in 1924 as a news reporter proved a turning point in his life. At the dinner, Stokley made contacts that led to his employment by Science Service in Washington, District of Columbia, the following year. In 1927, Stokley visited planetariums in Berlin and Jena, Germany, returning with the conviction that he would one day become a planetarium director. Although he had risen to the position of astronomical editor with Science Service in 1931, his earlier ambition was realized when he was appointed director of the Franklin Institute's Fels Planetarium, the second Zeiss-equipped facility to be opened in the United States.

Stokley's programs at the Fels Planetarium at first emulated the cycle of topics devised by **Philip Fox** and **Maude Bennot** at Chicago's Adler Planetarium. However, Stokley's background as a science journalist gave him a sharper appreciation of audience tastes than research astronomers possessed. He was less inhibited about trying new and unconventional topics that were greeted with skepticism by scientific colleagues.

Within a few years, Stokley's original programs had ranked him as the most audacious of astronomical showmen. The techniques he pioneered later became widely adopted among other major American planetariums. For example, astronomers had adopted a theory of planetary conjunctions as the most likely explanation for the Star of Bethlehem. In 1933, Stokley developed a program entitled, "Skies of the First Christmas." Using the Zeiss projector, the succession of conjunctions between the planets Jupiter, Saturn, and Mars, during the period 7–6 BCE, could be accurately reproduced. Scriptural readings, recorded music, and lighting effects, including a crèche scene, were also employed in the program. Stokley's Christmas Star program proved remarkably successful, not only for audiences who witnessed his performances, but among later generations who were exposed to similar presentations at planetariums elsewhere. The Christmas Star became the most widely presented astronomical topic in planetariums, large or small.

In 1936, Stokley presented a program, "How Will the World End?" that earned him a reputation as the greatest showman in the field. After the Christmas Star, no other subject captured such media attention, was so widely copied by other planetariums, or drew such criticism from contemporary astronomers. Either intense heat (from a sudden flare-up) or freezing temperatures (following depletion of its nuclear fuel) were considered possible consequences when the Sun reached the end of its normal evolution. Some astronomers had predicted (incorrectly, it turned out) the possible disruption of our Moon, when tidal friction drew this satellite closer to the Earth. Finally, the remote chance of a collision, between an asteroid or comet and the Earth, was another scenario depicted by Stokely's "end of the world" program. His shows presaged motion picture adaptations of similar themes in the decades that followed.

Stokley then adapted a topic that originated in a 1938 program developed at New York's Hayden Planetarium. This was an imaginary "Trip to the Moon." He prepared visual effects that transformed the planetarium chamber into a space ship. For this, he sought the aid of Dick Calkins, creator of the *Buck Rogers* comic strip, who, it is said, personally designed the control panel that audiences saw on the projected navigator's bridge. While admitting that the rocket trip was "pure fantasy," Stokley defended its educational value on the basis of "absolute scientific knowledge." Through highly creative uses of audiovisual resources, Stokley demonstrated that planetariums could significantly aid popular understanding of astronomy, space travel, and many other scientific wonders.

The 1936 conference of the American Association of Museums gave Stokley an opportunity to report on the economic circumstances under which America's four Zeiss planetariums were administered. It marked the first occasion on which comparative accounts of the incomes, expenditures, and attendances at those facilities were openly discussed. This address highlighted Stokley's prominence within the American planetarium community, along with growing interest among museum professionals in planetariums as public educational institutions.

A close professional relationship developed between Stokley and Hayden Planetarium director, G. Clyde Fisher. The Philadelphia and New York planetariums functioned effectively as a dyad, with frequent exchanges made of program materials and guest lecture appearances. By comparison, larger personal as well as geographic distances separated the directors of the Chicago and Los Angeles planetariums (both professional astronomers) from their eastern colleagues and prevented the formation of a viable professional association.

During this period, Stokley played other roles in the popularization of astronomy. He was invited to witness the pouring of the first 200-in. mirror blank for the Palomar Mountain telescope. In 1935, he had the honor of presenting the first planetarium demonstration ever witnessed by physicist **Albert Einstein**. Stokley observed five total solar eclipses, including that of 8 June 1937 from a cruise ship in the Pacific Ocean.

Stokley served as consultant to, and was later chosen director of, Pittsburgh's Buhl Planetarium and Institute of Popular Science, which opened in 1939. His appointment there was short-lived, however, as he resigned the following year, and withdrew altogether from the planetarium community to became chief publicist for General Electric in Schenectady, New York. He remained in that position from 1941 to 1956. Stokley then joined the faculty of Michigan State University in 1956 as an associate professor of journalism and astronomy, serving in that capacity until his retirement in 1969.

Stokley wrote seven books and many papers in both popular and technical journals. Although his career in planetariums lasted but one decade, he was never far removed from astronomy, and in 1961 the second of his two books on that topic was published. In 1949, Stokley received an honorary Sc.D. degree from Wagner College, Staten Island, New York.

In 1933, Stokley married Susan A. Doughton. The couple had two children, a daughter Marcia and son Donald.

Jordan D. Marché, II

Selected References

Anon. (Dec. 1936). "Who's Who in the Franklin Institute: James Stokley." *Institute News*: 4.

Marché II, Jordan D. (2005). *Theaters of Time and Space: American Planetaria, 1930–1970*. New Brunswick, New Jersey: Rutgers University Press.

Menke, David H. (1987). "James Stokley and the Fels Planetarium." *Planetarian* 16, no. 3: 39–43.

Stokley, James (1936). *Stars and Telescopes*. New York: Harper and Brothers.

——— (1937). "Planetarium Operation." *Scientific Monthly* 45: 307–316.

——— (1961). *Atoms to Galaxies: An Introduction to Modern Astronomy*. New York: Ronald Press.

Stone, Edward James

Born **London, England, 28 February 1831**
Died **Oxford, England, 9 May 1897**

Edward Stone served as Her Majesty's Astronomer at the Royal Observatory, Cape of Good Hope (1870–1879), where he compiled data on more than 12,000 stars for the *Cape Catalogue*. Stone was the son of Edward Stone, a London businessman. Educated at home until the age of twenty, he first attended King's College, London, but then transferred to Queen's College, Cambridge, from which he graduated as fifth wrangler in 1859. Stone was subsequently offered a fellowship from the college.

In 1860, Stone succeeded Reverend Robert Main as chief assistant at the Royal Greenwich Observatory. Over the coming decade, he accomplished a number of duties under director **George Airy**. One of his principal investigations concerned a reanalysis of data from the 1769 transit of Venus, with a view of deriving the most accurate value of the solar parallax (giving the Earth's true distance from the Sun). For this work, Stone was awarded the Gold Medal of the Royal Astronomical Society (1869). In 1866, he married Grace Tuckett; the couple had four children.

Stone was appointed to the Royal Observatory (1870), where he succeeded **Thomas Maclear**. His goal was to prepare a catalog of the positions of all Southern Hemisphere stars down to the seventh visual magnitude from observations made with a transit circle. In addition, he undertook the reduction of extensive observations taken by Maclear. Stone had few paid assistants for this task and was occasionally helped by his wife. While at the Cape, he observed the 1874 transit of Venus, along with the first (16 April 1874) of several total solar eclipses that he would witness during his lifetime.

Having completed his observations for the star catalog in 1879, Stone returned to Oxford, England, where he was appointed the Radcliffe Observer at the University's Observatory (following the death of Main). Stone's publication of the *Cape Catalogue*, containing the positions of 12,441 stars, earned him the Lalande Prize of the Paris Académie des sciences (1881). During the remainder of his career, Stone also published the *Radcliffe Catalogue* of 6,424 northern star positions. He was responsible for coordinating the observations of the 1882 transit of Venus, and for their reduction, from which he derived a value of 8.832 arc seconds for the solar parallax. Stone also attempted, but with only limited success, to employ the technique of minor planet observations suggested by **David Gill** as an independent measurement of the solar parallax. This triangulation method had to await the discovery of the minor planet (433) Eros, which comes closer to the Earth.

Stone served as honorary secretary (1866–1870) and president (1882–1884) of the Royal Astronomical Society and was likewise elected a fellow of the Royal Society of London. He was awarded an honorary doctorate in natural philosophy from the University of Padua (1892).

Jordan D. Marché, II

Selected References

Anon. (1897). "Edward James Stone, Radcliffe Observer." *Astronomische Nachrichten* 143: 295–296.

E. D. (1898). "Edward James Stone." *Monthly Notices of the Royal Astronomical Society* 58: 143–151.

Stone, Edward James (1881). *Catalogue of 12,441 stars, for the Epoch 1880; from Observations Made at the Royal Observatory, Cape of Good Hope, during the Years 1871 to 1879*. London: G. E. Eyre and W. Spottiswoode.

——— (1894). *Catalogue of 6424 Stars for the Epoch 1890. Formed from Observations Made at the Radcliffe Observatory, Oxford, during the Years 1880–1893*. Oxford: J. Parker.

Turner, Herbert Hall (1897). "Edward James Stone." *Observatory* 20: 234–237.

Stone, Ormond

Born **Pekin, Illinois, USA, 11 January 1847**
Died **near Manassas, Virginia, USA, 17 January 1933**

Ormond Stone – educator, observatory director, and discoverer of many double stars—was the son of Elijah (a Methodist minister) and Sophia (*née* Creighton) Stone. He grew up excelling in mathematics. After his family moved to Chicago, Illinois, Stone visited the Dearborn Observatory and began a program of studies under director **Truman Safford**. Stone attended, but did not graduate from, the University of Chicago (*circa* 1867–1870); he was later awarded an A.M. degree (1875). He married Catherine Flagler in 1871 and, following her death in 1914, married Mary Florence Brennan in 1915.

Stone's teaching career began while he was still attending the University of Chicago; he served as an instructor at Racine College in Wisconsin (1867–1868) and at the Northwestern Female College at Evanston, Illinois (1869). In 1870, Stone was offered an assistantship at the United States Naval Observatory, which he held until 1875. His work drew admiration from **Simon Newcomb**, who recommended him for the position of director of the Cincinnati Observatory, which Stone held from 1875 to 1882.

Under Stone's direction, the object glass of Cincinnati's 11-in. Merz and Mahler refractor was refigured by **Alvan Clark**. Stone then observed comets and launched a successful program of discovering new southern double stars, of which some 44 of those were recorded in **Robert Aitken**'s *New General Catalogue of Double Stars* (1932). Carleton College astronomer **William Payne** gained a summer's experience in observatory practice under Stone. This opportunity may have sparked Payne's ambition to revive former Cincinnati Observatory director **Ormsby Mitchel**'s periodical, *The Sidereal Messenger*.

Stone was appointed director of the University of Virginia at Charlottesville's Leander McCormick Observatory in 1882. It housed several telescopes, the largest being a 26-in. refractor. Over the next 30 years, he carried out a visual observing program covering a wide range of topics, including the discovery and photometry of nebulae, observations of planetary satellites, variable and double star measurements, and the study of comets.

Stone was a member of several professional societies, including the American Association for the Advancement of Science, and the Washington, Virginia, and Wisconsin academies of sciences. At the University of Virginia, he taught courses in astronomy, established its Philosophical Society, and in 1884 founded the journal, *Annals of Mathematics*, that is still published under the same name. More than 30 Vanderbilt Fellows studied under Stone, including **Heber Curtis**; many went on to successful careers in science. Stone participated in three solar eclipse expeditions, in the years 1869, 1878, and 1900. He published more than 100 papers.

Stone retired from the McCormick Observatory in 1912 to his farm in Virginia, where he continued to be active in the local and state affairs. He died after being struck by an automobile.

Scott W. Teare

Selected References

Olivier, Charles P. (1933). "Ormond Stone." *Popular Astronomy* 41: 295–298.

Stone, Ormond (1879). "Micrometrical Measurements of 1054 Double Stars, Observed with the 11 Inch Refractor from January 1, 1878, to September 1, 1879." *Publications of the Cincinnati Observatory*, no. 5.

——— (1887). "Second List of Nebulas Observed at the Leander McCormick Observatory, and Supposed to be New." *Astronomical Journal* 7 (1887): 57–61.

Zund, Joseph D. (1999). "Stone, Ormand." In [sic] *American National Biography*, edited by John A. Garraty and Mark C. Carnes. Vol. 20, pp. 867–868. New York: Oxford University Press.

Stoney, George Johnstone

Born **Oakley Park, (Co. Offaly), Ireland, 15 February 1826**
Died **London, England, 5 July 1911**

George Johnstone Stoney was a mathematical physicist with a very wide range of interests. Though most of his working life was taken up with university administration, he made fundamental discoveries in physics and astronomy. He is best remembered for giving the name "electron" to the smallest possible quantity of electric charge.

Stoney was the elder son of George and Anne (*née* Blood) Stoney. The family's rural property in County Offaly greatly depreciated in value after the Napoleonic wars and had to be sold at the time of the Irish Famine (1846–1848). The family moved to Dublin where George and his brother Bindon entered Trinity College, earning their fees by tutoring other students. Both graduated with distinction, George in 1848 and Bindon in 1850. On completing his studies in physics and mathematics, Stoney became the first astronomical assistant to **William Parsons**, the Third Earl of Rosse, spending 2½ years at Parsonstown (Birr), from July 1848 to August 1850 and from August to December 1852. In addition to observing nebulae (most of which turned out to be galaxies) with the great 6-ft. reflector, Stoney served as tutor to Lord Rosse's children.

While at Parsonstown, Stoney prepared for a fellowship in Trinity College. He applied for it in 1852, taking second place and thereby winning the Madden Prize, which was worth about £300. As Stoney could not afford to try again for the fellowship, Lord Rosse used his influence to have him appointed to the Chair of Natural Philosophy at Queen's College, Galway. Stoney remained 5 years in Galway and then became secretary to Queen's University, which brought him back to Dublin in 1857.

As a university administrator, Stoney devoted himself enthusiastically to improving the effectiveness of the provincial colleges in Belfast, Cork, and Galway. It was therefore a great blow to him when the Queen's University was dissolved in 1882 and its place was taken by the Royal University, which had the power of conferring degrees purely by examination. The Irish government frequently consulted Stoney on educational matters, and he was for many years superintendent of civil service examinations in Ireland.

Stoney played a very active part in the affairs of the Royal Dublin Society, serving as honorary secretary from 1871 to 1881 and as vice president from 1881 to 1911. During Stoney's tenure, the society underwent profound changes. It handed over its great collections to the government and received capital to pursue its scientific functions and to improve Irish agriculture. Stoney's own research work was usually communicated first to the society and then reported in its publications and in Royal Society journals. Stoney and his gifted nephew, **George FitzGerald**, played central roles in the society's scientific meetings and discussions.

In 1863, Stoney married his cousin, Margaret Sophia (*née* Stoney); the couple had two sons and three daughters. In spite of the death of his wife in 1872, followed by two severe illnesses of his own (smallpox in 1875 and typhoid in 1877), and his heavy load of administrative duties, he still managed to carry out scientific research, often rising at five in the morning to write or to experiment before going to his office. In 1893, Stoney left Dublin to live in London, in order to give his daughters the opportunity of a university education, which was denied to them at that time in Dublin. On retiring, he developed the lines of research that he had not fully explored while he was occupied with his administrative duties.

One of the main themes of Stoney's research was his interest in the kinetic theory of gases. In a paper published in 1858, he showed that Boyle's law is contrary to the view that the particles of a gas are at rest or that a gas can be a continuous, homogeneous substance. Ten years later, he estimated the number of molecules in a given volume of gas at normal temperature and pressure, independently of a similar estimate by Amadeo Avogadro. In 1868, Stoney first considered the limitations of planetary atmospheres. He correctly explained the absence of hydrogen and helium in the Earth's atmosphere and the absence of an atmosphere on the Moon in terms of the concept of escape velocity. Stoney was also the first to suggest that rotation of the vanes of a Crooke's radiometer arose from the unsymmetrical impacts of molecules contained within the glass envelope.

Stoney introduced the word "electron" (from the Greek word for amber) into the scientific vocabulary. In a paper read before the British Association for the Advancement of Science in 1874, he pointed out "an absolute unit of quantity of electricity exists in that amount of it which attends each chemical bond or valency." He proposed that this quantity should be regarded as the fundamental unit of electricity and suggested the name "electron" in 1881. In the same paper, Stoney proposed the adoption of a system of natural units of mass, length, and time, based on the gravitational constant, the velocity of light, and the electric charge. In 1899, physicist Max Planck proposed a similar set of units that are of significance in cosmology today.

From 1896 onward, Stoney wrote a series of papers concerning the Leonid meteor showers. Together with A. M. W. Downing, he showed in principle how meteor storms could be predicted. More recently, these ideas have been successfully developed to predict the behavior of individual dust trails within the broader streams.

Stoney wrote extensively on the optical theory of microscopes and telescopes, using his concept of spherical wavelets. In 1868, he considered how periodic motions of electrons within atoms could give rise to spectral lines. In later work on the origin of atomic spectra, Stoney proposed that electrons described elliptical orbits in molecules and used this idea to explain double and triple lines in gas spectra.

While primarily a theorist, Stoney was also a practical man. He invented a novel form of heliostat that could be constructed more readily than contemporary instruments of French design. Stoney was keenly interested in music, both scientifically and artistically. By persuading the Royal Dublin Society to hold chamber music concerts, he much enhanced musical culture in Dublin.

Stoney received many honors and distinctions during his life. Perhaps the one he valued most highly was the award of the first Boyle Medal from the Royal Dublin Society in 1899. The medal was instituted to commemorate Irishman Robert Boyle's role in founding the Royal Society of London. Stoney was elected to the Royal Society in 1861, and served as vice president (1898–1899) and on its council (1898–1900). He was a member of the Royal Irish Academy and a fellow of the Royal Astronomical Society. Stoney regularly attended the meetings of the British Association for the Advancement of Science and was president of Section A at the 1879 meeting in Sheffield. He was a visitor to the Royal Observatory at Greenwich and to the Royal Institution, and a foreign member of the United States National Academy of Sciences, and the American Philosophical Society. Stoney received honorary doctorates from Queen's University in Ireland (1879) and the University of Dublin (1902). A lunar crater at 55°.3S, 156°.1W has been named in his honor.

Ian Elliott

Selected References

Anon. (1912). "George Johnstone Stoney." *Monthly Notices of the Royal Astronomical Society* 72: 253–255.

Ball, Robert Stawell (1911). "Dr. G. Johnstone Stoney, F.R.S." *Observatory* 34: 287–290.

Joly, John (1912). "George Johnstone Stoney, 1826–1911." *Proceedings of the Royal Society of London* A 86: xx–xxxv.

Kelham, Brian B. (1976). "Stoney, George Johnstone." In *Dictionary of Scientific Biography*, edited by Charles Coulston Gillispie. Vol. 13, p. 82. New York: Charles Scribner's Sons.

O'Hara, James G. (1975). "George Johnstone Stoney, F.R.S., and the Concept of the Electron." *Notes and Records of the Royal Society of London* 29: 265–276.

Owen, W. B. (1920). "Stoney, George Johnstone." In *Dictionary of National Biography, Supplement, January 1901–December 1911*, edited by Sir Sidney Lee, pp. 429–431. London: Oxford University Press.

Storer, Arthur

Born **Lincolnshire, England, 1642**
Died **Calvert County, Maryland, (USA), 1686**

As a boy, Arthur Storer was a playmate of **Isaac Newton**. He observed comet 1P/1682 Q1 (Halley) for Newton after immigrating to America.

Selected Reference

Thrower, Norman J. W. (1990). *Standing on the Shoulders of Giants: A Longer View of Newton and Halley*. Berkeley: University of California Press.

Störmer, Fredrik Carl Mülertz

Born **Skien, (Norway), 3 September 1874**
Died **Blindern, Norway, 13 August 1957**

Carl Störmer made substantial contributions to the understanding of polar auroral displays, from both theoretical and empirical viewpoints. His findings also had wider application to the study of cosmic rays.

Störmer was the son of Georg Ludvig Störmer, a pharmacist, and Henriette Mülertz. He attended the national university at Christiania (now Oslo) from 1892 to 1897. Störmer was awarded a *candidatus realium* (graduate) degree in the following year, and then offered a 5-year research fellowship, which allowed him to conduct advanced studies at the Sorbonne (Paris) and Göttingen University. In 1900, he married Ada Clauson; the couple had five children.

Störmer was appointed professor of pure mathematics at the University of Oslo in 1903; he occupied this post for forty three years, until his retirement in 1946. There, his colleague, physicist **Kristian Birkeland**, introduced him to the nature of cathode rays and their behavior in the presence of magnetic fields. Through experiments in which a magnetized sphere was bombarded with cathode rays under vacuum conditions, Birkeland and Störmer were able to simulate a number of phenomena relating to auroral displays. Störmer then undertook detailed analyses of the paths of charged particles in magnetic fields, including numerical integration of differential equations (long before the advent of electronic computation). In 1907, he described one such pathway in which a charged particle becomes entrapped within a dipole converging magnetic field. Although little recognized at the time, Störmer's mathematical solution received dramatic confirmation 50 years later, with **James Van Allen**'s discovery of radiation belts surrounding Earth. These were identified by the United States Explorer I satellite that was launched in 1958 during the International Geophysical Year [IGY].

In 1909, Störmer began intensive photographic studies of the aurorae, using parallactic photography along baselines as large as 27 km. By these means, he established the heights in Earth's atmosphere over which auroral displays occur. His photographic archives eventually encompassed more than 40,000 images. Störmer published a *Photographic Atlas of Auroral Forms* (1930). Results from his career-long research were brought together in his textbook, *The Polar Aurora* (1955).

Störmer was a research associate at Mount Wilson Observatory in 1912. He was appointed chairman of the auroral committee of the International Association of Terrestrial Magnetism and Electricity within the International Union of Geodesy and Geophysics [IUGG], and also chosen president of the auroral committee of the Second International Polar Year (1932/1933). For his auroral research, Störmer was awarded the Janssen Medal of the Paris Académie des sciences (1922). He also received honorary doctorates from Oxford University (which invited him to deliver its 1947 Halley Lecture), the University of Copenhagen, and the Sorbonne. A coeditor of the journal *Acta Mathematica* beginning in 1906, Störmer was elected president of the International Congress of Mathematicians (1936).

Jordan D. Marché, II

Selected References

Chapman, Sydney (1958). "Fredrik Carl Mülertz Störmer." *Biographical Memoirs of Fellows of the Royal Society* 4: 257–279.

Störmer, Carl (1955). *The Polar Aurora*. Oxford: Clarendon Press.

Stoyko, Nicolas

Born **Odessa, (Ukraine), 2 May 1894**
Died **Menton, Alpes-Maritimes, France, 14 September 1976**

Nicolas Stoyko is chiefly remembered for his contributions to the precision measurement of astronomical time and its distribution through the Bureau International de l'Heure [BIH] in Paris, France, which he directed from 1945 to 1964. Stoyko was a student at the University of Novorossia (Odessa). From 1914 to 1916, he was an unpaid trainee at the Odessa Observatory under Aleksandr Orlov, a specialist in polar motion studies. Stoyko received his bachelor's degree in 1916. Soon mobilized into the Russian army (1916–1918), he was certified as *agrégé de mathématiques* in 1920. Unable to obtain a position in his native land, Stoyko immigrated to Bulgaria and taught at a boys' school in Pleven. But following a military *coup d'état* (1923), Stoyko moved to France. Upon a recommendation from his former instructor Orlov, he traveled to the Paris Observatory. In 1924, Stoyko was given a post at the BIH where he spent the remainder of his career.

The BIH had been established in 1919 at the Paris Observatory. Its missions were to centralize astronomical time determinations; to assure the accuracy of all time signal receptions; to distribute a *heure provisoire* (provisional time); and, through data analysis, to publish the *heures définitives* (definitive time), as those quantities were then called. After several years of freelance work and study, Stoyko became *pour ordre et à titre étranger* and made *aide-astronome* (positions available for noncitizens). He became a naturalized French citizen in 1930, under the name of Stoyko, and obtained the title of *docteur d'État français* after defending his thesis, entitled *La mesure du temps et les problèmes qui s'y rattachent* (Measurement of time and related problems, 1931).

In his roles at the Paris Observatory and the BIH, Stoyko contributed numerous astronomical observations, reportedly as many as 100,000. He collaborated with Armand Lambert who was arrested for being a Jew in 1943 and executed at Auchwitz in the following year. In 1945, Stoyko was appointed "*Chef du service horaire*" at the Paris Observatory and *Chef des services* (the nominal director) at the BIH.

During his 40-year career, Stoyko conducted the BIH's longitude campaigns (1926 and 1933) to measure secular changes in the rotation period of the Earth. His work involved improvements to the timekeepers and pendulum clocks (with their constant-pressure cases), along with studies concerning the propagation of radio waves. His investigations into the causes of polar motion led to the creation, under his responsibility, of the Service International Rapide des Latitudes [SIR]. Stoyko's name is associated with the seasonal variations of the Earth's rotation, a phenomenon he disclosed in 1937. Difficult to predict, this effect eventually led to abandonment of the day (= rotation period of the Earth) as the basic unit of time. In the early 1960s, Stoyko introduced an atomic clock into the BIH, which later provided the official time scale established in 1972. When he retired in 1964, astronomers were no longer the only "masters of time."

Appointed *astronome titulaire* in 1946, Stoyko was elected a corresponding member of the Bureau des longitudes and *Chevalier de la Légion d'honneur* in 1952. He was also honored in foreign countries, for example, as a member of the Academy of Technical Sciences in Warsaw (1938–1976). His collaborator and spouse, Anna, reports that during Stoyko's lifetime, he published nearly 300 papers.

Jacques Lévy

Alternate name

Stoiko-Radilenko, Nicolas

Selected Reference

Stoyko, Nicolas (1959). "Proceedings of the Symposium on the Rotation of the Earth and Atomic Time Standards: Part II. The Rotation of the Earth: Variations periodiques et al atoires de la rotation de la Terre." *Astronomical Journal* 64: 99 .

Strand, Kaj Aage Gunnar

Born **Hellerup, Denmark, 27 February 1907**
Died **Washington, District of Columbia, USA, 31 October 2000**

Danish–American astrometrist Kaj Strand pioneered the use of reflecting telescopes for the measurement of parallax and orbits of visual binaries. The United States Naval Observatory, under his direction, completed the first extensive infrared survey (2 μm) in 1969, which found about 3,000 sources.

Strand was the son of Viggo Peter and Constance (*née* Malmgren) Strand; he married Emilie Rashevsky on 10 June 1949, and they had two daughters, Kristin Ragna and Constance Vibeke. Strand received his BA and M.Sc. degrees from the University of Copenhagen in 1931 and his Ph.D. from the same university in 1938 for work done with **Ejnar Hertzsprung** while both were at Leiden, the Netherlands. He became a geodesist with the Royal Geodetic Institute, Copenhagen, 1931. In 1933, Strand was appointed assistant to the director of the University Observatory, Leiden, the Netherlands, where he remained until 1938, working with Hertzsprung. Strand joined the faculty of Swarthmore College, Pennsylvania, USA, in 1938 and remained there until 1946, except for a break for war service in the US Army and US Army Air Force, in which he served with the rank of captain, training special aircrews. Strand became an associate professor at the Yerkes Observatory, University of Chicago, in 1946/1947, thereafter research associate, serving also as professor of astronomy, Northwestern University and director of Dearborn University from 1947 to 1958. In that year, he was appointed director of the Division of Astrometry and Astrophysics at the United States Naval Observatory, becoming the observatory's scientific director in 1963 until his retirement in 1977. Strand also served as a consultant to, among other agencies, the National Aeronautics and Space Administration, the National Science Foundation, the Office of Naval Research, and the National Bureau of Standards.

Upon retirement, Strand received the Navy's Distinguished Civilian Award. He was elected a member of the Royal Danish Academy of Sciences and Letters in 1965. Strand was president of the International Astronomical Union's commission on double stars from 1964 to 1967.

Strand's work with Hertzsprung during his years at Leiden, photographic observations of double stars, influenced him throughout his working life, both in his methods of working and in the topics he selected for research, namely, parallax determinations and basic data on the masses and luminosities of stars and then interpretation of these data, which are of fundamental importance to modern astrophysics. At Swarthmore College, where he worked with **Peter van de Kamp** and his team at the Sproul Observatory, Strand continued in the study of visual binaries, working particularly on the detection of invisible (possibly planetary) companions of those objects. Although many such companions claimed as discoveries by the Sproul group have not withstood further scrutiny, Strand's companion to one of the components of 61 Cygni (now believed to be B) has not entirely been ruled out. At Swarthmore, Strand also pioneered the use of coarse diffraction gratings for accurate measurements of the relative positions of visual binary stars where one star is much brighter than the other.

Astrometry had traditionally been carried out with refracting telescopes of long focal length. Strand became convinced that a large astrometric reflector could provide accurate parallaxes for many intrinsically faint but nearby stars; he first publicly proposed this at a conference he organized at Northwestern University in 1953. When Strand moved to the Naval Observatory, he had the opportunity to create such a telescope. A 61-in. (1.55-m) reflector was commissioned and built at the observatory's station in Flagstaff, Arizona. It came into operation in the spring of 1964 and has continued to produce astrometric measurements of high precision, including a large number of parallaxes. Results from this telescope, which is now known as the Strand astrometric telescope, bear comparison with those obtained so far from space. While he was at the Naval Observatory, Strand also established its first Southern Hemisphere station at El Leoncito, Argentina, where a 7-in. (18-cm) transit telescope was in operation from 1966 to 1973, contributing to the Southern Reference Star Program. He also instituted a long-term program of photographic observations of double stars at the observatory, which he ran from 1958 until 1981.

Alan H. Batten

Selected References

Dick, Steven J. (2001). "Kaj Aage Strand, 1907-2000." *Bulletin of the American Astronomical Society* 33: 1584–1585.

Strand, K. Aa. (1943). "61 Cygni as a Triple System." *Publications of the Astronomical Society of the Pacific* 55: 29–32.

——— (1943). "The Orbital Motion of 61 Cygni." *Proceedings of the American Philosophical Society* 86: 364–367.

——— (1944). "The Astrometric Study of Unseen Companions in Double Stars." *Astronomical Journal* 51: 12–13. (For a more popular account see *The Sky* 5, no. 7 (1941): 6–8.)

——— (ed.) (1963). *Basic Astronomical Data*. Vol. 3 of *Stars and Stellar Systems*. Chicago: University of Chicago Press.

——— (1964). "The New 61-inch Astrometric Reflector." *Sky & Telescope* 27, no 4: 204–209, 232–233. (For a more technical account see Part 1 of the *Publications of the U.S. Naval Observatory* 20 [1971]. Parts 2 and 3 [1967 and 1970 respectively] of the same volume, mostly by other authors, supplement that latter account.) (Catalogs of the parallaxes measured by the instrument have appeared in subsequent volumes of the same *Publications*.)

——— (1977). "Hertzsprung's Contributions to the HR Digaram." In *Memory of Henry Norris Russell*, edited by A. G. Davis Philip and David H. DeVorkin, pp. 55–60. Albany, New York: Dudley Observatory.

Stratton, Frederick John Marrian

Born **Birmingham, England, 16 October 1881**
Died **Cambridge, England, 2 September 1960**

Frederick Stratton excelled as a teacher, administrator, and international organizer of astronomy and was instrumental in holding together the International Council of Scientific Unions through the difficult years of World War II. His influence on the progress of astronomy as a worldwide enterprise, through his students and international relationships, was perhaps as great as or greater than that of any previous or subsequent astronomer.

The son of Stephen Samuel and Mary Jane (*née* Marrian) Stratton, Frederick never married and had no descendants. He was educated at King Edward VI Grammar School, Five Ways, Birmingham, and Mason College (later to become the University of Birmingham), before proceeding, early in the 20th century, to Gonville and Caius College, Cambridge, with which he was to be associated for the rest of his life. In 1904, the year that **Arthur Eddington** was senior wrangler in the mathematics *tripos*, Stratton was third wrangler. In 1905 Stratton held the Isaac Newton Studentship, and in 1906 he was Smith's Prizeman.

Stratton took some time to find his particular niche in astronomy. His first paper (his essay for the Smith's Prize) was a purely mathematical discussion of the effect of tidal forces on the obliquities of the rotational axes of the planets. After publishing this, he joined the staff of the Cambridge University Observatory, then under Sir **Robert Ball**, and completed a study of the proper motions of faint stars under the direction of **Arthur Hinks**. Stratton then became interested in solar physics and stellar spectroscopy and transferred to the Solar Physics Observatory, recently moved from Kensington to Cambridge. Hugh Frank Newall (1857–1944), the director of that observatory, appointed Stratton as assistant director in 1913, after he had published two papers on the spectrum of Nova Geminorum 1912.

Stratton's astronomical career was interrupted by World War I. He rose to the rank of lieutenant-colonel in the Royal Corps of Signals, and was decorated both by Britain and France, receiving the Distinguished Service Order [DSO], *Croix de Chevalier*, and the *Légion d'honneur*. Stratton returned to the Signal Corps during World War II serving mainly in Special Duties in connection with radio security. He traveled widely in that role, later served as deputy scientific advisor to the Army Council, and was generally known as Colonel Stratton thereafter.

When he returned to Cambridge after World War I, Stratton was appointed tutor at Caius College and relinquished his formal appointments at the Solar Physics Observatory. In those days, the tutor of a Cambridge college was well placed to influence appointments in his own field. Stratton's influence can be measured by the fact that, at one time, the Astronomer Royal, the Astronomer Royal for Scotland, and Her Majesty's Astronomer at the Cape were all graduates of Caius College!

Despite his responsibilities in his college, Stratton maintained a considerable research output. In 1925, he published the book *Astronomical Physics*, one of the earliest textbooks on astrophysics. Although that book is long out of date, one of its appendices remained useful as long as prism spectrographs were in frequent use; it showed how to compute the Hartmann constants necessary for reducing stellar spectrograms.

Novae and solar physics became Stratton's prime interests. In 1926 he went on an eclipse expedition to Sumatra. Stratton led or took part in several other eclipse expeditions but, because of bad luck with the weather, none was as successful as 1926.

In 1928, Stratton succeeded Newall as professor of astronomy and director of the Solar Physics Observatory. The appearance of Nova Herculis in 1934 provided him with another opportunity to study the nova phenomenon spectroscopically; in collaboration with W. H. Manning he produced an atlas of the changing spectrum of that nova. At about the same time Stratton wrote an article on novae for the *Handbuch der Astrophysik*, a multivolume compendium of astronomy that continued to be useful well into the 1960s. Serious stellar spectroscopy rose with the advent of photography in the second half of the 19th century. Spectroscopic observations of novae and of the solar chromosphere during eclipses were, of course, possible only on the rare occasions that novae appeared or total solar eclipses occurred. The early 20th century was, therefore, a time when the groundwork of these two fields of study was laid, groundwork to which Stratton contributed his share.

Stratton frequently stressed the importance of international cooperation in astronomy; his record of service to both national and international organizations shows that he practiced what he preached. A side of his work not generally known was his quiet assistance of astronomers who were victims of political persecution in their own countries.

The esteem in which Stratton was held by the international community is shown in the first two volumes of *Vistas in Astronomy* which, under the editorship of Arthur Beer (1900–1980), were conceived as a tribute to Stratton on the occasion of his 70th birthday; only later did *Vistas* become a continuing serial publication. At that time too, minor planet (1560) was named Strattonia in his honor – a rarer tribute then than it has since become. Stratton was a prominent member of the Cambridge Unitarian Church, of which he was chairman for more than 50 years. Like many physical scientists of his generation, Stratton also had an interest in what are now called paranormal phenomena, and was president of the Society for Psychical Research from 1953 to 1955.

A dedicated member of the Royal Astronomical Society [RAS], Stratton served on the RAS Council for more than 40 years, being elected successively as treasurer (1923–1927), president (1933–1935), and foreign secretary (1945–1955), and for several terms as vice president. He was general secretary of the International Astronomical Union from 1925 to 1935 and was president of Commission 38 (Exchange of Astronomers) for an unusual three terms from 1948 to 1958. Stratton was general secretary of the International Council of Scientific Unions from 1937 to 1952 and also general secretary of the British Association for the Advancement of Science from 1930 to 1935. In 1947, he was elected a fellow of the Royal Society and of the Institute of Coimbra (Portugal). In 1929 Stratton was made an Officer of the Order of the British Empire [OBE], and he was an honorary or corresponding member of several academies and the recipient of a number of honorary doctorates.

Alan H. Batten

Selected References

Chadwick, Sir James (1961). "Frederick John Marrian Stratton." *Biographical Memoirs of Fellows of the Royal Society* 7: 281–293.

McCrea, W. H. (1982). "F. J. M. Stratton, DSO, OBE, TD, FRS, 1881–1960." *Quarterly Journal of the Royal Astronomical Society* 23: 358–362.

Stratton, F. J. M. (1925). *Astronomical Physics*. London: Methuen and Co.

——— (1936). "Novae." In *Handbuch der Astrophysik*. Vol. 7, *Ergänzungsband*, pp. 671–684. Berlin: J. Springer.

——— (1949). *History of the Cambridge Observatories*. Cambridge: Cambridge University Press.

Stratton, F. J. M. and W. H. Manning (1934). *Atlas of Spectra of Nova Herculis*. Cambridge: Solar Physics Observatory.

Streete, Thomas

Born **probably Cork, Ireland, 15 March 1622**
Died **Westminster, (London), England, 27 August 1689**

Thomas Streete was an observational astronomer, a publisher of ephemerides, and introduced, through his writings, **Johannes Kepler**'s laws of planetary motion to **Isaac Newton**. Streete was employed in London as a clerk in the Excise Office under Elias Ashmole. He had contacts with Gresham College, but little seems to be known about his education. He knew a number of the leading astronomers in England and abroad, and often assisted them in observations. Streete was careless about citing his sources, which led to accusations of plagiarism. Still, he published highly regarded ephemerides, worked on the problem of determining longitude at sea, and was engaged in the resurvey of London after the Great Fire of 1666.

Streete was very highly regarded in his own day as an astronomical observer. His tables, even if not described as the best, are regularly cited by Newton in the *Principia*. In 1661, Streete published *Astronomia Carolina*, which provided a list of apparent planetary diameters at their mean distance from us and which disseminated Kepler's first and third laws. In the mid-1660s Newton took nearly *verbatim* notes on Streete's book, which contains the first statement of Kepler's laws that Newton is likely to have seen. Streete presented the third law as exact for the sidereal periods of the planets, which were well determined even in the 17th century, if a value for our distance from the Sun was assumed. Newton's note suggested that he accepted Streete's procedure at that time, developing doubts about the accuracy of the third law only later.

Newton knew that Streete, among others, thought that an equant construction for the planetary elliptical orbits could be explained by quasi-Cartesian vortices. Newton did not learn of the second law (planetary radius vectors sweeping out equal areas in equal times) then, from Streete's treatise, nor did he know of it as early as 1661.

Streete's insistence on the exactitude of Kepler's third law was not based on his own observations but on those made by **Jeremiah Horrocks** and on the latter's value for our distance from the Sun. By 1669 Newton began to worry about the accuracy of Kepler's third law, while he also entertained the hypothesis of vortices. We may conclude that Newton's path to universal gravitation was aided in part by his struggles with the accuracy of Kepler's laws and with the failure of vortices to confirm Kepler's second law. It was Streete among others who had provided Newton with the materials and problems that would eventually lead him to combine inertia with gravitational forces to derive Kepler's three laws.

André Goddu

Selected Reference

Wilson, Curtis (1989). *Astronomy from Kepler to Newton*. London: Variorum Reprints.

Strömberg, Gustav

Born **Gothenburg, Sweden, 16 December 1882**
Died **Pasadena, California, USA, 30 January 1962**

Gustav Strömberg was educated at Gothenburg, Kiel, Stockholm, and Lund universities. From 1906 to 1913, he was an assistant at the Stockholm Observatory. In 1917, he went to the United States and joined the staff of Mount Wilson Observatory.

Strömberg's first important work was on the luminosity of the long-period variable stars . His work on the radial motions of stars and nebulae led to his striking discovery, announced in 1923, of the "asymmetry of stellar motions" explicable in the Lindblad–Oort theory of galactic rotation, enunciated soon afterward.

Strömberg also attempted to correlate radial velocities of nebulae, measured by **Vesto Slipher**, with estimates of their distances, in about 1925. This was before Edwin Hubble established the red-shift-distance relation. Strömberg's version included the possibility of negative velocities, so as to include the globular clusters.

Selected Reference

Strömberg, Gustav (1946). "The Motions of the Stars Within 20 Parsecs of the Sun." *Astrophysical Journal*. 104: 12.

Strömgren, Bengt Georg Daniel

Born **Gothenburg, Sweden, 21 January 1908**
Died **Copenhagen, Denmark, 4 July 1987**

Danish astronomer Bengt Strömgren is strongly associated with the concept of the Strömgren sphere, the idea that a star with a certain ultraviolet luminosity can keep a certain mass of diffuse hydrogen gas ionized, thus accounting for the sizes and shapes of the ionized (H II) regions around young, massive stars and the planetary nebulae around old, very hot stars. He was born into a Swedish family, that of astronomer **Elis Strömgren** and dentist Hedvig Strömgren. His father was appointed professor of astronomy and director of the observatory at the University of Copenhagen in 1907, and the son naturally became Danish. Bengt Strömgren married in 1931, and one of their two daughters is the distinguished biologist Nina Strömgren Allen.

As a result of his father's position, Strömgren began observing in 1919; a catalog incorporating some of his measurements was published when he was 17. He received an M.Sc. from Copenhagen University in 1927 and a Ph.D. in 1929 for work on the determination of comet orbits partially carried out with **Karl Kustner**. During the next few years in Copenhagen, Strömgren worked with the physicists at the Institute for Theoretical Physics (usually called the Niels Bohr Institute after its director, **Niels Bohr**), acquiring a familiarity with general relativity, quantum mechanics, and spectroscopy, and working on problems in stellar structure. Among the visitors he interacted with was **Cecilia Payne-Gaposchkin**, whose 1925 thesis had shown that stellar atmospheres have hydrogen as their dominant constituent. In 1932 Strömgren reached the conclusion that hydrogen was abūndant also in the interior of stars. He also showed that some of the heavy elements in the Sun have the same relative elemental abundances as they have in meteorites, thus tying stellar astronomy to the work of **Viktor Goldschmidt** on abundances in the meteorites.

Strömgren spent the years 1936–1938 at the University of Chicago as assistant, then associate professor of astronomy, working with **Subrahmanyan Chandrasekhar** and **Otto Struve** on problems of stellar structure and composition. He returned to Copenhagen in 1938 as professor of astronomy and was appointed director of the observatory in 1940 upon his father's retirement. This retirement was marked by an outstanding *Festschrift*, which included a paper by young Strömgren on the structure of the solar atmosphere, incorporating the work of **Rupert Wildt** on the importance of the negative hydrogen ion H^- for solar opacity. His classic paper on ionization of interstellar hydrogen dates from 1939. During the war years, he worked in relative isolation and with limited resources at Copenhagen Observatory on stellar atmospheres, geometrical optics, and calculation of tables useful in both these fields.

After World War II and several short visits to the United States, Strömgren returned to Chicago as professor of astronomy in 1951, and from 1952 to 1957 was also the director of Yerkes and McDonald observatories, before becoming professor of astrophysics at the Institute for Advanced Study in Princeton. His best-known work from this period was in photoelectric photometry in a system he invented (called Strömgren colors) to determine the temperatures, gravitational fields, and compositions of a wide range of kinds of stars. Strömgren colors were incorporated into the new Copenhagen Observatory system when he returned there in 1967.

Meanwhile, more of Strömgren's attention went into scientific administration. He was general-secretary of the International Astronomical Union during 1948 to 1952 and its president during the period 1970–1973. He was elected president of the American Astronomical Society, but his term was curtailed by his return to Denmark, and president-elect **Albert Whitford**, with whom Strömgren had worked on photoelectric colors at Lick Observatory, took over a year early. Strömgren was active in the development of Kitt Peak National Observatory while in the United States and the European Southern Observatory after his return to Denmark. He received medals from the Franklin Institute, the Astronomical Society of the Pacific, the French Academy of Sciences, and the Royal Astronomical Society (London). In 1967, Strömgren was selected as the outstanding scientist of Denmark, entitling him and his family to residence in the Carlsberg Mansion, formerly occupied by Bohr.

Helge Kragh

Selected References

Lang, Kenneth R. and Owen Gingerich (eds.) (1979). *A Source Book in Astronomy and Astrophysics, 1900–1975*. Cambridge, Massachusetts: Harvard University Press, pp. 588–592. (For a reprint of Strömgren's 1939 paper, "The Physical State of Interstellar Hydrogen.")

Rebsdorf, Simon O. (2003). "Bengt Strömgren: Growing ūp with astronomy, (1908–1932)." *Journal for the History of Astronomy* 34: 171–199.

Strömgren, Bengt (1983). "Scientists I Have Known and Some Astronomical Problems I Have Met." *Annual Review of Astronomy and Astrophysics* 21: 1–11.

Strömgren, Svante Elis

Born **Hälsingborg, Sweden, 31 May 1870**
Died **Copenhagen, Denmark, 5 April 1947**

Early in the 20th century, an often asked question about comets was whether they originate beyond the Solar System – perhaps in other stellar systems. Elis Strömgren endeavored to calculate the eccentricity of comet orbits backward in time, taking planetary perturbations into account. None of the comets examined by Strömgren appeared to have an initial hyperbolic orbit.

At the Copenhagen Observatory, Strömgren's program to communicate timely astronomical information to colleagues in other countries evolved into the International Astronomical Union's Central Bureau for Astronomical Telegrams (transferred to Harvard College Observatory when Denmark fell to the Nazis at the outset of World War II). Strömgren was a colleague of **Carl Burrau** and the father of astronomer **Bengt Strömgren**.

Selected Reference

Lundmark, Knut (1948). "Svante Elis Strömgren." *Monthly Notices of the Royal Astronomical Society* 108: 37.

Stroobant, Paul-Henri

Born **Ixelles, Belgium, 11 April 1868**
Died **Brussels, Belgium, 15 July 1936**

Paul Stroobant directed the Royal Observatory at Uccle, Belgium, and contributed to a number of astronomical specialties, including the study and discovery of minor planets. Before he was twenty years old, he observed the bright comets of 1882 (C/1882 RI) and 1885 and became a voluntary assistant at the Royal Observatory. Stroobant earned a doctorate in mathematics and physics from the University of Brussels in 1889. During the following year, he studied at the Sorbonne, Paris and the Observatoire de Paris.

In 1891, Stroobant returned to Belgium and joined the staff of the Royal Observatory as assistant astronomer. He was to spend the remainder of his career there, becoming in succession assistant director (1918) and director (1925), succeeding Georges Lecointe. He retired only a few weeks before his death. Stroobant also served as professor of astronomy at the University of Brussels after 1896. He presided over its faculty of sciences between 1906 and 1909.

Among Stroobant's most important work was his statistical investigation of the minor planets, based upon more than 800 known objects. From this sample, he estimated the existence of more than 100,000 asteroids brighter than 20th magnitude and calculated their total expected mass.

Stroobant conducted a number of other investigations. These included dynamical studies of the satellites of Saturn, the "personal equation" in meridian circle observations (systematic errors in measuring stellar positions that vary from one observer to another), the direction in space of the Sun's motion, and the dynamics and distribution of stars and clusters in the Milky Way Galaxy.

As the observatory's centennial celebration (1935) drew near, Stroobant ordered and installed newer photographic and spectroscopic equipment, particularly relevant to continued research on minor planets. As a result, **Eugène Delporte** discovered a number of new minor planets at the observatory, including two of particular importance. Minor planets (1221) Amor and (2101) Adonis make relatively close approaches to the Earth.

Stroobant presided over the Belgian National Committee on Astronomy and was elected president of the International Astronomical Union's [IAU] Committee on Bibliography. His textbook, *Précis d'Astronomie*, passed through two editions (1903, 1933). Between 1907 and 1920, Stroobant issued the *Annuaire de l'Observatoire Royal de Belgique*, containing reviews of current astronomical research. He was awarded the Lalande Prize of the Paris Académie des sciences in 1921 and made a Commander of the Legion of Honor. Today, the Paul and Marie Stroobant Prize of the Royal Academy of Belgium honors both the subject of this sketch and his wife.

Jordan D. Marché, II

Selected References

Anon. (1937). "Paul Henri Stroobant." *Monthly Notices of the Royal Astronomical Society* 97: 289–290.
Cox, J. F. (1936). "Paul-Henri Stroobant" (in French). *Astronomische Nachrichten* 260: 175–176.
F. D. R. (1936). "Paul Stroobant." *Observatory* 59: 349–352.

Struve, Friedrich Georg Wilhelm

Born **Altona, (Hamburg, Germany), 15 April 1793**
Died **Saint Petersburg, Russia, 23 November 1864**

Wilhelm Struve (as he was usually known) was the founding director of the Pulkovo Observatory and codiscoverer of stellar parallax. Through his many descendants, he created a family dynasty of Russian-born astronomers across the span of four generations. Struve, the son of Jakob and Maria Emerentia (*née* Wiese) Struve, was educated at the Christianeum in Altona, the gymnasium at which his father was rector, and later at the University of Dorpat (now Tartu) in Estonia (then part of the Russian empire). He studied classical philology and graduated in 1810 but switched to the study of astronomy and received both master's and doctor's degrees in 1813. Struve married Emilie Wall in 1815; the couple had 12 children, the third of whom (and oldest to survive to adulthood) was the astronomer **Otto Wilhelm Struve**, who became his father's successor. After Emilie's death in 1834, Wilhelm Struve married Johanna Bartels, with whom he had another six children. Struve was elected an associate of the Royal Astronomical Society in 1823 and received its Gold Medal in 1826. He became a corresponding member of the Imperial Academy of Sciences (Saint Petersburg) in 1827 and a full member in 1832. Struve was also elected a foreign member of the Royal Society of London (1827).

Struve's childhood and adolescence coincided with the Napoleonic wars; the troubled nature of those times led to his being sent to the comparative safety of Dorpat University, rather than a nearby one such as Kiel or Göttingen. He was attracted to astronomy by the physicist Georg Parrot, who encouraged him to pursue graduate studies in that subject. The university's observatory had just been completed, but its only telescope (a small transit instrument) remained in its packing cases because the professor of astronomy, **Johann Huth**, was too ill to install it. Struve, largely unsupervised, succeeded in installing the instrument and using it to determine the longitude and latitude of the new observatory, which earned him his advanced degrees. He was then appointed extraordinary professor of astronomy. Struve became ordinary professor and director of the observatory in 1820, following Huth's death.

At that time, an astronomer's duties included surveying and geodetic work. Struve was active in a survey of Livland (much of modern Estonia and Latvia) in the years 1816–1818. This task led him to conceive the measurement of an arc of the meridian through Dorpat and stretching from Hammerfest in Norway, to the mouth of the Danube near the modern border between Ukraine and Romania. Spanning more than 25°, this measurement offered a significant contribution to knowledge of the dimensions of the Earth and was to occupy Struve until illness prevented him from undertaking any further scientific work in 1858. In addition, he directed an expedition to measure the difference in levels between the Black Sea and Caspian Sea, although the fieldwork was done by others. Struve also planned to measure a parallel of latitude from the west coast of Ireland to the Ural Mountains, but illness prevented him from doing so.

In 1820, Struve had visited Munich for the purpose of ordering instruments to be used in his geodetic surveys. During this visit, he saw the large refractor of 9 (Paris)-inches aperture that **Joseph von Fraunhofer** was constructing. Struve was able to persuade Dorpat University to purchase this instrument, which determined the course of much of his astronomical career. The "Great Refractor," as it came to be called, was the largest astronomical telescope of its day and arrived at Dorpat in late 1824. It was promptly assembled by Struve, even though Fraunhofer had forgotten to send detailed instructions for its installation. Struve used it principally for making a census of double stars visible in the northern sky, of which he had already published a preliminary catalog. His research marked the first systematic study of these objects since their gravitational binding had been confirmed by **William Herschel**. The results of Struve's survey were presented in the *Catalogus Novus Stellarum Duplicium et Multiplicium* (1827). The catalog contains more than 3,000 pairs, most of which were Struve's own discoveries. Even before its publication, Struve received the Gold Medal of the Royal Astronomical Society.

His success with the Great Refractor, along with his geodetic work, brought him to the attention of the Tsar Nicholas I, who invited Struve to supervise the construction of a major new observatory (Pulkovo) of which he became the first director in 1839. In that year, he and his family left Dorpat and took up residence at Pulkovo, just outside Saint Petersburg. Much of his time in Pulkovo was spent in reducing his measurements of the arc of the meridian and in determining the constant of aberration from observations made with the prime-vertical telescope in Pulkovo. Although Struve equipped the observatory with an even greater refractor of 15-in. aperture, he rarely used that instrument himself and left it mainly to his son, Otto Wilhelm.

The elder Struve also brought one important piece of unfinished business with him – a determination of the parallax of Vega (α Lyrae). While still at Dorpat, Struve had published an important work, usually known by its abbreviated title of *Mensurae Micrometricae* (1837), a collection of his micrometer measurements of double stars made with the Dorpat refractor that is still of value today. In this work, he presented a preliminary value of the parallax of Vega, as determined by comparing its position throughout the year with respect to its much fainter (and presumably more distant) "companion". His result was in good agreement with the modern value, but its uncertainty was sufficiently large that Struve himself did not regard it as definitive. His newer measurements of Vega were not reduced or published until 1840, after **Friedrich Bessel** had published his determination of the parallax of 61 Cygni. Undoubtedly, Bessel's finding provided the first convincing value for a stellar parallax, but Struve shares with him and **Thomas Henderson** the credit for demonstrating almost simultaneously that the measurement of stellar parallax was within reach.

In 1847, Struve published *Études d'astronomie stellaire*, which might be described as one of the earliest textbooks on stellar statistics. It was necessarily tentative because only a few stellar parallaxes had yet been successfully measured. This book tried to extend Herschel's work on the "construction of the heavens." Controversial in its day, Struve's text remains significant today for containing one of the first suggestions that starlight was absorbed by the presence of an interstellar medium. He correctly argued that the absorption of starlight must be considered in any attempt to measure the distribution of stars in space. Struve attempted quantitatively to estimate the amount of absorption and came up with a value of roughly the same order of magnitude as that found nearly a century later by **Robert Trumpler**. Astronomers of Struve's day, however, were unwilling to consider this possibility, and only Trumpler's work eventually convinced them of the reality of the general interstellar absorption.

Apart from his own achievements in astronomy, Struve is remarkable for having created a dynasty of astronomers that included one of his sons, two grandsons, two great-grandsons, and (briefly) one great-great-grandson. Taken together, the Struve family established a record of achievement in astronomy that is rivaled only by the Herschels and the Cassinis.

Struve likewise had considerable talents as a bibliographer. He equipped Pulkovo Observatory with a first-class library, for which he provided the catalog. The library was further enriched by the purchase of **Heinrich Olbers**'s personal collection of books. These works survived the observatory's destruction during World War II, although much of the Olbers collection was damaged or destroyed by a fire in 1997.

As late as 1857, Struve was still completing his report on the measurement of the arc of the meridian. Overwork was taking its toll, and he planned a "rest-cure" in Europe during which he began negotiations for his project to measure the parallel of latitude. Deteriorating health cut short this venture, and Struve became seriously ill in 1858. Although he recovered enough to enjoy a few more years, his capacity for scientific work was never regained. He resigned in 1862 as director of the Pulkovo Observatory, where his son Otto Wilhelm had been the director in all but name for several years. Struve retired to the city of Saint Petersburg.

Alan H. Batten

Alternate name

Struve, Vasily Yakovlevich

Selected References

Batten, Alan H. (1988). *Resolute and Undertaking Characters: The Lives of Wilhelm and Otto Struve*. Dordrecht: D. Reidel.

Hirshfeld, Alan W. (2001). *Parallax: The Race to Measure the Cosmos*. New York: W. H. Freeman.

Pritchard, C. (1865). "Friedrich Georg Wilhelm Struve." *Monthly Notices of the Royal Astronomical Society* 25: 83–98.

Sokolovskaya, Z. K. (1976). "Struve, Friedrich Georg Wilhelm." In *Dictionary of Scientific Biography*, edited by Charles Coulston Gillispie. Vol. 13, pp. 108–113. New York: Charles Scribner's Sons.

Struve, Georg Otto Hermann

Born **Tsarskoye Selo, Russia, 29 December 1886**
Died **Berlin, Germany, 10 June 1933**

An expert on the Solar System and especially the planet Saturn, Georg Struve was the son of **Karl Hermann Struve** and Olga (*née* Struve) Struve. When Georg was born, his father was an adjunct astronomer at the Pulkovo Observatory. Georg was taken to eastern Prussia when his father became director of the Albertus University Observatory at Königsberg in 1895. Struve attended the humanistically oriented Königsberg Wilhelms-Gymnasium from which he graduated in 1905. He then studied mathematics and astronomy at the universities of Heidelberg and Berlin, where one of his teachers was Julius Bauschinger.

While a student (1908), Struve assisted **Paul Guthnick** in his pioneering astrophotometric observations. He received his Ph.D. in 1910 for an investigation of the orbital motion of minor planet (2) Pallas. Struve dedicated this work to the memory of his grandfather, **Otto Wilhelm Struve**. He married Marie Mock in 1912; the couple had two sons.

During 1911/1912, Struve worked as an assistant in the observatories located at Bonn, Berlin–Babelsberg (under his father's direction), and at Hamburg–Bergedorf. He was then appointed an astronomer in the Wilhelmshafen Naval Observatory (1913–1919), where he was placed in charge of its chronometers and compasses. Using the observatory's 4.8-in. Repsold meridian circle, Struve measured the positions of Saturn and its satellites, sharing an interest in that subject with his father. He likewise observed the positions of more than 500 stars in order to derive their proper motions. Struve returned to the Berlin–Babelsberg Observatory in 1919, where he regularly used its 26-in. refractor for visual observations. He later held the post of professor there until his death.

Starting in 1917, Struve published some ten papers on Saturn, its satellites, and rings, which included a new determination of the planet's equatorial plane, the orbits of its satellites, the periodic disappearance of its rings when seen edge-on, and comparisons of visual and photographic observations of the planet's satellites. Struve observed the eclipses of Jupiter's satellites, measured the diameter of Venus through an application of the theory of contrasts, studied the outer planets and their satellites, and observed the opposition of minor planet (433) Eros in 1930/1931 to improve knowledge of distances within the Solar System. He supplemented his own observational data with that collected during visits to the Johannesburg Station of the Yale University Observatory, South Africa, and to the Lick and Yerkes Observatories in the United States.

Politically, Struve was an active member of the Deutschnationale Volkspartei, which fought unsuccessfully against the Nazis and attempted a restoration of the Hohenzollern monarchy. As a result of the complex political situation in the early 1930s, Struve suffered a nervous breakdown, experienced a pulmonary embolism, and died suddenly.

Victor K. Abalakin

Selected References

Guthnick, Paul (1934). "Georg Struve." *Astronomische Nachrichten* 251: 47–48.

Sokolovskaya, Z. K. (1976). "Struve, Georg Otto Hermann." In *Dictionary of Scientific Biography*, edited by Charles Coulston Gillispie. Vol. 13, p. 113. New York: Charles Scribner's Sons.

Struve, Otto (1933). "Georg Struve (1886–1933)." *Publications of the Astronomical Society of the Pacific* 45: 289–291.

Struve, Gustav Wilhelm Ludwig

Born **Pulkovo, Russia, 1 November 1858**
Died **Simferopol, Crimea, (Ukraine), 4 November 1920**

An expert on lunar occultations and stellar positions, Ludwig (as he was usually known) Struve was the son of **Otto Wilhelm Struve** and the younger brother of **Karl Hermann Struve**. He completed gymnasium studies at Vyborg in 1876 and entered Dorpat University from which he graduated in 1880. He then moved back to Pulkovo and worked part-time at the observatory, which his father directed. One of his earliest published papers concerned the double star η Cassiopeiae. In 1883, Struve defended his magister's thesis on the star Procyon (α Canis Minoris). Afterward, he was sent abroad to further his scientific education (1883–1885) and worked at the observatories of Bonn, Milan, and Leipzig.

In 1885, Struve took part in the general meeting of the Deutsche Astronomische Gesellschaft (German Astronomical Society) held in Geneva, Switzerland, and visited observatories located at Paris, Greenwich, Leiden, and Potsdam. He returned briefly to Pulkovo before obtaining a position in 1886 at the Dorpat University Observatory. Struve was married to Elsa Elisabeth Grohmann; the couple had four children.

While at Dorpat, Struve devoted himself to the determination of positions and proper motions of stars. He collaborated with the Astronomische Gesellschaft (essentially an international society) in compiling a catalog of stellar positions for the Dorpat zone (from +70° to +75° declination) of the *Astronomische Gesellschaft Katalog*. Struve's doctoral degree was awarded in 1887 for his detailed comparison of stellar positions gleaned from the Bradley–Auwers catalog and the Pulkovo catalogs of 1845 and 1855.

In 1887, Struve estimated the angular rotation rate of the Galaxy to be −0.41 ± 0.42 arc seconds per century, under the assumption of a rigid body rotation. (The modern value of this parameter is −0.58 arc seconds per century at the Sun's distance from the galactic center.) That year, he also participated in Pulkovo's expedition to observe a total solar eclipse from the location of Smolensk. Between 1884 and 1888, Struve observed occultations of stars during total lunar eclipses for the purpose of determining the Moon's precise radius. For these results, published in 1893, he was awarded the first prize of the Imperial Russian Astronomical Society. In 1910, he also received the Society's Glasenapp Prize for his treatment of occultations over the past two decades. Professor Theodore Wittram wrote of Struve (1915) that he "should be considered as the most competent scholar in this field."

Struve moved to Kharkov University in 1894. He was the first professor extraordinarius and later full professor of astronomy and geodesy from 1897 to 1919. He also directed the university's observatory and from 1912 to 1919 was dean of the Faculty of the Physical and Mathematical Sciences. Together with N. N. Yevdokimov and B. I. Kudrevich, Struve observed the positions of selected zodiacal stars used for deriving the positions of minor planet (433) Eros, and of circumpolar stars from +79° declination to the celestial pole. These observations were applied to new determinations of the constant of precession and of the direction of motion of the Solar System. He took an active part in several geodetic projects, including the leveling work by which the Kharkov Observatory was included in the Russian Vertical Control Network, and he conducted measurements with the Rebeur–Pashwitz horizontal pendula.

In 1919, under mounting pressure from political events in post-revolutionary Russia, Struve and his family fled to Simferopol, Crimea. There, he obtained an academic position at the newly founded Tauride University. But grave misfortunes followed him, including the death of his son Werner and a younger daughter. Struve himself died suddenly while attending a meeting of the Tauride Learned Association, where he had gone to present a paper on Nova Cygni 1920. The new star was independently discovered and observed by his son, **Otto Struve**, and strongly influenced the latter's astronomical career.

Victor K. Abalakin

Alternate name

Struve, Ludwig Ottovich

Selected References

Courvoisier, L. (1921). "Todesanzeige: Ludwig Struve." *Astronomische Nachrichten* 212: 351–352.

Sokolovskaya, Z. K. (1976). "Struve, Gustav Wilhelm Ludwig." In *Dictionary of Scientific Biography*, edited by Charles Coulston Gillispie. Vol 13, pp. 113–114. New York: Charles Scribner's Sons.

Struve, Hermann Ottovich

➔ **Struve, Karl Hermann**

Struve, Karl Hermann

Born **Pulkovo, Russia, 3 October 1854**
Died **Kurort Herrenalb, (Baden-Württemberg), Germany, 12 August 1920**

A specialist in optics and planetary satellites, Hermann Struve (as he was usually known) was the son of **Otto Wilhelm Struve** and the elder brother of **Gustav Struve**. He was first educated at the gymnasia of Karlsruhe, Germany, and Vyborg, Russia. After passing his final exams at Revel, Russia (now Tallin, Estonia), he enrolled at the University of Dorpat in 1872, where he studied mathematics and physics. In 1874, Struve took part in Bengt Hasselberg's Pulkovo expedition to eastern Siberia and the port of Possiet to make observations of the transit of Venus. This experience interrupted his studies for almost a year. He graduated from Dorpat University in 1877 and returned to Pulkovo as a part-time astronomer under his father's direction.

To continue his education, Struve traveled to Paris, Strasbourg, Berlin, and Graz. At Berlin, his tutors were **Hermann von Helmholtz**, **Gustav Kirchhoff**, and Karl Weierstrass. While in Graz, he began a thesis under Ludwig Boltzmann's guidance on the problem of Fresnel interference and the diffraction of light, which was completed in 1881. Thereafter, he was awarded his magister's degree *cum lauda* from Dorpat University. In the following year, Struve received his doctoral degree in mathematics, also from Dorpat University, for the elaboration of a new theory of diffraction phenomena, which he tested with apparatus of his own design and construction.

Upon returning to Pulkovo as an adjunct astronomer, Struve conducted precise observations of Saturn's satellites, especially Iapetus and Titan, first with the observatory's 15-in. and, after 1885, with its 30-in. refractors. Upon his father's retirement, Struve served as senior astronomer from 1890 to 1895. During this time, he investigated the dynamics of Neptune, Mars, and Jupiter, observed the positions of planetary satellites, measured double stars, and published works on theoretical optics. He married Olga Struve, the daughter of his father's cousin, in 1885. The couple had two sons, one of whom, Georg Struve, became an astronomer.

To be nearer to his father at Karlsruhe, Germany, Struve accepted the directorship of the Albertus University Observatory at Königsberg in 1895. There, he published his most important work (1898), which provided a complete list of the basic constants of motion related to Saturn's ring and satellite system. Among Struve's discoveries was the recognition of libration motions of two satellite pairs: Mimas-Tethys and Enceladus-Dione.

Struve was awarded the Damoiseau Prize by the Paris Académie des sciences (1897) and the Gold Medal of the Royal Astronomical Society (1903). In 1904, he became director of the Berlin–Babelsberg Observatory and, from 1913 until his death, directed the Neu-Babelsberg Astrophysical Observatory that he helped to found.

Victor K. Abalakin

Alternate name

Struve, Hermann Ottovich

Selected References

Courvoisier, L. (1920). "Hermann Struve." *Astronomische Nachrichten* 212: 33–38.

F. W. D. (1921). "Karl Hermann Struve." *Monthly Notices of the Royal Astronomical Society* 81: 270–272.

Sokolovskaya, Z. K. (1976). "Struve, Karl Hermann." In *Dictionary of Scientific Biography*, edited by Charles Coulston Gillispie. Vol. 13, pp. 114–115. New York: Charles Scribner's Sons.

Struve, Ludwig Ottovich

➔ **Struve, Gustav Wilhelm Ludwig**

Struve, Otto

Born **Kharkov, (Ukraine), 12 August 1897**
Died **Berkeley, California, USA, 6 April 1963**

Russian–American stellar astronomer Otto Struve contributed to our understanding of the spectra of stars and nebulae, binary stars, the interstellar medium, and stellar structure and evolution. He was the great-grandson of **Friedrich Struve,** the grandson of **Otto Wilhelm Struve**, the nephew of **Karl Struve**, and the son of **Gustav Struve**, professor of astronomy at Kharkov University, and his wife Elizabeth.

Otto Struve had begun studies in astronomy at Kharkov University but enlisted in the Imperial Russian Army in 1916. At the end of World War I, he returned to his studies, completing a diploma (BS) in 1919 and then rejoined the army as a lieutenant in the White Russian forces opposing the revolution. When that cause was lost, he and many others fled to Turkey, where Struve attempted to make contact with members of his immediate family (few of whom survived). He got in touch with a German aunt who notified astronomers at the University of Chicago of his survival and circumstances. **Edwin Frost** was able to offer him transport and a position at Chicago and Yerkes, where he completed a Ph.D. in 1923. Struve continued collaboration with Russian astronomers throughout his career, including **Grigory Shain** on stellar rotation and carbon stars, **Boris Gerasimovich** on the interstellar medium (a very important investigation that nearly led to a pre-World-War-II recalibration of the cosmic distance scale), and K. F. Ogorodnikov (on β Cephei stars), though he said later that his inspiration to a career in astrophysics had come from the work of **Henry Norris Russell**.

Struve was appointed an instructor in astrophysics at Yerkes Observatory following his Ph.D. in 1924 and in 1927 became an assistant professor at Yerkes as well as a citizen of the United States. A hard worker, even a driven man, he became an associate professor in 1930, assistant director of the observatory in 1931, and director in 1932. In 1947, he became chairman and honorary director.

In 1950, in declining health, Struve moved to the University of California at Berkeley as head of the astronomy department. He returned home to Yerkes in 1959 and, in the same year, accepted the position of director of the new National Radio Astronomy Observatory in Green Bank, West Virginia. This move was a great surprise to most, for Struve had no experience at all in the field and, at that time, radio astronomy was considered by many to be outside "real" astronomy, a matter best left to amateurs and electrical engineers. Struve knew what he was doing, however, and he gave instant respectability to the discipline, which has flourished ever since.

By 1962, Struve resigned due to a further deterioration of his health, and took dual positions at Princeton University's Institute for Advanced Study and the California Institute of Technology. He died at 65, worn out by his labors. Struve and his wife, the former Mary Martha Lanning, had no children, and so with them the Struve dynasty came to an end.

From 1932 until 1947, Struve was editor of the *Astrophysical Journal*, then as now the premier publication in the field of astronomy. This was a demanding task (at least half time for most people), yet it hardly slowed down his astronomical research. In this position, Struve established and maintained a reputation for rigor combined with openness and fair mindedness.

While at Yerkes, Struve played a key role in the construction of what was then the world's second largest telescope. By 1932 the University of Texas had come into control of an $800,000 bequest of banker William J. McDonald, Paris, Texas, to build an observatory; however, Texas had no astronomy department. Meanwhile, Struve was frustrated by the lack of a large, modern telescope at Yerkes and the large proportion of cloudy nights there. Learning of the McDonald fund, he brokered an agreement whereby the University of Texas would build and own the observatory while the University of Chicago would operate it and pay all ongoing expenses. By the spring of 1939 an 82-in. reflector was in operation at the W. J. McDonald Observatory on Mount Locke in Trans-Pecos, Texas, where the skies are dark and for the most part clear. Struve was director at McDonald until 1947, and during his tenure dealt with the difficult problems involved in keeping it (and, of course, Yerkes as well) operating through World War II and its aftermath.

Despite his onerous administrative and editorial duties, Struve was first and foremost an observational astronomer, with stellar spectroscopy his *forte*. He particularly used spectra of very high dispersion for his era, which could be obtained only with telescopes of large aperture. Struve was not primarily a theoretician. Though he attempted to interpret his observations whenever possible, he made it a point to publish them even when they apparently defied explanation; the case of the massive, luminous, interacting binary β Lyrae is a prime example.

Starting his career by studying the radial velocities of more or less normal spectroscopic binaries, Struve was soon drawn to a lifelong fascination with unusual ones and, indeed, peculiar and even bizarre stars of all types, including variables. Many (but not all!) of his binary-star observations could be explained by invoking the presence of gas streams that transferred material from one star to the other.

Struve worked at a time when atomic physics was coming of age and, often by working at high dispersion, was able to apply and extend many recent advances. For example, he demonstrated the existence and influence of the Stark effect, rapid axial rotation, and turbulence on stellar spectral lines.

In another area, Struve demonstrated that absorption lines of ionized calcium in the spectra of distant stars were due to diffuse interstellar gas clouds. This led him to a general study of the interstellar medium and diffuse emission and reflection nebulae and, in turn, to his detection, with **Jesse Greenstein** of faint hydrogen emission throughout our Galaxy, using a specially designed nebular spectrograph at McDonald.

Struve's work on stellar rotation and gas streaming inspired him to speculate in the late 1940s on their possible roles in the evolution of stars as they age. His conclusion required major revision in light of future understanding of nuclear reactions in stars and post-main-sequence evolution as pioneered by **Martin Schwarzschild**.

Struve served the American Astronomical Society as vice president (1941) and president (1946–1949) and the International Astronomical Union [IAU] as vice president (1948–1952) and president (1952–1955). During the 1955 IAU General Assembly in Dublin, he was able to arrange a compromise whereby the union would meet in the Soviet Union (Moscow) in 1958 and in the United States (Berkeley) in 1961, thereby enabling international cooperation to continue in astronomy where it had ceased in almost all other activities. He was elected to the academies of science in the United States, the United Kingdom, Belgium, the Netherlands, and several other countries, received the Gold Medal of the Royal Astronomical Society, and was awarded nine honorary doctorates.

Struve was a prodigious author, with six books and more than 400 technical papers to his credit, as well as dozens of book reviews and popular articles. **Albrecht Unsöld**'s 1963 obituary notice in *Mitteilungen der Astronomischen Gesellschaft* (pp. 11–22) contains an exhaustive list of Struve's scientific publications, most of which appeared in the *Astrophysical Journal*. In addition, there are two long series of articles on a general level that appeared in *Popular Astronomy* from 1924 to 1951 and in *Sky & Telescope* from 1946 to 1963.

Of Struve's books, the most widely read was *Stellar Evolution: An Exploration from the Observatory* (Princeton University Press, 1950). His speculations in this area were not widely accepted at the time and soon became completely outmoded. However, the observational data presented there are still valid and as fascinating as ever, while the work has great historical value as an illustration of the way one astronomer was thinking at the time when evolution of stars was first seriously considered.

The 82-in. reflecting telescope at McDonald Observatory is named in Struve's honor.

Ronald A. Schorn

Selected References

Cowling, T. G. (1964). "Otto Struve." *Biographical Memoirs of Fellows of the Royal Society* 10: 283–304.

Evans, D. and J. Mulholland (1986). *Big and Bright: A History of the McDonald Observatory*. Austin: University of Texas Press. (The McDonald years are discussed from a Texan viewpoint.)

Goldberg, Leo (1964). "Otto Struve." *Quarterly Journal of the Royal Astronomical Society* 5: 284–290.

Milne, Edward A. (1944). "Address Delivered by the President, Professor E. A. Milne, on the Award of the Gold Medal to Professor Otto Struve." *Monthly Notices of the Royal Astronomical Society* 104: 112–120.

Osterbrock, Donald E. (1997). *Yerkes Observatory, 1892–1950: The Birth, Near Death, and Resurrection of a Scientific Research Institution*. Chicago: University of Chicago Press. (Covers Struve's career at the University of Chicago.)

Payne-Gaposchkin, Cecilia (1963). "Otto Struve as an Astrophysicist." *Sky & Telescope* 25, no. 6: 308–310. (Perhaps the best general survey of Struve's work.)

Struve, Otto and Velta Zebergs (1962). *Astronomy of the 20th Century*. New York: Macmillan.

Struve, Otto Wilhelm

Born **Dorpat (Tartu, Estonia), 7 May 1819**
Died **Karlsruhe, Germany, 14 April 1905**

Otto Wilhelm Struve's career spanned 50 years at the Pulkovo Observatory, where he succeeded his father as director from 1862 to 1889. His own work concerned astrometric topics like double stars and precession, but he supported the early development of astrophysics at Pulkovo.

Otto Wilhelm was the third son of **Friedrich Struve** and Emilie (*née* Wall) Struve. Otto Wilhelm received much of his early education at home before proceeding to the University of Dorpat (Tartu), where he also studied under his father. He graduated in 1839, shortly before the family relocated to Pulkovo, where he became one of four associate astronomers. In 1842, he married Emilie Dyrssen, with whom he had ten children, including **Karl Hermann Struve** and **Gustav Struve**, both of whom became astronomers. After Emilie's death in 1868, Otto Wilhelm married Emma Jankowsky in 1871, with whom he had one daughter, Eva.

It is unclear when Struve chose to become an astronomer. His two older brothers died prematurely, leaving him the eldest survivor of his parents' 12 children. He became deputy director of the Pulkovo Observatory in 1854 and assumed added duties after his father's incapacitating illness (1858), being officially appointed director in 1862. Struve's increasing involvement in its administration doubtlessly interfered with his own scientific productivity. He remained director until his retirement in 1889. Struve was elected an associate of the Royal Astronomical Society (1848) and received its Gold Medal in 1850. He was also elected a corresponding member of the Imperial Academy of Sciences (Saint Petersburg) in 1852 and made a full member in 1861. Struve was elected a foreign member of the Royal Society of London (1873) and an honorary member of the United States National Academy of Sciences (1883).

Struve continued his family's tradition of careful work in positional astronomy and in the discovery and measurement of double stars. He had a large share of the time on the new 15-in. refractor at Pulkovo (then the world's largest telescope) and added to his father's discoveries of binary stars. In particular, he discovered the binary nature of δ Equulei, a star that once held the record as the visual binary with the shortest known orbital period. Struve attempted to estimate personal errors by his construction and measurement of artificial double stars.

Struve used the 15-in. telescope for other kinds of observations, too. In 1852, he was perhaps the last person to observe the separated fragments of comet 3P/Biela, as it receded from the Sun. Struve assisted in the observatory's geodetic work, especially its determination of the longitude difference between Altona and Greenwich (1846). This task, which he undertook with Wilhelm Döllen (later his brother-in-law), formed the second half of a longitude determination undertaken between the Pulkovo and Greenwich Observatories. His father had supervised the earlier measurements. After his father became ill, Struve completed publication of the *Arc du Meridien*.

Struve's contemporaries considered his most important work to be new determinations of the constant of precession and of the solar motion. These were published by the Saint Petersburg Academy (1841) and the work awarded the Gold Medal of the Royal Astronomical Society. Astronomer Royal and society president **George Airy** gave a detailed description and analysis of Struve's work at the time of this award.

Despite a reputation to the contrary, Struve encouraged some of the earliest astrophysical observations made at Pulkovo. For example, he attempted spectroscopic observations of the aurora in 1868.

Increasingly, however, Struve became the observatory's administrator and enabled other members of his staff to do the scientific work. He coordinated Russian observations of the 1874 transit of Venus and facilitated an American expedition to the neighborhood of Vladivostok. Struve concluded that further observations of the Venus transit were unlikely to lead to an improved value of the solar parallax and did not organize any official Russian expedition in 1882.

Struve was very active in affairs of the Deutsche Astronomische Gesellschaft (German Astronomical Society), the only organization of its kind that then tried to operate on an international level. He was a prominent member of the International Metre Commission and tried, unsuccessfully, to get Imperial Russia to adopt the Gregorian calendar.

Although Struve was exclusively a visual observer, he was much impressed by the early photographic work of the French brothers **Paul** and **Prosper Henry**, and discussed the possibilities of astronomical photography in correspondence with **David Gill**. He became president of the Astrographic Congress held in Paris (1887) that initiated the photographic *Carte du Ciel* project. His letters, however, show that Struve remained ambivalent about its probable success. Events were to prove him correct, at least after his death, but his initial skepticism contributed to a feeling of isolation from the astronomical community that only increased with time.

Struve's final gift to Pulkovo was the construction of a 30-in. refractor. While the observatory's 15-in. telescope had once been the world's largest, it had long since lost that distinction. In 1878, he was authorized to make inquiries about the cost of a 30-in. refractor, an instrument sufficiently large then to restore Pulkovo's preeminence. United States Naval Observatory director **Simon Newcomb** influenced him to consider letting Alvan Clark & Sons make the telescope. This decision led to Struve's twice visiting the United States of America.

The building of the large refractor and dome encountered a number of delays, but the telescope arrived at Pulkovo in the summer of 1884 and was finally installed in the winter of 1884–1885. Struve made some use of the telescope at the beginning, but concluded that he was no longer strong enough to handle so large an instrument. Its chief user became his son Karl Hermann. The telescope was dismantled during World War II and its dome was destroyed during the siege of Leningrad. It has never been rebuilt.

Struve considered retirement on several occasions, partly because he found himself at odds with the Imperial Academy, which administered the Pulkovo Observatory. He remained in office at the tsar's request until the observatory's 50th anniversary had been celebrated. He then retired to Germany, eventually settling in Karlsruhe. After his second wife's death in 1902, Struve became increasingly infirm.

Alan H. Batten

Alternate name

Struve, Otton Vasilievich

Selected References

Anon. (1906). "Otto Wilhelm von Struve." *Monthly Notices of the Royal Astronomical Society* 66: 179–180.

Batten, Alan H. (1988). *Resolute and Undertaking Characters: The Lives of Wilhelm and Otto Struve*. Dordrecht: D. Reidel.

Nyrén, M. (1906). "Otto Wilhelm Struve." *Popular Astronomy* 14: 352–368. (English translation from *Vierteljahrsschrift der Astronomischen Gesellschaft* [1905], by Isabella Watson.)

Sokolovskaya, Z. K. (1976). "Struve, Otto Wilhelm." In *Dictionary of Scientific Biography*, edited by Charles Coulston Gillispie. Vol. 13, pp. 120–121. New York: Charles Scribner's Sons.

Struve, Otton Vasilievich

➔ **Struve, Otto Wilhelm**

Struve, Vasily Yakovlevich

➔ **Struve, Friedrich Georg Wilhelm**

Stukeley, William

Born **Holbeach, Lincolnshire, England, *circa* 1687**
Died **London, England, 3 March 1765**

Physician William Stukeley made studies of Stonehenge and thus foreshadowed the development of archaeoastronomy.

Stukeley, the eldest son in a family of four boys and a girl, was a man of wide interests, and was one of the first antiquaries to value ancient monuments and to show concern about their survival. He was educated at what is now Corpus Christi College, Cambridge, matriculating in 1704 at the age of 17. Although his interest in ancient relics began at Cambridge, Stukeley's investigations really got under way when he was in his 20s and practicing medicine in Lincolnshire. From 1718 to 1725, a period now regarded as the most significant of his archaeological career, Stukeley was very active in fieldwork. He was also a student of solar eclipses, observing the eclipse of 22 April 1715 while in practice in Boston, Lincolnshire; later he also observed the total solar eclipse of 11 May 1724 and the annular eclipse of 1 April 1764.

Stukeley was the first secretary of the Society of Antiquaries, which he helped to found in 1717, and was elected a fellow of the Royal Society in 1718. By the 1730s he was in Holy Orders, producing works of half-religious, half-antiquarian conjecture. Stukeley published, among other writings, *Itinerarium Curiosum* (1724), *Stonehenge, a Temple Restored to the British Druids* (1740), and *Avebury, a Temple of the British Druids* (1743). In 1747 he accepted the rectory of Saint George the Martyr, Bloomsbury, where in 1764 he observed the annular eclipse of the Sun.

In general Stukeley's fieldwork is reliable, but he could be wayward in his judgments. For instance, in 1757 he published as a genuine work of Richard of Cirencester (an authentic figure), Charles

Bertram's forgery *De Situ Britanniae*, which purported to be a history of Roman Britain. Stukeley's measurements of the stone circles at Avebury and Stonehenge were accurate and are still useful. However, he attributed the construction of the latter to the druids, and suggested without evidence that they had used it for esoteric purposes. But even though his ideas became increasingly eccentric, Stukeley was still widely acknowledged as the only antiquary in England interested in the pre-Roman period. Hence, in spite of his elaborate druidical fantasies, Stukeley more than anyone before him correctly placed the great stone circles in a prehistoric context, and unwittingly assisted in laying the foundations of archaeoastronomy.

Richard Baum

Selected Reference

Piggott, Stuart (1950). *William Stukeley, An Eighteenth Century Antiquary*. New York: Thames and Hudson. (Revised and enlarged, 1985.)

Su Song

Born **Tong'an, (Fujian), China, 1020**
Died **Runzhou (Zhenjiang, Jiangsu), China, 1101**

Su Song was a Chinese astronomer and pharmacologist in the Northern Song dynasty. His public name was Zirong. In 1042, Su Song passed the imperial examinations for government service. In 1086, he was ordered to investigate existing armillary spheres. He examined them and discussed armillary spheres with Han Gonglian, who had made a model of a water-driven armillary sphere. In 1087, Su Song started a project to make a new water-driven armillary sphere with Han Gonglian and others. After making a small model in 1088 and presenting a large model in 1089, Su Song completed the water-driven armillary sphere *cum* celestial globe called *Shuiyun yixiang tai* (tower of water-driven instrument) in 1092. He also composed the *Xin yixiang fayao* (Outline of the method for a new instrument), which is a detailed monograph on this instrument.

The invention of the armillary sphere and celestial globe in China is usually attributed to the time of the former Han dynasty. The inventor of the armillary sphere is said to have been Luoxia Hong, who contributed to the calendar reform (104 BCE) of the Taichu era. At that time, the armillary sphere was used to determine the equatorial system of coordinates. Probably only right ascension was measured initially and the north polar distance added later. At the beginning of the later Han dynasty, the concept of the ecliptic was established, and the first official instrument with an ecliptic circle is said to have been made in the year 103. The celebrated astronomer **Zhang Heng** made an armillary sphere and a celestial globe. The latter is said to have been the first attempt to rotate the celestial globe using waterpower.

The definite form of the armillary sphere was established by **Li Chunfeng** of the Tang dynasty in 633. His instrument contained three sets of rings: The outer set consisted of meridian, horizon, and equatorial rings; the middle set consisted of the rings for the equator, ecliptic, and the orbit of the Moon; while the inner set consisted of the polar axis and a sighting tube. **Yixing** and Liang Lingzan made an armillary sphere around 724. They also made a water-driven celestial globe in 725. The technology of the latter was improved by **Zhang Sixun** of the Song dynasty, who constructed a water-driven (mercury-driven in winter) celestial globe in 979. This was a predecessor of Su Song's instrument. Some armillary spheres were made in the Song dynasty, one of which was constructed by the polymath Shen Gua in 1074, who also wrote monographs on the armillary sphere, water clock, and gnomon. It may be mentioned here that Shen Gua and his colleague Wei Pu also contributed to the development of the calendar.

The technology to produce constant water flow was highly developed in China along with the development of the water clock. The water clock is said to have already been used in the Spring and Autumn and Warring States periods (770–221 BCE). The earliest extant water clocks in China are from the former (western) Han dynasty (206 BCE–8 AD). They are the simple outflow type of water clock. Of course, these simple water clocks cannot achieve a constant water flow, since the water pressure diminishes as the water volume decreases, so that the water flow also diminishes. The first attempt to achieve a constant water flow was made by Zhang Heng in the later (eastern) Han dynasty (25–220). He made an inflow-type of water clock with a double reservoir. Water was supplied by the upper reservoir, and the water level and water flow of the lower reservoir did not decrease much. In the Tang dynasty (618–907), Lü Cai made an inflow type of water clock with fourfold reservoir (where the upper three reservoirs are used to supply water) in the 7th century.

In the northern Song dynasty (960–1127), Yan Su made a kind of ultimate water clock in 1030. In this instrument, water was oversupplied by the upper reservoir to the lower reservoir, the water overflowing through a tube attached to the lower reservoir so that the water level of the lower reservoir is always at the height of the tube. This device was also utilized by Su Song. Actually, a simple siphon cannot achieve a constant water flow, as the water level of the acceptor increases and the difference between the water level of the reservoir and that of the acceptor decreases. In the water clock of Han Zhongtong, made in 1162, water from the reservoir siphon was accepted by a funnel, and then went to the acceptor. In this instrument, the above-mentioned defect was solved.

Su Song's *Shuiyun yixiang tai* was a huge clock tower. It consisted of three stories. The upper story on the roof was for an armillary sphere, the middle story was for a celestial globe, and the lower story was for mechanical devices to rotate the armillary sphere and the celestial globe (and also to move figures to indicate the time, and to signal the time). It had an escapement in order to control the movement.

Su Song's monograph on this instrument (*Xin yixiang fayao)* has a set of five star maps for drawing the celestial globe. They are clearly written, beautiful star maps (one map of circumpolar stars, two maps of non-circumpolar stars expanded around the equator, and a pair of maps of the Northern and Southern Hemispheres).

Alternate name

Su Sung

Selected References

Bo Shuren and Cai Jingfeng (1992). "Su Song." In *Zhongguo gudai kexue jia zhuanji* (Biographies of scientists in ancient China), edited by Du Shiran. Vol. 1, pp. 480–496. Beijing: Kexue chubanshe (Science Publishers).

Hua Tongxu (1991). *Zhongguo louke* (Chinese water clocks). Hefei: Anhui kexue jishu chubanshe (Science and Technology Publishing House of Anhui). (For the development of water clocks in China.)
Liu Xianzhou (1962). *Zhongguo jixie gongcheng faming shi* (History of mechanical engineering in China). Beijing Kexue chubanshe (Science Publishing House). (Pioneering work on the history of mechanical engineering in China, in which the contribution of Su Song is also discussed.)
Li Zhichao (1997). *Shuiyun yixiang zhi* (History of water-driven instruments), *Zhongguo kexue jishu daxue chubanshe*. Hefei: University of Science and Technology of China Press. (Includes a study of the instrument of Su Song containing the full text of *Xin yixiang fayao*.)
Needham, Joseph, with the collaboration of Wang Ling (1959). *Science and Civilisation in China*. Vol. 3, *Mathematics and the Sciences of the Heavens and the Earth*. Cambridge: Cambridge University Press. (For Chinese astronomy in general, including the contribution of Su Song.)
Needham, Joseph, Wang Ling, and Derek J. Price (1960). *Heavenly Clockwork*. Cambridge: Cambridge University Press. 2nd ed. 1989. (A detailed study of the instrument of Su Song.)
Ruan Yuan (1764–1846) (1799). *Chouren zhuan* (Biographies of astronomers). Reprint, Taipei: Shijie shuju. (See Vol. 20 for some classical accounts of Su Song.)
Su Song. *Xin yixiang fayao*. (A relatively popular reprint is the *Shoushange* series edition [1844 AD] included in the *Guoxue jiben congshu* series [1968] and also the *Renren wenku* series [1969] reprinted by Taiwan Commercial Press.)
Tuotuo (1314–1355 AD) *et al* (eds.) (1345 AD). *Song shi* (History of the Song dynasty.) (See Vol. 340 for the official biography of Su Song.)
Wang Zhenduo (1989). *Keji kaogu luncong* (Papers in technical archaeology, in Chinese). Beijing: Wenwu chubanshe (Cultural Relics Publishing House).
Yamada Keiji and Tsuchiya Hideo (1997). *Fukugen Suiun gishō dai* (Reconstruction of the tower of water-driven instrument). Tokyo: Shinyōsha. (A detailed study of the instrument of Su Song written in Japanese.)
Yan Zhongqi and Su Kefu (1993). *Su Song nianpu* (Chronological biography of Su Song). Changchun: Beifang funü ertong chubanshe.
Yi Shitong (1998). "Su Song." In *Zhongguo kexue jishu shi, Renwu juan* (A History of science and technology in China, Biographical Volume), edited by Jin Qiupeng, pp. 329–338. Beijing: Kexue chubanshe (Science Publishers).
Zeng Zhao. "*Su Sikong muzhiming*" (Epitaph presented to Sikong Su). In *Qufu ji*. Vol. 3. Included in the *Wenyuange* edition of *Siku quanshu* (The four treasuries of traditional literature compiled in the late 18th century.) (Reprint, Taiwan Commercial Press, 1983–1986. Vol. 1101, pp. 380–385).

Su Sung

➲ **Su Song**

Suárez, Buenaventura

Born **Santa Fe, Argentina, 3 September 1678**
Died **Paraguay, 24 August 1750**

The first native-born astronomer in the Western Hemisphere was Father Buenaventura Suárez, S. J. Assisted by Guarani Indians, Suárez constructed the telescopes for his mission observatory himself (including the grinding of lenses from native crystal) in the 1730s. His Southern Hemispheric observations included those of eclipses, comets, and occultations of Galilean satellites. Late in life, the Jesuits equipped Suárez with European-made telescopes. He wrote an almanac published in Spain and corresponded with **Anders Celsius**, **Joseph Delisle**, and other Old World astronomers.

Selected Reference

Toche-Boggino, Alexis (2000). "Buenaventura Suárez: The Pioneer Astronomer of Paraguay." *Journal of Astronomical History and Heritage* 3, no. 2: 159.

Suess, Hans Eduard

Born **Vienna, (Austria), 16 December 1909**
Died **San Diego, California, USA, 20 September 1993**

Austrian–American chemist and physicist Hans Suess, together with **Harold Urey**, compiled the table of abundances of the elements and isotopes in the Solar System that guided Alastair G. W. Cameron and E. Margaret Burbidge, Geoffrey R. Burbidge, **William Fowler**, and **Fred Hoyle** to understanding the origins of the elements, primarily from nucleosynthesis in stars. Suess was the son of Franz E. Suess (1867–1941), professor of geology at the University of Vienna and grandson of Eduard Suess (1831–1914), author of an important early work in geochemistry, *The Face of the Earth*, who had held the same position. Hans Suess's scientific interests were shaped by this background, and he received a Ph.D. in chemistry from the University of Vienna in 1935. After postdoctoral work in chemical institutes at the Swiss Technical University [ETH] and the University of Vienna, he accepted a position at the University of Hamburg in 1938.

During World War II, Suess worked primarily on the chemistry of deuterium, potentially important for nuclear fission reactors, and made occasional trips as a scientific advisor to the heavy water plant at Vemork, Norway, which had been occupied by the Germans in 1940 and was destroyed by Allied bombs in 1943. During the war years, he also became interested in the structure of nuclei and origin of the elements, realizing the importance of "magic numbers" of protons and neutrons, which made the nuclides containing those numbers more abundant and more stable than their neighbors. Suess was associated with J. Hans D. Jensen (1907–1973) in formulating a shell model of the nucleus (analogous to the model of electron shells due to **Niels Bohr**), for which Jensen shared the 1963 Nobel Prize with Maria Goeppert Mayer, who had worked out the same structure independently.

Suess moved to the University of Chicago to work with Harold Urey in 1950 and went on to the United States Geological Survey in Washington State, USA, the next year. There he set up a laboratory to do carbon-14 dating of organic materials dating from roughly the past 50,000 years. Suess developed a new technique in which carbon samples were processed to gaseous acetylene for measurement, and used it to establish the chronology of the end of the last Ice Age in the Northern Hemisphere and to show that dilution of atmospheric carbon dioxide by the burning of fossil fuels could make samples look older than they really are. This is sometimes called the Suess effect.

Roger Revelle invited Suess to join the Scripps Institute of Oceanography at the University of California, San Diego, in 1955, where he established a new radiocarbon lab and served as professor of geochemistry from 1958 to1977. An important early result was the calibration of carbon-14 dates against accurate counts of annual rings in trunks of very old trees. Deviations of about 10% in either direction around the average correlation were initially dubbed Suess wiggles, but are now understood to reflect changes in the speed and density of the solar wind flowing past the Earth and, therefore, in the intensity of cosmic rays that can penetrate that wind to produce carbon-14 in the Earth's upper atmosphere, and which may affect the climate.

The paper on "Abundances of the Elements" by Suess and Urey came in 1956. It included not just elemental abundances but also the relative amounts of the various stable isotopes of each element (ten for tin, but beryllium only one). The abundance numbers for elements that readily form solids came primarily from certain classes of primitive meteorites, and the numbers for gases like hydrogen and nitrogen from the Sun. A current compilation of abundances shows remarkably few and remarkably small differences from their results.

During his years at San Diego, Suess acted as a consultant to the International Atomic Energy Agency in Vienna. Suess received his Dr. *Habilitation* from Hamburg in 1938, an honorary degree from Queen's University (Belfast), and awards from the Guggenheim Foundation, the Alexander von Humboldt Foundation, and the Geochemical Society. He was a member of the national academies of science of the United States and Austria.

Fathi Habashi

Selected Reference

Arnold, James R., Kurt Marti, and Mark H. Thiemens (1993). "Hans E. Suess." In *In Memoriam*. Berkeley: University of California Press.

Ṣūfī: Abū al-Ḥusayn ʿAbd al-Raḥmān ibn ʿUmar al-Ṣūfī

Born **Rayy (near Tehran, Iran), 903**
Died **986**

Ṣūfī spent his life as an astronomer in Iran, in close relation to the regional rulers of the Buyid dynasty. The most important of his several astronomical and other works was the *Book on the Constellations* (*circa* 964). In it he gave a description of the 48 Ptolemaic constellations, based on the Arabic translations of **Ptolemy**'s *Almagest*, with detailed critique for each of the 1,025 stars in Ptolemy's star catalog, based on his own observations. Two drawings of each constellation were added, one "as seen in the sky," and one "as seen on the (celestial) globe."

The book became very influential both in the Orient and in Europe. Its text and nomenclature were taken up by many later authors, such as the encyclopedist Qazwīnī (died: 1283) and the Timurid Prince and astronomer **Ulugh Beg** in the star catalog of his astronomical handbook (epoch: 1437). For centuries, Arabic–Islamic astronomers followed the forms of the constellation figures as drawn in Ṣūfī's book, in written works and on instruments (celestial globes).

In Europe, Ṣūfī's book was not among the many scientific Arabic works that were translated into Latin between the late 10th and the 13th centuries. Nevertheless, its contents became known there and exerted considerable influence in several instances. King **Alfonso X** of Castile (reigned: 1252–1284) had a free recension of the book, with constellation drawings, included in his multivolume astronomical handbook, *Libros del saber de astronomia*; an Italian translation of this appeared in 1341. Perhaps also in the 13th century, a text corpus was compiled in Sicily, where drawings of the 48 constellations from Ṣūfī's book were combined with Ptolemy's star catalog (in the Latin translation of **Gerard of Cremona** from the Arabic) and extracts from some other astronomical and astrological texts (the so called Ṣūfī Latinus corpus, of which eight manuscripts are known today). In 1515, two maps of the Northern and Southern Celestial Hemispheres were printed in Nuremberg after woodcuts made by **Albrecht Dürer**. One of four portraits of important astronomers added by Dürer to the map of the Northern Hemisphere is an imaginary portrait of Ṣūfī (here called Azophi, with a medieval Latin spelling). In the 1530s, the German astronomer **Peter Apian** somehow made use of Ṣūfī's book, mentioned some old Arabic asterisms, and even converted them into drawn constellation figures on a star map. Ṣūfī's stellar nomenclature – in Arabic script – was also used on a celestial globe by J. A. Colom (*circa* 1635) and on the "King's globe" (1681–1683) by V. Coronelli. In 1665, Thomas Hyde published in Oxford an edition of **Ulugh Beg**'s star catalog; in the accompanying commentary he amply quoted from Ṣūfī's book. From here, **Giuseppe Piazzi** picked up around 100 Arabic star names, which he added to the 1814 edition of his Palermo star catalog, thereby introducing them into modern astronomy. Ṣūfī's name (in its medieval Latinized form, Azophi) was given by **Giovanni Riccioli** (1651) to one of the craters on the Moon.

Paul Kunitzsch

Selected References

Al-Ṣūfī, ʿAbd al-Raḥmān ibn ʿUmar (1986). *Kitāb Ṣuwar al-Kawākib.* Frankfurt am Main.(Facsimile of Oxford, Bodleian Library MS. Marsh 144.)

Kunitzsch, Paul (1986). "The Astronomer Abū 'l-Ḥusayn al-Ṣūfī and His Book on the Constellations." *Zeitschrift für Geschichte der Arabisch–Islamischen Wissenschaften* 3: 56–81. (Reprinted in Kunitzsch, *The Arabs and the Stars*, XI. Northampton: Variorum, 1989.)

Schjellerup, H. C. F. C. (trans.) (1874). *Description des étoiles fixes.* St. Pétersbourg. (French translation of Ṣūfī's text.)

Sulaymān ibn ʿIṣma: Abū Dāwūd Sulaymān ibn ʿIṣma al-Samarqandī

Flourished **Samarqand, (Uzbekistan), second half of the 9th century**

Much of our information on Sulaymān ibn ʿIṣma comes from the remarks of **Bīrūnī**. According to Bīrūnī, Sulaymān made observations in Balkh (Afghanistan) in 888–890 for determining the obliquity of the ecliptic. For this purpose, he used a mural quadrant (*libna*) provided with an alidade, the diameter of the quadrant being about 8 cubits (*dhirāʿ*), approximately 4 m. He found the meridian solar altitude at the winter solstice to be 29° 46′ and at the summer solstice 76° 54′. From this he determined that the obliquity of the ecliptic was 23° 34′, 1 min less than the result of **Battānī**. Bīrūnī also tells us of Sulaymān's determination of the length of Spring and Summer, and attributes to Sulaymān a *zīj* (astronomical handbook) dealing with the Sun and Moon (*Zīj al-nayyirayn*), as well as a book on the construction of an instrument for determining the visibility of the crescent (*Qānūn* II, p. 654). **Nasawī** claims that Sulaymān also wrote a commentary on the *Almagest*.

Finally, Sulaymān composed a commentary on the tenth book of Euclid's *Elements*, which is still extant.

Giuseppe Bezza

Selected References

Al-Bīrūnī, Abū al-Rayḥān Muḥammad b. Aḥmad (1954–1956). *al-Qānūn al-Masʿūdī*. Hyderabad.

Ali, Jamil (trans.) (1967). *The Determination of the Coordinates of Cities: Al-Bīrūnī's* Taḥdīd al-Amākin. Beirut: American University of Beirut.

Kennedy, E. S. (1973). *A Commentary upon Bīrūnī's Kitāb* Taḥdīd al-Amākin. Beirut: American University of Beirut.

Rosenfeld, B. A. and Ekmeleddin Ihsanoğlu (2003). *Mathematicians, Astronomers, and Other Scholars of Islamic Civilization and Their Works (7th – 19th c.)*. Istanbul: IRCICA, p. 78.

Sayılı, Aydın (1960). *The Observatory in Islam*. Ankara: Turkish Historical Society.

Schirmer, Oskar (1926/1927). "Studien zur Astronomie der Araber." *Sitzungsberichte der Physikalisch-Medizinische Sozietät in Erlangen* 58–59: 33–88.

Sezgin, Fuat (1978). *Geschichte des arabischen Schrifttums*. Vol. 6, *Astronomie*, p. 170. Leiden: E. J. Brill.

Steinschneider, M. (1870). "Zur Geschichte der übersetzung aus dem Indischen in's Arabische und ihres Einflusses auf die arabische Literatur." *Zeitschrift der deutschen morgenländischen Gesellschaft* 24: 325–392.

Sundman, Karl Frithiof

Born **Kaskinen, (Finland), 25 October 1873**
Died **Helsinki, Finland, 28 September 1949**

Karl Sundman is most widely remembered for his analytic solution to the so called three-body problem, and for his design of an analog computer, which was planned to perform the power series calculations needed for modeling planetary perturbations. He was the son of custom-house officer Johan Frithiof Sundman and Adolfina Fredrika Rosenqvist. His parents attempted to train him as a fisherman, but the boy was interested in academic learning, and prepared privately for admission to the Imperial Alexander-University at Helsinki. There, he studied mathematics and physical sciences (1893–1897) and also assisted in the bureau for stellar photography at the local astronomical observatory. From 1897 to 1899, Sundman studied at the Pulkovo Observatory, where he examined the orbital motions of the minor planets. His doctoral dissertation (1901) addressed the perturbations of minor planets having a mean motion twice that of Jupiter.

In 1902, Sundman was appointed *Privatdozent* (lecturer) in astronomy at Helsinki. He also conducted postdoctoral studies (1903–1906) in Germany and Paris, France. In 1907, Sundman was appointed extraordinary professor of astronomy, and in 1918 full professor and director of the Helsinki Observatory. He retired from that post in 1941.

During his directorship, the research program of the observatory (led by Ragnar Furuhjelm) concentrated on completion of the Helsinki Zone of the International *Carte du Ciel* Astrographic Chart and Catalogue begun by Sundman's predecessor, **Anders Donner**. Sundman accepted this responsibility as a matter of course, even though his own interests were directed toward celestial mechanics. Sundman was an unassuming scholar; he founded no school and had scarcely any followers who continued his work.

Sundman is best known for his theoretical solution to the three-body problem. Here, one considers three mass points having fixed masses, along with known initial positions and velocities, which attract one another according to Newton's law. The long history of the problem began with **Isaac Newton** himself. In 1772, **Joseph Lagrange** obtained five restricted solutions in closed form (Lagrangian points L1 through L5). By around 1890, it was generally accepted that a more complete solution must be sought in the form of a power series or expansion.

King Oscar II of Sweden offered a prize for the solution of some unsolved mathematical problems; one of them was the three-body problem. The prize was awarded in 1889 to **Jules Poincaré** for his pathbreaking study that initiated a new approach to celestial mechanics. Yet, Poincaré did not solve the problem posed. Within this context, Sundman began his own investigation, and he succeeded in providing a solution to the problem in two papers published by the Finnish Society of Sciences (1907, 1909). A more widely known summary was published in the journal *Acta Mathematica* (1912).

Sundman's work gave in principle an algorithm for calculating the power series representing the motions of the bodies. This algorithm, however, is so complicated that the series cannot in practice be used to compute the positions of the bodies or even to obtain a qualitative knowledge of their behaviors. Yet, the significance of Sundman's result arises from the fact that, after centuries of debate as to whether or not an analytic solution existed, he showed that it does exist for all initial values that provide a nonvanishing angular momentum to the system.

Sundman treated also the special case where the condition on the angular momentum is not fulfilled. In that case, a simultaneous collision of all three bodies becomes possible. The case of more than three bodies gives rise to complications that cannot be treated using Sundman's constructions.

Sundman's other works gave computationally more accessible treatments to the problem of perturbations. These included an article that he contributed to the *Encyclopädie der mathematischen Wissenschaften* (1915). Sundman also insisted that it was possible to construct a machine capable of performing (to a satisfactory approximation) the calculations needed for the perturbations. To achieve this end, he designed an analog computer and explained its principles in another 1915 paper. In his later years, he tried to realize these ideas, but his attempts came to naught. After the rise of digital electronic computers, Sundman's ideas retained only a historical value.

Sundman was in many respects a conservative scientist. He never accepted **Albert Einstein**'s general theory of relativity but speculated about nonrelativistic explanations for the anomalous perihelion advance of Mercury. This attitude was not uncommon among traditional astronomers of his generation.

Sundman was a member of the Finnish Society of Sciences, a foreign member of the Royal Swedish Academy of Sciences, and a member of the editorial board of the *Acta Mathematatica*. His work concerning the three-body problem earned him the 1913 de Pontécoulant Prize of the French Academy of Sciences. Sundman married twice – first to Edith Rosa Maria Anderson, and then to Fanny Alexandra Janhunen. The couples had no children.

Raimo Lehti

Selected References

Barrow-Green, June (1997). *Poincaré and the Three Body Problem*. Providence, Rhode Island: American Mathematical Society and London Mathematical Society.

Järnefelt, Gustav (1950). "Karl F. Sundman In Memoriam." *Acta Mathematica* 83, nos. 3–4: i–vi.

Lehti, Raimo (2001). "Karl Frithiof Sundman, forskare i himmelsmekanik, I–II." *Normat: Nordisk matematisk tidskrift* 49: 7–20, 71–88.

Schalén, C. (1976). "Sundman, Karl Frithiof." In *Dictionary of Scientific Biography*, edited by Charles Coulston Gillispie. Vol. 13, pp. 153–154. New York: Charles Scribner's Sons.

Suyūṭī: Abū al-Faḍl ʿAbd al-Raḥmān Jalāl al-Dīn al-Suyūṭī

Born **Cairo, (Egypt), 1445**
Died **1505**

Suyūṭī wrote an important work on "religious" astronomy, whose sources derived from the traditions of the Prophet. Born into a family engaged in religious scholarship and holding administrative offices, he became the most prolific authors in all of Islamic literature. His father was a preacher, taught Shāfiʿī religious law, and acted as a deputy judge (*qāḍī*). He died prematurely when his son was only 5 years old, but he had made financial arrangements that allowed Suyūṭī to pursue a path of scholarship through the guardianship and aid of his father's friends and students. Suyūṭī commenced his studies at an early age, with the study of Islamic religious sciences under various teachers. This included the study of *ḥadīth* (statements and actions of the Prophet Muḥammad and his companions as recorded by his contemporaries and collated into collections by later authors), some rudimentary arithmetic for the solution of problems of inheritance, and probably the study of rudimentary timekeeping (*mīqāt*) and traditional medicine. At the young age of 18, he assumed his father's former position of teaching religious law at the Shaykhū mosque and provided juridical consultative opinions. Soon afterward in 1467, Suyūṭī reinitiated the study of *ḥadīth* at the mosque of Ibn Ṭulūn. He was appointed to teach *ḥadīth* at the prestigious Shaykhūniyya *madrasa* (religious college) in 1472 and then was given a royal appointment by the Mamlūk Sultan Qā'it Bāy (reigned: 1468–1495) to the directorship of the Baybarsiyya *khānqāh* (Ṣūfī lodge) in 1486. Suyūṭī's personality and convictions resulted in controversy and polemics with contemporary scholars as well as officials among the ruling Mamluks. He withdrew from public life in 1501, following a conflict over the finances of the Baybarsiyya *khānqāh* and spent the rest of his days editing and revising his works.

Suyūṭī wrote over 500 works that primarily focus on topics and issues in the Islamic religious and the Arabic linguistic disciplines. Two of his works deal with astronomy and medicine. His interest in astronomy, however, was not in what we or his contemporaries would call scientific, *i. e.*, related to the pre-Islamic astronomical heritage that had been transmitted in the 8th and 9th centuries. Rather his interest in astronomy lay in the discussion of celestial objects and phenomena as found in the corpus of literature and activity, which comprises *ḥadīth*. As such, his *al-Hay'a al-saniyya fī al-hay'a al-sunniyya* (*The radiant cosmology: On sunnī cosmology*) is a religiously oriented account of "cosmology," that is to say, celestial and terrestrial entities from the perspective of *ḥadīth*, or more precisely the *ḥadīth* corpus which, in Suyūṭī's view, reflects the position of the Sunnī community as laid out by Sunnī religious scholars. In the introduction of the *Radiant Cosmology*, Suyūṭī states,

"This is a book on cosmology (*ʿilm al-hay'a*), which I have compiled from the traditions (*al-athār*) and have appended it with reports [by earlier narrators] (*akhbār*) so that those with intelligence may find delight and those with vision may reflect. I have titled it *The Radiant Cosmology: On Sunnı Cosmology*."

On the one hand, Suyūṭī wanted to inform his readers about Sunnī cosmology, as it was discussed in traditions and reports of earlier narrators. On the other hand, Suyūṭī's choice of the term cosmology (*hay'a*) for his religious enterprise was novel. The astronomers had utilized the term *hay'a* since the 9th century to signify the configuration of the celestial orbs. Thus the term *ʿilm al-hay'a* was used to signify the discipline of "astronomy." Suyūṭī's appropriation of the terms *hay'a* and *ʿilm al-hay'a* for his enterprise indicates a conscious attempt to present an alternative religious cosmology, that is to say an "Islamic cosmology," to replace the "scientific" cosmology of the astronomers. In his *Autobiography*, Suyūṭī is quite explicit regarding his views on science:

I do not occupy myself [with] logic and the philosophical disciplines (*ʿulūm al-falsafa*) because they are forbidden, and even if they were permissible, I would not prefer them to the religious disciplines.

During this period, astronomy, and other sciences, certainly fell under the classification of "philosophical disciplines." Suyūṭī and other religious scholars regarded them with suspicion for, in their view, these disciplines ultimately derived from pre-Islamic sources. Suyūṭī regarded his sources, in contrast, to be the unimpeachable views of religious scholars from earlier generations. Just as they had provided the material for the sound formulation of Islamic Law that governed all aspects of life, including the proper practice of rituals, the sound understanding of the text of the Qur'ān, and so forth, only they could provide the basis for a sound "Islamic" cosmology, that is to say the cosmology for Muslims who follow the path of tradition and orthodoxy (*i. e.*, the Sunnīs). He held similar views regarding medicine.

The subjects that Suyūṭī treats in the *Radiant Cosmology* comprise the Divine Throne (ʿ*arsh*), the Divine Footstool (*kursī*), the Tablet (*lawḥ*), and the Pen (*qalam*), which are entities mentioned in the Qurʾān, as well as the seven heavens and seven Earths, Sun, Moon, stars, night, day, hours, water and winds, clouds and rain, thunder, lightning, thunderbolt, Milky Way, rainbow, earthquakes, mountains, seas, and River Nile. Suyūṭī's approach to these subjects is apparent in his chapter headings, which refer to reports of the views of selected earlier authorities regarding these "cosmological" entities. As such, the *Radiant Cosmology* preserves the views of these earlier religious authorities whose works are lost to us.

Alnoor Dhanani

Selected References

Heinen, Anton M. (1982). *Islamic Cosmology: A Study of as-Suyūṭī's* al-Hay'a as-sanīya fī l-hay'a as-sunnīya *with Critical Edition, Translation, and Commentary*. Beirut: F. Steiner Verlag.

Ragep, F. J. (1993). *Naṣīr al-Dīn al-Ṭūsī's* Memoir on Astronomy *(al-Tadhkira fī* ʿ*ilm al-hay'a)*. 2 Vols. New York: Springer-Verlag.

Sartain, E. M. (1975). *Jalāl al-Dīn al-Suyūṭī: Biography and Background*. 2 Vols. Cambridge: Cambridge University Press.

Swan, William

Born **Edinburgh, Scotland, 13 March 1818**
Died **Helensburgh, (Strathclyde), Scotland, 1 March 1894**

William Swan is perhaps best known for his pioneering identification of features due to carbon compounds in the spectra of comets now called Swan bands and seen in stars and other sources. From 1850 to 1852 he was mathematical master in the Free Church of Scotland Normal School, and in 1853 was appointed teacher of mathematics, natural philosophy, and navigation in the Scottish Naval and Military Academy. He pursued scientific studies and published papers in *The Philosophical Magazine*, among others. Swan served as professor of natural philosophy at the United College of Saint Andrews University, Scotland, from 1859 to 1880, retiring due to ill health.

Swan was said to have been inquisitive, witty, and intelligent, with a fierce intolerance of fraud; he once described a wooden clock as "ferociously coarse and useless." Like many of his Victorian contemporaries, Swan had multiple interests, enjoyed broad literary and musical friendships, was a deacon of his church, and remained perpetually curious. His work in spectroscopy was meticulous.

Swan performed significant work on the flame spectra of carbon and hydrocarbon compounds. He insisted on the need for very pure samples to ensure that analysis would be fruitful. He was also one of the first (1856) to use a collimator in his spectroscope, which significantly improved the efficiency of his instrument.

Swan first associated several bright emission features in the spectra of comets with identical emission bands observed in candle flames in 1857. The Swan bands originate from the carbon molecule C_2; three prominent emission heads occur at wavelengths of 5636, 5165, and 4737 Å. These are formed when sunlight excites diatomic carbon in the tail of a comet. The carbon then fluoresces and emits light at these discrete wavelengths. The Swan bands give the comet's gaseous (or ion) tail its characteristic blue–green color.

Swan's 1856 attempt to extend **Joseph von Fraunhofer**'s study of solar absorption lines to the spectra of stars was not successful. He was, however, able to observe the spectrum of Mars and remarked that its appearance was "more brilliant than I anticipated." He is also credited with invention of the Swan prism photometer, used for measuring the brightnesses of stars, and establishing that "red protuberances" (prominences) seen in total solar eclipses arise from the Sun and not the Moon. Swan additionally made important suggestions for the improvement of lighthouse lenses.

Swan compiled a detailed inventory of the scientific and historic instruments preserved in the Department of Natural Philosophy at Saint Andrews. Among the instruments from his catalog that still exist are an armillary sphere, a Gregorian telescope, and several astrolabes.

Swan was awarded the Gold Medal of the Royal Scottish Society of Arts for his work in spectroscopy in 1883. He was also the recipient of honorary LLDs from Edinburgh University (1869) and Saint Andrews University (1886), and was a member of the Royal Society of Edinburgh.

Portions of Swan's correspondence are preserved at the Saint Andrews University Library.

Katherine Haramundanis

Selected References

Anon. (3 March 1894). "Obituary". *St. Andrews Citizen*.

Clerke, Agnes M. (1902). *A Popular History of Astronomy during the Nineteenth Century*. 4th ed. London: Adam and Charles Black.

Galbraith, James L. (1910). *A Scotch Professor of the Old School*. Glasgow: James MacLehose.

Hearnshaw, J. B. (1986). *The Analysis of Starlight: One Hundred and Fifty Years of Astronomical Spectroscopy*. Cambridge: Cambridge University Press.

Knott, C. G. (1902). "On Swan's Prism Photometer, commonly called Lummer and Brodhun's Photometer." *Proceedings of the Royal Society of Edinburgh* 23: 12–14.

Pannekoek, A. (1961). *A History of Astronomy*. London: George Allen and Unwin.

Poggendorff, J. C. (1898). "Swan." In *Biographisch-literarisches Handwörterbuch*. Vol. 2, pp. 1315–1316. Leipzig: Johann Ambrosius Barth.

Swan, William (1884). *Address, chiefly on the Graphic Arts, at the Annual Meeting of the Royal Scottish Society of Arts, 12 November 1883*. Edinburgh: Neill.

Swedenborg, Emanuel

Born **Stockholm, Sweden, 29 January 1688**
Died **London, England, 29 March 1772**

Best known for his religious writings, Emanuel Swedenborg was active as a scientist, engineer, statesman, and philosopher. In astronomy, he was the first to propose that the Solar System originated

from a swirling nebula. Swedenborg published treatises on nearly every scientific and philosophical issue of his age.

Upon graduation from the University of Uppsala, Swedenborg traveled abroad for 5 years, visiting England, Holland, France, and Germany, one of many such trips throughout his life. In 1710 he began studies in England. Swedenborg's studies might have ended before they began, as the impetuous young man was nearly hanged for breaking the plague quarantine of his vessel. He was an agent for his home university, with instructions to acquire whatever books and scientific instruments would improve its collection, and to make detailed reports about scientific ideas, techniques, and instruments.

Swedenborg met some of the leading intellectuals of the day. Astronomer Royal **John Flamsteed** inspired him to devise his own solution to the longitude problem. Swedenborg took careful notes about the instruments at Greenwich: His dream, never realized, was to establish an observatory in Sweden. Swedenborg became particularly close to **Edmond Halley**, who acquainted him with the problem of the lunar motion, and he spent several months in Oxford to be near his mentor. Although initially enthusiastic about Newtonianism, he found **Isaac Newton**'s reticence toward inquiring into the ultimate causes of phenomena (*hypotheses non fingo*) unsatisfying. In his *magnum opus*, also called *Principia* (1734), Swedenborg presents his own natural philosophy.

On his return home in 1715, Swedenborg launched Sweden's first scientific journal, *Daedalus Hyperboreus*, which featured new inventions, theories, and scientific discussion. He was appointed by the king as assessor to the Board of Mines, a position that he occupied (making substantial improvements to the mining industry) until his retirement in 1747. Swedenborg was active in research, sketching many inventions, (including a submarine, a flying machine, a new type of siphon, and an air pump) and making fruitful investigation into human perception and the brain. After 1747 he devoted the remainder of his life to religious writing.

While Swedenborg's contributions to neuroanatomy are better known, he also published astronomical and cosmological treatises. Two of these will be considered here: a solution to the longitude problem (1721), and a theory of the origin of the Solar System (1734).

Longitude determination was the outstanding practical issue of the time. The lunar distance method, proposed by **Johann Werner** in 1514, was in widespread use: It employs the Moon as a clock as it moves against the fixed stars. Werner's method was impractical at the time, since there existed neither sufficiently accurate star tables or instruments, nor a correct lunar theory. In 1714, the British Board of Longitude offered £20,000 for a practical and accurate method for reckoning longitude at sea. After extensive discussion with Flamsteed and Halley, Swedenborg published *Methodus nova inveniendi longitudines locorum terra marique ope lunae* (Amsterdam, 1721). The board rejected it, since his solution did not meet the requirements for the prize – he provided no tables for lunar and stellar positions. When compared to the 1765 prize-winning solution, which involved the combined efforts of several men including **Tobias Mayer**, Flamsteed, and **Leonhard Euler**, it is clear that Swedenborg's was not a complete solution.

Swedenborg's "nebular hypothesis" for the origin of the Solar System, described in his *Principia rerum naturalium* ... (Dresden, 1734), anticipated the cosmological theories of **Georges Leclerc** (Comte de Buffon; 1749), **Immanuel Kant** (1775), and **Pierre de Laplace** (1796). Swedenborg's conception was derived from his philosophy of "like-partedness," the idea that every entity is recursively composed of smaller, homologous versions of itself, and that it likewise forms a component part of a larger entity. For the Solar System, Swedenborg proposed that the Sun had developed a dense surface layer that was forced outward by the centrifugal force of its rotation, into the equatorial plane of the solar rotation. Continuing its outward motion, this ring eventually thinned and broke apart into smaller bodies that formed the planets and smaller bodies. Swedenborg's theory, unlike those of Buffon, Kant, and Laplace, is primarily based on an *a priori* conception rather than empirical investigation.

Glen M. Cooper

Selected References

Arrhenius, Svante (1908). "Emanuel Swedenborg as a Cosmologist." Introduction to *Emanuel Swedenborg: Opera quaedam aut inedita aut obsolete de rebus naturalibus*. Vol. 2, *Cosmologica*, edited by Alfred H. Stroh, pp. xxiii–xxxv. Stockholm. (Also contains the Latin text of the *Principia*.)

Benz, Ernst (2002). *Emanuel Swedenborg: Visionary Savant in the Age of Reason*, translated by Nicholas Goodrick-Clarke. West Chester, Pennsylvania: Swedenborg Foundation.

Koke, Steve (1988). "The Search for a Religious Cosmology." In *Emanuel Swedenborg: A Continuing Vision: A Pictorial Biography and Anthology of Essays and Poetry*, edited by Robin Larsen, pp. 457–466. New York: Swedenborg Foundation.

Swift, Lewis

Born **Clarkson, New York, USA, 29 February 1820**
Died **Marathon, New York, USA, 5 January 1913**

Lewis Swift discovered 13 comets and 1,248 previously uncataloged nebulae, which placed him after only **William Herschel** in the number of nebulae he discovered visually. His report of the hypothetical planet Vulcan was eventually discredited.

The son of Lewis and Ann (*née* Forbes) Swift, the younger Lewis was born into a distinguished family. His father was a general in the local militia, while his grandfather had served in the Revolutionary War in general Putnam's personal guard. An earlier ancestor had emigrated from England to Massachusetts in 1630. Swift's father farmed, and during the winter made farming implements; as a young man Swift also showed mechanical ingenuity. His life was changed when, at age 13, he fractured his hip during a farming accident and was unable to continue his work on the family farm. Instead – incapacitated for farm work – young Swift gained time for study. He trudged the 2 miles to the local school on crutches and laid the rudiments of his education. That same year (1983) he was awestruck by the great Leonid meteor storm, and 2 years later by the apparition of comet 1P/Halley. For a decade or more after completing the schooling available to him in Clarkson, Swift traveled as an itinerant science lecturer.

After his marriage in 1850 to Lucretia Hunt, Swift became a country storekeeper at Hunt's Corner, New York. Like **Edward Barnard**, he was stimulated in his incipient astronomical interest by reading the works of **Thomas Dick** and fashioned a 3-in. Spencer lens into a first small telescope. Inspired by the dramatic orations of **Ormsby Mitchel** in nearby Rochester during 1857 and 1858, when the Spencer lens was accidentally broken, Swift replaced that primitive first telescope with a 4½-in. Henry Fitz refractor. Two years later, Swift discovered his first comet, which had also been discovered independently by **Horace Tuttle** at Harvard College. The comet, 109P/1862 O1 (Swift–Tuttle) proved to be periodic; it returns to perihelion once every 134 years. Later in the 1860s, the Italian astronomer **Giovanni Schiaparelli** showed that comet 109P/Swift–Tuttle left debris in its orbit, which the Earth encounters each year as the Perseid meteor shower – the famous August meteors.

After Lucretia died in 1862, Swift married Caroline Topping of Long Island, New York, in 1864; they had one son, Edward. In 1872, Swift moved to Rochester, New York, to open a hardware store. From the flat roof above a nearby cider mill where he set up his telescope, Swift began discovering comets with considerable regularity. He found new comets in 1877 (C/1877 G2), 1878 (C/1878 N1), and 1879 (C/1879 M1), receiving Gold Medals from the Vienna Observatory for each of these discoveries.

One of Swift's most famous observations occurred during the total solar eclipse of 1878. From his observing station at Denver, Colorado, Swift reported observing Vulcan, a hypothetical planet between the Sun and Mercury. Vulcan had been postulated by the French mathematical astronomer **Urbain Le Verrier** to explain the advance of Mercury's perihelion. Swift's observations were taken seriously at the time, in part because University of Michigan astronomer **James Watson** also reported observing Vulcan from his eclipse site near Rawlins, Wyoming. Vulcan remained a controversial subject among astronomers for four decades. In 1916, **Albert Einstein** published his general theory of relativity, which satisfactorily accounted for the motions of Mercury for which Vulcan had been invoked as a cause.

Swift's work attracted the interest of Hulbert Harrington Warner, a patent-medicine vendor whose *Safe Remedy* promoted good bowel hygiene, a late-Victorian obsession. Warner endowed an observatory at the corner of East Avenue and Arnold Park in Rochester. The Warner Observatory was equipped, with public donations, with a 16-in. Clark refractor, then the fourth largest in the United States. Swift was placed in charge of the observatory. He gave public demonstrations of the telescope, served as clearinghouse for claims to Warner's prize for new comet discoveries, and charted new nebulae – most of them now known to be galaxies. The latter turned up by the hundreds in areas of the sky like Draco, which was located far from the Milky Way. Neither Swift nor anyone else knew what the nebulae were; they remained shrouded in mystery, and, as he told a meeting of the American Academy for the Advancement of Science in 1884, "this is, and for ages to come must be, true." However, Swift and others, notably **Stéphane Javelle** of France, were discovering so many new nebulae that **Johann Dreyer** abandoned his efforts to update his New General Catalogue and started the Index Catalogue to accommodate the new discoveries.

Swift found conditions in upstate New York – the snowbelt – less than ideal for observing; most portraits of him show him wearing a woolen cap. With city electrification, the skies over Rochester became less and less favorable. Swift managed to discover a new comet in 1892 (C/1892 E1), which became the brightest since 1883; it brightened to 3rd magnitude, and its intricate tail was captured in wide-angle photographs taken at Lick Observatory by **Edward Emerson Barnard**. After his business failed in the financial panic of 1893, Warner abandoned Rochester and moved to Minnesota, where he retired to an obscure old age.

With no support in Rochester, and after considering a number of offers, Swift joined Civil War balloonist and businessman Thadeus Sobieski Constantine Lowe (1832–1913) in setting up an observatory built around the 16-in. refractor at Echo Mountain, in California's Sierra Madre range. At the Lowe Observatory, Swift – dubbed the "Columbus of the Skies" by residents of nearby Pasadena and now in his 70s – continued to discover uncataloged nebulae. He discovered his last comet, his 13th, in 1899 (C/1899 E1), when he was almost 80. His son, Edward, also discovered a comet at the Lowe Observatory in 1894 (C/1894 W1).

The Lowe Observatory narrowly escaped a wildfire in 1900. By then Swift's eyesight was beginning to fail. At last in 1904 – even as **George Hale** arrived in California to develop the new Mount Wilson Observatory on another mountain, 3,000 ft. higher in the Sierra Madres – Swift retired, and returned to upstate New York to live in obscurity with his daughter. He died, all but forgotten by earlier generations of astronomers, among whom he had once been prominent. In his own words: "So much for fame!"

Swift was elected a fellow of the Royal Astronomical Society [RAS] and received an honorary Ph.D. from the University of Rochester in 1879. The French Academy of Sciences awarded Swift their Lalande Silver Medal and Prize in 1881, while the RAS selected him to be the first recipient of their Jackson-Gwilt Medal and Gift in 1897.

William Sheehan

Selected References

Bates, Ralph and Blake McKelvey (1947). "Lewis Swift, The Rochester Astronomer." *Rochester History* 9, no. 1: 1–19.

Baum, Richard and William Sheehan (1996). *In Search of the Planet Vulcan*. New York: Plenum Trade.

Seims, Charles (1976). *Mount Lowe, the Railway in the Clouds*. San Marino, California: Golden West Books.

Sheehan, William (1995). *The Immortal Fire Within: The Life and Work of Edward Emerson Barnard*. Cambridge: Cambridge University Press.

Swift, Lewis (1887). *History and Work of The Warner Observatory, Rochester, New York: 1883–1886*. Vol. 1. Rochester, New York: Democrat and Chronicle Book and Job Print.

Wlasuk, Peter T. (1996). "'So much for fame!': The Story of Lewis Swift." *Quarterly Journal of the Royal Astronomical Society* 37: 683–707.

Swings, Polydore [Pol] Ferdinand Felix

Born **Ransart, Belgium, 24 September 1906**
Died **Eseneux, Belgium, 28 October 1983**

Belgian spectroscopist Polydore Swings codiscovered the first interstellar molecule and identified a number of other molecules and radicals in the spectra of nebulae and comets. He was educated at the prestigious Athénée Charleroi, where an early interest in astronomy was stirred by the books of **Camille Flammarion**, and studied celestial mechanics under Marcel Dehalu at the University of Liège, earning a first degree in 1927. Swings spent the next 2 years in France, taking courses at the Sorbonne, the Collège de France, and the Institute d'Optique and working at the Observatoire de Paris in Meudon, where he learned the techniques of astronomical spectroscopy.

Swings returned to Liège in 1928 and, apart from brief visits abroad, spent the rest of his academic career there. Delahu and he set up a laboratory to carry out spectroscopy, since the wavelengths to be expected from many molecules (neutral and ionized) were not known or calculable at the time. He investigated the molecule S_2 on a visit to the Institut de Physique at Warsaw University, receiving a second degree from Liège in 1931 for this work. Swings focused on the phenomena of predissociation (in which a molecule is excited to a state with more energy than is required to unbind it) and fluorescence (in which an atom or molecule is excited by ultraviolet light in a line or continuum and emits a specific wavelength of visible light as it cascades back to its ground state), which he applied to a number of astronomical contexts over the years.

In 1931, Swings began a collaboration with **Otto Struve** observing and interpreting stellar and nebular spectra, and pushing the available data into ultraviolet wavelengths using the new quartz spectrograph at the (also new) McDonald Observatory. That collaboration became closer when the Swings family came to the United States for a sabbatical in 1939, remaining for the duration because Germany soon occupied Belgium. Pol and Christiane Swings's son Jean-Pierre, also an astronomer and eventually general secretary of the International Astronomical Union, was born in the USA. They spent time at Yerkes, McDonald, and Lick observatories, and Swings built spectrographs for the United States Navy at Lick. The most cited work by Struve and Swings dates from this period. They were able to interpret the spectra of stars with very extended envelopes, like P Cygni, in terms of absorption in the stellar photosphere plus emission from the extended gas, with the strengths of some lines greatly enhanced by fluorescence when the gas was moving at just the right velocity to have, for instance, a strong line of hydrogen or helium fall at the wavelength needed to excite some particular state. They also determined the expansion velocities of nova explosions, and showed that the hottest gas expanded most slowly.

In 1937, Swings and Leon Rosenfeld (1904–1974, a younger associate of **Niels Bohr**) identified a sharp 4303.3 Å astronomical line observed by **Theordore Dunham** that same year with a laboratory transition of the CH molecule. Since the line is a "stationary" one (*i. e.*, its wavelength does not shift back and forth with the stellar lines in the spectra of binary stars), it must be produced by gas that is not associated with the stars themselves. Swings and Rosenfeld therefore added a molecular component to the interstellar atomic gas discovered in 1904 by **Johannes Hartmann**.

During and after the war, Swings turned his attention increasingly to the study of cometary spectra, for which ultraviolet spectra are particularly important. He and **Andrew McKellar** found that a spectral feature at 4050 Å that they had seen in stars with carbon-rich atmospheres was also present in many comets, and they suggested it might be produced by a carbon molecule of more than two atoms, then a radical idea in astronomy. They were vindicated when German–Canadian spectroscopist **Gerhard Herzberg** and others identified C_3 or carbazone molecular features with the ones, now called the Swings band, in the stars and comets. Swings and his colleagues, first in the United States and later back in Belgium, gradually added CH^+, OH^+, CO_2^+, CN, CH, OH, and NH_2 to the cometary inventory and concluded that most of these must come from stable "parent" molecules of H_2O, CO_2, NH_3, and CH_4 in the ices of cometary nuclei. A 1956 Atlas of Representative Cometary Spectra summarized this work and long remained a standard. The name Swings effect is given to the fluorescence of ultraviolet CN features whose intensity changes as comets move toward and away from the Sun and Doppler shifts swing particular transitions in and out of resonance with the exciting solar line.

Swings established a long-running series of summer conferences at Liège, whose published proceedings were authoritative sources on stellar astronomy for many years. He was president of the International Astronomical Union from 1964 to 1967 and was one of about a dozen European astronomers who actively collaborated to establish the European Space Research Organization (now European Space Agency) and the European Southern Observatory. Swings received the highest Belgian scientific honor, the Prix Franqui, and the Royal Academy of Belgium now awards a prize in his and his wife's names to a promising young astrophysicist every 4 years.

Peter Wlasuk

Selected References

Garstang, R. H. (1986). "Pol Swings." *Quarterly Journal of the Royal Astronomical Society* 27: 305–308.

Ledoux, P. (1984). "In Memoriam: Pol Swings." *Astrophysics and Space Science* 102: 1–2.

Swope, Henrietta Hill

Born **Saint Louis, Missouri, USA, 26 October 1902**
Died **Pasadena, California, USA, 24 November 1980**

American variable star astronomer Henrietta Swope is remembered for the discovery and period determinations of a very large number of RR Lyrae, Cepheid, and other variables, the later ones on plates taken by **Walter Baade**.

Swope was the daughter of Gerard Swope (president of General Electric and director of the National Broadcasting Company) and Mary Dayton Hill. She became interested in astronomy while attending lectures by Margaret Harwood at Maria Mitchell Observatory on Nantucket, Massachusetts, and received a BA from Barnard College (Columbia) in 1925 and an MA from Radcliffe College (Harvard) in 1928.

Swope was appointed to an assistantship (partly financed by her father) at Harvard College Observatory in 1928 to continue her work with **Harlow Shapley** on galactic variable stars. Her work was of great precision; of 35 periods for RR Lyrae stars that she published in one 1929 paper, 34 remain definitive, and she worked on about 1,600 variables between 1927 and 1942. Her father's money also permitted the hiring of an assistant for her, **Dorrit Hoffleit**.

With the outbreak of World War II, Swope moved to the radiation laboratory at the Massachusetts Institute of Technology, where she helped to develop LORAN navigation tables with Fletcher G. Watson (1912–1997), and then in 1943 to the hydrographic office of the United States Department of the Navy, where she remained as a mathematician until 1947.

Her non-reappointment at Harvard after the war led to a position, first, as associate astronomer at Barnard College and Columbia (1947–1952), and then as assistant astronomer at Mount Wilson and Palomar observatories (1962–1967), where Swope was first Baade's assistant, and, after his death, his scientific executor. She completed the work on Cepheid variables that led to a new determination of the distance to the Andromeda Nebula (M31) of 2.2 million light years or 675,000 parsecs in 1962, very close to the best modern value. For several years, in the early 1960s, Swope appeared in the annual official photographs of the Mount Wilson–Palomar scientific staff, the only woman to do so in roughly the first 80 years of the observatories' existence. She held the title research fellow from her retirement in 1967 until shortly before her death.

Swope received the Annie J. Cannon Award of the American Astronomical Society in 1968 (the last one given as a lifetime achievement award) and is eponymized *via* the minor planet (2168) Swope. Her gift to the Carnegie Institution of Washington (former owners of Mount Wilson Observatory) made possible the initial development of the Las Campanas Observatory in Chile and the establishment of the first 40-in. Swope telescope there, though she herself never observed at either Mount Wilson or Palomar Mountain, under the policies then in force.

Katherine Haramundanis

Selected References

Rossiter, Margaret W. (1995). *Women Scientists in America: Before Affirmative Action, 1940–1972.* Baltimore: Johns Hopkins University Press.

Swope, Henrietta H. (3 August 1977). Oral History Interview with David DeVorkin. Niels Bohr Library, American Institute of Physics, College Park, Maryland.

Synesius of Cyrene

Born **Cyrene (near Darnah, Libya), *circa* 365–370**
Died **Ptolemaïs (near Al Marj, Libya), *circa* 413**

Synesius of Cyrene figured prominently in the literary, philosophical, scientific, and religious culture of the Greek east of Late Antiquity, playing a leading role on the contemporary political and historical stage. He also wrote a description of an astrolabe and encouraged the study of the heavens in order to know the divine.

Synesius's birth date remains conjectural. He spent his youth in Cyrenaica, receiving a classically grounded education typical of the landed aristocracy. Sometime after 390 he began the study of philosophy, mathematics, and the sciences in Alexandria with the Neoplatonist **Hypatia**, with whom he maintained close personal contact throughout his life. After several years in Alexandria, Synesius returned home, but soon traveled to Constantinople on diplomatic business on behalf of his province.

Synesius remained in the imperial capital until 402. While involved in the affairs of the court of Emperor Arcadius, he may have begun his association with Christianity. He later married a Christian and settled again in Cyrenaica, although he visited Alexandria a number of times in subsequent years, continuing his contact with Hypatia and others in her intellectual circle. Little is known about Synesius's other activities until 410, when he was appointed Bishop of Ptolemaïs, a post he held until his death around 413.

Synesius represents the blending of pagan Hellenism with the burgeoning Christian culture of late Roman times. Moreover, his wide range of interests, from the purely intellectual and practical to the mystical and even the magical, is evidence of the intellectual landscape of his day. Synesius's substantial extant written corpus consists of poems, hymns, homilies, and a variety of essays on philosophy and politics. In addition, a large collection of letters gives an especially intimate portrait of Synesius's life, of his intellectual pursuits, and of his relationship with his contemporaries, especially Hypatia.

A number of Synesius's works touch upon astronomical matters and cosmology. In some of the hymns (most notably nos. three and four) appears the Neoplatonist belief that the visible and invisible Universe is in itself divine, and several passages praise the beauty and majesty of the cosmos. Along with other philosophical topics, this same mystical engagement with the heavens appears in Synesius's letter to Paeonius (*Ad Paeonium de dono, circa* 397/398) that accompanied the gift of a silver astrolabe. To prompt in that imperial official an appreciation for astronomy, Synesius says that a knowledge of the divine is attainable through the study and practice of observing the skies. He also includes an epigram of **Ptolemy** that expresses a deep reverence for the visible Universe and his own verse composition on the benefits of using the instrument.

Additionally, though somewhat confused and at times vague, his technical description of the astrolabe itself, which Synesius calls only the *organon* ("instrument"), constitutes a contribution to its history and to an understanding of the general astronomical knowledge of the day. For example, Synesius famously ascribes to **Hipparchus** a knowledge of the astrolabe and assertes that Ptolemy had used such a device to determine the hours of the night. He also credits his teacher Hypatia with instructing him how to construct the instrument and claims that he himself had added the final details to perfect it, saying that he wrote a (now lost) treatise on the topic. He discusses actual design features of the object, including the placement of the celestial circles and the number of stars shown. Synesius also seems to have represented the celestial sphere as projected onto a shallow concave surface, thereby modifying a true astrolabe's stereographic projection of the celestial dome onto a flat plane.

Nevertheless, Synesius does not seem actually to have contributed anything to the theory or design of the astrolabe itself. In fact, his instrument may actually have been a kind of a celestial map incorporating only some features of an astrolabe.

John M. McMahon

Selected References

Bregman, Jay (1982). *Synesius of Cyrene, Philosopher–Bishop*. Berkeley: University of California Press. (Concentrates on philosophical, political, and religious dimensions.)

Cameron, Alan and Jacqueline Long (1993). *Barbarians and Politics at the Court of Arcadius*. Berkeley: University of California Press. (The most comprehensive biographical and analytical treatment of Synesius yet to appear.)

Dzielska, Maria (1995). *Hypatia of Alexandria*. Cambridge, Massachusetts: Harvard University Press. (For informative discussions of Synesius's relationship to Hypatia and her philosophical and scientific circle.)

Fitzgerald, Augustine (1926). *The Letters of Synesius of Cyrene*. London: Oxford University Press. (Reprint, 1980.)

——— (1930). *The Essays and Hymns of Synesius of Cyrene*. London: Oxford University Press. (Reprint, 1983.)

Garzya, A. (1979). *Synesii Cyrenensis epistolae*. Rome: Typis officinae polygraphicae.

Migne, J. P. (1864). *Patrologia Graeca*. Vol. 66. Paris: Migne (self-published). (For Synesius's complete works in the original Greek.)

Neugebauer, Otto (1949). "The Early History of the Astrolabe." *Isis 40*: 240–256. (Reprinted in Neugebauer, *Astronomy and History: Selected Essays*. New York: Springer-Verlag, pp. 278–294.) (A comprehensive explanation of Synesius's astrolabe in the *Ad Paeonium de dono*.)

——— (1975). *A History of Ancient Mathematical Astronomy*. 3 pts. New York: Springer-Verlag, pt. 2, pp. 876–877. (For a critique of Synesius's scientific abilities.)

Pando, José C. (1940). "The Life and Times of Synesius of Cyrene as Revealed in His Works." Ph.D. diss., Catholic University of America: Washington, D.C.: Catholic University of American Press. (Extracts of descriptive information from all of his works.)

Terzaghi, N. (1939). *Synesii Cyrenensis hymni*. Rome: Typis Officianae Polygraphicae.

——— (1944). *Synesii Cyrenensis opuscula*. Rome.

T

Ṭabarī: Abū Jaʿfar Muḥammad ibn Ayyūb al-Ḥāsib al-Ṭabarī

Flourished **(Iran), 1092–1108**

Ṭabari lived in Iran under the Saljūqs, probably in Āmul, and is the author of two independent treatises in Persian on the astrolabe as well as several other books, including two on arithmetic. Although several modern studies place him in the 13th century, he must have lived earlier based upon manuscript sources and since he is mentioned by al-Bayhaqī in his *Tatimmat ṣiwān al-ḥikma* (1164).

The first astrolabe treatise is known under the title *Shish faṣl* (Six chapters [on the knowledge of the astrolabe]), the oldest manuscript copy dating to 1176–1177. It was probably composed at the request of some students and is arranged in a question–answer (q/a) format:

(1) On the parts of the astrolabe and their names (60 q/a);
(2) On lines, figures, inscriptions, and circles on the astrolabe (77 q/a);
(3) On knowing the functioning of the back part of the astrolabe (49 q/a);
(4) On knowing the functioning of the face side of the astrolabe (136 q/a);
(5) On knowing how to check the exactness of the astrolabe (28 q/a);
(6) On the use of the astrolabe for land-surveying/measuring (*misāḥa*) (17 q/a).

The second treatise is a shorter and simplified version of the "six chapters" and is arranged simply into 104 entries. Entitled *ʿAmal wa-alqāb* ([On the] functions and names [of the astrolabe]), the oldest extant manuscript copy is dated 1162. It was written for a certain nobleman of his time, who is named in two copies as Abū al-Fatḥ Dawlatshāh ibn Sulaymān. In the beginning of the treatise, Ṭabari states that there are three types of astrolabes: (1) spherical (*kurī)* used in earlier times; (2) circular (*dawrī*), used in the time of the author, who describes it as circular and flat; and (3) boat-shaped/*navicula* (King, 1999, p. 352) (*zawraqī*), which is nevertheless described by Ṭabari as a "hemisphere, like a cup," who indicates that it was used in pre-Islamic Iran and that the astrolabe was called in Pahlavi "the cup that mirrors the world" (*jām-i jahān-namā*). This passage is not found in **Abū Rayḥān al-Bīrūnī's** *Tafhīm* and is quoted here as a curiosity. Ṭabarī's two treatises on the astrolabe are among the oldest extant Persian texts on the subject. For a detailed study of their content, they should be compared to the earlier chapter on the astrolabe in Bīrūnī's *Tafhīm* and to the later *Bist bāb dar maʿrifat-i usṭurlāb* (20 chapters on knowledge on the astrolabe) by **Naṣīr al-Dīn al-Ṭūsī**.

Živa Vesel

Selected References

King, David A. (1999). *World-Maps for Finding the Direction and Distance to Mecca*. Leiden: E. J. Brill.

Lazard, G. (1969). "A quelle époque a vécu l'astronome Moḥammad b. Ayyūb Ṭabari?" In *Yādnāmah-i Īrānī-i Mīnūrskī*, edited by Mujtabá Mīnuvī and Īraj Afshār, pp. 1–8. Tehran: 1347 H. Sh.

Maddison, F. (1997). "Observatoires portatifs." In *Histoire des sciences arabes*, edited by Roshdi Rashed. Vol. 1, pp. 139–172. Paris: Seuil.

Pingree, D. (1987). "Asṭorlāb." In *Encyclopaedia Iranica*, edited by Ehsan Yarshater. Vol. 2, pp. 853–857. Winona Lake, Indiana: Eisenbrauns.

Ṭabarī, Muḥammad ibn Ayyūb. (1993). *"Maʿifat-i usṭurlāb" maʿrūf bih "Shish faṣl" bih ḍamīmah-i "ʿAmal wa al-alqāb,"* edited by M. Amīn Riyāḥī. Tehran: 1371 H. Sh.

Tacchini, Pietro

Born **Modena, (Italy), 21 March 1838**
Died **Spilamberto, (Emilia-Romagna), Italy, 24 March 1905**

Astrophysicist, meteorologist, and seismologist, Pietro Tacchini distinguished himself as one of the fathers of solar astrophysics, inventor of one of the first sunspot classifications, editor of the oldest astrophysics review, first observer of the details of Venus' atmosphere spectrum, deviser of the first experiments of synchronization of astronomical observations, and organizer of scientific projects and institutions, both national and international.

Born the son of Bartolomeo Tacchini and Giuseppina Selmi, Pietro graduated *cum laude* in engineering during autumn 1857 at the Modena Archiginnasio. Tacchini was noticed for his talent by **Giuseppe Bianchi**, director of the Modena Observatory, who wanted to make him a good assistant. To allow him to learn astronomy, in April 1858, Bianchi (thanks to a scholarship bestowed by the Duke of Modena) sent Tacchini to the Padua Observatory where he served his apprenticeship under the guidance of **Giovanni Santini** and Virgilio Trettenero. In September 1859, Tacchini was designated *ad interim* director of Modena Observatory by the dictator Luigi Carlo Farini. Tacchini thus succeeded Bianchi who had suddenly resigned for political reasons. During his time in Modena, Tacchini continued the astronomical observations and corresponded with **Giovanni Schiaparelli**, director of the Brera Observatory, and **Angelo Secchi,** S.J., director of the Collegio Romano Observatory.

In 1863, following Schiaparelli's advice, Tacchini became assistant astronomer of the Palermo Observatory. After designing the refractor room and mounting, in 1865, the Merz equatorial 25-cm telescope (requested by Domenico Ragona to substitute for the Troughton used by **Giuseppe Piazzi**), Tacchini did his principal studies on solar meteorology consisting of a series of observations of solar photosphere and chromosphere, especially the faculae, prominences, and sunspots. He proposed classifications for these phenomena in 1871. (His sunspot classifications were based on connections with terrestrial magnetism.) Thanks to these studies, Palermo became one of the capitals of solar astrophysics. Tacchini was also one of the principal observers of southern stars and of seven among the most important contemporary solar eclipses. One of these was on 22 December 1870; its totality path passed through Sicily. Another was in 1883, which he observed from the Caroline Islands; this observation permitted Tacchini to note the calcium white prominences, different from the hydrogen red ones. In these years, together with Secchi, Arminio Nobile, and Emanuele Fergola, he made the first experiments in synchronization of astronomical observations of the solar limb by using a telegraph. (With Nobile, Tacchini measured the difference in longitude between Palermo and Naples.)

His success in solar spectroscopy and the need to monitor solar activity led Tacchini to found, in 1871 with Secchi and **Lorenzo Respighi**, the Società degli Spettroscopisti Italiani, the oldest professional society specifically devoted to astrophysics. Among its members were Nobile and **Giuseppe Lorenzoni**. From 1872 onward, Tacchini, as president of the Society, launched in Palermo the publication of its official journal, the *Memorie della Società degli Spettroscopisti Italiani* (now *Memorie della Società Astronomica Italiana*), an internationally distributed review that is considered the oldest one on astrophysics still in print.

In 1874, Tacchini was asked to organize the Italian astronomical expedition to Muddapur, India, to observe the passage of Venus across the solar disk on 8/9 December. The expedition, organized by Tacchini, Alessandro Dorna, and **Antonio Abetti**, observed, for the first time, the details of Venus's spectrum (**Joseph von Fraunhofer**'s lines C and B), thus confirming the existence of an atmosphere. The expedition also validated the use of spectroscopic observations to determine the exact instant of limb contact. During the trip to India, needing a low latitude observatory for winter solar observations, Tacchini founded the spectroscopic Calcutta Astronomical Observatory, at Saint Xavier College (directed by Eugène Lafont, S.J.)

From 1874 on, Tacchini, in order to reorganize astronomical research in Italy, promoted a reform project for astronomical observatories, accepted by the Italian government in 1876. He proposed to divide them into three classes: true astronomical observatories (in Florence, Naples, Milan, and Palermo), university observatories, and meteorological observatories.

In 1879, after Secchi's death, Tacchini was called to Rome as director of the Collegio Romano Observatory. In Rome, Tacchini also directed the new R. Ufficio Centrale di Meteorologia (and di Geodinamica, from 1887).

In 1880, the Bellini Observatory was founded. Promoted by Tacchini, it sat at a high altitude (2,940 m) on the Etna volcano to reduce atmospheric effects. It was equipped with a Merz 33-cm refractor. Despite the ideal altitude, ash emissions from the volcano hindered long-duration observations, forcing Tacchini to found the downtown Catania Astrophysical Observatory, financed by the government. This was necessary in order to be able to participate in the *Carte du Ciel* international enterprise promoted by the French Academy of Sciences and lasting more than 50 years. (This observatory was the only Italian participant.)

In 1890, at Catania, Tacchini helped found the first Italian astrophysics chair, assigned to **Annibale Riccò**, director of the Palermo Observatory.

In 1893, Tacchini was invited to the World Congress on Astronomy and Astro-Physics by **George Hale** who saw in him a master of organization. They collaborated to publish the *Astrophysical Journal*, from 1895, of which Tacchini was an associate editor.

Tacchini's work in astrophysics earned him the Rumford Medal of the London Royal Society in 1888 and the Prix Janssen of the Paris Académie des sciences in 1892. He was a fellow of the Accademia dei Lincei and of the Accademia Nazionale delle Scienze, and a foreign member of the Royal Astronomical Society and of the Royal Society.

Leonardo Gariboldi

Selected References

Abetti, Giorgio (1938). "Celebrazione del primo centenario della nascita di Pietro Tacchini." *Coelum* 9: 1–8.

——— (1976). "Tacchini, Pietro." In *Dictionary of Scientific Biography*, edited by Charles Coulston Gillispie. Vol. 13, pp. 232–233. New York: Charles Scribner's Sons.

Buffoni, Letizia, Edoardo Proverbio, and Pasquale Tucci (2000). *Pietro Tacchini: Lettere al Padre Angelo Secchi (1861–1877).* Milan: Università degli Studi-Pontificia Università Gregoriana.

Chinnici, Ileana (1993). "Pietro Tacchini: Ingegnere, Astrofisico, Meteorologo: Una prima ricostruzione biografica." Ph.D. diss., Università di Palermo.

——— (1997). "La Società degli Spettroscopisti Italiani e la fondazione di 'The Astrophysical Journal' nelle lettere di G. E. Hale a P. Tacchini." In *Atti del XVI Congresso nazionale di Storia della Fisica e dell'Astronomia*, edited by Pasquale Tucci, pp. 299–321. Como: Centro di Cultura Scientifica "A. Volta"

Chinnici, Ileana (ed.) (1994). "Il fondo 'Tacchini' dell'Ufficio Centrale di Ecologia Agraria in Roma." *Quaderni P. RI. ST. EM.* 7.

Foderà Serio, Giorgia and Ileana Chinnici (1997). *L'Osservatorio Astronomico di Palermo: La storia, e gli strumenti.* Palermo: Flaccovio.

Foderà Serio, Giorgia and Donatella Randazzo (1997). *Astronomi Italiani dall'Unità d'Italia ai nostri giorni: Un primo elenco.* Florence: Società Astronomica Italiana editore.

Millosevich, Elia (1905). "Pietro Tacchini" (in Italian). *Astronomische Nachrichten* 168: 15–16.

Riccò, Annibale (1905). "Pietro Tacchini." *Memorie della Società degli Spettroscopisti Italiani* 34: 85.

Tadahide

➤ **Nishikawa, Joken**

Takahashi, Yoshitoki

Born **Osaka, Japan, 1764**
Died **Asakusa, Edo (Tokyo), Japan, 1804**

Yoshitoki Takahashi was a leading scholar and advocate of calendrical reform in Japanese Edo culture, based on the importation of Western scientific methods. His father, Tokujiro Takahashi, was a lower-class official in Osaka. Considered a prodigy, the young Takahashi developed an early interest in mathematics. In 1779, he took up his father's profession and pursued intellectual interests in his spare time. Even if a man was of poor means in Edo Japan (1603–1867), he could still advance in social status if he had intellectual talent. Takahashi certainly had such talent, and he began to study with the noted calendar scholar **Goryu Asada** around 1787.

Takahashi found common interests with the wealthy merchant and instrument maker Shigetomi Hazama, who was also a student of Asada. They developed a long relationship of mixed cordiality and rivalry. With encouragement from Asada, Takahashi continued his work in mathematics and observational techniques related to the construction of an accurate lunar calendar. The work of Asada and his students came at a time when information about advances in Europe only trickled into Japan. Teachers and students at the Asada School studied the *Li-shiang K'ao-ch'eng Hou-pien* (a Sequel to the Compendium of Calendrical Science and Astronomy), edited by Ignatius Kögler, that included **Johannes Kepler**'s theory of elliptical orbits and **Jean-Dominique Cassini**'s work on the motions of the Sun and Moon. While the young student did not understand the full implications of celestial mechanics contained within the work, Takahashi found particularly fertile ground in its mathematics and empirically verifiable concepts.

As theoretical ideas matured, the Asada School developed a reputation that received notice from the *Tokugawa* shogunate in Edo. Seeing the accuracy gained by using advanced techniques, the shogunate felt it necessary to reform the current *Horyaku Reki* calendar. Asada was asked to join the calendar reform project but, on account of illness, he recommended Takahashi and Hazama instead. Both went to Edo, and Takahashi was assigned to the Tenmongata (Bureau of Astronomy) in 1795. Plans for calendar reform were completed in 1797, and a new calendar, called *Kansei Reki*, was officially placed into operation in 1798.

An accurate lunar calendar is dependent upon precise determinations of terrestrial latitudes and longitudes. Takahashi felt that accurate locations of Japanese cities should be acquired with the increased precision of astronomical instruments. Assisted by his own student, **Tadataka Ino**, a national survey under Takahashi's guidance (with the aid of Hazama) was conducted. This effort culminated in a large collection of maps of Japan used well into the 20th century.

Many have considered Takahashi to be a consummate theorist, but in an atmosphere of abject pragmatism, his gifts certainly found their most significant outlet in the development of more accurate methods of calendar construction. Working with secondary sources pertaining to western scientific developments, and not always understanding the full implications of such reports, he was able to adapt and apply modern methods to his own efforts in ingenious ways. For example, Takahashi was able to calculate the length of the tropical year and the synodic month within several decimal places of their currently accepted values.

Takahashi unhesitatingly tackled ideas from the west and even worked out an epicyclic theory for trepidation, misguided as the concept was. He began a translation from a Dutch version of **Joseph de Lalande**'s *Traité d'Astronomie* and compiled a multivolume work, entitled *Rarande Rekisho Kanken* (A Review of Lalande's *Astronomie*). Unable to finish the full translation, the work was completed by his second son. Takahashi was a prime influence in Ino's work related to precision measurements of latitude and longitude. Along with Asada, Takahashi was perhaps most influential in showing the viability of using western astronomical methods for calculating the movements of celestial bodies at a time of officially sanctioned superstition and indifference. At the time of his death, he was trying to develop rigorous techniques for accurately calculating planetary movements along with predictions of lunar and solar eclipses.

When Takahashi passed away, his eldest son, Kageyasu, succeeded him in the Bureau of Astronomy. The noted Shibukawa family later adopted his second son, Kagesuke. Both men continued in their father's footsteps.

Steven L. Renshaw and *Saori Ihara*

Selected References

Nakayama, Shigeru (1969). *A History of Japanese Astronomy: Chinese Background and Western Impact*. Cambridge, Massachusetts: Harvard University Press.

Sugimoto, Masayoshi and David L. Swain (1978). *Science and Culture in Traditional Japan*. Cambridge, Massachusetts: MIT Press.

Watanabe, Toshio (1986–1987). *Kinsei Nihon Tenmongakushi* (A Modern History of Astronomy in Japan). 2 Vols. Tokyo: Koseisha Koseikaku.

Tang-Jo-Wang

➤ **Schall von Bell, Johann Adam**

Taqī al-Dīn Abū Bakr Muḥammad ibn Zayn al-Dīn Maʿrūf al-Dimashqī al-Ḥanafī

Born **Damascus, (Syria), 14 June 1526**
Died **Istanbul, (Turkey), 1585**

Taqī al-Dīn was the founder and the director of the Istanbul Observatory and worked in the fields of mathematics, astronomy, optics, and mechanics. He made various astronomical instruments and was the first astronomer to use an automatic–mechanical clock for his astronomical observations. He advanced the arithmetic of decimal fractions and used them in the calculation of astronomical tables.

Taqī al-Din began his studies, as was normal, with the basic religious sciences and Arabic. Later on, he continued his religious studies and studied the mathematical sciences with scholars in Damascus and Egypt, including most significantly his father. It is probable that Taqī al-Dīn's teacher in mathematics was Shihāb al-Dīn al-Ghazzī whereas the one in astronomy was **Muḥammad ibn Abī al-Fatḥ al-Ṣūfī**. Taqī al-Dīn himself states in several of the forewords to his books that he was particularly interested in the mathematical sciences during his education.

Taqī al-Dīn, after completing his education, taught for a short while at various *madrasas* (schools) in Damascus. He, together with his father Maʿrūf Afandī, came to Istanbul around the year 1550 where he benefited from his association with a number of prominent scholars. Taqī al-Dīn would shortly return to Egypt where he spent most of the next 20 years. A brief trip back to Istanbul, also around 1550, brought him into the company of the Grand Vizier Samīz ʿAlī Pasha, who allowed him to use his private library and clock collection. Taqī al-Dīn would benefit from this association when ʿAlī Pasha was appointed governor of Egypt, where he held positions as a teacher and judge (*qāḍī*) in Egypt. Encouraged to deal with mathematics and astronomy by a grandson of **ʿAlī Qūshjī**, who collected and gave Taqī al-Dīn works by his grandfather, by **Jamshīd al-Kāshī**, and by **Qāḍīzāde**, as well as various observation instruments, Taqī al-Dīn undertook a serious pursuit of astronomy and mathematics. While a judge in Tinnīn, Egypt, he made astronomical observations by means of an astronomical instrument that he mounted in a well that was 25-m deep.

Taqī al-Dīn returned to Istanbul in 1570 and was appointed head astronomer (*Müneccimbası*) by Sultan Selīm II upon the death of **Muṣṭafā ibn ʿAlī al-Muwaqqit** in 1571. He continued his observations in a building situated on a height overlooking Tophane or in Galata Tower and gained the support of several high officials. This led to an imperial edict by Sultan Murad III in early 1579 to build an observatory, which was located on a height overlooking Tophane where the French palace is located today. Important astronomical books and instruments were collected there. Little is known about the size, shape, and so on, but we do have magnificent depictions of the scholars at work and of the astronomical instruments in use (in *Ālāt-i raṣadiyya li-Zij-i Shāhinshāhiyya* [Istanbul Univesity, TY, MS 1993] and in ʿAlāʾ al-Dīn Manṣūr al-Shīrāzī's *Shāhinshahnāme* [Istanbul University, TY, MS 1404]). Apart from the observatory building, we hear of a well called *çah-i raṣad* that was also used by Taqī al-Dīn. Unfortunately the observatory did not last long. Due to political reasons, as well as Taqī al-Dīn's incorrect astrological prognostications, it was demolished by the state on 22 January 1580.

Taqī al-Dīn's most important work in astronomy is entitled *Sidrat muntahā al-afkār fī malakūt al-falak al-dawwār* (= *al-Zīj al-Shāhinshāhī*). This work was prepared according to the results of the observations in Egypt and Istanbul in order to correct and complete *Zīj-i Ulugh Beg*, a project originally conceived in Egypt and furthered by the building of the Istanbul Observatory. In the first 40 pages of the work, Taqī al-Dīn deals with trigonometric calculation. This is followed by discussions of astronomical clocks, heavenly circles, and so forth. In the following parts, he treats observational instruments and their use, the observations of lunar and solar motions, and trigonometric functions calculated according to sexagesimal. As was normal in the Islamic astronomical tradition, Taqī al-Dīn used trigonometric functions such as sine, cosine, tangent, and cotangent rather than chords. Following the work done at the Samarqand Observatory, he developed a new method to find the exact value of sin 1°, which Jamshīd al-Kāshī had put into the form of an equation of third degree. Additionally, Taqī al-Dīn employed the method of "three observation points," which he was the first to use for calculating solar parameters; apparently **Tycho Brahe** was aware of his work. For determining the longitudes and latitudes of the fixed stars, he used Venus, Aldebaran, and α Virginis (Spica), which are near the ecliptic (rather than the Moon), as reference stars. As a result of his observations, he found the eccentricity of the Sun to be 2° 0′ and the annual motion of apogee 63″. Taqī al-Dīn's values turn out to be more precise than those of **Nicolaus Copernicus** and Brahe. This provides evidence for the precision of Taqī al-Dīn's methods of observation and calculation. It is thus a pity that the destruction of the observatory meant that Taqī al-Dīn was unable to complete his observation program. Indeed in the absence of a conclusion to this *Zīj*, it can probably be concluded that the book was never completed.

Taqī al-Dīn's second most important work on astronomy is a *zīj* entitled *Jarīdat al-durar wa kharīdat al-fikar*. In this work, for the first time we find the use of decimal fractions in trigonometric functions. He also prepared tangent and cotangent tables. Moreover, in this *zīj*, as in another of his *zījes* entitled *Tashīl zīj al-aʿshāriyya al-shāhinshāhiyya*, Taqī al-Dīn gave the parts of degree of curves and angles in decimal fractions and carried out the calculations accordingly. Excluding the table of fixed stars, all the astronomical tables in this *zīj* were prepared using decimal fractions.

In addition, Taqī al-Dīn has some other astronomical works of secondary importance. One of them is *Dustūr al-tarjīḥ li-qawāʿid al-tasṭīḥ*, which is about the projection of a sphere onto a plane as well as other topics in geometry. Another of his works is *Rayḥānat al-rūḥ fī rasm al-sāʿātʿalā mustawī al-suṭūḥ*, which deals with sundials drawn on marble surfaces and their features. This book was commented upon by his student Sirāj al-Dīn ʿUmar ibn Muḥammad al-Fāriskūrī (died: 1610) under the title *Nafḥ al-fuyūḥ bi-sharḥ rayḥānat al-rūḥ*; the commentary was translated into Turkish by an unknown writer in the beginning of the 17th century.

In addition to his 20 books on astronomy, Taqī al-Dīn wrote one book on medicine and zoology, three on physics-mechanics, and five on mathematics. He has a monograph entitled *Risāla fī ʿamal al-mīzān al-ṭabīʿī* on the specific gravity of substances and **Archimedes**' hydrostatic experiments. All of his books are in Arabic.

Taqī al-Dīn's works on physics and mechanics, besides being interesting in their own right, also have connections with astronomy. In 1559 while in Nablus, he wrote his *al-Kawākib al-durriyya fī waḍʿ al-bankāmāt al-dawriyya*, which dealt with mechanical-automatic clocks for the first time in the Islamic and Ottoman world. In the foreword, Taqī al-Dīn mentions that he benefited from using Samiz ʿAlī Pasha's private library and his collection of European mechanical clocks. In this work, Taqī al-Dīn discusses various mechanical clocks from a geometrical–mechanical perspective. His second book on mechanics is the one he wrote when he was 26, *al-Ṭuruq al-saniyya fī al-ālāt al-rūḥāniyya*. In this work, Taqī al-Dīn focuses on the geometrical-mechanical structure of clocks previously examined by the **Banū Mūsā** and Abū al-ʿIzz al-Jazarī. In the field of physics and optics, Taqī al-Dīn wrote *Nawr ḥadīqat al-abṣar wa-nūr ḥaqīqat al-Anẓar*, which dealt with the structure of light, its diffusion and global refraction, and the relation between light and color.

In his mathematical treatises, Taqī al-Dīn dealt with various aspects of trigonometry, geometry, algebra, and arithmetic. In the latter, he carried on the work of Kāshī in developing the arithmetic of decimal fractions both theoretically and practically.

Taqī al-Dīn was a successor to the great school of Samarqand and, following the lead of ʿAlī Qūshjī, tended toward a more purely mathematical approach in his scientific work that was beginning to abandon Aristotelian physics and metaphysics. Taqī al-Din's most significant achievement in the history of Islamic and Ottoman astronomy is his foundation of the Istanbul Observatory and his activities there. Besides using established instruments and techniques, he developed a number of new ones as well, including his use of the automatic–mechanical clock. Carrying on the work of his Islamic predecessors, Taqī al-Dīn's application of decimal fractions to trigonometry and astronomy stands as another important contribution to astronomy and mathematics.

İhsan Fazlıoğlu

Selected References

Demir, Remzi (2000). *Takiyüddin'de Matematik ve Astronomi*. Ankara: Aṭaturk Kūthur Merkeri Yainlari.

İhsanoğlu, Ekmeleddin *et al.* (1997). *Osmanlı Astronomi Literatürü Tarihi (OALT)* (History of astronomy literature during the Ottoman period). Vol. 1, pp. 199–217 (no. 96). Istanbul: IRCICA.

——— (1999). *Osmanlı Matematik Literatürü Tarihi* (OMLT). (History of mathematical literature during the Ottoman period). Vol. 1, pp. 83–87 (no. 47). Istanbul: IRCICA.

Mordtmann, J. H. (1923). "Das Observatorium des Taqî ed-Dîn zu Pera." *Der Islam* 13: 82–96.

Sayılı, Aydın (1960). *The Observatory in Islam*. Ankara: Turkish Historical Society, pp. 289–305.

Tekeli, Sevim (1958). "Nasirüddin, Takiyüddin ve Tycho Brahe'nin Rasat Aletlerinin Mukayesesi." *Ankara Üniversitesi Dil ve Tarih-Coğrafya Fakültesi Dergisi 16, nos. 3–* 4: 301–353.

——— (1966). *16'ıncı Asırda Osmanlılar'da Saat ve Takiyüddin'in "Mekanik Saat Konstrüksüyonuna Dair En Parlak Yıldızlar" Adlı Eseri*. Ankara (Turkish–English–Arabic text.): Ankara ūniversitesi, Dil, Tarih-Coğrafya Fakultesi Yayinlare

——— (1986). "Onaltıncı Yüzyıl Trigonometri Çalısmaları Üzerine Bir Arastırma: Copernicus ve Takiyuddin." *Erdem* 2, no. 4: 219–272.

Tekeli, Sevim, (ed.) (1960). "Alat el-Rasadiyye li Zic-i Sehinsahiyye." *İslâm Tetkikleri Enstitüsü Dergisi* 3, pt.1/ 2: 1–30.

Topdemir, Hüseyin Gazi (1999). *Takiyüddin'in Optik Kitabı*. Ankara: Aṭaturk Kūthur Merkeri Yayinlari.

Tarde, Jean

Born **La Roque-Gageac near Sarlat, (Dordogne), France, 1561 or 1562**
Died **Sarlat, (Dordogne), France, 1636**

Jean Tarde was an early French Copernican and student of sunspots. The social status of Tarde's family is uncertain, although probably bourgeois. At his death, Tarde left behind a considerable estate, and his family remained in high positions in the bourgeois community. Tarde received his doctorate in law from the University of Cahors and then continued his studies at the University of Paris. He was ordained a priest and assigned to the parish of Carves, near Belvès. Soon after, he was promoted to canon theologian of Sarlat's cathedral. The bishop of Sarlat, Louis de Salignac, wishing to ascertain the effects of France's religious wars on the diocese, appointed Tarde as vicar general in 1594, commissioning him to map out the diocese. In 1599, Henri IV appointed him an *almoner* (royal chaplain), for which Tarde received a pension.

Tarde mapped out the neighboring diocese at the request of the bishop of Cahors in 1606, employing a quadrant fitted with a compass needle and attached to a sundial. The bishop's interest in the quadrant prompted Tarde to publish *Les usages du quadrant à l'esguille aymantée* (1621), dedicated to the bishop. His cartographic work demonstrated his interest in geometry, drawing, drafting, and the practical uses of instruments – skills he found useful in his astronomical work.

In November 1614, Tarde visited **Galileo Galilei** in Florence (his second of two trips to Italy). According to Tarde's diary, he flattered Galilei and told him that he had read his *Siderius nuncius*, and wanted to know the latest observations made by the famed Italian. Galilei told him about several observations including his sighting of two stars (the rings) around Saturn, the phases of Venus, and the spots on the face of the Sun, remarking that since others using telescopes had also seen these spots, they could not be illusions. Galilei also told him that the spots took 14 days to move across the Sun's face and that their movements were similar to those of Mercury and Venus. When Tarde asked Galilei how to build a telescope, Galilei claimed he did not know how it worked and instead referred Tarde to **Johannes Kepler**'s book on optics, promising also to send him a lens in Rome. While Tarde never received the lens, he did visit with Galilei twice more before returning to France in February of 1615.

After his return, Tarde established an observatory to observe sunspots, an activity he pursued for 5 years. Based on his previous conversations with Galilei and his own observations, Tarde published his *Borbonia sidera* in 1620. In 1622 he published and dedicated to Louis XIII a French translation entitled *Les astres de Borbon*, with which was bound Tarde's theoretical work on the telescope, *Telescopium, seu demonstrationes opticae....* In *Les astres de Borbon*, Tarde defended the validity of telescopic data and described his own instrument for observing the Sun – a *caverne obscure* (*camera obscura*), the device invented by **Christoph Scheiner**, in which a telescope projected the image of the Sun onto a white surface in a darkened room.

Tarde rejected (as had Scheiner and Galilei) the notion that sunspots were illusions or that they were located within the Earth's atmosphere, claiming also that they were not spots on the Sun or comets. He concluded that sunspots were distinct bodies (planets, or as he called them, *étoiles errantes*) that orbited the Sun, and, following Galilei's precedent with the satellites of Jupiter, he named them the Bourbon stars after the French royal family. Tarde's theory was based on his firm belief that the Sun was the seat of God and could not be blemished or tarnished. Although he criticized the Aristotelians, Tarde did not break entirely free from that tradition and was determined to fit the new observational data into his own framework. In his discussions of the Sun, Tarde also addressed heliocentrism. Citing **Pythagoras** and **Nicolaus Copernicus**, he argued in favor of heliocentrism, indicating that it was an easier and more convenient system as far as astronomical calculations were concerned. But Tarde did not fully commit himself to heliocentrism either because of the church's prohibition of Copernicanism or because heliocentrism had not been proven yet.

Galilei raised several objections to Tarde's theory of *étoiles errantes* that the latter refuted. The one point that Tarde conceded to Galilei was retrograde motion – the spots on the Sun do not exhibit retrograde motion as the superior planets do. To address this argument, Tarde complained that the weather had interfered with his observations and that, in any case, retrograde motion was too difficult to detect for these "planets" in any case, because of their proximity to the Sun. **Pierre Gassendi** also criticized Tarde's theory in a 1625 letter to Galilei, claiming that the lack of a periodic return of the "planets" was proof that they were not circling the Sun. However, in 1633, Gassendi wrote to **Nicolas de Peiresc** that he was still waiting for more evidence in support of Tarde's theory that might persuade him more fully. The dispute faded away soon after Tarde's death.

In the mid-19th century, **Urbain Le Verrier** argued that his calculations indicated that there was a planet between the orbit of Mercury and the Sun. This led to a revival of interest in Tarde, whose work was scrutinized (including by one of Tarde's descendants, Gabriel Tarde) to find evidence in support of Le Verrier's theory. This attempt failed, however, when Le Verrier's theory was proven false. Nevertheless, late-19th-century interest in Tarde's work motivated Gabriel Tarde and Gaston de Gérard to publish in 1887 Tarde's unpublished history of Sarlat, *Les Chroniques de Jean Tarde*.

Voula Saridakis

Alternate name

Tardeus

Selected References

Ariew, Roger (2001). "The Initial Response to Galileo's Lunar Observations." *Studies in History and Philosophy of Science* A 32: 571–581.

Baumgartner, Frederic J. (1987). "Sunspots or Sun's Planets: Jean Tarde and the Sunspot Controversy of the Early Seventeenth Century." *Journal for the History of Astronomy* 18: 44–54.

Darricau, Raymond (1986). "Jean Tarde, chanoine théologal de Sarlat, et la science de son temps (1561–1636)." *Revue française d'histoire du livre* 51: 337–340.

Dujarric-Descombes, Albert (1882). "Recherches sur les historiens du Périgord au XVIIe siècle: Jean Tarde." *Bulletin de la Société historique et archéologique du Périgord* 9: 371–412, 489–497.

Favaro, Antonio (1887). "Di Giovanni Tarde e di una sua visita a Galileo dal 12 al 15 novembre 1614." *Bullettino di bibliografia e storia delle scienze matematiche e fisiche* 20: 345–374.

Rosen, Edward (1976). "Tarde, Jean." In *Dictionary of Scientific Biography*, edited by Charles Coulston Gillispie. Vol. 13, pp. 256–257. New York: Charles Scribner's Sons.

Tarde, Gabriel. "Observations au sujet des *Astres de Borbon* du chanoine Tarde." *Bulletin de la Société historique et archéologique du Périgord* 4 (1877): 169–173; 9 (1882): 391.

Tarde, Jean (1984). *A la rencontre de Galilée: Deux voyages en Italie*. Preface and notes by François Moureau. Geneva: Slatkine.

Tardeus

➔ **Tarde, Jean**

Taylor, Geoffrey Ingram

Born **Saint John's Wood, (London), England, 7 March 1886**
Died **Cambridge, England, 27 June 1975**

Geoffrey Ingram Taylor is recognized within astronomy for the description of Taylor columns (rotating, rising fluid structures, of which the Great Red Spot on Jupiter may be an example) and for development of the theory of the Rayleigh–Taylor instability, in which a dense fluid, held up by the pressure of a less dense one underneath, rather suddenly exchanges positions with the less dense fluid, in a swirl of eddies and fingers. Supernova explosions and a range of other astronomical events and sources show evidence of the phenomenon. His work also has found applications in meteorology, oceanography, engineering, and aeronautics.

Taylor's father was Edward Ingram Taylor. His mother was Margaret (*née* Boole) Taylor, daughter of George Boole, the mathematician. Taylor grew up in an atmosphere conducive to an appreciation of science under the influence of his parents and the other Boole daughters, respected scientists in their own right. Indeed, at the age of 12, he and a friend constructed an X-ray generator, only 2 years after Röntgen had first discovered the ray's existence. Taylor's mechanical skills and ingenuity were also on display as a teenager, when he managed not only to build a sailboat in his bedroom, but also to get it out of his window and sail it on the Thames. Starting in 1905, Taylor attended Trinity College at Cambridge, studying mathematics and physics. Upon graduation, he received a scholarship for postgraduate research, and worked with J. J. Thomson on the interference of low-intensity beams of photons. This led to his first published paper, in 1909. (The last appeared in 1973.) Taylor remained at Cambridge for most of his career, latterly as a fellow of Trinity College and a Royal Society Professor at the Cavendish Laboratory.

Taylor quickly shifted his interest from pure physics to fluid mechanics, and his second published paper, in 1910, studied the structure of shock waves and garnered him a Smith's Prize at Cambridge. In 1911, he received a readership in dynamical meteorology and delved into an analysis of small-scale processes, including momentum and heat transfer in response to turbulent fluctuations. Taylor found that turbulent velocity distributions were isotopic except near the ground, in contrast to prevailing theory of the time. In response to the sinking of the Titanic, the British government sponsored a study of the distribution of icebergs in the North Atlantic in 1913, and Taylor was named meteorologist of the research ship, *Scotia*. This allowed him to study transfer properties on a much larger scale. His analysis of the results put forward the concept of a mixing length for turbulent diffusion.

Taylor was also very interested in stability of turbulent systems, including his famous work on steady flow between concentric circular cylinders. He investigated the motion of objects in a rotating liquid. He discovered that where Coriolis effects are dominant, Taylor columns form, of which the Red Spot of Jupiter is considered to be a potential example. Taylor subsequently published several phenomenological papers on turbulence, culminating in the 1930s with an empirically testable understanding of the statistical properties of turbulence and a determination of the energy spectrum of turbulent motion.

Taylor was always interested in practical problems, and was active in war research in World War I and World War II. During the former, he studied shafts under torsion to build better airplane propeller shafts. In the latter, he worked on the Manhattan Project on shock waves and saw the first nuclear explosion at Alamogordo. Even after his retirement in 1951, and for the next 20 years, Taylor continued to investigate new problems in fluid mechanics, including how small organisms swim, electrohydrodynamics, and dynamics of thin sheets of liquid. Several other important discoveries in physics of fluids and solids bear his name, including the Taylor–Proudman theorem (one of whose consequences is that rotation of a fluid inhibits convection), Taylor–Couette instabilities, Taylor dislocations (in crystals, a topic he worked on intermittently from 1934 onward), and the Taylor dispersion relation (describing the relationship between frequency and wavelength in unstable, incompressible flows). It might seem surprising that his work on the behavior of a fluid trapped between a solid cylinder and the inside of a cylindrical chamber is still commonly cited. It is because there remain even now surprisingly few exact results in the description of convective energy transport in fluids.

Taylor was famous both for his ability to find the important problems to work on and for his ingenuity in finding simple and elegant experiments to test his predictions. Among his honors were fellowships in the Royal Society, the United States National Academy of Sciences, and the American Philosophical Society, knighthood in 1944, and admission to the United Kingdom Order of Merit in 1969. George Batchelor and Sir Brian Pippard, both from Cambridge, were among Taylor's scientific heirs.

Michael Fosmire

Selected References

Batchelor, George K. (1996). *The Life and Legacy of G. I. Taylor*. London: Cambridge University Press. (This biography includes appendices on the honors he received and a complete bibliography of his published papers.)

Pippard, Sir Brian A. (1975). "Sir Geoffrey Taylor." *Physics Today* 28, no. 9: 67.

Taylor, Geoffrey I. (1958–1971). *The Scientific Papers of Sir Geoffrey Ingram Taylor*, edited by G. K. Batchelor. 4 Vols. London: Cambridge University Press.

Tebbutt, John

Born **Windsor, New South Wales, (Australia), 25 May 1834**
Died **Windsor, New South Wales, Australia, 29 November 1916**

During the last 3 decades of the 19th century, John Tebbutt became Australia's leading astronomer, amateur or professional, and made valuable meteorological contributions. Although he was offered the Sydney Observatory directorship in 1862, he decided to remain an amateur astronomer and successfully combined astronomy and farming for the rest of his life.

The son of John and Virginia (*née* Saunders) Tebbutt, the younger John attended local church schools where he excelled academically. At age 15, he began working full-time on his father's farm on the outskirts of Windsor; he inherited the farm in 1870. On 8 September 1857, Tebbutt married Jane Pendergast; they had one son and six daughters.

Tebbutt was largely self-taught as an astronomer. During the 1850s he used his naked eye, a marine telescope, and a sextant to observe sunspots, aurorae, meteors, lunar eclipses and occultations, Jupiter's satellites, comets, and the variable star Algol. Tebbutt taught himself the mathematics required to compute comet orbits, and began publishing astronomical reports in the Sydney newspapers.

On 13 May 1861, Tebbutt detected a faint nebulous object in Eridanus. Comet C/1861 J1, or the Great Comet of 1861, developed into one of the most magnificent comets of the century, featuring a tail more than 100° in length at its prime. This discovery inspired Tebbutt to purchase an 8.3-cm Jones refractor. In 1863, he constructed a simple observatory that eventually housed this refractor, a 5.3-cm transit telescope from Sydney scientific instrumentmaker, Angelo Tornaghi, and a chronometer. Tornaghi also made a ring micrometer for the refractor. A full set of meteorological instruments completed the observatory's instrumentation. Over the years, Tebbutt arranged the construction of three new Windsor Observatory buildings (1874, 1879, and 1894) and the addition of 11.4-cm Cooke and 20.3-cm Grubb equatorial refractors and a 7.6-cm Cooke transit telescope. New micrometers and chronometers also were acquired.

Although the Windsor Observatory was modest by international standards, Tebbutt more than compensated for this with his dedication and enthusiasm. Between 1863 and his semiretirement in 1903, Tebbutt conducted an amazing range of observational programs. He continued intermittent observations up to 1915.

Comets were without doubt Tebbutt's favorite observational targets. Between 1853 and 1912 he observed 50 different periodic and nonperiodic comets. He obtained micrometric positions for many of these comets on every possible clear night. Tebbutt's longest series of observations (103 nights) for any one comet was on C/1898 L1 (Coddington–Pauly). Tebbutt observed comet 2P/Encke on eight different returns and is credited with its recovery on three occasions. He also searched successfully for new comets, discovering two great comets, C/1861 J1 (the first comet for which astronomers, including Tebbutt, forecast the passage of the Earth through the tail), and C/1881 K1, from which important advances in astrophysical knowledge of comets were gained.

Tebbutt made important contributions to variable-star astronomy, including a detailed light-curve of η Carinae between 1854 and 1898 that revealed its minor outburst of the 1880s. His observations of the Mira type variable R Carinae from 1880 to 1898 allowed a precise determination of its period. He discovered Nova V728 Scorpii in 1862.

Tebbutt successfully observed the 1874 transit of Venus, but overcast skies prevented him from recording the 1882 event. In addition, between 1866 and 1899, he observed four transits of Mercury, six partial solar eclipses, and five lunar eclipses. Between 1877 and 1915, Tebbutt made many micrometric measures of 133 different double stars, and recorded accurate positions of 23 different asteroids. His measures of asteroids and comets were prized by orbit computers as they were frequently either the earliest or the last measures available for an apparition. Through Tebbutt's painstaking lunar occultation work, the Windsor Observatory became one of Australia's fundamental geodetic reference points.

Tebbutt maintained a time service for Windsor and operated a meteorological station. He provided Sydney Observatory with monthly weather reports from 1863 through to 1898 when a curtailed meteorological program was adopted. (These continued up to the time of his death.) Tebbutt supplied Sydney and Windsor-district newspapers with meteorological data on a regular basis.

Almost single-handedly Tebbutt carried out the time-consuming reduction of his astronomical and meteorological observations, and communicated his results to colleagues through Australian and international journals. In all he wrote two books and 323 different papers, some of which appeared in more than one journal. His *Annual Reports* of the Windsor Observatory were produced as booklets from 1888 to 1903 (inclusive). Tebbutt wrote two chapters for books by other authors, eight meteorological monographs, a number of booklets on the relationship between astronomy and religion, and hundreds of newspaper articles to popularize astronomy, a phenomenal output for a one-man observatory.

Tebbutt was a member of the Philosophical (later the Royal) Society of New South Wales from 1861 (and a stalwart of its short-lived Astronomy Section) and a fellow of the Royal Astronomical Society [RAS] from 1873. When the New South Wales Branch of the British Astronomical Association was founded in 1895, Tebbutt was elected its inaugural president.

Although an amateur, Tebbutt quickly gained an international reputation as an astronomer, and from 1869 his Windsor Observatory was included in the *Nautical Almanac's* listing of world observatories. In 1867 the government presented him with a silver medal, and in 1905 he was awarded the Jackson-Gwilt Medal and Gift by the RAS. In 1973 the International Astronomical Union arranged for lunar crater Picard G to be renamed Tebbutt. In 1984 Tebbutt's portrait was featured on a new Australian $100 bank note. Two surviving Windsor Observatory buildings were refurbished as a museum of astronomy in 1989 by his grandson, John Halley Tebbutt.

Tebbutt's status as Australia's leading astronomer led eventually to a bitter feud with Sydney Observatory director, **Henry Chamberlain Russell** that only ended in 1907 with Russell's death. Despite modest equipment, Tebbutt was able to make valuable contributions to observational astronomy. He played an important role in the development of Australian astronomical groups and societies and, more than any other 19th-century Australian astronomer, helped popularize astronomy. Tebbutt was a remarkable scientist, running what Joseph Ashbrook has likened to "1/4 a one-man Greenwich Observatory in the Southern Hemisphere."

Wayne Orchiston

Selected References

Ashbrook, Joseph (1984). "John Tebbutt, His Observatory, and a Probable Nova." In *The Astronomical Scrapbook*, edited by Leif J. Robinson, pp. 66–71. Cambridge, Massachusetts: Sky Publishing Corp.

Orchiston, Wayne (1982). "Illuminating Incidents in Antipodean Astronomy: John Tebbutt and the Abortive Australian Association of Comet Observers." *Journal of the Astronomical Society of Victoria* 35: 70–83.

——— (1988). "Illuminating Incidents in Antipodean Astronomy: John Tebbutt and the Sydney Observatory Directorship of 1862." *Australian Journal of Astronomy* 2: 149–158.

——— (1997). "The Role of the Amateur in Popularising Astronomy: An Australian Case Study." *Australian Journal of Astronomy* 7: 33–66.

——— (2000). "John Tebbutt of Windsor, New South Wales: A Pioneer Southern Hemisphere Variable Star Observer." *Irish Astronomical Journal* 27: 47–54.

——— (2001). "'Sentinel of the Southern Heavens': The Windsor Observatory of John Tebbutt." *Journal of the Antique Telescope Society* 21: 11–23.

——— (2002). "Tebbutt vs Russell: Passion, Power and Politics in Nineteenth Century Australian Astronomy." In *History of Oriental Astronomy*, edited by S. M. Razaullah Ansari, pp. 169–201. Dordrecht: Kluwer Academic Publishers.

——— (2004). "John Tebbutt and Observational Astronomy at Windsor Observatory." *Journal of the British Astronomical Association* 114: 141–145.

Tebbutt, J. (1866). "On the progress and present state of astronomical science in New South Wales." In *Catalogue of the Natural and Industrial Products of New South Wales. Forwarded to the Paris Universal Exhibition of 1867, by the New South Wales Exhibition Commissioners*, pp. 55–64. Sydney: Government Printer.

——— (1887). *History and Description of Mr Tebbutt's Observatory*. Sydney: printed for the author.

——— (1908). *Astronomical Memoirs*. Sydney: Printed for the author. (Reprinted by the Hawkesbury Shire Council, with additional material, in 1986.)

Turner, H. H. (1918). "John Tebbutt." *Monthly Notices of the Royal Astronomical Society* 78: 252–255.

Wood, Harley W. (1976). "Tebbutt, John (1834–1916)." In *Australian Dictionary of Biography*. Vol. 6, *1851–1890*, pp. 251–252. Melbourne: Melbourne University Press.

Teller, Edward [Ede]

Born **Budapest, (Hungary), 15 January 1908**
Died **Palo Alto, California, USA, 9 September 2003**

Hungarian–American nuclear physicist Edward Teller collaborated with **George Gamow** in studying the rules for beta decay and applications of astrophysics to controlled thermonuclear reactions.

Teller was the son of Miksa Teller and Ilona Deutsch. His marriage to Augustzta Maria Harkanyi produced three children – Paul, Susan, and Wendy, the last of whom is coauthor with him of an autobiography.

Teller studied at the Institute of Technology in Karlsruhe, Germany (1926–1928) and earned a Ph.D. in physics at the University of Leipzig in 1930. He was a teacher and researcher in various universities (Leipzig: 1929–1931; Göttingen: 1931–1933; Copenhagen: 1934 [where he first met Gamow at the Neils Bohr Instutute]; and London, 1934/1935); then in 1935 he emigrated to the United States where he became a naturalized American citizen in 1941. Teller was professor of physics at various universities: first George Washington (at the time of his collaboration with Gamov), 1935–1941; Columbia, 1941–1942; Chicago, 1946–1952; and California, 1953–1975. He also held office in various research institutions: Los Alamos National Laboratory, assistant director, 1949–1952; University of California, Radiation Laboratory, staff member, 1952–1953; Lawrence Livermore Radiation Laboratory, associate director then director, 1954–1975; and the Hoover Institute for the Study of War, Revolution, and Peace, Stanford University, senior researcher, 1975–2003.

Teller was *honoris causa* doctor of more than a dozen universities including Yale (1954) and George Washington (1960). He was recipient of the Joseph Priestley Memorial Award (1957), Albert Einstein Award (1958), Research Institute of America Living History Award (1958), Thomas E. White Award (1962), Enrico Fermi Award (1962), Robins Prize (1963), Arvey Prize (1975), and the United States Presidential Medal of Freedom (2003). He was a member of the American Academy of Arts and Sciences and honorary member of the Hungarian Academy of Sciences (1990).

Teller retained some interest in astrophysics even after issues of war and politics dominated his life. In 1949, Teller wrote with Maria Goeppert Mayer (Nobel Prize in Physics in 1963 for formulation of the nuclear shell model) on the possibility of accounting for the abundances of the elements in the Universe in terms of the fragmentation of an enormous mass of neutrons.

In 1939 Teller and fellow Hungarian physicist Leo Szilárd convinced **Albert Einstein** to join them in writing a letter to US president Franklin Roosevelt on the necessity to develop American nuclear weapons. Teller worked on the Manhattan Project, a military program for constructing an atomic bomb for wartime use. He was the first to study thermonuclear reactions in connection with weapons. After completion of the atomic bomb, his scientific interest turned to the more powerful fusion bomb. In the University of California's Livermore Radiation Laboratory, he worked on the hydrogen bomb. In 1954 Teller testified against his former Los Alamos colleague **Julius Robert Oppenheimer** who opposed the production of the hydrogen bomb and was accused of being a communist sympathizer.

Teller made major contributions to spectroscopy of polyatomic molecules and the theory of atomic nucleus. He is the author (or coauthor) of over a dozen books, mostly dealing with the problems of nuclear energy and defense.

László Szabados

Selected References

Brown, Harold and Michael May (2004). "Edward Teller in the Public Arena." *Physics Today* 57, no. 8: 51–53.

Libbey, Stephen B. and Morton S. Weiss (2004). "Edward Teller's Scientific Life." *Physics Today* 57, no. 8: 45–50.

Mark, Hans and Lowell Wood (eds.) (1988). *Energy in Physics, War and Peace: A Festschrift Celebrating Edward Teller's 80th Birthday*. Dordrecht: Kluwer.

Teller, Edward and Judith L. Shoolery (2001). *Memoirs: A Twentieth-Century Journey in Science and Politics*. Oxford: Perseus Books.

Tempel, Ernst Wilhelm Leberecht

Born **Nieder-kunnersdorf, (Sachsen, Germany), 4 December 1821**
Died **Arcetri near Florence, Italy, 16 March 1889**

Comet and asteroid seeker Ernst Tempel grew up on the parental farm and was intended for that occupation. With the support of a local schoolmaster, early on he taught himself drawing and gained knowledge of the sky. Tempel spent time as an apprentice for a lithographer in Dresden. He then spent 3 years in Copenhagen, Denmark, practicing this craft. Around 1850, Tempel went to Italy and in the employ of scientists prepared especially fine botanical drawings.

In 1856, Tempel worked for several months in Marseilles, France, at the observatory directed by Jean Valz. Similarly, he spent a short term in Italy assisting at the Bologna Observatory of **Lorenzo Respighi**.

In 1858, Tempel lived in Venice from where his wife, Marianna Gambini, came. The practical experience from Marseilles and Bologna in celestial observations sparked in Tempel the wish to possess a telescope. He acquired a 108-cm refractor from Munich, which enabled magnifications of 24, 40, and 300×. With it in 1859 he discovered comet C/1859 G1 (Tempel). He also confirmed the existence of nebulosity about the Pleiades (the Merope Nebula).

Tempel returned in 1860 to Marseilles. Valz had offered him a formal position as an assistant. There Tempel discovered minor planet (64) Angelina in 1861 and, four days later, minor planet (65) Cybele. For both of these discoveries, he received the Lalande Prize of the French Academy of Sciences.

Tempel observed jointly with Valz the 18 July 1860 total solar eclipse, in southern Spain. But after Valz's retirement in September 1861, Tempel left the observatory, as he found Valz's successor (Charles Simon, formerly mathematics teacher at the secondary school in Algiers) to be professionally incompetent.

Tempel worked again as a lithographer and returned to being an amateur astronomer. At his private observatory on the rue de Pythagore, Marseilles, he found further asteroids between 1861 and 1868 for a total of five.

Yet Tempel was dedicated to finding comets. Altogether he would discover 21 new comets and recover eight others. As a "comet hunter," Tempel can be considered the peer of the likes of **Jean Pons**, **William Brooks**, or **Edward Barnard.**

In the progress of the 1871 Franco-Prussian War, Tempel was expelled from France and went to Milan, Italy. There, as a respected amateur, he was engaged by **Giovanni Schiaparelli** at the Brera Observatory. Tempel continued his comet discoveries and very detailed drawings of celestial bodies.

After the death of **Giovani Donati**, Schiaparelli recommended Tempel as his successor, and so Tempel became director of the Arcetri Observatory in Florence. Here he found insufficient equipment. However, Tempel was accustomed to working under simple conditions and was able to use the opportunity to make additional discoveries. Most notable among these was the modern codiscovery of the Great Red Spot on Jupiter.

Minor Planet 3808 is named Tempel.

Jürgen Hamel

Selected References

Clausnitzer, Lutz (1989). *Wilhelm Tempel und seine kosmischen Entdeckungen.* Archenhold-Sternwarte, Vorträge und Schriften, no. 70. Berlin-Treptow.

Dreyer, J. L. E. (1890). "Ernst Wilhelm Leberecht Tempel." *Monthly Notices of the Royal Astronomical Society* 5 (1890): 179.

Tennant, James Francis

Born **Calcutta, (India), 10 January 1829**
Died **London, England, 6 March 1915**

After observing the 1874 transit of Venus from India, British Army officer James Tennant became an authority on the notorious Black Drop effect. Tennant and **John Herschel, Jr.** (also in India) had previously observed the 1868 total solar eclipse, using spectroscopes.

Selected Reference

Anon. (1916). "Lieutenant-General James Francis Tennant." *Monthly Notices of the Royal Astronomical Society*. 76: 272.

Terby, François Joseph Charles

Born **Louvain, Belgium, 8 August 1846**
Died **Louvain, Belgium, 20 March 1911**

François Terby was an important contributor to visual studies of Mars and Jupiter as well as to the history of those observations.

The son of music professor François Pierre Terby, François Joseph was educated at the Collège de Josephites in Louvain, and later at the University of Louvain, receiving degrees in philosophy, letters, and a Ph.D. in natural sciences. For several years he taught at the University of Louvain, and at the Collège Communal, and then as a professor of physics, chemistry, and mechanics at the École Industrielle, both also in Louvain. In about 1871, Terby gave up teaching to pursue observational astronomy as a full-time avocation in his private observatory.

Terby observed with a 3½-in. Secretan refractor, and after 1885 with an 8-in. Grubb equatorial refractor, concentrating on the planets, with especially noteworthy results on both Mars and Jupiter. He had begun observing Mars in 1864 and continued his observations of the red planet throughout his life. In addition to observing Mars, Terby decided to collect and discuss all prior observations of that planet in one volume. In the course of his effort to collect the observations, Terby performed a valuable service in tracking down the papers of **Johann Schröter,** which he preserved from loss by recovering them from one of Schröter's nephews. Terby wrote a memoir on Schröter's *Aerographische Fragmente* and on the Mars observations by every other known observer of Mars back to the 1636 observations of **Francisco Fontana** in his *Aérographie*, a valuable compendium published in 1875. Terby was awarded a prize by the Brussels Academy for that effort. Unfortunately, in his own observations, Terby succumbed to the all too common fate of Mars observers in that period. After **Giovanni Schiaparelli** published his observations of *canali* (canals) on the red planet, Terby looked for using Schiaparelli's map and imagined that he found them.

On Jupiter, Terby was among the first observers to record the appearance of the Red Spot, though his recognition of its uniqueness in his drawing was only retrospective after the announcement of its discovery by **Carr Pritchett**. By 1876, Terby was recognized well enough for his Jovian work that the Royal Astronomical Society appointed him to the first committee ever organized to systematically study Jupiter. That committee included **William Huggins**, **Edward Knobel**, James Ludovic Lindsay (Lord Lindsay, 1847–1913), **Wilhelm Lohse**, **William Parsons** (Third Earl of Rosse), **Arthur Ranyard,** and **Thomas Webb.**

Venus, and Saturn also attracted Terby's observational attention. Moreover he recorded many observations of meteors, comets, and the aurorae, and studied the lunar rills.

Terby was honored by election to the Royal Academy of Belgium in 1891, and as an officer of the Order of Leopold in 1907.

Thomas R. Williams

Selected References

Anon. (1911). "Obituary." *Observatory* 34: 204.

E. B. K. (1912). "François Joseph Charles Terby." *Monthly Notices of the Royal Astronomical Society* 72: 257–258.

Hockey, Thomas (1999). *Galileo's Planet*. Bristol: Institute of Physics, p. 129.
Sheehan, William P. (1996). *The Planet Mars: A History of Observation and Discovery*. Tucson: University of Arizona Press, pp. 38, 84–85.
Terby, F. J. C. (1875). "Aérographie, ou étude comparative des observations faites sur l'aspect physiquede la planète Mars depuis Fontana (1636) jusqu'á nous jours (1873)." *Mémoires des savants étrangers de l'Académie Royale des Sciences de Belgique* 39.
——— (1880). "Observations de la tache rouge de Jupiter." *Bulletins de l'Académie de Belgique*, 2nd ser., 49: 210.
W. T. L. (1911). "Dr. Terby." *Journal of the British Astronomical Association* 21: 328.

Terrentius

Schreck, Johann

Terrenz, Jean

Schreck, Johann

Tezkireci Köse Ibrāhīm

Flourished **Szigetvár, (Hungary), 17th century**

Tezkireci, an Ottoman astronomer and bureaucrat who settled in Istanbul, is known for having translated the French astronomer Noel Durret's (died: *circa* 1650) work entitled *Nouvelle théorie des planètes* from French into Arabic; this was the first book in Ottoman scientific literature to have been translated from a European language. The work, which was printed in Paris in 1635, was translated sometime between 1660 and 1664 and appeared under the title *Sajanjal al-aflāk fī ghāyat al-idrāk* (The mirror of the orbs with the utmost perception). In addition to containing astronomical tables, it was the first work in the Ottoman world to discuss the Copernican system and **Tycho Brahe**'s model of the Universe. The book also included the first diagrams illustrating those systems.

A bureaucrat charged with writing official memoranda, Tezkireci found the time to occupy himself with astronomy. There is little other information about his life except what we can discern from his translated book. In the introduction, Tezkireci reports that when he first showed the translated work to the chief astronomer *(başmüneccim)* Müneccimek Şekîbî Mehmed Çelebi (died: 1667) in Istanbul, Müneccimek at first disapproved saying that "Europeans have many vanities similar to this one." But eventually Müneccimek came to appreciate the work after Tezkireci Köse prepared an ephemeris based on the French tables, and Müneccimek saw that it was in conformity with **Ulugh Beg**'s *Zīj* (astronomical handbook with tables). Müneccimek copied the work for himself and bestowed upon the translator a benefaction, saying, "You saved me from suspicion. Now I have full confidence in our *zījes*."

In 1663 Tezkireci Köse again worked on the translation during his time with the Ottoman army at the winter quarters in Belgrade, this time with the encouragement of the *Kâdîasker* (chief judge) Ünsî Efendi (died: 1664). Tezkireci recalculated all the solar, lunar, and planetary mean motions of the *zīj* (originally compiled according to the meridian of Paris) and used the sexagesimal system; Tezkireci further abbreviated the tables and arranged them according to the signs of the zodiac *(abrāj)*. He presented a copy of the work to *Kâdîasker* Ünsî Efendi.

Later, Tezkireci Köse would translate most of the introduction of the work from Arabic into Turkish, leaving a few explanations in Arabic. This became the final form of the work. In the introduction, after a brief account of the history of astronomy, Tezkireci presents explanations, arranged in 24 subchapters *(ta'līm)*, which are followed by tables. In 1683, Cezmî Efendi (died: 1692), a judge in Belgrade, found a copy of the *Sajanjal* that had probably been given to Ünsî Efendi, and prepared another edition of the work.

From the introduction to the *Sajanjal*, we learn from Tezkireci that he had written another work about which he states: "For the proofs I compiled a different and new treatise *(risāla)*, containing all operations that are easier [to use] than the *Almagest*, as well as compiled a work for ephemerides that are used internationally and that are more graceful and succinct than all [others]" (Istanbul, Kandilli Observatory Library, MS 403, fol. 2a).

Mustafa Kaçar

Selected References

İhsanoğlu, Ekmeleddin (1992). "The Introduction of Western Science to the Ottoman World: A Case Study of Modern Astronomy (1660–1860)." In *Transfer of Modern Science and Technology to the Muslim World*, edited by Ekmeleddin İhsanoğlu, pp. 67–120, esp. 69–76. Istanbul: IRCICA. (Reprinted in İhsanoğlu, *Science, Technology and Learning in the Ottoman Empire* Aldershot: Ashgate, 2004, article II.)
İhsanoğlu, Ekmeleddin *et al.* (1997). *Osmanlı Astronomi Literatürü Tarihi (OALT)* (History of astronomy literature during the Ottoman period). 2 Vols. Istanbul: IRCICA.

Thābit ibn Qurra

Born **near Ḥarrān, upper Mesopotamia, (Turkey), *circa* 830**
Died **Baghdad, (Iraq), 18 February 901**

As a member of the **Banū Mūsā** circle of scholars in 9th-century Baghdad, Thābit ibn Qurra contributed significantly to the development of astronomy and other sciences through his translations and commentaries of Greek and Hellenistic works and through his original treatises. Notable astronomical contributions include a translation of **Ptolemy**'s *Almagest* and treatises on the motion of the Sun and the Moon. More generally, Thābit's significance lies in the influence of his work on the development of the exact sciences in Islam.

Thābit was a member of the Sabian religious sect. His heritage was steeped in traditions of Hellenistic culture and pagan veneration of the stars. This background, and in particular, his knowledge of Greek and Arabic, made him an attractive prospect for inclusion in one particular community of scholars – the Banū Mūsā and their circle in Baghdad. Thābit seems to have been asked to join this circle by a family member, the mathematician **Muḥammad ibn Mūsā ibn Shākir,** who recognized his talents and potential.

Thābit remained mainly in Baghdad, becoming a noted translator, physician, and renowned scholar in a variety of disciplines. As in the case of his mentors and teachers, Thābit was part of a family tradition of scholarly activity, with son Sinān ibn Thābit and grandson **Ibrāhīm ibn Sinān Thābit ibn Qurra** also making contributions to medicine and the exact sciences.

Thābit is credited with dozens of treatises, covering a wide range of fields and topics. While some were written in his native Syriac, most were composed in Arabic. Thabit was trilingual, a skill that enabled him to play a key role in the translation movement of 9th-century Baghdad. He translated works from both Syriac and Greek into Arabic, creating Arabic versions of important Hellenistic and Greek writings. Several of Thābit's Arabic translations are the only extant versions of important ancient works.

A large percentage of Thābit's corpus is devoted to mathematics. This includes translations of Books V–VII of **Apollonius**'s *On Conics* and **Archimedes**' *Lemmata* and *On Triangles*. His work in mathematics also includes original treatises, with contributions in the many areas of geometry and number theory. His original contributions include proofs of the Pythagorean theorem, a proof of Menelaus's theorem, proofs of Euclid's fifth postulate, and work on composite ratios.

Thābit's achievements in astronomy are closely linked to his work in mathematics. The application of his mathematical work (*e. g.*, his theories of composite ratios) to the examination and development of Ptolemaic astronomy, as Morelon emphasizes, helped establish a tradition of mathematical astronomy in Islamic culture. Discussion of Thābit's ideas is found in the work of later astronomers, including **Khāzinī** and **Naṣīr al-Dīn al-Ṭūsī**.

Thābit's revision of **Ḥunayn ibn Isḥāq**'s translation of the *Almagest* survives in manuscript. In addition, something less than a dozen astronomical treatises by Thābit have survived, about a fourth of the number he is credited with composing. Two of these present the basics of Ptolemaic astronomy, including the structure of the cosmos according to Ptolemy's *Planetary Hypotheses*, a work whose Arabic translation Thābit revised. In the other extant treatises, Thābit addresses the problem of the unequal motion of the Sun, the motion of the Moon, the determination of crescent visibility, and the theory of sundials.

Two treatises traditionally attributed to Thābit are almost certainly not by him. One of these that survives only in Latin translation is *De motu octave spere* (On the motion of the eighth sphere); the misattribution may be due to the fact that a related treatise was written by his grandson Ibrāhīm Ibn Sinān. The author of *De motu* addresses a type of problem that astronomers in the centuries following Ptolemy have all had to confront – changes in astronomical parameters as a consequence of elapsed time. A new model for the precession of the equinoxes is presented in order to account for such changes. Two time-related changes that this model addresses are the increase in the rate of precession and the decrease in the value of the obliquity of the equinox since the time of the *Almagest*. In addition, a theory of oscillation or periodicity of these motions ("trepidation") is proposed.

The other misattributed treatise deals with the solar year. The author of this work attempts to show why adopting a sidereal year is preferable to accepting Ptolemy's tropical year as the basic time-unit for solar motion.

In addition to his works in mathematical astronomy, Thābit also wrote on philosophical and cosmological topics, questioning some of the fundamentals of the Aristotelian cosmos. He rejected **Aristotle**'s concept of the essence as immobile, a position Rosenfeld and Grigorian suggest is in keeping with his anti-Aristotelian stance of allowing the use of motion in mathematics. Thābit also wrote important treatises related to Archimedean problems in statics and mechanics.

Thābit's efforts provided a foundation for continuing work in the investigation and reformation of Ptolemaic astronomy. His life is illustrative of the fact that individuals from a wide range of backgrounds and religions contributed to the flourishing of sciences like astronomy in Islamic culture.

JoAnn Palmeri

Selected References

Carmody, Francis J. (1960). *The Astronomical Works of Thabit b. Qurra*. Berkeley: University of California Press.

Goldstein, Bernard R. (1964). "On the Theory of Trepidation." *Centaurus* 10: 232–247.

Morelon, Régis (1994). "Ṯābit b. Qurra and Arab Astronomy in the 9th Century." *Arabic Sciences and Philosophy* 4: 111–139.

——— (1996). "Eastern Arabic Astronomy between the Eighth and the Eleventh Centuries." In *Encyclopedia of the History of Arabic Science*, edited by Roshdi Rashed. Vol. 1, pp. 20–57. London: Routledge.

Neugebauer, Otto (1962). Review of *The Astronomical Works of Thabit b. Qurra*, by Francis Carmody. *Speculum* 37: 99–103.

Ragep, F. J. (1993). *Naṣīr al-Dīn al-Ṭūsī's* Memoir on Astronomy *(al-Tadhkira fī ʿilm al-hay'a)*. 2 Vols. New York: Springer-Verlag. (For a discussion of trepidation and the authorship of *De motu octave spere*.)

Rosenfeld, B. A. and A. T. Grigorian (1976). "Thābit ibn Qurra." In *Dictionary of Scientific Biography*, edited by Charles Coulson Gillispie. Vol. 13, pp. 288–195. New York: Charles Scribner's Sons.

Thābit ibn Qurra (1987). *Oeuvres d'astronomie*, edited by Régis Morelon. Paris: Les Belles Lettres.

Thackeray, Andrew David

Born **Chelsea, (London), England, 19 June 1910**
Died **Sutherland, (Western Cape), South Africa, 21 February 1978**

British-South African stellar astronomer Andrew Thackeray (always called David) was noted for his discovery with **Adriaan Wesselink** in 1952 of RR Lyrae stars in the Small Magellanic Cloud. This, in turn, demonstrated that the distance scale until then in use (which went back to **Harlow Shapley**) was too small by a factor of two, and the Universe therefore a factor of at least two older than had previously been supposed, removing most of the contradiction between the apparent age of the cosmos and the best estimates of solar and stellar ages.

Thackeray was the son of a classical scholar and a nephew of the solar physicist **John Evershed**. He was educated at Eton (1924–1929) and King's College Cambridge (1929–1933), where he studied mathematics. He started research work at the Solar Physics Laboratory, Cambridge during 1932–1934. From 1934 to 1936 Thackeray held a Commonwealth Fund Fellowship in Astrophysics and worked at the Mount Wilson Observatory in California, USA. There he studied the emission spectra of Mira variables and showed that fluorescence mechanisms similar to those then just discovered by **Ira Bowen** could explain some of the lines. This was the subject of his Cambridge Ph.D. dissertation in 1937. Thackeray served as chief assistant at the Solar Observatory in Cambridge from 1937 to 1948, except for a period of wartime service as an ambulance driver in Italy.

In 1948 Thackeray became chief assistant of the Radcliffe Observatory in Pretoria, then just completed. He became its director in 1950 and stayed in the post until the observatory was closed in 1974. His last years were spent as an honorary professor at the University of Cape Town. Thackeray was returning from an observing run with the Radcliffe 1.9-m reflector, which had in the meantime been moved to Sutherland, when he was killed in a freak accident.

On arrival in Pretoria in 1948, Thackeray found that he had almost unlimited observing time on the new telescope but rather minimal instrumentation to work with. He corresponded with **Walter Baade** about suitable programs to pursue and decided to investigate the globular-like clusters of the Magellanic Clouds as a priority. At the International Astronomical Union General Assembly in Rome in 1952 Baade announced that he had failed to detect RR Lyrae variables in M31 with the 200-in. telescope and that, therefore, it had to be further away than formerly believed. Immediately after he had finished speaking, Thackeray announced that he and Wesselink had actually detected RR Lyrae variables in NGC 121 in the Small Magellanic Cloud. Thus the absolute magnitudes of RR Lyraes and Cepheids could now be compared. The RR Lyraes were found to be 1.5 magnitudes fainter than predicted from the prevailing Cepheid distance scale, which therefore had to be wrong.

Another important campaign conducted by Thackeray with his collaborators Michael Feast and Wesselink was an investigation of the brightest stars in the Magellanic Clouds, which showed observationally the limits of stellar luminosity.

Thackeray's main area of expertise was stellar spectroscopy, and he took particular interest in the spectroscopy and long-term behavior of the eclipsing symbiotic variable AR Pav, the luminous galactic variable η Carinae, and the old nova RR Tel. He discovered the polarization of the halo of η Carinae, and his spectroscopic study of RR Tel is widely quoted.

As the director of the Radcliffe Observatory, Thackeray created an atmosphere where administrative matters were kept firmly at a minimal level and scientific discussion was very much encouraged. He never saw the necessity for seeking publicity, a fact that perhaps led to a lack of appreciation of his work in official circles. He had very wide interests in natural phenomena, and his publication list numbers about 300 items. He was an editor of *The Observatory* during 1938–1942. A fellow of the Royal Astronomical Society from 1933 onward, Thackeray was made an associate just days before his death. He was president of the Astronomical Society of Southern Africa in 1951–1952.

In 1944, Thackeray married Mary Rowlands, and they had four children.

Ian S. Glass

Selected References

Feast, Michael W. (1978). "Andrew David Thackeray 1910–1978. A Bibliography." *Monthly Notes of the Astronomical Society of Southern Africa* 37: 49–62.

——— (1979). "Andrew David Thackeray." *Quarterly Journal of the Royal Astronomical Society* 20: 216–220.

——— (2000). "Stellar Populations and the Distance Scale: The Baade–Thackeray Correspondence." *Journal of the History of Astronomy* 31: 29–36.

Warner, Brian (1978). "Andrew David Thackeray (Obituary)." *Monthly Notes of the Astronomical Society of Southern Africa* 37: 20–22.

Thaddaeus Hagecius

➔ **Hájek z Hájku, Tadeá**

Thales of Miletus

Born ***circa*** **625 BCE**
Died ***circa*** **547 BCE**

Thales was credited by **Aristotle** with founding Ionian natural philosophy. His fame as an astronomer is based more specifically on his purported prediction of a solar eclipse, an achievement that marks for some historians the beginning of western astronomical science.

One of the major intellectual traditions within pre-Socratic science between 600 and 400 BCE was that established and developed by the Milesians (after the city of Miletus; also Ionians, after the region, the present-day Turkish coast of Asia Minor). Of the new Greek communities that sprang up in Greece itself and across the Aegean Sea in Asia Minor, the most prosperous was Miletus. Now but lonely ruins inland from the coast because the river and harbor silted up long ago, Miletus was, in its time, the richest city in the Greek world.

One objective of Ionian science or philosophy – the two were not separate disciplines at this time – seems to have been to search for a basic substance or substances, which persist throughout all changes. The Ionians were more interested in cosmogony (the creation of the world) than in cosmology (the structure and evolution of the world).

According to Aristotle, writing more than two centuries after the fact, Ionian philosophers thought that matter or principles in the form of matter were the principle of all things:

Most of the first philosophers thought that principles in the form of matter were the only principles of all things: For the original source of all existing things, that from which a thing first comes-into-being and into which it is finally destroyed, the substance persisting but changing in its qualities, this they declare is the element and first principle of existing things, and for this reason they consider that there is no absolute coming-to-be or passing away, on the ground that such a nature is always preserved for there must be some natural substance, either one or more than one, from which the other things come-into-being, while it is preserved. Over the number, however, and the form of this kind of principle they do not all agree; but Thales, the founder of this type of philosophy, says that it is water (and therefore declared that the earth is on water), perhaps taking this supposition from seeing the nurture of all things to be moist, and the warm itself coming-to-be from this.

Thales was reputed to be the wisest of the seven wise men or sages of Greece. Asked what was difficult, he answered "to know thyself." Asked what was easy, he answered "to give advice." Supposedly, Thales was the first mathematician to demonstrate that a circle is bisected by its diameter, that the angle of a semicircle is a right angle, and that angles at the base of an isosceles triangle are equal. An acerbic scholar has noted, however, that "inevitably there accumulated round the name of Thales, as that of Pythagoras (the two often being confused), a number of anecdotes of varying degrees of plausibility and of no historical worth whatsoever."

Thales is also credited with predicting a solar eclipse. According to the Greek historian Herodotus, writing in the 5th century BCE, more than a century after Thales:

In the sixth year of the war, which they [the Medes and the Lydians] had carried on with equal fortunes, an engagement took place in which it turned out that when the battle was in progress the day suddenly became night. This alteration of the day Thales the Milesian foretold to the Ionians, setting as its limit this year in which the change actually occurred.

Either the warring parties took the eclipse of the Sun as a sign to cease fighting, or they were eager for any reason to cease and found the eclipse a convenient excuse.

Astronomical calculations indicate a total solar eclipse on 28 May 585 BCE at the place of the battle in northern Turkey, thus lending credence to Herodotus's history. Subsequent discussions have centered less on the credibility of the tradition itself and more on what methods Thales could have used to predict the solar eclipse. From a study of the periodic recurrence of solar eclipses, Thales might possibly have predicted a slightly later eclipse but taken credit for the 585 eclipse. It is not certain that the eclipse reported by Herodotus is the eclipse of 585; eclipses of 582 and 581 have been pointed to as other possibilities, though they were not total over Asia Minor. Also, some scholars dismiss the whole eclipse prediction legend as more myth than historic truth.

Another legend involving Thales has him providing a practical justification for the study of philosophy. This time Aristotle is the source, in his *Politics*:

For when they reproached him [Thales] because of his poverty, as though philosophy were no use, it is said that, having observed through his study of the heavenly bodies that there would be a large olive-crop, he raised a little capital while it was still winter, and paid deposits on all the olive presses in Miletus and Chios, hiring them cheaply because no one bid against him. When the appropriate time came there was a sudden rush of requests for the presses; he then hired them out on his own terms and so made a large profit, thus demonstrating that it is easy for philosophers to be rich, if they wish, but that it is not in this that they are interested.

Thus did Thales demonstrate that philosophers could be rich in conventional monetary terms if they wished. The philosopher's true wealth, however, is not measured in money; it is found in the pleasure derived from intellectual endeavor. In eschewing a myopic pursuit of wealth, Thales demonstrated his wisdom again. Still, Thales was not always practical, as **Plato** noted in his *Thaetetus*: .

Theodorus, a witty and attractive Thracian servant-girl, is said to have mocked Thales for falling into a well while he was observing the stars and gazing upward; declaring that he was eager to know the things in the sky, but that what was behind him and just by his feet escaped his notice.

Norriss S. Hetherington

Selected References

Aaboe, Asger (1972). "Remarks on the Theoretical Treatment of Eclipses in Antiquity." *Journal for the History of Astronomy* 3: 105–118. (For a general discussion of ancient understanding of eclipses.)

Dicks, D. R. (1959). "Thales." *Classical Quarterly* 53, n.s., 9: 294–309.

Hartner, Willy (1969). "Eclipse Periods and Thales' Prediction of a Solar Eclipse: Historic Truth and Modern Myth." *Centaurus* 14: 60–71. (Reprinted in Hartner, *Oriens-Occidens*, Vol. 2, pp. 86–97. Hildesheim: Georg Olms, 1984. [On Thales's purported eclipse prediction.])

Hetherington, Norriss S. (1987). *Ancient Astronomy and Civilization*. Tucson: Pachart Publishing House.

——— (1993). "Early Greek Cosmology." In *Encyclopedia of Cosmology: Historical, Philosophical, and Scientific Foundations of Modern Cosmology*. New York: Garland, pp. 183–188.

——— (1993). "The Presocratics." In *Cosmology: Historical, Literary, Philosophical, Religious, and Scientific Perspectives*. New York: Garland, pp. 53–66.

——— (1997). "Early Greek Cosmology: A Historiographical Review." *Culture and Cosmos* 1: 10–33.

Kirk, G. S., J. E. Raven, and M. Schofield (1983). *The Presocratic Philosophers: A Critical History with a Selection of Texts*. 2nd ed. Cambridge: Cambridge University Press. (The most useful and convenient collection of the raw materials for reconstructions and analyses of Presocratic philosophy.)

Lloyd, G. E. R. (1970). *Early Greek Science: Thales to Aristotle*. New York: W. W. Norton and Co.

Mosshammer, Alden A. (1981). "Thales' Eclipse." *Transactions of the American Philological Association* 111: 145–155. (For doubts concerning Thales' ability to predict a solar eclipse.)

Panchenko, Dmitri (1994). "Thales's Prediction of a Solar Eclipse." *Journal for the History of Astronomy* 25: 275–288.

Stephenson, F. Richard and Louay J. Fatoohi (1997). "Thales's Prediction of a Solar Eclipse." *Journal for the History of Astronomy* 28: 279–282.

Theodosius of Bithynia

Born **Bithynia, (Anatolia, Turkey),** ***circa*** **160 BCE**
Died ***circa*** **90 BCE**

Theodosius compiled a three-volume text on spherical geometry, which was much used in the Middle Ages and Renaissance.

Euclid's *Elements of Geometry*, while no minor work, does not treat spherical geometry. Astronomers throughout the Hellenistic world thus hungered for a text on this important subject. This appetite was satisfied by Theodosius' three-volume *Spherics*. *Spherics* attracts little compliment from modern writers: The mathematician T. Heath, (1921) judged Theodosius "simply a laborious compiler," for "there was practically nothing original in his work." **Otto Neugebauer** observed that Theodosius failed to recognize the significance of the great-circle triangle, that his theorems seldom treat more than what is obvious, and that its Euclidean rigor is but cosmetic: his "proofs" do little more than reword the conjectures, and seldom does Theodosius admit his assumptions. Writers as early as **Pappus** commented that the *Spherics* had a very theoretical tone, that (in contrast to a competing text by **Apollonius**) it hardly ever indicated where in astronomy the mathematics might be applied. Yet the *Spherics* has proven useful enough to endure nearly as long as Euclid's *Elements:* The Greek manuscript was translated into Arabic in the 10th century, and thence into Latin by **Gerard of Cremona** (and perhaps **Campanus of Norara**) in the 12th century. A Latin edition was printed in 1518, soon followed by **Johannes Vögelin**'s much-improved translation of 1529. The *Spherics* inspired **Christoph Clavius** to produce a new Latin translation and commentary in 1586, of which an English translation appeared in 1721. Latin translations were published also by other influential thinkers such as Jean Pena (1558, including the first printing of the Greek text), **Francesco Maurolico** (1558), and Isaac Barrow (1675). Boring and unoriginal *Spherics* may have been, but useless and disregarded it was not.

Theodosius' other astronomical works include two books *On Days and Nights* and a 12-theorem book *On Habitations*. These works, which have survived, discuss how views of the stars and the lengths of night and day depend on the observer's location on the Earth, and which parts of the Earth have habitable climates. *On Days and Nights* treats the daily passage of the Sun, with a view to determining the conditions under which the solstice occurs on the meridian, and when equinoctial night and day are truly equal. One interesting conclusion is that, if the year is equal to an irrational number of days, then the stellar phases will show no annual periodicity.

Theodosius is credited also with a work on astrology (containing material important to astronomy), and a commentary on **Archimedes**' *Mechanics*, both of which are lost. Some fragments survive from his *Description of Houses*, which treats problems in architecture. On the practical front, **Vitruvius** credits Theodosius with inventing a universal sundial, but of this we know no details.

Alistair Kwan

Selected References

Adler, A. (ed.) (1928–1938). *Suidae Lexicon*. 5 Vols. Stuttgart: B. G. Teubner. (Reprint, 1967–1971.) (This extensive encyclopedia on Mediterranean antiquity, compiled in the 10th century from numerous older sources, enumerates five men named Theodosius, the first of whom is the mathematician. Theodosius is often called "of Tripolis," which is widely blamed on confusion in the *Suda*. But the *Suda* in fact describes the Tripolitan Theodosius in a clearly separate entry.)

Heath, Sir Thomas L. (1921). *A History of Greek Mathematics*. 2 Vols. Oxford: Clarendon Press. (The *Spherics* is outlined in Vol. 2, pp. 245–253.)

Irby- Massie, Georgia L. and Paul T. Keyser (2002). *Greek Science of the Hellenistic Era: A Sourcebook*. London: Routledge.

Lorch, Richard (1996). "The Transmission of Theodosius' *Sphaerica*." In *Mathematische Probleme im Mittelalter: Der lateinische und arabische Sprachbereich*, edited by Menso Folkerts, pp. 159–183. Wiesbaden: Harrassowitz. (Traces the Greek–Arabic–Latin transmission of the *Spherica*.)

Neugebauer, Otto (1975). *A History of Ancient Mathematical Astronomy*. 3 pts. New York: Springer-Verlag (See "Spherical Astronomy before Menelaus," pp. 748–771, esp. 755–768.

Strabo. *Geography*. 5.467. (Records Theodosius the mathematician and his sons, in his list of wise men from Bithynia.)

Theodosius (1927). *De habitationibus; De diebus et noctibus*. Edited by Rudolf Fecht. Berlin: Weidmannsche Buchhandlung, 1927. (Neugebauer warns against Fecht's "absurd" diagrams which betray a lack of fundamental understanding needed to properly edit the text.)

Theodosius (1927). *Sphaerica*, Greek and Latin text edited by J. L. Heiberg. Berlin. Weidmannsche Buchhandlung.

Theodosius (1959). *Les sphériques de Theodose de Tripoli*. Translation with introduction and notes by Paul ver Eecke. Paris: Blanchard.

Theodosius (1529). *Spherics*. Translated by Johannes Vögelin. Vienna: Joannes Singrenius. (Reprinted, New York: Readex Microprint, 1986.)

Vitruvius. *De architectura*. 9.9. (Names Theodosius as coinventor of a sundial "for all climates," *i. e.*, for any latitude.)

Theon of Alexandria

Born **Alexandria, (Egypt),** ***circa*** **335**
Died **Alexandria, (Egypt),** ***circa*** **400**

The details of the life of Theon, Greco–Egyptian mathematician, astronomer, and teacher of late antique Alexandria, are speculative and derive primarily from later accounts that are frequently confused or inaccurate. His predictions and observances of the solar and lunar eclipses of 364, however, establish that he was an active scholar at that time; similarly, he is said to have reached maturity during the latter two decades of the 4th century. A pagan, Theon served as the last member of the *Mouseion* at Alexandria and devoted himself especially to the study of older Greek religious practices and beliefs. Though he seems not to have actually taught philosophy, he was regarded as a philosopher in some later sources. He worked with associates such as his older contemporary **Pappus**, the mathematicians Orgines and Eulalius, and his student Epiphanius. Importantly, Theon was also the father of the Neoplatonist philosopher and mathematician **Hypatia** and was her closest associate and collaborator. His death is thought to have occurred before she was killed by a Christian mob in 415.

Theon's surviving works all ultimately derive from his activities as a professor, being chiefly commentaries on and explications of the works of earlier authors in the fields of mathematics and astronomy. In the case of the former, Theon produced reworked editions of several of Euclid's treatises: a highly influential edition of the *Elements*, along with an edition of the *Data* (on the basics of geometrical figures), and of the *Optics*, subsequently identified as student notes taken down during lectures. He also produced an edition of the Pseudo-Euclidean work on visual reflection, the *Catoptrics*. In general, Theon's mathematical contribution is slight, but he does offer insights into the Greek sexagesimal system as it was applied in calculation.

Theon's astronomical contributions are more significant, and his commentaries on the two major works of **Ptolemy** are partially extant. The more extensive of these is that on the *Almagest*, originally written in 13 books but now missing book 11 and most of book 5. The commentary itself is a reworking of Theon's own lectures and thus has been criticized as being merely a scholastic exercise. Its value, however, lies in its incorporation of information from lost works on which Theon relied, including that of Pappus. Theon also wrote two commentaries on Ptolemy's *Handy Tables*, claiming that he was the first to do so. The *Great Commentary* is fragmentary, with slightly more than three of its original five books now extant (books 1–3, the beginning of book 4). In describing how to use the Ptolemaic computations, it also explains the reasoning and calculations behind them, thus repeating some of the information in the commentary on the *Almagest.* There is some indication that Hypatia may have revised book 3. The *Little Commentary* on the *Handy Tables* survives complete in one book and is consciously directed at students limited in their geometrical and mathematical preparation. In it Theon discusses the theory later known as "trepidation," the variability of the rate of precession. It is thought that in the 7th century the Syrian **Severus Sebokht** used Theon's *Little Commentary* in conjunction with the *Handy Tables.*

Theon also wrote a now lost work on the astrolabe, called in the 10th century *Suda Lexicon* (Treatise on the small astrolabe); Arabic sources also attribute to him a work on the instrument. Significantly, while Theon clearly did not invent the astrolabe, his work served as the most important link transmitting its theoretical concepts from the Greek to the Islamic world and thence to Europe. Indeed, Sebokht's work on the astrolabe, which is still extant, was drawn from and preserves material from Theon's lost treatise.

Theon's name is also associated with a number of other lost works, including tracts on planetary movements, on the star Sirius, and on other natural occurrences. He seems also to have written commentaries on esoteric religious and magical texts as well as to have composed poetry, one extant example of which is in praise of Ptolemy.

John M. McMahon

Selected References

Cameron, Alan (1990). "Isidore of Miletus and Hypatia: On the Editing of Mathematical Texts." *Greek, Roman and Byzantine Studies* 31: 103–127. (For Hypatia's collaboration with Theon and the revision of his *Great Commentary*.)

Dzielska, Maria (1995). *Hypatia of Alexandria.* Cambridge, Massachusetts: Harvard University Press, pp. 66–77. (For a more recent treatment of Theon's life and his relationship to his daughter and other contemporaries.)

Nau, F. N. (1899). "Le traité sur l'astrolabe plan de Sévère Sabokt." *Journal asiatique,* 9th ser., 13: 56–101, 238–303. (For Severus Sebokht's *Treatise on the Astrolabe.* An English version of it [from Nau's French] is in Robert T. Gunther, *The Astrolabes of the World,* Vol. 1: *The Eastern Astrolabes.* Oxford: University Press, 1932, pp. 82–103.)

Neugebauer, Otto (1949). "The Early History of the Astrolabe." *Isis* 40: 240–256. (Reprinted in Neugebauer, *Astronomy and History: Selected Essays.* New York: Springer-Verlag, 1983, pp. 278–294 [for a discussion of Theon's lost work on the astrolabe].)

Pingree, David (1994). "The Teaching of the *Almagest* in Late Antiquity." In *The Sciences in Greco-Roman Society*, edited by Timothy D. Barnes. Edmonton: Academic Printing and Publishing, pp. 73–98. (For the survival of Theon's commentary on the *Almagest*.)

Rome, A. (1931, 1936, 1943). *Commentaires de Pappus et de Théon d'Alexandrie sur* l'Almageste. Bks. 1–3. Vatican City: Billiotheca apostolica vaticana. (For Theon's commentary on the *Almagest*.)

Tihon, Anne (1985, 1991, 1999). *Le 'Petit commentaire' de Théon d'Alexandrie aux tables faciles de Ptolémée.* Vatican City, 1978; *Le "Great Commentaire" aux tables faciles de Ptolémée.* Bks. 1–4. Vatican City. (Recent editions, with French translation, of Theon's commentaries on the *Handy Tables*.)

Toomer, G. J. (1976). "Theon of Alexandria." In *Dictionary of Scientific Biography*, edited by Charles Coulston Gillispie. Vol. 13, pp. 321–325. New York: Charles Scribner's Sons. (An encyclopedic treatment of Theon, with extensive bibliography on all aspects of his life and works.)

Theon of Smyrna

Born ***circa* 70**
Died ***circa* 135**

Known as "Theon the mathematician" by **Ptolemy**, "the old Theon" by **Theon of Alexandria**, and "Theon the Platonist" by **Proclus**, Theon is best remembered as the author of a handbook on Pythagorean harmony, and for several widely cited observations of Mercury and Venus. Recorded by Ptolemy, these observations (127, 129, 130, and 132) were later used to determine the maximum elongation of the inferior planets. They also confirm Theon's flourish dates; an extant stone bust fixes his death before 140. Nothing certain is known of his life, education, and related writings. Theon's most influential extant writing, *Theonis Smyrnaei Platonici Eorum, quae in Mathematicis ad Platonis lectionem utilia sunt, Expositio*, was first translated from Greek to Latin by **Ismaël Boulliau**, the noted French astronomer (Theon of Smyrna the Platonist, exposition of the works on mathematics useful for reading Plato, Paris 1644, Bks I and II).

Now consisting of three extant books (Arithmetic, Music, and Astronomy) Theon's *Exposition* was designed to introduce general readers to the Cosmic Harmony that binds the mathematical and natural worlds, thus opening a path (if not a royal road) to understanding **Plato**'s philosophy. It is not an original or technically demanding work. Instead, it moves simply but elegantly from number to arithmetic, geometry, music, and astronomy to the Harmony of the World. In practice, Theon begins by defining numbers, prime and geometrical, finally focusing on more sophisticated ratios, proportional ratios, and progressions. By tradition, the principal value of the *Exposition* is not its originality but its use as a historical source concerning ancient writers and lost texts.

Yet Theon's treatise can no longer be viewed simply as an ancient text. Seldom discussed, Boulliau's translation of the *Exposition* marked something of a modern revival in the wake of **Johannes Kepler**'s harmonic conjectures. Pythagorean speculations about the Harmonies of the World were of particular interest, especially the kinds of numbers: odd and even, prime and composite, square and oblong, circular and spherical, pyramidal and perfect. In Book II, Theon's treatment of music addressed not only the mathematical relations between intervals but the role of ratios in eccentric and epicyclic constructions. The resonance with Kepler is obvious. But if Theon's main interest was Cosmic Harmony, Part Three draws together his central themes. Echoing Adrastus, Theon begins with a systematic introduction to the elements of astronomy, the ordering of the planets, their retrograde motions (discussing epicyclic and homocentric models), and finally, the problem of planetary distances, which Theon plays skillfully against musical intervals of the octave. More generally, the "harmony of the world" is rooted in number and proportion, which govern all bodies and movements. Theon's concluding discussion of planetary motion on the surface of geometrical solids – on parallel circles and in spirals – suggests possibilities later explored and exploited by the New Science.

Theon's other works on astronomy and mathematics are lost, among them commentaries on Ptolemy and Plato's *Republic*, and reportedly, a history of Plato's ancestry. Principal manuscripts of Theon's *Exposition* (Greek) are found in Paris, Venice, Florence, Naples, and Rome; reports suggest an Arabic version has been recently discovered. Published editions of the *Exposition* have appeared piecemeal over the last three centuries. Following Boulliau's edition of Book I (Arithmetic) is J. J. de Gelder (Book I, Greek & Latin, Leiden 1827). Book II (Music) has not been retranslated into Latin; Book III (Astronomy) is in an edition by T. H. Martin (Greek & Latin, 1849; Gröningen 1971). The first complete Greek edition is Eduard Hiller (Books I–III, Greek, Leipzig 1878). The first complete translation was J. Dupuis (Books I–III, Greek and French, Paris 1892) and most recently, Joëlle Delattre (Greek–French). No complete English translation (from Greek or Latin) exists.

Robert Alan Hatch

Selected References

Boulliau, Ismaël (ed. and trans.) (1644). *Theonis Smyrnaei Platonici*. Paris: Amherst.

Heath, Sir Thomas L. (1921). *A History of Greek Mathematics*. 2 Vols. Oxford: Clarendon Press. (Reprint, New York: Dover, 1981.)

Neugebauer, Otto (1975). *A History of Ancient Mathematical Astronomy*. 3 pts. New York: Springer-Verlag.

Vedova, G. C. (1951). "Notes on Theon of Smyrna." *American Mathematical Monthly* 58: 675–683.

Theophrastus

Born **Eresos, Lesbos (Lésvos Greece), 372/371 BCE or 371/370 BCE**

Died **possibly Athens, (Greece), 288/287 BCE or 287/286 BCE**

Theophrastus wrote numerous works on topics ranging from formal logic to practical legislation and from ethics to meteorology.

Tyrtamus, son of Melantas, was his full name; "Theophrastus" was a nickname. The ancient tradition that he studied at **Plato**'s Academy (and there met **Aristotle**) may be unreliable. He certainly accompanied Aristotle in his travels and researches from 347 BCE onward, becoming a member of Aristotle's school in the Lyceum when it was founded in 335 BCE, and head of the school when Aristotle went into exile in 323 BCE.

The contribution to knowledge for which Theophrastus is best known, and the one in which his major works survive, is his establishing of botany as a formal science. Other areas of his activity are known through some shorter surviving treatises and through secondhand reports and quotations.

Theophrastus's contributions to astronomy can be considered under five heads:

(1) The explanation of the motion of each planet in the theory of **Eudoxus** and **Callippus** involved several spheres in addition to that actually carrying the planet. Aristotle's conversion of this mathematical theory into a physical one involved the addition of further counteracting spheres to cancel out the motion of each planet so that it would not affect those below it. We are informed that Theophrastus applied the terms "starless" to the former and "carrying round in the opposite direction" to the latter, and the implication is that he invented these terms. It is a reasonable though not certain supposition that he also accepted Aristotle's theory itself, but see (3) below.

(2) In the surviving short treatise now known as his *Metaphysics* (not its original title), Theophrastus raises serious objections to Aristotle's theory that the circular movement of the heavenly spheres is caused by their desire for God, the Unmoved Mover. It seems likely, though this is disputed, that Theophrastus rejected the theory of the Unmoved Mover altogether. It does however seem that he retained the view that the heavenly bodies themselves are living beings.

(3) In Aristotle's view the heavens are composed of an element, *aether*, unique to them and distinct from the four elements (earth, water, air, and fire) found in the terrestrial region. Whether Theophrastus retained this view is disputed. In sections 5–6 of his treatise *On Fire*, in the course of a complex and somewhat inconclusive discussion of the status of fire as an element and its relation to heat, he considers the possibility that the Sun itself is hot, which in Aristotle's theory it is not. (According to Aristotle, its heating effect is caused by friction between the *aether* and the air beneath it.) But ancient reports of Theophrastus's views suggest that he retained Aristotle's theory of *aether* as a distinct element, and it is not easy to see how he could have accepted Aristotle's theory of the heavenly spheres (earlier, (1)) if he did not do so. It is possible that his views changed during his career; but, as is often the case, the tentative and exploratory nature of his discussions (where they are preserved in their original form at all) means that we cannot be sure.

(4) Like Aristotle, Theophrastus regarded such phenomena as comets, because of their irregular nature, as occurring in the region near the Earth and as part of the study of meteorology rather than of astronomy. He followed Aristotle in linking comets with wind and drought; an alleged connection with earthquakes is less certain. The treatise *On Weather-Signs*, which includes reference to astronomical signs of the seasons and of the weather, is not a genuine work of Theophrastus in its present form, though the astronomical signs probably derive from him.

(5) The date at which the Greeks developed an interest in astrology (in the modern sense of the term, rather than in its earlier Greek usage for what we now call astronomy) is disputed. Theophrastus is recorded as having referred to the Chaldaeans foretelling the fortunes of individuals from the heavenly bodies;

this, if genuine, would be one of the earliest explicit references to astrology in Greek.

R. W. Sharples

Alternate name

Tyrtamus

Selected References

Amigues, Suzanne (trans.) (1988–2006). *Recherches sur les plantes*, by Theophrastus. Paris: Les Belles Lettres.

Coutant, Victor (ed. and trans.) (1971). *Theophrastus, De igne: A Post-Aristotelian View of the Nature of Fire*. Assen: Royal Vangorcum. (For English translation of *On Fire*.)

Einarson, Benedict and George K. K. Link (trans.) (1976–1990). *De causis plantarum*, by Theophrastus. Loeb Classical Library, nos. 471, 474, and 475. Cambridge, Massachusetts: Harvard University Press.

Fortenbaugh, William W. and Dimitri Gutas (1992). *Theophrastus: His Psychological, Doxographical, and Scientific Writings*. New Brunswick, New Jersey: Transaction Publishers, pp. 307–345. (For sources of the spurious *On Weather-Signs*, which is included as an appendix in Hort.)

Fortenbaugh, William W. *et al.* (eds.) (1992). *Theophrastus of Eresus: Sources for His Life, Writings, Thought and Influence*. Leiden: Brill. (Evidence for Theophrastus' lost works is collected herein; texts relating to astronomy are in Vol. 1, pp. 318–333 [texts 158–168] and to comets and astrology in Vol. 1, pp. 362–365 [texts 193–194]. These texts and related issues are discussed in the accompanying commentary volume by Sharples (1998).

Hort, Sir Arthur (trans.) (1916). *Enquiry into Plants*, by Theophrastus. Loeb Classical Library, nos. 70 and 79. London: W. Heinemann. (English translation, superseded by Amigues.)

Sharples, R. W. (1998). *Theophrastus of Eresus: Sources for His Life, Writings, Thought and Influence, Commentary.* Volume 3.1, *Sources on Physics*. Leiden: Brill, pp. 85–112, 160–163.

Theophrastus. *Metaphysics*, translated by W. D. Ross and F. H. Fobes. Oxford: Clarendon Press, 1929; *Metaphysics*, translated by Marlein van Raalte. Leiden: E. J. Brill, 1993; *Métaphysique*, translated into French by Andre Laks and Glenn W. Most. Paris: Les Belles Lettres, 1993.

Thiele, Thorvald Nicolai

Born **Copenhagen, Denmark, 24 December 1838**
Died **Copenhagen, Denmark, 26 September 1910**

Thorvald Thiele was an observatory director and mathematician. He was the son of Just Thiele and Hanne Aagesen. In 1875, he was appointed professor of astronomy and director of the Copenhagen University Observatory, positions he held until 1907. Apart from his works in astronomy, Thiele contributed to statistics and the actuarial sciences. Thiele, who had bad eyesight, mostly worked in celestial mechanics. In his analyses of double star systems he developed what became known as the Thiele–Innes method. His most original work was in the mathematical theory of observations, published in 1903 as *Theory of Observations*.

Helge Kragh

Selected Reference

Lauritzen, Steffen L. (2002). *Thiele: Pioneer in Statistics*. Oxford: Oxford University Press.

Third Earl of Rosse

➲ **Parsons, William**

Thollon, Louis

Born **Ambronay, Ain, France, 2 May 1829**
Died **Nice, France, 8 April 1887**

Louis Thollon was a specialist in high-dispersion spectroscopy who worked in Paris in the physical laboratories of the Sorbonne, the École Normale, and the Collège de France, and from 1879 at the observatories of San Remo, Italy and Nice. Thollon was awarded the Prix Lalande by the Académie des sciences in 1885 for his large map of the solar spectrum, published posthumously in 1890.

Inspired by the large-scale lithographic spectrum atlases of **Anders Ångström** and **Alfred Cornu**, Thollon set out to sketch, with the best possible likeness, the "physiognomy" of each group of lines in the Fraunhofer spectrum. For this program, a maximum increase in the resolving power of his spectroscope was crucial. In the late 1870s he worked on fine-tuning his direct-vision spectroscope by means of an intricate multi-prism arrangement to minimize the total angle of deviation while maximizing the dispersion. Thollon's first arrangement from 1878, which employed eight glass prisms, attained an angle of 30° between the Fraunhofer lines B and H. Sixfold enlargement of this spectrum produced a total length of over 1 m.

In that same year, Thollon changed the prism combination to two of crown-glass and one fluid-filled, containing a mixture of ether and carbon bisulphide, carefully prepared to have the same index of refraction as the glass prisms. The prisms were positioned face to face without any air gap and were of the same index of refraction, so there was considerably less loss of light due to reflection. When his Parisian optician and precision instrument maker Léon Louis Laurent assembled two such block prisms into a prototype direct-vision spectroscope, Thollon realized what enormous dispersions were now within grasp: 12 ft. between the two D lines, equivalent to a prism chain of 16 carbon-bisulphide prisms or 30 glass prisms with a refractive index of 1.63. This instrument enabled him to verify by experiment the Doppler shift caused by the solar rotation in spectra from the solar limb. He also developed a device that allowed him to record the details of the thus generated spectra far more efficiently.

This work had been conducted at the Sorbonne, the École Normale, and the Collège de France. The next stage, mapping the full visible portion of the solar spectrum at this unprecedented dispersion, could not be carried out in Paris due to weather conditions.

For the greater part of 1879 Thollon worked in the private observatory of Prince Nicolas d'Oldenbourg in San Remo. Given clear skies and his ingeniously contrived apparatus, Thollon finished in under 3 months a 10–15 m-long drawing of the solar spectrum displaying approximately 4,000 spectrum lines between A and H. This map was presented to the Académie des sciences in late 1879 but was never published because, meanwhile, Thollon had embarked on an even more ambitious mapping project at the newly founded Observatoire de Nice. He had already had some dealings with that observatory (still under construction at the time) as scientific consultant for its spectroscopic equipment. This prestigious site, with which France hoped to compete against the new observatories in Potsdam and Strasbourg, was where Thollon set his four prisms, two of them of a high-dispersion carbon-bisulphide type, in a special mounting that guided the incident light twice through each prism.

This arrangement yielded a dispersion of 70° (or 30 mm on the map) between the two yellow sodium D lines, the optimum attainable with prism chains because of the significant loss in light intensity to each prism along the optical path. Only the big Rowland concave gratings (and later interferometric measurements) could achieve even higher dispersions. Thollon secured a constant ambient temperature for his strongly temperature-dependent bisulphide-of-carbon fluid prisms, by having water circulate within the table on which the spectroscope was mounted, and additionally by enclosing the whole instrumental setup within a double-walled metal box suspended from the ceiling. In a half-decade of assiduous labor, Thollon mapped 3,448 lines in roughly one half of the optical part of the spectrum from the visible red to the middle green (from 7600 Å to 5100 Å). Unfortunately, he died before his *magnum opus* was finished, which was eventually published in 1890 at the expense of Raphaël Bischoffsheim, the financier of the Nice Observatory.

Increased resolution of his map was not the only feature with which Thollon tried to improve upon his forerunners. He gave a fourfold representation of the solar spectrum: With the Sun near the horizon; at 30°; in a normal and dry atmosphere; and finally, omitting the atmospheric lines altogether. Of the 3,202 dark spectrum lines listed in Thollon's accompanying table, 2,090 were of purely solar origin, 866 of atmospheric absorption lines, and 246 were labeled "mixtes." It was this feature that resulted in the continued use of his map, well after **Henry Rowland**'s photographic maps of the normal spectrum of 1886 and 1888 became available. The latter had dispersions of between 4.8 and 2.4 Å per centimeter, depicted nearly 20,000 Fraunhofer lines, and were generally considered to be far superior to all foregoing lithographs.

It is significant that both Cornu and Thollon were appreciated by their colleagues for their unique combination of scientific and artistic talents. Thollon's beautiful drawings of the spectrum, which cost him 6 years of labor, plus another 3 for his engraver to transfer onto steel plates, marks the climax of the tradition of spectrum portraiture, which sought not only to plot the precise location of each spectral line, but also to portray its character.

Klaus Hentschel

Selected References

Anon. (1887). "Dr. L. *Thollon.*" *Observatory* 10: 207.

Guébhard, Adrien (1879). "Le spectroscope et ses derniers perfectionnements." *La Nature* 7, 2nd semester: 223–226.

Hentschel, Klaus (2002). *Mapping the Spectrum: Techniques of Visual Representation in Research and Teaching*. Oxford: Oxford University Press.

——— (2002). "Spectroscopic Portraiture." *Annals of Science* 59: 57–82.

Janssen, Pierre Jules César (1887). "La mort de M. Thollon." *Comptes rendus de l'Académie des sciences* 104: 1047–1048.

Spée, Eugène (1899). *Région b–f du spectre solaire*. Brussels: Polleunis and Centerick.

Thom, Alexander

Born **Mains Farm near Carradale, (Strathcyle), Scotland, 26 March 1894**

Died **Fort William, (Highland), Scotland, 7 November 1985**

Alexander Thom was pivotal in the development of the discipline of archaeoastronomy.

Alexander Thom's father was a dairy farmer and his mother, the daughter of a Glasgow muslin manufacturer. After school in Ayrshire, Alexander studied civil engineering at the Royal Technical College, Glasgow and was awarded its associateship in 1914. He then studied engineering at the University of Glasgow, gaining a B.Sc. in 1915. Thom married Jeanie Boyd Kirkwood in 1917, and they had three children: Archibald, Alan, and Beryl.

For a few years Thom worked for various engineering and aeronautical firms in Glasgow. In 1921 he returned to the University of Glasgow as a lecturer, where he remained until 1939, gaining a Ph.D. in 1926 and a D.Sc. in 1929. During World War II Thom worked at the Royal Aircraft Establishment, Farnborough. In 1945 he was appointed professor of engineering science at the University of Oxford and fellow of Brasenose College, and held those positions until his retirement in 1961. Thom received several honorary degrees: an MA from the University of Oxford in 1945, an LLD from the University of Glasgow in 1960, and an LLD from the University of Strathclyde in 1976. He was a fellow of the Royal Astronomical Society, elected in 1957, and a member of the British Astronomical Association.

Thom's professional career was in engineering. Most of his professional work in this field was concerned with developing techniques for the numerical solution of the partial differential equations that arise in fluid mechanics. These techniques were applied to various types of fluid flow and are summarized in a book written late in his career. Though they predate the introduction of electronic computers, they proved readily adaptable for use with these new devices.

However, Thom was interested in astronomy from an early age. (Indeed, his first scientific paper, published while he was still an undergraduate, was on an astronomical topic.) Another long-standing passion was sailing, and it was a visit to the stone circle at Callanish on the Isle of Lewis, during a sailing trip in 1933, that prompted the investigations into ancient astronomy for which Thom is best known.

Starting in 1933 and continuing for the next 40-odd years, Thom made the first extensive and accurate surveys of the standing stones and stone circles found in Britain and Brittany. Such "megaliths" are common in Britain, Ireland, and Brittany, and are usually considered to have been erected between 4000 and 1500 BCE, during the Neolithic (New Stone Age) and early to middle Bronze Age. Thom surveyed several hundred megaliths, a task of considerable difficulty as many of the sites are in remote and inhospitable locations. This achievement is all the more remarkable in that it was pursued independently of his career in engineering and was initially carried out during vacations. However, following his retirement in 1961, Thom was able to work full-time on his astronomical studies. He published his first paper on this work in 1954 and his first book in 1967. Several members of Thom's family contributed to the work, and in particular his later books were written with his eldest son Archibald (who also followed a career in engineering fluid mechanics).

Thom drew three broad conclusions from the analysis of his surveys. These conclusions are separate and stand independently of each other. First, many of the rings were constructed in shapes other than simple circles. Second, they were constructed using a standard unit of measurement, which he called the "megalithic yard." Finally, many of the sites had significant astronomical alignments (hence the term "megalithic astronomy" often applied to this field), perhaps pointing to the rising and setting positions of the Sun on the summer or winter solstice. Many of the alignments proposed by Thom were remarkably precise and would imply considerable sophistication on the part of the builders of the monuments.

The reaction to Thom's ideas was mixed. Among astronomers it was broadly favorable, though there was debate about individual alignments and concern about the statistical significance of the results. The ideas were less well received by archaeologists, perhaps because Thom made no attempt to reconcile his proposals with the generally accepted understanding of the period; indeed he disclaimed any expertise in archaeology. Many archaeologists caviled at the idea that the preliterate societies thought to have erected the megaliths could have mastered the relatively advanced mathematics and astronomy implied by Thom's results. Similarly, the suggested use of a standard megalithic yard over an extended geographic area and for a protracted period of time seemed ill-matched to the conventional view of a shifting patchwork of local tribal cultures.

Before Thom it was known that some stone circles were not accurately circular. However, he was the first to show that the rings were deliberately constructed with a noncircular shape, rather than being a poor attempt at a circle, though the precise geometrical constructions that he proposed are not now widely accepted. Similarly, the idea of the megaliths having astronomical alignments predates Thom's work; it goes back to at least Sir **Norman Lockyer**, if not to the antiquarian **William Stukeley**'s writing in 1740. However, Thom recognized that astronomical alignments are common in these monuments. Precise alignments of the sort that he proposed are no longer widely accepted, but it is believed that many megalithic sites incorporate astronomical symbolism and some approximate alignments, mostly lunar, but also some solar.

In addition to attracting the attention of astronomers and archaeologists, Thom's work aroused considerable public interest. Popular accounts appeared in numerous magazine articles and were featured in several balanced and well-produced programs broadcast on British television.

The real significance of Thom's work is that he recognized the importance of extensive, accurate, and systematic surveys of megalithic monuments, and amassed a large body of accurate data on several hundred such sites. Modern studies of the astronomy of megalithic monuments date from his work.

A. Clive Davenhall

Selected References

Atkinson, R. J. C. (1986). "Alexander Thom." *Journal for the History of Astronomy* 17: 73–75.

Heggie, D. C. (1987). "Alexander Thom." *Quarterly Journal of the Royal Astronomical Society* 28: 178–182.

Ruggles, C. L. N. (ed.) (1988) *Records in Stone – Papers in Memory of Alexander Thom*. Cambridge: Cambridge University Press. (Includes a complete bibliography and two personal reminiscences.)

——— (1999). *Astronomy in Prehistoric Britain and Ireland*. New Haven: Yale University Press. (For a recent appraisal of Thom's astronomical contribution; see esp. Chap. 2, pp. 49–67.)

Thom, Alexander (1961). *Field Computations in Engineering and Physics*. Written with C. J. Apelt. London: Van Nostrand. (Thom's work on fluid mechanics is described herein.)

——— (1967). *Megalithic Sites in Britain*. Oxford: Clarendon Press.

——— (1971). *Megalithic Lunar Observatories*. Oxford: Clarendon Press.

Thom, Alexander and A. S. Thom (1978). *Megalithic Remains in Britain and Brittany*. Oxford: Clarendon Press. (Coauthored with his son.)

Thom, Alexander, A. S. Thom, and A. Burl (1980). *Megalithic Rings*. BAR British Series, 81. Oxford: British Archaeological Reports.

——— (1990). *Stone Rows and Standing Stones*. 2 Vols. BAR International Series, 560. Oxford: British Archaeological Reports. (Published posthumously.)

Thome, John [Juan] Macon

Born **Palmyra, Pennsylvania, USA, 22 August 1843**
Died **Córdoba, Argentina, 27 September 1908**

Córdoba Observatory director John Thome played a dominant role in compiling the most extensive and widely used visual catalogs of Southern Hemispheric stars. Thome was admitted to Lehigh University in 1866, and was awarded the degree of C.E. (civil engineer) in 1870. Thome was one of several assistants recruited by **Benjamin Gould** to accompany him to Argentina and establish that country's first national observatory at Córdoba. Having gained the support of Argentina's newly elected president, Domingo F. Sarmiento, Gould wished to extend the precise mapping of Southern Hemisphere stars to the same level of accuracy accomplished by the northern *Bonner Durchmusterung* star catalog of **Friedrich Argelander**.

Thome and Gould arrived at Córdoba at the end of September 1870, before construction of the Observatory had begun. (Dedication took place 24 October 1871.) The team's first project was to prepare an atlas of all stars visible to the unaided eye, of which Thome obtained the bulk of the observations. This work appeared as *Uranometria Argentina* (1879), and would earn a Gold Medal from the Royal Astronomical Society, in 1883. Despite Thome's lack of formalized training or experience in astronomy, "his subsequent career and the excellent work ... accomplished at Córdoba Observatory show[ed] the wisdom of Dr. Gould's selection."

The remainder of Gould's plan called for precise measurements of fainter stars in successive zones and publication of the corresponding catalogs. As Gould's senior assistant, Thome increasingly assumed a greater part of this task. During Gould's absences in 1874, 1880, and 1883, Thome served as acting director. Following Gould's permanent return to the United States in 1885, Thome became the Observatory's second director and held this post until his death.

Thome's career of service to Argentina, to the Observatory, and the world's astronomical community, was manifested in his realization of Gould's vision. Thome supervised publication of the *Córdoba Zone Catalogues* (starting 1884) and the *Argentine General Catalogue* (starting 1886), which culminated in the *Córdoba Durchmusterung* (starting 1890). Those four volumes produced under Thome's guidance contain the positions and brightnesses of 630,000 stars, compiled from some 1.8 million observations.

During the severe economic depression of the 1890's, Thome faced extremely difficult circumstances. Collapse of the Argentinian economy and bankruptcy of the government made it nearly impossible to retain sufficiently skilled workers. Anti-American sentiments arose during the Spanish–American War, further hindering his institution's tasks. To an extent now impossible to measure, Thome received valuable assistance from his wife during this critical period.

Thome's enormous labors did not go unnoticed. In 1888, he was awarded an honorary doctorate by Lehigh University. The *Córdoba Durchmusterung*, however, was only completed by Thome's successor, **Charles Perrine**. Both this catalog and its counterpart, the *Bonner Durchmusterung*, form the basis of star catalogs used by astronomers to the present day. Thome was elected a foreign associate of the Royal Astronomical Society on 10 November 1899. In 1901, he was awarded the Lalande Prize of the French Academy of Sciences in recognition of his achievements.

With the return of improved economic and political conditions in 1900, Thome attended the International Astronomical Congress held in Paris, as the delegate from Argentina. There, he offered the services of Córdoba Observatory toward completion of the *Carte du Ciel* Astrographic Chart and Catalogue, a planned photographic atlas of the heavens. Córdoba adopted that portion of the sky that was originally assigned to the La Plata Observatory at Buenos Aires, Argentina. New equipment was ordered, and the work was begun under Thome's tenure, although the massive *Carte du Ciel* itself was never completed.

Apart from his preparation of the star catalogs, Thome conducted observations of minor planets, comets, variable stars, and the 1874 transit of Venus. Between 1877 and 1906, he served as the United States vice consul at Córdoba, although this title carried little responsibility. While Thome spent the majority of his adult life in South America, he never became a naturalized citizen, but instead retained allegiance to the United States.

Durruty Jesús de Alba Martínez

Selected References

Hodge, John E. (1971). "Juan M. Thome, Argentine Astronomer from the Quaker State." *Journal of Inter-American Studies and World Affairs* 23, no. 2: 215–229.

Holden, Edward S. (1892). "The National Observatory of the Argentine Republic." *Publications of the Astronomical Society of the Pacific* 4: 25–27.

Knobel, E. B. (1909). "John Macon Thome." *Monthly Notices of the Royal Astronomical Society* 69: 255–257.

Sersic, J. L. (1971). "The First Century of Cordoba Observatory." *Sky & Telescope* 42, no. 6: 347–350.

Thome, John M. (1904). "Report on the Work of the Argentine National Observatory, 1904." *Monthly Notices of the Royal Astronomical Society* 64: 807–812.

Thomson, George Paget

Born **Cambridge, England, 3 May 1892**
Died **Cambridge, England, 10 September 1975**

British experimental physicist Sir George Paget Thomson shared the 1937 Nobel Prize in Physics with Clinton J. Davisson for the discovery of electron diffraction, which demonstrated that subatomic particles also have wave properties, as part of the wave–particle dualism of quantum mechanics. He also chaired the committee that persuaded the British government, early in World War II, to become involved in the development of nuclear weapons.

George Thomson was the only son of the famous Joseph John Thomson (1856–1940; Nobel Prize in Physics 1906) and his wife Rose Paget, and he grew up in the privileged environment of the Cambridge elite. Thomson received his early education at the Perse School in Cambridge. He entered Trinity College in Cambridge in 1910, and in 1914 he earned first-class honors in mathematics and physics. Thomson took up postgraduate work under his father at the Cavendish Laboratory and became a mathematical lecturer at Corpus Christi College, Cambridge (until 1922).

When Britain entered World War I, Thomson immediately enlisted and was on the frontlines in France as a second lieutenant in the Royal West Surrey Regiment by November 1914. He was returned to England in 1915 as part of a group at the Royal Aircraft Factory working on the stability and performance of aircraft at Farnborough. **Francis Aston** was also part of this group. Thomson's first textbook, *Applied Aerodynamics* (1919), was the outgrowth of this experience. He also served as an advisor to the Air Ministry during the latter part of World War II.

After returning to Cambridge, Thomson resumed research on the behavior of electrical discharges in gases. In 1922 he became professor of natural philosophy (physics) at the University of Aberdeen, and (after a stay in 1929/1930 at Cornell University) in 1930 he was appointed professor at Imperial College, London. In 1952, Thomson became master of Corpus Christi College, Cambridge; from this position he retired in 1962.

Thomson was married and had four children. He was elected a fellow of the Royal Society in 1930, and, among other honors, he received the Hughes Medal (1939) and the Royal Medal (1949) of that society.

At the Oxford meeting of the British Association for the Advancement of Science in 1926 one main point of discussion was Louis de Broglie's (1892–1987) wave theory of matter. This stimulated Thomson to adapt certain experiments, which he had undertaken on scattering of (positive) anode rays, for testing this theory. He (together with Alexander Reid) passed electrons through a thin celluloid foil onto a photographic plate behind the foil, and the plate revealed a diffraction pattern, thus indicating the wavelike behavior of electrons. Experiments with thin metallic foils validated this result.

At the same time Clinton J. Davisson (1881–1958) – who had taken part in the 1926 conference, too – together with Lester H. Germer (1896–1971) at Bell Telephone Laboratories independently came to the same result from a similar experiment. In 1937 Thomson and Davisson shared the Nobel Prize for Physics for the experimental discovery of the diffraction of electrons by crystals. While Thomson in 1926/1927 had shown the wavelike character of electrons, his father, about 30 years previously, had demonstrated their particle character.

During the 1930s, Thomson moved his interest more and more to nuclear physics. With J. A. Saxton he looked for artificial radioactivity from positron bombardment, and together with another group of coworkers he studied the velocity distribution of slow neutrons. Thomson's last original work was on special cosmic-ray effects.

As World War II approached, Thomson was among the first to appreciate the significance of multiple slow neutrons coming out of natural uranium samples, and he urged the British government to buy up the Belgian stock of uranium residue. (They did not.) In April 1940, after the memorandum from Rudolf Peierls and Otto Frisch made clear to all knowledgeable readers that chain reactions in uranium would be possible, Thomson was appointed chair of the Maud Committee to investigate the implications. They reported in July 1941 that a superbomb could be made of istopically separated uranium-235. At this point, James Chadwick (discoverer of the neutron) was put in charge of the project and Thomson was sent to head the British Scientific Office in Canada. He returned to the Air Ministry in 1943.

After the war, Thomson became interested in the peaceful application of thermonuclear fusion. During his last years he was an engaged and passionate organizer and popularizer of science.

Horst Kant

Selected References

Feather, Norman (1975). "Sir George Thomson." *Physics Today* 28, no. 12: 66–67.

Moon, Parry B. (1977). "George Paget Thomson." *Biographical Memoirs of Fellows of the Royal Society* 23: 529–556.

Thomson, G. P. "Experiments on the Diffraction of Cathode Rays." Parts 1–3. *Proceedings of the Royal Society of London* A 117 (1928): 600–609; 119 (1928): 651–663; 125 (1929): 352–370.

——— (1961). *The Inspiration of Science*. London: Oxford University Press.

——— (1965). *J. J. Thomson and the Cavendish Laboratory in His Day*. Garden City, New York: Doubleday.

——— (1968). "The Early History of Electron Diffraction." *Contemporary Physics* 9: 1–15.

Thomson, G. P. and W. Cochrane (1939). *Theory and Practice of Electron Diffraction*. London: Macmillan and Co.

Thomson, G. P. and A. Reid (1927). "Diffraction of Cathode Rays by a Thin Film." *Nature* 119: 890.

Thomson, William

Born **Belfast, (Northern Ireland), 26 June 1824**
Died **Netherhall near Largs, (Strathclyde), Scotland, 17 December 1907**

William Thomson calculated the age of the Earth from its cooling rate, and concluded that it was too short to fit with Charles Darwin's theories of evolution. As Lord Kelvin, he is commemorated in the Kelvin temperature scale and the Kelvin–Helmholtz time scale.

Thomson's father, James Thomson, held the chair of Mathematics at Glasgow University. His mother died when William was 6 years old. William learned mathematics from his father, and became adept in that field at a very young age. He was admitted to Glasgow University at age 10, and began what we would now consider university-level work at 14.

Thomson won a medal from the University of Glasgow when he was 15, for an essay entitled "Essay on the Figure of the Earth." This essay contained many important ideas that he returned to repeatedly in his later career. Thomson was strongly influenced by the French mathematical approach to physical science, including the works by Joseph Fourier, Augustin Fresnel, **Adrien Legendre**, **Pierre de Laplace**, and **Joseph Lagrange**.

At the age of 17, in 1841, Thomson entered Cambridge University and published his first paper, on Fourier series. Papers on heat and electricity followed later in his undergraduate career. He graduated in 1845, becoming second wrangler in the mathematical *tripos* of 1845. Thomson was elected a fellow of Peterhouse College, Cambridge.

Thomson studied in Paris in the physical laboratory of Victor Regnault and had deep discussions with **Jean Biot**, Augustin Cauchy, Joseph Liouville, François Sturm, and J. B. Dumas. In 1846 he was elected to the chair of natural philosophy at Glasgow, with the help of his father.

Thomson collaborated closely with **George Stokes** on the theory of heat and its relation to the theory of fluids. In 1848, he proposed the absolute scale of temperature. The Kelvin scale of temperature derives its name from the title that Thomson was given by the British government in 1892: Baron Kelvin of Largs.

In 1852 Thomson observed the Joule–Thomson effect, namely the decrease in the temperature of gas when it expands in a vacuum. James Joule influenced Thomson's ideas, which were developed into a dynamical theory of heat that became the foundation for what we now know as statistical mechanics.

He was then led into the study of electricity and magnetism, and his ideas became the foundation on which **James Maxwell** built his remarkable new theory of electromagnetism. However, Thomson developed his own ideas differently, and diverged from Maxwell's viewpoint, is not accepting the existence of the displacement current.

Thomson was knighted in 1866 for work on the transatlantic cable connection; he had invented a very sensitive mirror galvanometer. He published more than 600 papers. Thomson was elected to the Royal Society in 1851 and was its president from 1890 to 1895.

In applying thermodynamics to cosmogony, Thomson foresaw the heat death of the Universe. He had an interest in the age of the Sun, and assumed its radiant energy came from the gravitational potential of matter that had fallen into it. Thomson estimated the Sun's age at 50 million years. The ideas he put forward were closely related to those earlier expressed by **Julius Meyer** and **John Waterston**, some of whose work he had seen.

David Jefferies

Alternate names

Baron Kelvin of Largs
Lord Kelvin

Selected References

Buchwald, Jed Z. (1976). "Thomson, Sir William." In *Dictionary of Scientific Biography*, edited by Charles Coulston Gillispie. Vol. 13, pp. 374–388. New York: Charles Scribner's Sons.

Garnett, William and Hugo Munro Ross (1911). "Kelvin, William Thomson." In *Encyclopaedia Britannica*. 11th ed. Vol. 15, pp. 721–723. New York: Encyclopaedia Britannica.

Russell, Alexander (1938). *Lord Kelvin*. London: Blackie and Son.

Smith, Crosbie and M. Norton Wise (1989). *Energy and Empire: A Biographical Study of Lord Kelvin*. Cambridge: Cambridge University Press.

Young, A. P. (1944). *Lord Kelvin, Physicist, Mathematician, Engineer*. London: Longmans, Green.

Tikhov, Gavril Adrianovich

Born **Smolevichi near Minsk, (Belarus), 1 May 1875**
Died **25 January 1960**

Pulkovo Observatory's Gavril Tikhov (and Meudon's **Charles Nordmann**) "demonstrated" that light of different wavelengths, propagating from the same astronomical source, arrives at different times. The demise of this theory, the Nordmann–Tikhov effect, is explicated by Hendrik van de Hulst in "Nanohertz Astronomy" (Sullivan, 1984).

Selected Reference

Sullivan III, W. T. (ed.) (1984). *The Early Years of Radio Astronomy: Reflections Fifty Years after Jansky's Discovery*. Cambridge: Cambridge University Press.

Timocharis

Flourished **first half of third century BCE**

Timocharis was one of the first astronomers in the school of Alexandria, and was a near-contemporary of **Aristarchus** of Samos. Little is known of him, except that he made astronomical observations during the approximate period 295 to 270 BCE. He may have founded a school, but if so then **Aristyllus** is the only member whose name is known.

Timocharis and Aristyllus are usually considered to have compiled the first true catalog of the fixed stars, in which stars are identified by numerical measurements of their positions. (In earlier lists, stars had been identified by descriptions of their locations, typically with respect to other stars and constellations.) The catalog is not extant. Indeed, while Aristyllus and Timocharis certainly amassed a set of numerical observations of star positions, it is not, strictly speaking, known whether these observations were assembled into a catalog or table. Probably fewer than a hundred stars were observed, and the positions were reputedly of low accuracy. Observations by Timocharis or Aristyllus survive in **Ptolemy**'s *Almagest* for some 18 stars.

The observations of Timocharis and Aristyllus were practically the only historical measurements of the positions of the fixed stars available to **Hipparchus**, who used them in combination with his own observations to discover the precession of the equinoxes. This discovery is probably the most important use to which they were put. However, much later **Edmond Halley** also used the observations of Timocharis, among others, to demonstrate the existence of proper motion.

Timocharis observed the planets as well as the fixed stars, and two observations that he made of Venus are preserved in the *Almagest*.

In *De Pythiae oraculis* (402 F) Plutarch includes Timocharis in a list of astronomers who wrote in prose. However, most of the information about him comes from Ptolemy's *Almagest*, particularly the discussion of precession and its discovery by Hipparchus.

A. Clive Davenhall

Selected Reference

Ptolemy (1984). *Ptolemy's* Almagest translated and annotated by G. J. Toomer. New York: Springer-Verlag.

Tisserand, François-Félix

Born **Nuits-Saint-Georges, Côte d'Or, France, 13 January 1845**
Died **Paris, France, 20 October 1896**

François-Félix Tisserand, known by the first name Félix, was born in the Burgundy region known for growing good wine. Together with a brother, he was the son of a cooper who died when Félix was still young. Tisserand entered the École Normale Supérieure when he was 18, graduating first in his class. At that time **Urbain Le Verrier** was the director of the Observatoire de Paris. Le Verrier recruited the bright Tisserand to find mistakes in **Charles Delaunay**'s lunar theory.

After passing his *Doctorat d'Etat* in 1868, Tisserand showed particular interest and value in the field of celestial mechanics. On the other hand, he was participating in astronomical expeditions such as the solar eclipse of 1868 and the transits of Venus in 1874 and 1882.

In 1873, Tisserand was appointed director of the Observatoire de Toulouse, reorganizing and reequipping it in such a way that it became a valuable astronomical center. In 1878 he was asked to teach at the Sorbonne, Paris, in celestial mechanics. Tisserand soon became a specialist in the last subject, establishing what is called the "critère de Tisserand," employed to know if a comet is new or if it is a returning object. He became director of the Observatoire de Paris in 1892, following his predecessor, admiral **Ernest Mouchez**, in particular with regard to the international enterprise, the *Carte du Ciel*.

Since the celebrated *Principia* by **Isaac Newton** and the *Mécanique céleste* by **Pierre de Laplace** one century earlier, nothing of great value had been published in the general field of celestial mechanics during the 19th century. The *Traité de mécanique céleste*, rigorous and clearly written by Tisserand, was published in four volumes, the last one appearing the year he died. Volumes I and II of his *Traité* were reedited in 1960 as was the complete set in 1990. After reading Newton and Laplace, specialists must read Tisserand to best understand **Jules Henri Poincaré** and **Albert Einstein**.

Tisserand was married twice. His first spouse died soon after their daughter was born. He had two more daughters from his second marriage.

In 1878, Tisserand succeeded Le Verrier as a member of the Académie des sciences. He was appointed as a member of the Bureau des longitudes in 1878.

Suzanne Débarbat

Selected References

McLaughlin, William I. and Sylvia L. Miller (2004). "The Shadow Effect and the Case of Félix Tisserand." *American Scientist* 92, no. 3: 262–267.

Tisserand F. (1889–1894). *Traité ole mécanique céleste*. Paris: Gauthier–Villars (4 vol.).

Titius [Tietz], Johann Daniel

Born **Konitz, (Chojnice, Poland), 2 January 1729**
Died **Wittenberg, (Germany), 11 December 1796**

Johann Titius was a German physicist, astronomer, and biologist, known for the discovery of the so called Titius–Bode law. He studied at the University of Leipzig and became professor of physics at Wittemberg in 1756.

His work in physics (especially thermometry), biology, and mineralogy being totally forgotten, Titius is remembered for his law on planetary distances, also known as Bode's law or the Titius–Bode law. This law, which is in fact more an empirical rule, was formulated by Titius as follows: If the distance Sun–Saturn is taken as 100, then Mercury is found at a distance of 4 from the Sun, Venus at 4 + 3 = 7, the Earth at 4 + 6 = 10, Mars at 4 + 12 = 16 (no planet being found at the distance 4 + 24 = 28), Jupiter at 4 + 48 = 52, and Saturn 4 + 96 = 100.

Titius formulated his discovery as a brief remark in his own 1766 adaptation in German of the then well-known book *Contemplations de la nature* by the Swiss philosopher Charles Bonnet. It remained unnoticed until **Johann Bode**, a much better-known astronomer, formulated the same scheme in 1772. Although Bode's note was nearly identical to Titius's, he did not refer to the older source, and it was not before 1823 – Titius being long since dead and forgotten – that Bode admitted to have read it. Both Titius and Bode were convinced that the Creator could not have left the space between Mars and Jupiter empty, but – and this is the only difference between the two theories – Bode clearly presumed the existence of an unknown planet in that sector, while Titius only speculated that this space was filled by yet undiscovered satellites of Mars and Jupiter.

Titius acknowledged to having been inspired by the German rationalist philosopher Christian Wolff, who had tried before to formulate some numerical regularity in the distance between the planets, and by the German mathematician and astronomer **Johann Lambert**, who, in his *Cosmologische Briefe* (1761), had asked if perhaps some planets were missing in the vast space between Mars and Jupiter.

Tim Trachet

Selected Reference

Nieto, Michael Martin (1972). *The Titius–Bode Law of Planetary Distances: Its History and Theory*. Oxford: Pergamon Press.

Todd, Charles

Born **Islington, (London), England, 7 July 1826**
Died **Adelaide, South Australia, Australia, 29 January 1910**

Sir Charles Todd established the Adelaide Observatory and directed its astronomical work for over 51 years. He is probably best remembered for overseeing the construction of the Australian transcontinental telegraph line from Adelaide to Darwin, completed in 1872. In addition, Todd established a network of weather stations and laid the foundations of the Australian Bureau of Meteorology.

The elder son of George Todd, a grocer and tea merchant who lived at Greenwich in London, Charles, at the age of 15, found employment at the Royal Observatory in Greenwich as a computer under **George Airy**, the Astronomer Royal. He held this post until the end of 1847, being engaged on lunar reductions. In 1848, Todd was appointed assistant astronomer at the Cambridge University Observatory and was put in charge of the Northumberland Equatorial. He made observations of the newly discovered planet Neptune and took a daguerreotype of the Moon, one of the earliest attempts at astronomical photography. In addition, Todd helped to determine the difference in longitude between Cambridge and Greenwich by means of the electric telegraph.

In 1854, Todd was recalled to Greenwich to take charge of the new time signal apparatus. In this position he was responsible for the transmission of time signals throughout England and for the dropping of time balls. In early 1855, Airy recommended Todd for the new posts of superintendent of telegraphs in the South Australia colony and director of Adelaide Observatory. Todd was appointed to those positions on 10 February, and married Alice Gillam (*née* Bell) in April; they arrived in Australia in November 1855. Todd immediately started taking meteorological observations, made plans for a telegraph network, and supervised construction of an observatory.

Despite all his other duties, Todd maintained a keen interest in astronomy and fully executed his duties as government astronomer. Completed in 1860, the Adelaide Observatory was equipped at first with only a 42-in. focal length transit instrument on loan from the government of Victoria. Todd observed the transits of Mercury in November 1861 and November 1868 with a Dollond refractor of 2¼-in. aperture. In 1868 he cooperated with the government astronomers in Victoria and New South Wales in the determination of a more accurate 141st meridian, later adopted as the common boundary between South Australia and New South Wales.

Todd observed the 1874 transit of Venus with a new 8-in. aperture Cooke equatorial equipped with a spectroscope and micrometer. The Cooke telescope was subsequently used for extensive observations of the motions of Jupiter's satellites, for the positions of comets, and for studying features of the planets. A 6-in. transit circle by Troughton and Simms was installed later, and a program of regular observations was begun in 1891. A founding member of the Astronomical Society of South Australia in 1892, Todd served as its first president until his death. When consulted by the government of Western Australia in 1895 about establishing a new observatory, Todd chose a site on Mount Eliza on the outskirts of Perth. He observed the annular eclipse of the Sun in March 1905 and the partial solar eclipse in February 1906.

Under Todd's supervision, eventually as postmaster general of Australia managing both the postal and telegraphic services, a telegraphic network extending over 5,000 miles of mostly unmapped territory was completed between 1856 and 1872. The network linked major cities in Australia to England *via* a transoceanic cable from Darwin through Java to Singapore; communication with England opened in October 1872. Todd achieved similar progress in weather observation and forecasting by equipping and training postal supervisors at 357 meteorological stations and 2,575 rainfall stations in Australia and New Zealand. Todd provided weather forecasts and pioneered in the production of weather maps.

Todd was made a companion of the Order of Saint Michael and Saint George in 1872 for his work on the continental telegraph system. He was elected a fellow of the Royal Society in 1889 and received a knighthood in 1893 for his public services. Todd held an honorary MA from the University of Cambridge and was a fellow of the Royal Meteorological Society and the Society of Electrical Engineers in addition to being a fellow of the Royal Astronomical Society.

In his later years Todd ruled his departments as a "benevolent autocrat," respected both by his employers and by his employees. He was so highly regarded that the South Australian parliament deliberately delayed compulsory retirement laws until the esteemed octogenarian retired of his own volition. Todd retired as postmaster general in 1905 and as government astronomer at the end of 1906, having served for over 51 years in the latter office.

Todd died of gangrene and was buried in Adelaide. He was survived by one son, Dr. C. E. Todd and four daughters, Lady Todd and his eldest son having predeceased him. In 1898, his daughter Gwendoline married (Sir) William Bragg, professor of mathematics and physics at the University of Adelaide.

David W. Dewhirst

Selected References

Edwards, P. G. (2004). "Charles Todd's Observations of the Transits of Venus." *Journal of Astronomical History and Heritage* 7, no. 1: 1–7.

Gibbs, W. J. (June 1998). "The Origins of Australian Meteorology." *Metarch Papers*, no. 12. Bureau of Meteorology.

McCarthy, Gavan (2001). "The Overland Telegraph and Undersea Cable: Australia's First Electronic Information Network." *Australasian Science* 22 : 46.

Symes, G. W. (1976). "Todd, Sir Charles (1826–1910)." In *Australian Dictionary of Biography*. Vol. 6, 1851–1890, pp. 280–282. Melbourne: Melbourne University Press.

Todd, Charles (1877). "Observations of the Phenomena of Jupiter's Satellites at the Observatory, Adelaide, and Notes on the Physical Appearance of the Planet." *Monthly Notices of the Royal Astronomical Society* 37: 284–300; see 284–285.

——— (1878). "Observations at the Adelaide Observatory." *Monthly Notices of the Royal Astronomical Society* 39: 1–22.

——— (1894). "Meteorological Work in Australia: A Review." In *Report of the Fifth Meeting of the Australasian Association for the Advancement of Science.*

Todd, David Peck

Born **Lake Ridge, New York, USA, 19 March 1855**
Died **Lynchburg, Virginia, USA, 1 June 1939**

David P. Todd was a versatile astronomer with a career that covered over 50 years. His base of operation was Amherst College, but he is probably best known for his many expeditions to observe astronomical phenomena all over the world.

After growing up on a New York farm, Todd received bachelor's (1875) and master's (1878) degrees from Amherst College. (His honorary doctorate was from Washington and Jefferson College.) Todd became professor of astronomy and director of the observatory at Amherst in 1881 after a brief stay at the United States Naval Observatory (USNO), working under **Simon Newcomb**.

Todd married Mable Loomis in 1879. (She is better known as the editor of Emily Dickinson's poetry.) The couple had one daughter, Millicent.

Todd participated in the 1978 Naval Observatory eclipse expedition to Texas, and led many other eclipse expeditions. He reduced the USNO's 1874 transit of Venus plates and was in charge of the Lick Observatory heliometric observations of the transit in 1882. (Neither data set resulted in a satisfactory solar parallax.)

Todd became a close friend and associate of **Percival Lowell,** who financed and accompanied him on a 1900 eclipse expedition to

Tripoli. Lowell also sponsored his trip to observe Mars from Chili, and sent **Earl Slipher** along as his assistant. Todd became a proponent of attempts to contact Mars *via* radio. He was institutionalized in 1922.

Todd was an active member of national and international astronomical societies, and the author of several books on astronomy. He contributed many articles to leading astronomical journals and magazines.

Henry L. Giclas

Selected Reference

Sheehan, William and Anthony Misch (2004). "Ménage à Trois: David Peck Todd, Mabel Loomis Todd, Austin Dickinson, and the 1882 Transit of Venus." *Journal for the History of Astronomy* 35: 123–134.

Tolman, Richard Chace

Born **West Newton, Massachusetts, USA, 4 March 1881**
Died **Pasadena, California, USA, 5 September 1948**

American chemist and mathematical physicist Richard Tolman was a pioneer in applying the ideas of general relativity and of thermodynamics to the large-scale structure of the Universe. He received a BS from the Massachusetts Institute of Technology [MIT] in 1903, and, after a year of study and work in an industrial chemistry laboratory in Germany, entered the MIT graduate program in the chemistry laboratory of Arthur Amos Noyes, receiving a Ph.D. in 1910 with a thesis on measurement of the electromotive force in rotating, conducting solutions. In 1909, Tolman and Gilbert N. Lewis (a fellow chemist, and also a native of West Newton, working in Noyes's lab) wrote the first American exposition on special relativity.

Tolman held academic positions at the universities of Michigan (1910/1911), Cincinnati (1911/1912), California (1912–1916), and Illinois (a full professorship of physical chemistry, 1916–1918). He worked largely in laboratory chemistry and chemical kinetics, coming to the conclusion that thermodynamic equilibrium was not usually a good description of the results of chemical reactions.

After the United States entered World War I, Tolman took a position with the Chemical Warfare Service. Noyes felt strongly that scientific involvement with government operations should continue after the war and arranged a position for Tolman as associate director and then director (1919–1922) of a laboratory working on fixation of nitrogen for explosives and fertilizers. Noyes moved on to the California Institute of Technology (Caltech) and brought Tolman there in 1922 as professor of physical chemistry and mathematical physics, where most of the work we remember him for was done. Tolman remained associated with Caltech the rest of his life, though from 1940 onward he held a variety of advisory wartime positions as vice chairman of the National Defense Research Council, scientific advisor to General Leslie Groves on the atomic bomb project, and United States advisor to the United Nations Atomic Energy Commission (1946 until his death).

In 1924 Tolman married psychologist Ruth Sherman whom he had met through his brother, Edward Chace Tolman (a psychologist and also member of the National Academy of Sciences, to which Richard Tolman was elected in 1923).

Following a suggestion from **Otto Stern**, Tolman began trying to apply the ideas of thermodynamics to the totality of a static universe. He quickly concluded that the observed ratio of helium to hydrogen could not have been achieved in equilibrium at any temperature. (The 1:3 ratio by mass observed is now understood as a product of the very nonequilibrium expansion of the early Universe.) By the time of his 1934 book, *Relativity, Thermodynamics, and Cosmology*, he had, of course, incorporated the evidence for an expanding universe presented in 1929 by **Edwin Hubble**. In 1935, Hubble and Tolman published a single joint paper on how one might use observations of the numbers of galaxies as a function of apparent brightness and of apparent brightness versus redshift as tests of whether the Universe would expand forever or be pulled back into contraction by the gravitational force of the total amount of matter. Tolman's own conclusion was that the data were not sufficient to make any reliable statement about the long-term behavior of the Universe, though he regarded the present expansion as well established. He took the possibility of an oscillating universe quite seriously.

Tolman considered a number of different ways of writing down equations to describe the behavior of space and time within general relativity. One of these proved to be particularly appropriate for the surroundings of a very compact object. In 1939, he made some progress in calculating the structure of such objects, and the almost simultaneous paper by **Julius Robert Oppenheimer** and his student **George Volkoff** successfully established the distribution of density and pressure expected inside a neutron star, if general relativity and quantum mechanics are the only relevant physics. The part common to these papers is generally called the Tolman–Oppenheimer–Volkoff equation of state.

In addition to his academy membership, Tolman received an honorary D.Sc. from Princeton University (1942) and was the 1932 J. Willard Gibbs Lecturer of the American Mathematical Society. His 1938 textbook on statistical mechanics remains in print and in use after 65 years.

George Gale

Selected References

Gale, George and John Urani (1999). "Milne, Bondi and the 'Second Way' to Cosmology." In *The Expanding Worlds of General Relativity*, edited by Hubert Goenner *et al.*, pp. 343–375. Boston: Birkhauser. (Describes Tolman's views about philosophy and cosmology.)

Hubble, Edwin and Richard, C. Tolman (1935). "Two Methods of Investigating the Nature of the Nebular Red-Shift." *Astrophysical Journal* 82: 302–337.

Kirkwood, John G., Oliver R. Wulf, and P. S. Epstein (1952). "Richard Chace Tolman." *Biographical Memoirs, National Academy of Sciences* 27: 139–153.

Kragh, Helge (1996). *Cosmology and Controversy: The Historical Development of Two Theories of the Universe*. Princeton, New Jersey: Princeton University Press. (This is the best general source on modern cosmology.)

North, John David (1994). *The Norton History of Astronomy and Cosmology*. New York: W. W. Norton.

Tolman, Richard C. (1932). "Models of the Physical Universe." *Science* 75: 367–373.

——— (1934). *Relativity, Thermodynamics and Cosmology*. Oxford Clarendon Press.

——— (1938). *The Principles of Statistical Mechanics*. Oxford: Clarendon Press. (Reprint, New York: Dover, 1987.)

Tombaugh, Clyde William

Born **Streator, Illinois, USA, 4 February 1906**
Died **Las Cruces, New Mexico, USA, 17 January 1997**

Clyde Tombaugh was the only person in the 20th century to discover an object classified as a major planet orbiting the Sun.

Tombaugh was the son of Muron and Della Tombaugh. The family moved to a farm near Burdett, Kansas in 1922 where Clyde attended high school. At the time of his death he was the only remaining staff member who had been present at Lowell Observatory, Flagstaff, Arizona, when Pluto was discovered.

At age 14, in a school assigned Autobiography, Tombaugh recounted his first experience observing the night sky with a telescope, when his uncle Leon Tombaugh showed him some stars and the planet Mars. He went on to say, "On New Year's Day [1919], my uncle gave me the telescope and said it was mine .... I was then the happiest boy in the world."

Tombaugh's early attempts at telescope making are documented in his letters to Napoleon Carreau, an optician from Wichita. Most enlightening are his comments in a 17 May 1926 letter to Carreau. He laments about problems, pointed out by Carreau, with an 8-in. mirror (his first) and explains in zealous detail about what he thinks are the reasons for its imperfect figure. Impressed with the quality of Tombaugh's subsequent work and attention to detail, Carreau later offered him a job.

At about the same time, Tombaugh sent his drawings of Mars and Jupiter to Lowell Observatory, where the director, **Vesto Slipher**, along with the trustee Roger Lowell Putnam, had decided to resume the late **Percival Lowell**'s search for Planet X. Tombaugh's letters and drawings captured the attention of the Lowell administrators, and they hired the young man, fresh from the farm in Kansas.

At Lowell Observatory in 1929, Tombaugh had the privilege of helping to bring a new telescope on line and developing an observing procedure for seeking a distant planet. The new instrument was the 13-in. Lawrence Lowell refractor named after the Harvard University president who financed the project. The lens, designed and partially finished by amateur astronomer **Joel Metcalf** before his death, arrived from Clark & Sons in January 1929, as did Tombaugh, who started work on 15 January. In a sense, Tombaugh and the telescope grew up together with Tombaugh helping to solve minor and major problems under the tutelage of Slipher, **Carl Lampland,** and Stanley Sykes.

Although Lowell had predicted a new planet based on orbital positions of Neptune and Uranus, Tombaugh's search had been underway for 10 months before he found the planet Pluto on plates exposed on 23 and 29 January and blinked on 18 February 1930. The discovery was announced on 13 March. Pluto was disappointingly faint, however. Various suggestions put forth at the time could not dispel the suspicion that its mass was far less than that assumed in Lowell's predictions.

The University of Kansas awarded Tombaugh the Edwin Emory Slosson 4-year Scholarship in 1931; he began his studies at Kansas in 1932. He would receive BA and MS degrees from Kansas in 1936 and 1939, and eventually an honorary D.Sc. from Northern Arizona State College in 1960.

Patricia Edson, her mother, and brothers James and Alden, moved to Lawrence, Kansas from Kansas City in 1932. Their house had several extra rooms, and brother James arranged to have Tombaugh move in as one of several boarders in the fall of 1933. Patricia enrolled at the University of Kansas that same fall as a philosophy major and, of course, met Tombaugh. They married at the end of the spring semester in 1934. That summer the couple drove to Flagstaff on the new Route 66. Daughter Annette was born in 1940 and son Alden in 1945.

After graduation from college, Clyde returned to Lowell and continued the search for other bodies in the outer Solar System. That photographic program, centered on the ecliptic, covered 70% of the celestial sphere. As Tombaugh pointed out, this study would have found planets brighter than magnitude 16.5, and, had there been an Earth-sized planet within 100 Astronomical Units, it would have been detected. Although the general public reveres a discovery, the null result from this systematic study is far more important. It has shaped the thinking of the astronomical community for more than 50 years, and it has influenced both our models of Solar System formation and our search for planetary systems around other stars. By-products of the study included the discovery of a cataclysmic variable (TV Corvis), six star clusters, two comets, observations of a number of asteroids and clusters of galaxies, and the discovery of one supercluster.

In 1943, Tombaugh was invited to teach physics and later navigation in a navy program at Arizona State Teacher's College (now the University of Northern Arizona). Toward the end of World War II the large number of veterans with delayed educations taxed US colleges. Clyde was contacted in 1944 by Fredrick Leonard at the

University of California at Los Angeles, and he spent a year teaching astronomy there. These activities interrupted his work in planetary astronomy, and when Tombaugh returned to Lowell hoping to continue, he found that he was out of a job!

James Edson facilitated Tombaugh's next move in 1946, helping him to launch his new career in rocket science at White Sands Proving Grounds in New Mexico. There, Tombaugh developed tracking systems to determine flight paths and rocket characteristics and helped establish the serious business of missile warfare.

Tombaugh returned briefly to Lowell in 1952. In a letter to Roger L. Putnam on 27 February 1952, he reports:

> In talking over several things, Dr. V. M. Slipher asked me if we could spend several weeks at the observatory this summer. He is desirous of repeating the 13-inch plates for [a] stellar proper motion survey, and wanted me to blink several pairs to ascertain what such a harvest would yield.

Although a brief survey was carried out in May 1952 with positive results, Tombaugh decided to stay in New Mexico, and the proper-motion study was completed by Henry Giclas.

From 1953 to 1958 Tombaugh directed a search for natural Earth-orbiting debris, and in 1955 he moved his operation to New Mexico State University [NMSU]. (This program was carried out initially at Lowell and later in Ecuador.) The telescope drive rate was adjusted to match the motion of objects orbiting at assumed distances from the Earth. Resulting exposures contained reference star trails and assured maximum exposure of faint sources. Again a systematic search yielded a null result. This time the result was welcome; near space was not hostile to manned activity.

At NMSU, Tombaugh assembled a team that provided systematic sets of planetary images that were used to support the Mariner, Viking, and Voyager missions. He was also instrumental in designing and obtaining funding for a 24-in. telescope for the university's Tortugas Mountain Observatory. It captured its first images in 1967, and carried out National Aeronautics and Space Administration [NASA] mission-supported research for almost 30 years. Bradford Smith, team member on the Mars Mariner and Viking missions, and principal investigator for the Voyager imaging team, began his career with Tombaugh's group.

In the mid-1960s, Tombaugh teamed up with two forward-thinking individuals on the NMSU campus. One was a recent aeronautical engineering/astronomy Ph.D. from the University of Arizona, W. L. Reitmeyer, and the other was research vice president William O'Donnell. Together they established a new Ph.D. granting Department of Astronomy in 1970.

From 1955 to retirement in 1973, Tombaugh taught geology and astronomy classes. His commitment was contagious, and his interest and dedication to public education did not flag as he entered retirement. Tombaugh continued to be a strong influence on students and made an amazing effort to satisfy the demands of the public. He came into his office regularly for 20 years after his retirement, providing never-to-be-forgotten experiences for students and faculty. In 1980, in collaboration with Patrick Moore, Tombaugh published his version of the discovery of Pluto, *Out of the Darkness: The Planet Pluto*. He also continued his interest in planetary astronomy, in particular NASA Mars projects and later, potential Pluto missions.

Reacting to lagging professional opportunities for young scientists, Clyde committed Patricia and himself to raising funds to establish the Clyde W. Tombaugh Fellowship at NMSU. Assisted by professor Bernard McNamara and biographer David Levy, he toured the United States and Canada from 1985 to 1990, presenting public lectures and raising funds in 53 locations in both countries.

Tombaugh published dozens of papers about the discovery of Pluto, research on observing techniques, and observations of planets, especially Mars and Jupiter. He also wrote hundreds of letters to lovers of astronomy of all ages.

Tombaugh received the Royal Astronomical Society Jackson-Gwilt medal in 1931. Other awards include the University of Kansas's Distinguished Service Citation in 1966 and Distinguished Public Service medals from NASA and the American Institute of Aeronautics and Astronautics, both in 1980. The Clyde W. Tombaugh Elementary School in Las Cruces proudly bears his name.

Tombaugh was a warm and stimulating friend who encouraged inventive thinking and perseverance. When he was not inflicting puns on his friends and colleagues, he was enthusiastically following new developments. Tombaugh was the only true "Plutocrat" of our time.

Books listed later by Hoyt and Levy provide the most outstanding and scholarly histories of the events before, during, and after the discovery. A detailed account of the natural Earth satellite search is given in the Final Technical Report, copies of which are available through the Rio Grande Historical Collections at NMSU. Information about Tombaugh's letters and other writings, including his age-14 autobiography, can be obtained from the same source at NMSU.

Herbert Beebe

Selected References

Hoyt, William Graves (1976). *Lowell and Mars*. Tucson: University of Arizona Press.

——— (1980). *Planets X and Pluto*. Tucson: University of Arizona Press.

Kuiper, G. and B. Middlehurst (eds.) (1961). *The Solar System*. Chicago: University of Chicago Press.

Levy, David H. (1991). *Clyde Tombaugh: Discoverer of Planet Pluto*. Tucson: University of Arizona Press.

Putnam, W. L. (1994). *The Explorers of Mars Hill*. West Kennebunk, Maine: Phoenix.

Shapley, Harlow (ed.) (1960). *Source Book in Astronomy, 1900–1950*. Cambridge, Massachusetts: Harvard University Press.

Tombaugh, Clyde W. (1960). "Reminiscences of the Discovery of Pluto." *Sky & Telescope* 19, no. 5: 264–270.

Tombaugh, Clyde W. and P. Moore (1980). *Out of the Darkness: The Planet Pluto*. Harrisburg, Pennsylvania: Stackpole.

Tomaugh, Clyde W., J. C. Robinson, B. A. Smith, and A. S. Murrell (1959). *The Search for Small Natural Earth Satellites: Final Technical Report*. Las Cruces: New Mexico State University Physical Science Laboratory.

Torricelli, Evangelista

Born **Faenza, (Emilia-Romagna, Italy), 15 October 1608**
Died **Florence, (Italy), 25 October 1647**

Evangelista Torricelli is best known for his researches on geometry, hydrodynamics, and motion of weights. He was a pupil of **Benedetto Castelli**, who was in turn a former pupil of **Galileo Galilei**. Torricelli was able to express the principle of inertia in a clear and modern form: "When acting forces are absent, the motion is rectilinear with constant velocity." He observed Jupiter and noted the colored bands parallel to the Equator of the planet. His most important contribution to astronomy was his ability as a telescope maker, particularly for his excellent lenses. In a letter to Galilei, written on behalf of Castelli, he introduced himself saying that he had studied **Ptolemy**, **Tycho Brahe**, **Johannes Kepler**, and **Nicolaus Copernicus**, and his studies had convinced him to accept the Copernican system. He was the first in Rome to have made a careful study of Galilei's *Dialogo sopra i massimi sistemi*.

Margherita Hack

Selected Reference

Loria, Gino and Giuseppe Vassura (eds.) (1919–1944). *Opere di Evangelista Torricelli*. 4 Vols. Faenza: Montanari.

Toscanelli dal Pozzo, Paolo

Born **Florence, (Italy), 1397**
Died **Florence, (Italy), May 1482**

Paolo Toscanelli's astronomical significance hinges primarily on his comet observations.

Toscanelli was the second son of Domenico, a physician, and Biagia Mei. His family was one of the richest medical families in Florence. Toscanelli never married and, apart from short periods of time he spent outside Florence, he lived in the household of his father and later of his brother and his nephews.

Our knowledge of Toscanelli's life and works is limited to a few documents, and his fame as one of the greatest personalities of the 15th century is mainly attested by the eulogies of his contemporaries. He was astronomer, mathematician, and physician, knew Greek, and owned an important collection of Greek and Latin manuscripts.

Nicholas Krebs (Nicholas of Cusa) reported he knew Toscanelli in Padua, where Nicholas attended the university from 1417 to 1424; therefore, we can argue that Toscanelli studied there in the same years. They remained in contact for life. Nicholas dedicated two mathematical works on the squaring of the circle to Toscanelli, and Toscanelli attended Nicholas at his deathbed in 1464. In 1425 he came back from Padua with his brother Piero, and they enrolled in the College of Florentine Physicians.

Two records testify to Toscanelli's public roles: In January 1442 he consulted about the construction of the dome of Florence Cathedral, Santa Maria del Fiore, and in September 1453 he consulted with the rulers of Florence as an astrologer. Toscanelli was a friend of famous artists like F. Brunelleschi and L. B. Alberti, and participated in the humanistic circles that flourished in Florence during the 15th century.

In Rome, Toscanelli also knew **Johann Müller** (Regiomontanus), who praised him as a mathematician and as an astronomer. In 1464, Müller reported that L. B. Alberti and Toscanelli made astronomical observations for the determination of the obliquity of the ecliptic.

Toscanelli was involved in the revival of the studies of geography and cartography and played an important role in the preparation of Christopher Columbus's voyages to America. He knew the Portuguese canon Fernao Martins in Florence and remained in contact with him after the latter's return to Lisbon. In 1474, Toscanelli sent Martins a letter describing a new route to the Far East, together with a map to clarify his theory. The letter and the map were later copied or directly sent to Columbus, inspiring his navigations.

Toscanelli was associated with the installation of the first meridian inside a church. A tradition states that it was made in 1468, but a document from 1475 recorded the payment for a gnomon to be installed in the lantern of Brunelleschi's dome in the Cathedral of Santa Maria del Fiore in Florence, to observe the Sun and determine the date of the summer solstice.

Toscanelli's only surviving works are contained in a few handwritten sheets discovered and published in the 19th century. There are notes and drawings on the observations of six comets, some mathematical computations, astronomical tables, lists of geographical places, a map grid, and two horoscopes.

The records of cometary observations are the most important of his surviving works. Toscanelli observed comet C/1433 R1 for 6 weeks, C/1449 Y1 for 7 weeks, 1P/1456 K1 (Halley) for a month, C/1457 A1 for several days, C/1457 L1 for 8 weeks, and C/1472 Y1 for nearly 3 weeks. In the 1890s, the Italian astronomer **Giovanni Celoria** was able to compute the orbital elements from an analysis of Toscanelli's documents. An evolution in his interests and his methods of representation of the observations has been indicated by J. Jervis after a more careful analysis of the manuscript. Initially in 1433, Toscanelli was most interested in the shape of the tail, probably to determine its astrological significance. In the observations of 1449/1450 he seemed more interested in a precise determination of the head of the comet, while in 1456, Toscanelli recorded the positions in longitude and latitude, suggesting he had begun to use an instrument like a torquetum or an armillary sphere. In 1472, probably due to his advanced age, he gave only verbal descriptions of cometary positions.

Toscanelli's notes represent the first example of observations of celestial phenomena over a long period of time, and of maps used as an integral part of precise measurements.

Giancarlo Truffa

Selected References

Apfelstadt, Eric (1992). "Christopher Columbus, Paolo dal Pozzo Toscanelli and Fernão de Roriz: New Evidence for a Florentine Connection." *Nuncius* 7, no. 2: 69–80.

Barker, Peter and Bernard R. Goldstein (1988). "The Role of Comets in the Copernican Revolution." *Studies in History and Philosophy of Science* 19: 299–319.

Celoria, G. (1921). *Sulle osservazioni di comete fatte da Paolo dal Pozzo Toscanelli e sui lavori suoi astronomici in generale*. Milan. Pubblicazioni del R. Osservatorio astronomico di Brera in Milano, 55. (First published in Uzielli, 1894.) (See pp. 308–385.)

Gallelli, C. (1993). "Paolo dal Pozzo Toscanelli." In *Il mondo di Vespucci e Verrazzano a cura di Leonardo Rombai*, pp. 71–92. Florence: Olschki.

Garin, E. (1967). *Ritratti di umanisti*. Florence, pp. 41–67. English translation by Victor A. Velen and Elizabeth Velen. *In Portraits from the Quattrocento*. New York: Harper and Row, 1972.

Heilbron, J. L. (1999). *The Sun in the Church*. Cambridge, Massachusetts: Harvard University Press, pp. 70–71.

Jervis, Jane L. (1985). *Cometary Theory in Fifteenth-Century Europe. Studia Copernicana*, Vol. 26. Dordrecht: D. Reidel, pp. 43–85, 162–169.
Kronk, Gary W. (1999). *Cometography*. Vol. 1, pp. 268–269, 271–273, 275–279, 286–289. Cambridge: Cambridge University Press.
Park, Katherine (1985). *Doctors and Medicine in Early Renaissance Florence*. Princeton, New Jersey: Princeton University Press, pp. 226–233.
Settle, T. B. (1978). "Dating the Toscanelli's Meridian in Santa Maria del Fiore." *Annali dell'Istituto e Museo di storia della scienza di Firenze* 3, pt. 2: 69–70.
Toscanelli dal Pozzo, Paolo. Florenze, Biblioteca Nazionale Centrale, Magliabechiano XI, 121, ff. 280r–280v with two horoscopes for 1448 and 1449.
——— Florenze, Biblioteca Nazionale Centrale, Banco Rari 30, 15 ff. excerpted from Magl. XI, 121.
Uzielli, G. (1894). *La vita e i tempi di Paolo Dal Pozzo Toscanelli: Ricerche e studi di Gustavo Uzielli: Con un capitolo (VI) sui lavori astronomici del Toscanelli di Giovanni Celoria*. Rome.
Yeomans, Donald K. (1991). *Comets: A Chronological History of Observation, Science, Myth, and Folklore*. New York: John Wiley and Sons, pp. 24–25, 406–408, 410.

Tousey, Richard

Born **Somerville, Massachusetts, USA, 18 May 1908**
Died **Prince Georges County, Maryland, USA, 15 April 1997**

American space physicist Richard Tousey headed the group whose spectrograph, flown on a captured German V-2 rocket, made the first-ever detection of ultraviolet radiation (to which the atmosphere is opaque) from an astronomical object, the Sun. Tousey was the son of Coleman and Adella Richards (*neé* Hill) Tousey, and married Ruth Lowe in 1932. She died in 1994; they had one daughter.

Tousey earned an AB (*summa cum laude*, with election to Phi Beta Kappa) from Tufts University in 1928, and an AM (1929) and Ph.D. (1933) from Harvard University. The latter was completed with a thesis called "An Apparatus for the Measurement of Reflecting Powers with Application to Fluorite at 1216 Å" carried out under **Theodore Lyman**, discoverer of the 1216 Å "Lyman-α" line of hydrogen gas. Tousey held fellowships and a teaching position at Harvard (1929–1936) followed by a faculty position at Tufts University (1936–1941). He came to the Naval Research Laboratory [NRL] in 1941 at the invitation of its director of research **Edward Hulbert** and was soon head of the instrument section of the optics division, working on wartime optics-related research. Tousey became head of the micro-wave branch in 1945 and the rocket spectroscopy branch in 1958, retiring in 1978 but continuing to work on the spectrum of the Sun with Charles Brown and **Charlotte Moore Sitterly**.

Early in 1946, Tousey and his colleagues built a grating spectrograph suitable for use in captured German V2 rockets (replaced later with Aerobes). Not a single identifiable piece of the spectrograph could be retrieved from the first rocket's impact crater near White Sands, New Mexico. For the second try, the spectograph went into the rocket tail, flew on 10 October 1946, and returned the very first spectrum of solar radiation in the spectral region called the rocket ultraviolet. It was several years before they were able to push the sensitivity of their detectors far enough to record wavelengths as short as Lyman α, but about 1950, Tousey could say that he had found in the Sun the important hydrogen feature that his advisor had found in the laboratory. Meanwhile, another group, also within Hulbert's empire at NRL, developed an electronic detector for still shorter wavelengths, and the 1949 discovery of solar X-rays is largely associated with the name of **Herbert Friedman**, later also director of research at NRL.

Tousey's group also developed the first rocket-borne coronographs, permitting observations of the corona and study of its variability outside eclipses. Measurements of the ultraviolet absorption as the rockets rose and returned also provided the first determination of the vertical distribution of ozone in the Earth's atmosphere. Their detectors on the Orbiting Solar Observatory number 7 (1971–1974) discovered the phenomenon called coronal mass ejections, and recorded the sequence of events as material is blown off the Sun.

Tousey's greatest success in his own view was the design, building, and flying of two instruments that flew with astronauts on the Skylab mission for 4 months in 1974. One recorded extreme ultraviolet (170–630 Å) images with very high spatial resolution, showing rapid changes in the Sun's upper atmosphere, connected with sunspots, flares, and mass ejections. The other was a spectrograph with very good wavelength resolution. This led to the identification of a large number of features in the solar spectrum, and so to a better understanding of the temperature and density structure of the chromosphere and corona and of their compositions (which differ importantly from that of the optical photosphere). In support of this work, the laboratory also conducted a number of studies of the ultraviolet properties of optical materials. Nighttime rocket launches led to the first direct measurements of the nighttime airglow, which is reemission of solar energy stored in the gases of the upper atmosphere.

Tousey was elected to membership in the United States National Academy of Sciences (1960) and received its Henry Draper Medal (1963), along with medals, awards, and lectureships from astronomical and optical societies in the United States, Britain, and France. In addition to memberships in societies connected with spectroscopy, geophysics, astronomy, and physics, he was active in groups connected with his long-standing interests in birds (on which he coauthored a field list for the District of Columbia area), American silver, and music.

Eugene F. Milone

Selected References

Anon. (1961). "Richard Tousey: Frederic Ives Medalist for 1960." *Journal of the Optical Society of America* 51: 379–383.
Baum, William A. (2002) "Richard Tousey." *Biographical Memoirs, National Academy of Sciences* 81: 341–355.
Hufbauer, Karl (1991). *Exploring the Sun: Solar Science since Galileo*. Baltimore: Johns Hopkins University Press, Chap. 4, "Solar Physics as a Beneficiary of War 1939–1957".
Koomen, M. J., W. R. Hunter, and N. R. Sheeley, Jr. (1997). "Richard Tousey, 1908–1997." *Bulletin of the American Astronomical Society* 29: 1494–1495.
Moe, O. Kjeldseth and E. F. Milone (1978). "Limb Darkening 1945–3245 Å for the Quiet Sun from *Skylab* Data." *Astrophysical Journal* 226: 301–314.
Tousey, Richard (1958). "Rocket Measurements of the Night Airglow." *Annales de géophysique* 14: 186–195.
——— (1961). "Solar Spectroscopy in the Far Ultraviolet." *Journal of the Optical Society of America* 51: 384–395.

——— (1962). "Techniques and Results of Extraterrestrial Radiation Studies from the Ultraviolet to X-Rays." In *Space Age Astronomy*, edited by A. J. Deutsch and W. B. Klemperer, pp. 104–114. New York: Academic Press.
——— (1967). "Some Results of Twenty Years of Optical Space Research." *Astrophysical Journal* 149: 239–252.
——— (1990). "The Development of Solar Space Astronomy." *Bulletin of the American Astronomical Society* 22: 814. (Paper abstract.)
Tousey, Richard (17 September 1981). Oral History Interview with David DeVorkin. Niels Bohr Library, American Institute of Physics, College Park, Maryland.
Tousey, R., E. F. Milone, W. Palm Schneider, J. D. Purcell, and S. G. Tilford (1974). *An Atlas of the Solar Ultraviolet Spectrum between 2226 and 2992 Angstroms*. Washington, DC: Naval Research Laboratory.

Triesnecker, Franz [Francis] de Paula von

Born **Kirchberg, (Austria), 2 April 1745**
Died **Vienna, (Austria), 29 January 1817**

Franz de Paula von Triesnecker was well known in his times for his mathematical and astronomical skills that he used in his work to determine the exact location of many central European geographical positions. Triesnecker became a member of the Society of Jesus at the age of 16. He studied philosophy at Vienna and mathematics at Tyrnau. After teaching mathematics and philosophy at Jesuit institutions for a few years, Triesnecker entered the university at Graz to study theology and completed his doctorate of philosophy. He received his ordination soon after his graduation.

In 1782, Triesnecker was appointed as an assistant to **Miksa Höll** (Maximilian Hell), the director of the Imperial Observatory in Vienna. Triesnecker's duties included serving as an assistant editor of the *Ephemerides Astronomicae ad meridianem Vindobonensem* from 1782 until he became the editor in 1792. With the death of Hell in April 1792, Triesnecker was appointed director of the observatory. Triesnecker then served as both the editor on the annual ephemerides and as the director until his retirement in 1806. He shared editorial duties with his fellow Austrian astronomer Johann Tobias Bürg. Between the years 1787 and 1806, Triesnecker annually published his *Tabulae Mercurii, Martis, Veneris, Solares*, along with most of his micrometrical observations of the Moon, planets, Sun, and hundreds of stellar positions.

Triesnecker's numerous treatises deal mainly with astronomy and geography. In 1798, Triesnecker published an occultation table in the *Allgemeine Georgraphische Ephemerdin* giving the geographical position of the Buda Observatory (Ofen). The table was based on observations of two solar eclipses and his own observations of 12 stellar occultations by the Moon.

Triesnecker's major published works include: *Veränderliche Schicksate dreyer merkwürdiger Längenbestimmungen von Pekin, Amsterdam und Regensburg* (1802, 1804), *Astronomische Beobachtungen an verschiedenen Sternwarten von 1805 (–1815), herausgegeben von F. Triesnecker* (1805–1815), and *Über die Ungewissheit einiger astronomischen Fixpunkte bei der Entwerfung einer Karte von Persien und der asiatischen Türkey* (1802). Triesnecker's work also appeared in such publications as **János von Zach**'s *Monatliche Correspondenz zur Befoerderung der Erd- und-Himmels-Kunde*, **Johann Bode**'s *Astronomische Jahrbuch*, the *Commentarii Societatis Regiae Scientiarum Gottingensis*, and the *Transactions of the Royal Society of Bohemia*. In 1802, he published his monumental work on calculating the motions of the Moon in *Novae motuum lunarium tabulae*.

In geography Triesnecker determined or corrected the longitude and latitude of various places from the best available data. He completed Georg Ignaz von Metzburg's triangulation of lower Austria. Triesnecker's data formed the basis for the production of a new map of Austria, and assisted him with the triangulation of Galicia.

Triesnecker became a corresponding member of the Russian Academy of Sciences in Saint Petersburg on 5 February 1812. He was also a member of the scientific societies in Breslau, Göttingen, Munich, and Prague. Triesnecker was honored by the International Astronomical Union with the naming of a nearside lunar crater (latitude 4.°2 N, longitude 3.°6 E) in 1935 and a system of rilles, named Rimae Triesnecker, in 1964.

Robert A. Garfinkle

Selected References

Günther (1894). "Triesnecker: Franz." In *Allgemeine Deutsche Biographie*. Vol. 38, pp. 607–608. Leipzig: Duncker und Humblot.
Stein, J. (1913). "Triesnecker." In *Catholic Encyclopedia*, edited by Charles G. Habermann *et al.* Vol. 15, pp. 44–45. New York: Encyclopedia Press.

Trouvelot, Étienne-Lêopold

Born **Guyencourt, Aisne, France, 26 December 1827**
Died **Meudon near Paris, France, 22 April 1895**

Étienne Trouvelot spent the busiest and most productive part of his life in the United States using the large refracting telescopes at Harvard College Observatory, the University of Virginia, and the United States Naval Observatory [USNO] to make original drawings of celestial phenomena. He published over 50 papers on astronomy, and made a huge number of evocative drawings of astronomical phenomena.

Little is known of Trouvelot's early life. It seems he dabbled in politics and had Republican tendencies. It is therefore possible he either fled or was exiled from France when Louis Napoleon rose to power after the *coup d'etat* of 1852. Whatever the situation, the events of that period are unconnected with astronomy.

Trouvelot immigrated to the United States, and in 1855 arrived in Massachusetts, supporting himself and his family as an artist. His leanings toward the natural sciences led him to join the Boston Natural History Society, and between 1868 and 1876 he contributed several papers on entomology to its publication. Trouvelot also became acquainted with Louis Agassiz. In 1860 he moved to Medhurst, a suburb of Boston, where he experimented in silkworm production. Not satisfied with the output of the native *Polyphemus* silkworm, Trouvelot went to Europe and sometime later in 1868 or early 1869 returned with eggs of the gypsy moth. Unfortunately, he either ignored or overlooked the possibility of accidental escape. The inevitable happened, and attempts to eradicate the consequent infestation proved ineffective; the subsequent defoliation of forests in the Northeastern United States is the fact for which Trouvelot is mainly ill-remembered.

Meanwhile, and perhaps fortunately, astronomy had caught Trouvelot's interest. In 1870 a number of quite spectacular auroral displays piqued his artistic sensibilities. His skillful renditions came to the notice of **Joseph Winlock**, director of the Harvard College Observatory, who in 1872 invited him to join the observatory staff. Observing regularly with the 15-in. refractor, Trouvelot produced numerous sketches for the series *Astronomical Engravings from the Observatory of Harvard College*, which comprised 35 plates, many in color.

In 1875, the year Trouvelot announced his discovery of veiled sunspots, he resolved to prepare a series of highly detailed drawings of celestial objects as they appear to experienced observers through large telescopes. Coincidentally that same year, he was offered the use of the USNO 26-in. Clark refractor. In September 1875 Trouvelot prepared an exquisite rendition of Saturn with the Washington instrument and started a magnificent drawing of the Great Nebula in Orion. In succeeding years he worked on a great number of drawings at various observatories including the 26-in. Clark refractor at the Leander McCormick Observatory of the University of Virginia. From those that were complete by 1876, Trouvelot selected 15 pastels as representative of his best work for display at the Philadelphia Centennial Exposition. The set, entitled *Astronomical Drawings*, was reproduced as chromolithographs and published in 1882. In 1878 he and his son George observed the total eclipse of the Sun on 29 July from Wyoming Territory.

In 1876, Trouvelot initiated the work with which his name will always be associated, a systematic study of the planets. In 1881 he published a major paper on Jupiter in the *Proceedings of the American Academy of Arts and Sciences*. In contributions to *The Observatory*, the *Comptes rendus* de l'Académie des Sciences, and *L'Astronomie*, Trouvelot considered aspects as diverse as white spots on Venus, the Great Red Spot of Jupiter, variations in the rings of Saturn, the apparent duplicity of the shadow of Jupiter's largest satellite Ganymede, various phenomena of Mars, and enigmatic appearances on the Moon. He also attempted, apparently unsuccessfully, to construct a lunar map 60 in. in diameter based on observations with a 6-in. refractor at Cambridge.

Trouvelot returned to France in 1882, and joined the leading solar expert **Pierre Janssen** at Meudon, Paris, where he indulged his fascination with the Sun. Trouvelot witnessed many spectacular prominences, including two massive eruptions in June 1885 (estimated to reach a height of 480,000 km) and another on 17 June 1891. He accompanied Janssen to the Caroline Islands to observe the total solar eclipse of 1883, and with **Johann Palisa** undertook a fruitless search for intramercurial planets. His last major publication, "Observations sur les Planètes Vénus et Mercure," communicated to the Société Astronomique de France in 1892, is perhaps his most important contribution. It is based on a twofold series of observations of the two planets, first at Cambridge, Massachusetts, 1876–1882, then at Meudon, 1882–1891, amounting to a total of 744 observations and 285 drawings. It is a landmark work of more than historical interest, and describes in great detail what the telescope reveals of these bodies.

Trouvelot was honored by the French Academy of Sciences with the award of their Valz Prize. A crater on the Moon is named in Trouvelot's honor.

Richard Baum

Selected References

Anon. (1895). "Étienne-Lêopold Trouvelot." *Astrophysical Journal* 2: 166–167.
Anon. (1895). "Obituary." *Observatory* 18 : 245–246.

Trumpler, Robert Julius

Born **Zürich, Switzerland, 2 October 1886**
Died **Oakland, California, USA, 10 September 1956**

Swiss–American statistical and observational astronomer Robert Trumpler is best known as the person who, in 1930, provided definitive evidence for systematic absorption and scattering of starlight by dust in the plane of the Milky Way, settling an issue that had been debated for decades, not least by **Jacobus Kapteyn**. Trumpler also made important contributions to statistical astronomy, the study of the motions of the stars in the Milky Way.

Third of ten children in a large family of businessmen and manufacturers, Trumpler's early interest in business was transformed into a career in science after a short apprenticeship in a Zürich bank. In his student days he also was an alpinist and skier. After graduation from the Gymnasium first in his class, Trumpler entered the University of Zürich, later transferring to Göttingen University where he received the Ph.D. *magna cum laude* in 1910. A 1913 meeting of the Astronomische Gesellschaft in Hamburg provided opportunities to meet American astronomers, including **Frank Schlesinger**, who invited him to Allegheny Observatory. Though called for military service in 1914, Trumpler accepted the offered assistantship in 1915 and shortly thereafter traveled to the United States.

In 1919, Trumpler was invited to Lick Observatory and appointed assistant astronomer in 1920; the following year he became a naturalized United States citizen. Among his early observational projects, he made a photographic-visual survey of Mars at its opposition in 1924. His map of the planet sketched many of the features now more fully understood from later photographs. Trumpler also observed Eros at its opposition in 1931 as part of the campaign of the International Astronomical Union to measure the solar parallax.

In 1930, with the publication of his catalog of star clusters in his paper "Preliminary Results on Distances, Dimensions, and Distribution of Open Star Clusters," Trumpler showed that interstellar absorption was real; his meticulous observations enabled him to demonstrate its effects conclusively. His observations showed that the apparent linear diameters of more distant clusters of all types, based on their H–R diagrams, were larger than the diameters of nearby clusters of the same types. Trumpler's further analysis using diameters based on central concentrations and brightnesses gave the same results. However, this conclusion did not make sense physically. His meticulous observations and exhaustive analysis enabled him to eliminate the possibility that cluster diameters did increase with distance and conclude that the discrepancy was caused by interstellar absorption of about one magnitude per kilopersec, close to the modern value.

It took some years for the implications of Trumpler's work on absorption to be fully integrated into astronomical knowledge. He himself perceived the meaning as that both the Sun-centered Kapteyn universe and the very differently centered, larger galaxy of **Harlow Shapley** could be right. During World War II, **Henri Mineur** recalibrated the galactic-distance scale based on Cepheid variable stars using Trumpler's absorption, but the result did not diffuse out of France fast enough to prevent great surprise on the part of **Walter Baade** when he turned the new 200-in. telescope toward the Andromeda Galaxy and did not see the RR Lyrae stars in it, beginning about 1950. Great also was the surprise of most of the people who first heard about Trumpler's conclusion at the 1952 Rome General Assembly of the International Astronomical Union. Incorporating Trumpler's numbers roughly doubled the distances to external galaxies and so doubled the best estimate of the age of the Universe.

Trumpler had earlier contributed his remarkable observational skills to a test of **Albert Einstein**'s general theory of relativity. Einstein had predicted that starlight passing close to the Sun's limb would be bent by 1.745 arcsec. Trumpler assisted **William Campbell** (director of the Lick Observatory) during the total solar eclipse of 1922 at Wallal, Australia, observing many more stars than did earlier expeditions. They confirmed Einstein's prediction with a result of 1.75 arc sec (±0.09 arc sec), more accurate and confirmatory than previous observations.

In his early professional career while at the Swiss Geodetic Survey, Trumpler determined the longitudes of the Swiss observatories, and at Allegheny Observatory he published the parallaxes of several stars, the proper motion of Nova Aquila, and the relative motions of stars in the Pleiades. He also began a classification of star clusters that would later be the basis for his work in interstellar absorption. In his later career, with H. F. Weaver, Trumpler wrote the text *Statistical Astronomy*, which became a classic in the field.

Trumpler was an inspired teacher, fostering the development of many astronomers in his classes at the University of California at Berkeley. Even in retirement he continued to add to his catalog of clusters. His observations of certain O stars in clusters indicated anomalously high masses for some of them – data that have yet to be explained.

Trumpler served the American Astronomical Society as a councilor and the Astronomical Society of the Pacific as president in 1932 and 1949. The latter named its Trumpler Prize (given for an outstanding Ph.D. dissertation each year) for him. He was elected to the United States National Academy of Sciences (1932) and the American Academy of Arts and Sciences.

Katherine Haramundanis

Selected References

Anon. (1999). "Trumpler, Robert Julius." In *A Dictionary of Scientists*. Oxford: Oxford University Press.

Pannekoek, A. (1961). *A History of Astronomy*. London: George Allen and Unwin.

Struve, Otto and Velta Zebergs (1962). *Astronomy of the 20th Century*. New York: Macmillan.

Weaver, Harold F. (2000). "Robert Julius Trumpler." *Biographical Memoirs, National Academy of Sciences* 78: 277–297.

Tserasky [Tzeraskii], Vitol'd [Witold] Karlovich

Born **Sluck, (Belarus), 9 May 1849**
Died **Moscow, (Russia), 29 May 1925**

Vitol'd Tserasky was a prominent astronomer of Polish–Lithuanian descent who became director of the Moscow University Observatory (1890–1916). He was a pioneer in both the applications of photometry and photography to astronomical research. But due to the specific conditions of Tsarist Russia, Tserasky's ancestry led to complications in his scientific career.

Tserasky graduated from Moscow University in 1871, and after a number of successive appointments at the observatory, succeeded **Fedor Bredikhin** as director. Tserasky's doctoral dissertation (1887) was awarded for his construction of an apparatus (the Zöllner–Tserasky photometer) that employed an artificial star to permit the accurate measurement of starlight. In 1889, he was appointed professor of astronomy at Moscow University and taught there until his retirement in 1916. In 1914, Tserasky was elected a corresponding member of the Saint Petersburg Academy of Sciences.

Tserasky is chiefly remembered for his work in stellar and solar photometry, especially for an original measurement of the apparent stellar magnitude of the Sun (−26.5). He made the first determination of the lower limit of the Sun's surface temperature and is credited with the discovery of noctilucent clouds in the Earth's upper atmosphere. Among his preoccupations was the refinement of astronomical instruments.

At the Moscow Observatory, Tserasky's tenure as director began when all personnel "could be seated just on a single divan." Between 1891 and 1903, he performed numerous technical upgrades and completely refurbished the observatory with modern instruments (including a wide field, short-focus astrograph) for astrophysical research. This transition marked the beginning of the observatory's later growth and prosperity.

In 1895, Tserasky initiated a campaign for systematic photographic observations of variable stars. There, his wife Lydia Petrovna Tseraskaya (1855–1931) became an assistant and discovered 219 variable stars with the astrograph. Ultimately, this project created the foundation of the Moscow "glass library" of photographic plates. Continued by **Sergei Blazhko** and other Moscow astronomers, this plate collection is now among the richest in the world.

A crater on the Moon's farside has been named for Tserasky. In the former Soviet Union, however, his name was overshadowed by that of his successor, **Pavel Sternberg**, who was an active revolutionary.

Alexander A. Gurshtein

Alternate name

Ceraski, Vitol'd [Witold] Karlovich

Selected References

Anon. (1986). "Tserasky, Vitol'd Karlovich." In *Astronomy: Biograficheskii spravochnik* (Astronomers: A Biographical Handbook), edited by I. G. Kolchinskii, A. A. Korsun', and M. G. Rodriges, pp. 359–360. 2nd ed. Kiev: Naukova dumka.

Blažko, S. (1925). "W. Ceraski" (in French). *Astronomische Nachrichten* 225: 111–112.

Kulikovsky, P. G. (1976). "Tserasky (or Cerasky), Vitold Karlovich." In *Dictionary of Scientific Biography*, edited by Charles Coulston Gillispie. Vol. 13, pp. 481–482. New York: Charles Scribner's Sons.

Tsu Ch'ung-Chih

➜ **Zu Chongzhi**

Ṭufayl

➜ **Ibn Ṭufayl: Abū Bakr Muḥammad ibn ʿAbd al-Malik ibn Muḥammad ibn Muḥammad ibn Ṭufayl al-Qaysī**

Turner, Herbert Hall

Born **Leeds, England, 13 August 1861**
Died **Stockholm, Sweden, 20 August 1930**

Besides making fundamental contributions to the disciplines of astronomy and seismology, Herbert Turner was a leading advocate of international cooperation in science. He was the eldest son of John and Isabella Turner. From Leeds Modern School and then Clifton College, Bristol, Turner won a scholarship to Trinity College, Cambridge, and matriculated there in 1879. In 1882, he finished as second wrangler in the finals of the mathematical *tripos*; the following year, he was awarded the Smith's Prize. After a year as fellow of his college, Turner became chief assistant at the Royal Greenwich Observatory (1884) directed by **William Christie**. He was elected a fellow of the Royal Astronomical Society in 1885 and a fellow of the Royal Society (London) in 1896. He served the British Association for the Advancement of Science as general secretary from 1913 to 1922, and received a number of British and foreign honorary degrees. In 1899, Turner married Agnes Whyte of Blackheath; they had one daughter. Both survived him.

At the Royal Observatory, Turner became heavily involved in planning the institution's contribution to the international *Carte du Ciel* Astrographic Chart and Catalogue Project, inaugurated in 1887. This massive undertaking, which was never fully completed, sought the cooperation of 18 worldwide observatories to exploit dry-plate photography for the purpose of producing a photographic map and positional catalog of all stars down to about 11th magnitude.

Turner was responsible for two crucial technical innovations. First, he devised a rectilinear coordinate system (the method of "standard coordinates") that made it possible to use simple linear equations to reduce a star's apparent to true position, while correcting for errors of position on the photographic plate. Turner's methodology allowed ordinary photography to replace that part of the traditional work in positional astronomy that had been conducted with meridian circle instruments. Turner's second technical innovation was the invention of an eyepiece scale for plate measurements. This made it possible to employ semiskilled labor in the plate reductions, which in turn enabled many smaller observatories to participate in the enterprise. Turner brought nearly a quarter of the Astrographic Catalogue to completion. These achievements were recognized by the award of the Bruce Gold Medal of the Astronomical Society of the Pacific in 1927.

In 1893, Turner was elected Savilian Professor of Astronomy at Oxford University, with a fellowship at New College. Turner was an obvious choice because his predecessor, **Charles Pritchard**, had already committed the University Observatory to the *Carte du Ciel*. A recently deceased benefactor enabled the observatory to acquire a 13-in. astrographic telescope (mounted in 1893), which served as its principal instrument.

During his lengthy career at Oxford, Turner became a leading figure in the Royal Astronomical Society, serving on its council for 43 years, as foreign secretary from 1919 to 1930, and its president from 1903 to 1905. Along with **John Dreyer**, Turner coedited the centennial *History of the Royal Astronomical Society* (1923) and furnished its chapter on the society's formative decade (1820–1830).

Turner traveled to the principal observatories in the United States and was especially well informed about developments in astrophysics. He was himself a keen observer of solar eclipses, although Oxford lacked the necessary resources to develop such research. The proximity of the privately owned Radcliffe Observatory, completely refurbished in 1902 with an 18/24-in. double refractor, made it impossible for Turner to obtain new equipment for his own observatory.

From his analysis of the results obtained by different observatories, Turner devised methods of deriving stellar magnitudes from the measured diameters of their images, and coined the term *parsec* to denote a distance corresponding to an angular measurement. In 1919, the new International Astronomical Union [IAU] elected Turner as president of the commission overseeing the Astrographic Catalogue. Yet, unable to secure further instrumental and other resources from his university for astronomy, he was increasingly drawn into the field of seismology.

One of Turner's professional characteristics was his belief that no good measurements should ever be lost to science. One consequence was that he brought the work of four British variable star observers to publication. Another of Turner's characteristics was an aptitude for identifying periodicities, and an evolving conviction that multidisciplinary studies might facilitate the deduction of physical processes in variable stars, terrestrial earthquakes, tides, and meteorology. He published tables that made it possible to use harmonic analysis to model stellar variability. These and related beliefs led Turner in 1913 to supervise the coordination and reform of his late friend John Milne's worldwide network of seismological reporting stations. The Royal Astronomical Society was already a forum for the emerging science of geophysics, and, in 1919, Turner became a prime mover in its Council's formation of a Geophysical Committee. In 1920, Turner was appointed its secretary while the IAU had likewise elected him as first president of its Commission on Seismology. Although seismology deflected Turner from astronomy after 1913 and especially after 1919, he knew that he had developed a unique resource.

In the history of seismology, Turner is remembered for four achievements: two of an administrative nature and two on a scientific/technical basis. His two administrative achievements were:

(1) that he kept Milne's international network for earthquake reporting in operation through World War I; and
(2) he effected the crucial transition to collating data by seismic *event*, rather than by reporting station, and this made useful analysis possible.

His first theoretical achievement was that he developed the Zoppritz–Turner tables for more successful locating of earthquakes. Using this refined tool, his discovery of deep-focus earthquakes in 1922 represented a great step forward. That same year, Turner founded the journal, *International Seismological Survey*, at the observatory, and had it accepted as the international publication of seismic events. Between 1919 and 1924, he was a leading promoter of broadcasting a world time service in order to facilitate the accurate correlation of earthquake reports. Turner was one of the first to suggest that the Earth had a liquid core, and his tables were the basis for Sir **Harold Jeffreys**'s and David Bullen's work that proved that hypothesis.

Turner's second technical achievement was that, through use of *International Seismological Survey* data, he was the first (in 1930) to map volcanic and earthquake events together as the "ring of fire" around the Pacific Rim. Turner innovated the way that seismology is done today, while the *International Seismological Survey* (now based at Newbury in Berkshire) and his discovery of deep-focus earthquakes are his legacy.

After 1902, a lobby to economize the observatory's budget (on account of the nearby Radcliffe Observatory) led to complex politics that Turner handled with insufficient care and increasing resentment. Through his achievements in seismology, and his tireless committee work, he managed to keep the increasingly obsolete observatory within the international front rank during a very difficult period for both sciences. Not an observer but a mathematician and a manager who believed in international cooperation to advance science, Turner was a big man with an outgoing personality who encouraged young and amateur talent better than he coped with university politics. But, overwork, and the worry caused by an over-stretched budget, along with attacks by a faculty colleague who sought to undermine Turner's new cooperation with the Radcliffe Observatory, led to a cerebral hemorrhage while presiding at an IAU meeting on seismology in Stockholm. He died in the Sabbatsberg Hospital.

Roger D. Hutchins

Selected References

Aitken, Robert Grant (1930). "Herbert Hall Turner." *Publications of the Astronomical Society of the Pacific* 42: 277–280.

Bellamy, F. A. and Ethel F. B. Bellamy (1931). *Herbert Hall Turner: A Notice of His Seismological Work*. Isle of Wight: County Press.

Bullen, K. E. (1976). "Turner, Herbert Hall." In *Dictionary of Scientific Biography*, edited by Charles Coulston Gillispie. Vol. 13, pp. 500–501. New York: Charles Scribner's Sons.

Cannon, Annie J. (1931). "Herbert Hall Turner." *Popular Astronomy* 39: 59–66.

Hollis, H. P. (1930). "Turner." *Observatory* 53: 290–296.

Hutchins, Roger D. (forthcoming, 2007). *British University Observatories, ca. 1800–1939*. Ashgate: Ashgate Publishing Ltd., Aldershot, Hants.

Plummer, H. C. (1931). "Professor H. H. Turner." *Monthly Notices of the Royal Astronomical Society* 91: 321–334.

Ṭūsī: Abū Jaʿfar Muḥammad ibn Muḥammad ibn al-Ḥasan Naṣīr al-Dīn al-Ṭūsī

Born **Ṭūs, (northeast Iran), 17 February 1201**
Died **Baghdad, (Iraq), 25 June 1274**

Naṣīr al-Dīn al-Ṭūsī's major scientific writings in astronomy, in which he worked to reform Ptolemaic theoretical astronomy, had an enormous influence upon late medieval Islamic astronomy as well as the work of early-modern European astronomers, including **Nicholas Copernicus.** Ṭūsī wrote over 150 works, in Arabic and Persian, that dealt with the ancient mathematical sciences, the

Greek philosophical tradition, and the religious sciences (law [*fiqh*], dialectical theology [*kalām*], and Sufism). He thereby acquired the honorific titles of *khwāja* (distinguished scholar and teacher), *ustādh al-bashar* (teacher of mankind), and *al-muʿallim al-thālith* (the third teacher, the first two being **Aristotle** and **Fārābī**). In addition, Ṭūsī was the director of the first major astronomical observatory, which was located in Marāgha (Iran).

Ṭūsī was born into a family of Imāmī (Twelver) Shīʿa. His education began first at home; both Ṭūsī's father and his uncles were scholars who encouraged him to pursue *al-ʿulūm al-sharʿiyya* (the Islamic religious sciences) as well as the *ʿulūm al-awā'il* (the rational sciences of the ancients). He studied the branches of philosophy (*ḥikma*) and especially mathematics in Ṭūs, but eventually traveled to Nīshāpūr (after 1213) in order to continue his education in the ancient sciences, medicine, and philosophy with several noted scholars; among the things he studied were the works of **Ibn Sīnā**, who became an important formative influence. Ṭūsī then traveled to Iraq where his studies included legal theory; in Mosul (sometime between 1223 and 1232), one of his teachers was Kamāl al-Dīn ibn Yūnus (died: 1242), a legal scholar who was also renowned for his expertise in astronomy and mathematics.

In the early 1230s, after completing his education, Ṭūsī found patrons at the Ismāʿīlī courts in eastern Iran; he eventually relocated to Alamūt, the Ismāʿīlī capital, and witnessed its fall to the Mongols in 1256. Ṭūsī then served under the Mongols as an advisor to Īlkhānid ruler Hūlāgū Khan, becoming court astrologer as well as minister of religious endowments (*awqāf*). One major outcome was that Ṭūsī oversaw the construction of an astronomical observatory and its instruments in Marāgha, the Mongol headquarters in Azerbaijan, and he became its first director. The Marāgha Observatory also comprised a library and school. It was one of the most ambitious scientific institutions established up to that time and may be considered the first full-scale observatory. It attracted many famous and talented scientists and students from the Islamic world and even from as far away as China. The observatory lasted only about 50 years, but its intellectual legacy would have repercussions from China to Europe for centuries to come. Indeed, it is said that **Ulugh Beg**'s childhood memory of visiting the remnants of the Marāgha Observatory as a youth contributed to his decision to build the Samarqand Observatory. Mughal observatories in India, such as those built by **Jai Singh** in the 18th century, clearly show the influence of these earlier observatories, and it has been suggested that **Tycho Brahe** might have been influenced by them as well. In 1274 Ṭūsī left Marāgha with a group of his students for Baghdad.

Ṭūsī's writings are both synthetic and original. His recensions (*taḥārīr*) of Greek and early Islamic scientific works, which included his original commentaries, became the standard in a variety of disciplines. These works included Euclid's *Elements*, **Ptolemy**'s *Almagest*, and the so called *mutawassiṭāt* (the "Intermediate Books" that were to be studied between Euclid's *Elements* and Ptolemy's *Almagest*) with treatises by Euclid, **Theodosius**, **Hypsicles**, **Autolycus**, **Aristarchus**, **Archimedes**, **Menelaus**, **Thābit ibn Qurra**, and the **Banū Mūsā**. In mathematics, Ṭūsī published a sophisticated "proof" of Euclid's parallels postulate that was important for the development of non-Euclidian geometry, and he treated trigonometry as a discipline independent of astronomy, which was in many ways similar to what was accomplished later in Europe by **Johonn Müller** (Regiomontanus). Other important and influential works include books on logic, ethics, and a famous commentary on a philosophical work of Ibn Sīnā.

In astronomy, Ṭūsī wrote several treatises on practical astronomy (*taqwīm*), instruments, astrology, and cosmography/theoretical astronomy (*ʿilm al-hay'a*). He also compiled a major astronomical handbook (in Persian) entitled *Zīj-i Īlkhānī* for his Mongol patrons in Marāgha. Virtually all these works were the subject of commentaries and supercommentaries, and many of his Persian works were translated into Arabic. They were influential for centuries, some still being used into the 20th century.

Ṭūsī's work in practical astronomy, as well as his *Zīj-i īlkhānī*, were not particularly original or innovative. This was not the case with his work in planetary theory. There he sought to rid the Ptolemaic system of its inconsistencies, in particular its violations of the fundamental principle of uniform circular motion in the heavens. Ṭūsī set forth an astronomical device (now known as the Ṭūsī-couple) that consisted of two circles, the smaller of which was internally tangent to the other that was twice as large. The smaller rotated twice as fast as the larger and in the opposite direction. Ṭūsī was able to prove that a given point on the smaller sphere would oscillate along a straight line. By incorporating this device into his lunar and planetary models, Ṭūsī reproduced Ptolemaic accuracy while preserving uniform circular motion. A second version of this couple could produce (approximately) oscillation on a great circle arc, allowing Ṭūsī to deal with irregularities in Ptolemy's latitude theories and lunar model.

These models were first found in Ṭūsī's Persian treatise *Ḥall-i mushkilāt-i Muʿīniyya* (Solution of the difficulties in the *Muʿīniyya*), written for his Ismāʿīlī patrons, and were further developed and incorporated years later in his famous Arabic work *al-Tadhkira fī ʿilm al-hay'a* (Memoir on astronomy), composed during his years with the Mongols. Ṭūsī's devices are of major significance for several reasons. First, they produced models that adhered to both physical and mathematical requirements; the two versions of the Ṭūsī couple, from the perspective of mathematical astronomy, allowed for a separation of the effect of distance of the planet from its speed (which had been tied together in the Ptolemaic models). Ṭūsī was thus able, for example, to circumvent Ptolemy's reliance on a circular motion to produce a rectilinear, latitudinal effect. Second, Ṭūsī's new models greatly encouraged and influenced the work of Islamic astronomers, such as his student **Quṭb al-Dīn al-Shīrāzī** and **Ibn al-Shāṭir** (14th century) as well as the work of early-modern European astronomers such as Copernicus. The Ṭūsī couple also appears in Sanskrit and Byzantine texts.

Ṭūsī also influenced his astronomical and cosmological successors with his discussion of the Earth's motion. Although he remained committed to a geocentric universe, Ṭūsī criticized Ptolemy's reliance on observational proofs to demonstrate the Earth's stasis, noting that such proofs were not decisive. Recent research has revealed a striking similarity between Ṭūsī's arguments and those of Copernicus.

Ṭūsī was committed to pursing knowledge in all its forms, and he tried to reconcile the intellectual traditions of late Greek Antiquity with his Islamic faith. As was the case with many Islamic scientists, he held that the certitude of the exact mathematical sciences, especially astronomy and pure mathematics, was a means toward understanding God's creation.

F. Jamil Ragep

Selected References

Al-Ṭūsī, Naṣīr al-Dīn (1998). *Contemplation and Action (Risālah-i Sayr wa sulūk)*, edited and translated by S. J. Badakhchani. London: I. B. Tauris.

Kusuba, Takanori and David Pingree (2002). *Arabic Astronomy in Sanskrit: Al-Birjandī on* Tadhkira II, *Chapter 11 and Its Sanskrit Translation*. Leiden: E. J. Brill. (For the translation of a part of a commentary on Ṭūsī's *Tadhkira* into Sanskrit.)

Ragep, F. Jamil (1987). "The Two Versions of the Ṭūsī Couple." In *From Deferent to Equant: A Volume of Studies in the History of Science in the Ancient and Medieval Near East in Honor of E. S. Kennedy*, edited by David A. King and George Saliba, pp. 329–356. Annals of the New York Academy of Sciences, Vol. 500. New York: New York Academy of Sciences.

——— (1993). *Naṣīr al-Dīn al-Ṭūsī's* Memoir on Astronomy *(al-Tadhkira fī ʿilm al-hayʾa)*. 2 Vols. New York: Springer-Verlag.

——— (2000). "The Persian Context of the Ṭūsī Couple." In *Naṣīr al-Dīn al-Ṭūsī: Philosophe et savant du XIIIe siècle*, edited by N. Pourjavady and Ž. Vesel, pp. 113–130. Tehran: Institut français de recherche en Iran/Presses universitaires d'Iran.

——— (2000). "Al-Ṭūsī, Naṣīr al-Dīn: As scientist." In *Encyclopaedia of Islam*. 2nd ed. Vol. 10, pp. 750–752. Leiden: E. J. Brill.

——— (2001). "Ṭūsī and Copernicus: The Earth's Motion in Context." *Science in Context* 14: 145–163.

——— (2004). "Copernicus and His Islamic Predecessors: Some Historical Remarks." *Filozofski vestnik* 25: 125–142.

Riḍawī, M. M. (1976). *Aḥwāl wa-āthār… Naṣīr al-Dīn*. Tehran: Farhang Iran.

Rosenfeld, B. A. and Ekmeleddin Ihsanoğlu (2003). *Mathematicians, Astronomers, and Other Scholars of Islamic Civilization and Their Works (7th–19th c.)*. Istanbul: IRCICA, pp. 211–219.

Sayılı A. (1960). *The Observatory in Islam*. Ankara: Turkish Historical Society. (On the Marāgha Observatory, see pp. 187–223.)

Tuttle, Horace Parnell

Born **Newfield, Maine, USA, 24 March 1839**
Died **Falls Church, Virginia, USA, 1893**

As colorful an assistant astronomer as any that have ever served at Harvard College Observatory [HCO], Horace Tuttle made independent discoveries of eight comets of which six bear his name, and discovered two asteroids.

The son of Moses and Mary (*née* Merrow) Tuttle of Newfield, Horace followed his older brother, Charles Wesley Tuttle (1829–1881) in taking an unpaid position at HCO in 1857. At the time, **William Bond** was still director of the observatory, and other members of the extraordinarily talented staff – paid or unpaid – included **George Bond**, **Asaph Hall**, and Sidney Coolidge.

While Charles Wesley Tuttle and Coolidge specialized in detailed observations of the Saturnian ring system, in part motivated by the effort to obtain data supporting Bond's theory of the fluid composition of the rings, Horace swept the skies for comets. In the year following his arrival at Harvard, Tuttle made independent discoveries of four comets, three of which bear his name. For this remarkable *fête*, Tuttle received, at the age of only 21, the Lalande Prize of the French Academie of Sciences. He also shared in the discovery of two of the most celebrated of all periodic comets: 109P/1862 O1 (Swift–Tuttle) and 55P/1865 Y1 (Tempel Tuttle), the parent comets of the Perseid and Leonid meteor showers, respectively. The period of comet 109P/Swift–Tuttle is 134 years, so that after its discovery during the Civil War, it was not seen again until 1992. It was then recovered by a Japanese amateur astronomer, Tsuruhiko Kiuchi. Its next return will not occur until 2126.

In addition to finding comets 55P and 109P, Tuttle discovered comets 8P/1858 L1 (Tuttle), 41P/1858 J1 (Tuttle–Giacobini–Kresák), C/1858 R1 (Tuttle), and C/1861 Y1 (Tuttle), and he made independent discoveries of C/1858 L1 (Donati), C/1859 G1 (Tempel), and the great comet C/1860 M1. Tuttle also discovered minor planets (66) Maja on 9 April 1861 and (73) Klytia on 7 April 1862.

Tuttle left Harvard in 1862 to join the Union army, serving with the 44th Massachusetts Infantry. His Harvard Observatory colleague Coolidge also joined the war effort, and would die in battle at Chickamauga. After 9 months in the infantry, Tuttle received an appointment as an accounting paymaster in the navy, and in 1864 made observations of comet C/1864 N1 (Tempel) from the deck of the ironclad *U. S. S. Catskill*. He played a role in the capture of the English blockade runner *Deer* shortly before the end of the war. After the war, Tuttle returned to Harvard and was awarded an honorary MA degree. It was during this return visit, in January 1866, that he independently identified 55P/1865 Y1 (Tempel–Tuttle), more than 3 weeks after it had first been located by **Ernst Tempel** at the Marseilles Observatory. This comet had actually been seen by Chinese astronomers as long ago as 1366, as noted by British comet orbit computer **John Hind**.

Unfortunately, Tuttle's descent was almost as rapid as his ascent. When the navy account books were audited after the war, a substantial shortfall was noted in Tuttle's figures. On one occasion, Tuttle was found to have illegally cashed a naval bill of exchange and claimed the lion's share had been stolen from him. He was eventually found guilty of embezzlement and "scandalous conduct tending to the destruction of good morals," and in 1875 was discharged from the navy. In spite of the cloud over his reputation, he returned to science in various roles for the United States Geographical and Geological Survey – among other things, he assisted in surveying the boundary between Wyoming and the Dakota Territory – and served for a time at the United States Naval Observatory, where in 1888, he independently discovered comet C/1888 U1 (the official credit for which went to **Edward Barnard** at the Lick Observatory). Certainly Tuttle never fulfilled his youthful promise. When he died he was forgotten and almost penniless and was laid to rest in a pauper's grave in Falls Church.

William Sheehan

Selected References

Kronk, Gary W. (2003). *Cometography: A Catalogue of Comets*. Vol. 2, *1800–1899*. Cambridge: Cambridge University Press.

Vsekhsvyatskii, S. K. (1964). *Physical Characteristics of Comets*. Jerusalem: Israel Program for Scientific Translations.

Yeomans, Donald K. (1991). "Comet Hunter, Civil War Hero, and Embezzler." In *Comets: A Chronological History of Observation, Science, Myth, and Folklore*, pp. 238–239. New York: John Wiley and Sons.

Tyrtamus

Theophrastus

U

ˁUbaydī: Jalāl al-Dīn Faḍl Allāh al-ˁUbaydī

Died **1350**

There is little information about the identity and life of ˁUbaydī. According to a recorded note in one of his works, he was the student of **Quṭb al-Dīn al-Shīrāzī** (Istanbul, Süleymaniye, Nuri Efendi MS 149/2, 16b–81b). Since copies of his astronomical works are extant in Turkish manuscript libraries, it is assumed that he was educated, and that he studied in Anatolia.

ˁUbaydī's work represents a continuation of the tradition of studying mathematics and astronomy at the Marāgha School and Marāgha Observatory as well as ideas put forth by **Ibn al-Haytham**. He was particularly interested in *ˁilm al-hay'a* (theoretical astronomy) and wrote a commentary on **Maḥmūd al-Jaghmīnī**'s *al-Mulakhkhaṣ fī ˁilm al-hay'a al-basīṭa*. ˁUbaydī informs us in the preface that he wrote the commentary in 3 days at the request of some professors and students. There are at least 20 extant copies of the commentary in Turkish manuscript libraries.

ˁUbaydī wrote another important astronomical work, in February 1328, entitled *Bayān al-Tadhkira wa-tibyān al-tabṣira*, which was a commentary on **Naṣīr al-Dīn al-Ṭūsī**'s *al-Tadhkira fī ˁilm al-hay'a*. "*Al-tabṣira*" in the title refers to **Muḥammad al-Kharaqī**'s *al-Tabṣira fī ˁilm al-hay'a*. There are at least four extant copies of this work in Turkey.

One often finds copies of both works bound together; this was probably intentional since their contents complement one another. ˁUbaydī's two commentaries need to be examined more closely; only then will their place within the tradition of *ˁilm al-hay'a* be established. We do know, however, that ˁUbaydī's teaching of the subject in various schools certainly contributed toward making the tradition more widespread.

İhsan Fazlıoğlu

Selected References

Brockelmann, Carl (1943). *Geschichte der arabischen Litteratur*. 2nd ed. Vol. 1, p. 624. Leiden: E. J. Brill.

İzgi, Cevat (1997). *Osmanlı Medreselerinde İlim*. 2 Vols. Vol. 1, pp. 389, 401–402. Istanbul: iz Yayinlare.

Karatay, Fehmi Edhem (1966). *Topkapı Sarayı Müzesi Kütüphanesi Arapça Yazmalar Kataloğu*. Vol. 3, p. 755 (no. 7058), pp. 763–764 (no. 7083). Istanbul: TopKapi sarayi Muzesi Yayinlare.

Kātib Čelebī (1943). *Kashf al-Ẓunūn ˁan asāmī al-kutub wa-'l-funūn*. Vol. 2, col. 1819. Istanbul: Mulli Egition Bahantigi Yayinlare.

Ragep, F. J. (1993). *Naṣīr al-Dīn al-Ṭūsī's* Memoir on Astronomy *(al-Tadhkira fī ˁilm al-hay'a)*. 2 Vols. Vol. 1, pp. 61–62. New York: Springer Verlag.

Ulugh Beg: Muḥammad Ṭaraghāy ibn Shāhrukh ibn Tīmūr

Born **Sulṭāniyya, (Iran), 22 March 1394**
Died **near Samarqand, (Uzbekistan), 27 October 1449**

Ulugh Beg (Turkish for "great prince") was governor of Transoxiana and Turkestan and, during the last 2 years of his life, Timurid Sultan. However, he is mostly remembered as a patron of mathematics and astronomy. In Samarqand, he founded a school and the famous astronomical observatory, where the most extensive observations of planets and fixed stars at any Islamic observatory were made. Ulugh Beg is associated with a Persian astronomical handbook (*zīj*) that stands out for the accuracy with which its tables were computed.

Ulugh Beg was the first-born son of Shāhrukh (youngest son of the infamous conqueror Tīmūr or Tamerlane) and his first wife Gawharshād. He was raised at the court of his grandfather and, at the age of 10, was married to his cousin Agha Bīkī, whose mother was a direct descendent of Chingiz Khan. Thus Ulugh Beg could use the epithet Gūrgān, "royal son-in-law," which had originally been used for Chingiz's son-in-law.

In the years after Tīmūr's death in 1405, Ulugh Beg became governor of Turkestan and Transoxiana, the most important cities of which were the cultural centers Samarqand and Bukhara. Although not completely divorced from affairs of state, he is better known for his interest in religion, architecture, arts, and sciences, which were fostered by the Mongols as well as by the Timurids. Ulugh Beg is said to have spoken Arabic, Persian, Turkish, Mongolian, and some Chinese. He had a thorough knowledge of Arabic syntax and also wrote poetry. Although he honored Turkic–Mongolian customs, he also knew the Quran by heart, including commentaries and citations. Ulugh Beg was also a passionate hunter.

By 1411, Ulugh Beg had developed a lively interest in mathematics and astronomy, which may have been aroused by a visit in his childhood to the remnants of the Marāgha Observatory that had been directed by **Ṭūsī**. In 1417, he founded in Samarqand a *madrasa* (religious school or college) that can still be seen on the Registan Square. At this institution, unlike other *madrasas*, mathematics and astronomy were among the most important subjects taught. The most prominent teacher was **Qāḍīzāde al-Rūmī**, who was joined somewhat later by **Kāshī**.

Two extant letters by Kāshī to his father in Kāshān make clear that Ulugh Beg was personally involved in the appointment of scholars and that he was frequently present, and actively participated, in seminars, where he displayed a good knowledge of mathematical and astronomical topics. Kāshī relates how Ulugh Beg performed complicated astronomical calculations while riding on horseback. Anecdotes from other sources show that Ulugh Beg, like many other Muslim rulers, believed in astrology and fortune-telling. He appears as a person who very much respected the scholars he appointed, and whose main objective was to reach scientific truth.

In 1420, Ulugh Beg founded his famous astronomical observatory on a rocky hill outside the city of Samarqand. Its circular main building, beautifully decorated with glazed tiles and marble plates, had a diameter of about 46 m and three stories reaching a height of approximately 30 m above ground level. The north–south axis of the main building was occupied by a huge sextant with a radius of 40 m (called Fakhrī sextant after that of **Khujandī**). On the scale of this instrument, which partially lay in an underground slit with a width of half a meter, 70 cm corresponded to 1° of arc, so that the solar position could be read off with a precision of 5″. On the flat roof of the main building various smaller instruments could be placed, such as an armillary sphere, a parallactic ruler, and a triquetrum. Among other instruments known to have been used in Samarqand are astrolabes, quadrants, and sine and versed sine instruments.

Although Ulugh Beg was the director of the Samarqand Observatory, Kāshī was in charge of observations until his death in 1429, after which he was succeeded by Qāḍīzāde, who died after 1440. The observational program was completed by **Qūshjī**, who had studied in Kirmān (southeastern Iran) before returning to Samarqand. The results of the observations made under Ulugh Beg include the measurement of the obliquity of the ecliptic as 23° 30′17″ (the actual value at the time was 23° 30′48″) and that of the latitude of Samarqand as 39° 37′33″ N. (modern value: 39° 40′). Furthermore, most of the planetary eccentricities and epicyclic radii were newly determined, and the longitudes and latitudes of the more than 1,000 stars in **Ptolemy**'s star catalogue were verified and corrected. Precession was found to amount to 51.4″ per year (corresponding to 1° in little more than 70 years; the actual value is 50.2″ per year).

The observatory of Ulugh Beg stayed in operation for little more than 30 years. It was finally destroyed in the 16th century and completely covered by earth in the course of time. In 1908, archaeologist V. L. Vyatkin recovered the underground part of the Fakhrī sextant, consisting of two parallel walls faced with marble and the section of the scale between 80° and 57° of solar altitude. Ulugh Beg's observatory exerted a large influence on the huge masonry instruments built by **Jai Singh** in five Indian cities (most importantly Jaipur and Delhi) in the 18th century, more than 100 years after the invention of the telescope.

The main work with which Ulugh Beg is associated is an astronomical handbook with tables in Persian, variously called *Zīj-i Ulugh Beg*, *Zīj-i Jadīd-i Sulṭānī*, or *Zīj-i Gūrgānī*. In the introduction, Ulugh Beg acknowledges the collaboration of Qāḍīzāde, Kāshī, and Qūshjī, who were undoubtedly responsible for the underlying observations as well as the computation of the tables. The *Zīj* is in many respects a standard Ptolemaic work without any adjustments to the planetary models. It consists of four chapters dealing with chronology, trigonometry and spherical astronomy, planetary positions, and astrology, respectively. The instructions for the use of the tables, which were edited and translated into French by L. Sédillot in the middle of the 19th century, are clear but very brief and do not even include examples of the various calculations.

Thus, the most significant part of Ulugh Beg's *Zīj* lies in the observations and computations underlying the tables. Most impressively, the sine table, covering 18 pages in the manuscript copies, displays the sine to five sexagesimal places (corresponding to nine decimals) for every arc minute from 0° to 87° and to six sexagesimal places (11 decimals) between 87° and 90°. All independently calculated values for multiples of 5′ are correct to the precision given, whereas the intermediate values, calculated by means of quadratic interpolation, contain incidental errors of at most two units. Also most of the planetary tables in the *Zīj* were calculated to a higher precision than before. New types of tables were added that simplified the calculation of planetary positions. Ulugh Beg's star catalog for the year 1437 represents the only large-scale observations of star coordinates made in the Islamic realm in the medieval period. (Most other catalogs simply adjusted Ptolemy's ecliptic coordinates for precession or were limited to a relatively small number of stars.)

Ulugh Beg's *Zīj* was highly influential and continued to be used in the Islamic world until the 19th century. It was soon translated into Arabic by Yaḥyā ibn ʿAlī al-Rifāʿī and into Turkish by ʿAbd al-Raḥmān ʿUthmān. Reworkings for various localities were made in Persian, Arabic, and Hebrew by scholars such as ʿImād al-Dīn ibn Jamāl al-Bukhārī (Bukhara), **Ibn Abī al-Fatḥ al-Ṣūfī** (Cairo), Mullā Chānd ibn Bahāʾ al-Dīn and Farīd al-Dīn al-Dihlawī (both Delhi), and Sanjaq Dār and Husayn Qusʿa (Tunis). Commentaries to the *Zīj* were written by Qūshjī, **Mīram Chelebī**, **Bīrjandī**, and many others. Hundreds of manuscript copies of the Persian original of Ulugh Beg's *Zīj* are extant in libraries all over the world. Already in 17th-century England, various parts of the *Zīj* were published in edition and/or translation.

Little is known about other works of Ulugh Beg. A marginal note by him in the India Office manuscript of Kāshī's *Khāqānī Zīj* presents a clever improvement of a spherical astronomical calculation. A *Risāla fī istikhrāj jayb daraja wāḥida* (Treatise on the extraction of the sine of 1°) has been attributed to Ulugh Beg on the basis of a citation in Bīrjandī, although most manuscripts of this work mention Qāḍīzāde as the author. Aligarh Muslim University Library lists a treatise *Risāla-yi Ulugh Beg* that is yet to be inspected. Finally, an astrolabe now preserved in Copenhagen and made in 1426/1427 by Muḥammad ibn Jaʿfar al-Kirmānī, who is known to have worked at the observatory in Samarqand, was originally dedicated to Ulugh Beg.

In 1447, Ulugh Beg succeeded his father Shāhrukh as sultan of the Timurid empire. However, he was killed on the order of his son ˁAbd al-Laṭīf. An investigation of Tīmūr's mausoleum by Soviet scholars in the 1940s showed that Ulugh Beg was buried as a martyr in accordance with *Sharīˁa* (Islamic law), *i. e.*, fully clothed in a sarcophagus.

Benno van Dalen

Alternate name

Gūrgān

Selected References

Bagheri, Mohammad (1997). "A Newly Found Letter of Al-Kāshī on Scientific Life in Samarkand." *Historia Mathematica* 24: 241–256.

Barthold, V. V. (1958). *Four Studies on the History of Central Asia.* Vol. 2, *Ulugh-Beg*. Leiden: E. J. Brill.

Kary-Niiazov, T. N. (1950). *Astronomicheskaya shkola Ulugbeka* (The astronomical school of Ulugh Beg) (in Russian). Moscow: Akademia Nauk SSSR. (Second enlarged edition in Kary-Niiazov, *Izbrannye trudy* [Collected works]. Vol. 6. Tashkent: FAN, 1967.)

——— (1976). "Ulugh Beg." In *Dictionary of Scientific Biography*, edited by Charles Coulston Gillispie. Vol. 13, pp. 535–537. New York: Charles Scribner's Sons.

Kennedy, E. S. (1956). "A Survey of Islamic Astronomical Tables." *Transactions of the American Philosophical Society*, n.s., 46, pt. 2: 121–177, esp. 125–126 and 166–167. (Reprint, Philadelphia: American Philosophical Society, 1989.)

——— (1960). "A Letter of Jamshīd al-Kāshī to His Father: Scientific Research and Personalities at a Fifteenth Century Court." *Orientalia* 29: 191–213. (Reprinted in E. S. Kennedy, et al., *Studies in the Islamic Exact Sciences*, edited by David A. King and Mary Helen Kennedy. Beirut: American University of Beirut, 1983, pp. 722–744.)

——— (1998). "Ulugh Beg as Scientist." Chapter 10 in *Astronomy and Astrology in the Medieval Islamic World*. Aldershot: Ashgate. (Describes the marginal note by Ulugh Beg in a manuscript of Kāshī's *Zīj*.)

Knobel, Edward Ball (1917). *Ulugh Beg's Catalogue of Stars: Revised from All Persian Manuscripts Existing in Great Britain, with a Vocabulary of Persian and Arabic Words*. Washington: Carnegie Institution of Washington.

Krisciunas, Kevin (1992). "The Legacy of Ulugh Beg." In *Central Asian Monuments*, edited by H. P. Paksoy, pp. 95–103. Istanbul: Isis. (Includes a bibliography of all publications of parts of the *Zīj* of Ulugh Beg.)

——— (1993). "A More Complete Analysis of the Errors in Ulugh Beg's Star Catalogue." *Journal for the History of Astronomy* 24: 269–280.

Kunitzsch, Paul (1998). "The Astronomer al-Ṣūfī as a Source for Uluġ Beg's Star Catalogue." In *La science dans le monde iranien ā l'époque islamique*, edited by Ž. Vesel, H. Beikbaghban, and B. Thierry de Crussol des Epesse, pp. 41–47. Tehran: Institut français de recherche en Iran.

Manz, Beatrice F. (2000). "Ulugh Beg." In *Encyclopaedia of Islam*. 2nd ed. Vol. 10, pp. 812–814. Leiden: E. J. Brill.

Rosenfeld, B. A. and Jan P. Hogendijk (2002/2003). "A Mathematical Treatise Written in the Samarqand Observatory of Ulugh Beg." *Zeitschrift für Geschichte der Arabisch-Islamischen Wissenschaften* 15: 25–65.

Sayılı, Aydın (1960). *The Observatory in Islam*. Ankara: Turkish Historical Society, esp. pp. 260–289.

———(1960). *Ghiyâth al-Dîn al Kâshî's Letter on Ulugh Beg and the Scientific Activity in Samarqand*. Ankara: Turkish Historical Society.

Schoy, Carl (1927). *Die trigonometrischen Lehren des persischen Astronomen Abū 'l-Rayḥān Muḥammed ibn Aḥmed al-Bīrūnī dargestellt nach al-Qānūn al-Masˁūdī*. Hanover: Lafaire. (Reprinted in Schoy, *Beiträge zur arabisch-islamischen Mathematik und Astronomie*, edited by Fuat Sezgin, Vol. 2, pp. 629–746. Frankfurt am Main: Institute for the History of Arabic-Islamic Science, 1988.) (Includes an edition of parts of Ulugh Beg's sine and tangent tables.)

Sédillot, Louis P. E. Amélie (1847). *Prolégomènes des tables astronomiques d'Oloug-Beg. Publiés avec notes et variantes et précédés d'une introduction*. Paris: Firmin Didot.

——— (1853). *Prolégomènes des tables astronomiques d'Oloug-Beg. Traduction et commentaire*. Paris: Firmin Didot.

Shevchenko, Mikhail Yu (1990). "An Analysis of Errors in the Star Catalogues of Ptolemy and Ulugh Beg." *Journal for the History of Astronomy* 21: 187–201.

Umawī: Abū ˁAlī al-Ḥasan ibn ˁAlī ibn Khalaf al-Umawī

Born **Cordova, (Spain), 1120**
Died **Seville, (Spain), 1205/1206**

Abū al-Ḥasan al-Umawī, known as al-Khaṭīb (the preacher), was an expert in the Islamic religious sciences and the Arabic language. He wrote a number of treatises among which there are two on Arabic ethnoastronomy: *Kitāb al-Luʾluʾ al-manẓūm fī ma ˁrifat al-awqāt bi-'l-nujūm* (Book of the pearl in the necklace on the knowledge of time by means of the stars) and *Kitāb al-Anwāʾ* (Book about the *Anwāʾ*). The book belongs to a genre that aims to compile astronomical and meteorological materials from traditional Arabic lore inside the framework of the *anwāʾ*, periods of 13 days defined by the risings and settings of certain asterisms (lunar mansions) located along the lunar ecliptic, which account for the complete solar year. Umawī's main source is the *Kitāb al-Anwāʾ wa-'l azmina* by another Cordovan, Ibn ˁĀṣim (died: 1013), who had compiled materials taken from philologists of eastern Islam from the 8th century onward.

As a religious scholar, Umawī expanded on and completed Ibn ˁĀṣim's chapters on the procedures of Arabic folk astronomy that could help determine the times of prayers (*mīqāt*) or find the direction of Mecca (*qibla*). The treatise contains a method for determining night hours based upon the appearance of the asterisms of the *anwāʾ* system – this chapter seems to be related with Umawī's other astronomical treatise mentioned above, two series of lengths of shadows cast by a gnomon to determine prayer times (one of them written in a numerical notation, the *Rūmī* ciphers, found only in Andalusia and north Africa), and a long fragment on the possibility of observing Canopus (*Suhayl*) from Muslim Spain, a star used to determine the direction of Mecca. The author seems to be aware of more sophisticated forms of astronomy as he mentions two unusual sundials, the *mīzān fazārī* and the *mukḥūla*.

There are two possible reasons for Umawī's interest in continuing a tradition that by his time was two centuries old: First, the rulers of the period, the Almohads, used to train their sons in the observation of the asterisms of the *anwāʾ* system; and second, the Almohad mosques, unlike those built by their predecessors, the Almoravids, were often directed toward the rising of Canopus. About a century later, this treatise was used by the famous Moroccan astronomer **Ibn al-Bannāʾ** as a source for his *Kitāb fī al-anwāʾ* (Book on the *anwāʾ*). Only the second treatise has come down to us, albeit in fragmentary form (preserved in El Escorial Library, MS 941).

Miquel Forcada

Alternate name
al-Khatīb al-Umawī al-Qurṭubī

Selected References
Forcada, Miquel (1990). "*Mīqāt* en los calendarios andalusíes." *Al-Qanṭar* 11: 59–69.

——— (1992). "Les sources andalouses du calendrier d'Ibn al-Bannā' de Marrakech." In *Actas del II Coloquio Hispano-Marroquí de Ciencias Históricas: Historia, ciencia y sociedad*, pp. 183–196. Madrid: Agencia Española de Cooperación Internacional.

——— (1993). *El Kitāb al-anwā' wa-l-azmina - al-qawl fī l-šuhūr de Ibn ʿĀṣim (Tratado sobre los anwā' y los tiempos – capítulo sobre los meses)*. Fuentes Arabico-Hispanas, 15. Madrid: Consejo Superior de Investigaciones Cientifícas.

——— (1994). "Esquemes d'ombres per determinar el moment de les pregàries en llibres d'*anwā'* i calendaris d'al-Andalus." In *Actes de les I Trobades d'Història de la Ciència i de la Tècnica,* edited by J. M. Camarasa, H. Mielgo, and A. Roca, pp. 107–118. Barcelona: Societat Catalana de Física-Secció de Ciència i Tècnica de l'Institut Menorquí d'Estudis-Societat Catalana d'Història de la Ciència i de la Tècnica.

——— (1998). "Books of *Anwā'* in al-Andalus." In *The Formation of al-Andalus*, Part 2: *Language, Religion, Culture and the Sciences*, edited by Maribel Fierro and Julio Samsó, pp. 305–328. Aldershot: Ashgate.

Ibn al-Abbār (1887). *Al-Takmila li-Kitāb al-Ṣila*, edited by F. Codera. Madrid: Biblioteca Arabico-Hispana (V–VI), Biography no. 46.

Rius, Mònica (2000). *La alquibla en al-Andalus y al-Magrib al-Aqṣà*. Barcelona: Institut "Millàs Vallicrosa" d'Història de la Ciència Árab.

Unsöld, Albrecht

Born **Bolheim near Tübingen, (Baden-Württenberg), Germany, 20 April 1905**
Died **Kiel, Germany, 23 September 1995**

German stellar astrophysicist Albrecht Unsöld made major theoretical and observational contributions to the detailed analysis of stellar spectra resulting in accurate measurements of temperatures, densities, and compositions of the Sun and stars. He was the son of a minister, and had begun reading and writing about atomic physics as early as age 14 when he sent a manuscript to Arnold Summerfeld in Munich, Germany. Unsöld received enough encouragement that he left the University of Tübingen very shortly after Gymnasium to study in Munich, where he received a Ph.D. in 1927 with a thesis on quantum mechanics of atomic structure carried out under Summerfeld. His own students in due course included a number of other German stellar astrophysicists holding university professorships: Karl-Heinz Bohm, Kurt Hunger, Gerhard Traving, Volker Weidemann, Bodo Baschek, and Dieter Reimers. He married Dr. Liselotte Kuhnert, a biologist, in 1934, and they had four children.

After brief stays at Potsdam Observatory, Munich, Mount Wilson Observatory (1928–1929 on a Rockefeller Fellowship), and Hamburg University (1930–1932), Unsöld was appointed professor at Kiel University (the youngest such appointment recorded in Germany) in 1932 and remained there for the rest of his career, officially retiring in 1973 but remaining active in the astronomical community until the death of his wife in 1990. He obtained his *Habilitation* from Munich in 1929 with a paper interpreting the Balmer hydrogen lines in the solar spectrum.

Specific contributions to the analysis of conditions in the Sun and stars included the recognition that absorption lines are broadened by thermal Doppler effects, the Stark effect, the close approach of atoms to each other, and the natural lifetimes of the excited states. This enabled Unsöld to combine strong and weak lines of a given element to learn the temperature of the layers producing the lines and the total number of atoms responsible, hence the abundance of that element. He was also a pioneer in understanding the relationship between convection in the solar atmosphere and the granulation of the level we see. His 1937 monograph *Physics of Stellar Atmospheres* made his methods available to the entire community, where they were widely used.

A fortunate visit to McDonald Observatory enabled Unsöld to take some high-resolution stellar spectra in 1939, which provided material for him and his students to work on during World War II. Unsöld and his close friend and colleague W. Lochte-Holtgreven, a plasma physicist, were drafted to work on meteorology during the war. Proximity to Kiel enabled Unsöld to save the library of **Heinrich Schumacher** (founder of the oldest astronomical journal, *Astronomische Nachrichten*) when the observatory buildings were bombed.

Unsöld edited the main West German publication in the field, *Zeitschrift für Astrophysik*, until its 1969 merger with other European journals to form *Astronomy and Astrophysics*. He had, as it happened, written the first paper ever published in the *Zeitschrift* (founded in 1930).

After the war, Unsöld became the first dean of the faculty in the reopened University of Kiel and later served as rector. He opened a small radio astronomy observatory in connection with the optical observatory and continued to collaborate with students on theory of stellar atmospheres. The second 1955 edition of *Physics of Stellar Atmospheres* included many of the new developments like mixing-length theory and model atmospheres obtained by numerical methods. His 1967 text *Der Neue Kosmos* was intended to hark back to **Alexander von Humboldt**'s *Kosmos* of 1859 and to introduce a new generation to modern astrophysics.

Unsöld's own scientific interests broadened to include the origin of the chemical elements as well as their abundances in the Sun and stars, and at this point he began to diverge from the majority of the community. He was not convinced that synthesis in stars was important, and instead favored bulk production in "little bangs" at the beginning of galactic history. His last book in 1981 on evolution dealt not just with cosmology and biology, but also with psychology, art, and religion.

Unsöld received honorary degrees from Utrecht, Edinburgh, and Munich universities as well as the Bruce Medal of the Astronomical Society of the Pacific, a Gold Medal and the Darwin Lectureship of the Royal Astronomical Society (London), and memberships in a number of honorary organizations. He was both a violinist and a painter in watercolors, continuing these beyond the time he ceased interacting with other scientists.

Christian Theis

Selected References
Baschek, Bodo (1996). "Albrecht Unsöld." *Mitteilung Astronomische Gesselschaft*, no. 79: 11–15.

Jeffreys, Harold (1957). "The President's Address on the Award of the Gold Medal to Professor Albrecht Unsöld." *Monthly Notices of the Royal Astronomical Society* 117: 344–346.
Weidemann, Volker (1996). "Albert Unsöld (1905–1995)." *Publications of the Astronomical Society of the Pacific* 108: 553–555.
Wilson, Olin C. (1956). "The Award of the Bruce Gold Medal to Dr. Albrecht Unsöld." *Publications of the Astronomical Society of the Pacific* 68: 89–92.

ˁUrḍī: Mu'ayyad (al-Milla wa-) al-Dīn (Mu'ayyad ibn Barīk [Burayk]) al-ˁUrḍī (al-ˁĀmirī al-Dimashqī)

Born **probably ˁUrḍ between Palmyra and Ruṣāfa, (Syria), *circa* 1200**
Died **Marāgha, (Iran), *circa* 1266**

ˁUrḍī was one of the major figures of Islamic astronomy in the 13th century, and participated in a number of important scientific innovations and developments. Sometime before 1239, ˁUrḍī moved to Damascus, where he worked as an engineer, a teacher of geometry. and, possibly, of astronomy as well. In 1252/1253, as he says in his *Risālat al-Raṣd*, he built the socalled perfect instrument for al-Malik al-Manṣūr, Lord of Ḥimṣ. In the 1250s **Naṣīr al-Dīn al-Ṭūsī** asked him to come to Marāgha in Azerbaijān (now in northwest Iran) to help in the building of an observatory under the patronage of the Mongol ruler Hülegü. The observatory, one of the most important ever built in the Islamic world and arguably the first full-scale observatory in the modern sense, was founded in 1259, and ˁUrḍī arrived in Marāgha in that year (or shortly before). He took part in building the observatory outside the city and erected special devices and water wheels to raise the water to the observatory hill; he also participated in the construction of a mosque and a special building for Hülegü's residence. At the observatory in Marāgha, ˁUrḍī probably joined in the observations for Ṭūsī's *Īlkhānī Zīj* and was mentioned in this treatise. Though a noted astronomer and instrument maker, his participation in the observatory was limited to its early years, in as much as he constructed instruments there only before 1261/1262. Several instruments for which ˁUrḍī tells us he prepared models were actually seen by later visitors, further suggesting that he was not the only instrument maker at Marāgha. His son Muḥammad, also a member of the observatory staff, made a copy of his father's *Kitāb al-Hay'a* and constructed a celestial globe, now preserved in Dresden, which was used at the observatory. ˁUrḍī, as well as Ṭūsī, was a member of the so called School of Marāgha, which also included **Quṭb al-Dīn al-Shīrāzi** and a number of other astronomers.

Urḍī's *Risāla fī Kayfiyyat al-arṣād* (or simply *Risālat al-Raṣd*) is a rich and informative treatise on observational instruments, preserved in a unique manuscript in Paris. Some of the instruments mentioned in this treatise were well known, others were invented by ˁUrḍī himself. The treatise mentions the instruments built before and up to 1261/1262. The introduction describes the determination of the meridian by means of an "Indian circle." ˁUrḍī tells us the place and time of the erection of the instruments, and he also outlines his relationship to Ṭūsī. The following instruments are mentioned: a mural quadrant, that seems to be used in general for altitude measurement, as well as for a careful determination of the latitude of Marāgha and the obliquity of the ecliptic; an armillary sphere for the measurement of the ecliptic longitude and latitude; a solstitial armilla for the determination of the obliquity of the ecliptic; an equinoctial armilla for the determination of the entry of the Sun into the equatorial plane and the path of the Sun at the equinoxes; a so called dioptrical ruler of **Hipparchus** for the measurements of the apparent diameters of the Sun and the Moon and the observation of eclipses; an azimuth ring for the determination of the altitude and the azimuth; and several other rulers and instruments, such as the "perfect instrument" for the measurement of the azimuth. ˁUrḍī ends with a critique of the parallactical ruler described by **Ptolemy**. As for the size of the instruments, ˁUrḍī remarks that the instruments should be as large as possible to have the required division of the scales.

ˁUrḍī's *Kitāb al-Hay'a*, written sometime before ˁUrḍī reached Marāgha in 1259, is a work on theoretical astronomy that includes a critique of Ptolemy's *Almagest* and his *Planetary Hypotheses*. There exist two versions of ˁUrḍī's treatise: an earlier one compiled sometime between 1235 and 1245 and a later version in which he edited whole chapters of his original text to make the arguments more consistent. In the *Kitāb al-Hay'a* ˁUrḍī introduces the reader to Ptolemaic astronomy and then explains the difficulties arising from some of Ptolemy's methods and techniques. He then presents his own astronomical models as an alternative. For ˁUrḍī, as well as for other astronomers of the so called School of Marāgha, the main problem in Ptolemaic astronomy was the lack of consistency between the mathematical models and the principles of natural philosophy. Examples occurred in the prosneusis point for the Moon, the deferent in the lunar model, the equant in the model for the superior planets, the inconsistencies in the planetary distances, and the inclination and deviation of the spheres of Mercury and Venus that were meant to account for latitude. In ˁUrḍī's opinion, these inconsistencies violated the essential consistency between the theoretical mathematical models and the accepted natural and physical axioms. ˁUrḍī held to the basic principles of Greek astronomy, especially the circular and uniform motion of the heavenly bodies, and the Earth as the unmovable center of the Universe; he also appreciated the validity of the Ptolemaic planetary observations as quoted in the *Almagest*. But he objected to the mathematical models that Ptolemy had devised to describe the motions of the planets. ˁUrḍī tried to find astronomical models that would preserve Ptolemy's observations, and which would also be consistent mathematically as well as physically. To this end, he devised the ˁUrḍī lemma, a developed form of the theorem by **Apollonius** that transformed eccentric models into epicyclic ones. ˁUrḍī stated that if we construct two equal lines on the same side of any straight line so that they make two equal angles with that straight line, be they corresponding or interior, and if their endpoints are connected, then the line resulting from connecting them will be parallel to the line upon which they were erected (*Kitāb al-Hay'a*, p. 220). The new technique of bisecting the Ptolemaic eccentricity allowed him to preserve Ptolemy's deferent, while preserving the uniform, circular motions of the celestial orbs that revolve on their own centers, thus avoiding the apparent contradictions in Ptolemy's model. ˁUrḍī's *Kitāb al-Hay'a* was written within a tradition of astronomical literature that was critical of Ptolemy, but it apparently did not depend upon the work of Ṭūsī, who also presented alternative models in several of his works (many of which were based upon the Ṭūsī couple that transforms circular motion into linear motion). ʿUrḍī's work was quoted by **Ibn al-Shāṭir**, and influenced **Bar Hebraeus** and Quṭb al-Dīn al-Shīrāzī. Furthermore, there are many similarities to **Nicholaus Copernicus**'s work. ʿUrḍī's technical alternative to Ptolemy's model for the upper planets is essentially the same as that in Copernicus's *De revolutionibus*.

ʿUrḍī also wrote some minor treatises: a commentary on **Kharaqī**'s astronomical treatise *Kitāb al-Tabṣira fī ʿilm al-hayʾa*, that closely follows Kharaqī's wording (extant in a unique manuscript in Madrid); a supplement to a problem in the *Almagest*, probably preserved in Mashhad and Ankara; a short treatise on the determination of the solar eccentricity, preserved in Ankara and Istanbul; and a *Risālat al-ʿAmal fī al-kura al-kāmila* on the armillary sphere, mentioned in ʿUrḍī's *Risālat al-Raṣd* as well as in his *Kitāb al-Hayʾa*, which seems no longer extant. In addition, ʿUrḍī himself, or his son, copied in 1252/1253 the recension of the *Almagest* by Ṭūsī, which is preserved in Cairo.

Both of ʿUrḍī's main works, the *Risālat al-Raṣd* and the *Kitāb al-Hayʾa*, are characterized by improvement and refinement. On the one hand, he tried to make precise instruments – some standard, others of his own invention – that would result in the best observations possible. The *Risālat al-Raṣd* gives the reader a rare insight into the equipment of a medieval Islamic observatory. On the other hand, he attempted to make the Ptolemaic astronomy more consistent by developing new and highly sophisticated planetary theories, some of them mathematically identical to Copernicus's non-Ptolemaic models.

Petra G. Schmidl

Selected References

Drechsler, Adolph (1873). *Der Arabische Himmels-Globus angefertigt 1279 zu Maragha von Muhammed bin Muwajid Elardhi zugehörig dem Königl. Mathematisch-physikalischen Salon zu Dresden*. Dresden: Königl. Hofbuchhandlung von Hermann Burdach. (2nd edition reprinted in Sezgin, *Astronomische Instrumente*, Vol. 4, pp. 215–241; (reprinted in Sezgin, *School of Marāgha*, Vol. 1, pp. 261–289; and reprinted in Sezgin, *Astronomical Instruments*, Vol. 4, pp. 215–241.) (On the celestial globe made by ʿUrḍī's son.)

Frank, Josef (1929). "Review of 'Hugo J. Seemann, *Die Instrumente der Sternwarte zu Marâgha nach den Mitteilungen von al-ʿUrḍī*.'" *Zeitschrift für Instrumentenkunde* 49: 356–367. (Reprinted in Sezgin, *Astronomische Instrumente*, Vol. 6, pp. 130–141; reprinted in Sezgin, *School of Marāgha*, Vol. 2, pp. 194–205; and reprinted in Sezgin, *Astronomical Instruments*, Vol. 6, pp. 130–141.) (On ʿUrḍī's *Risālat al-Raṣd*.)

Saliba, George (1979). "The First Non-Ptolemaic Astronomy at the Maraghah School." *Isis* 70: 571–576. (Reprinted in Saliba, *History of Arabic Astronomy*, pp. 113–118.)

——— (1979). "The Original Source of Quṭb al-Dīn al-Shīrāzī's Planetary Model." *Journal for the History of Arabic Science* 3: 3–18. (Reprinted in Saliba, *History of Arabic Astronomy*, pp. 119–134.)

——— (1985). "The Determination of the Solar Eccentricity and Apogee According to Mu'ayyad al-Dīn al-ʿUrḍī (d. 1266 A.D.). " *Zeitschrift für die Geschichte der Arabisch-Islamischen Wissenschaften* 2: 47–67. (Reprinted in Saliba, *History of Arabic Astronomy*, pp. 187–207.)

——— (1987). "The Height of the Atmosphere According to Mu'ayyad al-Dīn al-ʿUrḍī, Quṭb al-Dīn al-Shīrāzī, and Ibn Muʿādh." In *From Deferent to Equant: A Volume of Studies in the History of Science in the Ancient and Medieval Near East in Honor of E. S. Kennedy*, edited by David A. King and George Saliba, pp. 445–465. *Annals of the New York Academy of Sciences*. Vol. 500. New York: New York Academy of Sciences.

——— (1987). "The Rôle of Maragha in the Development of Islamic Astronomy: A Scientific Revolution before the Renaissance." *Revue de synthèse* 108: 361–373. (Reprinted in Saliba, *History of Arabic Astronomy*, pp. 245–257.)

——— (1989). "A Medieval Arabic Reform of the Ptolemaic Lunar Model." *Journal for the History of Astronomy* 20: 157–164. (Reprinted in Saliba, *History of Arabic Astronomy*, pp. 135–142.)

——— (1990): *The Astronomical Work of Mu'ayyad al-Dīn al-ʿUrḍī (Kitāb al-Hay'a): A Thirteenth Century Reform of Ptolemaic Astronomy* (in Arabic with English introduction). Bayrūt: Markaz Dirāsat al-Waḥda al-ʿarabiyya. (On ʿUrḍī's *Kitāb al-Hay'a*.)

——— (1991). "The Astronomical Tradition of Marāgha: A Historical Survey and Prospects for Future Research." *Arabic Sciences and Astronomy* 1: 67–99. (Reprinted in Saliba, *History of Arabic Astronomy*, pp. 258–290.)

——— (1994). *A History of Arabic Astronomy: Planetary Theories during the Golden Age of Islam*. New York: New York University Press.

Sayılı, Aydın (1960). *The Observatory in Islam*. Ankara: Turkish Historical Society.

Seemann, Hugo J. (1928). "Die Instrumente der Sternwarte zu Marâgha nach den Mitteilungen von al ʿUrḍī." *Sitzungsberichte der Physikalisch-medizinischen Sozietät zu Erlangen* 60: 15–126. (Reprinted in Sezgin, *School of Marāgha*. Vol. 2, pp. 81–192 and reprinted in Sezgin, *Astronomical Instruments*. Vol. 6, pp. 17–129.)

Sezgin, Fuat (ed.) (1990/91). *Astronomische Instrumente in orientalistischen* Studien. Vol 4 and 6. Frankfurt: Institute for the History of Arabic-Islamic Science.

Sezgin, Fuat (ed.) (1998). *The School of Marāgha and Its Achievements*. Vols. 1 and 2. *Islamic Mathematics and Astronomy*. Vols. 50 and 51. Frankfurt.

——— (1998). *Astronomical Instruments and Observatories in the Islamic World*. Vols. 4 and 6. *Islamic Mathematics and Astronomy*. Vols. 88 and 90. Frankfurt.

Tekeli, Sevim (1970). "*'The Article on the Quality of Observations' by al-ʿUrḍī*." Araztima 8: 157–169.

Urey, Harold Clayton

Born **Walkerton, Indiana, USA, 29 April 1893**
Died **San Diego County, California, USA, 5 January 1981**

American chemist Harold C. Urey received the 1934 Nobel Prize in Chemistry for his discovery of heavy hydrogen (deuterium), which proved to be of enormous importance in understanding both energy generation in the Sun and stars and conditions in the early Universe, where all the deuterium that exists today was produced. Astronomers also remember him for a definitive table of the abundances of the elements, compiled in collaboration with **Hans Suess**, which guided the modern understanding of nucleosynthesis in stars. (See **William Fowler** and **Fred Hoyle**.) In addition, Urey suggested a way of imitating atmospheric conditions of the early Earth with a laboratory experiment, carried out in 1953 by Stanley Miller (but called a Urey atmosphere experiment), which demonstrated that simple molecules like methane, ammonia, and water could form amino acids and other biologically important molecules under natural conditions.

Urey was the son of Reverend Samuel Clayton Urey and entered the University of Montana in 1914, receiving a BS in zoology in 1917. After a few years work at Barrett Chemical Company in Maryland and as an instructor at the University of Montana, he entered graduate school at the University of California (Berkeley) in 1921, receiving a Ph.D. in chemistry in 1923 for work with Gilbert Newton Lewis. Urey spent the following year at **Niel Bohr**'s Institute in Copenhagen on a Scandinavian–American fellowship. During and after his stay in Copenhagen, he held a research associateship at Johns Hopkins University (1924–1929). Urey was appointed associate professor of chemistry at Columbia University, moving up to professor in 1934, and serving as executive officer of his department from 1939 to 1942. During the war years Urey also directed a Columbia laboratory connected with the atomic bomb project.

Urey moved to a professorship at the Institute of Nuclear Studies at the University of Chicago in 1952 and held the Ryerson Professorship there from 1952 to 1958. During his Columbia and Chicago years he also spent various periods as visiting professor or endowed lecturer in England, Israel, and at a number of American universities, and was editor of the *Journal of Chemical Physics* from 1933 to 1940. On reaching normal retirement age, Urey was appointed to one of the first professorships-at-large at the University of California [UC], meaning that his tenure resided across the whole system. He chose, however, to locate at the relatively new UC San Diego campus in La Jolla. Quite remarkably for one of his distinction, Urey enjoyed teaching introductory chemistry, and his department recalled him to active duty to do so long after his official retirement, indeed until a few years before his death, when he could no longer easily climb the steps to the lecture room podium.

Urey's early research at Columbia concerned absorption spectra and structure of molecules. In 1931 he devised a method for evaporating a large quantity of liquid hydrogen down to a milliliter, which would automatically concentrate any heavy component (the lighter atoms or molecules tend to evaporate more easily), at a time when the concept of an isotope was fairly new. This led to the discovery of deuterium. He and E. W. Washburn evolved the electrolytic method for isotope separation, and Urey was able to investigate the properties of deuterium and measure its abundance, which he found to be about 2 parts in 10,000 in ordinary water.

In his postwar career in Chicago, Urey turned almost fully to geochemistry and cosmochemistry. His work with the heavy isotope oxygen-18 enabled him to devise a method for estimating ocean temperatures as far back as 180 million years ago. Again, light things evaporate more easily, so the ratio of O^{18} to O^{16} in fossil seashells is an indicator of the water temperature in which they lived. The isotope work in turn led Urey to the study of the relative abundances of the chemical elements on Earth and in meteorites and toward the development of a theory of the formation of the elements in stars, put forward soon after by E. Margaret Burbidge, Geoffrey R. Burbidge, Fowler, and Hoyle (as well as by Alastair G. W. Cameron in Canada). Consideration of the compositions of the Earth's early atmosphere and its implications for the origin of life was also a product of his Chicago years.

After 1958, Urey became a scientific advisor to the space program. One of his conclusions was that billions of years of impacts of small meteorites on the surface of the Moon, chipping off bits from the lunar rocks, would probably have covered the lunar surface with a thick layer of quicksand-like dust, a major hazard for any landing on the surface. Luckily this proved not to be the case. Thomas Gold of Cornell University was right in his expectation that the dust would be firmly compacted and able to support both spacecraft and astronauts, while taking clear impressions of landing gear and boots.

Urey's major books include *Atoms, Molecules and Quanta* (two volumes, 1930) with Arthur E. Ruark; *The Planets, Their Origin and Development* (1952); and *Isotopes and Cosmic Chemistry* (1962) with H. Craig, G. L. Miller, and Gerald J. Wasserburg.

In addition to the Nobel Prize, Urey was the recipient of 23 honorary degrees from universities stretched across the United States and Europe, from Athens to Wayne State. He was awarded medals and other prizes by the Franklin Institute, the University of Paris, the United States government (National Medal of Science), the United States National Academy of Sciences, the American Chemical Society, and the Royal Astronomical Society (London), among others. Urey was elected to honorary membership in scientific academies in Belgium, Portugal, Great Britain, India, Ireland, France, and Sweden. He married in 1936, and he and his wife had four children, one of whom, John Clayton Urey, is a biochemical geneticist.

Fathi Habashi

Selected References

Asimov, I. (1982). *Biographical Encyclopedia of Science and Technology*. 2nd ed. New York: Doubleday.

Doel, Ronald Edmund (1996). *Solar System Astronomy in America: Communities, Patronage and Interdisciplinary Science, 1920–1960*. Cambridge: Cambridge University Press.

James, L. K. (ed.) (1993). *Nobel Laureates in Chemistry 1901–1992*. Washington, DC: American Chemical Society, pp. 211–215

ˁUṭārid: ˁUṭārid ibn Muḥammad al-Ḥāsib

Flourished **9th century**

ˁUṭārid ibn Muḥammad is sometimes also referred to as al-Kātib (the scribe), but the usual appellation, al-Ḥāsib (the arithmetician), is more appropriate. Little is known of his life. Ibn al-Nadīm tells us that he was an arithmetician and an astrologer (*al-munajjim*) as well as a man of excellence and learning. From Ibn al-Nadīm we also know the titles of five books by ˁUṭārid:

(1) *Kitāb al-Jafr al-hindī* (Book on Indian divination), which may have dealt with divination based upon letters of the alphabet or perhaps meteorological predictions;
(2) *Kitāb al-ˁAmal bi-'l-asṭurlāb* (Book on using the astrolabe);
(3) *Kitāb al-ˁAmal bi-dhāt al-ḥalaq* (Book on using the armillary sphere);
(4) *Kitāb Tarkīb al-aflāk* (Book on the arrangement of the heavens); and
(5) *Kitāb al-Marāyā al-muḥriqa* (Book on burning mirrors).

There is also a report that ˁ**Abd al-Raḥman al-Ṣūfī** saw a book of ˁUṭārid (in latter's own handwriting) about the 48 constellations. In addition, both **Bīrūnī** and **Sijzī** attribute to ˁUṭārid a *Kitāb al-Miḥna al-munajjim* (Book on examining astrologers), a work specifically for testing the skills of astrologers. A text with a similar subject is by **Qabīṣī**. None of the above mentioned works attributed to ˁUṭārid are extant.

Of ˁUṭārid's works, only two have reached us. One is an astrological work entitled *Sirr al-asrār* (Secret of secrets) or *al-Asrār al-samāwiyya* (The secrets of the heavens), and also known as *Fuṣūl li-ˁUṭārid al-Ḥāsib fī al-asrār al-samāwiyya.* One can find excerpts in **Majrīṭī**'s *Ghāyat al-ḥakīm* that deal with the astrological topic of elections (*ihktiyārāt*). The other is *Kitāb al-Jawāhir wa-'l-aḥjār* (Book on the properties of stones), perhaps the earliest work of its kind in Arabic, which follows the so called Lapidary of **Aristotle**.

Giuseppe Bezza

Selected References

Ibn al-Nadīm (1970). *The Fihrist of al-Nadīm: A Tenth-Century Survey of Muslim Culture,* edited and translated by Bayard Dodge. 2 Vols. New York: Columbia University Press, p. 658.

Rosenfeld, B. A. and Ekmeleddin Ihsanoğlu (2003). *Mathematicians, Astronomers, and Other Scholars of Islamic Civilization and Their Works (7th–19th c.).* Istanbul: IRCICA, p. 92.

Saidan, Ahmad (1977). "Kitāb tasṭīḥ al-ṣuwar wa-tabṭīḥ al-kuwar li-Abī al-Rayḥān al-Bīrūnī." *Dirāsāt majalla ˁilmiyya* (University of Jordan) 4: 7–22.

Sezgin, Fuat (1974). *Geschichte des arabischen Schrifttums.* Vol. 5, *Mathematik*: 254; Vol. 6, *Astronomie* (1978): 161; Vol. 7, *Astrologie – Meteorologie und Verwandtes* (1979): 137. Leiden: E. J. Brill.

Schjellerup, H. C. F. C (trans.) (1874). *Description des étoiles fixes.* Saint Petersburg.

Steinschneider, M. (1871). "Intorno ad alcuni passi d'opere del medio evo relativi alla calamita." *Bullettino di bibliografia e di storia delle scienze matematiche e fisiche* 4: 257–298.

V

Väisälä, Yrjö

Born **Kontiolahti, Utra, (Finland), 6 September 1891**
Died **Rymättylä, Finland, 21 July 1971**

Yrjö Väisälä was influential in practical astronomy and the related fields of optics and geodesy. He developed several optical designs including a reflecting telescope with a spherical-aberration correction plate that he designed independently of **Bernhard Schmidt**. Väisälä discovered three comets and, with his students, over 800 asteroids. He was also a founding member of the amateur astronomers' association URSA, intended to promote astronomical education.

Väisälä's parents, Johannes and Emma Birgitta Veisell, had seven children, three of whom became well-known scientists. Yrjö's brother Kalle was a mathematician; another brother Vilho was a meteorologist and the founder of the Vaisala Company, which continues to manufacture scientific instruments.

Yrjö Väisälä obtained his Ph.D. in physics (optics) in Helsinki in 1922. After working briefly in the Geodetic Institute, he became the professor of physics at Turku University in 1922 and remained there for his entire career, serving as professor of astronomy from 1928 until his retirement in 1965.

Väisälä was the founder of the Turku University Observatory. Together with his students, Väisälä found six new comets, three of which bear his name, and over 800 asteroids. After his death, one of the asteroids was later identified by his student Liisi Oterma as a comet and is now known as comet 139/P Väisälä/Oterma. A record of 141 new asteroid discoveries was obtained in 1942 when the disturbing city lights had to be dimmed due to the war.

To facilitate his work Väisälä developed new methods. In his double-point method a photographic plate is exposed twice with an interval of approximately half an hour, while pointing the telescope in slightly different directions. On such a plate all stars appear as two dots close to each other while the dots produced by moving asteroids are more widely separated.

Before modern computers, the work of orbit calculation from observational data was very laborious. For this work Väisälä developed a new method, which was much faster to use than the earlier methods.

Väisälä also founded the Tuorla Research Center outside Turku. In Tuorla he built an optical laboratory, which has been developed further and is currently used for designing and making optics for telescopes and satellites. It is unfortunate that Väisälä published very little, and even many of his published results appeared in domestic series only. Väisälä's handwritten notes from 1924 indicate that he understood the principles of several modern telescope designs, Ritchey-Chrétien, Maksutov, and Schmidt camera, several years before they were "officially" invented. However, he thought these inventions were too simple for publishing. In an ordinary Schmidt camera, which is a telescope used for photographing large areas of the sky, the focal surface is not plane but curved. Väisälä built several Schmidt cameras of his own design with an additional lens to flatten the focal surface; such telescopes are sometimes called Schmidt–Väisälä telescopes. He also experimented with a multiple mirror telescope in which six spherical mirrors were grouped in a circle around a seventh mirror. Light was reflected to a common focus and spherical aberration corrected with a Schmidt-like corrector lens. In concept, the telescope was a precursor to the larger multiple mirror telescopes constructed several decades later.

Väisälä made several other inventions in the fields of physics, geodesy, and astronomy. His interferometer has been used to accurately measure lengths of baselines for geodetic surveys in several countries. He also found a method to make standard quartz meters of exactly the same length.

In 1913 Väisälä married Martta Johanna Levanto; they had four children: Marja, Aune, Veikko, and Vuokko. In 1951 he was elected to the Finnish Academy.

Hannu Karttunen

Selected References

Jacchia, Luigi (1978). "Forefathers of the MMT." *Sky & Telescope* 55, no. 2: 100–102.

Niemi, A. (ed.) (1991). *Yrjö Väisälä – Tuorlan taikuri* (Yrjö Väisälä – The magician of Tuorla). Fysiikan kustannus Oy. (A collection of articles in Finnish by relatives, coworkers, and students. Although not a biography, this is the only work containing detailed information on Väisälä's life.)

Oja, Heikki and Tapio Markkanen (1984). "A Celebration of Finnish Astronomy." *Sky & Telescope* 68, no. 6: 503–505.

Väisälä, Yrjö (1936). "Über Spiegelteleskope mit großem Gesichtsfeld." (On mirror telescopes with a large field of view.) *Astronomische Nachrichten* 259: 197–204. (Description of the Schmidt–Väisälä telescope.)

——— "Beobachtungen von Kleinen Planeten." (Observations of minor planets.) *Astronomische Nachrichten* 267: 115–118; 269 (1939): 176–179; 271 (1940): 35–40; 272 (1942): 231–235; 274 (1943): 41–47. (Notes on new asteroids found at the observatory of the University of Turku.)

——— (1939). "Eine Einfache Methode der Bahnbestimmung." (A simple method for orbit determination.) *Mitteilung der Sternwarte der Universität Turku*, no. 1. Annales Academiae Scientiarum Fennicae, ser A., 52, no. 2. Helsinki. (This article explains the fast method for determining the orbit from the double-point observations.)

Van Albada, Gale Bruno

Born **Amsterdam, the Netherlands, 28 March 1911**
Died **Amsterdam, the Netherlands, 18 December 1972**

Gale van Albada was well known for his work on the evolution of clusters of galaxies, the theory of formation of stellar associations and double stars, and for observations and orbit determinations of binary stellar systems.

Van Albada earned a Ph.D. in astrophysics at the University of Amsterdam as a student of **Herman Zanstra** and **Antonie Pannekoek** in 1945. His dissertation was a study of line intensities in stellar spectra. After a postdoctoral fellowship at the Warner and Swasey Observatories of Case Institute, Cleveland, Ohio, during which he worked on near-infrared spectra of late type stars with Jason John Nassau, van Albada was appointed director of the Bosscha Observatory at Lembang, Java, Netherlands East Indies. He arrived at the height of the revolutions that resulted in the independence of Indonesia in December 1949.

Van Albada revived the Bosscha Observatory during his 10 years there. Acquisition of a Schmidt telescope with an objective prism for the Bosscha Observatory permitted van Albada and his students to secure photometric and spectroscopic observations of many variable stars. Van Albada and his students also conducted an active program of photographic observations of double stars using the observatory's 23.6-in. refractor. From 1951 onwards, van Albada promoted astronomy very efficiently through his teaching at Bandung University. By the time he left the country, a new generation of astronomers was ready to take over astrophysical research in Indonesia.

In 1959, Leiden University welcomed van Albada home to the Netherlands for a short time until he was appointed director of research at the University of Amsterdam after the retirement of Zanstra. At the time of his death, van Albada was the director of the Astronomical Institute of the University of Amsterdam, where he also held a chair of professor of astronomy. Van Albada was married to Elsa van Dien, who was also well-known as an astronomer.

Léo Houziaux

Selected References

Anon. (1973). "Dutch Stellar Astronomer." *Sky & Telescope* 45, no. 3: 160.

Oort, J. H. (1973). "In memoriam Prof. Dr G. van Albada, 28 March 1911–18 December 1972." *Hemel en Dampkring* 71: 47–48.

van Albada, G. B. (1958). "Photographic Measures of Double Stars from Plates Obtained with the 60 cm Refractor." *Annals of the Bosscha Observatory Lembang* 9, pt. 2.

——— (1962). "Distribution of Galaxies in Space." In *Problems of Extra-galactic Research: I.A.U. Symposium No. 15, August 10–12, 1961*, edited by G. C. McVittie, pp. 411–428. New York: Macmillan.

——— (1962). "Gravitational Evolution of Clusters of Galaxies, with Consideration of the Complete Velocity Distribution." *Bulletin of the Astronomical Institutes of the Netherlands* 16: 172–177.

——— (1963). "Simple Expressions for Observable Quantities in Some World Models." *Bulletin of the Astronomical Institutes of the Netherlands* 17: 127–131.

Van Allen, James Alfred

Born **Mount Pleasant, Iowa, USA, 7 September 1914**
Died **Iowa City, Iowa, USA, 9 August 2006**

American space scientist James Van Allen is sometimes called the father of space science, because he developed and provided instruments for the first US satellites that resulted in the discovery

of radiation zones encircling the Earth, subsequently named Van Allen belts.

Van Allen was the son of Alfred Morns and Alma Olney Van Allen, and was educated at Iowa Wesleyan College (BS in physics, in 1935, working with Thomas Poulter in physics and Delbert Wobbe in chemistry) and at the University of Iowa (MS: 1936; Ph.D. in physics: 1939). He married Abigail Fithian Halsey in 1945 and had five children.

After receiving his doctorate, Van Allen moved to the city of Washington, in September of 1939 under a Carnegie Fellowship, to work in the Department of Terrestrial Magnetism [DTM] of the Carnegie Institution of Washington, then directed by Merle Tuve. After the outbreak of World War II, the DTM became involved in fission research following the announcement of deuteron-induced fission of the uranium nucleus. Initially, Van Allen worked with the DTM Van de Graff accelerator studying the photodissociation of the deuteron. Soon, however, he became involved in the prewar effort to improve the technical capabilities of the United States navy, specifically through the development of a "proximity fuse" that would explode upon approaching a target. By the spring of 1942, the fuse work was transferred to the Applied Physics Laboratory [APL], which was established under the auspices of Johns Hopkins University to develop the concept and introduce it to the navy. By November 1942, Van Allen was commissioned as a lieutenant in the United States Naval Reserve and sailed to the Pacific with a supply of proximity fuses for use by the navy's antiaircraft guns.

As it turned out, building rugged electronics to be shot out of a gun with enormous force was excellent preparation for a number of subsequent postwar tasks undertaken by Van Allen and his team. At the end of the war, they performed high-altitude experiments with captured German V-2 rockets to measure cosmic radiation and solar ultraviolet spectra and to take pictures of Earth's curved horizon, among other things.

At the end of 1950, Van Allen moved to Iowa to become professor and head of the Department of Physics at the university where he had obtained his Ph.D. There, he continued his high-altitude research, but now with the domestic Aerobee rocket that he helped develop while at APL. This set the foundations for involvement in the forthcoming entry of the United States into the space era. The V-2 Rocket Panel, set up after the war to oversee high-altitude research, evolved into the Upper Atmosphere Rocket Research Panel and later into the Rocket and Satellite Research Panel with Van Allen as a principal driver and eventually as chair. This panel organized US participation in the 1957/1958 International Geophysical Year [IGY], and panel members actively promoted the use of scientific satellites around Earth as part of the IGY and spearheaded the science base of the National Aeronautics and Space Administration [NASA], which also was established in 1958.

The implications of a brief stay at Princeton University during this period are mentioned in the article on **Martin Schwarzschild**.

By 1956, Van Allen had already submitted a formal proposal to the IGY for an instrument to be included in the payload of an Earth satellite. The objective would be to map the global distribution of cosmic rays, a goal that he and his collaborators had been pursuing with balloons and rockets for more than a decade. The proposed instrument was included in the payload of the first flight of the Army's Jupiter C rocket, launched on 31 January 1958. This first US satellite, named Explorer I, worked well, and the Iowa Geiger–Mueller tube's high counting rates turned out not to be at all consistent with the modest intensities expected from high-altitude cosmic rays. These measurements were later combined with those of Explorer III, which carried a tape recorder and was capable of providing global coverage. The two data sets resulted in the discovery of the radiation belts that bear Van Allen's name, huge regions of space populated mostly by protons and electrons trapped in the Earth's magnetic field with energies extending to millions of electron volts for the electrons and hundreds of millions for the protons (*i. e.*, sufficient energies to penetrate several inches of steel). The public announcement was made at the National Academy of Sciences on 1 May 1958. Significantly, confirmation of this key discovery by Van Allen and colleagues was provided later that month by the launch of Sputnik III. The Iowa instrument "factory" continued to provide radiation detectors for subsequent satellites (Explorer IV and Pioneers I and III, all in 1958) that provided a large body of data establishing the radial dimensions of the region containing trapped radiation, subsequently named the "magnetosphere."

By late 1958, NASA had taken over most space research in the United States. Under NASA support, Iowa instruments and entire satellites were built and flown, with some data reception through makeshift antennas near Iowa City. Seminal discoveries were made that illuminated plasma processes in every region of Earth's magnetosphere.

In the meantime there was incessant curiosity about potential magnetospheres around other planets and the properties of the interplanetary medium in between. The National Academy of Sciences had now established the Space Science Board, chaired by Lloyd Berkner, with Van Allen as an influential member. Soon, plans were made for missions to Venus and Mars, and Van Allen's radiation detectors were included in both payloads.

Mariner II passed by Venus on 14 December 1962 but did not detect any radiation, suggesting the absence of an extended, Earth-like, magnetosphere. The plasma detector on Mariner II, however, did establish firmly the presence of the solar wind, a continuous stream of plasma flowing radially outward from the Sun, and delineated its properties for the first time. Similarly, Mariner IV passed by Mars on 15 July 1965 and established an upper limit to the Martian magnetic moment as 0.001 of the Earth. Nevertheless, the Mariner IV mission made several discoveries en route, including electron emissions in solar flares and an 8-month record of solar x-ray flare activity.

This "null" result in finding magnetospheres at Earth's neighbors only accelerated the search for such regions in the outer planets, the prime candidate being Jupiter, a known radio emitter since the 1950s. Van Allen's Lunar and Planetary Missions Board soon recommended that NASA proceed with two spacecraft to investigate the region around Jupiter. Pioneers 10 and 11 (launched in 1972 and 1973, respectively) carried a set of Iowa's Geiger–Müller radiation detectors along with other instruments. They encountered Jupiter in 1973 and 1974 and established the presence of a huge magnetosphere with high-energy/high-intensity radiation belts and a magnetic field easily 10 times that of Earth. The Pioneer 11 trajectory was deflected using Jupiter's gravity to direct the spacecraft toward an encounter with Saturn in 1979, where it measured the latter planet's magnetic field and magnetosphere.

Even before the two Pioneers were launched, the Van Allen Board had recommended that NASA take advantage of a rare alignment of the outer planets to plan and launch what eventually became the Voyager 1 and 2 missions. These spacecraft were to perform comprehensive investigations not only of Jupiter and Saturn, but also Uranus and Neptune. The Voyager missions became a spectacular success and "rewrote the book" on humanity's knowledge of the outer planets.

Van Allen's fascination with Jupiter did not stop with the Voyager missions. He chaired a working group in the mid-1970s that recommended a mission to orbit Jupiter and drop a probe into its atmosphere. This mission resulted in the Galileo spacecraft, launched in 1989 and orbiting Jupiter in 1995.

In addition to his purely scientific contributions, Van Allen was a strong exponent of the view that most space research can be accomplished by robotic spacecraft and automated equipment, while human crews are of limited utility and are only needed for laboratory-type experiments in low gravity conditions. In this vein, he was an ardent opponent of manned spaceflight, particularly the space shuttle and space station, from the inception of each program.

In addition to his research work, Van Allen was an educator at both the undergraduate and graduate level, a leader in science policy-making at the national level, and an eloquent spokesman on scientific issues at various public fora.

Van Allen's legacy includes the supervision of 34 Ph.D. dissertations and 45 MS theses, some 260 published papers between 1967 and 2002, and a number of books and public and technical lectures throughout the world. He was the recipient of 15 honorary degrees, the United States National Medal of Science (1987), the Crafoord Prize of the Royal Swedish Academy of Science (1989), and a number of other honors and awards.

Stamatios Kimigis

Selected References

Van Allen, James A. (1983). *Origin of Magnetosphere Physics.* Washington, DC: Smithsonian Institution Press.

——— (1990). "What is a Space Scientist? An Autobiographical Example." *Annual Review of Earth and Planetary Sciences* 18: 1–26.

——— (1951–1987). Records of James Van Allen. Main Library Archives, University of Iowa, Iowa City, Iowa.

Van Biesbroeck, Georges-Achille

Born **Ghent, Belgium, 21 January 1880**
Died **Tucson, Arizona, USA, 23 February 1974**

Georges Van Biesbroeck, the leading observer of close double stars in America during his lifetime, was renowned for his long hours at the telescope spanning nearly 70 years. He retired as a University of Chicago faculty member at age 65, only to begin a second career in astronomy that ended with his death.

Though his father was an artist, Van Biesbroeck himself studied civil engineering at the University of Ghent. Indeed, he was employed for 6 years (1902–1908) by the Brussels Department of Roads and Bridges. Even so, Van Biesbroeck was fascinated by astronomy from an early age; he studied the subject in college. He volunteered as an observer at the Royal Observatory (Uccle, near Brussels) and spent time at the Royal Observatory (Greenwich) and various observatories in Germany (including Heidelberg, working with **Maximilian Wolf**, and Potsdam, working with **Johannes Hartmann** and **Karl Müller**), before returning to Uccle as an adjunct astronomer in 1908. It was from Belgium that he observed an annular eclipse in 1912, presumably kindling his later interest in eclipses, which he observed whenever possible.

In June 1914 Van Biesbroeck, still an adjunct astronomer at Uccle, was contacted by **Edwin Frost**, director of the Yerkes Observatory. Frost sought an expression of interest from Van Biesbroeck regarding a potential opening for an astronomer with micrometer experience and an interest in double-star work. Faced with the certain need to replace **Sherwin Burnham** when he retired, Frost was also concerned about the eventual retirement of **Edward Barnard**, already ill with late-onset diabetes. Van Biesbroeck accepted Frost's offer and came to Yerkes as a visiting professor to work with the 40-in. refractor on a temporary assignment. He arrived in Williams Bay, Wisconsin, USA in late August 1915. Van Biesbroeck spent 10 months at Yerkes during which he performed a variety of tasks to Frost's satisfaction and appeared to fit in well with the Yerkes staff, but then returned to his family in war-torn Belgium. When Frost offered him a permanent assignment, in spite of the continuing dangers during World War I, Van Biesbroeck took his family on a harrowing emigration through neutral The Netherlands and traveled to the United States by way of Norway, arriving in Williams Bay in 1917.

In addition to his double star observing at Yerkes, Van Biesbroeck routinely measured the positions of comets and asteroids and made observations of variable stars depending on the instruments available for his use. He was also, in effect, the resident engineer at Yerkes and participated in the design and construction of many new instruments. When the University of Chicago and the University of Texas agreed to jointly develop and operate the McDonald Observatory near Fort Davis, Texas, in the 1930s, Van Biesbroeck tested and approved many of the optical systems. He also designed and constructed a complete mounting for a 20-in. Schmidt telescope for which the optics were never completed.

Though he became professor emeritus at the University of Chicago in 1945, Van Biesbroeck's observations at the Yerkes and MacDonald observatories continued without interruption. At an age when many men stay close to home, in 1947, 1948, and 1952 he traveled to Brazil, Korea, and Sudan, respectively, to explore the relativistic deflection of starlight during eclipses of the Sun. In between these years (1949/1950), Van Biesbroeck undertook a 6-month mission for the Belgian government to conduct an observatory site survey in the remote Congo.

In 1963, the Yerkes Observatory director advised Van Biesbroeck that, in consideration of his age and the perceived dangers that using the telescope entailed, he could no longer use the 40-in. refractor. Incensed at this denial of his continued capability, Van Biesbroeck accepted an offer from **Gerard Kuiper** to join the staff of the University of Arizona's Lunar and Planetary Laboratory, where he remained active as an astronomer until the end of his life. He traveled to Perry, Florida, at the age of 90 to observe the solar eclipse of 7 March 1970 but was clouded out. It was his last attempt to observe an eclipse.

All told, Van Biesbroeck amassed a body of approximately 600 publications. These covered his observations and computed orbits for double stars, asteroids, comets, and planetary satellites. His orbit of Nereid resulted in the determination of a more accurate mass for Neptune. He was the discoverer of 11 asteroids and three comets (C/1925 W1, C/1935 Q1, and 53P/1954 R1). Van Biesbroeck was credited with the discovery (in 1944) of the least luminous star recognized at the time: a companion of BD +04° 4048 (Van Biesbroeck's star).

Like so many of his contemporaries, Van Biesbroeck's life in astronomy was a quest for ever-increasing aperture. He started out as an observer of close binary stars with a 15-in. refractor at Uccle. At Yerkes he made use of the 40-in. refractor, or the 24-in. reflector if the former was unavailable, continuing in the tradition of Burnham and Barnard. In fact, it was Van Biesbroeck and Barnard's niece Mary Calvert who edited Barnard's unpublished work after his death. At MacDonald, Van Biesbroeck used the 82-in. reflector which he helped design. He continued to receive observing time on large (for the time) telescopes into his 10th decade. In all, Van Biesbroeck made thousands of double star measurements including those of stars he had discovered.

In his lifetime Van Biesbroeck was awarded the Gold Medal of the Royal Danish Society of Sciences, the Valz Prize of the Paris Academy of Science, the Donohoe Comet Medal of the Astronomical Society of the Pacific (twice), the Franklin L. Burr prize of the National Geographic Society, honorary membership in the Royal Astronomical Society of Canada (one of only 15), and an honorary doctorate from the University of Brussels (1935). He was a fellow of the Royal Astronomical Society. Minor planet (1781) was named for Van Biesbroeck by the International Astronomical Union.

Van Biesbroeck's wife Julia died before him. He was survived by two daughters and a son.

Thomas Hockey and *Thomas R. Williams*

Selected References

Hardie, Robert H. (1974). "Georges van Biesbroeck, 1870–1974." *Journal of the Royal Astronomical Society of Canada* 68: 202–204.

Heard, J. F. (1974). "Some Recollections of Mr. Van B." *National Newsletter* (Toronto, Canada): L23–L24.

Roemer, Elizabeth (1974). "George A. Van Biesbroeck (1880–1974)." *Icarus* 23: 134–135.

Van Biesbroeck, George (1960). *Measurements of Double Stars.* Chicago: University of Chicago. (*Publications of the Yerkes Observatory*, the last of four catalogues of double star measures published by Yerkes between 1927 and 1960, totaling some 670 pages of data.)

Worley, Charles E. (1974). "The Prince of Observers." *Sky & Telescope* 47, no. 6: 370–372.

Van de Kamp, Peter [Piet]

Born **Kampen, the Netherlands, 26 December 1901**
Died **Amsterdam, the Netherlands, 18 May 1995**

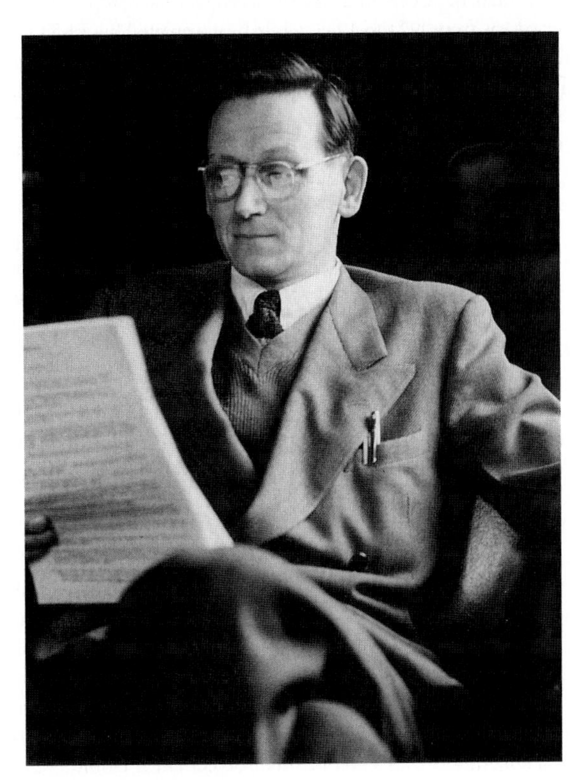

Dutch–American astrometrist Peter van de Kamp is remembered for an extended series of searches for low-mass companions to nearby stars, which greatly increased our knowledge of binary systems, but found none of the extrasolar-system planets he had hoped for. He was the son of Lubbertag and Egelina (*nee* Van der Wal) van de Kamp. At an early stage Piet, or Peter as he became known in the United States, was exposed to music through his father who played the organ in a local church and had a piano at home. In fact, when Peter enrolled at the University of Utrecht, the Netherlands, working with A. A. Nijland, in 1918, he had to choose between music and astronomy. He decided the latter would lead to a better career. Van de Kamp graduated from the University of Utrecht in 1920 and received a doctorate in 1922. After working as an assistant at the Kapteyn Astronomical Laboratory, Groningen, from 1922 to 1923, he came to the United States, becoming a naturalized citizen in 1942.

In the 1920s and 1930s, Samuel A. Mitchell brought together a group of astronomers of varied interests at the University of Virginia, Charlottesville. Besides van de Kamp, who arrived in 1923, there were A. N. Vyssotsky, a specialist in stellar motions; Emma T. R. Williams, with interests in the same field, who would marry Vyssotsky; **Harold Alden**, interested in stellar parallaxes and a future director of Virginia's Leander McCormick Observatory; **Rupert Wildt**, an astrophysicist who would move to Yale; plus M. Kovalenko and D. Reuyl. Many of these people formed the nucleus of the Observatory Mountain Orchestra with van de Kamp as conductor. From such a beginning the Charlottesville Symphony Orchestra grew.

Van de Kamp served as an associate astronomer at the University of Virginia's Leander McCormick Observatory, from 1923 to 1924; then he accepted a Martin Kellogg Fellowship at the Lick Observatory, from 1924 to 1925, while he completed his Ph.D. at the University of California at Berkeley. He was awarded a similar degree, from Groningen University, the following year, for a thesis on systematic errors in the proper motions of stars in the Boss catalog, carried out under **Pieter van Rhijn**. Van de Kamp appears to have been the first Dutch astronomer to earn a Ph.D. in the United States.

Van de Kamp returned to Virginia as an instructor in 1925 and was promoted to assistant professor in 1928. He held this post until in 1937 he was invited to Swarthmore College, Swarthmore, Pennsylvania, as an associate professor and director of the Sproul Observatory. He was promoted to professor in 1940, a rank Van de Kamp held until his retirement in 1972 when he was appointed research professor.

Van de Kamp was an able and popular teacher. He occupied a number of visiting professorships at Harvard University (1936), at Wesleyan University (1956), and at the New School for Social Research (1944–1962). In addition he was a Fulbright Professor at the University of Paris (1949), and at Amsterdam University (1972). Van de Kamp was awarded the President and Visitor's Prize of the University of Virginia for 3 years (1927, 1937, and 1938). In addition he won a Glover Award from Dickinson College (1961), Swarthmore's Nason Award (1963), and the Gold Medal of Philadelphia's Rittenhouse Society (1965).

An early concern of van de Kamp was the effect of interstellar absorption on distances in various directions within the Galaxy. Such absorption causes a star or other object like a globular cluster to appear fainter than it actually is and thus at a greater distance. In applying his results to an estimate of the distance to the galactic center, van de Kamp lowered **Harlow Shapley**'s original estimate by some 4,000 parsecs (about 13,000 light years) to 12,000 parsecs.

During his tenure at Swarthmore, van de Kamp became recognized internationally for his research in astrometry with long-focus telescopes such as the 60-cm refractor of Sproul Observatory. Here he expanded his research program to include a study of nearby stars. One of the goals of this project was to discover low-mass companions and, potentially, to find out if other stars were accompanied by planets. At that time, a half century ago, the question of the existence of other planetary systems was not the popular question that it is today, nor was the technology to pursue this question as refined as it is now. Experience had shown that "unseen" companions of visible stars could be detected from their mutual gravitational attraction, which would cause a wobble in the visible object's track across the plane of the sky; its proper motion, the change in its angular position over a period of time, would not be a straight line. Companions to Sirius and Procyon had been detected in this manner by **Friedrich Bessel**.

In 1917 the American astronomer **Edward Barnard** found that a faint, nearby red star in Ophiuchus had a very large proper motion – in fact the largest ever measured. This object was from that time on called Barnard's Star. From the analysis of data accumulated from 1916 through 1962 at the Sproul Observatory, van de Kamp, in 1963, announced that observed deviations of this star's motion could be attributed to a planet moving about it in a period of 24 years. His final analysis (1982) indicated that Barnard's star had planets with masses 0.7 and 0.5 times that of Jupiter in orbits with periods of 12 and 20 years, respectively. However, by then, other experts had become skeptical, blaming the results on instrumental effects. Most recently a study of images of Barnard's star compiled at the Leander McCormick Observatory from 1969 to 1998 showed no sign of planets.

The research program that van de Kamp initiated has led to the discovery of a number of low-mass stellar companions and has increased our knowledge of stellar duplicity in the Galaxy. From his studies of known binary stars, he contributed additional data on stellar masses, one of the basic parameters of the physical Universe.

In 1954 van de Kamp took a leave of absence from Swarthmore for the academic year to become the first program director for astronomy in the National Science Foundation, a position he held through the summer of 1955. One of his innovations was to solicit research proposals from all members of the American Astronomical Society. He also became involved with a panel, chaired by **Robert McMath**, that would lead to the eventual creation of the Association of Universities for Research in Astronomy [AURA], Incorporated, and the Kitt Peak National Observatory. Subsequently van de Kamp became a director-at-large of the AURA board. For his research on binary stars, van de Kamp was named president of Commission on Double Stars of the International Astronomical Union, 1958–1964. He also became a member of the United States National Committee.

Van de Kamp was a member of several other organizations including the American Astronomical Society, the Astronomical Society of the Pacific, Sigma Xi, Phi Beta Kappa, and the Royal Dutch Academy. He married Olga Ptrushoka in 1947. Toward the end of World War II he became a member of the Alsos Mission, a group charged by the Office of Scientific Research and Development with studying the progress of science and technology – notably atomic bombs, radar, and guided missiles – in Germany under the Nazis.

Throughout his life van de Kamp felt that astronomy was a marvelous synthesis of art and science. He played the viola, violin, and piano. In addition to the amateur orchestra that he had directed at the University of Virginia he became the conductor of the Swarthmore College Orchestra. He loved to play ragtime music and was fond of the blues. Among his own compositions was *Blackout Blues*, to commemorate the great northeastern blackout of November 1965.

An oral-history interview with van de Kamp may be found at the Niels Bohr Library of the American Institute of Physics in College Park, Maryland. His research correspondence along with other papers have been deposited at the Library of the United States Naval Observatory in Washington.

George S. Mumford

Selected References

Fredrick, Laurence W. (1996). "Peter van de Kamp (1901–1995)." *Publications of the Astronomical Society of the Pacific* 108: 556–559.

Lippincott, Sarah Lee (1995). "Peter van de Kamp, 1901–1995." *Bulletin of the American Astronomical Society* 27: 1483–1484.

van de Kamp, Peter (1962). "Astrometry with Long-Focus Telescopes." In *Astronomical Techniques*, edited by W. A. Hiltner, pp. 487–536. Vol. 2 of *Stars and Stellar Systems*. Chicago: University of Chicago Press.

——— (1967). *Principles of Astrometry, with Special Emphasis on Long-Focus Photographic Astronomy*. San Francisco: W. H. Freeman.

——— (1982). *Stellar Paths: Photographic Astrometry with Long-Focus Instruments*. Dordrecht: D. Reidel.

——— (1986). *Dark Companions of Stars: Astrometric Commentary on the Lower End of the Main Sequence*. Dordrecht: D. Reidel. (Reprinted from *Space Science Reviews* 43, nos. 3/4 (1986): 211–327.)

Van de Sande Bakhuyzen [Bakhuysen], Hendrik Gerard [Hendrikus Gerardus]

Born **The Hague, the Netherlands, 2 April 1838**
Died **Leiden, the Netherlands, 8 January 1923**

Dutch astronomer Hendrikus van de Sande Bakhuyzen, primarily known for precise measurements of positions of stars, was the son of a successful landscape painter, also named Hendricus. After attending the local gymnasium (a classical secondary school), Bakhuyzen began his studies in 1855 at the Delft Polytechnic, where in 1859 he took a degree in civil engineering. He subsequently matriculated at the University of Leiden, where he came under the powerful influence of the astronomer **Frederik Kaiser**. In 1863 Bakhuyzen obtained his doctorate with a dissertation on the flexure of the recently acquired meridian circle at the Leiden Observatory.

For a year Bakhuyzen worked at the new, but poorly staffed, Leiden Observatory as an unpaid observer. When his prospects did not improve, he decided to accept a post as teacher, first at the gymnasium in The Hague, then, 2 years later, at the newly instituted Hogere Burgerschool (technical secondary school) in Utrecht. In 1867 he was appointed professor of applied physics at Delft University. Two years later he married Geertruida van Vollenhoven of Rotterdam. They had a son and two daughters. When Kaiser died in 1872, Bakhuyzen succeeded him at the University of Leiden as professor of astronomy and director of the observatory. He held this position until his retirement in 1908. During this entire period he was assisted by his brother Ernst Frederik (1848–1918), who held the position of first observer. Ernst would eventually succeed Hendrick as director of the observatory.

Bakhuyzen continued the tradition established by Kaiser: the exact determination of stellar positions. The principal instrument in the observatory was the Pistor and Martins meridian circle, and up to 1919 meridian astronomy remained the main domain of activity at the observatory. The program included observations of 84 circumpolar stars and, subsequently, of 303 southern stars, all selected by the Astronomische Gesellschaft as reference stars. These measurements were intended as a contribution to **Arthur von Auwers**' Fundamental Catalog for the zones of the Astronomische Gesellschaft. Unfortunately, Bakhuyzen's high demands on the reduction of the data, and the small number of (human) computers at the observatory, delayed the publication of the observations, thereby preventing their inclusion in the Catalog.

All of Bakhuyzen's research testifies to his penchant for precision. In the spirit of his predecessor he meticulously investigated all possible instrumental and personal errors. In 1879, he was the first to provide solid evidence for the influence of the magnitude of stars on the personal equation. To such ends he had designed an apparatus to determine the personal equation by the observation of moving artificial stars. To improve precision in the observation of stellar occultations by the Moon, he also investigated the effect of the brightness of a point on the time of perception of its sudden appearance and disappearance (a somewhat different phenomenon). (The personal equation describes the systematic differences in the positions of the same stars determined by different observers using the same equipment, it arises from systematic differences in estimating the time at which a star crosses a hair-like mark in the field of the telescope.)

Atmospheric refraction was a favorite subject of the Bakhuyzen brothers. Between 1879 and 1898 the Leiden Observatory determined the zenith distances of two zones of stars, one near the zenith of the Cape of Good Hope, the other near the zenith of Leiden. These stars were also observed at the Cape, and comparison of the conclusions resulted in a correction to the adopted value of the constant of refraction. In 1907 Bakhuyzen published his research on the temperature distribution in the atmosphere, based on measurements made during balloon ascents, and the resulting effect on atmospheric refraction.

Other investigations that testify to his predilection for precision include Bakhuyzen's redetermination in 1885 of the rotation period of the planet Mars. Using all previous observations, including those by **Christiaan Huygens**, **Johann Schröter**, and Kaiser, Bakhuyzen derived the rotation period with a probable error of about one part per million.

Much of Bakhuyzen's research concerned geodetical questions. He served as president of the Dutch National Geodetic Committee from 1882 onwards, and as *secrétaire perpétuel* of the International Geodetic Association from 1900. In the former capacity he supervised the precision leveling, as well as the triangulation, of the Netherlands. The committee also conducted tidal research in Dutch ports. In this connection Bakhuyzen investigated the change of sea level at Amsterdam and Den Helder from 1700 to 1910. He used the results to discuss the variation of latitude as well as the secular motion of the soil. Bakhuyzen's "compensation" of all European determinations of longitude, published in 1894, became the main source for the longitudes of European observatories as given in the national ephemerides.

Bakhuyzen participated in several national and international committees. From 1888 to 1910 he served as president of the Royal Dutch Academy of Sciences. He represented the academy at the triennial meetings of the International Association of Academies from 1904 to 1913. Bakhuyzen was a member of the (German) Astronomische Gesellschaft, serving as its vice president from 1889 to 1896. He was a correspondent of the Institut de France and a foreign member of the Royal Astronomical Society, the Italian Academia dei Lincei, and the Royal Belgian Academy. Bakhuyzen also took a prominent share in the international conferences of the *Carte du Ciel* Project.

Frans van Lunteren

Selected References

de Sitter, W. (1924). "Hendrik Gerard van de Sande Bakhuyzen." *Monthly Notices of the Royal Astronomical Society* 84: 226–230.

——— (1933). *Short History of the Observatory of the University at Leiden, 1633–1933*. Haarlem: J. Enschede en zonen.

van de Sande Bakhuyzen, H. G. (1879). "Beobachtungen zur Bestimmung der Änderung des persönlichen Durchgangsfehlers mit der Helligkeit der Sterne, angestellt in Leiden." *Vierteljahrsschrift der Astronomischen Gesellschaft* 14: 408–414.

——— (1879). "Apparat zur Bestimmung des absoluten persönlichen Fehlers bei Durchgangsbeobachtungen an Meridian-Instrumenten." *Vierteljahrsschrift der astronomischen Gesellschaft* 14: 414–418.

——— (1885). "The Rotation-Period of Mars." *Nature* 33: 153.

——— (1889). 'Über die Bestimmung der persönlichen Gleichung bei Passagebeobachtungen." *Vierteljahrsschrift der Astronomischen Gesellschaft* 24: 249–250.

——— (1893). "Résultats d'une compensation du réseau des longitudes déterminées depuis 1860 en Europe, en Algérie et en quelques stations Asie." *Astronomische Nachrichten* 134: 153–160.

——— (1908). "Réfraction astronomique d'après une distribution de la temperature atmospherique déduite de sondages en ballons." *Archives Néerlandaises des sciences exactes et naturelles* 14: 342–355.

——— (1913). "Die Änderung der Mehreshöhe und ihre Beziehung zur Pohlhöhenschwankung." *Vierteljahrsschrift der Astronomischen Gesellschaft* 48: 218–221.

van Herk, G. and H. Kleibrink (1983). *De Leidse Sterrewacht*. Leiden.

Went, F. A. F. C. (1923). "Hendricus Gerardus van de Sande Bakhuyzen." *Verslag van de gewone vergadering der wisen natuurkundige afdeeling der Koninklijke Akademie van Wetenschappen te Amsterdam* 32: 4–7.

Van den Bos, Willem Hendrik

Born **Rotterdam, the Netherlands, 25 September 1896**
Died **Johannesburg, South Africa, 30 March 1974**

During his career of over 30 years in South Africa, Willem van den Bos discovered approximately 3,000 new visual double stars and likewise computed the orbits of almost 100 double stars, carrying the work well into his retirement until forced to stop by declining health.

Van den Bos entered the University of Leiden in 1913, but his studies were interrupted by war. He rose to the rank of lieutenant in the Netherlands Coast Artillery. Afterwards, Van den Bos resumed his studies at a time when such illustrious personalities as **Albert Einstein**, Paul Ehrenfest, **Hendrik Lorentz**, **Willem de Sitter**, and **Ejnar Hertzsprung** were faculty members. In 1921, he became a member of the Leiden Observatory staff. His 1925 doctoral dissertation was on the subject of binary stars. **Robert Innes**, director of the Union (later Republic) Observatory, Johannesburg, South Africa, offered Van den Bos a short-term research position to utilize the 26.5-in. refractor, construction of which was to be completed in 1927. After 3 years, Van den Bos was appointed chief assistant by Innes's successor, H. E. Wood, and eventually succeeded him as director in 1941, retaining this position until his retirement in 1956.

Van den Bos compiled a card catalog of southern double stars that was embodied in one volume of the *Lick Observatory Publications* Series. This comprehensive account presented measurements of more than 64,000 binary stars. Van den Bos himself supplied that portion of the catalog listing binaries south of declination –19°. He later updated the catalog when appointed a visiting astronomer at the Lick Observatory, Mount Hamilton, California. His paper on binary-star orbit computation in the University of Chicago's *Stars and Stellar Systems* (1962) series became a classic reference work.

As an aid to observational astronomers and orbit computers, Van den Bos discussed, in 1958, techniques for obtaining good double-star measurements and listed sound etiquette for reporting such observations. He also discussed the relative merits of refractors versus reflectors for double-star observations. For the presentation of results from orbital calculations, Van den Bos advocated the system laid down by the International Astronomical Union [IAU] in 1935.

In connection with the formation of the South African Astronomical Observatory [SAAO] in the early 1970s, Van den Bos and **William Finsen** expressed considerable reservations toward the plan to relocate the principal observing facilities to Sutherland and to establish the SAAO's headquarters at Cape Town. Loss of the dedicated Union/Republic Observatory refractor was a serious blow to fundamental double star research, in the Southern Hemisphere.

Van den Bos was elected president of IAU Commission 26 (Double Stars) from 1938 to 1952, the longest serving president of this commission. He was the recipient of the Gold Medal of the Royal Danish Academy, and the Gill Medal of the Astronomical Society of Southern Africa.

John McFarland

Selected References

Finsen, W. S. (1974). "Double Star Observer Extraordinary." *Sky & Telescope* 48, no. 1: 24–25.

——— (1974). "Willem Hendrik van den Bos." *Monthly Notes of the Astronomical Society of Southern Africa* 33: 60–61.

Hirst, W. P. (1974). "Dr. W. H. van den Bos – a Tribute." *Monthly Notes of the Astronomical Society of Southern Africa* 33: 61.

Muller, P. and P. Baize (1974). "G. Van Biesbroeck, et W. H. van den Bos." *L'Astronomie* 88: 305, 309–311.

Van den Hove, Maarten

Born **Delft, The Netherlands, 1605**
Died **Leiden, The Netherlands, 17 August 1639**

Martinus Hortensius' significant contributions were related to the dissemination of the Copernican theory and finding the angular diameter of the Sun. He corresponded with **René Descartes**, **Marin Mersenne**, **Pierre Gassendi**, **Christiaan Huygens**, **Nicolas de Peiresc**, **and Wilhelm Schickard**.

Hortensius was named Maarten van den Hove after his mother's father; his father's family name was van Swaanswijk. He was Isaac Beeckman's student at Dordrecht and attended the Latin School of Rotterdam from December 1621 to 1627, after which he left the Dordrecht area. Hortensius then became **Willebrord Snel**'s pupil, but was not enrolled at the university. He later enrolled at Leiden and Ghent to study mathematics and theology from 1628 to 1630.

Hortensius was a self-taught astronomer who at first followed the lead of **Tycho Brahe**. He made several observations of the Sun with Beeckman in Dordrecht. Beeckman introduced him to **Philip Lansbergen** in Middelburg. Hortensius later translated Lansbergen's *Commentations in motum terrae* (1630). In the preface, Hortensius mentioned that Lansbergen had told him of several mistakes made by Brahe; he also discussed **Johannes Kepler**'s views regarding the distance of the Sun. This work of Lansbergen was challenged by Libert Froidmond in Louvain and **Jean Morin** in Paris, while defended by his son **Jacob Lansbergen**. The preface in particular provoked a response from Kepler concerning the measurement of the diameter of the Sun, *Ephemeris ad annum 1624*. Although Kepler had died, Hortensius, on the advice of Lansbergen, gave a public response to Kepler in *Responsio ad additiunculam D. J. Kepleri praefixam Ephemeridi ejus in annum 1625* (1631) and dedicated it to Abraham van der Meer, Councillor in the States-General of Holland.

Hortensius then wrote *Dissertatio de Mercurio in sole viso et venere invisa, instituta cum clarissimo ac doctissimo Viro D. Petro Gassendo* (1633). About this time he was also actively involved with improving the telescope. In March 1634 Hortensius was asked to lecture in mathematics at the newly established Amsterdam Atheneum. In May 1634 he accepted the position with a speech, *De dignitate et utilitale matheseos*. Hortensius later taught a course on the beginnings of astronomy. It was during this time that he translated Willem Blaeu's *Guilielmi Blaeu Institutio astronomica de usu globorum et sphaerorum coelestium ac terrestrium* (1634).

Hortensius was appointed full professor in the Copernican theory in July 1635. He lectured on optics and dedicated these lectures, *De oculo ejusque praestantia* (1635), to the Polish nobleman Hyacinthus de Rozdrazew Rozdrazewsky. Soon afterward Hortensius taught courses in navigation. Throughout this period he traveled several times to Leiden, Delft, and The Hague. In November 1636, he was nominated to a commission that was formed to discuss with **Galileo Galilei** his system of longitudinal determination by observing Jupiter's satellites. On 20 August 1637 1,000 guilders was paid to acquire the necessary instruments to conduct this research. Several letters were exchanged by Hortensius with Hugo de Groot and Elia Diodati as well as with Galilei, who on 5 September 1637, provided information regarding the telescope, the astronomical almanac, and the ephemerides.

In 1639 the University of Leiden appointed Hortensius full professor. He became ill soon after his appointment and died, leaving behind him a son. He was buried at Delft.

Unpublished letters of Hortensius and Peiresc are in the library of Inguimbert in Carpentras and also in the Bibliothèque nationale in Paris. His correspondence with Schickard is in the Stuttgart Library, and that with Constantine Huygens is in the Royal Academy of Amsterdam.

Suhasini Kumar

Alternate name

Martinus Hortensius [Ortensius]

Selected Reference

de Waard, C. (1911). "Hortensius." *Nieuw Nederlandsch biografisch woordenboek*. Vol. 1, cols. 1160–1164. Leiden.

Van Lansbergen, Philip

➔ **Lansbergen, Philip**

Van Maanen, Adriaan

Born **Sneek, Friesland, The Netherlands, 31 March 1884**
Died **Pasadena, California, USA, 26 January 1946**

Dutch–American observational astronomer Adriaan van Maanen is, sadly, remembered almost entirely for a mistake, measurements that he thought indicated that material was flowing out along the arms of the spiral nebulae, or that the nebulae were rotating, at speeds such that they could not possibly be located outside of the Milky Way Galaxy. This prolonged the debate around 1920 about whether other galaxies even existed.

Van Maanen received his degrees from the University of Utrecht: B.A. (1906), M.A. (1909), and Ph.D. (1911), the last with a thesis on "The Proper Motions of 1418 Stars in and near the Clusters h and Chi Persei," written in Dutch but with a translation in English financed by a wealthy relative. Although **Albert Nijland**, professor of astronomy at Utrecht, was his official advisor, most of van Maanen's work was actually done at the University of Groningen (1908–1910) under the influence of **Jacobus Kapteyn**.

It was Kapteyn who arranged for van Maanen to spend a year, 1910/1911, at Yerkes Observatory as a volunteer astronomer, financed by that same wealthy relative, Mr. K. Blokhuis, the owner of the Gasworks at Haarlem, the Netherlands. **George Hale** then appointed van Maanen to a staff position at Mount Wilson Observatory, preferring him to either **Ejnar Hertzsprung** or **Pieter van Rhijn** as successor to **Philip Fath** (discoverer of emission lines in the cores of a few spiral galaxies now called Seyfert galaxies). Van Maanen thus became the first of a large number of Dutch astronomers to pursue careers in the United States, frequently organized by Kapteyn, and later by van Rhijn and **Willem de Sitter**. Others who appear in this book include **Willem Luyten**, **Peter van de Kamp**, **Dirk Brouwer**, **Bart Bok**, **Gerard Kuiper**, and **Adriaan Wesselink**.

At Yerkes, van Maanen continued his work on stellar parallaxes and proper motions, studying stars in and around the Orion Nebula. When he first arrived at Mount Wilson, there was no apparatus appropriate for such work there, so he participated in the program on stellar and solar spectroscopy. Van Maanen concluded that the average magnetic field of the Sun was quite large. He did this on the basis of measurements of Zeeman broadening of spectral features that must have been subject to some systematic error comparable with his later one in spiral nebulae. This is because the actual average field is, at most, 10% of what van Maanen deduced.

In 1914, the Cassegrain focus at the 60-in. telescope became available, and van Maanen resumed astrometric work, which had traditionally been done only with very long-focal-length refracting telescopes. He undoubtedly established that astrometry with reflectors was possible, measuring his 500th parallax in 1945. Van Maanen also concluded that the number of stars near the Sun as a function of real brightness peaks near absolute magnitude 10.3, only a factor 100 fainter than the Sun. There is indeed such a peak, but a factor of 10 fainter than this.

In 1917, van Maanen reported a star with a parallax of 0.24 arc sec (*i. e.*, only a little more than 4 parsecs away), with apparent magnitude 12 and the color of a G star like the Sun. With a bit of arithmetic, one can see that this means a brightness of 0.02% that of the Sun and, with the same color (temperature), that the star must have a radius only about 1% that of the Sun, about the same as that of the Earth. Now called van Maanen's star, this was the second white dwarf discovered (following Sirius B by **Walter Adams**), and it remains unique in having strong features due to iron in its spectrum, and little or no hydrogen. (Helium would be invisible at its temperature near 6000 K.)

Between 1916 and 1923, van Maanen sought to measure proper motions in several spiral nebulae, using plates taken both before and during his tenure with several different cameras and telescopes. He devised, and had constructed for the purpose, a stereocomparator, which allowed the observer to look at one plate with each eye, so that anything that had moved would seem to jump out of the image. The motions van Maanen reported, about 0.02 arcsec/year, would have corresponded to physical speeds of at least 10,000 km/s for any location that would permit the spirals to be separate galaxies comparable in size with the Milky Way. This would have implied enormous masses. **Harlow Shapley**, then a colleague and close friend, automatically believed the results, and his opposition to the existence of external galaxies, voiced at the debate with **Heber Curtis**, was based partly on van Maanen's work.

In 1923, when the discovery of Cepheid variables in a couple of spirals had been made by **Edwin Hubble** (but before this was widely announced), **Knut Lundmark** remeasured the plates and found no evidence for proper motions. Van Maanen never entirely recanted. Hubble repeated the measurements again in 1935 with the same plates and comparator and found no motion, while van Maanen's own remeasurements merely halved his previous result. The comparator itself was used successfully by others, though it retained a sign instructing astronomers to consult van Maanen before touching it, long after his death. Computer reprocessing of the numbers he recorded found no error in the arithmetic. Thus no one has ever fully understood what went wrong, but it was to do with how van Maanen perceived the images on the plates. It is perhaps significant that, during that period and long after, he was the only one at Mount Wilson working primarily in astrometry. The others were spectroscopists and interpreters of images. In any case, his reputation never fully recovered.

Van Maanen maintained memberships in the International Astronomical Union, in astronomical societies in England, France, Germany, and the United States, and in at least five Dutch scientific organizations, but seems never to have been elected to office or to have received any special recognition from them. He never married, but was a social person, active in the Valley Hunt Club, Young Mens' Christian Association, and the local chamber music society (as an organizer, not a performer).

Adriaan Blaauw

Selected References

Hetherington, Norriss S. and Ronald S. Brasher (1992). "Walter S. Adams and the Imposed Settlement between Edwin Hubble and Adriaan van Maanen." *Journal for the History of Astronomy* 23: 53–56.

Joy, Alfred H. (1946). "Adriaan van Maanen, 1884–1946." *Popular Astronomy* 54: 107–110.

Seares, Frederick H. (1946). "Adriaan van Maanen, 1884–1946." *Publications of the Astronomical Society of the Pacific* 58: 89–103. (On implications for the discovery of external galaxies.)

Van Rhijn, Pieter Johannes

Born **Gouda, The Netherlands, 24 March 1886**
Died **Groningen, The Netherlands, 9 May 1960**

Dutch statistical astronomer Pieter van Rhijn made an important determination, called the van Rhijn luminosity function, of the numbers of stars in the solar neighborhood as a function of their absolute brightness. He was the son of Cornelis H. van Rhijn, a professor of theology at Groningen University, and Aletta J.F. Kruijt. Van Rhijn and his wife, *née* Regnera L. G. C. de Bie, had one son and one daughter.

Van Rhijn received his Ph.D. from the University of Groningen in 1915 for work carried out under the direction of **Jacobus Kapteyn** and partly done at the Mount Wilson Observatory

between 1912 and 1914. He was appointed professor of astronomy at Groningen in 1921.

A collaborator, and successor of J. C. Kapteyn in the chair of astronomy at the University of Groningen and the directorate of the Kapteyn Astronomical Laboratory, van Rhijn contributed mainly to research on the stellar content and the interstellar matter of the Galaxy. Kapteyn had initiated and obtained wide international collaboration for his Plan of Selected Areas, aimed at exploring the structure and size of the Galaxy. Van Rhijn saw it as one of his primary tasks to realize this project and promote its extension. He encouraged and coordinated observational projects at observatories elsewhere and conducted extensive projects for the measurement of photographic plates at the Kapteyn Laboratory for both the study of proper motions and the measurement of stellar brightness. Among the main partners in these projects were the Harvard and Mount Wilson observatories in the United States and the Hamburg Observatory in Germany.

Besides these projects van Rhijn mainly worked on two aspects of galactic research: the distribution of stellar luminosity and the properties of the interstellar matter in the Galaxy. The first one resulted in the so called van Rhijn luminosity function that has for many years been the standard reference. It describes the frequency distribution of the absolute magnitudes – a measure of the star's intrinsic brightness – per unit volume in the solar neighborhood and is derived from careful analyses of the statistical distributions of stellar proper motions and apparent magnitudes. It was one of the starting points for the subsequent studies by other authors of the Initial Luminosity Functions [ILF] (also called "birth-luminosity function"). The ILF is then a guide to the distributions of masses of stars when they form, an important factor in the evolution of galaxies and their composition. The observed luminosity function determined from nearby stars can also be used to analyze counts of stars at larger distances to determine how stellar populations change with position in the Galaxy.

In his thesis of 1915, "Derivation of the Change of Colour with Distance and Apparent Magnitude," van Rhijn searched for a possible increase in the reddening of the apparent colors of the stars with increasing distance as evidence for the existence of interstellar matter, but the result was not conclusive. In later years, after the introduction of photoelectric measurements of stellar colors, he returned more successfully to this subject. Using the data on spectral classification for the Selected Areas that resulted from the joint project with the Hamburg Observatory, he determined the local density distribution separately for stars of different spectral classes.

During the occupation of the Netherlands in World War II (1940–1945) van Rhijn was hospitalized with tuberculosis but recovered fully, retiring only in 1957. He served as president of the Commission on Selected Areas (32) of the International Astronomical Union (1932–1958, one of the longest tenures of a single individual as Commission president) and as rector magnificus (vice chancellor) of Groningen University (1939–1940). Van Rhijn was elected a foreign associate of the Royal Astronomical Society (London), and was knighted by Queen Juliana of the Netherlands in 1956.

A nearly complete set of van Rhijn's publications, as well as many of his notebooks for lectures and research, are kept in the archives of the Kapteyn Institute of Groningen University.

Adriaan Blaauw

Selected References

Blaauw, A. (1984). "P. J. Van Rhijn." In *Biografisch woordenboek van Nederland* (Biographical dictionary for the Netherlands). Vol. 20.

Blaauw, A. and J. J. Raimond. "P. J. van Rhijn 1886–1960" (in Dutch). *Hemel and Dampkring* 58: 137–139.

Bok, Bart J. (1960). "Two Famous Dutch Astronomers." *Sky & Telescope* 20, no. 2: 74–76.

Macpherson, Hector (1933). *Makers of Astronomy*. Chap. 8. Oxford: Clarendon Press.

Pickering, Edward C., J. C. Kapteyn, and P. J. van Rhijn (1923–1924). "Durchmusterung of Selected Areas." *Annals of the Astronomical Observatory of Harvard College* 102—103.

Schwassmann, A., P. J. van Rhijn, and L. Plaut (1935–1953). *Bergedorfer Spektral-Durchmusterung*. Bergedorf, Germany.

van Rhijn, P. J. (1965). "Classical Methods for the Determination of Luminosity Function and Density Distribution." In *Galactic Structure*, edited by A. Blaauw and M. Schmidt, pp. 27–39. Vol. 5 of *Stars and Stellar Systems*. Chicago: University of Chicago Press. (Written shortly before his death.)

van Rhijn, P. J. and A. Schwassmann (1935). "Die Dichteverteilung der Sterne in höheren galacktischen Breiten" (Stellar density distribution at high galactic latitudes). *Zeitschrift für Astrophysik* 10: 161–187.

van Rhijn, P. J., F. H. Seares, and J. C. Kapteyn (1930). *Mount Wilson Catalogue of Photographic Magnitudes in Selected Areas 1–139*. Washington: Carnegie Institute of Washinton. (See also *Third Report on the Progress of the Plan of Selected Areas* [1923], and later reports, up to 1958, in the *Transactions of the International Astronomical Union*.)

Varāhamihira

Born **possibly Kapitthaka, (India), 505**
Died **Avanti, (Ujjain, Madhya Pradesh, India), 587**

Varāhamihira's three major works are *Pancha Siddhantika*, *Brihat Samhita*, and *Brihat Jataka*. The first of these summarizes the five important astronomical schools current at his time, the second is an encyclopedia, and the third is an astrological text. The *Pancha Siddhantika* presents the theories of the Paulisha, Romaka, Vasishtha, Paitamaha, and Surya Siddhantas. The *Brihat Samhita* has chapters on astronomy, geography, the calendar, meteorology, botany, agriculture, economics, engineering, zoology, and so on. Varāhamihira's main importance lies in his review of older astronomical theories, especially the Surya Siddhanta, and the description in his encyclopedia of the popular knowledge that existed in India during his time.

A. Vagişwari

Selected Reference

Sarma, K. V. (1993). *Panca-Siddhantika of Varahamihira*. Madras: PPST Foundation.

Vaucouleurs, Gérard Henri de

Born **Paris, France, 25 April 1918**
Died **Austin, Texas, USA, 7 October 1995**

Gérard de Vaucouleurs surveyed, classified, and cataloged galaxies in the Northern Hemisphere and Southern Hemisphere, and created a complex Galaxy classification system that differed from **Edwin Hubble**'s system and was more useful scientifically.

Vaucouleurs grew up in Paris. Nothing is known of his father; he adopted his mother's maiden name. Vaucouleurs showed a boyhood interest in astronomy, observing the Moon with a borrowed nautical telescope from the balcony of his family's apartment. With a telescope later given to him by his mother, he became an expert visual observer, timing lunar occultations and mapping the visible appearances of the planets. With his home-made map of Martian surface features, the young amateur astronomer measured the rotation rate of Mars with an accuracy unsurpassed until the 1960s Mariner spacecraft missions to the planet made further improvements possible.

Vaucouleurs studied astronomy, physics, mathematics, optics, photography, and spectroscopy at the University of Paris where he received an undergraduate degree in 1939. He joined amateur astronomer **Julien Péridier** at the latter's Le Houga Observatory to continue his planetary work, but the onset of World War II forced the closing of the observatory a few months later. Vaucouleurs served in the French army artillery for 19 months before the French capitulation in May 1941. He then returned to Le Houga, where in addition to continuing his lifelong studies of Mars, he took up double stars and variable stars. As a research scholar at the Sorbonne Physics Research Laboratory (1943–1945) and the Institut d'Astrophysique (1945–1950), Vaucouleurs earned a doctoral degree in 1949, defending a dissertation on molecular (Rayleigh) scattering and depolarization of light in liquids and gases. In 1944, he married Antoinette Pietra, an accomplished astronomer in her own right, who for the next 43 years would collaborate with him in many of his astronomical researches.

The Vaucouleurs immigrated to England where he was on the staff of the University of London's Mill Hill Observatory. In 1951 the couple moved to Australia when Gérard was awarded a fellowship to do planetary research at the Australian National University's Mount Stromlo Observatory. There he earned a second doctor of science degree, for research in molecular physics, optics, photography, astronomy, and astrophysics. During this time Vaucouleurs also served as observer-in-charge of the Yale–Columbia Southern Station in Canberra.

At Mount Stromlo, Vaucouleurs took advantage of the fact that the southern sky remained relatively unexplored by large telescopes or photography; little was known about the thousands of galaxies visible from the Southern Hemisphere. From 1952 to 1956 Vaucouleurs surveyed bright southern galaxies, and reobserved the 1,300 galaxies in the Shapley–Ames Catalog, measuring the brightnesses and radial velocities of hundreds of galaxies and determining their distances. Using his measurements, he mapped clusters of galaxies that aggregated to form what Vaucouleurs called the "Local Supercluster." In 1953 Vaucouleurs pointed out that most of the bright galaxies were distributed along a relatively narrow belt at an angle of roughly 90° to the plane of the Milky Way. He interpreted this distribution as the perspective effect of looking edge-on at a great disk of galaxies tens of millions of light-years across. At the time, few astronomers took Vaucouleurs's Local Supercluster model seriously. However, it is now generally accepted that galaxies are distributed in great sheets or bubbles of large-scale structure separated by large voids in intergalactic space.

From 1953 to 1956 Vaucouleurs made a detailed study of the Magellanic Clouds with **Frank Kerr**, discovering that the Milky Way's neighbors, previously regarded as irregular and chaotic in their form and motions, actually showed spiral structure and rotated. These studies resulted in the first accurate determination of the masses of the clouds; Vaucouleurs classified both clouds as barred spirals with a specific type of asymmetry. Vaucouleurs was also the first astronomer to classify the Milky Way as a barred spiral galaxy.

In 1957, Vaucouleurs came to the United States, where he would live for the rest of his life. After brief stints at the Lowell Observatory in Flagstaff, Arizona, and the Harvard College Observatory, Vaucouleurs joined the faculty of the University of Texas, Austin, in 1960. He became a naturalized citizen in 1962, and was named a full professor in 1965. At McDonald Observatory near Fort Davis, Texas, Vaucouleurs took photographs and made photometric and spectroscopic measurements of thousands of galaxies. For his survey, Vaucouleurs also designed and built a Fabry–Perot interferometer. With his data, he mapped the shape of the visible Universe with unprecedented accuracy, separating spiral galaxies according to their two major morphological components – the central bulge and the disk.

With his wife Antoinette and other collaborators, Vaucouleurs was an indefatigable cataloger of galaxies, publishing three *Reference Catalogues* of bright galaxies in 1964, 1976, and 1991 (the last published after Antoinette's death in 1987), and other valuable galaxy cataloges including databases of those objects that appeared in the southern sky. His catalogs were not mere compilations or lists, but contained much original data on angular sizes, magnitudes, colors, and radial velocities (redshifts) that Vaucouleurs gathered himself with the world's largest telescopes. He also applied a critical eye to data from other sources incorporated into his galaxy catalogs, carefully weighing it for its relative reliability.

Vaucouleurs devised a complex alternative to Hubble's scheme for classifying galaxies based on their morphologies. Vaucouleurs used many different parameters including the averaged surface brightness of the galaxy, its photometric brightness at different wavelengths, the ratio of the galaxy's HI content to its magnitude, and the ratio of a galaxy's central bulge to its disk. He developed formulae relating galaxies' angular dimensions to their luminosity profiles, discovering the "$r^{1/4}$" law that empirically defines the surface brightness distribution for elliptical galaxies.

Intensely interested in the cosmic distance scale – the absolute distances separating galaxies and clusters of galaxies in the Universe – Vaucouleurs questioned and revised the standard distance indicators and developed many new ones. He believed in spreading the risks by averaging the effects of many different distance indicators to cancel out systematic and statistical errors. Vaucouleurs' measurements of Galaxy distances led him to continue supporting a H_0 value near 100 km s^{-1} Mpc^{-1} (the majority view between about 1955 and 1970) after others had adopted values of 50–75. He was a proponent of hierarchical or fractal structure in the Universe, and of a non-zero cosmological constant, so that his large value of H was not in contradiction with the best estimates of the best estimates of the age of the Universe. His work and that of other proponents of the "short distance scale" (H_0 values in the range of 80 to 100 km s^{-1} Mpc^{-1}) stood in sharp contrast to the work of Allan Rex Sandage (born: 1926), Gustav Tammann, and others who favored values of H_0 nearer 50 km s^{-1} Mpc^{-1}, the "long distance scale." The precise value of H_0 remains somewhat uncertain in 2006, but seems to fall between 57 (Sandage) and 71 (Hubble Space Telescope key project).

Vaucouleurs received the Royal Astronomical Society's 1980 Herschel Medal and the American Astronomical Society's 1988 Henry Norris Russell Prize and lectureship. In 1986, he was elected a member of the United States National Academy of Sciences.

Vaucouleurs was survived by his second wife, the former Elysabeth Bardavid of Paris, France, whom he had known for a number of years and married in 1988.

Peter Wlasuk

Selected References

Burbidge, E. Margaret (2002). "Gérard de Vaucouleurs." *Biographical Memoirs, National Academy of Sciences* 82: 99–113.

Buta, Ronald (1996). "Gerard Henri de Vaucouleurs, 1918–1995." *Bulletin of the American Astronomical Society* 28: 1449–1450.

Capaccioli, M. and H. G. Corwin, Jr. (1989). *Gerard and Antoinette de Vaucouleurs: A Life for Astronomy*. Singapore: World Scientific Press.

Vaucouleurs, Gérard de (1953). *A Revision of the Harvard Survey of Bright Galaxies*. Canberra: Australian National University.

Vaucouleurs, Gérard de and F. J. Kerr (1955). "Rotation and other Motions of the Magellanic Clouds." *Australian Journal of Physics* 8: 508–522.

Vaucouleurs, Gérard de, A. de Vaucouleurs, H. G. Corwin, Jr., R. J. Buta, and G. Patural (1991). *Third Reference Catalogue of Bright Galaxies*. New York: Springer-Verlag.

Verbiest, Ferdinand

Born **Pittem, (Belgium), 1623**
Died **Beijing, China, 28 January 1688**

Ferdinand Verbiest was a prominent Jesuit mathematician and astronomer who served the Chinese emperor. Born in West Flanders into a family of landed gentry, Verbiest received his secondary education in Bruges and Kortrijk and after a brief stay at the Arts Faculty of Louvain University (1640/1641), became a Jesuit novice of the Flemish–Belgian province in Mechlin (1641–1643). He returned to Louvain to complete his philosophical studies at the local Jesuit College, and finally obtained his degree in 1645. During a year's theological studies in Rome (1652–1653) at the celebrated Roman College, he made the acquaintance of the renowned scholar and mathematician, **Athanasius Kircher**.

In the second half of 1655 Verbiest received official permission to join the missionary group that was to set out for China under the leadership of M. Martini. When the group was forced to spend a year in Portugal, Verbiest was sent to Coimbra to teach mathematics, an indication that he was already developing a reputation as a mathematician. From Portugal he maintained his contacts with Kircher in Rome and managed to obtain some of Kircher's recent works. The group finally set sail and reached Macao, the gateway to the Chinese mainland, on 17 July 1658. Verbiest's practical schooling in astronomy apparently continued during the voyage, with Martini as his teacher. In Macao he acquired his first knowledge of Chinese by reading the Four Confucian Classics.

Verbiest was first sent to Hsi-an-fu in the Shensi province where, despite his still deficient command of the Chinese language, he performed the work of a simple missionary with great enthusiasm. On 26 February 1660 **Johann Adam Schall von Bell**, with the endorsement of the Shun-chih emperor, called him to Beijing. Schall von Bell saw Verbiest as a possible successor to the very important position Schall von Bell held as head of the Astronomical Bureau, where he had the ultimate responsibility for the annual calendar. Schall von Bell undoubtedly chose Verbiest because of his fame as a mathematician. That his reputation as a mathematician had already penetrated into China at this early stage of the mission is explicitly attested by various correspondence in those years between Beijing and Rome. Upon his arrival in Beijing on 9 June 1660, Verbiest was received by Schall von Bell in the Hsi-t'ang residence where he was initiated and prepared for his future task as head of the Astronomical Bureau, primarily of its calendar section. Also in these initial years Verbiest was active in more areas than specifically required for the calendar. As early as 1661 and 1662 he assisted Schall von Bell in the difficult task of hanging an enormous clock in the Beijing Bell Tower with the aid of pulleys, and he had produced a series of drawings for new astronomical instruments based on the system of **Tycho Brahe**.

In 1664 Yang Kuang-hsien launched an outright attack against the western influences within the Chinese Astronomical Bureau, in particular against the unquestioned application by the Jesuits since 1644 of the "western rule" in the establishment of the annual calendar. After several failures Yang Kuang-hsien, taking advantage of the altered climate since the death of the Shun-chih emperor (1661) and of the anti-western mood under the so called Oboi regency (1661–1669), seized his opportunity. A legal complaint was filed, and the Jesuits were arrested, put on trial, and convicted. During these particularly trying times Verbiest displayed great devotion towards his half-paralyzed mentor, Schall von Bell, who died on 15 August 1666.

After the conversion of the initial death sentence to house arrest, and after the banishment of 25 Jesuits to Canton, Verbiest with several Jesuit companions spent 4 years (1665 to 1669) under surveillance in the Tung-Vang residence. During this time he managed to show his skill at forecasting solar eclipses, and he occupied himself with the construction of mechanical devices, with experiments in steam-powered vehicles, with tests of gnomons, and with meteorological observations. Here we see Verbiest, under house arrest, preparing himself as the many-sided engineer and mathematician, who from 1669 would serve the emperor in a wide variety of mathematical and mechanical disciplines. By his prediction and verification of planet positions and by his successful tests of gnomons, Verbiest convinced the emperor of his thorough command of western astronomical science. He thus managed to have the "western rule" of calendar making reinstated, and himself along with it.

Verbiest was soon appointed by the Ministry of Rites to the position of prefect of the Astronomical Bureau, a title he was to hold until the end of his life. From the position that office provided him, he would henceforth develop an uncommonly busy and varied activity. The services he rendered to the emperor, to some Chinese officials, and to China, in fact, made him indispensable, a situation accordingly recognized in the form of distinctions of all kinds that quite rapidly elevated him to the second rank of the mandarin hierarchy in 1676.

A survey of the writings of that period also reveals Verbiest's missionary activity, particularly in the form of several treatises on points of religious belief. His principal contribution in this area, however, consisted in maintaining the emperor's friendship by performing his manifold tasks, thus protecting the mission in the rest of China. In addition he was active as a diplomat, for example, in the Sino-Russian negotiations concerning the Amour frontier. By designating a competent successor, as well as by attracting French Jesuits in 1685, Verbiest also ascertained the survival of the mission. His death prevented him from witnessing the realization of two important aims for which he himself had laid the groundwork: the treaty of Nerchinsk on 28 January 1689, establishing peace between China and Russia along the Amour River, and the K'ang-hsi edict of 1692, guaranteeing a far-reaching toleration for Catholicism.

George V. Coyne

Selected References

Golvers, Noël (1993). *The* Astronomia Europea *of Ferdinand Verbiest*. Nettetal: Steyler Verlag.

Harris, S. J. (1988). "Jesuit Ideology and Jesuit Science: Scientific Activity in the Society of Jesus, 1540–1773." Ph.D. diss., University of Wisconsin.

Ronan, Charles E. and Bonnie B. C. Oh (eds.) (1988). *East Meets West: The Jesuits in China, 1582–1773*. Chicago: University of Chicago Press.

Rowbotham, Arnold H. (1942). *Missionary and Mandarin: The Jesuits at the Court of China*. Berkeley: University of California Press.

Very, Frank Washington

Born **Salem, Massachusetts, USA, 12 February 1852**
Died **Massachusetts, USA, 23 November 1927**

Frank Very was an 1873 graduate of the Massachusetts Institute of Technology. He was employed at, or associated with, successively, the Allegheny, Lowell, and Westwood observatories. Very assisted two of the best-known figures in late-19th-century American astronomy: **Samuel Langley** and **Percival Lowell**. He was recognized in his own right for investigating the lunar albedo bolometrically.

Selected Reference

Ogden, J. Gordon (1928). "Frank W. Very." *Popular Astronomy* 36: 391–397.

Vespucci, Amerigo

Born **Florence, (Italy), 18 March 1454**
Died **Seville, Spain, 22 February 1512**

New World navigator Amerigo Vespucci was born into one of the leading mercantile, aristocratic families of Florence. He demonstrated in 1501 and 1502 that the lands discovered in 1492 by Christopher Columbus were continental in extent, ascertained a longitude by means of a lunar occultation of the planet Mars, and gave both findings wide publicity.

Vespucci studied under a student of the Italian astronomer **Paolo Toscanelli**, and enjoyed the support of an established mercantile network – providing supplies and services for Columbus himself in the process. In 1499 and 1502 Vespucci used the *Ephemerides* of **Johann Müller** (Regiomontanus) the leading European authority, to determine his longitude.

Attempting to explain to a general audience the navigational problems in determining longitude at sea, Vespucci wrote:

> I found nothing better to do than to watch for and take observations at night of the conjunction of one planet with another, and especially of the conjunction of the moon with the other planets. . . . One night, the twenty-third of August, 1499, there was a conjunction of the moon with Mars, which according to the almanac [for the city of Ferrara] was to occur at midnight or a half hour before. I found that when the moon rose an hour and a half after sunset, the planet had passed that position in the east. (quoted in Boorstin, 1991, p. 357.)

The difference in longitude between the two locations was then the same as the difference in the observed times of this event.

Vespucci was included in a Spanish expedition of 1499 and served as a leader of an expedition to the coast of Brazil sponsored by King Manuel I of Portugal, in 1501 and 1502. Vespucci's accounts of his travels, written in evocative, sometimes poetic terms, won a wide readership. The community of Florence ordered the illumination of the Vespucci mansion in celebration of his announcement, following the 1501/1502 voyage, that the western lands were an entire continent. Latin, French, German, and Dutch translations of his *Mundus Novus* appeared rapidly, and in many reprintings. A printed atlas with an edition of Vespucci's letters appeared in 1507, containing a map labeling the New World "America."

Michael Meo

Selected References

Apelstadt, Eric (1992). "Christopher Columbus, Paolo dal Pozzo Toscanelli and Fernão de Roriz: New Evidence for a Florentine Connection. " *Nuncius* 7, no. 2: 69–80.

Arciniegas, Germán (1955). *Amerigo and the New World: The Life and Times of Amerigo Vespucci* translated by H. De Onís. New York: Alfred A. Knopf.

Boorstin, Daniel J. (1991). *The Discoverers*. Vol. 1. 2nd ed. New York: Harry N. Abrams. (Devotes a chapter to Vespucci's career, pp. 354–368, in a gracefully written general account of the exploration of the New World.)

Pohl, Frederick J. (1966). *Amerigo Vespucci: Pilot Major.* New York: Octagon Books. 1st ed. 1944. (Historical conflict has raged for centuries over the authenticity of Vespucci's printed letters as distinct from three letters found since the mid-18th century in Florentine archives and accepted by all scholars as genuine; the originals of *Mundus Novus* and of the 1507 letter have never been found. Pohl rejects the printed letters as forgeries and Arciniegas attacks the many arguments of Alberto Magnaghi, published in 1926, which Pohl terms "not seriously challenged.")

Heilbron, J. L. (1999). *The Sun in the Church.* Cambridge, Massachusetts: Harvard University Press.

Herbermann, Charles George (ed.) (1907). *The* Cosmographiae Introductio *of Martin Waldseemüller.* New York: United States Catholic Historical Society, Monograph 4. (For a translation of the atlas and letter of 1507.)

Morison, Samuel Eliot (1962). *Admiral of the Ocean Sea: A Life of Christopher Columbus*. 2 Vols. New York: Time Reading Program, Time Incorporated. 1st ed. 1942. (While not the most recent, this biography remains standard.)

Thorndike, Lynn (1923–1958). *A History of Magic and Experimental Science.* 8 Vols. New York: Columbia University Press.

Vico, Francesco de

Born **Macerata, (Marche, Italy), 19 May 1805**
Died **London, England, 15 November 1848**

Comet hunter father Francesco de Vico became a Jesuit in 1823 and joined the faculty of the Collegio Romano. In 1833 he began his studies in astronomy and started his observations of comets 1P/Halley and 3D/Biela. In 1838 de Vico became director of the observatory in the Collegio Romano. After making planetary observations (Mimas, Enceladus, Saturn, and Venus) he discovered seven comets between 1844 and 1847. Apart from his work in astronomy, he was active in helping the sick during the 1837 cholera epidemic. The unrest during 1848 forced de Vico to leave Rome and go to the United States as a refugee.

Mariafortuna Pietroluongo

Selected References

Anon. (1849). "Francis de Vico." *Monthly Notices of the Royal Astronomical Society* 9: 65–66.

Anon. (1847–1848). "Obituary" (in Italian). *Atti Pontif. Acc. Di Sc. Dei Nuovi Lincei* 1: 172.

Green, Dan (1995). "The Re-discovery of Comet 122P/de Vico. " *International Comet Quarterly* 17: 161.

Secchi, A. (1851). *Ragguaglio intorno alla vita ed ai lavori di padre F. de Vico* Mem. Oss. dell'Univ. Gregoriana in Collegio Romano dell'anno 1850. Rome.

Sommervogel, Carlos (1898). *Bibliothèque de la Compagnie de Jesus.* Vol. 8, cols. 641–644. Brussels: Société Belge de Libraire.

Stein, G. (1949). *F. de Vico ed i suoi contributi alle scienze astronomiche.* La Civiltà Cattolica, 2, pp. 190 and 314.

Vinci, Leonardo da

Born **Florence, (Italy), 15 April 1452**
Died **Cloux, near Amboise, (Indre-et-Loire), France, 2 May 1519**

Artist Leonardo da Vinci, of Florence and Milan, produced three naked-eye renderings of the Moon. He thought of the Earth's satellite as a nonuniform mirror (likely watery with some land) reflecting the light of the Sun. Knowing that a smooth sea would result in a specular reflection, da Vinci imagined white-capped waves breaking on lunar shores. He also correctly identified "earthshine" as being produced by sunlight reflected from the Earth to the Moon.

Selected Reference

Reaves, Gibson and Carlo Pedretti (1987). "Leonardo Da Vinci's Drawings of the Surface Features of the Moon." *Journal for the History of Astronomy* 18: 55–58.

Virdung, Johann

Born **Hassfurt, (Bavaria, Germany), 14 March 1463**
Died **Heidelberg, (Germany), 1538/1539**

Johann Virdung was one of the most influential astrologers around 1500. Virdung, about whose childhood and youth nothing is known, began his studies in 1481 in Leipzig, and continued them in Krakow, where he attended (*inter alia*) the lectures by **Albertus de Bruzewo** and Johannes von Glogau. He returned to Leipzig as "Johannes Johannis baccalaureus Cracovensis," where he obtained his master's degree in 1491. The following year Virdung moved to Heidelberg, where he gave lectures on medicine, mathematics, and astronomy at the university and entered the service of the Electoral Palatinate court. From 1521 until his death he ran the court dispensary; and around 1529/1530 he was conferred a medical doctorate. In 1487, his first (or the earliest so far known) of his prognostications appeared in Leipzig; his bibliography includes at least 80, generally minor, astrological works in German and Latin, but also medical books.

Virdung compiled numerous almanacs, annual astrological forecasts, and interpretations of special celestial events. His annual prognostications followed the traditional pattern: the determination of the ruling planet for the year, the prediction of weather conditions, the prospects for the harvest, and the determination of the time for carrying out medical procedures, as well as the interpretation of eclipses. Particularly widespread were Virdung's writings about the Great Conjunction of 1524, which appeared in Latin in 1521 and in both 1522 and 1523 in German, and in which he endorsed the general predictions for multiple disasters in that year. These prophecies on the eve of the great German Peasants' War reflected the tense social situation in the first-third of the 15th century. Virdung used astrological prophecies in the hope that reforms would be introduced into the church, that serious shortcomings in the Church would cease, and that neither the nobles nor the "common folk" would rebel against the clergy.

Virdung's great standing in the scholarly world is shown by the fact that he was asked, on behalf of Heidelberg University, to present a report to the Lateran Council (1512–1517) regarding the planned calendar reform (which was, however, not implemented).

Virdung's lesser astrological works were extremely popular, which is partly shown by the numerous editions in Latin, German, and Low German, printed in Leipzig, Nuremberg, Lübeck, Oppenheim, Stendal, and Strasbourg with the same work often appearing in several places. After Virdung's death, some of his predictions were reproduced in various collections of astrological prophecies. His observations of optical phenomena (haloes that appeared around the Sun and Moon) show that apart from Virdung's interpretations of them as miraculous, he had genuine powers of observation.

Of his major works, Virdung's *Tabulae resolutae* must be mentioned. This was published in 1542, after the author's death, by his pupil Jacobus Curio. The foreword contains important information about Virdung. There was also an iatromathematical work on the astrological basis for medicine, *Novae medicinae methodus* (1532, second edition, 1533), in which he complains that doctors disregard astrology, and that he hoped to give doctors reliable assistance with this book.

Jürgen Hamel

Selected References

Bruckner, Ursula (1969). "Wenzel Faber von Budweis oder Johannes Virdung?" *Beiträge zur Inkunabelkunde*, 3rd ser., 4: 123–140.

Steinmetz, Max (1986). "Johann Virdung von Hassfurt, sein Leben und seine astrologischen Flügschriften." In *Astrologi Hallucinati: Stars and the End of the World in Luther's Time*, edited by Paola Zambelli, pp. 196–214. Berlin: Walter de Gruyter.

Thorndike, Lynn (1943). "Another Virdung Manuscript." *Isis* 34: 291–293.

Virgil [Vergil]

Born **Andes, Cisalpine Gaul, (Lombardy, Italy), 15 October 70 BCE**
Died **Brundisium (Brindisi, Italy), 20 September 19 BCE**

The Roman poet Virgil adapted and incorporated traditional themes and earlier literary treatments of astronomy into his own works. According to ancient accounts and evidence from his poems themselves, Virgil, the most illustrious of Latin poets, was born in the small village of Andes near Mantua in the Po Valley region of Italy. Details of his family circumstances and childhood are uncertain, but he seems to have been educated at Cremona and later at Milan before going to Rome. He also spent time in Naples, where he associated with the Epicurean philosophical community there. In 42 BCE after the victory of Octavian (later the emperor Augustus) in the civil war that followed the assassination of Julius Caesar, Virgil's family property was confiscated to settle veterans but subsequently restored. After the publication of his earliest poems (*circa* 38 BCE) Virgil was invited to join the circle of Maecenas, an intimate friend, and advisor to Octavian. Thereafter he, received official support for his poetry, and in turn generally promoted the policies of the regime in his works. He associated with the leading literary and political figures of his day until his death. From the outset, Virgil's literary influence has been enormous.

Of a number of works ascribed to Virgil by commentators ancient and modern, only three, all written in epic hexameters, can be deemed genuine. The earliest of these works are the *Eclogues*, a collection of 10 individual poems styled after the pastorals of the Greek poet Theocritus. The four books of the *Georgics*, published in 29 BCE, comprise a structurally complex and multifaceted poem, ostensibly modeled on **Hesiod**, whom Virgil acknowledges (*Geo.* 2.176), but drawing inspiration as well from **Aratus** and a host of other Greek and Roman writers. It is superficially about the order and unity of the agricultural world, but in reality poses unsettling questions about cosmic and societal ambiguity and ambivalence. Virgil's final work, left unfinished at his death and edited for publication at the behest of Augustus, is the *Aeneid,* the Roman national epic, consciously suffused by its author with Homeric details, references, and literary echoes. The *Eclogues* and the *Aeneid* make general reference to the workings of the heavens in a number of passages; in the *Georgics,* especially Books 1 and 2, Virgil focuses directly on both the traditional and the literary functions of astronomy. Although each work differs markedly in its general theme and content, taken together they reveal the poet's awareness of the skies and a studied familiarity with the literary heritage of astronomical writing.

Virgil acknowledges that for the scientific aspects of astronomy, other peoples, specifically the Greeks, are better suited than the Romans (*Aen.* 6.849-51). In Virgil's view, too, the stars represent one element of the opposition between the two cultures. Yet he honors the Greek intellectual achievement by naming the Alexandrian astronomer **Conon** and by alluding to another, probably **Eudoxus** (*Ecl.* 3.40-42), even though for the Romans astronomy was primarily of practical value in its application to the agricultural year and, later, for navigation. A comprehensive Virgilian cosmogony appears in *Eclogue* 6, where the satyr Silenus describes the beginnings of things (31-40) in language suggestive of **Lucretius**'s Epicureanism. Virgil's recounting of the themes sung by the bard Iopas (*Aen.* 1.742–6), incorporating as they do the mention of commonly known constellations and celestial occurrences, represents his awareness of larger cosmic concerns.

Even in a primarily literary environment, much astronomical material is evident in Virgil. For example, he identifies five celestial zones and their terrestrial counterparts (*Geo.* 1.231–58), closely following several earlier writers, especially **Eratosthenes**. The description likewise presupposes a geocentric universe (242–3). The poet also describes the Sun's passage through the 12 signs of the zodiac (231–2), although he names only six anywhere in his works. Virgil values a general knowledge of the heavens (*Geo.* 1.1; 2.1; 2.475–82) and also recognizes the ancient observances of the risings, settings, and daily motions of the heavenly bodies (*Ecl.* 9.49, *Aen.* 3.515), including both the morning star, Lucifer (*Ecl.* 8.17; *Geo.* 3.324), and its evening counterpart, Vesper (*Ecl.* 6.86; *Geo.* 1.251; 3.337). The movements of the planets, the Sun, and the Moon provide a source of insight (*e. g., Geo.* 1.337; *Aen.* 1.742), while lunar and solar weather signs form an important part of the *Georgics* (1.351–463). Significant also are eclipses (*Geo.* 1.467; 2.478; *Aen.* 1.742), which the poet relates to terrestrial affairs, and comets (*Geo.* 1.365–67, 488; *Aen.* 5. 527–28; 6.693–4; 10.272), which had portentous status in antiquity. The famous Sidus Iulium, a comet that marked the ascent of Julius Caesar as a divinity after his death, appears in Virgil's early works (*Ecl.* 9.46–9) as a source of abundance and later figures prominently on the pro-Augustan iconography depicted on the shield of Aeneas (*Aen.* 8.681).

The stars themselves seem to generate both knowledge and wonder, as when the pilot Palinurus is depicted gazing upward at them (*Aen.* 5.25, 853). As might be expected in a body of work drawing extensively on an earlier epic tradition, Virgil refers to the constellations of **Homer** and Hesiod on several occasions: the Great Bear (Arctos), the Pleiades, and the Hyades (*Geo.* 1.138; cf. 1.246, 4.231–5); the two Bears (Triones) (*Aen.* 1.744; 3.516); and Orion (*Aen.* 1.535, 3. 517, 4.52, 7.719). The stars Arcturus (*Geo.* 1.68, 204; *Aen.* 1.744, 3.516) and Sirius (*Geo.* 4.425; *Aen.* 3.141. 10.273) are mentioned as well. Other constellations, primarily from Aratus and agricultural works, appear in the farmer's calendar section of the *Georgics* (1.204–30) and in its description of the great Celestial Sphere (1.231–58). Here Virgil describes how Draco, the Haedi, Boötes, Corona Borealis, Canis Major, Libra, and Taurus serve as astronomical markers for farming tasks. Elsewhere, he recognizes several additional zodiacal constellations, including Cancer (*Ecl.* 10.68), Aquarius (*Geo.* 3.304), Virgo (*Geo.* 1.33), Scorpius (*Geo.* 1.35), and perhaps Pisces (*Geo.* 4.234). The Virgilian creation of Libra from the claws (Chelae) of the Scorpion (*Geo.* 1.33–35) reveals the wedding of the astronomical and literary spheres to contemporary politics in its praise of Augustus, initiating a general tendency of linking Roman rulers to the heavens that would continue to develop in imperial culture. A number of passages suggest traces of astrological tendencies in Virgil's works as well, particularly as evident in the phrase "stars aware of fate" (*Aen.* 4.519–20, 9.429) and in references to mortals who possess a special insight into the heavens (*Aen.* 3.359–60, 10.176). In later ages, of course, Virgil himself was considered a seer and magician, and his works were consulted as an oracular source of knowledge of the future.

John M. McMahon

Alternate name

Vergilius Maro, Publius

Selected References

Aujac, G. (1984–1991). "Astronomie." In *Enciclopedia Virgiliana*. Vol. 1, pp. 382–385. Rome: Istituto della Enciclopedia italiana.

Braund, S. M. (1997). "Virgil and the Cosmos: Religious and Philosophical Ideas." In *The Cambridge Companion to Virgil*, edited by Charles Martindale, pp. 204–221, esp. 207–210. Cambridge: Cambridge University Press. (For Virgilian cosmology in literary contexts.)

d'Hérouville, P. (1940). *L'astronomie de Virgile*. Paris: Les Belles Letters. (The standard treatment of Virgilian astronomy.)

Ramsey, John T. and A. Lewis Licht (1997). *The Comet of 44 B. C. and Caesar's Funeral Games*. Atlanta: Scholars Press, esp. Chap. 4, pp. 61–94 and Chap. 7, pp. 135–153. (For the *Sidus Iulium*.)

Virgil (1971). *The Aeneid of Virgil*, a verse translation by Allen Mandelbaum. Berkeley: University of California Press. *The Aeneid*, translated by R. Fitzgerald. New York: Random House, 1983.

——— (1982). *The Georgics*, translated by Robert Wells. Manchester: Carcanet New Press. *Georgics*. 2 Vols, edited by Richard F. Thomas. Cambridge: Cambridge University Press, 1988 (Latin text and English commentary); *Georgics*. 2 Vols, edited by R. A. B. Mynors. Oxford: Clarendon Press, 1990. (Latin text and English commentary)
——— (1997). *Vergil's Eclogues*, translated by Barbara Hughes Fowler. Chapel Hill: University of North Carolina Press.

Vergilius Maro, Publius

➔ **Virgil [Vergil]**

Vitruvius, Marcus

Born **Fundi (Fondi, Campania, Italy), *circa* 85 BCE**
Died ***circa* 15 BCE**

Vitruvius is best known for his writings on architecture, but he also wrote on astronomy, describing the understanding of his time and its usefulness in making sundials.

Vitruvius was born amid the death throes of the Roman Republic to an old South Italian family prominent at Fundi (midway between Rome and Naples). Trained in architecture, as a young man he served in the army corps of engineers under Caesar, first in Gaul (known service includes Larignum 56 BCE and Marseilles 48 BCE), then in North Africa (at Zama in 46 BCE). After his general's assassination, Vitruvius joined the troops of Octavian (the future Augustus), on active duty as an artillery engineer; by 33 BCE he was an aqueduct official.

Vitruvius wrote one known work, a handbook (*institutio*) in ten books on "architecture" – that is civil engineering – from selection of a city site through design and construction to maintenance and defense. He gives extensive theoretical justifications for each precept, devoting over two-thirds of Book 9 to astronomy and astrology, as the basis for constructing sundials. He wrote for many years around 25 BCE, in his old age and during peacetime.

Vitruvius offers no original astronomy, but summarizes contemporary belief and preserves some otherwise lost astronomy. Short shrift but no polemics is given to astrology (recently popular). Constellations visible in Rome are described from a star map based on **Aratus**' poem (possibly using **Hipparchus**' commentary). Annual solar motion causes seasonal phenomena, especially variable day length, with equinoxes and solstices occurring at 8° of Aries, Leo, Libra, and Capricorn. Vitruvius gives two explanations of lunar phases: **Aristarchus** of Samos thought moonlight was reflected sunlight, the phases being explained by geometry, while **Berôssus** of Babylonia claimed the luminous lunar hemisphere is attracted by sunlight which rotates the Moon. The Vitruvian universe is standard for the era, described in mechanical terms: the heavens rotate about the Earth on pin-like poles beyond the stars, round which wheel rims roll as on a lathe (*tornus*). In orbits contrary to the stars, the Moon, Mercury, Venus, the Sun, Mars, Jupiter, and Saturn (the standard order) "wander" from west to east, one above another "as if on a staircase."

For the planets outside the Sun, Vitruvius gives the most accurate ancient solar periods: Mars 683 days (4 days less than the modern value), Jupiter 11 years, 313 days (2 days less), and Saturn 29 years, 160 days (7 days less). He explains their apparent retrogradations through an alleged greater attraction by solar rays at greater distances. Vitruvius offers a heliocentric model for Mercury and Venus in which the Sun's rays serve as a center that those planets "crown," their varying speeds being explained by their varying distance from the attractive Sun.

Vitruvius bases his architectural theory primarily upon Hermogenes of Alabanda (*circa* 160 BCE) and earlier Greeks, attributing authority to Antiquity (*e. g.*, Aratus, Aristarchus, Berôssus, and Ktesibius – all 3rd century BCE), the latest astronomer cited being Hipparchus. Since the partially heliocentric theory presumes epicycles, it postdates **Apollonius**, and probably Hipparchus (whom **Ptolemy** alleges attempted no planetary theory). One may suspect the heliocentrist **Seleukus** or perhaps the neopythagorean astronomer Apollonius of Mundos, who theorized that comets were long-period planets on elongated orbits.

Paul T. Keyser

Alternate name

Pollio, Marcus

Selected References

Aicher, Peter J. (1999). "Vitruvius." In *Dictionary of Literary Biography*. Vol. 211. *Ancient Roman Writers*, edited by Ward W. Briggs, pp. 366–370. Detroit: Gale Group.
Baldwin, Barry (1990). "The Date, Identity, and Career of Vitruvius." *Latomus* 4: 425–434.
Granger, Frank (ed. and trans.) (1931–1934). *On Architecture*, by Vitruvius. 2 Vols. London: W. Heinemann. (Translation with facing Latin.)
Russo, Lucio (1994). "The Astronomy of Hipparchus and His Time: A Study Based on Pre-Ptolemaic Sources." *Vistas in Astronomy* 38: 207–248, esp. 225–230. (caveat lector.)

Vogel, Hermann Carl

Born **Leipzig, (Germany), 3 April 1841**
Died **Potsdam, Germany, 13 August 1907**

German observational astronomer and spectroscopist Hermann Vogel was the first person to recognize a spectroscopic binary, that is, a pair of stars that reveal their mutual orbit through changes of their velocities along the line of sight *via* a changing Doppler shift of their spectral features. He was the sixth child of a Leipzig educator, Carl Christophy Vogel, and grew up in a family that valued intellectual pursuits. His older brother Eduard's friendship with the director of the Leipzig Observatory acquainted him with astronomy early on in his life. In 1860 Vogel attended the Dresden Polytechnic School to prepare for a career in technology. However, during his studies, his parents died, leaving him without financial support. He worked himself out of debt, partly through teaching the new technology of photography, and returned to Leipzig in 1863 to study natural science and work as an assistant at the

university's observatory, working under **Karl Bruhns**. Vogel earned his doctorate in 1868 for observations of the positions of nebulae and star clusters and a historical survey of nebular observations.

In 1870, Vogel was given the directorship of the private observatory of F. G. von Bulow in Bothkamp. There Vogel began a research program of spectroscopic observations of the Sun, stars, planets, nebulae, and even lightning and aurorae. He was one of the first professional astronomers to use photography to enhance the accuracy of observations. His paper on the spectra of planets won a prize from the Copenhagen Academy of Sciences and was regarded as an authoritative work for decades after its publication. Based on his spectral observations of stellar bodies, Vogel proposed modifications to **Angelo Secchi**'s classification scheme for stellar spectra. Utilizing Doppler shifts of spectra to infer velocities, he observed the rotation of the Sun, establishing that the photosphere rotates at the same rate as previous sunspot observations indicated.

Intrigued by the attempts of **William Huggins** to determine the line-of-sight or radial velocity of fixed stars, Vogel took up the problem and started making visual observations of a few stars. Around this time, the Prussian state decided to build an observatory in Potsdam, and Vogel was asked to be an observer and to help plan the equipment needs of the new observatory. The Potsdam Astrophysical Observatory was finished in 1879, and Vogel was appointed director in 1882, a position he held until his death. With the new facilities available to him Vogel was able to make real progress on his studies, and in 1892, he published the first accurate observations of the radial velocities of 51 stars.

One of the most striking achievements of the radial velocity study was the discovery of spectroscopic binaries. These are pairs of stars too close together to be seen as separate images. Vogel showed that Algol and Spica are binaries. For the former, this explained the periodic brightness fluctuations first found in 1669, and it confirmed an explanation put forward in 1783 by **Charles Goodricke**, that one star was passing in front of the other. At nearly the same time, **Edward Pickering** at Harvard College Observatory announced that Mizar and β Aurigae were also spectroscopic binaries, the latter having been recognized by **Antonia Maury**. About half of all stars are binaries, and the fraction detectable as such by spectroscopic measurements increases with the accuracy of the velocity measurements. The combination of the velocity measurements and eclipse light curve for Algol enabled Vogel and **Julius Scheiner** to make the first estimate of the mass of a body outside the Solar System.

As time went on, Vogel spent increasing amounts of time on organizational duties and securing upgrades to observational equipment. His management of the Potsdam Astrophysical Observatory helped to make it an important and productive center for research, and he was highly regarded for getting the most out of the equipment he had. Among his many honors and memberships in learned societies, the following stand out: In 1892 Vogel was elected to the Berlin Academy of Sciences, in 1893 he won the Gold Medal of the Royal Astronomical Society, and in 1906 he won the Bruce Medal of the Astronomical Society of the Pacific. Vogel died after several years of failing health.

Michael Fosmire

Selected References

Frost, Edwin B. (1908). "Hermann Carl Vogel." *Astrophysical Journal* 27: 1–11.

Townley, Sidney D. (1907). "Address of the Retiring President of the Society, in Awarding the Bruce Medal to Geheimer Ober-Reg. Rath Professor Dr. Hermann Carl Vogel." *Publications of the Astronomical Society of the Pacific* 18: 101–110.

Vögelin, Johannes

Flourished **(Austria), 16th century**

Viennese astronomer Johannes Vögelin attempted to measure the parallax for the comet of 1532 (c/1532 R1).

Alternate name

Vogelinus

Selected Reference

Hellman, C. Doris (1944). *The Comet of 1577: Its Place in the History of Astronomy*. New York: Columbia University Press, 1944. (Reprint, New York: AMS Press, 1971.)

Vogelinus

➜ **Vögelin, Johannes**

Vogt, Heinrich

Born **Gau-Algesheim, (Rheinland-Pfalz), Germany, 5 October 1890**
Died **Heidelberg, (Germany), 23 January 1968**

Heinrich Vogt and collaborators solved problems of the internal structure of stars, origin of spiral arms in galaxies, and the redshift in cosmology.

Vogt was born in a family of farmers headed by Philipp and Margarete Vogt, in a small town near Mainz am Rhein. The Gymnasium in Mainz and the university in Heidelberg prepared him well for the career of a professional astronomer. Soon after beginning university studies (1911) he started routine observations at the Heidelberg-Königstuhl Observatory under supervision of the director **Maximilian Wolf**, who appointed him as his assistant in the next year. Vogt was attracted more to theory than to observations. After an interruption to his studies due to World War I he graduated in April 1919.

Vogt's thesis dealt with the interpretation of observed properties of stars, including limb darkening, reflection of light in binaries, and distortions of the shapes of binary stars by their companions. His subsequent work at the observatory on the photometry of variable stars, comets, novae, and star clusters (especially h and χ Persei) resulted in his appointment to an associate professorship at the university.

Vogt continued theoretical investigations on the internal structure of stars, communicating with **James Jeans**, **Arthur Eddington**, and **Edward Milne**, who became his close friend. Vogt concluded in 1926 that the internal structure of a star was completely fixed by its mass and chemical composition. This is called the Vogt–Russell theorem in Europe and the Russell–Vogt theorem in the United States. It remains true for the evolution of a star of homogeneous composition, but the theorem fails when, for instance, the star develops an exhausted helium core. In the same year, Vogt obtained positions as assistant professor at Heidelberg University and observer at the observatory.

Vogt spent 3 years (1929–1932) in Jena as full professor and director of the Jena Observatory. For the first time he had freedom to develop the institute according to his plans. Soon his group of collaborators and students received international recognition in theoretical astrophysics.

After the death of Wolf in 1932, Vogt returned to Heidelberg and became director of the observatory and professor at the university. He refused an invitation to the Potsdam Astrophysical Observatory and continued his theoretical investigations of stellar interiors even though he was forced to manage a lot of organizational work at both institutions. After the end of World War II, in 1946, Vogt resigned the directorship at the observatory but increased his teaching activities in Heidelberg and Stuttgart (as visiting professor). He also began to write popular books on astronomy and cosmology.

Vogt was a member of the academies in Heidelberg, Berlin, and Halle.

Martin Solc

Selected References

Bohrmann, A. (1968). "Heinrich Vogt." *Mitteilungen der Astronomischen Gesellschaft*, no. 24: 7–10.

Kollnig-Schattschneider, Erika (1970). "Heinrich Vogt." *Astronomische Nachrichten* 292: 45–46.

Vogt, Heinrich (1919). "To the Theory of Algol-type Variables." Ph.D. thesis, Heidelberg.

——— (1927). "Der innere Aufbau und die Entwicklung der Sterne." *Ergebnisse der exakten Naturwissenschaften* 6: 1–26.

——— (1943). *Aufbau und Entwicklung der Sterne*. Probleme der kosmischen Physik, Vol. 24. Leipzig: Akademische Verlagsgesellschaft.

——— (1946). *Die Spiralnebel*. Heidelberg: C. Winter.

——— (1949). *Der Bau des Weltalls*. Stuttgart: C. E. Schwab.

——— (1951). *Kosmos und Gott*. Heidelberg: F. H. Kerle.

——— (1960). *Aussergalaktische Sternsysteme und die Struktur der Welt im Grossem*. Leipzig: Akademische Verlagsgesellschaft.

——— (1961). *Die Struktur des Kosmos als Ganzes*. Berlin: Morus-Verlag.

——— (1964). *Das Sein in der Sicht des Naturforschers*. Berlin: Morus-Verlag.

Volkoff, George Michael

Born **Moscow, Russia, 23 February 1914**
Died **Vancouver, British Columbia, Canada, 24 April 2000**

Working with **Julius Robert Oppenheimer**, in 1939, Russian–Canadian physicist George Volkoff calculated properties of a star supported by the pressure of neutron degeneracy. Such an object eventually would be called a neutron star.

Selected Reference

Oppenheimer, J. R. and G. M. Volkoff (1939). "On Massive Neutron Cores." *Physical Review* 55: 374–381.

Vorontsov-Veliaminov [-Velyaminov], Boris Aleksandrovich

Born **Dnipropetrovs'k, (Ukraine), 14 February 1904**
Died **Moscow, Russia, 27 January 1994**

Boris Vorontsov-Veliaminov was a prolific researcher, devoted pedagogue, and writer on a variety of astronomical subjects. He was the author of a standard astronomy textbook for high school students that, over many decades, passed through numerous editions. His books crafted for general readers attracted several generations of future astronomers. Starting in 1947, Vorontsov-Veliaminov was the only Soviet astronomer who was elected a corresponding member of the Soviet Academy for Pedagogical Science. His disciples at various levels in the former Soviet Union were almost numberless. Like **Nicolas Flammarion** in France, Vorontsov-Veliaminov in the Soviet Union was the foremost of astronomical promoters during his lifetime.

Vorontsov-Veliaminov graduated from Moscow University in 1925, where he conducted research throughout his life at the Sternberg State Astronomical Institute. He was appointed a professor of astronomy in 1934. His scientific contributions spanned a range of astrophysical topics, including "early-type" stars, diffuse and planetary nebulae, novae, and irregular galaxies.

Independently of **Robert Trumpler**, Vorontsov-Veliaminov argued (1929/1930) for the absorption of starlight in the Galaxy, although he did not provide a value for the amount of absorption (which Trumpler soon supplied). Nonetheless, his skepticism toward models of the Galaxy that neglected interstellar absorption was later vindicated.

With a team of collaborators, Vorontsov-Veliaminov published two atlases and catalogs depicting several hundred interacting galaxies (1959 and 1970). Under his guidance, detailed morphological descriptions of some 32,000 galaxies were published between 1962 and 1974. Vorontsov-Veliaminov argued, incorrectly, that magnetic fields, rather than gravitational (tidal) interactions, were chiefly responsible for the filaments and "tails" observed.

A very gentle person, Vorontsov-Veliaminov tried to do his best as an administrator when necessary. He was imprisoned for a short time under Josef Stalin's regime. Vorontsov-Veliaminov wrote extensively on the history of astronomy in Russia and the USSR and composed a memoir on astronomy in Moscow after the 1917 Bolshevik Revolution. He was awarded the Bredikhin Prize of the Academy of Sciences in 1962.

Alexander A. Gurshtein

Selected References

Anon. (1986). "Vorontsov-Veliaminov, Boris Aleksandrovich." In *Astronomy: Biograficheskii spravochnik* (Astronomers: A Biographical Handbook), edited by I. G. Kolchinskii, A. A. Korsun', and M. G. Rodriges, pp. 67–68. 2nd ed. Kiev: Naukova dumka.

Vorontsov-Veliaminov, B. A. (1948). *Gazovye tumannosti i novye zvezdy* (Gaseous Nebulae and New Stars). Moscow: Izd-vo Akademii nauk SSSR.

Vorontsov-Veliaminov, B. A. and A. A. Krasnogorskaya. (1962–1974). *Morfologicheskii katalog galaktik* (A Morphological Atlas of Galaxies). Moscow: Izd-vo Moskovskogo Universiteta.

——— (1969). *Astronomical Problems: An Introductory Course in Astronomy*. Oxford: Pergamon Press.

——— (1987). *Extragalactic Astronomy*. Chur, Switzerland: Harwood Academic Publishers.

Voytekhovich, Marian Albertovich

➤ **Kovalsky, Marian Albertovich**

W

Wābkanawī: Shams al-Munajjim [Shams al-Dīn] Muḥammad ibn ʿAlī Khwāja al-Wābkanawī [Wābkanawī]

Flourished **(Iran), early 14th century**

Wābkanawī is the author of the important astronomical handbook *al-Zīj al-muḥaqqaq*, which contains valuable historical information on lost earlier works and is one of only two *zīj*es known to be based on the observations carried out at the famous observatory at Marāgha.

Wābkanawī presumably hailed from the village Wābkana (or Wābakna) nearly 20 km from the important cultural center of Bukhara (now in Uzbekistan). Hardly anything is known about his life, and the available information about his astronomical career derives mainly from his astronomical handbook with tables, *al-Zīj al-muḥaqqaq al-sulṭāni ʿalā uṣūl al-raṣad al-Īlkhānī* (The correct *zīj* for the sultan based on the principles of the Īlkhān observations). From the introduction to this work it appears that Wābkanawī made observations during a period of 40 years, presumably at the famous observatory in Marāgha in northwestern Iran, which had been founded by Hülegü Khān at the instigation of **Ṭūsī** in 1258. However, Wābkanawī was also involved in the reform of the Malikī calendar ordered by Maḥmūd Ghāzān Khān (reigned: 1295–1304), who had an observatory built in Tabrīz. It is therefore possible that Wābkanawī spent part of his career in Marāgha and part of it in Tabrīz.

The *Zīj* of Wābkanawī is extant in four or five manuscript copies, of which no. 2694 of the Aya Sofia Library in Istanbul is the most complete. The work is written in Persian even though the title given above (found on f. 4a of the Aya Sofia manuscript) is in Arabic. Wābkanawī started working on the *Zīj* under Öljeytü Khān (reigned: 1304–1316) and finally dedicated it to Abū Saʿīd (reigned: 1316–1335). It consists of five treatises (*maqāla*s) dealing in a very extensive way with all the standard topics of *zīj*es, in particular chronology, planetary positions and eclipses, spherical astronomy, and timekeeping.

Only scattered parts of the work have been studied. The introduction is important because it mentions a number of earlier *zīj*es that are nonextant and not known from earlier sources; these include, in particular, the six *zīj*es of al-Fahhād.

The chronological chapter of the *Zīj* describes the reform of the Malikī or Jalālī calendar carried out on the order of Maḥmūd Ghāzān Khān in 1302. The original calendar had been adopted by the Seljuk Sultan Malikshāh I in 1079. Wābkanawī and various other astronomers appointed by Ghāzan Khān modified the exact definition of the beginning of the year (*i. e.*, the day of the vernal equinox), adopted a new epoch called "Khānī," and introduced the use of Turkish month names. Wābkanawī writes that he adopted the new calendar in his *Zīj*, although he uses the year 188 Malikshāh (1266) as epoch, possibly in order to cover the dates of observations made at Marāgha. Wābkanawī also presents an extensive explanation of the Chinese–Uighur calendar that was introduced into Iran by the Mongols and first described in the *Īlkhānī Zīj* of Naṣīr al-Dīn al-Ṭūsī.

The present author has made a cursory analysis of the planetary tables in *al-Zīj al-muḥaqqaq*. The mean motions were shown to have been derived from those in the *Adwār al-anwār*, the latest of the three *zīj*es by **Ibn Abī al-Shukr al-Maghribī** and known to be based on the extensive observational program carried out by that astronomer at Marāgha. Most of Wābkanawī's tables for the planetary equations were simply copied from the *Adwār*.

A work by Wābkanawī on the astrolabe, the *Kitāb-i maʿrifat-i usṭurlāb-i shamālī* (On the northern astrolabe), likewise in Persian, is extant in a manuscript in the library of the Topkapı Saray Museum in Istanbul. It consists of two chapters: one on the parts of the astrolabe and one on the operations with it. An Arabic fragment by Wābkanawī on the difference in setting times of the Sun and the Moon is extant in Cairo.

Benno van Dalen

Selected References

Dalen, Benno van, E. S. Kennedy, and Mustafa K. Saiyid (1997). "The Chinese-Uighur Calendar in Ṭūsī's Zīj-i Īlkhānī." *Zeitschrift für Geschichte der arabisch–islamischen Wissenschaften* 11: 111–152.

Kennedy, E. S. (1956). "A Survey of Islamic Astronomical Tables." *Transactions of the American Philosophical Society*, n.s., 46, pt. 2: 123–177, esp. p. 130. (Reprint, Philadelphia: American Philosophical Society, 1989.)

Krause, Max (1936). "Stambuler Handschriften islamischer Mathematiker." *Quellen und Studien zur Geschichte der Mathematik, Astronomie und Physik, Abteilung B, Studien* 3: 437–532, esp. pp. 518–519.

Saliba, George (1983). "An Observational Notebook of a Thirteenth-Century Astronomer." *Isis* 74: 388–401. (Provides proof that the *Adwār al-anwār* of Ibn Abī al-Shukr al-Maghribī is based on the observations carried out at Marāgha.)

Sayılı, Aydın (1960). *The Observatory in Islam*. Ankara: Turkish Historical Society.

Storey, C. A. (1958). *Persian Literature*. Vol. 2, pt. 1. A. *Mathematics*. B. *Weights and Measures*. C. *Astronomy and Astrology*. D. *Geography*. London: Luzac and Co., esp. p. 65.

Wachmann, Arno Arthur

Born **Hamburg, Germany, 8 March 1902**
Died **Hamburg, (Germany), 24 July 1990**

German observational astronomer Arno Wachmann is most likely to be recognized for his discovery of several comets, jointly with **Arnold Schwassmann**, although most of his work was in stellar statistics and variability. Wachmann studied astronomy in Kiel and obtained his Ph.D. degree under **Carl Wirtz** in 1926. His dissertation concerned the proper motions of some 8,800 stars. After a short time at the Remeis Observatory in Bamberg, Germany, Wachmann returned to Hamburg in July 1927, remaining there until his retirement on 31 March 1969. His position there began as a scientific assistant to Schwassmann, and he then became a scientific advisor, department chief, and finally, a senior observer. In 1961, he was appointed an honorary professor at the University of Hamburg. As a result of an invitation by Walter J. Miller, Wachmann worked at the Fordham Astronomical Laboratory in New York, USA, in 1958 and again in 1961/1962.

Much of Wachmann's work concerned an international program to investigate the statistics of stars over the entire sky. Together with Schwassmann, he worked on the Bergdorf spectral survey whereby the brightnesses, colors, spectral types, and proper motions of stars were observed over various sky regions with particular attention given to certain calibration fields, the Selected Areas of **Jacobus Kapteyn**. It was during these stellar surveys that Schwassmann and Wachmann discovered four comets. These comets, three of which are periodic, were 29P/1927 V1 (Schwassmann–Wachmann), 31P/1929 B1 (Schwassmann–Wachmann), 73P/1930 J1 (Schwassmann–Wachmann), and C/1930 D1 (Peltier–Schwassmann–Wachmann). Comet 29P, in an unusually low-eccentricity orbit beyond Jupiter, is well known for its frequent, bright outbursts. 73P disintegrated almost entirely in 2006.

For more than four decades, Wachmann discovered and monitored the light fluctuations of variable stars, first photographically and then using photoelectric techniques. In 1939, he discovered the peculiar star, FU Orionis, a variable star that became a prototype of a subgroup of T Tauri stars that were young, low-mass stars with a slow but very large brightness rise. Now called FUORs, these stars are thought to experience occasional episodes of rapid accretion of material from disks surrounding them. In 1954, Wachmann published a monograph on FU Orionis. (He was a member of the Commission on Variable Stars of the International Astronomical Union.)

During his professional career, Wachmann discovered and observed several minor planets. One minor planet, (1704) Wachmann, honors his professional achievements.

Donald K. Yeomans

Selected Reference

Haug, Ulrich (1991). "Arno Arthur Wachmann, 1902–1990." *Mitteilungen der Astronomischen Gesellschaft* 74: 5–6.

Walcher of Malvern

Flourished **England, *circa* 1100**

Walcher was a prior at the abbey of Malvern. He produced an ecclesiastical (lunar) calendar, benchmarked to his observation of a lunar eclipse on 18 October 1092.

Selected Reference

Lindberg, David C. (1978). *Science in the Middle Ages*. Chicago: University of Chicago Press.

Waldmeier, Max

Born **Olten, Switzerland, 18 April 1912**
Died **Zürich, Switzerland, 26 September 2000**

A custodian and compiler of **Johann Wolf**'s Zürich sunspot number, Swiss solar physicist Max Waldmeier created the nine sunspot classifications, A–J.

Selected Reference

Izenman, Alan Julian (1983). "J. R. Wolf and H. A. Wolfer: An Historical Note on the Zurich Sunspot Relative Numbers." *Journal of the Royal Statistical Society* 146: 311.

Wales, William

Born **Warmfield, Yorkshire, England, February 1734**
Died **London, England, 29 December 1798**

William Wales was an observational astronomer and teacher, best known for serving as one of the two astronomers on James Cook's second voyage to the South Seas. Born of humble parents, little is known of Wales's childhood except that he showed an early aptitude for mathematics. In 1765 he married Mary Green, youngest sister of **Charles Green** of the Royal Observatory, Greenwich, and was outlived by his wife and five of his children.

In 1766, Wales was commissioned by the Royal Observatory to carry out computations for the *Nautical Almanac* of 1767, and soon after was selected as one of the British observers for the 1769 transit of Venus expeditions. He and **Joseph Dymond** were forced to overwinter at the Prince of Wales's Fort in Hudson Bay, Canada, before successfully observing all four transit contacts on 3 June. They subsequently published their results in the *Philosophical Transactions of the Royal Society*. In common with other 18th-century observers, Wales and Dymond noted the contact-timing problems associated with the notorious "black drop" effect.

Following his Canadian escapade, Wales was appointed as astronomer on the *Resolution* for Cook's second voyage to the South Seas. His role was to make astronomical observations in order to determine latitude and longitude for navigational purposes, and equally importantly to monitor the accuracy of the Larcum Kendall K1 chronometer and one of the three chronometers manufactured by John Arnold. The K1 was a faithful copy of Harrison's famous prize-winning timekeeper. In addition to performing ongoing astronomical observations associated with navigation and timekeeping, Wales also observed solar and lunar eclipses and documented a number of aurorae, and together with William Bayly (astronomer on the second vessel, the *Adventure*) set about establishing the precise latitude and longitude of Queen Charlotte Sound, New Zealand, Cook's favorite Pacific revictualling center. After the voyage, Wales and Bayly wrote up their work, and *The Original Astronomical Observations. Made in the Course of a Voyage towards the South Pole* … was published in 1777.

But this did not end Wales's involvement with the Cook voyages, for in 1778 he was charged with producing the official astronomical account of the first voyage, a task fraught with difficulty given the death of the expedition's astronomer (his brother-in-law Green) on the voyage and the deplorable state of his papers. Compounding matters was the directive to include data from earlier South Sea voyages. As a result, *Astronomical Observations. Made in the Voyages which were Undertaken by Order of His Present Majesty, for Making Discoveries in the Southern Hemisphere* … only came off the press in 1788.

Part of the reason for the delay in preparing this volume lay in the fact that Wales was in full-time employment, for after returning from the South Seas in 1775 he had been appointed master of the Royal Mathematical School at Christ's Hospital in London. This novel academic institution was founded in 1673 in order to train budding ships' officers and others in the principles of mathematics and navigation. Over the years, Wales built the school into a great naval academy, and published two new editions of Robertson's *Elements of Navigation*, a book of his own on aspects of navigation (which also ran to two editions), and a volume on British demography.

In late 1776 Wales was elected a fellow of the Royal Society, a considerable honor at the time, and in 1795 he was appointed secretary of the Board of Longitude. By the time he died, he had made important contributions to 18th-century astronomy, navigation, and education. Professor **Thomas Hornsby**'s figure for the Astronomical Unit derived from Wales's 1769 transit of Venus observations and those of other British observers vary little from the currently adopted value. Wales's volumes on Cook voyage astronomy and on navigation are a testament to scholarship. His evaluation of the chronometers during the second voyage led eventually to their widespread adoption by the Royal Navy and through his post at Christ's Hospital a generation of boys learned the rudiments of mathematics and navigation.

Wayne Orchiston

Selected References

Carlyle, Edward I. (1921–1922). "Wales, William." In *Dictionary of National Biography*, edited by Sir Leslie Stephen and Sir Sidney Lee. Vol. 20, pp. 490–491. London: Oxford University Press.

Moore, P. (1981). "William Wales." In *The Journal of H.M.S. Resolution, 1772–1775*, edited by J. Cook, pp. 86–93. Guildford: Genesis Publications.

Orchiston, Wayne (1998). *Nautical Astronomy in New Zealand: The Voyages of James Cook*. Occasional Papers, no. 1. Wellington, New Zealand: Carter Observatory.

Orchiston, Wayne and D. Howse (1998). "From Transit of Venus to Teaching Navigation: The Work of William Wales." *Astronomy and Geophysics* 39, no. 6: 21–24. (Reprinted, with modifications, in *Journal of Navigation* 53 (2000): 156–166.)

Walker, Arthur Geoffrey

Born **Watford, Hertfordshire, England, 17 July 1909**
Died **Sussex, England, 31 March 2001**

British mathematician Arthur Geoffrey Walker is remembered within cosmology for the formulation of the Robertson–Walker metric, a very general description of the four-dimensional structure of a homogeneous, isotropic space-time, applicable to the Universe as a whole.

Walker received an M.A. from Oxford University and a Ph.D. from the University of Edinburgh. He was appointed a lecturer in mathematics at Imperial College, London, in 1935, moved to Liverpool University (1936–1947) and to the chair of mathematics at the University of Sheffield, and finally returned to Liverpool University, from where he retired in 1974. Walker was a fellow of the Royal Society of London and of the Royal Society of Edinburgh, and received prizes from the Royal Society of Engineers and the London Mathematical Society.

Walker's research focused on geometry, especially Reimannian and other manifolds. In 1935/1936, he (and, independently, **Howard Robertson**) recognized that the solutions to **Albert Einstein**'s equations of general relativity, published earlier by **Wilhelm de Sitter** and **Georges Lemaître**, embodied a way of looking at the geometry of space-time that could be generalized to apply to a wider range of theories of gravity and cosmological models for any homogeneous, isotropic universe. The metric is still in general use, and a theory of gravity that cannot be put into metric form (this or some other) is automatically somewhat suspect.

Douglas Scott

Selected Reference

Walker, Arthur G. (1935). "On Reimannian Spaces with Spherical Symmetry about a Line, and the Conditions of Isotropy in General Relativity." *Quarterly Journal of Mathematics* 6: 81–93.

Walker, Sears Cook

Born **Wilmington, Massachusetts, USA, 28 March 1805**
Died **Cincinnati, Ohio, USA, 30 January 1853**

Sears Walker, a leading American mathematical astronomer, founded one of the first major research observatories in the United States and calculated a precise orbit for the newly discovered planet Neptune. He also headed the United States Coast Survey's pioneering development of longitude determinations using the telegraph, a technique that dominated geodesy worldwide in the 19th and into the early 20th century.

Walker was born near Boston, of farmers Benjamin Walker and Susanna Cook. As a child, his intellectual precocity and retentive memory were the wonder of the village and a worry to his mother, who tried to discourage his studiousness in favor of outdoor activity – to little avail; Walker was what today would be termed a workaholic, and was plagued by lifelong fragile health and obesity.

In 1825 Walker graduated from Harvard College, remarkably apt at acquiring languages and mastering Latin, Greek, Italian, and German. He taught school in Boston for 2 years before moving to Philadelphia, Pennsylvania (1827), his home for the next two decades. There Walker continued teaching for 8 years, spending his free time studying medicine, natural history, geology, mineralogy, physics, chemistry, and astronomy. One of his significant astronomical publications during this time was a set of tables (1834) for the parallax of the Moon that he had calculated for the latitude of Philadelphia, which reduced the time required for computing the phases of a lunar occultation to under half an hour. He also acquired a small Dollond refractor, a 20-in. focal-length transit instrument, and an astronomical clock, and began astronomical observations, eventually devoting all his leisure hours to astronomy alone.

In 1836, Walker gave up teaching to become an actuary for the Pennsylvania Company for the Insurance of Lives and Granting Annuities (a position analogous to that held by **Nathaniel Bowditch** in Boston around the same time), where he remained until 1845. In 1837, the Central High School in Philadelphia was founded, with $5,000 allocated to erect an astronomical observatory for the school. Walker, who in 1839 became the observatory's director, advised the founders to order the main instrument from Merz and Mahler in Munich, Bavaria, rather than from England or France. The 6-in. refractor (with a focal length of 8 ft.), installed in 1840, was one of the two largest telescopes in the United States from 1840 to 1843, and introduced the superior craftsmanship of the German makers to American astronomers. In the hands of Walker and his astronomer half brother E. Otis Kendall (who succeeded him as director), this high-school observatory became an acclaimed research institution, noted on both sides of the Atlantic for its comprehensive timings of lunar culminations and occultations for determining geographical longitudes and latitudes.

Although previously well-to-do, by 1845, unfortunate investments left Walker destitute at the age of 40. Fortunately, he was invited to join the US Naval Observatory, moving to Washington in 1846. That fall, the observatory's director **Matthew Maury** ordered Walker to calculate an orbit for the planet Neptune, recently discovered by **Johann Galle** in Berlin near the position for a hypothetical planet predicted almost simultaneously by **Urbain Le Verrier** of Paris and **John Adams** of Cambridge University. In early 1847, Walker discovered that on 8 and 10 May 1795, Neptune had been observed as a fixed star by **Joseph de Lalande**, director of the Paris Observatory, and had been so recorded in Lalande's 1802 catalog of 47,000 stars, *Historie céleste française*. From that additional position, Walker was able (in part based on gravitational perturbations calculated by Harvard mathematician **Benjamin Peirce**) to calculate an orbit for Neptune that accurately predicted the planet's future positions – an achievement that won him international renown.

Meanwhile, unhappy under Maury, Walker left the Naval Observatory and was promptly hired by **Alexander Bache**, superintendent of the Coast Survey, the agency charged with accurately charting all US shorelines. There, Walker took charge of determining longitudes by the new technology of the telegraph, a position he held until his death.

Because of the rotation of the Earth, a difference in longitude is equivalent to a difference in local time (as measured by the stars). But there was no direct way to compare clocks that were hundreds of miles apart – until 1844, with the advent of the telegraph and its seemingly instantaneous signal-transmission time. As soon as telegraph lines reached a city, Walker traveled to local observatories to determine longitude differences between them and other key points, rapidly tying into one geodetic system points ranging from Charleston, South Carolina, to Cincinnati, Ohio, to Halifax, Nova Scotia, Canada. The telegraphic method of determining longitudes – soon dubbed the American method – became standard in geodesy until displaced by radio techniques in the 1920s.

By pure serendipity, Walker's work with the American method led to a crucial scientific observation. In January 1849, during the first experiment with permanently recording the local times of star transits over four observatories of differing longitudes, Walker noticed that the transit times recorded at each observatory compared with the time signals from one central clock differed depending on the distance of the observatory from the recording apparatus. In short, he stumbled onto the discovery that an electromagnetic signal was not instantaneous (the accepted view) but had a finite and measurable velocity through the circuit of about a tenth the speed of light in a vacuum.

In August 1851 Walker was afflicted by mild paralysis that deprived him of the use of one hand for several days. Despite pleas from friends and physician, he kept his usual schedule of working dawn to dusk. In 1852, showing symptoms of mental illness, he spent several months recuperating at two asylums, even there still working ceaselessly to refine the ephemeris of Neptune. During a visit to his brother Timothy, the never-married Walker died from fever.

Trudy E. Bell

Selected References

Bartky, Ian R. (2000). *Selling the True Time: Nineteenth-Century Timekeeping in America*. Stanford, California: Stanford University Press.

Bell, Trudy E. (2002). "The Victorian Global Positioning System." *The Bent* 93(2): 14–21.

Gould, Jr., Benjamin Apthorp (1854). *An Address in Commemoration of Sears Cook Walker, Delivered Before the American Association for the Advancement of Science*, April 29, 1854. Cambridge, Massachusetts: Joseph Lovering.

H. L. S. [Hamilton Lamphere Smith], "Obituary [Sears Cook Walker]" *Annals of Science* 1 (1853): 110.

Hubbell, John G. and Robert W. Smith (1992). "Neptune in America: Negotiating a Discovery" *Journal for the History of Astronomy* 23: 261–291.

Jones, Bessie Zaban and Lyle Gifford Boyd (1971). *The Harvard College Observatory: The First Four Directorships, 1839–1919*. Cambridge: Harvard University Press.

Loomis, Elias (1856). *The Recent Progress of Astronomy; Especially in the United States*. New York: Harper & Brothers. Includes a detailed write-up of the Philadelphia High-School Observatory and an entire chapter chronicling the early development of the American method of determining longitudes.

Rothenberg, Marc (1983). "Observers and Theoreticians: Astronomy at the Naval Observatory, 1845–1861." In *Sky with Ocean Joined: Proceedings of the Sesquicentennial Symposia of the U.S. Naval Observatory*, December 5 and 8, 1980, edited by Steven J. Dick and LeRoy E. Doggett, pp. 29–43. Washington, DC: US Naval Observatory.

Williams, Frances Leigh, *Matthew Fontaine Maury: Scientist of the Sea*. New Brunswick, New Jersey: Rutgers University Press, 1963.

Wallace, Alfred Russel

Born **Usk, Monmouthshire, Wales, 8 January 1823**
Died **Wimborne near Bournemouth, Dorset, England, 7 November 1913**

Although Alfred Russel Wallace made significant contributions to astronomy, he is best known as a central figure in the emergence of the fields of evolutionary biology and biogeography.

As the eighth of nine children of Thomas Vere and Mary Anne (*née* Greenell) Wallace, poor but middle-class English parents, Alfred Wallace led a rather ordinary life until his midteens. At that time, while working as a surveyor in western England and Wales, he began to take an amateur's interest in natural history. In the early 1840s Wallace became involved in local mechanics' institutes as a lecturer, curator, and possibly librarian. In 1844 he took a position as a master at Leicester School, where he incidentally met another famous naturalist-to-be, Henry Walter Bates. The two would eventually decide to turn professional as natural history collectors. In 1848 they voyaged to the Amazon region, where in the following years they were quite successful collecting specimens. Wallace returned to England in 1852 when his health deteriorated; on the way he narrowly escaped death when his ship caught fire and sank, but he lost some 2 years of collections in the disaster. Undaunted, Wallace set off for the Far East 18 months later to reimmerse himself in collecting activities.

Wallace's name is now inextricably linked with the Malay Archipelago, where 8 years of fieldwork (1854–1862) secured for him a reputation among future generations as history's greatest tropical naturalist. While there he thought out the theory of natural selection; the famous essay on the subject he sent to Charles Darwin is now well known to have propelled the latter into finally committing his own ideas to paper in *On the Origin of Species* in 1859. Over the same period Wallace made fundamental contributions to the study of biotic distribution patterns, and is now regarded as the father of the science of zoogeography. Wallace returned to England in 1862, thereafter settling down to a long career of study and writing.

Wallace's contributions to astronomy are overshadowed by his fame in other natural sciences, but his thoughts and writings on astronomical topics were and still are influential, and in some areas he may be regarded as an important pioneer. Although lacking even a secondary-school education, he developed a firm grasp of basic scientific principles, and later was particularly brilliant at marshaling evidence and drawing conclusions. Wallace's attention was drawn to astronomy during his early surveying days, when practical geodetical matters were of daily concern. He developed a talent for cartography, a skill he would exercise during his Amazon travels by producing one of the first reliable maps of the course of the Rio Negro.

In 1865, after his return from the Malay Archipelago, Wallace became embroiled in a public discussion on the shape of the Earth. While discussing this incident in his 1905 autobiography *My Life* he produced a nontechnical explanation of the derivation of latitude that geographer Yi Fu Tuan would later describe as never having been surpassed in clarity. Wallace's fascination with geodesy culminated in 1870 when he devised the celebrated Bedford Canal experiment, an attempt to silence the claims of a particularly outspoken advocate of a flat Earth.

In the 1860s Wallace also became interested in **James Croll**'s ideas about the possible astronomical causes of the glacial epochs. Wallace adopted some of Croll's theory of climate change as related to eccentricity of the Earth's orbit and precessional movement of its axis, but added his own twist by examining possible synergistic interplays between astronomical and climato-geographical forces. His fully developed theory along these lines – the first of its kind – was presented as the opening sections of the book *Island Life* in 1880.

In 1896 Wallace was invited to Switzerland to give a lecture on scientific progress; the research he did for this lecture and in 1898 for a related book, *The Wonderful Century*, rekindled his interest in astronomy, and he soon took up the subject again. Adopting **William Whewell**'s position on the plurality of worlds and relying on his thorough review of the recent astronomical literature, Wallace attempted to make the argument that the Earth and Solar System are located at the very center of the Universe. Further, he argued that, on a consideration of the physical improbabilities involved, ours is probably the only existing world inhabited by advanced creatures. This position was first advanced by Wallace in an essay published in early 1903, but later that year he produced a much-expanded discussion in the book *Man's Place in the Universe*, which drew both much attention and much criticism.

A few years later Wallace was drawn into the discussion surrounding **Percival Lowell**'s sensational view that the planet Mars is inhabited by advanced beings. In 1907 Wallace published a short book, *Is Mars Habitable?* that devastatingly criticized the range of problems inherent in Lowell's position. The discussion remained close to principles of basic science with Wallace surmising that the Red Planet's surface must be desertlike and devoid of higher life forms. He was able to accurately deduce its likely surface temperatures and albedo, and to suggest that its polar caps are probably frozen carbon dioxide rather than frozen water.

The astronomical writings Wallace produced over the last decade of his life reflect an unusually flexible worldview: one scientific enough to address questions bearing on proximate causalities, yet philosophical enough to find a place for final causes. Although he has sometimes been accused of theistic leanings, he strictly rejected the notion of a reality operating on first causes and therefore, in spite of all his spiritualist beliefs, was in no sense a creationist. Still, he did believe there was purpose exhibited by natural structure and its programs of change. In examining this matter scientifically in the context of astronomy Wallace became perhaps the first important purveyor of what has come to be known as the anthropic principle. With this philosophical perspective it is all the more interesting that his most important contribution to the progress of astronomy was a methodological one: his analytical approach to the study of planetary atmospheres and surfaces toward the end of assessing their potential for life-sponsoring conditions. For this latter work he may justly be regarded as a founding father of the science of astrobiology.

Wallace's career, especially after 1862, was characterized by frequent public controversy, for in addition to his natural science interests Wallace was also a vocal and demonstrative spiritualist, land nationalizer, antivaccinationist, and socialist. In April 1866 at the age of 43, he married Ann Mitten, the 20-year-old daughter of the English botanist William Mitten. They had three children, two of whom survived to adulthood. By the time of his death, Wallace was well honored: He was a fellow of the Royal Society and received the society's Royal Medal (1868), its Darwin Medal (1890), and its Copley Medal (1908). Among many other honors including two honorary doctorates, he was the first recipient of the Darwin–Wallace Medal of the Linnean Society of London (1908) and the Order of the British Empire.

Charles H. Smith

Selected References

Balashov, Yuri V. (1991). "Resource Letter AP-1: The Anthropic Principle." *American Journal of Physics* 59: 1069–1076.

Dick, Steven J. (1996). *The Biological Universe: The Twentieth-Century Extraterrestrial Life Debate and the Limits of Science*. Cambridge: Cambridge University Press.

Fichman, Martin (2004). *An Elusive Victorian: The Evolution of Alfred Russel Wallace*. Chicago: University of Chicago Press.
Garwood, Christine (2001). "Alfred Russel Wallace and the Flat Earth Controversy." *Endeavour* 25, no. 4: 139-143.
Heffernan, William C. (1978). "The Singularity of Our Inhabited World: William Whewell and A. R. Wallace in Dissent." *Journal of the History of Ideas* 39: 81–100.
Tipler, Frank J. (1981). "A Brief History of the Extraterrestrial Intelligence Concept." *Quarterly Journal of the Royal Astronomical Society* 22: 133–145.
Tuan, Yi-Fu (1963). "Latitude and Alfred Russel Wallace." *Journal of Geography* 62: 258–261.
Wallace, Alfred Russel (19 May 1866) "Is the Earth an Oblate or a Prolate Spheroid?" *Reader* 7, no. 177: 497b-c.
——— (1898). *The Wonderful Century: Its Successes and Its Failures*. London: Swan Sonnenschein. (New revised edition, 1903 with the subtitle *The Age of New Ideas in Science and Invention*.)
——— (1903). *Man's Place in the Universe: A Study of the Results of Scientific Research in Relation to the Unity or Plurality of Worlds*. London: Chapman and Hall. (4th ed., London: G. Bell and Sons, 1904.)
——— (1905). *My Life; A Record of Events and Opinions*. 2 Vols. London: Chapman and Hall. (New condensed and revised edition, 1908.)
——— *Is Mars Habitable? A Critical Examination of Professor Percival Lowell's Book Mars and Its Canals, With an Alternative Explanation*. London: Macmillan, 1907; New York: Macmillan, 1908.

Wallis, John

Born **Ashford, Kent, England, 23 November 1616**
Died **Oxford, England, 28 October 1703**

Aristarchus expert John Wallis was primarily a mathematician, and should be considered one of the inventors of analytic geometry. He lived during the English Revolution, and Oxford fell into the portion of England that sided with the Parliamentarians. Wallis decoded some messages from Royalists that came into the hands of the Parliamentarians. He was later accused of having decoded the personal letters of King Charles himself, a charge that Wallis adamantly denied. In his old age, Wallis taught what he knew of cryptography to his grandson, William Blencowe, though by then, Wallis admitted, the new French methods of encryption were too complicated to break by the means used by Wallis.

In 1649, Wallis became Savilian Professor of Geometry at Oxford, more likely because of his support for the Parliamentarians than for his mathematical ability. However, he soon proved that, political appointment or not, he well deserved the chair. In 1663, he was elected a fellow of the Royal Society.

Through the use of conjecture and interpolation, he was able to obtain an infinite product expansion for π, and had a considerable influence on **Isaac Newton**'s mathematical development. Wallis also played an important role in the development of analytic geometry and was among the first to consider curves defined purely by an algebraic equation.

Wallis's main contribution to astronomy was his publication and annotation of the Greek text *On the Sizes and Distances of the Sun and Moon* by Aristarchus. Aristarchus was the first to put forward a heliocentric model of the planetary system, and **Nicolaus Copernicus** used Aristarchus's work to support his own. Latin and Arabic translations of Aristarchus's writings were widely available, but Wallis's 1688 version was the first printed edition of the Greek text. Wallis based his work on two copies, one made by **Henry Savile** from a copy in the Vatican, and the second a Greek manuscript in the possession of Edward Bernard, the Savilian Professor of Astronomy at Oxford.

Jeff Suzuki

Selected References

Heath, Sir Thomas L. (1913). *Aristarchus of Samos, the Ancient Copernicus*. Oxford: Clarendon Press. (Reprint, New York: Dover, 1981.)
Scott, J. F. (1938). *The Mathematical Work of John Wallis*. London: Taylor and Francis.

Walther, Bernard [Bernhard]

Born **Memmingen, (Bavaria, Germany), 1430**
Died **Nuremberg, (Germany), 15 June 1504**

Bernard Walther was, after the death of **Johann Müller** (Regiomontanus), the leading astronomer of his time. Though unpublished in his lifetime, Walther's 30-year series of astronomical observations established a new approach to observational astronomy based on the integrity of the instrument with due concerns for accuracy, and awareness and recording of conditions attendant to observations. Walther's observations were later used to advantage by **Nicolaus Copernicus**,

Johannes Kepler, and **Tycho Brahe**, and for the first time demonstrated the value of an extended series of observations for testing theories.

Walther was a Nuremberg merchant who may have helped to fund the observing program of Regiomontanus. Though such patronage has not been proven, Walther proclaimed he was a student of Regiomontanus. Walther and Regiomontanus often observed together. Like Regiomontanus before him, Walther noted discrepancies for the positions of the planets and circumstances of eclipses published in Regiomantanus' almanacs, which were based on the *Alfonsine Tables*. The understanding of these important discrepancies and improvement of astronomical tables became the rationale for their observing program of planetary positions and eclipse timings.

Walther carried on the procedures of astronomical observation after the death of his teacher. He safeguarded much of Regiomontanus' estate, preserving for eventual publication many manuscripts involving both astronomical data and theory. Strangely though, Walther published neither manuscripts from Regiomontanus nor his own data. He apparently withheld the observations and manuscripts that he had inherited from other astronomers during his own lifetime.

Walther modified his Nuremberg house so that he could make nighttime observations of the sky. The Walther residence was evidently Nuremberg's first astronomical observatory. The famous artist **Albrecht Dürer** later occupied the house; it remained standing until it was destroyed in World War II.

After Regiomontanus's death, Walther bought his mentor's brass observing instruments, and continued to make many astronomical observations. These instruments included a mechanical clock, an armillary sphere, and astrolabes.

Regiomontanus suggested using a geared clock for timing astronomical events. This suggestion spurred Walther to use such a clock in some of his astronomical observations beginning in 1484. Though clocks at that time were not very reliable, and he evidently did not use one all the time, Walther is one of the first astronomers to do so, though his use of a clock was possibly preceded by John Abramius several decades earlier. An analysis of the timings of six different eclipses by Walther and Regiomontanus has shown that they attained an accuracy of about 7″.

Another Walther contribution to improved observing practice was that he recorded sky conditions and estimated the reliability of each astronomical observation. Walther's example set a precedent that would be followed in the 16th century by others who read his notes after they were published with the observations in 1544 by **Johannes Schöner**.

Walther's positional observations were unusually accurate. His long series of many hundreds of observations of the Sun, planets, and at least one comet made an impact on later observers and analysts of astronomical data. For example, Schöner evidently obtained the papers of Walther and Regiomontanus sometime after Walther's death and may have provided them to Copernicus who used some of Walther's observations in his *De revolutionibus*. In discussing the Moon's motion in his 1604 *Optical Part of Astronomy*, Kepler used the example of Walther's 1482 observation of a lunar occultation of Saturn. Brahe also alluded to Walther often, and used his data to evaluate atmospheric refraction as well as competing theories of solar motion. Walther was a careful enough observer that he discovered atmospheric refraction for objects low in the sky before learning in astronomy treatises that earlier observers also had done so.

Little else is known about Walther, other than that he was part of a circle of humanists in Nuremberg and that he knew Greek. A surviving will indicates that he was a man of modest means.

Daniel W. E. Green and *Thomas R. Williams*

Selected References

Beaver, Donald Deb. (1970). "Bernard Walther: Innovator in Astronomical Observation." *Journal for the History of Astronomy* 1: 39–43.

Donahue, W. H. (trans.) (2000). *Johannes Kepler: Optics*. Sante Fe: Green Lion Press, p. 411.

Eirich, Raimund (1987). "Bernhard Walther (1430–1504) und seine Familie." *Mitteilungen des Vereins für Geschichte der Stadt Nürnberg* 74: 77–128.

Kremer, Richard L. (1980). "Bernard Walther's Astronomical Observations." *Journal for the History of Astronomy* 11: 174–191.

——— (1981). "The Use of Bernard Walther's Astronomical Observations: Theory and Observation in Early Modern Astronomy." *Journal for the History of Astronomy* 12: 124–132.

Pilz, K. (1970). "Bernhard Walther und seine astronomischen Beobachtungsstände." *Mitteilungen des Vereins für Geschichte der Stadt Nürnberg* 57: 176–188.

Schoener, J. (ed.) (1544). "*Ioannis de Monteregio et Barnardi Waltheri eivs discipvli ad solem obseruationes*" and "Ioannis de Monteregio, Georgii Pevrbachii, Bernardi Waltheri, ac aliorum, Eclipsium, Cometarum, Planetarum ac Fixarum obseruationes." In *Scripta clarissimi mathematici* ... Nuremberg: Ioannem Montanum & Ulricum Neuber. (Contains typographical errors in the printing of Walther's observations.)

Snell, W. (ed.) (1618). *Coeli & siderum in eo errantium Observationes Hassiacae Lantgravii ... Quibus accesserunt Ioannis Regiomontani & Bernardi Walteri Observationes Noribergiacae*. Leiden: Justum Colsterum. (Repeats typographical errors in the printing of Walther's observations in this work.)

Steele, John M. and F. Richard Stephenson (1998). "Eclipse Observations Made by Regiomontanus and Walther." *Journal for the History of Astronomy* 29: 331–344.

Zinner, Ernst (1990). *Regiomontanus: His Life and Work*, translated by Ezra Brown, pp.138 ff. Amsterdam: Elsevier Science Publications.

Wang Xun

Born **(Hebei), China, 1235**
Died **China, 1281**

Wang Xun was a Chinese mathematician and astronomer in the Yuan dynasty (1260–1367). He learned to read at 6 years of age and became interested in mathematics and nature. Wang Xun went to Zijing Mountain in Hebei to study with Liu Bingzhong (1216–1274), a well-known scholar in the Yuan dynasty. At the same time, **Guo Shoujing** (1231–1316) was also one of Liu's students. Under the guidance of Liu, Wang Xun devoted himself to the study of mathematics and astronomy, which laid the foundation for his future astronomical research. After finishing his study at Zijing Mountain, Wang Xun was recommended by his teacher to Khublai Khan (1215–1294) and was appointed tutor to the crown prince. In 1261/1262 he was promoted twice to higher office. Meanwhile, he became a well-known mathematician.

In 1276, acting on the earlier advice of Liu Bingzhong, Khublai Khan decided to reform the calendar after the existing calendar was found to be inaccurate. He set up a new astronomical and historiographical bureau, the Taishiju, to undertake the work. Wang Xun was appointed supervisor of the academic activities in the bureau; Guo Shoujing was also transferred there from the water conservancy department. The Taishiju developed into the Taishiyuan, the Imperial Bureau of Astronomy and Calendrics (1279). Wang Xun therefore was promoted to Taishiling, the highest academic post in the bureau, and Guo Shoujing became his deputy. Of course, there were other higher officials, such as **Xu Heng**, who were in charge of the administration of the institute.

To fulfill the task of calendrical reform, an observatory was established in the capital, Dadu (the present Beijing), and several new astronomical instruments were designed and built. Meanwhile, a large-scale research program was initiated. The program consisted of astronomical observation, nationwide surveying of the altitude of the North Celestial Pole (equivalent to geographical latitudes), measurement of gnomon-shadow lengths at the summer solstice, studies of the history of calendars, and a great deal of calculation. Several scholars took part in the program, sharing the work and cooperating with one another. Wang Xun took on the task of the calculations, Guo Shoujing designed and made instruments and astronomical observations, Xu Heng studied the history of astronomy, and Duan Zhen constructed the observatory. The result of the program was the completion of a new calendar in 1281 called the Shoushi Calendar.

As a mathematician, astronomer, and supervisor-general of the whole program, Wang Xun made a significant contribution to the development of the Chinese calendar and astronomy. Worn out from constant overwork and the death of his family members, Wang died.

Guo Shirong

Selected References

Bai Shangshu and Li Di (1985). *Shisan shiji Zhongguo shuxuejia Wang Xun* (Wang Xun: A Chinese mathematician of the 13th century). In *Zhongguo shuxueshi lunwenji* (Collection of papers on the history of mathematics in China, Series 1). Jinan: Shandong Educational Press, pp. 37–51.

Su Tianjue (1294–1352). *Yuanchao mingchen shilue* (Sketches of eminent officials of the Yuan Dynasty). Vol. 9.

Yuan shi. *(History of the Yuan Dynasty).* Compiled 1369–1370. (For biography of Wang Xun.)

Ward, Isaac W.

Born **Belfast, (Northern, Ireland), 13 September 1834**
Died **Drumbeg, (Northern, Ireland), 11 October 1916**

Belfast amateur Isaac Ward spied a "new" star on 19 August 1885. This object, appearing in the Great Andromeda Nebula (M31), was the first recorded extragalactic supernova (SN 1885A). The following evening it was also observed by **Ernst Hartwig**.

Selected Reference

Beesley, David E. (1985). "Isaac Ward and S Andromedae." *Irish Astronomical Journal* 17: 98–102.

Ward, Seth

Born **Aspenden, Hertfordshire, England, April 1617**
Died **London, England, 6 January 1689**

Seth Ward was a mathematician, astronomer, clergyman, and controversialist. Baptized on 5 April 1617, he was the second son of John Ward, an attorney, and Martha (*née* Dalton) Ward, mother of six. After taking degrees at Cambridge (BA: 1637; MA: 1640) Ward was appointed Savilian Chair in Astronomy at Oxford (1649–1661), having been nominated by his ousted predecessor, **John Greaves**. Previous Savilian professors—**John Bainbridge** and Greaves—taught **Ptolemy**; Ward was the first to lecture on the Copernican Systems. A founding fellow of the Royal Society, Ward produced several short works in mathematics before becoming Bishop of Exeter (1662) and Salisbury (1667). When Ward finally left Oxford and academic life (1662), the Savilian Chair was taken by **Christopher Wren**.

Ward's career in science was marked by controversy. The first dispute, now known as the Webster–Ward debate, focused on the role of the New Science in the university curriculum. In his *Vindiciae academiarum* (Oxford, 1654), Ward opposed claims of John Webster and positions taken by Thomas Hobbes, arguing that the mathematical sciences were faring well in England and that university reform was unnecessary. Ward later attacked Hobbes' materialist philosophy in his *Thomae Hobbii philosophiam exercitatio epistolica* (Oxford, 1656).

But Ward's defining contribution to astronomy – the so called "simple elliptical hypothesis" – emerged from the Boulliau–Ward debate. This dispute focused on Kepler's laws, or more precisely, alternatives to **Johannes Kepler**'s first two planetary rules, now known as the ellipse and area laws. Significantly, deep cosmological concerns formed the core of the debate, issues first framed by **Ismaël Boulliau** in his influential *Astronomia philolaïca* (Paris, 1645). Here Boulliau made startling claims against Kepler's now-famous cosmological contributions. Calling Kepler "ingenious" and "sagacious," as well as a "mediocre geometer," Boulliau rejected Kepler's celestial physics as "mere figments" and dismissed his demonstrations as "a-geometric." In place of Kepler's *anima mortrix* and "magnetic fibers" Boulliau argued it was more natural to assume that the planets were self-moved. In place of Kepler's cumbersome geometry (involving trial and error approximation) Boulliau proposed a "direct solution" based on mean motion.

Here Ward opposed Boulliau's assumptions, methods, and conclusions. Prompted by Sir Paul Neile, Ward published two treatises attacking Boulliau's geometrical procedures. In his *Inquisitio brevis* (1653) Ward claimed to produce a more accurate elliptical theory, the "simple elliptical hypothesis," which had in fact been known earlier to Kepler, **Albert Curtz**, and Boulliau himself. Focusing strictly on mathematical methods, Ward never questioned the empirical accuracy of Boulliau's theory but insisted (incorrectly) that his results were more accurate. Three years later Ward published his *Astronomia geometrica* (1656), which provided a more polished account but which continued to ignore empirical factors. Finally, in the following year, Boulliau responded with his *Astronomia philolaica fundamenta clarius explicata* (Paris, 1657). After acknowledging his error (noted in his *Astronomia philolaïca*), Boulliau demonstrated that Ward had mistakenly identified the conical hypothesis with his alternative "simple elliptical" model, that is, an ellipse where the empty (non-solar) focus served as an equant point. To demonstrate the difference, Boulliau cleverly argued that if Ward's hypothesis were applied empirically to the planet Mars, it would result in a maximum error of over 7′ in heliocentric longitude – not the 2.5′ calculated from the conical hypothesis. While his "modified elliptical hypothesis" (1657) seemed to win the day – it surpassed the accuracy of Kepler's "area law" for Mars – Boulliau's cosmological principles failed to excite focused attention.

But the technical debate continued. Over the next three decades the great debate on the "problem of the planets" was largely reduced to quibbles about "saving the appearances." But there was wide agreement about planetary orbits. By the 1660s the elliptical path was commonplace. Thanks to Boulliau's "English Lieutenants" (**Jamy Shakerley,** John Newton, **Vincent Wing**, **Nicolaus Kauffman** (Mercator), and **Thomas Streete**), the modified elliptical hypothesis had been continually refined. Yet in retrospect this technical success was also marked by failure. Debates of the 1660s and 1670s failed to illuminate the critical relationship between computational simplicity and cosmological explanation. If this lapse marks a "retrograde step," it was not a singular *faux pas*. In any case, by the 1670s the "problem of the planets" had became an English affair, and by tradition, that decade marks the uneasy divide between "post-Keplerian" and "pre-Newtonian" astronomy. In the end, by the time of **Isaac Newton**'s *Principia* (1686), Ward had left a large legacy of sermons but nothing new in science. Most of Ward's papers are lost – reportedly his cook used them for kindling and as doilies for cooling potpies.

Robert Alan Hatch

Selected References

Aubrey, John (1949). *Aubrey's Brief Lives*, edited by Oliver Lawson Dick. London: Secker and Warburg.

Pope, Walter (1697). *The Life of Seth: Lord Bishop of Salisbury*. (Reprint, edited by J. B. Bamborough Oxford: Basil Blackwell, 1961.)

Wilson, Curtis A. (1970). "From Kepler's Laws, So-called, to Universal Gravitation: Empirical Factors." *Archive for the History of the Exact Sciences* 6: 89–170.

Wargentin, Pehr Wilhelm

Born **Sunne Prästgard, Sweden, 11 September 1717**
Died **Stockholm, Sweden, 13 December 1783**

Pehr Wargentin, secretary general of the Royal Swedish Academy of Sciences from 1749 to 1783, was a renowned Swedish astronomer and statistician who devoted much of his scientific energy to the study of the motion of Jupiter's satellites. Pehr was the son of Wilhelm Wargentin, vicar of Sunne Parish, and Christina Aroselius. He married Christina Magdalena Raan in 1756. Together, they had three daughters and remained together until Christina's death in 1769.

Wargentin became interested in astronomy as a youth and, after receiving his early education at Frösö Trivialskola, he began his studies at Uppsala University on 30 January 1735. There he was introduced to professor **Anders Celcius**, who led Wargentin to become interested in the study of the motion of Jupiter's satellites. On 12 December 1741, Wargentin defended his dissertation entitled *De satillitibus Iovis*, a treatise on the history and motions of these satellites and developed *Tabulae pro eclipsibus satellitym Iovis*, a table for finding the known moons of Jupiter, which caught the attention of the academic world. In 1746 he presented the dissertation *De incremento astronomiae ab ineunte hoc seculo* and was appointed to the position of senior lecturer at the university. Wargentin was elected to the Swedish Academy of Sciences in 1748 and was appointed to the philosophy faculty at Uppsala University in the same year.

In 1753 Wargentin became the first director of the Stockholm Observatory and arrived there on 12 April. The observatory opened on 20 September 1753, and again he devoted his time to studies of the motion of Jupiter's satellites. His continued interest in this area is understandable, because one of the most important problems of the day was the development of a method for determining the "longitude of places," particularly for shipboard navigation. At that time there was considerable support for using the motions of the Jovian moons as a means of determining time for conversion to a measure of terrestrial longitude. In addition, Wargentin's studies also included observations of the opposition of Mars in 1751 and observing the transits of Venus in the 1760s when it twice passed between the Earth and the Sun. The quality of his work was well recognized and resulted in his election to fellow of the Royal Society in 1764 as well as his becoming a member of the French Academy of Sciences.

Wargentin's contributions to science went beyond astronomy. He held the position of secretary general of the Royal Swedish Academy of Sciences for more than 30 years, and he headed the government agency Tabellverket, which was established to compile population records and was the forerunner of Statistics Sweden. The collection of population records in Sweden had its beginnings with a 1686 law that required the Church of Sweden to keep records of people who moved in and out of each parish. Through Wargentin's efforts in compiling these records he is considered to have laid the groundwork for modern Swedish population statistics.

Scott W. Teare and *M. Colleen Gino*

Selected References

Debus, Allen G. (ed.) (1968). "Wargentin, Pehr Vilhelm." In *World Who's Who in Science*. 1758. Chicago: Marquis Who's Who.

Nordenmark, N. V. E. (1939). "Pehr Wilhelm Wargentin" (in Swedish). Uppsala: Almquist and Wiksells.

Wargentin, P. W. (1776). "A Letter from Mr. Wargentin ... to the Rev. Mr. Maskelyne ... containing an Essay on a new Method of determining the Longitude of Places, from Observations of the Eclipses of Jupiter's Satellites." *Philosophical Transactions* 56: 278–286.

Wassenius [Vassenius], Birger

Born **Mankärr near Vänersborg, Sweden, 26 September 1687**
Died **Mankärr near Vänersborg, Sweden, 11 January 1771**

Birger Wassenius is best known for his early descriptions of solar prominences and earthshine. Wassenius's father, Jonas Wassenius, was yeoman of the guard under Carl XI, and his mother was Märta Torstendotter. Meager family means kept him in Vänersborg schools until 1712, when he began studying mathematics, physics, and astronomy in Uppsala, obtaining a philosophy degree in 1722. While a student Wassenius made a living by regularly journeying to the countryside to tutor students, and he is said to have built his own astronomical instruments out of wood and moose antlers. Wassenius married Ebba Regina Spalk, and from 1735 onward, under the patronage of Göteborg Bishop Erik Benzelius, he held lectureship positions at the gymnasium in Göteborg. In 1751 he retired to Mankärr.

Wassenius's claim to astronomical fame rests primarily on his description of solar prominences and earthshine, both of which he noted while observing the total solar eclipse of 13 May 1733 from the vicinity of Göteborg. Wassenius's writings represent one of the earliest, unambiguous descriptions of solar prominences. He also first noted the existence of earthshine, the faint illumination of the lunar disk at totality due to sunlight reflected from Earth. Wassenius suggested that prominences were clouds in the Moon's atmosphere, an idea that was convincingly refuted only much later, via sequences of photographs taken by **Warren de la Rue** and **Angelo Secchi** at the 18 July 1860 eclipse, in Spain. Wassenius also wrote yearly almanacs from 1724 to 1748, and had a lifelong interest in geophysical phenomena.

Paul Charbonneau

Selected References

Hofberg, H. (1906). *Svenskt biografiskt handlexikon*, p. 700. (For a short biography [in Swedish] of Birger Wassenius.)

Vassenio, Birgero (1733). "Observatio Eclipsis Solis totalis cum mora facta Gothoburgi Sveciae." *Philosophical Transactions* 38: 134–135. (The most readily accessible original description of Wassenius's 1733 eclipse observations is this short account [in Latin] he communicated to the Royal Society.)

Waterston, John James

Born **Edinburgh, Scotland, 1811**
Died **Edinburgh, Scotland, 18 June 1883**

John Waterston, a technically trained and competent amateur Victorian scientist, was a pioneer (with John Herapath) of the kinetic theory of gases and, along with **Julius Mayer**, one of the pre-discover of the Kelvin-Helmholtz theory of solar energy. His grandfather, William Waterston, was the founder of a successful business for the manufacture of sealing wax, while his grandmother, Catherine (*née* Sandeman, of the well-known port wine merchants) was within a line of religious intellectuals known as Glasites or Sandemanians, a Scottish sect that valued independence of thought and which included Michael Faraday among its active followers. John, the sixth of the nine children of George Waterston and Jane Blair, had a privileged childhood in a culturally rich and emotionally happy family, which valued education and intellectual achievement. Several of his family contemporaries were successful as military men, bankers, and merchants.

After attending the Edinburgh High School, a leading school in Scotland, Waterston became apprenticed to a civil engineering firm, while attending the University of Edinburgh. He studied mathematics and physics with Sir John Leslie, famous as a mathematician and physicist interested in the heat absorption and reflection properties of various materials. Leslie trained a series of distinguished scientists. In 1832 Waterston moved to London, to the firm of James Walker, a leading civil engineer and president of the Institution of Civil Engineers. After 3 further years in engineering, Waterston sought employment that might provide him the leisure and time to follow his scientific interests, and took a post in the Hydrography Department of the Admiralty. This course of action was prompted by his family's belief that one should separate one's main interests in life from the means by which one earned a living. In 1839 he obtained a post at a substantial salary as naval instructor at the Bombay Academy of the East India Company. Although he had published a scientific paper before leaving Edinburgh (an attempt at a mechanical explanation of gravity with obvious connections to his later work on the kinetic theory of gases), his main scientific work, which encompassed biology and psychology as well as physics and astronomy, was accomplished in India.

Navigation and gun laying were Waterston's instructional responsibilities with the naval cadets, so his interests extended into astronomy as well as physiology, physics, and chemistry. His papers on astronomical topics, starting in 1842, included a navigational discourse on how to find the latitude, time, and azimuth at a given position; an interesting simple graphical method of predicting lunar occultations; several papers on comets (in one of which Waterston came very close to describing the effect of solar wind on cometary tail formation and morphology); and an interesting treatise on observations of solar radiation. Most of these papers were communicated to the Royal Astronomical Society [RAS] in London and printed in their *Monthly Notices*. In 1852, Waterston became a life member of the RAS.

Waterston had become interested in the kinetic theory of gases after trying to calculate the age of the Sun. He realized that chemical processes were insufficient to account for the energy released by the Sun, and postulated that kinetic energy released by the impact of meteors was a possible source of the needed additional energy. Waterston wrote two notes, and then a full paper, setting out his ideas on a kinetic theory of gases. In the paper, Waterston laid out an elaboration on his theory that heat in a gas was a function of the *vis viva* or kinetic energy of the molecules in the gas. His statistical approach included the concept of mean free path and other details that anticipated the work of Rudolf Clausius and **James Maxwell** by about 20 years. He showed how the empirical laws of Boyle and Charles could be derived theoretically from kinetic theory, and that through it, the laws of Avogadro and of Dulong and Petit, would follow directly.

In 1845, Waterston's paper was communicated to the Royal Society in London, where it was read but rejected for publication by the society's referees. As was the practice, Waterston's manuscript was not returned to him, but instead became the property of the Royal Society and was retained in the archives. Waterston had not made a copy of this complex manuscript and was unable to reconstruct it in sufficient detail to submit it for publication elsewhere. By the time he realized it would not be published by the Royal Society, his interest had passed on to the physical chemistry of liquids and gases. Although Waterston later referred to his paper on the kinetic theory of gases several times in oral presentations and other publications, especially one on acoustics and the speed of sound, he never published the ideas presented in this seminal paper.

Waterston returned to Scotland from India in 1857, apparently in frustration over his problems with getting his other scientific work published. In Edinburgh, later in Inverness, and then again in Edinburgh, he continued work that was essentially in physical chemistry, working on the properties of gases and liquids at their interface and exploring the properties of the triple point, surface tension, and capillarity. During his time in India Waterston had suffered a heatstroke while making measurements of solar radiant energy, and was thereafter subject to periodic spells of disequilibrium and dizziness. It is likely that such a spell accounted for his mysterious disappearance during a walk along the Edinburgh waterfront; authorities at the time assumed he fell in the water and drowned, although his body was never found.

While working on acoustics, Lord Rayleigh found Waterston's manuscript in 1891, and recognized its importance in the light of the later theories of Maxwell and Clausius. He immediately had Waterston's manuscript published in the *Transactions of the Royal Society*. In his introduction to the paper, Lord Rayleigh commented that Waterston's paper represented "... an immense advance in the direction of the now generally received theory. The omission to publish it at the time was a misfortune which probably retarded the development of the subject by ten or fifteen years." The original rejection of Waterston's paper may be understood in the context of the prevailing theories of heat, despite the fact that he had previously published papers in the RAS *Monthly Notices*. In mitigation of the Royal Society decision, Lord Rayleigh and others have pointed out that the referees were acting in good faith and on their best understanding of the state of the science at the time, and had acted properly, though unfortunately, in the matter.

David Jefferies

Acknowledgment

The author acknowledges the previous work of J. S. Haldane.

Selected References

Haldane, J. S. (1928). "Memoir of J. J. Waterston." In *The Collected Scientific Papers of John James Waterston*, pp. xiii–lxviii. Edinburgh: Oliver and Boyd.

——— (1928). *Liquids and Gases, a Contribution to Molecular Physics*. Edinburgh: Oliver and Boyd.

Turner, Herbert Hall (1923). "The Decade 1820–1830." In *History of the Royal Astronomical Society, 1820–1920*, edited by J. L. E. Dreyer and H. H. Turner, pp. 1–49, esp. pp. 27–28. London: Royal Astronomical Society. (Reprint, Oxford: Blackwell Scientific Publications, 1987.)

Waterston, John James (1892). "On the Physics of Media that are Composed of Free and Perfectly Elastic Molecules in a State of Motion." Introduction by Lord Rayleigh. *Philosophical Transactions of the Royal Society of London* A 183: 1–79. (Read 5 March 1846.)

Watson, James Craig

Born **near Fingal, (Ontario, Canada), 28 January 1838**
Died **Madison, Wisconsin, USA, 22 November 1880**

As a professor and observatory director at the universities of Michigan and Wisconsin, James Watson was an important figure in 19th-century American astronomy. His astronomical career was largely dedicated to visual observations and orbit calculations of solar-system objects. He discovered 22 asteroids, but his search for the hypothetical intra-Mercurial planet Vulcan proved fruitless.

Watson was the eldest son of William Watson, a farmer, schoolmaster, and factory worker, and Rebecca (*née* Bacon) Watson. William's father was one of the original settlers in 1811 in

the region of colonial Canada where James was born and raised. Though James's grandfather prospered, his father struggled to support his wife and four children. Hoping for better prospects in the United States, the family moved to Michigan in 1850, but in fact James had to work with his father in various menial jobs and in a factory. The mechanical skills he picked up there proved valuable later in his career. He learned mathematics, Greek, and Latin mainly on his own and in his spare time. This helped him gain admission to the University of Michigan at the young age of 15. Watson was a precocious youngster who seemed to master any subject that he encountered.

The circumstances that led Watson to astronomy were fortuitous. In 1852 the University of Michigan at Ann Arbor appointed its first president, Henry Philip Tappan. Determined that science should be prominent in the curriculum, Tappan wanted an observatory to be part of the university. With the financial backing of generous donors in Detroit, he was able to make his dream in that regard a reality. In 1853, as Watson was beginning his studies at the university, Tappan was in Germany finding out firsthand about the education system there and how best to equip an observatory. At Berlin, Tappan met the famed astronomer **Johann Encke** and his assistant **Franz Brünnow**. On their advice, Tappan ordered a Pistor & Martins meridian circle and clock made by Tiede, which were installed in Ann Arbor the following year. A Fitz 13.6-in. refractor was put into service in 1857. Watson was to use these instruments to great advantage in the years ahead. Tappan recruited Brünnow himself to be the first director of the observatory. Few students were able to benefit from Brünnow's demanding courses in spherical and practical astronomy, but those that did became his intellectual descendants forming what has been called the Ann Arbor School of Astronomy. Watson was foremost among them.

Even before Watson received his A.B. degree in 1857 his first scientific contribution was published – the elements of comet C/1857 D1 (d'Arrest). It was his first of over 60 papers giving either elements or observations of comets or asteroids. His command of celestial mechanics and his great facility in rapid computation made him a natural for this sort of work. While pursuing his master's degree, Watson assisted in the observatory of the University of Michigan and in 1859 took charge of it while Brünnow was away for a year as interim director of the Dudley Observatory in Albany, New York. On Brünnow's return in 1860, Watson was assigned to the chair of physics and was temporarily diverted from observing. That year he wrote a popular book on comets, married Annette Waite, and began extramural work reducing observations for **Benjamin Gould**.

When, at age 25, Watson became professor of astronomy and director of the observatory following Brünnow's resignation, he began his astronomical career in earnest, lecturing, computing, observing, and discovering the first of his 22 asteroids. His prolific success in this field earned him many honors. Nowadays Watson is more likely to be remembered for his textbook *Theoretical Astronomy* published in 1868 and reprinted as recently as 1981.

Watson was elected to the National Academy of Sciences [NAS] in 1868, and under their auspices in 1874/1875 he and his wife made a round-the-world trip culminating in the successful observation of the transit of Venus in China. The NAS appointed Watson to three solar-eclipse expeditions, one to Iowa in 1869, another to Sicily the following year, and the last to Wyoming in 1878.

It was on the latter occasion, during totality, that Watson was convinced that he had discovered two planets close to the Sun. The actual discovery of an intra-Mercurial planet would not only have made Watson a celebrity but would have been seen as a triumphant vindication of **Urbain Le Verrier**'s theoretical proposal of 1859 that such a planet could explain the anomalous advance of Mercury's perihelion. As it turned out, his observations (and similar though discordant ones by **Lewis Swift**) were never explained. The phenomenon that Le Verrier sought to explain was eventually accounted for in 1915 by **Albert Einstein** in his theory of general relativity. Watson, however, was always convinced of the legitimacy of his observations and hoped to confirm his work. This quest was a factor in his decision to accept a position as director of the new Washburn Observatory in Madison, Wisconsin.

Attracted by the promise of better equipment, Watson and his wife and mother moved to Madison in 1879. Much of the design and supervision of the observatory's construction fell on his shoulders. At his own expense, he also began work on an underground solar telescope, which, by means of a heliostat, would allow him to search for planets very close to the Sun. All these responsibilities took their toll, and Watson died of intestinal inflammation, leaving his widow and no children.

Watson's hardship in his early years affected him throughout his life. He seemed to be continually seeking financial opportunities beyond the observatory. He was an insurance agent, and later an actuary; he was in the stationery business, selling photos and books, and he became president of the Ann Arbor Printing and Publishing Company. Consequently, even though he died young, Watson left a considerable estate. Most of it, over $18,000, went to the NAS to continue work on the asteroids he had discovered; **Simon Newcomb** was among his executors. The fact that Watson was brilliant and that he learned and earned a great deal without outside help made him ambitious and confident. Except for Vulcan, his achievements were genuine and were recognized by his colleagues at home and abroad with honorary degrees and awards, including the Lalande Prize in 1870.

Watson's papers are found mostly in the University of Michigan Archives, Bentley Historical Library, Ann Arbor; in the Archives of the Department of Astronomy, University of Wisconsin; and in the Library of the Wisconsin Historical Society, Madison.

Peter Broughton

Selected References

Broughton, P. (1996). "James Craig Watson (1838–1880)." *Journal of the Royal Astronomical Society of Canada* 90: 74–81.

Comstock, G. C. (1895). "Memoir of James Craig Watson." *Biographical Memoirs, National Academy of Sciences* 3: 43–57.

Plotkin, H. (1980). "Henry Tappan, Franz Brünnow and the Founding of the Ann Arbor School of Astronomers, 1852–1863." *Annals of Science* 37: 287–302.

Whitesell, Patricia S. (1998). *A Creation of His Own: Tappan's Detroit Observatory.* Ann Arbor: Bentley Historical Library, University of Michigan.

——— (2003). "Detroit Observatory: Nineteenth-Century Training Ground for Astronomers." *Journal of Astronomical History and Heritage* 6, no. 2: 69–106.

Watts, Chester Burleigh

Born **Winchester, Indiana, USA, 27 October 1889**
Died **Annandale, Virginia, USA, 17 July 1971**

Chester Watts contributed important scientific and technological improvements to meridian astrometry and to our knowledge of the topography of the Moon in the marginal or libration zones.

Watts was born to Joseph and Ada (*née* Irey) Watts. His father was a railway postal clerk while his mother was, for a short time, a schoolteacher. Watts became interested in astronomy at an early age and constructed a working sextant with which he took his first astronomical measurements. From July 1911 to September 1914 he worked at the United States Naval Observatory [USNO] as a miscellaneous computer, returning to Indiana to marry Ada Williams and complete his BA degree with distinction in astronomy at Indiana University in 1915. Watts then resumed his employment at the USNO and spent his entire career there. For the first 4 years he worked in the USNO Time Service Division. In 1919 he was transferred to the Six-Inch Transit Circle Division and was appointed director of that division in 1934.

His imaginative abilities with machinery and creative approach to experimentation led Watts into a remarkable career in observational astronomy. He brought marked improvements to the USNO's Six-Inch Transit Circle, including automation of the traveling thread micrometer and photographic recording of the circle and micrometer readings together with an automated measurement of these films. Those improvements were applied by Watts to his design of the Seven-Inch USNO Transit Circle for which construction began in 1947; it was placed in service in 1955. These improvements resulted in significant advances in the accuracy of stellar positions and in other astronomical measurements dependent on those fundamental catalogs.

In 1942, Watts suggested the USNO carry out a survey of the marginal zone of the Moon. The purpose was to produce the most detailed map of the lunar limb to date, as well as a precise datum from which to reference lunar topography. The project was delayed by World War II. From 1946 until 1963 Watts struggled with a number of scientific problems inherent in the project, but world politics intervened to give the project a higher priority and bring additional resources to bear.

The United States was in the midst of the Cold War with the Soviet Union for which the premise of mutually assured destruction in the event of a war resulted in substantial strategic preparation. One problem the US military had in launching intercontinental ballistic missiles was that they could not be sure the missiles would hit their targets on another continent. The launch points and target zones were on different mapping datums that would cause errors in targeting.

The Army Map Service was given the mission to link the various continental datums, a problem for which astronomical measurements offered the main solutions available at the time. Specifically, precise timing of transient astronomical events that could be predicted and observed from both continents could be used for that purpose. Astronomical events that would lend themselves to such a project include total solar eclipses in which the appearance and disappearance of Bailey's beads were timed accurately. (See **Edward Halbach**.) Another type of suitable event was a grazing occultation observed from both datums. However, for both the solar eclipse and the grazing occultation observations, accurate charts of the lunar marginal areas, the areas that appear and disappear according to the ever-changing libration of the Moon, would have to be available to increase the accuracy of the datum observations. The Army Map Service provided support to speed the USNO project along. The Lunar Limb Charts (now known as the Watts Limb Charts) were finally published in 1963 as "The Marginal Zone of the Moon," Volume 17 of the *Astronomical Papers of the American Ephemeris*.

The Watts Charts were derived from photographs taken at three locations – 571 photographs at Washington, USA, between 1947 and 1956; 247 photographs taken at Johannesburg, South Africa, between 1927 and 1952; and 49 photographs taken at Flagstaff, Arizona, USA, between 1927 and 1928. The primary telescope used was the Naval Observatory's horizontal refractor. This refractor is of 12.2-m focal length, 12.7-cm aperture (a 5-in. *f*/96), and was built in 1873 by **Alvan Clark**.

The most significant feature of the Watts Charts is their reference to a spherical datum, actually two spherical datums, each with a specific purpose in the use of the charts. The finished product of this work was a series of 1,800 libration frames and a few pages of what became known as the "P & D Charts" for a total of 951 pages of data. Use of the Watts Charts originally involved graphical interpolations of the exact points of interest on the lunar surface. More recently, however, interpolation of the Watts Charts has been computerized by the International Occultation Timing Association [IOTA]. Data reduction of observational data is now remarkably simplified. Watts's finished work has led to highly refined occultation work, leading to other useful information as well, for example the discovery of new double stars from the precise manner in which the light of the star disappears when occulted on a precisely known surface of the Moon. Lunar occultation data have also been used to understand the cloud structures around young T-Tauri stars in the Taurus and Ophiuchus Dark Clouds, and to refine the map of the center of the Milky Way Galaxy.

Watts worked quietly on his demanding projects without seeking publicity for himself. It was the trademark of the work of Watts that he accomplished difficult and demanding work of great benefit to succeeding generations of astronomers without a great deal of recognition for himself. However, honors of various types came his way due to the effect his work had on the astronomical community. He was awarded the honorary degree of doctor of sciences by his *alma mater*, Indiana University, in 1953. Watts received the James Craig Watson Medal from the National Academy of Sciences in 1956. He earned the Federal Distinguished Civilian Service Award in 1959, which was also the year of his retirement from the USNO. Watts served as president of International Astronomical Union Commissions on Positional Astronomy and on the Motion and Figure of the Moon through the 1950s as well as president and chairman of the Astronomy Section of the American Association for the Advancement of Science.

Richard P. Wilds

Selected References

Ashbrook, Joseph (1964). "C. B. Watts and the Marginal Zone of the Moon." *Sky & Telescope* 27, no. 2: 94–95.

——— (1971). "Positional Astronomer." *Sky & Telescope* 42, no. 3: 131.

Dunham, David W. (1971). "Geodetic Applications of Grazing Occultations." In *Highlights of Astronomy*, edited by C. de Jager, pp. 592–600. Dordrecht: Reidel.

Gray, Mike (1992). *Angle of Attack: Harrison Storms and the Race to the Moon*. New York: W. W. Norton and Co.

Scott, F. P. (1972). "Chester Burleigh Watts." *Quarterly Journal of the Royal Astronomical Society* 13: 110–112.

——— (1973). "In Memoriam, C. B. Watts." *Moon* 6, nos. 3– 4: 233–234.

Stevens, Berton (1975). "The Automatic Computer Lunar Profile Plotting Program (ACLPPP)." *Occultation Newsletter* 1, no. 5.

Watts, C. B. (1963). "C. B. Watts Recording His Recollections of the Naval Observatory." Washington, DC: US Naval Observatory Library.

——— (1963). *The Marginal Zone of the Moon*. American Ephemeris and Nautical Almanac Astronomical Papers, no. 17. Washington, DC: US Naval Observatory.

Webb, Thomas William

Born **Tretire cum Michaelchurch, (Hereford and Worcester, England), 14 December 1806**

Died **Hardwick, Oxfordshire, England, 19 May 1885**

In many ways the patron saint of British amateur astronomers, Thomas Webb influenced amateurs more broadly around the world through his popular guide to telescopic astronomy, which has been updated and remained in print for over a century. He was the only son of a clergyman, Reverend John Webb, rector of Tretire cum Michaelchurch in the county of Hereford. The elder Webb was a sound classical scholar, and an eminent authority on Norman French, who was frequently called upon to give evidence in Courts of Law on the interpretation of early documents. But he was more particularly devoted to researches on the history of the west of England during the Civil War, and for the greater part of his long life – he lived to the ripe old age of 93 – John Webb was preoccupied with preparing a history of Herefordshire during the Civil War. But his dread of inaccuracy or precipitate thought slowed the work, which was left to his son to complete many years later.

Thomas's mother died when he was still a child, and he was educated entirely by his father. Brought up among books and manuscripts, Webb became precise, studious, and mature for his years. While he showed some aptitude for experimental science, especially electricity, his father insisted that he become a classical scholar. Thus Webb was dissuaded from studying Euclid; otherwise it is almost certain that he would have become a mathematician. His efforts at Latin and Greek, though carried out with diligence, did not progress to his father's satisfaction. In 1826 Webb entered Oxford University as a gentleman commoner at Magdalen Hall. He took his degree in 1829, with second-class honors in mathematics. Webb might have done better had he applied himself more earnestly, for he later admitted that he spent most of his time in the "idleness" of desultory reading in the library.

On leaving Oxford University, Webb was licensed to the curacy of Pencoyd where he remained for 2 years. He took holy orders in August 1831 and became a minor canon of Gloucester Cathedral. Ten years later Webb returned to Tretire and worked as curate under his father. Only in 1852 did he attain to the living of Hardwick, a large but thinly populated parish near the Welsh border. While diligent in the pursuit of his clerical duties – he set out early each afternoon with a knapsack on his back to visit the more remote members of his parish – Webb never lost his early interest in science.

Proficient in German, Webb read **Wilhelm Beer** and **Johann von Mädler** in the original and became the most reliable source of information about German selenographic studies, then the standard, in the English-speaking world. At Gloucester he began routine observations of the Sun, Moon, and planets with a rather modest instrument, a 3.7-in. achromatic refractor by the English maker Tulley. On the recommendation of his close friend and advisor, **William Rutter Dawes**, in 1859 Webb obtained a 5.5-in. refractor by **Alvan Clark**. This was followed by an 8- and then a 9.3-in. Newtonian reflector, both by George With, during the 1860s.

Webb immediately realized that there was still much useful work to be done, especially on the Moon. He wrote:

> A little experience showed that Beer and Mädler have not represented all that may sometimes be seen with a good common telescope. My own opportunities, even when limited to a 3 7/10-inch aperture, satisfied me, not only how much remains to be done, but how much a little willing perseverance might do.

What fascinated Webb above all else was the exciting possibility that the surface of the Moon was still in active condition. His first essay, "On the Lunar Volcanos," had been presented to the British Association as far back as 1838, but not until 1859 did he follow it up with "Notice of Traces of Eruptive Action in the Moon." Here Webb pointed out that while astronomers were generally agreed as to the cessation of such action on a large scale, "this would not necessarily infer the impossibility, or even improbability, of minor eruptions, which might still continue to result from a diminished but not wholly extinguished force." Webb's project of documenting lunar change, which had been all but banished since the publication of Beer and Mädler's treatise *Der Mond* (The Moon) in 1837, set the tone for much of the selenographical work by British amateurs for the next century, preoccupied with seeming minor changes as evidence of the Moon's ongoing geological activity, along with the resurgence of the possibility that the Moon was a living world, "habitable," as Webb expressed the hope, "in some way of its own."

Webb left four notebooks recording thousands of solar, lunar, planetary, and stellar observations. His meticulously detailed notes, all in delightful prose and exquisite calligraphy, were accompanied by hundreds of sketches. Webb made many contributions to **John Birmingham**'s *Red Star Catalog* and also discovered a number of variable stars.

Webb's influential book *Celestial Objects for Common Telescopes* appeared in 1859, the same year as Darwin's *On the Origin of Species*. Bringing a quarter of a century of Webb's experience to the aid of the beginner, it contained comprehensive discussions of telescopes and observing techniques, as well as exhaustive descriptions of the lunar, planetary, and stellar objects accessible to modest instruments, accompanied by historical accounts of the previous astronomers who had viewed the objects under discussion. It was an instant classic, and during his lifetime it passed through several revised and expanded editions with chapters on the Moon updated by **Thomas Elger**. Following Webb's death his friend, *protégé*, and executor, Reverend **Thomas Espin** would edit the fifth and sixth editions. A later edition, edited by **Margaret Mayall**, was published as late as 1962 and is still widely read today.

Thomas Dobbins and *William Sheehan*

Selected References

Both, Ernst E. (1961). *A History of Lunar Studies.* Buffalo, New York: Buffalo Museum of Science.

Chapman, Allan (1998). *The Victorian Amateur Astronomer: Independent Astronomical Research in Britain, 1820–1920.* Chichester: John Wiley and Sons.

Ranyard, Arthur Cowper (1886). "Thomas William Webb." *Monthly Notices of the Royal Astronomical Society* 46: 198–201.

Sheehan, William P. and Thomas A. Dobbins (2001). *Epic Moon: A History of Lunar Exploration in the Age of the Telescope.* Richmond, Virginia: Willmann-Bell.

Webb, Thomas William (1859). *Celestial Objects for Common Telescopes.* London: Longman, Green, Longman and Roberts.

——— (1859). "Notice of Traces of Eruptive Action in the Moon." *Monthly Notices of the Royal Astronomical Society* 19: 234–235.

——— (1864). "On Certain Suspected Changes in the Lunar Surface." *Monthly Notices of the Royal Astronomical Society* 24: 201–206.

Weigel, Erhard

Born **Weiden, (Bavaria, Germany), 16 December 1625**
Died **Jena, (Germany), 21 March 1699**

Erhard Weigel was a professor of mathematics at the University of Jena, who attempted to rename the classical constellation patterns, replacing them with figures from modern European heraldry. Thus, "Ursa Major," became "The Elephant of Denmark," and "Cygnus" became "The Ruta and Swords of Saxony." Weigel's uranography has not survived him.

Selected Reference

Brown, Basil (1932). *Astronomical Atlases, Maps, and Charts: An Historical and General Guide*. London: Search Publishing Co.

Weinek, László [Ladislaus]

Born **Buda, (Budapest, Hungary, 13 February 1848**
Died **Prague, (Czech Republic), 12 November 1913**

László Weinek directed the German University Observatory in Prague (after 1883) and is chiefly remembered for his extensive drawings and photographs of lunar-surface features. Weinek's father was a government official. Although his family was of German origin living in Hungary, Weinek declared himself to be Hungarian. After graduating from secondary school in Buda, he studied mathematics, physics, and astronomy at the University of Vienna, where he was awarded a doctorate in 1870. A state-sponsored scholarship enabled him to continue postdoctoral studies at Berlin and Leipzig, Germany.

As a young astronomer, Weinek was involved in the Saxonian geodetic work of latitude and longitude determination (1872/1873). He undertook a pioneering role by introducing photographic techniques into astronomy. After completing a study of the measurement errors associated with photographic astrometry, Weinek was chosen to participate in the German expedition to observe the 1874 transit of Venus from Kerguelen Island in the Indian Ocean. He obtained 60 good-quality photographs of the event. After returning to Leipzig Observatory, Weinek analyzed the photographic plates of his own and others' expeditions, with a view toward deriving a more precise value for the Astronomical Unit. He then took part in the astrometric measurements for the Leipzig zone of the *Astronomische Gesellschaft Katalog*. For these contributions, Weinek was elected a member of the Hungarian Academy of Sciences (1879).

Unable to find employment as an astronomer in Hungary, Weinek accepted a position as professor of astronomy at the German University in Prague (1883). Those duties included the directorship of the university's Observatory. There, he became a leading investigator of lunar topographic features by virtue of his hundreds of detailed drawings and from analysis of photographs of the Moon taken at the Lick Obervatory and the Paris Observatory. Weinek's combined treatment of the drawings and photographs brought to light many unrecorded features on the lunar surface. Starting in 1889, he also participated in the international efforts to observe the effects of polar motion by astrometric means. A crater on the Moon has been named for him.

László Szabados

Selected References

Münzel, Gisela (2001). "Ladislaus Weinek (1848–1913)." In *Beitäge zur Astronomiegeschichte*, edited by Wolfgang R. Dick and Jürgen Hamel. Vol. 4, pp. 127–166. *Acta Historica Astronomiae*, Vol. 13. Frankfurt am Main: Harri Deutsch.

Scheller, A. (1913). "Todesanzeige: Ladislaus Weinek." *Astronomische Nachrichten* 196: 323–324.

Weiss, Edmund

Born **Freiwaldau (Jeseník, Czech Republic), 25 August 1837**
Died **Vienna, (Austria), 30 June 1917**

Edmund Weiss directed the University of Vienna Observatory from 1878 to 1910; his expertise in celestial mechanics demonstrated the connections between two periodic comets and their associated meteor showers. Weiss and his twin brother Gustav Adolf (who became a professor of botany at Prague) were raised in England. His father, a well-known physician, was the head of a hydropathic establishment near Richmond. Weiss's knowledge of the English language was likely helpful in shaping his career as an astronomer. But after his father died prematurely, Weiss returned to his homeland and attended the Gymnasium at Troppau, Austrian Silesia, from 1847 to 1855. He then studied mathematics, astronomy, and physics at the University of Vienna. In 1858, he was appointed an assistant at the Vienna Observatory, where **Karl von Littrow** served as director. Littrow came to appreciate Weiss's mathematical abilities and his skill and diligence as an observer. Weiss was awarded his Ph.D. in 1860.

In 1867, **Otto Wilhelm Struve** offered Weiss a position at the Pulkovo Observatory near Saint Petersburg, Russia, but he declined the invitation. Two years later, he received an offer from the Geodetic Institute in Berlin to undertake latitude and longitude determinations. Weiss also refused the latter offer and was rewarded with an *extraordinarius* professorship at the University of Vienna. He was appointed a full professor there in 1875. Weiss married Adelaide Fenzl in 1872; the couple had seven children.

During the 1870s, Viennese astronomers began construction of a new, modern observatory on the hills of Währing beyond the city limits. Weiss was sent to Great Britain and the United States to visit the leading observatories and telescope makers. He later published a report on his impressions in the proceedings of the Deutsche Astronomische Gesellschaft (German Astronomical Society). His visitations led to the purchase of a 27-in. refracting telescope manufactured by Sir **Howard Grubb** of Dublin (at that time the largest refractor in the world) and a smaller 12-in. refractor by Alvan Clark & Sons of Cambridge, Massachusetts, USA.

As Littrow's health declined, Weiss assumed more of the responsibility for planning and supervising the observatory's construction. After Littrow's death in 1878, Weiss was appointed director of the new Vienna Observatory. He retained this position until his retirement in 1910. During his directorship, Weiss equipped the Vienna Observatory with two more instruments: an equatorial Coudé telescope donated by the Viennese Baron Albert von Rothschild, and a "standard astrograph," which was chiefly used for the positional determinations of minor planets and comets.

A majority of Weiss's research publications dealt with orbital determinations and the calculation of ephemerides of comets and minor planets. He perfected new methods for finding the improved orbits of those bodies. Weiss investigated the orbits of meteors and demonstrated an association between the Lyrids and comet C/1861 G1 (Thatcher) and between the Andromedids and comet 3D/Biela. From these associations, he developed the accepted view that meteors are the disintegration products of comets.

Weiss maintained a special interest in traveling to observe unusual astronomical phenomena. He witnessed solar eclipses in Greece

(1861), in Dalmatia (1867), in Aden (1868), and in Tunis (1870). He likewise observed the 1874 transit of Venus from Jassy, and the expected great meteor shower of the Leonids in 1899 from Delhi.

On account of his interest in solar physics and eclipses, Weiss regularly attended the meetings of the International Union for Cooperation in Solar Research (a forerunner of the International Astronomical Union), which was founded by **George Hale** in 1904. Here, his knowledge of the English language was of considerable help to him. Weiss was also appointed to the Council of the Astronomische Gesellschaft from 1881 to 1915, and was a vice president from 1896 to 1913. He held the office of president at the Austrian commission of the International Commission for the Measurement of the Earth. Weiss was made a full member (1883) of the Vienna Academy of Sciences and was awarded an honorary doctorate from the University of Dublin. In 1889, he was appointed a member of the Permanent International Committee of the *Carte du Ciel* project.

At Vienna, Weiss had a high reputation as a lecturer and was aware of the need to interest the public in matters astronomical. He wrote popular articles and published an astronomical calendar that was distributed by the Vienna Observatory. Weiss was also responsible for revising and publishing the seventh and eighth editions of **Johann von Littrow**'s *Wunder des Himmels* (Wonder of the heavens), a well-known textbook, which was first printed in 1834.

Anneliese Schnell

Selected References

Crommelin, Andrew Claude De La Cherois (1918). "Edmund Weiss." *Monthly Notices of the Royal Astronomical Society* 78: 258–259.

Ferrari d'Occhieppo, Konradin (1976). "Weiss, Edmund." In *Dictionary of Scientific Biography*, edited by Charles Coulston Gillispie. Vol. 14, pp. 242–243. New York: Charles Scribner's Sons.

Hepperger, Josef (1917). "Todesanzeig: Edmund Weiss." *Astronomische Nachrichten* 204: 431–432.

Weizsäcker, Carl Friedrich von

Born **Kiel, Germany, 28 June 1912**

German theoretical physicist C. F. von Weizsäcker discovered (at about the same time as **Hans Bethe**) the commonest cycle of nuclear reactions that produces energy by converting hydrogen to helium in massive stars. He also revitalized the nebular or Kant–Laplace hypothesis for the origin of the Solar System.

Born into a family of statesmen, theologians, and scientists, Weizsäcker attended schools in eight different countries before beginning the study of physics at the University of Leipzig and obtaining his D.Phil. in 1933 for work with Werner Heisenberg. His diplomat father Ernst von Weizsäcker was sentenced to jail by the Allies at Nuremberg for service under the Nazis, and his brother Richard served as President of the Federal Republic of Germany from 1984 to 1994.

Carl von Weizsäcker initially worked as assistant to Heisenberg at Leipzig, and moved to the Kaiser Wilhelm Institute für Physik in 1936 as a research physicist, becoming *Privatdozent* (lecturer) the next year and lecturing at the University of Berlin. He worked partly with Lise Meitner and Otto Hahn, and his attention turned to applications of nuclear physics to astrophysics. The 1932 discovery of the neutron and its relationship *via* the weak interaction to the proton and neutron enabled Weizsäcker to go beyond the ideas of **Fritz Houtermans** and **Robert Atkinson** and to propose how a nucleus of carbon might act as a catalyst, capturing four protons in succession, dropping off a helium, and going through the same process many times. He published a couple of papers in 1937/1938 on the transformation of elements in stellar interiors.

In 1942, Weizsäcker was appointed associate professor of theoretical physics at the University of Strasbourg. Here he devised a more sophisticated and realistic version of the nebular hypothesis of **Immanuel Kant** and **Pierre de Laplace** for the origin of the Solar System.

Von Weizsäcker argued that the original dust cloud out of which the Solar System was formed would experience turbulence and break up into a number of smaller vortices and eddies. These vortices fell into gradually larger systems with increasing distance. At the boundaries between sets of vortices, conditions were supposed to be suitable for planets to form from the continuous aggregation of progressively larger bodies. His theory was able to resolve the problem of low angular momentum of the Sun and to provide a physical explanation of the Bode–Titius law of planetary distances. While Weizsäcker's theory was not able to resolve all the questions about planetary formation, his work directed a fresh stream of thought into this field and attracted a lot of attention from scientists in the field. Modifications and additions were later proposed by Dirk ter Haar, **Hannes Alfvén**, and **Fred Hoyle**.

In 1944, Weizsäcker returned to the Kaiser Wilhelm Institute and turned to a study of the nuclear reactions that take place in the fission of uranium and the problems of constructing a self-sustaining reactor. In April 1945, he was arrested by Allied forces along with other high-ranking German scientists. During the 8 months they were interned by the British government at Farm Hall, Godmanchester, their conversations were recorded. These form part of the historical record on how far Germany had advanced toward construction of a fission bomb.

After his return to Germany, Weizsäcker joined the staff of the Max Planck Institute for Physics at Göttingen in 1945 and remained there until 1957. During this period he also held an honorary professorship at the University of Göttingen. Then, in 1957, Weizsäcker accepted a position of professor of philosophy at the University of Hamburg. This decision was indicative of his intention to spend more time in thinking, and writing, and with matters of religion and philosophy, while holding an honorary chair at the University of Munich. During the 1960s, he was very active in the peace movement and became a strong spokesman for nuclear disarmament. He relinquished the Hamburg position in 1969 to take up the directorship of the Max Planck Institute on the Preconditions of Human Life in the Modern World in Starnberg in 1970. Weizsäcker retired from this position in 1980 having reached the age of 68. Since 1980 he has been emeritus scientific member (*Mitglied*) of the Max Planck Institute, Munich.

Weizsäcker married Gundalena Wille on 30 March 1937. Their children are Carl Christian, Ernest-Ulrich, Elizabeth (Mrs. Raiser), and Heinrich.

He has received many honors, including the Max Planck Medal in 1957 and 1966, the Goethe Prize (City of Frankfurt, 1958), the Order of Merit for Sciences and Arts (1961), the Arnold Raymond Prize for Physics (1965), the Erasmus Prize (1961), the Templeton Prize for Progress in Religion, and several others. He has also received honorary degrees from several universities.

Weizsäcker is a member of numerous societies. Amongst these are the Max-Planck-Gesselschaft, Deustsche Akademie der Naturforscher Leopoldina (Halle), Deutsche Akademie für Sprache und Dichtung, Göttinger Akademie der Wissenschaften, Sachsische Akademie der Wissenschaften (Leipzig), Österreichische Akademie der Wissenschaften (Wien), and Deutsche Physikalische Gessellschaft.

Y. P. Varshni

Selected References

Gamow, G. and J. A. Hynek (1945). "A New Theory by C. F. von Weizsäcker of the Origins of the Planetary System." *Astrophysical Journal* 101: 249–254.

Narins, Brigham (ed.) (2001). *Notable Scientists from 1900 to the Present.* Farmington Hills, Michigan: Gale Group.

Tango, Gerardo G. (1991). "Von Weizsäcker Finalizes His Quantitative Theory of Planetary Formation." In *Great Events from History II.* Vol. 3, *1931–1952,* edited by Frank N. Magill, pp. 1208–1212. Pasadena: Salem Press.

Weizsäcker, C. F. von (1935). "Zur Theorie der Kernmassen" (On the theory of nuclear masses). *Zeitschrift für Physik* 96: 431–458.

——— "Über Elementumwandlungen im Innern der Sterne" (On transformation of the elements in stellar interiors). Parts 1 and 2. *Physikalische Zeitschrift* 38 (1937): 176–191; 39 (1938): 633–646.

——— (1943). "Über die Entstehung des Planetensystems" (Origin of the planetary system). *Zeitschrift für Astrophysik* 22: 319–355.

——— Oral History Interviews, 9 July 1963, 18 April 1978, 15 February 1984. Niels Bohr Library, American Institute of Physics, College Park, Maryland.

——— (1978). Letter to Karl Hufbauer. Niels Bohr Library, American Institute of Physics, College Park, Maryland.

——— History of physics manuscript biography. Collection S-Z (1901–1989). Niels Bohr Library, American Institute of Physics, College Park, Maryland.

Wendelen, Govaart [Gottfried, Godefried]

Born **Herck-la-Ville, (Belgium), 6 June 1580**
Died **Ghent, (Belgium), 24 October 1667**

Godefried Wendelen demonstrated the uniformity of the precession of the equinoxes. He corrected numerous geographical longitudes and was the first to point out a solar parallax of 14 s. Wendelen also claimed to have discovered Kepler's third law, 8 years before Kepler himself did so.

Wendelen was the son of Nicolas Wendelen and his second wife, Elisabeth Corneli. After having received his first education in Herck-la-Ville, Wendelen went to the Jesuit college at Tournai in 1595 and afterward to Louvain to continue his studies. He subsequently traveled to Marseilles (1599) and Rome (1600). In Digne, Wendelen taught mathematics for a year (1601), where he met **Nicolas de Peiresc** and **Pierre Gassendi**. From September 1604 until April 1612, he tutored the two sons of André Arnaud, Seigneur de Miravail and lieutenant general of the Seneschal's court of Forcalquier. In 1612 Wendelen returned to his birthplace, where he became the supervisor of the local Latin school. In December 1619 he became subdean in Mechelen, and he was ordained priest in Brussels. In June 1620 Wendelen was nominated priest of Geetbets. When he was ordained priest in Herck-la-Ville in 1633, he returned to his birthplace, where he lived until 1648. In that year he became an official at the cathedral in Tournai. Wendelen retired in 1658 and lived the remainder of his life in Ghent.

Wendelen was a convinced Copernican, and was praised by **Galileo Galilei** and **René Descartes. Isaac Newton** included him among the 71 authors cited in the *Principia* (1687). Wendelen corresponded with his younger contemporaries **Marin Mersenne**, Pierre Gassendi, and Constantijn Huygens.

Wendelen tackled the problem of the variation of the obliquity of the ecliptic in his *Loxias, seu de obliquitate solis* (1626). It was traditionally thought that the obliquity had decreased from **Eratosthenes** until **Georg Peurbach**, and that it had increased from Peurbach until Wendelen. Wendelen examined the errors in the observations at hand and completed the available data by adding his own observations. He focused in particular on the atmospherical refraction, the solar parallax (which he found to be 1′ by adopting **Aristarchus**' method and using the telescope), and a correction of **Ptolemy**'s inaccurate latitude for Alexandria (which Wendelen improved by relying on information provided by navigators). On the basis of the new parameters Wendelen recalculated all the data, and concluded that the obliquity had been decreasing from **Thales** onward; he also provided a table to calculate the obliquity at any time.

In 1629, Wendelen published the first part of his *De deluvio*, a work that he would never finish. He took the seven floods that had occurred thus far in the history of mankind as his point of departure to reflect on the history of the Earth. In the last chapter, dealing with cosmology, the heliocentric theory was openly present. Wendelen moreover affirmed that he had observed sunspots from 1601 onward, *i. e.*, long before Galilei. He considered a comet to be a mass of fire that is ejected by the Sun and describes an elliptical path.

In his *Eclipses lunares ab anno 1573 ad 1644 observatae* (1644) Wendelen described in detail his observations of eclipses and the advantages they offer, especially for correcting geographical longitudes. His analysis of eclipses showed that the solar parallax only amounts to 14.656″ instead of 1′. The distance between the Sun and the Earth is fixed at 14,656 Earth radii.

Wendelen's Copernicanism clearly surfaced in his *De caussis naturalibus pluviae purpureae Bruxellensis* (1647), a work written on the occasion of the red rain observed by the Capuchins of Brussels. In a later edition, parts of Wendelen's correspondence with Chifflet, Cornelius Giselberti Plempius, **Pierre Gassendi**, and others were appended to the original text. From one of his letters to Gassendi, Wendelen's sympathy for **Johannes Kepler** becomes apparent, even though he thought that the lunar motion described an oval, rather than an elliptical path.

The *Teratologia cometica* (1652) discussed the comets of 1607, 1618, and 1652. Wendelen resumed his theory that comets are ejected by the Sun, similar to the fiery masses that are ejected by volcanoes on the Earth. Instead of being elliptical, however, the comets' path was described as being conchoidal. The heliocentric theory is present in this book, where Wendelen considered the Earth (called the *tertium corpus*) to be one of the planets describing an elliptical path around the Sun.

In 1658, Wendelen resumed a method that he had used before, in his *Luminarcani Arcanorum caelestium Lampas* (1643). This, Wendelen's last published work, is a small treatise containing

12 astronomical propositions that are presented in the form of anagrams. It contains Kepler's third law, which Wendelen possibly discovered independently in 1610. Either way, Wendelen certainly applied it for the first time to Jupiter's satellites. In the preface to this work, Wendelen expressed his belief that eclipses entail predictions for the future. This means that the demand for exact knowledge goes hand in hand with the belief in a celestial influence on earthly phenomena.

Additional work by Wendelen can be found in his manuscripts that are preserved at Brussels and Bruges, and in his correspondence with Pierre Gassendi, Marin Mersenne, Constantijn Huygens, and others.

Fernand Hallyn and *Cindy Lammens*

Alternate name

Godefridus Wendelinus

Selected References

Godeaux, Lucien (1938). *Biographie nationale publiée par l'Académie royale de Belgique* 27: 180–184.

Hallyn, Fernand (1998). "La cosmologie de Gemma Frisius à Wendelen." In *Histoire des sciences en Belgique de l'Antiquité à 1815*, edited by Robert Halleux, pp. 145–168. Brussels: Crédit Communal.

Huygen, F. (1980). *Govaart Wendelen: Herk-de-Stad, 1580–1667*. Gemeente, Herk-de-Stad.

Jacques, Emile (1983). "Les dernières années de Godefroid Wendelen (Wendelinus) († 1667)." *Lias* 10: 253–271.

Pelseneer, Jean (1976). "Wendelin (Vendelinus), Gottfried." In *Dictionary of Scientific Biography*, edited by Charles Coulston Gillispie. Vol. 14, pp. 254–255. New York: Charles Scribner's Sons.

Silversyser, Florent (1932). "Les autographes inédits de Wendelin à la Bibliothèque de Bruges." *Annales de la Société scientifique de Bruxelles* 52: 99–268. (This work contains Wendelen's manuscript treatise *Motus fixarum; Motus Solis; Motus Lunae; De systemate mundano* [MS 526 D] and MS 526 A of the Bruges Library.)

——— "Godefroid, Wendelen: sa vie, son ambiance et ses travaux (1580–1567)." *Bulletin de l'Institut Archéologique liégeois* 58 (1934): 89–158; 60 (1936): 137–190.

——— (1935). "Un astronome belge du 17e siècle: Godefroid Wendelen, ce en quoi il fut intéressant et précurseur en sciences." In *2e Congrès national des sciences*. Brussels, pp. 89–93.

Wendelen, G. (1626). *Loxias, seu de Obliquitate solis diatriba in qua zodiaci ab aequatore declinatio hactenus ignorata tandem eruitur et in canonem suum refertur*. Antwerp: Hieronymus Verdussius.

——— (1629). *De diluvio liber primus*. Antwerp.

——— (1632). *Aries seu Aurei Velleris encomium*. Antwerp: Balthasar Moretus.

——— (1637). *De Tetracty Pythagorae dissertatio epistolica ad Erycium Puteanum*. Louvain: C. Coenestesius and G. Lipsius.

——— (1640). *Leges salicae illustratae: illarum natale solum demonstratum, cum glossario Salico vocum Aduaticarum*. Antwerp: Balthasar Moretus.

——— (1643). *Lampas tetraluxnoj, quatuor obvelata hexametris quae totidem velut umbrae sunt, quatuor anagrammatis revelata quae totidem lumina, omnibus orbis terrarum mathematicis ac physicis traditur*. Brussels: Joan Mommartius.

——— (1644). *Eclipses lunares ab anno 1573 ad 1644 observatae quibus Tabulae Atlanticae superstruuntur earumque idea proponitur*. Antwerp: Hieronymus Verdussius.

——— (1647). *De caussis naturalibus pluviae purpureae Bruxellensis clarorum virorum judicia*. Brussels: J. Mommartus.

——— (1647). *Pluvia purpurea Bruxellensis*. Paris: L. de Heuqueville.

——— (1652). *Teratologia cometica occasione anni vulgaris aevae*.

——— (1658). *Sphinx et Oedipus seu Lampas dodekaluxnoj Duodecim obvelata Hexametris quae totidem umbrae sunt duodecim revelata Anagrammatis Quae sunt Lumina totidem omnibus orbis terrarum mathematicis ac physicis seram posteritatem traditur*. Tournai: Adrianus Quinqué.

Werner, Johannes

Born **Nuremberg, (Germany), 14 February 1468**
Died **Nuremberg, (Germany), probably March to June 1522**

Navigator Johannes Werner started his studies in Nuremberg and continued at the University of Ingolstadt (enrolled: 1484); from the very beginning he showed an inclination toward the exact sciences. In 1490, he was appointed as chaplain in Herzogenaurach, Germany, but he spent the years 1493–1497 studying in Rome. Werner entered the career of a priest, but besides studies of theology he substantially improved his knowledge of astronomy, mathematics, geography, and the Greek language, profiting from discussions with many learned Italian men. After his return to Germany in 1498, he settled finally at the Saint Johannis Church in Nuremberg, where he remained until his death.

In Nuremberg, Werner had frequent contacts with many scholars, including **Albrecht Dürer**, who sometimes needed advice on problems of mathematics and geometry. Werner also earned recognition abroad, and in 1503 he was invited to the emperor's court in Vienna. Later, **Johannes Stabius**, mathematician and historiographer at the Vienna court, prepared an edition of works on geography, which contained the writings of Werner (1514). Other texts of Werner were published separately, some remained in manuscripts, and some were lost.

In astronomy Werner continued the practical work of **Johann Müller** (Regiomontanus). For example, he refined the Jacob's staff (radius), an instrument consisting of two or three wooden rules with scales, which Müller used for measuring the angular distances between stars or other celestial bodies.

An invention of Werner's own was a nomographical tool called the "Meteoroscop" for easy numerical solving of spherical triangles, without the need of tables of trigonometric functions. It consisted of a metal plate with a pointer, divided into four quadrants. In two of them were circles corresponding to coordinate lines on the celestial sphere in the stereographic projection; another two with scales served to determine the sines, as is described in detail in the text *De Meteoroscopis*. The clock on the church in Herzogenaurach and sundials in Rosstal belong to the same category of Werner's handmade instruments.

In order to facilitate numerical computations using trigonometric functions, Werner derived the cosine formula and for the first time also the formula (originally in another, slightly cumbersome notation):

$$2(\sin)a \cdot (\sin)b = \cos(a - b) - \cos(a + b).$$

which was later used (before the invention of logarithms) for replacing multiplication by addition (*e.g.*, by **Tycho Brahe** and **Paul Wittich**). His treatise on the motion of the eight spheres and the statement that precession is an irregular motion drew severe criticism both from **Nicolaus Copernicus** and from Brahe. This treatise appeared in the collection of prints from 1522.

Werner's main achievement was his method of determining the difference of geographical longitudes between two places, later referred to as "the method of lunar distances," which became the principal method used in navigation at sea before reliable nautical chronometers came into use at the end of the 18th century. If the measured angular distances between the Moon and selected bright stars along the ecliptic are compared with the values in tables of lunar motion, which had been computed in advance for the time of a reference meridian, then the difference of the local time and the time at the reference meridian equals the difference of longitudes between the observing place and the reference meridian. This method is explained in comments to **Ptolemy**'s book on geography (contained in the collection of 1514), together with two other methods. One method, less precise, was based on measuring the lunar parallaxes from both places; according to the other, the difference in longitudes was equal to the difference of local times of both observers at the beginning or end of a lunar eclipse. The method of lunar distances triggered an effort to express the sophisticated lunar motion by means of mathematics, which gave strong impetus for the development both of mathematical analysis and celestial mechanics.

Martin Solc

Selected References

Doppelmayr, J. G. (1730). *Historische Nachricht von den Nürnbergischen Mathematicis und Künstlern*. Nuremberg.

Kressel, H. (1963–1964). "Hans Werner. Der gelehrte Pfarrherr von St. Johannis. Der Freund und wissenschaftliche Lehrmeister Albrecht Dürers." *Mitteilungen des Vereins für Geschichte der Stadt Nürnberg* 52: 287–304.

Zinner, E. (1934). "Die fränkische Sternkunde im 11. bis 16. Jahrhundert." *Bericht der naturforschenden Gesellschaft in Bamberg* 27: 111–113.

Wesselink, Adriaan Jan

Born **Hellevoetsluis, the Netherlands, 7 April 1909**
Died **New Haven, Connecticut, USA, 12 January 1995**

Adriaan Wesselink developed new instruments and techniques for the photometric observation of stars, provided powerful methods for the determination of stellar radii, and offered a significant correction to the cosmic distance scale. He was the son of Jan Hendrik Wesselink, a physician, and his wife Adriane Marina Nicolette Stok, a surgical nurse. Wesselink was admitted to the University of Utrecht and studied under **Ejnar Hertzsprung** and **Willem de Sitter**. From 1929 to 1946, he held an assistantship at the Leiden Observatory. His doctoral thesis, completed under Hertzsprung's supervision in 1937, examined the light curve of the eclipsing binary star, SZ Camelopardalis. Wesselink employed an objective grating, which provided multiple diffraction images with precisely known photometric ratios, to obtain the star's high-resolution light curve. In 1936, he and two colleagues applied another differential technique, involving size-graded mirrors, to the Sun (in deep partial eclipse), to derive a more accurate value for its limb-darkening function.

In 1939, Wesselink introduced the notion of relaxation oscillations to account for the asymmetries in the light curves of Cepheid variable stars. More importantly, he developed, independently of **Walter Baade**, a new method for determining the radius of a pulsating variable star. This technique was first demonstrated on the star δ Cephei in 1946. Here, Wesselink compared simultaneous measures of the star's color index and its radial velocity. He reasoned that if the pulsation hypothesis was correct, then brightness differences having identical color indices could be due only to changes in size. He then computed the star's mean radius (38 solar radii) and its variation over the pulsation cycle. In this way, Wesselink demonstrated the validity of the pulsation hypothesis with minimal assumptions, mainly, that a single-valued relation exists between the surface brightness and the color index only for the star being analyzed.

During World War II, Wesselink was one of a small group who remained in the Netherlands to look after the Leiden Observatory. In 1943, he married Jeanette van Gogh; the couple had three children.

After the war, Wesselink was appointed by **Jan Oort** to supervise the work at Leiden's southern field station, the Union Observatory in Johannesburg, South Africa (1946–1950). Stronger research opportunities attracted him to the nearby Radcliffe Observatory in Pretoria, South Africa, where he became chief assistant and adjunct director (1950–1964). Working with **Andrew Thackeray** and later with Michael Feast, Wesselink contributed to studies of the B stars of the southern Milky Way. More importantly, this trio detected RR Lyrae stars in the Magellanic Clouds, thereby confirming expansion of the distance scale of the Universe and solving a host of related problems. In the process, Wesselink demonstrated the existence of a significant amount of interstellar dust in the Small Magellanic Cloud.

In South Africa, Wesselink mastered techniques of photoelectric photometry and learned of a new device to observe variable stars differentially: the Walraven photometer (after Théodore Walraven). This instrument may be classified as a "two-star" photometer, the analog output of which was related to the brightness difference between two stars. The device was an antecedent of later chopping, gated, and pulse-counting versions constructed at the University of Calgary in the early 1980s under the title "Rapid Alternate Detection Systems" [RADS].

In 1964/1965, Wesselink became a research associate in the Astronomy Department at Yale University and, from 1966 to 1977, executive director of the Yale–Columbia Southern Observatory at El Leoncito, Argentina. His work at Yale brought him back to his earlier interests in color indices and surface brightnesses. He derived a general relation between these two quantities from stars the radii of which were precisely known through occultations or interferometry; this was widely recognized as a further important contribution.

Although he taught many undergraduates while at Yale, Wesselink had but two Ph.D. students: Carol Ann Williams, a celestial mechanics student; and E. F. Milone, the writer of this essay. Both went on to academic careers. Wesselink retired from Yale in 1977, but maintained an interest in astronomy even after suffering his first stroke. On occasion, he was invited to offer his perspectives at meetings held on stellar pulsation.

A quiet, unassuming person, Wesselink received but few awards during his lifetime, beyond recognition for the eponymous Baade-Wesselink method. He was elected a fellow of the Royal Astronomical Society and served as president of the Astronomical Society of South Africa. He enjoyed a brief turn in the spotlight in 1969 when his 1948 theory of a shallow layer of dust on the Moon's surface, deduced from thermal measurements and conductivity considerations, was proven accurate by the image of astronaut Neil Armstrong's footprint on the Moon.

Wesselink was a resourceful investigator. Once, when he wanted to observe an extremely faint object for which an exposure of many hours was required, he persuaded the manager of the city of Pretoria to arrange an extensive power blackout during the exposure (with the exception of the observatory!), thus darkening the sky sufficiently to produce the intended result.

Wesselink loved a good joke. He once told with glee about the time two well-known astronomers visited him in South Africa. His children came running up and announced that two astronomers, "Egg and Sandwich" (Eggen and Sandage), had arrived.

Taped interviews of Wesselink, obtained in 1977 and 1978 by D. H. Devorkin, may be found at the Niels Bohr Center for the history of Physics, American Institute of Physics.

Eugene F. Milone

Acknowledgement

The author would like to acknowledge discussions with Micheal Feast, Jan Wesselink, William van Altena (and contributions from Dorrit Hoffleit, Carlos Lopez, and Carol Ann Williams), Adriaan Blaauw, and, much earlier, Leendert Binnerdijk.

Selected References

Blaauw, Adriaan and Michael Feast (1996). "Adriaan Wesselink (1909–1995)." *Quarterly Journal of the Royal Astronomical Society* 37: 95–97.

Evans, David S. (1995). "Adriaan Jan Wesselink, 1909–1995." *Bulletin of the American Astronomical Society* 27: 1486–1488.

Wesselink, A. J. (1939). "Stellar Variability and Relaxation Oscillations." *Astrophysical Journal* 89: 659–668.

Wesselink, A. J. (1941). " *A study of SZ Camelopardalis*". *Annalen van de Sternwacht to Leiden* 17: part 3.

——— (1946). "The Observations of Brightness, Colour, and Radial Velocity of delta Cephei and the Pulsation Hypothesis." *Bulletin of the Astronomical Institutes of the Netherlands* 10: 91.

——— (1947). "A Redetermination of the Mean Radius of delta Cephei." *Bulletin of the Astronomical Institutes of the Netherlands* 10: 256–258.

——— (1964). "Astronomical Objective Gratings." *Applied Optics* 3: 889-893.

——— (1969). "Surface Brightness in the UBV System with Applications of M_V and Dimensions of Stars." *Monthly Notices of the Royal Astronomical Society*, 144: 297–311.

Weyl, (Claus Hugo) Hermann

Born **Elmshorn, (Schleswig-Holstein), Germany, 9 November 1885**

Died **Zürich, Switzerland, 8 December 1955**

German–Swiss–American mathematician Hermann Weyl appears to have been the first to write the general-relativistic equations of cosmology in a form in which the galaxies could be regarded as staying at fixed coordinate locations (in what are now called comoving coordinates) while the cosmic expansion was carried by a single function out in front of the rest of the equations. He also gave his name to an important mathematical entity called the Weyl tensor.

The son of Anna Dieck and Ludwig Weyl, a bank director, Weyl was educated at the Gymnasium at Altona (site of what was then a major observatory), whose headmaster sent him on to work at Göttingen with David Hilbert, a relative of the headmaster's. Weyl completed a doctorate and postdoctoral *Habilitation* in 1910 with a dissertation on a particular form of differential equations. He remained at Göttingen and continued his mathematics research there until 1913, working on various aspects of functional analysis.

That year was marked by three major changes in Weyl's life. He married Helene (Hella) Joseph, a translator of Spanish literature. His first book, *Die Idee der Riemannische Flachen*, dealing with Riemannian surfaces, an area of mathematics he later applied in relativity theory, was published. He was offered an opportunity to remain at Göttingen and assume the chair being vacated by the retiring **Felix Klein**, but chose instead to accept an appointment at the Eidgenossische Technische Hochschule [ETH] (Swiss Federal Institute of Technology) in Zürich.

At the ETH Weyl met **Albert Einstein**, who by then was working on general relativity, and welcomed the opportunity to work with him for approximately a year. In particular, Weyl found himself strongly interested in the application of tensor calculus to the representation of spacetime, which served to clarify some of Einstein's concepts. However, as a German citizen, with World War I under way, he was drafted into the German army in 1915. In 1916, at the urgent request of the neutral Swiss government, he was discharged and returned to the ETH.

In 1918 Weyl's second book, *Raum, Zeit, Materie*, later translated as *Space, Time, Matter*, was published, a comprehensive exposition of relativity theory in terms of differential geometry. It went through five editions in rapid succession, the fifth bearing the date 1923.

During 1918/1919, Weyl attempted to derive a unified field theory for electromagnetism and gravitation. Though he abandoned that effort as unsuccessful, his concept of gauge invariance remained fundamental for cosmology. Moreover, Weyl was able to relate gauge invariance to charge conservation.

Over the next decade, Weyl alternated his attention and work among basic mathematics, including its philosophical foundations, mathematical cosmology, and the mathematical foundations of quantum theory. In 1926, as the "new" quantum theory emerged, he was in regular communication with **Erwin Schrödinger** at the ETH and the physicists at the Göttingen school, notably Max Born, Werner Heisenberg, and **Pascual Jordan**. Weyl's third book, *Gruppentheorie und Quantenmechanik*, appeared in 1928, followed by an expanded second edition in 1931.

Weyl's contributions to physics are germane to cosmology in a broad contemporary view. In the 1931 edition of the group theory book he included an analysis of Paul Dirac's relativistic equation for the electron, showing that the positive particle its solutions allowed could not be the proton because the equation constrained its mass to be identical to that of the electron. The positron, thus predicted, was found experimentally by **Carl Anderson** in 1932.

In 1927, Weyl had analyzed Wolfgang Pauli's hypothetical neutrino, and derived an equation then thought to be inadequate because it lacked the expected symmetry. Only after the experimental discovery in 1957 of the nonconservation of parity, by Chien-Shiung Wu and her collaborators, was it realized that Weyl's equation had predicted it, decades before the hypothesis published by T. D. Lee and C. N. Wang. The equivalence of mathematical symmetry and physical conservation had been established by Emmy Noether in 1919, but the predictive power of Noether's theorem was not fully realized for many years.

Over the same years, Weyl's work on cosmology continued and progressed. In the early 1920s he showed that the redshifted galactic spectra reported by **Vesto Slipher** required an expanding Universe rather than the stationary models postulated by Einstein and by **Willem de Sitter**, although de Sitter's allowed for redshifts. However, Weyl's postulate on the redshifts led to his model showing all geodesic world lines diverging from a common point as an origin, never intersecting in finite space-time, and projected back as converging toward negative infinity in the past. He thus postulated that all galaxies had a common origin in the very distant past, and that expansion was therefore required. His postulate also enabled him to define cosmic time. Weyl's paper, "Redshift and Cosmology," appeared in 1930.

That year the Weyls and their two sons, Fritz Joachim (also a mathematician) and Michael, moved back to Germany. Weyl had at last accepted an appointment to the faculty at Göttingen, following Hilbert's retirement. However, as the political situation in Germany became increasingly ominous, the Weyls chose to immigrate to the United States in 1934 while it was still possible to do so openly. Weyl was then appointed to the faculty of the Institute for Advanced Study at Princeton University, where Einstein had preceded him. The Weyls were familiar with Princeton since he had spent the year 1928/1929 as a visiting professor at the university.

At the institute, Weyl was able to resume research in several areas of mathematical physics, including spinor theory. His last book, other than revisions and reminiscences, was *The Classical Groups: Their Invariants and Representations*, which was published in 1939.

Weyl became an American citizen in 1939. In 1940 he was elected to the US National Academy of Sciences. He was also a foreign member of the Royal Society (London) and a corresponding member of the Paris Académie des sciences. During World War II, Weyl interrupted his research and carried out analyses on fluid dynamics and shock waves pertinent to national needs. This applied work as well as his basic research led to papers that appeared throughout the 1940s, including one in 1944 pertinent to relativity theory.

In 1948, Hella Weyl died; in 1950, Hermann married Ellen Lohnstein Bär, a sculptor. He retired from the institute in 1951, and marked that occasion with a retrospective book, *Symmetry*, published in 1952. Thereafter, the Weyls divided their time between Princeton and Zürich. The revised and translated version of his first book, *The Concept of a Riemannian Surface*, appeared in 1955. That year, Weyl learned that he had overstayed the time allowed for naturalized American citizens to remain out of the country, and was barred from returning to the United States until a legal exception could be arranged. Unfortunately, he died of a heart attack while still in Zürich.

Weyl's contributions to mathematics, physics, cosmology, and philosophy comprise 150 papers and books. Roger Penrose declared him to be "the greatest mathematician of [the 20th] century."

Frieda A. Stahl

Selected References

Dieudonné, J. (1976). "Weyl, Hermann." In *Dictionary of Scientific Biography*, edited by Charles Coulston Gillispie. Vol. 14, pp. 281–285. New York: Charles Scribner's Sons.

Drucker, Thomas (1995). "Hermann Weyl." In *Notable Twentieth-Century Scientists*, edited by Emily J. McMurry, pp. 2161–2164. Detroit: Gale Research.

Hetherington, Norriss S. (ed.) (1993). *Encyclopedia of Cosmology*. New York: Garland, pp. 571–572.

Kragh, Helge (1996). *Cosmology and Controversy: The Historical Development of Two Theories of the Universe*. Princeton: Princeton University Press.

Weyl, Hermann (1930). "Redshift and Relativistic Cosmology." *Philosophical Magazine*, ser. 7, 9: 936–943.

——— (1949). *Philosophy of Mathematics and Natural Science*, translated by Olaf Helmer. Princeton, New Jersey: Princeton University Press.

——— (1952). *Space – Time – Matter*, translated by Henry L. Brose. New York: Dover.

——— (1952). *Symmetry*. Princeton, New Jersey: Princeton University Press.

Zund, Joseph D. (1999). "Weyl, Hermann." In *American National Biography*, edited by John A. Garraty and Mark C. Carnes. Vol. 23, pp. 101–103. New York: Oxford University Press.

Wharten, George

Born **1617**
Died **1681**

English mathematician George Wharten compiled "royalist" (Copernican) almanacs starting in 1641.

Selected Reference

Capp, Bernard (1979). *English Almanacs 1500–1800: Astrology and the Popular Press*. Ithaca, New York: Cornell University Press.

Wheeler, John Archibald

Born **Jacksonville, Florida, USA, 9 July 1911**

American theoretical physicist John A. Wheeler is very often cited for having coined the phrase "black hole," but his most lasting impact has been as the founder and mentor of an American school

of scientists who have applied the ideas of general relativity to problems in astrophysics. Corresponding groups in Russia (the former USSR) and England (United Kingdom) are associated with the names of **Yakov Zel'dovich** and Dennis Sciama.

Wheeler is the son of librarians Joseph Lewis and Mabel Archibald Wheeler. His marriage to Janette Hegner in 1935 produced three children.

In 1927, Wheeler enrolled at Johns Hopkins University in a program that led directly to a Ph.D. without intermediate degrees. He received his doctoral degree in 1933 at age 21 with a thesis on the theory of absorption and scattering of light by helium, under the direction of Karl Ferdinand Herzfeld. Wheeler used a National Research Council fellowship to spend the years 1933–1935 working with Gregory Breit at New York University and with **Niels Bohr** in Copenhagen. He was appointed to a faculty position at the University of Chapel Hill, North Carolina, USA, in 1935 and joined the faculty of Princeton University in 1938. During World War II, Wheeler worked on the atomic-bomb project, at the Metallurgy Lab at the University of Chicago (1942/1943), and as a consultant to E. I. du Pont de Nemours & Company, at the Hanford reactor site (1944/1945).

Returning to Princeton, Wheeler was promoted to associate and then full professor of physics and was named to the Joseph Henry chair in 1966, from which he retired in 1976. He then became director of a center for theoretical physics at the University of Texas at Austin for another decade, and retired back to Princeton as professor Emeritus in 1986.

Most of the work for which Wheeler is known was carried out at Princeton—much of it in collaboration with students. His early research (1933–1955) centered on atomic nuclei, elementary particles, and the interaction of radiation with matter. Before going to Princeton, he contributed to the theory of the scattering of light by light (with Breit, in 1934, making a prediction that was finally confirmed experimentally in 1997), to a way of describing quantum-mechanical processes called the S-matrix (1937), and to a first theory of nuclear rotation (in 1938, with **Edward Teller**). Wheeler's famous work on the theory of nuclear fission was completed at Princeton (in 1939, with Bohr, very soon after the discovery of fission).

Wheeler's best-known Ph.D. student was Richard P. Feynman, who completed his Princeton doctorate in1942. During World War II, Wheeler and Feynman were occasionally able to find time at Los Alamos to work together, resulting in two notable papers published in 1945 and 1949 on "action at a distance" (electrodynamics without fields). They showed that the one-way flow of time observed in electromagnetic phenomena results from the enormous amount of matter in the Universe, and that in a hypothetical universe containing only few particles, time would observably run in both directions, with the future affecting the past.

Other Wheeler contributions included a study of the "universal" weak interaction (in 1949, with Jayme Tiomno), a way of looking at the structure of atomic nuclei that accounted for their sometimes very nonspherical shapes (in 1953, with David Hill), the use of muons to probe nuclear properties (1953), and a theory of scattering useful in molecular as well as nuclear physics (in 1959, with Kenneth Ford).

In 1952, Wheeler requested and was granted approval to offer the first-ever course in relativity theory at Princeton University. This triggered an interest in gravitation and relativity that dominated most of his remaining career. He began by pushing the theory to its limits, far from the tiny effects that had initially validated it (the advance of the perihelion of Mercury, the deflection of starlight by the Sun, and the gravitational redshift of solar radiation). Wheeler studied, for example, the possibility of radiation so intense that it held itself together gravitationally. This "geon" (1955) he later showed to be unstable. But in some manifestation—electromagnetic waves, gravitational waves, or neutrinos—he suggested, geons could have had transitory existences, making them potential sources of gravitational radiation.

Wheeler's attention then turned to gravitational collapse, which had been studied as early as 1939 by **J. Robert Oppenheimer** and colleagues. With Tullio Regge, he showed that a Schwarzschild singularity could be a stable entity (1957). Even in his first paper on relativity (1955), Wheeler already was exploring links between gravitation and quantum theory, and soon (1957) he introduced the idea of "quantum foam," pointing out the significance of what he called the "Planck length" and "Planck time" as measures of the scale of quantum fluctuations in spacetime.

Working with Charles Misner on an "already unified field theory"—physics built on curved empty space only (1957)—Wheeler studied and named the "wormhole," a hypothetical conduit between remote points of a multiply connected space–time geometry (leading others to speculate about the wormhole as a mechanism of time travel). He and Misner went on to suggest the idea of "charge without charge"—that is, the ends of wormholes disgorging and engorging lines of electric field to give the appearance of sources and sinks of field.

With various students, Wheeler explored all the possible fates of cold, baryonic matter (1958). He likes to say that he fought against the idea that matter can collapse to a singularity, looking for and hoping to find a rescue from this fate in quantum theory and elementary-particle physics. When Wheeler concluded that nothing could stop the collapse of a sufficiently massive body after its fuel is exhausted, he gave the resulting entity a name, "black hole" (coined in 1967 and first used by him in print in 1968). He then became a leader in working out the properties of black holes. To encapsulate the idea that a black hole is characterized by only its mass, charge, and spin, he introduced the phrase, "a black hole has no hair." In the monograph *Gravitation* (published in 1973 and still in print), Wheeler, Misner, and Kip Thorne laid out the theory of gravitational physics in more than 1,200 pages, much of the content developed by Wheeler with Misner, Thorne, and other students.

In the realm of quantum theory, Wheeler put forward the idea of a "delayed choice" experiment in which a decision to check whether a photon followed one of two possible paths, or followed both paths at once, is made *after* the photon is well on its way through the apparatus. The 1978 conclusion (that the counter intuitive predictions of quantum theory will be borne out) was tested in a laboratory experiment by Carroll Alley and others in 1984.

Throughout his career, Wheeler has served the scientific community and the United States government in many ways and received many honors. With **Lyman Spitzer**, he founded Project Matterhorn in Princeton and contributed to the development of thermonuclear weapons (1951-1953). (Matterhorn later evolved into the Princeton Plasma Physics Laboratory.) An outgrowth of his 1958 "Project 137" was the "Jasons," a group of informal advisors to the government on technical aspects of military matters. Wheeler was president of the American Physical Society (1966) and served as an advisor to Oak Ridge National Lab, the Los Alamos Scientific Lab, the Battelle Memorial Institute, and a number of other organizations. He holds honorary degrees from a dozen institutions in several countries. Wheeler received the National Medal of Science in 1971 and the Wolf Prize in Physics in Israel in 1997. He is a member of the United States National Academy of Sciences and the American Philosophical Society, among other professional and honorary organizations.

Kenneth W. Ford

Selected References

Ciufolini, Ignazio and John Wheeler (1995). *Gravitation and Inertia*. Princeton, New Jersey: Princeton University Press.

Ferris, Timothy (1998). *The Whole Shebang: A State of the Universe(s) Report*. New York: Simon and Schuster.

Harrison, B. Kent, Kip S. Thorne, Masami Wakano, and John Wheeler (1965). *Gravitation Theory and Gravitational Collapse*. Chicago: University of Chicago Press.

Misner, Charles W., Kip S. Thorne, and John Wheeler (1973). *Gravitation*. San Francisco: W. H. Freeman.

Taylor, Edwin F. and John Wheeler (1966). *Spacetime Physics*. San Francisco: W. H. Freeman. (An undergraduate textbook.)

——— (2000). *Exploring Black Holes: Introduction to General Relativity*. San Francisco: Addison Wesley Longman.

Wheeler, John (1955). "Geons." *Physical Review* 97: 511–536.

——— (1957). "On the Nature of Quantum Geometrodynamics." *Annals of Physics* 2: 604–614.

——— (1962). *Geometrodynamics*. New York: Academic Press. (For the first of his books, Wheeler chose for the title a word he had coined to emphasize that the physics of gravity is the dynamics of geometry.)

——— (1990). *A Journey into Gravity and Spacetime*. New York: Scientific American Library.

——— (1994). *At Home in the Universe*. New York: American Institute of Physics.

Wheeler, John and Niels Bohr. (1939). "The Mechanism of Nuclear Fission." *Physical Review* 56 (1939): 426-–450.

Wheeler, John and Richard Feynman. (1949). "Classical Electrodynamics in Terms of Direct Interparticle Action." *Reviews of Modern Physics* 21 (1949): 425–433.

Wheeler, John and Kenneth Ford (1998). *Geons, Black Holes and Quantum Foam: A Life in Physics*. New York: W. W. Norton. (Wheeler's autobiography)

Wheeler, John and Charles Misner. (1957). "Classical Physics as Geometry: Gravitation, Electromagnetism, Unquantized Charge, and Mass as Properties of Curved Empty Space." *Annals of Physics* 2 (1957): 525–603.

Wheeler, John and Tullio Regge (1957). "Stability of a Schwarzschild Singularity." *Physical Review* 108: 1063–1069.

Wheeler, John, Martin Rees, and Remo Ruffini (1974). *Black Holes, Gravitational Waves, and Cosmology*. New York: Gordon and Breech.

Whewell, William

Born **Lancaster, Lancashire, England, 24 May 1794**
Died **Cambridge, England, 6 March 1866**

William Whewell was a philosopher of science and a central figure in early Victorian science and mathematics whose astronomical work focused upon tides. Born the eldest son of a carpenter, he attended the grammar school at Heversham, Westmorland, and then entered Trinity College, Cambridge, graduating second wrangler. He became a fellow of the college in 1817, taking his M.A. degree in 1819 and his D.D. degree in 1844.

By the first quarter of the 19th century, French mathematicians, applying analytical methods to Newtonian physics, had established a supremacy over British mathematicians. In 1819, the year Whewell helped form the Cambridge Philosophical Society, he published his first textbook, *An Elementary Treatise on Mechanics*, the first English text on applied mathematics that consistently used continental symbols and became the standard for undergraduates at Cambridge. With his second textbook, *A Treatise on Dynamics* (1823), Whewell became a leading agent for French analytical methods in Britain. He went on to hold professorships first in mineralogy (1828), then moral philosophy (1838); ultimately he accepted the mastership of Trinity College, Cambridge (1841), an appointment he held until his death.

At Cambridge, Whewell developed one of the foremost mathematical curricula in history. A zealous and prolific researcher, he published significant works in experimental physics, crystallography, mineralogy, physical astronomy, science education, architecture, poetry, and religion, along with a bewildering number of more popular reviews, lectures, and sermons. He was the inventor of the self-registering anemometer, and the originator of many new scientific terms, including "ion," "cathode," "Eocene," "Miocene," "physicist," and "scientist." Whewell is best known for his multivolume *History of the Inductive Sciences* (1837) and his equally impressive *Philosophy of the Inductive Sciences* (1840), both unrivaled in their day. These works helped to define what "science" was in the early Victorian era, an important period in the professionalization of the sciences.

Whewell's work in history and philosophy, and his own researches in physical astronomy, were intimately linked; in the mid-1830s, Whewell composed his *History*, outlined his *Philosophy*, and published his most extensive tidal researches. Physical astronomy, the "queen of sciences" according to Whewell, had reached a state of maturity that no other science could emulate. He referred to it as the only complete science, and it was central to both his

History and his *Philosophy*. Whewell laid out a complete philosophy of scientific methodology. His *History* focused on the gradual ascension of scientific knowledge from facts, to phenomenological laws, and finally to causal laws. Each science began with a "prelude" in which a mass of unconnected facts predominated. The act of "colligation" by the scientist brought about an "inductive epoch" where a useful theory was formed through the creative role of the scientist. A "sequel" followed where the successful theory was refined and applied. Whewell's historical analysis of physical science provided the basis for his philosophy of science.

Deeply influenced by **Immanuel Kant**, Whewell, like Kant, emphasized the creative role of the mind and the need for bold unifying conjectures that far surpassed the empirical evidence. Because a "boldness and license of guessing" was a necessary aspect of all progress in science, Whewell also believed the scientist must be equally prepared for testing each hypothesis. Though a correct theory should be able to account for all of the observed facts and predict new ones, the true test of a scientific hypothesis came when it explained "cases of a kind different from those which were contemplated in [its] formation." According to Whewell, cases in which "inductions from classes of facts altogether different have thus jumped together," a peculiar feature he termed "consilience of induction," belonged only to the best-established theories in the history of science.

Whewell's own research in physical astronomy was in what he termed "tidology." Between 1833 and 1850, he wrote 14 major papers on the study of the tides, along with numerous shorter essays. By following the analogy of physical astronomy, his model science, and **Johannes Kepler**, his model scientist, Whewell sought to have masses of observations made around the globe to determine the phenomenological laws of the tides. He followed two major lines of research: The first advanced the earlier work of John William Lubbock and entailed an analysis of long-term observations to determine the tidal constants at the major ports in Great Britain, including the establishment of each port and the effects of the parallax and declination of the Sun and Moon. His second line of research was unique and entailed an analysis of short-term but simultaneous observations along the entire coast of Great Britain, and eventually Europe and America. In July 1835, Whewell organized a "great tide experiment" where the tides were measured every 15 minutes for a fortnight at over 650 tidal stations in nine countries, including Great Britain, France, and the United States. He used these simultaneous measurements to draw a map of "co-tidal lines" to determine the motion of the tides across the ocean.

Whewell's work on the tides was modestly successful. He combined his method of analyzing long-term observations with simultaneous short-term measurements in a unique fashion to determine the course of the tide around the British coast. He determined the empirical laws for the parallax and declination of the Moon and Sun, and quite correctly noted the importance of the diurnal inequality – his prize analysis – for any future theory. Along with **John Herschel**, Whewell pioneered the graphical representation of data and its use in theoretical investigations. He used his unique "graphical method of curves" throughout his tidal studies, and, in turn, used his tidal researches as an explanation of the process of data reduction and analysis in his *Philosophy*. Thus, though Herschel had laid out the graphical method in 1833, it was Whewell who explained it for the first time in combination with other methods of data analysis, such as the method of residues, and popularized its use through the pages of his *Philosophy*. He received the Royal Society's Royal Medal for his efforts in 1838.

Whewell held many titles, including fellow of the Royal Society of London and the Royal Astronomical Society, and honorary membership in numerous foreign societies.

Michael S. Reidy and *Malcolm R. Forster*

Selected References

Butts, Robert E. (1976). "Whewell, William." In *Dictionary of Scientific Biography*, edited by Charles Coulston Gillispie. Vol. 14, pp. 292–295. New York: Charles Scribner's Sons.

Fisch, Menachem (1991). *William Whewell, Philosopher of Science*. Oxford: Clarendon Press.

Yeo, Richard (1993). *Defining Science: William Whewell, Natural Knowledge, and Public Debate in Early Victorian Britain*. Cambridge: Cambridge University Press.

Whipple, Fred Lawrence

Born **Red Oak, Iowa, USA, 5 November 1906**
Died **Cambridge, Massachusetts, USA, 30 August 2004**

American cometary astronomer Fred Whipple formulated the modern theory of the structure of comet nuclei, the "dirty snowball" or "dirty iceberg," in which a matrix of frozen water, carbon dioxide, and other ices embeds solids including interstellar dust grains. The son of Henry Lawrence Whipple and Celestia (*née* McFarland) Whipple, he spent his first 14 years on an Iowa farm. After the family moved to Long Beach, California, he graduated from high school there. Whipple enrolled as a mathematics major at Occidental College, switching after one year to an astronomy major at the University of California [UC] at Los Angeles where he received an AB in 1927. His graduate work at UC Berkeley and Lick Observatory was supported by a teaching assistantship; he received a Ph.D. in 1931 with a thesis on radial velocity curves of Cepheid variables under the direction of **Armin Leuschner**. The next year, Whipple accepted a position at the Harvard College Observatory [HCO] under the direction of **Harlow Shapley**, and remained associated with the various astronomical institutions in Cambridge, Massachusetts, thereafter, first as director of the outlying Oak Ridge Station (1932–1937), then as a faculty member of HCO (1937–1977 with secondment to the Office of Scientific Research and Development 1942–1945), retiring as Phillips Professor of Astronomy in 1977. Whipple also served a term (1949–1958) as chair of the Harvard astronomy department. He concurrently helped arrange the relocation of the Smithsonian Astrophysical Observatory [SAO] to Cambridge in 1955, serving first as its director and then, after an institutional merger in 1973, as director of Harvard–Smithsonian Center for Astrophysics until 1977.

Whipple married Dorothy Woods in 1928, the marriage ending in divorce in 1935 after the birth of a son. His second marriage to Babette Frances Samuelson in 1946 produced two daughters.

Soon after the discovery of Pluto in 1930 by **Clyde Tombaugh**, Whipple collaborated with another graduate student to attempt to determine its orbit from a very limited stretch of data. One of their several possible orbits turned out to be essentially correct. During the period 1936–1941, he collaborated with **Cecilia Payne-Gaposchkin** on interpretation of the spectrum of Nova

Hercules and an attempt to understand the then completely mysterious spectra of supernovae, focusing on the very bright 1937 event now known to have been a Type Ia. Their conclusion that the dominant contributions must be very greatly broadened lines of common elements was essentially correct. Whipple's estimate of the total light production of supernovae, allowing for changes in the extragalactic distance scale since 1939, also came close to the modern value.

Virtually all of the rest of Whipple's work concentrated on meteors, comets and dust in the Solar System. His war work had two parts: designing strips of aluminum-foil chaff to be deployed from Allied aircraft and confuse enemy radar, and devising a thin absorbing shield to protect high-flying aircraft from meteorite impacts. Spacecraft later carried such Whipple shields.

Whipple began observing meteors by photographing them from several sites to get three-dimensional trajectories. He soon verified his underlying assumption that most of them share cometary kinematics. His meteor spectra showed silicon and iron, and Whipple concluded that comets should also contain heavy elements in dust form as well as come from a remote ring of progenitors outside the orbit of Pluto, as suggested in 1932 by **Ernst Öpik** and later refined by **Jan Oort** for whom the comet cloud is named. The conclusion that meteors are cometary debris sheds light on a long-standing problem, apparent deviations of cometary orbits, including that of the most famous, 1P/Halley, from precise periodicity. Ejecting the material that becomes meteoroids and then meteors if they intersect with the Earth's atmosphere causes an "equal and opposite force" on the comet, which can change its orbit.

By 1949 Whipple had concluded that comets consisted of a solid nucleus of frozen water, methane, carbon dioxide, carbon monoxide, and ammonia with embedded dust particles of silicates and hydrocarbons. As the nucleus approaches the Sun, the ices volatilize sequentially, and the gases and dust form a coma and two tails responsible for the standard appearance, while jets of ejecta can change the orbit. He further refined the model over the next decade.

As director of SAO, Whipple expanded its mission from effects of solar radiation on the Earth to include the small bodies of the Solar System and extra-solar-system astronomy. He participated in the early years of the space program, developing cameras and tracking teams to follow artificial satellites and continued with the program throughout his career. He became a member of the scientific team of the National Aeronautics and Space Administration's [NASA] CONTOUR (Comet Nucleus Tour) mission in 1998. (The mission failed at the stage when the main engine should have ignited to take it out of near Earth orbit.)

Whipple served on advisory panels for NASA, the National Science Foundation, and the US Army, US Navy, and US Air Force; maintained memberships in a number of professional organizations; held a vice presidency of the American Astronomical Society (1962–1964); and served as president of the commission of the International Astronomical Union that deals with the physical nature of comets and other small bodies. He wrote popular books as well as more than 150 research publications and was a devotee of science fiction. Whipple held a number of honorary degrees; received medals and other honors from the US National Academy of Sciences, the American Academy of Arts and Sciences, the Royal Astronomical Society (London), the Astronomical Society of the Pacific, the American Association for the Advancement of Science, the Meteoritical Society, and the US government; and has a minor planet named for him. Between 1932 and 1942, Whipple also discovered six comets (on Harvard photographic plates) that are named for him, including 36P/Whipple.

Frieda A. Stahl and Virginia Trimble

Selected References

Bullock, Raymond E. (1995). "Fred Lawrence Whipple." In *Notable Twentieth-Century Scientists*, edited by Emily J. McMurry, pp. 2167–2170. Detroit: Gale Research.

Calder, Nigel (1981). *The Comet Is Coming*. New York: Viking Press.

Field, George B. and A. G. W. Cameron (eds.) (1975). *The Dusty Universe*. New York: Neale Watson for the Smithsonian Astrophysical Observatory.

Levy, David H. (2002). "Dr. Comet at 95." *Sky & Telescope* 103, no. 1: 89–90.

Whipple, Fred L. "A Comet Model." Parts 1–3. *Astrophysical Journal* 111 (1950): 375–394; 113 (1951): 464–474; 121 (1955): 750–770.

——— (1964). *History of the Solar System*. Cambridge, Massachusetts: Smithsonian Institution Astrophysical Observatory.

——— (1968). *Earth, Moon, and Planets*. 3rd ed. Cambridge: Harvard University Press.

——— (1981). *Orbiting the Sun: Planets and Satellites of the Solar System*. Cambridge, Massachusetts: Harvard University Press.

Whipple, Fred L. and Daniel W. E. Green (1985). *The Mystery of Comets*. Washington, DC: Smithsonian Institution Press.

Whipple, Fred L. and Walter F. Huebner (1990). *Physics and Chemistry of Comets*. Berlin: Spring-Verlag.

Yeomans, Donald K. (2004). "Fred Lawrence Whipple, 1906–2004." *Bulletin of the American Astronomical Society* 36: 1688–1690.

Whiston, William

Born **Norton-Juxta-Twycross, Leicestershire, England, 9 December 1667**
Died **Lyndon, Leicestershire, England, 22 August 1752**

William Whiston popularized Newtonian physics and astronomy, which he incorporated into his own cosmogony that also reached wide audiences. He was born to Josiah and Katherine Whiston. Whiston's father, the rector of Norton, intended his son to become a clergyman. A year after his father's death in 1685, William matriculated at Clare Hall, Cambridge. In 1693, Whiston became a senior fellow of Clare and was ordained in the Church of England. At Cambridge, he continued his study of mathematics and Cartesian mechanical philosophy, but he was converted shortly afterward to the physics that **Isaac Newton** presented in his *Principia* (1687).

After meeting Newton in 1694, Whiston published his first book, *A New Theory of the Earth*, with a dedication to Newton. Whiston's cosmogony, the first book-length popularization of Newtonian physics and astronomy, was an attempt to correct Thomas Burnet's *Sacred Theory of the Earth*, which had used Cartesian physics to explicate the biblical accounts of creation and the Flood. Whiston employed Newtonian physics and astronomy to account for physical mechanisms used by God to create the Solar System and bring about the Noachic Deluge and also to describe the final apocalyptic conflagration. The chief mechanism came from Whiston's providentialist, Newtonian cometography. Whiston proposed that planets were comets captured by the Sun's gravitational pull. The most striking feature of the *New Theory* is its catastrophist model of Earth history; for example, the close passage of a comet accounts for the diurnal rotation of the Earth and the distortion of the Earth's orbit from a circular to elliptical shape. Whiston attributed the geological strata and buried marine fossils to the Noachic Flood, brought about by a later close passage of a comet that distended the Earth's crust, causing it to break open and release the subterranean waters. The rain described in the Genesis account of the Flood he attributed to escaping vapors from the comet, the tail of which he believed the Earth passed through during the Deluge. Whiston predicted that the gravitational attraction of yet another comet would pull the Earth out of its solar orbit in the future, leaving the Earth to travel freely through the Universe. Whiston answered some of his critics in his *A Vindication of the New Theory* (1698) and *A Second Defence of the New Theory* (1700).

Whiston became rector of Lowestoft cum Kessingland on the Suffolk coast in 1698, but returned to Cambridge in early 1701 after being asked to lecture as Newton's deputy. This appointment may have been, at least in part, the result of the *New Theory*, which Whiston claimed had been viewed with favor by Newton. Whiston evidently impressed the electors: He was elected Lucasian Professor in May 1702, shortly after Newton's resignation in late 1701.

Whiston lectured on astronomy, mathematical physics, and ancient eclipses. Instrumental in the election of **Roger Cotes** to the Plumian Professorship of Astronomy and Experimental Philosophy, Whiston went on to collaborate with Cotes in Cambridge's first experimental lecture course, which began in May 1707. While he was Lucasian Professor, Whiston published Newton's lectures on algebra (*Arithmetica universalis*, 1707). Unlike his immediate predecessor, Whiston attempted to reach his undergraduates with his lectures and textbooks, including various editions of Euclid's *Elements*.

After Whiston became aware of Newton's antitrinitarian heresy, he began in 1708 to preach antitrinitarian views openly, much to the consternation of his Cambridge colleagues. Characteristically, Whiston refused to be dissuaded by friends who warned him away from such a legally dangerous path; on 30 October 1710 the college heads expelled him from Cambridge and his professorship. Newton remained silent through Whiston's trial by the Convocation of Clergy that followed, and by 1714 broke with his quondam disciple completely.

With only the meager revenues from a small farm to support his growing family, Whiston moved to London and set himself up as a private mathematics tutor. By 1712, he began public lectures on experiments and formed a partnership with the instrument-maker Francis Hauksbee, Jr., whose shop, a few doors down from the Royal Society, provided the venue. In collaboration with engraver and instrument-maker John Senex, Whiston published in 1712 a much-copied chart of the Solar System illustrated with the paths of 21 comets. The solar eclipses of 1715 and 1724 provided the enterprising Whiston with further opportunities to secure income from astronomy; he delivered lectures on these events and, along with **Edmond Halley**, produced some of the earliest eclipse charts. Whiston continued to bring together his theological and astronomical interests in his *Astronomical Principles of Religion, Natural and Reveal'd*. Written in accessible prose and published with engravings of the Solar System, this book served as an effective popularization of both the new astronomy and natural theology.

Although Newton blocked his nomination, Whiston made several appearances as a nonmember at meetings of the Royal Society, including presentations of magnetic experiments in 1720 and 1722. In 1734, he made a presentation to the society on a reflecting telescope he had invented. Whiston continued to devote energy to the longitude problem and the mapping of the coast of Britain in the 1730s and 1740s, resulting in *An Exact Trigonometrical Survey of the British Channel* (1745).

Whiston made few original contributions to astronomy aside from his cometographical theories, but he played a pivotal role in the dissemination of Newton's work through his popularizing program. A major luminary in the catastrophist tradition of Earth history, Whiston's *New Theory* helped foster mechanical explanations (including impact theory) to describe the origin and development of the Solar System. Although his own schemes proved unsuccessful, Whiston played a pivotal lobbying role in the creation of the Board of Longitude and the Longitude Act. Whiston's influence can also be detected in the work of his grandson, astronomer and meteorologist Thomas Barker (1722–1809), whose *An Account of the Discoveries Concerning Comets, with the Way to Find Their Orbits* (1757) played a minor but important role in the history of cometography.

Stephen D. Snobelen

Selected References

Farrell, Maureen (1981). *William Whiston*. New York: Arno Press.

Force, James E. (1985). *William Whiston: Honest Newtonian*. Cambridge: Cambridge University Press.

Howarth, Richard J. (2003). "Fitting Geomagnetic Fields before the Invention of Least Squares: II. William Whiston's Isoclinic Maps of Southern England (1719 and 1721)." *Annals of Science* 60: 63–84.

Whiston, William (1696). *A New Theory of the Earth, from its Original, to the Consummation of All Things*. London: R. Roberts. (A second edition of the *New theory* appeared in 1708, followed by third [1722], fourth [1725], fifth [1737], and sixth [1755] editions. A German translation appeared in 1713 and a French epitome in 1718. Buffon summarized the main principles of the book in the first edition of his *Histoire naturelle* [1749], Vol. 1.)

——— (1707). *Prælectiones astronomicæ*. Cambridge: Cambridge University Press.

——— (1712). *A Scheme of the Solar System*. London. J. Senex.

——— (1715). *Astronomical Lectures*. London. J. Senex and W. Taylor.

——— (1716). *Sir Isaac Newton's Mathematick Philosophy more easily demonstrated*. London. J. Senex and W. Taylor.

——— (1717). *Astronomical Principles of Religion, Natural and Reveal'd*. London. J. Senex and W. Taylor.

Whiston, William and Francis Hauksbee, Jr. (1714). *A Course of Mechanical, Optical, Hydrostatical, and Pneumatical Experiments*. London.

Whitehead, Alfred North

Born **Ramsgate, Kent, England, 15 February 1861**
Died **Cambridge, Massachusetts, USA, 30 December 1947**

Alfred Whitehead was a leading mathematician and philosopher of the 20th century, whose works addressed theoretical physics and cosmology. Whitehead's early education was obtained at Sherborne in Dorset, a school founded in the 8th century. In 1880, he was admitted to Trinity College, Cambridge University, where two centuries earlier, **Isaac Newton** had laid down what he thought were the fundamental laws of the Universe. Upon his graduation in 1884, Whitehead was elected a fellow of his college. Although he held strong interests both in mathematics and philosophy, he chose the former because, as he said, "Mathematics must be studied; philosophy should be discussed." One of Whitehead's most important contributions to both disciplines was achieved through his collaboration with former student Bertrand Russell, on their three-volume *Principia Mathematica* (1910–1913), in which the pair took on the gargantuan task of translating all of mathematics into logic. In 1890, Whitehead married Evelyn Willoughby Wade; the couple had three children.

Whitehead resigned his post at Cambridge in 1910 and relocated to London. The following year, he obtained a position at University College but in 1914 was appointed to the chair of applied mathematics at the Imperial College of Science and Technology.

In 1924 (at the age of 63), Whitehead left England to join the philosophy department at Harvard University. He had, in years previous (*e. g., The Concept of Nature*, 1920), already turned his attention to the conceptual analysis typical of philosophers of science. At Harvard, however, he began work on his own great system of *process philosophy*, a theory according to which all things, even atoms – which Newton had conceived as being most real – are but intellectual abstractions that have no mind-independent existence. This aspect of Whitehead's philosophy resembles the idealist views of George Berkeley and Josiah Royce but especially William James's radical empiricism and, even more so, the phenomenalism of physicist Ernst Mach.

The main difference is that Whitehead conceives of the cosmos in terms of fundamental units of existence, instead of the inert atoms represented in the tradition of **Leucippus**, **Democritus**, and **Epicurus**. According to Whitehead's viewpoint, the ultimate atomic constituents of the Universe exist, like Leibnizian monads, as *processes* derived in relation to, and out of, the "now" of consciousness. But this is not to say that they are merely phenomenal or representational. As actual entities, that is, "actual occasions," they are not subject to the sort of mind–body problem as conceived in Cartesian dualism. The Universe according to Whitehead is one insubstantial substance that exists in a chaotic sort of Heraclitan perpetual flux. What immediately appears here and now is real; beyond that is nothing. The Universe, which exists without any static substances, must therefore be understood without the use of any static concepts typical of science and philosophy. The cosmos as a whole must be understood as an interconnected network of individually independent, but mutually complementary, *events*.

This revolutionary aspect of Whitehead's philosophy, though still poorly understood and not widely accepted today, was well ahead of its time. Events, as conceived by Whitehead, are themselves spatiotemporal unities; they exist as actual extensions, and they are what give rise from within the cosmic flux to individual organisms capable of being aware of themselves and of others. What we define as consciousness, he argues, consists of the relationships between events; and, more significantly, every entity consists of all its active relations with all others in a cosmic synchronicity. This Whitehead calls "prehensive occasion." And perhaps most significant of all, Whitehead's fundamental (*i. e.*, process) units of existence do not persist with identity over time. They have no permanent identity, and no history; they exist in a perpetual process of becoming. That is, the annihilation of one set of entities is, itself, the result of the creation of the Universe moving on to the next momentary birth in which each event loses its uniqueness, preserving thereby nothing but the flow of process.

The upshot of this rather extraordinary aspect of Whitehead's cosmology is that since the Universe exists in virtual flux, it cannot be completely understood: not ever, not by anyone. According to Whitehead, the single greatest error made by scientists and philosophers has been the mistaking of intellectual abstractions for actual entities, or what he calls "the fallacy of misplaced concreteness." Moreover, the truly permanent aspects of the Universe do not exist within the realm of *actuality* but only within the realm of *possibility*, and it is the possibilities themselves – not the momentary actualities – that constitute the "eternal objects" of the Universe. The virtual flux of creation and annihilation makes possible the existence of the Universe as a whole which, *necessarily*, is incomplete and unknowable, except as momentary influxes into the process of existence.

Anyone who recognizes in Whitehead's cosmic architecture certain similarities with quantum mechanics and quantum cosmology should not be surprised that he was singularly unimpressed with **Albert Einstein**'s general theory of relativity. He wrote:

> … in the 1880's … nearly everything was supposed to be known about physics that could be known – except a few spots, such as electromagnetic phenomena, which remained (or so it was thought) to be co-ordinated with the Newtonian principles. But, for the rest, physics was supposed to be nearly a closed subject. … By the middle of the 1890's there were a few tremors, a light shiver as of all not being quite secure, but no one sensed what was coming. By 1900 the Newtonian physics were demolished, done for! [This] had a profound effect on me; I have been fooled once, and I'll be damned if I'll be fooled again! … There is no more reason to suppose that Einstein's relativity is anything final, than Newton's *Principia*. The danger is dogmatic thought; it plays the devil with religion, and science is not immune from it.

Daniel Kolak

Selected References

Barker, William A. *et al.* (1976). "Whitehead, Alfred North." In *Dictionary of Scientific Biography*, edited by Charles Coulston Gillispie. Vol. 14, pp. 302–310. New York: Charles Scribner's Sons.

Sherburne, Donald W. (1999). "Whitehead, Alfred North." In *American National Biography*, edited by John A. Garraty and Mark C. Carnes. Vol. 23, pp. 256–259. New York: Oxford University Press.

Whitehead, Alfred North (1929). *Process and Reality: An Essay on Cosmology*. New York: Macmillan.

Whittaker, Edmund T. (1959). "Whitehead, Alfred North." In *Dictionary of National Biography, 1941–1950*, edited by L. G. Wickham Legg and E. T. Williams, pp. 952–954. London: Oxford University Press.

Whitford, Albert Edward

Born **Milton, Wisconsin, USA, 22 October 1905**
Died **Madison, Wisconsin, USA, 28 March 2002**

American photometrist Albert Whitford coordinated the first of five (to date) decadal reports laying out the future course of astronomy in the United States (*Ground-Based Astronomy: A Ten-Year Program*, generally called the Whitford Report and published in 1964). He is also eponymized in the Stebbins–Whitford effect, which might have been early evidence for the evolution of galaxies with the age of the Universe but was in fact an error in allowing for the effects of stars with strong hydrogen features on the spectra of the galaxies of which they are a part. As director of the Lick Observatory, he was instrumental in bringing its 120-in. telescope project into successful operation.

Whitford received a BA in physics in 1926 at Milton College, with which his family had long-standing connections, and his Ph.D. in physics in 1932 at the University of Wisconsin for work under Charles Mendenhall and Julian Mack on a problem in atomic spectra. This problem required him to acquire skills in laboratory instrumentation including vacuum technology and the measurement of very small electric currents. Jobs in physics not being plentiful in 1932, Whitford gladly accepted an assistantship at the Wisconsin and Washburn Observatory under its then director, **Joel Stebbins,** who was attempting to improve the technology of photoelectric photometry with a potassium hydride cell and a quartz-fiber electrometer. Whitford succeeded in attaching a vacuum-tube amplifier to the cell and encasing both the photoelectric cell and the amplifier in a vacuum chamber to reduce noise from cosmic-ray ionization. He spent the years 1933–1935 on a National Research Council Fellowship at the California Institute of Technology and at the Mount Wilson Observatory, working still with Stebbins, who observed a good deal at Mount Wilson, and also working on atomic spectra with **Ira Bowen**. During this period, Whitford developed from an instrument builder to a full-fledged observer, learning photographic spectroscopy from **Alfred Joy** and observing the absorption and reddening of stars at the North Galactic Pole with Stebbins.

Whitford returned to Wisconsin in 1935, interrupting his tenure there from 1941 to 1946 for work on radar at the Massachusetts Institute of Technology under Lee DuBridge, pushing the technology to shorter wavelengths that could resolve, for instance, the periscope of U-boats. Whitford returned to the University of Wisconsin as an associate professor in 1946, where teaching duties required that he acquire the broad knowledge of astronomy that had not been part of his own student career but that so much characterized his work thereafter. Whitford succeeded Stebbins as director of Washburn in 1948 and continued to observe at the Mount Wilson Observatory and Lick Observatory on Mount Hamilton. In 1956, he persuaded the Wisconsin administration to provide funding for a 36-in. Cassegrain photometric telescope on a dark site outside the city, considerably improving both the research and teaching opportunities in Madison, but he left in 1958 to become director of the Lick Observatory. There, Whitford supervised the completion of the 120-in. reflector (then the second-largest telescope in the world after the 200-in. on the Palomar Mountain) and played an active role in the development of its cameras, drawing on the lab skills he had been developing all his life. Whitford also played an important role in the founding of what became the Kitt Peak National Observatory, available to all American astronomers on a scientifically competitive basis.

Through the years, Whitford continued to improve the techniques of photometry, extending the standard wavebands from purely visible light into the ultraviolet and infrared as far as the Earth's atmosphere would permit. He used this to observe distant, hot, bright stars and thus trace out both the distribution of dust in the Milky Way and the properties of the grains that make it up and redden starlight. The best known of his students at Wisconsin was the late Olin Eggen, also a photometrist. After resigning the Lick directorship in 1968, Whitford again took up active undergraduate and graduate teaching. His students from this period (including Jack Baldwin, David Burstein, Alan Dressler, and David Soderblom) and from his postretirement years (including Michael Rich and David Terdrup) have branched out into an enormous range of kinds of galactic, extragalactic, and stellar astronomy. Whitford's last major research effort focused on attempting to determine whether the stars in the central bulge of our own galaxy are as rich in heavy elements as those found in elliptical galaxies and large bulges.

Whitford was elected to the National Academy of Sciences (1954) and the American Academy of Arts and Sciences, received the Henry Norris Russell Lectureship of the American Astronomical Society, and the Bruce Medal of the Astronomical Society of the Pacific (1986 and 1996). He served the community as vice president (1965–1967) and president (1967–1970) of the American Astronomical Society.

Michael Rich

Selected References

Kraft, Robert P. (2002). "Albert Edward Whitford, 1905–2002." *Bulletin of the American Astronomical Society* 34: 1387–1390.

Osterbrock, Donald E. (2003). "Albert Edward Whitford." *Physics Today* 56, no. 1: 67–68.

Rich, R. Michael and Donald M. Terndrup (1997). "Bulges of Galaxies: A Celebration of the 90th Birthday of Albert Whitford." *Publications of the Astronomical Society of the Pacific* 109: 571–583.

Whitford, Albert E. (1986). "A Half-Century of Astronomy." *Annual Review of Astronomy and Astrophysics* 24: 1–22.

Whiting, Sarah Frances

Born **Wyoming, New York, USA, 23 August 1846**
Died **Wilbraham, Massachusetts, USA, 12 September 1927**

With an abiding conviction that astronomy, in keeping with all science, should be taught as astronomy is done, in both laboratory and observatory settings, Sarah Whiting established the first such program in laboratory instruction for women in the United States at Wellesley College, and devoted her career to innovative teaching of astronomical principles and techniques. Whiting was one of two daughters of Elizabeth Lee Comstock Whiting and Joel Whiting. Sarah's father traced his American lineage back to a Mayflower passenger and other early settlers. He was a graduate of Hamilton College, and served successively as a teacher and a principal at several secondary-level academies in upper New York State. He guided Sarah's educational development during her childhood, tutoring her in Latin, Greek, and mathematics, as well as physical sciences, in all of which she was precocious. She often accompanied him to his school and helped him set up demonstrations for his "natural philosophy" classes.

This educational environment predisposed Sarah to seek her own career as a teacher. She attended Ingham University, a collegiate-level academy in Le Roy, New York, chartered in 1857 by the New York legislature to grant 4-year degrees to women. She received the AB degree in 1864, at the age of 18, and taught there initially. Employed subsequently at the Brooklyn Heights Seminary, a secondary school for girls, Whiting taught both classics and mathematics for approximately a decade.

Whiting's interests were not limited to those disciplines, however, and she frequently attended science lectures in New York City, many of them devoted to recent discoveries in astronomy, physics, photography, and spectroscopy. **Annie Cannon**'s obituary quoted Whiting: "I was really started on my scientific career by some lectures, brilliantly illustrated.... These had fascinated me with the application of the spectroscope to astronomy." A lecture on the 1869 solar eclipse furthered Whiting's interest in astronomy. In 1876, Henry Durant sought out Whiting for the faculty of Wellesley College, which he and his wife Pauline had founded the previous year for the express purpose of educating women. Durant specified that women would constitute the faculty and administration, including the president. Functioning as a trustee as well as a benefactor, Durant set the policy and developed a Wellesley curriculum in which science and mathematics were required disciplines on par with the humanities.

Durant was acquainted with **Edward Pickering** at the Massachusetts Institute of Technology [MIT]. Pickering established the first instructional physics laboratories in the United States. Following Pickering's example, Wellesley science courses incorporated a strong laboratory component. At Durant's request, Pickering allowed Whiting to study at MIT for 2 years as a guest, since women were not then admitted as students, to prepare herself for laboratory instruction in physics. From 1876 to 1878, Whiting taught mathematics at Wellesley while commuting to MIT 4 days a week. At MIT she studied the equipment as well as experimental procedures before ordering materials for her Wellesley laboratories.

In 1878, the fifth-floor attic of Wellesley's College Hall was adapted for use as a physics laboratory. Whiting showed a high degree of mechanical aptitude in the assembly, with Durant's assistance, of apparatus that had arrived in parts. She is credited with establishing the second undergraduate physics instructional laboratory program in the United States, the first for women.

After Whiting began teaching astronomy in 1879, she remained in contact with Pickering, who was by then director of the Harvard Observatory. She devised astronomical laboratory exercises that could be performed in daytime, to prepare for as well as supplement night observations. She taught astronomy for 20 years using only a 4-in. Browning portable telescope, a celestial globe, an ephemeris, photographs, and star charts.

Whiting was tireless at bringing phenomena to students and students to phenomena. Cannon, in the obituary she wrote for her former professor, commented, "She let no striking astronomical event pass without arousing the attention of the whole college," citing as examples a bright comet in 1882 (C/1882 R1) and, later that year, the transit of Venus. (In 1882, Cannon was a junior at Wellesley; in 1895 she returned to Wellesley as Whiting's teaching assistant.)

In 1898, Stephen V. C. White offered a 12-in. Fitz/Clark telescope for sale. Whiting had taken many of her Brooklyn Academy students to White's observatory to observe through this telescope and admired its optical quality. During a carefully planned dinner, Whiting impressed a wealthy Wellesley trustee, Mrs. John C. Whitin, with the college's need for an observatory. Whitin agreed to endow an observatory including the Fitz/Clark telescope. Dedicated in 1900, and expanded in 1906, the Whitin Observatory included a residence for Sarah Whiting, its first director. She shared the living quarters with her sister Elizabeth, an administrative employee of the college. The Whiting sisters were frequent and popular hostesses to students, alumnae, faculty, and college visitors at the observatory. Whiting was devoutly religious and strongly in favor of Prohibition, and in accord with the rigid expectations of her era, she remained unmarried.

In earlier years Whiting had expressed a feeling of stress for "being the only woman in places where women were not expected." Yet she was not deterred from her inquiries, visits, travels, and guest enrollments in advanced courses at both American and European institutions. Committed to the precept that good teaching arises from good knowledge, though she did no research herself throughout her career, Whiting vigilantly kept abreast of research results and instrumental innovations and kept her teaching updated. Whiting's curriculum in astronomy incorporated more physics than did other contemporaneous programs designed for women. Building on her experience in teaching both subjects, she included class and laboratory work on spectroscopy and photometry as well as instructional segments on variable stars and sunspots. Winifred Edgerton, the first woman to receive a Columbia Ph.D. in astronomy (1886), found that her undergraduate preparation in celestial mechanics under Whiting was at a level fully equivalent to graduate work at Columbia. By 1905, Wellesley was able to offer a master's degree in astronomy. Whiting used her sabbatical leaves for trips to Europe, during which she was welcomed at the laboratories of noted scientists, mostly in the British Isles but occasionally on the Continent. In 1896 she traveled to Scotland, where she studied in Edinburgh with Sir Peter Guthrie Tait.

Whiting spent extensive time with the astrophysical pioneers Sir **William Huggins** and Lady **Margaret Huggins**, noted for their work in stellar spectra, so it is not surprising that Whiting emphasized stellar spectroscopy in her teaching. She took special pleasure in her association with Lady Huggins. After the death of Sir William in 1910 Lady Huggins bequeathed astronomical equipment and artifacts from their Tulse Hill home and observatory to the Whitin Observatory. Whiting described these items in an article published in *Popular Astronomy*.

In 1912 Whiting relinquished her chairmanship in physics, but continued as director of the observatory until 1916. In anticipation of retirement, she mentored Louise McDowell, her former student in the class of 1898, who returned to Wellesley from Cornell in 1909 with a Ph.D. and an established record in physics research. McDowell became chairman of the physics department when Whiting stepped down. After retiring from the college in 1916, Sarah and Elizabeth moved to their final home in Wilbraham, Massachusetts. Sarah died of atherosclerosis and kidney disease.

A member of the American Astronomical Society and the American Physical Society, Whiting was honored for her innovative teaching at several stages of her career. In 1883 she was one of the first six women elected to fellow status in the American Association for the Advancement of Science, and the only one in astronomy or physics. In 1905, Tufts University awarded her an honorary Sc.D. degree.

Frieda A. Stahl

Selected References

Cannon, Annie Jump (1927). "Sarah Frances Whiting." *Popular Astronomy* 35: 539–545.

Glasscock, Jean (ed.) (1975). *Wellesley College 1875–1975: A Century of Women.* Wellesley, Massachusetts: Wellesley College.

Guernsey, Janet B. (1983). "The Lady Wanted to Purchase a Wheatstone Bridge: Sarah Frances Whiting and Her Successor." In *Making Contributions: An Historical Overview of Women's Role in Physics*, edited by Barbara Lotze. College Park, Maryland: American Association of Physics Teachers.

Hoffleit, Dorrit (1994). *The Education of American Women Astronomers before 1960.* Cambridge, Massachusetts: American Association of Variable Star Observers.

Lankford, John (1997). *American Astronomy: Community, Careers, and Power, 1859–1940.* Chicago: University of Chicago Press. (See particularly Chap. 9, which Lankford devotes to women in astronomy.)

Palmieri, Patricia Ann (1995). *In Adamless Eden: The Community of Women Faculty at Wellesley.* New Haven: Yale University Press.

Rossiter, Margaret W. (1982). *Women Scientists in America: Struggles and Strategies to 1940.* Baltimore: Johns Hopkins University Press.

Whiting, Sarah Frances (1912). *Daytime and Evening Exercises in Astronomy.* Boston: Ginn.

——— (1914). "Priceless Accessions to Whitin Observatory." *Popular Astronomy* 22: 487–492.

——— "History of the Physics Department at Wellesley College, 1878–1912." History Folder, Physics Department Collection. Wellesley College Archives, Wellesley, Massachusetts.

——— "Whiting, Sarah Frances: The First Woman to Found and Develop Two Departments of Science, Physics and Astronomy, in a College of High Rank." Folder labeled "S. F. Whiting's Notes for Speeches and Addresses." Wellesley College Archives, Wellesley, Massachusetts. (Autobiographical manuscript in the handwriting of Sarah Frances Whiting.)

Whitrow, Gerald James

Born **Kimmeridge, Dorset, England, 9 June 1912**
Died **London, England, 2 June 2000**

Gerald Whitrow published over 100 papers and many books, mostly on the subject of time.

Whitrow entered Christ Church, Oxford, graduating with a double first-class degree in 1933. He remained to carry out research with **Edward Milne** on kinematical relativity, and received his D.Phil. in 1939. Whitrow was a mathematics lecturer at Christ Church from 1936 until 1940, when he had to join the Ministry of Supply as a scientific officer doing war research. In 1945, he went to Imperial College, University of London, where he was successively an assistant lecturer and a lecturer in mathematics. Whitrow was promoted to reader in Applied Mathematics in 1951. In 1972 he received a personal chair in the History and Applications of Mathematics.

Perhaps the most important of Whitrow's books was *The Natural Philosophy of Time* (1960). He showed that time could be studied independently of its magnitude. Other books included *The Structure of the Universe* (1949) and (with H. Bondi, W. B. Bonnor, and **Raymond Lyttleton**) *Rival Theories of Cosmology* (1960). The latter was written at the time of the debate between the Big Bang and steady-state theories of the Universe. Whitrow's historical work included a paper on **Robert Hooke**.

Whitrow played an important part in many societies, libraries, and archives. He was president of both the British Society for the History of Science and the British Society for the Philosophy of Science, and was a founding member and first president of the British Society for the History of Mathematics.

Roy H. Garstang

Selected Reference

James, Frank A. J. L. (2001). "Gerald James Whitrow, 1912–2000." *Astronomy and Geophysics* 42, no. 2: 35–36.

Widmanstätten, Aloys [Alois] Joseph Franz Xaver von

Born **Graz, (Austria), 13 July 1754**
Died **Vienna, (Austria), 10 June 1849**

Austrian printer and meteoriticist Aloys von Widmanstätten has his name firmly associated with the patterns of lines seen in iron and iron–nickel meteorites when they are sectioned and etched. The family traced its ancestry to Georg Widmanstätten, a devoted Catholic and successful printer in Bavaria. He was knighted in 1548 by Emperor Karl V. The noble name was transferred to Johann Andrea Karl Beckh, husband of Susanne Widmanstätten and granddaughter of Georg, because her brother Ferdinand, who was Mayor and City Judge of Graz, had no son to inherit the title.

Johann and Susanne von Widmanstätten were the parents of Aloys, and their printing business was successful until Johann died in 1765, leaving it to his 11-year-old son. Aloys studied natural science at the University of Graz and, in 1806, sold his inherited business and moved to Vienna.

In Vienna, Widmanstätten took over the position of director of the newly founded Imperial Technical Museum (Fabriksproduktenkabinetts). According to Dr. Heinrich Thurn, the present director of the museum, there are no photographs available of Aloys von Widmanstätten. He was told that Widmanstätten considered himself so ugly that he did not want to be photographed. Widmanstätten remained a bachelor all his life and died at his home in Vienna (Spiegelgasse 25) at the age of 95.

Widmanstätten was 11 years in the service of the Imperial palace, from 1806 to 1817. It was during this period that Emperor Franz (1768–1835) sent him in 1808 to study the meteorite that fell in 1751 at Agram (the present Zagreb, Croatia, at that time part of the Austrian Empire). This was a particularly auspicious time in meteoritics, because the witnessed falls of meteorites at Wold Cottage, Yorkshire, England (December 1795) and L'Aigle, France (April 1803) had finally convinced most of the European scholarly community that stones truly do fall from the sky.

Because of his printing background, Widmanstätten was used to experimenting with printer's ink. He polished the Agram meteorite, etched it, inked the surface, and made a print on paper. When he noted the characteristic pattern, he studied other meteorite samples in the collection. In 1810, Widmanstätten examined a Siberian specimen and a Mexican specimen sent to the emperor from Berlin by the German chemist Martin Klaproth (1743–1817). In 1812, he examined the large meteorite from Elbogen in Bohemia, and finally in 1815 he examined a piece of a Carpathian meteorite. It was his coworker and successor as the curator of the Meteorite Collection, Carl von Schreibers, who first published the prints as a supplement to a book on meteorites.

German physicist **Ernst Chladni** (1756–1827), who started the study of meteorites, visited Vienna in the spring of 1812 and witnessed the printing of meteoritic patterns. He included in his book *Feuer Meteore* (Vienna, 1819) a number of lithographed illustrations of whole and sectioned meteorites made directly from the etched surface. The Widmanstätten pattern was explained by the mineralogist Gustav Tschermak von Seysenegg (1836–1927) in Vienna.

Iron meteorites are pieces of once molten metallic cores in asteroids that were subsequently eroded and fragmented by impacts after slow cooling. Depending on their nickel content, they are classified into three categories: <6% Ni, 6–14% Ni, and >14% Ni. Iron meteorites containing 6–14% nickel are the most common. Besides their nickel content, iron meteorites are distinguished from other iron metallic finds by the typical Widmanstätten structure. The formation of such a pattern is characteristic of systems consisting of two solid metallic alloys of different composition – in the present case with low and high nickel contents, respectively – that have separated slowly from the molten state upon cooling. The presence of large crystals in iron meteorites is also regarded as evidence of slow changes occurring in the metal over a long period of time.

Fathi Habashi

Acknowledgment

The author acknowledges with thanks the information supplied by Dr. Heinrich Thurn and Prof. Gero Kurat from Vienna.

Selected References

Anon. (1887). "Widmanstätter (Bekh-Widmanstätter) Alois Joseph Franz Xaver von." *Biographishes Lexikon des Kaisertums Österreich*. Vienna.

Habashi, F. (1995). "Aloys von Widmanstätten and the Story of Meteorites." *Bulletin of the Canadian Institute of Mining and Metallurgy* 88, no. 987: 61–66.

——— (1998). "Meteorites: History, Mineralogy, and Metallurgy." *Interdisciplinary Science Review* 23, no. 1: 71–81.

Schützenhofer, V. (1947). "Aloys von Widmanstätten." *Blätter für Technikgeschichte* (Vienna): 34–44.

Wildt, Rupert

Born **Munich, Germany, 25 June 1905**
Died **Orleans, Massachusetts, USA, 9 January 1976**

German–American theoretical astrophysicist Rupert Wildt solved a long-standing problem in the analysis of stellar atmospheres by recognizing that hydrogen atoms with an extra electron (H^- ions) play an essential role in controlling the flow of radiation through the outer layers of the Sun and other cool stars. The son of Gero and Hertha Wildt, he received a Ph.D. in 1927 from the University of Berlin with a thesis on color photography. His first positions were at the Observatory of Bonn (1928–1929) and the University of Göttingen (1930–1935).

In 1931, while at the University of Göttingen as part of the group headed by **Hans Kienle**, Wildt recognized that broad absorption features in the spectra of the giant planets, which had been discovered nearly 10 years earlier by **Vesto Slipher**, were produced by molecules of methane and ammonia absorbing in higher harmonics. This was confirmed by **Theodore Dunham** at the Mount Wilson Observatory soon after. Wildt immigrated to the United States in 1935, holding positions at the Mount Wilson Observatory (1935–1956 as a Rockefeller Fellow), Harvard University (visiting lecturer 1936), the Institute for Advanced Study and Princeton University (1936–1942), and the University of Virginia (assistant professor, 1942–1946). His tenure at Virginia was interrupted by war service (1944–1946) with the Office of Scientific Research and Development.

Wildt's best-known discovery was made at Princeton and published in 1939. Work by **Edward Milne** in the 1920s had established that in order to understand the light we see coming from the Sun and other cool stars it was necessary to postulate that the atmospheric gases were more opaque to light between about 4,000 and 9,000 Å than could be accounted for by adding up all the lines and ionization edges of known atoms and ions. Wildt identified the missing opacity source as H^-, hydrogen atoms with an extra electron attached just tightly enough that a 9,000 Å photon can remove it. The source of the extra electrons was necessarily metal atoms, accounting for the extra redness of metal-rich stars.

Yale University offered Wildt an assistant professorship in 1947, and he moved up to associate professor in 1949, professor in 1958, and acting department chair (1966–1968). He remained there until his retirement in 1973, apart from visiting professorships at Hamburg (1951), the National University of Mexico (1963), the universities of California, Berkeley (1956), and Göttingen (really an honorary professorship emeritus from 1960). When Yale joined the Association of Universities for Research in Astronomy [AURA] (which operated the national observatories at Kitt Peak and Cerro Tololo), Wildt became its representative on the board and was president of AURA from 1965 to 1968 and from 1971 to 1972, and chairman of the AURA board from 1972 almost until his death.

As early as 1938, Wildt had concluded that Jupiter and the other giant planets must have hydrogen as their most abundant element, probably hydrogen in the solid state (which has not yet quite been observed on Earth). His Yale student, Wendell DeMarcus, completed a 1951 thesis on the internal structure of Jupiter, based on the assumption that its chemical composition was the same as that of the Sun, which is roughly right, and that most of the hydrogen would be solid, which is still under discussion. Wildt's best-known Yale student is probably the planetary system dynamicist Myron Lecar of Harvard. Late in his career, Wildt studied the flash (chromospheric) spectrum of the Sun, attempting to use the maximum height at which various lines can be seen to map out coronal temperature and density. He also, before 1957, embarked on an effort to calculate the atmospheres of stars without making the assumption called Local Thermodynamic Equilibrium [LTE] (roughly, the assumption that the radiation field can be described by the same temperature as the kinetic energy of the atoms). That non-LTE calculations are essential is now universally recognized, though Wildt's work is not much cited.

Recognition on the whole came late. Wildt received the Bronze Medal of the University of Liège in 1962 and the Eddington Medal of the Royal Astronomical Society (London) 1966 for his work on H^-, and felt that his discovery of methane and ammonia in the giant planets had not received adequate credit. He was survived by his wife, the former Katherine Eldridge, and their two children.

Gary A. Wegner

Selected References

Anon. (1976). "Wildt, Rupert." In *Who Was Who in America*. Vol. 6, p. 438. Chicago: Marquis Who's Who.

Anon. (1976). "Yale Astrophysicist." *Sky & Telescope* 51, no. 3: 156.

DeMarcus, W. C. (1977). "Rupert Wildt (1905–1976) Gewidmet." *Icarus* 30: 441–445.

Hoffleit, D. (1992). *Astronomy at Yale, 1701–1968*. Memoirs of the Connecticut Academy of Arts and Sciences, Vol. 23. New Haven: Connecticut Academy of Arts and Sciences, pp. 156–168.

Unsöld, A. (1976). "Rupert Wildt." *Quarterly Journal of the Royal Astronomical Society* 17: 523–524.

Wildt, R. (1931). "Ultrarote Absorptionsbanden in den Spektren der groβen Planeten." *Naturwissenschaften* 19: 109–110.

——— (1938). "On the State of Matter in the Interior of the Planets." *Astrophysical Journal* 87: 508–516.

——— (1939). "Negative Ions of Hydrogen and the Opacity of Stellar Atmospheres." *Astrophysical Journal* 90: 611–620.

Wilhelm IV

Born **Kassel, (Hessen, Germany), 24 June 1532**
Died **Kassel, (Hessen, Germany), 20 August 1592**

Wilhelm IV built the first observatory in modern Europe with excellent instruments and staff, producing superior stellar catalogs. Wilhelm was the son of the Landgrave Philipp of Hessen (called the Magnaminous), who introduced the Reformation into Hesse and who became the leader of the Protestant princes. Wilhelm first obtained an education from private tutors at the court of Kassel, as well as, during the year 1546/1547, at the Gymnasium in Strasbourg founded by Johann Sturm. Already in childhood, he developed remarkable intellectual capabilities.

His interests in astronomy appeared to awaken early and were influenced by **Peter Apian**'s brilliant work, *Astronomicum Caesareum* (1540). Under the influence of this book, Wilhelm's interests developed in two directions. First, his fascination with the graphic modeling of revolving disks as a representation of the movement of stars (volvelles) led later to Wilhelm's contracting Eberhard Baldewein to build a planetary clock with a mechanism that is equally a work of art and a milestone in clock making. Wilhelm later had **Jost Bürgi** construct self-moving globes for him.

Wilhelm's second interest was in finding the precise locations of stars that were listed in the catalogs of his time. His first catalog, containing the locations of 58 stars observed with a torquetum, was the result of work during 1560–1563. He later refused to use a torquetum because its overhanging part did not provide enough stability. Observations during 1567/1568 resulted in a second catalog of 58 stars. This time the positions measured directly by Wilhelm himself were compared with those in contemporary catalogs. Both catalogs demonstrated an accuracy that is a clear advance over available star catalogs and represents an early example of the systematic observation of heavenly bodies.

Besides these observations, Wilhelm's systematic measurements of the locations of the Sun began in 1559, along with the observations of comets in 1558 and 1577, and the supernova of 1572 (SN B Cas) – to mention just a few of his publications. In addition to the large torquetum, Wilhelm used a wooden quadrant, as well as a carefully constructed metal azimuth quadrant. For these and other relatively large instruments, he erected a special structure at the Kassel palace as the first permanent observatory in modern Europe. When Wilhelm had to assume the regency of the Hesse–Kassel landgraviate after his father's death in 1567, only sporadic amounts of his time remained for astronomy. At times during the previous period, Wilhelm had brought scholars to his court to support him in his work. From around 1558, Andreas Schöner, son of **Johannes Schöner**, stayed in Kassel and was interested in both the early observations and the calculations involved in the planetary clock built by Baldewein and others. Baldewin also manufactured various observational instruments, such as a precision operating clock (the so-called minute clock), and participated in the observations. In addition to other observers at his observatory, Wilhelm collaborated with Victorinus Schönfeld, professor of mathematics in Marburg, and also held counsel, upon special occasions, with scholars such as **Caspar Peucer** (on the occasion of the supernova in 1572). In April 1575, **Tycho Brahe** visited the Kassel Observatory and received important support for his "Uranienburg," a facility soon to be built on Hven Island. By means of a diplomatic mission, Wilhelm arranged for support from the Danish king for Brahe's project.

On 25 July1579, Bürgi acquired his position as the landgrave's clock maker. Bürgi, coming from the Toggenburg region of Switzerland, constructed an excellent observational clock with a precision of movement unknown until then. With this device and with changes to the scales and instrument's sight settings, Bürgi, in collaboration with **Christoph Rothman** (who was working in Kassel around 1585), provided a remarkable improvement in the accuracy of the observations undertaken in Kassel.

Wilhelm recognized the need to compile a star catalog based on his own precise observations, to be used as the foundation for a reformed astronomy, one that took planetary motion into account when plotting the positions of planets alongside those of stars. The resulting star catalog was finally constructed on the basis of Rothmann's observations and represented, with its precise positions, a new quality in fixing the locations of stars. The reform of astronomy would, after Wilhelm, proceed on the basis of **Nicolaus Copernicus**' heliocentric system, Wilhelm being one of the first scholars to adopt it.

Already at the time of his observations of the supernova in 1572, Wilhelm concluded, on account of the absence of a possible parallax for this star, that it could not be found in the region below the Moon but in the realm of the stars. This realization was, along with Brahe's observations, one of the first indications that stars were mutable, a possibility that contradicted the principles of Aristotelian physics. Wilhelm further clarified this premise, especially during the appearance of the 1585 comet (C/1585 T1).

In the discussions surrounding Pope Gregory XIII's calendar reform, Wilhelm occupied a central position among the Protestant princes. He was involved in an exchange of opinions focusing above all on the fact that the reform should be rejected if it was merely an attempt to combine Catholic and Protestant church doctrine.

In all the work undertaken at the Kassel Observatory, Wilhelm proved himself not only to be accurately informed but always in the position to devise concrete observational tasks. Even in his extensive correspondence with Brahe, Wilhelm always appears as an astronomer of equal standing.

Important records of Wilhelm's observations can be found in the Kassel State Library, including the star catalogs from 1559 to 1563 and from 1566 to 1567. His scientific correspondence is currently found in the Hessen State Archives in Marburg and in the Saxon Main State Archives in Dresden.

Jürgen Hamel

Alternate name

Landgrave of Hessen-Kassel

Selected References

Brahe, Tycho (1596). *Epistolarum astronomicarum libri*. 1596. Vol. 6 of *Opera Omni*, edited by J. L. E. Dreyer. Copenhagen, 1919. (For correspondence with Brahe.)

Hamel, Jürgen (1999). "Die Kalenderreform Papst Gregors XIII. von 1582 und ihre Durchsetzung (unter besonderer Berüecksichtigung der Landgrafschaft Hessen–Kassel)." In *Geburt der Zeit: Eine Geschichte der Bilder und Begriffe*, edited by Hans Ottomeyer *et al.*, pp. 292–301. Wolfratshausen.

——— *Die astronomischen Forschungen in Kassel unter Wilhelm IV: Mit einer Teiledition der deutschen Übersetzung des Hauptwerkes von Copernicus um 1586* (The astronomical research in Kassel under Wilhelm IV: With a partial edition of the German translation of the main work of Copernicus c.1586). (2nd rev. ed. *Acta Historica Astronomiae*, Vol. 2. Frankfurt am Main: Harri Deutsch, 2002.)

Justi, K. W. (1817). "Züge aus dem Leben des Hessischen Landgrafen Wilhelm's IV, des Weisen." *Für müssige Stunden* 2: 135–226.

Leopold, J. H. (1986). *Astronomen, Sterne, Geräte: Landgraf Wilhelm IV. und seine sich selbst bewegenden Globen*. Munich: Callwey-Verlag.

Mackensen, Ludolf von (1988). *Die erste Sternwarte Europas mit ihren Instrumenten und Uhren*. Munich: Callwey-Verlag.

Zach, Franz Xaver von. (1805). "Landgraf Wilhelm IV." *Monatliche Correspondenz zur Befoerderung der Erd- und Himmelskunde* 12: 267–302.

Wilkins, Hugh Percival

Born **Carmarthen, Wales, 4 December 1896**
Died **Bexleyheath, (London), England, 23 January 1960**

As an amateur astronomer, Hugh Percival Wilkins, as he was known, specialized in selenography and was a leading visual observer of the Moon in the middle of the 20th century. Wilkins was educated at Carmarthen Grammar School. He early on showed a marked aptitude for mechanics, and after service in the British Army during World War I, became a practical engineer in Llanelly, South Wales. He married in 1931, and subsequently moved to Kent, in the south of England, settling first in Barnehurst, then after World War II in Bexleyheath. Wilkins gave up practical engineering in 1941 and joined the Ministry of Supply. He remained a government official until 1959, when he retired with the intention of devoting the rest of his life to astronomical research. Within days, though, he suffered a heart attack. In spite of confident hopes of a full recovery, he had a relapse and died soon afterward.

Though deeply interested in geology, Wilkins's chief passion was astronomy. This developed early and led first to youthful experiments in telescope making and mirror grinding. Around 1909, Wilkins took up serious observation, notably of the planets, with the telescope he had made. His passion was so intense that he took a small spyglass with him into the trenches during the war.

His main preoccupation, however, was mapping the Moon, especially its then poorly known limb regions. In this respect, he was following a long tradition of outstanding British amateur lunar observers including **William Birt**, **Edmund Nevill**, **Thomas Elger**, and **Walter Goodacre**. Wilkins produced three maps of the visible hemisphere. He published his first map to a scale of 60 in. to the Moon's diameter in 1924. A second, 100 in. in diameter, was completed in 1932, and a third, of 300-in. diameter, appeared in 1946. The latter was revised for a third edition in 1951, and yet another revision was issued in 1954. Even so, he still was dissatisfied, and at the time of his death yet another map was planned.

Sadly, Wilkins's enthusiasm for the subject outweighed his abilities as a cartographer. His maps, which represent a prodigious effort, testify to his industry and dedication, yet are so crowded and unreliable as to be of historical interest only.

In 1946 the Lunar Section of the British Astronomical Association [BAA] was virtually moribund. Wilkins was entrusted with its reinvigoration. Ten years later, when he resigned his directorship, it had been transformed into a useful and enthusiastic organization with over 100 participating observers. The BAA Lunar Section served as a model for the burgeoning US-based Association of Lunar and Planetary Observers, and similar groups across the world. Following his resignation, Wilkins founded the International Lunar Society, becoming its first president and then its director of research. He also served on the council of the British Interplanetary Society.

Wilkins was an excellent lecturer, and broadcast frequently on radio and television. He also undertook a lecture tour of North America. Wilkins produced several books on popular astronomy, and contributed chapters and essays to multiauthored works. In recognition of his contributions to astronomy, the University of Barcelona conferred an honorary doctorate upon him in 1953.

Richard Baum

Selected References

McKim, Richard (ed.) (1990). *The British Astronomical Association: The Second Fifty Years*. London: British Astronomical Association.

Moore, Patrick (1960). "Hugh Percival Wilkins; 1896–1960." *Journal of the International Lunar Society* 1, no. 6: 138–139.

——— (1960). "Hugh Percival Wilkins." *Journal of the British Astronomical Association* 70: 237–238.

Wilkins, H. Percy (1937). "The Lunar Mare 'X.'" *Journal of the British Astronomical Association* 48: 80–82.

——— (1947). "Tenth Report of the Lunar Section." *Memoirs of the British Astronomical Association* 36, pt. 1.

——— (1950). "Eleventh Report of the Lunar Section." *Memoirs of the British Astronomical Association* 36, pt. 3.

——— (1954). *Our Moon*. London: Muller.

Wilkins, H. Percy and Patrick Moore (1955). *The Moon; A Complete Description of the Surface of the Moon, Containing the 300-inch Wilkins Lunar Map*. New York: Macmillan.

Wilkins, John

Born **Fawsley, Northamptonshire, England, 1614**
Died **Chester, England, 16 November 1672**

John Wilkins's popular astronomical works, *The Discovery of a World in the Moone* (1638) and *A Discourse Concerning a New Planet* (1640), appearing when Wilkins was only in his mid-20s, became "the most influential English defense of Copernican astronomy in the second half of the 17th century." Though not scientifically original, they transmitted the insights and the excitement of **Galileo Galilei** and **Johannes Kepler** to a vernacular English readership in an engaging, speculative style that laid the foundation for the genre of science fiction.

Wilkins received a classical education at Oxford in the school of Edward Sylvester and was such an outstanding student that he matriculated at Oxford University in 1627, at the age of 13. After receiving his BA in 1631 he became a clergyman and served, among other roles, as warden of Wadham College (1648–1659), master of Trinity College, Cambridge (1659–1660), and Bishop of Chester (1668–1672). Wilkins's theological position was unclear, but he was sufficiently adaptable ecclesiastically and politically to flourish both before and after the Restoration (May 1660). In 1656 Wilkins married Robina French, the widowed youngest sister of Oliver Cromwell.

Wilkins chaired the November 1660 meeting of the newly founded Royal Society at which it was resolved to petition King Charles II for a charter. Wilkins's lifelong sociable encouragement of science earned him praise as "a great preserver and promoter of experimental philosophy."

Before Wilkins's publication of *The Discovery*, serious popular attention to or even awareness of the Copernican model in England was rare. For this reason, Wilkins was careful in introducing the most shocking, counterintuitive tenet of Copernicanism, namely, that the Earth moves and is therefore a "new planet." Although the main focus of *The Discovery* is lunar, Wilkins explicitly draws the inference of "other worlds" theory that follows from Copernican cosmology: If the Earth is a planet, then perhaps the planets (including the Moon) may be conceived to be Earths.

The Discovery strove to overcome resistance to Copernicanism that results from literalistic interpretation of certain passages of Scripture, from the sheer novelty of the idea of a moving Earth, and from traditional notions regarding the physical uniqueness and mutability of the Earth, in contrast to the pristine realms beyond. This work's main inspiration at the imaginative level appears to be Kepler's posthumous *Somnium* (1634). At the physical level, however, it hews closely to Galilei's *Sidereus Nuncius*. Thus Wilkins demonstrates the Earth-like, mountainous character of the Moon, and offers a similarly Galilean account of how both objects reflect the Sun's light mutually.

This simultaneous poetic and physical domestication of the Moon then opens the way to its being imagined as another world, another place of habitation (not in the older dominant sense of "world" as universe).

Moreover, if light may travel from Earth to the Moon, so perhaps can other things. In the revised edition of *The Discovery* in 1640, Wilkins suggested how "our posterity" might "find out a conveyance to this other world." The possibility, and difficulty, of space flight led Wilkins to speculate on the nature of gravity. For the difficulty seemed less if, as Wilkins argued, that "natural vigour whereby the earth does attract dense bodies unto it, is less efficacious at a distance." An account of precisely how much less this "natural vigour" grows with distance was offered later by **Isaac Newton**. But Wilkins's both imaginative and practical struggle to conceptualize human space travel helped to place the issue of gravity squarely on the agenda of physical theory.

A further problem for space travel that Wilkins helped to remove from the popular imagination was the idea of the crystalline spheres, which post-Copernican developments of astronomical thinking in both **Tycho Brahe** and Kepler had already effectively eliminated from scientific consideration. Robert Burton, in his contemporary *The Anatomy of Melancholy*, saw clearly what the solid spheres' removal implied for possible human exploration: "If the heavens then be penetrable, … it were not amiss in this aerial progress to make wings and fly up." Following Galilei, Wilkins treated "the heavens or stars" as "of a material substance." And he built into this physical approach to astronomy a satirical denial of the materiality – indeed the reality – of the supposed crystalline spheres, "this astronomical fiction," as he called them.

Wilkins offered his vernacular audience touches of satire, patient logical and arithmetical refutation of objections against Copernicanism, simple diagrams and explanations, unwavering piety, and undeniable poetic and rhetorical charm. For his countrymen, these qualities combined to diminish the strangeness of the new astronomy and to open its vistas in a moderate, unthreatening way. Wilkins thus contributed to both a buffering and a kindling of the spirit of science, as well as of science fiction. He articulated, in a word, new scientific and cosmic prospects.

Wilkins's impact also extended to other literatures beyond English. Both *The Discovery* and *A Discourse* were translated into French by Jean de la Montagne under the title *Le monde dans la lune* (Rouen, 1656). This edition is said in part to have inspired Cyrano de Bergerac's *Histoire comique des états et empires de la Lune*, 1656, and, most influentially, **Bernard de Fontenelle**'s *Entretiens sur la pluralité des mondes*, 1686. It is probable too that Wilkins helped stimulate the "other worlds" speculations of **Christiaan Huygens**, whose *Cosmotheoros* was published in Latin in 1698 and translated into English in the same year under the title *The Celestial Worlds Discover'd*. Wilkins's first German translator was **Johann Doppelmayer**, whose edition appeared in 1713 as *Johannis Wilkins ... Vertheidigter Copernicus, oder, Curioser und gründlicher Beweiss der copernicanischen Grundsätze*.

Dennis Danielson

Selected References

Crowther, J. G. (1960). *Founders of British Science*. London: Cresset Press.

McColley, Grant (1938). "The Ross-Wilkins Controversy." *Annals of Science* 3: 153–189.

Shapiro, Barbara J. (1969). *John Wilkins, 1614–1672: An Intellectual Biography*. Berkeley: University of California Press.

Wilkins, John (1708). *The Mathematical and Philosophical Works of the Right Rev. John Wilkins*. London. (Reprint, London: Frank Cass, 1970.)

——— (1973). *The Discovery of a World in the Moone*. Delmar, New York: Scholars' Facsimiles and Reprints.

William of [Guillaume de] Conches

Born **Conches, (Eure), France, *circa* 1100**
Died ***circa* 1154**

William was a philosopher, theologian, and astronomer, who published a survey of contemporary astronomical knowledge in the 12th century. He studied in Chartres, where he was a pupil of Bernard at Chartres (Bernardus Carnotensis). Later he was a teacher at Chartres and Paris, and he acted as Bernard's successor, whose school at Chartres he represented. From 1144 to 1149 William lived in the court of Geoffroi le Bel, and at the end of his life he found refuge in the court of the Duke Geoffroi V. Plantagenet. William also worked as a tutor of Geoffroi's son Henri II, future duke of Normandy and future king of England.

Interested in natural science, cosmology, and philosophical questions, William belongs among the early scholars who examined the biblical cosmological opinions not only from the traditional "literary" approach of exegesis of the Holy Scriptures, but also from the "scientific" point of view. For example, he refused the interpretation of the Venerable **Bede** and others on "supracelestial waters."

William's main work, Encyclopaedia *Philosophia Mundi*, contains among other things a survey of contemporary astronomical knowledge. It was published as a work of *Honorius Augustodunensis* (*Honorius von Autun*) in PL 172, and as a work of Beda Venerabilis in PL 90. Among William's other works, the dialogue *Dragmaticon philosophiae* (edited Strassburg, 1567; reprinted Frankfurt, 1967) stands out. He wrote the following commentaries: on **Boethius**' *De consolatione philosophiae*, on **Macrobius**' *Commentarii in Somnium Scipionis*, and on **Plato**'s *Timaios* (edited E. Jeauneau, *Glossae super Platonem*, Paris, 1965). His commentary on **Martianus Capella**'s *Encyclopaedia of Septem Artes Liberales* (the seven liberal arts) contains a basic knowledge of the later *trivium* and *quadrivium*, including astronomy. The glosses on Iuvenalis are perhaps also written by him (edited B. Wilson, Paris, 1980).

Alena Hadravová and *Petr Hadrava*

Alternate name

Guilelmus de Conchis

Selected References

Grant, Edward (1994). *Planets, Stars, and Orbs: The Medieval Cosmos, 1200–1687*. Cambridge: Cambridge University Press.

Gregory, Tullio (1955). *Anima Mundi: La filosofia di Guglielmo di Conches e la scuola di Chartres*. Florence: G. C. Sansoni.

Lemay, Helen Rodnite (1977). "Science and Theology at Chartres: the Case of the Supracelestial Waters." *British Journal for the History of Science* 10: 226–236.

Lerner, Michel-Pierre (1996–1997). *Le monde des sphères*. 2 Vols. Paris: Les Belles Lettres.

Migne, J. P. (1844–1864). *Patrologia Latina*. Paris.

Nauta, Lodi (1999). *Guillelmi de Conchis Glosae Super Boetium*. Corpus Christianorum, Continuatio Mediaevalis 158. Turnhout: Brepols.

Ronca, Italo (1997). *Guillelmi de Conchis Dragmation*. Corpus Christianorum, Continuatio Mediaevalis 152. Turnhout: Brepols.

William of Moerbeke

Born **Brabant, (Belgium), *circa* 1215**
Died **Corinth, (Greece), *circa* 1286**

William of Moerbeke translated **Aristotle** and **Archimedes** to Latin from Greek (as opposed to Arabic).

Selected Reference

Clagett, Marshall (ed.) (1964). *Archimedes in the Middle Ages*. Madison: University of Wisconsin Press.

William of [Guillaume de] Saint-Cloud

Flourished **Saint-Cloud, France, *circa* 1290**

Virtually nothing is known about calendricist William of Saint-Cloud's life other than the fact that he lived in Saint-Cloud, France, around 1290. Knowledge of his existence appears to stem largely from a report he gave (possibly more than one, as he is known for two different discoveries around the same time) in an almanac published in about 1290. William was probably a part of the School of Paris – Saint-Cloud is very close to Paris – and was most likely associated with the Church in some way.

William of Saint-Cloud is known for two significant discoveries, both dating from roughly the same time (possibly both reported in the same almanac in 1290). The first item that he discussed is the impairment of the eyes while viewing eclipses for too long. The eclipse in question is that of 4 June 1285. (Some reports indicate 5 June, but this is erroneous.) The manuscript, which was dated 5 years later, reported some observers experiencing near blindness for several hours or even several days in some cases. William followed the observational report by suggesting the use of a *camera obscura* – essentially the pinhole projection method still used today to safely view an eclipse. He also reported on the power lenses and mirrors.

William's other great accomplishment, around the same time, was his calculation of the obliquity of the ecliptic and the time of the vernal equinox from the Sun's position at the solstice. His observations brought out the inaccuracies in the Tables of Toledo developed by **Zārqalī**.

Ian T. Durham

Selected Reference

Duhem, Pierre (1911). "Physics, History of." In *Catholic Encyclopedia*, edited by Charles G. Habermann *et al*. Vol. 12, pp. 47–67. New York: Encyclopedia Press, 1911.

Williams, Arthur Stanley

Born **Brighton, (East Sussex), England, 1861**
Died **Feock, Cornwall, England, 1938**

Stanley Williams was one of the notable English amateurs in the late 19th and early 20th centuries who collectively established much of our knowledge about Jupiter's atmospheric dynamics. Specifically, Williams invented the terminology that has been adopted universally to describe the Jovian atmospheric zones and belts. He also perfected the technique for making central-meridian transit timings of various features within those belts to determine rotation periods and drift rates. His work firmly established the existence of stable atmospheric currents in various latitudes. Williams also discovered a number of variable stars and determined useful light curves and periods for many of these stars.

Unfortunately, little seems to be known of Williams's family, early life, and education. Professionally he was a solicitor, though he retired at an early age to devote his life to astronomy.

Although most of Williams's astronomical observations were made with a Calver equatorial reflecting telescope of only 6.5-in. aperture, he was able to see and recognize spots on Jupiter that others would have found difficult even with larger telescopes. He pioneered a method of measuring the longitudes (and hence drift rates) of Jovian spots by visually timing transits. His method of eye-estimates brought a withering attack from **George Hough**, who championed micrometer measurements of longitudes. The two had a robust argument about the relative merits of these techniques in the pages of the *Monthly Notices of the Royal Astronomical Society* in 1904 and 1905. In spite of Hough's attack, Williams and the other British amateurs continued making visual transit timings. Their work reliably established the pattern of atmospheric currents on Jupiter that is now known to be permanent. It is undoubtedly true that Hough's micrometer methods produced more accurate individual measurements of Jovian longitudes. Nevertheless, such measures were time-consuming and could not be applied to the large number of smaller and fainter objects observed productively by the British amateurs because the micrometer wires obscured these objects. The method of visual central-meridian transit timings remained a staple technique of planetary observation throughout the twentieth century.

Williams's own very detailed reports for 1880 were published privately as *Zenographical Fragments* volumes I and II. In the *Monthly Notices* in 1896 he reviewed all observations to date and made the first systematic listing of nine atmospheric currents on Jupiter.

In his analysis of colors on Jupiter, Williams combined his own observations (of over nearly 50 years from 1878 to 1936) with historical research. He argued that there was a periodic alternation in color between the major equatorial belts. However, like many proposed periodicities on Jupiter, his theory failed to hold up over a longer epoch.

Williams's other major astronomical activity was observing variable stars. While on a visit to Australia in 1885/1886, he engaged in systematic visual photometry of the brighter stars in the southern skies, eventually publishing his results as *A Catalogue of the Magnitudes of 1081 Stars Lying between -30° Declination and the South Pole*. In the course of his extended study, Williams detected the variability of V Puppis with a period of 1.45 days, the first of his many discoveries and period determinations for variable stars. Williams was the first British astronomer to use photography in searching for variable stars beginning in 1899. He was only able to continue this work for about 4 years because of ill health; after World War I he could not afford the cost of photographic plates. At a time when relatively few variable stars were known, Williams discovered over 50 variables, including the irregular variable RX Andromedae, Y Lyrae, Y Aurigae, YZ Aurigae, and the very short-period eclipsing binary WY Tauri.

Williams's photographic search for variable stars produced one other result of great importance by chance. A plate exposed for 1 hour and 16 min on the night of 20 February 1901 covered the area in which **Thomas Anderson** discovered Nova Persei 1901 at magnitude 2.7 only 28 hours later. Williams's plate did not contain an image of Nova Persei, confirming that the nova had brightened enormously in the intervening period. Williams later made important observations of Nova Persei 1901.

Williams was born near the sea and had always been a keen sailor. In 1920 he won the Challenge Cup for what was then a notable single-handed voyage to Vigo, Spain, and back. In his retirement he lived on a barge in Cornwall with his observatory on the shore nearby. His last Jupiter observation was a month before his death. In 1923, the Royal Astronomical Society awarded Williams its Jackson-Gwilt Medal recognizing his contributions to both planetary and variable star astronomy.

John Rogers and *Roy H. Garstang*

Selected References

Anon (1939). "Arthur Stanley Williams." *Journal of the British Astronomical Association* 49: 359–362.

Clerke, Agnes M. (1893). *A Popular History of Astronomy during the Nineteenth Century*. 3rd ed. London: Adam and Charles Black.

Garstang, Roy H. (1961). "Variable Star Observations of Stanley Williams." *Quarterly Journal of the Royal Astronomical Society* 2: 24–35.

Lankford, John (1979). "Amateur versus Professional: The Transatlantic Debate over the Measurement of Jovian Longitude." *Journal of the British Astronomical Association* 89: 574–582.

——— (1981). "Amateurs versus Professionals: The Controversy over Telescope Size in Late Victorian Science." *Isis* 72: 11–28.

Phillips, Theodore Evelyn Reese (1939). "Arthur Stanley Williams." *Monthly Notices of the Royal Astronomical Society* 99: 313–316.

Williams, Evan Gwyn

Born **London, England, 13 October 1905**
Died **Pretoria, South Africa, 31 May 1940**

Though he showed considerable early promise in stellar spectroscopy, Gwyn Williams's career was cut tragically short by complications arising from surgery in 1940. The elder son of Christopher Williams, an artist and portrait painter, Williams was educated at the Froebel Educational Institute School, London, England, and the King Alfred School, Hampstead, London, England. He played cricket and hockey and was active in the Boy Scout movement. In 1924 Williams entered Trinity College at the University of Cambridge where he studied natural sciences. He continued to play hockey but joined several student societies including the University Mountaineering Club, chess club, and Cambridge Photographic Club. After graduating in 1927 Williams continued his astronomical studies at University College, London, until 1928 when he also became a fellow of the Royal Astronomical Society.

Williams left London for Cambridge in 1928 where he worked at the Solar Physics Observatory. In 1929, he obtained an Isaac Newton Studentship and was granted a Commonwealth Fund Fellowship in 1931. The fellowship enabled him to study for 2 years at the California Institute of Technology and the Mount Wilson Observatory. His work was primarily on B type stars, whose spectrum predominantly contains absorption lines of hydrogen and neutral helium. On Williams's return from the United States, he continued at Cambridge as an Isaac Newton Student until 1936 when he again obtained an assistant position at the Solar Physics Observatory.

In his stellar spectrophotometry, Williams measured absorption-line intensities and developed a classification system for B type stars. He made a special study of the spectra from Nova Herculis 1934. In 1936 Williams traveled to Omsk in Siberia, USSR, where he successfully identified new lines in the Paschen series of hydrogen from the solar corona during a total solar eclipse.

After his marriage to Fiona Lauder in April 1937, Williams became first junior observer at the Radcliffe Observatory in Pretoria, South Africa. Originally founded in Oxford, England, in the late 18th century, the observatory had been managed by the trustees of the Radcliffe estate. In the early 1930s, the trustees took the imaginative, and at the time controversial, decision to move the observatory to South Africa and furnish it with a 74-in. reflector. The large telescope would be invaluable for studies of the distant and highly luminous stars of types O and B in the southern Milky Way. The importance of those stars to theories of stellar evolution was already understood, and Williams was emerging as an expert in this area.

The turret and the telescope mounting were well advanced in South Africa by 1938, but work on the primary mirror was delayed in England by the onset of war in September 1939. In the interim, Williams devoted himself to three-color photometry of B type stars, Cepheid variables, and other objects of interest using the 7-in. finder for the large telescope. Williams's sudden death from complications following emergency surgery was a great loss to the Radcliffe Observatory. It was only after the war in 1948 that the 74-in. mirror was installed and the telescope placed in productive service by **David Evans**.

Williams wrote over 20 astronomical articles, published mainly in the *The Monthly Notices of the Royal Astronomical Society*. His early death was a great loss to astronomy.

David W. Dewhirst and *Mark Hurn*

Selected References

Dunham, Jr., Theodore, (1956)." Methods in Stellar Spectroscopy." *Vistas in Astronomy* 2: 1223–1283, see p. 1250.

Knox-Shaw, H. (1941). "Evan Gwyn Williams." *Monthly Notices of the Royal Astronomical Society* 101: 141–142.

Stratton, F. J. M. (1940). "Gwyn Williams." *Observatory* 63: 213–214.

Wilsing, Johannes Moritz Daniel

Born **Berlin, (Germany), 8 September 1856**
Died **Potsdam, Germany, 23 December 1943**

Germany astronomer Johannes Wilsing was an early spectrophotometrist. The son of Eduard Wilsing and Clara Hitzig Wilsing, he studied at Göttingen and received his doctorate from the University of Berlin. Wilsing joined the Astrophysical Observatory in Potsdam in 1881, was promoted to observer in 1893 and chief observer in 1898, and retired from Potsdam in 1921.

Before 1900, Wilsing worked on the rotational period of the Sun (trying to explain the variation with latitude), on the density of the Earth, on the parallax of the star 61 Cygni (one of the first ever measured, by **Friedrich Bessel**, in 1838), and on the classification of stellar spectra. In 1896, together with **Julius Scheiner**, Wilsing attempted to detect radio emission from the Sun. They failed because the equipment was not sensitive enough; such emission was not seen until the work of **James Hey** during World War II. Later, Wilsing's work included spectrophotometric observations of stars, and, together with Scheiner and (later) W. Munch, the determination of the effective temperatures of 109 stars. In 1907/1908, Wilsing and Scheiner extended their spectroscopic work to determine the composition of the surface of the Moon. Further work covered the determination of the diameters of stars from their colors and brightnesses, the laws of blackbody radiation, and investigations of refractor objectives with **Johannes Hartmann**.

A lunar crater is named for Wilsing.

Christof A. Plicht

Selected References

Ludendorff, H. J. (1936). "Wilsing zum 80. Geburtstage." *Die Sterne* 16: 202.

Poggendorff, J. C. "Wilsing." *Biographisch–literarisches Handwörterbuch*. Vol. 4 (1904): 1645–1646; Vol. 5 (1926): 1376–1377; Vol. 6 (1940): 2896. Leipzig and Berlin. (For lists of his works.)

Wilsing, J. (1880). "Ueber den Einfluss von Luftdruck und Warme auf die Pendelbewegung." Ph.D. thesis, Berlin.

Wilson, Albert George

Born **Houston, Texas, USA, 28 July 1918**

Albert Wilson exposed and located dwarf galaxies on more than 1,000 Palomar Observatory Sky Survey [POSS] plates. Wilson succeeded **Vesto Slipher** as director of the Lowell Observatory.

Selected Reference

Putnam, William Lowell (1994). *The Explorers of Mars Hill: A Centennial History of Lowell Observatory, 1894–1994*. West Kennebunk, Maine: Phoenix Publishing.

Wilson, Alexander

Born **Saint Andrews, Scotland, 1714**
Died **Edinburgh, Scotland, 16 October 1786**

Alexander Wilson suggested that stars may be prevented from falling together under mutual gravitational attraction by their motion around a distant common center of gravity, but he is best known today for his observation that sunspots at the limb of the visible surface of the Sun appear to be saucer-shaped indentations, known as the Wilson effect.

Wilson was the younger son of Patrick and Clara (*née* Fairfoul) Wilson. His father, the town clerk of Saint Andrews, Scotland, died when he was comparatively young, leaving Alexander to be raised primarily by his mother. After graduation with an MA from Saint Andrews University in 1733, he apprenticed to a surgeon and apothecary in Glasgow and then moved to London to take charge of an apothecary's shop. In 1742, Wilson conceived of a new method of casting printer's type and returned to Saint Andrews where he established a typefoundry. In 1744 he moved the enlarged typefoundry to near Glasgow, and provided type for Mssrs. Foulis, printers of University of Glasgow publications, especially for their editions of the Greek classics. Wilson married Jean Sharp in 1752.

As a person of rather broad experimental interests and interested in astronomy since he was a student, Wilson began making reflecting telescopes in the late 1740s. In 1756, Glasgow University received a legacy of valuable astronomical instruments from Dr. A. Macfarlane of Jamaica, and built an observatory to house these instruments. In 1760 King George II appointed Wilson as the first professor of practical astronomy and observer at the University of Glasgow. Wilson published several astronomical papers in the *Philosophical Transactions of the Royal Society* dealing with the 1769 transit of Venus, the eclipses of Jupiter's first satellite, and the cross-wires in eyepieces.

Wilson's best-known paper, which dealt with the nature of sunspots and what became known as the Wilson effect, appeared in the *Philosophical Transactions* in 1774. The paper was based on his observations of a very large sunspot in November 1769. Wilson noticed that just before the Sun's rotation carried the spot beyond the Sun's limb, the apparent width of the penumbra on the side of the spot remote from the limb was less than that on the side closest to the limb. When the spot later reappeared on the other limb of the Sun, the narrowing of the penumbra was obvious on the opposite side of the spot, but the spot regained its usual appearance as it moved away from the limb. Although this effect had been remarked upon by such earlier observers as **Christoph Scheiner**, **Philippe de la Hire**, **Jacques Cassini**, Leonard Rost of Nuremberg, and Pastor Schülen of Essingen, Wilson was the first to analyze the observation in a geometrical sense. He accounted for the observed saucer shape of the spot as an effect of perspective. He extended his reasoning to make a novel interpretation of the nature of the Sun. Arguing that the Sun was a dark body surrounded by a luminous atmosphere, Wilson suggested that sunspots might be funnel-shaped holes in the Sun's atmosphere, and that the umbra of the sunspot was the hole at the bottom of the funnel near the surface of the darker solid body. Although **Joseph de Lalande** challenged Wilson's description, the idea was supported by **William Herschel** and was only displaced by spectroscopic studies a century later.

Reacting to **Isaac Newton**'s question as to why the stars do not all fall together under the force of their gravity, Wilson suggested that this might be because they are in periodic rotation around some distant center of gravitation. A letter conveying this speculation was acknowledged by William Herschel in the latter's paper on the structure of the Universe.

Wilson was interested in the variation of the temperature with altitude in the Earth's atmosphere, which he studied using thermometers mounted on kites. He also worked on methods of determining the specific gravity of liquids and solids. Wilson resigned his position in 1784, and his second son, Patrick, succeeded him as a professor. One of the founders of the Royal Society of Edinburgh, Wilson received a Gold Medal from the Royal Danish Academy, and an honorary MD degree from Saint Andrews University. A lunar crater is named for him.

Roy H. Garstang

Selected References

Clerke, Agnes M. (1893). *A Popular History of Astronomy during the Nineteenth Century*. 3rd ed. London: Adam and Charles Black, pp. 64, 192.

Garstang, R. H. (1964). "Alexander Wilson, 1714–1786." *Journal of the British Astronomical Association* 74: 201–203.

Herschel, William (1783). "On the Proper Motion of the Sun and Solar System." *Philosophical Transactions* 73: 247–283.

Stronach, George (1921–1922). "Wilson, Alexander." In *Dictionary of National Biography*, edited by Sir Leslie Stephen and Sir Sidney Lee. Vol. 21, pp. 545–546. London: Oxford University Press.

Wilson, Alexander (1774). "Observations on the Solar Spots." *Philosophical Transactions* 64: 1–30.

Wilson, Patrick (1824). "Biographical Account of Alexander Wilson." *Transactions of the Royal Society of Edinburgh* 10: 279–297.

Wilson, Herbert Couper

Born **Lewiston, Minnesota, USA, 28 October 1858**
Died **Northfield, Minnesota, USA, 9 March 1940**

Herbert Couper Wilson enjoyed a long and successful career as a teacher, administrator, and editor of two important journals in American astronomy.

Wilson's long relationship with Carleton College began at age 14, when he entered the Carleton Preparatory Department, eventually matriculating to the college itself at age 17. A successful student of mathematics and astronomy under **William Payne**, Wilson served for a year as principal at a public school in Jaynesville, Minnesota, after receiving his B.A. from Carleton in 1879. He then returned to astronomy as a graduate student at the University of Cincinnati under **Ormand Stone.** Wilson served as *astronomer pro tempore* in charge of the Cincinnati Observatory until 1884 when **Jermain Porter** arrived to replace Stone. In 1886, Wilson earned the first Ph.D. granted by Cincinnati for "six faithful years of work" on double stars, comets, and asteroids. During his work at Cincinnati, Wilson met and married Mary B. Nichols. They had three daughters and one son, **Ralph Elmer Wilson**; Ralph would earn a Ph.D. in astronomy and serve professionally at the Lick and Mount Wilson observatories.

After completing his work at Cincinnati, Wilson worked as a computer for the Transit of Venus Commission at the US Naval Observatory in Washington, while he unsuccessfully sought positions at the Washburn Observatory, University of Wisconsin, and the Lick Observatory, University of California. In the fall of 1887, he joined the faculty at Carleton as associate professor and stayed there for the remainder of his career.

At the time of his return, Payne was still publishing *The Sidereal Messenger*, and Wilson became a frequent contributor to that journal. When Payne and **George Hale** joined forces to publish *Astronomy & Astro-Physics*, Wilson became an associate editor of that journal. He remained in that capacity when Payne stopped publishing *Astronomy & Astro-Physics* and introduced a new journal, *Popular Astronomy*, the third astronomical journal to be published at Carleton College under his editorship. When Payne retired in 1909, Wilson became the editor of *Popular Astronomy* and continued in that capacity until his own retirement in 1926.

Wilson was a favorite teacher on campus, particularly well known for his lanternslide lecture on a "Trip to the Moon." He organized the Carleton Observatory's first off-campus expedition in 1889 to view and photograph a total solar eclipse in California. Wilson developed his expertise with cameras while a student at the Lookout Mountain Observatory in Cincinnati. His plates of nebulae, planets, and variable stars were often requested by other astronomers for study and illustration, especially his photograph of nebulosity surrounding stars in the Pleiades cluster. Wilson founded the *Publications of the Goodsell Observatory*, in which some of his photos of nebulae, sunspots, asteroids, and the spectra of the Sun's corona are presented.

Wilson served as the Carleton College dean of faculty as well as chairman of the Mathematics and Astronomy Department for many years, and shared responsibility, as a member of a committee of three, for the general administration of the college during a 2-year period in which the college was without a president.

Thomas R. Williams

Selected References

Gingrich, Curvin H. (1940). "Herbert Couper Wilson, 1858–1940." *Popular Astronomy* 48: 231–240.

Greene, Mark (1988). *A Science Not Earthbound: A Brief History of Astronomy at Carleton College*. Northfield, Minnesota: Carleton College, pp. 10–21.

Lankford, John (1997). *American Astronomy: Community, Careers, and Power, 1859–1940*. Chicago: University of Chicago Press, p. 157.

Osterbrock, Donald E. (1995). "Founded in 1895 by George E. Hale and James E. Keeler: The *Astrophysical Journal* Centennial." *Astrophysical Journal* 438: 1–7.

Wilson, Latimer James

Born **Nashville, Tennessee, USA, 1 December 1878**
Died **Nashville, Tennessee, USA, 17 May 1948**

Latimer Wilson excelled as an amateur planetary observer, producing excellent sketches and photographs over a career that spanned four decades. His mother had described the thrill of looking through telescopes with the youthful **Edward Barnard** at the Vanderbilt University Observatory to her young son while introducing him to the night skies. In 1908, Wilson took up astronomy with a homemade refractor using a single 4-in. lens; he fashioned a 10-in. Newtonian reflector in 1910, and in 1912 completed the 12-in. Newtonian reflector that was to be his primary instrument for the remainder of his life.

By 1913 the quality of Wilson's planetary sketches attracted the attention of a young Frederick Charles Leonard (1896–1960), who appointed Wilson director of the Society for Practical Astronomy's [SPA] Planetary Section. At the 1915 meeting of the society at the University of Chicago, Wilson was elected president of the SPA. With **Forrest Moulton** and George N. Saegmueller of Bausch and Lomb, Wilson planned for the third annual SPA meeting which was to be held in Rochester, New York. However, the society collapsed when World War I diverted interest from astronomy and financial pressures arose after Leonard matriculated at the University of Chicago and his parents dropped their support of the society.

Wilson continued to observe, sketching and photographing the planets, especially Mars, and publishing his results from time to time in *The English Mechanic*, *Knowledge*, and *Popular Astronomy*. His observing interests expanded to include meteors; **Charles Olivier** appointed him a regional director for the American Meteor Society. Wilson joined the American Astronomical Society at the invitation of **Philip Fox**, and soon thereafter was made a member of the Sociétié Astronomique de France by **Nicholas Flammarion**. From 1919 to 1922 Wilson was employed as an editor by *Popular Science Magazine*, and for a time in the 1930s he served as the director of the Chattanooga Observatory. In 1935 Wilson became a member of the Amateur Astronomers Association of America [AAAA]; his observations were accorded much discussion in their journal *Amateur Astronomy* by AAAA Planetary Section director, Edwin P. Martz (1918–1966). Through Martz, Wilson's observations were included in **Gérard de Vaucouleurs**' published works on Mars. Wilson's careful observation detected bright flashes in the southern polar cap of Mars on 30 May 1937, a phenomenon also observed in 1890 by **Percival Lowell** and **William Pickering**, and in 2001 by a number of members of the Association of Lunar and Planetary Observers.

Thomas R. Williams

Selected References

Scanlon, Leo J. (1938). "Latimer James Wilson." *Amateur Astronomy* 4, no. 1: 1, 12.
Vaughn, Frank R. (1948). "Latimer J. Wilson (1878–1948)." *Strolling Astronomer* 2, no. 10: 1–2.

Wilson, Olin Chaddock, Jr.

Born **San Francisco, California, USA, 13 January 1909**
Died **West Lafayette, Indiana, USA, 14 July 1994**

American stellar spectroscopist Olin Chaddock Wilson, Jr. – the suffix disappeared in the 1940s – discovered the first set of stellar activity cycles analogous to the solar sunspot cycle and is eponymized in the Wilson–Bappu effect, a correlation between the absolute brightness of cool stars and the strength of the emission-line cores in strong absorption lines due to hydrogen and calcium.

By the time Wilson earned the BA degree in physics from the University of California, Berkeley, in 1930, the 60-in. telescope no longer stood as the largest in the world: It had been surpassed by the mighty 100-in. Hooker telescope on Mount Wilson, which saw first light on 2 November 1917.

Wilson received the first Ph.D. awarded in astronomy by the California Institute of Technology, in 1934, for work on a comparison of the Paschen and Balmer series of hydrogen lines in stellar spectra, done with **Paul Merrill** (the paper appearing as Merrill and Wilson). Wilson had been hired as assistant astronomer at the Mount Wilson Observatory on 1 July 1931, but was not related to Benjamin "Don Benito" Wilson, after whom the mountain is named. His thesis observations rested on spectrograms from the 100-in. telescope. Wilson detailed the shapes of higher principle quantum number members of the Balmer and Paschen series in luminous A and B-spectral type stars, including the emission lines of P Cygni and γ Cassiopeia.

Wilson became a full staff astronomer at the Mount Wilson Observatory in 1936. Working generally at bright lunar phase between the great dark-phase observations of **Edwin Hubble** and **Milton Humason**, Wilson continued to illuminate stellar phenomena through spectra obtained with the 100-in. telescope. Wilson studied Wolf–Rayet objects, and discovered the first binary member of its class. He investigated the expansion of planetary nebulae, atmospheric inhomogeneities in the cool supergiant, ζ Aurigae, and radial velocities of interstellar neutral sodium and singly ionized calcium lines.

Wilson spent the war years working with Charles Lauritsen on the California Institute of Technology rocket project, where he met a fellow worker named Katherine Johnson. They married in 1943.

With M. K. V. Bappu, Wilson found that the width of the chromospheric emission cores of the singly ionized calcium H and K Fraunhofer lines in a cool star accurately marks its luminosity. Wilson also found that the intensity of the H and K emission cores decreases with age, and later greatly extended that conclusion in collaboration with Sir **Richard van der Riet Woolley**.

The hallmark of chromospheric activity on the Sun is its 11-year periodicity. Wilson was struck in the 1930s by the **George Hale–Seth Nicholson** work on calcium K-line spectroheliograms showing that the solar cycle was prominent in disk-averaged calcium emission. Wilson hypothesized that the fluxes of the H and K emission cores of other lower main-sequence stars, whose disks were unresolved, might display measurable variations similar to the solar cycle. Wilson early obtained spectrograms at the 100-in. telescope; after World War II he revisited the project only to conclude that photographic plates were too insensitive to record the subtle cycles.

By 1966 a more-sensitive photometric system at the 100-in. telescope encouraged Wilson to survey approximately 100 stars on or near the lower main sequence. With monthly measurements accumulated over 10 years, Wilson discovered that stellar chromospheric activity includes the property of interannual variation. Three patterns are observed in lower main-sequence stars: *cyclic*, with a period of several years to a decade, resembling the solar cycle; *variable*, with either multiple periods or nonperiodic variability; and *flat*, with no measurable variability. In general, stars with highly variable records are relatively young, while the cyclic and flat records may be two different phases of centuries-long variability, as recorded for the Sun, for example, in comparing the present, pronounced cycle of the Sun to its phase of diminished variability during the Maunder minimum (*circa* 1640–1720).

After Wilson reached compulsory retirement age in 1974, the activity-cycle project continued under Arthur Vaughn, who had constructed the chromospheric spectrograph for it, and is now in its 36th year of recording chromospheric variations in lower main-sequence stars under the leadership of the author of this sketch. At Wilson's prompting, the survey now includes evolved stars.

Wilson was awarded the Russell Lectureship of the American Astronomical Society in 1977 – his lecture dealt with stellar cycles – and the Bruce Medal of the Astronomical Society of the Pacific in 1984. He served as president of the Astronomical Society of the Pacific in the 1950s.

A monument erected in Wilson's memory is found on the southeastern side of the 100-in. telescope dome on Mount Wilson.

Sallie Baliunas

Selected Reference

Abt, Helmut A. (2002). "Olin Chaddock Wilson." *Biographical Memoirs, National Academy of Sciences* 82: 353–371.

Preston, George W. (1995). "Olin C. Wilson (1909–1994)." *Publications of the Astronomical Society of the Pacific* 107: 97–103.

Wilson, Olin C. (1963). "Lithium in a Main Sequence Star." *Publications of the Astronomical Society of the Pacific* 75: 62–64.

——— (1976). "Chromospheric Variations in Main Sequence Stars." In *Basic Mechanisms of Solar Activity*, edited by Vaclav Bumba and Josip Kleczek, pp. 447–452. IAU Symposium No. 71. Dordrecht: D. Reidel.

——— (1982). "Photoelectric Measures of Chromospheric H and K and Hε in Giant Stars." *Astrophysical Journal* 257: 179–192.

Wilson, Ralph Elmer

Born **Cincinnati, Ohio, USA, 14 April 1886**
Died **Corona del Mar, California, USA, 25 March 1960**

American observational astronomer Ralph E. Wilson made his greatest impact with an extensive catalog of radial velocities published in 1953 after his retirement. Wilson's catalog remained useful into the modern era of precision stellar measurements. He was the son of **Herbert Wilson**, long-term director of the Goodsell Observatory of Carleton College in Northfield, Minnesota, and received his AB from Carleton College in 1906. His Ph.D. came in 1910 from the University of Virginia for a thesis on positions of stars in the Orion Nebula. After a year as acting director at Goodsell, he was appointed assistant astronomer at the Lick Observatory from 1911 to 1913, where he worked on **William Campbell**'s bright-star radial-velocity program. Following this, Wilson spent the next 5 years as assistant astronomer at Lick Observatory's Southern Station in Santiago, Chile. Employing the 36-in. reflector there, he pursued a similar project on the radial velocities of bright stars. He later extended this research to include radial velocities of planetary nebulae in the Magellanic Clouds. After obtaining the necessary spectrograms, Wilson found that their radial velocities were high, leading him to suspect that the Magellanic Clouds might be external to the Galaxy and could be closely connected with the spiral galaxies.

Following war service in 1918 as an aeronautical engineer with the Bureau of Aircraft Production, Wilson transferred to the Dudley Observatory in Albany, New York, where he remained for the next 20 years. Wilson's work included fundamental studies of the Galaxy in general and also star clusters, principally by meridianal proper-motion measurements. His results helped to delineate the motion of the Sun through the Galaxy. Wilson carried out early investigations on the rotation of the Galaxy and the Sun's galactocentric distance.

It was at the Dudley Observatory that Wilson completed the five-volume *General Catalogue of 33,342 Stars* initiated by **Benjamin Boss**. This catalog also remained useful for many decades. The Carnegie Institution was gradually transferring its support for astronomy from the Dudley Observatory to Mount Wilson, and Ralph Wilson moved there in 1938, retiring in 1951. The agreement of names is a coincidence but must have prompted a good deal of joking.

While at Dudley in the early 1920s, Wilson had conducted a vigorous investigation into the zero-point of the period–luminosity [PL] relation of Cepheid variables. After moving to the Mount Wilson Observatory in 1938, he revisited this problem by combining proper-motion and radial-velocity measurements to investigate the space motions of different types of stellar objects, in particular red variables and long-period variables, deriving an independent value for the zero-point of the Cepheid PL relationship. Wilson's photographic PL curve and zero-point essentially confirmed **Harlow Shapley**'s values. However, research published by **Edwin Hubble** in 1932 indicated that the absolute magnitudes derived for the globular clusters in the Andromeda nebula were on average about 1.5 magnitudes too faint in comparison with their galactic counterparts.

The problem with the galactic and extragalactic distance scales had two pieces. First, Shapley had neglected the possibility of interstellar absorption of starlight by dust. This was discovered in 1930 by **Robert Trumpler** but not incorporated into redetermination of the distance scales until after World War II. Second, it turns out that there are two kinds of Cepheid variables associated with young and old stars, with the young ones being about twice as bright (1.5 magnitudes) as the latter. The two had been confused, making it seem as if external galaxies, based on the young Cepheids in their disks, were much closer than they really were – and would have been found to be if it had been possible to observe the old Cepheids in their globular clusters. This was not fully sorted out until 1952 when **Walter Baade** announced, and **David Thackeray** immediately confirmed, an approximate doubling of all distances outside the Milky Way. Further expansions occurred later, and Shapley's scale inside the Milky Way eventually proved to be somewhat too large.

Wilson's retirement from the Mount Wilson Observatory in May 1951 was marked by a session at a meeting of the Astronomical Society of the Pacific on "The Radial-Velocity Programs of the Pacific Coast Observatories," featuring a contribution by Wilson himself. Concluding an exemplary scientific career, Wilson published his famous *General Catalogue of Stellar Radial Velocities* in 1953.

Wilson's other achievements included serving as editor and associate editor of *Popular Astronomy* (1910–1914), associate editor of the *Astronomical Journal* (1929–1949), and president of the Astronomical Society of the Pacific (1946). He received a Gold Medal from the Danish Academy in 1926 and was a member of the US National Academy of Sciences.

Ralph Wilson died after a protracted illness and was survived by his wife Mary Adelaide (*née* Macdonald) and their son, Herbert Ralph.

John McFarland

Selected References

Anon (1960). "Ralph E. Wilson." *Publications of the Astronomical Society of the Pacific* 72: 434.

Baade, W. (1956). "The Period–Luminosity Relation of the Cepheids." *Publications of the Astronomical Society of the Pacific* 68: 5–16.

Boss, Benjamin (1937). *General Catalogue of 33342 Stars for the Epoch 1950.* Five Vols. Washington, DC: Carnegie Institution of Washington.

Hubble, Edwin (1932). "Nebulous Objects in Messier 31 Provisionally Identified as Globular Clusters." *Astrophysical Journal* 76: 44–69.

Joy, A. H. (1962). "Ralph Elmer Wilson." *Biographical Memoirs, National Academy of Sciences* 36: 314–329.

Russell, John A. (1951). "Los Angeles Meeting of the Astronomical Society of the Pacific." *Publications of the Astronomical Society of the Pacific* 63: 168–171. (The four invited contributions comprising the symposium were: Walter S. Adams, "Stellar Radial-Velocity Programs of the Mount Wilson Observatory," pp. 183–190; G. H. Herbig, "Stellar Radial-Velocity Programs of the Lick Observatory," pp. 191–199; R. M. Petrie, "Stellar Radial-Velocity Programs of the Dominion Astrophysical Observatory," pp. 215–222; and Ralph E. Wilson, "A New General Catalogue of Radial Velocities," pp. 223–231.)

Shapley, Harlow (1930). *Star Clusters. Harvard Observatory Monographs,* no. 2. New York: McGraw-Hill.

Wielen, R. *et al.* (2001). "The Combination of Ground-based Astrometric Compilation Catalogues with the HIPPARCOS Catalogue. II. Long-term predictions and Short-term Predictions." *Astronomy and Astrophysics* 368: 298–310.

Wilson, Ralph E. (1910). "New Positions of the Stars in the Huyghenian Region of the Great Nebula of Orion, from Observations Made at the Leander McCormick Observatory." Ph.D. diss., University of Virginia.

——— (1915). "Note on the Radial Velocities of Six Nebulae in the Magellanic Clouds." *Publications of the Astronomical Society of the Pacific* 27: 86–87.

——— (1923). "The Proper-Motions and Mean Parallax of the Cepheid Variables." *Astronomical Journal* 35: 35–44.

——— (1939). "The Zero Point of the Period–Luminosity Curve." *Astrophysical Journal* 89: 218–243.

——— (1953). *General Catalogue of Stellar Radial Velocities.* Washington, DC: Carnegie Institution of Washington.

Wing, Vincent

Born **North Luffenham, (Leicestershire), England, 19 April 1619**

Died **North Luffenham, (Leicestershire), England, 30 September 1668**

Vincent Wing was a surveyor, an astrologer, an almanac compiler, and a prolific writer of astronomical works. Wing, whose father was a small landowner, was apparently self-educated, having learned at an early age, by his own exertions, some Latin, Greek, and mathematics. He lived throughout his life in or near North Luffenham.

Wing's first publication was an *Almanack* (1641) that included solar-eclipse data computed for English coordinates for the years 1641 to 1654. His *Urania Practica* (1649), written with the assistance of William Leybourn, was the first English book to describe the fundamental principles of computational astronomy and provided tables for calculating the times of lunar and solar eclipses. Based on **Philip Lansbergen**'s *Tabulae motuum coelestium perpetuae* (1632), and Ptolemaic in spirit, the book was criticized for various alleged errors by the English astronomer **Jeremy Shakerley** in the latter's *The Anatomy of Urania Practica* (1649) and quickly defended by Wing in his own *Ens fictum Shakerlaei: or the Annihilation of Mr. Jeremie Shakerley, His In-artificial Anatomy of Urania Practica* (1649).

In his *Harmonicum Coeleste: Or the Coelestial Harmony of the Visible World* (1651), Wing added tables to calculate the positions of the planets Mercury, Venus, Mars, Jupiter, and Saturn. From these tables he derived *An Ephemerides of the Coelestiall Motions for 1652 to 1658* (1652). The dispute with Shakerley may have influenced Wing's conversion to Copernicanism, which was evident in his *Astronomia Britannica* (1652). Wing published revised planetary tables in his *Astronomia instaurata: Or a New and Comprehensive Restauration of Astronomy* (1656), from which he derived *An Ephemerides of the Coelestiall Motions* for 1659 to 1671 (1658). Wing's two sets of *Ephemerides*, in the view of **John Flamsteed**, were the "exactest" to be had during this period. In the last few years of his life Wing was engaged in a heated dispute with the English astronomer **Thomas Streete** regarding the accuracy of the latter's *Astronomia Carolina* (1661); the disagreement centered on the magnitude of the horizontal parallax of the Sun.

Wing's annual almanac, *Olympia Domata*, had sales averaging 50,000 copies per year, and it was continued by various family descendants until 1805. Wing also published *Geodates Practicus: Or the Art of Surveying* (1664).

Craig B. Waff

Selected References

Applebaum, Wilbur (1976). "Wing, Vincent." In *Dictionary of Scientific Biography*, edited by Charles Coulston Gillispie. Vol. 14, pp. 446–447. New York: Charles Scribner's Sons.

Kelly, John T. (1991). *Practical Astronomy during the Seventeenth Century: Almanac-Makers in America and England.* New York: Garland.

Winlock, Joseph

Born **Shelby County, Kentucky, USA, 6 February 1826**

Died **Cambridge, Massachusetts, USA, 11 June 1875**

Joseph Winlock, a mathematical astronomer, was twice superintendent of the *American Ephemeris* before becoming the third director of the Harvard College Observatory. There he upgraded the observatory's equipment, expanded its research programs into the New Astronomy of astrophysics, and invented the photoheliograph.

Winlock's grandfather, a surveyor, participated in the convention that framed Kentucky's constitution, and both his grandfather and father had distinguished military careers during the War of 1812. Winlock was educated in his home state. His mathematical prowess was so evident that immediately upon his graduation from Shelby College in 1845, he was offered an appointment there as professor of mathematics and astronomy. Winlock spent his first savings on a set of the *Astronomische Nachrichten*, then the world's foremost astronomical journal; to gain enough fluency to read it, he arose daily before dawn to speak German with a laborer on his father's farm.

By all accounts, Winlock was rescued from frontier obscurity by attending the fifth meeting of the American Association for the Advancement of Science, held in Cincinnati, Ohio, in 1851, where he met Harvard astronomer and mathematician **Benjamin Peirce**. That contact led in 1852 to Winlock's joining the corps of calculators for the *American Ephemeris and Nautical Almanac*, which was then headquartered in Cambridge, Massachusetts. He remained there for 5 years, meeting and marrying Isabel Lane (1856), from whom he eventually had six children.

In 1857, Winlock was appointed professor of mathematics at the US Naval Observatory in Washington. The next year, however, he was made superintendent of the *Ephemeris and Almanac*, and returned to Cambridge. In 1859, he became head of the mathematical department of the US Naval Academy in Annapolis, Maryland. But when the Civil War removed the academy to Newport, Rhode Island in 1861, Winlock returned to Cambridge and his old superintendent's position. Over his intermittent 11 year service with the ephemeris office, he became known for his carefully prepared tables of Mercury, and was one of the original founding members of the National Academy of Sciences (1863). In 1866, Winlock made his last move – this time locally within Cambridge – to become director of the Harvard College Observatory and Philips Professor of Astronomy; 2 years later, he concurrently became professor of geodesy at Harvard's Lawrence and Mining Schools.

Winlock inherited an institution with aging equipment, small endowment, inadequate staff, and a huge backlog of unpublished raw observations from his two predecessors, father and son astronomers **William Bond** and **George Bond**. As director, Winlock made it his priorities to modernize the instrumentation, get the massive research of the Bonds into print, and turn the observatory into an efficient research center with a secure financial base. In so doing, Winlock revealed significant talent as an inventor, fund-raiser, and administrator.

Because the observatory's original meridian circle had suffered damage during its transportation from Europe, the main 15-in. Merz and Mahler refractor was often pressed into service as a substitute. To free the great telescope for more suitable research, Winlock solicited more than $12,000 in donations to purchase a brand new meridian circle; the new circle, mounted in 1870, was customized to his own specifications (among them shortening the piers and sealing the bearings from dust under glass) – improvements adopted by later observatories. Under his 9-year directorship, the observatory also acquired an auxiliary 7-ft. Clark equatorial, a number of clocks and chronometers, a Russian "broken" transit, self-recording meteorological instruments, and several spectroscopes. These last Winlock acquired to expand Harvard College Observatory's research beyond traditional positional astronomy and into the fledgling field of astrophysics; he himself used the spectroscopes during total solar eclipses to study the solar corona.

Winlock also proved to be an optical innovator. In 1869, Winlock led an expedition to Kentucky to observe the total solar eclipse of 7 August. Determined to photograph the Sun's corona – something not yet captured on film – Winlock rejected the then-standard method of eyepiece projection, instead placing the photographic plates at his telescope lens's prime focus. Although his images were thus very small – the Sun's disk was only 0.75 in. in diameter – Winlock's photographs not only revealed the corona, but also showed that it extended farther from the Sun than astronomers had realized. To attain larger images at the total solar eclipse on 22 December 1870 in Spain, Winlock invented a horizontal telescope using a lens 4 in. in diameter having a focal length of 40 ft. The telescope lens, a heliostat (an unsilvered plane mirror for reflecting the Sun into the lens), and camera were mounted on separate piers; daylight was excluded by a tube disconnected from them all. Winlock's design pioneered what later became known as a photoheliograph – a very long horizontal telescope that served as the centerpiece of many late-19th-century eclipse expeditions. (Some later astronomers contested his priority.) From 1870 on, Winlock's 4-in. horizontal telescope was used for daily solar observations as well as for photographing the transit of Venus in 1874.

When Winlock took over the Harvard Observatory, its annual operating budget was $200, a sum meager even in its time. For the eclipse of 1869, to stretch an additional $500 allocated by the Harvard Corporation for his 10-man expedition, Winlock pioneered the method of requesting free rail transportation for astronomical observers and equipment. In 1871, he followed the lead of **Samuel Langley**, director of the Allegheny Observatory in Pittsburgh, Pennsylvania, and began charging for the accurate time signals the Harvard Observatory had been telegraphing for free to New England railroads, jewelers, hotel managers, and other customers. By 1875, the observatory's average annual income from the time service was about $2,400 (and later peaked at about $3,000).

Winlock died of a mysterious illness quite unexpectedly. A man unusually laconic in conversation, he also wrote unusually few papers. Both his untimely death and his emphasis in publishing his predecessors' zone catalogs of stars, solar drawings, and aurora observations in the *Annals of the Astronomical Observatory of Harvard College* resulted in much of his own research not being printed during his lifetime. Thus today, Winlock's original early contributions to astronomical photography, photometry, and spectroscopy are less recognized than his faithful stewardship of the Harvard College Observatory.

Trudy E. Bell

Selected References

Annon. (1875). "Joseph Winlock," *American Journal of Science* 110: 159–160.

Bailey, Solon I. (1931). *The History and Work of Harvard Observatory, 1839 to 1927,* New York and London: McGraw-Hill Book Company, Inc.

Bartky, Ian R. (2000). *Selling the True Time: Nineteenth-Century Timekeeping in America*. Stanford, California: Stanford University Press.

Jones, Bessie Zaban and Lyle Gifford Boyd (1971). *The Harvard College Observatory: The First Four Directorships, 1839–1919*. Cambridge, Massachusetts: Harvard University Press.

Lovering, Joseph (1877). "Memoir of Joseph Winlock." *Biographical Memoirs, National Academy of Sciences* 1: 329–343.

Rothenberg, Marc (1999). "Joseph Winlock." In *American National Biography*, edited by John A. Garraty and Mark C. Carnes. Vol. 23, pp. 637–638. New York : Oxford University Press.

Winnecke, Friedrich August Theodor

Born **Gross-Heere near Hildesheim, (Niedersachsen, Germany), 5 February 1835**
Died **Strasbourg, (Bas-Rhin, France), 2 December 1897**

August Winnecke was an outstanding observational astronomer. He was the son of the local clergyman, Heinrich F. L. Winnecke (1803–1852), and his wife Dorette (*née* Quensell), who died some days after giving birth to August. Two aunts, sisters of his father, took care of the household and the child for the first 5 years. In 1840 the father retired and sent his son first to school in Gittelde, where relatives accommodated him. Later, August was sent to Hoya to attend high school and moved to Hanover in 1850 to prepare for university, which he started in 1853 at Göttingen, studying astronomy.

This science had fascinated Winnecke since he had moved to Hanover, where he used a small telescope for his first observations. In Göttingen he replaced this instrument with a 3-in. comet seeker, made by the workshop of Merz in Munich. During his observations in a garden, Winnecke met a young high-school pupil who later would also be a well-known astronomer: **Arthur Auwers. Carl Gauss**, who had given up lecturing due to his age, helped Winnecke with his studies and even sent his first papers to the *Astronomische Nachrichten* to have them published.

In the fall of 1854 Winnecke moved to Berlin for further studies. Several letters, published in the *Astronomische Nachrichten* during the following 2 years, show that he was active during night and day, either at the telescope or at the table reducing the observations he had made. Here he also discovered his first comet C/1854 Y1 (Winnecke–Dien). In later years he found 12 more comets, 5 of them as codiscoverer. On 7 August 1856 Winnecke received his doctorate with a thesis *de stella η Coronae borealis duplici*, which he dedicated to **Johann Encke**.

After finishing his studies in Berlin, Winnecke moved to Bonn to work with **Friedrich Argelander**. Here he extended his knowledge of practical astronomy, reflected in his published papers in the *Astronomical Nachrichten*, Volumes 45–48. His main subjects were the 6-in. heliometer, which he tested thoroughly, and parallax observations on the star Lalande 21185 and a planetary nebula with the aforementioned instrument. In addition he acquired positional data for the stars of the Praesepe cluster.

In the fall of 1857 **Friedrich Struve** visited Bonn and met Winnecke, whom he invited to work at the well-equipped main observatory in Pulkovo, Russia. Winnecke accepted and arrived there in July 1858. His observations between September 1858 and October 1864, made with the meridian circle manufactured by Repsold, represent the majority of the data for the first Pulkovo catalog of stars. Another major work was the observations of Mars during the 1868 opposition, executed by several observatories worldwide, which were based on a suggestion by Winnecke and led to an improved value for the solar parallax as given by Encke 40 years earlier. Here one of Winnecke's favorite sayings bore fruit: "Man muss an allem Zweifeln" (One has to question everything).

Additional work included observations of comet C/1858 L1 (Donati) with the 7-in. heliometer and, together with **Otto W. Struve** and Frederico A. Oom, the total solar eclipse on 18 July 1860 in Spain. Winnecke got acquainted with Sir **George Airy** on that occasion and met him again in Greenwich in 1864, where Winnecke was sent to test **James Bradley**'s instruments. Winnecke's papers, published again in the *Astronomische Nachrichten*, Volumes 49–66, report of his widespread interests, such as observing comets, nebulae, and variable stars.

Winnecke was soon promoted to elder astronomer and then to Vice Director. Soon after his marriage to Otto Struve's niece, Hedwig Dell, in May 1864, he had to take care of the director's duties when Otto Struve fell ill. Winnecke himself became ill in the fall of the same year. This and the stress of these additional duties may have led to a mental illness from which he did not recover easily. So he resigned from his post in December 1865 and went to Bonn, seeking the help of a Dr. Hertz. Winnecke recovered within a year and moved to Karlsruhe, Germany, where the climate was much milder than in Pulkovo.

He started observing again, first with his old comet seeker and then with a 5-in. instrument made by Reinfelder and Hertel. The collection of instruments was extended when Winnecke received the Berlin heliometer for further tests in preparation for the upcoming Venus transit in 1874. During the 5 years in Karlsruhe, he found four comets and observed mainly variable stars. In 1869 he was elected to the board of the Astronomische Gesellschaft and held that post for 12 years.

In 1872 Winnecke was invited to build an observatory at the new University of Strasbourg. His tasks there included, besides constructing and supervising the erection of the new buildings (which were completed in 1880), observing nebulae at the old observatory and teaching at the university. In addition he led the preparations for the expeditions to observe the Venus transit and computed a lot of the results. The new 18-in. refractor arrived in 14 crates on 6 August and was tested by Winnecke and erected by Repsold between 6 and 30 November. On 21 November Winnecke celebrated the move to his new home in Strasbourg with his wife, two sons, and three daughters.

On 13 January 1881, in the weeks when the equipment of the observatory was moved to the new buildings, Winnecke's son Fritz had a fatal accident on a frozen lake. After a while Winnecke returned to work, but suffered from a relapse of his mental illness a year later. Before he moved again to Bonn in February 1882 to seek help, he was offered the post of professor in Munich to succeed **John Lamont**. Instead he stayed in Strasbourg, where he was elected headmaster in the early weeks of 1882. Winnecke never recovered. He was buried in Strasbourg.

Numerous comets (including 7P/Pons–Winnecke) bear his name, and minor planet (207) Hedda was named in honor of Winnecke's wife.

Christof A. Plicht

Selected References

Auwers, A. (1897). "Todes-Anzeige: August Freidrich Theodor Winnecke." *Astronomische Nachrichten* 145: 161–166.

Berberich, A. (1898). "Obituary". *Naturwissenschaftliche* 13.

Hartwig, E. (1898). "Obituary". *Vierteljahresschrift der Astronomischen Gesellschaft* 33.

Schmadel, Lutz D. (1999). *Dictionary of Minor Planet Names*. 4th ed. New York: Springer-Verlag.

Winthrop, John

Born **Boston, Massachusetts, (USA), 19 December 1714**
Died **Cambridge, Massachusetts, USA, 3 May 1779**

More than any other colonial, John Winthrop was responsible for transplanting Newtonian physics to the American colonies. Winthrop was one of 16 children born to Adam and Anne Winthrop. He was educated at the Boston Latin School and then at Harvard College, where he graduated in 1732. In 1738, he was appointed as the second Hollis Professor of Mathematics and Natural Philosophy at Harvard, succeeding Isaac Greenword. Winthop held this position until his death. He married Rebecca Townsend in 1746, and then 3 years after her death in 1756, Winthrop married the widow Hannah Fayerweather, who survived him. In 1766, he was elected as a fellow of the Royal Society and 3 years later, he became a member of the American Philosophical Society. Winthrop received honorary LL.D. degrees both from the University of Edinburgh (1771) and from Harvard (1773). He was the great-grandnephew of his namesake, John Winthrop. The earlier Winthrop was one of the founding members of the Royal Society of London and Governor of Connecticut from 1660 until his death in 1676.

Along with his Yale College contemporary Thomas Clap, Winthrop is given credit for first introducing Newtonian physics and calculus to college students in the English colonies. Winthrop had charge of Harvard's collection of scientific instruments, and after a disastrous fire on 24 January 1764, he set about replenishing the collection with the help of his family's influential friends, including John Hancock and Benjamin Franklin. More than a dozen of his scientific publications appeared in the *Philosophical Transactions of the Royal Society* and covered such topics as earthquakes, weather, the Mercury transits of 1740, 1743, and 1769, the return of comet 1P/1758 Y1 (Halley), and the Venus transits of 1761 and 1769. With regard to earthquakes, he attributed the effects to a wave of the Earth with vertical and horizontal components. He used the Mercury transit observations to help determine the longitude difference between Greenwich, England, and Cambridge, Massachusetts. The Venus transit observations were part of an international campaign to determine the solar parallax and hence the absolute scale of the Solar System. Winthrop successfully observed the Venus transit of 6 June 1761 from Saint John's, Newfoundland, the only useful set of observations from North America. From Cambridge, Massachusetts, he also observed the Venus transit of 3 June 1769. Winthrop's two lectures on the return of comet 1P/Halley showed that he was abreast of the ideas on comets presented by **Edmond Halley** and **Isaac Newton**.

Donald K. Yeomans

Selected References

Cohen, I. Bernard (1950). *Some Early Tools of American Science*. New York: Russell and Russell.

Stearns, Raymond Phineas (1970). *Science in the British Colonies of America*. Urbana: University of Illinois Press.

Turner, G. L'E. (1976). "Winthrop, John." In *Dictionary of Scientific Biography*, edited by Charles Coulston Gillispie. Vol. 14, pp. 452–453. New York: Charles Scribner's Sons.

Wirtanen, Carl Alvar

Born **Kenosha, Wisconsin, USA, 11 November 1910**
Died **Santa Cruz, California, USA, 7 March 1990**

Carl Wirtanen was noted for the carefulness and dedication to his work across a long astronomical career and is remembered for the Shane-Wirtanen counts of galaxies.

Wirtanen completed his undergraduate (B.S: 1936) and graduate (A.M: 1939) degrees at the University of Virginia and majored in astronomy, mathematics, and physics. Until 1941, he worked at the Leander McCormick Observatory of the University of Virginia, where he derived stellar distances from their trigonometric parallaxes. In October 1941, Wirtanen became an observing assistant at the Lick Observatory of the University of California. During World War II, he took part in ballistics research at the Naval Ordnance Test Station at China Lake, California.

In October 1946, Wirtanen returned to the Lick Observatory. For the next 32 years, he acted as an observer and research assistant in conjunction with the observatory's 51-cm Carnegie astrograph, a specialized photographic survey telescope. Like other Lick astronomers, Wirtanen lived on Mount Hamilton (with his wife, Edith, and their son and daughter) until the Lick staff moved their offices to the new Santa Cruz campus in 1966. He retired in 1978.

Wirtanen's major astronomical contributions were centered on programs using the Carnegie astrograph. This survey telescope was conceived by **William Wright** to measure the proper motions of the stars in our Galaxy by using background galaxies as the reference frame. Proper motions are the small angular displacements of stars that can only be detected by comparing photographs taken at widely separated epochs. The Lick Observatory proper-motion program was one of the first to determine "absolute" proper motions of stars, relative to the reference frame provided by some 50,000 faint galaxies. The program involved photographing a major part of the sky (north of declination −33°) in 1,246 exposures between 1947 and 1954. The work was shared equally between Wirtanen and Lick Observatory director **Charles Shane**. Wirtanen also participated in taking the plates during the second epoch (1968–1978) under the direction of Stanislavs Vasilevskis; he was the only astronomer involved in both epochs.

The first results (*Lick Northern Proper Motion Program: NPM1 Catalog*) for 149,000 stars in fields outside the Milky Way were published in 1993. The final catalog (*NPM2*), which includes 232,000 stars within the Milky Way fields, was published in 2004.

Starting in the 1950s, Shane and Wirtanen, independently and in duplicate, counted the number of galaxies in each of the 1.6 million fields on the proper-motion survey plates. Previously, galaxy counts had only been possible on the small fields obtained with large reflectors, or else were derived from inhomogeneous survey data. The Shane–Wirtanen counts first revealed the clustered distribution of galaxies over the entire northern sky, based on a homogeneous sample. These counts have been a valuable tool for astronomers. Twenty years later, they were used for correlation analysis and they provide information on galaxy distribution that is still of interest.

While studying the Lick survey plates, Wirtanen found five comets. In recognition of his discovery of the first four comets, he received the Donohue Comet Medal Awards of the Astronomical Society of the Pacific. One of these (46P/1948 A1) is of particular interest. Perturbations by Jupiter have changed this comet's orbit to one with a period of 5.46 years and a perihelion distance of 1.06 AU, making it accessible to investigation by spacecraft. The European Space Agency's International Rosetta Mission was originally expected to rendezvous within a kilometer of the nucleus of comet 46 P/Wirtanen in 2011. Unfortunately, the launch of this spacecraft was delayed so that another target had to be selected.

Wirtanen also discovered numerous minor planets. Two of these, (1965) Toro and (29075) 1950 DA, are on orbits that bring them unusually close to the Earth, making them useful for detailed observation; (29075) is of further interest because its rotation period (2.1 hours) is the second fastest known for its size. About the time of Wirtanen's retirement, the minor planet (2044) Wirt was named in his honor, employing the name by which he was generally known to family and colleagues alike.

During the 1960s, Wirtanen was closely involved with Thomas D. Kinman in a program with the astrograph to discover RR Lyrae variables. These variable stars have characteristic short periods and may be recognized at great distances. This program showed that the halo of our Galaxy (of which these variables are tracers) was more extended than previously thought. Besides taking plates for this survey, Wirtanen played the crucial role of "blinking" (or comparing) the plates to discover the variables.

A bibliography of Wirtanen's publications may be found in the Mary Lea Shane Archives of the Lick Observatory, Santa Cruz, California.

Thomas D. Kinman

Selected Reference

Klemola, Arnold Richard (1991). "Carl Alvar Wirtanen, 1910–1990." *Bulletin of the American Astronomical Society* 23: 1495–1496.

Wirtz, Carl Wilhelm

Born **Krefeld, (Nordrhein-Westfalen), Germany, 24 August 1876**

Died **Kiel, Germany, 18 February 1939**

Carl Wirtz, the astronomer, measured magnitudes and positions of nebulae, and made many observations of Solar System objects. He was among the first to study the redshift-magnitude and redshift-diameter diagrams of nebulae (galaxies), which is why Allan Sandage, the pupil and successor of **Edwin Hubble**, called Wirtz "the European Hubble without telescope."

Wirtz studied astronomy at Bonn University (1895–1898). After that he served as an assistant at the Wien-Ottakrieg Observatory (1899–1901), as a lecturer in the Hamburg School of Navigation (1901–1902), and as an astronomer–observer and professor at the Strasbourg Observatory (1901–1915). In 1905 he married Helene Borchardt. He served in the German army (1916–1918). In 1919 Wirtz was appointed as an extraordinary professor of the Kiel Observatory, and in the years 1934–1936 he served as the director of that observatory.

The last years of Wirtz' life were obscured by the political situation in Germany. Mrs. Wirtz was of Jewish origin, and under the Nazi regime the whole family fell into disgrace. In 1937 Carl Wirtz lost the right to teach.

At the Bonn Observatory, Wirtz measured the declinations of 487 stars. At the Strasbourg Observatory he participated in the observations of magnitudes and positions of 1,257 nebulae, and in data reduction. At Strasbourg, Wirtz also made observations of asteroids and planets. His observations of many comets are valuable, especially the long series of magnitude estimations. During the Kiel period Wirtz was one of the first astronomers to work in the field of extragalactic astronomy.

In his 1922 paper "A Note Concerning the Radial Motion of Spiral Nebulae," Wirtz put forward the first correlation between redshift (velocity) and distance (as determined from apparent magnitudes) that was roughly linear. Like the correlation published in 1929 by Hubble, it was based on velocities measured by **Vesto Slipher**. A diagram of velocities versus apparent diameters (another distance indicator) followed in 1924, along with a plot of the surface brightnesses of galaxies versus their diameters, and, in 1926 he published a luminosity function for galaxies. Because the distance indicators used by Wirtz were not as reliable as Hubble's Cepheid variables, his correlations were less tight and were not seen by the community as of great importance.

At the Fifth General Assembly of the International Astronomical Union in 1935, Wirtz presented the project "An Extragalactic Reference Frame for Stellar Proper Motion Measurements."

In 1912 Wirtz won the Lalande Prize of the Paris Academy, and he was honored by having a Martian crater named for him.

Mihkel Joeveer

Selected Reference

Theis, C., S. Deiters, C. Einsel, and F. Hohmann (1999). "Hans Rosenberg und Carl Wirtz." *Sterne und Weltraum* 38, no. 2: 126–130.

Witt, Carl Gustav

Born **Berlin, (Germany), 29 October 1866**
Died **Falkensee near Berlin, Germany, 3 January 1946**

Carl Witt discovered the minor planet (433) Eros, which makes relatively close approaches to the Earth.

Witt was born as the son of a teamster. In 1887, he passed the graduate examination of the Andreas-Realgymnasium (Berlin). For the next three years, Witt studied mathematics and physics at the University of Berlin, where he came under the influence of astronomers Friedrich Tietjen and Wilhelm Julius Foerster. But thereafter, Witt was offered a stenographer's position attached to the Reichstag (Germany's democratically elected Parliament). This occupation provided Witt's means of employment for many years; he gradually advanced to the head of the stenographer's bureau.

But Witt never lost his desire to become an astronomer, and increasingly devoted his spare time to studies of the heavens, especially minor planet and comet observations, which were aided by the development of astrophotography. On 13/14 August 1898, Witt exposed a photographic plate using the 12-in. telescope at the students' Urania Observatory that recorded the trail of a minor planet whose rapid motion indicated a close proximity to the Earth. This minor planet, later designated (433) Eros, was unusual because it was the first such object whose perihelion lay *inside* the orbit of Mars. For that reason, Eros subsequently was used to determine the value of the solar parallax (and the true dimension of the Astronomical Unit).

Witt's discovery of Eros provided an important spur to his astronomical studies and caused him to abandon his stenographic career. In 1905, Witt completed (and published) his doctoral dissertation on the orbital mechanics of Eros, which displayed the strengths of classic astrometry. The following year, he was awarded a Ph.D. by the University of Berlin. In 1908, he deduced a value of 8.803″ for the solar parallax. He was made a *Privatdozent* (a lecturer, paid by student fees) from 1909 to 1920 at the University of Berlin and was appointed director of the Urania Observatory. In 1913, that structure was relocated to Babelsberg, and Witt remained its director until resigning from that post in 1923. In 1902, Witt had married Martha Thiele; the couple had one son and two daughters.

Following his retirement, Witt took an increasing interest in the Berliner Mathematischen Gesellschaft. (Berlin Mathematical Society), and was eventually elected its president. He prepared ephemerides of Eros for its particularly close opposition of 1930/1931. These tables enabled more precise observations to be made, and a more accurate value of the solar parallax to be determined. Witt left unpublished a detailed study of the perturbations on Eros by the gravitational attraction of Jupiter.

In February 2001, the NEAR (Near-Earth Asteroid Rendezvous) Shoemaker spacecraft made a soft-landing on Witt's minor planet (433) Eros – the first such encounter ever attempted. This highly elongated, 33 × 13 km object is heavily cratered and has a rotation period of only 5.3 hours. Had Witt been alive on this occasion, he doubtlessly would have been thrilled to learn the true nature of this unusual asteroid discovered by him, more than 100 years earlier in Berlin.

Jordan D. Marché, II

Selected References

Cunningham, Clifford J. (1988). *Introduction to Asteroids: The Next Frontier.* Richmond, Virginia: Willmann-Bell, esp. pp. 10–11, 96.
Kahrstedt, Albrecht (1948). "G. Witt." *Astronomische Nachrichten* 276: 192.
Poggendorff, J. C. "Witt." In *Biographisch–literarisches Handwörtenbuch*. Vol. 5: (1926): 1384; Vol. 6 (1940): 2912; Vol. 7a (1962): 1043. Leipzig and Berlin.
Witt, Gustav (1905). *Untersuchung über die Bewegung des Planeten (433) Eros.* Berlin: Norddeutsche Buchdruckerei.

Wittich, Paul

Born **Breslau (Wroćlaw, Poland),** ***circa*** **1546**
Died **Vienna, (Austria), 9 January 1586**

Paul Wittich was one of many late-16th-century mathematicians who pursued the project of geometrically modifying the Copernican models of planetary motion to adapt them to a central Earth. He also studied the trigonometric problem of prosthaphaeresis.

Little is known about the family of Paul Wittich except that his uncle was the physician of Wroclaw Balthasar Sartorius, and that he was survived by a sister who inherited his books and his papers. The first known record of Wittich concerns his matriculation at the University of Leipzig in the summer of 1563. He later matriculated at Wittenberg, in June 1566, and at Frankfurt an der Oder, in 1576, but he is not known to have received degrees from these or any other institutions. Wittich seems to have preferred the lifestyle of the itinerant scholar: He wandered widely in between attending these universities, and indeed subsequently.

Wittich was a talented mathematician, and his contributions to astronomy derive from the combination of his mathematical interests and his peripatetic lifestyle. He wanted to transform the Copernican models of planetary motion so as to adapt them to the stability and centrality of the Earth. He also worked on prosthaphaeresis, a method of reducing problems involving the multiplication and division of trigonometric functions to ones of addition and subtraction. Wittich's progress in both of these fields is attested by annotations he made to the several copies of *De Revolutionibus* that he owned, and marginalia in further copies of **Nicolaus Copernicus**' text are among the sources that indicate the transmission of these ideas to others.

During the course of his travels, and the intermittent sojourns in his hometown of Wroćlaw, Wittich met many individuals actively interested in mathematics and astronomy with whom he collaborated or whom he instructed. His known contacts include the Altdorf professor Johannes Praetorius, the imperial physician **Tadeá Hájek z Hájku**, the Oxford mathematician **Henry Savile**, and the Scottish physicians **John Craig** and **Duncan Liddel**. (It is possible that Wittich's "discovery" of the first prosthaphaeretic identity was facilitated by Johannes Praetorius, who had come into contact with a manuscript by **Johannes Werner** that contained it.) In the late 1570s, Wittich communicated the prosthaphaeretic method to Craig, who later shared it with **John Napier**; he also divulged his work on planetary models to Savile in 1581. The most consequential of Wittich's collaborations, however, resulted from his visits to the two chief centers of astronomical endeavor in the late-16th century, the observatories of **Tycho Brahe** and Landgrave **Wilhelm IV** of Hesse.

In the autumn of 1580, Wittich visited Brahe's observatory in Denmark, revealing to him the first prosthaphaeretic identity and showing him his geometrical manipulations of the Copernican planetary models. Both were important to Brahe's astronomical project: Prosthaphaereois greatly simplified the task of reducing observational data, and by a slight alteration of Wittich's models, Brahe would arrive at the geoheliocentric scheme he promoted as the true system of the world. Brahe envisaged a long and fruitful collaboration with Wittich, but for reasons that are unclear, Wittich left Uraniborg after only a few months, deceiving the Danish astronomer with a promise to return.

In 1584, Wittich made his way to Kassel, where he worked with Wilhelm's mechanic **Jost Bürgi** in improving the instruments of the observatory of Kassel according to the design principles employed at Uraniborg. When Brahe learned of this collaboration, he was angered by the fact that Wittich had not credited him with these improved instrument designs; however, Brahe quickly came to appreciate the close agreement between the observational data produced at Uraniborg and Kassel that resulted. Wittich also designed an astrolabe for Landgrave Wilhelm, and he divulged to Bürgi the first prosthaphaeretic identity. Bürgi went on to discover a second, with proofs for both, and later showed these to **Nicholaus Bär** (Raimarus Ursus). As a consequence, Brahe's priority dispute with Bär over the invention of the geoheliocentric world-system was entangled with the quest to establish priority for himself and Wittich in the development of prosthaphaeresis.

Wittich also made observations, some of which he shared with other astronomers. However, both **Christoph Rothmann** and Brahe declared that Wittich was a poor observer, and in this respect was a better mathematician than astronomer.

Adam Mosley

Selected References

Gingerich, Owen (2002). *An Annotated Census of Copernicus'* De Revolutionibus *(Nuremberg, 1543 and Basel, 1566)*. Leiden: Brill. (For the most up-to-date account of Wittich's annotated copies of *De Revolutionibus* and their significance.)

Gingerich, Owen and R. Westman (1988). "The Wittich Connection: Conflict and Priority in Late Sixteenth-Century Cosmology." *Transactions of the American Philosophical Society* 78, pt. 7. (The most important work on Wittich to date.)

Goulding, R. (1995). "Henry Savile and the Tychonic World System." *Journal of the Warburg and Courtauld Institutes* 58: 152–179.

Mackensen, Ludolf von (ed.) (1988). *Die erste Sternwarte Europas mit ihren Instrumenten und Uhren*. Munich: Callwey Verlag, pp. 126–127. (The astrolabe Wittich designed for Wilhelm IV is described and illustrated herein.)

Thoren, Victor E. (1988). "Prosthaphaeresis Revisited." *Historia Mathematica* 15: 32–39.

Wolf, Charles-Joseph-Étienne

Born **Vorges, Aisne, France, 9 November 1827**
Died **Saint-Servan, Ille-et-Vilaine, France, 4 July 1918**

Charles Wolf and **Georges Rayet** have their names associated with a category of peculiar stars that they discovered at the Paris Observatory. Wolf came from an Alsatian family, the son of Pierre Frédéric Wolf, then lieutenant du Régiment des Chasseurs britanniques, and of Marie Josèphe (*dite* Louise) Pirachet. Several of Wolf's brothers likewise acquired positions in the field of teaching. Having passed the entrance exam, Wolf was admitted to the École Normale Supérieure in 1848 and studied physics; he graduated (*agrégé de physique*) in 1851. Wolf first taught at the Lycée of Nîmes and later at Metz, where he was married. His wife died prematurely, leaving him to raise their only child, a daughter.

Wolf's first research investigated capillarity as a function of temperature; for this work, he was awarded the degree of *Docteur d'État ès sciences de la Faculté de Paris* in 1856. He then transferred to the faculty of sciences at the University of Montpellier. There, he conducted experiments on the spectra of the alkali metals and found that these too displayed a variation with temperature. These findings were noticed by **Urbain Le Verrier**, director of the Paris Observatory, who offered Wolf a position as assistant astronomer in 1862.

First assigned to the Service Méridien, Wolf completed a study of the so called personal equation affecting meridian observations of stars. He also devised a direct-view spectroscope that was employed at the Service des Équatoriaux. A few years later, he was placed in charge of these instruments.

With his spectroscope, Wolf first observed some bright emission lines in the spectrum of a nova (T Coronae Borealis) on 20 May 1866. He then undertook a systematic search for similar lines in the spectra of other stars. In 1867, using **Jean-Bernard-Léon Foucault**'s 40-cm silvered-glass reflector, he and Rayet discovered three eighth-magnitude stars that are now known as Wolf–Rayet stars. Today, more than 100 such objects are known. Wolf–Rayet stars exhibit broad emission lines of helium, carbon, and nitrogen in their spectra. The outer layers of these highly evolved stars have been stripped away, revealing their exceptionally hot cores. Such stars are often found at the centers of planetary nebulae.

In 1869, Wolf explained the problem of the "black drop" effect, seen during transits of Mercury or Venus across the Sun's disk, as a purely instrumental consideration that arose from the contrast gradients of the two images. Wolf personally observed the 1874 and 1882 transits of Venus as tests of his ideas. Between 1873 and 1875, he prepared the most complete astrometric catalog of the Pleiades star cluster (achieved by visual means), giving the positions and magnitudes of more than 500 of its members. His results were superseded only by the development of astronomical photography. Wolf was made an assistant at the Sorbonne (1875), where he taught physical astronomy, and was a delegate to the Astrographic Congress (1887). He left the observatory in 1892 to devote full time to teaching.

Wolf undertook a historic study of the standard weights and measures housed in the collections of the Paris Observatory. In 1902, he published a definitive history of the Paris Observatory from its founding to the year 1793, drawn upon studies of original documents preserved in its archives and elsewhere.

Wolf retired in 1901 and returned to his native town but was forced to leave it when France was occupied during World War I. He was elected to the Paris Académie des sciences (1883) and later served as its president (1898).

Solange Grillot

Selected References

Anon. (1918). "Notice nécrologique." *L'Astronomie* 32: 255–256.

Bigourdan, G. (1919). "Charles Joseph Etienne Wolf." *Monthly Notices of the Royal Astronomical Society* 79: 235–236.

Lévy, Jacques R. (1976). "Wolf, Charles Joseph Étienne." In *Dictionary of Scientific Biography*, edited by Charles Coulston Gillispie. Vol. 14, pp. 479–480. New York: Charles Scribner's Sons.

Wolf, Charles Joseph Étienne (1902). *Histoire de l'Observatoire de Paris de sa fondation à 1793*. Paris: Gauthier-Villars.

Wolf, Charles Joseph Étienne and G. Rayet (1866). "Note sur deux étoiles." *Comptes rendus de l'Académie des sciences* 62: 1108–1109.

——— (1867). "Spectroscopie stellaire." *Comptes rendus de l'Académie des sciences* 65: 292–296.

Wolf, Johann Rudolf

Born **Fällanden, (Zürich), Switzerland, 7 July 1816**
Died **Zürich, Switzerland, 6 December 1893**

Johann Wolf is best known for his observations of sunspots and, in particular, his development of a formula for describing the number of observed sunspots. Wolf was born to Johannes Wolf (1768–1827), the fourth generation of Evangelical pastors in his family, and to Regula Gossweiler (1780–1867), a daughter of a Protestant minister. Wolf claimed to owe his scientific career in large part to his older brother, Johannes, who first announced his intention to carry on the family's religious tradition.

Wolf began his education at the Technological Institute in Zürich, but soon transferred to the newly founded Zürich University in 1833, where he studied until 1836, though he left without receiving a degree. He spent the following 2 years traveling to various universities and observatories across Europe. His most important steps on this journey were:

(1) an extended 18-month stay in Vienna, where he attended physics and astronomy lectures at the university and worked with **Johann von Littrow**;
(2) a 4-month stay in Berlin, where he rubbed shoulders with a number of established physicists and astronomers at the Berlin Observatory and the Academy of Sciences;
(3) a short but influential stay at the Göttingen Observatory, where he was introduced by **Karl Gauss** to contemporary geomagnetic theories and measurements; and,
(4) his visit to Gotha, where he became acquainted with the library collection and historical researches of **János von Zach**.

Wolf returned to Zürich at the end of 1838, and the following year began teaching mathematics, physics, and astronomy in Bern, where he became director of the local observatory in 1847. In 1855, he moved back to Zürich with a triple appointment as lecturer in mathematics at the Hochschule, extraordinary professor at the university, and professor of astronomy at what is now the Eidgenossische Technische Hochschule [ETH]. Wolf subsequently became the first director of the Federal Observatory inaugurated at the ETH in 1864, a position he held until his death. He was a member of the Swiss Naturforschenden Gesellschaft already in his student days, and over the years presided over its Bern and Zürich chapters. Wolf was also for a time director of the Swiss Meteorological Headquarters, and president of the Commission on Meteorology and Geodesy. He was elected a foreign member or associate of the Astronomische Gesellschaft Leipzig (Germany, 1850), the Società degli Spettroscopisti (Italy, 1859), the Royal Astronomical Society (England, 1864), and the Académie des sciences (Paris, 1885). He was granted an honorary doctorate by Bern University in 1852. Wolf never married, and after enjoying good health throughout his life, died after a short illness.

Wolf's astronomical interests ranged from comets to nebulae, but by far his most important contribution to astronomy was his historical reconstruction of solar activity based on sunspot numbers. His interest in such matters was fired by the observation of a particularly large and long-lived sunspot group in December 1847. Already aware of **Heinrich Schwabe**'s 1843 announcement of the sunspot cycle, Wolf embarked on his own sunspot-observing program. Using observatory records from across Europe, he began a program of historical researches aimed at extending sunspot cycle data prior to Schwabe's observations. In 1850, he introduced his relative sunspot number (R_Z), defined as

$$R_Z = k\,(10g + f),$$

where g is the number of sunspot groups observed on a given day, f the number of individual sunspots, and k a numerical scaling coefficient. Setting $k = 1$ for his own observations, Wolf assigned distinct k values to different observers, so that their numerical values for g and f would yield the same R_Z on common observing days. This simple rescaling procedure thus allowed him to put on the same numerical scale sunspot observations carried out by observers of widely varying ability and diligence, and using equally widely varying instruments and techniques.

By 1852, Wolf had revised Schwabe's 10-year cycle duration to an average value of $11^1/_9$ years, and offered evidence for significant variations in the cycle's duration, an anticorrelation between cycle amplitude and duration, and longer, secondary periodicities superimposed on the primary cycle. By 1868, he had extended his sunspot number reconstruction back to 1700. Wolf continued to revise his time series of sunspot numbers throughout his life, as more and more data became available to him. His successors in Zürich continued his work for nearly a century, with the Brussels Observatory carrying on the tradition since 1981. The Wolf sunspot number, as it is now called, remains to this day the classical (and most intensely studied) measure of solar activity.

In July 1852, Wolf was one of four researchers (along with **Edward Sabine** in England, **Johann von Lamont** in Germany, and **Jean Gautier** in Switzerland) to demonstrate independently and more or less simultaneously that a marked 11-year periodicity also appears in geomagnetic measurements. Wolf went on to discover the correlation between sunspot numbers and auroral records, also independently noted by American scientist **Elias Loomis**. Wolf continued to seek sunspot-related periodicities in various meteorological phenomena, but with inconclusive results.

Throughout his life, Wolf was very active in the Bern and Zürich chapters of the Swiss Naturforschenden Gesellschaft, and contributed numerous papers to the society's *Vierteljahrsschrift*, for which he also acted as editor for many years. Already in 1855, Wolf's wide-ranging interests and scholarship led to his appointment as the first ETH librarian, and it was largely through his initiative that the library's remarkable historical collections were assembled. Other technical interests of his included geodesy, surveying, number theory, and the empirical study of probability.

Wolf was an indefatigable worker and a prolific writer by any standards. In the course of his career, he authored some 258 articles or books in the fields of astronomy, meteorology, mathematics, surveying, and the history of science, culture, and religion. He also regularly penned articles and delivered lectures aimed at the general public. In 1856, Wolf inaugurated his solo astronomical journal, *Astronomische Mittheilungen*, adding up to 13 volumes by the year of his death, and in which he published results of his astronomical and historical researches.

Paul Charbonneau

Selected References

Anon. (1894). "Dr. Rudolf Wolf." *Monthly Notices of the Royal Astronomical Society* 54: 206–208.

Anon. (1993). *Vierteljahrsschrift der Naturforschenden Gesellschaft in Zürich* 138, no. 4. (This special issue focuses on Wolf's life and work.)

Hoyt, Douglas V. and Kenneth H. Schatten (1997). *The Role of the Sun in Climate Change*. Oxford: Oxford University Press, Chap. 2.

Jaeggli, A. (1968). *Die Berufung des Astronomen Joh. Rudolf Wolf nach Zürich 1855*. Zürich: Publication of the ETH Library, no. 11.

Lutstorf, H. T. (1993). *Professor Rudolf Wolf und seine Zeit*. Zürich: Publication of the ETH Library, no. 31.

Wolf, Rudolf (1856). *Taschenbuch für Mathematik, Physik, Geodäsie und Astronomie*. Bern: Haller

——— (1858–1861). *Biographien zur Culturgeschichte der Schweiz*. 4 Vols. Zurich: Orell, Füssli, & Co.

——— (1869–1872). *Handbuch der Mathematik, Physik, Geodäsie und Astronomie*. 2 Vols. Zurich: Friedrich Schulthess

——— (1877). *Geschichte der Astronomie*. Munich: R. Oldenbourg.

——— (1890–1893). *Handbuch der Astronomie: Ihrer Geschichte und Litteratur*. Zurich: F. Schulthess.

Wolf, Maximilian Franz Joseph Cornelius

Born **Heidelberg, (Germany), 21 June 1863**
Died **Heidelberg, Germany, 3 October 1932**

Max Wolf, considered a pioneer in astrophotography, observed many new nebulae both within the Milky Way and outside our Galaxy. He discovered more than 200 asteroids along with three comets that now bear his name.

Wolf was born to Franz Wolf and Elise Helwerth. As Wolf became interested in astronomy, his father, a physician, constructed a private observatory for him, which he used from 1885 until 1896. In 1884, when only 21 years old, he discovered comet 14P/Wolf. This discovery was remarkable because the object was first thought to be an asteroid.

Wolf received his Ph.D. from Heidelberg in 1888 working under Leo Königsberger. He then studied with **Hugo Gyldén** in Stockholm from 1888 to 1890. Wolf became *Privatdozent* (lecturer) in 1890 and served as professor of astrophysics and astronomy at the University of Heidelberg from 1893 to 1932. He prompted the building of a new observatory near Heidelberg at Königstuhl, and became its director. Wolf is now also known as the "Father of Heidelberg astronomy."

Wolf developed several new photographical methods for observational astronomy, and was the first astronomer to use time-lapse photography, useful, for example, in detecting asteroids. Wolf brought the "dry plate" technique to astronomy in 1880, and introduced the blink comparator in 1900 in conjunction with the Carl Zeiss optics company in Jena. Using a blink comparator, a microscope that optically superimposes two photographic plates onto the same viewing region by blinking between them so quickly that the two plates look like only one, an astronomer can compare two plates and easily find differences between them. The blink comparator turned out to be a valuable, useful astronomical tool, used in the discovery of Pluto by **Clyde Tombaugh** in 1930. While Wolf himself did not contribute to this discovery, he was able to locate the new planet on his older plates.

Wolf discovered more than 200 asteroids using various photographic techniques. The first, discovered in 1891 during a search for the minor planets (10) Hygiea and (30) Urania, was (323) Brucia, named by Wolf in honor of Catherine Wolfe Bruce, who had contributed $10,000 for one of his telescopes. Already by 1892, while overcoming difficulties with the optics, Wolf had found 17 new asteroids. In 1906, Wolf discovered (588) Achilles, the first of the so called Trojan minor planets, which orbit the Sun in low-eccentricity stable orbits with semi-major axes very close to that of Jupiter. These objects manifest the triangular three-body system analyzed and predicted theoretically by **Joseph Lagrange** in the 18th century.

Wolf was the first to observe comet 1P/Halley when it approached Earth in 1909. Halley's comet produced much excitement the following year because it was so close to the Earth that some expected the Earth would pass through its tail.

Wolf used wide-field photography to study the Milky Way. He discovered about 5,000 nebulae and galaxies and also found new stars, such as Wolf 359, an extremely faint star, the third closest to the Earth after Alpha Centauri 3 and Barnard's star. Though Wolf 359 is much too dim to be visible to the naked eye, Wolf was able to discover it with photographic techniques.

Wolf used statistical star counts to prove the existence of dark nebulae. Independently of American astronomer **Edward Barnard**, Wolf discovered that the dark "voids" in the Milky Way are in fact nebulae obscured by vast quantities of dust. In studying their spectral characteristics and distribution, he was among the first astronomers to show that spiral nebulae have absorption spectra typical of stars and thus differ from gaseous nebulae like planetary nebulae.

Around 1905, Wolf suggested building an observatory in the Southern Hemisphere, though it was not until 1930 that such plans were realized by Berlin and Breslau astronomers in Windhoek (Namibia). Observations there were stopped by World War II. In the 1950s, the European Southern Observatory [ESO] tested two sites in South Africa and South America, with a new observatory opening eventually in 1969 in northern Chile.

Wolf was a co-developer of the stereo comparator together with Carl Pulfrich from the Zeiss company. The stereo comparator consists of a pair of microscopes arranged so that one can see simultaneously two photographic plates of the same region taken at different times. Wolf seems to have experimented with such techniques as early as 1892, but without success. When Pulfrich approached him to adapt the technique from geodesy to astronomy, Wolf was delighted. A steady exchange of letters followed. Wolf and Pulfrich then worked together to analyze the rapidly growing accumulation of photographic plates. Tragically, Pulfrich lost one eye in 1906, preventing him from using the stereographic tool from then on.

Wolf also provided in 1912 suggestions for the idea of the modern planetarium, while advising on the new Deutsches Museum in Munich, Germany. Wolf was a gifted teacher who attracted students from all over the world. He was also highly esteemed by amateur astronomers, helping them out with pictures and slides. In 1930, Wolf became a Bruce Medalist, awarded each year by the directors of six observatories – three in the United States and three abroad. He received the Gold Medal of the Royal Astronomical Society in 1914.

Wolf was survived by his wife Gisela Wolf (*née* Merx), whom he married in 1897. She had assisted him often with his work at the blink comparator. In addition to the three comets, Wolf has a lunar crater, a star (Wolf 359), the minor planet (827) Wolfiana, and an irregular galaxy (Wolf–Lundmark–Melotte) named in his honor.

Oliver Knill

Selected References

Bok, Bart J. (1960). "Wolf's Method of Measuring Dark Nebulae." In *Source Book in Astronomy, 1900–1950*, edited by Harlow Shapley, pp. 285–288. Cambridge, Massachusetts: Harvard University Press.

Freiesleben, H. Christian (1962). *Max Wolf*. Grosse Naturforscher. Vol. 26. Stuttgart: Wissenschaftliche Verlagsgesellschaft.

——— (1976). "Maximilan Franz Joseph Cornelius." In *Dictionary of Scientific Biography*, edited by Charles Coulston Gillispie. Vol. 14, pp. 481–482. New York: Charles Scribner's Sons.

Wollaston, William Hyde

Born **East Dereham, Norfolk, England, 6 August 1766**
Died **London, England, 22 December 1828**

William Wollaston, an English scientist, physician, and inventor, was a gifted polymath who made important contributions to physiology, optics, mineralogy, and chemistry. Son of Francis and Althea (*née* Hyde) Wollaston, he was first educated at Charterhouse and then admitted to Caius College, Cambridge, where he studied botany. Upon his graduation in 1787, Wollaston pursued a medical degree and received his M.D. in 1793. He was subsequently elected to the Royal Society (London) and later served as its secretary (1804–1816).

In 1800, Wollaston abandoned his medical practice and began instead to pursue scientific research full time. He investigated a vast range of subjects, stretching from human physiology, chemistry, and metallurgy, to theoretical and experimental physics and astronomy. Wollaston identified bladder stones (composed of the first known amino acid) and their method of formation, advocated the use of meniscus lenses in eyeglasses, and made precise analyses of the human hearing mechanism. He discovered how to work platinum, tungsten, and other transition metals, and his original process (employing powder metallurgy) for rendering platinum malleable made him extremely wealthy. Wollaston used the money to further his research. He discovered two new metals, palladium and rhodium. The mineral wollastonite ($CaSiO_3$), widely used in ceramic products such as tiles and porcelain, is named after him. His improvements to the chemical battery were adopted for the rest of the century. Wollaston was one of the avatars of modern atomic theory; his anticipation of the three-dimensional geometrical arrangement of atoms paved the way for the work of John Dalton, Jacobus van't Hoff, and Joseph le Bel.

In 1802, Wollaston first reported that the solar spectrum was crossed by a series of dark lines; he erroneously took this to mean that there were only four primary colors in its spectrum. These dark lines were investigated a decade later by **Joseph von Fraunhofer**, after whom they are named, although they remained unexplained until chemists **Robert Bunsen** and **Gustav Kirchhoff** laid the foundations of observational astrophysics in 1859.

Much of Wollaston's research was prompted by his own development of new optical techniques and instruments. He invented the reflecting goniometer, a device used for measuring the angles between facets of crystalline minerals. This instrument had a notable impact on the sciences of crystallography and mineralogy. Wollaston also perfected a new type of sextant, the dip sector, which was used in early exploration near the Earth's North Pole. In 1807, he designed and built the *camera lucida*, containing a quadrilateral prism that aided the production of making sketches. His improvement of the *camera obscura* (using meniscus lenses) helped to inspire the "fixing" of landscape imagery onto a screen and the invention of photography. Shortly before his death, Wollaston left substantial sums to both the Royal and Geological Societies of London for the encouragement of scientific research.

Daniel Kolak

Selected References

Goodman, D. C. (1976). "Wollaston, William Hyde." In *Dictionary of Scientific Biography*, edited by Charles Coulston Gillispie. Vol. 14, pp. 486–494. New York: Charles Scribner's Sons.

Hartog, Philip J. and Charles H. Lees (1921–1922). "Wollaston, William Hyde." In *Dictionary of National Biography*, edited by Sir Leslie Stephen and Sir Sidney Lee. Vol. 21, pp. 782–787. Oxford: Oxford University Press.

Woltjer, Jan, Jr.

Born **Amsterdam, The Netherlands, 3 August 1891**
Died **Leiden, The Netherlands, 28 January 1946**

Dutch theoretical astronomer Jan Woltjer is recognized for a study of the motion of the Saturian satellite Hyperion that provided the first, and still best established, example of what is now called chaotic behavior in an astronomical system. He was the son of Jan Woltjer, Sr., a professor of classical languages at Amsterdam, and received his Ph.D. in 1918 for work on the theory of Hyperion at Leiden University under **Willem de Sitter**. He was a lector at Leiden University (a position which did not permit having formal Ph.D. students) for most of his career. He married Hillegonda Hester de Vries, and they had four children, all scholars: Anna (sociology), Margo (classical languages), Jan Juliaan (history), and Lodewijk (astronomy).

Following his Ph.D., Woltjer continued and completed his study of the complex motion of Hyperion, the seventh satellite of Saturn. Hyperion's orbit presented particular problems because of the strong influence of the very massive satellite Titan. Contrary to **Simon Newcomb**'s earlier skepticism, Woltjer showed that an expansion into powers of the eccentricity of Titan's orbit led to a successful ephemerid for Hyperion. More recently his theory has been taken up again and further elaborated by D. B. Taylor and others.

By the mid 1920s Woltjer's interests had shifted toward astrophysical problems. Among his contributions may be mentioned studies of opacities in stellar interiors, radiative transfer in moving media, and the dynamics of the solar chromosphere. However, Woltjer's most significant work pertained to the pulsation theory of Cepheid variable stars, where he was the first to obtain quantitative results on the excitation of overtone pulsations and their interaction with the fundamental mode. Woltjer succeeded in casting the equations of pulsation theory into a Hamiltonian form, so that the methods of celestial interaction between the fundamental mode and an overtone could limit the amplitude of the pulsations. This also permitted an understanding of the variables with multiple periods like RR Lyrae and ζ Geminorum.

The oscillation of a star would damp very quickly if there were no energy fed into it to compensate the damping. Woltjer developed an iterative procedure to deal with these "nonadiabatic" effects. To first order, the method consists of using the adiabatic equations to compute the flux variations and then utilizing these in the equations for the nonadiabatic temperature variations. From these, the energy input into the oscillation can be obtained. In principle this procedure can be repeated to obtain higher order corrections, but the first step is already quite complex. This method was later used by J. P. Cox and others to quantitatively demonstrate that the helium ionization zone could play an important role in maintaining the Cepheid pulsations. With the advent of powerful computers, direct integration of the hydrodynamic and radiative transfer equations became possible, and the Woltjer method has fallen into disuse.

Woltjer's work was characterized by a search for mathematical rigor, though it was also motivated by observational problems. Unfortunately, World War II, and his untimely death soon after, stopped his work in midstream.

Lodewijk Woltjer

Selected References

Brouwer, Dirk (1947). "Jan Woltjer, Jr." *Monthly Notices of the Royal Astronomical Society* 107: 59–60.

Rosseland, S. (1940). *The Pulsation Theory of Variable Stars*. Oxford: Clarendon Press. (Woltjer's Hamiltonian methods are described and extended herein.)

Woltjer, Jr., Jan (1928). "The Motion of Hyperion." *Annals of the Observatory at Leiden* 16, pt. 3: 1–139.(For elaboration of Woltjer's theory by D. B. Taylor, see "A Comparison of the Theory of the Motion of Hyperion with Observations Made During 1967–1982." *Astronomy and Astrophysics* 141 [1984]: 151–158.)

——— (1929). "The Ca^+ Chromosphere." *Bulletin of the Astronomical Institutes of the Netherlands* 5: 43–48. (See also "On the Diffusion of Monochromatic Radiation through a Medium Consisting of Plan – Parallel Layers in Relative Motion." *Bulletin of the Astronomical Institutes of the Netherlands* 7 [1934]: 217–226.) (For discussion of radiation pressure in the K-line of Ca^+ on the structure of the solar chromosphere, as well as the resulting more general problems of radiative transfer in moving media that became the subject of numerous papers by others in subsequent decades.)

——— (1935). "On Periodic Solutions in Adiabatic Star-Pulsations." *Monthly Notices of the Royal Astronomical Society* 95: 260–263. (Here the Hamiltonian formulation of the pulsation equations is represented.)

——— (1943). "On the Anomalous Phase Relation between First and Second Harmonic in the Radial Velocity of ζ Geminorum and Related Stars." *Bulletin of the Astronomical Institutes of the Netherlands* 9: 435–440. (For a study of the interactions of the three modes and an application to ζ Geminorum stars. This was further investigated in the posthumously published paper ", On Excitation and Maintenance of Secondary Oscillations in Pulsating Stars." *Bulletin of the Astronomical Institutes of the Netherlands* 10 (1946): 125–130.)

——— (1946). "On the Theory of Anadiabatic Star-Pulsations." *Bulletin of the Astronomical Institutes of the Netherlands* 10: 130–135. (In this paper the earlier studies of nonadiabatic pulsations were "continued, extended and amended." The latter theoretical framework, first developed in a 1936 paper, has been applied by subsequent authors to pulsations of realistic stellar models that became available in the 1950s. For a clear description by John P. Cox of the method and its application to Cepheids, see "A Preliminary Analysis of the Effectiveness of Second Helium Ionization in Inducing Cepheid Instability in Stars." *Astrophysical Journal* 132 [1960]: 594–626.)

Wood, Frank Bradshaw

Born **Jackson, Tennessee, USA, 21 December 1915**
Died **Gainesville, Florida, USA, 10 December 1997**

Frank Wood was a leading authority on photometry and eclipsing binary stars and provided leadership to the binary star subcommunity through the International Astronomical Union for a number of years. The son of Thomas Frank and Mary Bradshaw, he received his undergraduate education in Florida. He then earned a Ph.D. in astronomy from Princeton University as a student of **Raymond Dugan**, and then **Henry Norris Russell** when Dugan died, for a dissertation on eclipsing binary stars. Wood remained a specialist in close binary stars for the remainder of his career. He was the first to document that most close binary systems in which irregular period changes are observed are dynamically unstable with at least one member of the pair filling its Roche limit and likely losing mass. However, his physical explanation of the mass loss as occurring in jets directed along the axis of rotation was rejected by most theorists.

As department chair and director of the Flower Observatory and then Cook Observatory, Wood was responsible for the consolidation of these two facilities into one larger Flower and Cook Observatory in Paoli, Pennsylvania, more distant from Philadelphia city lights than either previous observatory. He married Elizabeth Hoar Pepper; they had four children.

Thomas R. Williams

Selected References

Kopal, Zdeněk (1959). *Close Binary Systems*. London: Chapman and Hall.

Sahade, Jorge (1958). "The Transfer of Mass in Close Binary Stars." *Astronomical Society of the Pacific Leaflet*, no. 344.

Sahade, Jorge and Kwan Yu Chen (1998). "Frank Bradshaw Wood, 1915–1997." *Bulletin of the American Astronomical Society* 30: 1469–1470.

Wood, Robert Williams

Born **Concord, Massachusetts, USA, 2 May 1868**
Died **Amityville, New York, USA, 11 August 1955**

Robert Wood, a brilliant experimentalist, contributed substantially to our physical understanding of the optical characteristics of gases, including those in magnetic fields; his work was critical to the evolution of astrophysical understanding through the application of the spectroscope to celestial objects. From his laboratory at Johns Hopkins University, Baltimore, Maryland, he supplied many high-precision ruled diffraction gratings that made their way into astronomical spectrographs. Wood's work in color photography led to the design of ultraviolet and infrared filters of importance to astronomy as well.

Wood, son of Robert and Lucy (*née* Davis) Wood, received his B.A. in chemistry from Harvard College in 1891. He then studied at Johns Hopkins University (1892), the University of Chicago (1892–1894), and the University of Berlin (1894–1896), where he became an assistant to physical chemist Wilhelm Ostwald and physicist Heinrich Rubens. Although Wood never earned a doctorate, he taught at the University of Wisconsin (1897–1901) before becoming professor of experimental physics (1901–1938) at the Johns Hopkins University, succeeding **Henry Rowland**. Wood was subsequently named professor emeritus and research professor (1938–1955). He married Gertrude Ames in 1892; the couple had four children.

A prolific inventor, Wood developed methods for thawing frozen water pipes by passing electric currents through them, frosting the insides of glass light bulbs, and introducing the so called Vienna method for detecting forged documents (using ultraviolet light). He was perhaps the first person to show animated films, and successfully demonstrated the principles of fisheye camera lenses and photography. Wood conducted pioneering investigations in the fields of ultrasound and biophysics, including the first study of the physiological effects of high-frequency sound waves.

Wood made a number of fundamental contributions to optics and spectroscopy. He greatly improved the efficiency of diffraction gratings with his design of the echelette grating, which allowed selection of a narrow range of wavelengths for detailed study. In 1897, he became the first person to observe the "field emission" of charged particles from a collector placed in an electric field. Wood's studies on the fluorescent properties of gases had profound implications for the theory of atomic structure. He was nominated in 1927 for the Nobel Prize in Physics by **Erwin Schrödinger**, but did not receive the award. Wood, however, received several honorary doctorates in his career.

Although the discovery of radiation beyond the visible spectrum and techniques of sensitizing photographic emulsions to record them predated Wood's activities, he was the first to make photographic filters that excluded visible wavelengths. He was also the first to capture ultraviolet fluorescence on film. Although infrared emulsions would not be commercially available until the 1930s, Wood published infrared landscape photographs taken with experimental films *circa* 1910. That same year, Wood undertook the first spectrophotometric investigation of the Moon, identifying localized chromatic differences in lunar soils. A region near the crater Aristarchus that appears dark in ultraviolet radiation is often referred to as Wood's spot. The Wood lamp for generating ultraviolet radiation, but commonly referred to as a "black light," bears his name.

In the early 1900s, Wood conducted experiments with rotating mercury mirrors. By applying various rotation speeds to pools of the liquid metal, Wood demonstrated the feasibility of turning them into paraboloidal mirrors for reflecting telescopes. He even hoped to find some substance that could be allowed to solidify while rotating, thereby saving most of the labor expended in constructing such telescope mirrors. Although not realized in Wood's day, this principle was later applied by the University of Arizona astronomer Roger Angel, whose rotating kilns enabled the production of giant (8 m) "spun-cast" glass telescope mirrors.

A highly visual thinker, Wood wrote in his textbook, *Physical Optics*, that he had "attempted to give, in as many instances as possible, a physical picture of the processes usually described by equations." Throughout his life, Wood retained a mischievous nature; his principal biographer, William Seabrook, styled Wood as a "small boy who … never grew up."

In 1912 Wood was elected to the US National Academy of Sciences, from which he received the Henry Draper Gold Medal for his contributions to astronomy. He was one of few foreigners elected as a member of the Royal Society of London, and received that society's Rumford Gold Medal for his achievements in physical optics. Wood served as president of the American Physical Society (1935). He authored *Physical Optics* (1905), a standard textbook on the subject for many years, along with *Researches in Physical Optics* (1913–1919), and *Supersonics, the Science of Inaudible Sounds* (1939), along with more than 220 scientific articles. Wood is also known for *How To Tell the Birds from the Flowers* (1907), a book of nonsense verses written to amuse his children, and *The Man Who Rocked the Earth* (1915), a science fiction novel coauthored with Arthur Train. He is commemorated by a 78-km diameter crater on the Moon. A collection of Wood's papers is held at the Niels Bohr Library of the American Institute of Physics.

Thomas A. Dobbins

Selected References

Dieke, G. H. (1956). "Robert Williams Wood." *Biographical Memoirs of Fellows of the Royal Society* 2: 327–345.

Hines, Terence (1999). "Wood, Robert Williams." In *American National Biography*, edited by John A. Garraty and Mark C. Carnes. Vol. 23, pp. 776–777. New York: Oxford University Press.

Lindsay, R. B. (1976). "Wood, Robert Williams." In *Dictionary of Scientific Biography*, edited by Charles Coulston Gillispie. Vol. 14, pp. 497–499. New York: Charles Scribner's Sons.

Seabrook, William (1941). *Doctor Wood: Modern Wizard of the Laboratory*. New York: Harcourt, Brace.

Wood, R. W. (1909). "The Mercury Paraboloid as a Reflecting Telescope." *Astrophysical Journal* 29: 164–176.

Woolley, Richard van der Riet

Born **Weymouth, England, 24 April 1906**
Died **Sutherland, South Africa, 24 December 1986**

British optical astronomer Richard Woolley is most happily remembered for his role in the postwar development of optical astronomy in Australia, Britain, and South Africa, and less happily for having firmly said that spaceflight would always be impossible, shortly before it became a reality. Woolley was the son of a rear admiral who had been stationed at Simonstown, near Cape Town, South Africa, and a South African mother (whence the name van der Riet), daughter of the resident magistrate there. Woolley began his education in England, but returned to South Africa when his father retired there, earning a B.Sc. in 1924 and an M.Sc. in 1925 at the University of Cape Town, and briefly holding a position as demonstrator in physics there.

Woolley returned to England, to Cambridge University in 1926, receiving a high-rank degree in mathematics in 1928. Encouraged to take an interest in astronomy by professor **Frederick Stratton**, Woolley began work on solar and stellar atmospheres with **Arthur Eddington** and received a Ph.D. in 1931, partly for work on the solar spectrum carried out at the Mount Wilson Observatory, California, USA, under a Commonwealth Fellowship.

After 2 years on an Isaac Newton Studentship at Cambridge, Woolley was appointed chief assistant at the Royal Greenwich Observatory, where he wrote the book *Eclipses of the Sun and Moon* with **Frank Dyson**, then the Astronomer Royal. After 2 more years in Cambridge (1937–1939) as assistant to Eddington, Woolley was appointed Commonwealth astronomer and director of the Commonwealth Solar Observatory in Canberra, Australia. Almost immediately, he put it on a war footing, devoting its optical expertise to the design of gun sights and the like.

After the war, Woolley arranged for the transfer of the observatory, renamed the Mount Stromlo Observatory, from the government to the newly founded Australian National University. His own work, meanwhile, had shown that the upper chromosphere of the Sun was hot, in collaboration with **Clabon Allen** (best known as the editor of several iterations of *Allen's Astrophysical Quantities*), though the mechanism they suggested (back-warming from the corona) cannot be the whole story. A collaboration with Douglas W. N. Stibbs resulted in an important book on stellar atmospheres in 1953. Woolley had arranged for the financing and construction of a 1.9-m telescope for Mount Stromlo, which was installed not long after he had returned to England in 1956 as Astronomer Royal and director of the Royal Greenwich Observatory [RGO].

Once again Woolley engaged in a great deal of hard work in the realm of scientific politics, connected with the transfer for RGO from its hopeless site near London to the only slightly less hopeless site near Brighton, Sussex (at Herstmonceux), and the construction and commission of the Isaac Newton telescope (a 98-in., eventually relocated, at least in parts, to La Palma in the Canary Islands, where it finally became a productive instrument). Woolley initiated the annual Herstmonceux Conferences in astronomy and the student summer courses as well as inaugurating new programs in photometry and dynamics of nearby stars. He collaborated with new staff member Olin J. Eggen and with **Olin Wilson** on determinations of stellar distances, ages, and velocities.

Woolley's next major initiative was the Anglo–Australian Telescope, a 4-m class instrument, to be sited under clear, southern skies but jointly owned and operated. His age-dictated retirement as Astronomer Royal in 1971 occurred before this project was completed.

Returning to South Africa, Woolley engaged yet again in observatory building. Three separate existing facilities, all short of funds and modern instrumentation, were merged to form the South African Astronomical Observatory at Sutherland under his directorship. There he encouraged work on abundances of the elements in stars and galaxies and on quasars. His own work continued after yet another retirement in 1976. It included exploitation of infrared light curves to improve the Baade–Wesselink method for determining brightnesses and distances to Cepheids (and other variable stars) and determination of the kinematics of the older RR Lyrae stars in the galactic halo.

Woolley served as a vice president of the International Astronomical Union (1952–1958) and president of the Royal Astronomical Society (1963–1965). He was elected to the Royal Society (London) in 1953, received a Sc.D. from Cambridge in 1951, and was knighted in 1963. His first two wives predeceased him, and he was survived by the third, Sheila Woolley.

Roy H. Garstang

Selected References

Lynden- Bell, D. (1987). "Professor Sir Richard Woolley, OBE, ScD, FRS, 1906–86." *Quarterly Journal of the Royal Astronomical Society* 28: 546–551.

McCrea, Sir William (1988). "Richard van der Riet Woolley." *Biographical Memoirs of Fellows of the Royal Society* 34: 921–982.

Wren, Christopher

Born **East Knoyle, Wiltshire, England, 20 October 1632**
Died **London, England, 25 February 1723**

Sir Christopher Wren, remembered mostly for his architecture, was a key figure in the nascent Royal Society of London. Wren was the son of Reverend Christopher and Mary Wren, a royalist family. In 1634 the elder Christopher was appointed dean of Windsor and registrar of the Order of the Garter. Christopher was tutored by his father, who had some knowledge of mathematics, modern science, and architecture, and then by Reverend William Holder, later a fellow of the Royal Society. He entered Westminster School in 1642; John Dryden and John Locke were fellow students. Westminster was one of the few schools to offer mathematics. In 1647 Wren went to live in the London home of physician Charles Scarborough, first as a patient and then as a sort of student assistant. Here he met a number of prominent scientists, some of them refugees from the Puritan stronghold of Cambridge. Wren translated a tract on sundials by William Oughtred into Latin. It was appended to the 1652 edition of *Clavis Mathematica*, and Oughtred praised him as a youth

who had already enriched astronomy and other sciences, prefiguring John Evelyn's famous words,"that miracle of a youth, Christopher Wren."

In 1649 Wren matriculated at Wadham College, Oxford, whose master, **John Wilkins**, became his mentor. Wren's talent and temperament led to his acceptance by his scientific seniors. Wren had an interest in instrumentation, a mechanical flair, and a bent for invention, which he used in his own research, and also in setting up apparatus for others. His steady hand in dissection and artistic talent were also shared. Wren was a member, along with Robert Boyle, **Seth Ward**, and **Robert Hooke** of the "Experimental Philosophical Club" formed by William Petty. Hooke praised Wren's pioneering work in microscopic illustration in *Micrographia*.

Wren's interests turned to astronomy and mathematics. Ward had established an observatory at Wadham with telescopes of 6-, 12-, and 22-ft. focal lengths, where they made joint observations. Wren also joined the amateur Sir Paul Neile in observations through the 35-ft. telescope on his estate. Wren and **John Wallis** collaborated on an 80-ft. telescope that could supposedly view the full face of the Moon. Wren earned his AB in 1651 and his MA in 1653; he was awarded a fellowship in All Souls College.

Wren was named professor of astronomy at Gresham College, London, in 1657, possibly through Oliver Cromwell's intervention. On Charles II's restoration in 1660, Wilkins and Ward lost their posts, but Wren, of a royalist family, was appointed to the Savilian Professorship of Astronomy in 1661. Both Oxford and Cambridge awarded him the Doctor of Civil Laws degree in the same year. Wren was frequently in Oxford during his Gresham years, and in London after assuming the Savilian chair, allowing him to attend meetings of what became the Royal Society; he was a charter fellow (July 1662) and president (1681–1683).

Wren undertook his first architectural assignments in the early 1660s, and was appointed by King Charles to the commission to restore the dilapidated Saint Paul's Cathedral. He continued to make astronomical observations with Hooke. During his 1665 trip to study advanced French architecture, his most frequent companions were the astronomers **Adrien Auzout** (whose observations of the 1664 comet (C/1664 W1) agreed with Wren's), Henri Justel, and **Pierre Petit**, savants who shared his interest in both science and architecture.

The Great Fire of 1666 opened the way for Wren's great work. He was appointed Surveyor General (royal architect) in 1669, but not until 1673 did he resign the Savilian professorship. King Charles conferred a knighthood also in 1673.

Wren married Faith Coghill in 1669. Their first son, Gilbert, died in infancy. But Christopher, Jr., lived to be his father's colleague, heir, and literary executor. The first Lady Wren died of smallpox in 1675. Wren remarried, to Jane Fitzwilliam, in 1677. Their daughter Jane was talented in music and art, but their son William was retarded. The second Lady Wren died in 1679, and Wren lived the rest of his long life as a widower.

Wren's architectural achievements are self-evident. He invented the English Baroque style. Saint Paul's Cathedral, London, is his masterpiece. Wren built parish churches, hospitals, academic buildings, and the Royal Observatory in Greenwich. In addition, he found time to serve as president, vice president, and member of Council for the Royal Society, on the Committee of the Hudson Bay Company, and for a couple of terms as a member of parliament.

Wren was singled out in Thomas Spratt's *History of the Royal Society* (1667) where contributions to refraction, theory of motion, the rings of Saturn, his lunar globe, and celestial mapping are mentioned. Wren was a Baconian experimentalist who seemed satisfied with a well-warranted hypothesis. Unlike **Isaac Newton**, he was disinclined to venture into comprehensive theory or writing definitive papers. Some of his scientific papers were extant in 1740 when Ward wrote his *Lives of the Professors of Gresham College*. They are now lost.

Wren's 1657 inaugural lecture at Gresham College, which survives, was considered a definitive statement of the experimental philosophy. A 1659 paper on **Johannes Kepler**'s second law of planetary motion was extremely helpful to English astronomers, most of whom accepted elliptical orbits, but did not understand them. Wren was the leading authority on lunar geography, apparently incorporating a micrometer in the eyepiece of his telescope to refine his measurements. The public product was a 10-in. globe showing the visible face of the Moon in relief, which he presented to King Charles in 1661. Less triumphant was Wren's work on Saturn. He hypothesized that the appearance of the planet was due to an elliptical corona. But the elegance of **Christiaan Huygen**'s ring hypothesis appealed to him. Wren endorsed it and dropped his own work (1659). Indeed his 1658 copper model may illustrate Huygen's view better than his own!

The Great Comet of 1664 provided the occasion for joint observations and intense discussion with Hooke. Wren accepted the popular notion that comets traveled in linear paths, while Hooke speculated about closed, possibly circular, orbits. When a second comet appeared in 1665 (C/1665 F1), probably the same one outbound after passing behind the Sun, Wren remarked it might be the same comet, but apparently took the notion no further.

In 1663 Wren constructed a double telescope with a measuring scale that would enable two observers to focus on the same object and more accurately estimate the distance. Wren built the Royal Observatory in 1675. The building itself has an observation room where smaller telescopes could be used, but the heroic instruments of the day would be suspended from booms in the yard.

In 1692 Wren was involved in a scheme to mount a 123-ft. telescope in a staircase at Saint Paul's, but it did not work. Wren also kept one of the west bell towers clear so that it could be used as an observatory.

At the time of Wren's death he was combining study of Scripture with efforts to solve the problem of determining longitude at sea by some astronomical method.

Christian E. Hauer, Jr.

Selected References

Bennett, James A. (1982). *The Mathematical Science of Christopher Wren*. Cambridge: Cambridge University Press.

Chambers, James (1998). *Christopher Wren*. Stroud, Gloucestershire: Sutton Publishing.

Gray, Ronald (1982). *Christopher Wren and St. Paul's Cathedral*. Minneapolis: Lerner Publications. (British ed., Cambridge, 1979.)

Hauer, Jr., Christian E. (ed.) (1997). *Christopher Wren and the Many Sides of Genius: Proceedings of a Christopher Wren Symposium*. Lewiston, New York: Edwin Mellen Press.

Little, Brian D. (1975). *Sir Christopher Wren: A Historical Biography*. London: Robert Hale.

Tenniswood, Adrian (2001). *His Invention So Fertile: A Life of Christopher Wren*. New York: Oxford University Press.

Wren, Jr., Christopher and Stephen Wren (1965). *Parentalia: Or, Memoirs of the Family of the Wrens*. (The "heirloom," or "interleaved" edition.) London: Gregg Press. (Original ed. 1750.)

Wright, Chauncey

Born **Northampton, Massachusetts, USA, 20 September 1830**
Died **Cambridge, Massachusetts, USA, 12 September 1875**

Inspired by reading Herbert Spencer, American philosopher Chauncey Wright imagined an evolutionary cosmogony of the Solar System in which planets form by "meteoric aggregation" and gradually spiral inward toward the Sun.

Selected Reference

Ryan, Frank X. (ed.) (2000). *The Evolutionary Philosophy of Chauncey Wright*. Bristol: Thoemmes Press.

Wright, Thomas

Born **Byers Green near Durham, England, 22 September 1711**
Died **Byers Green near Durham, England, 25 February 1786**

Thomas Wright, the third son of carpenter and yeoman John Wright, was largely self-taught in mathematics, astronomy, and navigation. He made a living in the 1730s by surveying aristocratic estates and teaching public and private courses in the physical sciences. In 1742, Wright declined a position as professor of navigation at the Imperial Academy of Sciences in Saint Petersburg, Russia. Since he never held a formal teaching position, Wright's primary influence came *via* his publications. He never married and was survived by a daughter.

Wright's lifelong preoccupation involved reconciling religious views with astronomical knowledge. Therefore, in order to understand his cosmological speculations, one must not impose modern expectations on Wright's ideas. Through the German philosopher **Immanuel Kant**, Wright became known, though mistakenly, as the originator of the explanation of the Milky Way as a disk-shaped system. In contrast, Wright emphasized the idea that the physical or gravitational center of the Universe must necessarily coincide with the moral or supernatural center. Hence, he insisted on a spherical system. Ironically, Wright's astronomical significance is that he was later credited, by Kant and later writers, as the father of the modern explanation of the Milky Way Galaxy, an idea that he did not develop.

Robinson M. Yost

Selected References

Gushee, Vera (1941). "Thomas Wright of Durham, Astronomer." *Isis* 33: 197–218.

Hoskin, Michael A. (1970). "The Cosmology of Thomas Wright of Durham." *Journal for the History of Astronomy* 1: 44–52. (Reprinted in *Stellar Astronomy: Historical Studies*. Chalfont St. Giles: Science History Publications, 1982.)

——— (1976). "Wright, Thomas." In *Dictionary of Scientific Biography*, edited by Charles Coulston Gillispie. Vol. 14, pp. 518–520. New York: Charles Scribner's Sons.

Hughes, Edward (1951). "The Early Journal of Thomas Wright of Durham." *Annals of Science* 7: 1–24.

Wright, Thomas (1740). *The Use of the Globes: or, The General Doctrine of the Sphere ... To which is added, A Synopsis of the Doctrine of Eclipses ....* London: printed for John Sehex.

——— (1742). *Clavis Coelestis. Being the Explication of a Diagram, Entitled a Synopsis of the Universe: or, The Visible World Epitomized*. London. (London: Dawsons of Pall Mall, 1967 reprint.)

——— (1750). *An original theory or new hypothesis of the universe, founded upon the laws of nature....* London: Macdonald & co, New York, American, Elsevier, 1971.

——— (1968). *Second or Singular Thoughts upon the Theory of the Universe*, Edited from the unpublished manuscript by Michael A. Hoskin. London: Dawsons.

Wright, William Hammond

Born **San Francisco, California, USA, 4 November 1871**
Died **San Jose, California, USA, 16 May 1959**

William Wright distinguished himself as a spectroscopist, planetary photographer, and observatory director in a career that spanned an early period of rapid growth in astrophysics. Wright's spectroscopic studies of novae and planetary nebulae measured many lines not previously detected, traced the evolution of gas shells in novae, and demonstrated the increase in energy levels in planetary nebulae as the central or progenitor star is approached. His photographic studies of the planets, especially Mars, applied new multicolor techniques that revealed characteristics not previously observed.

The son of Seldon Stuart and Joanna Maynard (*née* Shaw) Wright, William earned the BS degree in Civil Engineering at the University of California in 1893 and was a graduate student at the Universities of California and Chicago (the Yerkes Observatory) from 1894 to 1897. He was appointed assistant astronomer at the Lick Observatory in 1897.

In 1903, as a last minute replacement for **William Campbell**, Wright selected the site for and supervised the construction of the Lick Observatory's Southern Station near Santiago, Chile. After establishing the observational and data reduction procedures, he remained in Santiago for 3 highly productive years. Wright returned to the Lick Observatory in 1906 and was promoted to astronomer in 1908.

Some of Wright's early work was on novae; he observed the spectrum of Nova Geminorum 1912 (with Campbell), Nova Ophiuchi 1919, and at least eight other novae up to 1933. His observations highlighted the complexity of the phenomena occurring in such stellar explosions, and traced the evolution of the spectrum of the nova into one similar to that of a gaseous nebula as the event matured. Wright's work on the spectra of novae laid the foundations for our modern understanding of this stage in a star's life.

From 1912 to 1919, Wright also made spectroscopic observations of gaseous nebulae, photographing 70 emission lines in the region between 3,313 Å and 6,730 Å and determining accurate wavelengths for most of them. (Thirty of those lines had not been previously observed, and only a few had had well-determined wavelengths.) He showed that the nuclei of planetary nebulae have spectra like those of Wolf–Rayet stars and that the higher excitation emission lines are more intense in the inner portions of planetary nebulae.

Spectroscopic studies of stars in general became possible only toward the end of the 19th century. Neither gaseous nebulae nor novae were fully understood at the time when Wright began his career. **William Huggins** had observed emission-line spectra of what were then called *diffuse nebulae* in 1864, showing that some of these nebulae were gaseous clouds and not unresolved star clusters. Wright's work, by providing accurate wavelengths for the emission lines, helped to elucidate the physical conditions within these nebulae. His data were used by **Ira Bowen** for the crucial identification of these lines with forbidden transitions among energy levels of ionized oxygen, nitrogen, neon, and other elements.

From 1924 to 1927, Wright photographed the planets in six different colors from 3600 Å to 7600 Å. In his work, which extended further into both the ultraviolet and the red ends of the spectrum than had previously been possible, Wright used special emulsions prepared by the Eastman–Kodak chemist and amateur astronomer Charles Edward Kenneth Mees (1882–1960). Wright claimed the photographs showed the Martian atmosphere to be about 60-miles deep and that the polar caps were partly atmospheric phenomena. However, his conclusions in this regard were subjected to a polite but scathing criticism by **Donald Menzel**, who was in a postdoctoral fellowship at Ohio State University, but later worked with Wright at the Lick Observatory. Wright's composite photograph showing one half of Mars photographed in infrared light and the other half in ultraviolet showed clearly the larger apparent diameter of the ultraviolet image and was a favorite illustration in texts and popular books for some time.

Wright's final research, which he did not live to finish, was to work on using the extragalactic nebulae (as they were then called) as fixed reference points for the system of fundamental astronomical constants. His intent was to develop a reference system completely isolated from the inertial system of the Milky Way Galaxy to permit the most unambiguous possible measurement of stellar proper motions. Development of the project took place over a number of years and involved the design of a special 20-in. widefield astrographic telescope by **Frank Ross**. Construction of the telescope by J. W. Fecker was delayed by World War II, but Wright lived to be present during the first and last exposures in the first series of plates. The second series of plates were not exposed until two decades later, but the program began to make positive contributions in the 1960s, particularly in the form of the Shane-Wirtanen counts of galaxies.

Wright made many other contributions to instrumental design. In particular, when the original Mills spectrograph was replaced with the new Mills, Wright introduced several innovations that were quickly copied by the staffs of other observatories engaged in spectroscopic research with spectrographs at the Cassegrain focus. Principally, these innovations were: support of the spectrograph at both ends, to reduce flexure; enclosure of the spectrograph and its support network in an insulated and thermostatically controlled box to eliminate temperature changes during the night; and a means of impressing the comparison spectrum without interrupting the stellar exposure.

From 1935 to 1942, Wright was the director of the Lick Observatory. During his tenure as director, he is credited with strengthening the staff by recruiting younger astronomers, while at the same time opening up the observatory scientifically by inviting astronomers from other observatories and other nations such as **Ejnar Hertzsprung** and **Polydore Swings** for extended visits. These enhancements enriched observatory life and ensured the continued productivity of the Lick Observatory although it would be several decades before an instrument with an aperture larger than 36-in. would be available on Mount Hamilton.

Wright was elected to the US National Academy of Sciences in 1922 and received the academy's Henry Draper Medal in 1928. In that same year, he received the Janssen Medal of the Paris Academy of Sciences. Wright was elected an associate of the Royal Astronomical Society in 1915 and was awarded that society's Gold Medal in 1938. He received honorary degrees from Northwestern University (DSc 1929) and the University of California (LLD 1944). Wright married Elna Warren Leib on 8 October 1901; they had no children.

Wright's correspondence and personal papers are in the Mary Lee Shane Archives of the Lick Observatory, University of California at Santa Cruz.

Alan H. Batten

Selected References

Menzel, Donald H. (1926). "The Atmosphere of Mars." *Astrophysical Journal* 63: 48–59.

Merrill, Paul W. (1959). "William Hammond Wright, 1871–1959." *Publications of the Astronomical Society of the Pacific* 71: 305–306.

Shane, C. D. (1979). "William Hammond Wright." *Biographical Memoirs, National Academy of Sciences* 50: 377–396.

Spencer Jones, Harold (1938). "Address Delivered by the President, Dr. H. Spencer Jones, on the award of the Gold Medal to Dr. W. H. Wright." *Monthly Notices of the Royal Astronomical Society* 98: 358–374.

Wright, William Hammond (1907–1911). "D. O. Mills Expedition." *Publications of the Lick Observatory* 9. (Wright's account of the Lick southern expedition, with an introduction by W. W. Campbell.)

——— (1918). "The Wave-Lengths of the Nebular Lines and General Observations of the Spectra of the Gaseous Nebulae." *Publications of the Lick Observatory* 13, pt. 6: 191–268.

——— (1928). "On Photographs of the Brighter Planets by Light of Different Colours." *Monthly Notices of the Royal Astronomical Society* 88: 709–718. (George Darwin Lecture, 8 June 1928; Wright discusses his work on planetary photography, with reproductions of many of the photographs.)

——— (1940). "A Preliminary Account of the Spectrum of Nova Ophiuchi (1919)." *Publications of the Lick Observatory* 14, pt. 1: 3–26.

———(1940). "The Spectrum of Nova Geminorum (1912)." *Publications of the Lick Observatory* 14, pt. 2: 27–92.

Wrottesley, John

Born **near Wolverhampton, Staffordshire, England, 5 August 1798**
Died **Wrottesley Hall, Staffordshire, England, 28 July 1867**

John, Lord Wrottesley (Second Baron of Wrottesley) contributed to 19th-century astronomy as a persistent observer and as an effective administrator of scientific organizations. He was educated at Corpus Christi College, Oxford (BA: 1819; MA: 1823), and married Sophia Elizabeth, third daughter of Thomas Gifford in 1821. His primary career was as a lawyer.

From May 1831 to July 1835, Wrottesley observed the right ascensions of stars from the sixth to seventh magnitudes. For this accomplishment he was awarded the Royal Astronomical Society's Gold Medal in 1839. In 1842, Wrottesley began construction of an observatory near his home containing an achromatic refracting telescope of 10-ft. 9-in focal length. He communicated his observations to the Royal Society; his primary research, based upon an earlier suggestion of **John Herschel**, involved the determination of parallax for optical double stars. He later performed research on the statistical calculation of errors related to stellar observations.

Wrottesley perhaps made his greatest contributions to science as an administrator selected by his colleagues. Ten years of service as secretary of the Royal Astronomical Society [RAS] (1831–1841), which overlapped the presidencies of John Herschel, **George Airy**, and others better known than he, were succeeded by the presidency of the RAS (1841), of the Royal Society (1854–1857, to which he was elected in 1841), and of the British Association for the Advancement of Science.

Robinson M. Yost

Selected References

Carlyle, Edward I. (1921–1922). "Wrottesley, Sir John." In *Dictionary of National Biography*, edited by Sir Leslie Stephen and Sir Sidney Lee. Vol. 21, pp. 1082–1083. London: Oxford University Press.

Layton, David (1968). "Lord Wrottesley, F. R. S., Pioneer Statesman of Science." *Notes and Records of the Royal Society of London* 23: 230–246.

Wrottesley, John (1836). "A Catalogue of the Right Ascensions of 1318 Stars Contained in the Astronomical Society's Catalogue, being chiefly those of the 6th and 7th Magnitudes." *Memoirs of the Royal Astronomical Society* 10: 157–234.

——— (1861). "A Catalogue of the Positions and Distances of 398 Double Stars." *Memoirs of the Royal Astronomical Society* 29: 85–168.

Wurm, Karl

Born **Siegen, (Nordrhein-Westfalen), Germany, 21 July 1899**
Died **Rosenheim, Bavaria, (Germany), 16 February 1975**

German astronomer Karl Wurm contributed to measurements of the properties of diffuse gas around hot stars, particularly the measurement of the density of the shells around Be Stars, and of the temperatures of the stars at the centers of planetary nebulae, and to the analysis of the emission line spectra of comets.

Wurm studied at Bonn University between 1921 and 1927, receiving his Ph.D. for work with R. Mecke on problems of molecular physics. He spent the next 14 years at the astrophysical observatory at Potsdam, with an interval (1938/1939) as visiting professor at the University of Chicago. His work there with **Otto Struve** on helium lines in the spectra of gaseous nebulae was his most cited work.

In 1941 Wurm went to the Bergedorf Observatory near Hamburg as deputy observer, was promoted in 1943 to observer, and began lecturing at the Hamburg University in 1946. In 1950 he went to the Humboldt University in Berlin as visiting professor but returned to Hamburg as head observer after only a year. From 1961 onward he visited at Mount Hamilton as a Morrison Research Associate of the Lick Observatory.

Between 1958 and 1964 Wurm was president of the International Astronomical Union Commission 15, Physical Studies of Comets and Minor Planets. In 1954 he began a collaboration with the Astrophysical Observatory of Asiago, the University of Padova, which lasted until his death.

The list of Wurm's papers and articles contains several with spectroscopic topics, including work on planetary nebulae, comets, and stellar atmospheres. Beside contributions to textbooks (*e. g., Handbuch der Astrophysik*, Berlin 1930) he wrote two books, one on planetary nebulae and another on comets. In addition he published a *Monocromatic Atlas of the Orion Nebula* (26 sheets) with a later supplement of 20 sheets. Minor planet (1785) Wurm is named in his honor.

Christof A. Plicht

Selected References

Anon. (1930). *Handbuch der Astrophysik*. Vol. 3, pt. 2, *Grundlagen der Astrophysik*, edited by Walter Grotian *et al.* Berlin: J. Springer.

Rosino, L. (1975). "In Memoriam." *Astrophysics and Space Science* 35: 221–222.

Wurm, Karl (1951). *Die planetarischen Nebel*. Berlin: Akademie-Verlag.

——— (1954). *Die Kometen*. Berlin: Springer.

Wyse, Arthur Bambridge

Born **Blairstown, Ohio, USA, 25 June 1909**
Died **over the Atlantic Ocean, off the New Jersey coast, 8 June 1942**

Arthur Wyse was best known for his analysis of the spectra of novae, especially Nova Aquilae 1918, which, he demonstrated, expanded at a nearly constant rate for more than 20 years. Wyse was the son of Reverend Charles and Celia Wyse, and received his degrees from the College of Wooster (AB: 1918), the University of Michigan (AM: 1931), and the University of California, Berkeley (Ph.D.: 1934). His doctoral thesis, under **Heber Curtis** at the Lick Observatory, was a study of the light curves of eclipsing binaries. He briefly held a Martin Kellogg Fellowship at Rochester University after completing his Ph.D.

In l935, Wyse was one of the first two new staff appointments made at Lick by **William Wright** upon succeeding **Robert Aitken** as director. In 1938/1939 Wyse collaborated with **Ira Bowen**, then visiting the Lick Observatory, on the spectra of planetary and gaseous nebulae. Wyse continued this research on his own until late 1941, when he joined the US Naval Reserve as a lieutenant.

Although Wyse was primarily an observational astronomer, he was also an able theoretician, as shown by his analysis of limb darkening in eclipsing binaries, and of the mass distribution and dynamics of the spiral galaxies M31 and M33 (the latter with **Nicholas Mayall**). But his best-known work was his study of the spectra of novae, including v603 Aql (Nova Aquilae 1918), HR Lyrae (1919), and EL Aql (1927), using archival Lick plates and v368 Aql (1936), using his own observations. His work on planetary and gaseous nebulae showed that they have similar compositions to each other, and probably also to those of the Sun and stars. He published a notable catalog of 270 emission lines observed in the spectra of 10 nebulae in 1942. In addition to the collaborations mentioned earlier, he guided **Daniel Popper** through his Lick thesis on spectrophotometry of CP Lac (Nova Lacertae 1936).

Wyse died as the result of an airship accident while on leave-of-absence from the Lick Observatory in naval service during World War II. As a result, Lick lost a very able astronomer, with wide interests in astrophysics, in the early stages of his scientific career.

John Hearnshaw

Selected References

Bowen, I. S. and Arthur B. Wyse (1939). "The Spectra and Chemical Composition of the Gaseous Nebulae NGC 6572, 7027, 7662." *Lick Observatory Bulletin* 19, no. 495: 1–16.

Moore, J. H. (1942). "Arthur Bambridge Wyse, 1909–1942." *Publications of the Astronomical Society of the Pacific* 54: 171–175.

Wright, W. H. (1939). "Arthur Bambridge Wyse, 1909–1942." *Astrophysical Journal* 97: 89–92.

Wyse, Arthur B. (1939). "An Application of the Method of Least Squares to the Determination of the Photometric Elements of Eclipsing Binaries." *Lick Observatory Bulletin* 19, no. 496: 17–27.

——— (1942). "The Spectra of Ten Gaseous Nebulae." *Astrophysical Journal* 95: 356–385.

Wyse, Arthur B. and N. U. Mayall (1942). "Distribution of Mass in the Spiral Nebulae Messier 31 and Messier 33." *Astrophysical Journal* 95: 24–47.

Xenophanes of Colophon

Born **Colophon (near Selçuk, Turkey), *circa* 571 BCE**
Died **possibly (Sicily, Italy), *circa* 475 BCE**

Xenophanes' primary contribution to astronomy was in cosmology, and he is often remembered more as a theologian.

The dates of Xenophanes' life, particularly his birth, should not be taken as exact. There are some scholars who suggest an earlier date of 580 BCE for his birth. He was said to be the son of Dexias, and is considered the founder of the Eleatic School. It is known that early in his life he left Ionia for Sicily when the Persians took over Colophon. In Sicily, around 545 BCE, he worked at the court of Hiero, possibly as a wandering poet. He later departed for Magna Graecia (the Greek colonies in southern Italy), where he took up the profession of philosophy. Some scholarly sources suggest that he became an eminent Pythagorean scholar (though some dispute this), and thus it is likely that he spent some time in Crotona (Croton or Crotone), where **Pythagoras** had founded his religious school. It is also possible that he spent time in Siris, which had been colonized by Greeks from Colophon, his birthplace. Both Siris and Crotona were in Magna Graecia. Some (Robinson, 1968) suggest that he remained in Sicily until he died, but this is unlikely given his predilection for wandering as a poet. It is more likely that he died in Elea, based on the extent of the literature. His relationship with Pythagoras is something that is still not settled. Robinson claims Xenophanes predated Pythagoras in his philosophy (though the two were nearly the same age), but less reliable sources indicate otherwise.

Xenophanes made several important contributions to early physical and astronomical theories in addition to cosmology. He contributed to an understanding of the Earth in that he recognized that water was cycled from the sea to the clouds and from the clouds into rain, which cycled back to the sea. He also suggested a theory of the Sun, saying that the Sun actually came into being each day from small pieces of fire collected together. The Earth, Xenophanes said, was infinite and was not enclosed by air or by the heavens. He also said that there were innumerable suns and moons and that everything was made of earth. This last assertion does leave one to wonder whether he meant everything found on Earth, or if he included the heavenly bodies in the Earth; for, in the same passage, he mentions the development of the Sun from fire. He later said that the Sun and stars come from the clouds and that the Sun is actually made of ignited clouds. Rainbows also were supposedly made of clouds. In relation to his concept that there existed many suns and moons, Xenophanes developed a very abstract theory of eclipses.

Ian T. Durham

Selected References

Fairbanks, Arthur (1898). *The First Philosophers of Greece*. London: Kegan Paul, Trench, Trübner and Co.

Kirk, G. S. and J. E. Raven (1957). *The Presocratic Philosophers*. Cambridge: Cambridge University Press.

Robinson, J. M. (1968). *An Introduction to Early Greek Philosophy*. Boston: Houghton-Mifflin.

Ximenes, Leonardo

Born **Trapani, (Sicily, Italy), 1716**
Died **Florence, (Italy), 1786**

Beginning in 1756, Sicilian Jesuit Leonardo Ximenes used a wall hole in the Duomo (Cathedral) of Florence to project the Sun onto the marble floor below. Doing so, he experimentally determined the rate-of-change for the obliquity of the ecliptic.

Selected Reference

Heilbron, J. L. (1999). *The Sun in the Church*. Cambridge, Massachusetts: Harvard University Press.

Y

Yaḥyā ibn Abī Manṣūr: Abū ʿAlī Yaḥyā ibn Abī Manṣūr al-Munajjim

Flourished **Baghdad, (Iraq), *circa* 820**
Died **near Aleppo, (Syria), 830**

Yaḥyā ibn Abī Manṣūr was the senior astronomer/astrologer at the court of the ʿAbbāsid caliph **Ma'mūn**. He is well-known for his leading role in the earliest systematic astronomical observations in the Islamic world, which were carried out in Baghdad in 828–829, and for the astronomical handbook, *al-Zīj al-mumtaḥan*, that was written on the basis of these observations.

Yaḥyā was of Persian descent and originally named Bizīst, son of Fīrūzān. Since his father, Abū Manṣūr Abān, was an astrologer in the service of the second ʿAbbāsid caliph al-Manṣūr (754–775), we may assume that Yaḥyā spent his youth in Baghdad. His first known position was as an astrologer for al-Faḍl ibn Sahl, vizier of the Caliph Ma'mūn. After al-Faḍl was assassinated in February 818, Yaḥyā converted to Islam and adopted his Arabic name. He became a boon companion (Arabic: *nadīm*) of Ma'mūn, and is known to have made astrological predictions for the caliph on various occasions. He was also associated with the House of Wisdom and is mentioned as a teacher of the **Banū Mūsā**.

Ma'mūn strongly supported scientific activities, including the translation of Greek and Syriac scientific works into Arabic. In 828 and 829, he ordered astronomical observations to be carried out in the Shammāsiyya quarter of Baghdad with the purpose of verifying the parameters of the astronomical models of **Ptolemy** as found in his *Almagest* and *Handy Tables*. Yaḥyā became one of the most important persons involved in these observations together with **Jawharī**, **Sanad ibn ʿAlī**, and **Marwarrūdhī**.

The observational activities at Baghdad did not last for more than one and a half years. In that period basic observations of the Sun and the Moon were made, but a determination of all planetary parameters was not possible. Some specific values that were found are: 23° 33′ for the obliquity of the ecliptic (encountered only in the works of Yaḥyā and incidentally in those of his later contemporary **Ḥabash al-Ḥāsib**); a precession of the equinoxes of 1° in 66 Persian years (which may, however, have been influenced by Sasanian–Iranian measurements); a maximum solar equation of 1° 59′; and a maximum equation of center for Venus of 1° 59′. All four results constituted major improvements upon Ptolemy's outdated or incorrect values.

Yaḥya's name is associated with an astronomical handbook with tables dedicated to Ma'mūn. This work is known as *al-Zīj al-Ma'mūnī* or, more commonly, *al-Zīj al-mumtaḥan*, that is the *Verified Zīj* (Latin *Tabulae probatae*). A late recension of the *Zīj* is extant in the manuscript Escorial árabe 927, which contains, besides original material from Yaḥyā, numerous chapters, treatises, and tables of later date. In particular, we find material from the important 10th-century astronomers **Ibn al-Aʿlam**, **Būzjānī**, and **Kūshyār ibn Labbān**. Furthermore, there are various tables specifically intended for a geographical latitude of 36°, which corresponds to Mosul rather than to Baghdad. In 2004 the manuscript Leipzig Vollers 821 was recognized to be a recension of the *Mumtaḥan Zīj*. In some respects it is similar to the one in the Escorial library, but with fewer later additions. This copy has various insertions originating from **Battānī** and was apparently used in present-day southeastern Turkey.

Among the materials in the Escorial manuscript explicitly attributed to Yaḥyā are the tables for the lunar equation and the theory of solar eclipses. The latter is a typical mixture of Indian, Sasanian, and Hellenistic influences. The Ptolemaic table for the solar equation, which is also found in Ḥabash's *zīj*, may not be original, since a table of a more primitive nature is attributed to Yaḥyā in the 14th-century *Ashrafī Zīj*. Whereas the planetary equations were directly copied from the *Handy Tables*, the tables for the latitudes of the Moon and the planets are of a simple sinusoidal type and based on otherwise unknown parameters. A table with longitudes and latitudes of 24 fixed stars is indicated to be for the year 829 and derived from the observations made at Shammāsiyya.

It is not known with certainty whether the original *Mumtaḥan Zīj* was a work by Yaḥyā alone or a coproduction of the group of astronomers who were involved in the observations carried out on the order of Ma'mūn and who were referred to as *aṣḥāb al-mumtaḥan*, "authors of the verified (tables)." It is also possible that various of these astronomers wrote their own works with the title *Mumtaḥan Zīj*. Similarly, it is unclear what Ibn al-Nadīm (10th century), the earliest important biographer of Muslim scholars, meant by a "first" and "second" "copy" (Arabic: *nuskha*) of the work. In any case, the *Mumtaḥan Zīj* was very well-known and frequently quoted. **Thābit ibn Qurra** (second half of the 9th century) wrote a treatise on the differences between the *Mumtaḥan Zīj* and Ptolemy's astronomical tables, which is unfortunately lost.

Very little is known about other works by Yaḥyā. Ibn al-Nadīm mentions a *Maqāla fī ʿamal irtifāʿ suds sāʿa li-ʿarḍ Madīnat al-Salām* (Treatise on the determination of the altitude of [each] sixth of an hour for the latitude of Baghdad), as well as a *Kitābun yaḥtawī ʿalā arṣād lahu* (Book containing his observations) and *Rasāʾil ilā jamāʿa fī al-arṣād* (Letters to colleagues concerning observations). A small astrological work by Yaḥyā entitled *Kitāb al-rujūʿ wa-'l-hubūṭ* (Book on retrogradation and descent) is extant in the very late manuscript 173 of Kandilli Observatory in Istanbul. It appears that Yaḥyā was also involved in the measurement of 1° on the meridian that was carried out on the order of Ma'mūn in the Sinjār plain (in northern Iraq). On the other hand, both the book *Fī al-ibāna ʿan al-falak* and a set of measurements of the obliquity made at Marv (mentioned by **Bīrūnī** in his geographical masterwork *Taḥdīd*) have been incorrectly attributed to Yaḥyā by modern authors; in fact, they are associated with the Tahirid Governor of Khurāsān, Manṣūr ibn Ṭalḥa (*circa* 870).

Yaḥyā died in the early summer of 830 during the first of Ma'mūn's expeditions against Tarsus in Asia Minor. He was buried in Aleppo, where his tomb could still be seen in the 13th century. Thus the astronomical observations carried out during the years 831 and 832 at the monastery of Dayr Murrān on Mount Qāsiyūn near Damascus and headed by Marwarrūdhī took place after Yaḥyā's death. A number of Yaḥyā's descendants were also boon companions of the ʿAbbāsid caliphs and well-known scholars. One of his four sons, Abū al-Ḥasan ʿAlī (died: 888), collected a huge library for al-Fatḥ ibn Khāqān, secretary of caliph al-Mutawakkil (847–861), where, among others, the famous astrologer **Abū Maʿshar** is known to have studied. Yaḥyā's grandson Yaḥyā ibn ʿAlī was a famous theorist of music. His great-great-grandson Hārūn ibn ʿAlī (died: 987) was an able astronomer and likewise author of a *zīj*.

Benno van Dalen

Selected References

Al-Qifṭī, Jamāl al-Dīn (1903). *Ta'rīkh al-ḥukamā'*, edited by J. Lippert. Leipzig: Theodor Weicher.

Dalen, Benno van (1994). "A Table for the True Solar Longitude in the Jāmiʿ Zīj." In *Ad Radices: Festband zum fünfzigjährigen Bestehen des Instituts für Geschichte der Naturwissenschaften der Johann Wolfgang Goethe-Universität Frankfurt am Main*, edited by Anton von Gotstedter, pp. 171–190. Stuttgart: Franz Steiner.

_____ (2004). "A Second Manuscript of the *Mumtaḥan Zīj*." *Suhayl* 4: 9–44.

Fleischhammer, M. (1993). "Munadjdjim, Banu 'l-." In *Encyclopaedia of Islam*. 2nd ed. Vol. 7, pp. 558–561. Leiden: E. J. Brill.

Ibn al-Nadīm (1970). *The Fihrist of al-Nadīm: A Tenth-Century Survey of Muslim Culture*, edited and translated by Bayard Dodge. 2 Vols. New York: Columbia University Press. (This and the biographical dictionaries of Ibn Khallikān and Ibn al-Qifṭī provide all our information on Yaḥyā's life and relatives.)

Kennedy, E. S. (1956). "A Survey of Islamic Astronomical Tables." *Transactions of the American Philosophical Society*, n.s., 46, pt. 2: 121–177, esp. 132 and 145–147. (Reprint, Philadelphia: American Philosophical Society, 1989.)

_____ (1977). "The Solar Equation in the Zīj of Yaḥyā b. Abī Manṣūr." In *ΠΡΙΣΜΑΤΑ* (Prismata). *Naturwissenschaftsgeschichtliche Studien: Festschrift für Willy Hartner*, edited by Y. Maeyama and W. G. Saltzer, pp. 183–186. Wiesbaden: Franz Steiner. (Reprinted in E. S. Kennedy, *et al.*, *Studies*, pp. 136–139.)

Kennedy, E. S., *et al.* (1983). *Studies in the Islamic Exact Sciences*, edited by David A. King and Mary Helen Kennedy. Beirut: American University of Beirut.

Kennedy, E. S. and Nazim, Faris (1970). "The Solar Eclipse Technique of Yaḥyā b. Abī Manṣūr." *Journal for the History of Astronomy* 1: 20–38. (Reprinted in E. S. Kennedy, *et al.*, *Studies*, pp. 185–203.)

King, David A. (2000). "Too Many Cooks ... A New Account of the Earliest Muslim Geodetic Measurements." *Suhayl* 1: 207–241.

Salam, Hala and E. S. Kennedy (1967). "Solar and Lunar Tables in Early Islamic Astronomy." *Journal of the American Oriental Society* 87: 492–497. (Reprinted in E. S. Kennedy, *et al.*, *Studies*, pp. 108–113.)

Sayılı, Aydın (1960). *The Observatory in Islam*. Ankara: Turkish Historical Society, esp. pp. 50–87.

Sezgin, Fuat. *Geschichte des arabischen Schrifttums*. Vol.4, *Mathematik* (1974): 227; Vol. 6, *Astronomie* (1978): 136–137; Vol. 7, *Astrologie–Meteorologie und Verwandtes* (1979): 116. Leiden: E. J. Brill.

_____ (ed.) (1986). *The Verified Astronomical Tables for the Caliph al-Ma'mūn. Al-Zīj al-Ma'mūnī al-mumtaḥan by Yaḥyā ibn Abī Manṣūr*. Frankfurt am Main: Institute for the History of Arabic–Islamic Science. (Facsimile of the unique manuscript of Yaḥyā's *zīj*.)

Vernet, Juan (1956). "Las 'Tabulae Probatae.'" In *Homenaje a Millás-Vallicrosa*. Vol. 2, pp. 501–522. Barcelona: Consejo Superior de Investigaciones Científicas. (Reprinted in Vernet, *Estudios sobre historia de la ciencia medieval*, pp. 191–212. Barcelona: Universidad de Barcelona, 1979.)

_____ (1976). "Yaḥyā ibn Abī Manṣūr." In *Dictionary of Scientific Biography*, edited by Charles Coulston Gillispie, Vol. 14, pp. 537–538. New York: Charles Scribner's Sons.

Viladrich, Mercè (1988). "The Planetary Latitude Tables in the *Mumtaḥan Zīj*." *Journal for the History of Astronomy* 19: 257–268.

Yaʿqūb ibn Ṭāriq

Flourished **Baghdad, (Iraq), 8th to 9th century**

Yaʿqūb ibn Ṭāriq is known as a contemporary and collaborator of the 8th-century scholars in Baghdad (particularly **Fazārī**) who developed from Greek, Indian, and Iranian sources the basic structure of Arabic astronomy. Works ascribed by later authors to Yaʿqūb include the *Zīj maḥlūl fī al-Sindhind li-daraja daraja* (Astronomical tables in the *Sindhind* resolved for each degree), *Tarkīb al-aflāk* (Arrangement of the orbs), and *Kitāb al-ʿilal* (Rationales [of astronomical procedures]). He is also said to have written a *Taqṭīʿ kardajāt al-jayb* (Distribution of the *kardaja*s of the sine [sine values]), and *Mā irtafaʿa min qaws niṣf al-nahār* (Elevation along the arc of the meridian), which may be related to or incorporated within one of his more general works. An otherwise unknown astrological work entitled *Al-maqālāt* (Chapters) is also attributed to Yaʿqūb by one (unreliable) source. None of the above works is now extant, and only the first three are known in any detail from later writings.

Yaʿqūb's *zīj* (handbook with astronomical tables), like that of Fazārī, was apparently based on the Sanskrit original of the *Zīj al-Sindhind*, translated by them in Baghdad in the 770s. (A highly embroidered 12th-century account of Yaʿqūb's involvement in this translation is given by **Abraham ibn ʿEzra**.) Also like Fazārī's, the surviving fragments of Yaʿqūb's *zīj* are a heterogeneous mix from different traditions. For example, the mean motion parameters are Indian, as is the rule for visibility of the lunar crescent; the calendar is Persian;

and the Indian sunrise epoch for the civil day appears to have been converted to the Greek-inspired noon epoch by the simple expedient of moving the prime meridian 90° (or 1/4th day) eastward from the usual location of Arin (Ujjain).

The *Tarkīb al-aflāk* was an early work on the topic that became known as *hay'a* or cosmography (*i. e.*, the arrangement, sizes, and distances of the celestial orbs). Yaʿqūb's work apparently stated the orbital radii and sizes of the planets, as well as rules for determining accumulated time according to techniques in Sanskrit treatises. **Bīrūnī** in the 11th century mentioned the *Tarkīb* as the only Arabic source using the Indian cosmographic tradition (although at least some of the same values were known from other *zījes*); if his descriptions of some of Yaʿqūb's rules are accurate, Yaʿqūb did not always fully understand or correctly interpret the Indian procedures.

It is also from Bīrūnī that we derive our knowledge of the *Kitāb al-ʿilal*, an early representative of the genre of "rationales" or "causes" treatises that undertook to provide mathematical explanations of the computational rules in *zījes*. All of Bīrūnī's references to this work are contained in his *al-Ẓilāl* (On shadows), so they are limited to trigonometric procedures using gnomon shadows in calculations of time and location. By this time, evidently, Yaʿqūb's works were valued primarily for the information they provided about early influences from the Indian tradition, many of which were replaced in later Islamic astronomy by predominantly Ptolemaic techniques.

Kim Plofker

Selected References

Hogendijk, Jan P. (1988). "New Light on the Lunar Visibility Table of Yaʿqub ibn Ṭāriq." *Journal of Near Eastern Studies* 47: 95–104.

Kennedy, E. S. (1968). "The Lunar Visibility Theory of Yaʿqūb ibn Ṭāriq." *Journal of Near Eastern Studies* 27: 126–132.

Pingree, David (1968). "The Fragments of the Works of Yaʿqūb ibn Ṭāriq." *Journal of Near Eastern Studies* 27: 97–125.

______ (1976). "Yaʿqūb ibn Ṭāriq." In *Dictionary of Scientific Biography*, edited by Charles Coulston Gillispie. Vol. 14, p. 546. New York: Charles Scribner's Sons.

Sezgin, Fuat (1978). *Geschichte des arabischen Schrifttums*. Vol. 6, *Astronomie*, pp. 124–127. Leiden: E. J. Brill.

Yasuaki

➔ **Asada, Goryu**

Yativṛṣabha

Flourished **Prākrit, Jadivasaha, (India), 6th century**

Little is known about Yativṛṣabha. He was a Jain monk who studied under Ārya Maṅkṣu and Nāgahastin. He composed, along with other traditional Jain works, the *Tiloyapaṇṇattī* (in Sanskrit, *Trilokaprajñapti* or Knowledge on the three worlds), a work on Jain cosmography. This work describes the construction of the Universe expressed in specific numbers; for example, the diameter of the circular Jambu continent, upon which India is located, is 100,000 yojanas and its circumference is 316,227 yojanas, 3 krośas, 128 daṇḍas, 13 aṅgulas, 5 yavas, 1 yūkā, 1 ṛikṣā, 6 karmabhūmivālagras, 7 madhyabhogabhūmivālagras, 5 uttama bhogabhūmivālagras, 1 ratharenu, 3 trasareṇus, 2 sannāsannas, and 3 avasannāsannas, plus a remainder of 23213/105409. Yativṛṣabha also gives formulas for computing the circumference (C) and the area (A) of a circle having a diameter of d:

$$C=\sqrt{10d^2},\ A=C\cdot\frac{d}{4}$$

Setsuro Ikeyama

Selected References

Hayashi, Takao (1993). *Indo no Sūgaku* (Mathematics in India). Chūkō-shinsho 1155. Tokyo: Chūōkōron-sha.

Pingree, David (1976). "Yativṛṣabha." In *Dictionary of Scientific Biography*, edited by Charles Coulston Gillispie, Vol. 14, pp. 548–549. New York: Charles Scribner's Sons.

______ (1994). *Census of the Exact Sciences in Sanskrit*. Series A. Vol. 5, pp. 319a–320b. Philadelphia: American Philosophical Society.

Yativṛṣabha (1943–1951). *Tiloyapaḥṇattī*, edited by H. Jaina and A. N. Upādhyāya. Sholapura (2nd ed. of Vol. 1, Sholapur, 1956). New edition by V. Mātājī (Vols. 1 and 2 covering the first four chapters have appeared), Koṭā, 1984–1986.

Yavaneśvara

Flourished **(western India), 149/150**

Yavaneśvara translated a Greek astrological text (probably composed in Alexandria in the 1st half of the 2nd century BCE) into Sanskrit prose in 149/150 at Ujjayinī, the capital of the Western Kṣatrapas, during the reign of Rudradāman I. (Yavaneśvara, literally "lord of the Greeks," was probably a title for leaders of Greek merchants in Western India, *circa* 78–390, and not a proper name.) This translation, which is no longer extant, was versified and titled *Yavanajātaka* by **Sphujidhvaja** in 269/270. Verse 61 of Chapter 79 of this work runs:

> Yavaneśvara, who sees the truth coming from the brightness of the sun and speaks unblamable words, conveyed this treatise on horoscopy for the local authority in primitive words.

The work of Yavaneśvara became one of the major sources for Indian horoscopy.

Setsuro Ikeyama

Selected References

Pingree, David (1976). "Yavaneśvara." In *Dictionary of Scientific Biography*, edited by Charles Coulston Gillispie. Vol. 14, p. 549. New York: Charles Scribner's Sons.

______ (1981). *Jyotiḥśāstra*. Wiesbaden: Otto Harrassowitz, pp. 89, 109.

______ (1994). *Census of the Exact Sciences in Sanskrit*. Series A. Vol. 5, p. 330b. Philadelphia: American Philosophical Society.
______ (ed. and trans.) (1978). *The Yavanajātaka of Sphujidhvaja*. 2 Vols. Harvard Oriental Series, Vol. 48. Cambridge, Massachusetts: Harvard University Press.

Yixing

Born **Changle (Nanle, Henan), China or Julu (Hebei), China, 683**
Died **(Shaanxi), China, 727**

Yixing was a Chinese Buddhist monk and astronomer during the Tang dynasty. Yixing was his Buddhist name; his secular name was Zhang Sui.

In 717, Yixing received a call from Emperor Xuanzong, and he moved to Chang'an, then the capital. In 721, at the emperor's request Yixing started a project to make a new calendar. Yixing made an armillary sphere with his colleague Liang Lingzan around 724. From 724 onward Yixing conducted astronomical observations at several places all over China with his colleague Nangong Yue. In 725, Yixing made a water-driven celestial globe with Liang Lingzan. After these preparations, Yixing started to compile the new calendar, and completed the draft of the Dayan calendar in 727. As Yixing died the same year, Zhang Shui and Chen Xuanjing edited Yixing's draft, and the Dayan calendar was officially promulgated after 729.

During the Sui (581–618) and Tang (618–907) dynasties, several calendars were constructed. The Huangji calendar (600) of Liu Zhuo (544–610) was not officially used, but was an excellent calendar in which the inequalities corresponding to the equations of the centers of the Sun and the Moon and the precession of the equinoxes were all considered. In it second-order interpolation was used for the first time in China. The Linde calendar (665) of **Li Chunfeng** (602–670) is another well-known calendar of the time. (Li Chunfeng is also famous for his armillary sphere.) The Dayan calendar (727) of Yixing was one of the best calendars of the Tang dynasty. The Xuanming calendar (822) of Xu Ang is another famous one, in which the method of prediction of eclipses was improved. The Chongxuan calendar (892) of Bian Gang deserves note as well.

The Tang dynasty was the period when Indian astronomy was introduced to China. Some information on Indian astronomy might have reached China during the later Han dynasty. A Buddhist text containing knowledge of Indian astrology and astronomy, the *Śārdūlakarṇa-avadāna*, was translated into Chinese in the 3rd century during the Three Kingdoms period. During the Tang dynasty, a detailed work of Indian mathematical astronomy, the *Jiuzhi li* (Jiuzhi calendar; 718), was composed in Chinese by the Indian astronomer (resident in China since his grandfather's time) Qutan Xida (Chinese transliteration of Gotama-siddartha in Sanskrit), and was included in his ([*Da*]*Tang*) *Kaiyuan zhanjing*. In the 8th century, a Chinese version of Indian astrology, the *Xiuyao jing*, was composed in Chinese by Bukong, an Indian monk (whose Sanskrit name was Amoghavajra; 705–774). Amoghavajra was a disciple of Vajrabodhi, with whom Yixing also studied. Yixing certainly had knowledge of Indian astronomy, but made his Dayan calendar in Chinese traditional style.

Yixing and Liang Lingzan made an armillary sphere called *Huangdao youyi* (Instrument with a movable ecliptic circle) around 724. In this instrument, the ecliptic circle could be moved in accordance with the precession of the equinoxes. It also had a movable circle for the lunar orbit. With this instrument, Yixing observed stars, particularly the 28 lunar mansions, and (comparing with previous observations) measured the change of their polar distance and right ascension (*i. e.*, the change due to the precession of the equinoxes). Yixing and Liang Lingzan also made a water-driven celestial globe in 725. Besides the celestial globe itself, the device had two wooden figures that struck a drum and gong automatically.

From 724 to 725, Yixing and Nangong Yue conducted astronomical observations at 13 different places from about 51° N to about 18° N. They observed the altitude of the North Celestial Pole, the length of the gnomon shadow at solstices and equinoxes, and the length of daytime and nighttime at solstices.

In regard to astronomical theory, the Dayan calendar of Yixing is one of the best calendars from China. It has several features of significance: For example, the inequality corresponding to the equation of the center of the Sun was discovered by **Zhang Sixun** in the sixth century for the first time in China. For this inequality, Yixing gave the values for 24 seasonal nodes in a year, which were divided according to the Sun's angular movement. Here, Yixing used second-order interpolation with unequal steps of argument for the first time in China. For the inequality corresponding to the equation of the center of the Moon, which was discovered during the later Han dynasty in the 1st century, Yixing used the second-order interpolation with equal steps of argument invented by Liu Zhuo (542–608) during the Sui dynasty.

An attempt to predict lunar eclipses was first made in the Santong calendar of Liu Xin (died: 23) at the end of the former Han dynasty, and the basis of the standard system of the prediction of solar and lunar eclipses was established in the Jingchu calendar (237) of Yang Wei in the Three Kingdoms period. For the prediction of solar eclipses, Yixing considered the lunar parallax at different places. Although his method was not perfect, it was a big step forward. The method of predicting eclipses was further developed in the Xuanming calendar (822) of Xu Ang.

Yixing also improved the calculation of the positions of the five planets, and used a type of interpolation in which the third difference is used, although it was not interpolation of the third order.

Another Yixing contribution was a device to calculate the length of the gnomon shadow and the length of daytime and nighttime in different seasons at different places. For this purpose, Yixing made a table of the gnomon shadow for every Chinese degree (*du*), from 0 to 81, of the Sun's zenith distance. (One Chinese *du* is the angular distance on the celestial sphere through which the Sun moves in one day.) This Yixing table is the earliest tangent table in the world.

For the transformation of spherical coordinates, the graphical method on the celestial globe had been used since the later Han dynasty. An arithmetical method was started in the Huanji calendar (600) of Liuzhuo, and Yixing also used the arithmetical method. In this method, the difference between right ascension and polar longitude (longitude of the requisite hour circle on the ecliptic) was assumed to be a linear function in a quadrant, and the difference was given by a table.

The Dayan calendar of Yixing was introduced to Japan, and was officially used there from 746 to 857.

Alternate names

I-Hsing
Seng Yixing
Yixing Chanshi

Selected References

Anon. *Jiu Tang shu* (Old version of the standard history of the Tang dynasty). (The official biography of Yixing is included in the section of biographies of technicians, "*Fangji zhuan*," chap. 191.)

Beer, A. *et al.* (1961). "An 8th-Century Meridian Line: I-HSING's Chain of Gnomons and the Pre-history of the Metric System." *Vistas in Astronomy* 4: 3–28. (On Yixing and his colleagues' astronomical observations.)

Chen Meidong (1992). "Yixing." In *Zhongguo gudai kexue jia zhuanji* (Biographies of scientists in ancient China), edited by Du Shiran, Vol. 1, pp. 360–372. Beijing: Kexue chubanshe (Science Publishing House).

______ (1998). "Yixing." In *Zhongguo kexue jishu shi, Renwu juan* (A history of science and technology in China, biographical volume), edited by Jin Qiupeng, pp. 278–294. Beijing: Kexue chubanshe (Science Publishing House).

______ (1995). *Guli xintan* (New research on old calendars). Shenyang: Liaoning jiaoyu chubanshe (Liaoning Educational Publishing House).

Kūkai (Kōbō-daishi). *Shingon-fuhō-den (Ryaku-fuhō-den)* (Biographies of the successors of the Mantra sect of Buddhism, in classical Chinese). 821 AD. (For an epitaph of Yixing written by Emperor Xuanzong, [reign: 712–756]).

Li Di (1964). *Tang dai tianwenxue jia Zhang Sui (Yixing)* (Astronomer Zhang Sui [Yixing] of the Tang dynasty). Shanghai: Shanghai renmin chubanshe (People's Publishing House of Shanghai).

Liu Xu *et al.* (eds.) *Jiu Tang shu* (Old history of the Tang dynasty). (Reprint, Beijing: Zhonghua shuju, 1974.) (See Chap. 34 for the system of the Yixing's Dayan calendar.)

Needham, Joseph, with the collaboration of Wang Ling (1959). *Science and Civilisation in China*. Vol. 3, *Mathematics and the Sciences of the Heavens and the Earth*. Cambridge: Cambridge University Press. (For Chinese astronomy in general, including the contribution of Yixing.)

Ôhashi, Yukio (1994). "Zui-Tō jidai no hokan-hō no sanjutsu-teki kigen." (Arithmetical origin of Chinese interpolation of Sui and Tang periods, in Japanese.) *Kagakusi Kenkyu (Journal of History of Science, Japan)*, ser. 2, 33, 189: 15–24.

______ (1995). "Daien-reki no hokan-hō nit suite." (On the Interpolation used in the Dayan calendar, in Japanese.) *Kagakusi Kenkyu (Journal of History of Science)*, ser. 2, 34, 195: 170–176.

Ouyang Xium *et al.* (eds.) (1060 AD). *Xin Tang shu* (New history of the Tang dynasty). (Reprint, Beijing: Zhonghua shuju, 1974.) (See Chaps. 27–28 for the system of the Yixing's Dayan calendar. The *Xin Tang shu* also contains Yixing's theoretical exposition of the Dayan calendar. This is an important treatise on astronomy.)

Qu Anjing (1997). "Dayan li zhengqe hanshu biao de chonggou." (The reconstruction of Yixing's tangent table). *Ziran Kexueshi Yanjiu* (*Studies in the History of Natural Sciences*) 16, no. 3: 233–244. (A Japanese translation is in *Sūgakushi Kenkyū* [*Journal of History of Mathematics*] 153 [1997]: 18–29. An important paper on the tangent table of Yixing.)

Qu Anjing, Ji Zhigang, and Wang Rongbin (1994). *Zhongguo gudai shuli tianwenxue tanxi* (Research on mathematical astronomy in ancient China). Xi'an: Xibei daxue chubanshe (Northwest University Press). (Informative work on classical Chinese calendars, including the Dayan calendar.)

Ruan Yuan (1799). *Chouren zhuan* (Biographies of astronomers). (Reprint, Taipei: Shijie shuju, 1962.) (For classical accounts of Yixing, Chaps. 14–16.)

Swetz, Frank J. and Ang Tian Se (1984). "A Brief Chronological and Bibliographic Guide to the History of Chinese Mathematics." *Historia Mathematica* 11: 39–56.

Wang Yingwei (1998). *Zhongguo guli tongjie* (Expositions of Chinese old calendars). Shenyang: Liaoning jiaoyu chubanshe (Liaoning Educational Publishing House). (For a detailed commentary on Yixing's Dayan calendar.)

Zanning (919–1001) *Song gaoseng zhuan* (Biographies of eminent Buddhist monks compiled in the Song dynasty). (For the classical biography of Yixing as a Buddhist monk, Chap. 5.)

Yixing Chanshi

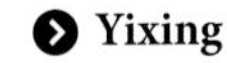 **Yixing**

Young, Anne Sewell

Born **Bloomington, Wisconsin, USA, 2 January 1871**
Died **Claremont, California, USA, 15 August 1961**

Anne Young, an outstanding teacher of astronomy, was one of the eight founders of the American Association of Variable Star Observers [AAVSO], and for many years contributed observations of variable stars and sunspots to that organization.

The niece of astronomer **Charles Young**, she received B.L. and M.S. degrees from Carleton College, and a Ph.D. in 1906 from Columbia. Recognized as an outstanding teacher, she taught astronomy for 3 years (1895–1898) at Whitman College, and then for 37 years (1899–1936) at Mount Holyoke College; among her students there who distinguished themselves in astronomy was **Helen Sawyer Hogg**.

Katherine Bracher

Selected Reference

Hogg, Helen Sawyer (1962). "Anne Sewell Young." *Quarterly Journal of the Royal Astronomical Society* 3: 355–357.

Young, Charles Augustus

Born **Hanover, New Hampshire, USA, 15 December 1834**
Died **Hanover, New Hampshire, USA, 3 January 1908**

Charles Young was a pioneer in solar physics who identified properties of the chromosphere.

Young's father, Ira Young, and grandfather, Ebenezer Adams, were both professors of mathematics and natural philosophy at Dartmouth College, Hanover, New Hampshire, where Young graduated in 1853. He then taught Latin and Greek for 2 years at Phillips Academy in Andover, Massachusetts. Enrolling at the Andover Theological Seminary, Young first contemplated missionary work, but in 1857 accepted a position at Western Reserve College in Hudson, Ohio as professor of natural philosophy and astronomy. That same year, he married Augusta S. Mixer; the couple later had three children. During the Civil War (1862), Young served for 4 months in the 85th regiment of Ohio volunteers. In 1866, he returned to Dartmouth College to assume the position held by his father. But when promised a much larger telescope (a 23-in. refractor) by Princeton University in 1877, Young accepted their offer and spent the remainder of his career as director of Princeton's Halsted Observatory.

Young worked chiefly in visual spectroscopy, and observed solar prominences without an eclipse, by using the spectroscope as a monochromator. During the 7 August 1869 total solar eclipse, Young, in collaboration with **William Harkness**, discovered a green line in the coronal spectrum without a known counterpart in terrestrial laboratory spectra. It took solar physicists over 60 years to realize that the green line belonged to a highly ionized state of iron, indicative of the million-degree temperature of the solar corona. During the 22 December 1870 eclipse, Young observed the "flash spectrum" of the chromosphere and explained its occurrence as due to a "reversing layer" above the Sun's photosphere. At that same eclipse, he also captured the first photograph of a solar prominence.

In 1876, Young used a grating spectroscope to make one of the earliest measurements of the Sun's rotation *via* the Doppler shift. He led other eclipse expeditions around the world (in 1878, 1887, and 1900) and to high mountain altitudes to make spectroscopic observations of the Sun's outer atmosphere. Many of these results were collected in his textbook, *The Sun* (1st edition, 1881), which influenced the next generation of American astrophysicists. Young's teaching and laboratory work turned the Princeton campus into a highly regarded training ground for future spectroscopists. His final scientific paper suggested that "atomic" (*i. e.*, nuclear) energy associated with radioactivity might one day explain the Sun's enormous energy production.

Young saw no conflict between scientific research and religious faith, regarding the "dignity of the human intellect" as the "offspring, and measurably the counterpart, of the Divine" (*Manual of Astronomy*). He was an effective and widely sought public speaker on science and astronomy. Young delivered the keynote address at the dedication of **George Hale**'s Kenwood Physical Observatory at Chicago in 1891. Notable students of Young (from Dartmouth and Princeton) included **Edwin Frost** and **Henry Norris Russell**. One of the most widely used textbooks of the early 20th century, written by Russell, **Raymond Dugan**, and **John Stewart**, was a revision of Young's *Manual of Astronomy*.

Young was awarded numerous honorary degrees and prizes, including the Janssen Medal of the French Academy of Sciences (1891). He served as president of the American Association for the Advancement of Science (1884) and was a member of the National Academy of Sciences, the American Philosophical Society, and the Royal Astronomical Society of Great Britain. Due to declining health, Young retired from Princeton in 1905 and moved back to his native Hanover.

Fathi Habashi

Selected References

Berendzen, Richard and Richard Hart (1976). "Young, Charles Augustus." In *Dictionary of Scienific Biography*, edited by Charles Coulston Gillispie. Vol. 14, pp. 557–558. New York: Charles Scribner's Sons.

DeVorkin, David H. (2000). *Henry Norris Russell: Dean of American Astronomers*. Princeton: Princeton University Press, esp. pp. 25–30.

Frost, Edwin B. (1909). "Charles Augustus Young." *Astrophysical Journal* 30: 323–338.

______ (1913). "Biographical Memoir of Charles Augustus Young." *Biographical Memoirs, National Academy of Sciences* 7: 89–114.

Russell, Henry Norris (1909). "Charles Augustus Young." *Monthly Notices of the Royal Astronomical Society* 69: 257–260.

Yūnus

➲ **Ibn Yūnus: Abū al-Ḥasan ʿAlī ibn ʿAbd al-Rahman ibn Aḥmad ibn Yūnus al-Ṣadafī**

Z

Zach, János Ferenc [Franz Xaver] von

Born **Pozsony (Bratislava, Slovakia), 13 June 1754**
Died **Paris, France, 2 September 1832**

Hungarian-born astronomer and geodetic surveyor, Baron János von Zach is best remembered for his organizational services. Zach was born to a noble family, son of József Zách and Klára Szontágh. He studied physics in Pest, Hungary, and finished his studies in the military academy in Vienna. In the second half of the 1770s, he taught mechanics at the University of Lemberg (now Lyio, Ukraine). When the university ceased operations, Zach moved to Paris (1780) and then to London (1783).

There, Zach made his acquaintance with several leading astronomers, including **Joseph de Lalande**, **Pierre de Laplace**, and **William Herschel**, as well as rich patrons of astronomy. With their influence, Zach was granted an astronomer's position by Duke Ernst II of Saxe-Gotha-Altenburg. Commissioned to plan and build a new observatory, he founded and became director of the Seeberg Observatory (near Gotha) from 1786 to 1804. Research began there in 1792 with instruments made by Jesse Ramsden. After the death of Ernst II in 1804, Zach was disgraced and left Gotha with Duchess Marie Charlotte Amalie (1751–1827), widow of Ernst II. They lived in various places and finally settled in Genova (Italy) in 1815. That year, he founded the Capodimonte Observatory in Naples, Italy. From 1827 on, Zach lived in Paris.

Zach's main astronomical contribution was the foundation of the first international astronomical periodical, the *Monatliche Correspondenz zur Beförderung der Erd- und Himmelskunde*, which was published in 28 volumes between 1800 and 1813. His previous journal, *Allgemeine Geographische Ephemeriden*, published between 1798 and 1799, covered both astronomy and geography. When Zach settled in Italy, he founded and edited a new journal, *Correspondence Astronomique*, which appeared in 13 volumes between 1818 and 1825. These journals enabled contemporary astronomers to distribute their observational results and newly developed mathematical methods in a very efficient way.

He regularly received guest astronomers in Seeberg, site of the first international meeting of astronomers, organized by Zach in 1798. He founded the first international association of astronomers, sometimes called the astronomical police, with the aim of launching an observational campaign for searching the planet missing between Mars and Jupiter, according to the Titius–Bode law. Members of this group were Ferdinand Adolf von Ende, **Johann Gildemeister**, **Karl Harding**, **Heinrich Olbers, Johann Schröter** (president), and Zach (secretary). Their Astronomische Gesellschaft is not identical with the still existing Astronomische Gesellschaft, founded in 1863. The first-discovered minor planet, (i) Ceres, was found by **Giuseppe Piazzi** just before Zach's group began their coordinated observations. Based on calculations by his pupil, **Carl Gauss**, Zach rediscovered the temporarily lost Ceres in December 1801, though this recovery is sometimes erroneously attributed to Olbers.

Zach's research activity also included observations of Mars during its opposition of 1790, and observations of the transits of Mercury in 1802 and 1805. He published corrected tables of solar motion (1792) and tables for aberration (1813).

Zach was a fellow of the Royal Society (1804) and an honorary member of the Hungarian Academy of Sciences (1832). A lunar crater and minor planet (999) Zachia are named for him.

László Szabados

Selected References

Brosche, Peter (2001). *Der Astronom der Herzogin: Leben und Werk von Franz Xaver Zach 1754–1832. Acta Historica Astronomiae*, Vol. 12. Frankfurt am Main: Harri Deutsch.

Brosche, Peter and M. Vargha (1984). "Briefe Franz Xaver von Zachs in sein Vaterland." *Publications of the Astronomy Department of Loránd Eötvös University*, No. 7. Budapest.

Zacut: Abraham ben Samuel Zacut

Born **Salamanca, (Spain), probably 1452**
Died **Damascus, (Syria), probably 1515**

Abraham Zacut was an important Jewish astronomer who contributed to observational astronomy, astronomical tables, and our historical knowledge of astronomy in Spain. Zacut came from a family originally from France; however, the evidence indicates that he was born in Salamanca, and spent his early years there as a pupil of Isaac Aboab, from whom he received the extensive knowledge that would later make him famous. Although professionally a doctor, Zacut's fame came from his works in astronomy.

During his lifetime, Zacut maintained relationships with several notable figures, including Don Juan de Zúñiga, the last Master of the Order of Alcántara (*Maestre de la Orden de Alcántara*), and the Bishop of Salamanca, Gonzalo de Vivero, to whom he dedicated his most famous astronomical work. When Bishop Vivero died in 1480, Zacut lost his protector in Salamanca and moved to the court of Don Juan de Zúñiga for whom he produced the following works: *Tratado breve de las influencias del cielo* (Short treatise on the influence of the heavens) and *De los eclipses del sol y la luna* (On solar and lunar eclipses). We know that Zacut was in Lisbon on 9 June 1493, working for Juan II of Portugal. It is logical to assume that he moved to this city when the Jews were expelled from Spain following the order of the Catholic kings in 1492. Zacut also worked for the brother of King Manuel I, who is said to have sought Zacut's advice for Vasco de Gama's trip around Africa, for which Zacut gave a favorable opinion. When in 1496 King Don Manuel ordered the Jews expelled from Portugal, Zacut fled Portugal and moved to Tunisia, where he was welcomed by a large Jewish colony. He lived in Carthage for several months, giving lessons in subjects for which his expertise was renowned. He eventually moved to the Ottoman lands and died, probably in 1515 although a death date of 1522 has also been suggested.

It is not clear whether or not Zacut taught at the University of Salamanca. However, he was in contact with and influenced some of the professors of astrology there. For example, Juan de Salaya, who was a professor of astrology from 1464 to 1469, translated Zacut's work titled *La Compilación Magna* (ha-*Hibbur ha-gadol* or The magnus compilation) from Hebrew into Spanish. The Latin translation, known as *Almanach Perpetuum*, was made by José Vizinho and first published in Leira in 1496. It became essential for the development of Spanish and Portuguese navigation at the end of the 15th century. The Spanish translation made Zacut famous due to its influence on his contemporaries.

La Compilación Magna was commissioned by Zacut's protector, Gonzalo de Vivero. Indeed the bishop left instructions regarding Zacut in his will as follows:

> ...to deliver to the Jew Abraham, astrologist, five hundred maravedises and ten measures of grain, and instructed that certain works which were in Romance, written by the mentioned Jew, should all be published in a volume together with his other books in his [*i. e.* the bishop's] church, because it is worthy to understand the tables made by the mentioned Jew.

This volume could be Incunable 176, presently kept at the Salamanca University Library, which contains the Spanish translation of *La Compilación Magna* that was dictated by Zacut to the translator Juan de Salaya.

La Compilación Magna is a collection of astronomical tables with rules (canons) that served several purposes. The tables were calculated for the meridian of Salamanca for the *radix* year 1473. The first part of the collection contains the rules in 19 chapters, a number Zacut uses because he considers it a golden number, following the indication of **Maimonides**. In those chapters he first analyzes the positions of the Moon and the Sun, their movements, circumstances, and eclipses, and then moves to the astrological houses and to the ascendant. He also provides the longitudes and latitudes of the main cities; finally he devotes a chapter to the fixed stars. In the second part of the canons, Zacut explains the circumstances of the other planets (Saturn, Jupiter, Mars, Venus, and Mercury) and devotes one chapter to the Jewish, Christian, Islamic, and Persian calendars. This second part comes to a close with Chapter 19, where he explains the movements of the seven planets and of the lunar node (Dragon Head). After the canons, he gives the tables for the material discussed in these 19 chapters. The structure of the tables is influenced by Jacob Poel (Bonet Bonjorn), an intermediary who connected the work of **Gersonides** and Zacut. More than 50 manuscripts are known of these tables, of which we should particularly note MS Sassoon 823 for its detailed "representation" of its catalog of stars. In addition to mentioning Bonet Bonjorn, whom the translator Salaya refers to by his Hebrew name Jacob Poel (Po ʿel meaning "the artisan"), Zacut mentions the Jewish scholar Yehuda ben Aser. There are also references to the tables and calendar of King **Alfonso X**.

The canons of the *Almanach perpetuum* also exist in another Spanish version that was made by the same José Vizinho who made the Latin translation. A copy of this Spanish version is kept in an *incunabulum* of the Colombin Library of Seville's Cathedral. It consists of 23 chapters dealing with the ascendants of the 12 houses, explanations of the positions of the Sun and the Moon and their eclipses, the places and movements of the planets, and a reference in the last chapter to an "animodar."

Zacut's empirical interests are indicated by his observation in 1474 of the Moon covering the star of the spike in Virgo's hand, when this constellation was approximately in the middle of the sky. Other astronomical observations attributed to him are an occultation of Venus by the Moon in July 1476, and a total solar eclipse in June 1478.

Cirilo Flórez Miguel

Selected References

Cantera Burgos, F. (1931). "Notas para la historia de la astronomía en la España medieval: El judio salmantino Abraham Zacut." *Revista de la Academia de Ciencias Exactas, Físicas y Naturales de Madrid* 27: 63–98.

Chabás, José and Bernard R. Goldstein (2000). *Astronomy in the Iberian Peninsula: Abraham Zacut and the Transition from Manuscript to Print.* Philadelphia: American Philosophical Society.

Flórez, Cirilo *et al.* (1989). *La ciencia del cielo.* Salamanca: CAJA DE AHORROS.

Romano, David (1992). *La ciencia hispanojudía.* Madrid: Mapfre.

Swerdlow, Noel M. (1977). "A Summary of the Derivation of the Parameters in the Commentariolus from the Alfonsine Tables with an Appendix on the Length of the Tropical Year in Abraham Zacuto's Almanach Perpetuum." *Centaurus* 21: 201–213.

Zanotti, Eustachio

Born **Bologna, Enilia-Romagna, (Italy), 27 November 1709**
Died **Bologna, Enilia-Romagna, (Italy), 15 May 1782**

Eustachio Zanotti was a versatile observer, professor, and observatory director in Bologna. The son of Gian Pietro Zanotti and Costanza Gambari, he came from a family known for its interest in the arts, humanities, and science. His early studies were at the Jesuit School. He was exposed to Bologna's most illustrious scientists, including the Manfredi and Beccari families, who were often guests at his house, and his uncle, Francesco Maria Zanotti, was president of the Istituto delle Scienze. Zanotti attended lessons at the institute and, under the guidance of **Eustachio Manfredi**, he became enthralled with astronomy and was appointed as Manfredi's assistant at the observatory in 1719.

On 22 August 1730 Zanotti graduated from the University of Bologna with a degree in philosophy. In 1738, after presenting an essay on the Newtonian theory of light, he embarked on his university career as professor of mechanics. That year, he discovered two comets to which he attributed a parabolic orbit. Following Manfredi's death the following year, Zanotti was appointed to the university chair of astronomy. Only a year before, the chair *ad Mathematicam*, established in 1569, was replaced by five other scientific disciplines, including astronomy. The program was also modified to permit the teaching of heliocentric theories (although it would not be until 16 April 1757 that the Sacred Congregation of the Index would permit the free circulation of such ideas).

Zanotti was also appointed professor of astronomy at the Istituto delle Scienze – then still independent of the university – in 1739. During his years of teaching, he continued to publish the *Ephemerides* started by Manfredi, compiling three volumes covering the years from 1751 to 1774. A fourth volume was published posthumously by his successor, Petronio Matteucci.

The modern instruments that Manfredi had ordered from London in 1738 were finally brought to Bologna in 1741: a mural quadrant with a radius of 1.2 m, a transit instrument with a focal length of 1 m, a movable quadrant, and a small reflecting telescope (built by English craftsman Jonathan Sisson and now exhibited at the Astronomical Museum of the Department of Astronomy at the University of Bologna, in the same rooms in which the astronomers used them). Because of the work required to restructure the room of meridian observations at the observatory and to put the instruments into operation and adjust them, they could not be used until 1749.

Zanotti worked with his assistants Giovanni Angelo Brunelli, who would later become mathematician to the king of Portugal, and Matteucci, conducting countless observations of the Sun, Moon, planets, and comets, and compiling a catalog of 446 stars, mostly in the zodiac. Their goal was to add to the knowledge of celestial motion and use lunar occultations to calculate more accurate terrestrial coordinates. This catalog – published by Zanotti in the reprint of Manfredi's *Introductio in Ephemerides* – also can be considered one of the first star catalogs drawn based on modern criteria. To calculate the positions of the stars, it considered not only the precession of the equinoxes but also the effects of annual aberration, discovered a short time before by **James Bradley** and confirmed by Manfredi. Moreover, the results were supplemented with a more accurate determination of the latitude of Bologna (estimated at 44° 29′ 54″, just 1.2″ more than the actual figure) and the γ point, or the intersection between the Equator and the ecliptic, corresponding to the vernal equinox.

Zanotti's other main observations include lunar occultation of stars, lunar eclipses, solar eclipses, numerous comets (including comet 1P/Halley in 1759), and the transits of Mercury and Venus across the solar disk. In 1750 the Académie des sciences invited him to participate in an international research project to measure the lunar parallax, and he provided some of the most accurate observations.

In 1760, Zanotti was moved to the chair of hydrometry, and the Bologna government asked him to oversee the construction of a number of navigable canals. In 1776, he restored the meridian line in the church of San Petronio, constructed by **Giovanni Cassini** in 1655. The meridian had lost its original position because the ground had sunk, and there were depressions in the floor caused by earthquakes. As a result, the gnomonic hole on the roof of the church had shifted. Zanotti described this restoration work in the book *La meridiana del Tempio di San Petronio rinnovata l'anno 1776*. He also published a treatise, *Trattato teorico – pratico di prospettiva*, which was studied widely during the period. It was reprinted in 1825 with Zanotti's biography as a foreword. The biography was written by one of his collaborators, Luigi Palcani Caccianemici, who asserted that Zanotti had also studied the variability of stellar brightness, although none of these observations are reported in his works.

In 1778, Zanotti took his uncle's place as the president of the Istituto delle Scienze. His epitaph can still be seen in the church of Santa Maria Maddalena.

Zanotti's manuscripts and astronomical logbooks are in the Historical Archive of the Department of Astronomy, University of Bologna.

Fabrizio Bònoli

Selected References

Baiada, E., F. Bònoli, and A. Braccesi (1985). "Astronomy in Bologna." *Museo della Specola – Catalog*. Bologna: Bologna University Press.

Bònoli, F. and E. Piliarvu (2001). *I lettori di astronomia presso lo Studio di Bologna dal XII al XX secolo*. Bologna: CLUEB.

Braccesi, A. and E. Baiada (1980). "Proseguendo sulla Specola di Bologna: Dagli studi del Manfredi sull'aberrazione al catalogo di stelle dello Zanotti." *Giornale di astronomia* 6: 5–29.

de Meis, S. (March 1999). "Alcune osservazioni astronomiche di Eustachio Manfredi e Vittorio Stancari a Bologna." *Giornale di astronomia* 25: 32–39.

de Meis, S. and A. Vitagliano (March 2001). "Due comete bolognesi: Manfredi-Stancari (1707) e Zanotti (1739)." *Giornale di astronomia* 27.

Zanstra, Herman

Born **Heerenveen, the Netherlands, 3 November 1894**
Died **Haarlem, the Netherlands, 9 October 1972**

Dutch astrophysicist Herman Zanstra devised the method that bears his name, for determining the temperatures of stars powering emission-line nebulae. He was educated at the technical college in Delft, graduating as a chemical engineer in 1917. After teaching at the same college and the Delft secondary school, Zanstra went to the University of Minnesota in 1921 to work as an instructor in physics, and obtained his Ph.D. from the university in 1923. In his thesis he investigated August Föppl's hypothesis that the angular momentum of the Universe about its center of mass is zero.

Zanstra then had a series of short appointments at universities in Chicago, Hamburg, and Pasadena (California, USA); the University of Washington; and Imperial College London. He spent the summer of 1927 at the Dominion Astrophysical Observatory in Victoria, Canada, returning to the Netherlands in 1931 as an assistant at the University of Amsterdam. Zanstra became a Radcliffe

Travelling Fellow in 1937, spending time in Oxford, England, and later at the Radcliffe Observatory in Pretoria, South Africa. He could not return to Europe because of the war and so taught physics at Howard College, Durban, from 1942 to 1946. In 1946 Zanstra was appointed professor of astronomy at the University of Amsterdam, where he remained until his retirement in 1961.

After completing his thesis, Zanstra began work on the excitation of gaseous nebulae. He realized that the primary mechanism for a hot star to excite a nebula is the photoionization of hydrogen and its subsequent recombination. He assumed that all stellar photons with wavelengths shorter than the Lyman limit would be absorbed and that the recombinations would give the Balmer lines and continuum. Each recombination would give one Balmer photon, so that measurement of the total Balmer emission would give a good estimate of the ultraviolet emission from the star, and this combined with the visible stellar radiation would give an estimate of the temperature of the star. The result was 34,000° K for O-type stars. Subsequently Zanstra studied planetary nebulae, and showed that their central stars had temperatures up to 150,000° K. In other works, Zanstra:

(1) suggested that the forbidden lines are excited by electron collisions;
(2) estimated the nebular distances and radii and their expansion velocities;
(3) modeled nebular expansion due to radiation pressure;
(4) suggested that lines and bands in cometary spectra are excited by resonance and fluorescence of the solar radiation in the cometary gases;
(5) made studies of Wolf–Rayet stars; and
(6) determined the density of a solar prominence.

He received the Gold Medal and George Darwin Lectureship of the Royal Astronomical Society in 1961. Minor planet (2945) was named for Zanstra.

Roy H. Garstang

Selected References

Garstang, R. H. (1973). *Mémoires de la Société royale des sciences de Liège*, ser. 6, Vol. 5, no. 11.

Osterbrock, Donald E. (2001). "Herman Zanstra, Donald H. Menzel, and the Zanstra Method of Nebular Astrophysics." *Journal for the History of Astronomy* 32: 93–108.

Plaskett, H. H. (1974). "Herman Zanstra." *Quarterly Journal of the Royal Astronomical Society* 15: 57–64.

Zanstra, H. (1961). "The Gaseous Nebula as a Quantum Counter." *Quarterly Journal of the Royal Astronomical Society* 2: 137–148.

Zarqālī: Abū Isḥāq Ibrāhīm ibn Yaḥyā al-Naqqāsh al-Tujībī al-Zarqālī

Died **Córdova, (Spain), 15 October 1100**

According to his biographer Isḥāq Israeli, Zarqālī was a renowned instrument maker in Toledo, where he taught himself astronomy. He worked for **Ṣāʿid al-Andalusī** and was a leading figure among Ṣāʿid's group of astronomers. An anonymous Egyptian 14th-century source (*Kanz al-yawāqīt*, Leiden Universiteitsbibliotheek, MS 468) quotes a passage from Ṣāʿid's lost work entitled *Ṭabaqāt al-ḥukamā'*, in which it is stated that Zarqālī constructed an astronomical instrument, called *al-zarqāla*, for al-Ma'mūn (1043–1075), the ruler of Toledo, in the year 1048/1049. It also says that Zarqālī wrote a treatise of 100 chapters on its use. Zarqālī left Toledo between 1081, the beginning of the reign of al-Qādir, and 1085, the date of the conquest of the city by Alfonso VI. He settled in Córdova, where he was protected by al-Muʿtamid ibn ʿAbbād (1069–1091), ruler of Seville.

There are many variations of the name of Zarqālī, known as Azarquiel in Latin. According to the *Ṭabaqāt al-umam* of Ṣāʿid al-Andalusī, he was known as *walad al-Zarqiyāl*, from whence came the Hispanicized form *Azarquiel*. The 13-century biographer al-Qifṭī maintains the expression *walad al-Zarqiyāl* in his *Akhbār al-ʿulamā" bi-akhbār al-ḥukamā'*. Other readings quoted in Andalusian sources are al-Zarqālluh, al-Zarqāl, or Ibn Zarqāl; readings such al-Zarqāla and al-Zarqālī (sometimes al-Zarqānī) seem to be classicized Eastern forms.

In his *Jāmi ʿal-mabādi' wa-'l-ghāyāt fī ʿilm al-mīqāt*, an encyclopedic work on astronomy, **Abū al-Ḥasan ʿAlī al-Marrākushī** (13th century) states that Zarqālī was making observations in Toledo in 1061. This testimony is confirmed by **Ibn al-Hā'im al-Ishbīlī** (flourished: 1204/1205) in his *al-Zīj al-kāmil fī al-taʿālīm*, who attributes to Zarqālī 25 years of solar observations and 37 years of observations of the Moon. Al-Qifṭī says that his observations were used by **Ibn al-Kammād**.

One can generally classify the contents of Zarqālī's work under four main categories: astronomical theory, astronomical tables, magic, and astronomical instruments.

The following four works by Zarqālī deal with astronomical theory: (1) There is a treatise on the motion of the fixed stars, written *circa* 1084/1085 and extant in Hebrew translation. It contains a study of three different trepidation models, in the third of which variable precession becomes independent of the oscillation of the obliquity of the ecliptic. (2) There is a lost work summarizing 25 years of solar observations, probably written *circa* 1075–1080. Its contents are known through secondary sources, both Arabic and Latin. The title was either *Fī sanat al-shams* (On the solar year) or *al-Risāla al-jāmiʿa fī al-shams* (A comprehensive epistle on the Sun). In this work Zarqālī established that the solar apogee had its own motion (of about 1° in 279 Julian years) and devised a solar model with variable eccentricity that became influential both in the Maghrib and in Latin Europe until the time of **Nicolaus Copernicus**. (3) There is an indirect reference to a theoretical work entitled *Maqāla fī ibṭāl al-ṭarīq allatī salaka-hā Baṭlīmūs fī istikhrāj al-buʿd al-abʿad li-ʿUṭārid* (On the invalidity of **Ptolemy**'s method to obtain the apogee of Mercury) mentioned by **Ibn Bājja**. (4) There is a reference in Ibn al-Hā'im's work to Zarqālī's lost writing (*bi-khaṭṭ yadi-hi*, in his own hand) describing a correction to the Ptolemaic lunar model. Ibn al-Hā'im understands this correction as a result of the displacement of the center of the lunar mean motion in longitude to a point on a straight line linking the center of the Earth with the solar apogee, and at a distance of 24′. This model met with some success, for we find the same correction in later Andalusian (Ibn al-Kammād) and Maghribī (**Ibn Isḥāq**, **Ibn al-Bannā'**) *zījes*, although restricted to the calculation of eclipses and the New Moon. It appears also in the Spanish canons of the first version of the *Alfonsine Tables* and in a

Provençal version of the tables of eclipses of **Gersonides**, although in these tables the amount is given as 29′ (either a copying error or a new estimation).

There are two works by Zarqālī dealing with astronomical tables: (1) The *Almanac* is preserved in Arabic, Latin, and in an Alfonsine translation. It is based on a Greek work calculated by a certain Awmātiyūs in the 3rd or 4th century, although the solar tables seem to be the result of the Toledan observations. Its purpose is to simplify the computation of planetary longitudes using Babylonian planetary cycles (*goal years*). (2) The *Toledan Tables* are known through a Latin translation. They seem to be the result of an adaptation of the best available astronomical material (*i. e.*, **Khwārizmī** and **Battānī**) to the coordinates of Toledo that was made by a team led by Ṣāʿid and in which Zarqālī seems to have been a prominent member. The mean-motion tables are original and are the result of observations. Ṣāʿid does not mention these tables although they had been completed before the writing of the *Ṭabaqāt* in 1068.

The only known magical work by Zarqālī is entitled *Risāla fī Ḥarakāt al-kawākib al-sayyāra wa-tadbīri-hi* (On the motions and influences of planets), which is a treatise on talismanic magic using magic squares to make talismans. It is preserved in two Arabic manuscripts, which contain two different versions of the text. There is also a third one summarized in a Latin translation.

Finally, Zarqālī has several works on astronomical instruments: (1) There is a treatise on the construction of the armillary sphere, which is preserved in an Alfonsine–Castilian translation. The original Arabic has not survived. (2) There are two treatises on the construction (*circa* 1080/1081) and use (*circa* 1081/1082) of the equatorium, dedicated to al-Muʿtamid. Zarqālī's equatorium differs from the earlier Andalusian model designed by **Ibn al-Samḥ** (*circa* 1025/1026) in that it is an independent instrument that represents all the planetary deferents and related circles on both sides of a single plate, while a second plate bears all the epicycles. Mercury's deferent is represented as an ellipse. (3) Marrākushī attributes to Zarqālī a sine quadrant with movable cursor (*majarra*), which is a graphic scale of solar declination with the solar longitude as argument. It is similar to the quadrant *vetustissimus*, although in this quadrant the argument used is the date of the Julian year. (4) There are two treatises on two variants of the same astronomical universal instrument (*al-ṣafīḥa al-mushtaraka li-jamīʿ al-ʿurūḍ*): A 100-chapter treatise on the use of the *ṣafīḥa* (plate), called the *zarqāliyya*, and another treatise of 60 chapters on the use of the *ṣafīḥa shakkāziyya*. In both instruments the stereographic equatorial projection of the standard astrolabe is replaced by a stereographic meridian projection onto the plane of the solstitial colure. In fact, it is a dual projection corresponding to each of the Celestial Hemispheres, one of which had its viewpoint at the beginning of Aries and the other at the beginning of Libra. The end result was obtained by superimposing the projection from Aries (turning it) onto the projection from Libra. The two variants of the *ṣafīḥa* differ slightly. The *zarqāliyya* has, on its face, a double grid of equatorial and ecliptical coordinates and a ruler horizon representing the horizontal ones. On its back, in addition to the features proper to the astrolabe, it shows an orthographic meridian projection of the sphere, a trigonometric quadrant, and a small circle (named "of the Moon") used to compute the geocentric distance of the Moon. The *shakkāziyya* is a simplification of the *zarqāliyya*, as Marrākushī states in his *Jāmiʿ*. On its front it bears a single grid of equatorial coordinates and a grid of ecliptical ones reduced to the ecliptic line and the circles of longitude marking the beginning of the zodiacal signs. The back of this kind of *ṣafīḥa* is the same as the back of the astrolabe. There is an Alfonsine translation of the treatise on the *zarqāliyya*, as well as several translations into Latin and Hebrew of the treatise on the *shakkāziyya*.

Roser Puig

Alternate name

Azarquiel

Selected References

Al-Marrākushī, Abū al-Ḥasan ʿAlī (1984). *Jāmi ʿal-mabādiʾ wa-'l-ghāyāt fī ʿilm al-miqāt*. Facsimile edition. Frankfurt am Main. Partially translated in J. J. Sédillot, *Traité des instruments astronomiques des arabes*, Paris, 1834–1835. (Reprint, Frankfurt, 1984); L. A. Sédillot, "Mémoire sur les instruments astronomiques des arabes," *Mémoires de l'Académie royale des inscriptions et belles-lettres de l'Institut de France* 1 (1894): 1–229. (Reprint, Frankfurt, 1989.)

Al-Qifṭī, Jamāl al-Dīn *Akhbār al-ʿulamāʾ bi-akhbār al-Ḥukamāʾ*. Beirut, n.d.

Boutelle, Marion (1967). "The Almanac of Azarquiel." *Centaurus* 12: 12–20.

Comes, Mercè (1991). *Ecuatorios andalusíes: Ibn al-Samḥ, al-Zarqālluh y Abū-l-Ṣalt*. Barcelona.

Goldstein, Bernard R. (1964). "On the Theory of Trepidation according to Thābit b. Qurra and al-Zarqāllu and Its Implications for Homocentric Planetary Theory." *Centaurus* 10: 232–247.

Ibn al-Abbār (1920). *Al-Takmila li-kitāb al-Ṣila*, edited by A. Bel and M. Ben Cheneb. Algiers.

Israeli, R. Isaac (1946–1948). *Liber Jesod olam seu Fundamentum mundi*, edited by B. Goldberg and L. Rosenkranz, with commentary by D. Cassel. Berlin.

King, David A. (1986). *A Survey of the Scientific Manuscripts in the Egyptian National Library*. Winona Lake, Indiana: Eisenbrauns.

______ (1997). "Shakkāziyya." In *Encyclopaedia of Islam*. 2nd ed. Vol. 9, pp. 251–253. Leiden: E. J. Brill.

Mercier, Raymond (1987). "Astronomical Tables in the Twelfth Century." In *Adelard of Bath: An English Scientist and Arabist of the Early Twelfth Century*, edited by Charles Burnett, pp. 87–118. London: Warburg Institute. (See pp. 104–112.)

Millás Vallicrosa, José María (1932). "La introducción del cuadrante con cursor en Europa." *Isis* 17: 218–258. (Reprinted in Millás Vallicrosa, *Estudios sobre historia de la ciencia española*. Barcelona, 1949.)

______ (1943–1950). *Estudios sobre Azarquiel*. Madrid–Granada.

Puig, Roser (1985). "Concerning the *ṣafīḥa shakkāziyya*." *Zeitschrift für Geschichte der arabisch–islamischen Wissenschaften* 2: 123–139.

______ (1987). *Los tratados de construcción y uso de la azafea de Azarquiel*. Madrid.

______ (2000). "The Theory of the Moon in the *Al-Zīj al-Kāmil fī-l-Taʿālīm* of Ibn al-Hāʾim (ca. 1205)." *Suhayl* 1: 71–99.

______ (1986). *Al-Šakkāziyya: Ibn al-Naqqāš al-Zarqālluh. Edición, traducción y estudio*. Barcelona.

Rico y Sinobas, Manuel (1863–1867). *Libros del saber de astronomía del rey D. Alfonso X de Castilla, copilados, anotados y comentados por Don Manuel Rico y Sinobas*. 5 Vols. Madrid.

Richter-Bernburg, Lutz (1987). "Ṣāʿid, the *Toledan Tables*, and Andalusī Science." In *From Deferent to Equant: A Volume of Studies in the History of Science in the Ancient and Medieval Near East in Honor of E. S. Kennedy*, edited by David A. King and George Saliba, pp. 373–401. *Annals of the New York Academy of Sciences*. Vol. 500. New York: New York Academy of Sciences.

Ṣāʿid al-Andalusī (1985). *Kitāb Ṭabaqāt al-umam*, edited by Ḥayāt Bū ʿAlwān. Beirut. (French translation with notes by Régis Blachère as *Livre des catégories des nations*. Paris: Larose, 1935.)

Samsó, Julio (1992). *Las ciencias de los antiguos en al-Andalus*. Madrid: Mapfre.
______ (1994). "Trepidation in al-Andalus in the 11th Century." In *Islamic Astronomy and Medieval Spain*, VIII. Aldershot: Variorum.
______ (1994). "Sobre el modelo de Azarquiel para determinar la oblicuidad de la eclíptica." In *Islamic Astronomy and Medieval Spain*, IX. Aldershot: Variorum.
______ (1994). "Ibn al-Bannā', Ibn Isḥāq and Ibn al-Zarqālluh's Solar Theory." In *Islamic Astronomy and Medieval Spain*, X. Aldershot: Variorum.
______ (2002). "Al-Zarḳālī." In *Encyclopaedia of Islam*. 2nd ed. Vol. 11, pp. 461–462. Leiden: E. J. Brill.
Samsó, Julio and Honorino Mielgo (1994). "Ibn al-Zarqālluh on Mercury." *Journal for the History of Astronomy* 25: 289–296.
Sesiano, Jacques (1996). *Un traité médiéval sur les carrés magiques: De l'arrangement harmonieux des nombres*. Lausanne: Presses polytechniques et universitaires romandes.
Toomer, G. J. (1968). "A Survey of the Toledan Tables." *Osiris* 15: 5–174.
______ (1969). "The Solar Theory of az-Zarqāl: A History of Errors." *Centaurus* 14: 306–336.
______ (1987). "The Solar Theory of az-Zarqāl: An Epilogue." In *From Deferent to Equant: A Volume of Studies in the History of Science in the Ancient and Medieval Near East in Honor of E. S. Kennedy*, edited by David A. King and George Saliba, pp. 513–519. *Annals of the New York Academy of Sciences*. Vol. 500. New York: New York Academy of Sciences.

Zeeman, Pieter

Born **Zonnemaire, the Netherlands, 25 May 1865**
Died **Amsterdam, the Netherlands, 9 October 1943**

Dutch physicist Pieter Zeeman made the laboratory discovery of the effect bearing his name, in which spectral lines emitted or absorbed by atoms in magnetic fields are slightly shifted in wavelength and polarized. He shared the 1902 Nobel Prize in Physics with **Hendrik Lorentz** who had immediately provided a theoretical explanation of the observation for "their researches into the influence of magnetism upon radiation phenomena."

Zeeman was educated at the University of Leiden in the laboratory directed by Keike Kamerlingh Onnes (Nobel Prize 1913 for his discovery of liquid helium), receiving his Ph.D. in 1893. He remained for several years as a *Privatdozent* and lecturer, before being appointed to a professorship at the University of Amsterdam and, in 1908, also as director of the Physical Institute there.

The critical experiments were done in Leiden in 1896, apparently over some considerable objection by Onnes. Zeeman had the good fortune to select neutral sodium gas for his investigation. It has a single active electron when in its ground state and so displayed what is now called a normal Zeeman pattern, splitting into two or three components in a magnetic field (with the amount of the split proportional to the strength of the field). When one looks along the direction of the field, one sees two components shifted in opposite directions, with oppositely directed circular polarization (the longitudinal components). Perpendicular to the field, one sees three components: two shifted components with linear polarization perpendicular to the field, and one undeviated, with polarization along the field direction. Lorentz was able to explain these results (and also the relative intensities and angular beaming of the components) in terms of radiation by individual electrons, discovered in 1897 by J. J. Thomson. Sodium was a lucky choice, because other elements, with several active electrons, can display a dozen or more shifted "anomalous" Zeeman components, which require the quantum mechanics of the 1920s for their explanation.

Zeeman's importance for astronomy lies in the application of his effect to the measurement of magnetic fields in the Sun, stars, and interstellar medium. In 1908, he guided **George Hale** in the interpretation of lines in the spectra of sunspots when they were observed near the center of the solar disk. The observed splitting into two components implied a magnetic field directed radially in the spot umbrae. Thus, Zeeman concluded, spots on the solar limb should show three components with appropriate polarizations. This turned out to be the case, with the field in sunspots ranging up to about 6,000 G in strength. Later, **Horace Babcock** searched for Zeeman broadening and polarization of spectral features produced in the atmospheres of other stars, finding strengths up to more than 30,000 G for some A-type stars. The strongest fields recognized from the Zeeman effect, nearly 1 billion gauss, are found in white dwarfs, and have been measured by **Jesse Greenstein** and others. Much weaker fields (of a thousandth of a gauss) in interstellar gas clouds also have been revealed by the Zeeman polarization of atomic and molecular features at radio wavelengths.

Zeeman's later work was in the general area of the propagation of electromagnetic radiation through various media in the presence of electric and magnetic fields. This included the topic of his Ph.D. dissertation, the Kerr effect, in which a liquid (whose molecules have been partially aligned by an applied electric field) transmits light of perpendicular polarizations at slightly different speeds. The polarizations get out of phase, and, with proper choice of parameters, light can pass through the Kerr cell only when the field is turned on.

Anne J. Kox

Selected References

Kox, A. J. (1997). "The Discovery of the Electron: II. The Zeeman Effect." *European Journal of Physics* 18: 139–144.
Zeeman, P. (1913). *Researches in Magneto-Optics*. London: Macmillan.

Zeipel, Edvard Hugo von

Born **Uppsala, Sweden, 8 February 1873**
Died **Uppsala, Sweden, 8 June 1959**

Swedish mathematical astronomer Edvard von Zeipel is renowned for von Zeipel's theorem and von Zeipel's paradox, closely related topics in the theory of rotating stars.

After finishing studies at Uppsala University in 1904, von Zeipel moved to Paris to increase his knowledge of celestial mechanics under the supervision of **Henri Poincaré**; subsequently he visited the Pulkovo Observatory for a long time. In 1904, he was appointed associate professor of astronomy at Uppsala University, and full professor in 1920. After 1911 he also held the position of the observer (*Observator Regius*) at Uppsala Observatory, until retirement in

1938. Von Zeipel was one of the founding personalities of the Swedish Astronomical Society and president during the years 1926 to 1935. In honor of him, minor planet (8870) bears his name.

The scientific interests of von Zeipel were concentrated on several topics, mainly theoretical in nature. In celestial mechanics, he developed a Hamiltonian formalism of short-term, long-term, and secular perturbations and applied it on the changes of orbits of periodic comets and minor planets including those with orbital elements like that of (108) Hecuba. The method followed an idea of Poincaré and is often referred to as the von Zeipel method.

In 1924, von Zeipel proved (in a literal, mathematical sense) that a star in rigid rotation (like the Earth) must have a very particular distribution of sources of energy such that the energy release would be negative in the outer parts. This was clearly not true. The problem was recognized a year later by **Arthur Eddington** and **Heinrich Vogt** and is sometimes called von Zeipel's paradox. The solution is that, if the distribution of energy release is not the required one, then it sets up currents that are primarily in planes through the axis of rotation (meridional circulation), which are important in gradually mixing stellar material.

Von Zeipel's theorem, also from 1924 and derived by **Edward Milne** as well, states that the local surface brightness of a point on a star is proportional to the local acceleration due to gravity, including the effects of rotation, tidal distortion by another star or planet, and so forth. This is still a useful guide in interpreting observations of stellar atmospheres.

The last topics of von Zeipel's work were the problems of stellar masses and distributions of stars within globular clusters.

Martin Solc

Selected References

Kienle, Hans (ed.) (1924). *Probleme der Astronomie*. Berlin: J. Springer. (Von Zeipel's theorem appeared in this festschrift for Hugo von Seeliger.)

Wallenquist, A. (1959). "Obituary" (in Swedish). *Popular astronomisk tidsskrift* 40: 143.

Zeipel, Edvard Hugo von (1904). "Recherches sur les solutions périodiques de la troisième sorte dans le problème de trois corps." Ph.D. thesis, Uppsala University.

______ (1921). "Recherches sur le mouvement des petites planèts." *Arkiv foer matematik, astronomi och fysik* 11, 12, 13. (The "Von Zeipel method" was developed in this series of publications.)

Zel'dovich, Yakov Borisovich

Born **Minsk, (Belarus), 8 March 1914**
Died **Moscow, (Russia), 2 December 1987**

Soviet theoretical physicist Yakov Zel'dovich made his mark in the astronomical world as founder of the Soviet school of astrophysics and cosmology, within which many ideas of current importance were formulated at the same time (between about 1955 and 1975) as they were being developed in the United States and Europe. Zel'dovich was born into a Jewish family of too high a social class to be initially eligible for college training in Stalin's USSR, and he began his career as a laboratory technician. Zel'dovich's great promise was recognized by more senior scientists, and he was then educated at the Institute of Physics and Technology in Leningrad and the Institute of Chemical Physics, obtaining the degree of Candidate of Sciences in 1936. He was later professor at Moscow State University and divisional head at the Sternberg Astronomical Institute.

Work on the Soviet nuclear program meant that Zel'dovich was 68 when he first traveled outside the Soviet sphere. Nevertheless, his international influence, particularly in cosmology, has been immense. He published around 500 papers, founded a whole school of scientific thought in the Soviet Union, and had close to 100 scientists who would consider themselves to have been his student.

Zel'dovich married several times. All six of his children and several of his grandchildren eventually earned Ph.D.s, largely in the physical sciences. His son Boris Yakovich Zel'dovich is a distinguished condensed-matter physicist.

Yakov Zel'dovich's early work was in chemical physics, including the theory of combustion and detonation, and contributed to the understanding of solid-fuel burning; there is a Zel'dovich number in the theory of combustion. He was also among the first to understand the importance of chain reactions in uranium, and became a leader in the Soviet nuclear program. By the late 1950s he had begun working on elementary particle physics. This led to early limits on particle properties using cosmological considerations.

From the early 1960s Zel'dovich was leading the Soviet efforts in relativistic astrophysics and cosmology. He worked on accretion onto black holes, suggesting in 1964 (simultaneously with Edwin Salpeter) that supermassive black holes could be the energy source of quasars.

With several younger colleagues, Zel'dovich wrote about the possibility of discovering neutron stars and black holes if they happened to be in binary systems, because the transfer of material from a normal star onto a compact one would result in the emission of X-rays. The key papers were published shortly before an improved position for the first extra-solar-system X-ray source, Sco X-1, permitted an optical identification and led a number of other theorists to similar considerations.

Zel'dovich was one of the first to point out the importance of using the Cosmic Microwave Background [CMB] to probe the early history of density perturbations. With Andrei Doroshkevich, Rashid Sunyaev, and others, he worked out the physics of the era of hydrogen recombination and early estimates of the microwave anisotropies. Zel'dovich also studied primordial nucleosynthesis and early Universe particle physics, including the behavior of neutrinos, quarks, and monopoles. He calculated how the CMB spectrum would be distorted by unstable particles, evaporation of mini black holes, annihilation of antimatter, *etc.* On top of this he investigated the origin of astronomical magnetic fields and dynamo theory, among other diverse topics.

Later Zel'dovich worked on the birth and spontaneous creation of the Universe, having some ideas that were forerunners of inflation, which he embraced in the 1980s. Zel'dovich also proposed (independently of Edward Harrison and James Peebles) a spectrum of initial perturbations that has equal power at all scales at horizon crossing. This popular assumption is often referred to as the Harrison–Zel'dovich spectrum.

Zel'dovich's work on the subsequent growth of such cosmological perturbations led to the Zel'dovich pancake picture, which he espoused in the 1970s. He invented a clever trick of using linear particle displacements to study nonlinear density enhancements up to the formation of these pancakes, a method which is termed the el'dovich approximation.

Zel'dovich's name is probably most commonly mentioned now in reference to the Sunyaev–Zel'dovich effect, which is the inverse

Compton scattering of CMB photons off hot electrons in clusters of galaxies. This effect, first described in the early 1970s, has become an important tool for understanding the physical properties of clusters, as well as for determining fundamental cosmological parameters.

Zel'dovich was elected a foreign member of the United States National Academy of Sciences [NAS] and of the Royal Society (London) as well as to membership in the Soviet Academy of Sciences. He received a number of prizes and medals from both Soviet and foreign organizations.

Zel'dovich was able to travel to the United States only once (to address the NAS) and died after suffering an unexpected heart attack, under circumstances that would probably not have been fatal in other countries. He was a widely read polymath even outside the sciences, quite unable to understand how scholars in the United States and Europe, with the right to read absolutely anything they wanted, could take so little advantage of their opportunities.

Douglas Scott

Selected References

Doroshkevich, A. G., Ya. B. Zel'dovich, and I. D. Novikov (1967). "The Origin of Galaxies in an Expanding Universe." *Soviet Astronomy-AJ* 11: 233–239.

Doroshkevich, A. G., Ya. B. Zel'dovich, and R. A. Sunyaev (1978). "Fluctuations of the Microwave Background Radiation in the Adiabatic and Entropic Theories of Galaxy Formation." *Soviet Astronomy* 22: 523–528.

Guseinov, O. Kh. and Ya. B. Zel'dovich (1966). "Collapsed Stars in Binary Systems." *Soviet Astronomy-AJ* 10: 251–253.

Ostriker, J. P., G. I. Barenblatt, and R. A. Sunyaev (eds.) (1993). *Selected Works of Yakov Borisovich Zel'dovich*. Vol. 2, *Particles, Nuclei and the Universe*. Princeton, New Jersey: Princeton University Press.

Sakharov, Andrei (1988). "A Man of Universal Interests." *Nature* 331: 671–672.

Sunyaev, R. A. and Ya. B. Zel'dovich (1972). "The Observations of Relic Radiation as a Test of the Nature of X-Ray Radiation from the Clusters of Galaxies." *Comments on Astrophysics and Space Physics* 4: 173–178.

Zel'dovich, Ya. B. (1964). "The Fate of a Star and the Evolution of Gravitational Energy upon Accretion." *Soviet Physics-Doklady* 9: 195–197.

——— (1970). "Gravitational Instability: An Approximate Theory for Large Density Perturbations." *Astronomy and Astrophysics* 5: 84–89.

——— (1992). *My Universe: Selected Reviews*. Chur, Switzerland: Harwood.

Zel'dovich, Ya. B. and I. D. Novikov (1970). "A Hypothesis for the Initial Spectrum of Perturbations in the Metric of the Friedmann Model Universe." *Soviet Astronomy-AJ* 13: 754–757.

——— (1983). *Structure and Evolution of the Universe*. 2 Vols. Chicago: University of Chicago Press.

Zhamaluding: Jamāl al-Dīn Muḥammad ibn Ṭāhir ibn Muḥammad al-Zaydī al-Bukhārī

Flourished **(Mongolia) and Beijing, China, *circa* 1255–1291**

The Muslim astronomer Zhamaluding (Chinese transliteration of Jamāl al-Dīn) was the first director of the Islamic Astronomical Bureau established in Beijing in 1271. He was involved in the compilation of a *zīj* (astronomical handbook with tables) in Persian, which was largely based upon newly observed planetary parameters and was translated into Chinese, under the title *Huihuilifa*, during the early Ming dynasty. Furthermore, Zhamaluding's name is associated with a "Geography of the Yuan empire," finished in 1291.

Most of the historical information concerning Zhamaluding stems from the official annals of the Yuan dynasty, the *Yuanshi*, and from the "Annals of the Yuan Office of Confidential Records and Books" (*Yuan bishujian zhi*, reprinted in the *Sikuquanshu*). It appears that Zhamaluding was in the service of the Mongol Great Khans from the 1250s onward. A certain Jamāl al-Dīn Muḥammad ibn Ṭāhir ibn Muḥammad al-Zaydī al-Bukhārī, hailing from the region of Bukhara in present-day Uzbekistan and presumably identical with Zhamaluding, is mentioned in the *Jāmiʿ al-tawārīkh* (World history) of the famous Persian historian Rashīd al-Dīn (died: 1317) as not having been capable of carrying out the construction of an astronomical observatory for Möngke Khan (1251–1259) in his capital Karakorum in central Mongolia. Möngke's successor Khubilai had already consulted Zhamaluding and other Muslim astronomers before he became the first emperor of the Yuan dynasty in 1264 and moved his capital to Beijing (Dadu). Three years later, Zhamaluding presented to Khubilai the *Wannianli* (Ten thousand-years calendar, presumably an Islamic *zīj*), which was for a short period distributed as an official calendar but is no longer extant. Furthermore, Zhamaluding offered models or depictions of seven astronomical instruments of Islamic type, namely an armillary sphere, a parallactic ruler, an instrument for determining the time of the equinoxes, a mural quadrant, a celestial and a terrestrial globe, and an astrolabe.

In 1271, Khubilai Khan founded an Islamic Astronomical Bureau with observatory, which was to operate parallel to the traditional Chinese bureau. He thus maintained the bureaucratic structure of the preceding Jin dynasty, but at the same time allowed Chinese observations and predictions to be checked against those of the highly respected Muslim astronomers. Zhamaluding became the first director of the Islamic Bureau and headed a staff of approximately 40 persons, including astronomers, teachers, and administrative personnel. Because, in particular during the 1260s, tens of thousands of Muslims had arrived in China, it need not surprise us that the staff included capable astronomers and that a large observational program could be carried out in order to redetermine most of the planetary parameters and to measure anew the longitudes and latitudes of hundreds of fixed stars. The Islamic Astronomical Bureau of Yuan China thus became one of the very few Islamic institutions where observations were carried out at such a large scale. Although the bureau was not abolished until 1656, its direct influence on Chinese astronomy was very limited and no Islamic methods were incorporated into the official calendar of the Yuan dynasty, the *Shoushili*, by **Guo Shoujing**.

Zhamaluding was also one of the directors of the imperial "Office for Confidential Records and Books" (*bishujian*), to which both astronomical bureaus were subordinate. The extant annals of this office contain a list of books and instruments present at the Islamic Observatory and in Zhamaluding's private library. From the Chinese transliterations of the book titles and brief descriptions, it can be seen that the following works were available: the *Almagest* of **Ptolemy**, the *Elements* of Euclid, the *Madkhal* (Introduction to astrology) by **Kūshyār ibn Labbān**, the *Stellar constellations* by **Ṣūfī**, *zījes*, and books on *hay'a* (cosmology) and the construction of instruments. The transliterations were clearly made from the Persian (rather than from the Arabic), as can be seen from certain grammatical elements and some small variations in terminology.

Zhamaluding was very probably the author of a *zīj* in Persian, or at least was associated with its compilation. The original of this work is lost, but a Chinese translation entitled *Huihuilifa* (Islamic calendar) has drawn the attention of Chinese scholars ever since its publication in the annals of the Ming dynasty (*Mingshi*) prepared in the late 17th century. The translation was made in 1383 by a Muslim astronomer, Ma-shayihei (possibly a *shaykh* who had assumed the common Chinese surname for Muslims, Ma), in cooperation with Chinese scholars. This project, which also included a translation of Kūshyār's *Madkhal*, was carried out at the Astronomical Bureau of the new capital Nanjing on the order of the first emperor of the Ming dynasty, Hong Wu.

In recent years the number of known sources from which the contents of Zhamaluding's original *zīj* may be reconstructed has significantly increased. A late-15th-century restoration of the Chinese translation by the vice director of the Astronomical Bureau in Nanjing, Bei Lin, as well as a Korean reworking made on the order of King Sejong (1419–1451), turned out to be more complete than the version published in the *Mingshi*. An Arabic *zīj* written in Tibet in 1366 by al-Sanjufīnī contains many tables taken directly from the *Huihuilifa* and others that were derived from that work. Al-Sanjufīnī's solar tables are said to be based on the "Jamālī observations," *i. e.*, probably, those carried out under Zhamaluding. A Persian–Arabic manuscript at the Oriental Institute in Saint Petersburg, Russia, which was clearly copied by someone who did not know Arabic or Persian very well, was presumably a working document of the Chinese translators, since it contains original tables for Beijing besides newly computed ones for Nanjing.

An investigation of all these sources has shown that Zhamaluding's original *zīj* contained planetary tables of standard Ptolemaic type, but based on mostly new values for the mean motions, eccentricities, and epicycle radii. For example, the solar mean motion in longitude as found in the *Huihuilifa* implies a length of the tropical year (in sexagesimal notation) of 365;14,31,55 days, one of the most accurate values hitherto found in Islamic sources (the actual year length in 1300 was approximately 365.242236, *i. e.*, 365;14,32,3 days). Zhamaluding's method for predicting solar and lunar eclipses appears to be a mixture of Islamic and Chinese methods. The origin of the star table in the *Huihuilifa*, which lists non-Ptolemaic longitudes, latitudes, and magnitudes of 277 stars near the ecliptic with Ptolemaic as well as Chinese star names, has not yet been completely clarified. The translators in the early Ming dynasty certainly made various modifications to this table, which they utilized for the calculation of so called encroachments (*lingfan*), *i. e.*, passings of the Moon and planets through stellar constellations, which were highly significant in Chinese astrology.

In 1286, undoubtedly as a senior scholar, Zhamaluding suggested to Khubilai a large-scale geographical survey of the Yuan empire. He became the head of an office especially established for this purpose and, since he did not speak Chinese, was provided with a personal translator. The result of the survey, the *Dayitongzhi* (Geography of the whole empire) in 755 volumes, was offered to the emperor in 1291 and finally printed in 1347. Unfortunately, only the introduction of this work is extant.

Benno van Dalen

Alternate name

Jamāl al-Dīn

Selected References

Chen Jiujin (1996). *Huihui tianwenxue shi yanjiu* (Investigations on the history of Muslim astronomy, in Chinese). Nanning: Guangxi kexue jishu chubanzhe.

Dalen, Benno van (1999). "Tables of Planetary Latitude in the *Huihui li* (II)." In *Current Perspectives in the History of Science in East Asia*, edited by Yung-Sik Kim and Francesca Bray, pp. 316–329. Seoul: Seoul National University Press.

______ (2000). "A Non-Ptolemaic Islamic Star Table in Chinese." In *Sic itur ad astra: Studien zur Geschichte der Mathematik und Naturwissenschaften. Festschrift für den Arabisten Paul Kunitzsch zum 70. Geburtstag*, edited by Menso Folkerts and Richard Lorch, pp. 147–176. Wiesbaden: Harrassowitz.

______ (2002). "Islamic and Chinese Astronomy under the Mongols: A Little-Known Case of Transmission." In *From China to Paris: 2000 Years Transmission of Mathematical Ideas*, edited by Yvonne Dold-Samplonius *et al.*, pp. 327–356. Stuttgart: Steiner.

______ (2002). "Islamic Astronomical Tables in China: The Sources for the Huihui li." In *History of Oriental Astronomy*, edited by S. M. Razaullah Ansari, pp. 19–31. Dordrecht: Kluwer Academic Publishers.

Hartner, Willy (1950). "The Astronomical Instruments of Cha-ma-lu-ting, Their Identification, and Their Relations to the Instruments of the Observatory of Marāgha." *Isis* 41: 184–194. (Reprinted in Hartner, *Oriens-Occidens*, edited by Gunter Kerstein *et al.*, pp. 215–226. Hildesheim: Georg Olms, 1968.)

Miyajima, Kazuhiko (1982). "Genshi tenmonshi kisai no isuramu tenmongiki ni tsuite" (New Identification of Islamic astronomical instruments described in the Yuan dynastical history, in Japanese). In *Tōyō no kagaku to gijutsu* (Science and skills in Asia: A festschrift for the 77th birthday of professor Yabuuti Kiyosi), pp. 407–427. Kyoto: Dohosha.

Shi Yunli (2003). "The Korean Adaptation of the Chinese-Islamic Astronomical Tables." *Archive for History of Exact Sciences* 57: 25–60.

Tasaka, Kōdō (1957). "An Aspect of Islam Culture Introduced into China." *Memoirs of the Research Department of the Tōyō Bunko* 16: 75–160.

Yabuuti, Kiyosi (1987). "The Influence of Islamic Astronomy in China." In *From Deferent to Equant: A Volume of Studies in the History of Science in the Ancient and Medieval Near East in Honor of E. S. Kennedy*, edited by David A. King and George Saliba, pp. 547–559, *Annals of the New York Academy of Sciences*. Vol. 500. New York: New York Academy of Sciences.

______ (1997). "Islamic Astronomy in China during the Yuan and Ming Dynasties" (translated and partially revised by Benno van Dalen). *Historia Scientiarum*, 2nd ser., 7: 11–43.

Yano, Michio (1999). "Tables of Planetary Latitude in the *Huihui li* (I)." In *Current Perspectives in the History of Science in East Asia*, edited by Yung-Sik Kim and Francesca Bray, pp. 307–315. Seoul: Seoul National University Press.

______ (2002). "The First Equation Table for Mercury in the *Huihui li*." In *History of Oriental Astronomy*, edited by S. M. Razaullah Ansari, pp. 33–43. Dordrecht: Kluwer Academic Publishers.

Zhang Heng

Born **Xi'e, (Henan), China, 78**
Died **Luoyang, (Henan), China, 139**

Zhang Heng (public name, Pingzi) was a Chinese astronomer and man of letters in the later (eastern) Han dynasty who first fully described the *huntian* or spherical-Earth cosmological model.

In his youth, Zhang Heng traveled to Chang'an (capital of the former Han dynasty) and Luoyang (capital of the later Han dynasty), and studied several subjects. He was appointed an imperial official by Emperor An (reigned: 106–125) and promoted to the post of *Taishiling* (director of the Bureau of Astronomy and Calendrics) shortly thereafter. He was reappointed *Taishiling* by Emperor Shun (reigned: 126–144).

The Chunqiu and Zhanguo ("Spring and Autumn" and "Warring States") periods (770–221 BCE) can be said to be the period of preparation of classical Chinese astronomy. In this era, the 28 lunar mansions were first established and the divisions of the tropical year (which finally became the 24 *qi*-nodes during or just before the early former Han dynasty) came into use. The naive cosmology in this period was the *tianyuan difang* theory, which described a circular heaven over a square Earth. This model developed into the *gaitian* theory of the former Han dynasty, according to which the upper heaven resembled the canopy of a carriage and the lower Earth an inverted bowl.

The former (western) Han dynasty (206 BCE–8) can be said to be the period of establishment of classical Chinese astronomy. The Taichu calendar, which established the standard style of the classical Chinese calendar, was devised in 104 BCE. Luoxia Hong, who was one of the major contributors to the establishment of the Taichu calendar, is said to have invented the *huntianyi* (or *hunyi*) or armillary sphere. At this time, only right ascension must have been measured to which north polar distances were added slightly later. The Taichu calendar of 104 BCE was developed into the Santong calendar of Liu Xin (died: 23) at the end of the former Han dynasty.

The invention of the armillary sphere must be connected with the development of the *huntian* theory of cosmology, in which heaven is considered to be spherical and the Earth is at its center. This theory was fully developed by Zhang Heng of the later Han dynasty, and subsequently became the orthodox theory in classical Chinese astronomy.

Classical Chinese astronomy continued to develop during the later (eastern) Han dynasty (25–220). A new later Han Sifen calendar was made in the year 85.

Also the armillary sphere continued to develop. Previously, during the former Han dynasty, the armillary sphere was only used to measure equatorial coordinates, and at that time, Gen Shouchang noticed that the movement in right ascension of the Sun and the Moon was not uniform. This inequality corresponds to the reduction to the Equator. At the beginning of the later Han dynasty, a nongovernment astronomer, **Fu An**, began to observe the movement of the Sun and Moon along the ecliptic, probably for the first time in China. Then, Li Fan and Su Tong discovered that the movement of the Moon is not uniform even if it is measured along the ecliptic. **Jia Kui** analyzed their discovery, and concluded in his report (92) that this is due to the real inequality of lunar motion caused by the varying distance of the Moon, and that the point on the lunar orbit where the Moon's speed is fastest revolves once in 9 years. This inequality evidently corresponds to the equation of the center of the Moon.

An instrument with an ecliptic circle was originally used by nongovernment astronomers. The first official instrument with an ecliptic circle is said to have been made in 103. Zhang Heng also made an armillary sphere and a celestial globe, as discussed later. The obliquity of the lunar orbit to the ecliptic was also discovered in the later Han dynasty. This fact and the inequality of lunar motion were taken into consideration in the Qianxiang calendar composed by Liu Hong in 206.

During the Han dynasty, there were three theories of cosmology, namely, the *gaitian* theory (where the hemispherical dome of heaven and Earth's surface are parallel), the *huntian* theory (where heaven is spherical), and the *xuanye* theory (where heaven is infinite). Of the three, the *huntian* theory became the orthodox theory. Zhang Heng fully developed the *huntian* theory, and composed the cosmological works, the *Lingxian* (Sublime constitution of the Universe) and the *Hunyi* (Commentary on the armillary sphere). The latter is sometimes called *Hunyizhu*.

In his *Hunyi*, Zhang Heng wrote that heaven is like the shell of a hen's egg, and the Earth is at its center like the yolk of an egg. Probably, the Earth was considered to be flat. As heaven was thought to be spherical, spherical coordinates could be set up. Chinese equatorial coordinates consisted of right ascension, for which the hour circles passing through the determinative stars of the 28 lunar mansions were used as datum lines, and the north polar distance. The angular distance was measured in terms of *du*, which is the angular distance on the celestial sphere through which the Sun moves in one day. (It may be noted here that the term *du* is now used to denote "degree" in modern Chinese.)

Some Chinese astronomers also used the polar longitude, that is, the longitude of the hour circle passing through the object on the ecliptic, for which datum lines are also the hour circles passing through the representative stars of the 28 lunar mansions. The conversion of polar longitude and right ascension was performed graphically on the celestial globe. In his *Hunyi*, Zhang Heng recorded a method to convert the right ascension of the Sun into longitude on the celestial globe.

According to his *Hunyi*, Zhang Heng constructed an armillary sphere called the *tongyi* (bronze instrument) for observational purposes, and a celestial globe called *xiaohun* (small sphere) for demonstration and graphical calculation. According to a later historical record (*Jin shu*), Zhang Heng's celestial globe was rotated by waterpower in a room, and its movements coincided precisely with the actual movement of the sky. Details of its construction are not recorded, but it was evidently the first water-driven celestial globe in China.

The water clock is said to have been used in the Chunqiu and Zhanguo periods (770–221 BCE), but how it was constructed is not recorded. Extant water clocks date back as far as the former Han dynasty and are simple outflow type. According to his fragmentary work *Loushuizhuan huntianyi zhi* (Construction of the water-driven armillary sphere), Zhang Heng built an inflow-type water clock with double reservoir. The double reservoir was intended to make the water flow constant. As water is supplied by the upper reservoir, the water level and water flow of the lower reservoir do not decrease much. This was the first attempt to make the water flow constant in China. Due to the different lengths of daytime and nighttime hours, different acceptors were used for day and night. The technique of making the water flow of the water clock constant was further developed later in China.

Zhang Heng's astronomical works, the *Lingxian*, the *Hunyi*, and the *Loushuizhuan huntianyi zhi*, exist as fragmentary quotations in later works, and are collected in compilations of such fragments, for example the *Quan Hou Han wen* (Complete collection of writings of the later Han dynasty), chapter 55 in the *Quan Shanggu sandai Qin Han Sanguo Liuchao Wen* (Complete collection of writings from high antiquity, the Three Dynasties, Qin, Han, Three Kingdoms, and Six Dynasties) of Yan Kejun (1762–1843), and the "*Tianwen lei*" (Works of astronomy) in the *Yuhanshanfang jiyi shu* of Ma Guohan (1794–1857).

Alternate name

Chang Heng

Selected References

Bo Shuren (1990). "Jinnianlai tianwenxue shi jie youguan Zhang Heng de ruogan zhenglun" (Some controversies about Zhang Heng in the recent research on the history of astronomy). *Zhongguo shi yanjiu dongtai* (Trends in the study of Chinese history) 3: 1–6. (A convenient review of some controversies about Zhang Heng.)

______ (1992). "Zhang Heng." In *Zhongguo gudai kexuejia zhuanji* (Biographies of scientists in ancient China), edited by Du Shiran, Vol. 1, pp. 72–95. Beijing: Kexue chubanshe (Science Publishing House).

Chen Meidong (1998). "Zhang Heng." In *Zhongguo kexue jishu shi, Renwu juan* (A history of science and technology in China, biographical volume), edited by Jin Qiupeng, pp. 65–81. Beijing: Kexue chubanshe (Science Publishing House).

Cullen, Christopher (1996). *Astronomy and Mathematics in Ancient China: The Zhou bi suan jing*. Cambridge: Cambridge University Press. (Discusses the text of the *gaitian* theory of cosmology.)

Eberhard, W. R. Müller (1936). "Contribution to the Astronomy of the San-kuo Period." *Monumenta Serica* 2, no. 2: 149–164. Also included in Wolfram Eberhard, *Sternkunde und Weltbild im alten China*. Taipei: Chinese Materials and Research Aids Service Center, 1970. (Discusses the *huntian* theory of cosmology in the Three Kingdoms period.)

Fan Ye (AD 398–445). "Zhang Heng zhuan" (Biography of Zhang Heng). Chapter 89 in *Hou Han shu* (Standard history of the later Han dynasty). (The official biography of Zhang Heng.)

Ho Peng Yoke (1966). *The Astronomical Chapters of the Chin Shu*. Paris: Moulton and Co. (The three schools of cosmology are described in the chapter on astronomy compiled by Li Chunfeng [602–670] in the *Jin shu* [History of the Jin dynasty], which has been translated into English in this work.)

Hua, Tongxu (1991). *Zhongguo louke* (Chinese water clocks). Hefei: Anhui kexue jishu chubanshe (Anhui Science and Technology Publishing House). (On the development of water clocks in China.)

Lai, Jiadu (1956). *Zhang Heng*. Shanghai: Shanghai renmin chubanshe (People's Publishing House of Shanghai).

Liu Yongping (ed.) (1996). *Kesheng Zhang Heng* (The scientific sage, Zhang Heng). Zhengzhou: Henan renmin chubanshe (People's Publishing House of Henan).

Needham, Joseph, with the collaboration of Wang Ling (1959). *Science and Civilisation in China*. Vol. 3, *Mathematics and the Sciences of the Heavens and the Earth*. Cambridge: Cambridge University Press. (For discussion of Chinese cosmologies, including that of Zhang Heng.)

Maspero, Henri (1929). "L'astronomie chinoise avant les Han. *T'oung Pao* 26: 267–356.

Nōda, Chūryō (1943). *Tōyō tenmonkaku-shi ronsō* (Papers on the history of Eastern astronomy). Tokyo: Kōsei-sha. (Reprint, 1989.) (Contains some important papers on the history of Chinese cosmology, including the contribution of Zhang Heng, in Japanese.)

Ôhashi, Yukio (1999). "Historical Significance of Mathematical Astronomy in Later-Han China." In *Current Perspectives in the History of Science in East Asia*, edited by Yung Sik Kim and Francasca Bray, pp. 259–263. Seoul: Seoul National University Press. (Although Zhang Heng is not mentioned, this may be consulted for the history of mathematical astronomy in the later Han dynasty.)

Ruan, Yuan (1799). *Chouren zhuan* (Biographies of astronomers). (Reprint, Taipei: Shijie shuju, 1962.) (See Chap. 3 for some classical accounts of Zhang Heng.)

Sun, Wenqing (1935). *Zhang Heng Nianpu* (Chronological biography of Zhang Heng). (Reprint, with revisions, Shanghai: Shangwu yinshu guan [Commercial Press], 1956.)

Zhang Pu. *Zhang Heng Ji* (Collected works of Zhang Heng). In *Han Wei Liuchao Baisan jia ji* (Chap. 14) by the Míng dynasty. Included in the *Wenyuange* edition of *Siku quanshu* (The four treasuries of traditional literature, compiled in the 18th century). (Reprint, Taiwan Commercial Press, 1983–1986, Chap. 1412.) (For a compilation of Zhang Heng's works.)

Zhang Sixun

Born **(Sichuan), China, 10th century**

Zhang Sixun was renowned as the maker of an important armillary. After successfully passing the astronomical examination at the Northern Song court in 978, Zhang Sixun was assigned the work of designing and making astronomical instruments at the Imperial Bureau of Astronomy and Calendrics in the capital Kaifeng. Because he had studied astronomical instruments before the examination, it was not long after his assignment that he designed and made an astronomical instrument (979). The instrument was named Taiping hunyi (Armillary sphere of the great peace) by Emperor Taizong. At the time Zhang Sixun was serving as a junior official in charge of administering astronomical instruments.

The Taiping Hunyi had the shape of a three-story building. The lower and middle stories contained devices to announce the hours, and in the upper one there was a celestial globe. The height of the instrument was about 4 m. The power for the machines in the instrument was supplied by flowing mercury. All the machines and devices were installed inside the building and could not be seen from the outside. At each quarter hour, three wooden puppets would come out of the lower building to announce the time by producing sounds. The puppet on the left shook small bells, the middle one beat a drum, and the one on the right rang a bell, after which they reentered the building. In the middle story, there were 12 images of deities with different appearances that would alternately move in and out on the hour to show the time.

On the celestial globe in the upper story, the Sun, Moon, and five planets (Venus, Jupiter, Mercury, Mars, and Saturn) were depicted. It is recorded that they could move along the surface of the globe, but we do not know how this was accomplished. The surface of the globe further showed the Chinese constellations, the Celestial Ecliptic, and the Celestial Equator. The globe rotated once a day. It is said that people could look up and watch it, but where they stood, is not mentioned.

Obviously, the Taiping Hunyi was a large astronomical instrument. It was highly praised in its time. The design is thought to have come from Zhang Sui's (also **Yixing**) Shuiyun hunyi (Water-driven armillary sphere). The Taiping hunyi was also the forerunner of the famous Shuiyun yixiang tai (Tower of the water-driven celestial globe) designed by **Su Song**.

Li Di

Selected References

Li Di (1984). "Zhang Sixun's Taiping hunyi." *Journal of Inner Mongolia Normal University* (Natural Science Edition) 2: 50–57.

Su Song (Song Dynasty). *Xin Yixiang Fayao* (Outline of a method for designing a new celestial globe). In *Congshu jicheng chubian*. (Reprint, Shanghai: Commercial Press, 1935–1937.)

Zinner, Ernst

Born **Goldberg (Slask, Poland), 2 February 1886**
Died **Planeg near Munich, (Germany), 20 August 1970**

German astronomer–historian Ernst Zinner is known primarily for the compilation of very extensive bibliographies of manuscripts and books about astronomy written from the late Middle Ages to the 18th century, and for his own monographs on the history of astronomy.

Zinner attended the Gymnasium in Legnica (over Breslau [Wroclaw]) and studied astronomy and mathematics in Jena and Munich, receiving his doctorate from Jena for work under O. Knopf on a classic problem in astrometry – the reduction of observed to real places of stars. Zinner continued study of astronomy at Lund, Sweden, with **Carl Charlier**, at Paris with **Jules Poincaré**, and at Heidelberg; moving to Bamberg in 1910, where he carried out a major project on historical light curves of variable stars and codiscovered short-period comet 21P/1913 U1 (Giacobini–Zinner). During World War I, Zinner served in the field weather service, followed by a term with the Bavarian Commission for the International Measurement of the Earth (analyzing gravity measurements made in Bavaria), and appointment as professor at Munich in 1924.

In 1926, Zinner was named director of the Bamberg Observatory, retaining the position until his retirement in 1953. His first project there included extensive tables of the apparent brightnesses of stars and of variable star light curves, the latter drawing on observations by **Friedrich Winnecke**, **Arno Wachmann**, and **Ernst Hartwig**. But Zinner's interests were turning ever more strongly to the history of astronomy, particularly the identification and indexing of manuscript sources and books published in German. He visited, as much as possible, the actual facilities holding such works throughout Europe. His published works include a 1925 index of manuscripts, a 1941 history and bibliography of German Renaissance astronomical literature, and a 1936 catalog of German and Dutch astronomical instruments built between the 11th and 18th centuries (and the instrument manufacturers who produced them). The library and museum collections Zinner recorded underwent considerable changes during and after World War II, and historians have not yet quite caught up in producing modern catalogs and bibliographies as extensive as his.

The most notable of Zinner's own historical monographs was the 1931 *History of Astronomy*. It stands out for its extensive discussion of the astronomy of the Jews, Persians, Indians, Chinese, and other East Asian peoples, as well as the preliterature astronomy of the Celts and early Slavs. Curiously, everything lying outside Greek, Roman, and Arabic astronomy appeared under the heading "astronomy of the Germans." This included the work of **Galileo Galilei**, **Isaac Newton**, and even American astronomers. Part of this was conformity to the dominant political views of German cultural and scientific preeminence at the times and places when and where Zinner was writing.

Zinner's studies of **Johann Müller** (Regiomontanus) led him to publish in 1937 a facsimile of the earlier scholar's calendar, which originally appeared in 1474, and, in the following year, to complete his greatest monograph, in which he fully appreciates the value of this influential 15th-century astronomy. Zinner's *Entstehung und Ausbreitung der coppernicanischen Lehre* (Origins and dissemination of Copernican teaching) appeared in 1943, and it is a work still worthy of the fullest consideration by every historian.

A significant portion of Zinner's scientific collection, including 2,700 books and editions of journals, manuscripts, autographs, portraits of scholars, and scientific images, was purchased by the San Diego State University and is kept as the Ernest Zinner collection. Another portion of his collection was donated to the Institute for the History of Physics in Frankfurt am Main.

Jürgen Hamel

Selected References

Anon. (1967). "Veroeffentlichungen 1907–1965 von Ernst Zinner." *Bericht der Naturforschende Gesellschaft, Bamberg* 41: 1–23.

Westman, Robert S. (1997). "Zinner, Copernicus, and the Nazis." *Journal for the History of Astronomy* 28: 259–270.

Hamel, Jürgen (1997). "Ernst Zinner (1886–1970) Quellekündenals Grundlage der Historiographie der Wissenschaften." *Nachrichtenblatt der Deutschen Gesellshaft für Geschichte der Medizin, Naturwissenschaften und Technik* 47: 164–169.

______ (ed.) (1987–1993). *Zentralkatalog alter astronomischer Drucke in den Bibliotheken der deuschen Bundesländer Mecklenburg-Vorpommern, Brandenburg, Berlin, Sachsen-Anhalt, Thüringen und Sachsen (bis 1700)*. Vols. 1–5. Publications of the Archenhold Observatory, Berlin-Treptow, 16–20. Berlin-Treptow, 1987–1993.

Schaldach, Karlheinz (1994). "Das Ernst Zinner-Archiv." In *Ad Radices: Festband zum fünfzigjährigen Bestehen des Instituts für Geschichte der Naturwissenschaften der Johann Wolfgang Goethe-Universität Frankfurt and Main*. Berlin, pp. 25–28.

Zinner, Ernst (1925). *Verzeichnis der astronomischen Handschriften des deutschen Kulturgebietes*. Munich: C. H. Beck.

______(1931). *Die Geschichte der Sternkunde von den ersten Anfängen bis zur Gegenwart*. Berlin: J. Springer.

______(1938). *Leben und Wirken des Johannes Müller von Königsberg, genannt Regiomontanus*. Munich: C. H. Beck. (2nd rev. ed., Osnabrück, 1975.)

______(1941). *Geschichte und Bibliographie der astronomischen Literatur in Deutschland zur Zeit der Renaissance*. Leipzig: K. W. Hiersemann. (2nd rev. ed. 1964.)

______(1956). *Deutsche und niederländische astronomische Instrumente des 11.–18. Jahrhunderts*. Munich: Beck. (2nd rev. ed. 1967.)

______(1943). *Entstehung und Ausbreitung der coppernicanischen Lehre*. Erlangen. (2nd rev. ed. and supplemented by Heribert M. Nobis and Felix Schmeidler, Munich: C. H. Beck, 1988.)

Zöllner, Johann Karl Friedrich

Born **Berlin, (Germany), 8 November 1834**
Died **Leipzig, Germany, 25 April 1882**

From an early age, Johann Zöllner displayed strong mechanical aptitude and fabricated various scientific instruments. Although Zöllner's father wished him to take over the management of his factory, the youth wanted no part of the business. In 1855, he began to study physics at the University of Berlin. Two years later, he enrolled at the University of Basel. By 1859, Zöllner had earned a Ph.D. for research on photometric problems. This work was later published as *Photometrische Untersuchungen* (1865).

Under Zöllner's direction, an instrument named the astrophotometer was constructed at Aarau. This device enabled a direct comparison to be made between an artificial star and a star observed through a telescope. Zöllner's methodology formed the basis of later research at the Potsdam Observatory and its *Photometrische Durchmusterung*.

Zöllner was appointed as professor of physical astronomy at the University of Leipzig in 1866. Another of his inventions, the reversion spectroscope, enabled the dispersion of spectral lines to be effectively doubled. This technique proved valuable in the measurements of Doppler shifts associated with the rotational velocities of sources. Zöllner undertook studies of the Sun's temperature and constitution, and of the nature of comets. He proposed a theory that comets vaporize while in close proximity to the Sun, described in his *Uber die Natur der Cometen* (1872). He also fabricated the horizontal pendulum, later used extensively in geophysical research.

Zöllner was elected to the Saxon Academy of Sciences at Leipzig and as an associate member of the Royal Astronomical Society in 1872. His later life was increasingly devoted to aspects of spiritualism, while his scientific productivity (and reputation) correspondingly declined.

Fathi Habash

Selected References

Anon. (1883). "Johann Karl Friedrich Zöllner." *Monthly Notices of the Royal Astronomical Society* 43 (1883): 185.

Herrmann, Dieter B. (1974). "Karl Friedrich Zöllner und die '*Potsdamer Durchmusterung*.'" *Sterne* 50: 170–180.

______ (1976). "Zöllner, Johann Karl Friedrich." In *Dictionary of Scientific Biography*, edited by Charles Coulston Gillispie, Vol. 14, pp. 627–630. New York: Charles Scribner's Sons.

Knott, Robert (1900). "Zöllner: Johann Karl Friedrich." In *Allgemeine Deutsche Biographie*. Vol. 45, pp. 426–428. Leipzig: Duncker and Humblot.

Staubermann, Klaus G. (1998). "Controlling Vision: The Photometry of Karl Friedrich Zöllner." Ph.D. diss., University of Cambridge.

Zu Chongzhi

Born **Fanyang, (Hebei), China, 429**
Died **500**

Zu Chongzhi (public name, Wenyuan) was a Chinese mathematician and astronomer of the Southern dynasties during the Northern and Southern dynasties period (420–589).

Zu Chongzhi served as an official during the Song (Liu-Song) dynasty (420–479) and southern Qí dynasty (479–502). He made a new calendar entitled the Daming calendar, and requested to use it officially in 462. He was strenuously opposed, and the calendar was not accepted. Zu Chongzhi's son Zu Gengzhi (or Zu Geng) was also a mathematician and astronomer. Thanks to Zu Gengzhi's efforts, the Daming calendar was officially used, beginning in 510.

It was during the eastern Jìn dynasty (317–420) that the precession of the equinoxes was discovered by Yu Xi (281–356; probably independently of **Hipparchus**). After the fall of eastern Jin in 420, the northern and southern dynasties period (420–589) began. In the Song (Liu-Song) dynasty (420–479), the first of the southern dynasties, an astronomer named He Chengtian (370–447) made an excellent calendar called the Yuanjia calendar. Before this, all Chinese calendars assumed that the length of a calendrical synodic month to be constant, although the inequality corresponding to the equation of the center of the Moon had been discovered in the 1st century during the later Han dynasty. He Chengtian tried to adjust the first day of a calendrical synodic month to the true conjunction of the Moon, corrected by the lunar inequality, and proposed a new calendar in 443. However, he was opposed by **Qian Lezhi**, the then director of the Imperial Bureau of Astronomy and Calendrics, and Yan Can, the deputy director. They felt that the new calendar was too complicated, although **Qian Lezhi** and others admitted that He Chengtian's calendar was quite accurate. He Chengtian modified his calendar and made the length of a calendrical synodic month constant. In 445, He Chengtian's Yuanjia calendar was finally accepted as an official calendar. This controversy shows how opinions could diverge regarding the difference between mathematical astronomy as a pure science and the civil calendar as applied technology.

Zu Chongzhi thought that He Chengtian had carried out a reform, but his calendar was still inaccurate. Zu Chongzhi then devised a more accurate calendar called the Daming calendar. However, he was strenuously opposed by the conservative Dai Faxing. (Mention may also be made here of **Zhang Sixun** of the northern dynasties, who discovered the inequality corresponding to the equation of the center of the Sun in the 6th century, probably independently of ancient Mediterranean astronomy.) Most Chinese calendars before Zu Chongzhi used the 19-year cycle for intercalation, during which 7 intercalary months are added. This cycle is called the *zhang*, and was already in use by the end of the "Warring States" period (475–221 BCE). Although this cycle is the same as the Greek Metonic cycle, the Chinese and Greek cycles are probably independent discoveries. The earliest Chinese calendar to abandon the 19-year cycle was the Xuanshi calendar (412) of Zhao Fei of the northern Liang dynasty (396–440), who used a 600-year cycle during which 221 intercalary months were added. The Daming calendar (462) of Zu Chongzhi was the second to abandon the 19-year cycle; it used a 391-year cycle during which 144 intercalary months were added.

Another important innovation of the Daming calendar of Zu Chongzhi is that it took into account the precession of the equinoxes. Precession had already been discovered by Yu Xi (281–356), but Zu Chongzhi was the first to accept it in a calendar. The Daming calendar also used a length of the nodical month 717,777/26,377 (= 27.212230…) days, which is quite accurate.

Zu Chongzhi further devised a method to determine the exact time of the winter solstice from observations of the midday gnomon shadow.

Zu Chongzhi also was a great mathematician. He calculated that the value of π lies between 3.1415926 and 3.1415927. He is said to have composed a high-quality mathematical work *Zhuishu*, which is not extant.

Alternate name

Tsu Ch'ung-chih

Selected References

Du Shiran (1992). "Zu Chongzhi." In *Zhongguo gudai kexuejia zhuanji* (Biographies of scientists in ancient China), edited by Du Shiran. Vol. 1, pp. 221–234. Beijing: Kexue chubanshe (Science Publishing House).

______(1998). "Zu Chongzhi." In *Zhongguo kexue jishu shi, Renwujuan* (A history of science and technology in China, biographical volume), edited by Jin Qiupeng, pp. 164–176. Beijing: Kexue chubanshe (Science Publishing House).

Kobori, Akira (1976). "Tsu Ch'ung-chih." In *Dictionary of Scientific Biography*, edited by Charles Coulston Gillispie. Vol. 13, pp. 484–485. New York: Charles Scribner's Sons. (A biography written from the mathematical point of view.)

Li Di (1977). *Zu Chongzhi*. Shanghai: Shanghai renmin chubanshe (People's Publishing House of Shanghai).

Li Yan and Du Shiran (1987). *Chinese Mathematics: A Concise History*, translated by John N. Crossley and Anthony W. C. Lun. Oxford: Clarendon Press. (For the history of mathematics in China, including the contribution of Zu Chongzhi.)

Li Yanshou (659). *Nanshi* (Standard history of the southern dynasties). (See Chap. 72 for an official biography of Zu Chongzhi.)

Needham, Joseph with the collaboration of Wang Ling (1969). *Science and Civilisation in China*. Vol. 3, *Mathematics and the Sciences of the Heavens and the Earth*. Cambridge: Cambridge University Press. (For Chinese mathematics and astronomy in general, including the contribution of Zu Chongzhi.)

Ruan Yuan (1799). *Chouren zhuan* (Biographies of astronomers). (Reprint, Taipei: Shijie shuju, 1962). (See Chap. 8 for classical accounts of Zu Chongzhi.)

Shen Yue (ed.) (441–513). *Song shu* (Standard history of the [Liu-] Song dynasty). (The Daming calendar of Zu Chongzhi is recorded in the chapter on calendrics in the *Song shu*, Chap. 13.)

Xiao Zixian (489–537 AD). *Nan Qi shu* (History of the southern Qi dynasty). (See section devoted to biographies of men of letters, *Wenxue zhuan*. For an official biography of Zu Chongzhi, see Chap. 52.)

Zucchi, Nicollo

Born **Parma, (Italy), 1596**
Died **Rome, (Italy), 1670**

Nicollo Zucchi taught mathematics at the Jesuit Roman College, did impressive research in optics, and was a well-known telescope maker. **Joseph de Lalande** speaks with great admiration of his contributions to the reflecting telescope. He designed an apparatus using a lens to observe the image focused by a concave mirror, thus providing an early version of the reflecting telescope. From his description of this apparatus in his *Optica*, other scientists such as **Isaac Newton** were able to make necessary improvements on this instrument. Using his reflecting telescope, Zucchi made a careful study of the spots on Mars (which had already been discovered), and from this data **Jacques Cassini** was able to discover the rotation rate of Mars.

Zucchi was held in such great esteem that he was sent as a papal legate to the court of Emperor Ferdinand II of Austria, where he met **Johannes Kepler**. One of the most touching of Kepler's letters was the dedication of his last book, *The Dream* (1634), which contains a long letter of gratitude to Jesuits **Paul Guldin** and Zucchi, who brought Kepler a telescope during his exile:

> To the very reverend Father Paul Guldin, priest of the Society of Jesus, venerable and learned man, beloved patron. There is hardly anyone at this time with whom I would rather discuss matters of astronomy than with you. Father Zucchi could not have entrusted this most remarkable gift – I speak of the telescope – to anyone whose effort in this connection pleases me more than yours.

A lunar crater is named to honor Zucchi: It is 64 km in diameter and is located at 61°.4 south latitude and 309°.7 east longitude.

Joseph F. MacDonnell

Selected References

Campedelli, Luigi (1976). "Zucchi, Niccolò." In *Dictionary of Scientific Biography*, edited by Charles Coulston Gillispie. Vol. 14, pp. 636–637. New York: Charles Scribner's Sons.

Sommervogel, Carolus (1890–1960). *Bibliothèque de la Compagnie de Jésus*. 12 Vols. Brussels: Société Belge de Libraire.

Zupi, Giovan Battista

Born **Catanzaro, (Calabria, Italy), 2 November 1589**
Died **Naples, (Italy), 26 August 1667**

Giovan Zupi was one of the earliest telescopic observers of Jupiter's bands. In the year 1610 Zupi became a Jesuit priest. After teaching humanities, philosophy, and theology, he was teacher of mathematics in the Jesuit College in Naples for 27 years.

Zupi was very active as an observer in collaboration with **Francesco Fontana**, who employed a telescope made of two convex lenses. Fontana, in his book *Novae Coelestium*, attributes to Zupi the early observation of what would become known as Jupiter's belts and zones. **Giovanni Riccioli** mentions in his *Almagestum Novum* Zupi's observations of Jupiter's bands and also many other observations of the planet Mercury. Riccioli included Zupi's name in his lunar map. There are no known books written by Zupi.

Juan Casanovas

Selected Reference

Gatto, Romano (1994). *Tra scienza e immaginazione: Le matematiche presso il collegio gesuitico napoletano (1552–1670)*. Florence: Leo Olschki.

Zwicky, Fritz

Born **Varna, Bulgaria, 14 February 1898**
Died **Pasadena, California, USA, 8 February 1974**

Swiss–American physicist and astrophysicist Fritz Zwicky is remembered as more or less the first:

(1) to point out the very large amounts of dark matter in rich clusters of galaxies;

(2) to show that gravitational lensing of one galaxy by another is much more likely than star–star lensing; and
(3) with **Walter Baade** to associate the supernova phenomenon with the formation of neutron stars and the acceleration of cosmic rays.

Zwicky was born into a family of Swiss merchants working abroad, and returned with them to the village of Mollis, in the home canton of Glarus (where he was eventually buried) in 1904. Though employed for nearly his entire career in the United States, he remained a Swiss citizen, returning home to vote, and remarking that "a naturalized citizen is always a second-class citizen." Zwicky was educated at the Eidgenossische Technische Hochschule [ETH] (Swiss Federal Institute of Technology), completing a diploma thesis (first degree) under mathematician **Herman Weyl** and a Ph.D. dissertation in theory of crystals under Peter Debye (winner of the 1936 Nobel Prize in Chemistry) and Paul Scherrer in 1922. Following a 3-year period as an assistant at the ETH, Zwicky moved in 1925 to the California Institute of Technology (Caltech), at the invitation of its president **Robert Millikan**, receiving a Rockefeller Fellowship from the International Education Board to support 2 years' work there.

Zwicky remained at Caltech the rest of his professional career as assistant (1927–1929), associate professor of physics (1929–1942), and professor of astrophysics (1942–1968), the first person to hold such a title there. Millikan had expected Zwicky to work on quantum theory of solids and liquids, and he indeed published in these areas, but his primary interests gradually turned to astrophysics, beginning with cosmic rays and ideas for how they might arise. In 1929, very shortly after the announcement of the expansion of the Universe by **Edwin Hubble**, Zwicky suggested that the data (a linear correlation between distance and redshift) might equally well be explained by "tired light," that is, the idea that photons simply become less energetic after traveling very long distances. This alternative was finally ruled out only about 60 years later, with the discovery of supernovae in distant galaxies and their time-dilated light curves, showing that the expansion is real. Zwicky himself remained suspicious of large redshifts, and carefully referred to "symbolic velocities," although much of his work in fact assumed the standard redshift–distance relation.

In 1933, Zwicky measured redshifts for galaxies in the Coma cluster, and found that they were not all quite the same. Instead, there was a spread in velocities of the cluster members, which required a very large mass for the cluster to hold it together. The effect was confirmed in 1937 by the discovery of a similarly large velocity spread in the Virgo cluster by **Sinclair Smith**. Zwicky's paper was published in a German–Swiss journal, so he spoke of *dunkle materie* rather than dark matter, and suggested that it might consist of some combination of small, faint galaxies and diffuse gas (perhaps molecular hydrogen). Modern work has shown that both of these are present (though the gas is very hot and ionized rather than molecular), but that an even larger quantity of mass is some other form of dark matter. Zwicky's results implied that the average mass of a galaxy must be much larger than that advocated by Hubble, and he originally suggested that gravitational lensing of one galaxy by another would be a good way to decide which was right, and that it would also allow study of galaxies too distant and faint to be seen otherwise. These 1937 ideas both proved to be correct, but with the first such lens identified only in 1975 and mass measurements and Zwicky telescopes in the 1990s.

Zwicky did live to see the confirmation of another of his seminal ideas. Baade had come to Mount Wilson Observatory in the early 1930s. He was interested in novae, and Zwicky had begun to think that these stars might be sources of cosmic rays. Together in 1933/1934 they put forward the ideas that a small subset of novae were actually much brighter supernovae – **Knut Lundmark** had said the same thing a year or two earlier—that the energy source was the collapse of a normal star to a neutron star, that some of the energy would go into accelerating cosmic rays, and that the Crab Nebula was an example of the remnant of such an event and that one should look for a neutron star in it. Because the neutron had been discovered only in 1932, these were remarkably prescient ideas, which were confirmed by the discovery of pulsars in 1968 and a pulsar in the Crab Nebula in 1969. The Russian theoretical physicist **Lev Landau** also conceived the neutron star idea, probably independently, but somewhat later, and the first serious calculations were done by **Julius Robert Oppenheimer** and **George Volkoff** in 1939. Zwicky was never convinced that an object too massive to form a stable neutron star would collapse to a black hole (as implied by work by Oppenheimer and **Hartland Snyder** the same year), and advocated a hierarchy of more compact objects, beginning with pygmy stars and object Hades beyond neutron stars.

Zwicky had begun deliberate searches for supernovae using a small camera mounted atop Robinson Laboratory at Caltech in 1934, soon after his 1932 marriage to Dorothy Vernon Gates, daughter of a wealthy California family. The marriage ended in divorce, but not before Gates had paid a large fraction of the cost of the first telescope erected on Palomar Mountain, an 18-in. Schmidt in 1936, with which Zwicky began finding new supernovae for systematic study. Both supernova searches and a desire for a wide field of view to study clusters of galaxies motivated Zwicky to be a strong supporter of the 48-in. Schmidt, which began operation at Palomar in 1948. He personally discovered 122 supernovae (more than half of those known at the time of his death). Images from the Palomar Observatory Schmidt Survey also yielded a six-volume catalog of clusters of galaxies, completed by Zwicky and several collaborators in 1968 and still very much in use. He also compiled, in the wake of the 1963 discovery of quasars, a catalog of compact galaxies and compact parts of galaxies, and noted their connection with Seyfert galaxies. His catalog, with **Milton Humason**, of high-latitude B stars (HZ objects) turned out to include both a variety of highly evolved stars and some quasars.

During and after World War II, Zwicky worked on rocketry and propulsion systems with Aerojet (later Aerojet General) Corporation, for which he received a high civilian award, the United States Medal of Freedom in 1949. He was for many years vice president of the International Academy of Astronautics, and many of his more than 50 patents were in rocketry, although his 1957 attempt to put a small mass into cislunar space (with a secondary firing of a projectile off a rocket as it ascended) was probably a failure.

Zwicky was, in fact, a very hands-on scientist, who developed not only telescopes but ways of handling photographs, for instance the subtraction of a negative of one image from a positive of another (in another color, taken at a different time, or in a different polarization) to reveal aspects of galaxies and nebulae that would otherwise have been missed. Others of his ideas were extremely theoretical, for instance the possibility of estimating the mass of the particle that carries the gravitational force (the graviton) from the nonexistence of structures in the Universe larger than, perhaps, 100 megaparsecs on the distance scale then in use. The absence of larger structure turns out to be correct, the finite mass of the graviton probably not.

Zwicky had a number of nonastronomical interests, including Alpine climbing, the rebuilding of ravished European libraries after World War II, and the housing of war orphans, through the Pestalozzi Foundation, whose board of trustees he chaired for a number of years. Reminiscences of Zwicky (invariably attempting to reproduce his Swiss–German accent—he is said to have spoken seven languages, all badly) point out that, late in life, he became inclined to mention that he had been working on some astronomical problems for many decades, and quote one or more of his uncomplimentary remarks about colleagues, most often "spherical bastards" (meaning from which ever way you look at them). On the other hand, his claim that a particular Mount Wilson colleague was color blind in his description of stars turned out, upon application of the Ishihara test, to be literally true. This irascibility (though he could be very kind as well) is probably responsible for the paucity of honors Zwicky received, despite his enormous accomplishments. Apart from the Medal of Freedom, these were limited to the 1948 Halley Lecture (in which he presented the concept of morphological astronomy) and the 1972 Gold Medal of the Royal Astronomical Society.

Zwicky was survived by his second wife, Margrit Zurcher, whom he met and married in Switzerland, and their three daughters.

Oliver Knill

Selected References

Arp, Halton (1974). "Fritz Zwicky." *Physics Today* 27, no. 6: 70–71.

Müller, Roland (1986). *Fritz Zwicky: Leben und Werk des grossen Schweizer Astrophysikers, Raketenforschers und Morphologen (1898–1974)*. Glarus: Verlag Baeschlin.

Payne-Gaposchkin, Cecilia (1974). "A Special Kind of Astronomer." *Sky & Telescope* 47, no. 5: 311–313.

Thorne, K. S. (1994). *Black Holes and Time Warps*. New York: W. W. Norton Co.

Wild, Paul (1989). "Fritz Zwicky." In *Morphological Cosmology*, edited by P. Flin and H. W. Duerbeck, pp. 391–398. Berlin: Springer-Verlag.

Zwicky, Fritz (1966). *Entdecken, Erfinden, Forschen im morphologischen Weltbild*. Munich: D. Knaur. (Reprint, Glarus: Verlag Baeschlin, 1989.)

______ (1971). *Jeder ein Genie*. Bern: H. Lang, 1971. (Reprint, 1992.)

General Bibliography

These are selected, recently published books on the history of astronomy (excluding textbooks). Consult them in addition to article, Selected References. Preference is given to those books written in–or translated into–English, the language of this encyclopedia. References to biographies, institutional histories, conference proceedings, and journal/compendium articles appear in individual entry bibliographies.

Ince, Martin. *Dictionary of Astronomy.* Fitzroy Dearborn, 1997.

GENERAL

Abetti, Giorgio. *The History of Astronomy.* H. Schuman, 1952.

Berry, Arthur. *A Short History of Astronomy from Earliest Times Through the Nineteenth Century*. Dover Publications, 1961.

Clerke, Agnes. *A Popular History of Astronomy During the Nineteenth Century.* A. and C. Black, 1902.

DeVorkin, David (editor). *Beyond Earth: Mapping the Universe*. National Geographic Society, 2002.

Ferris, Timothy. *Coming of Age in the Milky Way*. William Morrow & Company, 1988.

Grant, Robert. *History of Physical Astronomy From the Earliest Ages to the Middle of the Nineteenth Century*. Johnson Reprint Corporation, 1966.

Hoskin, Michael. *The Cambridge Illustrated History of Astronomy.* Cambridge University Press, 1997.

Hoyle, Fred. *Astronomy.* Crescent Books, 1962.

Lankford, John (editor). *History of Astronomy: An Encyclopedia*. Garland Publishing, 1997.

Motz, Lloyd and Weaver, Jefferson. *The Story of Astronomy.* Plenum Press, 1995.

North, John. *The Norton History of Astronomy and Cosmology.* W. W. Norton and Company, 1995.

Pannekoek, Antoine. *A History of Astronomy.* Dover Publications, 1989.

Tauber, Gerald E. *Man's Discovery of the Universe: A Pictorial History*. Crown Publishers, 1979.

Toulmin, Stephen and Goodfield, June. *The Fabric of the Heavens: The Development of Astronomy and Dynamics*. Harper and Row, 1961.

Wilson, Robert. *Astronomy through the Ages: The Story of the Human Attempt to Understand the Universe*. Princeton University Press, 1997.

ANCIENT ASTRONOMY

Bobrovnikoff, Nicholas. *Astronomy Before the Telescope*, vols. 1 & 2. Pachart Publishing House, 1984.

Dicks, D. R. *Early Greek Astronomy to Aristotle*. Cornell University Press, 1970.

Dreyer, J. L. E. *A History of Astronomy from Thales to Kepler.* Dover Publications, 1953.

Evans, James. *The History and Practice of Ancient Astronomy.* Oxford University Press, 1998.

Grant, Edward. *Planets, Stars, and Orbs: The Medieval Cosmos, 1200–1687*. Cambridge University Press, 1996.

Heath, Thomas. *Greek Astronomy.* Dover Publications, 1991.

Hetherington, Barry. *A Chronicle of Pre-Telescopic Astronomy.* John Wiley & Sons, 1996.

Hetherington, Norriss. *Ancient Astronomy and Civilization*. Packart Publishing House, 1987.

McCluskey, Stephen C. *Astronomies and Cultures in Early Medieval Europe*. Cambridge University Press, 1998.

O'Neil, William. *Early Astronomy: From Babylonia to Copernicus*. Sydney University Press, 1986.

Pedersen, Olaf. *Early Physics and Astronomy: A Historical Introduction* (Revised Edition). Cambridge University Press, 1993.

Thurston, Hugh. *Early Astronomy.* Springer Verlag, 1994.

Walker, Christopher (editor). *Astronomy Before the Telescope*. Saint Martin's Press, 1996.

RENAISSANCE ASTRONOMY

Crowe, Michael J. *Theories of the World from Antiquity to the Copernican Revolution*. Dover Publications, 1990.

Eleanor-Roos, Anna Marie. *Luminaries in the Natural World: The Sun and Moon in England, 1400–1720.* Peter Lang Publishing, 2001.

Jervis, Jane. *Cometary Theory in Fifteenth-Century Europe*. D. Reidel Publishing Company, 1985.

Koyré, Alexander. *The Astronomical Revolution: Copernicus – Kepler – Borrelli*. Dover Publications, 1992.

Kuhn, Thomas S. *The Copernican Revolution: Planetary Astronomy in the Development of Western Thought*. Harvard University Press, 1957.

Randles, W. G. L. *The Unmaking of the Medieval Christian Cosmos, 1500–1760: From Solid Heavens to Boundless Æther.* Ashgate Publishing Company, 1999.

Schofield, Christine. *Tychonic and Semi-tychonic World Systems.* Arno Press, 1981.

Taton, René and Wilson, Curtis (editors). *Planetary Astronomy from the Renaissance to the Rise of Astrophysics*, parts A & B. Cambridge University Press, 1989 & 1995.

van Helden, Albert. *Measuring the Universe: Cosmic Dimensions from Aristarchus to Halley*. University of Chicago Press, 1985.

Warner, Deborah. *The Sky Explored: Celestial Cartography 1500–1800.* Alan R. Liss, Inc., 1979.

ASTROPHYSICS

DeVorkin, David. *The History of Modern Astronomy and Astrophysics: A Selected, Annotated Bibliography*. Garland Publishing, 1982.

Friedlander, Michael W. *A Thin Cosmic Rain: Particles from Outer Space*. Harvard University Press, 2000.

Gingerich, Owen (editor). *Astrophysics and Twentieth-century Astronomy to 1950*. Cambridge University Press, 1984.

Herrmann, Dieter. *The History of Astronomy from Herschel to Hertzsprung*. Cambridge University Press, 1984.

Leverington, David. *A History of Astronomy: From 1890 to the Present*. Springer Verlag, 1995.

Leverington, David. *New Cosmic Horizons: Space Astronomy from the V2 to the Hubble Space Telescope*. Cambridge University Press, 2000.

Struve, Otto and Zebergs, Velta. *Astronomy of the 20th Century*. Macmillan, 1962.

THE MOON

Montgomery, Scott. *The Moon and the Western Imagination*. University of Arizona Press, 1999.

Sheehan, William and Dobbins, Thomas. *Epic Moon: A History of Lunar Exploration in the Age of the Telescope.* Willman-Bell, Inc., 2001.

Whitaker, Ewen. *Mapping and Naming the Moon: A History of Lunar Cartography and Nomenclature*. Cambridge University Press, 1999.

Wilhelms, Don E. *To A Rocky Moon: A Geologist's History of Lunar Exploration*. University of Arizona Press, 1993.

THE SUN

Brody, Judit. *The Enigma of Sunspots: A Story of Discovery and Scientific Revolution*. Floris Books, 2002.

Falck-Ytter, Harald; Lovgren, Torbjorn; and Alexander, Robin. *Aurora: The Northern Lights in Mythology, History, and Science*. Anthroposophic Press, 1999.

Hufbauer, Karl. *Exploring the Sun: Solar Science Since Galileo*. Johns Hopkins University Press, 1991.
Pang, Alex Soojung-Kim. *Empire and Sun: Victorian Solar Eclipse Expeditions*. Stanford University Press, 2002.
Steele, Duncan and Davies, Paul. *Eclipse: The Celestial Phenomenon that Changed the Course of History*. Henry John, 2001.

PLANETS

Baum, Richard and Sheehan, William. *In Search of Planet Vulcan: The Ghost in Newton's Clockwork Universe*. Plenum Press, 1997.
Hockey, Thomas. *Galileo's Planet: Observing Jupiter Before Photography*. Institute of Physics, 1996.
Hoyt, William. *Planets X and Pluto*. University of Arizona Press, 1980.
Schorn, Ronald. *Planetary Astronomy from Ancient Times to the Third Millennium*. Texas A & M University Press, 1998.
Sheehan, William. *Planets and Perception: Telescopic Views and Interpretations, 1609–1909*. University of Arizona Press, 1988.
Sheehan, William. *Worlds in the Sky*. University of Arizona Press, 1992.
Sheehan, William. *The Planet Mars: A History of Observation and Discovery*. University of Arizona Press, 1996.
Standage, Tom. *The Neptune File: A Story of Astronomical Rivalry and the Pioneers of Planet Hunting*. Walker & Company, 2000.

SMALL SOLAR-SYSTEM BODIES

Cunningham, Clifford J. *Introduction to Asteroids: The Next Frontier*. Willmann-Bell, 1987.
Hoyt, William Graves. *Coon Mountain Controversies: Meteor Crater and the Development of Impact Theory*. University of Arizona Press, 1987.
Littmann, Mark. *The Heavens on Fire: The Great Leonid Meteor Storms*. Cambridge University Press, 1998.
Peebles, Curtis. *Asteroids: A History*. Smithsonian Institution Press, 2000.
Schechner Genuth, Sara. *Comets, Popular Culture, and the Birth of Modern Cosmology*. Princeton University Press, 1997.
Yeomans, Donald K. *Comets: A Chronological History of Observation, Science, Myth and Folklore*. Wiley, 1991.

STELLAR ASTRONOMY

Harrison, Edward. *Darkness at Night: A Riddle of the Universe*. Harvard University Press, 1987.
Hetherington, Norriss. *Science and Objectivity: Episodes in the History of Astronomy*. Iowa State University Press, 1988.
Hirshfeld, Alan. *Parallax: The Race to Measure the Cosmos*. W. H. Freeman, 2001.
Jones, Kenneth Glyn. *The Search for the Nebulae*. Alpha Academic, 1975.
Murdin, Lesley. *Under Newton's Shadow: Astronomical Practices in the Seventeenth Century*. Hilger, 1985.
Paul, Erich. *The Milky Way Galaxy and Statistical Cosmology, 1890–1924*. Cambridge University Press, 1993.
Whitney, Charles A. *The Discovery of Our Galaxy*. Alfred A. Knopf, 1971.

COSMOLOGY

Berendzen, Richard; Hart, Richard; and Seeley, Daniel. *Man Discovers the Galaxies*. Science History Publications, 1976.
Crowe, Michael J. *Modern Theories of the Universe from Herschel to Hubble*. Dover Publications, 1994.
Danielson, Dennis Richard. *The Book of the Cosmos: Imagining the Universe from Heraclitus to Hawking*. Perseus Publishing, 2000.
Ferguson, Kitty. *Measuring the Universe*. Walker and Company, 1999.
Ferris, Timothy. *The Red Limit: The Search for the Edge of the Universe*. William Morrow, 1977.
Hetherington, Norriss S. *Cosmology: Historical, Literary, Philosophical, Religious, and Scientific Perspectives*. Garland, 1993.
Kragh, Helge. *Cosmology and Controversy*. Princeton University Press, 1996.
North, J. D. *The Measure of the Universe: A History of Modern Cosmology*. Clarendon Press, 1965.
Pecker, Jean-Claude. *Understanding the Heavens: 30 Centuries of Astronomical Ideas from Ancient Thinking to Modern Cosmology*. Springer Verlag, 2000.

COSMOGONY

Brush, Stephen G. *History of Modern Planetary Physics*, vols. 1, 2, & 3. Cambridge University Press. 1996.
Crowe, Michael. *The Extraterrestrial Life Debate, 1750–1900*. Cambridge University Press, 1986.
Dick, Steven J. *Plurality of Worlds: The Origins of the Extraterrestrial Life Debate from Democritus to Kant*. Cambridge University Press, 1982.
Dick, Steven J. *Life on Other Worlds: The 20th Century Extraterrestrial Life Debate*. Cambridge University Press, 1998.
Gribbin, John and Gribbin, Mary. *Ice Age*. Barnes and Noble, Inc., 2002.
Jaki, Stanley L. *Planets and Planetarians: A History of Theories of the Origin of Planetary Systems*. John Wiley and Sons, 1977. This book contains many whiggish sentiments, but nonetheless provides an important synthesis.

REGIONAL ASTRONOMY

Chapman, Allan. *The Victorian Amateur Astronomer: Independent Astronomical Research in Britain, 1820–1920*. Praxis Publishing, 1998.
Doel, Ronald. *Solar System Astronomy in America: Communities, Patronage, and Interdisciplinary Research, 1920–1960*. Cambridge University Press, 1996.
Evans, David. *Under Capricorn: A History of Southern Hemisphere Astronomy*. Institute of Physics Publishing, 1988.
Fauvel, John; Flood, Raymond; and Wilson, Robin (editors). *Möbius and his Band: Mathematics and Astronomy in Nineteenth Century Germany*. Oxford University Press, 1993.
Haynes, Raymond; Haynes, Roslynn; Malin, David; and McGee, Richard. *Explorers of the Southern Sky: A History of Australian Astronomy*. Cambridge University Press, 1996.
Ho Peng Yoke. *Modern Scholarship on the History of Chinese Astronomy*. Australian National University Press, 1977.
Hyde, Vicki. *Godzone Skies: Astronomy for New Zealanders*. Canterbury University Press, 1992.
Jarrell, Richard. *The Cold Light of Dawn: A History of Canadian Astronomy*. University of Toronto Press, 1988.
Lankford, John. *American Astronomy: Community, Careers, And Power, 1859–1940*. University of Chicago Press, 1997.
Levy, B. Barry. *Planets, Potions, and Parchments: Scientific Hebraica from the Dead Sea Scrolls to the Eighteenth Century*. McGill-Queen's University Press, 1990.
Moore, Patrick and Collins, Pete. *The Astronomy of Southern Africa*. Robert Hale & Company, 1977.
Nakayama, Shigeru. *A History of Japanese Astronomy: Chinese Background and Western Impact*. Harvard University Press, 1969.
Selin, Helaine and Sun Xiaochun (editors). *Astronomy Across Cultures: A History of Nonwestern Astronomy*. Kluwer Academic Publishers, 2000.
Sen, S. N. and Shukla, K. S. (editors). *History of Astronomy in India*. Indian National Science Academy, 1985.

ORGANIZATIONS

Blaauw, Adriaan. *History of the IAU: The Birth and First Half-Century of the International Astronomical Union*. Kluwer Academic Publishers, 1994.
DeVorkin, David (editor). *The American Astronomical Society's First Century*. American Astronomical Society, 1999.
Heck, André. *StarGuides 2001: A World-Wide Directory of Organizations in Astronomy, Related Space Sciences, and Other Related Fields*. Kluwer Academic Publishers, 2001.
Howell, Kenneth J. *God's Two Books: Copernican Cosmology and Biblical Interpretation in Early Modern Science*. University of Notre Dame Press, 2002.

Sullivan, Walter. *Assault on the Unknown: The International Geophysical Year.* McGraw-Hill, 1961.
Udías, Augustín. *Searching the Heavens and the Earth: The History of Jesuit Observatories.* Kluwer Academic Publishers, 2003.

ASTRONOMY EDUCATION

Marché, II, Jordan D. *Theaters of Time and Space: American Planetaria, 1930–1970.* Rutgers University Press, 2005.

Illustrations

BEA illustrations represent a sample of those available. We acknowledge the following sources.

Page 2, Abbe, Reproduced from *Biographical Memoirs, National Academy of Sciences* 8 (1919)

Page 8, Abney, Reproduced from *Proceedings of the Royal Society of London* A 99 (1921)

Page 14, Adams, Walter, Reproduced by permission of Yerkes Observatory

Page 20, Airy, Courtesy of History of Science Collections, University of Oklahoma Libraries

Page 21, Aitken, Reproduced by permission of the Mary Lea Shane Archives of the Lick Observatory, University of California at Santa Cruz

Page 23, Albert the Great, Courtesy of History of Science Collections, University of Oklahoma Libraries, Small Portraits Collection

Page 31, Alfvén, Courtesy of Alfvén Laboratory, Royal Institute of Technology, Stockholm

Page 51, Apian, Courtesy of History of Science Collections, University of Oklahoma Libraries, Small Portraits Collection

Page 54, Arago, Courtesy of History of Science Collections, University of Oklahoma Libraries, Large Portraits Collection

Page 56, Archimedes, Courtesy of History of Science Collections, University of Oklahoma Libraries

Page 58, Argelander, Reproduced by permission of Helsinki University Museum (A photograph of the portrait by Carl Peter Mazér in 1837)

Page 59, Argoli, Courtesy of History of Science Collections, University of Oklahoma Libraries

Page 68, Atkinson, Reproduced by permission of Indiana University Department of Astronomy and Indiana University Archives

Page 79, Bacon, Francis, Courtesy of History of Science Collections, University of Oklahoma Libraries

Page 82, Baillaud, Reproduced from *Popular Astronomy* 27, no. 9 (Nov. 1919)

Page 89, Ball, Courtesy of History of Science Collections, University of Oklahoma Libraries

Page 112, Berman, Reproduced by permission of the Mary Lea Shane Archives of the Lick Observatory, University of California at Santa Cruz and the Berman family

Page 116, Bessel, Courtesy of History of Science Collections, University of Oklahoma Libraries

Page 121, Bianchini, Courtesy of History of Science Collections, University of Oklahoma Libraries

Page 122, Bickerton, Reproduced from *Knowledge* 35, no. 522 (Jan. 1912)

Page 125, Biot, Jean-Baptiste, Reproduced by permission of Académie des sciences

Page 135, Blaauw, Courtesy of Adriaan Blaauw

Page 141, Bode, Courtesy of History of Science Collections, University of Oklahoma Libraries

Page 142, Böethius, Reproduced by permission of Glasgow University Library, Department of Special Collections (MS Hunter 374, folio 4^{R})

Page 145, Bok, Courtesy of Mrs. Joyce Ambruster

Page 148, Bond, William, Reproduced by permission of the Mary Lea Shane Archives of the Lick Observatory, University of California at Santa Cruz

Page 154, Boss, Lewis, Reproduced from *Popular Astronomy* 20, no. 9 (Nov. 1912)

Page 155, Bouillau, Courtesy of Robert A. Hatch

Page 161, Bradley, Courtesy of History of Science Collections, University of Oklahoma Libraries

Page 163, Brahe, Courtesy of History of Science Collections, University of Oklahoma Libraries, Small Portraits Collection

Page 166, Brashear, Reproduced from *Journal of the Royal Astronomical Society of Canada* 14, no. 5 (June 1920)

Page 172, Brorsen, Courtesy of Martin Solc

Page 178, Bruhns, Reproduced by permission of the Mary Lea Shane Archives of the Lick Observatory, University of California at Santa Cruz

Page 180, Bruno, Courtesy of History of Science Collections, University of Oklahoma Libraries

Page 182, Bunsen, Courtesy of History of Science Collections, University of Oklahoma Libraries, Small Portraits Collection

Page 186, Burnham, Reproduced by permission of the Mary Lea Shane Archives of the Lick Observatory, University of California at Santa Cruz

Page 195, Campbell, William, Reproduced by permission of the Mary Lea Shane Archives of the Lick Observatory, University of California at Santa Cruz

Page 198, Cannon, Reproduced from *New England Magazine* 6 (1892) (Cannon is at far right)

Page 205, Cassini, Giovanni, Courtesy of History of Science Collections, University of Oklahoma Libraries

Page 212, Cavendish, Courtesy of History of Science Collections, University of Oklahoma Libraries

Page 212, Cayley, Courtesy of History of Science Collections, University of Oklahoma Libraries

Page 218, Chamberlin, Reproduced by permission of Wisconsin Historical Society (Image WHi-2241)

Page 222, Chapman, Reproduced by permission of the Geophysical Institute, University of Alaska Fairbanks

Page 230, Chladni, Reproduced by permission of Staatsbibliothek zu Berlin

Page 232, Christie, Reproduced by permission of the Editors of *The Observatory* (Published previously in 1922 in vol. 45)

Page 241, Clerke, Reproduced from *Astrophysical Journal* 25, no. 3 (April 1907)

Page 245, Comrie, Reproduced by permission of J. K. Comrie

Page 247, Comstock, Reproduced by permission of the Mary Lea Shane Archives of the Lick Observatory, University of California at Santa Cruz

Page 248, Comte, Courtesy of History of Science Collections, University of Oklahoma Libraries, Small Portraits Collection

Page 249, Condamine, Courtesy of History of Science Collections, University of Oklahoma Libraries, Small Portraits Collection

Page 250, Cooper, Reproduced by permission of the Editors of *The Irish Astronomical Journal* (Published previously in 1998 in vol. 25)

Page 252, Copernicus, Courtesy of Owen Gingerich

Page 254, Cornu, Courtesy of History of Science Collections, University of Oklahoma Libraries, Small Portraits Collection

Page 262, Croll, Courtesy of History of Science Collections, University of Oklahoma Libraries

Page 266, Curtiss, Reproduced by permission of UM Department of Astronomy Records, Bentley Historical Library, University of Michigan

Page 270, d'Alembert, Courtesy of History of Science Collections, University of Oklahoma Libraries, Small Portraits Collection

Page 278, Darwin, Reproduced from *Popular Astronomy* 3, no. 10 (June 1896)

Page 280, Davis, Charles, Courtesy of United States Naval Observatory

Page 283, d'Azambuja, Reproduced by permission of Observatoire de Paris (Lucien d'Azambuja [left] and Henri Deslandres [right] about 1907 standing in front of the building at the Observatory of Meudon housing the spectroheliograph)

Page 285, Dee, Reproduced by permission of the Royal Astronomical Society

Page 293, Descartes, Courtesy of History of Science Collections, University of Oklahoma Libraries

Page 301, Divini, Reproduced by permission of "E. Divini", San Severino Marche (MC), Italy

Page 303, Dollond, John, Courtesy of M. Eugene Rudd

Page 305, Donner, Reproduced by permission of Helsinki University Museum (A photograph of the portrait by Eero Järnefelt in 1926)

Page 310, Draper, Henry, Reproduced by permission of Harvard College Observatory

Page 312, Dreyer, Reproduced by permission of Armagh Observatory

Page 317, Dunér, Reproduced from *Astrophysical Journal* 41, no. 2 (March 1915)

Page 325, Eddington, Reproduced by permission of the Astronomical Society of the Pacific

Page 333, Ellerman, Courtesy of The Observatories of the Carnegie Institution of Washington

Page 336, Ellison, Reproduced by permission of Mr. J. C. Liddell

Page 347, Euler, Reproduced by permission of öffentliche Kunstammulung Basel (Color portrait by Emmanuel Handmann, 1753)

Page 363, Federer, Reproduced by permission of *Sky & Telescope*

Page 364, Ferguson, Courtesy of History of Science Collections, University of Oklahoma Libraries

Page 365, Fernel, Courtesy of History of Science Collections, University of Oklahoma Libraries, Small Portraits Collection

Page 366, Ferrel, Courtesy of National Oceanic and Atmospheric Administration, NOAA Central Library

Page 369, FitzGerald, Courtesy of History of Science Collections, University of Oklahoma Libraries

Page 372, Flammarion, Courtesy of History of Science Collections, University of Oklahoma Libraries

Page 373, Flamsteed, Courtesy of History of Science Collections, University of Oklahoma Libraries, Large Portraits Collection

Page 376, Fontana, Courtesy of History of Science Collections, University of Oklahoma Libraries

Page 376, Fontenelle, Courtesy of History of Science Collections, University of Oklahoma Libraries

Page 378, Foucault, Courtesy of History of Science Collections, University of Oklahoma Libraries

Page 382, Fowler, William, Courtesy of Caltech (Fowler holding the National Medal of Science)

Page 385, Fracastoro, Courtesy of History of Science Collections, University of Oklahoma Libraries, Small Portraits Collection

Page 387, Franz, Courtesy of Leonard B. Abbey. Reproduced from *Vierteljahrsschrift der Astronomischen Gesellschaft* 49 (1914)

Page 388, Fraunhofer, Courtesy of History of Science Collections, University of Oklahoma Libraries, Small Portraits Collection

Page 390, Freundlich, Reproduced by permission of D. B. Herrmann from *The International Portrait Catalogue of the Archenhold Observatory* (Berlin: Archenhold Obs., 1984)

Page 393, Frisi, Courtesy of History of Science Collections, University of Oklahoma Libraries

Page 400, Galileo, Courtesy of History of Science Collections, University of Oklahoma Libraries, Small Portraits Collection

Page 402, Galle, Reproduced from *Journal of the Royal Astronomical Society of Canada* 4, no. 5 (Sept.-Oct. 1910)

Page 405, Gaposchkin, Courtesy of Katherine Haramundanis (Photograph taken in 1934)

Page 408, Gassendi, Courtesy of History of Science Collections, University of Oklahoma Libraries, Small Portraits Collection

Page 410, Gauss, Courtesy of History of Science Collections, University of Oklahoma Libraries, Small Portraits Collection

Page 420, Gill, Reproduced from *Proceedings of the Royal Society of London* A 91, no. 633 (Sept. 1915)

Page 430, Goodricke, Reproduced by permission of the Royal Astronomical Society

Page 431, Gore, Courtesy of Ian Elliott

Page 444, Grosseteste, Reproduced by permission of the British Library (Portrait from a fifteenth-century manuscript, MS Royal 6 E. V)

Page 452, Gyldén, Reproduced by permission of Helsinki University Observatory

Page 465, Halley, Courtesy of Donald Yeomans

Page 468, Hansen, Reproduced by permission of AIP Emilio Segrè Visual Archives, T.J.J. See Collection

Page 470, Harkness, Courtesy of United States Naval Observatory

Page 473, Hartmann, Courtesy of Niedersächsische Staats - und Universitätsbibliothek Göttingen

Page 476, Hatanaka, Photograph given to the authors by Kinuko Hatanaka

Page 478, Hegel, Courtesy of History of Science Collections, University of Oklahoma Libraries, Small Portraits Collection

Page 480, Helmholtz, Courtesy of History of Science Collections, University of Oklahoma Libraries

Page 483, Henry, Prosper-Mathieu, Reproduced from *Popular Astronomy* 11, no. 10 (Dec. 1903)

Page 491, Herschel, Caroline, Courtesy of History of Science Collections, University of Oklahoma Libraries

Page 493, Herschel, John, Reproduced by permission of the Master and Fellows of St. John's College, Cambridge (Portrait by H.W. Pickersgill)

Page 495, Herschel, William, Reproduced by permission of John Herschel-Shorland (A photograph of the portrait by John Russell)

Page 503, Hevelius, Courtesy of History of Science Collections, University of Oklahoma Libraries

Page 505, Higgs, Courtesy of Alan J. Bowden, Curator of Earth Sciences, Liverpool Museum

Page 507, Hill, Courtesy of United States Naval Observatory

Page 529, Hough, Courtesy of Northwestern University Archives

Page 535, Huggins, Margaret, Courtesy of History of Science Collections, University of Oklahoma Libraries

Page 536, Huggins, William, Courtesy of History of Science Collections, University of Oklahoma Libraries

Page 539, Humboldt, Courtesy of History of Science Collections, University of Oklahoma Libraries

Page 543, Huygens, Courtesy of History of Science Collections, University of Oklahoma Libraries

Page 576, Ino, Photograph by Steven Renshaw (Postage stamp released in 1995 celebrating Tadataka Ino and his mapping of Japan)

Page 594, Jeffreys, Reproduced by permission of Royal Astronomical Society (Presidential Portrait 59)

Page 602, Joy, Photograph originally appearing in the *Publications of the Astronomical Society of the Pacific* in an article by R. J. Trumpler (1950, *PASP*, 62, 33), copyright 1950, Astronomical Society of the Pacific; reproduced with permission of the Editors

Page 610, Kant, Courtesy of History of Science Collections, University of Oklahoma Libraries, Small Portraits Collection

Page 611, Kapteyn, Courtesy of Kapteyn Astronomical Institute of the University of Groningen, The Netherlands

Page 615, Keckermann, Courtesy of History of Science Collections, University of Oklahoma Libraries

Page 617, Keeler, Reproduced from *Astrophysical Journal* 12, no. 4 (Nov. 1900)

Page 620, Kepler, Courtesy of History of Science Collections, University of Oklahoma Libraries, Small Portraits Collection

Page 636, King, Reproduced from *Popular Astronomy* 24, no. 6 (June-July 1916)

Page 641, Kircher, Courtesy of History of Science Collections, University of Oklahoma Libraries, Small Portraits Collection

Page 642, Kirchhoff, Courtesy of History of Science Collections, University of Oklahoma Libraries

Page 647, Knorre, Courtesy of Mrs. Inga von Knorre

Page 649, Kolmogorov, Reproduced by permission of the Archives of the Russian Academy of Sciences

Page 657, Krieger, Reproduced from *Popular Astronomy* 22, no. 1 (Jan. 1914)

Page 668, Lalande, Reproduced by permission of photographer Simon Dumont (Photograph of statue of Lalande in Paris on Louvre façade, cour Napoléon)

Page 673, Lanczos, Reproduced by permission of North Carolina State University (Lanczos Collection, vol . VI, page 3-615)

Page 675, Langley, Reproduced from *Astrophysical Journal* 23, no. 4 (May 1906)

Page 677, Lansbergen, Philip, Courtesy of History of Science Collections, University of Oklahoma Libraries

Page 678, Laplace, Courtesy of History of Science Collections, University of Oklahoma Libraries

Page 683, Leavitt, Reproduced from *Popular Astronomy* 30, no. 4 (April 1922)

Page 684, Leclerc, Courtesy of History of Science Collections, University of Oklahoma Libraries, Small Portraits Collection

Page 687, Leibniz, Courtesy of History of Science Collections, University of Oklahoma Libraries, Large Portraits Collection

Page 689, Lemaître, Reproduced by permission of the Archives Lemaître, Université catholique de Louvain, Institute d'Astronomie et de Géophysique G. Lemaître, Louvain-la-Neuve, Belgium

Page 693, Leuschner, Reproduced by permission of the Mary Lea Shane Archives of the Lick Observatory, University of California at Santa Cruz

Page 696, Liais, Courtesy of History of Science Collections, University of Oklahoma Libraries

Page 702, Locke, Reproduced by permission of the Ohio Historical Society

Page 702, Lockyer, Reproduced from *Journal of the Royal Astronomical Society of Canada* 15, no. 2 (Feb. 1921)

Page 704, Lodge, Courtesy of History of Science Collections, University of Oklahoma Libraries

Page 705, Lohrmann, Courtesy of Niedersächisische Staats - und Universitätsbibliothek Göttingen

Page 710, Lowell, Reproduced by permission of Lowell Observatory

Page 721, Maclaurin, Courtesy of History of Science Collections, University of Oklahoma Libraries, Small Portraits Collection

Page 722, Maclear, Courtesy of Ian Elliott

Page 725, Magini, Courtesy of History of Science Collections, University of Oklahoma Libraries

Page 731, Malebranche, Courtesy of History of Science Collections, University of Oklahoma Libraries, Small Portraits Collection

Page 741, Maskelyn, Courtesy of History of Science Collections, University of Oklahoma Libraries, Small Portraits Collection

Page 748, Maupertuis, Reproduced by permission of Saint-Malo, Musée d'Histoire (Photograph by Michel Dupuis, Ville de Saint-Malo)

Page 754, Mayer, Julius, Courtesy of History of Science Collections, University of Oklahoma Libraries, Small Portraits Collection

Page 760, McMath, Reproduced by permission of Robert Raynolds McMath Collection, Bentley Historical Library, University of Michigan (McMath at observing desk, May 1948)

Page 764, Méchain, Courtesy of History of Science Collections, University of Oklahoma Libraries, Small Portraits Collection

Page 771, Merrill, Courtesy of The Observatories of the Carnegie Institution of Washington

Page 772, Mersenne, Courtesy of History of Science Collections, University of Oklahoma Libraries, Large Portraits Collection

Page 773, Messier, Courtesy of History of Science Collections, University of Oklahoma Libraries, Small Portraits Collection

Page 786, Minkowski, Hermann, Courtesy of History of Science Collections, University of Oklahoma Libraries

Page 791, Mitchell, Reproduced from *Popular Astronomy* 19, no. 7 (Sept. 1911)

Page 795, Molyneux, William, Reproduced by permission of Trinity College Dublin (A reproduction of the portrait by Robert Home)

Page 797, Monck, Reproduced by permission of the Mary Lea Shane Archives of the Lick Observatory, University of California at Santa Cruz

Page 802, Moore, Reproduced by permission of the Mary Lea Shane Archives of the Lick Observatory, University of California at Santa Cruz

Page 806, Morin, Courtesy of Robert A. Hatch

Page 814, Müller, Johann, Courtesy of History of Science Collections, University of Oklahoma Libraries

Page 820, Napier, Courtesy of History of Science Collections, University of Oklahoma Libraries

Page 821, Nasmyth, Courtesy of History of Science Collections, University of Oklahoma Libraries

Page 827, Newcomb, Courtesy of History of Science Collections, University of Oklahoma Libraries

Page 830, Newton, Isaac, Courtesy of History of Science Collections, University of Oklahoma Libraries, Large Portraits Collection

Page 848, Olbers, Courtesy of History of Science Collections, University of Oklahoma Libraries

Page 854, Oort, Courtesy of Adriaan Blaauw

Page 856, Öpik, Reproduced by permission of Armagh Observatory

Page 861, Oriani, Reproduced from *L'Astronomo Valtellinese e la Scoperta di Cerere* by permission of Luca Invernizzi, Alessandro Manara, and Peiro Sicoli (Fondazione Credito Valtellinese, 2001)

Page 868, Palitzsch, Courtesy of History of Science Collections, University of Oklahoma Libraries, Large Portraits Collection

Page 874, Parsons, Laurence, Reproduced by permission of The Earl of Rosse/Irish Picture Library

Page 876, Parsons, William, Reproduced by permission of The Earl of Rosse/Davison & Associates

Page 879, Payne-Gaposchkin, Courtesy of Katherine Haramundanis (*circa* 1920)

Page 883, Peary, Courtesy of History of Science Collections, University of Oklahoma Libraries

Page 885, Peek, Reproduced by permission of The British Astronomical Association

Page 893, Perrotin, Reproduced by permission of Observatoire de la Côte d'Azur Archives

Page 904, Piazzi, Courtesy of Luca Invernizzi. Reproduced from *L'Astronomo Valtellinese e la Scoperta di Cerere* by Luca Invernizzi, Alessandro Manara, and Peiro Sicoli (Fondazione Credito Valtenllinese, 2001)

Page 905, Picard, Reproduced from artist's depiction in *Le ciel* by Alphonse Berget (Paris, 1923)

Page 907, Pickering, Edward, Reproduced by permission of AIP Emilio Segrè Visual Archives

Page 911, Pişmiş, Reproduced by permission of photographer Juan Carlos Yustis

Page 917, Plaut, Courtesy of Adriaan Blaauw

Page 922, Pogson, Reproduced from *Popular Astronomy* 21, no. 8 (Oct. 1913)

Page 924, Poincaré, Courtesy of History of Science Collections, University of Oklahoma Libraries

Page 927, Popper, Reproduced by permission of the International Astronomical Union

Page 931, Poynting, Courtesy of History of Science Collections, University of Oklahoma Libraries

Page 934, Pritchett, Reproduced from *Popular Astronomy* 18, no. 6 (June-July 1910)

Page 938, Ptolemy, Courtesy of History of Science Collections, University of Oklahoma Libraries

Page 940, Purcell, Reproduced by permission of Physics Department, Harvard University

Page 957, Rayet, Reproduced from *Astrophysical Journal* 25, no. 1 (Jan. 1907)

Page 958, Reber, Reproduced by permission of National Radio Astronomy Observatory/AUI (Grote Reber pictured at the National Radio Astronomy Observatory at Green Bank, West Virginia, USA *circa* 1959)

Page 960, Recorde, Reproduced from *The American Mathematical Monthly* 28, no. 8-9 (1921)

Page 978, Roberts, Isaac, Reproduced by permission of the Mary Lea Shane Archives of the Lick Observatory, University of California at Santa Cruz

Page 981, Robinson, Reproduced by permission of Armagh Observatory

Page 990, Rowland, Courtesy of History of Science Collections, University of Oklahoma Libraries

Page 994, Runge, Reproduced by permission of the American Astronomical Society

Page 1012, Sampson, Reproduced by permission of the Royal Astronomical Society

Page 1018, Schaeberle, Reproduced by permission of Schaeberle Family Photograph Collection, Bentley Historical Library, University of Michigan

Page 1021, Scheiner, Julius, Reproduced from *Astrophysical Journal* 41, no. 1 (Jan. 1915)

Page 1022, Scheuchzer, Courtesy of History of Science Collections, University of Oklahoma Libraries

Page 1023, Schiaparelli, Courtesy of History of Science Collections, University of Oklahoma Libraries, Small Portraits Collection

Page 1024, Schickard, Courtesy of History of Science Collections, University of Oklahoma Libraries, Small Portraits Collection

Page 1030, Schönfeld, Courtesy of Christof A. Plicht

Page 1037, Schwarzschild, Karl, Reproduced by permission of Springer-Verlag (Karl Schwarzschild as a young man in Munich, 1900)

Page 1051, Shapley, Harlow, Reproduced by permission of AIP Emilie Segrè Visual Archives, Harlow Shapley Collection

Page 1074, Somerville, Reproduced by permission of the Royal Astronomical Society

Page 1075, Sorby, Reproduced by permission of the University of Sheffield

Page 1077, Spencer Jones, Reproduced by permission of the Royal Astronomical Society (Presidential Portrait 50)

Page 1081, Spörer, Reproduced by permission of the Astrophysikalisches Institut Potsdam

Page 1091, Stoeffler, Courtesy of History of Science Collections, University of Oklahoma Libraries, Small Portraits Collection

Page 1092, Stokes, Courtesy of History of Science Collections, University of Oklahoma Libraries

Page 1095, Stone, Ormond, Reproduced from *Popular Astronomy* 3, no. 9 (May 1896)

Page 1096, Stoney, Reproduced by permission of the Royal Dublin Society

Page 1100, Stratton, Reproduced by permission of the Royal Astronomical Society (Presidential Portrait 48)

Page 1102, Stromgren, Bengt, Reproduced by permission of the History of Science Department, University of Aarhus

Page 1108, Struve, Otto Wilhelm, From the collection of the late Nils Lindhagen (*circa* 1860)

Page 1116, Swedenborg, Courtesy of History of Science Collections, University of Oklahoma Libraries

Page 1119, Swope, Courtesy of Katherine Haramundanis

Page 1122, Tacchini, Reproduced from *Astrophysical Journal* 22, no. 1 (July 1905)

Page 1142, Thomson, Reproduced from *Proceedings of the Royal Society of London* A 81 (1908)

Page 1147, Tombaugh, Courtesy of photographer Thomas Hockey

Page 1152, Trumpler, Portrait originally appearing in the *Publications of the Astronomical Society of the Pacific* in an article by H. Weaver and P. Weaver (1957, *PASP*, 69, 304), copyright 1957, Astronomical Society of the Pacific; reproduced with permission of the Editors

Page 1167, Väisälä, Reproduced by permission of Tuorla Observatory, University of Turku

Page 1169, Van Allen, Reproduced by permission of The University of Iowa

Page 1171, Van de Kamp, Reproduced by permission of Friends Historical Library, Swarthmore College

Page 1177, Van Rhijn, Courtesy of the Kapteyn Astronomical Institute of the University of Groningen, The Netherlands

Page 1180, Vespucci, Courtesy of History of Science Collections, University of Oklahoma Libraries, Small Portraits Collection

Page 1182, Vinci, Courtesy of History of Science Collections, University of Oklahoma Libraries, Small Portraits Collection

Page 1185, Vogel, Reproduced from *Astrophysical Journal* 27, no. 1 (Jan. 1908)

Page 1190, Wachmann, Reproduced by permission of Hamburger Sternwarte

Page 1191, Wales, Courtesy of the Museum and Archive and Governors of Christ's Hospital

Page 1193, Wallace, Courtesy of History of Science Collections, University of Oklahoma Libraries

Page 1195, Wallis, Courtesy of History of Science Collections, University of Oklahoma Libraries

Page 1200, Watson, Reproduced by permission of the Mary Lea Shane Archives of the Lick Observatory, University of California at Santa Cruz

Page 1207, Werner, Courtesy of History of Science Collections, University of Oklahoma Libraries

Page 1211, Wheeler, Courtesy of Princeton University (Photograph by Robert Matthews, 1973)

Page 1223, Wilkins, Courtesy of History of Science Collections, University of Oklahoma Libraries

Page 1234, Wirtanen, Reproduced by permission of the Mary Lea Shane Archives of the Lick Observatory, University of California at Santa Cruz (Carl Wirtanen at the Carnegie Astrograph of the Lick Observatory, 29 March 1948)

Page 1244, Wren, Courtesy of History of Science Collections, University of Oklahoma Libraries

Page 1246, Wright, William, Reproduced by permission of the Mary Lea Shane Archives of the Lick Observatory, University of California at Santa Cruz

Page 1256, Young, Charles, Reproduced from *Astrophysical Journal* 30, no. 5 (Dec. 1909)

Entry Index

D

E

F

G

H

I

J

K

L

M

N

Q

R

S

T

U

V

W

X

Y

Z

Subject Index

B

D

E

F

G

H

N

O

P

S

T

U

Contributor Index

E

F

I

J

M

T

U

V

W

Y

Z